W9-AAI-398

VISIT AN ONLINE LIBRARY, 24 X 7

InfoTrac® College Edition with InfoMarks® is a university database with millions of articles from nearly 5,000 publications, including *Discover, Science, Science News, The New York Times,* and *USA Today.* This online resource, accessible through **1pass,** is a tremendous resource for doing homework or catching up on the news.* With InfoTrac, you have access to InfoMarks, which lets you set up a bibliography of stable links to articles, journals, and searches for help in writing papers. In addition, you can access InfoWrite, a set of online guides to grammar, writing research papers, and "critical thinking."

GET MORE STUDY SUPPORT AT THE BOOK COMPANION WEBSITE

Visit http://biology.brookscole.com/starr11 for other resources that complement what you study in this textbook. For instance, an electronic InfoMarks reader has a prime list of biology-related articles for each textbook chapter, selected for their easy-reading style and their relevance to your own life. Register for a login at **1pass,** and you will be able to access the articles from each chapter's "Suggested InfoTrac Readings" list with a click.

Register with 1pass™ and say goodbye to multiple passwords

Simplify your life with **1pass,*** a single point of access to the media resources described on these two pages. **1pass** provides you with the ability to access a wide range of media by using a single, personalized username and password—no more multiple URLs and log-in information to remember!

If you purchased a new book, you may already have **1pass** access to these resources. To register, see the enclosed **1pass** card for directions. If you did not purchase a new book, you still have the option to electronically purchase access by visiting http://1pass.thomson.com.

1PASS RESOURCES AVAILABLE WITH THIS TEXT:

- **BiologyNow,** including *How Do I Prepare?* and **vMentor**
- **InfoTrac® College Edition with InfoMarks®**

* Instructors: vMentor™ and InfoTrac College Edition are available for college and university adopters only. Other restrictions may apply.

BIOLOGY

THE UNITY AND DIVERSITY OF LIFE

ELEVENTH EDITION

BIOLOGY
THE UNITY AND DIVERSITY OF LIFE

ELEVENTH EDITION

CECIE STARR / RALPH TAGGART

Christine A. Evers

Lisa Starr

THOMSON

BROOKS/COLE

Australia • Brazil • Canada • Mexico • Singapore
Spain • United Kingdom • United States

PUBLISHER Jack C. Carey

VICE-PRESIDENT, EDITOR-IN-CHIEF Michelle Julet

SENIOR DEVELOPMENT EDITOR Peggy Williams

ASSOCIATE DEVELOPMENT EDITOR Suzannah Alexander

EDITORIAL ASSISTANT Chris Ziemba, Kristina Razmara

TECHNOLOGY PROJECT MANAGER Keli Amann

MARKETING MANAGER Ann Caven

MARKETING ASSISTANT Brian Smith

MARKETING COMMUNICATIONS MANAGER Nathaniel
Bergson-Michelson

PROJECT MANAGER, EDITORIAL PRODUCTION Andy Marinkovich

CREATIVE DIRECTOR Rob Hugel

PRINT BUYER Karen Hunt

PERMISSIONS EDITOR Joohee Lee, Sarah Harkrader

PRODUCTION SERVICE Grace Davidson & Associates

TEXT AND COVER DESIGN Gary Head

PHOTO RESEARCHER Myrna Engler

COPY EDITOR Kathleen Deselle

ILLUSTRATORS Gary Head, ScEYEnce Studios, Lisa Starr

COMPOSITOR Lachina Publishing Services

TEXT AND COVER PRINTER R.R. Donnelley/Willard

COVER IMAGE *Ko'olau Mountains on the windward side of
Oahu. This island and others of the Hawaiian Archipelago
are a natural laboratory for the study of evolution.*
Photographer Russ Lowgren

Printed in the United States of America
2 3 4 5 6 7 09 08 07 06

Library of Congress Control Number: 2005931271

Student Edition: ISBN 0-495-01599-7

For more information about our products, contact us at:
Thomson Learning Academic Resource Center
1-800-423-0563

For permission to use material from this text or product, submit
a request online at http://www.thomsonrights.com.
Any additional questions about permissions can be submitted
by e-mail to thomsonrights@thomson.com.

Thomson Higher Education
10 Davis Drive
Belmont, CA 94002-3098
USA

Asia (including India)
Thomson Learning
5 Shenton Way
#01-01 UIC Building
Singapore 068808

Australia/New Zealand
Thomson Learning Australia
102 Dodds Street
Southbank, Victoria 3006
Australia

Canada
Thomson Nelson
1120 Birchmount Road
Toronto, Ontario M1K 5G4

UK/Europe/Middle East/Africa
Thomson Learning
High Holborn House
50/51 Bedford Row
London WC1R 4LR
United Kingdom

CONTENTS IN BRIEF

DETAILED CONTENTS

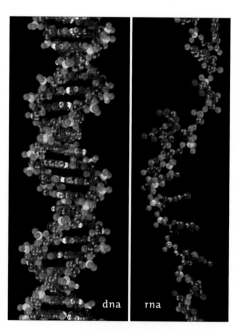

dna   rna

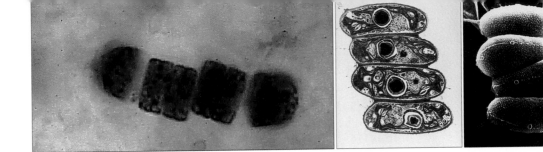

4 Cell Structure and Function

5 A Closer Look at Cell Membranes

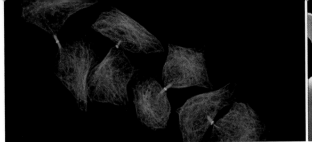

UNIT II PRINCIPLES OF INHERITANCE

9 How Cells Reproduce

10 Meiosis and Sexual Reproduction

11 Observing Patterns in Inherited Traits

12 Chromosomes and Human Inheritance

16 Studying and Manipulating Genomes

17 Evidence of Evolution

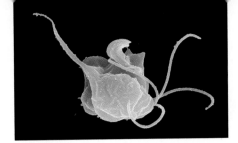

24 Fungi

25 Animal Evolution—The Invertebrates

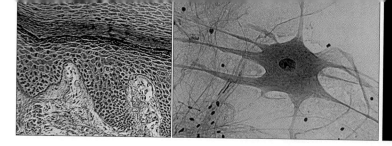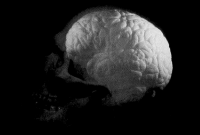

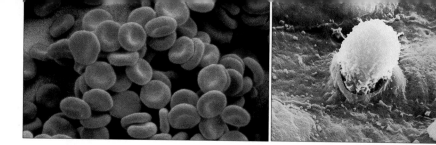
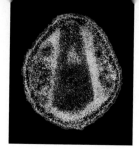

38 Circulation

39 Immunity

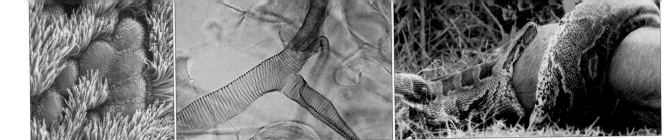

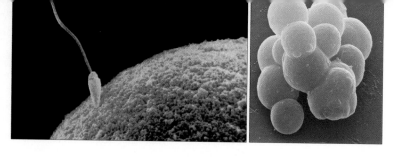

UNIT VII   PRINCIPLES OF ECOLOGY

45 Population Ecology

46 Community Structure and Biodiversity

47 Ecosystems

48 The Biosphere

49 Behavioral Ecology

IMPACTS, ISSUES *My Pheromones Made Me Do It* *898*

PREFACE

Biology's overriding paradigm is one of interpreting life's spectacular diversity as having evolved from simple molecular beginnings. For decades, researchers have been chipping away at the structural secrets of biological molecules. They are showing how different kinds are put together, how they function, and what happens when they mutate. They are casting light on how life originated, what happened over the past 3.8 billion years, and what the future may hold for humans and other organisms, individually and collectively.

This is profound stuff. *Yet introductory textbooks— including previous editions of this one—have not given students enough information to interpret for themselves the remarkable connection between molecular change, evolution, and their own lives.* We continue to rebuild this book in ways that clarify the connection. When students "get" the big picture of life, they become confident enough to think critically about the past, present, and future on their own. This is one of biology's greatest gifts.

MAKE THE DIRECT CONNECTION

New to this edition are connections that have direct and often controversial impact on our lives. What students learn today may help them make decisions tomorrow, in the voting booth as well as in their personal lives.

How Would You Vote? Each chapter starts with an essay on a current issue that relates to its content (see list at right). Essays are expanded in custom *videoclips* and exercises on the student website and in the Power-Lecture, a one-stop PowerPoint tool for instructors.

We ask students in each chapter, *How would you vote on research or an application related to this issue?* Then the exercise invites them to read a balanced selection of articles, pro and con, before voting. Students throughout the country are already voting *online*, and are accessing campuswide, statewide, and nationwide tallies. This interactive approach to issues reinforces the premise that individual actions can make a difference.

HELP STUDENTS TRACK CONNECTIONS

Linking Concepts Also new to this edition are links to concepts within and between chapters. Students have an easier time making connections when links guide them. Each chapter's opening spread has a section-by-section list of *key concepts*, each with a simple title. We repeat the titles at the top of appropriate text pages as reminders of the conceptual organization. A brief list of *links to earlier concepts* helps students assess concepts they should already know before they begin. For instance, before reading about neural function, they may wish to scan an earlier chapter's section on active transport. We repeat the linking icons in page margins to help students refer back to relevant sections.

Concept Spreads Judging from extensive feedback, *concept spreads* are immensely popular. Students report that they can easily study and master parts of a chapter

IMPACTS/ISSUES ESSAYS

1 State of the world, including bioterrorism
2 Toxic chemicals in and around the body
3 Methane hydrates, energy, and extinctions
4 What is a cell and can researchers make one?
5 Membrane defects and cystic fibrosis
6 Student binge drinking and liver damage
7 Why your life depends on plants
8 Disorders arising from abnormal mitochondria
9 Henrietta Lacks, HeLa cells, and cervical cancer
10 Why females and males?
11 Rose breeding, a big business
12 Genes and mental illness
13 Cloning—reprogramming the DNA
14 The bioterror agent ricin affects ribosomes
15 Genes linked with breast cancer
16 Malnutrition, biotechnology, and Golden Rice
17 Asteroid impacts and human evolution
18 Natural selection and the rise of super rats
19 Hawaiian honeycreepers and humans
20 Bioprospecting in extreme environments
21 West Nile virus and Alexander the Great
22 Protists as architects and pathogens
23 Evolutionary perspective on deforestation
24 Reindeer, reindeer herders, and lichens
25 New neuroactive drugs from snail venom
26 How people interpret and misinterpret the past
27 A cautionary tale from Easter Island
28 Homeostasis and a lethal case of heat stroke
29 Plants and the rise and fall of civilizations
30 Phytoremediation at military waste sites
31 No more pollinators? No more chocolate
32 Plant hormones and our food supply
33 Superman and stem cell research
34 Dropping dead with Ecstasy
35 National security, sonar, and the whales
36 Endocrine disrupters and deformed frogs
37 Andro, creatine, and pumped-up athletes
38 From a bulldog to automated defibrillators
39 AIDS and a generation of African orphans
40 Cigarette smoking, biologically speaking
41 From early hominids to hefty humans
42 Urine tests, past and present
43 The point of mammalian (and human) sexual behavior
44 Fertility drugs and a plethora of babies
45 St. Matthews Island, the accidental experiment
46 What can stop the fire ants?
47 Global warming and the disappearing bayous
48 El Niños, super surfing, and suffering seals
49 Pheromones and the "killer bees"

when they have a small window of time rather than wading into an overwhelming mass of information. We place all text, art, and evidence in support of each concept in one section that consists of two facing pages, at most. Each section starts at a numbered tab and ends with a blue-boxed *on-page summary*, which students can use to check whether they understand the main points before turning to the next concept spread.

Concept spreads offer teachers flexibility in assigning topics to fit course requirements. Those who spend less time on, say, photosynthesis might bypass details on the properties of light and ATP formation. All spreads are part of the chapter topic, but some offer added depth.

Read-Me-First Art *Read-me-first diagrams* are visual learning devices. Students can walk through them step-by-step as a preview of the text. The same art occurs in narrated, animated form online at BiologyNow and on the PowerLecture DVD. Repeated exposure to material in a variety of modalities accommodates diverse learning styles and reinforces knowledge of concepts.

Critical Thinking Like all textbooks at this level, we walk students through examples of problem solving and experiments throughout the book. We integrate many simple experiments in the text proper. *Focus on Science* essays highlight the detailed ones. The entries *Experiment, examples*, and *Test, examples* in the main index list our selections.

Each chapter ends with a number of *Critical Thinking* questions, some with art, and some more challenging than others. They invite students to think outside the memorization box. We include an *annotated scientific paper* as an example of how scientists work and then report their findings (Appendix IX).

We selectively use some chapter introductions, even a few whole chapters, to emphasize scientific methods. However, introductory students are not scientists, so we do not shoehorn *all* material into an experimental context. We cannot expect them to learn the language and processes of science by intuition alone; we help them build these skills in a paced way.

Connections essays, flagged by vertical red bands on page edges, invite students to step back and connect the dots between chapters and units. An *Epilogue* invites them to make the most sweeping connection of all.

WRITE CLEARLY AND SELECTIVELY

Students often complain that their textbooks are dry and boring. We write to engage them without glossing over the science, even through such formidable topics as the laws of thermodynamics (Section 6.1).

To keep book length manageable, we were selective about which topics to include so that we could allocate enough space to explain them clearly. For instance, most nonmajors simply do not want to memorize each catalytic step of crassulacean acid metabolism. They do want to learn about the basis of sex, and many women would like to know what will go on inside their body if they become pregnant. Sufficient information will help them make informed decisions on many biology-related

issues, such as STDs, fertility drugs, prenatal diagnoses, gene therapies, and stem cell research. Over the years, students have written to tell us that they did read such material closely even when it was not assigned.

Our choices for which topics to condense, expand, or delete reflect more than three decades of feedback from teachers of millions of students around the world.

OFFER STUDENTS EASY-TO-USE MEDIA TOOLS

With this edition, logging in to online assets is simpler than before. Students register their free **1pass access code** at http://1pass.thomson.com and then log in to access all resources outlined below. An access code is packaged in each new copy of the book and also is available through e-commerce.

BiologyNow™ For each text chapter, *BiologyNow* provides *animated tutorials* as well as guidance to other resources that can help students master the content. Responses to *diagnostic pretest questions* yield a student-driven, *personalized learning plan.* Answer a question incorrectly, and the plan lists text sections, figures, and chapter videos. The plan also gives links to relevant animation.

The *How Do I Prepare* section has tutorials in math, chemistry, graphing, and basic study skills. *vMentor* is an online service with a live tutor who interacts with students through voice communication, whiteboard, and text messaging.

The *post-test* can be used as self-assessment tool or submitted to an instructor. BiologyNow is built in iLrn, Thomson Learning's course management system, so that answers and results can be fed directly to an electronic gradebook. BiologyNow also can be integrated with WebCT and Blackboard, so students may log in through these systems as well.

InfoTrac College Edition™ 1pass also grants access to *InfoTrac College Edition*, an online database with more than 4,000 periodicals and close to 18 million articles. The articles are in full-text form and can be located easily and quickly with a *key word* search.

In BiologyNow, the *How Would You Vote* exercise for each chapter references specific InfoTrac articles and websites. As students vote on each issue, the site provides a running tally by campus, state, and the country. Instructors can assign the exercises from iLrn, or students can access them through the free website at:

http://biology.brookscole.com/starr11

This website also features more InfoTrac articles and web links, as well as *interactive flashcards* that define all of the book's boldface terms with pronunciation guides.

Also, an online *Issues and Resources Integrator* correlates chapter sections with applications, videos, InfoTrac articles, and websites. This guide is updated each semester.

Audiobook New college editions of the textbook can also include free access to the *Audiobook*. Students can either listen to narration online or download mp3 files for use on a portable mp3 player.

CONTENT REVISIONS

Previous adopters of *Biology: The Unity and Diversity of Life* who continue to use the book may wish to scan this list of refinements and updates in chapter content. Again, all chapters start with new essays on current issues that are more likely to engage students.

A New Tree of Life Thanks to the wealth of information from biochemistry and molecular biology, the tree of life is coming into sharp focus. Working closely with Walter Judd at the University of Florida, we became confident enough to create a simple but powerful graphic of the evolutionary connections, including the main archaean and protist groups. We present the tree in Section 19.7 as a prelude to the diversity unit. We repeat parts of it in the diversity chapters, as regional road maps. The first chapter does have a simplified classification system in which six kingdoms are subsumed into three domains. It is all that students will require until they reach the later units on evolutionary principles and biodiversity.

We use informal names in the tree and in much of the text. Our advisors and focus-group participants agree that it is more important for introductory students to think about evolutionary connections than ongoing refinements in taxonomy. For reference purposes, Appendix I still provides taxonomic names for major groups.

Energy and Life's Organization Throughout the book, we strengthened the concept of energy flow as the basis of life's levels of organization. Students generally do not find bioenergetics an endearing topic, but we have two new, nonthreatening sections that might engage them through the use of real-life examples (6.1, 6.2). This conceptual overview may help them make sense of metabolism. The photosynthesis chapter (7) has reorganized, streamlined text, new art (including an update on chloroplast structure), and good bioenergetic and evolutionary threads.

Evolution We continue to apply evolutionary theory all through the book. The early, simple overview in Section 1.4 is enough to get students thinking about the concept without bogging them down in details. As examples, we invite them to think about why sexual reproduction evolved (the new Chapter 10 introduction), a possible evolutionary connection between mitosis and meiosis (10.6), and past and potential future threats of methane hydrate deposits on the seafloor. We continue to expand on the connections between mutation, evolution, and development, or "evo-devo" (15.3, 17.8, 17.9, 18.1, 25.1, 25.2, 25.8, 26.4, 26.14, 32.6, 34.1, 39.1, and 43.5 are some examples). The macroevolution chapter (18) has a new essay and text section on the adaptive radiation and current extinctions of the Hawaiian honeycreepers.

Cell Communication We strengthened the art and text presentations of signaling pathways. Sections 22.12 and 25.1 can get students thinking about the evolutionary origins of cell communication. Section 28.5 introduces a generalized graphic of signal reception, transduction, and response that we repeat throughout the animal and plant anatomy units. Section 28.5 also has a specific example: apoptosis in the development of the human hand. We build on this introduction in the sections on neural function (34.1–34.5), sensory function (35.1), hormone action in animals (36.2) and in plants (32.3), immune function (39.4, 39.6, 39.7), urine formation (42.4), and embryonic development (43.2, 43.5).

Unit I Chapter 1: reorganized, simplified, with a new section on experimental tests. Chapter 2: more accessible introduction to electrons and energy levels. Chapter 3: revised sections on levels of protein structure. Chapter 4: new art and text on cell structures. Chapter 5: sections on diffusion, transport mechanisms, and osmosis revised for clarity. Chapter 6: new art, more accessible writing. Chapter 7: extensively revised, updated. Chapter 8: new art for the Krebs cycle and the electron transfer system.

Unit II Chapter 9: new introduction and new essay on cancer. Chapter 10: stronger evolutionary framework for explaining meiosis. Chapter 11: section on impact of crossing over on gene linkage moved to this chapter; new genetics problems on rose and watermelon hybrid experiments. Chapter 12: reorganization of sections on autosomal and X-linked inheritance. Chapter 13: new, updated essay on how cloning methods manipulate information in DNA. Chapter 14: subtle rewrites for clarity, updated alternative mRNA processing. Chapter 15: reorganized, starts with overview of control points; introduces *ABC* flowering model; new essay on effect of mutant *Drosophila* master genes on development. Chapter 16: major reorganization and rewrite, new sections on genomics, organ factories, and other research that will have enormous impact on our lives.

Unit III Chapter 17: now starts with an essay on mass extinctions as one measure of geologic time; introduces evidence for evolution (fossils, radiometric dating, biogeography and plate tectonics, and comparative morphology, embryology, and biochemistry); has new examples of evolutionary constraints on developmental patterns for plants and animals. Chapter 18: some new examples, as in introductory essay; revised section on genetic drift; ends with essay on adaptation (adaptation to what?). Chapter 19: revised text and art, some new examples for reproductive isolating mechanisms and speciation models; updated sections on taxonomy and systematics; new tree of life.

Unit IV Chapter 20: now a less detailed introduction and timeline for the diversity chapters; updated sections on origin of life and of organelles. Chapter 21: essay on West Nile virus; rewrite and updates on bacterial and archaean lineages; updated essay on infectious disease, including SARS and prion diseases. Chapter 22: basically a new chapter on "protists" with clarified phylogenies; has new essay on amoebozoans and the origin of cell communication and multicellularity. Chapter 23: starts and ends with essays on deforestation; new overview of plant origins and evolution; charophyes now included in plant clade; more on flowering plants and pollinators to set the stage for Section 32.3. Chapter 24: new essay on lichens; new material on chytrids, microsporidians, *Neurospora* life cycle, and endophytic fungi.

Chapter 25: major new, updated sections on animal origins and characteristics, including the basis of large, internally complex bodies; reordering of placozoans, flatworms, roundworms, other major groups; essay on cephalopod evolution. Chapter 26: *Archaeopteryx* essay; revised sections on chordate family tree and vertebrate evolutionary trends; aquatic origin of tetrapods with new art; focus essay on vanishing amphibians; sections on dinosaurs now integrated in this chapter; revised text and art on rise of amniotes; new text and art on adaptive radiation of mammals; updated text and art on primate and hominid evolutionary trends.

Chapter 27: now includes a section on environmental impact of human population growth; Rachel Carson essay; conservation biology and sustaining biodiversity, with Monteverde Cloud Forest as an example; ends on a more optimistic note with examples of what might be accomplished by thinking outside the box.

Introduction to Units V and VI New to this edition. Chapter 28: presents common challenges facing plants and animals at increasingly complex levels structural organization; introduces concept of homeostasis and requirements for gas exchange, internal transport, and cell communication, as listed earlier; animal examples of homeostasis are temperature regulation and oxygen deficiency at high altitudes; plant examples are systemic acquired resistance and leaf folding in response to shifts in environmental conditions.

Unit V Chapter 29: reorganized, rewritten introduction to plant tissues with new graphics and micrographs. Chapter 31: incorporates new essay on coevolution of flowering plants and pollinators, including text and photograph of the hawkmoth and the Christmas Star; additional diagrams, improved classification of fruits. Chapter 32: heavily revised, updated; auxin signaling pathway used as new example of cell communication; ends the unit with a new connections essay on global protein deficiency and quinoa research (earlier edition's essay on pesticides is now in Chapter 47 as part of an expanded essay on biomagnification).

Unit VI Chapter 33: new essay on stem cell research; new micrographs; tissue descriptions rewritten, a bit more on muscle tissue; chapter builds on Section 25.1 introduction to compartmentalization (division of labor) as emergent property of multicellularity; new essay on vertebrate skin. Chapter 34: all sections simplified and reorganized; opens with Ecstasy essay; new overview of invertebrate and central/peripheral vertebrate nervous systems; rewritten sections on neurotransmitters and neuromodulators; new section on neuroglia; updated essay on psychoactive drugs. Chapter 35: new essay on oceanic noise pollution. Chapter 36: new essay and text

on hormone disruptors, amphibians. and human sperm counts; major text reorganization (gland by gland); new essay on diabetes and impact of stress on health.

Chapter 37: new section on origin of invertebrate and vertebrate skeletons. Chapter 38: new text, art on cardiac conduction system. Chapter 39: fully revised chapter; more accessible, yet reflects current research and up-to-date concepts, including evolutionary and functional interrelationship between adaptive and innate immunity; easier-to-follow art; update on HIV/AIDS. Chapter 40: rewrite on countercurrent flow in fish gills; new text on controls over respiration. Chapter 41: opens with essay on hormones and appetite; updated human nutritional requirements. Chapter 42: major reorganization and new step-by-step graphics and text on urine formation; expanded emphasis on kidney disorders.

Chapter 43: frog life cycle now integrated with the overview of stages of development; sections on cleavage, pattern formation, and constraints on development have been clarified; new Critical Thinking question on RNA interference for interested students (page 769); updates on aging hypotheses. Chapter 44: text tightened with some reorganization (sections on fertility control and STDs now located after human reproduction and before development); updates on STDs, pregnancy guidelines, fetal alcohol syndrome; oxytocin secretion during labor a new example of positive feedback; essay on bioethics of interventions in fertility now at end of chapter.

Unit VII Chapter 45: unit now opens with essay on reindeer overshooting carrying capacity on St. Matthews Island; revised sections on limiting factors; updates on human population growth; demographic transition model now qualified. Chapter 46: rewrites and new art for mutualism, competitive interactions, predation models; new examples of parasitism; revised community structure models; updates on *Codium* aquarium strain, kudzu, rabbits in Australia; some additions to section on biogeographic patterns; Critical Thinking question on Wallace's line (page 841). Note that Chapter 50 sections of the previous edition are now rolled into Chapters 47 and 48. Chapter 47: global warming and bayous; revised essays on pesticides and biomagnification, energy flow, watershed experiments; global water crisis integrated in this chapter; carbon cycle updates; global warming updates; revised nitrogen cycle. Chapter 48: text on solar energy and wind energy now integrated in section on global air circulation patterns; essay on air pollution (ozone thinning, smog, acid rain, particulates); better historical look at biogeographic realms; new section on differences in sunlight, soils, and moisture, with desert soils and moisture differences as case study; new essay on desertification, including a model for feedback loops between desertification, climate change, and losses in biodiversity; expansion of freshwater provinces with a section on water pollution; new section on north central Pacific gyre as garbage dump. Chapter 49: update on genetics of courtship behavior in fruit flies and pair bonds in voles; new sections on primate social behavior and on biological aspects of human behavior.

ACKNOWLEDGMENTS

Once again I am thinking about all of the advisors and reviewers who helped shape the content of this book in significant ways. Bits and pieces of earlier critiques still flit through my mind, and they are making me realize that educators are symbionts of the highest order. Why else would they give so much of themselves, year after year, to the cooperative enterprise called education? The demands are great and the money is not. Somehow they all have become hardwired to contribute to the greater good. Each year when I interact with them, some new manifestation of their commitment manages to take my breath away; which is why I am always uncomfortable with calling this book "the Starr book." It is our book.

Dan Fairbanks, John Jackson, Walt Judd, and Bill Wischusen have been fearless about approaching this established book with a sledgehammer. They are aware that research directions change, students change, and textbooks must change with them. They dig out flaws in my thinking with needles. I admire them immensely.

Biology: The Unity and Diversity of Life wins awards for subsuming design and art in the service of pedagogy instead of slapping them on as afterthoughts. All it takes is abiding passion for biology and a compulsive urge to assess trends in education, keep up with research, write, create art, and finely lay out workable concept spreads, all at the same time. For some time now, I have been mentoring Lisa Starr and Christine Evers in the odd science of being multitasking authors. Lisa is educated in biochemistry, biotechnology, and computer graphics; Chris in animal behavior, evolutionary biology, and media technologies. With their combined talents and dedication, they have become indispensable partners.

Starting with Susan Badger, Michael Johnson, and Sean Wakely, Thomson Learning proved once again with this edition that it is one of the world's foremost publishers. Michelle Julet, what would I do without your laser focus, arm twisting, intelligence, and wit? Peggy Williams is Empress of the Team. So many of the improvements in this book are direct outcomes of her extraordinary talent for organizing and interpreting the results of class tests, workshops, and focus groups. Andy Marinkovich and dear Grace Davidson continue to take my world-class compulsivity in stride and still get the book out on time. Amazing. Gary Head, aka The Rock, suffers through being my designer, year in, year out. Thank goodness for Myrna Engler's devotion to the team and Suzannah Alexander's quiet competence. Ann Caven, Kathie Head, Laura Argento, Keli Amann, Marlene Veach, Karen Hunt, Chris Ziemba, Kristina Razmara, the professionals at Lachina—the list goes on, but no listing conveys how this team interacts to create something extraordinary.

Jack Carey, how long ago was it that you signed me to do eighteen books to keep me from writing for anybody else in this lifetime? I never did get around to the other fourteen, but probably you forgive me. Through our long partnership, we helped move textbook publishing in new directions, to the benefit of students around the world. Never would have done it without you, partner.

CECIE STARR, *November 2005*

MAJOR ADVISORS AND REVIEWERS

DANIEL FAIRBANKS *Brigham Young University*
JOHN D. JACKSON *North Hennepin Community College*
WALTER JUDD *University of Florida*
E. WILLIAM WISCHUSEN *Louisiana State University*

JOHN ALCOCK *Arizona State University*
CHARLOTTE BORGESON *University of Nevada*
DEBORAH C. CLARK *Middle Tennessee University*
MELANIE DEVORE *Georgia College and State University*
TOM GARRISON *Orange Coast College*
DAVID GOODIN *The Scripps Research Institute*
CHRISTOPHER GREGG *Louisiana State University*
PAUL E. HERTZ *Barnard College*
TIMOTHY JOHNSTON *Murray State University*
EUGENE KOZLOFF *University of Washington*
ELIZABETH LANDECKER-MOORE *Rowan University*
KAREN MESSLEY *Rock Valley College*
THOMAS L. ROST *University of California, Davis*
LAURALEE SHERWOOD *West Virginia University*
STEPHEN L. WOLFE *University of California, Davis*

CONTRIBUTORS: 2004–2005 WORKSHOPS AND REVIEWS

ANDERSON, DIANNE *San Diego City College*
ASMUS, STEVE *Centre College*
BLAUSTEIN, ANDREW R. *Oregon State University*
BROSSAY, LAURENT *Brown University*
CAPORALE, DIANE *University of Wisconsin-Stevens Point*
CAWTHORN, J. MICHELLE *Georgia Southern University*
COLLIER, ALEXANDER *Armstrong Atlantic State University*
CONWAY, ARTHUR F. *Randolph-Macon College*
DAVIS, GEORGE T. *Bloomsburg University*
DEGRAUW, EDWARD *Portland Community College*
DELANEY, CYNTHIA LEIGH *University of South Alabama*
DESAIX, JEAN *University of North Carolina*
DIERINGER, GREG *Northwest Missouri State University*
D'ORGEIX, CHRISTIAN *Virginia State University*
D'SILVA, JOSEPH G. *Norfolk State University*
FONG, APRIL *Portland Community College*
GARCIA, RIC A. *Clemson University*
HAIGH, GALE *McNeese State University*
HEARRON, MARY *Northeast Texas Community College*
HENDERSON, WILEY J. *Alabama A&M University*
HINTON, JULIANA *McNeese State University*
HUANG, SARA *Los Angeles Valley College*
HUFFMAN, DONNA *Calhoun Community College*
JONES, KEN *Dyersburg State Community College*
JULIAN, GLENNIS *University of Texas at Austin*
KROLL, WILLIAM *Loyola University, Chicago*
KURDZIEL, JOSEPHINE *University of Michigan*
LAMMERT, JOHN M. *Gustavus Adolphus College*
LAZOTTE, PAULINE *Valencia Community College*
MADTES, JR., PAUL *Mount Vernon Nazarene University*
MAJDI, BERNARD *Waycross College*
MATA, JUAN LUIS *University of Tennessee at Martin*
MCKEAN, HEATHER R. *Eastern Washington University*
METZ, TIMOTHY *Campbell University*
MOSS, ANTHONY *Auburn University*
NEUFELD, HOWARD S. *Appalachian State University*
NOLD, STEPHEN *University of Wisconsin*
ORR, CLIFTON *University of Arkansas at Pine Bluff*
PETERS, JOHN *College of Charleston*
PLUNKETT, JENNIE *San Jacinto College*
POMARICO, STEVEN M. *Louisiana State University*
RAINES, KIRSTEN *San Jacinto College*
REED, ROBERT N. *Southern Utah University*
RIBEIRO, WENDA *Thomas Nelson Community College*
RICHEY, MARGARET G. *Centre College*
ROBERTS, LAUREL *University of Pittsburgh*

ROMANO, FRANK A., III, *Jacksonville State University*
RUPPERT, ETTA *Clemson University*
SHOFNER, MARCIA *University of Maryland, College Park*
SIEVERT, GREG *Emporia State University*
SIMS, THOMAS L. *Northern Illinois University*
SONGER, STEPHANIE R. *Concord University*
SPRENKLE, AMY B. *Salem State College*
ST. CLAIR, LARRY *Brigham Young University*
TEMPLET, ALICE *Nicholls State University*
TURELL, MARSHA *Houston Community College*
WALSH, PAT *University of Delaware*
WILKINS, HEATHER DAWN *University of Tennessee at Martin*
WINDELSPECHT, MICHAEL *Appalachian State University*
WYGODA, MARK *McNeese State University*
ZAHN, MARTIN D. *Thomas Nelson Community College*
ZANIN, KATHY *The Citadel*

CONTRIBUTORS: INFLUENTIAL CLASS TESTS AND REVIEWS

ADAMS, DARYL *Minnesota State University, Mankato*
ANDERSON, DENNIS *Oklahoma City Community College*
BENDER, KRISTEN *California State University, Long Beach*
BOGGS, LISA *Southwestern Oklahoma State University*
BORGESON, CHARLOTTE *University of Nevada*
BOWER, SUSAN *Pasadena City College*
BOYD, KIMBERLY *Cabrini College*
BRICKMAN, PEGGY *University of Georgia*
BROWN, EVERT *Casper College*
BRYAN, DAVID W. *Cincinnati State College*
BURNETT, STEPHEN *Clayton College*
BUSS, WARREN *University of Northern Colorado*
CARTWRIGHT, PAULYN *University of Kansas*
CASE, TED *University of California, San Diego*
COLAVITO, MARY *Santa Monica College*
COOK, JERRY L. *Sam Houston State University*
DAVIS, JERRY *University of Wisconsin, LaCrosse*
DENGLER, NANCY *University of California, Davis*
DESAIX, JEAN *University of North Carolina*
DIBARTOLOMEIS, SUSAN *Millersville University of Pennsylvania*
DIEHL, FRED *University of Virginia*
DONALD-WHITNEY, CATHY *Collin County Community College*
DUWEL, PHILIP *University of South Carolina, Columbia*
EAKIN, DAVID *Eastern Kentucky University*
EBBS, STEPHEN *Southern Illinois University*
EDLIN, GORDON *University of Hawaii, Manoa*
ENDLER, JOHN *University of California, Santa Barbara*
ERWIN, CINDY *City College of San Francisco*
FOREMAN, KATHERINE *Moraine Valley Community College*
FOX, P. MICHAEL *SUNY College at Brockport*
GIBLIN, TARA *Stephens College*
GILLS, RICK *University of Wisconsin, La Crosse*
GREENE, CURTIS *Wayne State University*
GREGG, KATHERINE *West Virginia Wesleyan College*
HARLEY, JOHN *Eastern Kentucky University*
HARRIS, JAMES *Utah Valley Community College*
HELGESON, JEAN *Collin County Community College*
HESS, WILFORD M. *Brigham Young University*
HOUTMAN, ANNE *Cal State, Fullerton*
HUFFMAN, DAVID *Southwestern Texas University*
HUFFMAN, DONNA *Calhoun Community College*
INEICHER, GEORGIA *Hinds Community College*
JOHNSTON, TAYLOR *Michigan State University*
JUILLERAT, FLORENCE *Indiana University, Purdue University*
KENDRICK, BRYCE *University of Waterloo*
KETELES, KRISTEN *University of Central Arkansas*
KIRKPATRICK, LEE A. *Glendale Community College*
KREBS, CHARLES *University of British Columbia*
LANZA, JANET *University of Arkansas, Little Rock*
LEICHT, BRENDA *University of Iowa*
LOHMEIER, LYNNE *Mississippi Gulf Coast Community College*
LORING, DAVID *Johnson County Community College*
MACKLIN, MONICA *Northeastern State University*

MANN, ALAN *University of Pennsylvania*
MARTIN, KATHY *Central Connecticut State University*
MARTIN, TERRY *Kishwaukee College*
MASON, ROY B. *Mount San Jacinto College*
MATTHEWS, ROBERT *University of Georgia*
MAXWELL, JOYCE *California State University, Northridge*
McCLURE, JERRY *Miami University*
McNABB, ANN *Virginia Polytechnic Institute and State University*
MEIERS, SUSAN *Western Illinois University*
MEYER, DWIGHT H. *Queensborough Community College*
MICKLE, JAMES *North Carolina State University*
MILLER, G. TYLER *Wilmington, North Carolina*
MINOR, CHRISTINE V. *Clemson University*
MITCHELL, DENNIS M. *Troy University*
MONCAYO, ABELARDO C. *Ohio Northern University*
MOORE, IGNACIO *Virginia Tech*
MORRISON-SHETTLER, ALLISON *Georgia State University*
MORTON, DAVID *Frostburg State University*
NELSON, RILEY *Brigham Young University*
NICKLES, JON R. *University of Alaska, Anchorage*
NOLD, STEPHEN *University of Wisconsin- Stout*
PADGETT, DONALD *Bridgewater State College*
PENCOE, NANCY *State University of West Georgia*
PERRY, JAMES *University of Wisconsin, Center Fox Valley*
PITOCCHELLI, DR. JAY *Saint Anselm College*
PLETT, HAROLD *Fullerton College*
POLCYN, DAVID M. *California State University, San Bernardino*
PURCELL, JERRY *San Antonio College*
REID, BRUCE *Kean College of New Jersey*
RENFROE, MICHAEL *James Madison University*
REZNICK, DAVID *California State University, Fullerton*
RICKETT, JOHN *University of Arkansas, Little Rock*
ROHN, TROY *Boise State University*
ROIG, MATTIE *Broward Community College*
ROSE, GRIEG *West Valley College*
SANDIFORD, SHAMILI A. *College of Du Page*
SCHREIBER, FRED *California State University, Fresno*
SELLERS, LARRY *Louisiana Tech University*
SHAPIRO, HARRIET *San Diego State University*
SHONTZ, NANCY *Grand Valley State University*
SHOPPER, MARILYN *Johnson County Community College*
SIEMENS, DAVID *Black Hills State University*
SMITH, BRIAN *Black Hills State University*
SMITH, JERRY *St. Petersburg Junior College, Clearwater Campus*
STEINERT, KATHLEEN *Bellevue Community College*
SUMMERS, GERALD *University of Missouri*
SUNDBERG, MARSHALL D. *Emporia State University*
SVENSSON, PETER *West Valley College*
SWANSON, ROBERT *North Hennepin Community College*
SWEET, SAMUEL *University of California, Santa Barbara*
SZYMCZAK, LARRY J. *Chicago State University*
TAYLOR, JANE *Northern Virginia Community College*
TERHUNE, JERRY *Jefferson Community College, University of Kentucky*
TIZARD, IAN *Texas A&M University*
TRAYLER, BILL *California State University at Fresno*
TROUT, RICHARD E. *Oklahoma City Community College*
TURELL, MARSHA *Houston Community College*
TYSER, ROBIN *University of Wisconsin, LaCrosse*
VAJRAVELU, RANI *University of Central Florida*
VANDERGAST, AMY *San Diego State University*
VERHEY, STEVEN *Central Washington University*
VICKERS, TANYA *University of Utah*
VOGEL, THOMAS *Western Illinois University*
WARNER, MARGARET *Purdue University*
WEBB, JACQUELINE F. *Villanova University*
WELCH, NICOLE TURRILL *Middle Tennessee State University*
WELKIE, GEORGE W. *Utah State University*
WENDEROTH, MARY PAT *University of Washington*
WINICUR, SANDRA *Indiana University, South Bend*
WOLFE, LORNE *Georgia Southern University*
YONENAKA, SHANNA *San Francisco State University*
ZAYAITZ, ANNE *Kutztown University of Pennsylvania*

Introduction

Current configurations of the Earth's oceans and land masses—the geologic stage upon which life's drama continues to unfold. This composite satellite image reveals global energy use at night by the human population. Just as biological science does, it invites you to think more deeply about the world of life—and about our impact upon it.

1 INVITATION TO BIOLOGY

What Am I Doing Here?

Leaf through a newspaper on any given Sunday and you might get an uneasy feeling that the world is spinning out of control. There is a lot about the Middle East, where great civilizations have come and gone. You will not find much on the spectacular coral reefs of the surrounding seas, especially at the northern end of the Red Sea. Now the news is about oil and politics, bioterrorists and war.

Think back on the 1991 Persian Gulf conflict, when an Iraqi dictator ordered 460 million gallons of crude oil to be dumped into the Gulf and oil fields to be set on fire. Thick smoke blocked out sunlight, and black rain fell. Today, the human population in neighboring Kuwait shows a spike in cancer rates, which may have been caused by dangerous particles in the smoke. Similarly, New Yorkers who breathed in the dense, noxious dust that billowed through the air

after the 2001 World Trade Center attack are still reporting chronic health problems.

Besides terrorists, nature itself seems to have it in for us. Cholera, the flu, and SARS pose global threats. A long-term AIDS pandemic is unraveling the fabric of African societies. Monstrous storms, droughts, heat waves, and fires batter the land (Figure 1.1). Once-vast glaciers and the polar ice caps are melting fast, the atmosphere is warming up, and living conditions may change just about everywhere.

It is enough to make you throw down the paper and go sit in a park. It is enough to make you wish you lived in the good old days, when things were so much simpler.

Of course, read up on the good old days and you will find they weren't so good. Bioterrorists were around in 1346, when soldiers catapulted the corpses of bubonic plague

Figure 1.1 Biology is a way of thinking critically about life, one that helps us understand nature and our place in it. It starts with the premise that any aspect of nature—including this forest fire in Montana—has one or more underlying causes. To the right, an oil field burning out of control during the Persian Gulf War, an example of human impact on nature.

victims into a walled city under siege. Infected people and rats fled the city and helped fuel the Black Death, a plague that left 25 million dead all across Europe. Later, in 1918, the Spanish flu raced around the world and left between 30 million and 40 million people dead. Like today, many felt helpless in a world that seemed out of control.

What it boils down to is this: For at least a couple of million years, we humans and our immediate ancestors have been trying to make sense of the natural world. We observe it, we come up with ideas, and we test the ideas. However, the more pieces of the puzzle we fit together, the bigger the puzzle gets. We are now smart enough to know that it is almost overwhelmingly big.

You might choose to walk away from the challenge and simply let others tell you what to think. Or you might choose to develop your own understanding of the puzzle. Maybe you are interested in the pieces that affect your health, the food you eat, or your home or family. Maybe you simply find organisms and their environment fascinating. Regardless of the focus, the scientific study of life—*biology*—can deepen your perspective on the world.

Throughout this book, you will come across examples of how organisms are constructed, how they function, where they live, and what they do. The examples support concepts which, when taken together, convey what "life" is. This chapter is an overview of the basic concepts. It also sets the stage for forthcoming descriptions of scientific observations, experiments, and tests that help show how you can develop, modify, and refine your views of life.

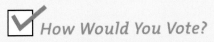

Watch the video online!

✓ How Would You Vote?

The warm seas of the Middle East support some of the world's most spectacular coral reef ecosystems. Should the United States provide funding to help preserve these reefs? See BiologyNow for details, then vote online.

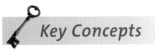

Key Concepts

LEVELS OF ORGANIZATION

The world of life has levels of organization that extend from atoms and molecules to the biosphere. The quality called "life" emerges at the level of cells. Section 1.1

LIFE'S UNDERLYING UNITY

The world of life shows unity, for all organisms are alike in key respects. They require inputs of energy and materials to function and maintain their complex organization. They work to keep their internal operating conditions within a tolerable range. They all sense and respond to conditions inside and outside themselves. They inherit DNA from parents, which gives them a capacity to survive and reproduce. Section 1.2

LIFE'S DIVERSITY

Millions of diverse kinds of organisms, or species, have appeared and disappeared over time. Each species is unique in some traits—that is, in some aspects of its body plan, functioning, and behavior. Section 1.3

EXPLAINING UNITY IN DIVERSITY

Theories of evolution, especially a major theory of evolution by way of natural selection, help explain the link between life's unity and diversity. The theories are a useful foundation for research in all fields of biological inquiry. Section 1.4

HOW WE KNOW

Biologists engage in systematic observations, hypotheses, predictions, and experimental tests in the outside world and in the laboratory. Well-designed tests can be repeated by others and yield the same results each time. Sections 1.5–1.7

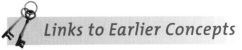

Links to Earlier Concepts

This book parallels nature's levels of organization, from atoms to the biosphere. Learning about the structure and function of atoms and molecules primes you to understand the structure of living cells. Learning about protein synthesis, active transport, and other processes that keep a single cell alive can help you understand how large organisms survive, because their trillions of living cells use the same processes. Knowing what it takes to survive can help you sense why and how organisms interact with one another and the environment.

At the start of each chapter, we will be reminding you of such structural and functional connections. Within chapters, you will come across keychain icons with cross-references that will link you to relevant sections in earlier chapters.

1.1 Levels of Organization in Nature

The world of life shows increasingly inclusive levels of organization. Take time to see how the levels connect to get a sense of how the topics of this book are organized and where they will take you.

MAKING SENSE OF THE WORLD

If we are interpreting the distant past correctly, the first humans lived in small bands that did not venture far from home. Safety, danger, and resources that did not extend beyond the immediate horizon comprised their world. Much later in time, human populations dispersed all around the globe. They soon had a lot more to observe, think about, and explain.

Today, even the far reaches of the known universe hold clues to our world. Scientists, clerics, farmers, astronauts, and anyone else who is of a mind to do so try to make sense of things. Interpretations differ, for no one person can be expert in everything learned so far or have foreknowledge of all that remains hidden. If you are reading this book, then you are starting to explore how a subset of scientists, the biologists, think about things, what they found out, and what they are up to now. Alternative ways of explaining the world are available in books for nonscience classes, such as those dealing with philosophy and religion. As you read this book, keep an open mind. Doing so can help you make more enlightened decisions about which explanations work for you.

A PATTERN IN BIOLOGICAL ORGANIZATION

What do we call the external world in its entirety? *Nature.* Biologists are interested in its forms of life, past and present. They have studied life all the way down to interacting atoms and all the way up to the impacts of organisms on a global scale. In doing so, they discovered a great pattern of organization.

That pattern starts at the level of atoms, which are the smallest units of nature's fundamental substances (Figure 1.2*a*). At the next level, atoms have combined into larger units called molecules (Figure 1.2*b*).

The pattern reaches the threshold of life as certain molecules are assembled as cells (Figure 1.2*c*). Complex

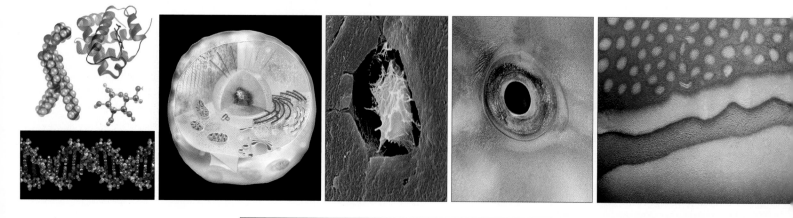

b molecule ⟶ **c** cell ⟶ **d** tissue ⟶ **e** organ ⟶ **f** organ system

molecule Two or more joined atoms of the same or different elements. "Molecules of life" are complex carbohydrates, lipids, proteins, DNA, and RNA. Only living cells now make them.

cell Smallest unit that can live and reproduce on its own or as part of a multicelled organism. It has an outer membrane, DNA, and other components.

tissue Organized array of cells and substances that are interacting in some task. Many cells (*white*) made this bone tissue from their own secretions.

organ Structural unit made of two or more tissues interacting in some task. A parrotfish eye is a sensory organ used in vision.

organ system Organs interacting physically, chemically, or both in some task. Parrotfish skin is an integumentary system with tissue layers, organs such as glands, and other parts.

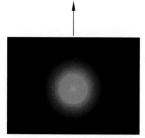

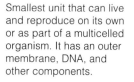

a atom

Elements are fundamental forms of matter. Atoms are the smallest units that retain an element's properties. Electrons, protons, and neutrons are its building blocks. This hydrogen atom's electron zips around a proton in a spherical volume of space.

Figure 1.2 *Animated!* Levels of organization in nature.

carbohydrates, complex fats and other lipids, proteins, DNA, and RNA—these are the "molecules of life." In nature, only living cells can make them. A **cell** is the smallest unit of life; it has a capacity to survive and reproduce on its own, given raw materials, an energy source, information encoded in its DNA, and suitable environmental conditions.

Each kind of organism, or **species**, consists of one or more cells. In multicelled species, cells form tissues, organs, and organ systems. Often there are trillions of specialized cells that interact directly or indirectly in the task of keeping the whole multicelled body alive. Figure 1.2*d–g* defines these levels of organization.

The population is the next level of organization. It is a group of the same species in some specified area. Fields of poppies in a valley or schools of fish in a lake are such groups (Figure 1.2*h*). The next level is the community. It includes all populations of all species in a specified area (Figure 1.2*i*). A community inside an underwater cave in the Red Sea, a forest in Argentina, or populations of tiny organisms that live, reproduce, and die quickly inside a flower are examples.

The next level of organization is the ecosystem, or a community interacting with its physical and chemical environment. The highest level, the biosphere, includes all regions of Earth's crust, waters, and atmosphere in which organisms live.

Bear in mind, life is more than the sum of its parts. At each successive level of organization, new properties emerge that are not inherent in any part by itself, as when living cells emerge from "lifeless" molecules. The interactions among parts generate **emergent properties**.

This book is a journey through the globe-spanning organization of life. Take a moment to study Figure 1.2. You can use it as a road map of where each part fits in the great scheme of nature.

Nature shows levels of organization, from the simple to the increasingly complex. Life's unique characteristics emerge as atoms and molecules interact and form cells. They extend from interactions among cells to populations, communities, ecosystems, and the biosphere.

g multicelled organism

Individual made of different types of cells. Cells of most multicelled organisms, including this Red Sea parrotfish, are organized as tissues, organs, and organ systems.

h population

Group of single-celled or multicelled individuals of the same species occupying a specified area. This is a fish population in the Red Sea.

i community

All populations of all species occupying a specified area. This is part of a coral reef in the Gulf of Aqaba at the northern end of the Red Sea.

j ecosystem

A community that is interacting with its physical environment. It has inputs and outputs of energy and materials. Reef ecosystems flourish in warm, clear seawater throughout the Middle East.

k the biosphere

All regions of Earth's waters, crust, and atmosphere that hold organisms. In the vast universe, Earth is a rare planet. Without its abundance of free-flowing water, there would be no life.

1.2 Overview of Life's Unity

"Life" is not easy to define. It is just too big, and it has been changing for 3.8 billion years! Even so, you can characterize it in terms of its unity and diversity. Here's the unity part: All living things require inputs of energy and materials; they sense and respond to change, as when they adjust conditions inside their body; and they reproduce with the help of DNA. But they differ in the details of their traits. That's the diversity part—variation in traits.

ENERGY AND LIFE'S ORGANIZATION

Cells, remember, are the smallest units that are alive. To stay alive, they get energy from the environment and convert it to forms that help them do work, such as constructing and organizing molecules into cell parts. **Energy** is the capacity for doing work. Whether a cell is free-living or a tiny bit of a multicelled organism, its organization would end without continuous inputs of energy. Higher levels of organization—populations, communities, and ecosystems—also would fall apart in the absence of energy inputs from the environment.

Producers make their food from simple materials in the environment. Plants and other photosynthetic types use sunlight energy to construct sugars from carbon dioxide and water molecules. They use the sugars as packets of energy and as building blocks for making complex carbohydrates, fats, and proteins. Animals and decomposers are **consumers**. They cannot make food; they eat producers and other organisms. Decomposers break down the remains of organisms to simpler raw materials, some of which become cycled back to the producers.

We are outlining a series of energy transfers from the environment, through producers, then on through consumers, and back to the environment. It is a one-way flow of energy, because all the energy that enters the world of life in a given interval eventually leaves it. Why? At each transfer step, a small amount escapes as an unorganized form of energy: heat. All organisms are participants in this continuous, directional flow of energy (Figure 1.3).

ORGANISMS SENSE AND RESPOND TO CHANGE

All organisms are alike in another way. They sense changes in their surroundings and make controlled, compensatory responses to them. They do so with the assistance of **receptors**. Receptors are molecules and structures that detect stimuli, which are specific kinds of energy. Different receptors can respond to different stimuli. A stimulus may be sunlight energy, chemical energy, or the mechanical energy of a bite (Figure 1.4).

Activated receptors trigger changes in the activities of organisms. As a simple example, after you finish eating an apple, sugars leave your small intestine and enter blood. Think of blood and the fluid around cells as the body's *internal* environment. The composition and volume of that fluid must be kept within a range that your cells can tolerate. Too much or too little sugar changes the composition of blood, as happens with diabetes and other medical problems. Normally, when there is too much sugar, your pancreas secretes more insulin. Most living cells in your body have receptors for this hormone. It stimulates cells to take up sugar.

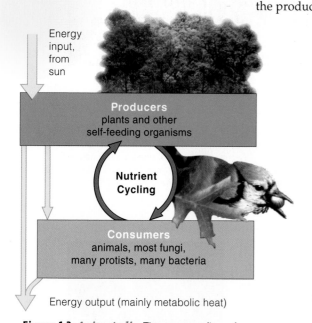

Energy input, from sun

Producers
plants and other self-feeding organisms

Nutrient Cycling

Consumers
animals, most fungi, many protists, many bacteria

Energy output (mainly metabolic heat)

Figure 1.3 *Animated!* The one-way flow of energy and cycling of materials in the world of life.

Figure 1.4 A roaring response to signals from pain receptors, activated by a lion cub flirting with disaster.

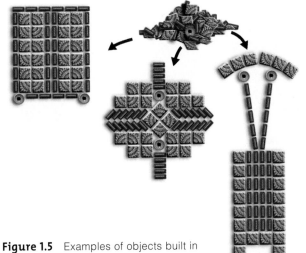

Figure 1.5 Examples of objects built in different ways from the same materials.

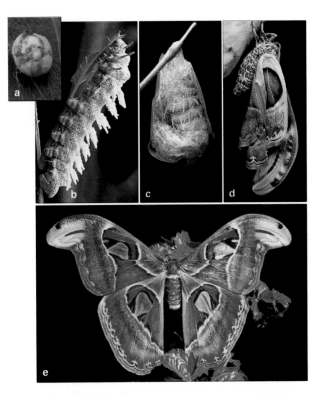

Figure 1.6 "The insect"—actually a series of stages of development guided largely by instructions in DNA. Here, a silkworm moth, from a fertilized egg (**a**), to a larval stage called a caterpillar (**b**), to a pupal stage (**c**), to the winged form of the adult (**d,e**).

When enough cells are doing so, the blood sugar level returns to the normal range.

As conditions in the internal environment change in potentially harmful ways, receptor-driven mechanisms kick in and return conditions to the state that cells can tolerate. **Homeostasis** is the name for this state, and it is a defining feature of life.

ORGANISMS GROW AND REPRODUCE

Organisms grow and reproduce based on information in **DNA**, a nucleic acid. DNA is *the* signature molecule of life. No chunk of granite or quartz has it.

DNA holds the information about building proteins from a few kinds of amino acids. Each protein has a particular amino acid sequence, which is the start of its particular shapes and properties.

By analogy, if you access suitable instructions and apply energy to the task, you might organize a pile of a few kinds of ceramic tiles into diverse patterns. Figure 1.5 has examples.

Protein-building information is vital for cell growth and reproduction. Proteins have many structural and functional roles. Important kinds function as enzymes, the cell's main worker molecules. With enzymes, cells build, split, and rearrange molecules exceedingly fast. Without enzymes, there could be no more complex carbohydrates, complex lipids, proteins, and nucleic acids. There could be no cells, no life.

In nature, an organism inherits DNA—the basis of its traits—from its parents. *Inheritance* is the acquisition of traits after parents transmit their DNA to offspring. Think about it. Why do baby storks look like storks and not like pelicans? Because they inherited stork DNA, which is different from pelican DNA.

Reproduction refers to actual mechanisms by which parents transmit DNA to offspring. For trees, humans, and other large organisms, the information in DNA is used in ways that guide growth and *development*—the transformation of the first cell of the new individual through orderly stages. The outcome is a multicelled adult, typically with tissues and organs (Figure 1.6).

Life's levels of organization start with a one-way flow of energy from the environment, through producers and consumers, then back to the environment.

Organisms interact through this one-way flow of energy and through a cycling of raw materials.

Organisms maintain their organization by sensing and responding to change. Many responses return conditions in the body's internal environment to a range that cells can tolerate, a state called homeostasis.

Organisms grow and reproduce based on information encoded in DNA, which they inherit from their parents.

1.3   If So Much Unity, Why So Many Species?

Although unity pervades the world of life, so does diversity. Organisms differ enormously in body form, the functions of their body parts, and behavior.

Superimposed on life's unity is tremendous diversity. How many species are with us today? Estimates range as high as 100 million. And 99.9 percent of all species that ever lived are extinct. So far, we have named approximately 1.8 million species.

For centuries, many scholars have been organizing information about life's diversity. Carolus Linnaeus, a naturalist, came up with the strategy of giving each species a two-part name. The first part designates the **genus** (plural, genera). A genus is a grouping of one or more species characterized by certain traits, at least one of which is unique to them. The second part of the name designates a particular species within the genus that has at least one trait no other species has.

Scarus gibbus is the formal name for the humphead parrotfish shown in Figure 1.2*g*. A different species in its genus is named *S. coelestinus* (midnight parrotfish). This example also shows that you can abbreviate the genus designation after you spell it out the first time.

Later, ever more inclusive groupings were devised, such as phylum (plural, phyla), order, kingdom, and domain. These are rankings of classification systems, which simply are ways to organize knowledge about relationships among species. Observable traits are still markers, but so is a richly expanding base of molecular evidence of descent from a shared ancestor.

Most biologists now favor a classification system having three domains: Bacteria, Archaea, and Eukarya (Figure 1.7 and Table 1.1). Protists, plants, fungi, and animals make up domain Eukarya (Figure 1.8).

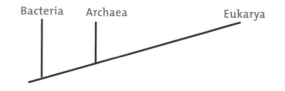

Figure 1.7   Three domains of life.

— Bacteria —

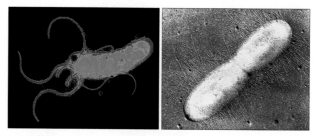

The most common prokaryotes; collectively, these single cells are the most metabolically diverse species on Earth.

— Archaea —

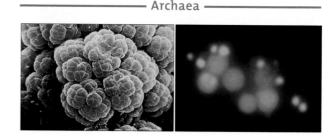

These prokaryotes are evolutionarily closer to eukaryotes than to bacteria. *Left*, a colony of methane-producing cells. *Right*, two species from a hydrothermal vent on the seafloor.

Figure 1.8   A few representatives of life's diversity.

Bacteria (singular, bacterium) and **archaeans** are *prokaryotic* cells. These single-celled organisms do not have a nucleus, which is a membrane-bound sac that, in all other species, encloses DNA. Of all organisms, they show the greatest metabolic diversity. Different species are producers and consumers in near-boiling water, frozen desert rocks, sulfur-clogged lakes, and other exceptionally harsh environments. Experimental evidence suggests that the first cells on Earth faced similarly hostile challenges to survival.

Structurally, the **protists** are the simplest organisms that are *eukaryotic*, which means their cells contain a nucleus. Different kinds are producers or consumers. Many are single cells, larger and more complex than prokaryotes. Some are tree-size, multicelled seaweeds. Actually, the protists are so diverse that they are being reclassified into a number of separate major lineages.

Cells of species we call fungi, plants, and animals are eukaryotic. Most **fungi** are multicelled, and not all of them form mushrooms (the reproductive structures for species that grow mostly underground). Many are decomposers. They secrete enzymes that digest food outside their body, then individual cells absorb the bits. Nearly all **plants** are multicelled and photosynthetic.

Table 1.1	Comparison of Life's Three Domains
Bacteria	Single cells, prokaryotic (no nucleus). Most ancient lineage.
Archaea	Single cells, prokaryotic. Evolutionarily closer to eukaryotes.
Eukarya	Eukaryotic cells (with a nucleus). Single-celled and multicelled species categorized as protists, plants, fungi, and animals.

Protists Single-celled and multicelled eukaryotic species that range from the microscopic to giant seaweeds. Many biologists are now viewing the "protists" as many major lineages.

Plants Generally, photosynthetic, multicelled eukaryotes, many with roots, stems, and leaves. Plants are the primary producers for ecosystems on land. Redwoods and flowering plants are examples.

Fungi Single-celled and multicelled eukaryotes; different kinds are decomposers, parasites, or pathogens. Without decomposers, communities would become buried in their own wastes.

Animals Multicelled eukaryotes that ingest tissues or juices of other organisms. Like this basilisk lizard, most actively move about during at least part of their life.

They make all of their own food by using sunlight as an energy source, and atoms of carbon dioxide and water as building blocks.

All **animals** are multicelled consumers that ingest tissues or juices of other organisms. Herbivores are grazers, carnivores eat meat, scavengers eat almost anything edible, and parasites pilfer nutrients from a host's tissues. All animals grow and develop through a series of stages. Most of them actively move about during at least part of their lives.

Pulling this information together, are you getting a sense of what it means when someone says that life shows unity *and* diversity?

On the basis of observable traits and molecular evidence of shared ancestry, we rank species in ever more inclusive groupings. The largest groupings are domains: archaea, bacteria, and eukarya (protists, fungi, plants, and animals).

1.4 An Evolutionary View of Diversity

How can organisms be so much alike and still show tremendous diversity? A theory of evolution by way of natural selection is one explanation.

Individuals of a species share certain traits, which are aspects of their physical form, function, and behavior. Rarely are individuals exactly alike; they differ in the details. Except for identical twins, for instance, the 6.4 billion individuals of our species (*Homo sapiens*) vary in height, hair color, and other traits.

Variation in most traits arises through **mutations**, or changes in DNA, which offspring inherit from their parents. Most mutations have neutral or bad effects. Some cause a trait to change in a way that makes one individual of a population better adapted than others to prevailing conditions. That is, its bearer might have an easier time securing food, a mate, and so on—so it has a better chance of reproducing and passing on the mutation to offspring. What is the outcome? Consider how a naturalist, Charles Darwin, expressed it:

First, a natural population tends to increase in size until its individuals compete more and more for food, shelter, and other dwindling environmental resources.

Second, those individuals differ from one another in the details of their shared, heritable traits.

Third, bearers of adaptive forms of traits are more likely to survive and reproduce, so those forms tend to become more common over successive generations. This outcome is called **natural selection**.

Consider how pigeons vary in feather color, size, and other traits (Figure 1.9). Say that pigeon breeders prefer black, curly-tipped feathers. They select captive birds having the darkest, curliest-tipped feathers and

WILD ROCK DOVE

let only those birds mate. In time, no birds in their captive populations have light, uncurly feathers. By culling through many traits, breeders have developed well over 300 varieties of domesticated pigeons.

Pigeon breeding is a case of *artificial* selection. One form of a trait is favored over others in an artificial environment under contrived, manipulated conditions. Darwin saw that breeding practices could be an easily understood model for *natural* selection, a favoring of some forms of a given trait over others in nature.

Just as breeders are "selective agents" that promote reproduction of certain pigeons, different agents act on the range of variation in the wild. Among them are pigeon-eating peregrine falcons, as in Figure 1.9. The swifter or better camouflaged pigeons are more likely to avoid falcons and live long enough to reproduce, compared with not-so-swift or too-flashy pigeons.

When different forms of a trait are becoming more or less common over successive generations, evolution is under way. In biology, **evolution** simply means that heritable change is occurring in a line of descent. Later chapters get into actual mechanisms of evolution. For now, it is enough to remember these preview points.

> *Body form, function, and behavior are mostly heritable traits. Different forms of a trait arise through DNA mutations. One may be more adaptive than others to prevailing conditions.*
>
> *Natural selection is an outcome of differences in survival and reproduction among individuals of a population that vary in one or more heritable traits. Evolution, or change in lines of descent, gives rise to life's diversity.*

Figure 1.9 Outcome of artificial selection: just a few of the hundreds of varieties of domesticated pigeons, all descended from captive populations of wild rock doves. At right, peregrine falcons are agents of natural selection in the wild.

1.5   The Nature of Biological Inquiry

The preceding sections introduced some big concepts. Consider approaching this or any other collection of "facts" with a critical attitude. "Why should I accept that they have merit?" The answer requires a look at how biologists make inferences about observations, then test their inferences against actual experience.

OBSERVATIONS, HYPOTHESES, AND TESTS

To get a sense of "how to do science," you might start with practices that are common in scientific research:

1. Observe some aspect of nature and research what others have found out about it, then frame a question or identify a problem related to your observation.

2. Develop a **hypothesis**: a testable explanation of the observed phenomenon or process.

3. Using the hypothesis as a guide, make a **prediction** —a statement of what you should find in nature if you were to go looking for it. This is often called the "if–then" process. *If* gravity does not pull objects toward Earth, *then* it should be possible to observe an apple falling up, not down, from a tree.

4. Devise ways to **test** the accuracy of predictions, as by making systematic observations, building models, and conducting experiments. **Models** are theoretical, detailed descriptions or analogies that might help us visualize an object or event that has not been, or cannot be, directly observed.

5. If your tests do not confirm the prediction, check to see what might have gone wrong. It may be that you overlooked a factor that had impact on the results. Or maybe the hypothesis is not a good one.

6. Repeat the tests or devise new ones—the more the better, because hypotheses that withstand many tests have a higher probability of being useful.

7. Objectively analyze and report the test results, as well as the conclusions you drew from them.

You might hear someone refer to these practices as "the scientific method," as if all scientists march to the drumbeat of an absolute, fixed procedure. They do not. Many observe and describe some aspect of nature and leave the hypothesizing to others. A few are lucky; they stumble onto information that they are not even looking for. Of course, it isn't always a matter of luck. Chance seems to favor a mind that has already been prepared, by education and experience, to recognize what the information might mean.

However, scientists do have something in common. It is a critical attitude about testing ideas in rigorous ways that are designed to disprove them.

Careful observations are a logical way to test the predictions that flow from a hypothesis. So are **experiments**, or tests carried out under controlled conditions that researchers manipulate. Such tests are carried out in nature and in laboratories, and they remove irrelevant factors that might skew the results. You will find two examples in the section to follow.

ABOUT THE WORD "THEORY"

Suppose a hypothesis has not been disproved after years of rigorous tests. Scientists use it to interpret more data or observations, which often involve more hypotheses. When a hypothesis meets these criteria, it may become accepted as a **scientific theory**.

You may hear people apply the word "theory" to a speculative idea, as in the phrase, "It's just a theory." However, a scientific theory differs from speculation in a big way. *After testing a scientific theory's predictive power many times and in many ways in the natural world, researchers have yet to find evidence that disproves it.* That is why the theory of evolution by natural selection is respected. It has been used successfully to explain a diverse number of questions about the natural world, such as how life diversified, how river dams can alter ecosystems, and why antibiotics can stop working.

Perhaps a well-tested theory is as close to the truth as we can get. For instance, after more than a century of many thousands of tests, Darwin's theory holds, with only minor modification. Yet we cannot prove it holds under all possible conditions, because doing so would take an infinite number of tests. We *can* say that a theory has a high probability of not being wrong. Even then, biologists keep on looking for information and devising tests that may disprove its premises. This willingness to modify even an entrenched theory is a strength of science, not a weakness.

Scientific inquiry into nature involves asking questions, formulating hypotheses, making predictions, testing predictions, and objectively reporting the results.

A scientific theory is a time-tested intellectual framework that is used to interpret a broad range of observations and data. Scientific theories remain open to rigorous tests, revision, and tentative acceptance or rejection.

1.6 The Power of Experimental Tests

Experiments are tests that can simplify observation in nature, because conditions under which observations are made can be controlled. Well-designed experiments test predictions about what you will find in nature when a hypothesis is correct—or won't find if it is wrong.

AN ASSUMPTION OF CAUSE AND EFFECT

A scientific experiment starts with this premise: *Any aspect of nature has an underlying material cause that can be tested through controlled experiments.* This premise sets science apart from faith in the supernatural ("beyond nature"). It means a hypothesis must be testable in the natural world in ways that might well disprove it.

Most aspects of nature are outcomes of interacting variables. A **variable** is a feature of an object or event that may differ over time or among the representatives of that object or event.

Researchers design experiments to test one variable at a time. They establish a **control group**, a standard used for comparison against one or more **experimental groups**. One type of control group is *identical* with an experimental group, but the test conditions are not. Those conditions differ by one variable. Another type of control group *differs* from an experimental group in one variable, but the test conditions are identical for both groups.

EXAMPLE OF AN EXPERIMENTAL DESIGN

In 1996 the FDA approved Olestra®, a type of synthetic fat replacement made from sugar and vegetable oil, as a food additive. Potato chips were the first Olestra-laced food product on the market in the United States. Controversy raged. After eating the chips, some people complained of bad cramps. Two years later, researchers at Johns Hopkins University designed an experiment. If Olestra causes gastrointestinal cramps, then people who eat food that contains Olestra are more likely to get cramps than people who do not.

To test this prediction, they used a Chicago theater as the "laboratory." They asked more than 1,100 people between ages thirteen and thirty-eight to watch a movie and eat their fill of potato chips. Each person got an unmarked bag that held thirteen ounces of chips. The individuals who received a bag of Olestra-laced potato chips were the experimental group. Individuals who got a bag of regular chips were the control group.

Afterward, researchers contacted the subjects and tabulated reports of gastrointestinal cramps. Of 563 people in the experimental group, 89 (or 15.8 percent) complained of problems. But so did 93 of 529 people

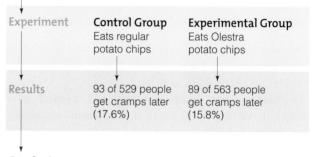

Hypothesis
Olestra® causes intestinal cramps.

Prediction
People who eat potato chips made with Olestra will be more likely to get intestinal cramps than those who eat potato chips made without Olestra.

Experiment	Control Group Eats regular potato chips	Experimental Group Eats Olestra potato chips
Results	93 of 529 people get cramps later (17.6%)	89 of 563 people get cramps later (15.8%)

Conclusion
Percentages are about equal. People who eat potato chips made with Olestra are just as likely to get intestinal cramps as those who eat potato chips made without Olestra. These results do not support the hypothesis.

Figure 1.10 *Animated!* Example of a typical sequence of steps taken in a scientific experiment.

(17.6 percent) in the control group, who had munched on regular chips! The experiment yielded no evidence that eating Olestra-laced potato chips, at least in one sitting, causes gastrointestinal problems (Figure 1.10).

EXAMPLE OF A FIELD EXPERIMENT

Many toxic or unpalatable species are vividly colored and often have distinct patterning. Predators learn to avoid individuals that display these visual cues after eating a few of them and getting sick.

In some cases, two bad-tasting species of butterflies resemble each other. **Mimicry** is a case of looking like something else and confusing predators (or prey). The naturalist Fritz Müller wondered why such a visual similarity persists. As he hypothesized, it may benefit both species. He knew that young birds catch and eat butterflies before learning to avoid eating bad-tasting ones. If both of the butterfly species are sampled, then each would lose fewer individuals to predatory birds.

Durrell Kapan, an evolutionary biologist, tested the hypothesis with *Heliconius cydno* in a rain forest. There are two forms of these bad-tasting butterflies. One has yellow markings on its wings, and the other does not (Figure 1.11a). Moreover, the yellow-marked butterfly resembles *H. eleuchia*, another bad-tasting species in a different part of the forest. The wings of *H. eleuchia* also have yellow markings (Figure 1.11b).

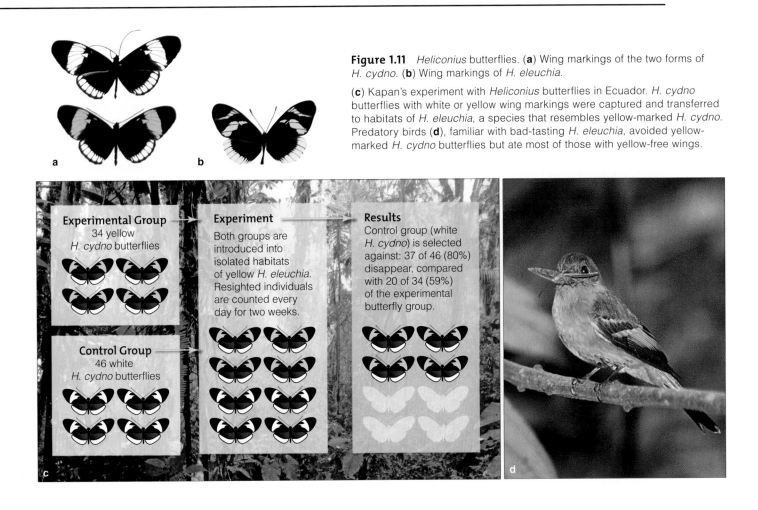

Figure 1.11 *Heliconius* butterflies. (**a**) Wing markings of the two forms of *H. cydno*. (**b**) Wing markings of *H. eleuchia*.

(**c**) Kapan's experiment with *Heliconius* butterflies in Ecuador. *H. cydno* butterflies with white or yellow wing markings were captured and transferred to habitats of *H. eleuchia*, a species that resembles yellow-marked *H. cydno*. Predatory birds (**d**), familiar with bad-tasting *H. eleuchia*, avoided yellow-marked *H. cydno* butterflies but ate most of those with yellow-free wings.

Experimental Group
34 yellow
H. cydno butterflies

Control Group
46 white
H. cydno butterflies

Experiment
Both groups are introduced into isolated habitats of yellow *H. eleuchia*. Resighted individuals are counted every day for two weeks.

Results
Control group (white *H. cydno*) is selected against: 37 of 46 (80%) disappear, compared with 20 of 34 (59%) of the experimental butterfly group.

Kapan predicted: If birds have learned to avoid *H. eleuchia* in their part of the forest, then they also will avoid imports of yellow-marked *H. cydno* butterflies. For the control group, Kapan used captured *H. cydno* butterflies with yellow-free wings. *H. cydno* butterflies with yellow-marked wings made up the experimental group. He released both of the groups into isolated *H. eleuchia* habitats. For two weeks, the approximate life span of the butterflies, he counted the survivors.

Kapan found that butterflies of the experimental group were less likely to survive in the new forest habitat. Because they did not display the visual cue recognized by the local birds—yellow markings—they were more likely to be eaten. The control group did better, as the test results in Figure 1.11c indicate. We can expect that predatory birds recognized this batch of imports as bad tasting and avoided them. Besides confirming Kapan's prediction, these test results also provide evidence of natural selection in action.

BIAS IN REPORTING RESULTS

Experimenters run a risk of interpreting data in terms of what they wish to prove or dismiss. That is why they prefer *quantitative* reports, with actual counts or some other precise measurements. With such reports, researchers get a chance to repeat an experimental test and check conclusions. Appendix IX gives an example.

This last point gets us back to the value of thinking critically. Scientists must keep asking themselves: *Will my observations or experiments show that a hypothesis is false?* They are expected to put aside pride or bias by testing ideas in ways that might prove them wrong. Even if someone won't do so, others will, for science works as a community that is both cooperative and competitive. Ideally, individuals will share their ideas, knowing that it is as useful to expose errors as it is to applaud insights. They can and often do change their mind when shown contradictory evidence.

> *Experiments simplify observations in nature by restricting a researcher's focus to one variable at a time. A variable is any feature of an object or event that may differ over time or among representatives of the object or event.*
>
> *Tests are based on the premise that any aspect of nature has one or more underlying causes. Scientific hypotheses can be tested in ways that might disprove them.*

1.7 The Limits of Science

Beyond the realm of scientific inquiry, some events are unexplained. Why do we exist, for what purpose? Why do we have to die at a particular moment? Such questions lead to *subjective* answers, which come from within, as an outcome of all the personal experiences and mental connections that shape our consciousness. People differ enormously in this regard, which is why subjective answers do not readily lend themselves to scientific analysis and experiments.

This is not to say subjective answers are without value. No human society can function for long unless its individuals share a commitment to standards for making judgments, even if they are subjective. Moral, aesthetic, philosophical, and economic standards vary from one society to the next. But they all guide people in deciding what is important and good, and what is not. All attempt to give meaning to what we do.

Every so often, scientists stir up controversy when they explain something that was thought to be beyond natural explanation, or belonging to the supernatural. This is often the case when a society's moral codes are interwoven with religious interpretations of the past. Exploring a long-standing view of the natural world from a scientific perspective might be misinterpreted as questioning morality, even though the two are not the same thing.

As one example, centuries ago in Europe, Nikolaus Copernicus studied the planets and decided that Earth circles the sun. Today this seems obvious. Back then, it was heresy. The prevailing belief was that the Creator made Earth—and, by extension, humans—as the fixed center of the universe. Galileo Galilei, another scholar, thought the Copernican model of the solar system was a good one and said so. He was forced to retract his statement, on his knees, and put Earth back as the fixed center of things. Word has it that he muttered, "Even so, it *does* move." Later on, Darwin's theory of evolution also ran up against prevailing beliefs.

Today, as then, society has sets of standards. Those standards might be questioned when a new, natural explanation runs counter to supernatural beliefs. This does not mean that scientists who raise questions are less moral, less lawful, less sensitive, or less caring than anyone else. It means a specific standard guides their work: *Explanations about nature must be testable in the external world, in ways that others can repeat.*

Science does not address subjective questions. The external world, not internal conviction, is the testing ground for the theories generated in science.

Summary

Section 1.1 Nature has increasingly inclusive levels of organization, with life emerging at the cellular level. All organisms consist of one or more cells. In most multicelled species, the cells are organized as tissues, organs, and organ systems. Individuals of the same species in a specified area form a population, and all populations in the same area form a community. An ecosystem is a community and its environment. The biosphere includes all regions of Earth's atmosphere, waters, and land that hold systems of life.

Distinctive properties emerge at each successive level of organization. It takes interactions among the parts to generate these emergent properties of life.

Biology⌇Now
Explore levels of biological organization with the interaction on BiologyNow.

Section 1.2 Life shows unity. All organisms require energy and raw materials from the environment to grow, maintain their organization, and reproduce, based on information encoded in DNA. All sense and respond to change. They work to counter changes in their internal environment so that conditions remain tolerable for cell activities, a state called homeostasis (Table 1.2).

Biology⌇Now
Using instructions with the animation on BiologyNow, view how different objects are assembled from the same materials. Also view energy flow and materials cycling.

Section 1.3 As in the past, species now show great diversity. Each species has unique aspects of body form, function, and behavior. Species are ranked in ever more inclusive groupings in classification systems, starting with a two-part name (genus and species name). One classification system assigns all species to three domains: Bacteria, Archaea, and Eukarya. Protists, plants, fungi, and animals make up Domain Eukarya.

Biology⌇Now
Explore the characteristics of the three domains of life with the interaction on BiologyNow.

Section 1.4 Life's diversity arises through mutation: change in the structure of DNA molecules. Mutations are the basis for variation in heritable traits, which are the traits that parents bestow on offspring. Such traits include most details of body form and function.

Individuals of a population differ in the details of their shared heritable traits. Variant forms of traits may affect the ability to survive and reproduce. The adaptive forms give their bearers a competitive edge, so they tend to become more common among successive generations; less adaptive traits become less common or are lost. Thus the traits that help define the population (and species) may change over successive generations; that is, the population may evolve. The outcome of differences in reproduction among individuals that differ in one or more heritable traits is called natural selection.

Biology ⑤ Now

Learn more about natural selection and evolution with InfoTrac readings on BiologyNow.

Read the InfoTrac article "Will We Keep Evolving?" Ian Tattersall, Time, April 2000.

Section 1.5 Scientific methods differ, but they all are based on the premise that any aspect of nature has one or more underlying causes. Researchers observe some object or event, form hypotheses (testable explanations about it), make predictions about what they can expect to find if the hypothesis is not wrong, and then test the predictions. Their tests may involve making more observations, building models, or doing experiments.

Scientists analyze and share test results. A hypothesis that does not hold up under repeated testing is modified or discarded. A scientific theory is a set of hypotheses that can be used to explain a broad range of observations and data. Many diverse tests have supported it.

Sections 1.6, 1.7 Supernatural explanations cannot be tested. Science deals only with aspects of nature that lend themselves to systematic observation, hypotheses, predictions, and experimental tests. Most aspects are outcomes of many interacting variables that differ among individuals and over time. A scientific experiment can simplify observations in nature and in the laboratory because the variables can be precisely manipulated and controlled. A scientist changes one variable at a time and observes what happens. A typical experiment is designed so that one or more experimental groups can be compared with a control group.

Self-Quiz

Answers in Appendix II

1. The smallest unit of life is the _____ .

2. _____ is required to maintain levels of biological organization, from cells to populations, communities, and even entire ecosystems.

3. _____ is a state in which the internal environment is being maintained within a tolerable range.

4. Researchers assign all species to one of three _____ .

5. DNA _____ .
   a. contains instructions for building proteins
   b. undergoes mutation
   c. is transmitted from parents to offspring
   d. all of the above

6. _____ is the acquisition of traits from parents who transmit their DNA to offspring.
   a. Reproduction   c. Homeostasis
   b. Development   d. Inheritance

7. Differences in heritable traits arise through _____ .

8. A trait is _____ if it improves an organism's ability to survive and reproduce in the prevailing environment.

9. A control group is _____ .
   a. the standard against which experimental groups can be compared
   b. the experiment that gives conclusive results

Table 1.2 Summary of Life's Characteristics

Shared characteristics that reflect life's unity

1. Life emerges at the level of cells. All organisms consist of one or more cells.

2. In nature, only organisms make complex carbohydrates and lipids, proteins, and nucleic acids. They all use these molecules of life as building blocks and energy sources, and for the preservation of heritable information.

3. Organisms require ongoing inputs of energy to maintain their complex organization. All obtain energy from the environment and convert it to forms that can be used for growth, survival, and reproduction.

4. Organisms sense and make controlled responses to conditions in their external and internal environments.

5. Organisms grow and reproduce based on heritable information in DNA.

6. The traits that define a population of organisms can change over the generations; the population can evolve.

Foundations for life's diversity

1. Mutations (heritable changes in the structure of DNA) give rise to variation in heritable traits, which are most details of body form, function, and behavior.

2. Diversity is the sum total of variations that accumulated in different lines of descent over the past 3.8 billion years, as by natural selection and other processes of evolution.

---

10. Match the terms with the most suitable description.

____ emergent properties
____ natural selection
____ scientific theory
____ hypothesis
____ prediction

a. statement of what you expect to find in nature based on hypotheses
b. testable explanation about what causes an event or aspect of nature
c. requires interaction of parts that make up a new level of organization
d. a time-tested, related set of hypotheses that explains a broad range of observations and data
e. outcome of differences in survival and reproduction among individuals of a population that differ in the details of one or more traits

Additional questions are available on **Biology ⑤ Now**™

Critical Thinking

1. Assess your bedroom. Is the bed made? Are the sheets clean? Are socks and underwear folded and put away? Are clothes strewn all over the floor? Now explain what the bedroom has in common with a living cell.

2. It is often said that only living things respond to the environment. Yet even a rock shows responsiveness, as when it yields to gravity's force and tumbles down a hill or changes its shape slowly under the repeated batterings of wind, rain, or tides. So how do living things differ from rocks in their responsiveness?

3. Witnesses in a court of law are asked to "swear to tell the truth, the whole truth, and nothing but the truth." What are some of the problems inherent in the question? Can you think of a better alternative?

a Natalie, blindfolded, randomly plucks a jelly bean from a jar of 120 green and 280 black jelly beans; a ratio of 30 to 70 percent.

b The jar is hidden before she removes her blindfold. She observes a single green jelly bean in her hand and assumes the jar holds only green jelly beans.

c Still blindfolded, Natalie randomly picks 50 jelly beans from the jar and ends up with 10 green and 40 black ones.

d The larger sample leads her to assume one-fifth of the jar's jelly beans are green and four-fifths are black (a ratio of 20 to 80). Her larger sample more closely approximates the jar's green-to-black ratio. The more times Natalie repeats the sampling, the greater the chance she will come close to knowing the actual ratio.

Figure 1.12 A simple demonstration of sampling error.

4. The Olestra potato chip experiment in Section 1.6 was a *double-blind* study: Neither the subjects of the experiment nor the researchers who made the follow-up phone calls knew which potato chips were in which bag. What are some of the challenges a researcher must consider when performing a double-blind study?

5. Suppose an outcome of some event has been observed to happen with great regularity. Can we predict that the same thing will always happen again? Not really, because there is no way for us to account for all of the possible variables that might affect the outcome. To illustrate this point, Garvin McCain and Erwin Segal offer a parable:

Once there was a highly intelligent turkey. The turkey lived in a pen, attended by a kind, thoughtful master. It had nothing to do but reflect on the world's wonders and regularities. It observed some major regularities.

Morning always started out with the sky turning light, followed by the clop, clop, clop of the master's footsteps, which was always followed by the appearance of delicious food. Other things varied—sometimes the morning was warm and sometimes cold—but food always followed footsteps. The sequence of events was so predictable that it eventually became the basis of the turkey's theory about the goodness of the world.

One morning, after more than 100 confirmations of the goodness theory, the turkey listened for the clop, clop, clop, heard it, and had its head chopped off.

Scientists understand that all well-tested theories about nature have a high probability of not being wrong. They realize, however, that any theory is subject to modification if and when contradictory information becomes available. The absence of absolute certainty has led some people to conclude that "facts are irrelevant—facts change." If that is so, should we just stop doing scientific research? Why or why not?

6. Many magazines are loaded with articles on exercise, diet, and many other health-related topics. Some authors recommend a specific diet or dietary supplement. What kinds of evidence should the articles include so that you can decide whether to accept the recommendations?

7. Rarely can experimenters observe all individuals of a group. They select subsets or samples of populations, events, and other aspects of nature. However, they must try to avoid bias, which means risking a test by using subsets that are not really representative of the whole. *Sampling error* can occur when estimates are based on a limited sample rather than the whole population (Figure 1.12). Test results are less likely to be distorted when a sampling is large and the test is repeated. Explain how sampling error could have affected results of the potato chip experiment described in Section 1.6 if the experimenters had not been careful.

8. In 1988 Dr. Randolph Byrd and his colleagues started a study of 393 patients admitted to the San Francisco General Hospital Coronary Care Unit. In the experiment, born-again Christian volunteers were asked to pray daily for a patient's rapid recovery and for prevention of complications and death.

None of the patients knew if he or she was being prayed for. None of the volunteers or patients knew each other. Byrd categorized how each patient fared in the hospital as "good," "intermediate," or "bad." He determined that patients who had been prayed for fared a little better than those who had not. His was the first experiment that had documented statistically significant results that seemed to support the prediction that prayer might have beneficial effects for seriously ill patients.

His published results engendered a storm of criticism, mostly from scientists who cited bias in the experimental design. For instance, Byrd had categorized the patients after the experiment was over. Think about how bias might play a role in interpreting medical data. Why do you suppose the experiment generated a heated response from many in the scientific community?

I | Principles of Cellular Life

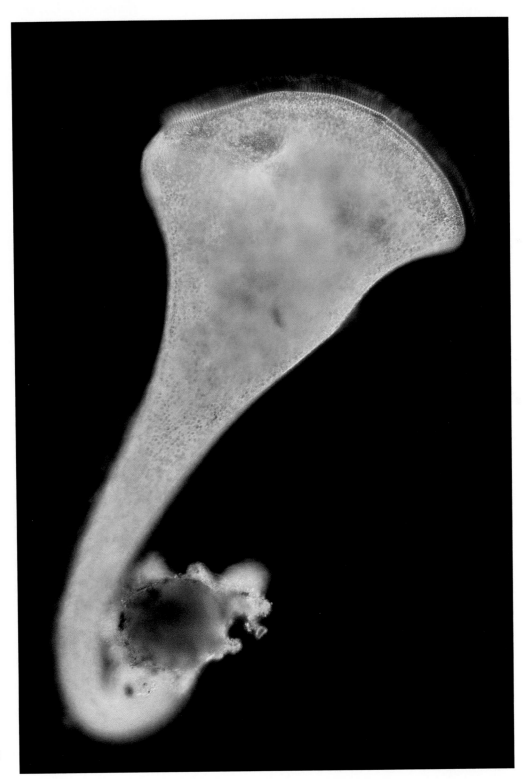

Staying alive means securing energy and raw materials from the environment. Shown here, a living cell of the genus Stentor. *This protist has hairlike projections around the opening to a cavity in its body, which is about two millimeters long. Its "hairs" of fused-together cilia beat the surrounding water. They create a current that wafts food into the cavity.*

2 LIFE'S CHEMICAL BASIS

What Are You Worth?

Hollywood thinks Leonardo DiCaprio is worth $20 million per movie, the Yankees think shortstop Alex Rodriguez is worth $217 million per decade, and the United States thinks the average teacher is worth $44,367 per year. Chemically, though, how much is the human body really worth (Figure 2.1a)?

Each of us is a collection of **elements**, or fundamental substances that each consist of only one kind of atom. An **atom** is the smallest unit of an element that still retains the element's properties. It occupies space, has mass, and cannot be broken down into something else, at least by everyday means.

Oxygen, carbon, hydrogen, nitrogen, and calcium are the main elements in organisms. Next are phosphorus, potassium, sulfur, sodium, and chlorine. There are a lot of *trace* elements, each making up less than 0.01 percent of the body's weight. Selenium and lead are examples.

Wait a minute! Selenium, lead, mercury, arsenic, and many other elements are toxic, right? So how can they be part of the collection? We're finding that trace amounts

Mass of Elements in a 70-Kilogram Human Body		Cost (Retail)
Oxygen	43.00 kilograms (kg)	$0.021739
Carbon	16.00 kg	6.400000
Hydrogen	7.00 kg	0.028315
Nitrogen	1.80 kg	9.706929
Calcium	1.00 kg	15.500000
Phosphorus	780.00 grams (g)	68.198594
Potassium	140.00 g	4.098737
Sulfur	140.00 g	0.011623
Sodium	100.00 g	2.287748
Chlorine	95.00 g	1.409496
Magnesium	19.00 g	0.444909
Iron	4.20 g	0.054600
Fluorine	2.60 g	7.917263
Zinc	2.30 g	0.088090
Silicon	1.00 g	0.370000
Rubidium	0.68 g	1.087153
Strontium	0.32 g	0.177237
Bromine	0.26 g	0.012858
Lead	0.12 g	0.003960
Copper	72.00 milligrams (mg)	0.012961
Aluminum	60.00 mg	0.246804
Cadmium	50.00 mg	0.010136
Cerium	40.00 mg	0.043120
Barium	22.00 mg	0.028776
Iodine	20.00 mg	0.094184
Tin	20.00 mg	0.005387
Titanium	20.00 mg	0.010920
Boron	18.00 mg	0.002172
Nickel	15.00 mg	0.031320
Selenium	15.00 mg	0.037949
Chromium	14.00 mg	0.003402
Manganese	12.00 mg	0.001526
Arsenic	7.00 mg	0.023576
Lithium	7.00 mg	0.024233
Cesium	6.00 mg	0.000016
Mercury	6.00 mg	0.004718
Germanium	5.00 mg	0.130435
Molybdenum	5.00 mg	0.001260
Cobalt	3.00 mg	0.001509
Antimony	2.00 mg	0.000243
Silver	2.00 mg	0.013600
Niobium	1.50 mg	0.000624
Zirconium	1.00 mg	0.000830
Lanthanum	0.80 mg	0.000566
Gallium	0.70 mg	0.003367
Tellurium	0.70 mg	0.000722
Yttrium	0.60 mg	0.005232
Bismuth	0.50 mg	0.000119
Thallium	0.50 mg	0.000894
Indium	0.40 mg	0.000600
Gold	0.20 mg	0.001975
Scandium	0.20 mg	0.058160
Tantalum	0.20 mg	0.001631
Vanadium	0.11 mg	0.000322
Thorium	0.10 mg	0.004948
Uranium	0.10 mg	0.000103
Samarium	50.00 micrograms (μg)	0.000118
Beryllium	36.00 μg	0.000218
Tungsten	20.00 μg	0.000007
Grand Total		$118.63

Figure 2.1 (**a**) What are you worth, chemically speaking? (**b**) Proportions of the most common elements in a human body, Earth's crust, and seawater. How are they similar? How do they differ?

a

Human		Earth's Crust		Seawater	
Oxygen	61.0%	Oxygen	46.0%	Oxygen	85.7%
Carbon	23.0	Silicon	27.0	Hydrogen	10.8
Hydrogen	10.0	Aluminum	8.2	Chlorine	2.0
Nitrogen	2.6	Iron	6.3	Sodium	1.1
Calcium	1.4	Calcium	5.0	Magnesium	0.1
Phosphorus	1.1	Magnesium	2.9	Sulfur	0.1
Potassium	0.2	Sodium	2.3	Calcium	0.04
Sulfur	0.2	Potassium	1.5	Potassium	0.03

b

of at least some of them have vital functions. For instance, even a little selenium is toxic, but *too* little can cause heart problems and thyroid disorders.

Superficially, then, the human body can be viewed as a balanced collection of elements. The amounts are worth no more than $118.63, and the kinds are not even unique; they occur in Earth's crust and even seawater (Figure 2.1*b*). However, the *proportions* of elements in humans and other organisms are unique relative to nonliving things. Look at all of that carbon, for instance! Also, you will never find a clod of dirt or a volume of seawater that comes close to the *structural and functional organization* of a living body. Assembling that collection of elements into an organized, operational body takes a fabulous molecular library (DNA), enzymes and other metabolic workers, and large, ongoing inputs of energy (just ask any pregnant woman).

Remember this when someone tries to say "chemistry" has nothing to do with you. It has everything to do with you. People, toothpaste, turkeys, refrigerators, jet fuel, health, disease, corsages, acid rain, nerve gas, old-growth forests—name any living or nonliving bit of the universe, and chemistry is part of it.

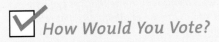

Watch the video online!

How Would You Vote?

Fluoride helps prevent tooth decay. But too much wrecks bones and teeth, and causes birth defects. A lot can kill you. Many communities in the United States add fluoride to their supply of drinking water. Do you want it in yours? See BiologyNow for details, then vote online.

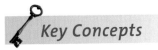

Key Concepts

ATOMS AND ELEMENTS

An element is a fundamental substance made of one type of atom. The atom is the smallest unit of an element that still retains the element's properties, and its building blocks are protons, electrons, and neutrons. Isotopes are atoms of an element that vary in the number of neutrons. Sections 2.1, 2.2

WHY ELECTRONS MATTER

Atoms acquire, share, and give up electrons. Whether one atom will bond with others depends on the number and arrangement of its electrons. Section 2.3

ATOMS BOND

The bonding behavior of biological molecules starts with the number and arrangement of electrons in each type of atom. Ionic, covalent, and hydrogen bonds are the main categories of bonds between atoms in biological molecules. Section 2.4

NO WATER, NO LIFE

Life originated in water and is adapted to its properties. Water has temperature-stabilizing effects. Many kinds of substances dissolve easily in it. Water also shows cohesion. Section 2.5

HYDROGEN IONS RULE

Life depends on precise controls over the formation, use, and buffering of hydrogen ions. Section 2.6

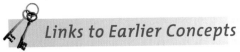

Links to Earlier Concepts

With this chapter, we start at the base of life's levels of organization, so take a moment to review the simple chart in Section 1.1. It all starts with atoms and energy. Life's organization requires tapping into a great one-way flow of energy and storing it in bonds between atoms (1.2).

The chapter also has a simple example of how the body's built-in mechanisms help return the internal environment to a homeostatic state when conditions shift beyond ranges that cells can tolerate (1.2).

2.1 Start With Atoms

LINK TO
SECTION
1.1

Know a bit about protons, neutrons, and electrons, and you have a clue to why the elements that make up the body behave as they do. Each element's unique properties start with the number of protons in its atoms.

An element, again, is a fundamental substance made of only one kind of atom. Atoms are built from three kinds of subatomic particles: protons, electrons, and neutrons. Each **proton** carries a positive *charge*, which is a defined amount of electricity. You can symbolize a proton as p^+. An atom's nucleus, or core region, holds one or more protons. Except for the hydrogen atom, it also holds **neutrons**, which carry no charge. Moving around the atomic nucleus are one or more **electrons**, which carry a negative charge (e^-). Figure 2.2 shows a few simple models for atomic structure.

The positive charge of one proton and the negative charge of one electron balance each other. Therefore, an atom that has the same number of electrons and protons has no net electrical charge.

Each element has a unique *atomic number*, which is the number of protons in the nucleus of its atoms. A hydrogen atom has one proton, so the atomic number is 1. For carbon, with six protons, it is 6.

Protons and neutrons contribute to an atom's mass. (Electrons are too tiny to do so.) We can assign each element a *mass number*, or the total number of protons *and* neutrons in the atomic nucleus. For carbon, with six protons and six neutrons, the mass number is 12.

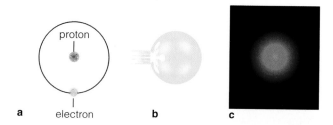

a electron   **b**   **c**

Figure 2.2 Different ways to represent atoms, using hydrogen (H) as the example. (**a**) The shell model, good for showing the number of electrons and their organization around the nucleus. (**b**) Ball models show the sizes of atoms relative to one another. (**c**) Electron density clouds are best at conveying the distribution of electrons around the nucleus.

Why bother with the number of electrons, protons, and neutrons? Knowing them can help you predict how each kind of element will behave under a variety of conditions inside and outside the body.

Elements were being classified in terms of chemical similarities long before their subatomic particles were discovered. In 1869, Dmitry Mendeleev, known more for his extravagant hair than his discoveries (he cut it only once a year), arranged the known elements in a repeating pattern, based on their chemical properties. By using gaps in this **periodic table of the elements**, Mendeleev correctly predicted the existence of many elements that had not yet been discovered.

Elements fall into order in the table according to their atomic number (Figure 2.3). All elements in each vertical column have the same number of electrons that are available for interaction with other atoms. As a result, they behave in similar ways. For example, helium, neon, radon, and other gases in the farthest right column of the periodic table are *inert* elements. Not one of the electrons in their atoms is available for chemical interactions. Such elements rarely do much; they occur mostly as solitary atoms.

You won't find all of the elements in nature. Those after atomic number 92 are extremely unstable. Some have been formed in exceedingly small quantities in laboratories—sometimes no more than a single atom —and they wink out of existence fast.

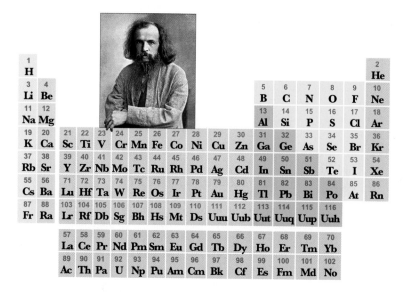

Figure 2.3 Periodic table of the elements and Dmitry Mendeleev, who created it. Some symbols for elements are abbreviations for their Latin names. For instance, Pb (lead) is short for *plumbum;* the word "plumbing" is related, because ancient Romans used lead to make their water pipes.

An element is a fundamental substance consisting of only one kind of atom. Atoms are the smallest units that retain an element's properties.

An atom consists of one or more positively charged protons, negatively charged electrons, and (except for hydrogen) neutrons. Whether any given atom will interact with others depends on how many electrons it has.

2.2 Putting Radioisotopes To Use

*All elements are defined by the number of protons in their atoms—but an element's atoms can differ in their number of neutrons. We call such atoms **isotopes** of the same element. Some are radioactive.*

In 1896, Henri Becquerel made a chance discovery. He had placed some uranium crystals in a desk drawer, next to a coin and metal screen on top of some sheets of opaque black paper. Underneath that paper was a photographic plate. A day later, the physicist used the film and developed it. Oddly, a negative image of the coin and the metal screen showed up. Becquerel hypothesized that energy radiating from the uranium salts had passed through the paper—which was impenetrable to light—and exposed the film around both metal objects.

As we now know, uranium has isotopes—fifteen of them. Most naturally occurring elements do. Carbon has three isotopes, nitrogen has two, and so on. A superscript number to the left of an element's symbol is the isotope's mass number. For instance, carbon's three natural isotopes are ^{12}C (carbon 12, the most common form, with six protons, six neutrons), ^{13}C (with six protons, seven neutrons), and ^{14}C (with six protons, eight neutrons).

Some isotopes are unstable, or radioactive. A radioactive isotope, or **radioisotope**, spontaneously emits energy in the form of subatomic particles and x-rays when its nucleus disintegrates. This process is called **radioactive decay**, and it can transform one element into another. As an example, ^{13}C and ^{14}C are radioisotopes of carbon. Each predictably decays with a particular amount of energy into a more stable product. After 5,700 years, about half of the atoms in a sample of ^{14}C will have turned into ^{13}N (nitrogen) atoms. Researchers use radioactive decay to estimate the age of rocks and biological remains, as Section 17.5 explains.

The different isotopes of an element are still the same element. For the most part, carbon is carbon, regardless of how many neutrons it has. Living systems use ^{12}C the same way as ^{14}C. Knowing this, researchers or clinicians who want to track a particular substance construct a **tracer**. Tracers are molecules in which a radioisotope has been substituted for a more stable isotope. They can be delivered into a cell or multicelled body, even into populations used in laboratory experiments. The energy from radioactive decay is like a shipping label. It helps researchers track the pathway or destination of a substance of interest with the help of radioactivity-detecting instruments.

For example, Melvin Calvin and his colleagues used a tracer to discover specific reaction steps of photosynthesis. They let growing plants take up a radioactive gas (carbon dioxide made with ^{14}C). By using radioactivity-detecting instruments, they tracked the carbon radioisotope through steps by which plants produce simple sugars and starches.

Radioisotopes also are used in medicine. *PET* (short for Positron-Emission Tomography) uses radioisotopes to study metabolism. Clinicians attach a radioisotope to glucose or another sugar. They inject this tracer into a patient, who is moved into a PET scanner (Figure 2.4a). Cells in different parts of the body absorb the tracer at different rates. The scanner detects radiation caused by energy from the decay of the radioisotope. That radiation is used to form an image on a monitor, as in Figure 2.4. Such images reveal variations and abnormalities in metabolic activity.

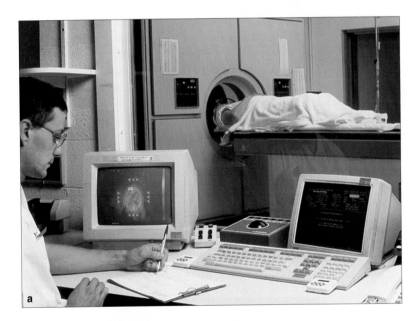

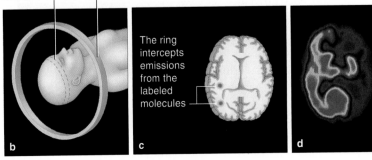

portion of the patient's body being scanned

detector ring inside the PET scanner

The ring intercepts emissions from the labeled molecules

b

c

d

Figure 2.4 *Animated!* (**a**) Patient whose brain is being probed in a PET scanner. (**b,c**) A ring of detectors intercepts radioactive emissions from tracers that had been injected into the patient. The body region of interest is scanned. Computers analyze and color-code the number of emissions from each location in the scanned region. Results are converted into digital images and displayed on computer screens.

(**d**) Different colors in a scan signify differences in metabolic activity. Cells in the left half of this brain absorbed and used labeled molecules at expected rates. Cells in the right half showed little activity. This patient has a neurological disorder.

2.3   What Happens When Atom Bonds With Atom?

LINK TO
SECTION
1.1

*Atoms acquire, share, and donate electrons. The atoms
of some elements do this quite easily; others do not.
Why is this so? To come up with an answer, look to the
number and arrangement of electrons in atoms.*

ELECTRONS AND ENERGY LEVELS

In our world, simple physics explains the motion of
an apple falling from a tree. Tiny electrons belong to a
strange world where everyday physics does not apply.
(If electrons were as big as apples, you would be 3.5
times taller than our solar system is wide.) Different
forces bring about the motion of electrons, which can
get from here to there without going in between!

We can calculate where an electron is, although not
exactly. The best we can do is say that it is somewhere
in a fuzzy cloud of probability density. Where it can
go in the cloud depends on how many other electrons
belong to the atom. The electrons become arranged in
orbitals, or volumes of space around the atomic
nucleus. Many orbitals, each with a characteristic
three-dimensional shape, are possible.

An atom has the same number of electrons as
protons. Most atoms have many electrons. How
are they all arranged, given that electrons repel
each other? Think of each atom as a multilevel
apartment building with lots of vacant rooms to
rent to electrons, and a nucleus in the basement.
Each "room" is one orbital, and it rents out to two
electrons at most. An orbital holding one electron
only has a vacancy; another electron can move in.

Each floor in that atomic apartment building
corresponds to an energy level. There is only one
room on the first floor (one orbital at the lowest
energy level, closest to the nucleus). It fills first.
For hydrogen, the simplest atom, a lone electron

occupies the room (Figure 2.5). Helium, with its two
electrons, has no vacancies at the first (lowest) energy
level. In larger atoms, more electrons rent the second-
floor rooms. If the second floor is filled, then more
electrons rent third-floor rooms, and so on. *Electrons
fill orbitals at successively higher energy levels.*

The farther an electron is from the basement (the
nucleus), the greater its energy. An electron in a first-
floor room can't move to the second or third floor, let
alone the penthouse, unless a boost of energy puts it
there. Suppose the electron absorbs just enough energy
from, say, sunlight, to get excited about moving up.
Move it does. If nothing fills that lower room, though,
the electron will quickly return to it, emitting extra
energy as it does. Later on, you will see how cells in
plants and in your eyes harness and use that energy.

FROM ATOMS TO MOLECULES

In shell models for atoms, nested "shells" correspond
to energy levels. They give us an easy way to check
for electron vacancies, as in Figure 2.6. Bear in mind,
atoms do not look like these flat diagrams. The shells
are not three-dimensional volumes of space, and they
certainly don't show the electron orbitals.

The atoms with vacancies in their outermost shell
tend to give up, acquire, or share electrons. Actually,
what we call **chemical bonds** are just a case of atoms
sharing their electrons with one another. An atom
with no vacancies rarely bonds with others. But the
most common atoms in organisms—such as oxygen,
carbon, hydrogen, nitrogen, and calcium—do have
vacancies in orbitals at their outermost energy level.
They tend to participate in bonds.

A **molecule** is simply two or more atoms of the
same or different elements joined in a chemical bond.

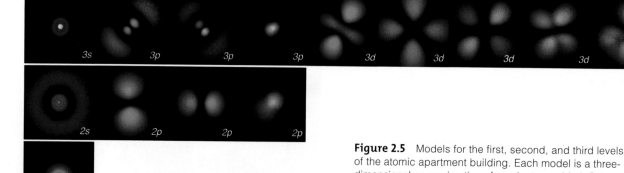

third
energy level
(second floor)

second
energy level
(first floor)

first
energy level
(closest to the
basement)

Figure 2.5 Models for the first, second, and third levels
of the atomic apartment building. Each model is a three-
dimensional approximation of an electron orbital. Colors
are most intense in locations where electrons are most
likely to be in any given instant. Orbitals farthest from the
nucleus have greater energy and are more complex.

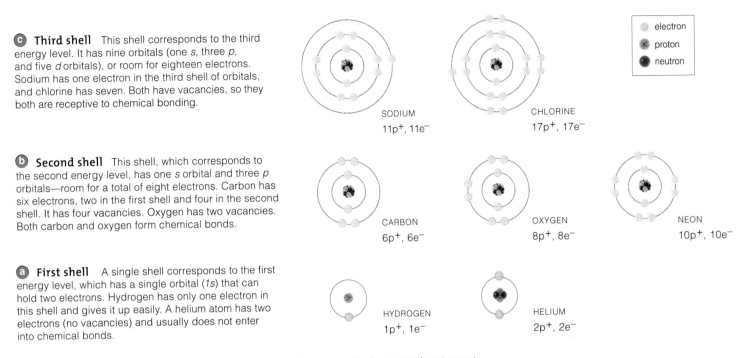

c **Third shell** This shell corresponds to the third energy level. It has nine orbitals (one *s*, three *p*, and five *d* orbitals), or room for eighteen electrons. Sodium has one electron in the third shell of orbitals, and chlorine has seven. Both have vacancies, so they both are receptive to chemical bonding.

SODIUM
11p⁺, 11e⁻

CHLORINE
17p⁺, 17e⁻

electron
proton
neutron

b **Second shell** This shell, which corresponds to the second energy level, has one *s* orbital and three *p* orbitals—room for a total of eight electrons. Carbon has six electrons, two in the first shell and four in the second shell. It has four vacancies. Oxygen has two vacancies. Both carbon and oxygen form chemical bonds.

CARBON
6p⁺, 6e⁻

OXYGEN
8p⁺, 8e⁻

NEON
10p⁺, 10e⁻

a **First shell** A single shell corresponds to the first energy level, which has a single orbital (*1s*) that can hold two electrons. Hydrogen has only one electron in this shell and gives it up easily. A helium atom has two electrons (no vacancies) and usually does not enter into chemical bonds.

HYDROGEN
1p⁺, 1e⁻

HELIUM
2p⁺, 2e⁻

Figure 2.6 *Animated!* Shell models, which help us visualize vacancies in an atom's outermost orbitals. Each circle, or shell, represents all orbitals at one energy level. Larger circles correspond to higher energy levels. Such models are highly simplified. A more realistic rendering would show electrons as fuzzy clouds of probability density about 10,000 times larger than the nucleus.

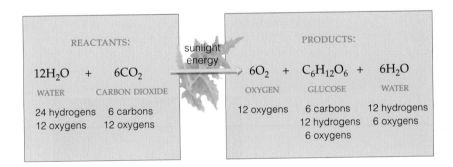

REACTANTS:

$12H_2O$ + $6CO_2$

WATER   CARBON DIOXIDE

24 hydrogens   6 carbons
12 oxygens   12 oxygens

sunlight energy

PRODUCTS:

$6O_2$ + $C_6H_{12}O_6$ + $6H_2O$

OXYGEN   GLUCOSE   WATER

12 oxygens   6 carbons   12 hydrogens
            12 hydrogens   6 oxygens
            6 oxygens

Figure 2.7 Chemical bookkeeping. We use formulas when writing out chemical equations, which represent reactions between atoms and molecules. Substances entering a reaction (reactants) are written to the left of a reaction arrow, and products to the right. How many molecules (or atoms) enter as reactants or form as products are indicated by a number that precedes their formula. *The same number of atoms that enter a reaction must be there at the end.* The atoms get shuffled around, but they never vanish. To be sure you wrote an equation correctly, count the atoms.

You can write a molecule's chemical composition as a formula, which uses symbols for the elements present and subscripts for the number of atoms of each kind of element (Figure 2.7). For example, one molecule of water has the chemical formula H_2O. The subscript number shows that there are two hydrogen (H) atoms for each oxygen (O) atom. If you have six molecules of water, then you would write $6H_2O$.

Compounds are molecules that consist of two or more different elements in proportions that never do vary. Water is an example. All water molecules have one oxygen atom bonded to two hydrogen atoms. The ones in rain clouds, the seas, a Siberian lake, flower petals, your bathtub, or anywhere else have twice as many hydrogen atoms as oxygen atoms. In a **mixture**, two or more substances intermingle without bonding.

For example, when you swirl sugar into water, you make a mixture. The proportions of elements in this mixture, or any other kind of mixture, can vary.

Electrons occupy orbitals, or defined volumes of space around an atom's nucleus. Successive orbitals correspond to levels of energy, which become higher with distance from the atomic nucleus.

One or at most two electrons can occupy any orbital. The atoms with vacancies in orbitals at their highest level tend to interact and form bonds with other atoms.

A molecule is two or more atoms joined in a chemical bond. In compounds, atoms of two or more elements are bonded together. A mixture consists of intermingled substances.

2.4 Major Bonds in Biological Molecules

Electrons of one type of atom interact with electrons of others in specific ways. Those interactions give rise to the distinctive properties of biological molecules.

ION FORMATION AND IONIC BONDING

An electron, recall, has a negative charge equal to a proton's positive charge. When an atom contains as many electrons as protons, these charges balance each other, so the atom has a net charge of zero. When an atom *gains* an extra electron, it acquires a net negative charge. When an atom *loses* an electron, it acquires a net positive charge. Either way, it has become an **ion**.

Consider: A chlorine atom has seven protons. It has seven electrons (one vacancy) in the third orbital level—which is most stable when filled with eight. This atom tends to attract an electron from someplace else. With that extra electron, it becomes a chloride ion (Cl^-), with a net negative charge.

Also consider: A sodium atom has eleven protons and eleven electrons. Its second orbital level is full of electrons, and only one electron is in the third orbital level. Giving up the one electron is easier than getting seven more. When it does so, the atom still has eleven protons. But now it has ten electrons. It has become a sodium ion, with a net positive charge (Na^+).

Remember that opposite charges attract each other. When a positively charged ion encounters a negatively charged ion, the two may associate closely with each other. A close association of ions is an **ionic bond**. For example, Figure 2.8*a* shows a crystal of table salt, or NaCl. In such crystals, ionic bonds hold ions of sodium and chloride in an orderly, cubic arrangement.

COVALENT BONDING

In an ionic bond, an atom that has lost one or more electrons associates with an atom that gained one or more electrons. What if both atoms have room for an extra electron? They can *share* one in a hybrid orbital that spans both atomic nuclei. The vacancy in each atom becomes filled with the shared electrons.

When atoms share two electrons, they are joined in a single **covalent bond** (Figure 2.8*b*). Such bonds are stable and are much stronger than ionic bonds.

We can represent covalent bonds as single lines in structural formulas, which show how the atoms of a molecule are physically arranged. A line between two atoms represents a pair of electrons that are being shared in a single covalent bond. To give examples of this bonding pattern, molecular hydrogen (H_2) has one covalent bond and can be written as H—H. Two

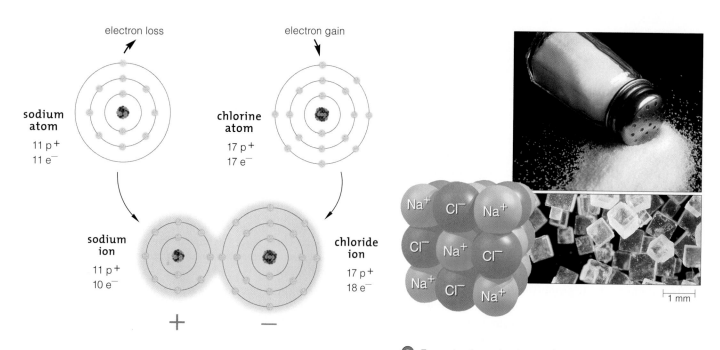

electron loss   electron gain

sodium atom
11 p+
11 e−

chlorine atom
17 p+
17 e−

sodium ion
11 p+
10 e−

chloride ion
17 p+
18 e−

+   −

a Example of ongoing interactions called ionic bonding. In each crystal of table salt, or NaCl, many sodium ions and chloride ions are staying close together because of the mutual attraction of their opposite charges.

Figure 2.8 *Animated!* Important bonds in biological molecules.

atoms share two electron pairs in a *double* covalent bond. Molecular oxygen (O=O) is like this. Others share three electron pairs in a *triple* covalent bond, as in molecular nitrogen (N≡N). Each time you take a breath, O_2 and N_2 molecules flow toward your lungs.

In a *nonpolar* covalent bond, two atoms are sharing electrons equally, so the molecule shows no difference in charge between the two "ends" of the bond. We find such bonds in molecular hydrogen (H_2), oxygen (O_2), and nitrogen (N_2).

In a *polar* covalent bond, two atoms do not share electrons equally. Why not? The atoms are of different elements, and one has more protons than the other. The one with the most protons exerts more of a pull on the electrons, so its end of the bond ends up with a slight negative charge. We say it is "electronegative." The atom at the other end of the bond ends up with a slight positive charge. For instance, a water molecule (H—O—H) has two polar covalent bonds. The oxygen atom carries a slight negative charge, and each of its two hydrogen atoms carries a slight positive charge.

HYDROGEN BONDING

A **hydrogen bond** is a weak attraction that has formed between a covalently bound hydrogen atom and an electronegative atom in a different molecule or in a different region of the same molecule.

Because hydrogen bonds are weak, they form and break easily. Collectively, however, many hydrogen bonds contribute to the properties of liquid water, as you will see next.

Hydrogen bonds also play important roles in the structure and function of biological molecules. They often form between different parts of large molecules that have folded over on themselves and hold them in particular shapes. Many of these bonds hold DNA's two nucleotide strands together. Figure 2.8c hints at the number of these interactions in DNA.

> Ions form when atoms acquire a net charge by gaining or losing electrons. Two ions of opposite charge attract each other. They can associate in an ionic bond.
>
> In a covalent bond, atoms share a pair of electrons. When atoms share the electrons equally, the bond is nonpolar. When the sharing is not equal, the bond is polar—slightly positive at one end, slightly negative at the other.
>
> In a hydrogen bond, a covalently bound hydrogen atom attracts a small, negatively charged atom in a different molecule or in a different region of the same molecule.

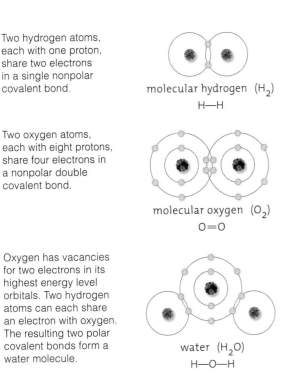

Two hydrogen atoms, each with one proton, share two electrons in a single nonpolar covalent bond.

molecular hydrogen (H_2)
H—H

Two oxygen atoms, each with eight protons, share four electrons in a nonpolar double covalent bond.

molecular oxygen (O_2)
O=O

Oxygen has vacancies for two electrons in its highest energy level orbitals. Two hydrogen atoms can each share an electron with oxygen. The resulting two polar covalent bonds form a water molecule.

water (H_2O)
H—O—H

b Covalent bonding. Each atom becomes more stable by sharing electron pairs in hybrid orbitals.

Two molecules interacting weakly in one H bond, which can form and break easily.

hydrogen bond

water molecule   ammonia molecule

H bonds helping to hold part of two large molecules together.

Many H bonds hold DNA's two strands together along their length. Individually each one is weak, but collectively they can stabilize DNA's large structure.

c Hydrogen bonds. Such bonds can form at a hydrogen atom that is already covalently bonded in a molecule. The atom's slight positive charge weakly attracts an atom with a slight negative charge that is already covalently bonded to something else. As shown, this can happen between one of the hydrogen atoms of a water molecule and the nitrogen atom of an ammonia molecule.

2.5 Water's Life-Giving Properties

No sprint through basic chemistry is complete unless it leads to the collection of molecules called water. Life originated in water. Organisms still live in it or they cart water around with them inside cells and tissue spaces. Many metabolic reactions use water. A cell's structure and shape absolutely depend on it.

POLARITY OF THE WATER MOLECULE

Figure 2.9*a* shows the structure of a water molecule. Two atoms of hydrogen have formed polar covalent bonds with an oxygen atom. The molecule has no net charge. Even so, the oxygen pulls the shared electrons more than the hydrogen atoms do. Thus, the molecule of water has a slightly negative "end" that is balanced out by its slightly positive "end."

The water molecule's polarity attracts other water molecules. Also, the polarity is so attractive to sugars and other polar molecules that hydrogen bonds form easily between them. That is why polar molecules are known as **hydrophilic** (water-loving) substances.

That same polarity repels oils and other nonpolar molecules, which are **hydrophobic** (water-dreading) substances. Shake a bottle filled with water and salad oil, then set it on a table. Soon, new hydrogen bonds replace the ones that the shaking broke. The reunited water molecules push out oil molecules, which cluster as oil droplets or as an oily film at the water's surface.

The same kinds of interactions proceed at the thin, oily membrane between the water inside and outside cells. Membrane organization—and life itself—starts with such hydrophilic and hydrophobic interactions. You will read about membrane structure in Chapter 5.

WATER'S TEMPERATURE-STABILIZING EFFECTS

Cells are mostly water, and they also release a lot of metabolic heat. The many hydrogen bonds in water keep cells from cooking in their own juices. How? All bonds vibrate nonstop, and they move more as they absorb heat. **Temperature** is a measure of molecular motion. Compared to most other fluids, water absorbs more heat energy before it gets measurably hotter. So water serves as a heat reservoir, and its temperature remains relatively stable. Over time, increases in heat step up the motion within water molecules. Before that happens, however, much of the heat will go into disrupting hydrogen bonds between molecules.

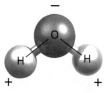

slight negative charge
on the oxygen atom

The + and – ends balance each other; the whole molecule carries no net charge, overall.

slight positive charge
a on the hydrogen atoms

Figure 2.9 *Animated!*   Water, a substance essential for life.

(**a**) Polarity of an individual water molecule.

(**b**) Hydrogen bonding pattern among water molecules in liquid water. Dashed lines signify hydrogen bonds, which break and re-form rapidly.

(**c**) Hydrogen bonding in ice. Below 0°C, every water molecule hydrogen-bonds with four others, in a rigid three-dimensional lattice. The molecules are farther apart, or less densely packed, than they are in liquid water. As a result, ice floats on water.

Thanks partly to rising levels of methane and other greenhouse gases that are contributing to global warming, the Arctic ice cap is melting. At current rates, it will be gone in fifty years. So will the polar bears. Already their season for hunting seals is shorter, bears are thinner, and they are giving birth to fewer cubs.

Figure 2.10 Spheres of hydration around two ions.

Figure 2.11 Examples of water's cohesion. (**a**) When a pebble hits liquid water and forces molecules away from the surface, the individual water molecules do not fly every which way. They stay together in droplets. Why? Countless hydrogen bonds exert a continuous inward pull on individual molecules at the surface.

(**b**) And just how does water rise to the very top of trees? Cohesion, and evaporation from leaves, pulls it upward.

a

b

When water temperature is stable, hydrogen bonds form as fast as they break. When water gets hotter, the increase in molecular motion can keep the bonds broken, so individual molecules at the water's surface can escape into air. By this process, **evaporation**, heat energy converts liquid water to gaseous form. The increased energy has overcome the attraction between water molecules, which break free. Water's surface temperature decreases during evaporation.

Evaporative water loss helps you and some other mammals cool off when you sweat on hot, dry days. Sweat, about 99 percent water, evaporates from skin.

Below 0°C, water molecules do not move enough to break their hydrogen bonds, so they become locked in the latticelike bonding pattern of ice (Figure 2.9c). Ice is less dense than water. During winter freezes, ice sheets may form near the surface of ponds, lakes, and streams. The ice "blanket" insulates the liquid water beneath it and helps protect many fishes, frogs, and other aquatic organisms against freezing.

WATER'S SOLVENT PROPERTIES

Water is an excellent *solvent*, meaning ions and polar molecules easily dissolve in it. A dissolved substance is known as a **solute**. In general, a substance is said to be *dissolved* after water molecules cluster around ions or molecules of it and keep them dispersed in fluid.

A clustering of water molecules around a solute is called a *sphere of hydration*. Such spheres form around any solute in cellular fluids, tree sap, blood, the fluid in your gut, and every other fluid associated with life. Watch it happen after you pour table salt (NaCl) into a cup of water. In time, the crystals of salt separate

into ions of sodium (Na^+) and chloride (Cl^-). Each Na^+ attracts the negative end of some water molecules even as Cl^- attracts the positive end of others (Figure 2.10). Spheres of hydration formed this way keep the ions dispersed in fluid.

WATER'S COHESION

Still another life-sustaining property of water is its cohesion. **Cohesion** means something is showing a capacity to resist rupturing when it is stretched, or placed under tension. You see its effect when a tossed pebble breaks the surface of a lake, a pond, or some other body of liquid water (Figure 2.11a). At or near the surface, uncountable numbers of hydrogen bonds are exerting a continuous, inward pull on individual molecules. Bonding creates a high surface tension.

Cohesion is working inside organisms, too. Plants, for example, absorb nutrient-laden water when they grow. Columns of liquid water rise inside pipelines of vascular tissues, which extend from roots to leaves. Water evaporates from leaves when molecules break free and diffuse into air (Figure 2.11b). The cohesive force of hydrogen bonds pulls replacements into the leaf cells, in ways explained in Section 30.3.

Being polar, water molecules hydrogen-bond to one another and to other polar (hydrophilic) substances. They tend to repel nonpolar (hydrophobic) substances.

The unique properties of liquid water make life possible. Water has temperature-stabilizing effects, cohesion, and a capacity to dissolve many substances easily.

2.6 Acids and Bases

LINK TO
SECTION
1.2

Ions dissolved in fluids inside and outside each living cell influence its structure and function. Among the most influential are hydrogen ions. They have far-reaching effects largely because they are chemically active and because there are so many of them.

THE pH SCALE

At any instant in liquid water, some water molecules split into ions of hydrogen (H^+) and hydroxide (OH^-). These ions are the basis of the **pH scale**. The scale is a way to measure the concentration of hydrogen ions in solutions such as seawater, blood, or sap. The greater the H^+ concentration, the lower the pH. Pure water (not rainwater or tap water) always has as many H^+ as OH^- ions. This state is neutrality, or pH 7.0 (Figure 2.12).

A decrease in pH by just one unit from neutrality corresponds to a tenfold increase in H^+ concentration, and an increase by one unit corresponds to a tenfold decrease in H^+ concentration. One way to get a sense of the range is to taste dissolved baking soda (pH 9), water (pH 7), and lemon juice (pH 2).

HOW DO ACIDS AND BASES DIFFER?

When dissolved in water, substances called **acids** *donate* hydrogen ions and **bases** *accept* hydrogen ions. *Acidic* solutions, such as lemon juice, gastric fluid, and coffee, release H^+; their pH is below 7. *Basic* solutions, such as seawater and egg white, easily combine with H^+. Basic solutions, which also are known as alkaline solutions, have a pH above 7.

Nearly all of life's chemistry occurs near pH 7. Most of your body's internal environment (tissue fluids and blood) is between pH 7.3 and 7.5. Seawater is more basic than body fluids of the organisms living in it.

Acids and bases can be weak or strong. The weak acids, such as carbonic acid (H_2CO_3), are stingy H^+ donors. Strong acids readily give up H^+ in water. An example is the hydrochloric acid that dissociates into H^+ and Cl^- inside your stomach. The H^+ makes your gastric fluid far more acidic, which in turn activates protein-digesting enzymes.

Too much HCl can cause an *acid stomach*. Antacids taken for this condition, including milk of magnesia, release OH^- ions that combine with H^+ to reduce the pH of stomach contents.

Actually, strong acids and bases can cause severe chemical burns. That is why we are supposed to read the labels on containers of ammonia, drain cleaner, and many other common household products. That is why we are not supposed to let a car battery's sulfuric acid drip on skin.

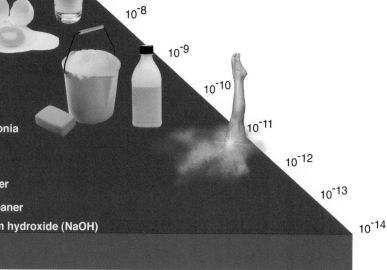

Figure 2.12 *Animated!* The pH scale, which represents the concentrations of hydrogen ions in a solution. Also shown are approximate pH values for some solutions. This pH scale ranges from 0 (the most acidic) to 14 (most basic). A change of 1 on the pH scale means a tenfold change in the H^+ concentration.

hydrochloric acid (HCl) — 10^0
gastric fluid (1.0-3.0) — 10^{-1}
lemon juice, some acid rain — 10^{-2}
vinegar, wine, beer, oranges — 10^{-3}
tomatoes — 10^{-4}
bananas
black coffee
bread — 10^{-5}
typical rainwater
urine (5.0-7.0) — 10^{-6}
milk (6.6)
pure water — ($H^+ = OH^-$) — 10^{-7}
blood (7.3-7.5)
egg white (8.0) — 10^{-8}
seawater (7.8-8.3)
baking soda
phosphate detergents — 10^{-9}
bleach, Tums
soapy solutions, milk of magnesia — 10^{-10}
household ammonia (10.5-11.9) — 10^{-11}
hair remover — 10^{-12}
oven cleaner — 10^{-13}
sodium hydroxide (NaOH) — 10^{-14}

Figure 2.13 Emissions of sulfur dioxide from a coal-burning power plant. Airborne pollutants such as sulfur dioxide dissolve in water vapor to form acidic solutions. They are a component of acid rain.

At high concentrations, strong acids or bases that enter an ecosystem can kill organisms. For instance, fossil fuel burning and nitrogen-containing fertilizers release strong acids that lower the pH of rainwater (Figure 2.13). Some regions are sensitive to this *acid rain*. Alterations in the chemical composition of soil and water harm fishes and other organisms in these regions. We return to this topic in Section 48.2.

SALTS AND WATER

A **salt** is any compound that dissolves easily in water and releases ions *other than* H^+ and OH^-. It commonly forms when an acid interacts with a base. For example:

$$HCl \text{ (acid)} + NaOH \text{ (base)} \rightleftharpoons NaCl \text{ (salt)} + H_2O$$

HYDROCHLORIC ACID   SODIUM HYDROXIDE   SODIUM CHLORIDE

NaCl, the salt product of this reaction, dissociates into sodium ions (H^+) and chloride ions (Cl^-) when it is dissolved in water. Many of the ions that are released when salts dissolve in fluid are important components of cellular processes. For example, ions of sodium, potassium, and calcium are essential for nerve and muscle cell functions. They also help plant cells take up water from soil.

BUFFERS AGAINST SHIFTS IN pH

Cells must respond fast to even slight shifts in pH, because excess H^+ or OH^- can alter how biological molecules function. Responses are rapid with **buffer systems**. Think of such a system as a dynamic chemical partnership between a weak acid and its salt. These two related chemicals work in equilibrium to counter slight shifts in pH. For example, if a small amount of a strong base enters a buffered fluid, the weak acid partner can neutralize excess OH^- ions by donating some H^+ ions to the solution.

Most body fluids are buffered. Why? Enzymes, receptors, and all other essential biological molecules work most efficiently within a narrow range of pH. Deviation from the range disrupts cellular processes.

Carbon dioxide, a by-product of many reactions, becomes part of a buffer system as it combines with water to form carbonic acid and bicarbonate. When the pH of human blood rises slightly, carbonic acid can neutralize the excess OH^- by releasing hydrogen ions, which combine with OH^- to form water:

$$OH^- + H_2CO_3 \rightarrow HCO_3^- + H_2O$$

CARBONIC ACID   BICARBONATE (salt)   WATER

When blood becomes more acidic, bicarbonate mops up excess H^+ and thus shifts the balance of the buffer system toward the acid:

$$HCO_3^- + H^+ \rightarrow H_2CO_3$$

BICARBONATE   CARBONIC ACID

Buffer systems can neutralize only so many excess ions. With even a slight excess above that point, the pH swings widely. When the blood pH (7.3–7.5) falls even to 7, buffering fails, and the consequences can be severe. An individual may fall into a *coma*, an often irreversible state of unconsciousness. This happens in *respiratory acidosis*. Carbon dioxide accumulates, too, much carbonic acid forms, and blood pH plummets. By contrast, when the blood pH increases even to 7.8, *tetany* may occur; skeletal muscles cannot be released from contraction. *Alkalosis* is a rise in blood pH that, if not reversed by medical treatment can be lethal.

Ions dissolved in fluids on the inside and outside of cells have key roles in cell function. Acidic substances release hydrogen ions, and basic substances accept them. Salts are compounds that release ions other than H^+ and OH^-.

Acid–base interactions help maintain pH, which is the H^+ concentration in a fluid. Buffer systems help maintain the body's acid–base balance at levels suitable for life.

Summary

Introduction   Chemistry can help us understand the composition and behavior of the substances that make up cells, organisms, and all components of the biosphere. Table 2.1 summarizes some key chemical terms that you will encounter throughout this book.

Section 2.1   All substances consist of one or more elements. Atoms are the smallest units that still retain

the element's properties. An uncharged atom consists of one or more positively charged protons, an equal number of negatively charged electrons, and (except for hydrogen) one or more neutrons, which carry no charge. Protons and neutrons occupy an atom's core region, or nucleus, and essentially account for its mass.

Section 2.2   Atoms of an element typically differ in the number of neutrons; they are isotopes. Radioisotopes are unstable, and their nucleus spontaneously decays.

Biology Now
Learn about how radioisotopes are used in a PET scan with the animation on BiologyNow.

Section 2.3   Whether one atom will interact with others depends on the number and arrangement of its electrons. Electrons occupy orbitals (volumes of space) around the atomic nucleus. The shell model for atomic structure is a diagram with successively larger circles, or shells, that keep track of all electrons in the orbitals at a given energy level.

When an atom has one or more vacancies in orbitals, it interacts with other atoms by donating, accepting, or sharing electrons (forming chemical bonds).

Biology Now
Use the animation and interaction on BiologyNow to investigate electron distribution and the shell model.

Section 2.4   Each chemical bond is an interaction between the electron structures of atoms. The main types are called ionic, covalent, and hydrogen bonds.

When an atom loses or gains one or more electrons, it becomes an ion, with a positive or a negative charge. In an ionic bond, a positive ion and a negative ion stay together by mutual attraction of their opposite charges.

Atoms often fill vacancies in their outermost orbitals by sharing one or more pairs of electrons. Two atoms share electrons equally in a *nonpolar* covalent bond. The sharing is unequal in a *polar* covalent bond, so the bond has a slight negative charge at one end and a slight positive charge at the other. The charges balance, so the participating atoms carry no net charge, overall.

In a hydrogen bond, a covalently bound hydrogen atom weakly attracts an electronegative atom that is bound in a different molecule or a different region of the same molecule.

Biology Now
Compare the types of chemical bonds in biological molecules using the animation on BiologyNow.

Section 2.5   Polar covalent bonds join three atoms in a water molecule (two hydrogen atoms and one oxygen). The polarity of the water molecule invites extensive hydrogen bonding between molecules in bodies of water. The polarity is the basis of hydrogen bonding, which gives liquid water a notable ability to resist temperature changes, to show internal cohesion, and to easily dissolve diverse polar or ionic substances. These properties of water help make life possible.

Biology Now
Explore the structure and properties of water with the animation on BiologyNow.

Table 2.1	Summary of Important Players in the Chemical Basis of Life
Element	Fundamental substance consisting of one kind of atom
Atom	Smallest unit of an element that still retains element's properties. Occupies space, has mass, and cannot be broken apart by ordinary physical or chemical means.
Proton (p^+)	Positively charged particle of the atomic nucleus
Electron (e^-)	Negatively charged particle that can occupy a volume of space (orbital) around the nucleus
Neutron	Uncharged particle of the atomic nucleus
Isotope	One of two or more forms of an element's atoms that differ in the number of neutrons in the nucleus
Radioisotope	An unstable isotope that emits particles and energy; has an unstable combination of protons and neutrons
Tracer	Molecule that incorporates one or more atoms of a radioisotope. Used with tracking devices to identify the movement or destination of the molecule or atom in a metabolic pathway, the body, or some other system
Ion	An atom that has gained or lost an electron and carries a positive or negative charge. A proton without an electron zipping around it is a hydrogen ion (H^+)
Molecule	Unit of matter in which two or more atoms of the same element, or different ones, are bonded together
Compound	Molecule of two or more different elements in unvarying proportions (e.g., water)
Mixture	Intermingling of two or more elements or compounds in proportions that usually vary
Solute	Any molecule or ion dissolved in some solvent
Hydrophilic substance	Polar molecule or molecular region that can readily dissolve in water
Hydrophobic substance	Nonpolar molecule or molecular region that strongly resists dissolving in water
Acid	Substance that releases H^+ when dissolved in water
Base	Substance that accepts H^+ when dissolved in water
Salt	Compound that releases ions other than H^+ or OH^- when dissolved in water

Section 2.6 The pH scale is used to measure the hydrogen ion (H^+) concentration of a solution. A typical pH range is from 0 (highest H^+ concentration; most acidic) to 14 (lowest H^+ concentration; the most basic or alkaline). At pH 7, or neutrality, H^+ and OH^- concentrations are equal.

Salts are compounds that dissolve easily in water and release ions other than H^+ and OH^-. Acids release H^+ ions in water. Bases combine with them. A buffer system is a dynamic chemical partnership between a weak acid or base and its salt. The two go back and forth donating and accepting ions to counter slight shifts in pH and thus maintain a favorable pH. Most biological processes operate within a narrow pH range.

Biology⑧Now
Investigate the pH of common solutions with the interaction on BiologyNow.

Figure 2.14 Laboratory of a typical alchemist.

Critical Thinking

1. Some molecules consist of atoms of a single element, but others are compounds. Explain which type of molecule you would expect to be more abundant in living things.

2. *Ozone* is a chemically active form of oxygen gas. High in Earth's atmosphere, a vast layer of it absorbs about 98 percent of the sun's harmful rays. Normally, oxygen gas consists of two oxygen atoms joined in a double nonpolar covalent bond: O=O. Ozone has three covalent bonds in this arrangement: O=O—O. It is highly reactive with a variety of substances, and it gives up an oxygen atom and releases gaseous oxygen (O=O). Using what you know about chemistry, explain why you think it is so reactive.

3. Some undiluted acids are less corrosive than when diluted with a little water. In fact, lab workers are told to wipe off splashes with a towel before washing. Explain.

4. Medieval scientists and philosophers called alchemists were predecessors of modern-day chemists (Figure 2.14). Many tried to transform lead (atomic number 82) into gold (atomic number 79). Why didn't they succeed?

5. David, an inquisitive three-year-old, poked his fingers into warm water in a metal pan on the stove and didn't sense anything hot. Then he touched the pan itself and got a nasty burn. Explain why water in a metal pan heats up far more slowly than the pan itself.

6. Why can water striders (Figure 2.15) and the basilisk lizard shown in Figure 1.8 walk on water?

7. Why do you think H^+ is often written as H_3O^+?

Self-Quiz

Answers in Appendix II

1. Is this statement false: Every type of atom consists of protons, neutrons, and electrons.

2. Electrons carry a _____ charge.
   a. positive   b. negative   c. zero

3. A(n) _____ is any molecule to which a radioisotope has been attached for research or diagnostic purposes.
   a. ion   b. isotope   c. element   d. tracer

4. Atoms share electrons unequally in a(n) _____ bond.
   a. ionic   c. polar covalent
   b. hydrogen   d. nonpolar covalent

5. In a hydrogen bond, a covalently bound hydrogen atom weakly attracts an _____ atom in a different molecule or a different region of the same molecule.
   a. electronegative   b. electropositive

6. Liquid water shows _____ .
   a. polarity   d. cohesion
   b. hydrogen-bonding capacity   e. b through d
   c. notable heat resistance   f. all of the above

7. Hydrogen ions (H^+) are _____ .
   a. the basis of pH values   d. dissolved in blood
   b. unbound protons   e. both a and b
   c. targets of certain buffers   f. a through d

8. When dissolved in water, a(n) _____ donates H^+, and a(n) _____ accepts H^+.

9. A(n) _____ is a dynamic chemical partnership between a weak acid and its salt.
   a. ionic bond   c. buffer system
   b. solute   d. solvent

10. Match the terms with their most suitable description.
   ____ trace element   a. atomic nucleus components
   ____ salt   b. two atoms sharing electrons
   ____ covalent   c. any polar molecule that readily
        bond   dissolves in water
   ____ hydrophilic   d. releases ions other than H^+ and
        substance OH^- when dissolved in water
   ____ protons,   e. makes up less than 0.001
        neutrons   percent of body weight

Additional questions are available on Biology⑧Now™

Figure 2.15 Water strider, not sinking.

3 MOLECULES OF LIFE

Science or the Supernatural?

About 2,000 years ago in the mountains of Greece, the oracle of Delphi made rambling, cryptic prophecies after inhaling sweet-smelling fumes that had collected in the sunken floor of her temple. She actually was babbling in a hydrocarbon-induced trance. We now know her temple was perched on intersecting, earthquake-prone faults. When the faults slipped, methane, ethane, and ethylene seeped out from the depths. All three gases are colorless hallucinogens.

Ancient Greeks thought Apollo spoke to them through the oracle; they believed in the supernatural. Scientists looked for a natural explanation, and they found carbon compounds behind her words. Why is their explanation more compelling? It started with tested information about the structure and effects of natural substances, and it was based on analysis of gaseous substances at the site.

All three gases consist only of carbon and hydrogen atoms; hence the name, hydrocarbons. Thanks to scientific inquiry, we now know a lot about them. Consider methane. It was present when Earth first formed. It is released when volcanoes erupt, when we burn wood or peat or fossil fuels, and when termites and cattle pass gas. Methane collects in the atmosphere and in ocean depths along the continental shelves. Methane also is one of the greenhouse gases and a contributing factor in global warming.

And methane all by itself may be big trouble. Long ago, organic remains of marine organisms sank to the bottom of the ocean. Today, a few kilometers below the sediments that slowly accumulated on top of them, the remains have become food for methane-producing archaeans. Collectively, their metabolic activity produces tremendous quantities of methane. All of that gas bubbles upward and seeps from the seafloor (Figure 3.1*b*). At these methane seeps, the low temperature and high water pressure "freeze" methane into icy methane hydrate.

There may be a thousand billion tons of frozen methane hydrate on the seafloor. It is the world's largest reservoir of natural gas, but we do not have a safe, efficient way to retrieve it. Why not? The icy crystals are unstable. They instantaneously fall apart into methane gas and liquid water as soon as the temperature goes up or the pressure goes down. It does not take much, only a few degrees.

Methane hydrate can disintegrate explosively. It can cause an irreversible chain reaction that may vaporize neighboring deposits on the seafloor. We see plenty of evidence of small methane hydrate explosions in the past that pockmarked the ocean floor. Immense explosions have caused underwater landslides that stretched, almost unbelievably, from one continent to another.

methane

Figure 3.1 *Left*, ruins of the Temple of Apollo, where hydrocarbon gases seep out from the ground. *Right*, microorganisms and bubbles of methane gas almost 230 meters (750 feet) below sea level in the Black Sea. The methane is produced by archaeans far beneath the seafloor, then seeps into deep ocean water.

Watch the video online!

Also consider this: The greatest of all mass extinctions occurred 250 million years ago and marked the end of the Permian period. All but about 5 percent of the species in the seas and about 70 percent of the known plants, insects, and other species on the land abruptly vanished. Scientists, who are not given to hyperbole, call it The Great Dying.

Chemical clues locked in fossils dating from that time point to a sharp spike in the atmospheric concentration of carbon dioxide—not just any carbon dioxide, but molecules that had been assembled by living things. Methane hydrate disintegrated abruptly, and in a gargantuan burp, millions of tons of methane exploded from the seafloor. Methane-eating bacteria converted nearly all of it to carbon dioxide —which displaced most of the oxygen in the seas and sky.

Too much carbon dioxide, too little oxygen. Imagine being transported abruptly to the top of Mount Everest and trying to jog in the "thin air," with its lower oxygen concentration. You would pass out and die. Before The Great Dying, free oxygen made up about 35 percent of the atmosphere. After the burp, its concentration plummeted to 12 percent. We can expect that most animals on land and in the seas suffocated.

The methane problem is closer than you might think. Not long ago, researchers found vast methane hydrate deposits 96 kilometers (60 miles) or so off the coast of Newport, Oregon, and off the Atlantic seaboard. What is to become of us if there is another methane burp?

In short, knowledge about lifeless substances can tell you a lot about life, including your own. It will serve you well when you turn your mind to just about any topic concerning the past, present, and future—from ancient myths, to health or disease, to forests, to physical and chemical conditions that affect life everywhere.

How Would You Vote?

Should we work toward developing the vast undersea methane deposits as an energy source, given that the environmental costs and risks to life are unknown? See BiologyNow for details, then vote online.

Key Concepts

NO CARBON, NO LIFE

We define cells partly by their capacity to assemble the organic compounds called complex carbohydrates and lipids, proteins, and nucleic acids. These large molecules of life have a backbone of carbon atoms, and functional groups attached to the backbone influence their properties. All are assembled from cellular pools of simple sugars, fatty acids, amino acids, and nucleotides. Sections 3.1, 3.2

CARBOHYDRATES

Carbohydrates are the most abundant biological molecules in nature. The simple sugars function as quick energy sources or transportable forms of energy. Complex carbohydrates are structural materials or energy reservoirs. Section 3.3

LIPIDS

Some kinds of complex lipids function as the body's energy reservoirs, others as structural components of cell membranes, as waterproofing or lubricating substances, and as signaling molecules. Section 3.4

PROTEINS

Structurally and functionally, proteins are the most diverse molecules of life. They include enzymes, structural materials, signaling molecules, and transporters. Sections 3.5, 3.6

NUCLEOTIDES AND NUCLEIC ACIDS

Both DNA and RNA are nucleic acids made of a few kinds of nucleotide subunits. They interact as the cell's system of storing, retrieving, and translating heritable information about building all the proteins necessary for life. Section 3.7

Links to Earlier Concepts

You are about to enter the next level of organization in nature, as represented by the molecules of life. Keep the big picture in mind by quickly scanning Section 1.1 once again.

You will be building on your understanding of how electrons are arranged in atoms (2.3) as well as the nature of covalent bonding and hydrogen bonding (2.4). Here again, you will be considering one of the consequences of mutation in DNA (1.4), this time with sickle-cell anemia as the example.

3.1 Molecules of Life—From Structure to Function

LINK TO
SECTIONS
1.1, 2.4

Under present-day conditions in nature, only living cells make complex carbohydrates, lipids, proteins, and nucleic acids. Different classes of these biological molecules are a cell's instant energy sources, structural materials, metabolic workers, cell-to-cell signals, and libraries and translators of hereditary information.

WHAT IS AN ORGANIC COMPOUND?

The molecules of life are **organic compounds**, which are defined as containing the element carbon and at least one hydrogen atom. The term is a holdover from a time when chemists thought "organic" substances were the ones made naturally in living organisms only, as opposed to the "inorganic" substances that formed abiotically. The term persists, although scientists now synthesize organic compounds in laboratories and have reason to believe that organic compounds were present on Earth before organisms were.

The **hydrocarbons** consist only of hydrogen atoms covalently bonded to carbon. Examples are gasoline and other fossil fuels. Like other organic compounds, each has a specific number of atoms that are arranged in specific ways. Each organic compound has one or more **functional groups**, which are particular atoms or clusters of atoms covalently bonded to carbon.

In this book we use the following color code for the main atoms of organic compounds:

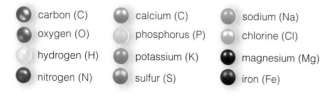

carbon (C)   calcium (C)   sodium (Na)
oxygen (O)   phosphorus (P)   chlorine (Cl)
hydrogen (H)   potassium (K)   magnesium (Mg)
nitrogen (N)   sulfur (S)   iron (Fe)

START WITH CARBON'S BONDING BEHAVIOR

Living things consist mainly of oxygen, hydrogen, and carbon (Figure 2.1). Their oxygen and hydrogen are primarily in the form of water. Put water aside, and carbon makes up more than half of what is left.

Carbon's importance to life starts with its versatile bonding behavior. *Each carbon atom can covalently bond with as many as four other atoms.* Such bonds, in which two atoms share one, two, or three pairs of electrons, are relatively stable. They join carbon atoms together as a backbone to which hydrogen, oxygen, and other elements are attached. In those configurations—in the arrangement of atoms and the distribution of electric charge—we find clues to how the different molecules of life will function and what their three-dimensional shapes will be.

WAYS TO REPRESENT ORGANIC COMPOUNDS

Methane is the simplest organic compound to think about. This colorless, odorless gas is present in the atmosphere, sea sediments, termite colonies, stagnant swamps, and stockyards. Its four hydrogen atoms are covalently bonded to one carbon atom (CH_4). You can use a ball-and-stick model to depict bond angles and show how the mass of this molecule or any other is distributed (in atomic nuclei). A space-filling model is better at conveying a molecule's size and surfaces:

structural formula   ball-and-stick   space-filling
for methane   model   model

Let's use a ball-and-stick model to depict an organic compound with six covalently bonded carbon atoms from which hydrogen *and* oxygen atoms project:

ball-and-stick model for the linear structure of glucose

This type of carbon backbone sometimes forms chains inside cells. But most of the time it coils back on itself, and its two ends connect to form a ring structure:

six-carbon ring structure of glucose that usually forms inside cells

We typically depict carbon ring structures in simpler ways. A flat structural model may show the carbons but not other atoms bonded to them. If an icon for the ring shows no atoms at all, it is understood that one carbon atom occupies each "corner" of the ring:

simplified structural formula for a six-carbon ring

icon for a six-carbon ring

Figure 3.2 shows ways to represent hemoglobin, a much larger molecule. You and all other vertebrates make this protein, which transports oxygen to tissues throughout your body. The ball-and-stick and space-filling models can give you an idea of this molecule's mass and structural complexity. But neither will tell you much about its oxygen-transporting function.

Now look at Figure 3.2c. This ribbon model shows how a hemoglobin molecule consists of four chains. As you will see later, each chain is a string of subunits called amino acids. Different regions of each chain are straight, folded, and coiled. For now, it is enough to know hemoglobin's three-dimensional shape includes four pockets, each containing a small cluster of atoms called a heme group. A heme group binds or releases oxygen in different body regions in response to how concentrated this gas is in different tissues.

More sophisticated models are now in use. Certain computer models, for instance, show local differences in electric charge across molecular surfaces. The areas color-coded, say, red on one molecule's surface might be attractive to a blue surface on another part of the same molecule or a different one (Figure 3.3).

Ultimately, such insights into the three-dimensional structure of molecules help us understand how cells and multicelled organisms function. For instance, virus particles can infect a cell when they dock at specific proteins located at the cell surface. Like Lego blocks, the proteins have ridges, clefts, and charged regions at their surface that can fit precisely into ridges, clefts, and charged regions of a protein at the surface of the virus. If a researcher can design a drug molecule that matches up with a viral protein and figure out how to deliver enough copies of it into a patient, then a lot of virus particles may be tricked into binding with the decoys instead of infecting body cells.

You will come across different kinds of molecular models throughout this book. In each case, the model selected gives you a glimpse into the structure and function of the molecule being described.

Carbohydrates, lipids, proteins, and nucleic acids are the main biological molecules—the organic compounds that only living cells assemble under present-day conditions in nature.

Organic compounds have diverse, three-dimensional shapes and functions that start with a carbon backbone and the bonding arrangements that arise from it.

Insights into the structure of molecules ultimately help us understand how cells, and multicelled organisms, function.

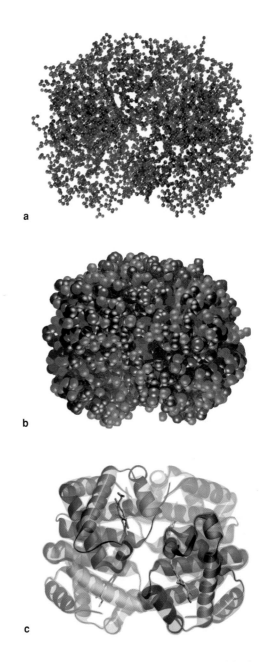

a

b

c

Figure 3.2 Visualizing the structure of hemoglobin, the oxygen-transporting molecule in red blood cells. (**a**) Ball-and-stick model, (**b**) space-filling model, and (**c**) ribbon model, with four heme groups (*red-orange*). Unlike the color coding for atoms, colors used for ribbon models and simple icons for complex molecules vary, depending on the context.

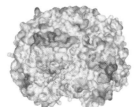

Figure 3.3 Model for the charged surface regions of a hemoglobin molecule. In this case, *blue* indicates positive charge and *red* indicates negative charge.

3.2 How Do Cells Build Organic Compounds?

LINK TO
SECTION 2.3

Before taking a run through the characteristics of the main biological molecules, get acquainted with their building blocks and how they are put together.

FOUR FAMILIES OF BUILDING BLOCKS

What is your favorite flower? Cells in the plant that made it turned carbon (from carbon dioxide), water, and the sun's energy into small organic compounds. The four main families of these small compounds are called simple sugars, fatty acids, amino acids, and nucleotides. Many kinds of molecules in each family contain two to thirty-six carbon atoms, at most.

Cells maintain and replenish pools of small organic compounds, which collectively account for about 10 percent of all organic material in a cell. They use up some molecules as ongoing sources of energy. They use others as individual subunits, or **monomers**, of the larger molecules necessary for their structure and functioning. The larger molecules—**polymers**—consist of three to millions of subunits that may or may not be identical. When they are broken apart, the released monomers might be used at once for energy, or they might reenter the cellular pools as free molecules.

A VARIETY OF FUNCTIONAL GROUPS

Functional groups, again, are lone atoms or clusters of atoms covalently bonded to carbon atoms of organic compounds. Each has specific chemical and physical properties that are consistent from one molecule to the next. How do such groups differ from hydrocarbon regions? They are more reactive. Important features of carbohydrates, lipids, proteins, and nucleic acids arise from the number, kind, and arrangement of functional groups, such as those shown in Figure 3.4.

Consider: Sugars in your diet belong to a class of organic compounds, the **alcohols**, which have one or more *hydroxyl* groups (—OH). Enzyme action can split

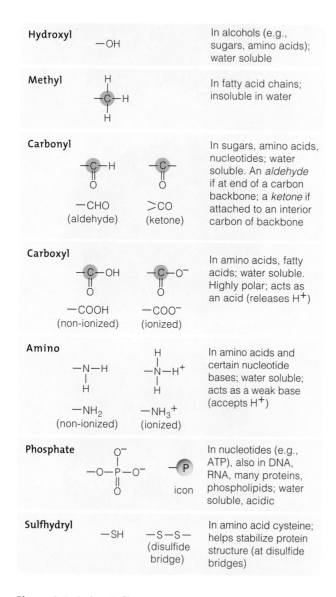

Hydroxyl	—OH	In alcohols (e.g., sugars, amino acids); water soluble
Methyl		In fatty acid chains; insoluble in water
Carbonyl	—CHO (aldehyde) / >CO (ketone)	In sugars, amino acids, nucleotides; water soluble. An *aldehyde* if at end of a carbon backbone; a *ketone* if attached to an interior carbon of backbone
Carboxyl	—COOH (non-ionized) / —COO⁻ (ionized)	In amino acids, fatty acids; water soluble. Highly polar; acts as an acid (releases H⁺)
Amino	—NH₂ (non-ionized) / —NH₃⁺ (ionized)	In amino acids and certain nucleotide bases; water soluble; acts as a weak base (accepts H⁺)
Phosphate	(P) icon	In nucleotides (e.g., ATP), also in DNA, RNA, many proteins, phospholipids; water soluble, acidic
Sulfhydryl	—SH / —S—S— (disulfide bridge)	In amino acid cysteine; helps stabilize protein structure (at disulfide bridges)

Figure 3.4 *Animated!* Common functional groups in biological molecules, with examples of their occurrences.

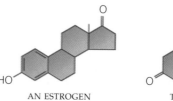

AN ESTROGEN   TESTOSTERONE

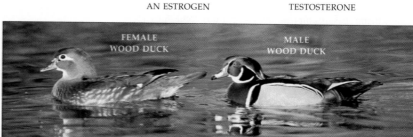

FEMALE WOOD DUCK   MALE WOOD DUCK

Figure 3.5 Observable differences in traits between the male and female wood duck (*Aix sponsa*). Two sex hormones govern the development of feather color and other traits that help males and females recognize each other and so promote reproductive success. Both hormones—testosterone and one of the estrogens—have the same carbon ring structure. They differ in the position of functional groups attached to the ring.

molecules or join them at such groups. Also, small alcohols dissolve swiftly because water molecules hydrogen-bond with them. Larger alcohols do not dissolve quickly because they have hydrocarbon chains, which are water insoluble. Such chains are also part of fatty acids, which is why lipids with fatty acid tails resist dissolving in water.

We find *carbonyl* groups—highly reactive and prone to electron transfers—in carbohydrates and fats, and *carboxyl* groups in amino acids and fatty acids. ATP activates other molecules by giving up *phosphate* groups. This group also combines with sugars to form the backbones of DNA and RNA. *Sulfhydryl* groups help stabilize many proteins.

How much can one functional group do? Look at a seemingly minor difference in the functional groups of two structurally similar sex hormones (Figure 3.5). Early on, an embryo of a wood duck, human, or any other vertebrate is neither male nor female. If it starts making testosterone (a hormone), a set of tubes and ducts will develop into male sex organs and later govern male traits. In the *absence* of testosterone, the ducts and tubes will develop into female sex organs. In that case, estrogens will guide the development of female traits.

FIVE CATEGORIES OF REACTIONS

So how do cells actually do the construction work? It will take more than one chapter to sketch out answers (and best guesses) to that question. For now, simply be aware that the reactions by which the cell builds, rearranges, and splits up organic compounds require more than energy inputs. They also require **enzymes**, a class of proteins that cause metabolic reactions to proceed much faster than they would on their own. Different enzymes mediate different reactions. In later chapters, you will come across specific examples of these five categories of reactions:

1. *Functional-group transfer*. One molecule gives up a functional group entirely, and a different molecule immediately accepts it.

2. *Electron transfer*. One or more electrons stripped from one molecule are donated to another molecule.

3. *Rearrangement*. Juggling of internal bonds converts one type of organic compound into another.

4. *Condensation*. Covalent bonds join two molecules into a larger molecule.

5. *Cleavage*. A molecule splits into two smaller ones.

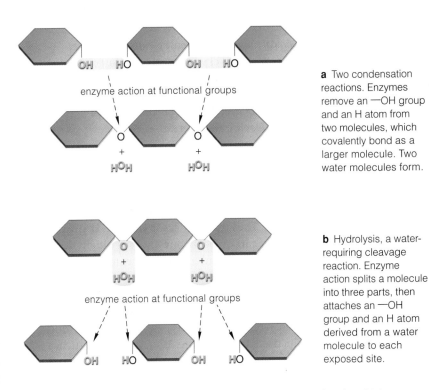

a Two condensation reactions. Enzymes remove an —OH group and an H atom from two molecules, which covalently bond as a larger molecule. Two water molecules form.

b Hydrolysis, a water-requiring cleavage reaction. Enzyme action splits a molecule into three parts, then attaches an —OH group and an H atom derived from a water molecule to each exposed site.

Figure 3.6 *Animated!* Examples of the metabolic reactions by which most biological molecules are synthesized, rearranged, or broken apart.

To get a sense of these cell activities, think about a **condensation reaction**. Enzymes split an —OH group from one molecule and an H atom from another, and a covalent bond forms at the exposed sites on both of the fragments. The discarded atoms often form water (Figure 3.6*a*). Starch and other large polymers form by way of repeated condensation reactions.

Another example: A type of cleavage reaction called **hydrolysis** is like condensation, but in reverse (Figure 3.6*b*). Enzymes split molecules at specific groups, then attach one —OH group and an H atom derived from a water molecule to the exposed sites. Cells can cleave polymers into smaller molecules when these are required for building blocks or for energy.

Cells build large molecules mainly from four families of small organic compounds called simple sugars, fatty acids, amino acids, and nucleotides.

Functional groups covalently bonded to carbon backbones add enormously to the structural and functional diversity of organic compounds, cells, and multicelled organisms.

Cells continually assemble, rearrange, and degrade organic compounds by enzyme-mediated reactions involving the transfer of functional groups or electrons, rearrangement of internal bonds, and a combining or splitting of molecules.

3.3   The Most Abundant Ones—Carbohydrates

Which biological molecules are most plentiful in nature? Carbohydrates. Most carbohydrates consist of carbon, hydrogen, and oxygen in a 1:2:1 ratio. Cells use them as structural materials and transportable or storable forms of energy. Monosaccharides, oligosaccharides, and polysaccharides are the main classes.

THE SIMPLE SUGARS

"Saccharide" is from a Greek word that means sugar. The *mono*saccharides (one sugar unit) are the simplest carbohydrates. They have at least two —OH groups bonded to their carbon backbone and one aldehyde or ketone group. Most dissolve easily in water. Common types have a backbone of five or six carbon atoms that tends to form a ring structure when dissolved.

Ribose and deoxyribose are the sugar monomers of RNA and DNA, respectively; each has five carbon atoms. Glucose has six (Figure 3.7*a*). Cells use glucose as an instant energy source, as a building block, and

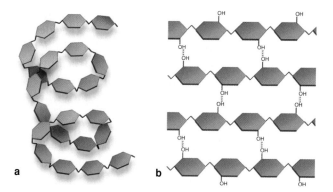

Figure 3.8   Bonding patterns for glucose units in (**a**) starch, and (**b**) cellulose. In amylose, a form of starch, a series of covalently bonded glucose units form a chain that coils. In cellulose, bonds form between glucose chains. The pattern stabilizes the chains, which can become tightly bundled.

as a precursor (parent molecule). For instance, glucose might be remodeled into vitamin C (a sugar acid) or into glycerol, an alcohol with three —OH groups.

SHORT-CHAIN CARBOHYDRATES

Unlike the simple sugars, an *oligo*saccharide is a short chain of covalently bonded sugar monomers. (*Oligo*– means a few.) The *di*saccharides consist of two sugar monomers. Lactose, a disaccharide in milk, consists of one glucose and one galactose unit. Sucrose, the most plentiful sugar in nature, has a glucose and a fructose unit (Figure 3.7*c*). Table sugar is sucrose extracted from sugarcane and sugar beets. Many proteins and lipids have oligosaccharide side chains. Later in the book, you will learn about oligosaccharide side chains that function in self-recognition, immunity, and other tasks. They are components of diverse molecules that are like docks and flags at the cell surface.

COMPLEX CARBOHYDRATES

The "complex" carbohydrates, or *poly*saccharides, are straight or branched chains of many sugar monomers —often hundreds or thousands. Different kinds have one or more types of monomers. The most common kinds are cellulose, starch, and glycogen. All three consist of glucose but differ in their properties (Figures 3.8 and 3.9). Why? The answer starts with differences in the covalent bonding patterns between their glucose units, which are joined together in chains.

In starch, the pattern of covalent bonding puts each glucose unit at an angle relative to the next unit in line. The chain ends up coiling like a spiral staircase

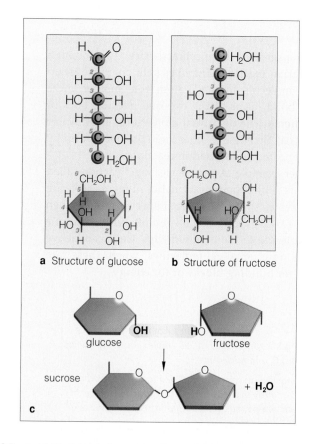

a Structure of glucose   **b** Structure of fructose

glucose   fructose

sucrose   + **H₂O**

Figure 3.7   (**a**,**b**) Straight-chain and ring forms of glucose and fructose. For reference purposes, carbon atoms of these simple sugars are numbered in sequence, starting at the end closest to the molecule's aldehyde or ketone group. (**c**) Condensation of two monosaccharides into a disaccharide.

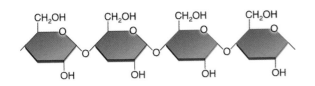

a Structure of amylose, a soluble form of starch. Cells inside tree leaves briefly store excess glucose monomers as starch grains in their chloroplasts, which are tiny, membrane-bound sacs that specialize in photosynthesis.

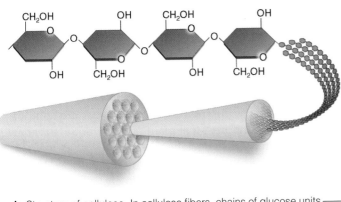

b Structure of cellulose. In cellulose fibers, chains of glucose units stretch side by side and hydrogen-bond at —OH groups. The many hydrogen bonds stabilize the chains in tight bundles that form long fibers. Few organisms produce enzymes that can digest this insoluble material. Cellulose is a structural component of plants and plant products, such as wood and cotton dresses.

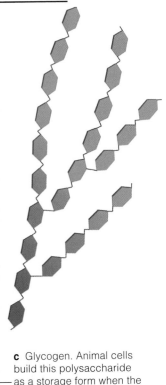

c Glycogen. Animal cells build this polysaccharide as a storage form when the body has excess glucose. It is especially abundant in the liver and muscles of highly active animals, including fishes and people.

Figure 3.9 Molecular structure of (**a**) starch, (**b**) cellulose, and (**c**) glycogen, and their typical locations in a few organisms. All three carbohydrates consist only of glucose units.

(Figure 3.8*a*). Many —OH groups project out from the coils, which makes the chains accessible for cleavage reactions. This is important. For example, plants store much of their photosynthetically produced glucose in the form of starch. When free glucose is in short supply, enzymes can quickly hydrolyze the starch.

In cellulose, glucose chains stretch side by side and hydrogen-bond to one another, as in Figure 3.8*b*. The bonding arrangement stabilizes the chains in a tightly bundled pattern, which can resist hydrolysis by most enzymes. Long fibers of cellulose are a structural part of plant cell walls (Figure 3.9*b*). Like the steel rods in reinforced concrete, these fibers are tough, insoluble, and resistant to weight loads and mechanical stress, as when stems are buffeted by strong winds.

In animals, glycogen is the sugar-storage equivalent of starch in plants (Figure 3.9*c*). Muscle and liver cells store a lot of it. When the sugar level in blood falls, liver cells degrade glycogen, and the released glucose enters the blood. Exercise strenuously but briefly, and muscle cells tap glycogen for a burst of energy.

Chitin is a modified polysaccharide, with nitrogen-containing groups attached to its glucose monomers.

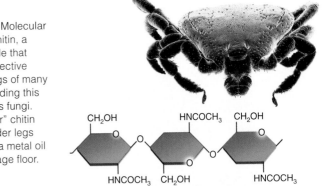

Figure 3.10 Molecular structure of chitin, a polysaccharide that occurs in protective body coverings of many animals, including this tick, as well as fungi. You may "hear" chitin when big spider legs clack across a metal oil pan on a garage floor.

Chitin strengthens the external skeleton and other hard parts of many animals, including crabs, earthworms, insects, spiders, and ticks of the sort shown in Figure 3.10. It also strengthens the cell walls of fungi.

Carbohydrates include simple sugars (such as glucose), oligosaccharides (such as sucrose), and polysaccharides (such as starch). Cells use some carbohydrates as structural materials, others as packets of instant energy, and others as transportable or storable forms of energy.

3.4 Greasy, Oily—Must Be Lipids

Lipids are greasy or oily to the touch. Cells use different lipids as energy reservoirs, structural materials, and signaling molecules. Fats, phospholipids, and waxes have fatty acid tails. Sterols have a backbone of four carbon rings.

FATS AND FATTY ACIDS

Being nonpolar hydrocarbons, **lipids** do not dissolve in water, but they mix with other nonpolar substances —for instance, as butter does in warm cream sauce.

Fats are lipids with one, two, or three fatty acids dangling like tails from a glycerol molecule. A **fatty acid** starts as a carboxyl group attached to a backbone of as many as thirty-six carbon atoms. Each carbon in the backbone has one, two, or three hydrogen atoms covalently bonded to it (Figure 3.11). *Unsaturated* fatty acids contain one or more double covalent bonds. The *saturated* fatty acids have single bonds only.

Weak interactions keep many saturated fatty acids tightly packed in animal fats. These fats are solid at room temperature. Most plant fats stay liquid at room temperature, as "vegetable oils." Their packing is not as stable because of rigid kinks in their fatty acid tails. That is why vegetable oils flow freely.

Neutral fats such as butter, lard, and vegetable oils are mostly **triglycerides**. Each has three fatty acid tails linked to one glycerol (Figure 3.12). Triglycerides are the most abundant lipids in your body and its richest reservoir of energy. Gram for gram, they yield more than twice as much energy as complex carbohydrates such as starches. All vertebrates store triglycerides as droplets in fat cells that make up adipose tissue.

Layers and patches of adipose tissue insulate the body and cushion some of its parts. Like many other

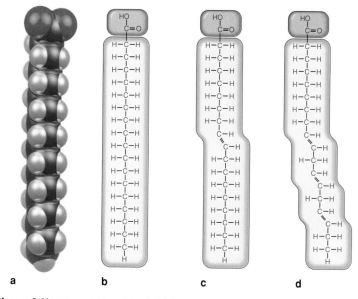

Figure 3.11 Three fatty acids. (**a,b**) Space-filling model and structural formula for stearic acid. The carbon backbone is fully saturated with hydrogen atoms. (**c**) Oleic acid, with a double bond in its backbone, is an unsaturated fatty acid. (**d**) Linolenic acid, also unsaturated, has three double bonds.

Figure 3.12 *Animated!* Condensation of (**a**) three fatty acids and one glycerol molecule into (**b**) a triglyceride. The photograph shows triglyceride-protected emperor penguins during an Antarctic blizzard.

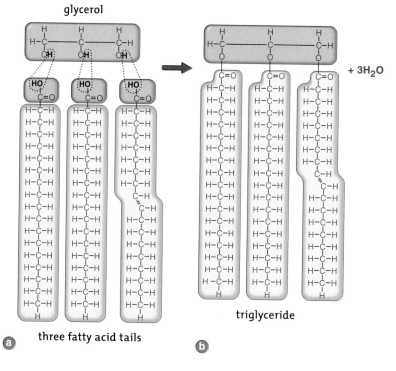

three fatty acid tails

triglyceride

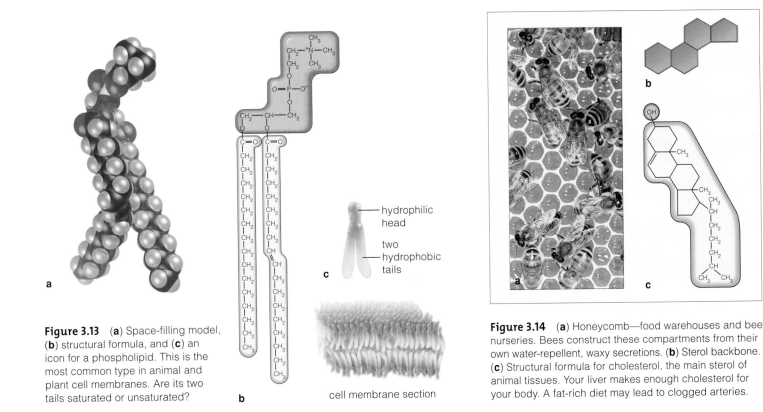

Figure 3.13 (**a**) Space-filling model, (**b**) structural formula, and (**c**) an icon for a phospholipid. This is the most common type in animal and plant cell membranes. Are its two tails saturated or unsaturated?

hydrophilic head

two hydrophobic tails

cell membrane section

Figure 3.14 (**a**) Honeycomb—food warehouses and bee nurseries. Bees construct these compartments from their own water-repellent, waxy secretions. (**b**) Sterol backbone. (**c**) Structural formula for cholesterol, the main sterol of animal tissues. Your liver makes enough cholesterol for your body. A fat-rich diet may lead to clogged arteries.

animals, penguins of the Antarctic can keep warm in extremely cold winter months thanks to a thick layer of triglycerides beneath their skin (Figure 3.12).

PHOSPHOLIPIDS

Phospholipids have a glycerol backbone, two nonpolar fatty acid tails, and a polar head (Figure 3.13). They are the main component of cell membranes, which consist of two layers of lipids. Phospholipid heads of one layer are dissolved in the cell's fluid interior, and phospholipid heads of the other layer are dissolved in the fluid surroundings. Sandwiched between the two are all of the hydrophobic tails. You will read about membrane structure and function in Chapter 5.

WAXES

Waxes have long-chain fatty acids tightly packed and bonded to long-chain alcohols or carbon rings. All have a firm consistency; all repel water. Surfaces of plants have a cuticle that contains waxes and another lipid, cutin. A plant cuticle restricts water loss and thwarts some parasites. Waxes also protect, lubricate, and lend pliability to skin and to hair. Birds secrete waxes, fats, and fatty acids that waterproof feathers. Bees use beeswax for honeycomb, which houses each new bee generation as well as honey (Figure 3.14*a*).

CHOLESTEROL AND OTHER STEROLS

Sterols are among the many lipids with no fatty acids. The sterols differ in the number, position, and type of their functional groups, but all have a rigid backbone of four fused-together carbon rings (Figure 3.14*b*).

Every eukaryotic cell membrane contains sterols. Cholesterol (Figure 3.14*c*) is the most common type in animal tissues. It is remodeled into compounds as diverse as bile salts, steroids, and vitamin D, which is required for strong bones and teeth. Bile salts play a part in fat digestion inside the small intestine. The steroids called sex hormones are essential for gamete formation and the development of secondary sexual traits. Such traits include the amount and distribution of hair in mammals, and feather color in birds.

Being largely hydrocarbon, lipids can intermingle with other nonpolar substances, but they resist dissolving in water.

Triglycerides, or neutral fats, have a glycerol head and three fatty acid tails. They are the major energy reservoirs. Phospholipids are the main component of cell membranes.

Sterols such as cholesterol are membrane components and precursors of steroid hormones and other compounds. Waxes are firm yet pliable components of water-repelling and lubricating substances.

3.5 Proteins—Diversity in Structure and Function

Of all large biological molecules, proteins are the most diverse. Some kinds speed reactions; others are the stuff of spider webs or feathers, bones, hair, and other body parts. Nutritious types abound in seeds and eggs. Many proteins move substances, help cells communicate, or defend against pathogens. Amazingly, cells assemble thousands of different proteins from only twenty kinds of amino acids.

An **amino acid** is a small organic compound with an amino group ($-NH_3^+$), a carboxyl group ($-COO^-$, the acid), a hydrogen atom, and one or more atoms called its R group. In most cases, these components are attached to the same carbon atom (Figure 3.15a). Appendix V shows all of the biological amino acids.

When a cell constructs a protein, it strings amino acids together, one after the other. Instructions coded

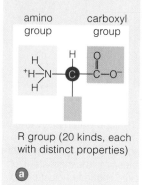

amino group   carboxyl group

R group (20 kinds, each with distinct properties)

a

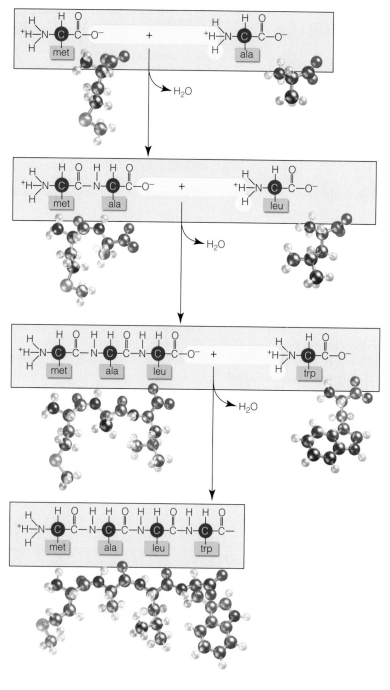

b Instructions encoded in DNA specify the order of amino acids to be joined in a polypeptide chain. The first amino acid is usually methionine (met). Alanine (ala) comes next in this example.

c In a condensation reaction, a peptide bond forms between the methionine and alanine. Leucine (leu) is next in line.

d A peptide bond forms between the alanine and the leucine. Tryptophan (trp) is next.

Figure 3.15 *Animated!*
(**a**) Generalized formula for amino acids. The *green* box highlights the R group, one of the side chains that include functional groups. Appendix V shows ball-and-stick models for twenty amino acids.

(**b–e**) Peptide bond formation during protein synthesis. Section 14.4 offers a closer look at protein synthesis.

e Part of the newly formed polypeptide chain. The sequence of amino acids in this part is met–ala–leu–trp. The reactions may continue until there are hundreds or thousands of amino acids in the chain.

Figure 3.16 The first three of four levels of protein structure. (**a**) Primary structure is a linear sequence of amino acids. (**b**) Many hydrogen bonds (dotted lines) along a polypeptide chain result in a helically coiled or sheetlike secondary structure. (**c**) Coils and sheets packed into stable domains represent a third structural level.

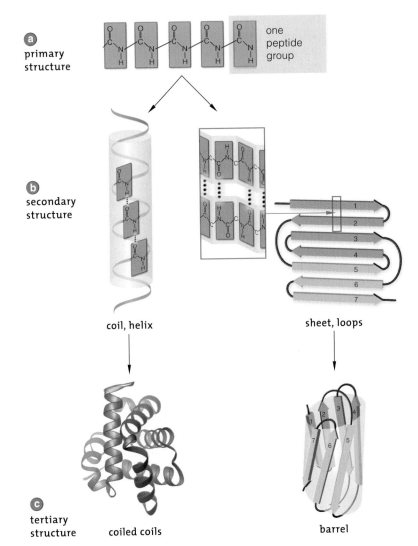

in DNA specify the order in which any of the twenty kinds of amino acids will occur. A *peptide* bond forms as a condensation reaction joins the amino group of one amino acid and the carboxyl group of the next in line (Figure 3.15*b–e*). Each **polypeptide chain** consists of three or more amino acids. The carbon backbone of this chain incorporates nitrogen atoms in this regular pattern: —N—C—C—N—C—C—.

A protein's sequence of amino acids is known as its *primary* structure (Figure 3.16*a*). Its *secondary* structure emerges as the chain twists, bends, loops, and folds. Hydrogen bonding between certain R groups makes some stretches of amino acids coil helically, a bit like a spiral staircase, or makes them form sheets or loops as in Figure 3.16*b*. Bear in mind, part of the primary structure for each type of protein is unique in certain respects, but you will come across similar patterns of coils, sheets, and loops among different proteins.

Much as an overly twisted rubber band coils back on itself, the coils, sheets, and loops of a protein fold up even more, into compact domains. A "domain" is a polypeptide chain or a part of it that has become organized as a structurally stable unit. This third level of organization is a protein's *tertiary* structure. The shape of domains and the charge distribution around that shape determines protein function. For instance, the barrel-shaped domains of some proteins function as tunnels through membranes (Figure 3.16*c*).

Many proteins are two or more polypeptide chains bonded together or associating intimately with one another. This is the fourth level of organization, or *quaternary* protein structure. Many enzymes and other proteins are globular, with several polypeptide chains folded into rounded shapes. Hemoglobin, described shortly, is a classic example of such a protein.

Protein structure doesn't stop here. Enzymes often attach short, linear, or branched oligosaccharides to a new polypeptide chain, making a *glyco*protein. Many glycoproteins occur at the cell surface or are secreted from cells. Lipids also get attached to many proteins. The cholesterol, triglycerides, and phospholipids that your body absorbs after a meal are transported about as components of *lipo*proteins.

Many proteins are fibrous, with polypeptide chains organized as strands or sheets. They contribute to cell shape and organization, and help cells and cell parts move about. Other fibrous proteins make up cartilage, hair, skin, and parts of muscles and brain cells.

A protein's primary structure is a sequence of covalently bonded amino acids that make up a polypeptide chain.

Local regions of a polypeptide chain become twisted and folded into helical coils, sheetlike arrays, and loops. These arrangements are the protein's secondary structure.

A polypeptide chain or parts of it become organized as structurally stable, compact, functional domains. Such domains are a protein's tertiary structure.

Many proteins show quaternary structure; they consist of two or more polypeptide chains.

A protein's shape and charge distribution around that shape dictate protein function.

3.6 Why Is Protein Structure So Important?

LINK TO
SECTION
1.4

Cells are good at making proteins that are just what their DNA specifies. But mistakes and mutations happen, and they may alter the protein's primary structure in bad ways. The consequences are sometimes far-reaching.

JUST ONE WRONG AMINO ACID . . .

Four tightly packed polypeptides called globins make up each hemoglobin molecule. Each globin chain is folded into a pocket that cradles a **heme** group, a large organic molecule with an iron atom at its center (Figure 3.17). Heme is an oxygen transporter. During its life span, each of the red blood cells in your body transports billions of oxygen molecules, all bound to the heme in globin molecules.

Globin comes in two slightly different forms, alpha and beta. Two of each form make up one hemoglobin molecule in adult humans. Glutamate is normally the sixth amino acid in the beta globin chain, but a DNA mutation sometimes puts a different amino acid—valine—in the chain's sixth position (Figure 3.18*b*). Unlike glutamate, which carries an overall negative charge, valine has no net charge. As a result of that one substitution, a tiny patch of the protein changes from polar to nonpolar—which in turn causes globin's behavior to change slightly. Hemoglobin that has this mutation in its beta chain is designated HbS.

. . . AND SICKLE-CELL ANEMIA MAY FOLLOW!

Every human inherits two genes for beta globin, one from each of two parents. (Genes are units of DNA that encode heritable traits.) Cells access both genes when they make beta globin. If one gene is normal and the other has the valine mutation, a person makes enough normal hemoglobin and can lead a relatively normal life. Someone who inherits two mutant genes can only make hemoglobin HbS. The outcome, *sickle-cell anemia*, is a severe genetic disorder.

As blood moves through lungs, hemoglobin in red blood cells binds oxygen and then gives it up in body regions where oxygen levels are low. After the oxygen is released, red blood cells quickly return to the lungs and pick up more. In the few moments when they have no bound oxygen, hemoglobin molecules clump together just a bit. However, HbS molecules do not form such clusters in regions where oxygen levels are low. They form large, stable, rod-shaped aggregates.

Red blood cells containing these aggregates become distorted into sickle shapes (Figure 3.18*c*). These cells clog tiny blood vessels and disrupt blood circulation. Tissues become oxygen-starved. Figure 3.18*d* lists the far-reaching effects of sickle-cell anemia.

PROTEINS UNDONE—DENATURATION

Environmental conditions, too, can skew a protein's functioning. Globin cradles heme, an enzyme speeds some reaction, a receptor transduces an energy signal. These proteins and others cannot function unless they stay coiled, folded, and packed in a precise way. Their shape depends on many hydrogen bonds and other interactions that heat, shifts in pH, or detergents can disrupt. At such times, polypeptide chains unwind and change shape in an event called **denaturation**.

Consider albumin, a protein in the white of an egg. When you cook eggs, the heat does not disrupt the covalent bonds of albumin's primary structure. But it destroys albumin's weaker hydrogen bonds, and so the protein unfolds. When the translucent egg white turns opaque, we know albumin has been altered. For a few proteins, denaturation might be reversed if and

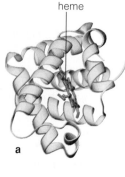

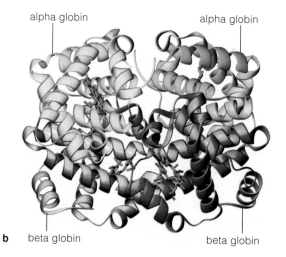

Figure 3.17 *Animated!*
(**a**) Globin, a coiled polypeptide chain. The chain cradles heme, a functional group that contains an iron atom. (**b**) Hemoglobin, an oxygen-transport protein in red blood cells. This is one of the proteins with quaternary structure. It consists of four globin molecules held together by hydrogen bonds. To help you distinguish among them, the two alpha globin chains are color-coded yellow and orange, and the two beta globins are color-coded blue and green.

heme

alpha globin   alpha globin

a

b   beta globin   beta globin

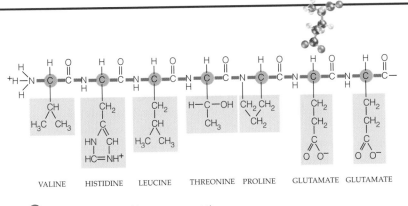

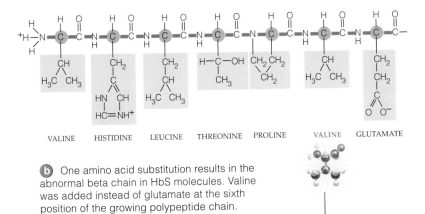

a Normal amino acid sequence at the start of a beta chain for hemoglobin.

VALINE   HISTIDINE   LEUCINE   THREONINE   PROLINE   GLUTAMATE   GLUTAMATE

b One amino acid substitution results in the abnormal beta chain in HbS molecules. Valine was added instead of glutamate at the sixth position of the growing polypeptide chain.

VALINE   HISTIDINE   LEUCINE   THREONINE   PROLINE   VALINE   GLUTAMATE

c Glutamate has an overall negative charge; valine has no net charge. This difference gives rise to a water-repellent, sticky patch on HbS molecules. They stick together because of that patch, forming rod-shaped clumps that distort normally rounded red blood cells into sickle shapes. (A sickle is a farm tool that has a crescent-shaped blade.)

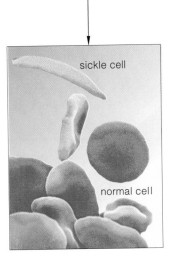

sickle cell

normal cell

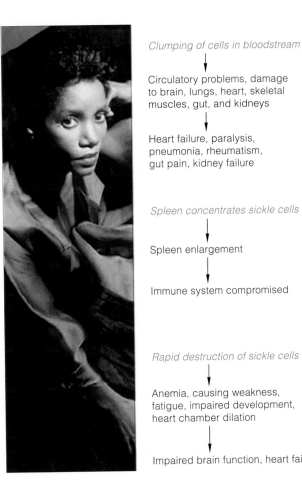

Clumping of cells in bloodstream

Circulatory problems, damage to brain, lungs, heart, skeletal muscles, gut, and kidneys

Heart failure, paralysis, pneumonia, rheumatism, gut pain, kidney failure

Spleen concentrates sickle cells

Spleen enlargement

Immune system compromised

Rapid destruction of sickle cells

Anemia, causing weakness, fatigue, impaired development, heart chamber dilation

Impaired brain function, heart failure

d Melba Moore, celebrity spokesperson for sickle-cell anemia organizations. *Right*, range of symptoms for a person with two mutated genes (Hb^S) for hemoglobin's beta chain.

Figure 3.18 *Animated!* Sickle-cell anemia's molecular basis and its symptoms. Section 18.6 explores evolutionary and ecological aspects of this genetic disorder.

when normal conditions return, but albumin isn't one of them. There is no way to uncook an egg.

What is the take-home lesson? *A protein's structure dictates its function.* Hemoglobin, hormones, enzymes, transporters—such proteins help us survive. Twists and folds in their polypeptide chains form anchors, or membrane-spanning barrels, or jaws that grip enemy agents in the body. Mutations can alter the chains enough to block or enhance an anchoring, transport, or defensive function. Sometimes the consequences are awful. Yet changes in sequences and functional domains also give rise to variation in traits—the raw material for evolution. *Learn about protein structure and function and you are on your way to comprehending life in its richly normal and abnormal expressions.*

> The structure of proteins dictates function. Mutations that alter a protein's structure sometimes have drastic consequences for its function, and for the health of organisms harboring them.

3.7 Nucleotides, DNA, and the RNAs

Certain small organic compounds called nucleotides are energy carriers, enzyme helpers, and messengers. Some are the building blocks for DNA and RNA. They are central to metabolism, survival, and reproduction.

Nucleotides have one sugar, at least one phosphate group, and one nitrogen-containing base. Deoxyribose or ribose is the sugar. Both sugars have a five-carbon ring structure; ribose has an oxygen atom attached to carbon 2 of the ring and deoxyribose does not. The bases have a single or double carbon ring structure.

The nucleotide **ATP** (adenosine triphosphate) has a row of three phosphate groups attached to its sugar (Figure 3.19). ATP can readily transfer the outermost phosphate group to many other molecules and make them reactive. Such transfers are vital for metabolism.

Other nucleotides have different metabolic roles. Some are **coenzymes**, necessary for enzyme function. They move electrons and hydrogen from one reaction site to another. NAD+ and FAD are major kinds.

Still other nucleotides act as chemical messengers within and between cells. Later in the book, you will read about one of these messengers, which is known as cAMP (cyclic adenosine monophosphate).

Certain nucleotides also function as monomers for single- and double-stranded molecules called **nucleic acids**. In such strands, a covalent bond forms between the sugar of one nucleotide and the phosphate group of the next (Figure 3.20). The nucleic acids DNA and RNA store and retrieve heritable information.

All cells start life and then maintain themselves with instructions in their double-stranded molecules of deoxyribonucleic acid, or **DNA**. This nucleic acid is made of four kinds of deoxyribonucleotides. Figure 3.20a shows their structural formulas. As you can see, the four differ only in their component base, which is adenine, guanine, thymine, or cytosine.

Figure 3.21 shows how hydrogen bonds between bases join the two strands along the length of a DNA molecule. Think of every "base pairing" as one rung of a ladder, and the two sugar–phosphate backbones as the ladder's two posts. The ladder twists and turns in a regular pattern, forming a double helical coil.

The sequence of bases in DNA encodes heritable information about all the proteins that give each new cell the potential to grow, maintain itself, and even to reproduce. Part of that sequence is unique for each species. Some parts are identical, or nearly so, among many species. We return to DNA's structure and its functions in Chapter 13.

Figure 3.19
The structural formula for an ATP molecule.

base (*blue*)

three phosphate groups

sugar (*red*)

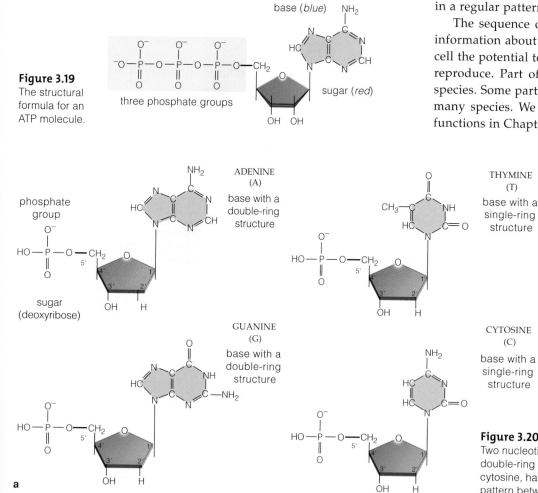

phosphate group

sugar (deoxyribose)

ADENINE (A)
base with a double-ring structure

GUANINE (G)
base with a double-ring structure

THYMINE (T)
base with a single-ring structure

CYTOSINE (C)
base with a single-ring structure

a

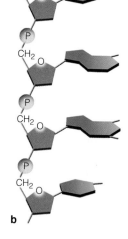

b

Figure 3.20 *Animated!* (**a**) Nucleotides of DNA. Two nucleotide bases, adenine and guanine, have a double-ring structure. The two others, thymine and cytosine, have a single-ring structure. (**b**) Bonding pattern between successive bases in nucleic acids.

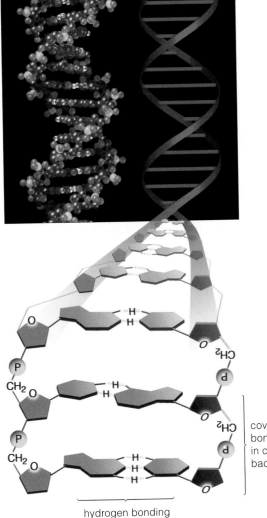

covalent bonding in carbon backbone

hydrogen bonding between bases

Figure 3.21 Models for the DNA molecule.

The **RNAs** (ribonucleic acids) have four kinds of ribonucleotide monomers. Unlike DNA, most RNAs are single strands, and one base is uracil instead of thymine. One type of RNA is a messenger that carries eukaryotic DNA's protein-building instructions out of the nucleus and into the cytoplasm, where they are translated into proteins by other RNAs. Chapter 14 returns to RNA and its role in protein synthesis.

Different nucleotides function as coenzymes, energy carriers such as ATP, chemical messengers, and building blocks for the nucleic acids DNA and the RNAs.

DNA consists of two nucleotide strands joined by hydrogen bonds and twisted as a double helix. Its nucleotide sequence encodes heritable protein-building information.

RNA usually is a single-stranded nucleic acid. Different RNAs have roles in processes by which a cell retrieves and uses genetic information in DNA to build proteins.

Summary

Section 3.1 Under present-day conditions in nature, only living cells can synthesize complex carbohydrates and lipids, proteins, and nucleic acids—the molecules of life. These molecules differ in their three-dimensional structure and function, starting with the carbon backbone and functional groups. Their structure affords clues to how cells, and multicelled organisms, function.

Biology⬡Now
Read the InfoTrac article "The Form Counts: Proteins, Fats, and Carbohydrates," Beatrice Trum, Consumer's Research Magazine, August 2001.

Section 3.2 All organic compounds have carbon and at least one hydrogen atom. Carbon atoms bond covalently with as many as four other atoms, often in long chains or rings. Functional groups attached to a carbon backbone influence an organic compound's properties. Enzyme-driven reactions synthesize all of the molecules of life from smaller organic molecules. Table 3.1 on the next page summarizes these compounds.

Biology⬡Now
Explore functional groups and view the animation of condensation and hydrolysis on BiologyNow.

Section 3.3 The main carbohydrates are simple sugars, oligosaccharides, and polysaccharides. Cells use carbohydrates as instant energy sources, transportable or storage forms of energy, and structural materials.

Section 3.4 Lipids are greasy or oily compounds that tend not to dissolve in water but mix easily with nonpolar compounds, such as other lipids. Neutral fats (triglycerides), phospholipids, waxes, and sterols are lipids. Cells use lipids as major sources of energy and as structural materials, as in cell membranes.

Biology⬡Now
Watch an animation showing how a triglyceride forms by condensation on BiologyNow.

Section 3.5 Structurally and functionally, proteins are the most diverse molecules of life. Their primary structure is a sequence of amino acids—a polypeptide chain. Such chains twist, coil, and bend into functional domains. Many proteins, including hemoglobin and most enzymes, consist of two or more chains. Certain aggregations of proteins form hair, muscle, connective tissue, and other body parts.

Biology⬡Now
Explore amino acid structure and learn about peptide bond formation with the animation on BiologyNow.
Read the InfoTrac article "Protein Folding and Misfolding," David Gossard, American Scientist, September 2002.

Section 3.6 A protein's overall structure determines its function. Sometimes a mutation in DNA results in an amino acid substitution that alters the protein's structure in ways that cause genetic diseases, including sickle-cell anemia. Weak bonds that hold a protein's

Table 3.1 Summary of the Main Organic Compounds in Living Things

Category	Main Subcategories	Some Examples and Their Functions	
CARBOHYDRATES . . . contain an aldehyde or a ketone group, and one or more hydroxyl groups	**Monosaccharides** (simple sugars) **Oligosaccharides** (short-chain carbohydrates) **Polysaccharides** (complex carbohydrates)	Glucose Sucrose (a disaccharide) Starch, glycogen Cellulose	Energy source Most common form of sugar; the form transported through plants Energy storage Structural roles
LIPIDS . . . are mainly hydrocarbon; generally do not dissolve in water but do dissolve in nonpolar substances, such as other lipids	**Lipids with fatty acids** *Glycerides:* Glycerol backbone with one, two, or three fatty acid tails *Phospholipids:* Glycerol backbone, phosphate group, one other polar group, and (often) two fatty acids *Waxes:* Alcohol with long-chain fatty acid tails **Lipids with no fatty acids** *Sterols:* Four carbon rings; the number, position, and type of functional groups differ among sterols	Fats (e.g., butter), oils (e.g., corn oil) Phosphatidylcholine Waxes in cutin Cholesterol	Energy storage Key component of cell membranes Conservation of water in plants Component of animal cell membranes; precursor of many steroids and vitamin D
PROTEINS . . . are one or more polypeptide chains, each with as many as several thousand covalently linked amino acids	**Fibrous proteins** Long strands or sheets of polypeptide chains; often tough, water-insoluble **Globular proteins** One or more polypeptide chains folded into globular shapes; many roles in cell activities	Keratin Collagen Enzymes Hemoglobin Insulin Antibodies	Structural component of hair, nails Structural component of bone Great increase in rates of reactions Oxygen transport Control of glucose metabolism Tissue defense
NUCLEIC ACIDS (AND NUCLEOTIDES) . . . are chains of units (or individual units) that each consist of a five-carbon sugar, phosphate, and a nitrogen-containing base	**Adenosine phosphates** **Nucleotide coenzymes** **Nucleic acids** Chains of nucleotides	ATP cAMP (Section 36.2) NAD^+, $NADP^+$, FAD DNA, RNAs	Energy carrier Messenger in hormone regulation Transfer of electrons, protons (H^+) from one reaction site to another Storage, transmission, translation of genetic information

shape can be disrupted by temperature, pH shifts, or exposure to detergent. Usually, the disruptions make the protein unfold permanently.

Biology Now

Learn more about hemoglobin structure and sickle-cell mutation by viewing the animation on BiologyNow.

Section 3.7 There are different kinds of nucleotides, but all consist of a sugar, a phosphate group, and a nitrogen-containing base. They have essential roles in metabolism, survival, and reproduction. ATP energizes many kinds of molecules by phosphate-group transfers. Other nucleotides function as coenzymes or chemical messengers. DNA and RNA are nucleic acids, each composed of four kinds of nucleotide subunits.

DNA's nucleotide bases encode information on the primary structure of all of the cell's proteins. Different kinds of RNA molecules interact with DNA and with one another in the translation of that information.

Biology Now

Explore DNA with the animation on BiologyNow.

Self-Quiz

Answers in Appendix II

1. Name the molecules of life and the families of small organic compounds from which they are built.

2. Each carbon atom can share pairs of electrons with as many as _____ other atoms.
   a. one   b. two   c. three   d. four

3. Sugars are a class of _____ , which have one or more _____ groups.
   a. proteins; amino   c. alcohols; hydroxyl
   b. acids; phosphate   d. carbohydrates; carboxyl

4. _____ is a simple sugar (a monosaccharide).
   a. Glucose   c. Ribose   e. both a and b
   b. Sucrose   d. Chitin   f. both a and c

5. The fatty acid tails of unsaturated fats incorporate one or more _____ .
   a. single covalent bonds   b. double covalent bonds

6. Sterols are among the many lipids with no _____ .
   a. saturation   c. hydrogens
   b. fatty acids   d. carbons

7. Which of the following is a class of molecules that encompasses all of the other molecules listed?
   a. triglycerides   c. waxes   e. lipids
   b. fatty acids   d. sterols   f. phospholipids

8. _____ are to proteins as _____ are to nucleic acids.
   a. Sugars; lipids   c. Amino acids; hydrogen bonds
   b. Sugars; proteins   d. Amino acids; nucleotides

9. A denatured protein has lost its _____ .
   a. hydrogen bonds   c. function
   b. shape   d. all of the above

10. Nucleotides occur in _____ .
    a. ATP   b. DNA   c. RNA   d. all are correct

11. Which of the following nucleotides is *not* found in DNA?
    a. adenine   b. uracil   c. thymine   d. guanine

12. Match the molecule with the most suitable description.
    ____ long sequence of amino acids   a. carbohydrate
    ____ energy carrier in cells   b. phospholipid
    ____ glycerol, fatty acids, phosphate c. polypeptide
    ____ two strands of nucleotides   d. DNA
    ____ one or more sugar monomers   e. ATP

Additional questions are available on Biology ⊜ Now™

Critical Thinking

1. In the following list, identify which is the carbohydrate, the fatty acid, the amino acid, and the polypeptide:

   a. $^+NH_3$—CHR—COO$^-$ c. (glycine)$_{20}$

   b. $C_6H_{12}O_6$ d. $CH_3(CH_2)_{16}COOH$

2. A clerk in a health-food store tells you that "natural" vitamin C extracts from rose hips are better than synthetic tablets of this vitamin. Given what you know about the structure of organic compounds, how would you respond?

3. It seems there are "good" and "bad" unsaturated fats. The double bonds of both put a bend in their fatty acid tails. The bend in *trans* fatty acid tails keeps them aligned in the same direction along their length. The bend in *cis* fatty acid tails makes them zigzag (Figure 3.22).

   Some *trans* fatty acids occur naturally in beef. But most form by industrial processes that solidify vegetable oils for margarine, shortening, and the like. These substances are widely used in prepared foods (such as cookies) and in french fries and other fast-food products. *Trans* fatty acids are linked to heart attacks. Speculate on why the body handles *cis* fatty acids better than *trans* fatty acids.

4. The shapes of a protein's domains often give us clues to functions. For example, Figure 3.23 is a model for one of the HLAs, a type of recognition protein perched above the surface of all vertebrate body cells. Certain cells of the immune system use HLAs to distinguish self (the body's own cells) from nonself. Each HLA has a jawlike region that can bind bits of an invader or some other threat. It thus alerts the immune defenders that the body has been invaded or otherwise threatened. Speculate on what may happen if a mutation makes the jawlike region misfold.

5. Cholesterol from food or synthesized in the liver is too hydrophobic to circulate in blood; complexes of protein and lipids ferry it around. Low density lipoprotein, or *LDL,* transports cholesterol out of the liver and into cells.

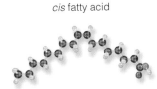

cis fatty acid

trans fatty acid

Figure 3.22 Maybe rethink the french fries?

High density lipoprotein, or *HDL,* ferries the cholesterol that is released from dead cells back to the liver.

High LDL levels are implicated in atherosclerosis, heart problems, and strokes. The main protein in LDL is called ApoA1. A mutant form of ApoA1 has the wrong amino acid (cysteine instead of arginine) at one place in its primary sequence. Carriers of this LDL mutation have very low levels of HDL, which is typically predictive of heart disease. Yet the carriers have no heart problems.

Some heart patients received injections of the mutant LDL, which acted like a drain cleaner. It quickly reduced the size of cholesterol deposits in the patients' arteries.

A few years from now, such a treatment may reverse years of damage. However, many researchers caution that a low-fat, low-cholesterol diet is still the best assurance of long-term health. Would you choose artery-cleansing treatments over a healthy diet?

where the molecule binds and displays "enemies" (*arrow*)

one of the chains spans the plasma membrane and anchors the molecule

Figure 3.23 From structure to function—a protein that helps your body defend itself against bacteria and other foreign agents. HLA-A2 has two polypeptide chains that are like jaws. Another protein anchors it to the plasma membrane.

4 CELL STRUCTURE AND FUNCTION

Animalcules and Cells Fill'd With Juices

Do you ever think of yourself as being close to 1/1,000 of a kilometer tall? Probably not. Yet that is how we refer to cells. We measure them in micrometers—in millionths of a millimeter, which is a thousandth of a meter, which is a thousandth of a kilometer. The bacterial cells in Figure 4.1 are a few micrometers "tall."

Before those cells were fixed on the head of a pin so someone could take their picture, they were the living descendants of a lineage far more ancient than yours. Their line of descent goes back so far they do not even have their DNA housed in a nucleus. They are prokaryotes, which are, at least structurally, the simplest cells of all. Your cells are eukaryotic. Somewhere in time, a nucleus developed in their single-celled ancestors, along with a variety of other internal compartments that help keep metabolic activities organized.

Nearly all cells are invisible to the naked eye. No one knew about them until the seventeenth century, when the first microscopes were being put together in Italy, then in France and England. These were not much to speak of. Galileo Galilei, for instance, simply used two glass lenses inside a cylinder, but the arrangement was good enough to reveal details of an insect's eyes. At midcentury, Robert Hooke focused a microscope on thinly sliced cork from a

mature tree and saw tiny compartments (Figure 4.2). He gave them the Latin name *cellulae*, meaning small rooms —hence the origin of the biological term "cell." Actually they were dead plant cell walls, which is what cork is made of, but Hooke did not think of them as being dead because neither he nor anyone else knew cells could be alive. He observed cells "fill'd with juices" in green plant tissues but didn't have a clue to what they were, either.

Given the simplicity of their instruments, it is amazing that the pioneers in microscopy observed as much as they did. Antoni van Leeuwenhoek, a Dutch shopkeeper, had exceptional skill in constructing lenses and possibly the keenest vision. By the late 1600s, he was spying on such wonders as sperm, protists, a bacterium, and "many very small animalcules, the motions of which were very pleasing to behold," in scrapings of tartar from his teeth.

In the 1820s, improved lenses brought cells into sharper focus. Robert Brown, a botanist, was the first to identify a plant cell nucleus. Later, the botanist Matthias Schleiden wondered if a plant cell develops as an independent unit even though it is part of the plant. By 1839, after years of studying animal tissues, the zoologist Theodor Schwann reported that cells and their products make up animals as well as plants. He also reported that cells have an individual

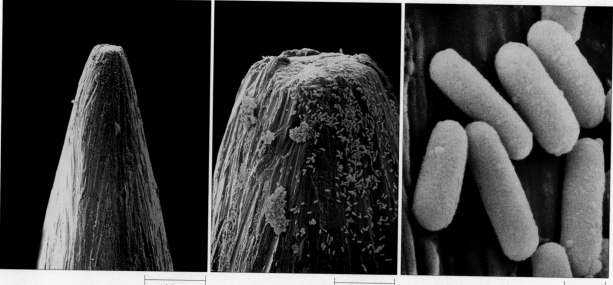

| 100 μm | 20 μm | 0.5 μm |

Figure 4.1 How small are cells? This example will give you an idea. Shown here at increasingly higher magnifications, a population of rod-shaped bacterial cells peppering the tip of a household pin.

Figure 4.2 Robert Hooke's microscope and his sketch of cell walls from cork tissue.

life of their own even when they are part of a multicelled body.

The physiologist Rudolf Virchow completed his own studies of a cell's growth and reproduction—that is, its division into daughter cells. Every cell, he decided, must come from a cell that already exists.

So microscopic analysis yielded three generalizations, which together constitute the **cell theory**. First, organisms consist of one or more cells. Second, the cell is the smallest unit of organization that still displays the properties of life. Third, the continuity of life arises directly from the growth and division of single cells.

This chapter introduces defining features of prokaryotic and eukaryotic cells. It is not meant for memorization. Read it simply to gain an overview of current understandings of cell structure and function. In later chapters, you might refer back to it as a road map through the details. With its images from microscopy, this chapter and others invite you into otherwise invisible worlds. Why bother to travel there? We are close to creating the simplest form of life in test tubes. That is something worth thinking about.

 How Would You Vote?

Researchers are modifying prokaryotes to identify what it takes to be alive. They are creating "new" organisms by removing genes from living cells, one at a time. What are the potential advantages or bioethical pitfalls of this kind of research? See BiologyNow for details, then vote online.

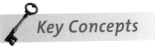 *Key Concepts*

WHAT ALL CELLS HAVE IN COMMON

A plasma membrane is a boundary between the interior of a cell and the surroundings, where the inputs and outputs of substances are controlled. Eukaryotic DNA is enclosed in a nucleus. Prokaryotic DNA is concentrated in a nucleoid. Cytoplasm is everything between the plasma membrane and the region of DNA. Section 4.1

MICROSCOPES

Microscopes employ rays of light or beams of electrons to reveal details at the cellular level of organization. They offer evidence of the cell theory: Each organism consists of one or more cells and their products, a cell has a capacity for independent life, and since the time of life's origin, all cells have arisen from cells that already exist. Section 4.2

PROKARYOTIC CELLS

Structurally, prokaryotic cells are the simplest and most ancient forms of life. Archaeans and bacteria are the only groups. Collectively, they show great metabolic diversity. Different kinds live in or on other organisms and in Earth's waters, soils, sediments, and rocky layers. Section 4.3

EUKARYOTIC CELLS

Organelles—small, membrane-bounded sacs—divide the interior of eukaryotic cells into functional compartments. All cells of protists, plants, fungi, and animals start out life with a nucleus; they are eukaryotic. They differ in the type and number of organelles, in cell structures, and in surface specializations. Sections 4.4–4.9

THE CYTOSKELETON

Arrays of a variety of protein filaments reinforce cell shape and keep its parts organized. Some filaments assemble and disassemble in dynamic ways that can move cells or their inner components to new locations. Sections 4.10–4.11

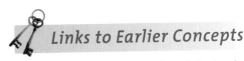 *Links to Earlier Concepts*

Look back on your road map through the levels of organization in nature (Section 1.1). With this chapter you arrive at the level of living cells. You will start to see how lipids are structurally organized as cell membranes (3.4), where DNA and RNA reside in cells (3.7), and where carbohydrates are built and broken apart (3.2–3.3). You will expand your view of how cell structure and function depend on proteins (3.5–3.6).

4.1 So What Is "A Cell"?

LINK TO
SECTION
3.4

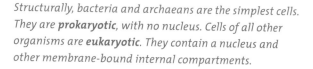

*Structurally, bacteria and archaeans are the simplest cells. They are **prokaryotic**, with no nucleus. Cells of all other organisms are **eukaryotic**. They contain a nucleus and other membrane-bound internal compartments.*

THE BASICS OF CELL STRUCTURE

The **cell** is the smallest unit with the properties of life: a capacity for metabolism, controlled responses to the environment, growth, and reproduction. Cells differ in size, shape, and activities, yet are all alike in three respects. They start out life with a plasma membrane, a region of DNA, and cytoplasm (Figure 4.3).

A **plasma membrane** defines the cell as a distinct entity. This thin, outer membrane separates metabolic activities from random events outside, but it does not isolate the cell interior. It is like a house with many doors that do not open for just anyone. Water, carbon dioxide, and oxygen enter and leave freely. Nutrients, ions, and other substances must be escorted.

In eukaryotic cells, the DNA occupies a **nucleus**, a membrane-bound, internal sac. In prokaryotic cells, it occupies a **nucleoid**, a region of the cytoplasm that is not enclosed in a membranous sac.

Cytoplasm is everything in between the plasma membrane and the region of DNA. It has a semifluid matrix and structural components that have roles in protein synthesis, energy conversions, and other vital tasks. For instance, cytoplasm holds many **ribosomes**, the molecular structures on which proteins are built.

PREVIEW OF CELL MEMBRANES

A **lipid bilayer** is a continuous, oily boundary that prevents the free passage of water-soluble substances across it (Figure 4.4). It is the structural basis of the plasma membrane and of various membranes inside

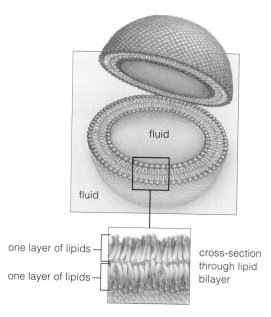

Figure 4.4 Simplified model for the lipid bilayer of all cell membranes. Remember the phospholipids (Section 3.4)? They are the most abundant lipids in cell membranes. Their hydrophobic tails are sandwiched between their hydrophilic heads, which are dissolved in cytoplasm on one side of the bilayer and in extracellular fluid on the other side.

eukaryotic cells. Membranes in the cytoplasm form channels or sacs that compartmentalize the tasks of transporting, synthesizing, modifying, stockpiling, or digesting substances.

Diverse proteins embedded in the lipid bilayer or positioned at one of its surfaces carry out most of the membrane functions (Figure 4.5). For instance, some proteins are channels and others are pumps across the bilayer. Others are receptors; they are like docks for hormones and other signaling molecules that trigger required changes in cell activities. You will read more about membrane proteins in chapters to follow.

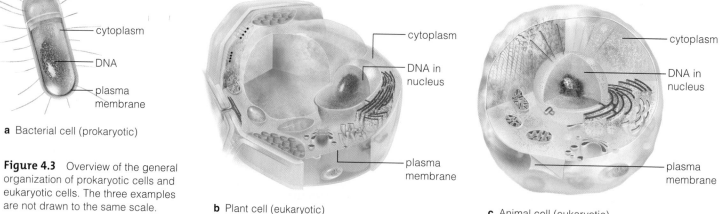

a Bacterial cell (prokaryotic)

Figure 4.3 Overview of the general organization of prokaryotic cells and eukaryotic cells. The three examples are not drawn to the same scale.

b Plant cell (eukaryotic)

c Animal cell (eukaryotic)

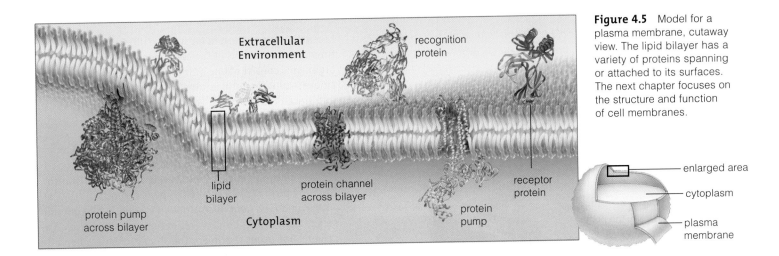

Figure 4.5 Model for a plasma membrane, cutaway view. The lipid bilayer has a variety of proteins spanning or attached to its surfaces. The next chapter focuses on the structure and function of cell membranes.

WHY AREN'T CELLS BIGGER?

Are any cells big enough to be seen without the help of a microscope? A few. They include "yolks" of bird eggs, cells in watermelon tissues, and amphibian and fish eggs. These cells can be large because they are not doing too much, metabolically speaking, at maturity. Most of their volume is a nutrient warehouse. If a cell has to perform many tasks, you can expect it to be too tiny to be seen by the unaided eye.

So why aren't all cells big? A physical relationship called the **surface-to-volume ratio** constrains increases in cell size. By this relationship, an object's volume increases with the cube of its diameter, but the surface area increases only with the square.

Apply this constraint to a round cell. As Figure 4.6 shows, *when a cell expands in diameter during growth, its volume increases faster than its surface area.* Suppose you could make a round cell increase four times in diameter. Its volume would increase 64 times (4^3), but its surface area would increase just 16 times (4^2). Each unit of its plasma membrane would now be required to service four times as much cytoplasm as before.

When the girth of any cell becomes too great, the inward flow of nutrients and outward flow of wastes will not be fast enough to keep up with the metabolic activity that keeps the cell alive. The outcome will be a dead cell.

Besides, a big, round cell also would have trouble moving materials through its cytoplasm. The random motion of molecules can distribute substances through tiny cells. When a cell is not tiny, you can expect it to be long or thin, or to have outfoldings or infoldings that increase its surface area relative to its volume. When a cell is smaller, narrower, or frilly surfaced, substances cross its surface and become distributed through the interior with greater efficiency.

Surface-to-volume constraints also shape the body plans of multicelled species. For example, small cells attach end to end in strandlike algae, so each interacts directly with its surroundings. Cells in your muscles are as long as the muscle itself, but each one is thin enough to efficiently exchange substances with fluids in the tissue surrounding them.

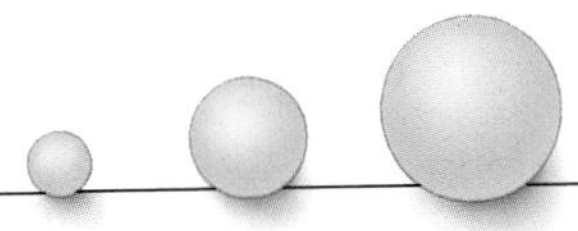

diameter (cm):	0.5	1.0	1.5
surface area (cm²):	0.79	3.14	7.07
volume (cm³):	0.06	0.52	1.77
surface-to-volume ratio:	13.17:1	6.04:1	3.99:1

Figure 4.6 *Animated!* One example of the surface-to-volume ratio. This physical relationship between increases in volume and surface area puts constraints on the sizes and shapes that are possible in cells.

All living cells have an outermost plasma membrane, an internal region called cytoplasm, and an internal region where DNA is concentrated.

Bacteria and archaeans are two groups of prokaryotic cells. Unlike eukaryotic cells, they do not have an abundance of organelles, particularly a nucleus.

Two layers of lipids are the structural framework for cell membranes. Proteins in the bilayer or positioned at one of its surfaces carry out diverse membrane functions.

A physical relationship called the surface-to-volume ratio constrains increases in cell size. The relationship also influences the shape of individual cells and the body plans of multicelled organisms.

4.2 How Do We "See" Cells?

Like their centuries-old forerunners, modern microscopes are our best windows on the cellular world.

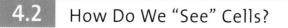

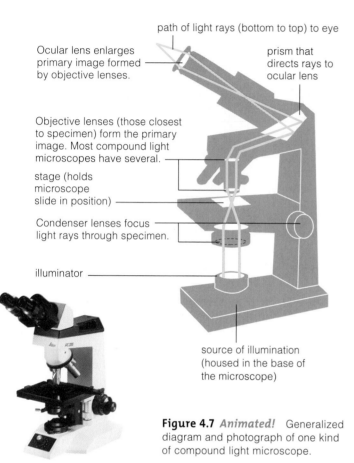

path of light rays (bottom to top) to eye

Ocular lens enlarges primary image formed by objective lenses.

prism that directs rays to ocular lens

Objective lenses (those closest to specimen) form the primary image. Most compound light microscopes have several.

stage (holds microscope slide in position)

Condenser lenses focus light rays through specimen.

illuminator

source of illumination (housed in the base of the microscope)

Figure 4.7 *Animated!* Generalized diagram and photograph of one kind of compound light microscope.

Research with modern microscopes still supports the three generalizations of the cell theory. In essence, all organisms consist of one or more cells, the cell is the smallest unit that still retains the characteristics of life, and since the time of life's origin, each new cell is descended from a cell that is already alive.

Like earlier instruments, many microscopes still use waves of light, and the light's wavelengths dictate their capacity to make images. Visualize a series of waves moving across an ocean. Each **wavelength** is the distance from the peak of one wave to the peak behind it.

In *compound light microscopes* (Figure 4.7), two or more sets of glass lenses bend waves of light passing through a cell or some other specimen. The shapes of lenses bend the waves at angles that disperse them in ways that form an enlarged image. *Micrographs* are simply photographs of images that emerge with the help of a microscope.

Cells become visible when they are thin enough for light to pass through them, but most cells are nearly colorless and look uniformly dense. Certain colored dyes can stain cells nonuniformly and make their component parts show up, but stains kill cells. Dead cells break down fast, which is why most cells are preserved before staining.

At present, the best light microscopes can enlarge cells about 2,000 times. Beyond that, cell structures appear larger but they are not clearer. Structures smaller than one-half of a wavelength of light are too small to resolve, so they cannot be distinguished.

Electron microscopes use magnetic lenses to bend and diffract beams of electrons, which cannot be diffracted through a glass lens. Electrons travel in wavelengths about 100,000 times shorter than those of visible light. Hence electron microscopes can resolve details that are 100,000 times smaller than you can see with a light microscope.

In *transmission* electron microscopes, electrons pass through a specimen and are used to make images of its internal details (Figure 4.8). *Scanning* electron microscopes direct a beam of electrons back and forth across a surface of a specimen, which has been given a thin metal coating. The metal responds by emitting electrons and x-rays, which can be converted into an image of the surface.

Figure 4.9 compares the resolving power of microscopes and of the human eye. Figure 4.10 compares the kinds of images that different microscopes offer.

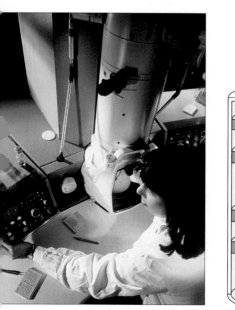

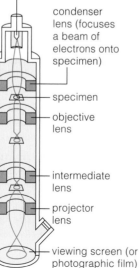

incoming electron beam

condenser lens (focuses a beam of electrons onto specimen)

specimen

objective lens

intermediate lens

projector lens

viewing screen (or photographic film)

Figure 4.8 *Animated!* Generalized diagram of an electron microscope. The photograph gives an idea of the lens diameters for a transmission electron microscope (TEM). When a beam of electrons from an electron gun moves down the microscope column, magnets focus them. With a transmission electron microscope, electrons pass through a thin slice of specimen and illuminate a fluorescent screen on a monitor. Shadows cast by the specimen's internal details appear, as in Figure 4.10c.

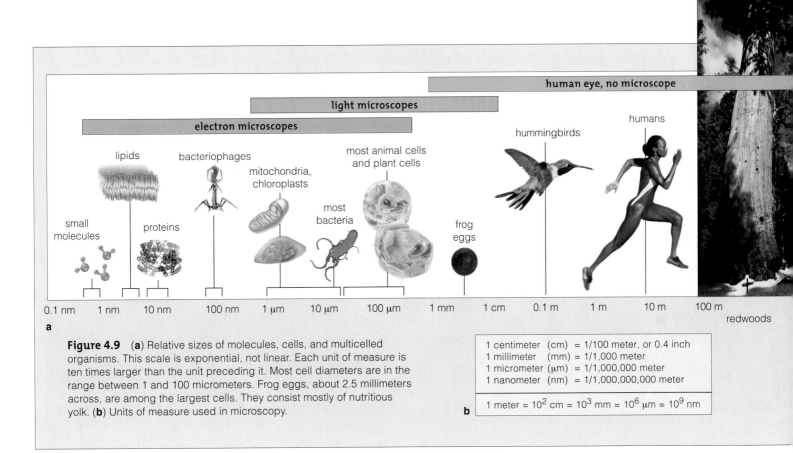

Figure 4.9 (**a**) Relative sizes of molecules, cells, and multicelled organisms. This scale is exponential, not linear. Each unit of measure is ten times larger than the unit preceding it. Most cell diameters are in the range between 1 and 100 micrometers. Frog eggs, about 2.5 millimeters across, are among the largest cells. They consist mostly of nutritious yolk. (**b**) Units of measure used in microscopy.

1 centimeter (cm)	= 1/100 meter, or 0.4 inch
1 millimeter (mm)	= 1/1,000 meter
1 micrometer (μm)	= 1/1,000,000 meter
1 nanometer (nm)	= 1/1,000,000,000 meter

$1 \text{ meter} = 10^2 \text{ cm} = 10^3 \text{ mm} = 10^6 \text{ μm} = 10^9 \text{ nm}$

b

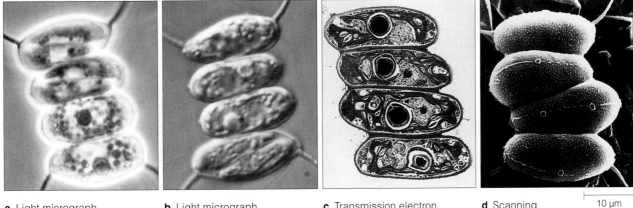

a Light micrograph (phase-contrast process)

b Light micrograph (Nomarski process)

c Transmission electron micrograph, thin section

d Scanning electron micrograph

10 μm

Figure 4.10 How different microscopes reveal different aspects of the same organism: a green alga (*Scenedesmus*). All four images are at the same magnification. (**a**,**b**) Light micrographs. (**c**) Transmission electron micrograph. (**d**) Scanning electron micrograph. A horizontal bar below a micrograph, as in (**d**), provides a visual reference for size. One micrometer (μm) is 1/1,000,000 of 1 meter. Using the scale bar, can you estimate the length and width of a *Scenedesmus* cell?

4.3 Introducing Prokaryotic Cells

LINKS TO
SECTIONS
1.1, 1.3, 3.5

The word prokaryote is taken to mean "before the nucleus." The name reminds us that bacteria and then archaeans originated before cells with a nucleus evolved.

Prokaryotes are the smallest known cells. As a group they are the most metabolically diverse forms of life on Earth. Different kinds exploit energy sources and raw materials in nearly all environments, including dry deserts, deep ocean sediments, and mountain ice.

We recognize two domains of prokaryotic cells—**Bacteria** and **Archaea** (Sections 1.3 and 19.5). Cells of both groups are alike in outward appearance and in size. However, the two groups differ in major ways.

Bacteria start making each new polypeptide chain with formylmethionine, a modified amino acid. Archaeans start a chain with methionine—as eukaryotic cells do (Section 3.5). Eukaryotic cells make many histones, a type of protein that structurally stabilizes the DNA. Archaeans make a few histones. Bacteria make a few histone-like proteins that stabilize the nucleoid.

Most prokaryotic cells are not much wider than one micrometer. The rod-shaped species are no more than a few micrometers long (Figures 4.11 and 4.12). Structurally, these are the simplest cells. A semirigid or rigid wall outside the plasma membrane imparts shape to most species. As Section 4.10 explains, arrays

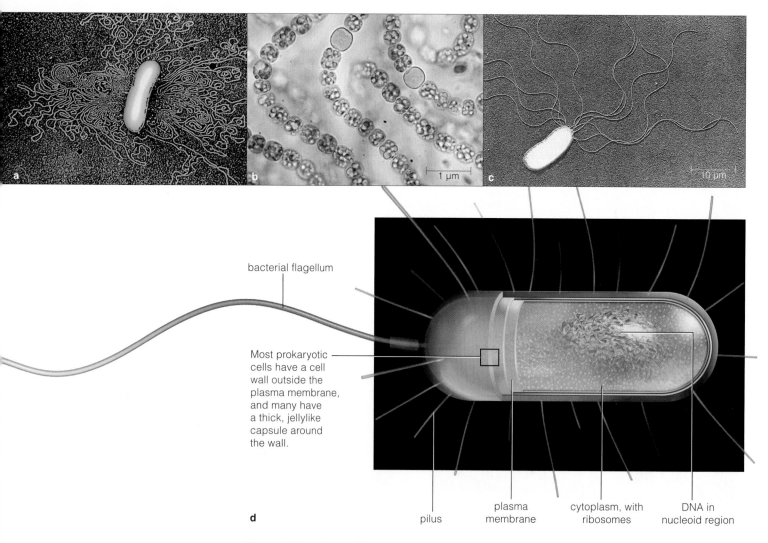

bacterial flagellum

Most prokaryotic cells have a cell wall outside the plasma membrane, and many have a thick, jellylike capsule around the wall.

d

pilus

plasma membrane

cytoplasm, with ribosomes

DNA in nucleoid region

1 µm

10 µm

Figure 4.11 *Animated!* (**a**) Micrograph of *Escherichia coli*. Researchers manipulated this bacterial cell to release its single, circular molecule of DNA. (**b**) Different bacterial species are shaped like balls, rods, or corkscrews. Ball-shaped cells of a photosynthetic bacterium (*Nostoc*) stick together in a thick, jellylike sheath of their own secretions. Chapter 21 offers more examples. (**c**) Like this *Pseudomonas marginalis* cell, many species have one or more bacterial flagella that propel the cell body through fluid environments. (**d**) Generalized sketch of a typical prokaryotic cell.

Figure 4.12 From Bitter Springs, Australia, fossilized bacteria dating to about 850 million years ago, in precambrian times. (**a**) A colonial form, most likely *Myxococcoides minor*. (**b**) Cells of a filamentous species (*Palaeolyngbya*).

(**c**) One of the structural adaptations seen among archaeans. Many of these prokaryotic species live in extremely hostile habitats, such as the ones thought to have prevailed when life originated. Most archaeans and some bacteria have a dense lattice of proteins that are anchored to the outer surface of their plasma membrane. In some species, the unique composition of the lattice may help the cell withstand extreme conditions in the environment. For instance, we find such lattices on archaeans living in near-boiling, mineral-rich water spewing from hydrothermal vents on the ocean floor.

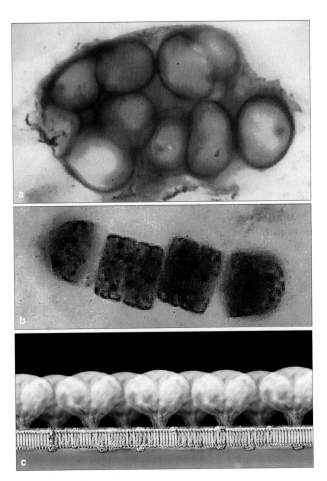

of protein filaments just under the plasma membrane reinforce the cell's shape.

Sticky polysaccharides often envelop bacterial cell walls. They let cells attach to such interesting surfaces as river rocks, teeth, and the vagina. Many disease-causing (pathogenic) bacteria have a thick protective capsule of jellylike polysaccharides around their wall.

Cell walls are permeable to dissolved substances, which cross them on the way to and from the plasma membrane. Some eukaryotic cells also have a wall, but it differs structurally from prokaryotic cell wall.

Many species have one or more bacterial flagella (singular, flagellum). These motile structures do not have a core of microtubules, as eukaryotic cells do (Section 4.11). They start at a tiny rotary motor in the plasma membrane and extend past the cell wall. They help move cells through fluid habitats, such as body fluids of host animals. Many bacteria also have pili (singular, pilus). These protein filaments help the cell cling to surfaces. A "sex" pilus latches on to another cell, then shortens. The attached cell is reeled in, and genetic material is transferred into it (Section 21.3).

As in eukaryotic cells, the plasma membrane of bacteria and archaeans selectively controls the flow of substances into and out of the cytoplasm. Its lipid bilayer bristles with protein channels, transporters, and receptors, and it incorporates built-in machinery for reactions. For example, the plasma membrane in photosynthetic bacterial species has organized arrays of proteins that capture light energy and convert it to bond energy of ATP, which helps build sugars.

The cytoplasm contains many ribosomes on which polypeptide chains are built. DNA is concentrated in an irregularly shaped region of cytoplasm called the nucleoid. Prokaryotic cells inherit one molecule of DNA that is circular, not linear. We call it a bacterial chromosome. The cytoplasm of some species also has plasmids: far smaller circles of DNA that carry just a

few genes. Typically, plasmid genes confer selective advantages, such as antibiotic resistance.

One more intriguing point: In cyanobacteria, part of the plasma membrane projects into the cytoplasm and is repeatedly folded back on itself. As it happens, pigments and other molecules of photosynthesis are embedded in the membrane—as they are in the inner membrane of chloroplasts. Is this a sign that ancient cyanobacteria were the forerunners of chloroplasts? Section 20.4 looks at this possibility. It is one aspect of a remarkable story about how prokaryotes gave rise to all protists, plants, fungi, and animals.

Bacteria and archaeans are two major groups of prokaryotic cells. These cells do not have a nucleus. Most have a cell wall around their plasma membrane. The wall is permeable, and it reinforces and imparts shape to the cell body.

Although structurally simple, prokaryotic cells as a group show the most metabolic diversity. Metabolic activities that are similar to ones occurring in eukaryotic organelles occur at the bacterial plasma membrane or in the cytoplasm.

4.7 Mitochondria

LINK TO
SECTION
3.7

Recall, from Section 3.7, that ATP is an energy carrier. It delivers energy, in the form of phosphate-group transfers, that drives reactions at sites throughout the cell. Without energy deliveries from ATP, cells could not grow, survive, or reproduce. ATP formation is essential for life.

The **mitochondrion** (plural, mitochondria) is a type of organelle that specializes in ATP formation. Reactions in this organelle help cells extract more energy from organic compounds than they can get by any other means. The reactions, called *aerobic* respiration, require free oxygen. With each breath, you take in oxygen mainly for mitochondria in your trillions of cells.

Typical mitochondria are 1 to 4 micrometers long; a few are 10 micrometers long. Some are branching. These organelles change shape, split in two, and fuse together. Figure 4.19 shows an example.

Each mitochondrion has an outer membrane and another one inside that most often is highly folded (Figure 4.19). The membrane arrangement creates two compartments. Hydrogen ions become stockpiled in the outer compartment. The ions then flow to the inner compartment in a controlled way. The energy inherent in the flow drives ATP formation.

No prokaryotic cells have mitochondria. Nearly all eukaryotic cells do. A single-celled yeast might have only one. Cells that have huge demands for energy may have a thousand or more. Skeletal muscle cells are one example. Liver cells, too, have a profusion of mitochondria. Take a closer look at Figure 4.14. That micrograph alone should tell you that the liver is an energy-demanding organ.

In size and biochemistry, mitochondria resemble bacteria. Like bacteria, they have their own DNA, and they divide on their own. They have some ribosomes. Did mitochondria evolve by way of endosymbiosis in ancient prokaryotic cells? By this theory, one cell was engulfed by another cell, or entered it as an internal parasite, but it escaped digestion. That cell and its descendants kept their plasma membrane intact and reproduced in the host. In time, they became protected, permanent residents. Structures and functions once required for independent life were no longer essential and were lost. Later descendants evolved into double-membraned mitochondria. We return to their possible endosymbiotic origins in Section 20.4.

All organisms require ATP, which carries energy (in the form of phosphate bonds) from one reaction site to another. ATP drives nearly all cell activities.

The organelles called mitochondria are the ATP-producing powerhouses of all eukaryotic cells.

Energy-releasing reactions proceed at the compartmented, internal membrane system of mitochondria. The reactions, which require oxygen, produce far more ATP than can be produced by any other cellular reaction.

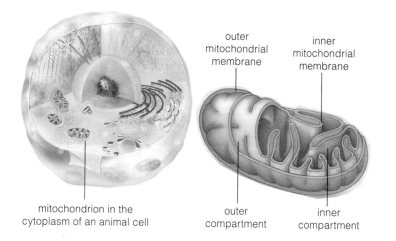

outer mitochondrial membrane

inner mitochondrial membrane

mitochondrion in the cytoplasm of an animal cell

outer compartment

inner compartment

Figure 4.19 Sketch and transmission electron micrograph, thin section, of a typical mitochondrion. This organelle specializes in forming large quantities of ATP, the main carrier of energy between reaction sites in cells. ATP formation in mitochondria cannot occur without free oxygen.

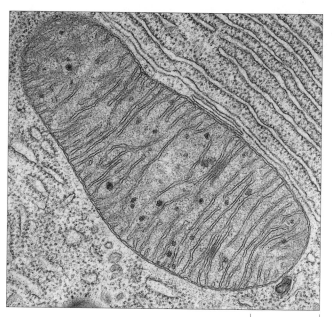

0.5 μm

4.8 Specialized Plant Organelles

Two kinds of organelles are prominent in many plant cells. They are plastids, such as chloroplasts, and central vacuoles.

CHLOROPLASTS AND OTHER PLASTIDS

Plastids are organelles that function in photosynthesis or storage in plants. Three types are common in plant tissues: chloroplasts, chromoplasts, and amyloplasts.

Chloroplasts are organelles that are specialized for photosynthesis. Most have oval or disk shapes. Two outer membranes enclose their semifluid interior, the stroma (Figure 4.20). In the stroma, a third membrane forms a single compartment that is commonly folded in intricate ways. Often the folds resemble a stack of flattened disks, called a granum (plural, grana).

Photosynthesis proceeds at the innermost, *thylakoid* membrane, which incorporates light-trapping pigments and other proteins. The most abundant photosynthetic pigments are chlorophylls, which reflect green light. Many kinds of accessory pigments assist chlorophylls in capturing light energy. The energy drives reactions in which ATP and an enzyme helper, NADPH, form. The ATP and NADPH are then used at sites in the stroma where sugars, starch, and other compounds are assembled. The new starch molecules may briefly accumulate in the stroma, as starch grains.

In many ways, chloroplasts are like photosynthetic bacteria. Like mitochondria, they may have evolved by endosymbiosis (Section 20.4).

Chromoplasts have no chlorophylls. They have an abundance of carotenoids, the source of red-to-yellow colors of many flowers, autumn leaves, ripe fruits, and carrots and other roots. The colors attract animals that pollinate the plants or disperse seeds.

Amyloplasts are pigment-free. They typically store starch grains. They are notably abundant in cells of stems, potato tubers (underground stems), and seeds.

CENTRAL VACUOLES

Many mature, living plant cells contain a fluid-filled **central vacuole**. This organelle stores amino acids, sugars, ions, and toxic wastes. Also, it expands during growth and increases fluid pressure on the pliable cell wall. The cell surface area is forced to increase, which favors absorption of water and other substances. Often the central vacuole takes up 50 to 90 percent of the cell's interior, with cytoplasm confined to a narrow zone in between this large organelle and the plasma membrane (Figures 4.15a and 4.20).

Photosynthetic cells of plants and many protists contain chloroplasts and other plastids that function in food production and storage.

Many plant cells have a central vacuole. When this storage vacuole enlarges during growth, cells are forced to enlarge, which increases the surface area available for absorption.

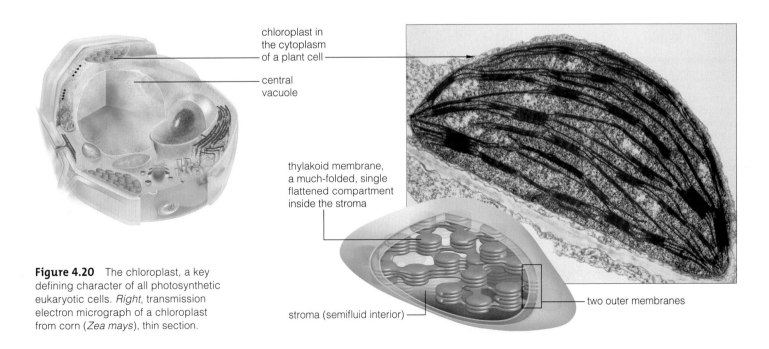

chloroplast in the cytoplasm of a plant cell

central vacuole

thylakoid membrane, a much-folded, single flattened compartment inside the stroma

stroma (semifluid interior)

two outer membranes

Figure 4.20 The chloroplast, a key defining character of all photosynthetic eukaryotic cells. *Right*, transmission electron micrograph of a chloroplast from corn (*Zea mays*), thin section.

4.9 Cell Surface Specializations

Turn now to a brief glimpse into some specialized surface structures of eukaryotic cells. Many of these architectural marvels are made primarily of cell secretions. Others are clusters of membrane proteins that connect neighboring cells, structurally and functionally.

EUKARYOTIC CELL WALLS

Single-celled eukaryotic species are directly exposed to the environment. Many have a **cell wall** around the plasma membrane. A cell wall protects and physically supports a cell, and imparts shape to it. The wall is porous, so water and solutes easily move to and from the plasma membrane. Any cell would die without these exchanges. Different plant cells form one or two walls around their plasma membrane. Cells of many protists and fungi have a wall. Animal cells do not.

Consider the growing parts of multicelled plants. New cells are secreting molecules of pectin and other gluelike polysaccharides, which form a matrix around them. They also secrete ropelike strands of cellulose molecules into the matrix. These materials make up the plant cell's **primary wall** (Figure 4.21). The sticky primary wall cements abutting cells together. Being thin and pliable, it allows the cell to enlarge under the pressure of incoming water.

Cells that have only a thin primary wall retain the capacity to divide or change shape as they grow and develop. Many types stop enlarging when they are mature. Such cells secrete material on the primary wall's inner surface. These deposits form a lignified, rigid **secondary wall** that reinforces cell shape (Figure 4.21*d*). The secondary wall deposits are extensive and contribute more to structural support.

In woody plants, up to 25 percent of the secondary wall is made of lignin. This organic compound makes plant parts more waterproof, less susceptible to plant-attacking organisms, and stronger.

At plant surfaces exposed to air, waxes and other cell secretions build up as a protective cuticle. This semitransparent surface covering limits water losses on hot, dry days (Figure 4.22*a*).

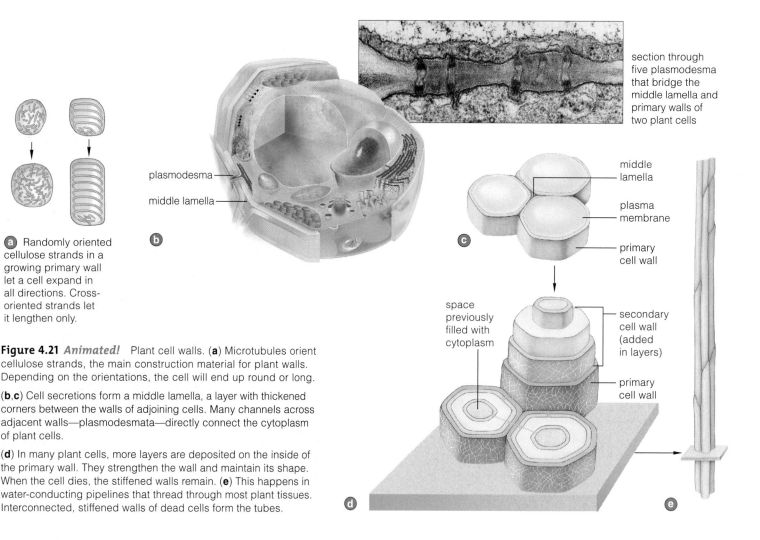

section through five plasmodesma that bridge the middle lamella and primary walls of two plant cells

plasmodesma

middle lamella

middle lamella

plasma membrane

primary cell wall

space previously filled with cytoplasm

secondary cell wall (added in layers)

primary cell wall

(a) Randomly oriented cellulose strands in a growing primary wall let a cell expand in all directions. Cross-oriented strands let it lengthen only.

Figure 4.21 *Animated!* Plant cell walls. (**a**) Microtubules orient cellulose strands, the main construction material for plant walls. Depending on the orientations, the cell will end up round or long.

(**b,c**) Cell secretions form a middle lamella, a layer with thickened corners between the walls of adjoining cells. Many channels across adjacent walls—plasmodesmata—directly connect the cytoplasm of plant cells.

(**d**) In many plant cells, more layers are deposited on the inside of the primary wall. They strengthen the wall and maintain its shape. When the cell dies, the stiffened walls remain. (**e**) This happens in water-conducting pipelines that thread through most plant tissues. Interconnected, stiffened walls of dead cells form the tubes.

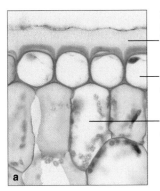

thick, waxy
cuticle at
leaf surface

cell of leaf
epidermis

photosynthetic
cell inside leaf

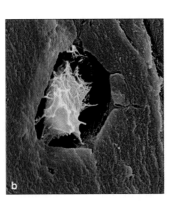

Figure 4.22 (**a**) Section through a plant cuticle, a surface covering made of cell secretions.

(**b**) A living cell imprisoned in hardened bone tissue, the stuff of vertebrate skeletons.

MATRIXES BETWEEN ANIMAL CELLS

Animal cells have no cell walls. Intervening between many of them are matrixes made of cell secretions and of materials absorbed from the surroundings. For example, the cartilage at the knobby ends of leg bones contains scattered cells and protein fibers embedded in a ground substance of firm polysaccharides. Living cells also secrete the extensive, hardened matrix that we call bone tissue (Figure 4.22*b*).

CELL JUNCTIONS

Even when a wall or some other structure imprisons a cell in its own secretions, the cell interacts with the outside world at its plasma membrane. In multicelled species, structures extend into neighboring cells or into a matrix. **Cell junctions** are molecular structures where a cell sends or receives signals or materials, or recognizes and glues itself to cells of the same type.

In plants, for instance, channels extend across the primary wall of adjacent living cells and interconnect the cytoplasm of both (Figure 4.21*b*). Each channel is a plasmodesma (plural, plasmodesmata). Substances flow quickly from cell to cell across these junctions.

In most tissues of animals, three types of cell-to-cell junctions are common (Figure 4.23). *Tight* junctions link the cells of most body tissues, including epithelia that line outer surfaces, internal cavities, and organs. These junctions seal abutting cells together so water-soluble substances cannot leak between them. That is why gastric fluid does not leak across the stomach lining and damage internal tissues. *Adhering* junctions occur in skin, the heart, and other organs subjected to continual stretching. At *gap* junctions, the cytoplasm of certain kinds of adjacent cells connect directly. Gap junctions function as open channels for a rapid flow of substances, most notably in heart muscle.

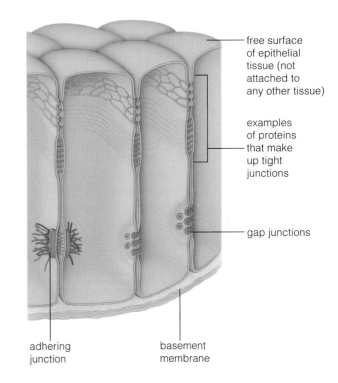

free surface of epithelial tissue (not attached to any other tissue)

examples of proteins that make up tight junctions

gap junctions

adhering junction

basement membrane

Figure 4.23 *Animated!* The three most common types of cell junctions in animal tissues.

A variety of protist, plant, and fungal cells have a porous wall that surrounds the plasma membrane.

Young plant cells have a thin primary wall pliable enough to permit expansion. Some mature cells also form a lignin-reinforced secondary wall that provides structural support.

Animal cells have no walls, but they and many other cells often secrete substances that form matrixes between cells. In multicelled species, different junctions commonly serve as structural and functional connections between cells.

4.10 Even Cells Have a Skeleton

What does the word "skeleton" mean to you? A collection of bones, such as a rib cage? That is one kind of skeleton in nature. However, anything that forms a structural framework is a skeleton—and cells, too, have one.

COMPONENTS OF THE CYTOSKELETON

The **cytoskeleton** of eukaryotic cells is an organized system of protein filaments that extends between the nucleus and plasma membrane. Different portions of it reinforce, organize, and move internal cell parts or the cell body. Many parts are permanent, and others form only at certain times in the life of a cell.

Microtubules and **microfilaments** are two classes of cytoskeletal elements in nearly all eukaryotic cells. Some cells also have ropelike **intermediate filaments** (Figures 4.24 and 4.25).

Microtubules Microtubules, the largest cytoskeletal elements, help keep organelles and cell structures in place or move them to new locations. For instance, before a cell divides, microtubules form a spindle that harnesses chromosomes and moves them about. Other microtubules move chloroplasts, vesicles, and other organelles through the cytoplasm.

In plant and animal cells, microtubules are hollow cylinders of tubulin monomers (Figure 4.24a). Tubulin consists of two chemically distinct polypeptide chains, each folded into a rounded shape. As a microtubule is being assembled, all of its monomers are oriented in the same direction. The assembly pattern puts slightly different chemical properties at opposite ends of the cylinder. Like bricks being stacked into a new wall, monomers join the cylinder's *plus* (fast-growing) end, which at first grows freely through the cytoplasm.

Microtubules of animal cells normally grow in all directions from small patches of dense material called centrosomes. Their *minus* (slow-growing) ends remain anchored in it. Microtubules can abruptly fall apart in controlled ways; they are not permanently stable.

In any given interval, some of the microtubules are being allowed to disassemble. Others get capped with proteins that stabilize them. For instance, microtubules in the advancing end of an amoeba stay intact when the cell is hot on a chemical trail to a potential meal. Nothing is stimulating microtubules to grow in that cell's trailing end, so they are allowed to fall apart.

Some plants make poisons that act on microtubules of plant-eating animals. By binding to tubulins, the poisons make microtubules fall apart and stop new ones from forming. For instance, the autumn crocus (*Colchicum autumnale*) makes colchicine (Figure 4.26a). The plant has an evolved insensitivity to colchicine, which can't bind well to its own tubulins. The western yew (*Taxus brevifolia*) makes the microtubule poison taxol (Figure 4.26b). A synthetic taxol stops the growth of certain tumors. It keeps microtubule spindles from forming, and thereby suppresses the uncontrolled cell divisions that form abnormal tissue masses. Taxol has the same effect on normal cells; it is just that tumor cells divide far more often than most of them.

Microfilaments Microfilaments are the thinnest of all cytoskeletal elements inside eukaryotic cells (Figure 4.24b). Two helically coiled polypeptide chains of actin monomers make up each filament.

Microfilaments are assembled and disassembled in controlled ways. Often they are organized in bundles or networks. As an example, just beneath the plasma membrane is a **cell cortex**: bundled up, crosslinked, and gel-like meshes of microfilaments. Some parts of the cortex reinforce a cell's shape. Others reconfigure the surface. For instance, during animal cell division, a ring of microfilaments around the cell's midsection contracts and pinches the cell in two. Microfilaments also anchor membrane proteins and are components of muscle contraction. In some large cells, *cytoplasmic streaming* gets under way as microfilament meshes loosen up. Local gel-like regions become more fluid and flow strongly, thereby redistributing substances and cell components through the cell interior.

Myosin and Other Accessory Proteins Genes for tubulin and actin were not drastically modified over time; they have been highly conserved in the DNA. All eukaryotic cells make similar forms of these monomers for microtubules and microfilaments. In spite of the structural uniformity, microtubules and microfilaments have different functions, thanks to other proteins that associate with them.

tubulin subunits

a   25 nm

actin subunit

b   5–7 nm

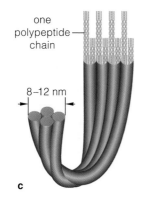

one polypeptide chain

8–12 nm

c

Figure 4.24 Structural arrangement of subunits in (**a**) microtubules, (**b**) microfilaments, and (**c**) one of the intermediate filaments.

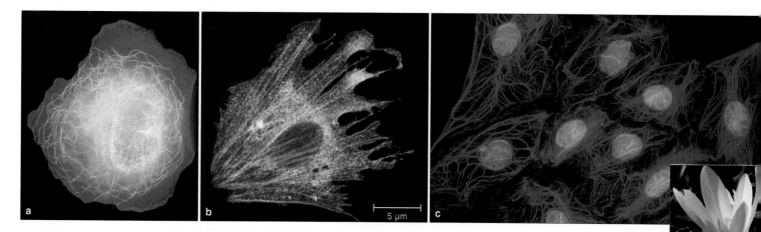

Figure 4.25 Cells stained to reveal the cytoskeleton. (**a**) Distribution of microtubules (*gold*) and actin filaments (*red*) in a pancreatic epithelial cell. This cell secretes bicarbonate, which functions in digestion by neutralizing stomach acid. The *blue* region is DNA. (**b**) Actin filaments (*red*) and accessory proteins (*green, blue*) help new epithelial cells such as this one migrate to proper locations in tissues as human embryos are developing. (**c**) Intermediate filaments of keratin (*red*) of cultured kangaroo rat cells. Each *blue*-stained organelle is a nucleus.

Figure 4.26
Two sources of microtubule poisons: (**a**) Autumn crocus, *Colchicum autumnale.* (**b**) Western yew, *Taxus brevifolia.*

For example, as you will read in the next section, kinesin and myosin are two kinds of *motor* proteins. Inputs of ATP energy make them move along tracks of cytoskeletal elements and put cell components in new locations. As another example, the *crosslinking* proteins interconnect microfilaments at the cell cortex.

Intermediate Filaments Intermediate filaments are between microtubules and microfilaments in size. They are eight to twelve nanometers wide and are the most stable elements of some cytoskeletons (Figure 4.25c). Six known groups strengthen and maintain the shape of cells or cell parts. For example, lamins help form a basketlike mesh that reinforces the nucleus. They anchor adjoining actin and myosin filaments as units of contraction inside muscle cells. Desmins and vimentins help hold these contractile units in position. Different cytokeratins reinforce cells that make nails, claws, horns, and hairs. Intermediate filaments may reinforce the nuclear envelope of all eukaryotic cells.

Different animal cells have different intermediate filaments in their cytoplasm. Because each type of cell contains one or at most two kinds, researchers can use these intermediate filaments to identify which type of cell it is. This typing is a useful tool in diagnosing the tissue origin of different forms of cancerous cells.

WHAT ABOUT PROKARYOTIC CELLS?

Unlike eukaryotic cells, bacteria and archaeans do not have a well-developed cytoskeleton. Until recently, we thought they did not have any cytoskeletal elements. However, reinforcing filaments have been identified in certain bacteria. What is more, the filaments are made of protein subunits that are a lot like tubulin and actin. Also, their repetitive pattern of assembly is similar to the assembly patterns for microtubules and microfilaments in eukaryotic cells.

For instance, just beneath the plasma membrane of rod-shaped bacterial cells, protein monomers (MreB) form filaments that help determine cell shape. These filaments, which are structurally like one of the actins, suggest that the eukaryotic cytoskeleton had its origin in ancestral prokaryotes.

A cytoskeleton is the basis of cell shape, internal structure, and movement. In eukaryotic cells, its components are microtubules, microfilaments, and in some cell types, intermediate filaments. Accessory proteins associate with these filaments and extend their range of functions.

Microtubules, cylinders of tubulin monomers, organize the cell interior and have roles in moving cell components.

Microfilaments consist of two helically coiled polypeptide chains of actin monomers. They form flexible, linear bundles and networks that reinforce or restructure the cell surface.

Intermediate filaments strengthen and maintain cell shapes. Some are present only in certain animal cells. Others may help reinforce the nuclear envelope of all eukaryotic cells.

Cytoskeletal elements similar to the microtubules and microfilaments have been identified in prokaryotic cells.

4.11 How Do Cells Move?

The skeleton of eukaryotic cells differs notably from your skeleton in a key respect. The cytoskeleton has elements that are not permanently rigid. At prescribed times, they assemble and disassemble.

MOVING ALONG WITH MOTOR PROTEINS

Think of a train station at the busiest holiday season, and you get an idea of what goes on in cells. Many of the cell's microtubules and microfilaments are like train tracks. The kinesins, dyneins, myosins, and other motor proteins function as the freight engines (Figure 4.27). Energy from ATP fuels the movement.

Some motor proteins move chromosomes. Others slide one microtubule over another; still others inch along tracks inside nerve cells that extend from your spine to your toes. Many engines are organized one after another, and each moves a vesicle partway along the track before giving it up to the next engine in line. From dawn to dusk, kinesins inside plant cells drag chloroplasts to new positions, the better to intercept light as the angle of the sun changes overhead.

Different kinds of myosins can move structures along microfilaments or slide one microfilament over another. As Sections 37.5 and 37.6 show, muscle cells form long fibers that are functionally divided into many contractile units along their length. Each unit has many parallel rows of microfilaments and myosin filaments. Myosin activated by ATP shortens the unit by sliding the microfilaments toward the unit's center. When all units shorten, the cell shortens; it contracts.

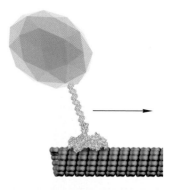

Figure 4.27 *Animated!* Kinesin (*brown*). This motor protein inches along a microtubule and drags a vesicle (*pink*) or some other cellular freight with it.

CILIA, FLAGELLA, AND FALSE FEET

Besides moving internal parts, many cells move their body or extend parts of it. Consider **flagella** (singular, flagellum) and **cilia** (singular, cilium). Both are motile structures that project from the surface of many types of cells. Both are completely sheathed by a membrane that is an extension of the plasma membrane.

Ciliated protists swim by beating their many cilia in synchrony. Cilia in certain airways to your lungs beat nonstop. Their coordinated movement sweeps out the airborne bacteria and particles that otherwise might cause disease (Figure 4.28a). Eukaryotic flagella usually are longer and not as profuse as cilia. Many single eukaryotic cells, such as sperm, swim with the help of whiplike flagella (Figure 4.28b).

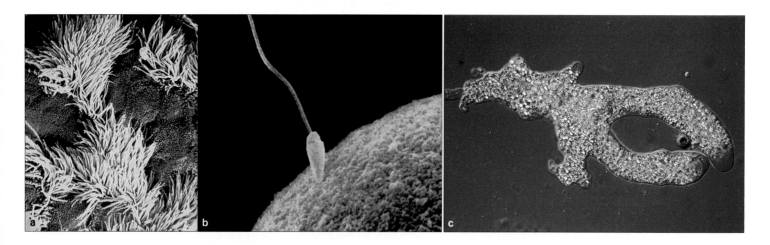

Figure 4.28 Cilia, a flagellum, and false feet. (**a**) Light micrograph of cilia (*gold*) projecting from the surface of some of the cells that line an airway to human lungs. (**b**) Scanning electron micrograph of a human sperm about to penetrate an egg. (**c**) Light micrograph of a predatory amoeba (*Chaos carolinense*) extending two pseudopods around a single-celled green alga (*Pandorina*).

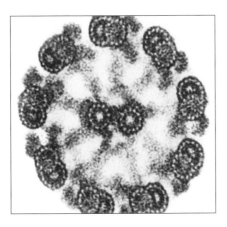

cross-section through one cilium

spokes, rings of connective system

central sheath

one central pair of microtubules

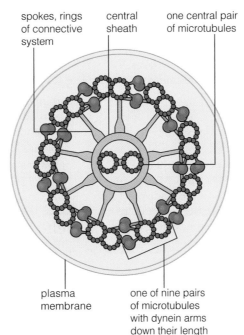

plasma membrane

one of nine pairs of microtubules with dynein arms down their length

microtubules near base of flagellum or cilium

basal body embedded in cytoplasm

plasma membrane

Figure 4.29 Internal organization of cilia and flagella. Both motile structures have a 9+2 array, an internal ring of nine pairs of microtubules around one pair at the core. Spokes and linking elements connected to the array stabilize it and keep it from slipping sideways, out of alignment.

Dynein arms projecting from microtubules in the ring incorporate ATPases. Repeated phosphate-group transfers from ATP cause them to bind briefly and reversibly to the microtubule pair in front of them. Each time, the arms force the pair to slide down a bit. The short, sliding strokes occur all around the ring, down the length of the microtubule, and it makes the motile structure bend.

Extending down the length of a cilium or flagellum is a *9+2 array*. Nine pairs of microtubules form a ring around a central pair, all stabilized by protein spokes and links. First a centrosome gives rise to a **centriole**. This barrel-shaped structure produces and organizes microtubules into the 9+2 array, then it remains below the finished array as a **basal body** (Figure 4.29).

Flagella and cilia move by a sliding mechanism. All pairs of microtubules extend the same distance into the motile structure's tip. Stubby dynein arms project from each pair in the outer ring. When ATP energizes them, the arms grab the microtubule pair in front of them, tilt in a short, downward stroke, then let go. As the bound pair slides down, its arms bind the pair in front of it, forcing it to slide down also— and so on around the ring. The microtubules cannot slide too far, but each *bends* a bit. Their sliding motion is converted to a bending motion.

As one more example, macrophages and amoebas form **pseudopods**, or "false feet." These temporary, irregular lobes bulge out from the cell. They move the cell and also engulf prey or some other target (Figure 4.28c). The pseudopods advance in a steady direction as microfilaments inside them are elongating. Motor proteins that are attached to the microfilaments are dragging the plasma membrane along with them in the direction of interest.

Cell contractions and migrations, chromosome movements, and other forms of cell movements arise by interactions among organized arrays of microtubules, microfilaments, and accessory proteins.

When energized by ATP, motor proteins move in specific directions, along tracks of microtubules and microfilaments. They deliver cell components to new locations.

Interactions among cytoskeletal elements bring about the movement of the motile structures called cilia and flagella, as well as the dynamic movements of pseudopods.

Summary

Section 4.1   Cells start life with a plasma membrane, cytoplasm that contains ribosomes and other structures, and DNA in a nucleus or nucleoid. Their membranes are a lipid bilayer with diverse kinds and numbers of proteins embedded in it or positioned at its surfaces. A physical relationship, the surface-to-volume ratio, constrains the sizes and shapes of cells.

Biology ⑤ Now
Use the interaction on BiologyNow to investigate the physical limits on cell size.

Section 4.2   Different microscopes use light or electrons to reveal cell shapes and structures. Microscopy reinforces the theory that all organisms are made of cells, that an individual cell has a capacity to live on its own, and that all cells now arise from preexisting cells.

Biology ⑤ Now
Learn how different types of microscopes function with the animation on BiologyNow.

Section 4.3   Bacteria and archaeans are prokaryotic; they have no nucleus (Table 4.2). They are structurally the simplest cells known, but collectively they show great metabolic diversity.

Biology ⑤ Now
View the animation about prokaryotic cell structure on BiologyNow.

Section 4.4   Organelles are membranous sacs that divide the interior of eukaryotic cells into functional compartments. Table 4.2 lists the major types.

Biology ⑤ Now
Introduce yourself to the major types of eukaryotic organelles with the interaction on BiologyNow.

Section 4.5   The nucleus keeps DNA molecules separated from metabolic reactions in the cytoplasm and controls access to a cell's hereditary information. It helps keep the DNA organized and easier to copy before a cell divides into daughter cells.

Biology ⑤ Now
Take a close-up look at the nuclear membrane with the animation on BiologyNow.

Section 4.6   In the endomembrane system's ER and Golgi bodies, new polypeptide chains are modified and lipids are assembled. Many proteins and lipids become part of membranes or packaged inside vesicles that function in transport, storage, and other cell activities.

Biology ⑤ Now
Follow a path through the endomembrane system with the animation on BiologyNow.

Section 4.7   The mitochondrion is the organelle that specializes in forming many ATP by aerobic respiration.

Section 4.8   The chloroplast is an organelle that specializes in photosynthesis.

Biology ⑤ Now
Look inside a chloroplast with the animation on BiologyNow.

Section 4.9   Most prokaryotic cells, cells of many protists and fungi, and all plant cells have a porous wall around their plasma membrane. In multicelled species, structural and functional connections link cells.

Biology ⑤ Now
Study the structure of cell walls and junctions with the animation on BiologyNow.

Sections 4.10, 4.11   Microtubules, microfilaments, and intermediate filaments make up a cytoskeleton in eukaryotic cells. They reinforce cell shapes, organize parts, and often move cells or structures, as by flagella.

Biology ⑤ Now
Learn more about cytoskeletal elements and their actions with the animation on BiologyNow.

Self-Quiz   *Answers in Appendix II*

1. Cell membranes consist mainly of a _____ .
   a. carbohydrate bilayer and proteins
   b. protein bilayer and phospholipids
   c. lipid bilayer and proteins

2. Identify the components of the cells shown in the two sketches at the bottom of the next page.

3. Organelles _____ .
   a. are membrane-bound compartments
   b. are typical of eukaryotic cells, not prokaryotic cells
   c. separate chemical reactions in time and space
   d. All of the above are features of organelles.

4. You will not observe _____ in animal cells.
   a. mitochondria   c. ribosomes
   b. a plasma membrane   d. a cell wall

5. Is this statement false: The plasma membrane is the outermost component of all cells. Explain your answer.

6. Unlike eukaryotic cells, prokaryotic cells _____ .
   a. lack a plasma membrane   c. have no nucleus
   b. have RNA, not DNA   d. all of the above

7. Match each cell component with its function.
   ____ mitochondrion   a. protein synthesis
   ____ chloroplast   b. initial modification of new
   ____ ribosome   polypeptide chains
   ____ rough ER   c. modification of new proteins;
   ____ Golgi body   sorting, shipping tasks
                         d. photosynthesis
                         e. formation of many ATP

Additional questions are available on **Biology ⑤ Now ™**

Critical Thinking

1. Why is it likely that you will never meet a two-ton amoeba on a sidewalk?

2. Your professor shows you an electron micrograph of a cell with many mitochondria, Golgi bodies, and a lot of rough ER. What kinds of cellular activities would require such an abundance of the three kinds of organelles?

3. *Kartagener syndrome* is a genetic disorder caused by a mutated form of the protein dynein. Affected people have chronically irritated sinuses, and mucus builds up in the

Table 4.2 Summary of Typical Components of Prokaryotic and Eukaryotic Cells

Cell Component	Function	Prokaryotic — Bacteria, Archaea	Eukaryotic — Protists	Eukaryotic — Fungi	Eukaryotic — Plants	Eukaryotic — Animals
Cell wall	Protection, structural support	✔*	✔*	✔	✔	None
Plasma membrane	Control of substances moving into and out of cell	✔	✔	✔	✔	✔
Nucleus	Physical separation and organization of DNA	None	✔	✔	✔	✔
DNA	Encoding of hereditary information	✔	✔	✔	✔	✔
RNA	Transcription, translation of DNA messages into polypeptide chains of specific proteins	✔	✔	✔	✔	✔
Nucleolus	Assembly of subunits of ribosomes	None	✔	✔	✔	✔
Ribosome	Protein synthesis	✔	✔	✔	✔	✔
Endoplasmic reticulum (ER)	Initial modification of many of the newly forming polypeptide chains of proteins; lipid synthesis	None	✔	✔	✔	✔
Golgi body	Final modification of proteins, lipids; sorting and packaging them for use inside cell or for export	None	✔	✔	✔	✔
Lysosome	Intracellular digestion	None	✔	✔*	✔*	✔
Mitochondrion	ATP formation	**	✔	✔	✔	✔
Photosynthetic pigments	Light–energy conversion	✔*	✔*	None	✔	None
Chloroplast	Photosynthesis; some starch storage	None	✔*	None	✔	None
Central vacuole	Increasing cell surface area; storage	None	None	✔*	✔	None
Bacterial flagellum	Locomotion through fluid surroundings	✔*	None	None	None	None
Flagellum or cilium with 9+2 microtubular array	Locomotion through or motion within fluid surroundings	None	✔*	✔*	✔*	✔
Complex cytoskeleton	Cell shape; internal organization; basis of cell movement and, in many cells, locomotion	Rudimentary***	✔*	✔*	✔*	✔

* Known to be present in cells of at least some groups.

** Many groups use oxygen-requiring (aerobic) pathways of ATP formation, but mitochondria are not involved.

*** Protein filaments form a simple scaffold that helps support the cell wall in at least some species.

airways to their lungs. Bacteria form huge populations in the thick mucus. Their metabolic by-products and the inflammation they trigger combine to damage tissues. Males affected by the syndrome can produce sperm, but they are infertile (Figure 4.30). Some have still become fathers with the help of a procedure that injects sperm cells directly into eggs. Explain how an abnormal dynein molecule could cause the observed effects.

4. As they grow and develop, many kinds of plant cells form a secondary wall on the *inner* surface of the primary wall that formed earlier. Speculate on the reason why the secondary wall does not form on the outside.

5. Reflect on Table 4.2. Notice how most prokaryotes, all plant cells, and many protist and fungal cells have walls, and that animal cells have none. Why do you suppose animal cells alone do not form walls?

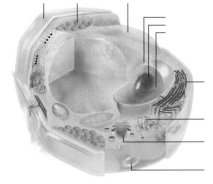

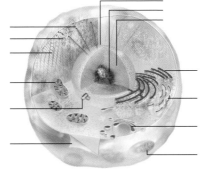

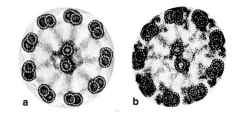

Figure 4.30 Cross-section through the flagellum of a sperm cell from (**a**) a male affected by Kartagener syndrome and (**b**) an unaffected male. Check the dynein arms projecting from the microtubule pairs.

One Bad Transporter and Cystic Fibrosis

Each living cell is engaged in risky business. Think of how it has to move something as ordinary as water in one direction or the other across its plasma membrane. If all goes well, it takes in or sends out water in just the right amounts—not too little, not too much. But who is to say life always goes well?

CFTR is one of the protein channels across the plasma membrane of epithelial cells. Sheets of these cells line sweat glands, airways and sinuses, and ducts in the digestive and reproductive systems. Chloride ions move through them, and water follows to form a thin film on the free surface of the linings. Mucus, which lubricates tissues and helps prevent infection, slides freely on the watery film.

Sometimes mutation changes how CFTR works. Not enough chloride and water reach the lining's free surface, so the film does not form. Mucus dries out and thickens. Among other things, it clogs ducts from the pancreas, so digestive enzymes cannot get to the small intestine where most food is digested and absorbed. Weight loss follows. Sweat glands secrete too much salt and alter the water–salt balance for the internal environment, which affects the heart and other organs. Males become sterile.

Problems also develop in airways to the lungs, where ciliated cells are supposed to sweep away bacteria and other particles stuck in mucus. Now the mucus makes cilia too sticky, and **biofilms** form. Biofilms are microbial populations anchored to one epithelial lining or another by stiff, sticky polysaccharides of their own making. They resist the body's defenses and antibiotics. *Pseudomonas aeruginosa,* the most efficient of the colonizers, cause low-grade infections that may last for years. Most patients can expect to live no longer than thirty years, at which time their lungs usually fail. At present there is no cure.

These symptoms—outcomes of mutation in the CFTR protein—characterize *cystic fibrosis* (CF), the most common fatal genetic disorder in the United States. More than 10 million people carry a mutant form of the gene. CF develops when they inherited a mutated gene from both parents. This happens in about 1 of every 3,300 live births (Figure 5.1).

CFTR is one of the ABC transporters in all prokaryotic and eukaryotic cells (Figure 5.2). Some of these proteins, including CFTR, are channels that let hydrophobic substances cross a membrane. Others pump substances across. By their action, some types affect what other membrane proteins are doing.

In all but 10 percent of CF patients, loss of a single amino acid during protein synthesis causes the disorder. Before a new CFTR protein is shipped to the plasma membrane, it is supposed to be modified in that endomembrane system you read about in Chapter 4. Copies of the mutant protein do enter the ER, but enzymes destroy 99 percent of them before they reach Golgi bodies. Thus few chloride channels reach their normal destinations.

Mutant CFTR may also contribute to the sinus problems of an estimated 30 million people in the United States

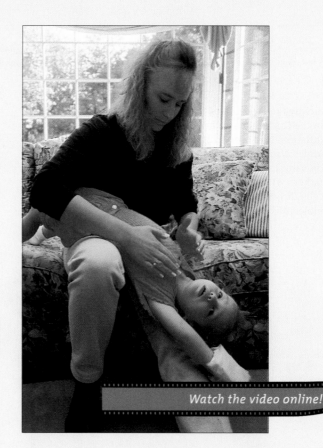

Watch the video online!

Figure 5.1 Child affected by cystic fibrosis, or CF, who each day endures chest thumps, back thumps, and repositionings to dislodge thick mucus that collects in airways to the lungs. Symptoms vary from one affected individual to the next, partly because the abnormal protein that causes CF has mutated in more than 500 ways. Environmental factors and a person's genetic makeup also affect the outcome.

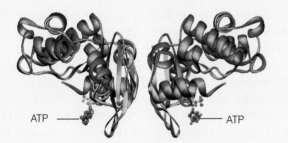

Figure 5.2 Model for part of an ABC transporter, a category of membrane proteins that includes CFTR. The parts shown here are ATP-driven motors that can widen an ion channel across the plasma membrane.

alone. In *sinusitis*, the linings of cavities inside the skull (around the nose) are chronically inflamed. In one study at Johns Hopkins University, researchers found a single copy of a mutant CFTR gene in 10 of 147 sinusitis patients. And they were only looking for 16 of more than 500 known mutant forms of the CFTR gene!

Think about it. A startling percentage of the human population can develop problems when the copies of even one kind of membrane protein don't work.

Your life depends on the functions of thousands of kinds of proteins and other molecules. Breathing, eating, moving, sleeping, crying, thinking—whatever you might be doing starts at the level of individual cells. And each cell functions properly only if it can be responsive to conditions in the microenvironments on both sides of its plasma membrane. Each eukaryotic cell also has to be responsive to conditions on both sides of its organelle membranes. *Cell membranes—these thin boundary layers make the difference between organization and chaos.*

How Would You Vote?

The ability to detect mutant genes that cause severe disorders raises bioethical questions. Should we encourage the mass screening of prospective parents for mutant genes that cause cystic fibrosis? Should society encourage women to give birth only if their child will not develop severe medical problems? See BiologyNow for details, then vote online.

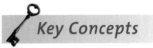

Key Concepts

MEMBRANE STRUCTURE AND FUNCTION

Cell membranes have a thin, oily, water-insoluble lipid bilayer that functions as a boundary between the outside environment and the cell interior.

The lipid bilayer consists primarily of phospholipids. Many diverse proteins are embedded in the bilayer or are positioned at one of its surfaces. The proteins carry out most membrane functions, such as transport across the bilayer and cell-to-cell recognition.   Sections 5.1, 5.2

DIFFUSION ACROSS MEMBRANES

Metabolism requires concentration gradients that drive the directional movements of substances. Cells have built-in mechanisms for increasing or decreasing water and solute concentrations across the plasma membrane and internal cell membranes.   Section 5.3

TRANSPORT ACROSS MEMBRANES

In passive transport, a solute crosses a membrane by diffusing through a channel inside a transport protein. In active transport, a different kind of transport protein pumps the solute across a membrane, against its concentration gradient. An input of energy, typically from ATP, jump-starts active transport.   Section 5.4

OSMOSIS

By a molecular behavior called osmosis, water diffuses across any selectively permeable membrane to a region where its concentration is lower.   Section 5.5

MEMBRANE TRAFFIC

Larger packets of substances and, in some cases, engulfed cells move across the plasma membrane by processes of endocytosis and exocytosis. Membrane cycling pathways extend from the plasma membrane to organelles of the endomembrane system.   Section 5.6

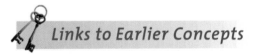

Links to Earlier Concepts

Reflect again on the road map in Section 1.1. Here you will see how complex lipids and proteins become organized in cell membranes (3.4, 4.1). Remember the different levels of protein organization? You will consider some examples of how protein structure translates into specific functions (3.6). You will be applying your knowledge of the properties of water molecules to the movement of water across membranes (2.5). You will see how the endomembrane system (4.6) helps cycle membranes.

5.1 Organization of Cell Membranes

LINKS TO
SECTIONS
3.4, 4.1

Cell membranes consist of a lipid bilayer in which many different kinds of proteins are embedded. The membrane is a continuous boundary layer that selectively controls the flow of substances across it.

REVISITING THE LIPID BILAYER

Think back on the phospholipids, the most abundant components of cell membranes (Section 3.4 and Figure 5.3*a*). Each has a phosphate-containing head and two fatty acid tails attached to one glycerol backbone. The head is hydrophilic, meaning it dissolves fast in water. The tails are hydrophobic; water repels them.

Immerse a lot of phospholipids in water, and they interact with water molecules and with one another until they spontaneously cluster into a sheet or film at the water's surface. Some line up as two layers, with all fatty acid tails sandwiched between the outward-facing hydrophilic heads. This is a **lipid bilayer**, the basic framework for cell membranes (Figure 5.3*c*).

THE FLUID MOSAIC MODEL

By the **fluid mosaic model**, every cell membrane has a mixed composition—or a *mosaic*—of phospholipids, glycolipids, sterols, and proteins. The lipids form an oily bilayer that serves as a barrier to water-soluble substances. Diverse proteins are either embedded in the bilayer or attached to one of its surfaces. They carry out most membrane functions.

The membrane is *fluid* because of interactions and motions of its components. The phospholipids differ in their heads and the length of their fatty acid tails. At least one of the tails is usually kinked, or unsaturated. Remember, an unsaturated fatty acid has one or more double covalent bonds in its carbon backbone; a fully saturated type has none. Also, most phospholipids drift sideways, spin around their long axis, and flex their tails, so they do not bunch up as a solid layer.

Figure 5.4 shows the fluid mosaic model. Section 5.2 is an overview of the membrane proteins that you will be reading about in many chapters to come.

DO MEMBRANE PROTEINS STAY PUT?

Some time ago, researchers figured out how to split a frozen plasma membrane down the middle of its bilayer. They found that proteins were not spread like a coat on the bilayer, as some had thought, but rather that many were embedded in it (Figure 5.5*a*). Were those proteins rigidly positioned in the membrane? No one knew until researchers designed an ingenious

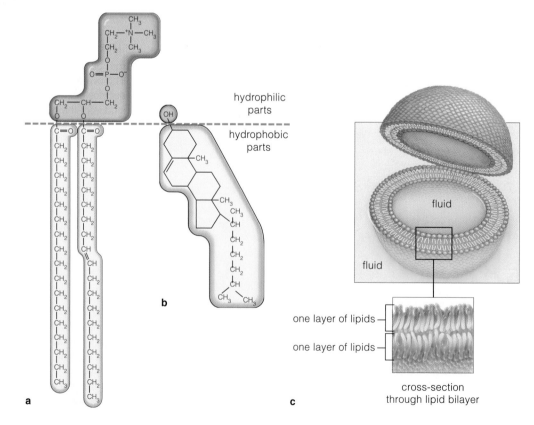

Figure 5.3 (**a**) Structural formula for phosphatidylcholine. This phospholipid is one of the most common molecules of animal cell membranes. *Orange* signifies its hydrophilic head, and *yellow*, its hydrophobic tails.

(**b**) Structural formula for cholesterol, the main sterol in animal tissues. Phytosterols are its equivalent in plant tissues.

(**c**) Spontaneous organization of lipid molecules into two layers (a bilayer structure). When immersed in liquid water, their hydrophobic tails become sandwiched between their hydrophilic heads, which dissolve in the water.

hydrophilic
parts

hydrophobic
parts

fluid

fluid

one layer of lipids

one layer of lipids

cross-section
through lipid bilayer

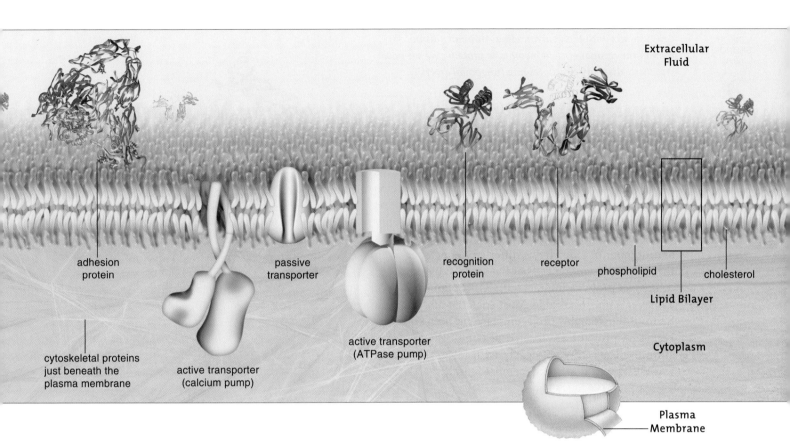

adhesion
protein

cytoskeletal proteins
just beneath the
plasma membrane

active transporter
(calcium pump)

passive
transporter

active transporter
(ATPase pump)

recognition
protein

receptor

phospholipid

cholesterol

Lipid Bilayer

Cytoplasm

Extracellular
Fluid

Plasma
Membrane

experiment. They induced an isolated human cell and an isolated mouse cell to fuse. The plasma membranes from the two species merged to form one continuous membrane in a new, hybrid cell. Most of the proteins mixed together in less than an hour (Figure 5.5*b*).

As we now know, many proteins are free to move laterally through the lipid bilayer, but others stay put. Some unite in complexes and do not move relative to one another. Receptors for acetylcholine, a signaling molecule, are like this. Cytoskeletal elements tether other proteins and restrict their lateral movements. For instance, a mesh of cross-lined spectrin proteins anchor glycophorin, a type of recognition protein, to the surface of all red blood cells. A transport protein that moves chloride one way and bicarbonate the other across the plasma membrane is similarly anchored.

All cell membranes consist of two layers of lipids—mainly phospholipids—and diverse proteins. Hydrophobic parts of the lipids are sandwiched between hydrophilic parts, which are dissolved in cytoplasmic fluid or in extracellular fluid.

All cell membranes have protein receptors, transporters, and enzymes. The plasma membrane also incorporates adhesion, communication, and recognition proteins.

Figure 5.4 *Animated!* Fluid mosaic model for the plasma membrane of an animal cell.

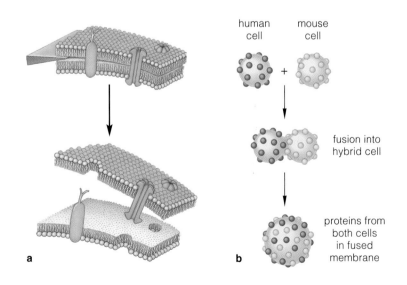

human
cell

mouse
cell

fusion into
hybrid cell

proteins from
both cells
in fused
membrane

a

b

Figure 5.5 *Animated!* Studying membranes. (**a**) Researchers split the two layers of a cell membrane's lipid bilayer apart, which revealed that proteins are embedded in the bilayer. (**b**) Result of an experiment in which plasma membranes from cells of two species were induced to fuse. Membrane proteins from both drifted laterally and became mixed.

5.2 Overview of the Membrane Proteins

LINK TO
SECTION
3.6

Cells interact with their surroundings through plasma membrane components. In membrane proteins, we see how structural diversity translates into functional diversity.

HOW ARE THE PROTEINS ORIENTED?

The fluid mosaic model is a good starting point for exploring membranes. But membranes differ in their composition and organization. Even the two surfaces of the same bilayer differ. For instance, many proteins (and lipids) of a plasma membrane have side chains of oligosaccharides and other carbohydrates, but only on the outward-facing surface (Figure 5.6). The kinds and number of side chains differ from one species to the next, even among cells of the same individual.

Integral proteins interact with hydrophobic parts of a bilayer's phospholipids. Most span the bilayer, with hydrophilic domains projecting beyond both surfaces. *Peripheral* proteins are located at one of the bilayer's surfaces. They interact weakly with integral proteins and with polar regions of membrane lipids.

WHAT ARE THEIR FUNCTIONS?

Figure 5.6 shows the main membrane proteins, lists their defining features, and gives some examples. The **transport proteins** either passively let specific solutes diffuse through a membrane-spanning channel in their interior or actively pump them through. Transporters are incorporated into all cell membranes.

The other proteins shown are typical of the plasma membrane. The **receptor proteins** bind extracellular substances, such as hormones, that can trigger change in cell activities. For example, certain enzymes control cell growth and division. They are switched on when somatotropin binds with receptors for it. Cells differ in their combinations of receptors.

Multicelled organisms have **recognition proteins** that are unique identity tags for each species; they are like molecular fingerprints. **Adhesion proteins** help cells of the same type locate each other and remain in the proper tissues. The **communication proteins** form channels that match up across the plasma membranes of two cells. They let signals and substances rapidly flow from the cytoplasm of one into the other.

> *All cell membranes have transporters that passively and actively assist water-soluble substances across the lipid bilayer. The plasma membrane, especially of multicelled species, has diverse receptors and proteins that function in self-recognition, adhesion, and communication.*

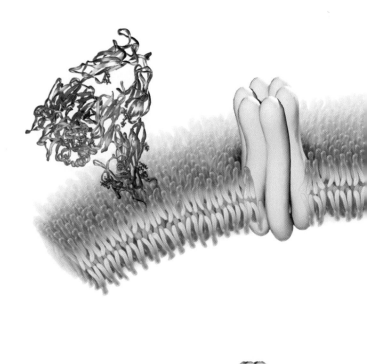

**Adhesion
Proteins**

These proteins are embedded in the plasma membrane. They help one cell adhere to another or to a protein, such as collagen, that is part of an extracellular matrix.

Integrins, including this one, relay signals across the cell membrane. Cadherins of one cell bind with identical cadherins in adjoining cells. Selectins, which hold cells together, are abundant in endothelium, the special lining of blood vessels and the heart.

**Communication
Proteins**

Communication proteins of one cell match up with identical proteins in the plasma membrane of an adjoining cell. Fingerlike projections of both intertwine in the space between the two cells. The result is a channel that directly connects the cytoplasm of both. Chemical and electrical signals flow fast through the channel.

This protein is one-half of a cardiac gap junction in heart muscle. The other half is in the lipid bilayer of another heart muscle cell (not shown) positioned above it. Signals flow so fast across such channels that heart muscle cells contract as a single functional unit.

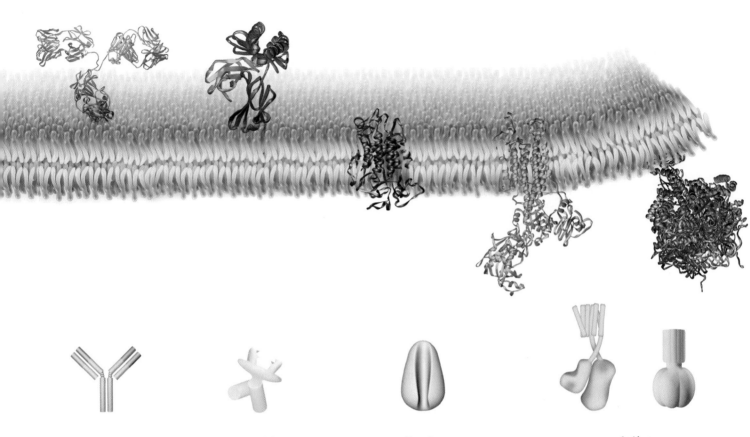

Receptor Proteins

Receptors embedded in a membrane are docks for hormones and other signaling molecules that may cause target cells to change their activities.

A signal might make a cell synthesize a certain protein, block or speed a reaction, secrete a substance, or get ready to divide.

Shown above, an antibody, a type of receptor made only by the type of white blood cell known as the B lymphocytes. These receptors are vital for all immune responses (Chapter 39).

Recognition Proteins

Certain glycoproteins (and glycolipids) project above the plasma membrane and identify a cell as *nonself* (foreign) or *self* (belonging to one's own body or a tissue).

Some, such as the HLAs (page 49), function in tissue defense. Foreign fragments bound to HLA sound the alarm for cells that defend the body. Other recognition proteins help cells stick to one another in tissues.

Passive Transporters

Passive transporters have a channel through their interior. Different kinds assist solutes or water simply by letting them diffuse through the channel, down concentration or electric gradients (Section 5.4). They do not require activation by energy inputs.

Shown here, GluT1; when its channel changes shape, glucose can cross a membrane. Aquaporins are open channels for water (page 89).

One cotransporter helps chloride and bicarbonate ions across a membrane at the same time, in opposite directions.

Ion-selective channels have molecular gates. Some gates open or close fast if a small molecule binds to them or if the charge distribution across the membrane shifts. Nerve and muscle cells have gated channels for sodium, calcium, potassium, and chloride ions.

Active Transporters

Active transport proteins pump a solute across the membrane to the side where it is more concentrated and less likely to move on its own. They require energy inputs to do this. Some are cotransporters that let one kind of solute flow passively "downhill" even as they pump a different kind "uphill."

Left, a calcium pump. Like the sodium–potassium pump, it is one of the ATPases.

Right, a type of ATPase that pumps H^+ through its interior channel, against gradients. It also can let H^+ diffuse back through the channel in a way that drives ATP synthesis. Hence its more precise name, ATP synthase (Chapters 7 and 8).

Figure 5.6 *Animated!* Major categories of membrane proteins. Included are simple icons and descriptions for membrane proteins that you will encounter in later chapters. The transporters span the lipid bilayer of all cell membranes. The other proteins shown are components of plasma membranes. Bear in mind, cell membranes also incorporate additional kinds of proteins, including some enzymes.

5.3 Diffusion, Membranes, and Metabolism

LINKS TO
SECTIONS
2.3, 2.5, 3.4

What determines whether a substance will move one way or another to and from a cell, across that cell's membranes, or through the cell itself? Diffusion down concentration gradients is part of the answer.

WHAT IS A CONCENTRATION GRADIENT?

A **concentration gradient** is a difference in the number per unit volume of molecules (or ions) of a substance between two adjoining regions. In the absence of other forces, the molecules move from a region where they are more concentrated to a region where they are not as concentrated. Why? Their inherent thermal energy keeps them in constant motion, so that they collide at random and bounce off one another millions of times each second. This happens more in regions where the molecules are most concentrated, and when you add it all up, the *net* movement is toward the region where they are not colliding and bouncing around as much. The molecules flow down their concentration gradient.

Diffusion is the name for the net movement of like molecules or ions down a concentration gradient. It is a factor in how substances move into, through, and out of cells. In multicelled species, it moves substances between body regions and between the body and its environment. For instance, when photosynthesis is going on in leaf cells, oxygen builds up and diffuses out of the cells and into air spaces in the leaf, where its concentration is lower. It then diffuses into the air outside the leaf, where its concentration is lower still.

Like other substances, oxygen tends to diffuse in a direction set by its *own* concentration gradient, not by gradients of other solutes. You can see the outcome by squeezing a drop of dye into water. The dye molecules diffuse slowly into the region where they are not as concentrated, and the water molecules move into the region where *they* are not as concentrated. Figure 5.7 shows simple examples of diffusion.

WHAT DETERMINES DIFFUSION RATES?

How fast a particular solute diffuses depends on the steepness of its concentration gradient, its size, the temperature, and electric or pressure gradients that may be present.

First, rates are high with steep gradients, because more molecules are moving out of a region of greater concentration compared with the number moving into it. Second, more heat energy makes molecules move faster and collide more often in warmer regions. Third, smaller molecules diffuse faster than large ones do.

Fourth, an electric gradient may alter the rate and direction of diffusion. An **electric gradient** is simply a difference in electric charge between adjoining regions. For example, each ion dissolved in fluids bathing a cell membrane contributes to a local electric charge. Opposite charges attract. Therefore, the fluid having more negative charge overall exerts the greatest pull on positively charged substances, such as sodium ions. Later chapters explain how many cell activities, such as ATP formation and the sending and receiving of signals in nervous systems, require the driving force of electric and concentration gradients.

Fifth, diffusion also may be affected by a **pressure gradient**. This is a difference in pressure exerted per unit volume (or area) between two adjoining regions.

MEMBRANE CROSSING MECHANISMS

Now think about the water bathing the surfaces of a cell membrane. Plenty of substances are dissolved in it, but the kinds and amounts close to its two surfaces

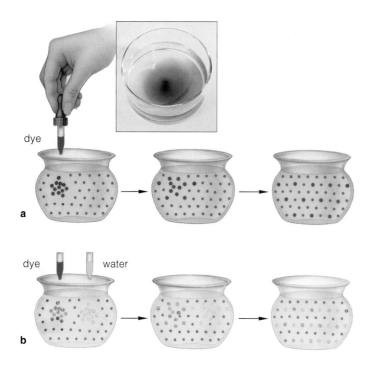

dye

a

dye   water

b

Figure 5.7 *Animated!* Two examples of diffusion. (**a**) A drop of dye enters a bowl of water. Gradually, the dye molecules become evenly dispersed through the molecules of water. (**b**) The same thing happens with the water molecules. Here, dye (*red*) and water (*yellow*) are added to the same bowl. Each substance will show a net movement down its own concentration gradient.

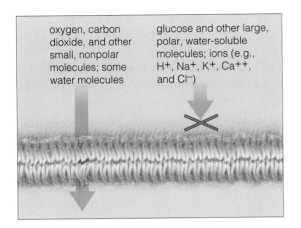

| oxygen, carbon dioxide, and other small, nonpolar molecules; some water molecules | glucose and other large, polar, water-soluble molecules; ions (e.g., H^+, Na^+, K^+, Ca^{++}, and Cl^-) |

Figure 5.8 *Animated!* Selective permeability of cell membranes. Small, nonpolar molecules and some water molecules cross the lipid bilayer. Ions and large, polar, water-soluble molecules and the water dissolving them cross with the help of transport proteins. Also, proteins called aquaporins specifically enhance the diffusion of water across the plasma membrane of certain cells.

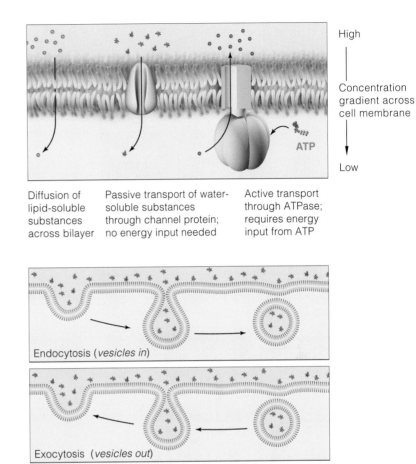

| Diffusion of lipid-soluble substances across bilayer | Passive transport of water-soluble substances through channel protein; no energy input needed | Active transport through ATPase; requires energy input from ATP |

High

Concentration gradient across cell membrane

Low

Endocytosis (*vesicles in*)

Exocytosis (*vesicles out*)

Figure 5.9 Overview of membrane crossing mechanisms.

differ. The membrane itself helps set up and maintain these differences. How? Its diverse lipid and protein components show **selective permeability**. They allow some substances but not others to enter and leave a cell. They also control when each substance can cross and how much crosses at a given time (Figure 5.8).

Membrane barriers and crossings are vital, because metabolism depends on the cell's capacity to increase, decrease, and maintain concentrations of substances required for reactions. That capacity also supplies the cell or organelles with raw materials, removes wastes, and maintains the cell volume and pH within ranges that favor reactions.

Lipids of a membrane's bilayer are mostly nonpolar, so they let small, nonpolar molecules such as O_2 and CO_2 slip across. Water molecules are polar, but some can slip through gaps that form when hydrophobic tails of lipids flex and bend (Section 5.1).

The lipid bilayer is impermeable to ions and large, polar molecules, including glucose. These substances cross a membrane by diffusing through the interior of transport proteins that span the bilayer. In many cells, proteins called aquaporins allow molecules of water to quickly cross the plasma membrane.

The passive transporters help specific solutes move down their concentration gradients but do not expend energy doing so. The mechanism, described shortly, is called *passive transport* or "facilitated" diffusion.

The active transporters help specific solutes diffuse across membranes, but they are not passive about it.

They move solutes against concentration and electric gradients, and they require an input of energy to do so. We call this mechanism *active transport*.

Other mechanisms move large particles into or out of cells. In *endocytosis* a vesicle forms around particles when a patch of plasma membrane sinks inward and seals back on itself. In *exocytosis*, a vesicle that formed in the cytoplasm fuses with the plasma membrane, so that its contents are released to the outside.

Before getting into these diverse mechanisms, you may wish to study the overview in Figure 5.9.

Diffusion is the net movement of molecules or ions of a substance into an adjoining region where they are not as concentrated. The steepness of such a concentration gradient as well as temperature, molecular size, and electric and pressure gradients affect diffusion rates.

Cellular mechanisms increase and decrease concentration gradients across cell membranes.

5.4 Working With and Against Gradients

LINK TO
SECTION
4.6

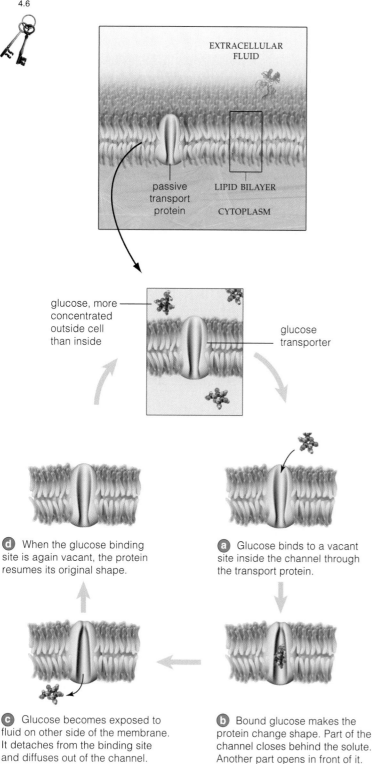

glucose, more
concentrated
outside cell
than inside

glucose
transporter

d When the glucose binding site is again vacant, the protein resumes its original shape.

a Glucose binds to a vacant site inside the channel through the transport protein.

c Glucose becomes exposed to fluid on other side of the membrane. It detaches from the binding site and diffuses out of the channel.

b Bound glucose makes the protein change shape. Part of the channel closes behind the solute. Another part opens in front of it.

Figure 5.10 *Animated!* Passive transport. This model shows one of the glucose transporters that span the plasma membrane. Glucose crosses in both directions. The *net* movement of this solute is down its concentration gradient.

Large, polar molecules and ions cannot diffuse across a lipid bilayer. They require the help of transport proteins.

Many kinds of solutes cross a membrane by diffusing through a channel or tunnel inside transport proteins. When one solute molecule or ion enters the channel and weakly binds to the protein, the protein's shape changes. The channel closes behind the solute and opens in front of it, which exposes the solute to fluid on the other side of the membrane. Now the solute is released; the binding site reverts to its original shape.

PASSIVE TRANSPORT

In **passive transport**, a concentration gradient, electric gradient, or both drive diffusion of a substance across a cell membrane, through the interior of a transport protein. The protein does not require an energy input to assist the directional movement. That is why this mechanism is also known as facilitated diffusion.

Some passive transporters are open channels; others open or close as conditions change. Figure 5.10 shows how a glucose transporter works. When one end of its channel is shut, the other is open and invites glucose in. The channel closes behind the glucose and opens in front of it, on the other side of the membrane.

The *net* direction of a solute's movement depends on how many of its molecules or ions are randomly colliding with the transporters. Encounters simply are more frequent on the side of the membrane where its concentration is greatest. The solute's *net* movement tends to be toward the side of the membrane where it is less concentrated.

If nothing else were going on, passive transport would continue until concentrations on both sides of a membrane were equal. However, other events affect the outcome. For example, the bloodstream moves glucose to all tissues. There, glucose transporters help molecules of glucose get into cells. But as fast as some glucose molecules are diffusing into the cells, others are being used as building blocks and energy sources. By *using* glucose, then, cells help maintain a gradient that favors the uptake of *more* glucose molecules.

ACTIVE TRANSPORT

Solute concentrations continually shift across the cell membrane. Living cells never stop expending energy to pump solutes into and out of their interior. With **active transport**, energy-driven protein motors help a particular kind of solute cross a cell membrane *against* its concentration gradient.

Only specific solutes can bind to functional groups that line the interior channel of an active transporter, which is activated by a phosphate group from an ATP molecule. The phosphate-group transfer changes the transporter's shape in a way that releases the solute on the other side of the membrane.

Figure 5.11 focuses on a **calcium pump**. This active transporter helps keep the concentration of calcium in a cell at least a thousand times lower than outside. What is so great about that? You will find out later, but for now think of how one of your muscle moves. The nervous system commands calcium ions to flood out from a specialized ER compartment that threads around muscle fibers inside the muscle. Calcium ions clear the way for trillions of motor proteins (myosins) to interact with actin filaments in ways that bring about contraction (Section 37.7). That muscle will go on contracting until staggering numbers of calcium pumps move those ions back inside the compartment, against their concentation gradient.

The **sodium–potassium pump** is a cotransporter that moves two kinds of ions in opposite directions. Sodium ions (Na^+) from the cytoplasm diffuse into the pump's channel and bind to functional groups. A phosphate-group transfer by ATP activates the pump, which changes shape. The change opens the channel on other side of the membrane, where Na^+ is released and potassium (K^+) diffuses in—down *its* gradient. The phosphate group is released. The channel closes behind the K^+, which is released to the other side of the membrane. As you will see, the nervous, digestive, and urinary systems of vertebrates cannot function without cellular pumps that respond to signals and to chemical changes (Sections 34.3, 41.5, and 42.3).

All cells incorporate membrane pumps. In Section 32.3, you will read about an H^+ pump that controls the transport of a hormone in growing plant parts.

Many membrane transport proteins act as open or gated channels across cell membranes. They undergo reversible changes in shape that assist solutes across the membrane.

In passive transport, a transporter allows a solute to cross a cell membrane simply by diffusing through its interior.

In active transport, the net diffusion of a specific solute is against its gradient. The transporter must be activated, usually by an energy input from ATP, which counters the force inherent in the gradient.

Passive and active transport continually help lower or raise gradients across a membrane, which helps the cell respond to signals and to chemical changes.

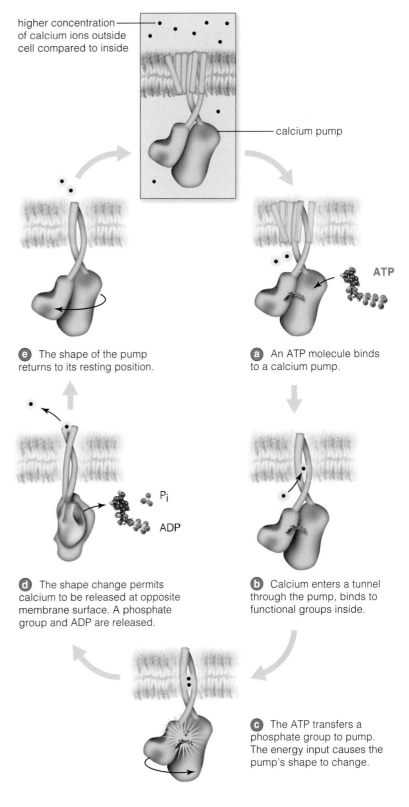

higher concentration of calcium ions outside cell compared to inside

calcium pump

e The shape of the pump returns to its resting position.

a An ATP molecule binds to a calcium pump.

ATP

d The shape change permits calcium to be released at opposite membrane surface. A phosphate group and ADP are released.

P_i

ADP

b Calcium enters a tunnel through the pump, binds to functional groups inside.

c The ATP transfers a phosphate group to pump. The energy input causes the pump's shape to change.

Figure 5.11 *Animated!* Active transport. This example uses a calcium pump that spans the plasma membrane. This sketch shows its channel for calcium ions. After two calcium ions bind to the pump, ATP transfers a phosphate group to it, thus providing energy that drives the movement of calcium *against* a concentration gradient across the cell membrane.

5.5 Which Way Will Water Move?

LINKS TO
SECTIONS
2.5, 4.8, 4.9

By far, more water diffuses across cell membranes than any other substance, so the main factors that influence its directional movement deserve special attention.

MOVEMENT OF WATER

Something as gentle as a running faucet or as mighty as Niagara Falls demonstrates **bulk flow**, or the mass movement of one or more substances in response to pressure, gravity, or another external force. Bulk flow accounts for some movement of water in multicelled organisms. A beating heart generates fluid pressure that pumps blood, which is mostly water. Sap flows inside tubes in trees, and this, too, is bulk flow.

What about the movement of water into and out of cells and organelles? If the concentration of water is not equal on both sides of a membrane, osmosis will probably occur. **Osmosis** is the diffusion of water across a selectively permeable membrane, to a region where the water concentration is lower.

You might be wondering: How can water —a liquid—be more or less concentrated? For the answer, you have to think of water in terms of its concentration relative to the amounts of solutes that may be dissolved in it. The greater the solute concentration, the lower the water concentration.

Visualize yourself pouring some glucose or another solute to a glass of water, so that you increase the volume of liquid. The glass has the same number of water molecules but in a greater volume of liquid.

Now visualize yourself using a membrane to divide the inside of another glass of water into two compartments. The membrane lets water but not glucose diffuse across it. Next, add glucose on one side of the membrane. Water follows its concentration gradient into the glucose solution until its concentration is the same on both sides of the membrane (Figure 5.12).

In cases of osmosis, "solute concentration" refers to the total number of molecules or ions in a volume of a solution. It does not matter whether the dissolved substance is glucose, urea, or anything else. The type of solute does not dictate water concentration.

EFFECTS OF TONICITY

Suppose you decide to test the statement that water tends to move into a region where solutes are more concentrated. You make three sacs from a membrane that water but not sucrose can cross, and fill each one with a solution that is 2 percent sucrose. You immerse the first sac in a liter of distilled water, the second sac in a solution that is 10 percent sucrose, and the third sac in a solution that is 2 percent sucrose.

In each experiment, tonicity dictates the extent and direction of water movement across the membrane, as Figure 5.13 shows. *Tonicity* refers to the relative solute concentrations of two fluids. When two fluids that are on opposing sides of a membrane differ in their solute concentrations, the **hypotonic solution** is the one with fewer solutes. The one having more solutes is a **hypertonic solution**. Water tends to diffuse from a hypotonic fluid into a hypertonic fluid. **Isotonic solutions** show no net osmotic movement.

Most cells have built-in mechanisms that counter shifts in tonicity. Red blood cells do not. Figure 5.13 shows what would happen to them if tonicity were to change. Normally, fluid in red blood cells is isotonic with tissue fluid. If the tissue fluid became hypotonic, too much water would diffuse into the cells, which would burst. If that tissue fluid became hypertonic, water would diffuse out, and the cells would shrivel.

EFFECTS OF FLUID PRESSURE

Most cells do not swell and burst from an influx of water by osmosis. For one thing, they can selectively transport solutes out. For another thing, the cells of plants and many protists, fungi, and bacteria have a

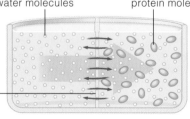

water molecules   protein molecules

semipermeable membrane between two compartments

Figure 5.12 Solute concentration gradients and osmosis. A membrane divides this container into two compartments. Water but not proteins can cross it. Pour 1 liter of water in the left compartment and 1 liter of a protein-rich solution in the right compartment. The proteins occupy some of the space available, and the net diffusion of water in this case is from left to right (large *gray* arrow).

wall that helps keep them from rupturing when they become turgid, or swollen with fluid.

In later chapters, you will see how osmosis affects the water and solutes inside plants and animals. For now, just think about the hypotonic and hypertonic solutions in Figure 5.14. Water molecules move back and forth until the water concentration is equal on both sides of a membrane that separates them. But the volume of the formerly hypertonic solution has now increased, because its solutes cannot diffuse out.

The same thing happens in plant cells, which tend to be hypertonic relative to soil water. When a young plant cell grows, water moves into it by osmosis and exerts fluid pressure on its primary wall (Section 4.9). Up to a point, this pliable wall expands under fluid pressure, and the cell increases in volume. Continued expansion ends when the wall shows enough resistance to stop the further inward movement of water.

Any volume of fluid exerts **hydrostatic pressure**, or *turgor* pressure, against the wall or membrane that contains it. The **osmotic pressure** of any fluid is one measure of the tendency of water to follow its water concentration gradient and move into that fluid. When hydrostatic pressure and osmotic pressure are equal in magnitude, osmosis stops completely.

Plant cells also are vulnerable to the loss of water, which can occur when soil dries or becomes too salty. Water stops diffusing in and starts diffusing out, so hydrostatic pressure falls and the cytoplasm shrinks.

> Osmosis is a net diffusion of water between two solutions that differ in solute concentration and are separated by a selectively permeable membrane. The greater the number of molecules and ions dissolved in a given amount of water, the lower the water concentration will be.
>
> Water tends to move osmotically to regions of greater solute concentration (from hypotonic to hypertonic solutions). There is no net diffusion between isotonic solutions.
>
> Fluid pressure that a solution exerts against a membrane or wall influences the osmotic movement of water.

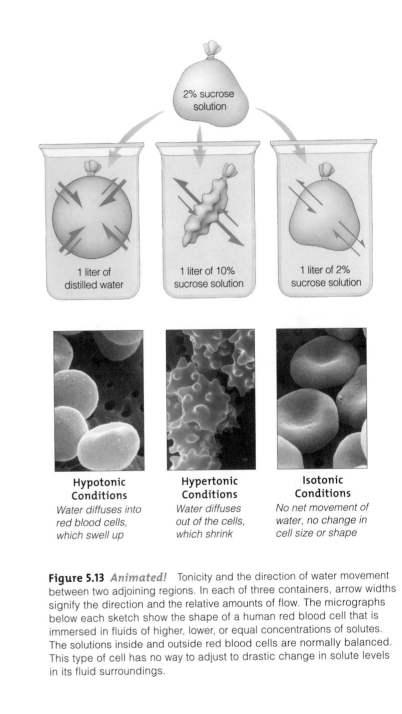

Hypotonic Conditions
Water diffuses into red blood cells, which swell up

Hypertonic Conditions
Water diffuses out of the cells, which shrink

Isotonic Conditions
No net movement of water, no change in cell size or shape

Figure 5.13 *Animated!* Tonicity and the direction of water movement between two adjoining regions. In each of three containers, arrow widths signify the direction and the relative amounts of flow. The micrographs below each sketch show the shape of a human red blood cell that is immersed in fluids of higher, lower, or equal concentrations of solutes. The solutions inside and outside red blood cells are normally balanced. This type of cell has no way to adjust to drastic change in solute levels in its fluid surroundings.

Figure 5.14 *Animated!* Experiment showing an increase in fluid volume as an outcome of osmosis. A selectively permeable membrane separates two compartments. Over time, the net diffusion will be the same in both directions across the membrane, but the fluid volume in the second compartment will be greater because there are more solute molecules in it.

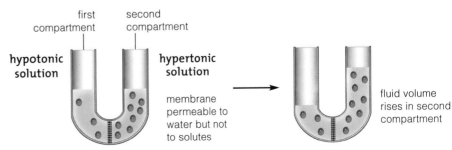

5.6 Membrane Traffic To and From the Cell Surface

LINKS TO
SECTIONS
4.6, 4.11

We leave this chapter with another look at exocytosis and endocytosis. By these mechanisms, vesicles move substances to and from the plasma membrane. Vesicles help the cell take in and expel materials in larger packets than transport proteins would be able to handle.

ENDOCYTOSIS AND EXOCYTOSIS

Think back on the lipid bilayer and how it minimizes the number of hydrophobic groups exposed to water. When the arrangement is disrupted—as when part of the plasma membrane or an organelle pinches off as a vesicle—the bilayer becomes self-sealing. Why? The disruption exposes too many of hydrophobic groups to the surroundings. When a patch of membrane is budding off, its phospholipids are being repelled by water on both sides of it. The water molecules "push" the phospholipids together, which rounds off the bud as a vesicle and also seals the rupture.

The lipid bilayer's self-sealing behavior is the basis of membrane traffic to and from a cell surface (Figure 5.15). That traffic moves in two directions.

By **endocytosis**, a small patch of plasma membrane balloons inward and pinches off inside the cytoplasm. It forms an endocytic vesicle that moves its contents to some organelle or stores them in a cytoplasmic region. By **exocytosis**, a vesicle moves to the cell surface, and then the protein-studded lipid bilayer of its membrane fuses with the plasma membrane. While this exocytic vesicle is losing its identity, its contents are released to the outside (Figure 5.15).

There are three endocytic pathways. With *receptor-mediated* endocytosis, a hormone, vitamin, mineral, or another substance binds to receptors at the plasma membrane. A slight depression, or pit, forms in the plasma membrane beneath the receptors. The pit sinks into the cytoplasm as hydrophobic interactions cause a vesicle to form (Figure 5.16).

Phagocytosis ("cell eating") is a common endocytic pathway. Phagocytes such as amoebas engulf microbes, food particles, or cellular debris. In multicelled species, macrophages and some other white blood cells engulf pathogenic viruses and bacteria, cancerous body cells, and other threats. Receptors play a different role in phagocytosis. When they bind to a specific substance, they cause microfilaments to become rearranged into a mesh just beneath the phagocyte's plasma membrane. The microfilaments contract and a bulging volume of cytoplasm is squeezed toward the cell periphery. The bulge, still enclosed in the plasma membrane, extends outward as a pseudopod (Section 4.11 and Figure 5.17).

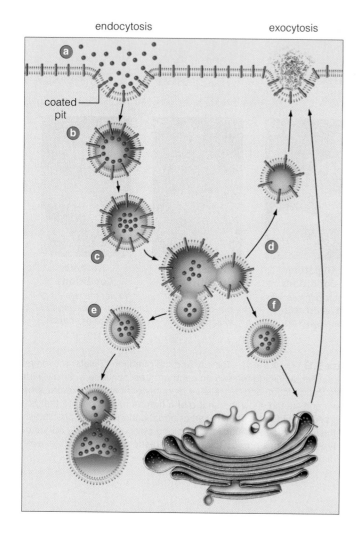

Figure 5.15 *Animated!* Endocytosis and exocytosis. This sketch starts with receptor-mediated endocytosis. (**a**) Molecules get concentrated inside coated pits at the plasma membrane. (**b**) The pits sink inward and become endocytic vesicles. (**c**) The vesicle contents are sorted and often released from receptors. (**d**) Many sorted molecules are cycled back to the plasma membrane. (**e,f**) Many others are delivered to lysosomes and stay there or are degraded. Still others are routed to spaces in the nuclear envelope and inside ER membranes, and others to Golgi bodies.

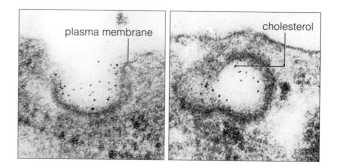

Figure 5.16 Endocytosis of cholesterol molecules.

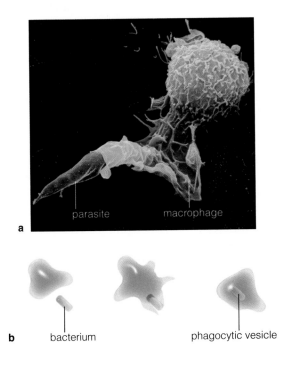

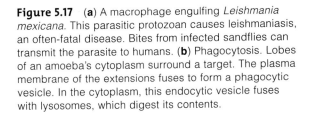

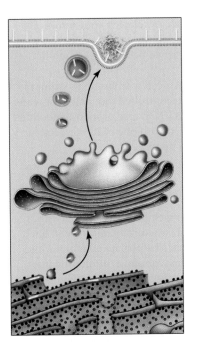

Figure 5.18 Example of how the asymmetric distribution of proteins, carbohydrates, and lipids in cell membranes originates. Proteins of the plasma membrane start out as new polypeptide chains, which become modified inside the channels of the ER and Golgi bodies. Many depart in vesicles that bud off, move to the plasma membrane, and fuse with it. The proteins inside automatically become oriented in the proper direction in the plasma membrane.

Figure 5.17 (**a**) A macrophage engulfing *Leishmania mexicana*. This parasitic protozoan causes leishmaniasis, an often-fatal disease. Bites from infected sandflies can transmit the parasite to humans. (**b**) Phagocytosis. Lobes of an amoeba's cytoplasm surround a target. The plasma membrane of the extensions fuses to form a phagocytic vesicle. In the cytoplasm, this endocytic vesicle fuses with lysosomes, which digest its contents.

Pseudopods flow completely around their target and then form a cytoplasmic vesicle. The vesicle sinks into the cytoplasm and fuses with lysosomes (Section 4.6). Lysosomal enzymes digest the vesicle's contents into fragments and smaller, reusable molecules.

Bulk-phase endocytosis is not as selective. A vesicle forms around a small volume of the extracellular fluid regardless of the kinds of substances dissolved in it.

MEMBRANE CYCLING

As long as a cell is alive, exocytosis and endocytosis are continually replacing and withdrawing patches of its plasma membrane, as in Figure 5.15. Apparently they do so at rates that maintain the total surface area of the plasma membrane. Steady losses in the form of endocytic membranes are balanced by replacements in the form of exocytic membranes.

For example, neurons release neurotransmitters in bursts of exocytosis. Each neurotransmitter is a type of signaling molecule that acts on neighboring cells.

An intense burst of endocytosis counterbalances each major burst of exocytosis.

The membranes are not shipped any which way. As an example, the composition and organization of the plasma membrane start inside the ER membranes, where many polypeptide chains become modified before being packaged and moved on to their final destinations (Section 4.6). Proteins that will become part of the plasma membrane are shipped in vesicles that fuse with a Golgi body. There, they are further modified, then sent off in other vesicles that fuse with the plasma membrane. As Figure 5.18 shows, fusion releases the proteins to the membrane surface that faces outside. There they will perform their functions.

Whereas transport proteins in a plasma membrane deal with ions and small molecules, exocytosis and endocytosis move large packets of materials in bulk across a plasma membrane.

By exocytosis, a cytoplasmic vesicle fuses with the plasma membrane, and its contents are released outside the cell.

By endocytosis, a small patch of plasma membrane sinks into the cytoplasm and pinches off as a vesicle. Membrane receptors often activate cytoskeletal elements that take part in endocytosis.

Phagocytosis is a form of endocytosis by which predatory amoebas engulf prey and certain white blood cells actively engulf tissue invaders, tissue debris, and cancer cells.

Summary

Section 5.1   Animal cell membranes consist mainly of phospholipids, along with glycolipids and sterols. The lipids are organized as a double layer, with all of their hydrophobic tails sandwiched between hydrophilic heads at both surfaces.

The lipid bilayer gives a cell membrane its primary structure and prevents uncontrolled movement of water-soluble substances across it. Diverse proteins embedded in the bilayer or associated with one of its surfaces carry out most membrane functions.

Biology Now
Learn about membrane structure and the experiments that elucidated it with the animation on BiologyNow.

Section 5.2   Each cell membrane associates with cytoplasmic proteins that structurally reinforce it. Each has receptors at its surface. The plasma membrane also contains adhesion proteins, communication proteins, recognition proteins, and diverse receptors (Figure 5.19). Differences in the number and types of proteins affect responsiveness to substances at the membrane, as well as cell metabolism, pH, and volume.

Water-soluble substances cross cell membranes by passing through the interior of transport proteins, which open to both sides of the membrane.

Receptor proteins bind extracellular substances, and binding triggers alterations in cell activities.

Recognition proteins are molecular fingerprints; they identify cells as being of a given type. Adhesion proteins help cells of tissues adhere to one another and to proteins of the extracellular matrix.

Communication junctions extend across the plasma membranes of adjoining cells; they let substances and signals travel swiftly from one into the other.

Biology Now
Use the animation on BiologyNow to familiarize yourself with the functions of receptor proteins.

Section 5.3   A concentration gradient is a difference in the number per unit volume of molecules (or ions) of a substance between two regions. The molecules tend to show a net movement down such a gradient, to the region where they are less concentrated. This behavior is called diffusion. The steepness of a concentration gradient, temperature, molecular size, and gradients in electrical charge and pressure influence diffusion rates.

Built-in cellular mechanisms work with and against gradients to move solutes across membranes.

Molecular oxygen, carbon dioxide, and other small, nonpolar molecules easily diffuse across a membrane's lipid bilayer. Ions and large, polar molecules such as glucose cross it through the interior of transport proteins that span the bilayer. Water molecules slip through gaps that briefly open in the bilayer. Aquaporins selectively assist water molecules across certain cell membranes.

Biology Now
Investigate diffusion across membranes with the interaction on BiologyNow.

Section 5.4   Many solutes cross membranes through transport proteins that act as open or gated channels or that reversibly change shape. Passive transport does not require energy input; a solute is free to follow its own concentration gradient across the membrane. Active transport requires an energy input from ATP to move a specific solute against its concentration gradient.

Biology Now
Compare the processes of passive and active transport, using the animation on BiologyNow.

Section 5.5   Osmosis is the diffusion of water across a selectively permeable membrane. The water molecules move down a water concentration gradient, which is influenced by solute concentrations and pressure.

Biology Now
Explore the effects of osmosis with the interaction and animation on BiologyNow.

Section 5.6   By exocytosis, a cytoplasmic vesicle fuses with the plasma membrane, and its contents are released outside. By endocytosis, a patch of plasma membrane forms a vesicle that sinks into the cytoplasm.

Biology Now
Use the animation on BiologyNow to discover how membrane components are cycled.

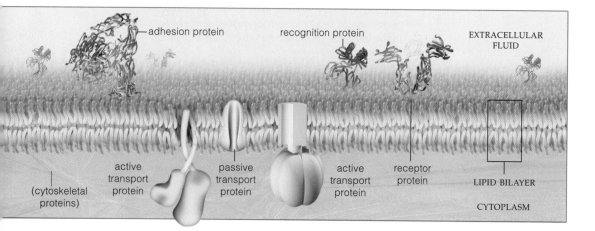

Figure 5.19   Summary of major types of membrane proteins.

Figure 5.20 Go ahead, name the mystery membrane mechanism.

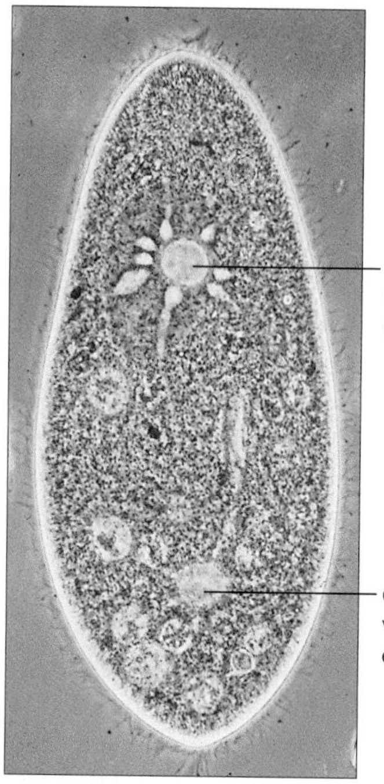

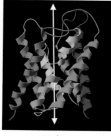

Figure 5.21 Light micrograph of one of the ciliated protozoans (*Paramecium*). This tiny single-celled body is crammed with diverse organelles, including contractile vacuoles.

— contractile vacuole filled

— contractile vacuole empty

Self-Quiz

Answers in Appendix II

1. Cell membranes consist mainly of a _____ .
   a. carbohydrate bilayer and proteins
   b. protein bilayer and phospholipids
   c. lipid bilayer and proteins

2. In a lipid bilayer, _____ of all of the lipid molecules are sandwiched between all of the _____ .
   a. hydrophilic tails; hydrophobic heads
   b. hydrophilic heads; hydrophilic tails
   c. hydrophobic tails; hydrophilic heads
   d. hydrophobic heads; hydrophilic tails

3. Most membrane functions are carried out by _____ .
   a. proteins   c. nucleic acids
   b. phospholipids   d. hormones

4. Plasma membranes incorporate _____ .
   a. transport proteins   c. recognition proteins
   b. adhesion proteins   d. all of the above

5. Diffusion is the movement of ions or molecules from one region to another where they are less concentrated. The rate of diffusion is affected by _____ .
   a. temperature   c. molecular size
   b. electrical gradients d. all of the above

6. _____ can readily diffuse across a lipid bilayer.
   a. Glucose   c. Carbon dioxide
   b. Oxygen   d. b and c

7. Some sodium ions cross a cell membrane through transport proteins that first must be activated by an energy boost. This is an example of _____ .
   a. passive transport   c. facilitated diffusion
   b. active transport   d. a and c

8. Immerse a living cell in a hypotonic solution, and water will tend to _____ .
   a. move into the cell   c. show no net movement
   b. move out of the cell d. move in by endocytosis

9. Vesicles form by way of _____ .
   a. membrane cycling   d. halitosis
   b. exocytosis   e. a through c
   c. phagocytosis   f. all of the above

10. Match the term with its most suitable description.
    ____ phagocytosis   a. molecular fingerprint
    ____ passive transport b. basis of diffusion
    ____ recognition protein c. big in membranes
    ____ active   d. one cell engulfs another
         transport   e. requires energy boost
    ____ phospholipid   f. docks for signals and
    ____ concentration   substances at cell surface
         gradient   g. no energy boost required
    ____ receptors   to move solutes

Additional questions are available on **Biology ⊜ Now**™

Critical Thinking

1. Is the white blood cell shown in Figure 5.20 disposing of a worn-out red blood cell by endocytosis, phagocytosis, or both?

2. Water moves osmotically into *Paramecium*, a single-celled aquatic protist. If unchecked, the influx would bloat the cell and rupture its plasma membrane, and the cell would die. An energy-requiring mechanism that involves contractile vacuoles expels excess water (Figure 5.21). Water enters the vacuole's tubelike extensions and collects inside. A full vacuole contracts and squirts water out of the cell through a pore. Are *Paramecium*'s surroundings hypotonic, hypertonic, or isotonic?

3. Water crosses cell membranes by diffusing past lipids that are jostling apart from one another in the bilayer. In many tissues, it also crosses faster through the interior channels of *aquaporins* (*white* arrow in Figure 5.22). As many as 3 billion water molecules per second flow through an aquaporin. Researchers already have found similar aquaporins in bacteria, plants, and insects.

Different aquaporins help different tissues respond to shifting conditions in the internal environment. They have roles in how the kidneys conserve or get rid of excess water. They play a part in producing and maintaining the fluid that bathes the spinal cord and brain, in producing saliva and tears, in keeping the lining of the lungs moist, and in keeping red blood cells from bursting or shriveling as the body's water–solute balance shifts.

If the gene for one of these water channels mutates, the outcome may be serious. Mutation in *aquaporin-0* results in cataracts, and mutation in *aquaporin-2* leads to a form of diabetes insipidus. Yet *aquaporin-1* seems less essential. In its absence, affected adults tend to produce unusually dilute urine but remain in good health as long as they drink plenty of water. Even so, affected individuals are rare. Speculate on the reasons why.

extracellular fluid

cytoplasm

Figure 5.22 Model for one of the four aquaporin subunits.

6 GROUND RULES OF METABOLISM

Alcohol, Enzymes, and Your Liver

The next time someone asks you to have a drink or two, or three, stop for a moment and think about the challenge confronting the cells that are supposed to keep a drinker alive—especially heavy drinkers. It makes little difference whether someone gulps down 12 ounces of beer, 5 ounces of wine, or 1–1/2 ounces of eighty-proof vodka. Each of these drinks has the same amount of "alcohol" or, more precisely, ethanol.

Ethanol molecules—CH_3CH_2OH—have water-soluble and fat-soluble components. Once they reach the stomach and the small intestine, they are quickly absorbed into the internal environment. The bloodstream transports more than 90 percent of ethanol's components to liver cells. There, enzymes speed their breakdown to a nontoxic form called acetate, or acetic acid. The liver has great numbers of alcohol-metabolizing enzymes, but they can detoxify only so much in a given hour.

One of the enzymes you will read about in this chapter is catalase, a foot soldier against toxins that can attack the body (Figure 6.1). Catalase assists another enzyme, alcohol dehydrogenase. When alcohol circulates through the liver, these enzymes convert it to acetaldehyde. The reactions cannot end there, because acetaldehyde becomes toxic when it accumulates in high concentrations. In healthy people at least, still another kind of enzyme speeds its breakdown to nontoxic forms.

Given the liver's central role in alcohol metabolism, habitually heavy drinkers gamble with alcohol-induced liver diseases. Over time, the capacity to tolerate alcohol diminishes because there are fewer and fewer liver cells —hence fewer enzymes—for detoxification.

What are some of the possible outcomes? A big one is *alcoholic hepatitis*, an all-too-common disease characterized by inflammation and destruction of liver tissue. Another disease, *alcoholic cirrhosis*, permanently scars the liver. In time, the liver stops working, with devastating effects.

The liver is the largest gland in the human body, and its activity impacts everything else. You would have a really hard time digesting and absorbing food without it. Your cells would have a hard time synthesizing and taking up carbohydrates, lipids, and proteins, and staying alive.

There is more to think about. The liver gets rid of a lot more toxic compounds than just acetaldehyde. It also makes certain plasma proteins that circulate freely in blood. In their absence, your body would not be able to defend itself well from attacks or to stop bleeding even from small cuts.

Figure 6.1 Something to think about—ribbon model for catalase, an enzyme that helps detoxify many substances that can damage the body, such as the alcohol in beer, martinis, and other drinks.

It would not be able to maintain the fluid volume of the internal environment that all of your cells depend upon.

Now think about a self-destructive behavior known as *binge drinking*. The idea is to consume large amounts of alcohol in a brief period. Binge drinking is now the most serious drug problem on campuses throughout the United States. Consider this finding from one study: Of nearly 17,600 students surveyed at 140 colleges and universities, 44 percent said they are caught in the culture of drinking. They report having five alcoholic drinks a day, on average.

Binge drinking can do far more than damage the liver. Put aside the related 500,000 injuries from accidents, the 70,000 cases of date rape, the 400,000 cases of (whoops) unprotected sex among students in an average year. Binge drinking can kill before you know what hit you. Drink too much, too fast, and you can abruptly end the beating of your heart.

With this example, we turn to **metabolism**, the cell's capacity to acquire energy and use it to build, degrade, store, and release substances in controlled ways. At times, the activities of your cells may be the last thing you want to think about. But they help define who you are and what you will become, liver and all.

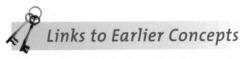

Watch the video online!

✔ How Would You Vote?

Some people have damaged their liver because they drank too much alcohol. Others have a diseased liver. There are not enough liver donors for all the people waiting for liver transplants. Should life-style be a factor in deciding who gets a transplant? See BiologyNow for details, then vote online.

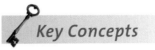

Key Concepts

THE NATURE OF ENERGY FLOW

Energy cannot be created or destroyed. It can only be converted from one form to another. Concentrated forms of energy tend to spread out, or disperse, spontaneously to less concentrated and less usable forms, such as dispersed heat. Collectively, chemical bonds resist this tendency and help organisms maintain their complex organization. Section 6.1

ENERGY CHANGES AND ATP

Metabolic reactions require an energy input before they can run spontaneously to completion. Some end with products that have more energy than the reactants. Others release more usable energy than the amount invested to start the reactions. ATP couples reactions that can release usable energy with reactions that require it. Section 6.2

ENERGY CHANGES AND ENZYMES

On their own, chemical reactions proceed too slowly to sustain life. Enzymes increase reaction rates enormously. They lower the amount of energy it takes to align reactive groups, destabilize electric charges, and break bonds so that products can form from reactants.

Temperature, pH, salinity, and other environmental factors influence enzyme activity. So does the availability of cofactors, or enzyme helpers. Sections 6.3, 6.4

THE NATURE OF METABOLISM

Metabolic pathways are enzyme-mediated sequences of reactions. By controlling enzymes that govern key steps in these pathways, cells build up, maintain, and decrease amounts of thousands of substances. ATP-forming pathways require redox reactions, or electron transfers. Section 6.5

FROM CONCEPT TO APPLICATION

We can simply absorb information about nature, including how enzymes work. We also can interpret information in novel ways that may have practical application. Section 6.6

Links to Earlier Concepts

Reflect again on the road map for life's organization (Section 1.1). Here you will gain insight into how organisms tap into a grand, one-way flow of energy to maintain that organization (1.2). You will start thinking about how cells use the chemical behavior of electrons and protons in ways that help make ATP (2.3). Remember what you learned about acids and bases (2.6)? Here you will see how pH influences enzyme activity.

6.1 Energy and Time's Arrow

LINK TO
SECTION
1.2

You know, almost without thinking about it, that your life does not stand still and will not start all over again. You have a sense of time's arrow—that everything we have observed and experienced, and all we expect will happen, goes forward. But <u>why</u> does it go forward?

WHAT IS ENERGY?

A dictionary definition of **energy** is "the capacity to do work," which doesn't say much. It is no more than a clue to two of the most sweeping laws of nature we humans have ever tried to wrap our minds around. In itself, the **first law of thermodynamics** seems simple enough: *Energy cannot be created or destroyed.*

Basically, this law deals with the *quantity* of energy in the universe. There is a finite amount distributed in different forms. However, one form may be converted to another.

One form, *potential* energy, is a capacity to do work because of something's location and the arrangement of its parts. While the skydivers shown in Figure 6.2 were still inside the plane, each had a store of potential energy because of their position above the ground. ATP and other molecules in their body had potential energy because of how their atoms were held together in particular arrangements by chemical bonds.

When the skydivers jumped, potential energy was transformed into *kinetic* energy, or energy of motion. ATP in muscle cells gave up some potential energy to molecules of contractile units and set them in motion. The motion of many thousands of muscle cells made whole muscles move. With each energy transfer from ATP, a bit of energy slipped off into the surroundings as *thermal* energy, or heat.

Figure 6.2 Forms of energy. How many can you identify?

The potential energy of molecules has its own name, *chemical* energy. It is measurable, as in kilocalories. A **kilocalorie** is the same as 1,000 calories, or the amount of energy it takes to heat 1,000 grams of water from 14.5°C to 15.5°C at standardized pressure.

THE ONE-WAY FLOW OF ENERGY

Now imagine all the energy changes going on inside the skydivers, between their hot bodies and molecules of air they are falling through, between the sun and those plants in the fields below them. Zoom out past the sun, past our solar system, and all the way to the boundary of the known universe, and you will pass staggeringly diverse energy conversions.

Amazingly, another law of thermodynamics helps explain every conversion by dealing with the *quality* of energy. It tells us why the sun will eventually burn out, why animals eat plants and one another, why nitroglycerin spontaneously explodes but you don't, why it rains, why skydivers fall down instead of up and splat if their parachutes don't open, and why your life can't start over again.

The **second law of thermodynamics** sounds simple enough: *Energy tends to flow from concentrated to less concentrated forms—so the cost of concentrating it in one area comes at a greater cost of energy dispersal or dilution somewhere else.* This tendency gives us our sense of "time's arrow." Skydivers in free fall do not have a concentrated form of energy that can get them back to the plane. A hot pan gives off heat as it cools, and that heat cannot diffuse back from the air to the pan. You can't rewind the winds flowing around the eye of a hurricane. You can't take back a scream.

The second law says only that concentrated energy tends to disperse spontaneously. It does not say when or how slowly or fast that might happen. In our own world, *the collective strength of chemical bonds resists the spontaneous direction of energy flow.* Think of the energy in uncountable numbers of chemical bonds between all the atoms making up the Rocky Mountains. Millions of years will pass before attacks by winds, rain, ice, and other assaults will break enough bonds to return the mountains to the seas. Or think of all the chemical bonds in your skin, heart, liver, and other body parts. The concentration of energy definable as "you" stays together as long as they do.

Entropy is the measure of how much and how far a concentrated form of energy has been dispersed after an energy change. It is not a measure of order versus disorder. In 2004, at one of the most active subduction zones on the seafloor, a major earthquake generated a

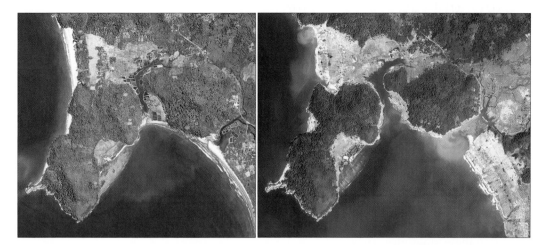

Figure 6.3 Aftermath of an appalling change in entropy. Gleebruk village, Indonesia, shown before and after a tsunami struck in December 2004. A major earthquake on the seafloor generated giant ocean waves—a form of concentrated mechanical energy that spread out after the waves hit land.

tsunami that spread across the Indian Ocean. Few of us missed the images of monstrous waves lashing at coasts all the way from Indonesia to Africa (Figure 6.3). Orderly rows of houses, hotels, and crops were ripped apart and washed away. But this horrific mess was not an increase in entropy; *it was its aftermath*. The earthquake had generated waves of pressure—a form of mechanical energy—in ocean water. There was a huge concentration of pressure that dissipated as the waves slammed against shorelines. The before/after difference in mechanical energy—from the original concentration on the seafloor to the last fingers of surf that fizzled away on land—was the entropy change.

The point is, *energy tends to flow in one direction, from concentrated to less concentrated forms*. The world of life is responsive to the flow. Photosynthetic cells of plants and other producers tap into a concentrated store of energy: light from the sun. They convert it to chemical bond energy in sugars and other organic compounds. Consumers as well as producers access that stored energy by breaking and rearranging chemical bonds. With each conversion, though, some energy is lost as heat, a dispersed form of energy that cells can't gather up again. The inevitable losses mean that the world of life must maintain its complex organization through ongoing replenishments of energy—which is being lost from someplace else (Section 1.2 and Figure 6.4).

Energy is the capacity to do work, and it cannot be created or destroyed. It can be converted from one form to another.

Energy concentrated in one place tends to spread out, or disperse, on its own. The collective strength of chemical bonds resists this spontaneous direction of energy flow.

Life continues as long as organisms tap into concentrated energy sources and use them to build complex molecules even as they continually lose energy in the form of heat.

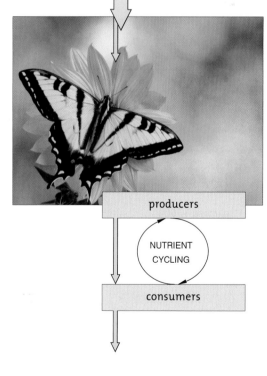

ENERGY LOST
Energy continually flows from the sun.

ENERGY GAINED

Sunlight energy reaches environments on Earth. Producers of nearly all ecosystems secure some and convert it to stored forms of energy. They and all other organisms convert stored energy to forms that can drive cellular work.

producers

NUTRIENT CYCLING

consumers

ENERGY LOST

With each conversion, there is a one-way flow of a bit of energy back to the environment. Nutrients cycle between producers and consumers.

Figure 6.4 A one-way flow of energy into an ecosystem compensates for the one-way flow of energy out of it.

6.2 Time's Arrow and the World of Life

LINKS TO
SECTIONS
1.5, 3.7, 5.4

Once the two laws of thermodynamics were just hypotheses. They became accepted as theories so powerful and wide-ranging that they became known as laws. Remember, the best theories—and laws—are supported by predictions based on them (Section 1.5). With respect to metabolism, the key prediction is this: It takes net inputs of energy to force small molecules to combine into larger ones, such as glucose, that are more concentrated forms of energy.

The molecules of life do not form spontaneously, and they are not that stable in the presence of free oxygen, which reacts with them and disrupts their structure. Cellular mechanisms safeguard these molecules. They control when and where energy changes will occur; that is, when energy will flow from one substance to another during chemical reactions.

You already know about some of the participants in metabolic reactions. Starting substances are called *reactants*. Substances formed before a reaction ends are *intermediates*, and those remaining are *products*. ATP

and other *energy carriers* activate enzymes and other molecules by phosphate-group transfers. *Enzymes* are catalysts; they speed specific reactions enormously. *Cofactors* are metal ions or coenzymes. They assist the enzymes by accepting and donating electrons, atoms, and functional groups. *Transport proteins* help solutes across membranes. Their action affects concentrations of substances, which in turn affects how, when, and whether a reaction can proceed.

ACTIVATION ENERGY—WHY THE WORLD DOESN'T GO UP IN SMOKE

When substances react, some of the chemical energy required to break bonds is conserved as new bonds form. Some energy is lost as heat, light, or both. Think of what happens after a spark from a campfire ignites tinder-dry plants. Plants are mostly cellulose—which has three reactive hydroxyl groups *in each one of many repeating units*: $C_6H_7O(OH)_3$. Once this reaction starts, it proceeds swiftly on its own. Most of the cellulose breaks down fully to carbon dioxide (CO_2) and water (H_2O), with the release of light and heat. Remember Figure 1.1? A single match can start a firestorm.

Why doesn't the world go up in flames on its own? It takes a boost of energy to overcome the strength of chemical bonds in reactants. The boost has to be big for some reactions but not much for others. Like other explosives, nitroglycerin has a lot of oxygen atoms. A hard shake or jarring is all it takes for nitroglycerin to start falling apart—explosively fast—into hot gases.

Each kind of reaction has a characteristic **activation energy**, the minimum amount of energy that can get the reaction to the point that it will run on its own. By controlling the energy inputs into reactions, cells are able to control when and how fast the reactions occur.

UP AND DOWN THE ENERGY HILLS

Let's see how this works when photosynthetic cells build glucose and other carbohydrates. An input of energy from the sun triggers two stages of reactions. Ultimately, six CO_2 and six H_2O molecules are used in the formation of one glucose molecule ($C_6H_{12}O_6$) and six oxygen molecules (O_2). Figure 6.5*a* is a simple way to think about these reactants and products.

The synthesis of glucose is an example of a series of reactions that just will not happen on their own. Why not? Carbon dioxide and water do have energy stored in their covalent bonds, but the bonds are so stable that it is as if they are at the base of an "energy hill." It takes inputs of energy to break those bonds

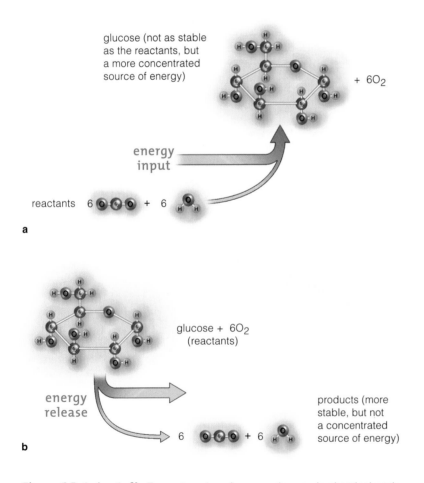

Figure 6.5 *Animated!* Two categories of energy changes in chemical work. (**a**) Endergonic reactions require a net input of energy. (**b**) Exergonic reactions end with a net release of usable energy.

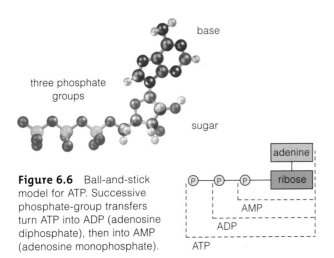

Figure 6.6 Ball-and-stick model for ATP. Successive phosphate-group transfers turn ATP into ADP (adenosine diphosphate), then into AMP (adenosine monophosphate).

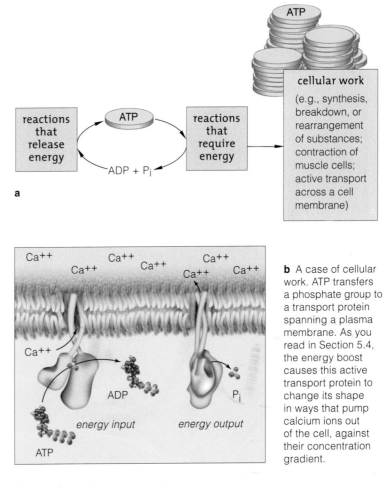

Figure 6.7 *Animated!* (**a**) ATP function. Recurring phosphate-group transfers convert ATP into ADP, and back to ATP. (**b**) Example of ATP-driven cellular work.

b A case of cellular work. ATP transfers a phosphate group to a transport protein spanning a plasma membrane. As you read in Section 5.4, the energy boost causes this active transport protein to change its shape in ways that pump calcium ions out of the cell, against their concentration gradient.

and convert them to something higher up on the hill. Any reactions that require a net input of energy are said to be *endergonic* (meaning energy in).

Glucose is a concentrated source of energy. When it is broken apart, energy is released. Cells capture some of the energy to do cellular work. Energy also ends up in covalent bonds of smaller, more stable products. With aerobic respiration, those products are six CO_2 and six H_2O molecules (Figure 6.5*b*).

Aerobic respiration releases energy bit by bit, with many conversion steps, so cells can capture some of it efficiently. This metabolic process is like a downhill run, from a concentrated form of energy (glucose) to less concentrated forms. Any reactions that end with a net release of energy are *exergonic* (meaning energy out). They, too, require an energy boost to get past the activation energy barrier. However, the net amount of energy released is more than the amount invested.

ATP—THE CELL'S ENERGY CURRENCY

It doesn't take a huge leap of the imagination to sense that cells stay alive by *coupling* reactions that require energy with reactions that release it. For nearly all metabolic reactions, adenosine triphosphate, or **ATP**, is the energy carrier, or coupling agent. All cells make this nucleotide, which consists of a five-carbon sugar (ribose), the base adenine, and three phosphate groups (Figure 6.6). ATP easily gives up phosphate groups to other molecules and thus primes them to react. Any phosphate-group transfer is called **phosphorylation**.

ATP is the currency in a cell's economy. Cells spend it in energy-requiring reactions and also invest it in energy-releasing reactions that help keep them alive. We use a cartoon coin to symbolize ATP (Figure 6.7*a*). Because ATP is the main energy carrier for so many reactions, you might infer—correctly—that cells have

ways of renewing it. When ATP gives up a phosphate group, ADP (adenosine diphosphate) forms. ATP can re-form when ADP binds to inorganic phosphate (P_i) or to a phosphate group that was split from a different molecule. Regenerating ATP by this **ATP/ADP cycle** helps drive most metabolic reactions (Figure 6.7*b*).

Activation energy is the minimum amount of energy required to get any reaction to the point where it will run spontaneously, with no further energy input. The amount differs for different reactions.

Reactions that build large organic compounds from smaller ones cannot run without a net input of energy. Reactions that degrade large molecules to smaller ones end with a net output of usable energy.

ATP, the main energy carrier in all cells, couples energy-releasing and energy-requiring reactions. It primes other molecules to react by phosphate-group transfers.

6.3 How Enzymes Make Substances React

If you left a cupful of glucose out in the open, years would pass before you would see its conversion to carbon dioxide and water. Yet that same conversion takes just a few seconds in your body. Enzymes make the difference.

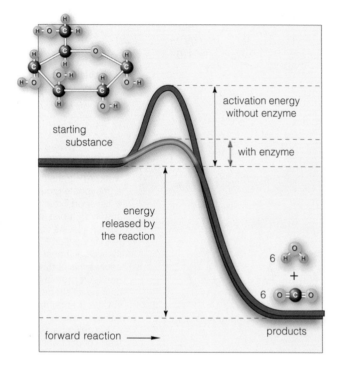

Enzymes are catalysts that can make reactions occur hundreds to millions of times faster than they would on their own. Enzyme molecules can work again and again; a reaction does not use them up or irreversibly alter them. Each type of enzyme chemically recognizes, binds, and alters specific reactants only. For instance, thrombin recognizes and cleaves only a peptide bond between arginine and glycine in a particular protein, thereby converting the protein into a factor that helps clot blood. Finally, nearly all enzymes are proteins. (A few kinds of RNAs also show enzymatic activity.)

Reactions cannot proceed until the reactants have a minimum amount of internal energy—the activation energy. Visualize activation energy as a barrier—a hill or brick wall (Figures 6.8 and 6.9). Enzymes lower it. How? *Compared with the surrounding environment, they offer a stable microenvironment, more favorable for reaction.*

Consider: Enzymes are far larger than **substrates**, another name for the reactants that bind to a specific enzyme. Their polypeptide chains are folded in ways that afford structural stability. Certain folds also form one or more chemically stable **active sites**: pockets or crevices where the substrates bind and where specific reactions can proceed rapidly and repeatedly.

Part of a substrate is complementary in shape, size, solubility, and charge to the active site. Because of the

Figure 6.8 *Animated!*
Activation energy: the minimum amount of internal energy that reactants must have before a reaction will run to products. An enzyme enhances the reaction rate by lowering the required amount; it lowers the energy hill.

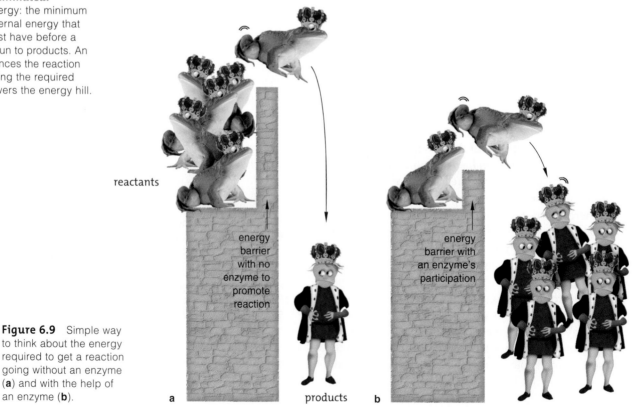

Figure 6.9 Simple way to think about the energy required to get a reaction going without an enzyme (**a**) and with the help of an enzyme (**b**).

one of four heme groups cradled in one of four polypeptide chains

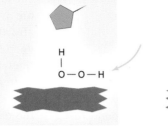

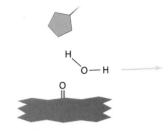

a Hydrogen peroxide (H_2O_2) enters a cavity in catalase. It is the substrate for a reaction aided by an iron molecule in a heme group (*red*).

b A hydrogen of the peroxide is attracted to histidine, an amino acid projecting into the cavity. One oxygen binds the iron.

c This binding destabilizes the peroxide bond, which breaks. Water (H_2O) forms. In a later reaction, another H_2O_2 will pull the oxygen from iron, which will then be free to act again.

Figure 6.10 *Animated!* How catalase works. This enzyme has four polypeptide chains and four heme groups (coded *red*).

fit, each enzyme chemically recognizes and binds its substrate among thousands of substances in cells.

Think back on the main types of enzyme-mediated reactions (Section 3.2). With *functional group transfers*, one molecule gives up a functional group to another. With *electron transfers*, one or more electrons stripped away from a molecule are donated elsewhere. With *rearrangements*, a juggling of internal bonds converts one kind of molecule to another. With *condensation*, two or more molecules become covalently bound into a larger molecule. Finally, with *cleavage* reactions, a larger molecule splits into smaller ones.

When we talk about activation energy, *we really are talking about the energy it takes to align reactive chemical groups, destabilize electric charges, and break bonds.* These events put a substrate at its **transition state**. Then, its bonds are at the breaking point, and the reaction can run easily to product (Figure 6.10).

The binding between an enzyme and its substrate is weak and temporary (that is why the reaction does not change the enzyme). However, energy is released when these weak bonds form. This "binding energy" stabilizes the transition state long enough to keep the enzyme and its substrate together for the reaction to be completed.

With enzymes, four mechanisms work alone or in combination to lower the activation energy and move substrates to the transition state:

Helping substrates get together. When they are at low concentrations, molecules of substrates rarely react. Binding at an active site is as effective as a localized boost in concentration, by as much as ten millionfold.

Orienting substrates in positions favoring reaction. On their own, substrates collide from random directions. By contrast, the weak but extensive bonds at an active site put reactive groups close together.

Shutting out water molecules. Because of its capacity to form hydrogen bonds so easily, water can interfere with the breaking and formation of chemical bonds in reactions. Certain active sites have an abundance of nonpolar amino acids with hydrophobic groups, which repel water and keep it away from the reactions.

Inducing a fit between enzyme and substrate. By the **induced-fit model**, a substrate is almost but not quite complementary to an active site. An enzyme restrains the substrate and stretches or squeezes it into a certain shape, often next to another molecule or to a reactive group. By optimizing the fit between them, it moves the substrate to the transition state.

On their own, chemical reactions occur too slowly to sustain life. Enzymes greatly increase reaction rates by lowering the activation energy. That is the minimum amount of energy required to align reactive groups, destabilize electric charges, and break bonds so that products can form from reactants.

In an enzyme's active site, substrates move to a transition state, when their bonds are at the breaking point and the reaction can run spontaneously to completion.

The transition state is reached by various mechanisms that concentrate and orient substrates, exclude water from the active site, and induce an optimal fit between the active site and substrate.

6.4   Enzymes Don't Work In a Vacuum

LINK TO
SECTION
2.6

Many factors influence what an enzyme molecule does at any given time or whether it is built in the first place. Here we highlight a few of the major factors.

CONTROLS OVER ENZYMES

What happens when one or another of the thousands of substances in cells becomes too abundant or scarce? Many controls over enzymes help cells respond fast by adjusting specific reactions. Feedback mechanisms can activate or inhibit enzymes in ways that conserve energy and resources. Cells produce what conditions require—no more, no less.

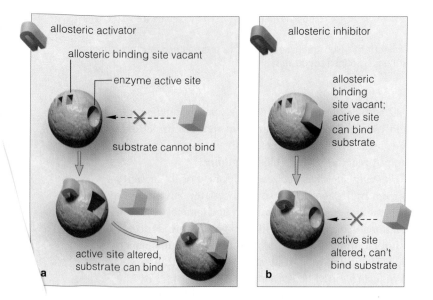

Figure **6.11** *Animated!*   Allosteric control over enzyme activity. (**a**) An active [site i]s unblocked when an activator binds to a vacant allosteric site. (**b**) An [active] site is blocked when an inhibitor binds to a vacant allosteric site.

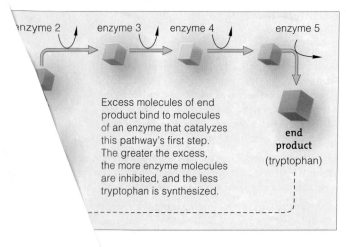

Excess molecules of end product bind to molecules of an enzyme that catalyzes this pathway's first step. The greater the excess, the more enzyme molecules are inhibited, and the less tryptophan is synthesized.

end product (tryptophan)

[F]eedback inhibition of a metabolic pathway. Five [enzymes in a sequ]ence to convert a substrate to tryptophan.

Controls maintain, lower, or raise concentrations of substances. They adjust how fast enzyme molecules are synthesized, and they activate or inhibit the ones already built. In multicelled species, enzyme controls keep individual cells functioning in ways that benefit the whole body.

In some cases, a molecule that acts as an activator or inhibitor can reversibly bind to an *allosteric* site on the enzyme, not to the active site (*allo*– other; *steric*, structure). Binding alters the enzyme's shape in a way that hides or exposes the active site (Figure 6.11).

Visualize a bacterial cell making tryptophan and other amino acids—the building blocks for proteins. Even when it has made enough proteins, tryptophan synthesis continues until its increasing concentration causes **feedback inhibition**. This means a change that results from a specific activity *shuts down the activity*.

A feedback loop starts and ends at many allosteric enzymes. In this case, unused tryptophan binds to an allosteric site on the first enzyme in the tryptophan biosynthesis pathway. Binding makes the active site change shape, so less tryptophan can be made (Figure 6.12). At times when not many tryptophan molecules are around, the allosteric sites are unbound. Thus the active sites remain functional, and the synthesis rate picks up. In such ways, feedback loops quickly adjust the concentrations of substances.

EFFECTS OF TEMPERATURE, pH, AND SALINITY

What an enzyme molecule actually does also depends on conditions in the environment. Temperature, pH, and salinity all have impact on it.

Temperature is a measure of molecular motion. As it rises, it boosts reaction rates both by increasing the likelihood that a substrate will bump into an enzyme and by raising a substrate molecule's internal energy. Remember, the more energy a reactant molecule has, the closer it gets to jumping that activation energy barrier and taking part in a reaction.

Above the range of temperatures that an enzyme can tolerate, weak bonds are broken. The shape of the enzyme changes, so substrates no longer can bind to the active site. The reaction rate falls sharply (Figure 6.13). Such declines typically occur with fevers above 44°C (112°F), which people usually cannot survive.

Also, remember how the pH of solutions can vary (Section 2.6)? In the human body, most enzymes work best at pH 6–8. For instance, trypsin is active in the small intestine (pH of about 8). The enzyme pepsin is one of the exceptions. This nonspecific protease can digest any protein. It is produced in inactive form and normally becomes activated only in gastric fluid, in the stomach. Gastric fluid is highly acidic, with a pH of 1–2. If activated pepsin were to leak out of the stomach, it would digest the proteins in your tissues instead of those in food. Figure 6.14 shows the effects of pH on pepsin and two other kinds of enzymes.

Also, most enzymes stop working effectively when the fluids in which they are dissolved are saltier or less salty than their range of tolerance. Too much or too little salt interferes with the hydrogen bonds that help hold an enzyme in its three-dimensional shape. By doing so, it inactivates the enzyme.

HELP FROM COFACTORS

Finally, don't forget the cofactors. These metal ions or coenzymes help at the active site of enzymes or taxi electrons, H^+, or functional groups to other reactions. **Coenzymes** are a class of organic compounds that may or may not have a vitamin component.

One or more metal ions assist nearly a third of all known enzymes. Metal ions easily give up and accept electrons. As part of coenzymes, they help products form by shifting electron arrangements in substrates or intermediates. That is what goes on at the hemes in catalase. Heme has an organic ring structure, with an iron atom at the center of the ring. As you saw in Figure 6.10, the iron atoms assist catalase in speeding the breakdown of hydrogen peroxide to water.

Like vitamin E, catalase is an **antioxidant**. It helps neutralize free radicals. *Free radicals*, or atoms with at least one unpaired electron, are leftovers of reactions. They attack the structure of DNA and other biological molecules. As we age, we make less and less catalase, so free radicals accumulate (Section 43.6).

Some coenzymes are tightly bound to an enzyme. Others, such as NAD^+ and $NADP^+$, can diffuse freely through the cytoplasm. Either way, they participate intimately in a metabolic reaction. Unlike enzymes, many become modified during the reaction, but they are regenerated elsewhere.

Controls over enzymes enhance or inhibit their activity. By doing so, they maintain, lower, or raise concentrations of many thousands of substances in coordinated ways.

Enzymes work best when the cellular environment stays within limited ranges of temperature, pH, and salinity. The actual ranges differ from one type of enzyme to the next.

Many enzymes are assisted by cofactors, which are specific metal ions or coenzymes.

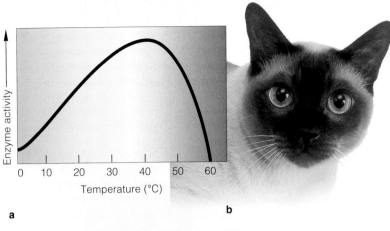

a   b

Figure 6.13 Enzymes and the environment. (**a**) How increases in temperature affect one enzyme's activity.

(**b**) The air temperature outside the body affects the fur color of Siamese cats. Epidermal cells that give rise to the cat's fur produce a brownish-black pigment, melanin. Tyrosinase, an enzyme in the melanin production pathway, is heat-sensitive in the Siamese. It becomes less active in warmer parts of the cat's body, which end up with less melanin, and lighter fur. Put this cat's feet in booties for a few weeks and its warm feet will become light.

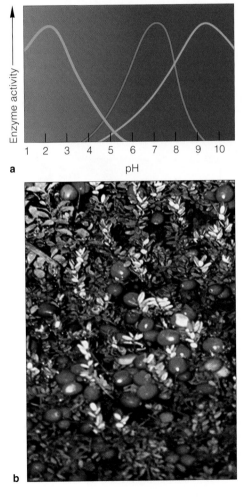

Figure 6.14 Enzymes and the environment. (**a**) How pH values affect three enzymes. The *blue* graph line shows the activity for pepsin. (**b**) Cranberry plants grow best in acidic bogs. Unlike most plants, they have no nitrate reductase. This enzyme converts nitrate (NO_3) found in most soils to metabolically useful ammonia (NH_3). In highly acidic soils, nitrogen is already in the form of ammonia (NH_4^+).

6.5 Metabolism—Organized, Enzyme-Mediated Reactions

LINK TO
SECTION
2.3

How cells use energy changes is one aspect of metabolism. Another is the concentration, conversion, and disposal of materials by energy-driven reactions. Most reactions in cells are part of stepwise metabolic pathways.

TYPES OF METABOLIC PATHWAYS

We have mentioned metabolic pathways in passing. Now let's formally define them. **Metabolic pathways** are enzyme-mediated sequences of reactions in cells. The *biosynthetic* (or anabolic) kinds require a net input of energy to produce glucose, starch, and other large molecules from small ones. Photosynthesis is the main biosynthetic pathway for the world of life (Figure 6.15).

Degradative (or catabolic) pathways are exergonic, overall, in that they end with a net release of usable energy. In degradative pathways, unstable molecules typically are broken down into smaller, more stable products, with the release of energy in forms that cells may use. Aerobic respiration is the main degradative pathway in the biosphere, and energy released during the reactions is used to form many ATP (Figure 6.15).

Many metabolic pathways are linear, a straight line from reactants to the products. In cyclic pathways, the last reaction regenerates the type of reactant molecule that is used in the first step of the reaction sequence. For instance, in the second stage of photosynthesis, a molecule known as RuBP is the entry point for cyclic reactions, and the cycle's last intermediate undergoes internal jugglings that convert it to RuBP. In branched pathways, reactants or intermediates are channeled into two or more different sequences of reactions.

THE DIRECTION OF METABOLIC REACTIONS

Bear in mind, metabolic reactions do not always run from reactants to products. They might start out in this "forward" direction. But most also run in reverse, with products being converted back to reactants.

Reversible reactions tend to run spontaneously toward **chemical equilibrium**, when the reaction rate is about the same in either direction. In most cases, the amounts of reactant and product molecules are not identical; they differ at that time (Figure 6.16). It is like a party where people drift between two rooms. The number in each room stays the same—say, thirty in one and ten in the other—even as individuals move back and forth.

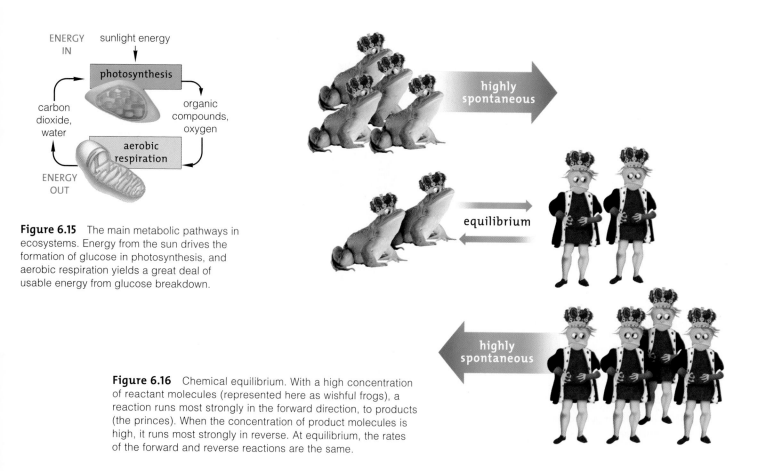

Figure 6.15 The main metabolic pathways in ecosystems. Energy from the sun drives the formation of glucose in photosynthesis, and aerobic respiration yields a great deal of usable energy from glucose breakdown.

Figure 6.16 Chemical equilibrium. With a high concentration of reactant molecules (represented here as wishful frogs), a reaction runs most strongly in the forward direction, to products (the princes). When the concentration of product molecules is high, it runs most strongly in reverse. At equilibrium, the rates of the forward and reverse reactions are the same.

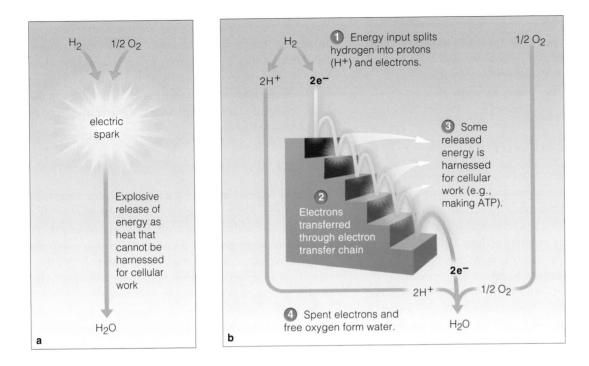

Figure 6.17 *Animated!* Uncontrolled versus controlled energy release. (**a**) Free hydrogen and oxygen exposed to an electric spark react and release energy all at once. (**b**) Electron transfer chains allow the same reaction to proceed in small, more manageable steps that can handle the released energy.

a (panel a labels)

H_2 $1/2 O_2$

electric spark

Explosive release of energy as heat that cannot be harnessed for cellular work

H_2O

b (panel b labels)

1 Energy input splits hydrogen into protons (H^+) and electrons.

H_2 $1/2 O_2$

$2H^+$ $2e^-$

3 Some released energy is harnessed for cellular work (e.g., making ATP).

2 Electrons transferred through electron transfer chain

$2e^-$

$2H^+$ $1/2 O_2$

4 Spent electrons and free oxygen form water.

H_2O

Why bother to think about this? *Each cell can bring about big changes in activities by controlling the enzymes that mediate a few steps of reversible metabolic pathways.*

For instance, when your cells need a quick bit of energy, they rapidly split glucose into two pyruvate molecules. They do so by a sequence of nine enzyme-mediated steps of a pathway called glycolysis. When glucose supplies are too low, cells quickly reverse this pathway and build glucose from pyruvate and other substances. How? Six steps of the pathway happen to be reversible, and the other three are bypassed. An input of energy from ATP drives the bypass reactions in the uphill (energetically unfavorable) direction.

What if cells did not have this reverse pathway? They would not be able to build glucose fast enough to compensate for episodes of starvation, when glucose supplies in blood become dangerously low.

REDOX REACTIONS IN THE MAIN PATHWAYS

You may be wondering: Why don't cells break down glucose all at once? Glucose, recall, is not as stable as the products of its full breakdown. Toss a cupful of glucose into a campfire, and its carbon and hydrogen atoms will explosively combine with oxygen in air. All of the released energy will be lost as heat.

Cells release energy efficiently by stepwise electron transfers, or **oxidation–reduction reactions**. In these "redox" reactions, one molecule gives up electrons (it is oxidized) and another gains them (it is reduced).

Commonly, hydrogen atoms are released at the same time. Remember, we represent free hydrogen atoms (naked protons) as H^+. Being attracted to the opposite charge of the electrons, H^+ tags along with them.

Start thinking about redox reactions, because they are central to photosynthesis and aerobic respiration. In the next two chapters, you will follow coenzymes as they pick up the electrons and H^+ stripped from substrates and then deliver them to **electron transfer chains**. Such chains are membrane-bound arrays of enzymes and other molecules that accept and give up electrons in sequence. Electrons are at a higher energy level when they enter a chain than when they leave.

Think of these electrons as descending a staircase and stingily losing a bit of energy at each step, as in Figure 6.17. In the case of photosynthesis and aerobic respiration, the stepwise electron transfers concentrate energy—in the form of H^+ electrochemical gradients—in ways that contribute to ATP formation.

Metabolic pathways are orderly, enzyme-mediated reaction sequences, some biosynthetic, others degradative.

Control over a key step of a metabolic pathway can bring about rapid shifts in cell activities.

Many aspects of metabolism involve electron transfers, or oxidation–reduction reactions. Electron transfer chains are important sites of energy exchange in both photosynthesis and aerobic respiration.

6.6 Light Up the Night—And the Lab

LINK TO SECTION 2.3

You can always think about organisms from a "gee-whiz-ain't-nature-grand" point of view. Or you can come up with novel ways to think about what they do and how they do it. The latter way of thinking puts you squarely in the camp of biologists. Here is a case in point.

ENZYMES OF BIOLUMINESCENCE

At night, in the warm waters of tropical seas or in the summer air above gardens and fields, you may catch sight of abrupt shimmerings or flashes of light. Many species, ranging from bacteria and algae to fishes to fireflies, flash with orange, yellow, yellow-green, or blue light. In seawater, great numbers of them often flash together with startling effect (Figure 6.18a).

All of the flashers emit light when enzymes called luciferases transduce chemical bond energy of certain molecules into light energy. Figure 6.18c shows the three-dimensional structure of the luciferase found in fireflies. Remember, electrons can be excited to higher energy levels with an input of energy. In this case, reactions start when ATP, in the presence of oxygen, transfers a phosphate group to luciferin. An array of electrons in this molecule become excited enough to enter reactions. At a certain reaction step, they release the extra energy in the form of *fluorescent* light. Any kind of destabilized molecule that reverts to a more stable configuration may emit such light. When the reactions take place in organisms, the emitted light is called **bioluminescence**.

A RESEARCH CONNECTION

People have been marveling over bioluminescence for a long time. Then biologists thought to borrow the genes for bioluminescence from fireflies and use them to make other organisms light up. Through methods of gene transfers, as sketched out in Chapter 16, they have now inserted copies of those genes into bacteria, plants, and mice (Figure 6.19).

In itself, making mice glow seems like a bizarre thing to do. However, some biologists immediately saw the potential for using bioluminescence genes as a diagnostic tool. For example, every year, 3 million people die from a lung disease caused by different strains of the bacterium *Mycobacterium tuberculosis*. No single antibiotic is effective against all the strains.

Figure 6.18 (**a**) Stirring up bioluminescent marine organisms near Vieques Island, Puerto Rico. As many as 5,000 free-living cells called dinoflagellates may be in each liter of water, and each flashes with blue light when agitated. (**b**) North American firefly (*Photinus pyralis*) emitting a flash from its light organ. Peroxisomes in this organ are packed with luciferase molecules. Firefly flashes help potential mates find each other in the dark. (**c**) Ribbon model for firefly luciferase. This enzyme catalyzes the reaction that releases light. Its single polypeptide chain is folded into multiple domains.

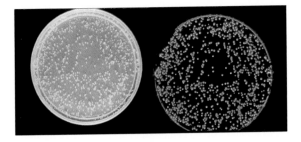

Figure 6.19 Colonies of bioluminescent bacteria in daylight (*left*) and glowing in a culture dish in the dark (*right*).

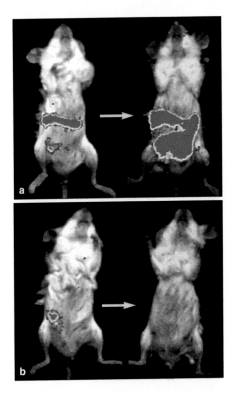

Figure 6.20 Utilizing bioluminescent bacterial cells to chart the location of infectious bacteria inside living laboratory mice and their spread through body tissues. (**a**) False-color images in this pair of photographs show how the infection spread in a control group that had not been given a dose of antibiotics. (**b**) This pair shows how antibiotics had killed most of the infectious bacterial cells.

If an infection has progressed to a dangerous stage, there is no time to waste, but there is no guarantee a treatment will be effective unless the particular strain causing a patient's infection is identified.

One way to do this is to take a sample of bacterial cells from the patient, then expose them to luciferase genes. In some cells, the genes get incorporated into the bacterial DNA. Those cells are isolated, and then colonies of their descendants are exposed to different antibiotics. When an antibiotic does *not* work, colonies glow; the cells are alive. When an antibiotic works, there is no glow; all of the bacterial cells are dead.

Christopher and Pamela Contag, two postdoctoral students at Stanford University, thought about using gene transfers to light up *Salmonella* cells in mice. As they knew, researchers who study viral or bacterial diseases had to infect dozens to hundreds of mice for experiments. Then they had to kill the mice and study their tissues to see whether infection had occurred. The practice was costly and tedious, and it meant the loss of a lot of infection-free experimental animals.

First the Contags approached a medical imaging researcher, David Benaron, with this hypothesis: If we make live, infectious bacteria bioluminescent, then flashes of light will shine through tissues of infected, living animals. As an early test of their prediction, the researchers put glowing *Salmonella* cells into a thawed chicken breast from a market. It glowed from inside.

The Contags transferred the bioluminescence genes into three *Salmonella* strains, which were injected into three experimental groups of laboratory mice. The Contags used a digital imaging camera to track and record whether infections developed in each group.

The first strain was weak; the mice fought off the infection in less than six days and did not glow. The second strain was not as weak but could not spread through the mouse body; it was localized. The third strain was dangerous. It spread rapidly through the entire mouse gut—which glowed.

Thus bioluminescent gene transfer, combined with imaging of enzyme activity, can track the course of infection. It is now used to evaluate the effectiveness of drugs in living organisms (Figure 6.20). It also may have use in gene therapy, which involves replacing one or more defective or cancer-causing genes in a patient with functional copies of the genes.

In short, people thought to use bioluminescence as visible evidence of metabolism—of the cell's capacity to acquire energy and use it to build, break apart, store, and release substances in controlled ways. Each flash reminds us that living cells are taking in energy-rich solutes, constructing membranes, storing things, replenishing enzymes, and checking out their DNA. A constant supply of energy drives these activities. The flashes remind us of how modern-day biologists are busy putting knowledge to use in practical ways.

Bioluminescence is an outcome of enzyme-mediated reactions that release energy as fluorescent light.

Biologists have transferred genes for bioluminescence, the luciferases, into a variety of organisms. The gene transfers have research applications and practical uses.

Summary

Section 6.1 Cells require energy for metabolism, or chemical work. The first law of thermodynamics tells us that energy cannot be created from scratch or destroyed. Energy can only be converted from one form to another. Examples are potential energy, such as that stored in chemical bonds, and kinetic energy (energy of motion).

The second law of thermodynamics tells us that energy tends to spread out or disperse spontaneously from concentrated to less concentrated forms, such as dispersed heat. Entropy is the measure of how much and how far a concentrated form of energy has been dispersed after an energy change.

The collective strength of the chemical bonds that hold together systems of life resists the spontaneous direction of energy flow. Those systems maintain their complex organization by being resupplied with energy lost from someplace else. Sunlight is the original source of energy for nearly all webs of life.

Section 6.2 Table 6.1 summarizes the functions of the key players in metabolism: reactants, intermediates, products, enzymes, cofactors, energy carriers, and transporters. The second law of thermodynamics lets us make this prediction about metabolism: It takes net inputs of energy to force stable molecules to combine into forms that are less stable but more concentrated sources of energy.

All reactants require a minimum amount of internal energy before they will enter into a reaction, although the amount differs among them. That amount is the activation energy for the reaction.

Some reactions are endergonic; they require a net energy input to run to completion. The formation of glucose from carbon dioxide and water is an example. Other reactions are exergonic; the net amount of energy they release is greater than the amount invested. Aerobic respiration is an example.

Cells couple reactions that require energy with other reactions that release energy. ATP is the main energy carrier between reaction sites. It jump-starts reactions by donating one or more of its phosphate groups to a reactant. It is regenerated when ADP binds to inorganic phosphate or to a phosphate group.

Biology Now
Learn about energy changes in chemical reactions and the role of ATP with the animation on BiologyNow.

Section 6.3 Enzymes are catalysts, which means they enormously enhance the rates of specific reactions. Nearly all are proteins that are much larger than their substrates (some RNAs also are catalytic). Folds in their polypeptide chains form active sites, or small clefts that create favorable microenvironments for reaction.

Enzymes speed reactions by lowering the activation energy. For each kind of reaction, that is the energy it takes to align reactive chemical groups, destabilize electric charges, and break chemical bonds. Enzymes move substrates faster to a transition state, when the reaction can proceed most easily to products.

Four mechanisms help get substrates to the transition state: Binding in an active site effectively boosts local concentrations of substrates, it orients substrates, it shuts out most or all water molecules that could interfere with the reaction, and it induces an optimum fit with the substrate that pulls it to the transition state.

Biology Now
Investigate how enzymes facilitate reactions with the animation and interaction on BiologyNow.

Section 6.4 Many factors influence enzyme action and, through it, metabolic pathways. Each type of enzyme functions best within a characteristic range of temperature, pH, and salinity. Controls over enzyme activity, including negative feedback mechanisms, influence the kinds and amounts of substances available in a given interval. Also, most enzymes require the assistance of cofactors. These metal ions and coenzymes help out at an active site by ferrying electrons, hydrogen ions, or functional groups to some other reaction site.

Biology Now
Observe mechanisms of enzyme control with the animation on BiologyNow.

Section 6.5 Cells modify concentrations of many thousands of substances, often by coordinating outputs of orderly, enzyme-mediated reaction sequences called metabolic pathways. The energy-requiring, *biosynthetic* pathways build large, unstable molecules from smaller, more stable ones. Photosynthesis is an example. The energy-releasing, *degradative* pathways break down large molecules to smaller products. Aerobic respiration is the main degradative pathway. Most metabolic reactions are reversible. Cells rapidly shift rates of metabolism by controlling a few steps of reversible pathways.

Table 6.1	Summary of the Main Participants in Metabolic Reactions
Reactant	Substance that enters a metabolic reaction or pathway; also called the substrate of a specific enzyme
Intermediate	Any substance that forms in a reaction or pathway, between the reactants and the end products
Product	Substance at the end of a reaction or pathway
Enzyme	A protein that greatly enhances reaction rates; a few RNAs also do this
Cofactor	Coenzyme (such as NAD^+) or metal ion; assists enzymes or move electrons, hydrogen, or functional groups to other reaction sites
Energy carrier	Mainly ATP; couples energy-releasing reactions with energy-requiring ones
Transport protein	Protein that passively assists or actively pumps specific solutes across a cell membrane

Cells release energy most efficiently by way of oxidation–reduction reactions, which simply are electron transfers. In both photosynthesis and aerobic respiration, redox reactions occur at electron transfer chains that are components of cell membranes.

Biology ⑤ Now
Compare the effects of controlled and uncontrolled energy release with the animation on BiologyNow.

Section 6.6 Bioluminescence is one outcome of enzyme-mediated reactions that release energy in the form of fluorescent light. The transfer of genes for bioluminescence into organisms has research and practical applications.

Biology ⑤ Now
Read the InfoTrac article, "An Enlightening Food Safety Tool," Lynn Petrak, The National Provisioner, October 2004.

Self-Quiz

Answers in Appendix II

1. _____ is life's primary source of energy.
   a. Food   b. Water   c. Sunlight   d. ATP

2. Energy _____ .
   a. cannot be created or destroyed
   b. can change from one form to another
   c. tends to flow spontaneously in one direction
   d. all of the above

3. Entropy is a measure of _____ .
   a. order versus disorder in a system
   b. how much and how far a form of energy has been dispersed after an energy change
   c. the forerunner of an energy change
   d. all of the above

4. Enzymes _____ .
   a. are proteins, except for a few RNAs
   b. lower the activation energy of a reaction
   c. are destroyed by the reactions they catalyze
   d. a and b

5. Enzyme function is influenced by _____ .
   a. changes in temperature
   b. changes in pH
   c. changes in salinity
   d. all of the above

6. Which of the following statements is *incorrect*? A metabolic pathway _____ .
   a. has an orderly sequence of reaction steps
   b. is mediated by only one enzyme that starts it
   c. may be biosynthetic or degradative, overall
   d. all of the above

7. Match the substance with its suitable description.
   ____ coenzyme or metal ion   a. reactant
   ____ adjusts gradients at membrane b. enzyme
   ____ substance entering a reaction c. cofactor
   ____ substance formed during   d. intermediate
        a reaction   e. product
   ____ substance at end of reaction   f. energy carrier
   ____ enhances reaction rate   g. transport
   ____ mainly ATP   protein

Additional questions are available on Biology ⑤ Now™

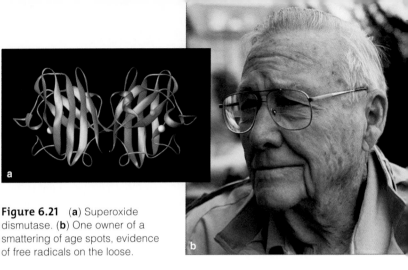

Figure 6.21 (**a**) Superoxide dismutase. (**b**) One owner of a smattering of age spots, evidence of free radicals on the loose.

Critical Thinking

1. State the law of thermodynamics that deals with the *quantity* of energy in the universe. State the law that deals with the *quality* of energy.

2. Cyanide, a toxic compound, binds irreversibly to an enzyme that is a component of electron transfer chains. The outcome is *cyanide poisoning*. Binding prevents the enzyme from donating electrons to a nearby acceptor molecule in the system. What effect will this have on ATP formation? From what you know of ATP's function, what effect will this have on a person's health?

3. Why does applying lemon juice to sliced apples keep them from turning brown?

4. One molecule of catalase can break down 6 million hydrogen peroxide molecules every minute. It is found in most organisms that live under aerobic conditions because hydrogen peroxide is toxic—cells must dispose of it fast or risk being damaged. Peroxide is catalase's substrate; but by a neat trick, catalase also can inactivate other toxins, including alcohol. Can you guess what the trick is?

5. Hydrogen peroxide bubbles if dribbled on an open cut but does not bubble on unbroken skin. Explain why.

6. *Free radicals* are unbound molecular fragments that have the wrong number of electrons. They form during many enzyme-catalyzed reactions, including the digestion of fats and amino acids. They slip out of electron transfer chains. They also form when x-rays and other kinds of ionizing radiation bombard water and other molecules. Free radicals are highly reactive. When they dock with a molecule, they alter its structure and function.

Superoxide dismutase (Figure 6.21*a*) is an enzyme that works with catalase to keep free radicals and hydrogen peroxide from accumulating in cells. As we age, cells make copies of enzymes in ever diminishing numbers, in altered form, or both. When this happens to superoxide dismutase and catalase, free radicals and hydrogen peroxide build up. Like loose cannonballs, they career through cells and blast away at the structural integrity of proteins, DNA, lipids, and other molecules. For instance, look at the "age spots" on an older person's skin (Figure 6.21*b*). Each spot is a mass of brownish-black pigments that have accumulated in skin cells. Do some research and identify other problems that arise when free radicals take over.

Sunlight and Survival

Think about the last bit of apple, lettuce, chicken, pizza, or any other food you put in your mouth. Where did it come from? Look past the refrigerator, the market or restaurant, and the farm. Look to plants, the starting point for nearly all of the food—the carbon-based compounds—you eat.

Plants are among the **autotrophs**, or "self-nourishing" organisms. Autotrophs get energy and carbon from the physical environment and use it to make their own food. Most bacteria, many protists, and all fungi and animals are like you; they cannot obtain energy and carbon from the physical environment. They are **heterotrophs**, which feed on autotrophs, one another, and organic wastes. *Hetero*– means other, as in "being nourished by others."

Plants are a type of *photo*autotroph. By the process of **photosynthesis**, they make sugars and other compounds by using sunlight as an energy source and carbon dioxide as their source of carbon. Each year, plants around the

world produce 220 billion tons of sugar, enough to make 300 quadrillion sugar cubes. That is a LOT of sugar. They also release great amounts of oxygen (Figure 7.1).

It wasn't always this way. The first prokaryotic cells on Earth were *chemo*autotrophs. Like the existing archaeans, they did not have the enzymes for complicated metabolic magic. They extracted energy and carbon from simple organic and inorganic compounds, such as methane and hydrogen sulfide, that happened to be around. Both gases were part of the chemical brew that made up the early atmosphere. Carbon dioxide also was present, but it takes special enzymes to harness it. There was little free oxygen.

Things did not change much for about a billion years. Then light-sensitive molecules evolved in a few lineages, which became the first *photo*autotrophs. Life had tapped into an immense supply of energy. Not long afterward, parts of the photosynthetic machinery became modified

Figure 7.1 Photosynthesis—the main pathway by which energy and carbon enter the web of life. This orchard of photosynthetic autotrophs is producing apples and oxygen at the Jerzy Boyz organic farm in Chelan, Washington.

in some photoautotrophs. Water molecules could now be split apart as a source of electrons for the reactions, and supplies of water were essentially unlimited. Over time, oxygen atoms released from uncountable water molecules diffused out of uncountable numbers of cells —and the world of life would never be the same.

Free oxygen reacts fast with metals, including metal ions that help enzymes. The reactions release free radicals which, as you know, are toxic to cells. So oxygen that had accumulated in the atmosphere put selection pressure on prokaryotic populations all over the world. Prokaryotes that could not neutralize toxic oxygen radicals vanished or were marginalized in muddy sediments, deep water, and other anaerobic (oxygen-free) habitats.

As you will read in this chapter, pathways that could detoxify the oxygen radicals evolved in some lineages. One pathway, aerobic respiration, lets cells *use* oxygen's reactive properties in highly beneficial ways.

Another bonus for life: As oxygen accumulated high in the atmosphere, many atoms combined to form ozone (O_3). An ozone layer formed and became a shield against lethal ultraviolet radiation from the sun. Life could now move out of the deep ocean, out from mud, out from under rocks, and diversify under the open sky.

As you read this chapter on photosynthesis, keep in mind that its emergence and its continuity are big reasons why *you* can exist, and read this book, and think about what it takes to stay alive.

 How Would You Vote?

The oxygen in Earth's atmosphere is a sure indicator that photosynthetic organisms flourish here. New technologies will allow astronomers in search of life to measure the oxygen content of the atmosphere of planets too far away for us to visit. Should public funds be used to continue this research? See BiologyNow for details, then vote online.

 Key Concepts

THE RAINBOW CATCHERS

A one-way flow of energy through the world of life starts after chlorophylls and other pigments absorb wavelengths of visible light from the sun's rays. In plants, some bacteria, and many protists, that energy ultimately drives the synthesis of glucose and other carbohydrates.   Sections 7.1, 7.2

OVERVIEW OF PHOTOSYNTHESIS

In plant cells and many protists, photosynthesis proceeds through two stages inside organelles called chloroplasts. At a membrane system in the chloroplast, the sun's energy is first converted to chemical energy. Then carbohydrates are synthesized in the chloroplast's semifluid matrix.   Section 7.3

MAKING ATP AND NADPH

In the first stage of photosynthesis, sunlight energy becomes converted to chemical bond energy in ATP. NADPH forms, and free oxygen escapes into the air.   Sections 7.4, 7.5

MAKING SUGARS

The second stage is the "synthesis" part of photosynthesis. Enzymes assemble sugars from atoms of carbon and oxygen obtained from carbon dioxide. The reactions use the ATP and NADPH that formed in the first stage of photosynthesis. The ATP delivers energy, and the NADPH delivers electrons and hydrogens to the reaction sites.   Sections 7.6, 7.7

GLOBAL IMPACTS OF AUTOTROPHS

The emergence of the world's main energy-releasing pathway, aerobic respiration, was an evolutionary consequence of photosynthesis—the world's main energy-acquiring pathway. Collectively, photoautotrophs and chemoautotrophs make the food that sustains all of life. They also have enormous impact on the global climate.   Section 7.8

 Links to Earlier Concepts

Before considering the chemical basis of photosynthesis, you may wish to review the nature of electron energy levels (Section 2.3), particularly how photons and electrons interact. You will be using your knowledge of carbohydrate structure (3.4), chloroplasts (4.8), active transport proteins (5.2, 5.4), and concentration gradients (5.3).

Remember the concepts of energy flow and the underlying organization of life (6.1 and 6.2)? They help explain how energy flows through photosynthesis reactions. You also will expand your understanding of how cells harvest energy through the operation of electron transfer chains (6.5).

7.1 Sunlight as an Energy Source

LINKS TO
SECTIONS
2.3, 6.1, 6.2

Remember how energy flows in one direction through the world of life? In nearly all cases, the flow starts when photoautotrophs intercept energy, in the form of wavelengths of visible light, from the sun.

PROPERTIES OF LIGHT

An understanding of photosynthesis requires a bit of knowledge of the properties of energy that radiates from the sun. That energy undulates across space in a manner analogous to the waves moving across a sea. The term **wavelength** refers to the horizontal distance between the crests of every two successive waves of radiant energy.

Although energy travels in waves, it has a particle-like quality. When absorbed, it can be measured as if it were organized in discrete packets, or **photons**. A photon consists of a fixed amount of energy. The least energetic photons travel in longer wavelengths, and the most energetic ones travel in shorter wavelengths.

Photoautotrophs only capture light of wavelengths between 380 and 750 nanometers. Humans and other organisms see light of these wavelengths as different colors, from deep violet through blue, green, yellow, orange, and red. Figure 7.2 shows where the spectrum of visible light fits in the **electromagnetic spectrum**—the range of all wavelengths of radiant energy, from shortest (gamma rays) to longest (radio waves).

Shorter wavelengths are energetic enough to alter or break chemical bonds in DNA and proteins. That is why UV (ultraviolet) light, x-rays, and gamma rays are a threat to all organisms. That is why early life evolved away from sunlight—deep in the ocean, or in sediments, or under rocks. Life did not move onto dry land until after the ozone layer formed high above Earth. The ozone layer absorbs much of the dangerous UV light (Sections 48.1 and 48.2).

Visible light of all wavelengths combined appears white. White light separates into its individual colors when it passes through a prism or water droplets in moisture-laden air. The prism or droplets bend light of longer wavelengths (yellow to red) more than they bend shorter wavelengths (violet to blue), the result being the band of colors we see in rainbows.

FROM SUNLIGHT TO PHOTOSYNTHESIS

Pigments are a class of molecules that absorb photons in particular wavelengths only. Certain kinds are the molecular bridges from sunlight to photosynthesis.

Photons that a specific pigment does not absorb are reflected by it or continue traveling right on through it. **Chlorophyll *a***, the major pigment in all but one group of photoautotrophs, absorbs violet and red light. It reflects green and yellow light, which is why plant parts with an abundance of chlorophylls appear green. *Accessory* pigments harvest additional wavelengths. The most common accessory pigment, **chlorophyll b**, reflects green and blue light. **Carotenoids** reflect red, orange, and yellow light. Besides their photosynthetic role, carotenoids impart color to many flowers, fruits, and vegetables. **Xanthophylls** reflect yellow, brown, blue or purple light. The **anthocyanins** reflect red and purple light, as they do in cherries and many flowers. In many deciduous plants, chlorophylls in green leaves mask accessory pigments until autumn (Figure 7.3*a*).

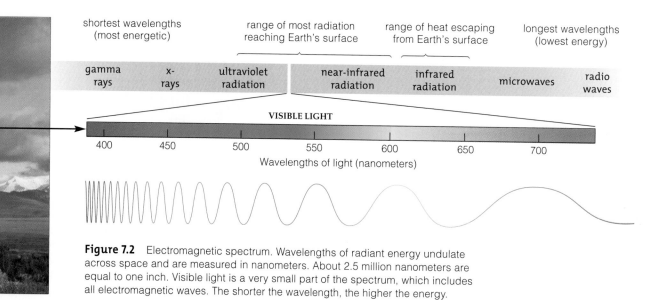

| shortest wavelengths (most energetic) | | range of most radiation reaching Earth's surface | range of heat escaping from Earth's surface | longest wavelengths (lowest energy) |

| gamma rays | x-rays | ultraviolet radiation | near-infrared radiation | infrared radiation | microwaves | radio waves |

VISIBLE LIGHT

400 450 500 550 600 650 700

Wavelengths of light (nanometers)

Figure 7.2 Electromagnetic spectrum. Wavelengths of radiant energy undulate across space and are measured in nanometers. About 2.5 million nanometers are equal to one inch. Visible light is a very small part of the spectrum, which includes all electromagnetic waves. The shorter the wavelength, the higher the energy.

a

The **phycobilins** reflect red or blue-green light. Red algae and cyanobacteria have notable amounts of these accessory pigments. A few bacteria of ancient lineages have unique pigments. Purple bacteriorhodopsin is the main kind in the archaean *Halobacterium halobium*.

Collectively, different photosynthetic pigments can absorb nearly all wavelengths across the spectrum of visible light. What happens next? You have to zoom into a pigment for the answer. As Figure 7.3b,c shows, a pigment molecule has at least one array of atoms in which single covalent bonds alternate with double covalent bonds. Remember electron orbitals (Section 2.3)? Electrons of these atoms share one orbital that spans the entire array. That array lets the pigment act like an antenna for receiving photon energy.

Each pigment absorbs light of specific wavelengths, which correspond to photon energy. Energy inputs, remember, boost electrons to higher energy levels. A photon is absorbed by a pigment only if it has exactly enough energy to boost an electron of the pigment's antenna region to a higher energy level.

An excited electron returns to a lower energy level almost immediately and emits its extra energy as heat or as a photon. As you will see shortly, that energy bounces back and forth like a fast volleyball among a team of photosynthetic pigments. It quickly reaches the team captain—a special chlorophyll that can *give up* excited electrons and so start the reactions.

Radiation from the sun travels in waves, which differ in length and energy content. We perceive visible light of different wavelengths as different colors and measure their energy content in packets called photons.

In plants, chlorophyll a and accessory pigments absorb specific wavelengths of visible light. They are molecular bridges between the sun's energy and photosynthesis.

Pigment molecules absorb photons at their arrays of alternating single and double covalent bonds. The arrays let the pigments act like energy-receiving antennas.

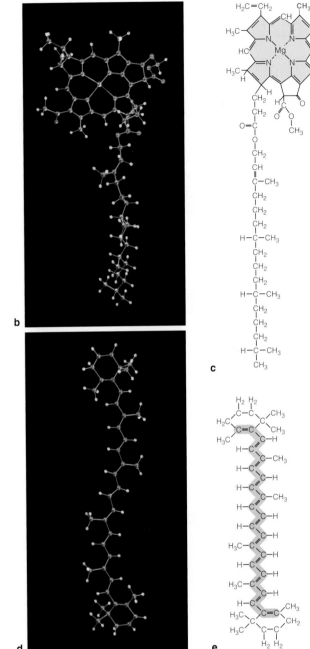

b

c

d

e

Figure 7.3 (**a**) Evidence of pigments in the changing leaves of autumn. In bright green leaves, photosynthetic cells continuously make chlorophyll, which masks accessory pigments. In autumn, chlorophyll synthesis lags behind its breakdown in many species. Accessory pigments then show through and give leaves characteristic red, orange, and yellow fall colors.

Ball-and-stick models and structural formulas for (**b,c**) chlorophyll *a* and (**d,e**) beta-carotene. The light-catching region of each pigment is tinted the specific color of light it transmits. Each pigment has a hydrocarbon backbone that readily dissolves in the lipid bilayer of cell membranes.

Chlorophylls *a* and *b* differ only in one functional group at the position shaded *red* ($-CH_3$ for chlorophyll *a* and $-COO^-$ for chlorophyll *b*). The light-catching portion is the flattened ring structure, which is similar to a heme (Section 3.1). It holds a magnesium atom instead of iron.

7.2 Harvesting the Rainbow

LINK TO SECTION 4,8

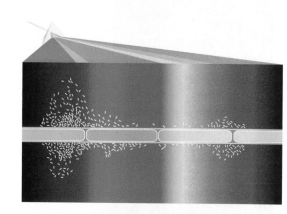

a Outcome of Engelmann's experiment

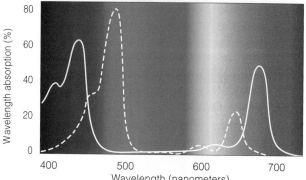

b Absorption spectra for chlorophyll *a* (solid graph line) and chlorophyll *b* (dashed line)

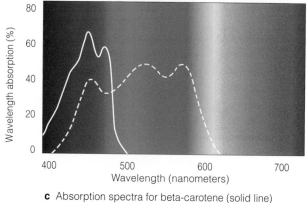

c Absorption spectra for beta-carotene (solid line) and one of the phycobilins (dashed line)

Different kinds of photosynthetic pigments work together. How efficient are these pigments at harvesting light of different wavelengths in the sun's rays?

At one time, people thought that plants used substances in soil to make food. By 1882, a few chemists had an idea that plants use sunlight, water, and something in the air. The botanist Wilhelm Theodor Engelmann wondered: *What parts of sunlight do plants favor?* He already knew that photosynthesis releases free oxygen. He came up with a hypothesis. If photosynthesis involves certain colors of light, then photosynthesizers will release more or less oxygen in response to different colors.

Engelmann also knew that certain bacteria use oxygen during aerobic respiration, and he predicted that they would gather in places where a photosynthetic organism was releasing the most oxygen. He directed a spectrum of visible light across a drop of water that contained bacterial cells (Figure 7.4a). The droplet also contained a strand of *Cladophora*, a photosynthetic alga (Figure 7.5).

Most of the bacterial cells gathered where violet and red light fell across the algal strand. More free oxygen had to be diffusing away from parts of the strand that were illuminated by the violet and red light—a sign that those colors are best at driving photosynthesis.

Engelmann did identify the wavelengths. But molecular biology was far in the future, so he did not know about the pigments that absorb the light.

Today, an **absorption spectrum** conveys how efficiently a given pigment absorbs light of different wavelengths. As Figure 7.4b shows, chlorophylls are best at absorbing red and violet light, but they transmit much of the yellow and green light. What if you combined absorption spectra for chlorophylls and all of the accessory pigments, including those in Figure 7.4b,c? You would see that, collectively, they respond to almost the full spectrum of wavelengths from the sun. They are efficient at what they do.

Figure 7.4 *Animated!* (**a**) One of the early photosynthesis experiments. W. T. Engelmann directed a ray of sunlight—broken into its component colors by a crystal prism—across a water droplet on a microscope slide. The droplet held an algal strand (*Cladophora*) and aerobic bacterial cells. As shown here, nearly all of the cells gathered under violet and red light, the most efficient wavelengths for photosynthesis.

(**b**,**c**) Later research revealed that all photosynthetic pigments combined absorb most wavelengths in the spectrum of visible light with remarkable efficiency. These graphs show absorption spectra for only four of many pigments: chlorophylls *a* and *b*, beta-carotene, and a phycobilin.

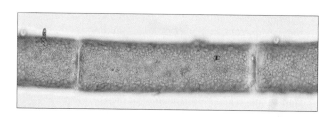

Figure 7.5 Light micrograph of the type of cells making up one algal strand.

7.3 Overview of Photosynthesis Reactions

Plants do something you never will do. They can make their own food from no more than light, water, and carbon dioxide.

Photosynthesis proceeds in two reaction stages. In the first stage—the **light-dependent reactions**—sunlight energy is converted to chemical bond energy of ATP. Water molecules are split, and typically the coenzyme NADP+ accepts the released hydrogen and electrons, thus becoming NADPH. The oxygen atoms released from water molecules escape into the surroundings.

The second stage, the **light-independent reactions**, runs on energy delivered by ATP. That energy drives the synthesis of glucose and other carbohydrates. The building blocks are the hydrogen atoms and electrons from NADPH, as well as carbon and oxygen atoms stripped from carbon dioxide and water.

Photosynthesis is often summarized this way:

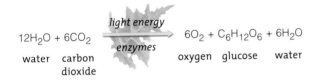

$$12H_2O + 6CO_2 \xrightarrow[\text{enzymes}]{\text{light energy}} 6O_2 + C_6H_{12}O_6 + 6H_2O$$

water   carbon   oxygen   glucose   water
     dioxide

We will focus on what goes on inside **chloroplasts**, the organelles of photosynthesis in plants and many protists. A chloroplast has two outer membranes that enclose a semifluid matrix called the **stroma**. A third membrane—the **thylakoid membrane**—is folded up inside the stroma. In many cells, it looks like stacks of flattened sacs (thylakoids) connected by channels. But the space inside all the sacs and channels forms one continuous compartment, as in Figure 7.6*b*. Sugars are synthesized outside the compartment, in the stroma.

As you will see in the next section, the thylakoid membrane is studded with pigments. Most pigments are packed together as light-harvesting complexes. A number of **photosystems**, or reaction centers, also are embedded in the membrane, and each is surrounded by hundreds of light-harvesting complexes that pass on energy to it. With enough energy, its electrons get excited. That excitation sets in motion the first stage of reactions, as sketched out in Figure 7.6*c*.

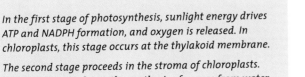

In the first stage of photosynthesis, sunlight energy drives ATP and NADPH formation, and oxygen is released. In chloroplasts, this stage occurs at the thylakoid membrane.

The second stage proceeds in the stroma of chloroplasts. Energy from ATP drives the synthesis of sugars from water and carbon dioxide.

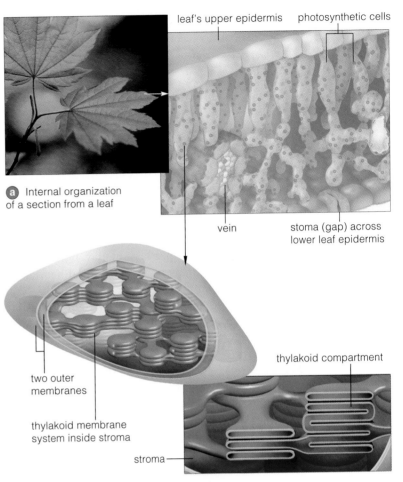

leaf's upper epidermis   photosynthetic cells

a Internal organization of a section from a leaf

vein   stoma (gap) across lower leaf epidermis

two outer membranes

thylakoid membrane system inside stroma

thylakoid compartment

stroma

b Cutaway view of a chloroplast inside the cytoplasm of one photosynthetic cell, and a close-up of its thylakoid compartment.

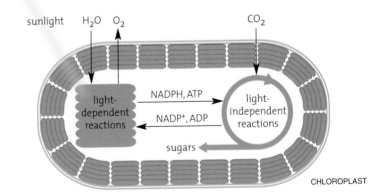

sunlight H_2O O_2 CO_2

light-dependent reactions → NADPH, ATP → light-independent reactions

NADP+, ADP

sugars

CHLOROPLAST

c Two stages of photosynthesis. The first stage depends on inputs of sunlight. It occurs at the thylakoid membrane system. ATP and NADPH form; free oxygen diffuses away. In the second stage, enzymes in the stroma catalyze the assembly of sugars. Energy from ATP starts the reactions. Building blocks are hydrogen atoms and electrons (from NADPH) and carbon atoms (from carbon dioxide).

Figure 7.6 *Animated!* Zooming in on sites of photosynthesis inside the leaf of a typical plant.

7.4 Light-Dependent Reactions

LINKS TO
SECTIONS
5.2, 6.5

In the first stage of photosynthesis, photons absorbed at photosystems drive ATP formation. Water molecules are split. Their oxygen diffuses away, but the coenzyme NADP⁺ picks up the released electrons and hydrogen.

WHAT HAPPENS TO THE ABSORBED ENERGY?

Visualize a lone photon as it collides with a pigment molecule. One of the pigment's electrons can absorb that photon's energy, which boosts the electron to a higher energy level. If nothing else were to happen, the electron would drop back to its unexcited state and lose the extra energy as a photon or as heat.

In the thylakoid membrane, however, energy that excited electrons give up is kept in play. Embedded in the membrane are many hundreds of light-harvesting complexes: circular clusterings of pigments and other proteins (Figure 7.7a). The pigments in light-harvesting complexes do not waste absorbed photons. Instead, their electrons hold on to photon energy by passing it back and forth, like a volleyball.

Energy released from one complex gets passed to another, which passes it on to another, and so on until the energy reaches a photosystem. Chloroplasts have two kinds of photosystems, *type I* and *type II*. The two have slightly different chlorophyll *a* molecules. Each contains other molecules, including different pigments. Hundreds of light-harvesting complexes surround it.

Look back on Figure 7.3, which shows the structure of chlorophyll. Two molecules of chlorophyll *a* are at the center of a photosystem. Their flat rings face each other so closely that the electrons in *both* rings are destabilized. When light-harvesting complexes pass on photon energy to a photosystem, electrons are popped right off of that special pair of chlorophylls.

The freed electrons immediately enter an electron transfer chain positioned next to the photosystem. As you know, **electron transfer chains** are components of cell membranes. Each is an orderly array of enzymes, coenzymes, and other proteins that transfer electrons step-by-step (Section 6.5). *The entry of electrons from a photosystem into an electron transfer chain is the first step in the light-dependent reactions*—in the conversion of photon energy to chemical energy for photosynthesis.

MAKING ATP AND NADPH

Figure 7.8 tracks electrons from a type II photosystem on through an electron transfer chain in the thylakoid membrane. As certain components of the chain accept and donate the electrons, they pick up hydrogen ions (H⁺) from the stroma and release them into the inner thylakoid compartment. They do so again and again. Soon, concentration and electric gradients are built up across the membrane, and the combined force of the gradients attracts H⁺ back toward the stroma.

But H⁺ cannot diffuse across the membrane's lipid bilayer. It can cross only through channels inside **ATP synthases**, a type of transport protein you read about in Section 5.2. In this case, the knoblike portion of the protein projects into the stroma. Ion flow through the channel makes the knob turn, which forces inorganic phosphate to become attached to an ADP molecule. In this way, ATP forms in the stroma.

As long as electrons flow through transfer chains, the cell can keep on producing ATP. But how are the electrons from photosystem II replaced? By a process called *photolysis*, new electrons are pulled away from water molecules, which then dissociate into hydrogen ions and molecular oxygen. The free oxygen diffuses out of the chloroplast, then out of the cell and into the surroundings. Hydrogen ions remain in the thylakoid compartment. They contribute to the concentration and electric gradients that drive ATP formation.

Figure 7.7 (**a**) Ringlike array of pigment molecules that intercept rays of sunlight coming from any direction. (**b**) One of the many photosystems (represented as a *green* sphere) embedded in a chloroplast's thylakoid membrane. Each photosystem collects energy from hundreds of light-harvesting complexes that surround it; only eight complexes are shown here.

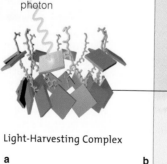

photon

Light-Harvesting Complex

a

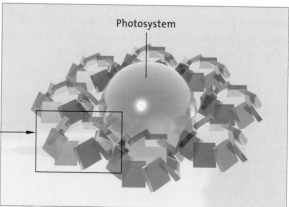

Photosystem

b

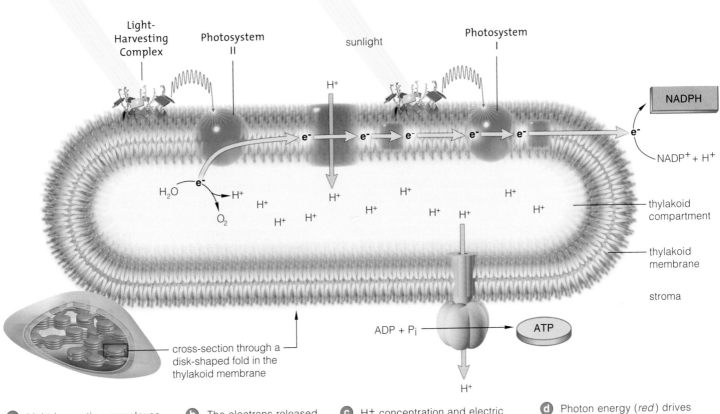

Light-Harvesting Complex | Photosystem II | sunlight | Photosystem I | NADPH

H$^+$

e$^-$ e$^-$ e$^-$ e$^-$ e$^-$ e$^-$ e$^-$

NADP$^+$ + H$^+$

H$_2$O e$^-$ H$^+$ H$^+$ H$^+$ H$^+$ H$^+$ thylakoid compartment

O$_2$ H$^+$ H$^+$ H$^+$ H$^+$ H$^+$ thylakoid membrane

stroma

cross-section through a disk-shaped fold in the thylakoid membrane

ADP + P$_i$ → ATP

H$^+$

a Light-harvesting complexes absorb photon energy (*red*), which drives electrons out of photosystem II. Replacement electrons are pulled from water molecules, which then split into oxygen and hydrogen ions (H$^+$). Oxygen leaves the cell as O$_2$.

b The electrons released from photosystem II enter an electron transfer chain (*brown*), which also moves H$^+$ from the stroma into the thylakoid compartment. The electrons continue on to photosystem I.

c H$^+$ concentration and electric gradients build up across the thylakoid membrane. The force of the gradients propels H$^+$ through ATP synthases. The flow causes this membrane protein to move in a way that forces the attachment of P$_i$ to ADP, thus forming ATP.

d Photon energy (*red*) drives electrons from photosystem I. An intermediary molecule (*brown*) adjacent to the photosystem in the membrane accepts and then transfers the electrons to NADP$^+$, which picks up H$^+$ at the same time and becomes NADPH.

Figure 7.8 *Animated!* How ATP and NADPH form during the first stage of photosynthesis. The drawing represents a cross-section through one of the disk-shaped folds of the thylakoid membrane. This entire sequence is called the *noncyclic* pathway of photosynthesis, because electrons that originally left photosystem II are not cycled back to it. They end up in NADPH.

But where do the electrons end up? After they pass through the electron transfer chain, they enter a type I photosystem where the light-harvesting complexes are volleying energy to a special pair of chlorophylls at the reaction center. The chlorophylls release electrons, which an intermediary molecule transfers to NADP$^+$. When this coenzyme accepts the electrons, it attracts hydrogen ions and thereby becomes NADPH.

We have been describing the *noncyclic* pathway of ATP formation in chloroplasts, so named because the electrons that leave photosystem II do not get cycled back to it; they end up in NADPH.

When too much NADPH forms, it accumulates in the stroma, so the photosystem II pathway backs up. At such times, photosystem I may run independently so that cells can continue to make ATP. It is a *cyclic* pathway of ATP formation, because the electrons that leave photosystem I get cycled back to it. Before they

return, they pass through an electron transfer chain that moves H$^+$ into the thylakoid compartment. The resulting H$^+$ gradients drive ATP formation, but no NADPH forms in this shorter pathway.

Two kinds of photosystems, type I and type II, are embedded in the thylakoid membrane. Hundreds of light-harvesting complexes surround and transfer photon energy to each one.

In a noncyclic pathway of photosynthesis, photon energy forces electrons out of photosystem II and on to an electron transfer chain, which sets up H$^+$ gradients that drive ATP formation. Electrons continue on through photosystem I and end up in a reduced coenzyme, NADPH.

ATP also can form by a cyclic pathway, in which electrons leave photosystem I and are cycled back to it. However, NADPH cannot form by this pathway.

7.5 Energy Flow in Photosynthesis

LINKS TO
SECTIONS
6.1–6.3

One of the recurring themes in biology is that organisms convert one form of energy to another in highly controlled ways. The energy exchanges during the light-dependent reactions, outlined in Figure 7.9, are a classic example.

The preceding section focused on a photosynthetic pathway that starts at photosystem II and ends with the formation of ATP and NADPH. However, a simpler pathway that was less energy efficient preceded it. When photoautotrophs first evolved, remember, they were anaerobic. Their light-dependent pathway of photosynthesis yielded ATP alone, and it still operates today.

Again, this set of reactions is said to be cyclic because excited electrons flow out of photosystem I, through an electron transfer chain, then back to photosystem I.

Later, the photosynthetic machinery in some kinds of photoautotrophs was remodeled. Photosystem II became part of it. That was the start of a combined sequence of reactions powerful enough to oxidize—that is, to strip hydrogen atoms from—water molecules. Free oxygen is a by-product of this pathway.

Remember, the combined pathway is noncyclic; the electrons that leave photosystem II are not returned to it. They end up in NADPH, which delivers them to the sugar factories in the stroma.

Today, some bacteria have only photosystem I. Others have only photosystem II. Cyanobacteria, plants, and all photosynthetic protists have photosystems of both types and carry out both cyclic and noncyclic pathways. Which pathway dominates depends on conditions at the time.

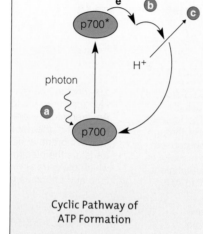

Cyclic Pathway of ATP Formation

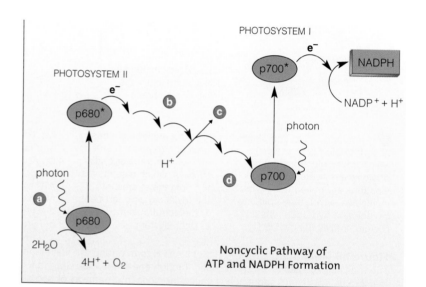

Noncyclic Pathway of ATP and NADPH Formation

ⓐ Photosystem I receives photon energy from a light-harvesting complex. It loses an electron.

ⓑ The electron passes from one molecule to another in an electron transfer chain that is embedded in the thylakoid membrane. It loses a little energy with each transfer, and ends up being reused by photosystem I (thus the pathway is considered "cyclic").

ⓒ Molecules in the transfer chain carry H⁺ across the thylakoid membrane into the inner compartment. Hydrogen ions accumulating in the compartment create an electrochemical gradient across the membrane that drives ATP synthesis, as shown in Figure 7.8.

ⓐ Photosystem II receives photon energy from a light-harvesting complex, then loses an electron. The electron moves through a different electron transfer chain. It loses a little energy with each transfer and ends up at photosystem I.

ⓑ Photosystem I receives photon energy from a light-harvesting complex, then loses an electron. Released electrons and hydrogen ions are used in the formation of NADPH from NADP⁺.

ⓒ As in the cyclic pathway, operation of the electron transfer chain pulls hydrogen ions into the thylakoid compartment. In this case, hydrogens released from dissociated water molecules also enter the compartment. The H⁺ concentration and electric gradient across the membrane are tapped for ATP formation (Figure 7.8).

ⓓ Electrons lost from photosystem I are replaced by the electrons lost from photosystem II. Electrons lost from photosystem II are replaced by electrons obtained from water. (Photolysis pulls water molecules apart into electrons, H⁺, and O₂.)

Figure 7.9 *Animated!* Using energy in the light-dependent reactions. The pair of chlorophyll *a* molecules at the center of photosystem I is designated p700. The pair in photosystem II is designated p680. The pairs respond most efficiently to wavelengths of 700 and 680 nanometers, respectively.

7.6 Light-Independent Reactions: The Sugar Factory

In the chloroplast's stroma, cyclic, enzyme-mediated reactions build sugars from hydrogen, carbon, and oxygen. These light-independent reactions run on energy that became conserved in ATP during the first stage of photosynthesis. NADPH that formed in the first stage donates the hydrogen and electrons. Plants get the carbon and oxygen from carbon dioxide (CO_2) in the air; algae get them from CO_2 dissolved in water.

The light-independent reactions proceed from carbon fixation, to PGAL formation, then RuBP regeneration. In **carbon fixation**, a carbon atom from CO_2 becomes attached to an organic compound. **Rubisco** (ribulose bisphosphate carboxylase/oxygenase) mediates this step in most plants. When it transfers the carbon to five-carbon RuBP (ribulose biphosphate), it opens the sugar factory—a series of enzyme-mediated reactions called the **Calvin–Benson cycle** (Figure 7.10).

The six-carbon intermediate that forms is unstable and splits at once into two PGA (phosphoglycerate) molecules, each with a three-carbon backbone (Figure 7.10a). Next, ATP energy and the reducing power of NADPH convert each PGA to a different three-carbon compound, PGAL (phosphoglyceraldehyde, or G3P).

How? ATP transfers a phosphate group to each PGA, and NADPH donates hydrogen and electrons to it.

Glucose, remember, has six carbon atoms. *Six CO_2 must be fixed and twelve PGAL must form to produce one glucose molecule and also to keep the Calvin–Benson cycle running.* Two PGAL combine to form one six-carbon glucose molecule with a phosphate group attached. The other ten PGAL undergo internal rearrangements in ways that regenerate RuBP (Figure 7.10c–f).

Most of the glucose is converted at once to sucrose or starch by other pathways that conclude the light-independent reactions. Sucrose is a transportable form of carbohydrate in plants. Excess glucose is converted to starch and briefly stored, as starch grains, in the stroma. Starch is converted to sucrose for export to leaves, stems, and roots. Plants can use photosynthetic products and intermediates as energy sources and as building blocks for all required organic compounds.

Driven by ATP energy, the light-independent reactions make sugars with hydrogen and electrons from NADPH, and with carbon and oxygen from carbon dioxide.

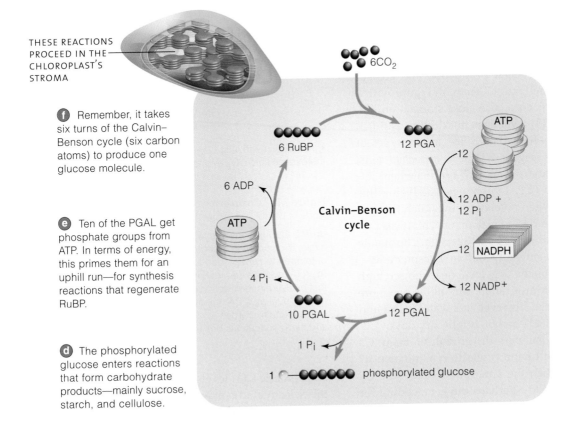

THESE REACTIONS PROCEED IN THE CHLOROPLAST'S STROMA

f Remember, it takes six turns of the Calvin–Benson cycle (six carbon atoms) to produce one glucose molecule.

e Ten of the PGAL get phosphate groups from ATP. In terms of energy, this primes them for an uphill run—for synthesis reactions that regenerate RuBP.

d The phosphorylated glucose enters reactions that form carbohydrate products—mainly sucrose, starch, and cellulose.

6 RuBP
6 ADP
ATP
4 Pi
10 PGAL
1 Pi
1 phosphorylated glucose

Calvin–Benson cycle

6CO₂

12 PGA
ATP
12
12 ADP + 12 Pi
12 NADPH
12 NADP+
12 PGAL

a CO_2 in air spaces inside a leaf diffuses into a photosynthetic cell. Six times, rubisco attaches a carbon atom of CO_2 to the RuBP that starts the Calvin–Benson cycle. Each time, the resulting intermediate splits to form two PGA molecules, for a total of twelve PGA.

b Each PGA molecule gets a phosphate group from ATP, plus hydrogen and electrons from NADPH. The resulting intermediate, PGAL, is thus primed for reaction.

c Two of the twelve PGAL molecules combine to form one molecule of glucose with an attached phosphate group.

Figure 7.10 *Animated!* Light-independent reactions of photosynthesis. The sketch is a summary of all six turns of the Calvin–Benson cycle and its product, one glucose molecule. *Brown* circles signify carbon atoms. Appendix VII details the reaction steps.

7.7   Different Plants, Different Carbon-Fixing Pathways

If sunlight intensity, air temperature, rainfall, and soil composition never varied, photosynthesis might be the same in all plants. But environments differ, and so do details of photosynthesis, as you can see by comparing what happens on hot, dry days when water is scarce.

C4 VERSUS C3 PLANTS

All plant surfaces exposed to air have a waxy, water-conserving cuticle. The only way for gases to diffuse into or out of a plant is at **stomata** (singular, stoma). These are tiny openings across the surface of leaves and green stems (Figure 7.11a). Stomata close on hot, dry days. Water stays inside the plant, but the CO_2 required for photosynthesis cannot diffuse in, and the O_2 by-product of photosynthesis cannot diffuse out.

That is why basswood, beans, peas, and many other plants do not grow well in hot, dry climates without steady irrigation. We call them **C3 plants**, because the *three*-carbon PGA is the first stable intermediate of the Calvin–Benson cycle. When their stomata are closed and the photosynthetic reactions are running, oxygen builds up in leaves and triggers a process that lowers a plant's sugar-making capacity. Remember rubisco, the enzyme that fixes carbon for the Calvin–Benson cycle? When O_2 levels rise, *photorespiration* dominates; rubisco attaches oxygen—not carbon—to RuBP. This reaction yields one molecule of PGA instead of two. The lower yield slows sugar production and growth of the plant. Compare Figure 7.11a with Figure 7.10.

C4 plants, such as corn, also close stomata on hot, dry days. But the CO_2 level does not decline as much because these plants fix carbon twice, in two types of photosynthetic cells (Figure 7.11b). In *mesophyll* cells, a four-carbon molecule, oxaloacetate, forms when CO_2 donates a carbon to PEP. The enzyme catalyzing this step will not use oxygen no matter how much there is. Oxaloacetate is converted to malate, which moves into *bundle-sheath cells* through plasmodesmata. The malate releases CO_2, which enters the Calvin–Benson cycle.

The C4 cycle keeps the CO_2 level near rubisco high enough to stop photorespiration. It requires one more ATP than the C3 cycle. However, less water is lost and more sugar can be made on hot, bright, dry days.

Photorespiration hampers the growth of many C3 plants. So why hasn't natural selection eliminated it? Rubisco evolved when the atmosphere held little O_2 and a great deal of CO_2. Perhaps the gene coding for rubisco's structure cannot mutate without disruptive effects on rubisco's primary role—carbon fixation.

Over the past 50 to 60 million years, the C4 cycle evolved independently in many lineages. Before then, atmospheric CO_2 levels were higher, so C3 plants had the selective advantage in hot climates. Which cycle will be most adaptive in the future? The CO_2 levels have been rising for decades and may double in the next fifty years. If so, C3 plants will yet again be at an advantage—and many vital crop plants may benefit.

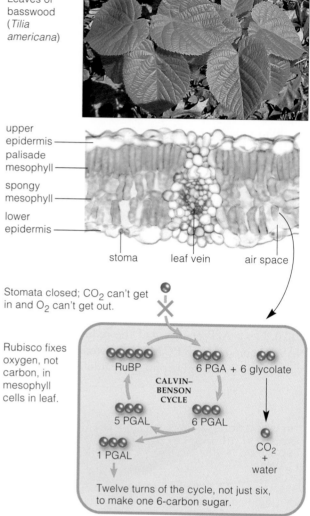

Leaves of basswood (*Tilia americana*)

upper epidermis
palisade mesophyll
spongy mesophyll
lower epidermis

stoma   leaf vein   air space

Stomata closed; CO_2 can't get in and O_2 can't get out.

Rubisco fixes oxygen, not carbon, in mesophyll cells in leaf.

RuBP   6 PGA + 6 glycolate

CALVIN–BENSON CYCLE

5 PGAL   6 PGAL

1 PGAL

CO_2 + water

Twelve turns of the cycle, not just six, to make one 6-carbon sugar.

a Carbon fixation in C3 plants during hot, dry weather, when there is too little CO_2 and too much O_2 in leaves.

Figure 7.11 Comparison of carbon-fixing adaptations in three kinds of plants that evolved in different environments.

(**a**) The Calvin–Benson cycle, which also is called the C3 cycle, is common in evergreens and many nonwoody plants of temperate zones, such as basswood and bluegrass. (**b**) A C4 cycle is common in grasses, corn, and other plants that evolved in the tropics and that fix CO_2 twice. (**c**) Prickly pear (*Opuntia*), a CAM plant. These plants, which open stomata and fix carbon at night, include orchids, pineapples, and many succulents besides cacti.

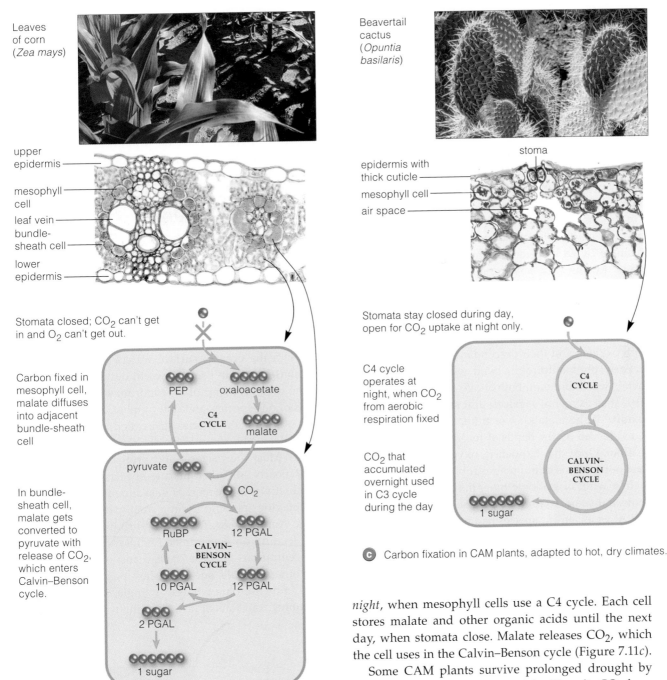

Leaves of corn (*Zea mays*)

upper epidermis

mesophyll cell

leaf vein

bundle-sheath cell

lower epidermis

Stomata closed; CO_2 can't get in and O_2 can't get out.

Carbon fixed in mesophyll cell, malate diffuses into adjacent bundle-sheath cell

PEP oxaloacetate

C4 CYCLE

malate

In bundle-sheath cell, malate gets converted to pyruvate with release of CO_2, which enters Calvin–Benson cycle.

pyruvate

CO_2

RuBP 12 PGAL

CALVIN–BENSON CYCLE

10 PGAL 12 PGAL

2 PGAL

1 sugar

b Carbon fixation in C4 plants during hot, dry weather, when there is too little CO_2 and too much O_2 in leaves.

Beavertail cactus (*Opuntia basilaris*)

stoma

epidermis with thick cuticle

mesophyll cell

air space

Stomata stay closed during day, open for CO_2 uptake at night only.

C4 cycle operates at night, when CO_2 from aerobic respiration fixed

CO_2 that accumulated overnight used in C3 cycle during the day

C4 CYCLE

CALVIN–BENSON CYCLE

1 sugar

c Carbon fixation in CAM plants, adapted to hot, dry climates.

night, when mesophyll cells use a C4 cycle. Each cell stores malate and other organic acids until the next day, when stomata close. Malate releases CO_2, which the cell uses in the Calvin–Benson cycle (Figure 7.11*c*).

Some CAM plants survive prolonged drought by keeping stomata shut even at night. They fix CO_2 from aerobic respiration. Not much forms, but it is enough to maintain low metabolic rates and very slow growth. Try growing cacti in mild climates, and you will see that they compete poorly with C3 and C4 plants.

C3 plants, C4 plants, and CAM plants respond differently to hot, dry conditions. At such times, stomata close to conserve water, and so photosynthetic cells must deal with too much oxygen and not enough carbon dioxide in leaves.

CAM PLANTS

We see a carbon-fixing adaptation to desert conditions in a cactus. This plant, a type of succulent, has juicy, water-storing tissues and thick surface layers that limit loss of water. It is one of many **CAM plants** (short for *C*rassulacean *A*cid *M*etabolism). A cactus will not open stomata on hot days; it opens them and fixes CO_2 *at*

7.8 Autotrophs and the Biosphere

We conclude this chapter by reflecting on the mind-boggling numbers of single-celled and multicelled photosynthesizers and other autotrophs. We find them on land and in the water provinces, and they profoundly influence the biosphere.

THE ENERGY CONNECTION

This chapter opened with a brief look at the origin of photosynthesis and its impact on the world of life. All organisms require ongoing supplies of energy and carbon-based compounds for growth and survival. Autotrophs get them from the physical environment. Energy-rich carbon compounds become concentrated in single-celled kinds and in the tissues of multicelled kinds. In this way, autotrophs become concentrated stores of food tempting to heterotrophs.

Autotrophs are more than carbon-rich food baskets for the biosphere. Early practitioners of the noncyclic pathway of photosynthesis enriched the atmosphere with oxygen, and their descendants still replenish it.

Early photoautotrophs lived when iron and other metals were abundant both above and below the seas. As fast as oxygen was released, it swiftly latched onto (oxidized) the metals. Over time, it rusted them out, as evidenced by the bands of red iron deposits on the seafloor. Once that happened, oxygen bubbled out of vast populations of photoautotrophs, unimpeded.

In a wink of geologic time, maybe a few hundred thousand years, oxygen levels rose in the seas and the sky. Most anaerobic species had no means of adapting to the change, and they perished in a mass extinction. Other chemoautotrophs that could not tolerate oxygen endured in seafloor sediments, hot springs, and other anaerobic habitats. Some still live near hydrothermal vents, where superheated water spews out from big fissures in the seafloor. Archaeans near the vents get hydrogen and electrons from hydrogen sulfide in the mineral-rich water. Chemoautotrophs live in oxygen-free soils, where they extract energy from nitrogen-rich wastes and remains of other organisms.

However, among some ancient species of bacteria, metabolic pathways became modified in ways that detoxified oxygen. Later on, a pathway that released energy from organic compounds became modified in ways that allowed it to *use* oxygen as a final electron acceptor. As a direct outcome of the selection pressure exerted by oxygen—a by-product of photosynthesis—aerobic respiration had evolved (Figure 7.12).

PASTURES OF THE SEAS

Today, aerobic species are all around us. Each spring, the renewed growth of photoautotrophs is evident as trees leaf out and fields turn green. At the same time, uncountable numbers of single-celled species drifting through the ocean's surface waters make a seasonal response. You can't see them without a microscope. In some regions, a cup of seawater may hold 24 million cells of one species, and that number does not include any other aquatic species suspended in the cup.

Collectively, these cells are the "pastures of the seas." Most are bacteria and protists that ultimately feed nearly all other marine species. Their primary productivity is the start of vast aquatic food webs.

Imagine zooming in on a small patch of "pasture" in an Antarctic sea. There, tiny shrimplike crustaceans are feeding on even tinier photosynthesizers. Dense concentrations of these crustaceans, or krill, are food for other animals, such as fishes, penguins, seabirds, and immense blue whales. A single, mature whale is straining four tons of krill from the water. Before they

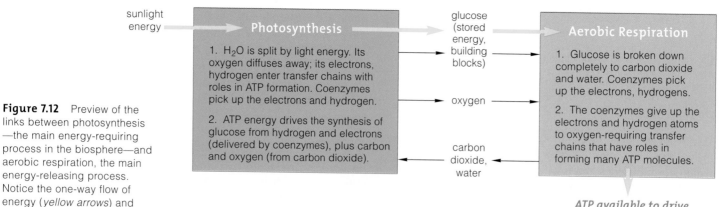

Figure 7.12 Preview of the links between photosynthesis —the main energy-requiring process in the biosphere—and aerobic respiration, the main energy-releasing process. Notice the one-way flow of energy (*yellow arrows*) and the cycling of materials.

sunlight energy

Photosynthesis

1. H_2O is split by light energy. Its oxygen diffuses away; its electrons, hydrogen enter transfer chains with roles in ATP formation. Coenzymes pick up the electrons and hydrogen.

2. ATP energy drives the synthesis of glucose from hydrogen and electrons (delivered by coenzymes), plus carbon and oxygen (from carbon dioxide).

glucose (stored energy, building blocks)

oxygen

carbon dioxide, water

Aerobic Respiration

1. Glucose is broken down completely to carbon dioxide and water. Coenzymes pick up the electrons, hydrogens.

2. The coenzymes give up the electrons and hydrogen atoms to oxygen-requiring transfer chains that have roles in forming many ATP molecules.

ATP available to drive nearly all cellular tasks

themselves became food for the whale, the four tons' worth of krill had munched their way through 1,200 tons of the pasture!

The pastures "bloom" in spring, when the seawater becomes warmer and greatly enriched with nutrients that currents churn up from the deep. The conditions favor huge increases in population sizes.

Until NASA gathered data from space satellites, we had no idea of the size and distribution of these marine pastures. Figure 7.13a shows the near-absence of photosynthetic activity one winter in the Atlantic Ocean. Figure 7.13b shows a springtime bloom that stretched from North Carolina all the way past Spain!

Collectively, these cells affect the global climate, because they deal with staggering numbers of gaseous reactant and product molecules. For instance, they sponge up nearly half of the carbon dioxide used in carbon fixation. Without them, atmospheric carbon dioxide would accumulate more rapidly and possibly accelerate global warming (Sections 47.9 and 47.10).

Although drastic global change is a real possibility, human activities release more carbon dioxide to the atmosphere than photoautotrophs can take up. Such activities include burning fossil fuels and setting fire to vast tracts of forests to clear land for farming.

There is more. Each day, tons of industrial wastes, raw sewage, and fertilizers in runoff from croplands enter the ocean and change its chemical composition. How long can we expect the marine photoautotrophs to function in this chemical brew? The answer may affect your life in more ways than one. It may affect populations and ecosystems throughout the world.

In sum, autotrophs exist in tremendous numbers. They nourish themselves and all other living things, and they are major players in the cycling of oxygen, nitrogen, phosphorus, and other elements all through the biosphere. Later chapters focus on their impact on the environment. In this unit, we turn next to major pathways by which all cells release the chemical bond energy that is stored in glucose and other biological molecules—the legacy of autotrophs everywhere.

Energy flow and the cycling of carbon and other nutrients through the biosphere starts with autotrophs.

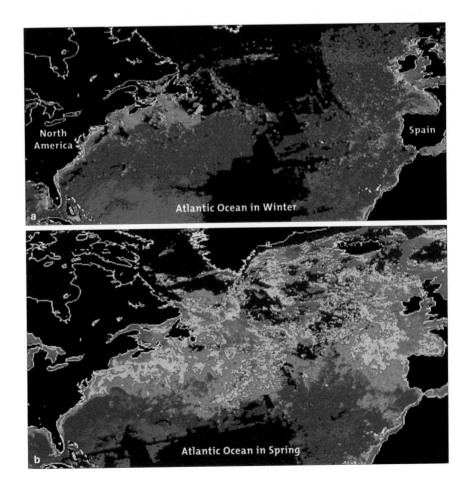

Figure 7.13 Two satellite images that convey the sheer magnitude of photosynthetic activity during springtime in the surface waters of the North Atlantic Ocean. Sensors in equipment launched with the satellite recorded concentrations of chlorophyll, which were greatest in regions coded *red*.

Take a deep breath while looking at these images. You just took in free oxygen that originated with some photoautotroph, somewhere in the world. Poison the autotrophs and how long will oxygen-dependent heterotrophs last?

Summary

Section 7.1 Photosynthesis runs on energy obtained when pigment molecules absorb wavelengths of visible light from the sun. Chlorophyll *a*, the main pigment, is best at absorbing violet and red wavelengths. Diverse photosynthetic pigments are accessory pigments. They form light-harvesting complexes that capture photons of particular wavelengths. Photons not captured are reflected as the characteristic color of each pigment.

Section 7.2 Collectively, photosynthetic pigments absorb most of the full range of wavelengths in the spectrum of visible light with impressive efficiency.

Section 7.3 Photosynthesis has two stages: light-dependent and light-independent reactions. Figure 7.14 and the following equation summarize the process:

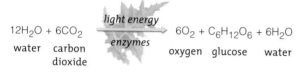

$$12H_2O + 6CO_2 \xrightarrow[\text{enzymes}]{\text{light energy}} 6O_2 + C_6H_{12}O_6 + 6H_2O$$

water   carbon   oxygen glucose   water
     dioxide

In chloroplasts, the light-dependent reactions occur at a thylakoid membrane that forms a single compartment in the semifluid interior (stroma).

Biology ⬧ Now
View the sites where photosynthesis takes place with the animation on BiologyNow.

Sections 7.4, 7.5 Accessory pigments arrayed in clusters in the thylakoid membrane absorb photons and pass energy to many photosystems. Light-dependent reactions use electrons released from photosystems in a noncyclic or a cyclic pathway of ATP formation.

In the noncyclic pathway, electrons are released from photosystem II and enter an electron transfer chain. Their flow through the chain causes hydrogen ions to accumulate in the thylakoid compartment. They flow on to photosystem I where photon absorption also causes the release of electrons. An intermediary molecule next to photosystem I accepts the electrons and transfers them to NADP+, which attracts hydrogen ions (H+) at the same time and becomes a reduced coenzyme, NADPH.

The electrons lost from photosystem II are replaced by way of photolysis—a reaction that pulls electrons from water molecules, with the release of H+ and O2.

In the cyclic pathway, electrons from photosystem I enter an electron transfer chain, then are cycled back to the same photosystem. NADPH does not form.

In both pathways, the H+ buildup in the thylakoid compartment forms concentration and electric gradients across the thylakoid membrane. H+ flows in response to the gradients, through ATP synthases. The flow causes P_i to be attached to ADP in the stroma, forming ATP.

Biology ⬧ Now
Review the pathways by which light energy is used to form ATP with the animation on BiologyNow.

Section 7.6 The light-independent reactions proceed in the stroma. In C3 plants, the enzyme rubisco attaches carbon from CO2 to RuBP to start the Calvin–Benson cycle. In this cyclic pathway, energy from ATP, carbon and oxygen from CO2, and hydrogen and electrons from NADPH are used to make phosphorylated glucose, which quickly enters reactions that form the products of photosynthesis (mainly sucrose, cellulose, and starch). It takes six turns of the Calvin–Benson cycle to fix the six CO2 required to make one glucose molecule.

Biology ⬧ Now
Read the InfoTrac article "Robust Plants' Secret? Rubisco Activase!" Marcia Wood, Agricultural Research, November 2002.

Section 7.7 Environments differ, and so do details of sugar production. On hot, dry days, plants conserve water by closing stomata, but O2 from photosynthesis cannot escape. In C3 plants, high O2/low CO2 levels cause the enzyme rubisco to use O2 in an alternate pathway that does not make as much sugar. In C4 plants, carbon fixation occurs in one cell type, and the carbon enters the Calvin–Benson cycle in a different cell type. CAM plants close stomata in the day and fix carbon at night.

Biology ⬧ Now
Read the InfoTrac article "Light of Our Lives," Norman Miller, Geographical, January 2001.

Section 7.8 Photoautotrophs and chemoautotrophs produce the food that sustains themselves and all other organisms. Also, staggering numbers of diverse aerobic and anaerobic autotrophs live in the seas as well as on land. They have impact on the global cycling of oxygen, carbon, nitrogen, and other substances, and the global climate. Human activities are having impact on them.

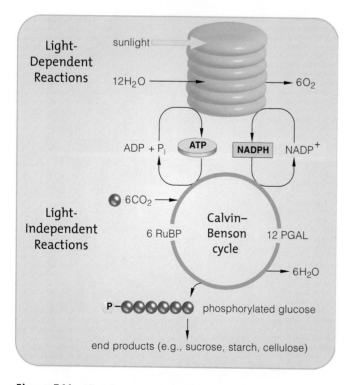

Figure 7.14 Visual summary of photosynthesis.

Figure 7.15 Leaves of *Elodea*, an aquatic plant.

Figure 7.16 (**a**) Red alga from a tropical reef. (**b**) Coastal green alga (*Codium*).

Self-Quiz

Answers in Appendix II

1. Photosynthetic autotrophs use _____ from the air as a carbon source and _____ as their energy source.

2. Chlorophyll *a* absorbs violet and red light, and it reflects light of _____ and _____ wavelengths.
   a. violet; red
   b. yellow; green
   c. green only
   d. white; orange

3. Light-*dependent* reactions in plants occur at the _____ .
   a. thylakoid membrane
   b. plasma membrane
   c. stroma
   d. cytoplasm

4. In the light-*dependent* reactions, _____ .
   a. carbon dioxide is fixed
   b. ATP and NADPH form
   c. CO_2 accepts electrons
   d. sugars form

5. What accumulates inside the thylakoid compartment during the light-*dependent* reactions?
   a. glucose   b. RuBP   c. hydrogen ions   d. CO_2

6. When a photosystem absorbs light, _____ .
   a. sugar phosphates are produced
   b. electrons are transferred to ATP
   c. RuBP accepts electrons
   d. light-dependent reactions begin

7. Light-*independent* reactions proceed in the _____ .
   a. cytoplasm   b. plasma membrane   c. stroma

8. The Calvin–Benson cycle starts when _____ .
   a. light is available
   b. carbon dioxide is attached to RuBP
   c. electrons leave photosystem II

9. What substance is *not* part of the Calvin–Benson cycle?
   a. ATP
   b. NADPH
   c. RuBP
   d. carotenoids
   e. O_2
   f. CO_2

10. Match each event with its most suitable description.
    ____ ATP formation only   a. rubisco required
    ____ CO_2 fixation   b. ATP, NADPH required
    ____ PGAL formation   c. electrons cycled back
                                  to photosystem II

Additional questions are available on **Biology ⋑ Now**™

Critical Thinking

1. About 200 years ago, Jan Baptista van Helmont did experiments on the nature of photosynthesis. He wanted to know where growing plants get the materials necessary for increases in size. He planted a tree seedling weighing 5 pounds in a barrel filled with 200 pounds of soil. He watered the tree regularly. Five years passed. Then van Helmont weighed the tree and the soil. The tree weighed 169 pounds, 3 ounces. The soil weighed 199 pounds, 14 ounces. Because the tree gained so much weight and the soil lost so little, he concluded the tree had gained all of its additional weight by absorbing water he had added to the barrel. Given what you know about biological molecules, why was he misguided? Knowing what you do about photosynthesis, what really happened?

2. A cat eats a bird, which earlier ate a caterpillar that chewed on a weed. Which organisms are autotrophs? Which are the heterotrophs?

3. Imagine walking through a garden of red, white, and blue petunias. Explain each of the colors in terms of which wavelengths of light the flower is absorbing.

4. Krishna exposes pea plants to a carbon radioisotope ($^{14}CO_2$), which they absorb. In which compound will the labeled carbon appear first if the plants are C3? C4?

5. While gazing into an aquarium, you observe bubbling from an aquatic plant (Figure 7.15). What is happening?

6. Only about eight classes of pigment molecules are known, but this limited group gets around in the world. For example, photoautotrophs make carotenoids, which move through food webs, as when tiny aquatic snails graze on green algae and then flamingos eat the snails. Flamingos modify the ingested carotenoids. Their cells split beta-carotene to form two molecules of vitamin A. This vitamin is the precursor of retinol, a visual pigment that transduces light into electric signals in eyes. Beta-carotene gets dissolved in fat reservoirs under the skin. Cells that give rise to bright pink feathers take it up.

   Select a similar organism and do some research to identify sources for pigments that color its surfaces.

7. Most pigments respond to only part of the rainbow of visible light. If acquiring energy is so vital, then why doesn't each kind of photosynthetic pigment absorb the whole spectrum? *Why isn't each one black?*

   If early photoautotrophs evolved in the seas, then so did their pigments. Ultraviolet and red wavelengths do not penetrate water as deeply as green and blue wavelengths do. Possibly natural selection favored the evolution of different pigments at different depths. Many relatives of the red alga in Figure 7.16*a* live deep in the sea. Some are nearly black. Green algae, such as the one in Figure 7.16*b*, live in shallow water. Their chlorophylls absorb red wavelengths, and accessory pigments harvest others. Some accessory pigments also function as shields against ultraviolet radiation.

   Speculate on how natural selection may have favored the evolution of different pigments at different depths, starting at hydrothermal vents. You might start at Richard Monasterky's article in *Science News* (September 7, 1996) on Cindy Lee Dover's work at hydrothermal vents.

When Mitochondria Spin Their Wheels

In the early 1960s, Swedish physician Rolf Luft mulled over some odd symptoms of a patient. The young woman felt weak and too hot all the time. Even on the coldest winter days she could not stop sweating, and her skin was always flushed. She was thin in spite of a huge appetite.

Luft inferred that his patient's symptoms pointed to a metabolic disorder. Her cells seemed to be spinning their wheels. They were active, but much of their activity was being dissipated as metabolic heat. So he ordered tests designed to detect her metabolic rates. The patient's oxygen consumption was the highest ever recorded!

Microscopic examination of a tissue sample from the patient's skeletal muscles revealed mitochondria, the cell's ATP-producing powerhouses. But there were far too many of them, and they were abnormally shaped. Other studies showed that the mitochondria were engaged in aerobic respiration—yet they were making very little ATP.

The disorder, now called *Luft's syndrome*, was the first to be linked directly to a defective organelle. By analogy, someone with this mitochondrial disorder functions like a city with half of its power plants shut down. Skeletal and heart muscles, the brain, and other hardworking body parts with the highest energy demands are hurt the most.

More than a hundred other mitochondrial disorders are now known. One, a heritable disease called *Friedreich's ataxia*, causes loss of coordination (ataxia), weak muscles, and visual problems. Many affected people die when they are young adults because of heart muscle irregularities. Figure 8.1 shows two affected individuals.

A mutant gene causes Friedreich's ataxia. Its abnormal protein product makes iron accumulate in mitochondria. Iron is vital for electron transfers that drive ATP formation, but too much favors the concentration of free radicals that can attack the structural integrity of all molecules of life.

Defective mitochondria also contribute to many age-related problems, including Type 1 diabetes, atherosclerosis, amyotrophic lateral sclerosis (Lou Gehrig's disease), as well as Parkinson's, Alzheimer's, and Huntington's diseases.

Clearly, human health depends on mitochondria that are structurally and functionally sound. However, zoom out from our characteristically human focus and you will find that every kind of animal, every kind of plant and fungus, and nearly every protist depend on them.

Remember how ATP forms at a membrane system inside chloroplasts? Similar events happen at a membrane system inside mitochondria. In both photosynthesis and aerobic respiration, electrons are released when an energy input breaks chemical bonds, and they are sent step-by-step through transfer chains in the membrane. Photosynthesis splits bonds in water molecules; aerobic respiration splits bonds in glucose and other organic compounds.

In mitochondria, too, energy harnessed as electrons go through transfer chains helps concentrate hydrogen ions on one side of the membrane. Here again, potential energy

Figure 8.1 Sister, brother, and broken mitochondria. Both of these individuals show symptoms of Friedreich's ataxia, a heritable genetic disorder that prevents them from making enough ATP to keep their body structurally and functionally sound. At age five, Leah started to lose her sense of balance and coordination. Six years later she was in a wheelchair and is now diabetic and partially deaf. Her brother Joshua was three when problems started. Eight years later he could not walk. He is now blind. Both sister and brother have heart problems; both had spinal fusion surgery. Special equipment allows them to attend school and work part-time. Leah is a professional model.

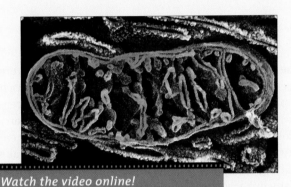

Watch the video online!

is briefly stored as concentration and electric gradients across that membrane. That stored energy is converted to the bond energy of ATP. And ATP is a transportable form of energy; it can jump-start nearly all of the metabolic reactions that keep cells, and organisms, going.

The point is, you already have a sense of how operation of electron transfer systems can concentrate energy in a form that can be tapped to make ATP. Even prokaryotic cells make ATP by operating simpler electron transfer systems, which are built into their plasma membrane. As you will see, the details differ from one group to the next. Yet even these variations do not obscure life's unity at the biochemical level.

How Would You Vote?

Developing new drugs is costly. There is little incentive for pharmaceutical companies to target ailments, such as Friedreich's ataxia, that affect relatively few individuals. Should the federal government allocate some funds to private companies that search for cures for diseases affecting a relatively small number of people? See BiologyNow for details, then vote online.

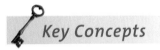

Key Concepts

THE MAIN ENERGY-RELEASING PATHWAYS

All organisms release chemical bond energy from glucose and other organic compounds to drive ATP formation. The main energy-releasing pathways all start in the cytoplasm. Only aerobic respiration, which uses free oxygen, ends in mitochondria. It has the greatest energy yield. Section 8.1

GLYCOLYSIS—FIRST STAGE OF THE PATHWAYS

Glycolysis is the first stage of aerobic respiration. It also is the first stage of anaerobic pathways, such as alcoholic and lactate fermentation. Enzymes partially break down glucose to pyruvate, and they can do so in the presence of oxygen or in its absence. Section 8.2

HOW AEROBIC RESPIRATION ENDS

Aerobic respiration continues through two more stages. In the second stage, pyruvate from glycolysis is broken down to carbon dioxide. Electrons and hydrogen that coenzymes picked up during the first two stages enter electron transfer chains. In the third stage, transfer chains set up conditions that favor ATP formation. Free oxygen accepts the spent electrons. Sections 8.3, 8.4

HOW ANAEROBIC PATHWAYS END

Fermentation also starts with glycolysis, but substances other than oxygen are the final electron acceptor. The net energy yield is always small. Section 8.5

WHAT IF GLUCOSE IS NOT AVAILABLE?

When required, molecules other than glucose can enter the aerobic pathway as alternative energy sources. Section 8.6

PERSPECTIVE AT UNIT'S END

We see evidence of life's unity in its molecular and cellular organization and in the utter dependence of all organisms on the one-way flow of energy. Section 8.7

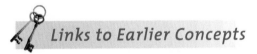

Links to Earlier Concepts

This chapter expands the picture of life's dependence on energy flow by showing how all organisms tap energy stored in glucose and convert it to transportable forms, ATP especially (Sections 6.1, 6.2). You may wish to review the structure of glucose (3.1, 3.3). You will see more examples of controlled energy release at electron transfer chains (6.4). You will reflect once more on global connections between photosynthesis and aerobic respiration (7.8).

8.1 Overview of Energy-Releasing Pathways

LINKS TO
SECTIONS
6.1, 6.2

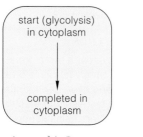

Plants make ATP during photosynthesis and use it to synthesize glucose and other carbohydrates. But all organisms, plants included, can make ATP by breaking down carbohydrates, lipids, and proteins.

Organisms stay alive only as long as they get more energy to replace the energy they use up (Section 6.1). Plants and all other photoautotrophs get energy from the sun; heterotrophs get energy by eating plants and one another. Regardless of its source, the energy must be converted to some form that can drive thousands of diverse life-sustaining reactions. Energy that becomes converted into chemical bond energy of adenosine triphosphate—ATP—serves that function.

COMPARISON OF THE MAIN TYPES OF ENERGY-RELEASING PATHWAYS

The first energy-releasing metabolic pathways were operating billions of years before Earth's oxygen-rich atmosphere evolved, so we can expect that they were *anaerobic*; the reactions did not use free oxygen. Many prokaryotes and protists still live in places where oxygen is absent or not always available. They make ATP by fermentation and other anaerobic pathways. Many eukaryotic cells still use fermentation, including skeletal muscle cells. However, the cells of nearly all eukaryotes extract energy efficiently from glucose by **aerobic respiration**, an oxygen-dependent pathway. Each breath you take provides your actively respiring cells with a fresh supply of oxygen.

Make note of this point: *In every cell, all of the main energy-releasing pathways start with the same reactions in the cytoplasm.* During the initial reactions, **glycolysis**, enzymes cleave and rearrange a glucose molecule into two molecules of **pyruvate**, an organic compound that has a three-carbon backbone.

After glycolysis, energy-releasing pathways differ. Only the aerobic pathway ends inside mitochondria. There, free oxygen accepts and removes electrons after they indirectly help the formation of ATP (Figure 8.2).

As you examine the energy-releasing pathways in sections to follow, keep in mind that enzymes catalyze each step, and intermediates formed at one step serve as substrates for the next enzyme in the pathway.

OVERVIEW OF AEROBIC RESPIRATION

Of all energy-releasing pathways, aerobic respiration gets the most ATP for each glucose molecule. Whereas anaerobic routes have a net yield of two ATP, aerobic respiration typically yields thirty-six or more. If you were a bacterium, you would not require much ATP. Being far larger, more complex, and highly active, you

start (glycolysis)
in cytoplasm

completed in
cytoplasm

Anaerobic Energy-
Releasing Pathways

start (glycolysis)
in cytoplasm

completed in
mitochondrion

Aerobic
Respiration

Figure 8.2 *Animated!* Where the main energy-releasing pathways of ATP formation start and end. Only aerobic respiration ends in mitochondria. This pathway alone delivers enough ATP to build and maintain big multicelled organisms, including redwoods and highly active animals, such as people and Canada geese.

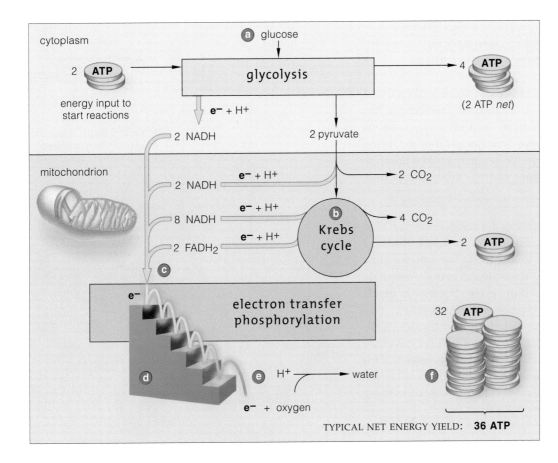

Figure 8.3 *Animated!* Overview of aerobic respiration. The reactions start in the cytoplasm, but they end inside mitochondria.

(**a**) In the first stage, glycolysis, enzymes partly break down glucose to pyruvate.

(**b**) In the second stage, enzymes break down pyruvate to carbon dioxide.

(**c**) NAD^+ and FAD pick up the electrons and hydrogen stripped from intermediates in both stages.

(**d**) The last stage is electron transfer phosphorylation. NADH and $FADH_2$, which are reduced coenzymes, deliver electrons to electron transfer chains. H^+ accompanies these electrons. Electron flow through the chains sets up H^+ gradients, which are tapped to make ATP.

(**e**) Oxygen accepts electrons at the end of the third stage, forming water.

(**f**) From start to finish, a typical net energy yield from a single glucose molecule is thirty-six ATP.

depend on the aerobic pathway's high yield. When a molecule of glucose is the starting material, aerobic respiration can be summarized this way:

$$C_6H_{12}O_6 + 6O_2 \longrightarrow 6CO_2 + 6H_2O$$

glucose   oxygen   carbon   water
                        dioxide

However, as you can see, the summary equation only tells us what the substances are at the start and finish of the pathway. In between are three reaction stages.

As you track the reactions, you will encounter two coenzymes, abbreviated **NAD⁺** (nicotinamide adenine dinucleotide) and **FAD** (flavin adenine dinucleotide). Both accept electrons and hydrogen derived from intermediates that form during glucose breakdown. Unbound hydrogen atoms are hydrogen ions (H^+), or naked protons. When the two coenzymes are carrying electrons and hydrogen, they are in a reduced form and may be abbreviated NADH and $FADH_2$.

Figure 8.3 is your overview of aerobic respiration. Glycolysis, again, is the first stage. The second stage is a cyclic pathway, the **Krebs cycle**. Enzymes break down pyruvate to carbon dioxide and water. These reactions release many electrons and hydrogen atoms.

Few ATP form during glycolysis or the Krebs cycle. The big energy harvest comes in the third stage, after reduced coenzymes give up electrons and hydrogen to electron transfer chains—the machinery of **electron transfer phosphorylation**. Operation of these chains sets up H^+ concentration and electric gradients. The gradients drive ATP formation at nearby membrane transport proteins. During this final stage, many ATP molecules form. It ends when free oxygen accepts the "spent" electrons from the last portion of the transfer chain. The oxygen picks up H^+ at the same time and thereby forms water, a by-product of the reactions.

Nearly all metabolic reactions run on energy released from glucose and other organic compounds. The main energy-releasing pathways start in the cytoplasm with glycolysis, a series of reactions that break down glucose to pyruvate.

Anaerobic pathways have a small net energy yield, typically two ATP for each glucose molecule metabolized.

Aerobic respiration, an oxygen-dependent pathway, runs to completion in mitochondria. From start (glycolysis) to finish, it typically has a net energy yield of thirty-six ATP.

8.2 Glycolysis—Glucose Breakdown Starts

LINKS TO
SECTIONS
2.4, 3.3

*Let's track what happens to a glucose molecule in the
first stage of aerobic respiration. Remember, these same
steps occur in anaerobic energy-releasing pathways.*

Any of several six-carbon sugars can be broken down
in glycolysis. Each glucose molecule, remember, has
six carbon, twelve hydrogen, and six oxygen atoms
(Section 3.3). The carbons form its backbone. During
glycolysis, this one molecule is partly broken down to
two molecules of pyruvate, a three-carbon compound:

glucose ⟶ ⓟ–glucose ⟶ 2 pyruvate

The initial steps of glycolysis are *energy-requiring*. One
ATP molecule primes glucose to rearrange itself by
donating a phosphate group to it. The intermediate
that forms, fructose-6-phosphate, accepts a phosphate
group from another ATP molecule. Thus, cells invest
two ATP to jump-start glycolysis (Figure 8.4*a*).

The resulting intermediate is split into one PGAL
(phosphoglyceraldehyde) and one DHAP, a molecule
that has the same number of atoms arranged a bit
differently. An enzyme can reversibly convert DHAP
into PGAL. It does so, and two PGAL molecules enter
the next reaction (Figure 8.4*b*).

In the first *energy-releasing* step of glycolysis, the
two PGAL are converted to intermediates that give
up a phosphate group to ADP, so two ATP form. In
later reactions, two more intermediates do the same
thing. Thus, four ATP have formed by **substrate-level
phosphorylation**. We define this metabolic event as a
direct transfer of a phosphate group from a substrate
of a reaction to another molecule—in this case, to ADP.

Meanwhile, the two PGAL give up electrons and
hydrogens to two NAD⁺, which becomes NADH.

Even though four ATP are now formed, remember
that two ATP were invested to start the reactions. The
net yield of glycolysis is two ATP and two NADH.

To summarize, glycolysis converts bond energy of
glucose to bond energy of ATP—a transportable form
of energy. The electrons and hydrogen stripped from
glucose and picked up by NAD⁺ can enter the next
stage of reactions. So can the products of glycolysis—
two pyruvate molecules.

> *Glycolysis is a series of reactions that partially break down
> glucose or other six-carbon sugars to two molecules of
> pyruvate. It takes two ATP to jump-start the reactions.*
>
> *Two NADH and four ATP form. However, when we subtract
> the two ATP required to start the reactions, the net energy
> yield of glycolysis is two ATP from one glucose molecule.*

Figure 8.4 *Animated!* Glycolysis. This first stage of the
main energy-releasing pathways occurs in the cytoplasm of
all prokaryotic and eukaryotic cells. Glucose is the reactant
in this example. Appendix VII gives the structural formulas of
reaction intermediates and products. Two pyruvate, two NADH,
and four ATP form in glycolysis. Cells invest two ATP to start
the reactions, however, so the *net* energy yield is two ATP.

Depending on the type of cell and environmental conditions,
the pyruvate may enter the second set of reactions of the
aerobic pathway, including the Krebs cycle. Or it may be
used in other reactions, such as those of fermentation.

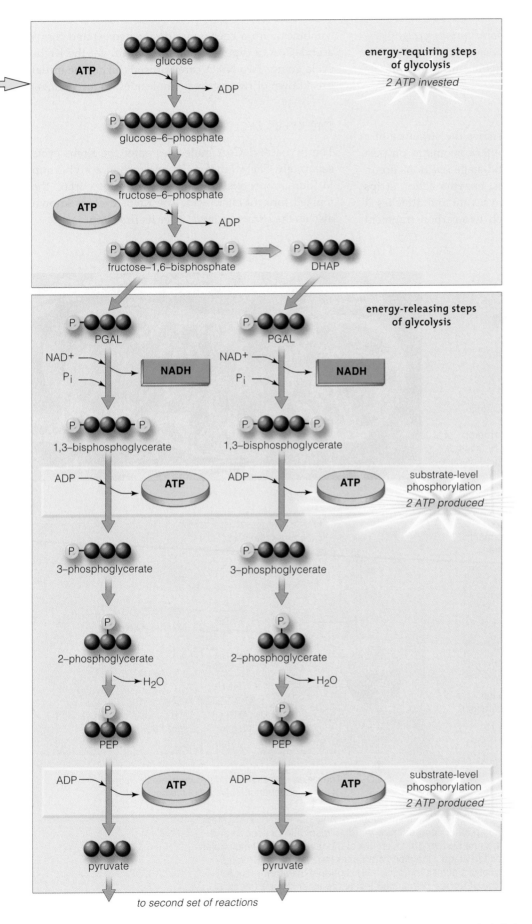

Track the six carbon atoms (*brown circles*) of glucose. Glycolysis requires an energy investment of two ATP:

a One ATP transfers a phosphate group to glucose, jump-starting the reactions.

b Another ATP transfers a phosphate group to an intermediate, causing it to split into two three-carbon compounds: PGAL and DHAP (dihydroxyacetone phosphate). Both have the same atoms, arranged differently, and they are interconvertible. But only PGAL can continue on in glycolysis. DHAP gets converted, so two PGAL are available for the next reaction.

c Two NADH form when each PGAL gives up two electrons and a hydrogen atom to NAD⁺.

d Two intermediates each transfer a phosphate group to ADP. *Thus, two ATP have formed by direct phosphate group transfers.* The original energy investment of two ATP is now paid off.

e Two more intermediates form. Each gives up one hydrogen atom and an —OH group. These combine as water. Two molecules called PEP form by these reactions.

f Each PEP transfers a phosphate group to ADP. *Once again, two ATP have formed by substrate-level phosphorylation.*

In sum, glycolysis has a net energy yield of two ATP for each glucose molecule. Two NADH also form during the reactions, and two molecules of pyruvate are the end products.

8.3 Second Stage of Aerobic Respiration

LINKS TO
SECTIONS
1.1, 2.4

The two pyruvate molecules that form during glycolysis may be completely dismantled in a mitochondrion. Many coenzymes pick up the released electrons and hydrogens.

ACETYL–CoA FORMATION

Start with Figure 8.5, which shows the structure of a typical mitochondrion. Figure 8.6a zooms in on part of the interior where the second-stage reactions occur. At the start of these reactions, enzyme action strips one carbon atom from each pyruvate and attaches it to oxygen, forming CO_2. Each two-carbon fragment combines with a coenzyme (designated A) and forms acetyl–CoA, a type of cofactor that can get the Krebs cycle going. Two NAD^+ are reduced during the initial breakdown of two pyruvate molecules (Figure 8.7a,b).

THE KREBS CYCLE

The two acetyl–CoA molecules enter the Krebs cycle separately. Each transfers its two-carbon acetyl group to four-carbon oxaloacetate, which forms citrate, the ionized form of citric acid. The Krebs cycle is known also as the *citric acid cycle*, after its first step.

mitochondrion   mitochondrion

Figure 8.5 Scanning electron micrograph of a mitochondrion, sliced crosswise. Remember, nearly all eukaryotic species, including plants and animals, contain these organelles.

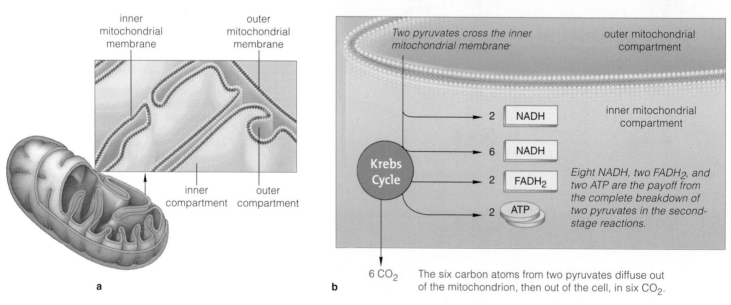

a

b

6 CO_2 The six carbon atoms from two pyruvates diffuse out of the mitochondrion, then out of the cell, in six CO_2.

Figure 8.6 *Animated!* (**a**) Functional zones of a mitochondrion. An inner membrane system divides the interior into an inner and an outer compartment. Aerobic respiration's second and third stages take place at this membrane system. (**b**) Overview of the number of ATP molecules and coenzymes that form in the second stage. Reactions start after the membrane proteins transport the two pyruvate from glycolysis across the outer mitochondrial membrane, then across the inner membrane. Both pyruvates are dismantled inside the inner compartment.

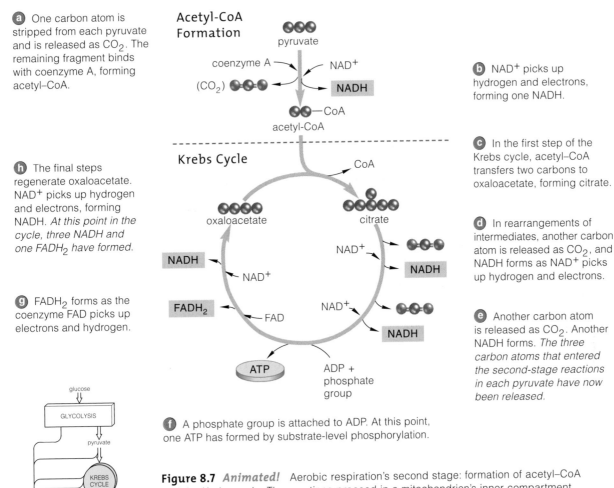

a One carbon atom is stripped from each pyruvate and is released as CO_2. The remaining fragment binds with coenzyme A, forming acetyl–CoA.

h The final steps regenerate oxaloacetate. NAD^+ picks up hydrogen and electrons, forming NADH. *At this point in the cycle, three NADH and one $FADH_2$ have formed.*

g $FADH_2$ forms as the coenzyme FAD picks up electrons and hydrogen.

Acetyl-CoA Formation

pyruvate

coenzyme A — → ← NAD^+

(CO_2) ← **NADH**

CoA

acetyl-CoA

Krebs Cycle

CoA

oxaloacetate

citrate

NAD^+ → **NADH**

NADH ← NAD^+

FADH$_2$ ← FAD

NAD^+ → **NADH**

ATP

ADP + phosphate group

b NAD^+ picks up hydrogen and electrons, forming one NADH.

c In the first step of the Krebs cycle, acetyl–CoA transfers two carbons to oxaloacetate, forming citrate.

d In rearrangements of intermediates, another carbon atom is released as CO_2, and NADH forms as NAD^+ picks up hydrogen and electrons.

e Another carbon atom is released as CO_2. Another NADH forms. *The three carbon atoms that entered the second-stage reactions in each pyruvate have now been released.*

f A phosphate group is attached to ADP. At this point, one ATP has formed by substrate-level phosphorylation.

glucose

GLYCOLYSIS

pyruvate

KREBS CYCLE

ELECTRON TRANSFER PHOSPHORYLATION

Figure 8.7 *Animated!* Aerobic respiration's second stage: formation of acetyl–CoA and the Krebs cycle. The reactions proceed in a mitochondrion's inner compartment. *It takes two turns of the cycle to break down the two pyruvates from glucose.* A total of two ATP, eight NADH, two $FADH_2$, and six CO_2 molecules form. Organisms release the CO_2 from the reactions into their surroundings. For details, see Appendix VII.

The second-stage reactions run twice, one for each pyruvate molecule. The remaining carbon atoms are released in the form of CO_2 (Figure 8.7*d,e*). Only two ATP form during the two turns, which does not add much to the small net yield from glycolysis. However, in addition to the two NAD^+ that were reduced when the acetyl–CoA formed, six more NAD^+ and two FAD molecules are reduced. With their cargo of electrons and hydrogen, these coenzymes—*eight NADH and two $FADH_2$*—are a big potential payoff for the cell. All of the electrons have potential energy, which coenzymes can deliver to the final reaction sites.

As you can see from Figure 8.7, a total of *six* carbon atoms (from two pyruvates) depart during the second stage of aerobic respiration, in six molecules of CO_2. Therefore, the glucose from glycolysis has lost all of its carbons; it has become fully oxidized.

A final note: Figure 8.7 is a simplified version of these second-stage reactions. For interested students, Figure B in Appendix VII offers more details.

Aerobic respiration's second stage starts after two pyruvate molecules from glycolysis move from the cytoplasm, across the outer and inner mitochondrial membranes, and then into the inner mitochondrial compartment.

During these reactions, pyruvate is converted to acetyl–CoA, which starts the Krebs cycle. Two ATP and ten coenzymes (eight NADH, two $FADH_2$) form. All of pyruvate's carbons depart, in the form of carbon dioxide.

Together with two coenzymes (NADH) that formed during glycolysis, the ten coenzymes from the second stage will deliver electrons and hydrogen to the third and final stage.

8.4 Third Stage of Aerobic Respiration—A Big Energy Payoff

LINKS TO
SECTIONS
5.2, 6.4

In the aerobic pathway's third stage, energy release goes into high gear. Coenzymes from the first two stages provide the hydrogen and electrons that drive the formation of many ATP. Electron transfer chains and ATP synthases function as the machinery.

ELECTRON TRANSFER PHOSPHORYLATION

The third stage starts as coenzymes donate electrons to electron transfer chains in the inner mitochondrial membrane (Figure 8.8a). The flow of electrons through the chains drives the attachment of phosphate to ADP molecules. That is what the name of the event, *electron transfer phosphorylation*, means.

Incremental energy release, recall, is more efficient than one burst of energy, nearly all of which would be lost as unusable heat (Section 6.4). As electrons flow through the chains, they transfer energy bit by bit, in tiny amounts, to molecules that briefly store it. The two NADH that formed in the cytoplasm (by glycolysis) cannot directly reach the ATP-forming machinery. They must give up their electrons and hydrogen to transport proteins, which shuttle them across the inner membrane, into the innermost compartment. There, NAD^+ or FAD

pick them up. The eight NADH and two $FADH_2$ from the second stage are already inside.

All of the coenzymes turn over electrons to transfer chains and at the same time give up hydrogen, which now has a positive charge (H^+). Again, electrons lose a bit of energy at each transfer through the chain. At three transfers, that energy drives the pumping of H^+ into the outer compartment (Figure 8.8b). There, many ions accumulate—which sets up concentration and electric gradients across the inner membrane.

H^+ cannot cross a lipid bilayer. Instead, it follows its gradients by flowing through ATP synthases (Section 5.2 and Figure 8.8c). The ion flow causes parts of the ATP synthase molecules to change their shape in a reversible way. The change promotes the attachment of unbound phosphate to ADP, thus forming ATP.

The last components of the electron transfer chains pass electrons to oxygen, which combines with H^+ and thereby forms water. *Oxygen is the final acceptor of electrons originally stripped from glucose.*

In oxygen-starved cells, the electrons have nowhere to go. The whole chain backs up with electrons all the way to NADPH, so no H^+ gradients form, and no ATP forms, either. Without oxygen, cells of complex organisms do not survive long. They cannot produce enough ATP to sustain life processes.

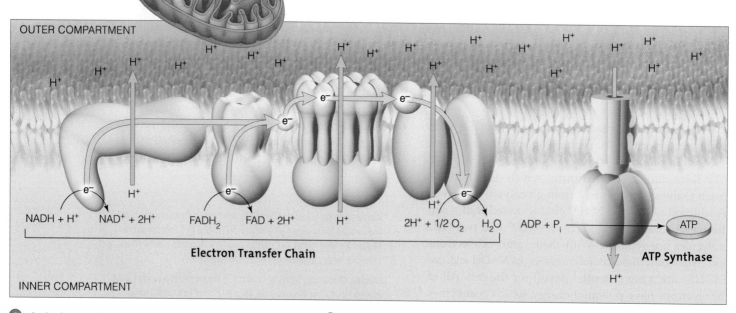

a At the inner mitochondrial membrane, NADH and $FADH_2$ give up electrons to transfer chains. When electrons are transferred through the chains, unbound hydrogen (H^+) is shuttled across the membrane to the outer compartment.

b Free oxygen is the final acceptor of electrons at the end of the transfer chain.

c H^+ concentration and electric gradients now exist across the membrane. H^+ follows the gradients through the interior of ATP synthases, to the inner compartment. The flow drives the formation of ATP from ADP and unbound phosphate (P_i).

Figure 8.8 Electron transfer phosphorylation, the third and final stage of aerobic respiration.

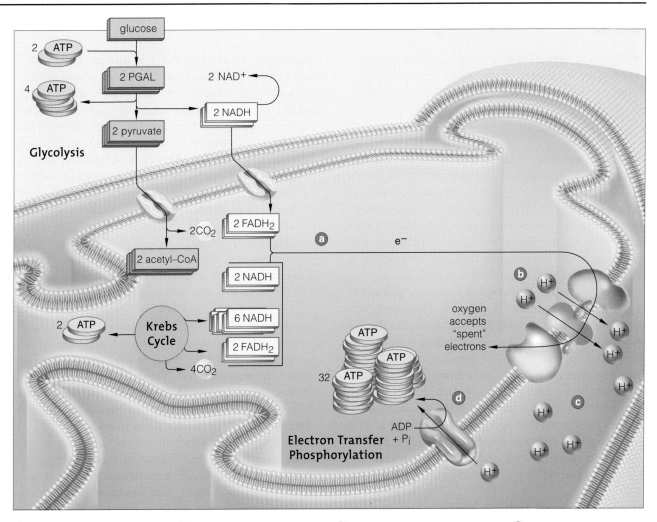

a Electrons and hydrogen from NADH and FADH₂ that formed during the first and second stages enter electron transfer chains.

b As electrons are being transferred through these chains, H⁺ ions are shuttled across the inner membrane, into the outer compartment.

c More H⁺ accumulates in the outer compartment than in the inner one. Chemical and electric gradients have been established across the inner membrane.

d Hydrogen ions follow the gradients through the interior of ATP synthases, driving ATP formation from ADP and phosphate (P$_i$).

Figure 8.9 *Animated!* Summary of the transfers of electrons and hydrogen from coenzymes involved in ATP formation in mitochondria.

SUMMING UP: THE ENERGY HARVEST

Thirty-two ATP typically form in the third stage. Add the four ATP from the earlier stages, and the net yield from a glucose molecule is *thirty-six ATP* (Figure 8.9). By contrast, anaerobic pathways may use up eighteen glucose molecules to get the same net yield.

The yield varies. First, reactant, intermediate, and product concentrations can change it. Second, the two NADH from glycolysis cannot enter a mitochondrion. They give up electrons and hydrogen to transport proteins in the outer mitochondrial membrane, which shuttle them across. NAD⁺ or FAD that are already inside accept them, forming NADH or FADH₂.

When NADH inside delivers electrons to a certain entry point into a transfer chain, enough H⁺ gets pumped across the membrane to make three ATP.

FADH₂ delivers them to a different entry point. Less H⁺ is pumped across, and only two ATP form.

In liver, heart, and kidney cells, the electrons and hydrogen enter the first entry point, and the energy harvest is thirty-eight ATP. More commonly, as in skeletal muscle and brain cells, they are transferred to FAD, so the harvest is thirty-six ATP.

In aerobic respiration's third stage, electrons from NADH and FADH₂ flow through transfer chains and H⁺ is shuttled across the mitochondrion's inner membrane, into the outer compartment. H⁺ concentration and electric gradients form across the inner membrane. H⁺ flows back outside through ATP synthases, which drives the formation of many ATP.

8.5 Anaerobic Energy-Releasing Pathways

LINK TO
SECTION
7.8

Unlike aerobic respiration, the anaerobic pathways do not use oxygen as the final acceptor of electrons. Their final steps have an important function: they regenerate NAD⁺.

FERMENTATION PATHWAYS

Fermenters are diverse. Many are protists and bacteria in marshes, bogs, mud, deep sea sediments, the animal gut, canned foods, sewage treatment ponds, and other oxygen-free places. When exposed to oxygen, some die. Bacteria that cause botulism are examples. Other fermenters are indifferent to oxygen's presence. Still other kinds use oxygen, but they also can switch to fermentation when oxygen becomes scarce.

Glycolysis is the first stage of fermentation, just as it is in aerobic respiration (Figure 8.4). Here again, two pyruvate, two NADH, and two ATP form. But the last steps do not degrade glucose to carbon dioxide and water. They get no more ATP beyond the small yield from glycolysis. *The final steps in fermentation pathways regenerate the essential coenzyme NAD⁺.*

Fermentation yields enough energy to sustain many single-celled anaerobic species. It helps some aerobic cells when oxygen levels are stressfully low. But it isn't enough to sustain large, multicelled organisms, this being why you will never see anaerobic elephants.

Alcoholic Fermentation   In **alcoholic fermentation**, the three-carbon backbone of two pyruvate molecules from glycolysis is split. The reactions result in two molecules of acetaldehyde (an intermediate with a two-carbon backbone), and two of carbon dioxide. Acetaldehyde accepts electrons and hydrogen from NADH to form ethyl alcohol, or ethanol (Figure 8.10).

Some yeasts are famous fermenters. Bakers mix *Saccharomyces cerevisiae* cells and sugar into dough. The bubbles of carbon dioxide that form expand the dough (make it rise). Oven heat forces bubbles out of the dough, and the alcohol product evaporates away.

Wild and cultivated strains of *Saccharomyces* help produce alcohol in wine. Crushed grapes are left in vats along with the yeast, which converts sugar in the juice to ethanol. Ethanol is toxic to microbes. When a fermenting brew's ethanol content nears 10 percent, yeast cells start to die, and fermentation ends.

Lactate Fermentation   With **lactate fermentation**, the NADH gives up electrons and hydrogen to the pyruvate. The transfer converts pyruvate to lactate, a three-carbon compound (Figure 8.11). You probably have heard of lactic acid, the non-ionized form of this compound. But lactate is by far the most common form in living cells, which is our focus here.

Lactate fermentation by *Lactobacillus* and certain other bacteria can spoil food, yet some species have commercial uses. Huge populations in vats of milk give us cheeses, yogurt, buttermilk, and other dairy products. Fermenters also help in curing meats and in pickling some fruits and vegetables, such as sauerkraut. Lactate is an acid; it gives these foods a sour taste.

a

b

c

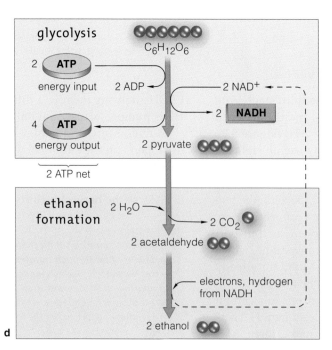

d

Figure 8.10 *Animated!* Steps of alcoholic fermentation. (**a**) Yeasts, which are single-celled fungi, use this anaerobic pathway to make ATP.

(**b**) A vintner examining the color and clarity of one fermentation product of *Saccharomyces*. Strains of this yeast live on the sugar-rich tissues of ripe grapes.

(**c**) Carbon dioxide released from cells of *S. cerevisiae* is making bread dough rise in this bakery.

(**d**) Alcoholic fermentation. The intermediate acetaldehyde functions as the final electron acceptor. The end product of the reactions is ethanol (ethyl alcohol).

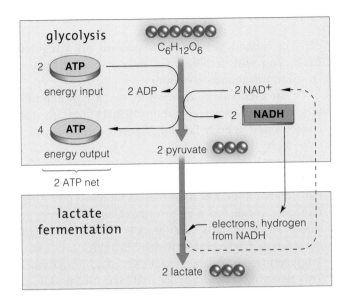

Figure 8.11 *Animated!* Steps of lactate fermentation. In this anaerobic pathway, the product (lactate) is the final acceptor of electrons originally stripped from glucose. These fermentation reactions have a net energy yield of two ATP (from glycolysis).

Figure 8.12 Sprinters, calling upon lactate fermentation in their muscles. The micrograph, a cross-section through part of a muscle, reveals three types of fibers. The lighter fibers sustain short, intense bursts of speed; they make ATP by lactate fermentation. The darker fibers contribute to endurance; they make ATP by aerobic respiration.

Lactate fermentation as well as aerobic respiration yields ATP for muscles that are partnered with bones. These skeletal muscles contain a mixture of cell types. Cells fused together inside *slow-twitch* muscle fibers support light, steady, prolonged activity, as during marathon runs or bird migrations. Cells of slow-twitch muscle fibers make ATP only by aerobic respiration, and they have many mitochondria. They are dark red because they hold large amounts of myoglobin. This pigment, which is related to hemoglobin, binds and stores oxygen for aerobic respiration (Figure 8.12).

By contrast, cells of pale *fast-twitch* muscle fibers have few mitochondria and no myoglobin. They use lactate fermentation to make ATP. They function when energy needs are immediate and intense, as during weight lifting or sprints (Figure 8.12). The pathway produces ATP quickly but not for very long; it does not support sustained activity. That is one reason you do not see chickens migrating. The flight muscles in a chicken contain mostly fast-twitch fibers, which make up the "white" breast meat.

Short bursts of flight evolved in the ancestors of chickens, perhaps as a way to flee from predators or improve agility during territorial battles. Chickens do walk and run; hence the "dark meat" (slow-twitch muscle) in their thighs and legs. Would you expect to see light or dark breast muscles in a migratory duck or in an albatross that skims ocean waves for months?

Section 37.5 offers more information on alternative energy pathways in skeletal and cardiac muscle cells.

ANAEROBIC ELECTRON TRANSFERS

Especially among prokaryotes, we see less common but more diverse energy-releasing pathways, some of which are topics of later chapters. Many assist in the global cycling of sulfur, nitrogen, and other elements. Collectively, the practitioners of these pathways affect nutrient availability in ecosystems everywhere.

Some bacteria and archaeans engage in **anaerobic electron transfers**. Electrons from organic compounds flow through transfer chains in the plasma membrane and H^+ flows out of the cell through ATP synthases. Inorganic compounds are often used as final electron acceptors. The net energy yield is variable but small.

Some anaerobic species in waterlogged soil give up electrons to sulfate, forming a putrid gas (hydrogen sulfide). Other anaerobes live in the nutrient-rich mud of some aquatic habitats. Still others are the basis of food webs at hydrothermal vents (Sections 7.8, 48.15).

In alcoholic fermentation, the final acceptor of electrons from glucose is acetaldehyde, a reaction intermediate. In lactate fermentation, it is pyruvate.

Both pathways have a net energy yield of two ATP, which forms in glycolysis. The remaining reactions regenerate the coenzyme NAD^+, without which glycolysis would stop.

Some bacteria and archaeans generate ATP by anaerobic electron transfers across the plasma membrane.

8.6 Alternative Energy Sources in the Body

LINKS TO
SECTIONS
3.3, 3.4

So far, you have looked at what happens after glucose molecules enter an energy-releasing pathway. Now start thinking about what cells can do when they have too much or too little glucose.

THE FATE OF GLUCOSE AT MEALTIME AND BETWEEN MEALS

What happens to glucose at mealtime? While you and all other mammals are eating, glucose and other small organic molecules are being absorbed across the gut lining, and your blood is transporting them through the body. The rising glucose concentration in blood prompts an organ, the pancreas, to secrete insulin. This hormone makes cells take up glucose faster.

Cells trap the incoming glucose by converting it to glucose–6–phosphate. This intermediate of glycolysis forms as ATP transfers a phosphate group to glucose (Figures 8.4 and 8.13). Phosphorylated glucose cannot be transported out of the cell.

When a cell takes in more glucose than it requires for energy, its ATP-forming machinery goes into high gear. Unless it is using ATP rapidly, the cytoplasmic concentration of ATP rises, and glucose–6–phosphate is diverted into a biosynthesis pathway. The glucose gets converted to glycogen, a storage polysaccharide in animal cells (Section 3.3). Liver cells and muscle cells especially favor the conversion. Together, these two types of cells maintain the body's largest stores of glycogen molecules.

Between meals, the blood level of glucose declines. If the decline were not countered, that would be bad news for the brain, your body's glucose hog. At any time, your brain is taking up more than two-thirds of the freely circulating glucose. Why? The brain's many hundreds of millions of nerve cells (neurons) use this sugar as their preferred energy source.

The pancreas responds to low glucose levels by secreting glucagon. This hormone causes liver cells to convert stored glycogen to glucose and send it back to the blood. Only liver cells do this; muscle cells will not give it up. The glucose level in blood rises, so brain cells keep on functioning. Thus, *hormones control whether your cells use free glucose as an energy source or tuck it away.*

Don't let this explanation lead you to believe that your cells store enormous amounts of glycogen. Glycogen makes up only 1 percent or so of the total energy reserves of the average adult's body—the energy equivalent of two cups of cooked pasta. Unless you eat on a regular basis, you will deplete your liver's small glycogen stores in less than twelve hours.

Of the total energy reserves in, say, a typical adult who eats well, 78 percent (about 10,000 kilocalories) is concentrated in body fat and 21 percent in proteins.

ENERGY FROM FATS

How does a human body access its fat reservoir? A fat molecule, recall, has a glycerol head and one, two, or three fatty acid tails (Section 3.4). The body stores most fats as triglycerides, which have three tails each. Triglycerides accumulate in fat cells of adipose tissue. This tissue is strategically located beneath the skin of buttocks and other body regions.

When the blood glucose level falls, triglycerides are tapped as an energy alternative. Enzymes in fat cells cleave bonds between glycerol and fatty acids, which both enter the blood. Enzymes in the liver convert the glycerol to PGAL. As you know, PGAL is one of the key intermediates in glycolysis (Figure 8.4). Nearly all cells of your body take up circulating fatty acids, and enzymes inside them cleave the fatty acid backbones. The fragments become converted to acetyl–CoA, which can enter the Krebs cycle.

Compared with glucose, a fatty acid tail has more carbon-bound hydrogen atoms, so it yields more ATP. Between meals or during steady, prolonged exercise, fatty acid conversions supply about half of the ATP that muscle, liver, and kidney cells require.

What happens if you eat too many carbohydrates? Aerobic respiration converts the glucose subunits to pyruvate, then to acetyl–CoA, which enters the Krebs cycle. When too much glucose is circulating through the body, acetyl–CoA is diverted to a pathway that synthesizes fatty acids. *Too much glucose ends up as fat.*

ENERGY FROM PROTEINS

Some enzymes in your digestive system split dietary proteins into their amino acid subunits, which are then absorbed into the bloodstream. Cells use amino acids to make proteins or other nitrogen-containing compounds. Even so, when you eat more protein than your body needs, amino acids are further degraded. Their $-NH_3^+$ group is pulled off, so ammonia (NH_3) forms. Depending on the types of amino acids, the leftover carbon backbones are split, and acetyl–CoA, pyruvate, or an intermediate of the Krebs cycle forms. Your cells can divert any of these organic compounds into the Krebs cycle (Figure 8.13).

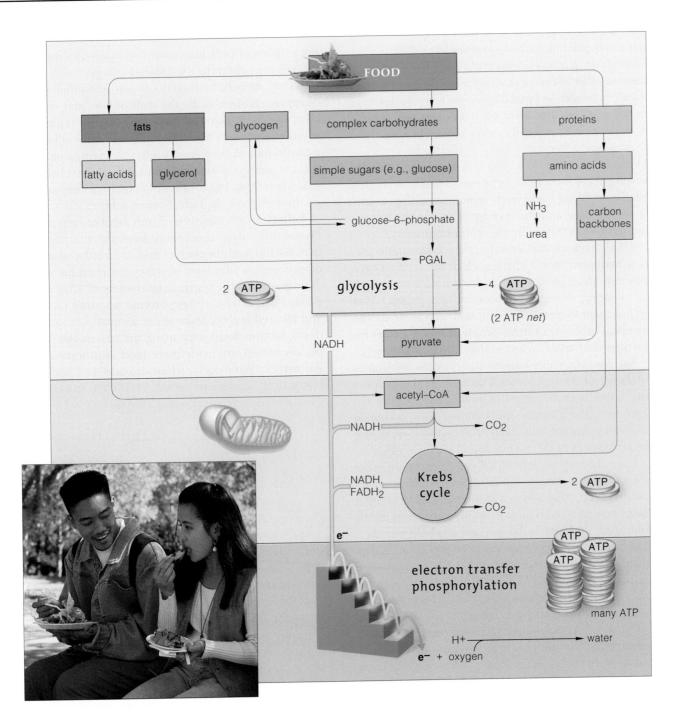

FOOD

| fats | glycogen | complex carbohydrates | proteins |

fats → fatty acids
fats → glycerol

complex carbohydrates → simple sugars (e.g., glucose)

proteins → amino acids

amino acids → NH_3 → urea

amino acids → carbon backbones

glucose–6–phosphate

PGAL

2 ATP →

glycolysis

→ 4 ATP

(2 ATP *net*)

NADH

pyruvate

acetyl–CoA

NADH → CO_2

NADH, $FADH_2$

Krebs cycle

→ 2 ATP

CO_2

e^-

electron transfer phosphorylation

ATP ATP ATP

many ATP

H+ → water

e^- + oxygen

As you can see, maintaining and accessing energy reserves is complicated business. Controlling the use of glucose is special because it is the fuel of choice for the brain. However, providing all of your cells with energy starts with the kinds of food you eat.

In humans and other mammals, the entrance of glucose or other organic compounds into an energy-releasing pathway depends on the kinds and proportions of carbohydrates, fats, and proteins in the diet.

Figure 8.13 *Animated!* Reaction sites where different organic compounds can enter the stages of aerobic respiration. The compounds shown are alternative energy sources in humans and other mammals. Notice how complex carbohydrates, fats, and proteins cannot enter the aerobic pathway directly. First the digestive system, then individual cells, must break apart these molecules to simpler compounds.

8.7 Reflections on Life's Unity

LINKS TO
SECTIONS
1.1, 2.4, 6.1, 7.8

In this unit, you traveled through many levels of life's organization. You have a sense that all organisms tap into a one-way flow of energy, that they alone make certain organic molecules, and that life emerges when molecules become organized and interact as units called cells. In short, you have a sense of life's molecular unity.

At this point in the book, you still may have difficulty sensing the connections between yourself—a highly intelligent being—and such remote-sounding events as energy flow and the cycling of carbon, hydrogen, and oxygen. Is this really the stuff of humanity?

Think back on the structure of a water molecule. Two hydrogen atoms sharing electrons with oxygen may not seem close to your daily life. Yet, through that sharing, water molecules show a polarity that invites them to hydrogen-bond with one another. The chemical behavior of three simple atoms is a start for the organization of lifeless matter into living things.

For now you can visualize other diverse molecules interspersed through water. The nonpolar kinds resist interaction with water; polar kinds dissolve in it. On their own, the phospholipids among them assemble into a two-layered film. Such lipid bilayers, recall, are the framework of cell membranes, hence all cells.

From the very beginning, the cell has been the basic *living* unit. The essence of life is not some mysterious force. It is organization and metabolic control. With a membrane to contain them, metabolic reactions *can* be controlled. With molecular mechanisms built into their membranes, cells can respond to energy changes and to shifts in solute concentrations in the environment. Response mechanisms operate by "telling" proteins —enzymes—when and what to build or tear down.

And it is not some mysterious force that creates proteins. DNA, the double-stranded treasurehouse of inheritance, has the chemical structure—*the chemical message*—that helps molecule reproduce molecule, one generation after the next. In your body, DNA strands

tell trillions of cells how countless molecules must be built or torn apart for their stored energy.

So yes, carbon, hydrogen, oxygen, and other atoms of organic molecules are the stuff of you, and us, and all of life. Yet it takes far more than organic molecules to complete the picture. Life continues only as long as a continuous flow of energy sustains its organization. It takes energy to assemble molecules into cells, cells into organisms, organisms into communities, and so on through the biosphere (Section 1.1).

Reflect on photosynthesis and aerobic respiration. Photosynthesizers use energy from the sun and raw materials to feed themselves and, indirectly, nearly all other forms of life. Long ago they enriched the whole atmosphere with oxygen, a leftover from a noncyclic pathway. That atmosphere exerted selection pressure and favored aerobic respiration, a novel way to break down food molecules by using the free oxygen. And photosynthesizers made more food with leftovers of the aerobic pathway—carbon dioxide and water. In this way the cycling of carbon, hydrogen, and oxygen through living things came full circle.

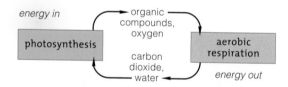

With few exceptions, infusions of energy from the sun sustain life's sweeping organization. And energy, remember, flows through time in one direction—from concentrated to less concentrated forms (Section 6.1). Only as long as more energy flows into the great web of life can life continue in all its rich expressions.

So life is no more *and no less* than a marvelously complex system for prolonging order. Sustained with energy transfusions from the sun, life continues by a capacity for self-reproduction. With energy and the hereditary codes of DNA, matter becomes organized, generation after generation. Even with the death of individuals, life elsewhere is prolonged. With each death, molecules are released and may be cycled as raw materials for new generations.

With this flow of energy and cycling of materials through time, each birth is affirmation of our ongoing capacity for organization, each death a renewal.

The diversity of life, and its continuity through time, arises from unity at the bioenergetic and molecular levels.

Summary

Section 8.1 All organisms, including photosynthetic types, access the chemical bond energy of glucose and other organic compounds, then use it to make ATP. They do so because ATP is a transportable form of energy that can jump-start nearly all metabolic reactions. They break apart compounds and release electrons and hydrogen, both of which have roles in ATP formation.

Glycolysis, the partial breakdown of one glucose to two pyruvate molecules, takes place in the cytoplasm of all cells. It is the first stage of all the main energy-releasing pathways, and it does not use free oxygen.

Anaerobic pathways end in the cytoplasm, and the net yield of ATP is small. An oxygen-requiring pathway called aerobic respiration continues in mitochondria. It releases far more usable energy from glucose.

Biology Now
Get an overview of aerobic respiration with the animation on BiologyNow.

Section 8.2 It takes two ATP molecules to jump-start glycolysis. One activates six-carbon glucose; the other primes an intermediate to split into two three-carbon molecules (PGAL). The two PGAL give up electrons and hydrogen to coenzymes (NAD$^+$), forming two NADH. Two intermediates each give up a phosphate group to ADP, forming two ATP. Two more intermediates do the same.

Thus, during glycolysis, four ATP form by substrate-level phosphorylation, but the *net* yield is two ATP (because two ATP had to be invested up front). The end products of glycolysis are two molecules of pyruvate, each with a three-carbon backbone.

Biology Now
Take a step-by-step journey through glycolysis with the animation on BiologyNow.

Section 8.3 The second and third stages of aerobic respiration get under way when the two pyruvates from glycolysis enter a mitochondrion.

The second stage consists of the Krebs cycle and a few preparatory steps before it. It takes two full turns of these cyclic reactions to break down both of the pyruvates.

Before the cycle, each pyruvate is stripped of a carbon atom (which departs in CO_2) as well as electrons and hydrogen (which NAD$^+$ picks up, forming NADH). A coenzyme (acetyl–CoA) picks up the two-carbon leftover.

The Krebs cycle starts when acetyl–CoA transfers the leftover to oxaloacetate, forming citrate. During stepwise rearrangements, intermediates give up two carbon atoms (which depart in CO_2) and electrons and hydrogen (which three NAD$^+$ and one FAD accept). An ATP forms.

In total, then, the second stage of aerobic respiration —including preparatory steps and two full turns of the Krebs cycle—results in the formation of six CO_2, two ATP, eight NADH, and two FADH$_2$.

Biology Now
Explore a mitochondrion and observe the reactions inside with the animation on BiologyNow.

Section 8.4 The third stage of aerobic respiration takes place at electron transfer chains and ATP synthases in the inner mitochondrial membrane.

The electron transfer chains accept electrons and hydrogen from NADH and FADH$_2$ that formed during the first two stages of the aerobic pathway. Electron flow through the chains causes H$^+$ to accumulate in the inner mitochondrial compartment, so H$^+$ concentration and electric gradients build up across the inner membrane.

H$^+$ follows its gradients and flows back to the outer mitochondrial compartment, through the interior of ATP synthases. This ion flow causes reversible changes in the shape of parts of the ATP synthase. Those changes force ADP to combine with unbound phosphate, thus forming ATP. This happens repeatedly, so many ATP molecules are produced.

Free oxygen picks up the electrons at the end of the transfer chains and combines with H$^+$, forming water.

Aerobic respiration has a typical net energy yield of thirty-six ATP for each glucose molecule metabolized.

Biology Now
Study how each step in aerobic respiration contributes to energy harvests with the animation on BiologyNow.

Section 8.5 Anaerobic energy-releasing pathways do not use free oxygen and occur only in cytoplasm. The net energy yield is small.

Fermentation pathways follow glycolysis and add no more to the two ATP net yield from glycolysis. They function only to regenerate NAD$^+$.

In alcoholic fermentation, the two pyruvates from glycolysis are converted to two acetaldehyde and two CO_2 molecules. When NADH transfers electrons and hydrogen to acetaldehyde, two ethanol (ethyl alcohol) molecules form and NAD$^+$ is regenerated.

In lactate fermentation, NAD$^+$ is regenerated when NADH gives up electrons and hydrogen to the two pyruvates. Two lactate molecules are end products.

Slow-twitch and fast-twitch skeletal muscle fibers support different levels of activity. Aerobic respiration and lactate fermentation occur in different fibers that make up these muscles.

Many prokaryotes make ATP by anaerobic electron transfers. They have electron transfer chains and ATP synthases in their plasma membrane. Sulfate or some other inorganic substance in the environment is often the final electron acceptor.

Biology Now
Compare alcoholic and lactate fermentation with the animation on BiologyNow.

Section 8.6 In the human body, simple sugars from carbohydrates, glycerol and fatty acids from fats, and carbon backbones of amino acids from proteins can enter the aerobic pathway as alternative energy sources.

Biology Now
Follow the breakdown of different organic molecules with the interaction on BiologyNow.

Section 8.7 Life's diversity and continuity arise from its unity at the bioenergetic and molecular levels.

Self-Quiz

Answers in Appendix II

1. Glycolysis starts and ends in the _____ .
   a. nucleus
   b. mitochondrion
   c. plasma membrane
   d. cytoplasm

2. Which of the following molecules does not form during glycolysis?
   a. NADH   b. pyruvate   c. FADH$_2$ d. ATP

3. Aerobic respiration is completed in the _____ .
   a. nucleus
   b. mitochondrion
   c. plasma membrane
   d. cytoplasm

4. In the third stage of aerobic respiration, _____ is the final acceptor of electrons from glucose.
   a. water   b. hydrogen   c. oxygen   d. NADH

5. Fill in the blanks in the diagram below.

6. In alcoholic fermentation, _____ is the final acceptor of electrons stripped from glucose.
   a. oxygen
   b. pyruvate
   c. acetaldehyde
   d. sulfate

7. Fermentation makes no more ATP beyond the small yield from glycolysis. The remaining reactions _____ .
   a. regenerate FAD
   b. regenerate NAD$^+$
   c. regenerate NADH
   d. regenerate FADH$_2$

8. In certain organisms and under certain conditions, _____ can be used as an energy alternative to glucose.
   a. fatty acids
   b. glycerol
   c. amino acids
   d. all of the above

9. Match the event with its most suitable description.
   ____ glycolysis
   ____ fermentation
   ____ Krebs cycle
   ____ electron transfer phosphorylation

   a. ATP, NADH, FADH$_2$, CO$_2$, and water form
   b. glucose to two pyruvates
   c. NAD$^+$ regenerated, two ATP net
   d. H$^+$ flows through ATP synthases

Additional questions are available on Biology Now™

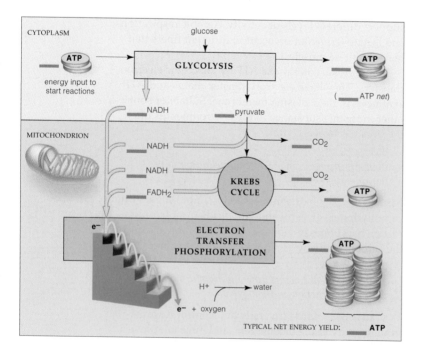

CYTOPLASM

glucose

ATP

GLYCOLYSIS

ATP

energy input to start reactions

(____ ATP *net*)

NADH   pyruvate

MITOCHONDRION

NADH   CO$_2$

NADH   CO$_2$

FADH$_2$ KREBS CYCLE   ATP

e$^-$ ELECTRON TRANSFER PHOSPHORYLATION   ATP

H$^+$ → water

e$^-$ + oxygen

TYPICAL NET ENERGY YIELD: ATP

Critical Thinking

1. Living cells of your body absolutely do not use their nucleic acids as alternative energy sources. Suggest why.

2. Suppose you start a body-building program. You are already eating plenty of carbohydrates. Now a qualified nutritionist recommends that you start a protein-rich diet that includes protein supplements. Speculate on how extra dietary proteins will be put to use, and in which tissues.

3. Each year, Canada geese lift off in precise formation from their northern breeding grounds. They head south to spend the winter months in warmer climates, then make the return trip in spring. As is the case for other migratory birds, their flight muscle cells are efficient at using fatty acids as an energy source. Remember, the carbon backbone of fatty acids can be cleaved into small fragments that can be converted to acetyl–CoA for entry into the Krebs cycle.

   Suppose a lesser Canada goose from Alaska's Point Barrow has been steadily flapping along for about three thousand kilometers and is approaching Klamath Falls, Oregon. It looks down and notices a rabbit sprinting from a coyote with a taste for rabbit.

   With a stunning burst of speed, the rabbit reaches the safety of its burrow.

   Which energy-releasing pathway predominated in muscle cells in the rabbit's legs? Why was the Canada goose relying on a different pathway for most of its journey? And why wouldn't the pathway of choice in goose flight muscle cells be much good for a rabbit making a mad dash from its enemy?

4. At high altitudes, oxygen levels are low. Mountain climbers risk altitude sickness, which is characterized by shortness of breath, weakness, dizziness, and confusion.

   Oddly, early symptoms of *cyanide poisoning* resemble altitude sickness. This highly toxic poison binds tightly to a cytochrome, the last molecule in mitochondrial electron transfer chains. When cyanide becomes bound to it, the cytochrome can't transfer electrons to the next component of the chain. Explain why cytochrome shutdown might cause the same symptoms as altitude sickness.

5. ATP forms in mitochondria. In warm-blooded animals, so does a lot of heat, which is circulated in ways that help control body temperature. Cells of *brown adipose tissue* make a protein that disrupts the formation of electron transfer chains in mitochondrial membranes. H$^+$ gradients are affected, so fewer ATP form; electrons in the transfer chains give up more of their energy as heat. Because of this, some researchers hypothesize that brown adipose tissue may not be like white adipose tissue, which is an energy (fat) reservoir. Brown adipose tissue may function in thermogenesis, or heat production.

   Mitochondria, recall, contain their own DNA, which may have mutated independently in human populations that evolved in the Arctic and in the hot tropics. If that is so, then mitochondrial function may be adapted to climate.

   How do you suppose such a mitochondrial adaptation might affect people living where the temperature range no longer correlates with their ancestral heritage? Would you expect people whose ancestors evolved in the Arctic to be more or less likely to put on a lot of weight than those whose ancestors lived in the tropics? See *Science*, January 9, 2004: 223–226 for more information.

II Principles of Inheritance

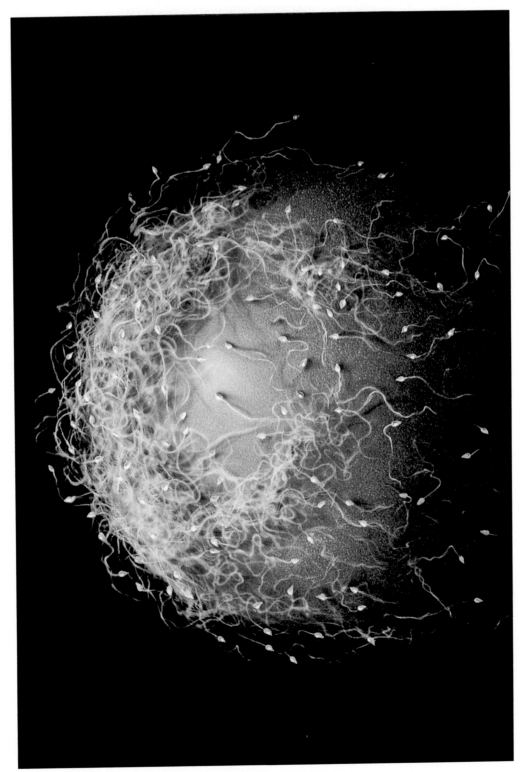

Human sperm, one of which will penetrate this mature egg and so set the stage for the development of a new individual in the image of its parents. This exquisite art is based on a scanning electron micrograph.

Henrietta's Immortal Cells

Each human starts out as a fertilized egg. By the time of birth, the body consists of about a trillion cells, all descended from that single cell. Even in an adult, billions of cells still divide every day and replace their damaged or worn-out predecessors.

In 1951, George and Margaret Gey of Johns Hopkins University were trying to develop a way to keep human cells dividing outside the body. An "immortal" cell lineage could help researchers study basic life processes as well as cancer and other diseases. Using cells to study cancer would be a better alternative than experimenting with patients and risking their already vulnerable lives.

For almost thirty years, the Geys tried to grow normal and diseased human cells. But they could not stop the cellular descendants from dying within a few weeks.

Mary Kubicek, a lab assistant, tried again and again to establish a self-perpetuating lineage of cultured human cancer cells. She was about to give up, but she prepared one last sample. She named them *HeLa* cells, a code for the first two letters of the patient's first and last names.

The HeLa cells began to divide. They divided again and again. Four days later, the researchers had to subdivide the cells into more culture tubes. The cell populations increased at a phenomenal rate; cells were dividing every twenty-four hours and coating the inside of the tubes within days.

Sadly, cancer cells in the patient were dividing just as frequently. Six months after she had been diagnosed with cancer, malignant cells had invaded tissues throughout her body. Two months after that, Henrietta Lacks, a young woman from Baltimore, was dead.

Although Henrietta passed away, her cells lived on in the Geys' laboratory (Figure 9.1). In time, HeLa cells were shipped to research laboratories all over the world. The Geys used HeLa cells to identify viral strains that cause polio, which at the time was epidemic. They developed the tissue culture techniques that were used to grow a vaccine. Other researchers used HeLa cells to investigate cancer, viral growth, protein synthesis, the effects of radiation on cells, and more. Some HeLa cells even traveled into space for experiments on the *Discoverer XVII* satellite. Even now, hundreds of important research projects move forward annually, thanks to Henrietta's immortal cells.

Figure 9.2 shows a photograph of Henrietta. She was only thirty-one when the runaway cell divisions killed her. Decades later, her legacy continues to help humans everywhere, through cellular descendants that are still dividing day after day.

Understanding cell division—and, ultimately, how new individuals are put together in the image of their parents—starts with answers to three questions. *First*, what kind of

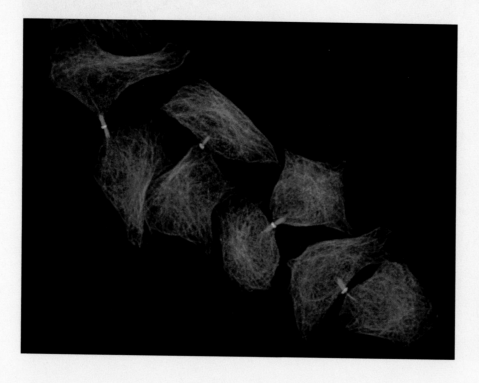

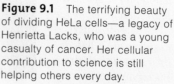

Figure 9.1 The terrifying beauty of dividing HeLa cells—a legacy of Henrietta Lacks, who was a young casualty of cancer. Her cellular contribution to science is still helping others every day.

Figure 9.2 Henrietta Lacks.

information guides inheritance? *Second,* how is information copied in a parent cell before being distributed to each daughter cell? *Third,* what kinds of mechanisms actually parcel out the information to daughter cells?

We will require more than one chapter to survey the nature of cell reproduction and other mechanisms of inheritance. In this chapter, we introduce the structures and mechanisms that cells use to reproduce.

Watch the video online!

 How Would You Vote?

It is illegal to sell your organs, but you can sell your cells, including eggs, sperm, and blood cells. HeLa cells are still being sold all over the world by cell culture firms. Should the family of Henrietta Lacks share in the profits? See BiologyNow for details, then vote online.

 **Key Concepts**

CHROMOSOMES AND DIVIDING CELLS

Individuals of a species have a characteristic number of chromosomes in their cells. Those chromosomes differ in length and shape, and they carry different parts of the cell's hereditary information. Division mechanisms parcel out the information to each daughter cell, along with enough cytoplasm to start up its own operation.   Section 9.1

WHERE MITOSIS FITS IN THE CELL CYCLE

A cell cycle starts when a daughter cell forms and ends when that cell completes its own division. A typical cycle goes through interphase, mitosis, and cytoplasmic division. In interphase, a cell increases its mass and number of components, and copies its DNA.   Section 9.2

STAGES OF MITOSIS

Mitosis divides the nucleus, not the cytoplasm. It has four continuous stages: prophase, metaphase, anaphase, and telophase. A microtubular spindle forms. It moves the cell's duplicated chromosomes into two parcels, which end up in two genetically identical nuclei.   Section 9.3

HOW THE CYTOPLASM DIVIDES

After nuclear division, the cytoplasm divides and typically puts a nucleus in each daughter cell. The cytoplasm of an animal cell is simply pinched in two. In plant cells, a cross-wall forms in the cytoplasm and divides it.   Section 9.4

THE CELL CYCLE AND CANCER

Built-in mechanisms monitor and control the timing and rate of cell division. On rare occasions, the surveillance mechanisms fail, and cell division becomes uncontrollable. Tumor formation and cancer are the outcome.   Section 9.5

Links to Earlier Concepts

Before reading, review the description about the changing appearance of chromosomes in the nucleus of eukaryotic cells (Section 4.6). You may wish to review the introduction to microtubules and the motor proteins associated with them (4.10, 4.11). Doing so will help you understand the nature of the mitotic spindle and the potential value of cancer research that is zeroing in on it. A look back at the walls of plant cells (4.9) will help give you a sense of why they cannot divide by pinching their cytoplasm in two parcels, as animal cells do.

9.1 Overview of Cell Division Mechanisms

LINK TO SECTION 4.5

*The continuity of life depends on **reproduction**. By this process, parents produce a new generation of cells or multicelled individuals like themselves. Cell division is the bridge between generations.*

A dividing cell faces a challenge. Each of its daughter cells must get information encoded in the parental DNA and enough cytoplasm to start up its own operation. DNA "tells" it which proteins to build. Some of the proteins are structural materials; others are enzymes that speed construction of organic compounds. If the cell does not inherit all of the required information, it will not be able to grow or function properly.

In addition, the parent cell's cytoplasm already has enzymes, organelles, and other metabolic machinery. When a daughter cell inherits what looks like a blob of cytoplasm, it actually is getting start-up machinery that will keep it running until it can use information in DNA for growing on its own.

MITOSIS, MEIOSIS, AND THE PROKARYOTES

Eukaryotic cells cannot simply split in two, because their DNA is housed in a single nucleus. They do split the cytoplasm into daughter cells, but not until *after* DNA has been copied and packaged into more than a single nucleus by way of mitosis or meiosis.

Mitosis is a nuclear division mechanism that occurs in *somatic* cells (body cells) of multicelled eukaryotes. It is the basis of increases in body size during growth, replacements of worn-out or dead cells, and tissue

repair. Many plants, animals, fungi, and single-celled protists also reproduce asexually, or make copies of themselves, by way of mitosis (Table 9.1).

Meiosis is a different nuclear division mechanism. It precedes the formation of gametes or spores, and it is the basis of sexual reproduction. The type of gametes known as sperm and eggs develop from *germ* cells, or immature reproductive cells. Spores form in the life cycle of many protists as well as plants and fungi.

As you will discover in this chapter and the next, meiosis and mitosis have much in common. Even so, their outcomes differ.

What about prokaryotes—bacteria and archaeans? All of these cells reproduce asexually by prokaryotic fission, an entirely different mechanism. We consider prokaryotic fission later, in Section 21.2.

KEY POINTS ABOUT CHROMOSOME STRUCTURE

In Section 4.5, you read that a eukaryotic chromosome is one double-stranded DNA molecule with a lot of proteins attached to it. Each eukaryotic species has a characteristic number of chromosomes inside its cells, and before a cell enters nuclear division, it duplicates every one of them. Each chromosome and its copy stay attached to each other as **sister chromatids** until late in the division process. Figure 9.3 is a simple way to think about what an unduplicated chromosome and a duplicated chromosome look like.

During an early stage of mitosis or meiosis, each duplicated chromosome coils back on itself again and

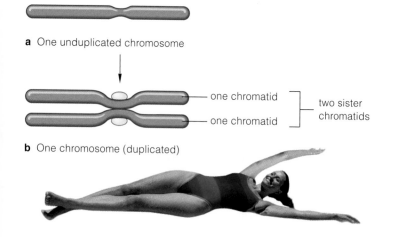

a One unduplicated chromosome

one chromatid ⎤
one chromatid ⎦ two sister chromatids

b One chromosome (duplicated)

Figure 9.3 A simple way to visualize a eukaryotic chromosome in the unduplicated state and duplicated state. Eukaryotic cells are duplicated before mitosis or meiosis. Each becomes two sister chromatids. Students sometimes have trouble visualizing which is which. Think of a chromatid as one arm and leg of a sunbather.

Table 9.1	Comparison of Cell Division Mechanisms
Mechanisms	Functions
Mitosis, cytoplasmic division	In *all* multicelled eukaryotes, the basis of the following three processes: 1. Increases in body size during growth 2. Replacement of dead or worn-out cells 3. Repair of damaged tissues In single-celled and many multicelled species, *also* the basis of asexual reproduction
Meiosis, cytoplasmic division	In single-celled and multicelled eukaryotes, the basis of sexual reproduction; precedes gamete formation or spore formation (Chapter 10)
Prokaryotic fission	In bacteria and archaeans, the basis of asexual reproduction (Section 21.2)

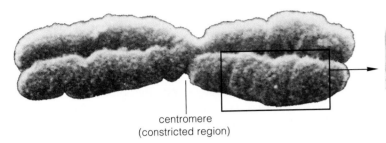

centromere
(constricted region)

a A duplicated chromosome in its most condensed form.

again, into a highly condensed form. Figure 9.4*a* gives an example of a duplicated human chromosome when it is most condensed. The orderly coiling arises from interactions between the DNA molecule and proteins associated with it. At regular intervals, the double-stranded DNA winds twice around tiny "spools" of proteins called **histones**. The repeating histone–DNA spools look like beads on a string when microscopists loosen them up (Figure 9.4*d*). Each of the "beads" is a **nucleosome**, which is the smallest unit of structural organization in eukaryotic chromosomes (Figure 9.4*e*).

While each duplicated chromosome is condensing, it becomes constricted at a predictable location along its length in a pronounced way. This constriction is a **centromere** (Figure 9.4*a*). The centromere's location is different for each type of chromosome and is one of its defining characteristics. During nuclear division, we find a kinetochore at the centromere of each chromatid. A kinetochore is a docking site for microtubules that will help move the chromatids to prescribed locations during nuclear division.

What is the point of this structural organization? The tight packaging probably keeps the chromosomes from getting tangled up while they are moved and sorted out into parcels *during* nuclear division. Also, *between* cell divisions, nucleosome packaging can be loosened in specific regions. In many cases, enzymes thereby gain access to the precise bits of hereditary information that a cell requires at that time.

When a cell divides, each daughter cell receives a required number of chromosomes and some cytoplasm. Eukaryotic cells divide their nucleus first, then the cytoplasm.

In eukaryotes, a nuclear division mechanism called mitosis is the basis of bodily growth, cell replacements, tissue repair, and often asexual reproduction.

Also in eukaryotes, a nuclear division mechanism called meiosis precedes the formation of gametes and, in many species, spores. Meiosis is the basis of sexual reproduction.

multiple levels of coiling
of DNA and proteins

b When a chromosome is most condensed, the proteins associated with it interact in ways that package loops of DNA, which are already coiled, into higher order levels of coiling.

fiber

c At a deeper level of structural organization, the chromosomal proteins and DNA are organized as a cylindrical fiber.

beads on
a string

d Immerse a chromosome in saltwater and it loosens to a beads-on-a-string organization. What appears to be a "string" is one DNA molecule. Each "bead" is a nucleosome.

DNA
double
helix

core of
histones

nucleosome

e A nucleosome consists of part of a DNA molecule looped twice around a core of histone proteins.

Figure 9.4 *Animated!* (**a**) Scanning electron micrograph of a duplicated human chromosome in its most condensed form. (**b,c**) Proteins package many loops of coiled DNA into the coiled array of a cylindrical fiber. (**d,e**) The smallest unit of structural organization is the nucleosome: part of a DNA molecule looped twice around a core of histone molecules. The transmission electron micrographs correspond to organizational levels (**c**) and (**d**).

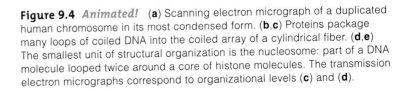

9.2 Introducing the Cell Cycle

Start by thinking of reproduction in terms of a cell's life, from the time it forms until it divides. This interval is not the same as a life cycle—the sequence of stages through which individuals of a species pass during their lifetime.

A **cell cycle** is a series of events from one cell division to the next (Figure 9.5). It starts when a new daughter cell forms by mitosis and cytoplasmic division. It ends when that cell divides. Mitosis, cytoplasmic division, and interphase constitute one turn of the cycle.

THE WONDER OF INTERPHASE

During **interphase**, the cell increases in mass, roughly doubles the number of components in its cytoplasm, and duplicates its DNA. For most cells, interphase is the longest portion of the cell cycle. Biologists divide it into three stages:

G1 Interval ("Gap") of cell growth and functioning before the onset of DNA replication

S Time of "Synthesis" (DNA replication)

G2 Second interval (Gap), after DNA replication when the cell prepares for division

G1, S, and G2 are code names for some events that are just amazing, considering how much DNA is packed in a nucleus. Remember, if you could stretch out all the DNA molecules from one of your somatic cells in a single line, they would extend past the fingertips of both outstretched arms. A line of all the DNA from one salamander cell would stretch about 540 feet.

The wonder is, enzymes and other proteins in cells *selectively* access, activate, and silence information in all that DNA. They also make base-by-base copies of every DNA molecule before cells divide. Most of this cellular work is completed in interphase.

G1, S, and G2 of interphase have distinct patterns of biosynthesis. Most of your cells remain in G1 while they are making proteins, carbohydrates, and lipids. Cells destined to divide enter S, when they copy their DNA as well as the proteins attached to it. During G2, they make proteins that will drive mitosis.

The length of the cycle is about the same for cells of the same type. However, it can differ from one cell type to the next. All of the neurons (nerve cells) inside your brain are stuck in G1 of interphase and normally will not divide again. Yet 2 million to 3 million cells in your red bone marrow are dividing every second. They give rise to new red blood cells, which will last only a few months before they wear out. As another example, when a new sea urchin is developing, the number of cells doubles every two hours.

Once S begins, DNA replication usually proceeds at a predictable rate and ends before a cell prepares to divide. The rate holds for all cells of a species, so you may well wonder if the cell cycle has built-in molecular brakes. It does. Apply the brakes that are supposed to work in G1, and the cycle stalls in G1. Lift the brakes, and the cell cycle runs to completion. Said another way, *control mechanisms govern the rate of cell division.*

Imagine a car losing its brakes just as it starts down a steep mountain road. As you will read later in the chapter, cancer begins this way. The crucial controls over the cell cycle are lost.

One more point: Stressful conditions often interrupt the cell cycle. For instance, when deprived of a vital nutrient, the free-living cells known as amoebas remain in interphase. They will remain there as long as they have not moved past a certain checkpoint. However, past that point, the cycle normally continues regardless of outside conditions, because built-in control mechanisms have lifted the brakes.

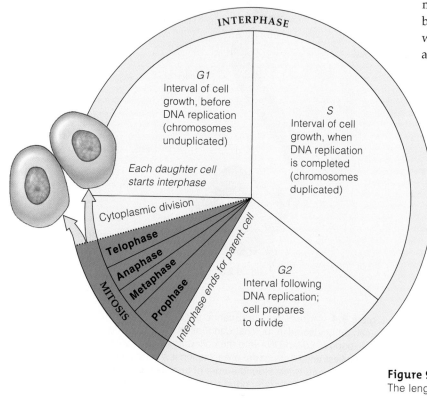

INTERPHASE

G1
Interval of cell growth, before DNA replication (chromosomes unduplicated)

S
Interval of cell growth, when DNA replication is completed (chromosomes duplicated)

Each daughter cell starts interphase

Cytoplasmic division

Telophase
Anaphase
Metaphase
Prophase
MITOSIS

Interphase ends for parent cell

G2
Interval following DNA replication; cell prepares to divide

Figure 9.5 *Animated!* Eukaryotic cell cycle, generalized. The length of each interval differs among different cell types.

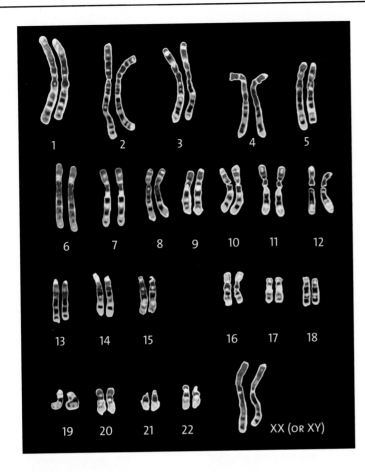

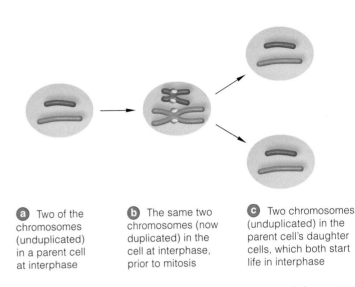

(a) Two of the chromosomes (unduplicated) in a parent cell at interphase

(b) The same two chromosomes (now duplicated) in the cell at interphase, prior to mitosis

(c) Two chromosomes (unduplicated) in the parent cell's daughter cells, which both start life in interphase

Figure 9.6 Preview of how mitosis maintains a parental chromosome number, one generation to the next. Human diploid cells have twenty-three pairs of metaphase chromosomes, for a total of forty-six (*left*). (The last ones in the lineup are a pair of sex chromosomes. In human females, they are two X chromosomes; in males, they are XY.)

The sketches above track what happens to just two of the forty-six. When all goes well, each time a human somatic cell undergoes mitosis and cytoplasmic division, daughter cells end up with an unduplicated set of twenty-three pairs of chromosomes. The icon below shows the bipolar mitotic spindle that helps bring about this outcome.

MITOSIS AND THE CHROMOSOME NUMBER

Mitosis follows G2, and it maintains the parent cell's chromosome number. The **chromosome number** is the sum of all chromosomes in cells of a given type. Body cells of gorillas and chimpanzees have 48, pea plants have 14, and humans have 46 (Figure 9.6).

Actually, your cells have a **diploid number** (*2n*) of chromosomes; there are two of each type. Those 46 are like volumes of two sets of books numbered from 1 to 23. You have two volumes of, say, chromosome 22—*a pair of them.* Except for one sex chromosome pairing (XY), both have the same length and shape, and carry the same hereditary information about the same traits.

Think of them as two sets of books on how to build a house. Your father gave you one set. Your mother had her own ideas about wiring, plumbing, and so on. She gave you an alternate edition on the same topics, but it says slightly different things about many of them.

With mitosis, a diploid parent cell can produce two diploid daughter cells. This doesn't mean each merely gets forty-six or forty-eight or fourteen chromosomes. If only the total mattered, then one cell might get, say, two pairs of chromosome 22 and no pairs whatsoever of chromosome 9. But neither cell could function like its parent *without two of each type of chromosome.*

Mitosis has four stages—*prophase, metaphase, anaphase,* and *telophase*—which require a **bipolar mitotic spindle**. This dynamic structure consists of microtubules that grow or shrink as tubulin subunits are added to or lost from their ends. Its microtubules extend from both spindle poles. Some overlap midway between the two poles, and others tether the duplicated chromosomes.

The next section will explain how the microtubules extending from one pole connect to one chromatid of each chromosome, and microtubules from the other pole connect to its sister. As you will see, the spindle moves the two chromatids apart, to opposite poles. The result is two complete sets of now-unduplicated chromosomes, one for each forthcoming daughter cell. Figure 9.6 is a preview of how mitosis maintains the parental chromosome number.

pole

pole

microtubule of bipolar spindle

Interphase, mitosis, and cytoplasmic division constitute one turn of the cell cycle. During interphase, a new cell increases its mass, doubles the number of its components, and duplicates its chromosomes. The cycle ends after the cell undergoes mitosis and then divides its cytoplasm.

9.3 A Closer Look at Mitosis

Focus now on a "typical" animal cell to see how mitosis can keep the chromosome number constant, division after division, from one cell generation to the next.

LINKS TO SECTIONS 4.10, 4.11

We know that a cell is in **prophase**, the first stage of mitosis, when its chromosomes become visible in light microscopes as threadlike forms. ("Mitosis" is from the Greek *mitos*, meaning thread.) Each chromosome was duplicated earlier, in interphase; each is two sister chromatids joined at the centromere. Now they twist and fold. By late prophase, they will be condensed in thicker, compact, rod-shaped forms (Figure 9.7a–c).

Also before prophase, two barrel-shaped centrioles and two centrosomes started duplicating themselves next to the nucleus. A centriole, recall, gives rise to a flagellum or cilium (Section 4.11). If you observe this structure, you can bet that flagellated or ciliated cells develop during the organism's life cycle.

In animal cells, each centriole helps organize one **centrosome**, a center where microtubules originate. In prophase, one set of the duplicated centrioles and centrosomes moves to the other side of the nucleus, then microtubules grow out of each centrosome. *These are the microtubules that form the bipolar spindle.*

During the transition from prophase to metaphase, the nuclear envelope breaks up completely into many tiny, flattened vesicles. The microtubules are now free to interact with chromosomes and with one another. Many dock at kinetochores; others keep on growing from centrosomes until they overlap midway between the two spindle poles. Remember the motor proteins associated with microtubules (Section 4.11)? Energy from ATP activates dyneins and kinesins, which then generate the force to assemble the mitotic spindle, and to bind and move the chromosomes.

Again, some microtubules extending from one pole tether one chromatid of each chromosome, and some from the opposite pole tether the sister chromatid. The opposing sets of microtubules engage in a tug-of-war. They add and lose tubulin subunits, so they grow and shrink until they are the same length. At that point, **metaphase**, all duplicated chromosomes are aligned midway between the spindle poles (*meta–*, midway). The alignment is crucial for the next stage of mitosis.

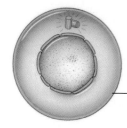

a Cell at Interphase

The cell duplicates its DNA, and prepares for nuclear division.

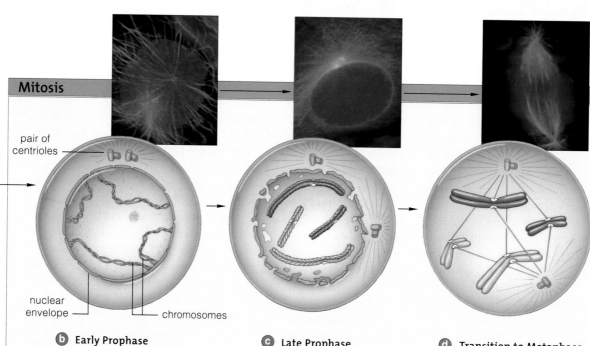

Figure 9.7
Animated! Mitosis. For clarity, these generalized sketches track only two pairs of chromosomes from a diploid (2n) animal cell. Cells of nearly all eukaryotic species have more pairs than this. The micrographs track a mouse cell through mitosis. The mouse cell's DNA is stained *blue*, and its microtubules are stained *green*.

Mitosis

pair of centrioles

nuclear envelope — chromosomes

b Early Prophase

Mitosis begins. The DNA and its associated proteins have started to condense. The two chromosomes color-coded *purple* were inherited from the female parent. The other two (*blue*) are their counterparts, inherited from a male parent.

c Late Prophase

The duplicated chromosomes continue to condense. New microtubules form. They move one of two pairs of centrioles and centrosomes to the opposite side of the nucleus. The nuclear envelope starts to break up.

d Transition to Metaphase

Now microtubules penetrate the nuclear region. Collectively, they form a bipolar spindle. Some tether one sister chromatid of each chromosome to one or the other spindle pole. Others overlap at the spindle equator.

At **anaphase**, sister chromatids of each chromosome are moved toward opposite spindle poles. How? Motor proteins attached to each chromatid's kinetochore are inching along microtubular tracks that lead to one or the other spindle pole. The microtubules themselves are shrinking at both ends even as motor proteins are dragging the chromatids with them. And so the sister chromatids end up at opposite poles (Figure 9.7f).

At the same time, the spindle poles themselves are being pushed farther apart! Different microtubules, the ones that overlap midway between the spindle poles, are ratcheting past one another. Motor proteins drive this interaction and push the poles apart.

One sister chromatid is a duplicate of the other. So once they detach from each other at anaphase, there are two separate chromosomes, one at each pole.

Telophase gets under way when one of each type of chromosome reaches a spindle pole. Two genetically identical clusters of chromosomes are now located at opposite "ends" of the cell. All of the chromosomes decondense and become threadlike. Vesicles derived from old nuclear envelope fuse and form patches of membrane around each cluster. Patch joins with patch until a new nuclear envelope encloses each cluster. And so two nuclei form (Figure 9.7g). In our example, the parent cell had a diploid number of chromosomes. So does each nucleus. Once two nuclei have formed, telophase is over—and so is mitosis.

Prior to mitosis, each chromosome in a cell's nucleus is duplicated, so it consists of two sister chromatids.

In prophase, chromosomes condense to rodlike forms, and microtubules form a bipolar spindle. The nuclear envelope breaks up. Some microtubules harness the chromosomes.

At metaphase, all duplicated chromosomes are aligned midway between the spindle's poles, at its equator.

At anaphase, microtubules move the sister chromatids of each chromosome apart, to opposite spindle poles.

At telophase, a new nuclear envelope forms around each of two clusters of decondensing chromosomes.

Thus two daughter nuclei have formed. Each has the same chromosome number as the parent cell's nucleus.

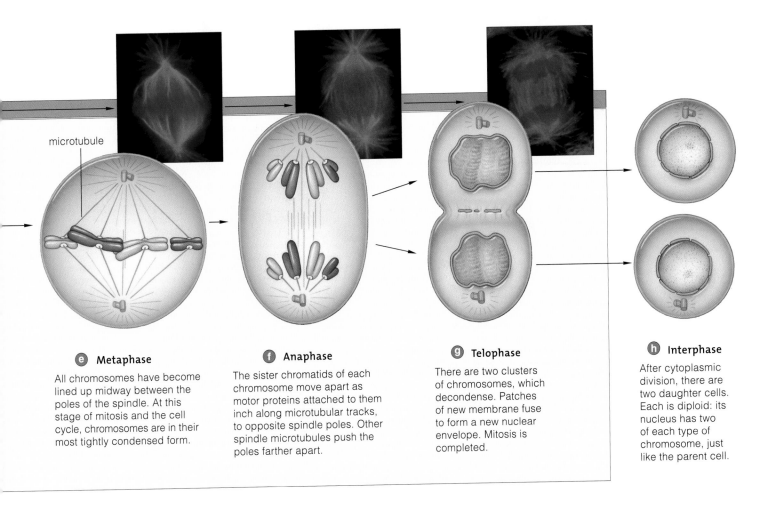

microtubule

ⓔ Metaphase

All chromosomes have become lined up midway between the poles of the spindle. At this stage of mitosis and the cell cycle, chromosomes are in their most tightly condensed form.

ⓕ Anaphase

The sister chromatids of each chromosome move apart as motor proteins attached to them inch along microtubular tracks, to opposite spindle poles. Other spindle microtubules push the poles farther apart.

ⓖ Telophase

There are two clusters of chromosomes, which decondense. Patches of new membrane fuse to form a new nuclear envelope. Mitosis is completed.

ⓗ Interphase

After cytoplasmic division, there are two daughter cells. Each is diploid: its nucleus has two of each type of chromosome, just like the parent cell.

9.4 Cytoplasmic Division Mechanisms

LINKS TO
SECTIONS
4.9, 4.11

In most cell types, the cytoplasm usually divides at some time between late anaphase and the end of telophase. The mechanism of cytoplasmic division—or, more formally, cytokinesis—differs among species.

HOW DO ANIMAL CELLS DIVIDE?

Dividing animal cells partition their cytoplasm by a **contractile ring mechanism**. Most often, the plasma membrane starts to sink inward as a thin indentation about halfway between the cell's poles (Figure 9.8*a*). This is a cleavage furrow, the first visible sign that the cytoplasm is dividing. It advances until it extends all around the cell. As it does so, it deepens along a plane corresponding to the former spindle's equator.

What is going on? Part of the cell cortex, that mesh of cytoskeletal elements under the plasma membrane,

is a ring of actin filaments organized as a thin band around the cell's midsection. The band is anchored to the plasma membrane. When energized by ATP, all the filaments contract and slide past one another in a way that shrinks the band diameter (compare Section 4.11). Being attached to the plasma membrane, the band drags it inward until the cytoplasm is pinched in two (Figures 9.8*a* and 9.9). Two daughter cells form this way. Each ends up with a nucleus and cytoplasm, enclosed within a plasma membrane.

HOW DO PLANT CELLS DIVIDE?

The contractile ring mechanism that works for animal cells could not work for plant cells. The contractile force could not pinch through plant cell walls, which are stiff with cellulose and often lignin. Microtubules

 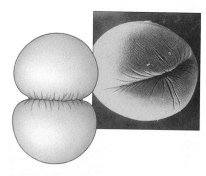 

1 Mitosis is completed, and the bipolar spindle is starting to disassemble.

2 At the former spindle equator, a ring of actin filaments attached to the plasma membrane contracts.

3 The diameter of the contractile ring continues to shrink and pull the cell surface inward.

4 The contractile mechanism continues to operate until the cytoplasm is partitioned.

a Contractile Ring Formation

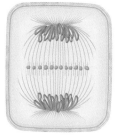 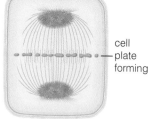 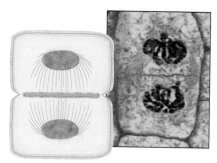 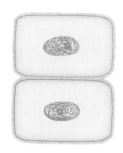

cell plate forming

1 The plane of division and of a future cross-wall was established by a band of microtubules and actin filaments that formed and broke up before mitosis. Vesicles cluster here when mitosis ends.

2 Vesicle membranes fuse. The wall material is sandwiched between two new membranes that lengthen along the plane of a newly forming cell plate.

3 Cellulose is deposited inside the sandwich. In time, these deposits will form two cell walls. Others will form the middle lamella between the walls and cement them together.

4 A cell plate grows at its margins until it fuses with the parent cell plasma membrane. The primary wall of growing plant cells is still thin. New material is deposited on it.

b Cell Plate Formation

Figure 9.8 *Animated!* Cytoplasmic division of an animal cell (**a**) and a plant cell (**b**).

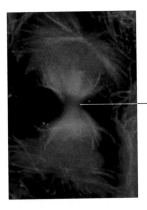

ring of microfilaments midway between the two spindle poles, in the same plane as the spindle equator

Figure 9.9 Micrograph capturing the contractile ring action inside an animal cell.

just beneath the plasma membrane help orient the fibers of cellulose in the wall. Before prophase, these microtubules disassemble and new ones assemble in a narrow band around the nucleus—a band that also includes actin filaments. As other microtubules form the bipolar spindle, the narrow band disappears, and an actin-depleted zone is left behind. *The zone marks the plane of cytoplasmic division* (Figure 9.8*b*).

Along that plane, tiny vesicles packed with wall-building materials from Golgi bodies fuse with one another. Together, deposits of these materials form a disk-shaped structure known as a cell plate. Deposits of cellulose accumulate at the plate. In time, they are thick enough to form a cross-wall through the cell. New plasma membrane extends across both sides of it. The wall grows until it bridges the cytoplasm and partitions the parent cell. This cytoplasmic division mechanism is known as **cell plate formation**.

APPRECIATE THE PROCESS!

Take a moment to look closely at your hands. Visualize the cells making up your palms, thumbs, and fingers. Now imagine the mitotic divisions that produced all of the cell generations that preceded them while you were developing, early on, inside your mother (Figure 9.10). And be grateful for the astonishing precision of mechanisms that led to their formation at prescribed times, in prescribed numbers, for the alternatives can be terrible indeed.

Why? Good health and survival itself depend on the proper timing and completion of cell cycle events. Some genetic disorders arise as a result of mistakes during the duplication or distribution of even one chromosome. In other cases, unchecked cell divisions often destroy surrounding tissues and, ultimately, the

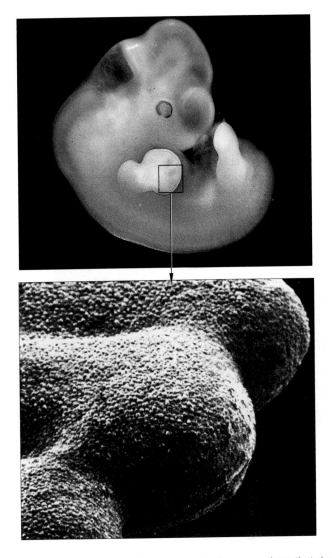

Figure 9.10 The paddlelike structure of a human embryo that develops into a hand by mitosis, cytoplasmic divisions, and other processes. The scanning electron micrograph shows individual cells.

individual. Such losses can start in body cells. They can start in the germ cells that give rise to sperm and eggs, although rarely. The last section of this chapter can give you a sense of the consequences.

After mitosis, a separate mechanism partitions the cytoplasm into two daughter cells, each with a nucleus.

A contractile ring mechanism partitions a dividing animal cell. A band of actin filaments around the cell midsection contracts and pinches the cytoplasm in two.

A mechanism called cell plate formation partitions plant cells. Golgi-derived vesicles deposit material at a plane of cytoplasmic division to form a cross-wall, which connects to the parent cell wall.

Summary

Section 9.1   By processes of reproduction, parents produce a new generation of individuals like themselves. Cell division is a bridge between generations. When a cell divides, its daughter cells each receive a required number of DNA molecules and some cytoplasm.

Only eukaryotic cells undergo mitosis, meiosis, or both. These nuclear division mechanisms partition the duplicated chromosomes of a parent cell into daughter nuclei. A separate mechanism divides the cytoplasm. Prokaryotic cells divide by a different mechanism.

Mitosis is the basis of multicellular growth, cell replacements, and tissue repair. Also, many singled-celled and multicelled species reproduce asexually by mitosis.

Meiosis, the basis of sexual reproduction, precedes the formation of gametes or spores.

A eukaryotic chromosome is a molecule of DNA and many histones and other proteins associated with it. The proteins structurally organize the chromosome and affect access to genes. The smallest unit of organization, the nucleosome, consists of a stretch of double-stranded DNA looped twice around a spool of histones.

When duplicated, the chromosome consists of two sister chromatids, each with a kinetochore (a docking site for microtubules). Until late in mitosis (or meiosis), the two remain attached at their centromere region.

Biology⟨ℰ⟩Now
Explore the structure of a chromosome with the animation on BiologyNow.

Section 9.2   Each cell cycle starts when a new cell forms, runs through interphase, and ends when that cell reproduces by nuclear and cytoplasmic division. A cell carries out most functions in interphase: it increases in mass, roughly doubles the number of its cytoplasmic components, then duplicates each of its chromosomes.

Biology⟨ℰ⟩Now
Investigate the stages of the cell cycle with the interaction on BiologyNow.

Section 9.3   The sum of all chromosomes in cells of a given type is the chromosome number. Human body cells have a diploid chromosome number of 46 (pairs of 23 types of chromosomes). Mitosis, which maintains the chromosome number, has four continuous stages:

Prophase. The duplicated, threadlike chromosomes start to condense. With the help of motor proteins, new microtubules start forming a bipolar mitotic spindle. The nuclear envelope starts to break apart. Some of the microtubules extending from one spindle pole tether one chromatid of each chromosome; others extending from the opposite pole tether the sister chromatid. Microtubules extending from both poles grow until they overlap at the spindle's midpoint.

Metaphase. At metaphase, all chromosomes have become aligned at the spindle's midpoint.

Anaphase. Sister chromatids detach from each other. The kinetochore of each drags it along microtubules, which are shortening at both ends. The microtubules that overlap ratchet past each other, pushing the spindle poles farther apart. Different motor proteins drive the movements. One of each type of parental chromosome ends up clustered together at each spindle pole.

Telophase. The chromosomes decondense to threadlike form. A new nuclear envelope forms around each cluster. Both nuclei have the parental chromosome number.

Fill in the blanks of the diagram below to check your understanding of the four stages of mitosis, and how it maintains the chromosome number.

Biology⟨ℰ⟩Now
Observe how mitosis occurs with the animation on BiologyNow.

Section 9.4   The mechanisms of cytoplasmic division differ. In animal cells, a microfilament ring that is part of the cell cortex contracts and pulls the cell surface inward until the cytoplasm is partitioned. In plant cells, a band of microtubules and microfilaments forms around the nucleus before mitosis starts. It marks the site where a cell plate will form from Golgi-derived material. The cell plate will enlarge and become a cross-wall that will partition the cytoplasm.

Biology⟨ℰ⟩Now
Compare the cytoplasmic division of plant and animal cells with the animation on BiologyNow.

Section 9.5   Checkpoint gene products are part of the controls over the cell cycle. Mutant checkpoint genes cause tumors by disrupting normal controls. Cancer is a multistep process involving altered cells that grow and divide abnormally. Malignant cells may metastasize, or break loose and colonize distant tissues.

Biology⟨ℰ⟩Now
See how cancers spread throughout a body with the animation on BiologyNow.

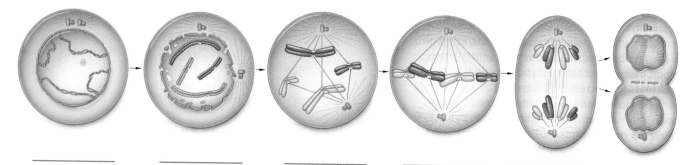

Self-Quiz

Answers in Appendix II

1. Mitosis and cytoplasmic division function in _____ .
   a. asexual reproduction of single-celled eukaryotes
   b. growth, tissue repair, often asexual reproduction
   c. gamete formation in prokaryotes
   d. both a and b

2. A duplicated chromosome has _____ chromatid(s).
   a. one   b. two   c. three   d. four

3. The basic unit that structurally organizes a eukaryotic chromosome is the _____ .
   a. higher order coiling   c. nucleosome
   b. bipolar mitotic spindle   d. microfilament

4. The chromosome number is _____ .
   a. the sum of all chromosomes in cells of a given type
   b. an identifiable feature of each species
   c. maintained by mitosis
   d. all of the above

5. A somatic cell having two of each type of chromosome has a(n) _____ chromosome number.
   a. diploid   b. haploid   c. tetraploid   d. abnormal

6. Interphase is the part of the cell cycle when _____ .
   a. a cell ceases to function
   b. a germ cell forms its spindle apparatus
   c. a cell grows and duplicates its DNA
   d. mitosis proceeds

7. After mitosis, the chromosome number of a daughter cell is _____ the parent cell's.
   a. the same as   c. rearranged compared to
   b. one-half   d. doubled compared to

8. Only _____ is not a stage of mitosis.
   a. prophase b. interphase c. metaphase d. anaphase

9. Match each stage with the events listed.
   _____ metaphase   a. sister chromatids move apart
   _____ prophase   b. chromosomes start to condense
   _____ telophase   c. daughter nuclei form
   _____ anaphase   d. all duplicated chromosomes are
                         aligned at the spindle equator

Additional questions are available on Biology Now™

Critical Thinking

1. Figure 9.15 shows a cell going through stages of mitosis. Notice the barrel-shaped spindle that is quite evident at anaphase. Also notice the dense array of short microtubules midway between the two clusters of chromosomes at telophase. From these clues, would you say that this is a plant cell or an animal cell?

2. Pacific yews (*Taxus brevifolius*) are among the slowest growing trees, which makes them vulnerable to extinction. People started stripping their bark and killing them when they heard that *taxol,* a chemical extracted from the bark, may work against breast and ovarian cancer. It takes bark from about six trees to treat one patient. Do some research and find out why taxol has potential as an anticancer drug and what has been done to protect the trees.

3. X-rays emitted from some radioisotopes damage DNA, especially in cells undergoing DNA replication. Humans

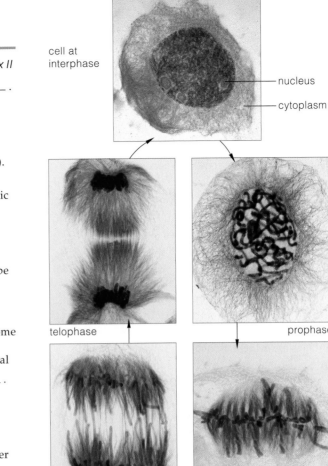

cell at interphase

nucleus

cytoplasm

Figure 9.15 Go ahead, name the mystery cell.

telophase

prophase

anaphase

metaphase

exposed to high levels of x-rays face *radiation poisoning.* Hair loss and a damaged gut lining are early symptoms. Speculate why. Also speculate on why radiation exposure is used as a therapy to treat some cancers.

4. Suppose you have a way to measure the amount of DNA in a single cell during the cell cycle. You first measure the amount at the G1 phase. At what points during the remainder of the cycle would you predict changes in the amount of DNA per cell?

5. The cervix is part of the uterus, a chamber in which embryos develop. The *Pap smear* is a screening procedure that can detect *cervical cancer* in its earliest stages.

   Treatments range from freezing precancerous cells or killing them with a laser beam to removal of the uterus (a hysterectomy). The treatments are more than 90 percent effective when this cancer is detected early. Survival chances plummet to less than 9 percent after it spreads.

   Most cervical cancers develop slowly. Unsafe sex increases the risk. A key risk factor is infection by human papillomaviruses (HPV), which cause genital warts. Viral genes coding for the tumor-inducing proteins get inserted into the DNA of cervical cells. Of one group of cervical cancer patients, 91 percent had been infected with HPV.

   Not all women request Pap smears. Many wrongly believe the procedure is costly. Many do not recognize the importance of abstinence or "safe" sex. Others don't want to think about whether they have cancer. Knowing about the cell cycle and cancer, what would you say to a woman who falls in one or more of these groups?

Why Sex?

Single-celled eukaryotes started to engage in sex many hundreds of millions of years ago, although no one knows how. An unsolved puzzle is why they did it at all.

Asexual reproduction by way of mitotic cell division is easier and faster. Evolutionarily speaking, one individual alone parcels out its DNA to offspring, which are just like their parent. Without sex, that one individual has all of its DNA represented in the new generation. The advantages are evident among most of the protists and fungi, which reproduce asexually most of the time. They quickly give rise to huge populations of cells just like themselves. The advantages are evident among many plants and many invertebrates, including corals, sea stars, and flatworms. Even after bits of these organisms bud or break off, or if the body splits in two, the parts grow into complete copies of the parent. How can the costs of sexual reproduction—such as all of the energy required to construct and use special mate-attracting body parts—beat that?

Sexual reproduction can be an alternative adaptation in changing environments. Consider the plant-sucking insects called aphids. In spring and summer, when plant juices are plentiful, a female aphid reproduces by parthenogenesis. In one day she can give birth to as many as five females, all from unfertilized eggs (Figure 10.1a). Aphid population sizes soar until autumn, when food dwindles. Males now form from eggs, aphids engage in sex, and large fertilized eggs are laid that can withstand winter conditions. Next spring, the eggs develop into asexually oriented females.

Alternative adaptations to the environment also may be why we find a few all-female species of fishes, reptiles, and birds—not mammals—in nature. Not content to let it go

at that, University of Tokyo researchers recently fused two mouse eggs in a test tube and made an embryo with no DNA from a male. The embryo developed into Kaguya, the world's first fatherless mammal (Figure 10.1b). The female mouse grew up, engaged in sex with a male mouse, and gave birth to offspring. But back to the big picture:

Sexual reproduction also has advantages when other organisms change. This is especially apparent when we consider the interactions between predators and prey, or between hosts and the parasites or pathogens that infect them. An intriguing idea, the Red Queen hypothesis, may explain the connection between these interactions and sexual reproduction.

In Lewis Carroll's book *Through the Looking Glass,* the Queen of Hearts tells Alice, "Now here, you see, it takes all the running you can do, to keep in the same place." When mutation introduces a better defense against a predator, parasite, or pathogen, we can comfortably predict that natural selection will favor it. However, we also can predict that selection will favor individual predators, parasites, or pathogens that have a novel means to overcome the new defense. The interacting species coevolve; each is running as fast as it can to keep up with the ongoing changes in the other. Talk about an evolutionary treadmill.

Applying the Red Queen hypothesis to our questions, sexual reproduction endures because individuals that practice it can come up with far more variety in heritable defenses compared to the ones that do not. Remember the chromosomes? Sexual reproducers typically have a diploid chromosome number; they inherit two of each type, from two parents. Their two sets of chromosomes

Figure 10.1 Reproductive moments. (**a**) Aphid giving birth. Like females of some other sexually reproducing species, this one reproduces asexually in spring but engages in sex before winter. (**b**) A fatherless mouse. (**c**) Poppy plant being helped by a beetle, which makes pollen deliveries for it. (**d**) Mealybugs mating.

generally hold information about the same traits, but the information about a given trait is not always *exactly* the same on both of them. Some of it might even be bad under prevailing conditions but might be useful in the future. As you will see shortly, meiosis and fertilization mix up information, so that a tremendous variety of novel traits is tried out among the offspring of each new generation. The capacity for rapid, adaptive responses to abiotic and biotic conditions may well be present somewhere in the expressed range of variation.

Asexual reproduction cannot shuffle information into novel combinations. It puts out the same versions of traits again and again into the environmental testing ground. Doing so works well enough—as long as the organism is already equipped to handle change.

With this chapter, we turn to mechanisms of sexual reproduction. Three interconnected events—meiosis, the formation of gametes, and fertilization—are hallmarks of this reproductive mode. The outcome is the production of offspring that display novel combinations of traits. As you will see throughout the book, that outcome has contributed immensely to the range of diversity, past and present.

Watch the video online!

☑ *How Would You Vote?*

Japanese researchers have successfully created a "fatherless" mouse that contains the genetic material from the eggs of two females. The mouse is healthy and fully fertile. Do you think researchers should be allowed to try the same process with human eggs? See BiologyNow for details, then vote online.

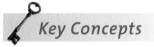

Key Concepts

SEXUAL VERSUS ASEXUAL REPRODUCTION

By asexual reproduction, one parent alone transmits genetic information to offspring. By sexual reproduction, offspring inherit novel combinations of information from more than one parent, because those parents typically differ in their alleles. Alleles are slightly different molecular forms of a gene that specify different versions of the same trait. Section 10.1

OVERVIEW OF MEIOSIS

Meiosis, a nuclear division mechanism, divides the parental chromosome number by half. It occurs only in cells set aside for sexual reproduction. Section 10.2

STAGES OF MEIOSIS

Meiosis sorts out a reproductive cell's chromosomes into four new nuclei. After it ends, gametes form by way of cytoplasmic division and other events. Section 10.3

CHROMOSOME RECOMBINATIONS AND SHUFFLINGS

During meiosis, each pair of chromosomes swaps segments and exchanges alleles. Also, one of each pair is randomly aligned for distribution into a new nucleus. Which ends up in a given gamete is a matter of chance. Chromosomes are shuffled again at fertilization. These events contribute to variation in traits among offspring. Section 10.4

SEXUAL REPRODUCTION IN THE LIFE CYCLES

In animals, gametes form by different mechanisms in males and females. In most plants, spore formation and other events intervene between meiosis and gamete formation. Spores store and protect hereditary information through times that predictably do not favor survival of offspring. Section 10.5

MITOSIS AND MEIOSIS COMPARED

Recent molecular evidence suggests that meiosis originated through mechanisms that already existed for mitosis and, before that, for repairing damaged DNA. Section 10.6

Links to Earlier Concepts

For this chapter, you will be drawing on your sense of the dynamic nature of microtubule assembly and disassembly (Sections 4.10, 4.11, 9.2). Be sure you have a clear picture of the structural organization of chromosomes (9.1) and that you can define chromosome number (9.2). Reflect on how a bipolar spindle made of microtubules moves chromosomes during nuclear division (9.3), and how the cytoplasm gets divided following nuclear division (9.4). You will be revisiting the checkpoint gene products that monitor and repair chromosomal DNA during the cell cycle (9.5).

LINKS TO
SECTIONS
9.1, 9.2

10.1 Introducing Alleles

Asexual reproduction produces genetically identical copies of a parent. Sexual reproduction introduces variation in the details of traits among offspring.

When an orchid or aphid reproduces by itself, what sort of offspring does it get? By the process of **asexual reproduction**, all offspring inherit the same number and kinds of genes from a single parent. **Genes** are sequences of chromosomal DNA. The genes for each species contain all the heritable information necessary to make new individuals. Rare mutations aside, then, asexually produced individuals can only be *clones*, or genetically identical copies of the parent.

Inheritance gets far more interesting with **sexual reproduction**, a process involving meiosis, formation of gametes, and fertilization—a union of two gametes. In most sexual reproducers, such as humans, the first cell of a new individual holds *pairs of genes* on pairs of chromosomes. Usually, one of each pair is maternal and the other paternal in origin (Figure 10.2).

If information in all pairs of genes were identical down to the last detail, sexual reproduction would also produce clones. Just imagine—you, every person you know, the entire human population might be a clone, with everybody looking alike. But the two genes of a pair might *not* be identical. Why not? The molecular structure of any gene can change permanently; it can mutate. So two genes that happen to be paired in an individual's cells may "say" slightly different things about a trait. Each unique molecular form of the same gene is called an **allele**.

Such tiny differences affect thousands of traits. For instance, whether your chin has a dimple depends on which pair of alleles you inherited at one chromosome location. One kind of allele at that location says "put a dimple in the chin." Another kind says "no dimple." Alleles are one reason why the individuals of sexually reproducing species do not all look alike. *With sexual reproduction, offspring inherit new combinations of alleles, which lead to variations in the details of their traits.*

This chapter gets into the cellular basis of sexual reproduction. More importantly, it starts you thinking about far-reaching effects of gene shufflings at certain stages of the process. The process introduces variations in traits among offspring that are typically acted upon by agents of natural selection. Thus, *variation in traits is a foundation for evolution.*

Figure 10.2 A maternal and a paternal chromosome. Any gene on one might be slightly different structurally than the same gene on the other.

Sexual reproduction introduces variation in traits by bestowing novel combinations of alleles on offspring.

10.2 What Meiosis Does

Meiosis is a nuclear division process that divides a parental chromosome number by half in specialized reproductive cells. Sexual reproduction will not work without it.

THINK "HOMOLOGUES"

Think back to the preceding chapter and its focus on mitotic cell division. Unlike mitosis, meiosis sorts out chromosomes into parcels not once but *twice*. Unlike mitosis, it is the first step leading to the formation of gametes. Male and female gametes—such as sperm and eggs—fuse to form a new individual. In most multicelled eukaryotes, cells that form in specialized reproductive structures or organs are the forerunners of gametes. Figure 10.3 gives three examples of where cells that give rise to gametes originate.

As you know, the **chromosome number** is the sum total of chromosomes in cells of a given type. If a cell has a **diploid number** ($2n$), it has a *pair* of each type of chromosome, often from two parents. Except for a pairing of nonidentical sex chromosomes, each pair has the same length, shape, and assortment of genes, and they line up with each other at meiosis. We call them **homologous chromosomes** (*hom–* means alike).

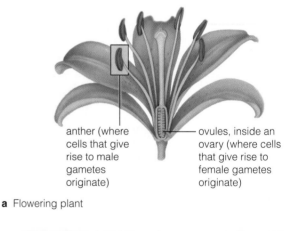

anther (where cells that give rise to male gametes originate)

ovules, inside an ovary (where cells that give rise to female gametes originate)

a Flowering plant

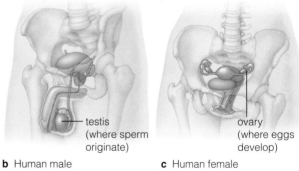

testis (where sperm originate)

ovary (where eggs develop)

b Human male

c Human female

Figure 10.3 Examples of reproductive organs, where cells that give rise to gametes originate.

The body cells of humans are diploid, with 23 + 23 homologous chromosomes (Figure 10.4). So are human germ cells that give rise to gametes. Following meiosis, every gamete normally gets 23 chromosomes—one of each type. Meiosis reduced the parental chromosome number by half, to a **haploid number** (*n*).

TWO DIVISIONS, NOT ONE

Bear in mind, meiosis *is* similar to mitosis in certain respects. As in mitosis, a germ cell duplicates its DNA in interphase. The two DNA molecules and associated proteins stay attached at the centromere, the notably constricted region along their length. For as long as they remain attached, we call them **sister chromatids**:

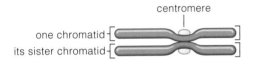

As in mitosis, the microtubules of a spindle apparatus move the chromosomes in prescribed directions.

With meiosis, however, *chromosomes go through <u>two</u> consecutive divisions that end with the formation of four haploid nuclei.* The germ cell does not enter interphase between the two nuclear divisions, which are known as meiosis I and meiosis II:

Interphase (DNA is replicated prior to meiosis I)	Meiosis I		*No* interphase (DNA is *not* replicated prior to meiosis II)	Meiosis II
	Prophase I			Prophase II
	Metaphase I			Metaphase II
	Anaphase I			Anaphase II
	Telophase I			Telophase II

In meoisis I, each duplicated chromosome aligns with its partner, *homologue to homologue.* After the two chromosomes of every pair have lined up with each other, they are moved apart:

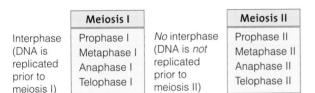

each homologue in the cell pairs with its partner

then the partners separate

The cytoplasm typically starts to divide at some point after each homologue detaches from its partner. The two daughter cells formed this way are haploid, with *one* of each type of chromosome. Don't forget, these chromosomes are still in the duplicated state.

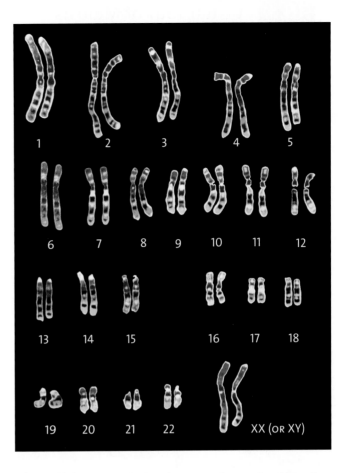

Figure 10.4 Another look at the twenty-three pairs of homologous human chromosomes. This example is from a human female, with two X chromosomes. Human males have a different pairing of sex chromosomes (XY). These chromosomes have been labeled with fluorescent markers.

Next, during meiosis II, *the two sister chromatids of each chromosome are separated from each other:*

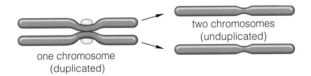

one chromosome (duplicated)

two chromosomes (unduplicated)

There are now four parcels of 23 chromosomes, and each has one chromosome of each type. New nuclear envelopes enclose them, as four nuclei. Typically the cytoplasm divides once more, so the outcome is four haploid (*n*) cells. Figure 10.5 on the next two pages puts these chromosomal movements in the context of the sequential stages of meiosis.

Meiosis, a nuclear division mechanism, reduces a parental cell's chromosome number by half—to a haploid number (n).

10.3 Visual Tour of Meiosis

Meiosis I

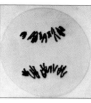
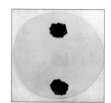

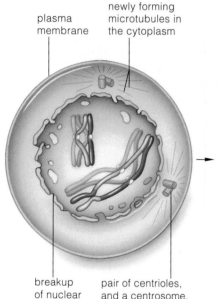

plasma
membrane

newly forming
microtubules in
the cytoplasm

spindle equator
(midway between
the two poles)

one pair of
homologous
chromosomes

breakup
of nuclear
envelope

pair of centrioles,
and a centrosome,
moving to opposite
sides of nucleus

ⓐ Prophase I

As prophase I begins, chromosomes become
visible as threadlike forms. Each pairs with its
homologue and usually swaps segments with it,
as indicated by the breaks in color in the large
chromosomes. Microtubules are forming a
bipolar spindle (Section 9.3). If two pairs of
centrioles are present, one pair is moved to
the opposite side of the nuclear envelope,
which is starting to break up.

ⓑ Metaphase I

Microtubules from one spindle
pole have tethered one of each type
of chromosome; microtubules from
the other pole have tethered its
homologue. By metaphase I, a
tug-of-war between the two sets
of microtubules has aligned all
chromosomes midway between
the poles.

ⓒ Anaphase I

Microtubules attached to each
chromosome shorten and move
it toward a spindle pole. Other
microtubules, which extend
from the poles and overlap at
the spindle equator, ratchet
past each other and push the
two poles farther apart. Motor
proteins drive the ratcheting.

ⓓ Telophase I

One of each type of
chromosome has
now arrived at the
spindle poles. For
most species, the
cytoplasm divides at
some point, forming
two haploid cells.
All chromosomes
are still duplicated.

Figure 10.5 *Animated!* Meiosis in one type of animal cell. This is a nuclear division mechanism.
It reduces the parental chromosome number in immature reproductive cells by half, to the haploid
number, for forthcoming gametes. To keep things simple, we track only two pairs of homologous
chromosomes. Maternal chromosomes are shaded *purple* and paternal chromosomes *blue*.

Of the four haploid cells that form by meiosis and cytoplasmic divisions, one or all may develop into
gametes and function in sexual reproduction. In plants, the cells that form may develop into spores,
a stage that precedes gamete formation in the life cycle.

Meiosis II

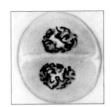

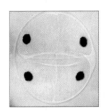

There is no DNA replication between the two nuclear divisions.

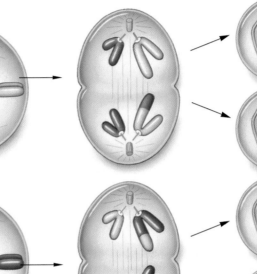

e Prophase II

A new bipolar spindle forms in each haploid cell. Microtubules have moved one member of the pair of centrioles to the opposite end of each cell. One chromatid of each chromosome becomes tethered to one spindle pole, and its sister chromatid becomes tethered to the opposite pole.

f Metaphase II

Microtubules from both spindle poles have assembled and disassembled in a tug-of-war that ended at metaphase II, when all chromosomes are positioned midway between the poles.

g Anaphase II

The attachment between sister chromatids of each chromosome breaks. Each is now a separate chromosome but is still tethered to microtubules, which move it toward a spindle pole. Other microtubules push the poles apart. A parcel of unduplicated chromosomes ends up near each pole. One of each type of chromosome is present in each parcel.

h Telophase II

In telophase II, four nuclei form as a new nuclear envelope encloses each cluster of chromosomes. After cytoplasmic division, each of the resulting daughter cells has a haploid (*n*) number of chromosomes.

10.4 How Meiosis Introduces Variations in Traits

As Sections 10.2 and 10.3 make clear, the basic function of meiosis is the reduction of a parental chromosome number by half. In evolutionary terms, two other functions are as important: Prophase I crossovers and the random alignment of chromosomes at metaphase I contribute greatly to the variation in traits among offspring.

The preceding section mentioned in passing that pairs of homologous chromosomes swap parts of themselves during prophase I. It also showed how a homologous chromosome becomes aligned with its partner during prophase I. Both events introduce new combinations of alleles into the gametes that form at some point *after* meiosis. Along with the chromosome shufflings that occur during fertilization, they contribute to variation in traits that occur among new generations of offspring in sexually reproducing species. Later in the book, you will explore how variation in traits has evolutionary and ecological consequences. We suggest that you read this section closely. It will serve you well later on.

CROSSING OVER IN PROPHASE I

Figure 10.6*a* is a simple sketch of a pair of duplicated chromosomes, early in prophase I of meiosis. Notice how they are in threadlike form. All chromosomes in a germ cell condense this way. When they do, each is drawn close to its homologue. The chromatids of one become stitched point by point along their length to the chromatids of the other, with little space between them. This tight, parallel orientation favors **crossing over**, a molecular interaction between a chromatid of one chromosome and a chromatid of the homologous partner. DNA strands break and seal in complex ways, but the outcome is that the two "nonsister" chromatids exchange corresponding segments; they swap genes.

a This maternal chromosome (*purple*) and paternal chromosome (*blue*) were duplicated in interphase. They appear in microscopes early in prophase I, when they are starting to condense to threadlike form. Sister chromatids of each chromosome are positioned so close together that they look like a single thread. (We pulled them apart a bit in this sketch so you can distinguish between them.)

b Each chromosome now becomes zippered up with its homologous partner, so all four chromatids are tightly aligned. If the two sex chromosomes have different forms (such as X paired with Y), they still get tightly aligned, but only in a tiny region at their ends.

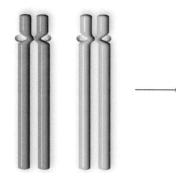 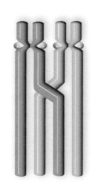  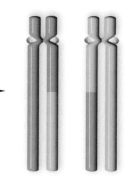

c Here is the simplest way to think about crossing over. However, don't forget that the chromosomes are not really rod-shaped during early prophase I. They are still condensing to threadlike form, and each is tightly aligned with its homologous partner.

d The intimate contact encourages crossovers at various intervals along the length of nonsister chromatids. Here we show the location of just one crossover.

e Nonsister chromatids exchange segments at the crossover site. They keep on condensing into thicker, rodlike forms. They will be fully unzipped from each other by metaphase I.

f Crossing over breaks up the old combinations of alleles and puts new ones together in homologous chromosomes. It mixes up maternal and paternal information about traits.

Figure 10.6 *Animated!* Key events of prophase I, the first stage of meiosis. For clarity, we show only one pair of homologous chromosomes and one crossover. More than one crossover usually occurs in each chromosome pair. *Blue* signifies a paternal chromosome, and *purple*, its maternal homologue.

Gene swapping would be pointless if each type of gene never varied. But remember, a gene can come in slightly different forms—alleles. You can predict that a number of the alleles on one chromosome will *not* be identical to their partner alleles on the homologous chromosome. Each crossover event is a chance to swap slightly different versions of heritable information on gene products.

We will look at the mechanism of crossing over in later chapters. For now, just remember this: *Crossing over leads to recombinations among genes of homologous chromosomes, and eventually to variation in traits among offspring.*

METAPHASE I ALIGNMENTS

Major shufflings of intact chromosomes start during the transition from prophase I to metaphase I. Suppose this is happening right now in one of your germ cells. Crossovers have already made genetic mosaics of the chromosomes, but put this aside in order to simplify tracking. Just call the twenty-three chromosomes you inherited from your mother the *maternal* chromosomes and the twenty-three you inherited from your father the *paternal* chromosomes.

At metaphase I, microtubules from both poles have now aligned all of the duplicated chromosomes at the spindle equator (Figure 10.5*b*). Have they tethered all maternal chromosomes to one pole and all paternal chromosomes to the other? Maybe, but probably not. When the microtubules were growing, they latched on to the first chromosome they contacted. Because the tethering was random, there is no particular pattern to the metaphase I positions of maternal and paternal chromosomes.

Now carry this thought one step further. During anaphase I, when a duplicated chromosome is moved away from its homologous partner, *either partner* can end up at either spindle pole.

Think of the possibilities while tracking just three pairs of homologues. By metaphase I, these three pairs may be arranged in any one of four possible positions (Figure 10.7). This means that eight combinations (2^3) are possible for forthcoming gametes.

Cells that give rise to human gametes have twenty-three pairs of homologous chromosomes, not three. Thus, every time a human sperm or egg forms, there is a total of *8,388,608* (or 2^{23}) possible combinations of maternal and paternal chromosomes! Moreover, in a sperm or an egg, many hundreds of alleles inherited from the mother might not "say" the exact same thing about hundreds of different traits as alleles inherited

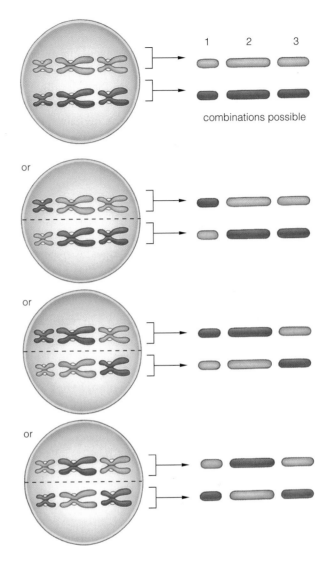

combinations possible

Figure 10.7 *Animated!* Possible outcomes for the random alignment of merely three pairs of homologous chromosomes at metaphase I. The three types of chromosomes are labeled 1, 2, and 3. With four alignments, eight combinations of maternal chromosomes (*purple*) and paternal chromosomes (*blue*) are possible in gametes.

from the father. Are you getting an idea of why such fascinating combinations of traits show up among the generations of your own family tree?

Crossing over, an interaction between a pair of homologous chromosomes, breaks up old combinations of alleles and puts new ones together during prophase I of meiosis.

The random tethering and subsequent positioning of each pair of maternal and paternal chromosomes at metaphase I lead to different combinations of maternal and paternal traits in each new generation.

10.5 From Gametes to Offspring

LINK TO
SECTION
9.4

What happens to the gametes that form after meiosis? Later chapters have specific examples. Here, simply focus on where they fit in the life cycles of plants and animals.

Gametes are not all the same in their details. Human sperm have one tail, opossum sperm have two, and roundworm sperm have none. Crayfish sperm look like pinwheels. Most eggs are microscopic in size, yet an ostrich egg inside its shell is as big as a football. A flowering plant's male gamete is just a sperm nucleus.

GAMETE FORMATION IN PLANTS

The life cycle of most plant species alternates between sporophyte and gametophyte stages. A *sporophyte* is a multicelled spore-producing body that makes sexual spores by way of meiosis (Figure 10.8a). In plants, each **spore** is a haploid reproductive cell that is not a gamete and that does not take part in fertilization. At some point, the spore undergoes mitotic cell divisions that give rise to a *gametophyte*. One or more gametes do form inside this multicelled haploid body.

Pine trees are examples of sporophytes, and their female gametophytes form on the scales of pinecones. Rose bushes and fuschias also are sporophytes, and gametophytes form inside their flowers. You will be focusing on plant life cycles in Chapters 23 and 32.

GAMETE FORMATION IN ANIMALS

In animals, diploid germ cells give rise to gametes. In a male reproductive system, a germ cell develops into a primary spermatocyte. This large, immature cell enters meiosis and cytoplasmic divisions. Four haploid cells result and develop into spermatids (Figure 10.9). Each

cell undergoes changes, such as the formation of a tail, and becomes a **sperm**, a type of mature male gamete.

In female animals, a germ cell becomes a primary oocyte, which is an immature egg. Unlike sperm, the primary oocyte increases in size and stockpiles many cytoplasmic components. In addition, its four daughter cells differ in size and function (Figure 10.10).

When the primary oocyte divides after meiosis I, one daughter cell—the secondary oocyte—gets nearly all of the cytoplasm. The other cell, a first polar body, is exceedingly small. Later, both of these haploid cells enter meiosis II, then cytoplasmic division. One of the secondary oocyte's daughter cells becomes the second polar body. The other daughter cell gets most of the cytoplasm and develops into a gamete. The mature female gamete is an ovum (plural, ova). An ovum also is known informally as an **egg**.

And so we have one egg. The three polar bodies that formed don't function as gametes and aren't rich in nutrients or plump with cytoplasm. In time they degenerate. But their formation assures that the egg will have a haploid chromosome number. Also, by getting most of the cytoplasm, the egg holds enough metabolic machinery to support early cell divisions of the new individual, as Chapters 43 and 44 explain.

MORE SHUFFLINGS AT FERTILIZATION

The chromosome number characteristic of the parents is restored at **fertilization**, a time when a female and male gamete unite and their haploid nuclei fuse. If meiosis did not precede fertilization, the chromosome number would double in each generation. Doublings would disrupt hereditary information, usually for the worse. Why? That information is like a fine-tuned set

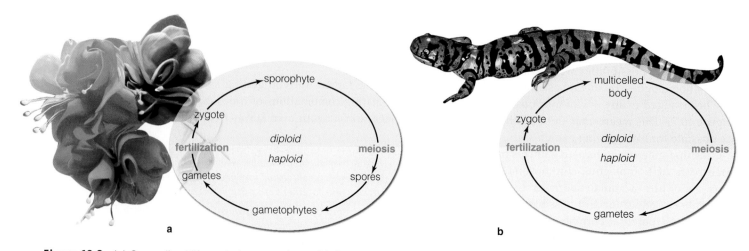

Figure 10.8 (a) Generalized life cycle for most plants. (b) Generalized life cycle for animals. The zygote is the first cell to form when the nuclei of two gametes fuse at fertilization.

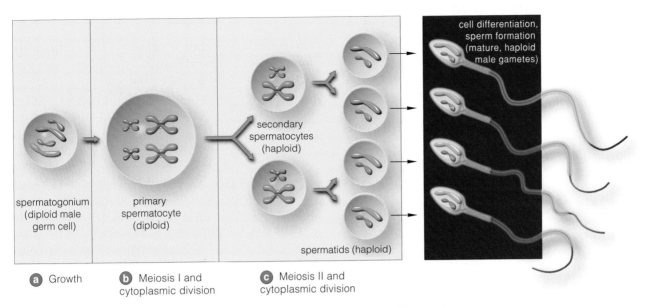

a Growth   **b** Meiosis I and cytoplasmic division   **c** Meiosis II and cytoplasmic division

Figure 10.9 *Animated!* Generalized sketch of sperm formation in animals. Figure 44.4 shows a specific example (how sperm form in human males).

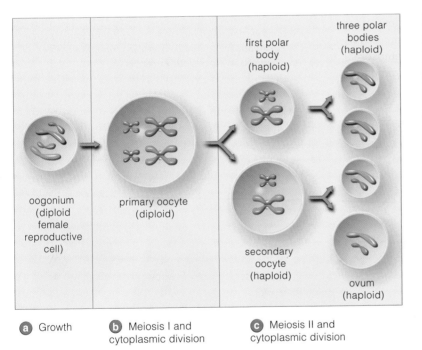

a Growth   **b** Meiosis I and cytoplasmic division   **c** Meiosis II and cytoplasmic division

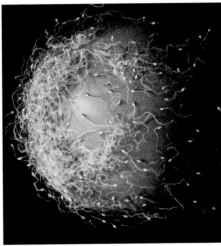

Figure 10.10 *Animated!* Animal egg formation. Eggs are far larger than sperm and larger than the three polar bodies. The painting above, based on a scanning electron micrograph, depicts human sperm surrounding an ovum.

of blueprints that must be followed exactly, page after page, to build a normal individual.

Fertilization also adds to variation among offspring. Reflect on the possibilities for humans alone. During prophase I, every human chromosome undergoes an average of two or three crossovers. In addition to the crossovers, random positioning of pairs of paternal and maternal chromosomes at metaphase I results in one of millions of possible chromosome combinations in each gamete. And of all male and female gametes

that form, *which* two actually get together is a matter of chance. The sheer number of combinations that can exist at fertilization is staggering!

> The distribution of random mixes of chromosomes into gametes, random metaphase chromosome alignments, and fertilization contribute to variation in traits of offspring.

10.6 Mitosis and Meiosis—An Ancestral Connection?

LINKS TO
SECTIONS
9.2, 9.5

This chapter opened with hypotheses about the survival advantages of asexual and sexual reproduction. It seems like a giant evolutionary step from producing clones to producing genetically varied offspring. But was it?

Figure 10.11 shows an obvious parallel between the four stages of mitosis and meiosis II. The same kind of bipolar spindle assorts duplicated chromosomes into parcels in very similar ways. Recent studies also reveal striking similarities at the molecular level.

In all organisms, from prokaryotes to mammals, certain genes code for proteins that can recognize and repair breaks in the double-stranded DNA molecules of chromosomes. Such damage, recall, is monitored by products of checkpoint genes while DNA is being replicated during the cell cycle (Sections 9.2 and 9.5). If they detect a problem, there is a pause in the cycle until it is repaired. Even in bacteria—the most ancient lineages on Earth—a mechanism exists that may well have been recruited for mitosis and meiosis.

Some highly conserved gene products often repair breaks and odd rearrangements in chromosomal DNA that occur during mitosis. They also put chromosomal DNA back together in prophase I, after homologous chromosomes exchange segments. This outcome—a form of genetic recombination—could have been part of the evolution of sexual reproduction.

Is *Giardia intestinalis* one model? This descendent of one of the earliest eukaryotic lineages does not have mitochondria, and it does not form a bipolar spindle during mitosis. This single-celled parasite has never been observed to reproduce sexually. Yet it has gene products that serve in meiosis in higher eukaryotes.

We invite you to think about these possibilities as you read later chapters in the book. We invite you to explore likely connections on your own. For instance, when you look at *Chlamydomonas*, a single-celled alga of freshwater habitats, mull over the fact that haploid *Chlamydomonas* cells reproduce asexually by mitotic cell division. But two cells of different mating strains also can function as *gametes*; they can fuse and form a diploid individual. Do such cells offer more clues to the origin of sexual reproduction? Maybe.

Recombination mechanisms that are vital for reproduction of eukaryotic cells might have evolved from DNA repair mechanisms in prokaryotic ancestors.

Meiosis I

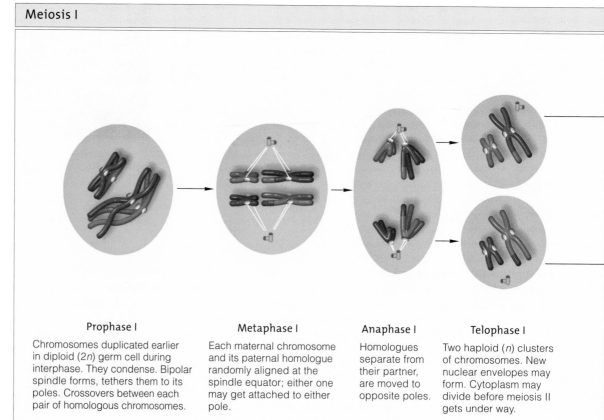

Figure 10.11 Comparative summary of key features of mitosis and meiosis, starting with a diploid cell. Only two paternal and two maternal chromosomes are shown. Both were duplicated in interphase, prior to nuclear division. Both use a bipolar spindle made of microtubules to sort out and move the chromosomes.

Mitosis maintains the parental chromosome number. Meiosis halves it, to the haploid number.

Mitotic cell division is the basis of asexual reproduction among eukaryotes. It is the basis of growth and tissue repair of multicelled eukaryotic species.

Meiotic cell division is a required step before the formation of gametes or sexual spores.

Prophase I
Chromosomes duplicated earlier in diploid (2n) germ cell during interphase. They condense. Bipolar spindle forms, tethers them to its poles. Crossovers between each pair of homologous chromosomes.

Metaphase I
Each maternal chromosome and its paternal homologue randomly aligned at the spindle equator; either one may get attached to either pole.

Anaphase I
Homologues separate from their partner, are moved to opposite poles.

Telophase I
Two haploid (n) clusters of chromosomes. New nuclear envelopes may form. Cytoplasm may divide before meiosis II gets under way.

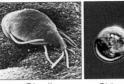

Giardia intestinalis

Chlamydomonas cells mating

Mitosis

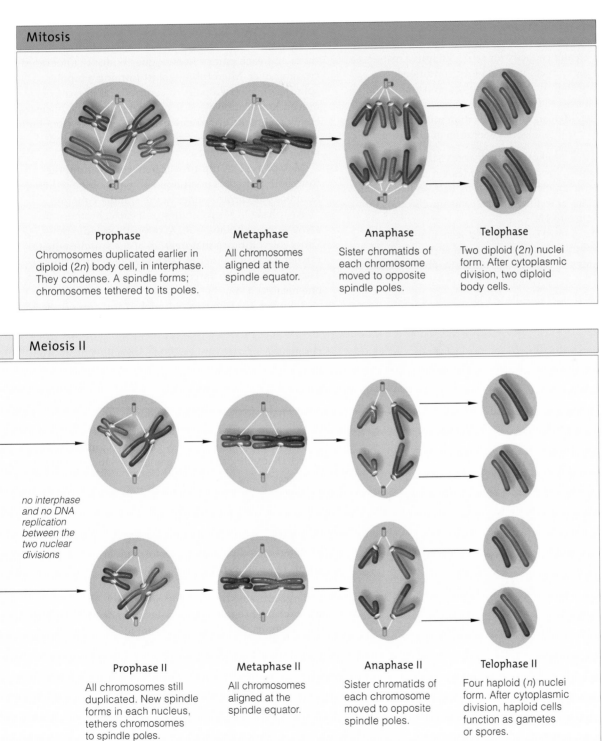

Prophase

Chromosomes duplicated earlier in diploid (2*n*) body cell, in interphase. They condense. A spindle forms; chromosomes tethered to its poles.

Metaphase

All chromosomes aligned at the spindle equator.

Anaphase

Sister chromatids of each chromosome moved to opposite spindle poles.

Telophase

Two diploid (2*n*) nuclei form. After cytoplasmic division, two diploid body cells.

Meiosis II

no interphase and no DNA replication between the two nuclear divisions

Prophase II

All chromosomes still duplicated. New spindle forms in each nucleus, tethers chromosomes to spindle poles.

Metaphase II

All chromosomes aligned at the spindle equator.

Anaphase II

Sister chromatids of each chromosome moved to opposite spindle poles.

Telophase II

Four haploid (*n*) nuclei form. After cytoplasmic division, haploid cells function as gametes or spores.

Summary

Section 10.1 Life cycles of eukaryotic species often have asexual and sexual phases.

Asexual reproduction by way of mitosis yields a clone, or offspring that are genetically the same as one parent. Compared with sexual modes, it is easier, requires less energy, and gives rise to huge populations in far less time.

Sexual reproduction involves two parents that engage in meiosis, gamete formation, and fertilization. It leads to novel allele combinations in offspring. Compared to asexual reproduction, the expressed range of variation offers a far greater capacity for rapid, adaptive response to novel changes in abiotic and biotic conditions.

Alleles are slightly different molecular forms of the same gene that specify different versions of the same gene product. Meiosis and fertilization mix up the alleles (and forms of traits) in each generation of offspring.

Section 10.2 Meiosis, a nuclear division process, precedes gamete formation. It divides the chromosome number characteristic of a species by half, so that fusion of two gametes at fertilization restores the chromosome number (Figure 10.12).

Offspring of most sexual reproducers inherit pairs of chromosomes, one from a maternal and one from a paternal parent. Except in individuals that have inherited

nonidentical sex chromosomes (e.g., X with Y), the pairs are homologous (alike); each pair has the same length, shape, and mostly the same gene sequence. All pairs interact at meiosis. Meiosis parcels out one chromosome of each type for forthcoming gametes.

Section 10.3 All chromosomes in a reproductive cell are duplicated in interphase, prior to meiosis. Meiosis sorts out duplicated chromosomes twice, in two divisions (meiosis I and II) that are not separated by interphase.

In meiosis I, the first nuclear division, homologous chromosomes are partitioned into two clusters, both with one of each type of chromosome.

Prophase I. Chromosomes condense into threadlike form, and each pair of homologues typically undergoes crossing over. Microtubules start forming a bipolar spindle. One of two pairs of centrioles, if present, is moved to the opposite side of the nucleus. The nuclear envelope breaks up, so microtubules growing from both spindle poles can penetrate the nuclear region and tether the chromosomes.

Metaphase I. A tug-of-war between microtubules from both poles has positioned all pairs of the tethered homologous chromosomes at the spindle equator.

Anaphase I. Microtubules pull each chromosome away from its homologue, to opposite spindle poles. Other microtubules that overlap at the spindle equator ratchet past each other to push the poles farther apart. There are now two parcels of duplicated chromosomes, one near each spindle pole.

Telophase I. Two haploid nuclei form around the parcels. Cytoplasmic division typically follows.

In meiosis II, the second nuclear division, the sister chromatids of each chromosome are pulled away from each other and partitioned into two clusters. This occurs in both haploid nuclei that formed in meiosis I. By the end of telophase II, there are four nuclei, each with a haploid chromosome number.

When the cytoplasm divides, there are four haploid cells. One or all may serve as gametes or, in plants, as spores that will give rise to gamete-producing bodies.

Biology ⑤Now
Explore what happens during each stage of meiosis with the animation on BiologyNow.

Section 10.4 Novel combinations of alleles and of maternal and paternal chromosomes arise through events in prophase I and metaphase I.

*Non*sister chromatids of homologous chromosomes undergo crossing over during prophase I. They break and exchange segments, so that each ends up with allelic combinations that were not present in either parent.

Maternal and paternal chromosomes get tethered randomly to one spindle pole or the other. Thus they are positioned at random when they are aligned at the spindle equator at metaphase I, so alleles of either one may end up in a new nucleus, then in a gamete.

Biology ⑤Now
Study how crossing over and metaphase I alignments affect allele combinations with the animation on BiologyNow.

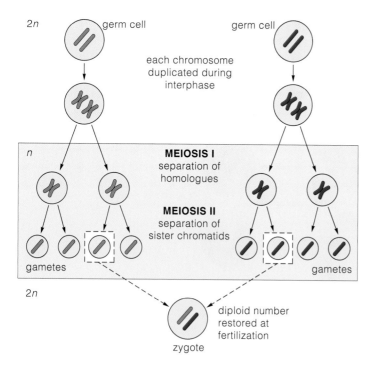

Figure 10.12 Summary of changes in chromosome number at different stages of sexual reproduction, using two diploid (2*n*) germ cells as the example. During two nuclear divisions, meiosis reduces the chromosome number by half (*n*). The union of haploid nuclei of two gametes at fertilization restores the diploid number.

Section 10.5 Life cycles of plants and animals have sexual phases. Sporophytes are a multicelled plant body that produces sexual spores. Such plant spores give rise to gametophytes, in which haploid gametes form.

In most animals, germ cells in reproductive organs give rise to sperm or eggs. Fusion of a sperm and egg nucleus at fertilization results in a zygote, the first cell of a new individual.

Biology⊘Now

Learn how gametes form with the animation on BiologyNow.

Section 10.6 Like mitosis, meiosis uses a bipolar spindle to move and sort duplicated chromosomes. But meiosis occurs only in sex cells and does not produce clones of the parent; it reduces the parental chromosome number by half. Crossing over and random alignments of different mixes of maternal and paternal chromosomes for distribution to gametes occur only in meiosis. These events, and the chance of any two gametes meeting at fertilization, contribute to enormous variation in traits among offspring.

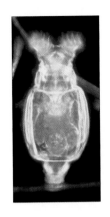

Figure 10.13 Bdelloid rotifer.

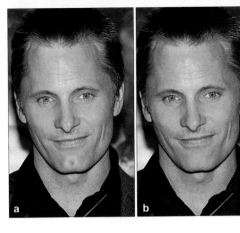

Figure 10.14 Viggo Mortensen (**a**) with and (**b**) without a chin dimple.

11. Match each term with its description.
_____ chromosome number   a. different molecular forms of the same gene
_____ alleles   b. none between meiosis I, II
_____ metaphase I   c. all chromosomes aligned at spindle equator
_____ interphase   d. sum total of all chromosomes in cells of a given type

Additional questions are available on Biology⊘Now™

Self-Quiz

Answers in Appendix II

1. Meiosis and cytoplasmic division function in _____ .
   a. asexual reproduction of single-celled eukaryotes
   b. growth, tissue repair, often asexual reproduction
   c. sexual reproduction
   d. both b and c

2. A duplicated chromosome has _____ chromatid(s).
   a. one   b. two   c. three   d. four

3. A somatic cell having two of each type of chromosome has a(n) _____ chromosome number.
   a. diploid   b. haploid   c. tetraploid   d. abnormal

4. Sexual reproduction requires _____ .
   a. meiosis   c. spore formation
   b. fertilization   d. a and b

5. Generally, a pair of homologous chromosomes _____ .
   a. carry the same genes   c. interact at meiosis
   b. are the same length, shape   d. all of the above

6. Meiosis _____ the parental chromosome number.
   a. doubles   b. halves   c. maintains   d. corrupts

7. Meiosis ends with the formation of _____ .
   a. two cells   c. eight cells
   b. two nuclei   d. four nuclei

8. The cell in the diagram below is in anaphase I rather than anaphase II. I know this because _____ .

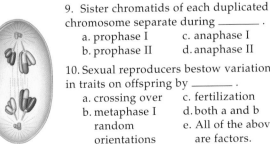

9. Sister chromatids of each duplicated chromosome separate during _____ .
   a. prophase I   c. anaphase I
   b. prophase II   d. anaphase II

10. Sexual reproducers bestow variation in traits on offspring by _____ .
   a. crossing over   c. fertilization
   b. metaphase I random orientations   d. both a and b
          e. All of the above are factors.

Critical Thinking

1. Why can you predict that meiosis will give rise to genetic variation between parent cells and daughter cells in fewer cell cycles than mitosis?

2. The bdelloid rotifer lineage started at least 40 million years ago (Figure 10.13). About 360 known species of these tiny animals live in many aquatic habitats worldwide. All are female. Do some research to identify conditions in the physical and biological environments to which they might be reproductively adapted.

3. Actor Viggo Mortensen inherited a gene that makes his chin dimple. Figure 10.14*b* shows what he might have looked like if he inherited a different form of that gene. What is the name for alternative forms of the same gene?

4. Assume you can measure the amount of DNA in the nucleus of a primary oocyte, and then in the nucleus of a primary spermatocyte. Each gives you a mass *m*. What mass of DNA would you expect to find in the nucleus of each mature gamete (egg and sperm) that forms after meiosis? What mass of DNA will be (1) in the nucleus of a zygote that forms at fertilization and (2) in that zygote's nucleus after the first DNA duplication?

5. The diploid chromosome number for the somatic cells of several eukaryotic species are listed at right. Write down the number of chromosomes that normally end up in gametes of each species. Then write what the number would be after three generations if meiosis did not occur before gamete formation.

Fruit fly, *Drosophila melanogaster*	8
Garden pea, *Pisum sativum*	14
Corn, *Zea mays*	20
Frog, *Rana pipiens*	26
Earthworm, *Lumbricus terrestris*	36
Human, *Homo sapiens*	46
Chimpanzee, *Pan troglodytes*	48
Amoeba, *Amoeba*	50
Horsetail, *Equisetum*	216

In Pursuit of a Better Rose

Researchers at Texas A&M and Clemson universities are breathing new life into *rose breeding*. People have been practicing this form of artificial selection for thousands of years. Starting with small, simple, five-petaled wild roses, they patiently cross-bred plants and in time were rewarded with a profusion of petals, fabulous fragrances, exquisite colors, and other compelling traits. Today, rose fanciers in thirty-six countries all over the world claim membership in the World Federation of Rose Societies. In any given year, people from all walks of life buy billions of dollars' worth of rosebuds and blooms. On Valentine's Day in the United States alone, 110 million cut roses are offered as symbols of love and romance. Roses are now big business.

Fossils in Colorado tell us that roses have been around for at least 40 million years. When rose breeding started, the ancestral stock had a diploid chromosome number— two sets of seven chromosomes. A great variety of cultivars now have four, seven, fourteen, even twenty-one sets of chromosomes! Within those chromosomes are genes that specify the size, number, and shape of petals and thorns, genes that deal with floral scents and colors, and genes that dictate whether plants bloom once or all year long. Other genes influence resistance to diseases and pests.

Unlike many wild roses, most of the cultivated varieties are susceptible to black spot, powdery mildew, and other diseases (Figure 11.1). The fungus that causes black spot is notably active in rainy, humid regions; the one that causes powdery mildew thrives in greenhouses. Fungicides work against pathogenic fungi, but they are costly, and many kinds also kill beneficial microorganisms.

Possibly a safer approach would be to cross-breed a wild plant known to have disease resistance with a plant known to be susceptible. However, traditional breeding practices are hit-or-miss, and they are tedious. Breeders have to wait for plants to form seeds, then plant the seeds, then observe whether any or all plants of the new generation, and the next, and the one after that show disease resistance.

Enter the new researchers. They are working to make genetic maps for all seven of the rose chromosomes. Just as road maps pinpoint cities along a highway, genetic maps can pinpoint where genes that influence specific traits are located along the length of chromosomes. By pinpointing a gene that influences a desired trait, breeders will be able to speed up their artificial selection practices.

For example, remember those radioisotopes described in Chapter 2? Researchers use them to make a DNA probe, a bit of radioactive DNA. They use it to test offspring from a cross for a specific DNA region—say, one near a gene that affects disease resistance. If the probe does not bind to the DNA of offspring, a breeder can assume the new plant has

Figure 11.1 One representative of a long history of artificial selection. Like most of the modern cultivars and unlike many wild roses, this one is vulnerable to black spot, a disease that results in the telltale destruction of leaves. Researchers are working to develop faster, more efficient ways to breed roses that have disease resistance and other desired traits.

not inherited the resistance gene and can try a new cross. Such *marker-assisted selection* is useful when several genes control a trait, as for disease resistance. A plant lineage that inherits all the genes may be the least vulnerable to attack.

In time, the maps being pieced together at Texas A&M, Clemson, and elsewhere will become consolidated into a permanent genetic map for roses. Its information will be retrieved for breeding programs. It also will be put to use for actual transfers of desirable genes into roses by way of biotechnology and genetic engineering. But these are cutting-edge topics that will not make much sense without in-depth knowledge of the structure and function of DNA, genes, and their protein products. We reserve them for later chapters in this unit. For now, start with something you already know about—the chromosomes and alleles introduced in the preceding chapter. This will be enough for you to follow the classical breeding practices that gave us our first glimpses of the principles of inheritance.

How Would You Vote?

The federal government helps support some agricultural extension programs that offer homeowners advice on gardens and ornamental plants. Do you consider this to be an appropriate use of government resources? See BiologyNow for details, then vote online.

Key Concepts

WHERE MODERN GENETICS STARTED

Gregor Mendel gathered the first experimental evidence of the genetic basis of inheritance: Each gene has a specific location on a chromosome. Organisms that have a diploid chromosome number have *pairs* of genes, at equivalent locations on pairs of homologous chromosomes. Alleles that are nonidentical may affect a trait differently. One allele is often dominant, in that its effect on a trait masks the effect of a recessive allele paired with it. Section 11.1

INSIGHTS FROM MONOHYBRID EXPERIMENTS

Some experiments yielded evidence of gene segregation: When one chromosome is separated from its homologous partner during meiosis, their pairs of alleles also separate and end up in different gametes. Section 11.2

INSIGHTS FROM DIHYBRID EXPERIMENTS

Other experiments yielded evidence of independent assortment: During meiosis, each pair of homologous chromosomes is sorted out for distribution into one gamete or another independently of how all of the other pairs of homologous chromosomes are assorted. Section 11.3

VARIATIONS IN GENE EXPRESSION

Not all traits have clearly dominant or recessive forms. One allele of a pair may be fully or partially dominant over its partner or codominant with it. Two or more gene pairs often influence the same trait, and some single genes influence many traits. The environment introduces more variation in gene expression. Sections 11.4–11.7

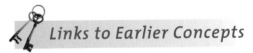

Links to Earlier Concepts

Before starting this chapter, review the definitions of genes, alleles, and diploid versus haploid chromosome numbers (Sections 10.1 and 10.2). As you read, you may wish to refer to the earlier introduction to natural selection (1.4) and to the visual road map for the stages of meiosis (10.3). You will be considering experimental evidence of two major topics that were introduced earlier—the effects that crossing over and metaphase I alignments have on inheritance (10.4).

11.1   Mendel, Pea Plants, and Inheritance Patterns

LINKS TO
SECTIONS
1.4, 10.1

We turn now to recurring inheritance patterns among humans and other sexually reproducing species. You already know meiosis halves the parental chromosome number, which is restored at fertilization. Here the story picks up with some observable outcomes of these events.

More than a century ago, people wondered about the basis of inheritance. Most had an idea that two parents contribute hereditary material to offspring, but few even suspected that it is organized in units, or genes.

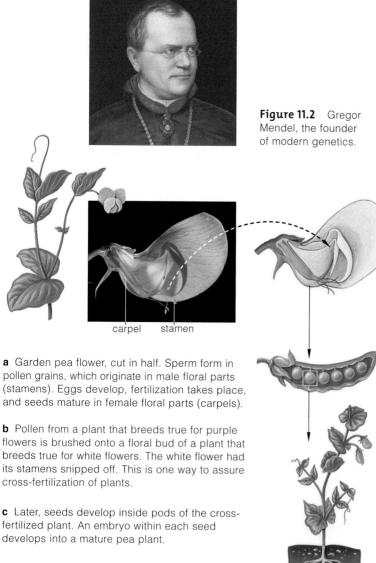

Figure 11.2 Gregor Mendel, the founder of modern genetics.

carpel   stamen

a Garden pea flower, cut in half. Sperm form in pollen grains, which originate in male floral parts (stamens). Eggs develop, fertilization takes place, and seeds mature in female floral parts (carpels).

b Pollen from a plant that breeds true for purple flowers is brushed onto a floral bud of a plant that breeds true for white flowers. The white flower had its stamens snipped off. This is one way to assure cross-fertilization of plants.

c Later, seeds develop inside pods of the cross-fertilized plant. An embryo within each seed develops into a mature pea plant.

d Each new plant's flower color is indirect but observable evidence that hereditary material has been transmitted from the parent plants.

Figure 11.3 *Animated!* Garden pea plant (*Pisum sativum*), which can self-fertilize or cross-fertilize. Experimenters can control the transfer of its hereditary material from one flower to another.

Rather, according to the prevailing view, hereditary material was fluid, with the fluids from both parents blending at fertilization like milk into coffee.

The idea of "blending inheritance" failed to explain the obvious. For example, many children who differ in eye color or hair color have the same two parents. If parental fluids blended, then the eye or hair color of children should be a blend of the parental colors. If neither parent had freckles, freckled children would never pop up. A white mare bred with a black stallion should consistently give birth to gray offspring, but as horse breeders knew, this was not always the case. Blending inheritance could scarcely explain much of the obvious variation in traits that people could see with their own eyes.

Even Charles Darwin accepted the blending notion until he and his cousin conducted experiments that disproved it. According to Darwin's theory of natural selection, individuals of a population show variation in traits. Over the generations, variations that help an individual survive and reproduce show up among more and more offspring, and less helpful variations become less frequent and might even disappear. Thus blending inheritance *seemed* to support the theory of natural selection. As it turned out, the idea of discrete units of information—genes—explain it better.

Even before Darwin presented his theory, someone was gathering evidence that eventually would help support it. A monk, Gregor Mendel (Figure 11.2), had already guessed that sperm and eggs carry distinct units of information about heritable traits. After he analyzed specific traits of pea plants, one generation after another, he found indirect but *observable* evidence of how parents transmit genes to offspring.

MENDEL'S EXPERIMENTAL APPROACH

Mendel spent most of his adult life in Brno, a city near Vienna that is now part of the Czech Republic. Yet he was not a man of narrow interests who accidentally stumbled onto dazzling principles.

Mendel's monastery was close to European capitals that were centers of scientific inquiry. Having been raised on a farm, he was keenly aware of agricultural principles and their applications. He kept abreast of literature on breeding experiments. He belonged to an agricultural society and won awards for developing improved varieties of vegetables and fruits. Shortly after entering the monastery, Mendel took courses in mathematics, physics, and botany at the University of Vienna. Few scholars of his time showed interest in both plant breeding *and* mathematics.

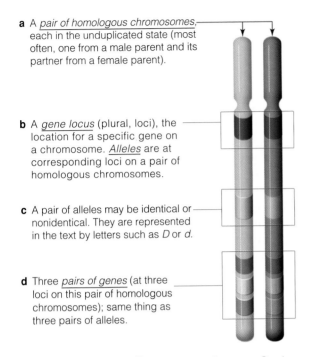

a A *pair of homologous chromosomes,* each in the unduplicated state (most often, one from a male parent and its partner from a female parent).

b A *gene locus* (plural, loci), the location for a specific gene on a chromosome. *Alleles* are at corresponding loci on a pair of homologous chromosomes.

c A pair of alleles may be identical or nonidentical. They are represented in the text by letters such as *D* or *d*.

d Three *pairs of genes* (at three loci on this pair of homologous chromosomes); same thing as three pairs of alleles.

Figure 11.4 *Animated!* A few genetic terms. Garden peas and other species with a diploid chromosome number have pairs of genes, on pairs of homologous chromosomes. Most genes come in slightly different molecular forms called alleles. Different alleles specify different versions of the same trait. An allele at any given location on a chromosome may or may not be identical to its partner on the homologous chromosome.

Shortly after his university training, Mendel started to study *Pisum sativum*, the garden pea plant (Figure 11.3). This plant is self-fertilizing. Its flowers produce both male and female gametes—call them sperm and eggs—that can come together and give rise to a new plant. One lineage of pea plants can "breed true" for certain traits. This means that successive generations will be like parents in one or more traits, as when all offspring grown from seeds of self-fertilized, white-flowered parent plants also have white flowers.

Pea plants also cross-fertilize when plant breeders transfer pollen from one plant to the flower of another plant. As Mendel knew, breeders open a floral bud of a plant that bred true for white flowers or some other trait and snip out its stamens. (Pollen grains, in which sperm develop, start forming in stamens.) The buds can be brushed with pollen from a plant that bred true for a *different* version of the trait—say, purple flowers.

As Mendel hypothesized, such clearly observable differences might help him track a given trait through many generations. If there were patterns to the trait's inheritance, *then those patterns might tell him something about heredity itself.*

TERMS USED IN MODERN GENETICS

In Mendel's time, no one knew about genes, meiosis, or chromosomes. As we follow his thinking, we will clarify the picture by substituting some modern terms used in inheritance studies, as stated here and in Figure 11.4:

1. **Genes** are units of information on heritable traits, which parents transmit to offspring. Each gene has a specific location (locus) in chromosomal DNA.

2. Cells with a **diploid** chromosome number ($2n$) have pairs of genes, on pairs of homologous chromosomes.

3. **Mutation** alters a gene's molecular structure and its message about a trait. It may cause a trait to change, as when a gene for flower color specifies yellow and a mutant form of the gene specifies white. All molecular forms of the same gene are known as **alleles**.

4. When offspring inherit a pair of *identical* alleles for a trait generation after generation, they typically are a true-breeding lineage. Offspring of a cross between two individuals that breed true for different forms of a trait are **hybrids**; each one has inherited *nonidentical* alleles for the trait.

5. A pair of identical alleles on a pair of homologous chromosomes is a *homozygous* condition. A pairing of nonidentical alleles is a *heterozygous* condition.

6. An allele is *dominant* when its effect on a trait masks the effect of any *recessive* allele paired with it. Capital letters signify dominant alleles, and lowercase letters signify recessive ones. *A* and *a* are examples.

7. Pulling this all together, a **homozygous dominant** individual has a pair of dominant alleles (*AA*) for the trait under study. A **homozygous recessive** individual has a pair of recessive alleles (*aa*), and a **heterozygous** individual has a pair of nonidentical alleles (*Aa*).

8. Two terms help keep the distinction clear between genes and the traits they specify. *Genotype* refers to the particular alleles that an individual carries. *Phenotype* refers to an individual's observable traits.

9. P stands for true-breeding parents, F_1 for the first-generation offspring, and F_2 for the second-generation offspring of self-fertilized or intercrossed F_1 individuals.

Mendel hypothesized that tracking clearly observable differences in forms of a given trait might reveal patterns of inheritance. He recognized patterns of dominance and recessiveness in certain traits, which later were connected with pairs of alleles on pairs of homologous chromosomes.

11.2 Mendel's Theory of Segregation

Mendel used monohybrid experiments to test a hypothesis: Pea plants inherit two "units" of information (genes) for a trait, one from each parent.

In **monohybrid experiments**, two homozygous parents differ in a trait that is governed by alleles of one gene. They are crossed to produce F_1 offspring that are all heterozygous ($AA \times aa \longrightarrow Aa$). Next, depending on the species, F_1 individuals are allowed to self-fertilize or mate in order to produce an F_2 generation.

MONOHYBRID EXPERIMENT PREDICTIONS

Mendel tracked seven traits for two generations. In one set of experiments, he crossed plants that bred true for purple *or* white flowers. All F_1 offspring had purple flowers, but in the next generation, some F_2 offspring had white flowers! So what was going on? Pea plants have pairs of homologous chromosomes. Assume one plant is homozygous dominant (AA) and another is homozygous recessive (aa) at the locus that governs flower color. Following meiosis, each sperm or egg that forms has only one of these alleles (Figure 11.5). Therefore, when a sperm fertilizes an egg, only one outcome is possible: $A + a \longrightarrow Aa$.

With his background in mathematics, Mendel knew about sampling error (Figure 1.12). He crossed seventy plants. He also counted and recorded the number of dominant and recessive forms of traits in thousands of offspring. On average, three of every four F_2 plants were dominant, and one was recessive (Figure 11.6).

The ratio hinted that fertilization is a chance event having a number of possible outcomes. Mendel knew about probability—*which applies to chance events and thus could help him predict possible outcomes of genetic crosses.* **Probability** means this: The chance that each outcome of an event will occur is proportional to the number of ways in which the outcome can be reached.

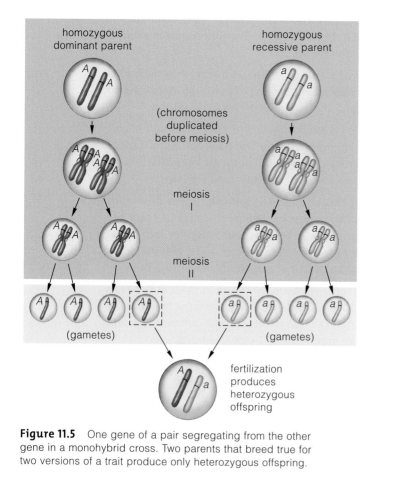

Figure 11.5 One gene of a pair segregating from the other gene in a monohybrid cross. Two parents that breed true for two versions of a trait produce only heterozygous offspring.

Figure 11.6 *Right,* from some of Mendel's monohybrid experiments with pea plants, counts of F_2 offspring having dominant or recessive hereditary "units" (alleles). On average, the 3:1 phenotypic ratio held for traits.

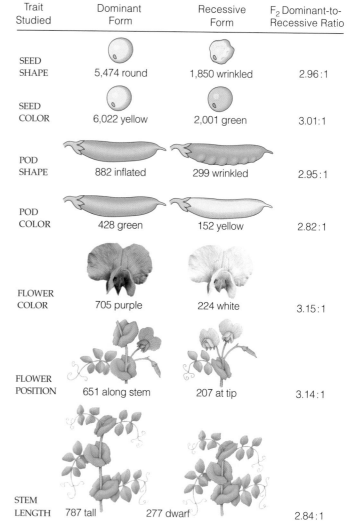

Trait Studied	Dominant Form	Recessive Form	F_2 Dominant-to-Recessive Ratio
SEED SHAPE	5,474 round	1,850 wrinkled	2.96:1
SEED COLOR	6,022 yellow	2,001 green	3.01:1
POD SHAPE	882 inflated	299 wrinkled	2.95:1
POD COLOR	428 green	152 yellow	2.82:1
FLOWER COLOR	705 purple	224 white	3.15:1
FLOWER POSITION	651 along stem	207 at tip	3.14:1
STEM LENGTH	787 tall	277 dwarf	2.84:1

A **Punnett-square method**, explained and applied in Figure 11.7, shows the possibilities. If half of a plant's sperm or eggs are *a* and half are *A*, then we can expect four outcomes with each fertilization:

POSSIBLE EVENT	PROBABLE OUTCOME
sperm *A* meets egg *A*	1/4 *AA* offspring
sperm *A* meets egg *a*	1/4 *Aa*
sperm *a* meets egg *A*	1/4 *Aa*
sperm *a* meets egg *a*	1/4 *aa*

Each F_2 plant has 3 chances in 4 of inheriting at least one dominant *A* allele (purple flowers). It has 1 chance in 4 of inheriting two recessive *a* alleles (white flowers). That is a probable phenotypic ratio of 3:1.

Mendel's observed ratios were not *exactly* 3:1. Yet he put aside the deviations. To understand why, flip a coin several times. As we all know, a coin is as likely to end up heads as tails. But often it ends up heads, or tails, several times in a row. If you flip the coin only a few times, the observed ratio might differ greatly from the predicted ratio of 1:1. Flip it many times, and you are more likely to approach the predicted ratio.

That is why Mendel used rules of probability and counted so many offspring. He minimized sampling error deviations in the observed results.

TESTCROSSES

Testcrosses supported Mendel's prediction. In such experimental tests, an organism shows dominance for a specified trait but its genotype may be unknown, so it is crossed with a homozygous recessive individual. The test results may reveal whether it is homozygous dominant or heterozygous.

For example, Mendel crossed F_1 purple-flowered plants with true-breeding white-flowered plants. If all were homozygous dominant, then F_2 offspring would all be purple flowered. If heterozygous, only about half would be. As it happened, about half of the testcross offspring had purple flowers (*Aa*) and half had white (*aa*). To predict outcomes of this testcross, construct a Punnett square.

The results from Mendel's monohybrid experiments became the basis of a theory of **segregation**, which we state here in modern terms:

> MENDEL'S THEORY OF SEGREGATION *Diploid cells have pairs of genes, on pairs of homologous chromosomes. The two genes of each pair are separated from each other during meiosis, so they end up in different gametes.*

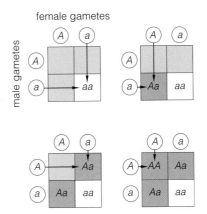

a Step-by-step construction of a Punnett square. Circles signify gametes. *A* and *a* signify a dominant and recessive allele, respectively. Possible genotypes among offspring are written in the squares.

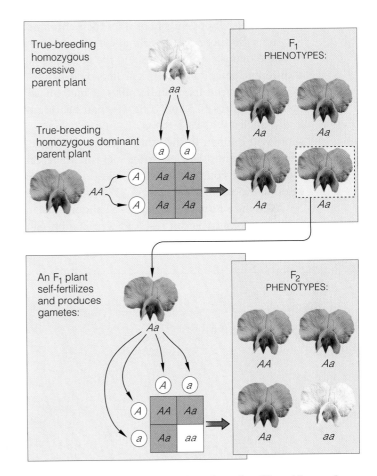

b Cross between two plants that breed true for different forms of a trait, followed by a monohybrid cross between their F_1 offspring.

Figure 11.7 *Animated!* (**a**) Punnett-square method of predicting probable outcomes of genetic crosses. (**b**) Results from one of Mendel's monohybrid experiments. On average, the ratio of dominant-to-recessive that showed up among second-generation (F_2) plants was 3:1.

11.3 Mendel's Theory of Independent Assortment

Mendel used dihybrid experiments to explain how two pairs of genes are sorted into gametes.

Dihybrid experiments start with a cross between true-breeding homozygous parents that differ in two traits governed by alleles of two genes. The F_1 offspring are all heterozygous for the alleles of both genes.

Let's duplicate one of Mendel's dihybrid crosses for flower color (alleles *A* or *a*) and for height (*B* or *b*):

True-breeding parents: *AABB* X *aabb*

Gametes: *AB AB* *ab ab*

F_1 hybrid offspring: *AaBb*

As Mendel would have predicted, F_1 offspring from this cross are all purple-flowered and tall (*AaBb*).

How will genes that control these traits assort in the F_1 plants? It depends in part on their chromosome locations. Suppose that the *Aa* alleles are on one pair of homologous chromosomes and the *Bb* alleles are on a different pair. Remember, chromosome pairs align midway between the spindle poles at metaphase I of meiosis (Figures 10.5 and 11.8). The pair bearing the *A* and *a* alleles will be tethered to opposite poles. The same will happen to the other chromosome pair that bears the *B* and *b* alleles. After meiosis, there can be four possible combinations of alleles in the sperm or eggs that form: 1/4 *AB*, 1/4 *Ab*, 1/4 *aB*, and 1/4 *ab*.

Given the alternative metaphase I alignments, many allelic combinations can result at fertilization. Simple

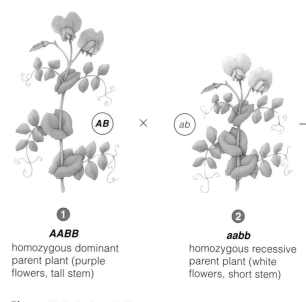

①

AABB

homozygous dominant parent plant (purple flowers, tall stem)

②

aabb

homozygous recessive parent plant (white flowers, short stem)

Figure 11.9 *Animated!* Results from one of Mendel's dihybrid experiments with the garden pea plant. The parent plants were true-breeding for different versions of two traits: flower color and plant height. *A* and *a* signify the dominant and recessive alleles for flower color. *B* and *b* signify dominant and recessive alleles for height. The Punnett square on the facing page shows all of the allelic combinations possible in the F_2 generation.

Adding up the corresponding F_2 phenotypes, we get:

☐ 9/16 or 9 purple-flowered, tall

☐ 3/16 or 3 purple-flowered, dwarf

☐ 3/16 or 3 white-flowered, tall

☐ 1/16 or 1 white-flowered, dwarf

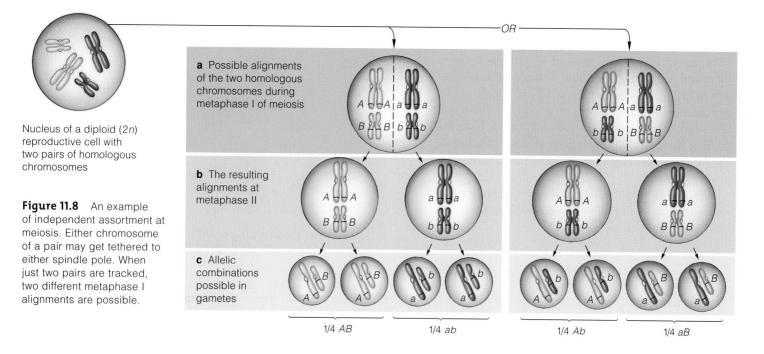

Nucleus of a diploid (2n) reproductive cell with two pairs of homologous chromosomes

Figure 11.8 An example of independent assortment at meiosis. Either chromosome of a pair may get tethered to either spindle pole. When just two pairs are tracked, two different metaphase I alignments are possible.

—OR—

a Possible alignments of the two homologous chromosomes during metaphase I of meiosis

b The resulting alignments at metaphase II

c Allelic combinations possible in gametes

1/4 *AB*   1/4 *ab*   1/4 *Ab*   1/4 *aB*

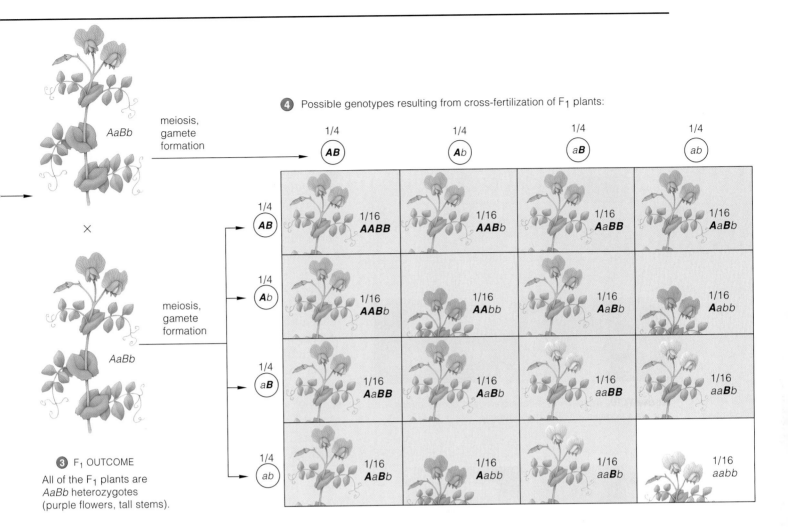

4 Possible genotypes resulting from cross-fertilization of F₁ plants:

	1/4 **AB**	1/4 **Ab**	1/4 **aB**	1/4 **ab**
1/4 **AB**	1/16 **AABB**	1/16 **AABb**	1/16 **AaBB**	1/16 **AaBb**
1/4 **Ab**	1/16 **AABb**	1/16 **AAbb**	1/16 **AaBb**	1/16 **Aabb**
1/4 **aB**	1/16 **AaBB**	1/16 **AaBb**	1/16 **aaBB**	1/16 **aaBb**
1/4 **ab**	1/16 **AaBb**	1/16 **Aabb**	1/16 **aaBb**	1/16 **aabb**

AaBb

meiosis, gamete formation

×

AaBb

meiosis, gamete formation

3 F₁ OUTCOME

All of the F₁ plants are *AaBb* heterozygotes (purple flowers, tall stems).

multiplication (four sperm types × four egg types) tells us that sixteen combinations of gametes are possible among F₂ offspring of a dihybrid cross (Figure 11.9).

Adding all possible phenotypes gives us a ratio of 9:3:3:1. We can expect to see 9/16 tall purple-flowered, 3/16 dwarf purple-flowered, 3/16 tall white-flowered, and 1/16 dwarf white-flowered F₂ plants. The results from the dihybrid experiment that Mendel reported were close to this ratio.

Mendel analyzed the numerical results from such experiments, but he did not know that seven pairs of homologous chromosomes carry a pea plant's "units" of inheritance. He could only hypothesize that two units for flower color were sorted out into gametes independently of the two units for height.

In time, his hypothesis became known as the theory of **independent assortment**. In modern terms, after meiosis ends, the genes on each pair of homologous chromosomes are sorted into gametes independently of how genes on other pairs of homologues are sorted out. Independent assortment and segregation give rise to genetic variation. In a monohybrid cross for one gene pair, three genotypes are possible: *AA*, *Aa*,

and *aa*. We represent this as 3^n, where n is the number of gene pairs. The more pairs, the more combinations are possible. If, say, the parents differ in twenty gene pairs, the number approaches 3.5 billion!

In 1866 Mendel published his work. Apparently his article was read by few and understood by no one. In 1871 he became monastery abbot, and his pioneering experiments ended. He died in 1884, never to know that his experiments would be the starting point for modern genetics. Mendel's theory of segregation still stands for most genes in most organisms: the units of hereditary material (genes) do retain their identity all through meiosis. However, his theory of independent assortment requires qualification, because the alleles of gene pairs do not *always* assort independently into gametes, as Section 11.5 explains.

MENDEL'S THEORY OF INDEPENDENT ASSORTMENT *As meiosis ends, genes on pairs of homologous chromosomes have been sorted out for distribution into one gamete or another, independently of gene pairs on other chromosomes.*

11.4 More Patterns Than Mendel Thought

LINKS TO
SECTIONS
4.6, 6.6

Mendel happened to focus on traits that have clearly dominant and recessive forms. However, expression of genes for some traits is not as straightforward.

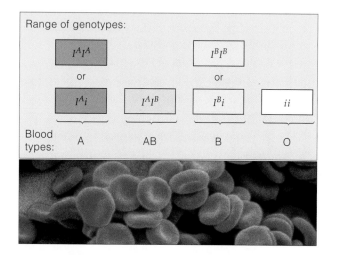

Range of genotypes:

Blood types: A   AB   B   O

Figure 11.10 *Animated!* Possible allelic combinations that are the basis for ABO blood typing.

homozygous parent   x   homozygous parent

All F₁ offspring heterozygous for flower color:

Cross two of the F₁ plants, and the F₂ offspring will show three phenotypes in a 1:2:1 ratio:

Figure 11.11 Incomplete dominance in heterozygous (*pink*) snapdragons, in which an allele that affects red pigment is paired with a "white" allele.

CODOMINANCE IN ABO BLOOD TYPES

In *codominance*, a pair of nonidentical alleles affecting two phenotypes are both expressed at the same time in heterozygotes. For example, red blood cells have a type of glycolipid at the plasma membrane that helps give them their unique identity. The glycolipid comes in slightly different forms. An analytical method, *ABO blood typing*, reveals which form a person has.

An enzyme dictates the glycolipid's final structure. Three alleles for this enzyme are present in all human populations. Two, I^A and I^B, are codominant when paired. (These superscripts represent two dominant alleles for the gene.) The third allele, *i*, is recessive when paired with I^A or I^B. The occurrence of three or more alleles for a single gene locus among individuals of a population is called a **multiple allele system**.

Each of these glycolipid molecules was assembled in the endomembrane system (Section 4.6). First, an oligosaccharide chain was attached to a lipid, then a series of sugars was attached to the chain. But alleles I^A and I^B specify different forms of the enzyme that attaches the last sugar. The two attach *different* sugars, which gives the glycolipid a different identity: A or B.

If you have $I^A I^A$ or $I^A i$, your blood is type A. With $I^B I^B$ or $I^B i$, it is type B. With codominant alleles $I^A I^B$, it is AB; you have both versions of the sugar-attaching enzyme. If you are (*ii*), the glycolipid molecules never did get a final sugar on the side chain, so your blood type is not A or B. It is O. Figure 11.10 is a simple way to think about these combinations.

INCOMPLETE DOMINANCE

In *incomplete* dominance, one allele of a pair is not fully dominant over its partner, so the heterozygote's phenotype is *somewhere between* the two homozygotes. Cross true-breeding red and white snapdragons and their F₁ offspring will be pink-flowered. Cross two F₁ plants and you can expect to see red, white, and *pink* flowers in a particular ratio (Figure 11.11). Why? Red snapdragons have two alleles that let them make a lot of molecules of a red pigment. White snapdragons have two mutant alleles and are pigment-free. Pink snapdragons have a "red" allele and a "white" allele; these genotypes have not "blended." Heterozygotes make enough pigment to color flowers pink, not red.

Two interacting gene pairs also can give rise to a phenotype that neither produces by itself. In chickens, interactions among alleles at the *R* and *P* gene loci specify walnut, rose, pea, and single combs, as shown in Figure 11.12.

EPISTASIS

Traits also arise through **epistasis**: interactions among products of two or more gene pairs. Two alleles might mask expression of another gene's alleles, and some expected phenotypes might not appear at all.

As an example, several gene pairs govern whether a Labrador retriever has black, yellow, or brown fur (Figure 11.13). Its coat color depends on how enzymes and other products of alleles at more than one gene locus make a dark pigment, melanin, and deposit it in tissues. Allele *B* (black) is dominant to *b* (brown). At a different locus, allele *E* promotes melanin deposition but two recessive alleles (*ee*) reduce it. In this case, fur appears yellow regardless of alleles at the *B* locus.

SINGLE GENES WITH A WIDE REACH

Alleles at one locus on a chromosome may affect two or more traits in good or bad ways, an outcome called **pleiotropy**. Many genetic disorders, including cystic fibrosis, sickle-cell anemia, and Marfan syndrome, are examples. *Marfan syndrome* arises from an autosomal dominant mutation of the gene for fibrillin, a protein in the most abundant, widespread vertebrate tissues —connective tissues. Thin, loose or crosslinked strands of fibrillin passively recoil after being stretched, as by the beating heart.

Altered fibrillin weakens the connective tissues in 1 of 10,000 men and women and puts the heart, blood vessels, skin, lungs, and eyes at risk. One mutation disrupts the synthesis of fibrillin 1, its secretion from cells, and its tissue deposition. It alters the structure and function of smooth muscle cells inside the wall of the aorta, a big vessel carrying blood out of the heart. Immune cells infiltrate and multiply inside the wall's lining. Calcium deposits accumulate and inflame the wall. Elastic fibers split into fragments. The aorta wall, thinned and weakened, can rupture abruptly during strenuous exercise. Until recent advances in medicine, Marfan syndrome killed most affected people before their fifties. Flo Hyman was one of them (Figure 11.14).

> An allele at a given gene locus may be fully dominant, incompletely dominant, or codominant with its partner on a homologous chromosome.
>
> Some gene products may interact with each other and influence the same trait through epistasis.
>
> A single gene's product may have pleiotropic effects, or positive or negative impact on two or more traits.

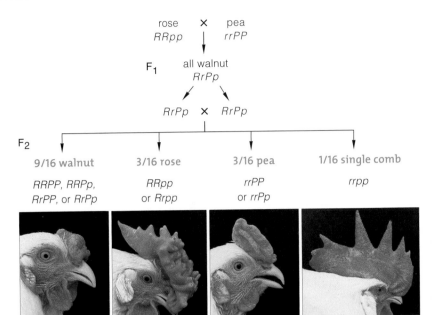

Figure 11.12 Polygenic inheritance in chickens. Interactions among alleles at two gene loci have variable effects on the comb on a chicken's head. The first cross is between a Wyandotte (rose comb) and a Brahma (pea comb).

a BLACK LABRADOR   **b** YELLOW LABRADOR   **c** CHOCOLATE LABRADOR

Figure 11.13 Coat color among Labrador retrievers. The trait arises through epistatic interactions among alleles of two genes.

Figure 11.14 Flo Hyman, left, captain of the United States volleyball team that won an Olympic silver medal in 1984. Two years later, at a game in Japan, she slid to the floor and died. A dime-sized weak spot in the wall of her aorta had burst. We know at least two affected college basketball stars also died abruptly as a result of Marfan syndrome.

11.5 Impact of Crossing Over on Inheritance

LINKS TO
SECTIONS
3.2, 10.3, 10.4

Crossing over between homologous chromosomes is one of the main pattern-busting events in inheritance.

We now know there are many genes on each type of autosome and sex chromosome. All the genes on one chromosome are called a **linkage group**. For instance, the fruit fly (*Drosophila melanogaster*) has four linkage groups, corresponding to its four pairs of homologous chromosomes. Indian corn (*Zea mays*) has ten linkage groups, corresponding to its ten pairs, and so on.

If genes on the same chromosome stayed together through meiosis, then there would be no surprising mixes of parental traits. You could expect parental phenotypes among, say, the F$_2$ offspring of dihybrid experiments to show up in a predictable ratio. As early experiments with fruit flies showed, however, that ratio was often predictably different for linked genes. In one dihybrid experiment, 17 percent of the F$_2$ offspring inherited a new combination of alleles that did not occur in either of their parents.

Many genes on the same chromosome do not stay linked through meiosis, but some stay together more often than others. Why? They are closer together on the chromosome, and so they are separated less often by crossing over. *The probability that crossing over will disrupt the linkage between any two genes is proportional to the distance between the two genes.*

If genes *A* and *B* are twice as far apart as genes *C* and *D* on a chromosome, then we can expect crossing over to disrupt the linkage between genes *A* and *B* more frequently than between the other two genes:

Two genes are very closely linked when the distance between them is small. Their combinations of alleles nearly always end up in the same gamete. Linkage is more vulnerable to crossing over when the distance between two gene loci is greater (Figure 11.15). When two loci are far apart, crossing over is so frequent that the genes assort independently into gametes.

Human gene linkages were identified by tracking DNA inheritance in families over the generations. One thing is clear from such studies: Crossovers are not rare. For most eukaryotes, meiosis cannot even be completed properly until at least one crossover occurs between each pair of homologous chromosomes.

All of the genes at different locations along the length of a chromosome belong to the same linkage group.

Crossing over between homologous chromosomes disrupts gene linkages and results in nonparental combinations of alleles in chromosomes.

The farther apart two genes are on a chromosome, the greater will be the frequency of crossing over and genetic recombination between them.

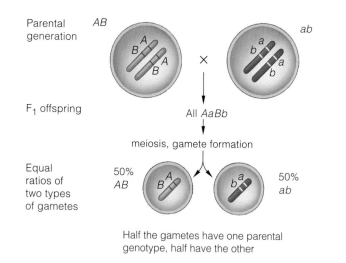

a Full linkage between two genes; no crossing over. Genes very close together along the length of the same chromosome typically stay together during gamete formation.

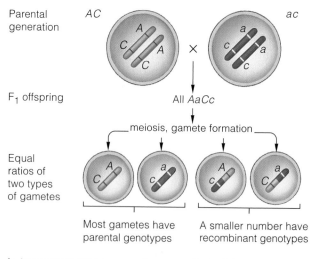

b Incomplete linkage; crossing over affected the outcome. Any two genes that are far apart along the length of a chromosome are more vulnerable to crossing over.

Figure 11.15 *Animated!* Examples of outcomes of crossing over between two gene loci: (**a**) full linkage and (**b**) incomplete linkage.

11.6 Genes and the Environment

The environment often contributes to variable gene expression among a population's individuals.

Possibly you have noticed a Himalayan rabbit's coat color. Like a Siamese cat, this mammal has dark hair in some parts of its body and lighter hair in others. The Himalayan rabbit is homozygous for the c^h allele of the gene specifying tyrosinase. Tyrosinase is one of the enzymes involved in melanin production. The c^h allele specifies a heat-sensitive form of this enzyme. This form is active only when the temperature around body cells is below 33°C, or 91°F.

When cells that give rise to this rabbit's hair grow under warmer conditions, they cannot make melanin, so hairs appear light. This happens in body regions that are massive enough to conserve a fair amount of metabolic heat. The ears and other slender extremities tend to lose metabolic heat faster, so they are cooler. Figure 11.16 shows one experiment that demonstrated how the environmental temperature can influence the production of melanin.

One classic experiment identified environmental effects on yarrow plants. These plants can grow from cuttings, so they are a useful experimental organism. Why? Cuttings from the same plant all have the same genotype, so experimenters can discount genes as a basis for differences that show up among them.

In this study, cuttings (clones) from each of several yarrow plants were grown at three elevations. The researchers periodically observed the growth of the plants in their habitats. They found that cuttings from the same parent plants grew differently at different altitudes. For example, cuttings from one plant grew tall at the lowest and the highest elevation, but a third cutting remained short at mid-elevation (Figure 11.17). Even though these plants were genetically identical, their phenotypes differed in different environments.

Similarly, plant a hydrangea in a garden and it may have pink flowers instead of the expected blue ones. Soil acidity affects the function of gene products that color hydrangea flowers.

What about humans? One of our genes codes for a transporter protein that moves serotonin across the plasma membrane of brain cells. This gene product has several effects, one of which is to counter anxiety and depression when traumatic events challenge us. For a long time, researchers have known that some people handle stress without getting too upset, while others spiral into a deep and lasting depression.

Mutation of the gene for the serotonin transporter compromises responses to stress. It is as if some of us are bicycling through life without an emotional helmet.

Figure 11.16 *Animated!* Observable effect of an environmental factor that alters gene expression. A Himalayan rabbit normally has black hair only on its long ears, nose, tail, and leg regions farthest from the body mass. In one experiment, a patch of a rabbit's white coat was removed and an icepack was placed over the hairless patch. Where the colder temperature had been maintained, the hairs that grew back were black.

Himalayan rabbits are homozygous for an allele that encodes a mutant version of tyrosinase, an enzyme required to make melanin. As described in the text, this allele encodes a heat-sensitive form of the enzyme, which functions only when air temperature is below about 33°C.

a Mature cutting at high elevation (3,060 meters above sea level)

b Mature cutting at mid-elevation (1,400 meters above sea level)

c Mature cutting at low elevation (30 meters above sea level)

Figure 11.17 Experiment demonstrating the impact of environmental conditions of three different habitats on phenotype in yarrow (*Achillea millefolium*). Cuttings from the same parent plant were grown in the same kind of soil at three different elevations.

Only when we take a fall does the phenotypic effect—depression—appear. Other genes also affect emotional states, but mutation of this particular gene reduces our capacity to snap out of it when bad things happen.

Variation in traits arises not only from gene mutations and interactions, but also in response to variations in environmental conditions that each individual faces.

11.7 Complex Variations in Traits

For most populations or species, individuals show rich variation for many of the same traits. Sometimes the phenotypes cannot be predicted, and most of the time they are part of a continuous range of variation.

REGARDING THE UNEXPECTED PHENOTYPE

Think back on Mendel's dihybrid crosses. Nearly all of the traits that he tracked occurred in predictable ratios because the two genes happened to be on different chromosomes or far apart on the same chromosome. They tended to segregate cleanly. Track two or more different pairs of genes—as Mendel did—and you might observe phenotypes that you would not have predicted at all. And not all of the variation is a result of tight linkage or crossing over.

As one example, *camptodactyly*, a rare abnormality, affects the shape and movement of fingers. Some of the people who carry a mutant allele for this heritable trait have immobile, bent fingers on both hands. Others have immobile, bent fingers on the left or right hand only. Fingers of still other people who have the mutant allele are not affected in any obvious way at all.

What causes such odd variation? Remember, most organic compounds are synthesized by a sequence of metabolic steps. *Different enzymes, each a gene product, control different steps.* One gene may have mutated in a number of ways. A gene product might be blocking some pathway or making it run nonstop or not long enough. Perhaps poor nutrition or some other variable factor in the individual's environment is influencing the activity of one of the pathway's enzymes. Such variable factors can introduce big or small variations even in otherwise expected phenotypes.

CONTINUOUS VARIATION IN POPULATIONS

Another point: Individuals of populations generally show a range of small differences in most traits. This feature of natural populations is known as **continuous variation**. It arises through **polygenic inheritance**, or the inheritance of multiple genes that affect the same trait. The distribution of all forms of a trait becomes more and more continuous when greater numbers of genes and environmental factors are involved.

Look in a mirror at your eye color. The colored part is the iris, a doughnut-shaped, pigmented structure just under the cornea (Figure 11.18). The color results from several gene products. Some products help make and distribute different kinds and amounts of melanins, which are similar to the light-absorbing pigment that affects coat color in mammals. Almost black irises have

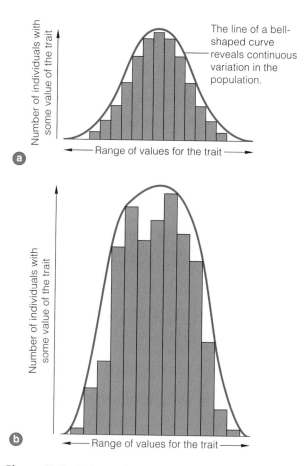

a The line of a bell-shaped curve reveals continuous variation in the population.

Number of individuals with some value of the trait

Range of values for the trait

b Range of values for the trait

Number of individuals with some value of the trait

Figure 11.19 *Animated!* Continuous variation. (**a**) A bar graph can reveal continuous variation in a population. The proportion of individuals in each category is plotted against the range of measured phenotypes. (**b**) The curved line above this particular set of bars is a real-life example of a bell-shaped curve that emerged for the population in Figure 11.20. It reflects continuous variation in body height, one of the traits that help characterize human populations.

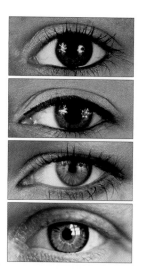

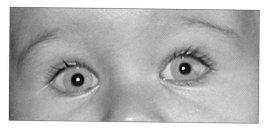

Figure 11.18 Sampling of the range of continuous variation in human eye color. Products of different gene pairs interact in making and distributing the pigment melanin, which helps color the iris. Small color differences arise from different combinations of alleles. The frequency distribution for the eye-color trait is continuous over a far larger range than this, from black to light blue.

dense melanin deposits, which can absorb most of the incoming light. Deposits are not as extensive in brown eyes, so some unabsorbed light is reflected out. Light brown or hazel eyes have even less melanin.

Green, gray, or blue eyes have lesser amounts of the pigments. Many or most of the blue wavelengths of light that enter the eyeball are simply reflected out.

How can you describe the continuous variation of some trait in a group? Divide the range of phenotypes for a trait—say, height—into measurable categories, such as numbers of inches. Next, do a count of how many individuals fall into each category; this will give you the relative frequencies of phenotypes across the range of measurable values. Finally, plot out the data as a bar chart, such as the one in Figure 11.19a.

In this figure, the shortest bars represent categories having the fewest individuals. The tallest bar signifies the category that has the most individuals. In this case, a graph line skirting the top of all of the bars will be a bell-shaped curve. Such **bell curves** are typical of

any trait showing continuous variation. Figure 11.19b is a bell curve based on real-life measurements at the University of Florida (Figure 11.20).

And so we conclude this chapter, which introduces heritable and environmental factors that give rise to great variation in traits. What is the take-home lesson? Simply this: An individual's phenotype is an outcome of complex interactions among its genes, enzymes and other gene products, and the environment. Chapter 18 will consider some of the evolutionary consequences.

Enzymes and other gene products control each step of most metabolic pathways. Mutations, interactions among genes, and environmental conditions typically affect one or more steps in ways that contribute to variation in phenotypes.

Individuals of populations or species show continuous variation—a range of small differences. Usually, the more genes and environmental factors that influence a trait, the more continuous the distribution of phenotypes.

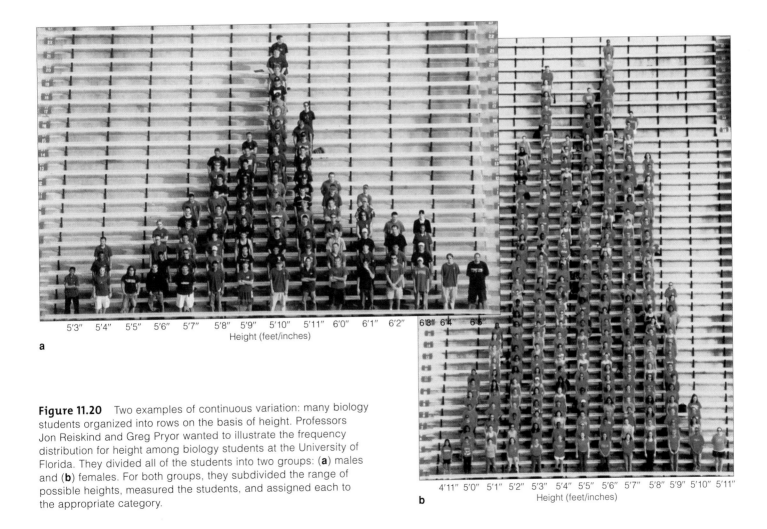

Figure 11.20 Two examples of continuous variation: many biology students organized into rows on the basis of height. Professors Jon Reiskind and Greg Pryor wanted to illustrate the frequency distribution for height among biology students at the University of Florida. They divided all of the students into two groups: (**a**) males and (**b**) females. For both groups, they subdivided the range of possible heights, measured the students, and assigned each to the appropriate category.

Summary

Section 11.1 Genes are heritable units of information about traits. Each gene has its own locus, or location, along the length of a particular chromosome. Different molecular forms of the same gene are known as alleles.

By experimenting with garden pea plants, Mendel was the first to gather evidence of patterns by which genes are transmitted from parents to offspring.

Offspring of a cross between two individuals that breed true for different forms of a trait are hybrids; each inherited nonidentical alleles for a trait being studied.

An individual with two dominant alleles for a trait (AA) is homozygous dominant. A homozygous recessive has two recessive alleles (aa). A heterozygote has two nonidentical alleles (Aa) for a trait. A dominant allele may mask the effect of a recessive allele partnered with it on the homologous chromosome.

Genotype refers to the particular alleles at any or all gene locations on an individual's chromosomes. *Phenotype* refers to an individual's observable traits.

Biology Now
Learn how Mendel crossed garden pea plants and the definitions of important genetic terms on BiologyNow.

Section 11.2 A cross between parents of different genotypes yields hybrid offspring. For monohybrid experiments, two parents that bred true for different forms of a trait produce F_1 heterozygotes that are identical for one pair of genes. Mendel's monohybrid experiments gave indirect evidence that some forms of a gene may be dominant over recessive forms.

All F_1 offspring of a parental cross $AA \times aa$ were Aa. Crosses between F_1 monohybrids resulted in these allelic combinations among the F_2 offspring:

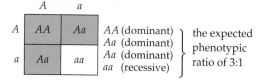

	A	a
A	AA	Aa
a	Aa	aa

AA (dominant)
Aa (dominant)
Aa (dominant)
aa (recessive)
} the expected phenotypic ratio of 3:1

Mendel's monohybrid experiment results led to a theory of segregation: Diploid organisms have pairs of genes, on pairs of homologous chromosomes. Genes of each pair segregate from each other at meiosis, so each gamete formed gets one or the other gene.

Biology Now
Carry out monohybrid experiments with the interaction on BiologyNow.

Section 11.3 Dihybrid experiments start with a cross between true-breeding heterozygous parents that differ for alleles of two genes ($AABB \times aabb$). All F_1 offspring are heterozygous for both genes ($AaBb$). In Mendel's dihybrid experiments, phenotypes of the F_2 offspring of F_1 hybrids were close to a 9:3:3:1 ratio:

9 dominant for both traits
3 dominant for A, recessive for b
3 dominant for B, recessive for a
1 recessive for both traits

His results support a theory of independent assortment: Before gamete formation, meiosis assorts gene pairs of homologous chromosomes independently of how gene pairs of all the other chromosomes are sorted. Random alignment of all pairs of homologous chromosomes at metaphase I is the basis of this outcome.

Biology Now
Observe the results of a dihybrid cross with the interaction on BiologyNow.

Section 11.4 Inheritance patterns are not always straightforward.

Some alleles are not fully dominant over their partner allele on the homologous chromosomes, and both are expressed at the same time. The phenotype that results from this allelic combination is somewhere between the two homozygous conditions.

Some alleles are codominant and are expressed at the same time in heterozygotes. An example occurs in the multiple allele system underlying ABO blood typing.

Also, products of one or more genes commonly interact in ways that influence the same trait, and a single gene may have effects on two or more traits.

Biology Now
Explore patterns of non-Mendelian inheritance with the interactions on BiologyNow.

Section 11.5 A linkage group consists of all genes along the length of one chromosome. Crossing over between pairs of homologous chromosomes disrupts expected inheritance patterns by breaking linkages. Its outcome is nonparental combinations of alleles in gametes. The farther apart two genes are on a chromosome, the greater will be the frequency of crossing over and genetic recombination between them.

Section 11.6 Environmental factors also can alter how genes are expressed in individuals of a population. An example is a difference in temperature that affects the activity of a heat-sensitive form of an enzyme—a gene product—that helps produce a coat color pigment.

Biology Now
See how the environment can affect phenotype with animation on BiologyNow.

Section 11.7 Gene interactions and environmental factors influence many enzymes differently among individuals, and many phenotypes result. They also contribute to small, incremental differences—a range of continuous variation—in a population.

Biology Now
Plot the continuous distribution of height for a class with the interaction on BiologyNow.

Self-Quiz   *Answers in Appendix II*

1. Alleles are _____ .
   a. different molecular forms of a gene
   b. different phenotypes
   c. self-fertilizing, true-breeding homozygotes

2. A heterozygote has a _____ for a trait being studied.
   a. pair of identical alleles
   b. pair of nonidentical alleles
   c. haploid condition, in genetic terms

3. The observable traits of an organism are its _____ .
   a. phenotype   c. genotype
   b. sociobiology   d. pedigree

4. Second-generation offspring of a cross between parents who are homozygous for different alleles are the _____ .
   a. F_1 generation   c. hybrid generation
   b. F_2 generation   d. none of the above

5. F_1 offspring of the cross $AA \times aa$ are _____ .
   a. all AA c. all Aa
   b. all aa d. 1/2 AA and 1/2 aa

6. Refer to Question 5. Assuming complete dominance, the F_2 generation will show a phenotypic ratio of _____ .
   a. 3:1   b. 9:1   c. 1:2:1   d. 9:3:3:1

7. Crosses between two dihybrid F_1 pea plants, which are offspring from a parental cross $AABB \times aabb$, result in F_2 phenotypic ratios close to _____ .
   a. 1:2:1   b. 3:1   c. 1:1:1:1   d. 9:3:3:1

8. The probability of a crossover occurring between two genes on the same chromosome is _____ .
   a. unrelated to the distance between them
   b. increased if they are close together
   c. increased if they are far apart

9. Two genes that are close together on the same chromosome are _____ .
   a. linked   c. homologous   e. all of the
   b. identical alleles d. autosomes   above

10. Match each example with the most suitable description.
   _____ dihybrid experiment   a. bb
   _____ monohybrid experiment   b. $AABB \times aabb$
   _____ homozygous condition   c. Aa
   _____ heterozygous condition   d. $Aa \times Aa$

Additional questions are available on **Biology 🌀 Now™**

Genetics Problems   *Answers in Appendix III*

1. A gene encodes the second enzyme in a melanin-synthesizing pathway. An individual who is homozygous for a recessive mutant allele of this gene cannot produce or deposit melanin in body tissues. *Albinism,* the absence of melanin, is the result.

   Humans and a number of other organisms can have this phenotype. Figure 11.21 shows two examples. In the following situations, what are the possible genotypes of the father, the mother, and their children?

   a. Both parents have normal phenotypes; some of their children are albino and others are unaffected.

   b. Both parents are albino and have albino children.

   c. The woman is unaffected, the man is albino, and they have one albino child and three unaffected children.

2. As rose breeders know, several alleles influence specific traits, such as long, symmetrical, urn-shaped buds, double flowers, glossy leaves, and resistance to mildew (Figure 11.22). Alleles of a single gene govern whether a plant will

Figure 11.21 Two albino organisms. By not posing his subjects as objects of ridicule, the photographer of human albinos is attempting to counter the notion that there is something inherently unbeautiful about them.

dominant   dominant

recessive   recessive

Figure 11.22 (**a**) Climbing rose and (**b**) shrub rose. (**c**) Globe-shaped buds versus (**d**) urn-shaped buds.

be a climber (dominant) or shrubby (recessive). When a true-breeding climber is crossed with a shrubby plant, all F_1 offspring are climbers. If an F_1 plant is crossed with a shrubby plant, about 50 percent of the offspring will be shrubby and 50 percent will be climbers. Using symbols A and a to represent the dominant and recessive alleles, make a Punnett-square diagram of the expected genotypic and phenotypic outcomes in the F_1 offspring and the offspring of the cross between an F_1 plant and a shrubby plant.

Figure 11.23 The Manx, a breed of cat that has no tail.

3. One gene has alleles *A* and *a*. Another has alleles *B* and *b*. For each of the following genotypes, what type(s) of gametes will form, assuming independent assortment during meiosis occurs?

    a. *AABB*   c. *Aabb*

    b. *AaBB*   d. *AaBb*

4. Refer to Problem 3. What will be the genotypes of offspring from the following matings? Indicate the frequencies of each genotype among them.

    a. *AABB* × *aaBB*   c. *AaBb* × *aabb*

    b. *AaBB* × *AABb*   d. *AaBb* × *AaBb*

5. Return to Problem 3. Assume you now study a third gene having alleles *C* and *c*. For each genotype listed, what type(s) of gametes will be produced, assuming that independent assortment occurs?

    a. *AABBCC*   c. *AaBBCc*

    b. *AaBBcc*   d. *AaBbCc*

6. Certain alleles are so essential for normal development that an individual who is homozygous recessive for a mutant form cannot survive. Such recessive, *lethal alleles* can be perpetuated in the population by heterozygotes.

    Consider the allele *Manx* (M^L) in cats. Homozygous cats ($M^L M^L$) die when they are still embryos inside the mother cat. In heterozygotes ($M^L M$), the spine develops abnormally. The cats end up with no tail (Figure 11.23).

    Two $M^L M$ cats mate. What is the probability that any one of their *surviving* kittens will be heterozygous?

7. In one experiment, Mendel crossed a pea plant that bred true for green pods with one that bred true for yellow pods. All the F_1 plants had green pods. Which form of the trait (green or yellow pods) is recessive? Explain how you arrived at your conclusion.

8. Mendel crossed a pea plant that produced plump and rounded seeds with a pea plant that produced wrinkled seeds. In the F_1 generation, all seeds were round. Mendel planted the F_1 seeds, which grew into plants that, when self-fertilized, produced 5,474 round seeds and 1,850 wrinkled seeds in the F_2 generation. The alleles that govern seed shape are designated *R* and *r*.

    a. What are the genotypes of the parents?

    b. What are the possible outcomes of a cross between a homozygous round-seeded plant and a wrinkle-seeded plant?

9. Mendel crossed a true-breeding tall, purple-flowered pea plant with a true-breeding dwarf, white-flowered plant. All F_1 plants were tall and had purple flowers. If an F_1 plant self-fertilizes, then what is the probability that a randomly selected F_2 offspring will be heterozygous for the genes specifying height and flower color?

10. Suppose you identify a new gene in mice. One of its alleles specifies white fur. A second allele specifies brown fur. You want to determine whether the relationship between the two alleles is one of simple dominance or incomplete dominance. What sorts of genetic crosses would give you the answer? What types of observations would you require to form conclusions?

11. In sweet pea plants, an allele for purple flowers (*P*) is dominant to an allele for red flowers (*p*). An allele for long pollen grains (*L*) is dominant to an allele for round pollen grains (*l*). Bateson and Punnett crossed a plant having purple flowers/long pollen grains with one having white flowers/round pollen grains. All F_1 offspring had purple flowers and long pollen grains. In the F_2 generation, the researchers observed the following phenotypes:

    296 purple flowers/long pollen grains

     19 purple flowers/round pollen grains

     27 red flowers/long pollen grains

     85 red flowers/round pollen grains

What is the best explanation for these results?

12. A dominant allele *W* confers black fur on guinea pigs. A guinea pig that is homozygous recessive (*ww*) has white fur. Fred would like to know whether his pet black-furred guinea pig is homozygous (*WW*) or heterozygous (*Ww*). How might he determine his pet's genotype?

13. Red-flowering snapdragons are homozygous for allele R^1. White-flowering snapdragons are homozygous for a different allele (R^2). Heterozygous plants (R^1R^2) bear pink flowers. What phenotypes should appear among first-generation offspring of the crosses listed? What are the expected proportions for each phenotype?

    a. $R^1R^1 \times R^1R^2$ c. $R^1R^2 \times R^1R^2$

    b. $R^1R^1 \times R^2R^2$ d. $R^1R^2 \times R^2R^2$

(In cases of incomplete dominance, alleles are usually designated by superscript numerals, as shown here, not by the uppercase letters for dominance and lowercase letters for recessiveness.)

For each cross, list which of these F_1 phenotypes show up as well as the proportion of each:

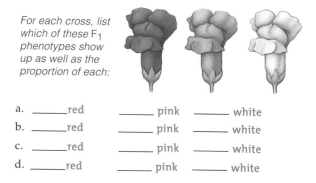

    a. _____red   _____ pink   _____ white

    b. _____red   _____ pink   _____ white

    c. _____red   _____ pink   _____ white

    d. _____red   _____ pink   _____ white

14. Two pairs of genes affect comb type in chickens (Figure 11.12), and they assort independently. When both are homozygous for recessive alleles, a chicken has a single comb. But a dominant allele of one gene, *P*, gives rise to a pea comb, and a dominant allele of the other gene (*R*) gives rise to a rose comb. An *epistatic* interaction occurs when a chicken has at least one of both dominant alleles, *P__ R __*, which gives rise to a walnut comb.

Predict the ratios resulting from a cross between two walnut-combed chickens that are heterozygous for both genes (*PpRr*) and list them below:

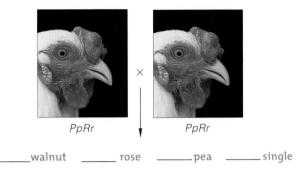

PpRr	*PpRr*

_____walnut _____rose _____pea _____single

15. As Section 3.6 explains, a single mutant allele gives rise to an abnormal form of hemoglobin (*Hb^S* instead of *Hb^A*). Homozygotes (*Hb^SHb^S*) develop the genetic disease sickle-cell anemia. Heterozygotes (*Hb^AHb^S*) show few obvious symptoms.

A couple who are both heterozygous for the *Hb^A* allele plan to have children. For *each* of the pregnancies, state the probability that this couple will have a child who is:
  a. homozygous for the *Hb^S* allele
  b. homozygous for the *Hb^A* allele
  c. heterozygous *Hb^AHb^S*

16. Watermelons (*Citrullus*) are important crops around the world (Figure 11.24). A single gene determines the density of green pigment that colors the rind, with solid light green (*g*) recessive to solid dark green (*G*). When a true-breeding plant having a dark-green rind is crossed with a plant having a light-green rind, what fraction of the dark-green F$_2$ offspring is expected to be heterozygous for this trait?

17. The rind of a watermelon that is homozygous for recessive allele *e* bursts, or splits explosively, when cut. Genotype *EE* results in a "nonexplosive" rind that is better for shipping watermelons to market. The rind of a watermelon that is homozygous for the recessive allele *f* has a furrowed surface. A furrowed rind has less market appeal than a smooth rind, which results from expression of dominant allele *F*.

For one testcross, a dihybrid plant that produces melons with a smooth, nonexplosive rind is crossed with a plant that produces melons with a furrowed, explosive rind. Make a Punnett square of the following results:

  118 smooth, nonexplosive
  112 smooth, explosive
  109 furrowed, nonexplosive
  121 furrowed, explosive

What is the smooth rind/furrowed rind ratio among the testcross offspring? What is the ratio of nonexplosive

Figure 11.24 A sampling of the variation in the rind characteristics of watermelon (*Citrullus*).

rind/explosive rind? Are the two gene loci assorting independently of each other?

18. Two pairs of genes determine kernel color in wheat plants. Alleles of one pair show incomplete dominance over the other pair. The product of allele A^1 at one locus produces enough pigment to add a dose of red color to the kernels, but that of allele A^2 does not. The product of allele B^1 at the second locus also adds a dose of red color to the kernels, but that of allele B^2 does not.

The chart shown below lists the numbers of different wheat kernel colors observed during a recent harvest, together with their corresponding genotypes. Using the information in this table, draw a graph showing the percentage of kernels in the wheat population that inherited each of the five kernel colors.

Explain why the kernel color in wheat plants shows a varied phenotypic distribution.

Genotype	Phenotype	Number Displaying the Trait	Percentage of Population
$A^1A^1B^1B^1$	Dark red	181	
$A^1A^1B^1B^2$ or $A^1A^2B^1B^1$	Red	360	
$A^1A^2B^1B^2$ or $A^1A^1B^2B^2$ or $A^2A^2B^1B^1$	Salmon	922	
$A^1A^2B^2B^2$ or $A^2A^2B^1B^2$	Pink	358	
$A^2A^2B^2B^2$	White	179	
Totals		2,000	

Strange Genes, Richly Tortured Minds

"This man is brilliant." That was the extent of a letter of recommendation from Richard Duffin, a mathematics professor at Carnegie Mellon University. Duffin wrote the line in 1948 on behalf of John Forbes Nash, Jr. (Figure 12.1). Nash was twenty years old at the time and applying for admission to Princeton University's graduate school.

Over the next decade, Nash made his reputation as one of America's foremost mathematicians. He was socially awkward, but so are many highly gifted people. Nash showed no symptoms of paranoid schizophrenia, a mental disorder that eventually debilitated him.

Full-blown symptoms emerged in his thirtieth year. Nash had to abandon his position at the Massachusetts Institute of Technology. Two decades passed before he was able to return to his pioneering work in mathematics.

Of every hundred people worldwide, one is affected by *schizophrenia*. This neurobiological disorder (NBD) is characterized by delusions, hallucinations, disorganized speech, and abnormal social behavior. As researchers know, exceptional creativity often accompanies schizophrenia. It also accompanies other NBDs, including autism, chronic depression, and bipolar disorder, which manifests itself as jarring swings in mood and social behavior.

Compared to the general population, highly intelligent individuals are *less* likely to develop NBDs—unless they also happen to be outside-the-box creative thinkers. Disturbingly creative writers alone are eighteen times more suicidal, ten times more likely to be depressed, and twenty times more likely to have bipolar disorder. Virginia Woolf's suicide after a prolonged mental breakdown is a tragic example.

We now have evidence that even emotionally healthy people who show creative brilliance have more personality traits in common with the mentally impaired than they do with individuals closer to the norm. For instance, they, too, are hypersensitive to environmental stimuli. Some may be on a razor's edge between mental stability and instability. Those who do go on to develop NBDs become part of a crowd that includes Socrates, Newton, Beethoven, Darwin, Lincoln, Poe, Dickens, Tolstoy, van Gogh, Freud, Churchill, Einstein, Picasso, Woolf, Hemingway, and Nash.

We have not yet identified all of the interactions among genes and the environment that might tip such individuals one way or the other. But we do know about several mutant genes that predispose them to develop NBDs.

Creatively gifted people, as well as those affected by NBDs, often turn up in the same family tree—which points

Figure 12.1 John Forbes Nash, Jr., a prodigy who solved problems that had baffled some of the greatest minds in mathematics. His early work in economic game theory won him a Nobel Prize. He is shown here at a premier of *A Beautiful Mind*, an award-winning film based on his battle with schizophrenia. His neural disorder places him in the ranks of other highly creative, distinguished, yet troubled individuals, including Abraham Lincoln, Virginia Woolf, and Pablo Picasso.

to a genetic basis for their special traits. Also, those affected by bipolar disorder and schizophrenia show altered gene expression in certain brain regions. Cells make too many or too few of the enzymes of electron transfer phosphorylation. Remember, this stage of aerobic respiration yields the bulk of the body's ATP. Does its disruption alter brain cells in ways that boost creativity but also invite illness? Perhaps.

With this intriguing connection, we invite you to reflect on how far you have come in this unit of the book. You first surveyed mitotic and meiotic cell divisions. You looked at how chromosomes and genes become shuffled during meiosis and then during fertilization. You also became acquainted with Gregor Mendel's discovery of major patterns of inheritance. This knowledge is your portal to the chromosomal basis of human inheritance.

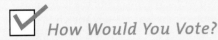

Watch the video online!

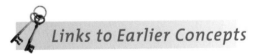

Key Concepts

AUTOSOMES AND SEX CHROMOSOMES

Sexually reproducing species have pairs of autosomes, which are chromosomes that are the same in length, shape, and which genes they carry. Nearly all animals also have a pair of sex chromosomes.

Karyotyping, a diagnostic tool, helps reveal changes in the structure or number of an individual's chromosomes. Section 12.1, 12.2

AUTOSOMAL INHERITANCE

Many alleles on autosomes are expressed in Mendelian patterns of simple dominance and recessiveness. Sections 12.3, 12.4

SEX-LINKED INHERITANCE

The pairing of sex chromosomes in human females (XX) differs from the pairing in males (XY). One of the genes on the Y chromosome dictates gender. Many alleles on the X chromosome are expressed in Mendelian patterns of simple dominance and recessiveness. Sections 12.5–12.7

CHANGES IN CHROMOSOME STRUCTURE

On rare occasions, a chromosome may undergo permanent change in its structure, as when a segment of it is deleted, duplicated, inverted, or translocated. Section 12.8

CHANGES IN CHROMOSOME NUMBER

Also on rare occasions, the parental number of autosomes or sex chromosomes changes. In humans, the change usually results in problems. Section 12.9

HUMAN GENETIC ANALYSIS AND OPTIONS

Various analytical and diagnostic procedures often reveal genetic disorders. Risks and benefits are associated with what individuals as well as society at large do with the information. Sections 12.10, 12.11

How Would You Vote?

Diagnostic tests for predisposition to neurobiological disorders will soon be available. Individuals might use knowledge of their susceptibility to modify choices in life-styles. Insurance companies and employers might also use that information to exclude predisposed but otherwise healthy individuals. Would you support legislation governing these tests? See BiologyNow for details, then vote online.

Links to Earlier Concepts

You will be drawing on your knowledge of chromosome structure (Sections 9.1, 9.3), meiosis (10.3, 10.4), and gamete formation (10.5). Be sure you understand dominance, recessiveness, and the homozygous and heterozygous conditions (11.1). Remember, environmental factors influence gene expression (11.6). Colchicine (4.10) will turn up again. So will glycolysis (8.2), this time in the context of a genetic disorder. You also will consider whether the hemoglobin family evolved after changes in chromosome structure (3.6).

12.1 Human Chromosomes

LINKS TO
SECTIONS
4.10, 9.1, 9.5, 10.3

You already know quite a bit about chromosomes and their roles in inheritance. Let's now focus on human autosomes and sex chromosomes.

Like nearly all animals, humans normally are male or female. Also like many species, they have a diploid chromosome number (2*n*), meaning that body cells have pairs of homologous chromosomes. Remember, all but one of the pairs are alike in their length, shape, and gene sequence. One member of the last pairing is a unique sex chromosome that is present in males or females, but not in both.

For instance, a diploid cell in a human female has two X chromosomes (XX). A diploid cell in a human male has one X and one Y chromosome (XY). This is a common inheritance pattern among mammals, fruit flies, and many other animals. It is not the only one, however. Among butterflies, moths, birds, and certain fishes, the males have two identical sex chromosomes and females do not.

Human X and Y chromosomes differ physically and in which genes they carry. Recall, from Section 10.3, that each pair of homologous chromosomes synapses (zippers together tightly) in prophase I of meiosis. An X chromosome and Y chromosome synapse in a small region along their length, but that is enough to allow the two to interact as homologues during meiosis.

Human X and Y chromosomes fall into the general category of **sex chromosomes**. As you will see later, when sex chromosomes are inherited in certain combinations, they dictate the gender of the new individual—that is, whether it will become a male or a female.

All of the other chromosomes in our body cells are the same in both sexes. We categorize them as **autosomes**.

The duplicated human chromosome shown in Figure 12.2 has a targeted band (*yellow*), an artistic way of introducing a key point: Molecular biology increased the power of diagnostic tools that were already in use to analyze chromosomes —as with fluorescent dyes that can label DNA regions linked to genetic disorders. In the next section, you will read about two of the diagnostic procedures.

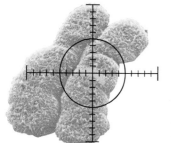

Figure 12.2 Long before the spectacular discoveries of molecular biology, researchers started identifying regions on chromosomes that probably held the genes responsible for certain genetic disorders.

> *Autosomes are pairs of chromosomes that are the same in males and females of a species. One other pairing, of sex chromosomes, differs between males and females.*

12.2 What Is Karyotyping?

With karyotyping, a diagnostic tool, images are constructed to analyze the structure and number of chromosomes in an individual's cells.

How do we know about an individual's autosomes and sex chromosomes? *Karyotyping* is one of the earliest diagnostic tools. A typical **karyotype** is a preparation of an individual's metaphase chromosomes, sorted out by length, shape, centromere location, and other defining features. Gross abnormalities in chromosome structure or an altered chromosome number can be pinpointed by comparing the individual's karyotype against a standard karyotype for the species.

Making a Karyotype   Human chromosomes are in their most condensed form and easiest to identify when a cell is at metaphase of mitosis (Sections 9.1 and 9.3). Technicians do not count on finding dividing cells in the body. They culture cells and induce mitosis artificially. They place a sample of cells, usually from blood, into a solution that stimulates growth and mitotic cell division. They add colchicine to the sample to arrest the cell cycle at metaphase. Colchicine, remember, is a poison that blocks spindle formation by preventing microtubules from forming (Section 4.10).

As Figure 12.3 explains, the cell culture is centrifuged to isolate all the metaphase cells. A hypotonic solution makes the cells swell, by way of osmosis, and move away from each other. The chromosomes inside them move away from each other, also. Then the cells are mounted on slides, fixed, and stained for microscopy.

Once the chromosomes are brought into focus, they are photographed. The photograph is cut with scissors or with a computer's cut-and-paste tools to separate the chromosomes. Then the chromosomes are lined up by size and shape, as in Figure 12.3*f*.

Spectral Karyotypes   *Spectral karyotyping*, a more recent diagnostic tool, uses a range of colored fluorescent dyes that bind to specific parts of chromosomes. Analysis of the resulting rainbow-hued karyotype often reveals abnormalities that would not otherwise be discernible.

Figure 12.4 shows a spectral karyotype. The Philadelphia chromosome in this karyotype, named after the city where someone discovered it, was the first chromosome to be specifically correlated with cancer—one of the leukemias. The Philadelphia chromosome was already known to be longer than human chromosome 9, which is its normal counterpart. But spectral karyotyping identified the extra length as a piece of chromosome 22.

By chance, both chromosomes broke inside a stem cell in bone marrow. Such cells give rise to blood cells. Enzymes reattached the pieces—but on the wrong chromosomes. You can identify the translocated parts in the Figure 12.4 karyotype. We will be returning to this type of change in the structure of chromosomes in Section 12.8.

Figure 12.3 *Animated!* Karyotyping, in which an image of metaphase chromosomes is cut apart. Individual chromosomes are aligned by their centromeres and arranged according to size, shape, and length.

(**a**) A sample of cells from an individual is put in a medium that stimulates cell growth and mitotic division. Colchicine is added to arrest the cell cycle at metaphase. (**b**) The culture is subjected to *centrifugation*, which works because cells have greater mass and density than the solution bathing them. A centrifuge's spinning force moves the cells farthest from the center of rotation, so they collect at the base of the centrifuge tubes.

(**c**) The culture medium is removed; a hypotonic solution is added. The cells swell, and chromosomes move apart. (**d**) The cells are mounted on a microscope slide and stained to make the chromosomes show up.

(**e**) A photograph of one cell's chromosomes is cut up and organized, as in the human karyotype in (**f**), which shows 22 pairs of autosomes and 1 pair of sex chromosomes—XX *or* XY. Scissors or computer tools do the cuts.

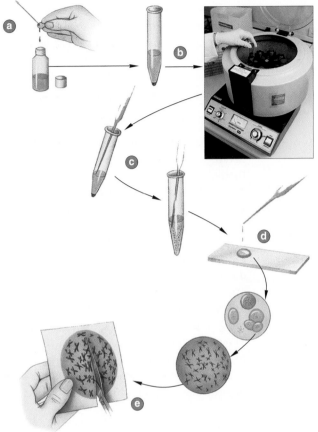

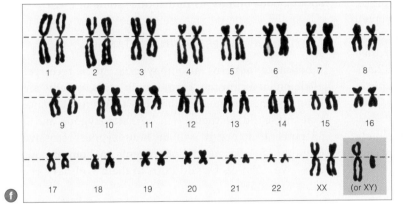

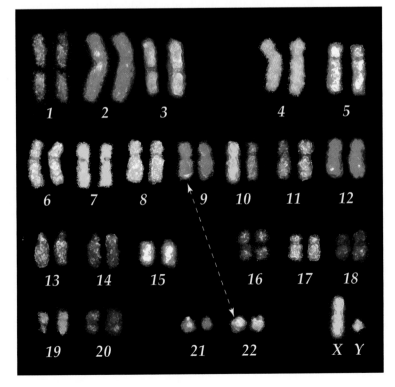

Figure 12.4 Image of a killer—the Philadelphia chromosome, as revealed by the artificial colors of spectral karyotyping. Its normal counterpart is human chromosome 9.

This chromosome exchanged a segment of itself with the nonhomologous chromosome 22. The broken end of chromosome 9 contained a gene that affects mitotic cell division. This gene fused with a DNA sequence in chromosome 22 that controls expression of another gene.

The fused gene is transcribed far more than it should be, and the cell cycle spins out of control (Section 9.5). The phenotypic outcome is *chronic myelogenous leukemia* (CML)—a rare form of leukemia in which the body produces far too many white blood cells. Uncontrolled divisions give rise to masses of malignant cells in bone tissues, where stem cells that give rise to white blood cells originate.

Chapter 12 Chromosomes and Human Inheritance 189

12.3 Examples of Autosomal Inheritance Patterns

LINKS TO
SECTIONS
8.2, 8.6

Most human traits arise from complex gene interactions, but many can be traced to autosomal dominant or recessive alleles that are inherited in simple Mendelian patterns. Some of these alleles cause genetic disorders.

AUTOSOMAL DOMINANT INHERITANCE

Figure 12.5a shows a typical inheritance pattern for an autosomal dominant allele. If one of the parents is heterozygous and the other homozygous, any child of theirs has a 50 percent chance of being heterozygous. The trait usually appears every generation. Why? The allele is expressed even in heterozygotes.

One autosomal condition, *achondroplasia*, affects 1 in 10,000 or so people. While they were still embryos, the cartilage model on which a skeleton is constructed did not form properly. Adults have abnormally short arms and legs relative to other body parts and they are only about four feet, four inches tall (Figure 12.5a). Most homozygotes die before or not long after birth. The allele does not affect the capacity of the survivors to grow and reproduce.

In *Huntington's disease*, the nervous system slowly deteriorates, and involuntary muscle action increases.

Symptoms often do not start until past age thirty, and those affected die during their forties or fifties. Many unknowingly transmit the mutant allele to children before then. The mutation causing the disorder alters a protein required for normal brain cell development. It is one of the *expansion* mutations, in which three nucleotides are repeated in series along the length of DNA. Hundreds of thousands of repeats occur within and between genes on human chromosomes, but this one (CAG) disrupts a gene product's function.

A few dominant alleles that cause severe problems persist in populations because expression of the allele may not interfere with reproduction, or affected people reproduce before the symptoms become severe. Also, spontaneous mutations reintroduce some of them.

AUTOSOMAL RECESSIVE INHERITANCE

Inheritance patterns also may point to a recessive allele on an autosome. First, if both of the parents are heterozygous for the allele, there is a 50 percent chance that any child of theirs will be heterozygous and a 25 percent chance it will be homozygous recessive (Figure 12.5b). Second, if both parents are homozygous recessive, then each child born to them will have the same condition.

Galactosemia is a heritable metabolic disorder that affects about 1 in every 100,000 newborns. This case of autosomal recessive inheritance involves alleles for an enzyme that helps digest the lactose in milk or milk products. The body normally converts lactose to glucose and galactose, then three enzymes convert the galactose to glucose–1–phosphate (Figure 12.6). This intermediate can enter glycolysis or be converted to glycogen (Sections 8.2 and 8.6). But galactosemics do not have functional copies for one of these three enzymes; they are homozygous recessive for a mutant

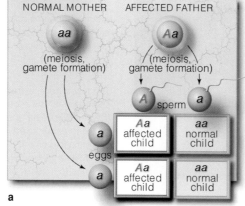

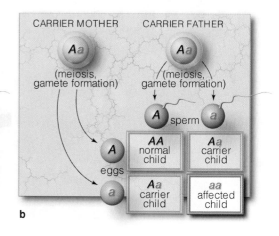

Figure 12.5 *Animated!* (**a**) Example of autosomal dominant inheritance. One dominant allele (*red*) is fully expressed in carriers. Achondroplasia, an autosomal dominant disorder, affects the three males shown above. At center, Verne Troyer (or Mini Me in the Mike Myers spy movies), stands two feet, eight inches tall.

(**b**) An autosomal recessive pattern. In this example, both of the parents are heterozygous carriers of the recessive allele (*red*).

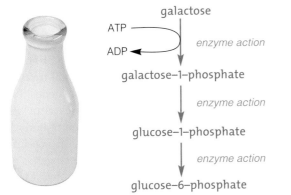

galactose

ATP

ADP

enzyme action

galactose–1–phosphate

enzyme action

glucose–1–phosphate

enzyme action

glucose–6–phosphate

Figure 12.6 How galactose is normally converted to a form that can enter the breakdown reactions of glycolysis. A mutation that affects the second enzyme in the conversion pathway gives rise to galactosemia.

allele that encodes it. Galactose–1–phosphate builds up to toxic levels in their body. High levels of this intermediate can be detected in urine. The excess leads to malnutrition, diarrhea, vomiting, and damage to the eyes, liver, and brain.

When they do not receive treatment, galactosemics typically die young. When they are quickly placed on a diet that excludes all dairy products, the symptoms may not be as severe.

WHAT ABOUT NEUROBIOLOGICAL DISORDERS?

Those human neurobiological disorders introduced at the start of the chapter do not follow simple patterns of Mendelian inheritance. In most cases, a lone gene does not give rise to depression, schizophrenia, or bipolar disorder. Still, it is useful to search for mutations that make some people more vulnerable, as long as we recognize that many genes and environmental factors contribute in individually small ways to the outcome.

For example, researchers who conducted extensive family studies and twin studies have predicted that mutant alleles in specific regions of autosomes 1, 3, 5, 6, 8, 11 through 15, 18, and 22 increase the chance of developing schizophrenia. Similarly, several mutant alleles have been reportedly linked to bipolar disorder and depression.

Some traits can be traced to dominant or recessive alleles on autosomes because they are inherited in simple Mendelian patterns. Certain alleles on these chromosomes give rise to genetic abnormalities and genetic disorders.

Sometimes textbook examples of the human condition seem a bit abstract, so take a moment to think about two boys who were too young to be old.

Imagine being ten years old with a mind trapped in a body that is getting a bit more shriveled, more frail—*old*—every day. You are barely tall enough to peer over the top of the kitchen counter. You weigh less than thirty-five pounds. Already you are bald and have a crinkled nose. Possibly you have a few more years to live. Would you, like Mickey Hays and Fransie Geringer, still be able to laugh?

On average, of every 8 million newborn humans, one will grow old far too soon. On one of its autosomes, that rare individual carries a mutant allele that gives rise to *Hutchinson–Gilford progeria syndrome*. While that new individual was still an embryo in its mother, billions of DNA replications and mitotic cell divisions distributed the information encoded in that gene to each newly formed body cell. Its legacy will be an accelerated rate of aging and a sharply reduced life span.

The mutation grossly disrupts gene interactions that are essential for growth and development. Observable symptoms start before age two. Skin that should be plump and resilient starts to thin. Skeletal muscles weaken. Limb bones that should lengthen and grow stronger soften. Premature baldness is inevitable (Figure 12.7). There are no documented cases of progeria running in families, so spontaneous mutation must be the cause. In one recent study, researchers examined twenty affected children. All the children carried a mutant gene that specifies lamin A, a structural protein that helps organize the nucleus.

Most progeriacs can expect to die in their early teens as a result of strokes or heart attacks. These final insults are brought on by a hardening of the wall of arteries, a condition typical of advanced age. Fransie was seventeen when he died. Before Mickey died at age twenty, he was the oldest living progeriac.

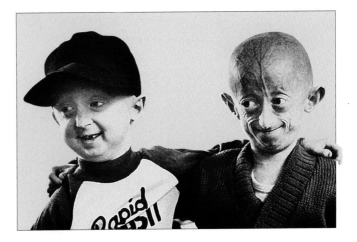

Figure 12.7 Two boys who met at a gathering of progeriacs at Disneyland, California, when they were not yet ten years old.

12.5 Sex Determination in Humans

*Expression of one of the genes on the Y chromosome—
that is what it takes to become a human male.*

Every normal egg produced by a human female has one X chromosome. Half of the sperm cells formed in a male carry an X chromosome, and half carry a Y. If an X-bearing sperm fertilizes an X-bearing egg, then the resulting zygote will develop into a female. If the sperm carries a Y chromosome, it will develop into a male (Figure 12.8a).

With only 255 genes, the human Y chromosome might seem relatively puny. But one of them is the *SRY* gene—which happens to be the master gene for male sex determination. Its expression in XY embryos triggers the formation of testes, or male gonads, as shown in Figure 12.8b. What do these primary male reproductive organs do? For one thing, some of their cells make testosterone, a sex hormone that controls the emergence of male secondary sexual traits.

An XX embryo has no Y chromosome, no *SRY* gene, and much less testosterone. Therefore, primary female reproductive organs—ovaries—form instead. Ovaries make estrogens and other sex hormones that govern the development of female secondary sexual traits.

The human X chromosome carries 1,141 genes. Like other chromosomes, it includes some genes associated with sexual traits, such as the distribution of body fat and hair. But most of its genes deal with *nonsexual* traits, such as blood-clotting functions. Such genes can be expressed in males as well as in females. Males, remember, also inherit one X chromosome.

Expression of the SRY gene on the human Y chromosome triggers testosterone synthesis, which makes a developing embryo become a male. In the absence of the Y chromosome (and the SRY gene), a developing embryo becomes a female.

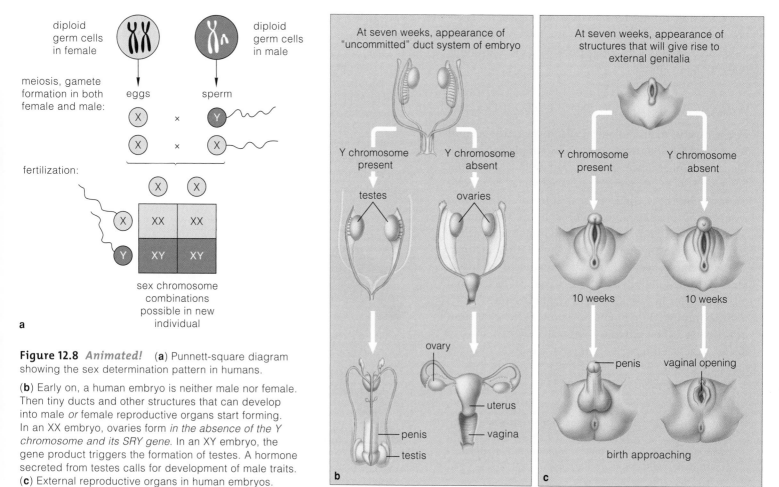

Figure 12.8 *Animated!* (**a**) Punnett-square diagram showing the sex determination pattern in humans.

(**b**) Early on, a human embryo is neither male nor female. Then tiny ducts and other structures that can develop into male *or* female reproductive organs start forming. In an XX embryo, ovaries form *in the absence of the Y chromosome and its SRY gene.* In an XY embryo, the gene product triggers the formation of testes. A hormone secreted from testes calls for development of male traits. (**c**) External reproductive organs in human embryos.

12.6 What Mendel Didn't Know: X-Linked Inheritance

After Mendel passed away in 1884, his paper on pea plants gathered dust in a hundred libraries. Then microscopists discovered chromosomes, and interest in the cellular basis of inheritance was rekindled. In 1900 researchers came across Mendel's paper while checking literature on genetic crosses. Their results confirmed what Mendel had already found out. Later, other researchers went on to discover something Mendel did not know about——genes on sex chromosomes.

By the early 1900s, researchers suspected that each gene has a specific location on a chromosome. Thomas Morgan and his coworkers confirmed it through their hybridization experiments with mutant forms of a fruit fly, *Drosophila melanogaster*. For instance, they found evidence that this fly's X chromosome has a gene for eye color and another gene for body color. They asked: Were the two genes linked on the same chromosome? That is, do they stay together during meiosis and end up in the same gamete?

Thomas Morgan, an embryologist, already had come across a relationship between sex determination and some *nonsexual* traits. For instance, human males and females both have blood-clotting factors. And yet, males are far more likely to develop hemophilia, which is a blood-clotting disorder. This sex-linked outcome was probably related to recessive forms of genes. But it was not like anything that Mendel identified in the results from his experimental crosses of pea plants. For pea plants, it made no difference which parent carried a recessive allele.

Morgan decided to study eye color and other nonsexual traits in *D. melanogaster*. This type of fruit fly has since become a favorite experimental organism. It can live in small bottles on nothing more expensive than a bit of agar, cornmeal, molasses, and yeast. A female lays hundreds of eggs in a few days, and her offspring reproduce in less than two weeks. Morgan knew that in a single year, he could use experimental tests to track observable traits for nearly thirty generations of thousands of fruit flies.

At first, all of the fruit flies in Morgan's bottles were wild type for eye color; they had brick-red eyes (Figure 12.9). Then Morgan got lucky. A gene that controls eye color mutated, and a white-eyed male appeared in a bottle, and Morgan quickly conducted **reciprocal crosses**. In the second of such paired crosses, a trait of each sex is reversed compared to the original cross to determine the role of parental sex on inheritance. In this case:

First cross: white-eyed male × red-eyed female
Second cross: red-eyed male   × white-eyed female

White-eyed males mated with homozygous red-eyed females. All of the F_1 offspring had red eyes, and after the F_1 individuals had mated with each other, some of their F_2 offspring were males with white eyes (Figure 12.9c).

Then Morgan allowed true-breeding red-eyed males to mate with white-eyed females. *Half* of the F_1 offspring of this cross turned out to be red-eyed females, and *half* were white-eyed males. Later, the phenotypes of F_2 offspring were 1/4 red-eyed females, 1/4 white-eyed females, 1/4 red-eyed males, and 1/4 white-eyed males. The results did not fit with the straightforward inheritance patterns of pea plants. Could Mendel's theories explain them?

They could if the locus of an eye-color gene were on a sex chromosome. But which one? Because females (XX) could be white-eyed, the recessive allele had to be on one of their X chromosomes. What if white-eyed males (XY) had the recessive allele on their X chromosome and their Y chromosome had no corresponding eye-color allele? In that case, they would have white eyes. They would have no dominant allele to mask the effect of the recessive one, as the Punnett-square diagram in Figure 12.9c shows.

And so Morgan's idea of an X-linked gene dovetailed with Mendel's concept of segregation. By proposing that a specific gene is located on an X chromosome but not on the Y chromosome, Morgan explained his reciprocal crosses. His experimental results matched predicted outcomes.

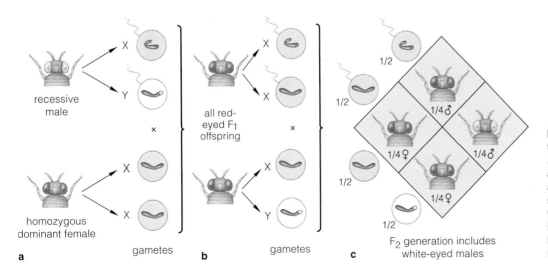

a recessive male × homozygous dominant female — gametes

b all red-eyed F_1 offspring — gametes

c F_2 generation includes white-eyed males

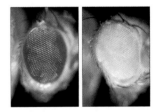

Figure 12.9 One of the experiments that pointed to sex-linked genes in *Drosophila melanogaster*. In this fruit fly, a wild-type allele specifies red eyes, and a mutant allele for the same locus specifies white eyes. *Wild-type* refers to a gene's most common form (either in nature or in standardized, laboratory-bred strains of a species) compared to less common, *mutant* alleles.

12.7 Examples of X-Linked Inheritance Patterns

LINK TO
SECTION
4.10

Alleles on an X chromosome give rise to phenotypes that also reflect simple Mendelian patterns of inheritance. Many of the recessive ones cause problems.

A recessive allele on an X chromosome often leaves certain clues when it causes a genetic disorder. First, more males than females are affected. Heterozygous females still have a dominant allele on their other X chromosome that masks the recessive allele's effects.

Males are not protected, because they inherit only one X chromosome along with one Y chromosome. Figure 12.10 reinforces this point. Second, a heterozygous female must be the bridge between an affected male and an affected grandson; an affected father cannot pass on the recessive allele to his son.

HEMOPHILIA A

Hemophilia A, a type of blood-clotting disorder, is one of the classic cases of X-linked recessive inheritance. Most of us have a functional clotting mechanism that quickly puts a stop to bleeding from minor injuries, in the manner explained in Section 38.9. The mechanism involves the synthesis of proteins that are products of genes on the X chromosome. Bleeding is prolonged in males who carry a mutant form of one of these X-linked genes. The affected males bruise easily, and the internal bleeding can cause problems in their muscles and joints.

This disorder affects 1 in 7,000 males, on average, but new mutations may account for a third of them. In heterozygous females, clotting time is close to normal. The disorder's frequency was relatively high among royal families of Europe and Russia in the nineteenth century. Figure 12.11 is a classic example of a pedigree for hemophilia A.

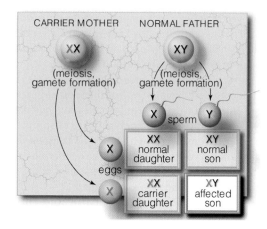

Figure 12.10 *Animated!* One pattern for X-linked recessive inheritance. In this case, the mother carries a recessive allele on one of her X chromosomes (*red*).

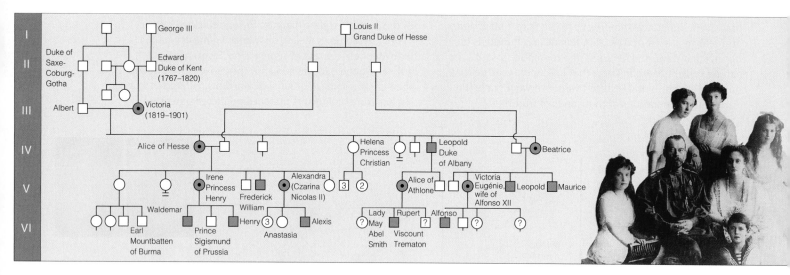

Figure 12.11 A classic case of X-linked recessive inheritance. This is a partial pedigree, or a chart of genetic connections, among descendants of Queen Victoria of England. It focuses on carriers and affected males who inherited the X-linked allele for hemophilia A (*white* circles and squares). At one time, the recessive allele was present in eighteen of Victoria's sixty-nine descendants, who sometimes intermarried. Of the Russian royal family members shown, the mother was a carrier. Through her obsession with the vulnerability of Crown Prince Alexis, a hemophiliac, she was sucked into political intrigue that helped trigger the Russian Revolution of 1917.

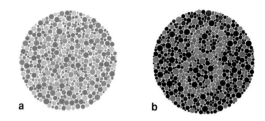

Figure 12.12 *Left,* what red–green color blindness means, using ripe red cherries on a green-leafed tree as an example. In this case, the perception of blues and yellows is normal, but the affected individual has difficulty distinguishing red from green.

Above, two of many Ishihara plates, which are standardized tests for different forms of color blindness. (**a**) You may have one form of red–green color blindness if you see the numeral "7" instead of "29" in this circle. (**b**) You may have another form if you see a "3" instead of an "8."

RED—GREEN COLOR BLINDNESS

The pattern of X-linked recessive inheritance shows up among individuals who have some degree of *color blindness.* The term refers to a range of conditions in which an individual cannot distinguish among some or all of the colors in the spectrum of visible light. Mutant gene products alter the structure and function of photoreceptors (light-sensitive receptors) in eyes.

Normally, humans can sense the differences among 150 colors. A person who is *red–green* color blind sees fewer than 25 because some or all of the receptors that respond to red and green wavelengths are weakened or absent. Other people confuse red and green colors. Still others see shades of gray instead of green but see blues and yellows quite well. Figure 12.12*a* represents this condition. Figure 12.12*b* is part of a standardized set of tests for color blindness.

Color blindness is more common in men, who are about twelve times more likely than women to develop the condition. Heterozygous women show symptoms as well. Can you explain why?

DUCHENNE MUSCULAR DYSTROPHY

Duchenne muscular dystrophy (DMD) is one of a group of X-linked recessive disorders characterized by rapid degeneration of muscles, starting early in life. About 1 in every 3,500 boys is affected.

The recessive allele encodes dystrophin. This is the protein that structurally supports fused-together cells in muscle fibers. It anchors much of the cell cortex to the plasma membrane (Section 4.10). In cases where

dystrophin is abnormal or absent, the cell cortex weakens and muscle cells die. The debris left behind in tissues triggers inflammation that becomes chronic.

Most individuals are diagnosed between ages three and seven. The progression of the disorder cannot be stopped. When the affected boy is about twelve years old, he will start using a wheelchair. His heart muscles will start to break down. Even with the best managed care, he usually will die before age twenty-five, most often as a result of respiratory failure.

Recently, researchers mapped all of the genes on the X chromosome. They discovered two things. First, only 5 percent of all of the genes we have reside on this sex chromosome. Second, the mutant alleles that cause or contribute to many known genetic disorders can occur at locations along this chromosome. More than 300 such connections have been identified.

> *Diverse recessive alleles on the human X chromosome are implicated in more than 300 genetic disorders.*
>
> *A heterozygous female may not show symptoms if she has a dominant allele on her other X chromosome, which masks the effect of the recessive allele. Males (XY) cannot transmit any X-linked allele to their sons.*

12.8 Heritable Changes in Chromosome Structure

LINK TO SECTION 10.3

On rare occasions, a chromosome's structure changes. Many of the alterations have severe or lethal outcomes.

MAIN CATEGORIES OF STRUCTURAL CHANGE

One or more changes in the physical structure of a chromosome may give rise to a genetic disorder or abnormality. Such changes are rare, but they do occur spontaneously in nature. Some also can be induced by exposure to certain chemicals or irradiation. Either way, the alteration may be detected by microscopic examination and karyotype analysis of cells during mitosis or meiosis. Four kinds of structural changes are chromosomal duplications, deletions, inversions, and translocations.

Duplication Even normal chromosomes have DNA sequences that are repeated two or more times. These are called **duplications**:

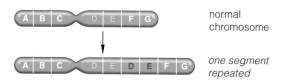

normal chromosome

one segment repeated

Duplications can occur through unequal crossovers at prophase I. Homologous chromosomes align side by side, but their DNA sequences misalign at some point along their length. The probability of this happening is greater in regions where DNA has long repeats of the same series of nucleotides. A stretch of DNA gets deleted from one chromosome and is spliced into the partner chromosome. Some duplications cause neural problems and physical abnormalities. As you will see, others apparently were important in the evolution of primates that were ancestral to humans.

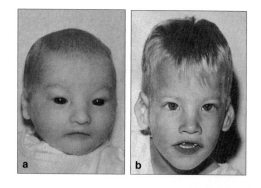

Figure 12.13 (**a**) A male infant who developed cri-du-chat syndrome. His ears are low on the side of the head relative to the eyes. (**b**) Same boy, four years later. By this age, affected humans stop making the mewing sounds typical of the syndrome.

Deletion A **deletion** is the loss of some portion of a chromosome, as by unequal crossovers, inversions, or chemical attacks:

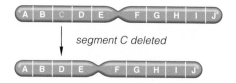

segment C deleted

In mammals, most deletions cause serious disorders or death. Missing or broken genes disrupt the body's growth, development, and metabolism. For instance, a tiny deletion from human chromosome 5 results in an abnormally shaped larynx and mental impairment. Crying infants sound like cats meowing (Figure 12.13). Hence the name of the disorder, *cri-du-chat* (cat-cry in French). Inversions and deletions often occur together as an outcome of unequal recombination events.

Inversion With an **inversion**, part of the sequence of DNA within the chromosome becomes oriented in the reverse direction, with no molecular loss:

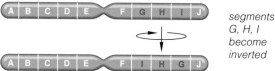

segments G, H, I become inverted

An inversion is not a problem for a carrier if it does not alter a crucial gene region. It can cause problems in meiosis. Chromosomes may mispair, and deletions may occur that can reduce the viability of gametes. Some individuals do not even know that they have an inverted chromosome region until a genetic disorder or abnormality surfaces in one or more children.

Translocation In Section 12.2, you came across a case of a broken part of one chromosome becoming attached to another chromosome. This type of change in chromosome structure is known as a **translocation**. Most translocations are reciprocal, in that both of the two chromosomes exchange broken parts:

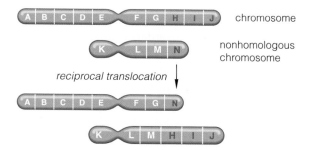

chromosome

nonhomologous chromosome

reciprocal translocation

Translocations often cause reduced fertility, because affected chromosomes have difficulty segregating in meiosis. Severe problems are rare, but they do arise. They include some sarcomas, lymphomas, myelomas, and leukemias.

DOES CHROMOSOME STRUCTURE EVOLVE?

Alterations in the structure of chromosomes generally are not good and may be selected against. Even so, many alterations with neutral effects have been built into the DNA of all species over evolutionary time.

Duplicates of genes could have bestowed adaptive advantages on descendants of their original bearers. Two or more copies of some gene means that one is free to mutate while the other continues to carry out its normal function. The slightly modified products of mutant genes can behave in slightly different or novel ways, some of which are beneficial.

Some duplications have proved adaptive. Reflect on the four globin chains of hemoglobin (Section 3.6). In humans and other primates, several genes for these polypeptide chains are strikingly similar. Apparently they evolved through duplications, mutations, and transpositions. They have slightly different molecular structures and slightly different capacities to bind and transport oxygen under a range of cellular conditions.

Alterations in chromosome structure might have contributed to the differences among closely related organisms, such as apes and humans. Eighteen of the twenty-three pairs of human chromosomes are almost identical with chimpanzee and gorilla chromosomes. The other five differ only at inverted and translocated regions. Figure 12.14 shows the striking similarities between some gibbon and human chromosomes that may have arisen by duplications and translocations.

To give one more example, human body cells have twenty-three pairs of chromosomes, but those of a chimpanzee, gorilla, or orangutan have twenty-four. Compare the banding patterns of these chromosomes. During human evolution, two chromosomes in an early ancestor fused, end to end, to form chromosome 2. In the fused region, researchers have discovered remnants of a telomere—the signature DNA sequence that caps the *ends* of all chromosomes (Figure 12.15).

A segment of a chromosome may be duplicated, deleted, inverted, or moved to a new location. Such changes can be harmful or lethal. Others have been conserved over time; they confer advantages or have had neutral effects.

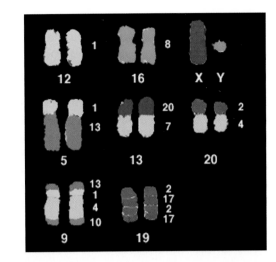
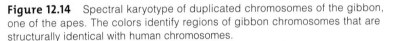

Figure 12.14 Spectral karyotype of duplicated chromosomes of the gibbon, one of the apes. The colors identify regions of gibbon chromosomes that are structurally identical with human chromosomes.

Top row: Chromosomes 12, 16, X, and Y are structurally the same in both primates. *Second row:* Translocations are present in gibbon chromosomes 5, 13, and 20, and they correspond to regions of human chromosomes 1, 13, 20, 7, 2, and 4.

Third row: Gibbon chromosome 9 corresponds to several human chromosome regions. In addition, duplications in gibbon chromosome 19 are present in human chromosomes 2 and 17.

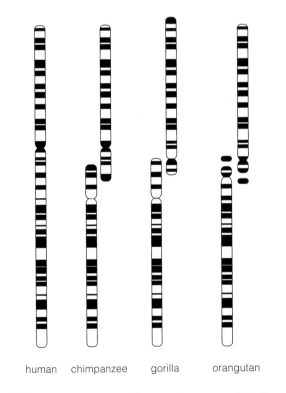

human   chimpanzee   gorilla   orangutan

Figure 12.15 Banding patterns of human chromosome 2 (*left*), compared with the patterns on two of the chromosomes in cells of the chimpanzee, gorilla, and orangutan. Such bands appear because different chromosome regions preferentially take up different kinds of stains. Their response to a given stain depends on their base composition and packing organization.

12.9 Heritable Changes in the Chromosome Number

LINKS TO
SECTIONS
9.3, 10.3

Occasionally, abnormal events occur before or during cell division, and gametes and new individuals end up with the wrong chromosome number. Consequences range from minor to lethal changes in form and function.

In **aneuploidy**, cells usually have one extra or one less chromosome. Autosomal aneuploidy is usually fatal for humans and is linked to most miscarriages. Aneuploidy typically arises through **nondisjunction**, whereby one or more pairs of chromosomes do not separate as they should during mitosis or meiosis. Figure 12.16 shows an example. In **polyploidy**, cells have three or more of each type of chromosome. Half of all species of flowering plants, some insects, fishes, and other animals are polyploid.

Such changes affect the chromosome number at fertilization. Suppose a normal gamete fuses with an $n+1$ gamete, with one extra chromosome. The new individual will be trisomic ($2n+1$), with three of one type of chromosome and two of every other type. Or what if an $n-1$ gamete and a normal n gamete fuse? In this case, the new individual will be monosomic, or $2n-1$. Mitotic divisions perpetuate such mistakes when an embryo is growing in size and developing.

AUTOSOMAL CHANGE AND DOWN SYNDROME

A few trisomics are born alive, but only trisomy 21 individuals reach adulthood. A newborn with three chromosomes 21 will develop *Down syndrome*. This autosomal disorder is the most frequent type of altered chromosome number in humans; it occurs once in every 800 to 1,000 births. It affects more than 350,000 people in the United States. Figure 12.16*a* shows a karyotype for a trisomic 21 female. About 95 percent of all cases arise through nondisjunction at meiosis. Affected individuals have upward-slanting eyes, a fold of skin that starts at the inner corner of each eye, a deep crease across each palm and foot sole, one (not two) horizontal furrows on their fifth fingers, and somewhat flattened facial features.

Not all of the outward symptoms develop in every individual. That said, trisomic 21 individuals do have moderate to severe mental impairment and heart problems. Also, their skeleton develops abnormally, so older children have shortened body parts, loose joints, and misaligned bones in hips, fingers, and toes. Their muscles and reflexes are weak. Motor skills, including speech, develop very slowly. With medical care, individuals can live for fifty-five years, on average.

The incidence of nondisjunction rises with increasing age of potential mothers (Figure 12.17). Nondisjunction might occur in the father, although less often. Trisomy 21 is just one of hundreds of conditions that can be detected through prenatal diagnosis (Section 12.11). With early training and medical intervention, individuals still can take part in normal activities. As a group, trisomics 21 tend to be cheerful and sociable.

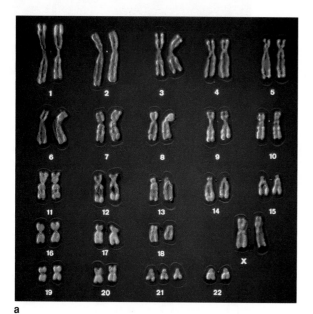

a

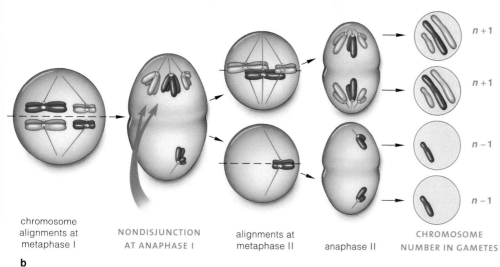

chromosome alignments at metaphase I

NONDISJUNCTION AT ANAPHASE I

alignments at metaphase II

anaphase II

CHROMOSOME NUMBER IN GAMETES

$n+1$

$n+1$

$n-1$

$n-1$

b

Figure 12.16 (**a**) A case of nondisjunction. This karyotype reveals the trisomic 21 condition of a human female. (**b**) One example of how nondisjunction arises. Of the two pairs of homologous chromosomes shown here, one fails to separate during anaphase I of meiosis. The chromosome number is altered in the gametes that form after meiosis.

CHANGE IN THE SEX CHROMOSOME NUMBER

Nondisjunction also causes most of the alterations in the number of X and Y chromosomes. The frequency of such changes is 1 in 400 live births. Usually, they lead to difficulties in learning and motor skills, such as speech, although problems can be so subtle that the underlying cause is not even diagnosed.

Female Sex Chromosome Abnormalities *Turner syndrome* individuals have an X chromosome and no corresponding X or Y chromosome (XO). About 1 in 2,500 to 10,000 newborn girls are XO. Nondisjunction originating with the father accounts for 75 percent of the cases. Yet cases are few, compared with other sex chromosome abnormalities. At least 98 percent of XO embryos may spontaneously abort early in pregnancy.

Despite the near lethality, XO survivors are not as disadvantaged as other aneuploids. On average, they are well proportioned, as shown here, but only four feet, eight inches tall. Most cannot make enough sex hormones; they do not have functional ovaries. The condition affects the development of secondary sexual traits, such as breast development. A few eggs form in the ovaries but degenerate by the time the girls are two years old.

A few females inherit three to five X chromosomes. An *XXX syndrome* occurs in about 1 of 1,000 live births. Adults are fertile. Except for slight learning difficulties, most fall in the normal range of social behavior.

Male Sex Chromosome Abnormalities About 1 of every 500 males has an XXY karyotype, with an extra chromosome inherited from the mother. Two-thirds of the cases are an outcome of nondisjunction at meiosis. Among the remainder, failure of the Y chromosome to separate at mitosis gave rise to a mosaic karyotype (XY in some cells and XXY in other cells).

The resulting *Klinefelter syndrome* develops during puberty. XXY males tend to be overweight and tall. The testes and the prostate gland usually are smaller than average. Many XXY males are within the normal range of intelligence, although some have short-term memory loss and learning disabilities. They make less testosterone and more estrogen than normal males, with feminizing effects. Sperm counts are low. Hair is sparse, the voice is pitched high, and the breasts are enlarged somewhat. When affected individuals enter

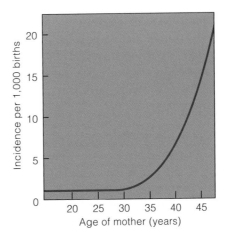

Figure 12.17 Relationship between the frequency of Down syndrome and mother's age at childbirth. The data are from a study of 1,119 affected children. The risk of having a trisomic 21 baby rises with the mother's age. This may seem odd, because about 80 percent of trisomic 21 individuals are born to women not yet thirty-five years old. But these women are in the age categories with the highest fertility rates, and they simply have more babies.

puberty, they can receive testosterone injections that can reverse the feminized traits.

About 1 in 500 to 1,000 males has one X and two Y chromosomes, an *XYY condition*. They tend to be taller than average, with mild mental impairment, but most fall in the normal phenotypic range. They were once thought to be genetically predisposed to a life of crime. This misguided view was based on a sampling error (too few cases of narrowly chosen groups, such as prison inmates) and were biased (the same researchers gathered karyotypes *and* personal histories). Fanning the stereotype was a report that a mass murderer of young nurses was XYY. He wasn't.

In 1976 a Danish geneticist reported results from his study of 4,139 tall males, all twenty-six years old, who had registered at their draft board. Besides their data from physical examinations and intelligence tests, the draft records offered clues to social and economic status, education, and criminal convictions, if there were any. Twelve of the males studied were XYY, which meant the "control group" had more than 4,000 males. The only finding was that mentally impaired, tall males who engage in criminal deeds are just more likely to get caught—irrespective of karyotype.

The majority of XXY, XXX, and XYY children may not even be diagnosed. Some are dismissed unfairly as being underachievers.

Nondisjunction in germ cells, gametes, or early embryonic cells changes the number of autosomes or the number of sex chromosomes. The change affects development and the resulting phenotypes.

Nondisjunction at meiosis causes most sex chromosome abnormalities, which typically lead to subtle difficulties with learning, and speech and other motor skills.

12.10 Human Genetic Analysis

Some organisms, including pea plants and fruit flies, are ideal for genetic analysis. They do not have very many chromosomes. They can grow and reproduce fast in small spaces, under controlled conditions. It does not take long to track a trait through many generations. Humans, however, are another story.

Unlike the flies in laboratory bottles, we humans live under variable conditions in diverse environments, and we live as long as the geneticists who study us. Most of us select our own mates and reproduce if and when we want to. Most families are not large, which means that there are not enough offspring available for researchers to make easy inferences.

Geneticists often gather information from several generations to increase the numbers for analysis. If a trait follows a simple Mendelian inheritance pattern, geneticists can be more confident about predicting the probability of its showing up again. The pattern also can be a clue to the past (Figure 12.18).

Such information is often displayed in **pedigrees**, or charts of genetic connections among individuals. Standardized methods, definitions, and symbols that

Figure 12.18 An intriguing pattern of inheritance. Eight percent of the men in Central Asia carry nearly identical Y chromosomes, which implies descent from a shared ancestor. If so, then 16 million males living between northeastern China and Afghanistan—close to 1 of every 200 men alive today—belong to a lineage that started with the warrior and notorious womanizer Genghis Khan. In time, his offspring ruled an empire that stretched from China all the way to Vienna.

represent different kinds of individuals are used to construct these charts. You already came across one in Section 12.7. Figures 12.19 and 12.20 are two more.

Those who analyze pedigrees rely on knowledge of probability and patterns of Mendelian inheritance that may yield clues to a trait. Such researchers have traced many genetic abnormalities and disorders to a dominant or recessive allele and often to its location on an autosome or a sex chromosome. Table 12.1 is a list of the ones used as examples in this book.

As individuals and as members of society, what do we do with the information? The next section gets into options. When considering them, keep in mind some important distinctions. First, a genetic *abnormality* is only a rare or uncommon version of a trait, as when a person is born with six digits on each hand or foot instead of the usual five. Whether you view such an abnormality as disfiguring or merely interesting is subjective only; there is nothing inherently life-threatening about it. By contrast, a genetic *disorder* is a heritable condition that sooner or later gives rise to mild to severe medical problems. Each

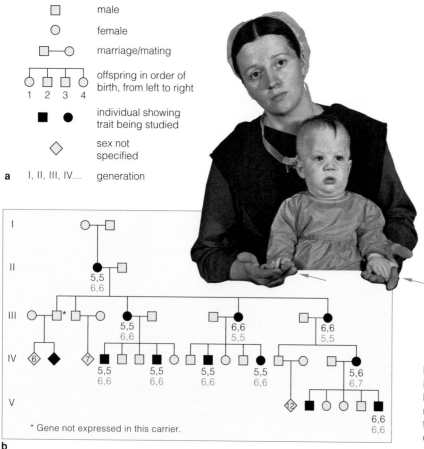

a I, II, III, IV... generation

□ male
○ female
□—○ marriage/mating
offspring in order of birth, from left to right
1 2 3 4
■ ● individual showing trait being studied
◇ sex not specified

*Gene not expressed in this carrier.

b

Figure 12.19 *Animated!* (**a**) Standardized symbols used in pedigrees. (**b**) A pedigree for *polydactyly*, characterized by extra fingers, toes, or both. *Black* numerals signify the number of fingers on each hand; *blue* numerals signify the number of toes on each foot. This condition recurs as one symptom of Ellis–van Creveld syndrome.

Table 12.1 Examples of Human Genetic Disorders and Genetic Abnormalities

Disorder or Abnormality	Main Symptoms	Disorder or Abnormality	Main Symptoms
Autosomal recessive inheritance		**X-linked recessive inheritance**	
Albinism	Absence of pigmentation	Androgen insensitivity syndrome	XY individual but having some female traits; sterility
Blue offspring	Bright blue skin coloration	Red–green color blindness	Inability to distinguish among some or all shades of red and green
Cystic fibrosis	Abnormal glandular secretions leading to tissue, organ damage		
Ellis–van Creveld syndrome	Extra fingers, toes, short limbs	Fragile X syndrome	Mental impairment
Fanconi anemia	Physical abnormalities, bone marrow failure	Hemophilia	Impaired blood-clotting ability
		Muscular dystrophies	Progressive loss of muscle function
Galactosemia	Brain, liver, eye damage	X-linked anhidrotic dysplasia	Mosaic skin (patches with or without sweat glands); other effects
Phenylketonuria (PKU)	Mental impairment		
Sickle-cell anemia	Adverse pleiotropic effects on organs throughout body	**Changes in chromosome structure**	
		Chronic myelogenous leukemia (CML)	Overproduction of white blood cells in bone marrow; organ malfunctions
Autosomal dominant inheritance		Cri-du-chat syndrome	Mental impairment; abnormally shaped larynx
Achondroplasia	One form of dwarfism		
Camptodactyly	Rigid, bent fingers	**Changes in chromosome number**	
Familial hypercholesterolemia	High cholesterol levels in blood; eventually clogged arteries	Down syndrome	Mental impairment; heart defects
Huntington's disease	Nervous system degenerates progressively, irreversibly	Turner syndrome	Sterility; abnormal ovaries, abnormal sexual traits
Marfan syndrome	Abnormal or no connective tissue	Klinefelter syndrome	Sterility; mild mental impairment
Polydactyly	Extra fingers, toes, or both	XXX syndrome	Minimal abnormalities
Progeria	Drastic premature aging	XYY condition	Mild mental impairment or no effect
Neurofibromatosis	Tumors of nervous system, skin		

genetic disorder is characterized by a specific set of symptoms—a **syndrome**.

One more point to keep in mind: Alleles that give rise to severe genetic disorders are generally rare in populations, because they put their bearers at risk. Why don't they disappear entirely? Rare mutations introduce new ones. In addition, in heterozygotes, a normal allele masks harmful effects that may result from expression of the mutant recessive allele. This means that heterozygotes can transmit harmful alleles to their offspring. The next section addresses how we may address the consequences.

Pedigree analysis may reveal simple patterns of Mendelian inheritance. From such patterns, specialists can infer the probability that offspring will inherit certain alleles.

A genetic abnormality is a rare or less common version of a heritable trait. A genetic disorder is a heritable condition that results in mild to severe medical problems.

Figure 12.20 Pedigree for Huntington's disease, a progressive degeneration of the nervous system. Researcher Nancy Wexler and her team constructed this extended family tree for nearly 10,000 Venezuelans. Their analysis of unaffected and affected individuals revealed that a dominant allele on human chromosome 4 is the culprit. Wexler has a special interest in the disease; it runs in her family.

12.11 Prospects in Human Genetics

With the arrival of their newborn, parents typically ask, "Is our baby all right?" Quite naturally, they want their baby to be free of genetic disorders, and most babies are. But what are the options when something goes wrong?

Many prospective parents have difficulty coming to terms with the possibility that a child of theirs might develop a severe genetic disorder. What are their options?

Phenotypic Treatments   Surgery, prescription drugs, hormone replacement therapy, and often dietary controls can minimize and in some cases eliminate the symptoms of many genetic disorders.

For instance, strict dietary controls work in cases of *phenylketonuria,* or PKU. Individuals affected by this genetic disorder are homozygous for a recessive allele on an autosome. They cannot make a functional form of an enzyme that catalyzes the conversion of the amino acid phenylalanine to tyrosine. Because the conversion is blocked, phenylalanine accumulates and is diverted into other metabolic pathways. The outcome is an impairment of brain function.

Affected people who restrict phenylalanine intake can lead essentially normal lives. They must avoid soft drinks and other products that are sweetened with aspartame, a compound that contains phenylalanine.

Genetic Screening   The idea behind genetic screening is to detect alleles that cause genetic disorders, provide information on reproductive risks, and help families who are already affected. Often, carriers or affected individuals are detected early enough to start countermeasures for minimizing the damage before symptoms develop.

A few large-scale screening programs are operational. Besides helping individuals, the information they generate is being used to estimate the prevalence and distribution of harmful alleles in populations. In the United States, for instance, most hospitals routinely screen newborns for PKU, so we now see fewer individuals with symptoms of the disorder.

There are social risks that must be considered. How would you feel if you were labeled as someone with "bad" alleles? Would the knowledge invite chronic anxiety? Would potential employers or insurance companies turn you down? How would you interact with an affected child that you brought into the world if you had known about the risk in advance? No easy answers here.

Prenatal Diagnosis   Doctors and clinicians commonly use methods of *prenatal diagnosis* to determine the sex of embryos or fetuses and to screen for more than 100 known genetic problems. *Prenatal* means before birth. *Embryo* is a term that applies until eight weeks after fertilization, after which the term *fetus* is appropriate.

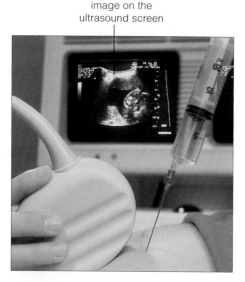

image on the ultrasound screen

Figure 12.21 *Animated!*   Amniocentesis, a prenatal diagnostic tool. A pregnant woman's doctor holds an ultrasound emitter against her abdomen while drawing a sample of amniotic fluid into a syringe. He monitors the path of the needle with an ultrasound screen, in the background. Then he directs the needle into the amniotic sac that holds the developing fetus and withdraws twenty milliliters or so of amniotic fluid. The fluid contains fetal cells and wastes that can be analyzed for genetic disorders.

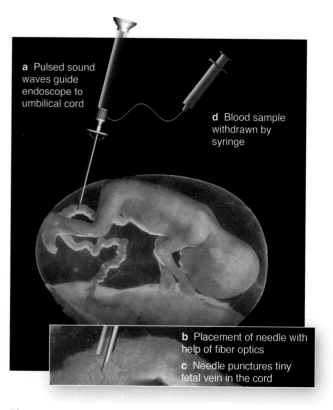

a Pulsed sound waves guide endoscope to umbilical cord

d Blood sample withdrawn by syringe

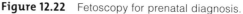

b Placement of needle with help of fiber optics

c Needle punctures tiny fetal vein in the cord

Figure 12.22   Fetoscopy for prenatal diagnosis.

Suppose a forty-five-year-old woman is pregnant and worries about Down syndrome. Between eight and twelve weeks after conception, she might opt for *amniocentesis* (Figure 12.21). By this diagnostic procedure, a clinician uses a syringe to withdraw a small sample of fluid from the amniotic cavity. The "cavity" is a fluid-filled sac, bounded by a membrane—the amnion—that encloses the fetus. The fetus normally sheds some cells into the fluid. Cells suspended in the fluid sample can be analyzed for many genetic disorders, including Down syndrome, cystic fibrosis, and sickle-cell anemia.

Chorionic villi sampling (CVS) is a similar diagnostic procedure. A clinician withdraws a few cells from the chorion, a membrane that surrounds the amnion and helps form the placenta. Unlike amniocentesis, however, CVS can be requested to find out information as early as eight weeks into pregnancy.

It is now possible to see a live, developing fetus with the aid of an endoscope, a fiber-optic device. In *fetoscopy*, sound waves are pulsed across the mother's uterus. Images of parts of the fetus, umbilical cord, or placenta show up on a computer screen that is connected to the endoscope (Figure 12.22). A sample of fetal blood is often drawn at the same time. This procedure can be used to diagnose many blood cell disorders, such as sickle-cell anemia and hemophilia.

There are risks to a fetus associated with all three procedures, including punctures or infections. Also, if the amnion does not reseal itself quickly, too much fluid can leak out of the amniotic cavity and endanger the fetus. Amniocentesis increases the risk of miscarriage by 1 to 2 percent. CVS may disrupt the placenta's development, which can cause missing or underdeveloped fingers and toes in 0.3 percent of newborns. Fetoscopy raises the risk of a miscarriage by 2 to 10 percent.

Genetic Counseling   Parents-to-be commonly ask genetic counselors to compare the risks associated with diagnostic procedures against the likelihood that their future child will be affected by a severe genetic disorder. At the time of counseling, they also should discuss the small overall risk (3 percent) that complications can affect *any* child during the birth process. They should talk about how old they are. The older either prospective parent is, the greater the risk may be.

As a case in point, suppose a first child or a close relative has a severe disorder. Genetic counselors come up with a program of diagnosis of parental genotypes, pedigrees, and genetic testing for known disorders. Using this information, counselors can predict risks for disorders in future children. They should remind prospective parents that the same risk usually applies to each pregnancy.

Regarding Abortion   What happens after prenatal diagnosis reveals a severe problem? Do prospective parents opt for an induced abortion? An *abortion* is an expulsion

Figure 12.23   Eight-cell and multicelled stages of human development.

of a pre-term embryo or fetus from the uterus. We can only say here that individuals must weigh awareness of the severity of the genetic disorder against their ethical and religious beliefs. Worse, today they must play out their personal tragedy on a larger stage that is dominated by a nationwide battle between highly vocal "pro-life" and "pro-choice" factions. We return to this volatile topic in Section 44.15, after explaining the stages of human embryonic development.

Preimplantation Diagnosis   This procedure relies on *in vitro fertilization*. Sperm and eggs from prospective parents are mixed in a sterile culture medium. One or more eggs may get fertilized. If this happens, mitotic cell divisions can turn an egg into a ball of eight cells within forty-eight hours (Figure 12.23).

According to one view, the tiny, free-floating ball is a pre-pregnancy stage. Like all of the unfertilized eggs that a woman's body discards monthly during her reproductive years, it has not attached to the uterus. All of its cells have the same genes. However, its cells are not yet committed to being specialized one way or another. Doctors carefully remove one of these undifferentiated cells and analyze its genes. If it has no detectable genetic defects, the ball is inserted into the uterus. The withdrawn cell will not be missed. Many of the resulting "test-tube babies" are born in good health.

Some couples who are at risk of passing on the alleles for cystic fibrosis, muscular dystrophy, or some other genetic disorder have opted for this procedure.

Summary

Section 12.1 Of twenty-three pairs of homologous chromosomes in human body cells, one is a pairing of sex chromosomes. The other chromosomes are called autosomes; in both sexes, they are the same length and shape, have the same centromere location, and carry the same genes along their length.

Section 12.2 In karyotyping, a diagnostic tool, an individual's metaphase chromosomes are prepared for microscopy, photographed, and arranged in sequence in a chart on the basis of their defining features.

Biology⊗Now
Learn how to create a karyotype with the animation on BiologyNow.

Sections 12.3, 12.4 Some dominant and recessive alleles on autosomes are inherited in simple Mendelian patterns that can be predictably connected with specific phenotypes. Some mutant forms of these alleles give rise to genetic abnormalities or genetic disorders.

Biology⊗Now
Investigate autosomal inheritance with the interaction on BiologyNow.

Sections 12.5, 12.6 Human females have identical sex chromosomes (XX) and males have nonidentical ones (XY). The *SRY* gene on the Y chromosome is the basis of sex determination. Its expression starts the synthesis of testosterone, a hormone that causes a human embryo to develop into a male. If an embryo has no Y chromosome (no *SRY* gene), it develops into a female.

Experiments with fruit flies yielded the first evidence that specific genes that give rise to nonsexual traits are located on the X chromosome.

Biology⊗Now
See how gender is determined in humans with the interaction on BiologyNow.

Section 12.7 Certain dominant and recessive alleles on the X chromosome are inherited in simple patterns. A number of alleles on the X chromosome contribute to more than 300 known genetic disorders. Males cannot transmit a recessive X-linked allele to their sons; an affected female must be the bridge of inheritance.

Biology⊗Now
Investigate X-linked inheritance with the interaction on BiologyNow.

Section 12.8 On rare occasions, a chromosome's physical structure undergoes abnormal alterations. Part of it may be duplicated, deleted, inverted, or moved to a new location (translocated) in the same chromosome or a different one.

Most alterations are harmful or lethal. Even so, many have accumulated in the chromosomes of all species over evolutionary time. Either they had neutral effects or they later proved to be useful. Many duplications, inversions, and translocations are built into primate chromosomes. They are strikingly similar among human, chimpanzee, gorilla, orangutan, and gibbon chromosomes, which is strong evidence of divergences from a common ancestor.

Section 12.9 The parental chromosome number can change permanently. Most often, this is an outcome of nondisjunction: the failure of one or more pairs of duplicated chromosomes to separate from each other, most often during meiosis.

Aneuploids have inherited one extra or one less chromosome than their parents. In the human population, trisomy 21, the most well-known form of aneuploidy, results in Down syndrome. Most human autosomal aneuploids die before birth.

Polyploids inherited three or more of each type of chromosome from their parents. About half of all flowering plants and some insects, fishes, and other animals are polyploid.

Changes in the number of sex chromosomes usually cause problems with learning and motor skills. Problems can be so subtle that the underlying cause may not be diagnosed, as among XXY, XXX, and XYY children.

Sections 12.10, 12.11 Traditionally, geneticists have constructed pedigrees, or charts of genetic connections among individuals, to estimate the probability that offspring will inherit a trait of interest. Phenotypic treatments, genetic screening, genetic counseling, prenatal diagnosis, and preimplantation diagnosis are options available for potential parents who are at risk of transmitting a harmful allele to offspring.

Biology⊗Now
Examine a human pedigree with the animation on BiologyNow.
Explore amniocentesis with the animation on BiologyNow.

Self-Quiz
Answers in Appendix II

1. The _____ of chromosomes in a cell are compared to construct karyotypes.
   a. length and shape   c. gene sequence
   b. centromere location   d. both a and b

2. The _____ determines gender in humans.
   a. X chromosome   c. *SRY* gene
   b. *Dll* gene   d. both b and c

3. If one parent is heterozygous for a dominant allele on an autosome and the other parent is homozygous, any child of theirs has a _____ chance of being heterozygous.
   a. 25 percent   c. 75 percent
   b. 50 percent   d. no chance; it will die

4. Expansion mutations occur _____ within and between genes in human chromosomes.
   a. only rarely   c. not at all
   b. frequently   d. only in multiples of ten

5. Galactosemia is a case of _____ inheritance.
   a. autosomal dominant   c. X-linked dominant
   b. autosomal recessive   d. X-linked recessive

6. Is this statement true or false: A son can inherit an X-linked recessive allele from his father.

7. Color blindness is a case of _____ inheritance.
   a. autosomal dominant   c. X-linked dominant
   b. autosomal recessive   d. X-linked recessive

8. A (An) _____ can alter chromosome structure.
   a. deletion   c. inversion   e. all of the
   b. duplication   d. translocation   above

9. Nondisjunction may occur during _____ .
   a. mitosis   c. fertilization
   b. meiosis   d. both a and b

10. Is this statement false: Body cells sometimes inherit three or more of each type of chromosome characteristic of the species, a condition called aneuploidy.

11. The karyotype for Klinefelter syndrome is _____ .
    a. XO   c. XXY
    b. XXX   d. XYY

12. A recognized set of symptoms that characterize a specific disorder is a _____ .
    a. syndrome   b. disease   c. pedigree

13. Match the chromosome terms appropriately.
    _____ polyploidy   a. number and defining
    _____ deletion   features of an individual's
    _____ nondisjunction   metaphase chromosomes
    _____ translocation   b. segment of a chromosome
    _____ karyotype   moves to a nonhomologous
    _____ aneuploidy   chromosome
                          c. extra chromosome sets
                          d. one outcome: gametes with
                             wrong chromosome number
                          e. a chromosome segment lost
                          f. change by one chromosome

Additional questions are available on Biology ⒺNow™

Genetics Problems
Answers in Appendix III

1. Human females are XX and males are XY.
   a. Does a male inherit the X from his mother or father?
   b. With respect to X-linked alleles, how many different types of gametes can a male produce?
   c. If a female is homozygous for an X-linked allele, how many types of gametes can she produce with respect to that allele?
   d. If a female is heterozygous for an X-linked allele, how many types of gametes might she produce with respect to that allele?

2. In Section 11.4, you read about a mutation that causes a serious genetic disorder, *Marfan syndrome*. A mutant allele responsible for the disorder follows a pattern of autosomal dominant inheritance. What is the chance that any child will inherit it if one parent does not carry the allele and the other is heterozygous for it?

3. Somatic cells of individuals with Down syndrome usually have an extra chromosome 21; they contain forty-seven chromosomes.
   a. At which stages of meiosis I and II could a mistake alter the chromosome number?
   b. A few individuals with Down syndrome have forty-six chromosomes, two of which are normal-appearing

Figure 12.24   A case of Klinefelter syndrome. Until his teenage years, Stefan was shy, reserved, and prone to rage for no apparent reason. Psychologists and doctors assumed he had learning disabilities that affected comprehension, auditory processing, memory, and abstract thinking. One told Stefan he was stupid and lazy, and would be lucky to graduate from high school. In time, Stefan was graduated from college with degrees in business administration and sports management. He never discussed his learning disabilities. Instead, he took pride in doing the work on his own and not being treated differently.

Stefan was twenty-five years old before laboratory tests as well as karyotyping revealed a 46XY/47XXY mosaic condition. That same year, he started a job as a software engineer. Having a full-time position helped him open doors to volunteer work with the Klinefelter syndrome network. During his volunteer work, he met his future fiancée, whose son also has the syndrome.

chromosomes 21 and a longer-than-normal chromosome 14. Speculate on how this chromosome abnormality may have arisen.

4. As you read earlier, *Duchenne muscular dystrophy* is a genetic disorder that arises through the expression of a recessive X-linked allele. Usually, symptoms start in childhood. Gradual, progressive loss of muscle function leads to death, usually by age twenty or so. Unlike color blindness, the disorder is nearly always restricted to males. Suggest why.

5. In the human population, mutation of two genes on the X chromosome causes two types of X-linked *hemophilia* (A and B). In a few cases, a woman is heterozygous for both mutant alleles (one on each of the X chromosomes). All of her sons should have either hemophilia A or B.
   However, on very rare occasions, one of these women gives birth to a son who does not have hemophilia, and his one X chromosome does not have either mutant allele. Explain how such an X chromosome could arise.

6. Does the phenotype indicated by red circles and squares in this pedigree show a Mendelian inheritance pattern that is autosomal dominant, autosomal recessive, or X-linked?

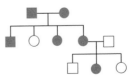

7. When it comes to acceptance of a genetic condition that is out of the ordinary, people tend to be subjective. As an example, consider the individual described in Figure 12.24. How would you have categorized him without knowing the genetic basis of his early behavior? How would you categorize him now in terms of what we as a society consider to be "ideal" phenotypes?

Goodbye, Dolly

In 1997 in Scotland, geneticist Ian Wilmut made headlines when he did not bother with the union of sperm and eggs to produce a new lamb. He wanted to make a genetic copy —a **clone**—of a fully grown sheep. He thought he could do so by slipping the nucleus of an adult's body cell into an unfertilized egg that had its own nucleus gently sucked out beforehand. He succeeded. One egg that his team modified developed into a cloned lamb, which they named Dolly.

Dolly grew up and later gave birth to six lambs of her own (Figure 13.1). Since then, researchers all over the world have been using adult DNA to make identical copies of other adult mammals. Mice, rabbits, pigs, cattle, goats, mules, deer, horses, and cats have all been cloned.

Sheep normally do not show symptoms of old age until they are about ten years old. By age five, Dolly had become arthritic and overweight. Less than a year later, an infection in her lungs proved irreversible and she was put to sleep.

Did Dolly develop health problems simply because she was a clone? Earlier studies of her telomeres had raised suspicions. Telomeres are short segments that cap the ends of chromosomes and stabilize them. They become shorter and shorter as an animal ages. When Dolly was only two years old, telomeres in some of her cells were as short as those of a six-year-old sheep—the exact age of the adult animal that was her genetic donor.

Using adult DNA to clone mammals is challenging. Most clones die before birth or shortly afterward. It took almost seven hundred attempts to get a clone of a guar, a wild ox on the endangered species list. Less than two days after his birth, he died of complications following an infection.

The clones that do survive often have health problems. Like Dolly, many become unusually overweight as they age. Other clones are exceptionally large from birth or have some enlarged organs. Cloned mice develop lung and liver problems, and almost all die prematurely. Cloned pigs have heart problems, they limp, and one never did develop a tail or, worse still, an anus.

Physically moving the DNA from an adult's cell into an egg stripped of its own nucleus is only part of the challenge. Most genes in a mature cell are inactive. To guide the reproductive process, they have to be reprogrammed or switched on in controlled ways. Apparently, not all of the genes in all clones are being properly activated.

Some people want to put a stop to adult cloning because the risk of bringing defective mammals into the world troubles them deeply. Other people want research into reprogramming DNA to continue. For instance, they point to patients who are desperate for organ transplants. People are already cloning pigs that were genetically modified to produce organs that human donors are less likely to reject. A few people on the fringes of bioethical common sense are toying with the idea of reprogramming DNA to clone an adult human.

Is DNA amazing? It certainly is. Are we amazing in our capacity to use and abuse its potential? You bet.

Figure 13.1 Dolly and one of her lambs. Dolly was the first mammal to be formed by way of adult DNA cloning. She awakened society to the goings-on of the molecular revolution by jarring our notions of what it takes to reproduce a complex animal.

Watch the video online!

Figure 13.2 Watson, Crick, and the model for DNA that brought about a revolution in molecular biology.

With this chapter, we move past the chromosomal basis of inheritance and turn to the investigations and models that led to our current understanding of DNA (Figure 13.2). The story is more than a march through the details of how its molecular structure encodes hereditary information. *It also is revealing of how ideas are generated in science.*

On the one hand, having a shot at fame and fortune quickens the pulse of men and women in any profession, and scientists are no exception. On the other hand, science proceeds as a community effort, with individuals sharing not only what they can explain but also what they do not understand. Even when an experiment fails to produce the anticipated results, it may turn up information that others can use or lead to questions that others can answer. Unexpected results, too, might be clues to something important about the natural world.

How Would You Vote?

Abnormal animals often form during animal cloning experiments, but cloning research may also result in new drugs and organ replacements for human patients. Should animal cloning be banned? See BiologyNow for details, then vote online.

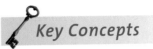

Key Concepts

DISCOVERY OF DNA'S FUNCTION

In all living cells, DNA molecules are storehouses of information that governs heritable traits. Section 13.1

THE DNA DOUBLE HELIX

DNA is a double-stranded molecule consisting of four kinds of nucleotides: adenine, thymine, guanine, and cytosine. The two strands coil together helically, like a spiral stairway.

Each nucleotide base of one strand is hydrogen-bonded to a base of the other strand. As a rule, adenine pairs with thymine, and guanine with cytosine.

The order in which one kind of base follows the next in a strand encodes heritable information. The DNA of each species has at least some unique base sequences that are not found in the DNA of any other species. Section 13.2

HOW CELLS DUPLICATE THEIR DNA

Before a cell divides, enzymes and other proteins replicate the DNA; they make a copy of it. Different kinds unwind the double helix and construct a new, complementary strand on the exposed bases of each parent strand. They do so according to the base-pairing rule for DNA.

Repair enzymes monitor mismatched base pairings and other changes in the DNA strands. Section 13.3

DNA AND THE CLONING CONTROVERSIES

What does it mean when we say that DNA holds heritable information? The answer hits home when the messages encoded in its base sequences are tapped to make an exact copy of an adult animal. Section 13.4

THE FRANKLIN FOOTNOTE

As in any profession, some were winners and some losers in the DNA chase. Some players might have taken less-than-noble shortcuts. Section 13.5

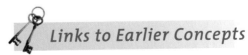

Links to Earlier Concepts

This chapter builds on your understanding of hydrogen bonding (Section 2.4), condensation reactions (3.2), and the earlier overview of DNA structure and function (3.7). Your knowledge of chromosomes, mitosis, and meiosis will help you understand the nuclear transfers that are part of cloning procedures (9.3, 10.3). Keep the image of the eight-cell stage of human development in mind when you read about embryo cloning, because the cells used are no more developed than this (12.11).

13.1 The Hunt for Fame, Fortune, and DNA

Why, in the spring of 1868, was Johann Miescher collecting cells from the pus of open wounds and, later, from sperm of a fish? This physician wanted to identify the chemical composition of the nucleus. Such cells have little cytoplasm, which makes it easier to isolate the nuclear material. In time he isolated an acidic compound that contains nitrogen and phosphorus. He had discovered what came to be known many years later as **deoxyribonucleic acid**, *or* **DNA***.*

EARLY AND PUZZLING CLUES

At the time Miescher made his discovery, no one knew much about the physical basis of inheritance. That is, *which substance encodes the information about reproducing parental traits in offspring?* Few researchers thought that DNA might hold the answer. For a long time, most were thinking PROTEINS! Because heritable traits are so diverse, they assumed that molecules of inheritance had to be structurally diverse, too. Proteins, they said, consist of unlimited combinations of twenty kinds of amino acids. Other molecules just seemed too simple.

Now fast-forward to 1928. An army medical officer, Frederick Griffith, wanted to develop a vaccine against the bacterium *Streptococcus pneumoniae*, a major cause of pneumonia. He did not succeed, but he isolated and cultured two strains that unexpectedly shed light on inheritance. The colonies of one strain had a rough surface appearance; colonies of the other appeared smooth. Griffith designated the strains *R* and *S*, and he used them in a series of experiments (Figure 13.3).

First, he injected mice with live *R* cells. The mice did not develop pneumonia. *The R strain was harmless.*

Second, he injected other mice with live *S* cells. The mice died. Blood samples from them teemed with live *S* cells. *The S strain was pathogenic; it caused the disease.*

Third, he killed *S* cells by exposing them to high temperature. *Mice injected with dead S cells did not die.*

Fourth, he mixed live *R* cells with heat-killed *S* cells. He injected them into mice. The mice died—*and blood samples drawn from them teemed with live S cells!*

What went on in the fourth experiment? Maybe heat-killed *S* cells in the mix were not really dead. But if that were so, then the mice injected with heat-killed *S* cells in experiment 3 would have died. Or maybe the harmless *R* cells had mutated into a killer strain. But if that were so, then the mice injected with the *R* cells only in experiment 1 would have died.

The simplest explanation was this: *Heat had killed the S cells but did not destroy their hereditary material— including the part that specified "how to cause infection."* Somehow, that material had been transferred from the dead *S* cells into living *R* cells, which put it to use.

Further tests made it clear that the transformation was permanently heritable. Even after a few hundred generations, *S* cell descendants were still infectious.

What was the hereditary material that caused the transformation? Scientists started looking in earnest, but most were still thinking PROTEINS!

Still, Griffith's results intrigued Oswald Avery, who began to transform harmless bacteria by mixing them with extracts of killed pathogenic cells. Avery asked: What part of the extracts caused the transformation? He found that adding protein-digesting enzymes to the extracts had no effect; cells were still transformed. However, adding a DNA-digesting enzyme to extracts prevented transformation. DNA was looking good.

CONFIRMATION OF DNA FUNCTION

By the 1950s, Max Delbrück, Alfred Hershey, Martha Chase, Salvador Luria, and other molecular sleuths were using viruses for experiments. These infectious particles hold information on substances required to make new virus particles. After viruses infect a host cell, their enzymes trick its metabolic machinery into synthesizing those substances. **Bacteriophages**, which only infect certain bacteria, were the viruses of choice for the early experiments.

As researchers knew, some bacteriophages consist only of DNA and a coat, probably of protein. Also, as

1 Mice injected with live cells of harmless strain *R*.

2 Mice injected with live cells of killer strain *S*.

3 Mice injected with heat-killed *S* cells.

4 Mice injected with live *R* cells *plus* heat-killed *S* cells.

Figure 13.3
Animated!
Summary of results from Fred Griffith's experiments with *Streptococcus pneumoniae* and laboratory mice.

Mice do not die. No live *R* cells in their blood.

Mice die. Live *S* cells in their blood.

Mice do not die. No live *S* cells in their blood.

Mice die. Live *S* cells in their blood.

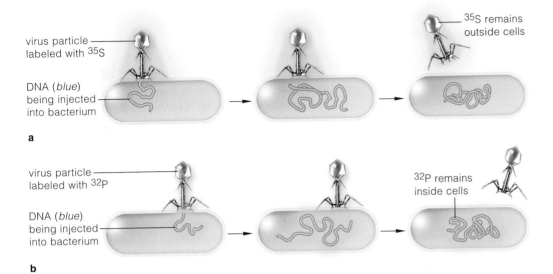

virus particle labeled with ³⁵S

DNA (*blue*) being injected into bacterium

a

³⁵S remains outside cells

virus particle labeled with ³²P

DNA (*blue*) being injected into bacterium

b

³²P remains inside cells

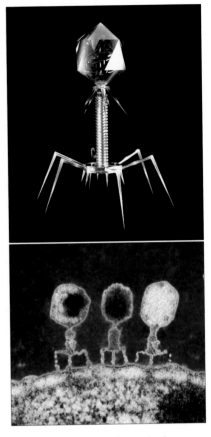

Figure 13.4 *Animated!* Example of the landmark experiments that tested whether genetic material resides in bacteriophage DNA, proteins, or both. Alfred Hershey and Martha Chase knew that sulfur (S) but not phosphorus (P) is a component of bacteriophage proteins. They also knew that phosphorus but not sulfur is a component of DNA.

(**a**) In one experiment, bacteria were grown in a culture medium with a tracer, the radioisotope ³⁵S. The cells used the ³⁵S when they built proteins. Bacteriophages infected the labeled cells, which started to make viral proteins. So the proteins of new virus particles became labeled with the ³⁵S. The labeled virus particles infected a new batch of unlabeled cells. The mixture was whirred in a kitchen blender. Whirring dislodged the viral coats from infected cells. Chemical analysis revealed the presence of labeled protein in the solution but only traces of it inside the cells.

(**b**) In another experiment, bacteriophages infected cells that had taken up the radioisotope ³²P. Later, the cells used ³²P when they built viral DNA. This labeled the DNA and new virus particles. The labeled viruses were used to infect bacteria in solution, then were dislodged from them. Most labeled viral DNA stayed in the cells—evidence that DNA is the genetic material of this virus.

c *Above*, model for a bacteriophage. *Below*, micrograph of virus particles injecting their DNA into an *E. coli* cell.

micrographs revealed, the coat remains on the *outer surface* of infected cells. Did viruses inject hereditary material only into cells? If so, then was the material protein, DNA, or both? Figure 13.4 outlines just two of many experiments that pointed to DNA.

Then Linus Pauling did something no one had done before. With his training in biochemistry, a talent for model building, and a dose of intuition, he deduced the structure of a protein—collagen. His discovery was electrifying. If someone could pry open the secrets of proteins, then why not DNA? And if DNA's structural details were deduced, would those details hold clues to how DNA functions in inheritance? *Someone could go down in history as having discovered the secret of life!*

ENTER WATSON AND CRICK

Scientists started to scramble after the prize. Among them were Francis H. Crick, a Cambridge University, researcher, and James Watson, a postdoctoral fellow recently arrived from Indiana University. They spent

hours arguing over everything they had read about DNA's size, shape, and bonding requirements. They fiddled with cardboard cutouts, and they badgered chemists to help them identify possible bonds they might have overlooked. They built models from thin bits of metal connected with wire "bonds."

In 1953, Watson and Crick built a model that fit all the pertinent biochemical rules and all the clues they had gleaned from other sources. They had discovered the structure of DNA. The molecule has breathtaking simplicity, and it helped Crick answer another riddle —*how life can show unity at the molecular level and still give rise to such spectacular diversity at the level of whole organisms.* Turn now to high points in the community effort that gave us these insights.

DNA functions as the cell's treasurehouse of inheritance. The cumulative efforts of many scientists, building on one another's work, resulted in the discovery of that function.

13.2 The Discovery of DNA's Structure

LINKS TO
SECTIONS
3.2, 3.7

Long before the bacteriophage studies were under way, biochemists knew that DNA contains only four kinds of nucleotides that are the building blocks of nucleic acids. But how were the nucleotides arranged in DNA?

DNA'S BUILDING BLOCKS

Recall, from Section 3.7, that a **nucleotide** in DNA has a five-carbon sugar (deoxyribose), a phosphate group, and one of the following nitrogen-containing bases:

adenine	guanine	thymine	cytosine
A	G	T	C

T and C are pyrimidines, with a backbone of carbon and nitrogen that forms a single ring. A and G are purines—larger, bulkier molecules having two rings. Overall, the four types of nucleotides have the same bonding pattern, as Figure 13.5 indicates.

By 1949, the biochemist Erwin Chargaff had shared with the scientific community two insights about the proportions of nucleotides in DNA. First, the amount of adenine relative to guanine differs among species. Second, the amounts of thymine and adenine in DNA are identical, and so are the amounts of cytosine and guanine. We may show this as A=T and G=C.

These symmetrical proportions had to mean something. As biochemists already knew, the nucleotides in DNA are joined to one another by way of condensation reactions that form long chains (Section 3.2). But how were the four kinds arranged in a chain, and in what order?

The first convincing clue to the actual arrangement emerged from Maurice Wilkins's research laboratory at Cambridge, England. Researcher Rosalind Franklin made exceptional **x-ray diffraction images** of DNA. Such images form after a beam of x-rays is directed at a molecule, which scatters the x-rays in a pattern that can be captured on film. The pattern consists only of dots and streaks; in itself, it is not the structure of the molecule. However, researchers can use it to calculate the positions of the molecule's atoms.

Before Franklin, researchers had been working with dehydrated DNA molecules. Franklin was the first to put DNA into a "wet" form—which is the form that occurs in cells—and make an exceptionally clear image of it. With that image, she painstakingly calculated that the DNA molecule is long and thin, and that it has a 2-nanometer diameter. She also found repeats of some molecular configuration every 0.34 nanometer along its length, and another repeat every 3.4 nanometers. These were crucial clues, but her part in the discovery process was downplayed until recently (Section 13.5).

What did the repeating variation in DNA mean? Could DNA be coiled along its length, like a circular stairway? Certainly Pauling thought so. After all, he had calculated that collagen is helically coiled. Like

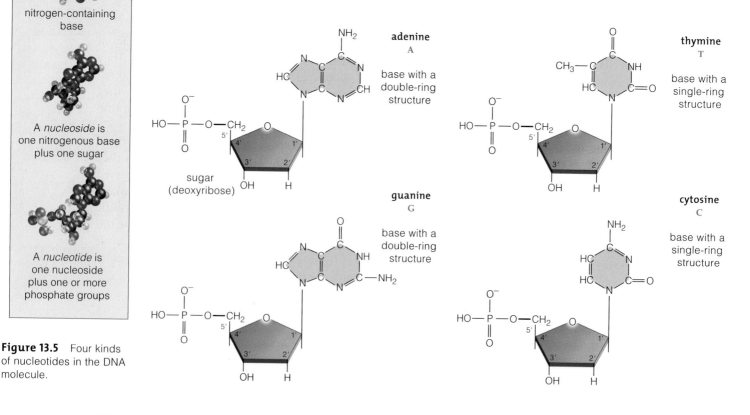

nitrogen-containing base

A *nucleoside* is one nitrogenous base plus one sugar

A *nucleotide* is one nucleoside plus one or more phosphate groups

Figure 13.5 Four kinds of nucleotides in the DNA molecule.

many others—including Wilkins, Watson, and Crick—he was thinking "helix." As Watson later wrote, "We thought, why not try it on DNA? We were worried that *Pauling* would say, why not try it on DNA? Certainly he was a clever man. He was a hero of mine. But we beat him at his own game. I still can't figure out why."

Pauling, it turned out, made a big chemical mistake. His model had all the negatively charged phosphate groups facing the interior of the DNA helix instead of facing outward. If they were that close together, they would repel each other too much to remain stable.

PATTERNS OF BASE PAIRING

Franklin filed away her image of wet DNA, but it still came to the attention of Watson and Crick. From all the clues that had accumulated, they perceived that DNA must consist of two strands of nucleotides, held together at their bases by hydrogen bonds (Figure 13.6). Such bonds form when the two strands run in opposing directions and twist to form a double helix. Only two kinds of base pairings typically form along the molecule's length: **A—T** and **G—C**.

This bonding pattern accommodates variation in the order of bases. For instance, a stretch of DNA from a rose, a human, or any other organism might be:

GGCCCCTTC — one base pair
CCGGGGAAG

or GCACCAATA or AAAAAAAAA
CGTGGTTAT TTTTTTTTT

All DNA molecules show the same bonding pattern. Many stretches of base sequences are the same in all of them. But some are unique for each species and even vary among individuals of a species! *The constancy in DNA's bonding pattern is the basis for life's unity—and variation in base sequences is the basis for life's diversity.*

Intriguingly, computer simulations show that if you want to pack a string into the least space, coil it into a helix. Was this space-saving advantage a factor in the molecular origin of the DNA double helix? Maybe.

> The pattern of base pairing between the two strands in DNA is constant for all species—A with T, and G with C. However, each species has a number of unique sequences of base pairs along the length of its DNA molecules.

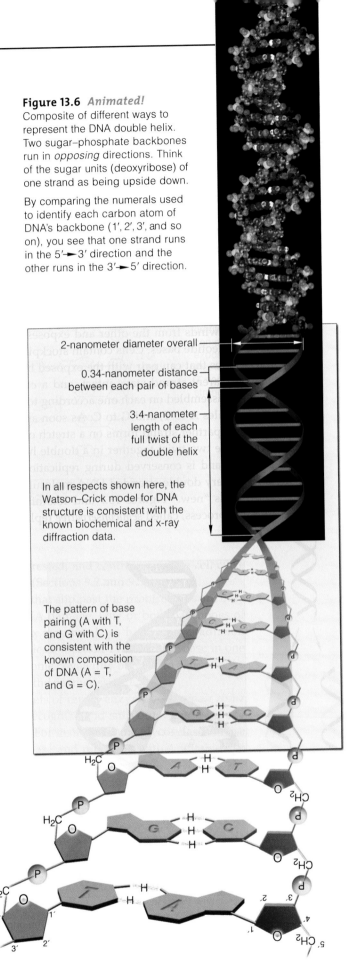

Figure 13.6 *Animated!*
Composite of different ways to represent the DNA double helix. Two sugar–phosphate backbones run in *opposing* directions. Think of the sugar units (deoxyribose) of one strand as being upside down.

By comparing the numerals used to identify each carbon atom of DNA's backbone (1', 2', 3', and so on), you see that one strand runs in the 5'→3' direction and the other runs in the 3'→5' direction.

2-nanometer diameter overall

0.34-nanometer distance between each pair of bases

3.4-nanometer length of each full twist of the double helix

In all respects shown here, the Watson–Crick model for DNA structure is consistent with the known biochemical and x-ray diffraction data.

The pattern of base pairing (A with T, and G with C) is consistent with the known composition of DNA (A = T, and G = C).

14.1 How Is RNA Transcribed From DNA?

LINKS TO
SECTIONS
3.7, 13.2, 13.3

*In **transcription,** the first step in protein synthesis, a sequence of nucleotide bases is exposed in an unwound region of a DNA strand. That sequence is the template upon which a single strand of RNA is assembled from adenine, cytosine, guanine, and uracil subunits.*

The chapter introduction may have left you with the impression that protein synthesis requires one class of RNA molecules. It actually requires three. When genes that specify proteins are transcribed, the outcome is **messenger RNA** (mRNA). *This is the only class of RNA that carries the protein-building codes.* **Ribosomal RNA** (rRNA) and **transfer RNA** (tRNA) are transcribed from different genes. The rRNA becomes a component of ribosomes, the structures in which polypeptide chains are assembled. The tRNA delivers amino acids one by one to ribosomes in the order specified by mRNA.

THE NATURE OF TRANSCRIPTION

An RNA molecule is almost but not quite like a single strand of DNA. It has four kinds of ribonucleotides, each with the five-carbon sugar ribose, one phosphate group, and one base. Three bases—adenine, cytosine, and guanine—are the same as those in DNA. In RNA, though, the fourth base is **uracil**, not thymine. Uracil, too, can pair with adenine, which means that a new RNA strand can base-pair with a DNA strand. Figure 14.2 is a simple way to think about this pairing.

Transcription *differs* from DNA replication in three respects. Only part of one DNA strand, not the whole molecule, is unwound and used as the template. The enzyme **RNA polymerase**, not DNA polymerase, adds ribonucleotides one at a time to the end of a growing strand of RNA. Also, transcription results in one free RNA strand, not a hydrogen-bonded double helix.

DNA contains many protein-coding regions. Each is transcribed separately, and each has its own START

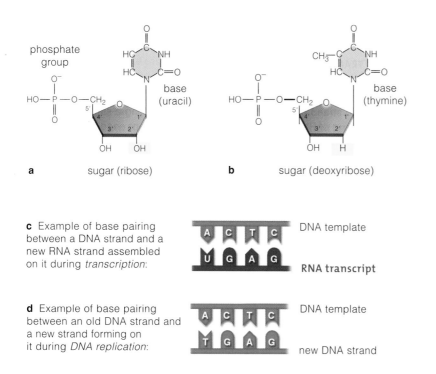

c Example of base pairing between a DNA strand and a new RNA strand assembled on it during *transcription*:

DNA template

RNA transcript

d Example of base pairing between an old DNA strand and a new strand forming on it during *DNA replication*:

DNA template

new DNA strand

Figure 14.2 (**a**) Uracil, one of four ribonucleotides in RNA. The other three—adenine, guanine, and cytosine—differ only in their bases. Uracil compared with (**b**) thymine, a DNA nucleotide. (**c**) Base pairing of DNA with RNA during transcription, compared with (**d**) base pairing during DNA replication.

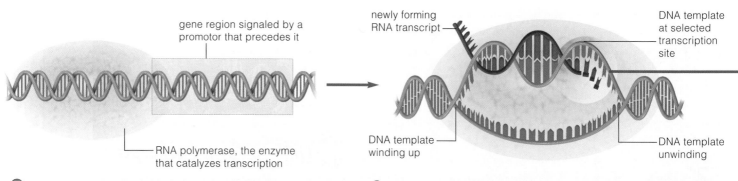

a RNA polymerase initiates transcription at a promoter in DNA. After binding to a promoter, RNA polymerases recognize a base sequence in DNA as a template for making a strand of RNA from free ribonucleotides, which have the bases adenine, cytosine, guanine, and uracil.

b All through transcription, the DNA double helix becomes unwound in front of the RNA polymerase. Short lengths of the newly forming RNA strand briefly wind up with its DNA template strand. New stretches of RNA unwind from the template (and the two DNA strands wind up again).

Figure 14.3 *Animated!* Gene transcription. By this process, an RNA molecule is assembled on a DNA template. (**a**) Gene region of DNA. The base sequence along one of DNA's two strands (not both) is used as the template. (**b–d**) Transcribing that region results in a molecule of RNA.

and STOP signal. A **promoter** is a START signal, a base sequence in DNA to which RNA polymerases bind and prepare for transcription. After binding, an RNA polymerase recognizes a gene region and moves along it. It uses the gene's base sequence as a template for covalently bonding free ribonucleotides together in a complementary sequence, as in Figure 14.3. When it reaches a sequence that signals "the end" of the gene region, the new RNA is released as a free transcript.

FINISHING TOUCHES ON THE mRNA TRANSCRIPTS

In eukaryotic cells, mRNA transcripts are modified before leaving the nucleus. Just as a dressmaker may snip off some threads or put bows on a dress before it leaves the shop, so do cells tailor their "pre-mRNA." For instance, some enzymes attach a modified guanine "cap" to the start of a pre-mRNA transcript. Others attach about 100 to 300 adenine ribonucleotides as a tail to the other end. Hence its name, poly-A tail.

Later, the pre-mRNA's cap will bind to a ribosome. Enzymes will nibble off the tail from the tip on back. Thus each tail's length dictates how long a particular protein-building message will last in the cytoplasm.

A transcript's message gets processed even before it leaves the nucleus. Eukaryotic genes contain **exons**: protein-coding base sequences that are interrupted by noncoding sequences, or **introns**. Both are transcribed, but all introns are snipped out before the transcript reaches the cytoplasm (Figure 14.4). Either all exons are retained in a mature mRNA transcript or some are

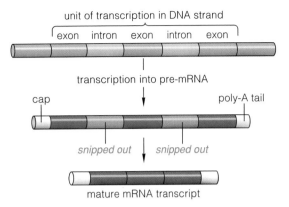

unit of transcription in DNA strand

exon intron exon intron exon

transcription into pre-mRNA

cap   poly-A tail

snipped out   *snipped out*

mature mRNA transcript

Figure 14.4
Animated! How pre-mRNA transcripts are processed into final form. Inside the nucleus, some or all introns are removed, and the transcript gets a cap and a tail.

removed and the rest are spliced together in various combinations. By this **alternative splicing**, one gene can specify two or more proteins that differ slightly in form and function! Cells use different combinations of exons at different times. Alternative splicing was once considered to be a rare event. However, it may occur in half (or all) genes of the human genome. It helps explain how human cells can make hundreds of thousands of proteins from only 21,500 or so genes.

In gene transcription, a sequence of exposed bases on one of the two strands of a DNA molecule serves as a template for synthesizing a complementary strand of RNA.

RNA polymerases assemble the RNA from four kinds of ribonucleotides that differ in their bases: A, U, C, and G.

Before leaving the nucleus, each new mRNA transcript, or pre-mRNA, undergoes modification into final form.

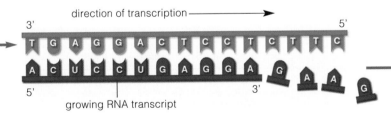

direction of transcription ⟶

3'   5'

T G A G G A C T C C T C T T C

A C U C C U G A G G A G A A G

5'   3'

growing RNA transcript

c What happened in the gene region? RNA polymerase catalyzed the covalent bonding of ribonucleotides to one another to form an RNA strand. The base sequence in the new strand is complementary to the exposed bases on the DNA as a template. Many other proteins assist in transcription; compare Section 13.3.

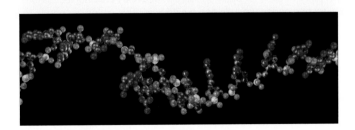

A C U C C U G A G G A G A A G

d At the end of the gene region, the last stretch of the new transcript is unwound and released from the DNA template. Shown below it is a model for a transcribed strand of RNA.

14.2 The Genetic Code

The correspondence between genes and proteins is encoded in protein-building "words" in mRNA transcripts. Three nucleotide bases make up each three-letter word.

Figure 14.5a shows a bit of mRNA transcribed from a DNA template. To translate it, you have to know how many letters (bases) make each word (amino acid). That is what Marshall Nirenberg, Philip Leder, Severo Ochoa, and Gobind Korana figured out. After mRNA has docked at a ribosome, its bases are "read" *three at a time*. The base triplets in mRNA are **codons**. Figure 14.5b shows how their sequence corresponds to the amino acid sequence in a growing polypeptide chain.

There are sixty-four different codons even though there are only twenty amino acids in proteins (Figure 14.6). Why so many? Think it through. If the codon were only one nucleotide, mRNA could specify only four kinds of amino acids. Codons of two nucleotides could code for sixteen kinds of amino acids—still not enough. Mixes of three nucleotides could code for sixty-four kinds—more than enough.

Certain codons actually do specify more than one kind of amino acid. For instance, both GAA and GAG specify glutamate. Also, in most species, the first AUG in the transcript is a START signal for translating "three-bases-at-a-time." It also means methionine is the first amino acid in all new polypeptide chains. UAA, UAG, and UGA do not specify any amino acid. They are STOP signals that block further additions of amino acids to a new chain.

The set of sixty-four different codons is the **genetic code**, and it has been highly conserved through time. Prokaryotes, a few organelles derived from them, and some protists of ancient lineages have a few slightly variant codons. For instance, a few unique codons give mitochondria their own "mitochondrial code." We can predict that they are outcomes of gene mutations that did not alter the mix of proteins in adverse ways. The near-universal use of the genetic code indicates that there is little tolerance for variation.

The genetic code is a set of sixty-four different codons, which are nucleotide bases in mRNA that are "read" in sets of three. Different codons (base triplets) specify different amino acids.

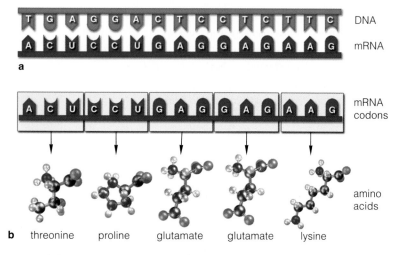

b   threonine   proline   glutamate   glutamate   lysine

Figure 14.5   Example of the correspondence between genes and proteins. (**a**) An mRNA transcript of a gene region of DNA. Three nucleotide bases, equaling one codon, specify one amino acid. This series of codons (base triplets) specifies the sequence of amino acids shown in (**b**).

Figure 14.6 *Animated!*   *Right*, the near-universal genetic code. Each codon in mRNA is a set of three ribonucleotide bases. Sixty-one of these base triplets encode specific amino acids. Three are signals that stop translation.

The *left* vertical column (*brown*) lists choices for the first base of a codon. The *top* horizontal row (*light tan*) lists the second choices. The *right* vertical column (*dark tan*) lists the third. To give three examples, reading left to right, the triplet U G G corresponds to tryptophan. Both U U U and U U C correspond to phenylalanine.

first base	second base				third base
	U	C	A	G	
U	phenylalanine	serine	tyrosine	cysteine	U
	phenylalanine	serine	tyrosine	cysteine	C
	leucine	serine	STOP	STOP	A
	leucine	serine	STOP	tryptophan	G
C	leucine	proline	histidine	arginine	U
	leucine	proline	histidine	arginine	C
	leucine	proline	glutamine	arginine	A
	leucine	proline	glutamine	arginine	G
A	isoleucine	threonine	asparagine	serine	U
	isoleucine	threonine	asparagine	serine	C
	isoleucine	threonine	lysine	arginine	A
	methionine (or START)	threonine	lysine	arginine	G
G	valine	alanine	aspartate	glycine	U
	valine	alanine	aspartate	glycine	C
	valine	alanine	glutamate	glycine	A
	valine	alanine	glutamate	glycine	G

14.3 The Other RNAs

Let's take stock. The codons in an mRNA transcript are the words in protein-building messages. Without translators, words that originated in DNA mean nothing; it takes the other two classes of RNA to synthesize proteins. Before getting into the mechanisms of translation, reflect on this overview of their structure and function.

Figure 14.7 shows the molecular structure for one of the tRNAs. All cells have pools of tRNAs and amino acids in their cytoplasm. Each tRNA has a molecular "hook," an attachment site for an amino acid. It has an **anticodon**, a ribonucleotide base triplet that can base-pair with a complementary codon in an mRNA transcript. When tRNAs bind to mRNA on a ribosome, the amino acid attached to each becomes positioned automatically in the order that the codons specify.

There are sixty-four codons but not as many kinds of tRNAs. How do tRNAs match up with more than one type of codon? According to base-pairing rules, adenine pairs with uracil, and cytosine with guanine. However, in codon–anticodon interactions, these rules

can loosen for the third base in a codon. This freedom in codon–anticodon pairing at a base is known as the "wobble effect." For example, AUU, AUC, and AUA specify isoleucine. All three codons can base-pair with one type of tRNA that hooks on to isoleucine.

Again, interactions between the tRNAs and mRNA take place at ribosomes. A ribosome has two subunits made of rRNA and structural proteins (Section 4.5 and Figure 14.8). In eukaryotic cells, they are built in the nucleus and moved to the cytoplasm. There, a large and small subunit converge as an intact, functional ribosome only when mRNA is to be translated.

LINK TO SECTION 4.5

Only mRNA carries DNA's protein-building instructions from the nucleus into the cytoplasm.

tRNAs deliver amino acids to ribosomes. Their anticodons base-pair with codons in the order specified by mRNA.

Polypeptide chains are built on ribosomes, each consisting of a large and small subunit made of rRNA and proteins.

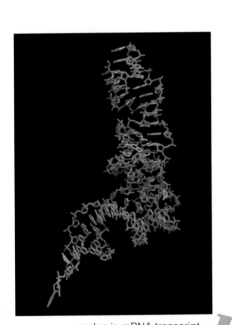

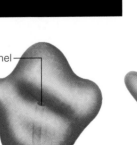

codon in mRNA transcript —
anticodon in tRNA —

amino acid

Figure 14.7 Model for a tRNA. The icon shown to the right is used in following illustrations. The "hook" at the lower end of this icon represents the binding site for a specific amino acid.

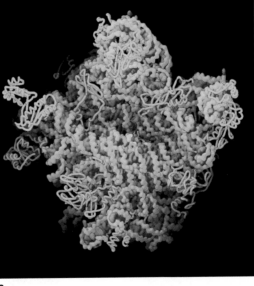

a

Figure 14.8 (**a**) Ribbon model for the large subunit of a bacterial ribosome. It has two rRNA molecules (*gray*) and thirty-one structural proteins (*gold*), which stabilize the structure. At one end of a tunnel through the large subunit, rRNA catalyzes polypeptide chain assembly. This is an ancient, highly conserved structure. Its role is so vital that the corresponding subunit of the eukaryotic ribosome, which is larger, may be similar in structure and function. (**b**) Model for the small and large subunits of a eukaryotic ribosome.

tunnel

small ribosomal subunit + large ribosomal subunit → intact ribosome

b

14.4 The Three Stages of Translation

LINKS TO
SECTIONS
3.5, 4.6

An mRNA transcript that encodes DNA's information about a protein enters an intact ribosome. There, its codons are translated into a polypeptide chain—a protein's primary structure (Section 3.5). Translation of the protein-building message proceeds through three continuous stages called initiation, elongation, and termination.

Only one kind of tRNA can start the *initiation* stage of translation. It alone has the anticodon UAC—which is complementary to the START codon of every mRNA transcript. The anticodon and codon meet up when this initiator tRNA binds to a small ribosomal subunit. Next, a large ribosomal subunit joins with the small subunit. Together, the initiator tRNA, the ribosome, and the mRNA transcript form an initiation complex (Figure 14.9a–c). The next stage can begin.

During the *elongation* stage, a polypeptide chain is synthesized while the mRNA passes between the two ribosomal subunits, a bit like a thread being moved through the eye of a needle. Many tRNA molecules deliver amino acids to the ribosome, and each binds to the mRNA in the order specified by their codons. One region of an rRNA molecule located at the center of the large ribosomal subunit is highly acidic, and it functions as an enzyme. It catalyzes the formation of peptide bonds between amino acids (Figure 14.9d–f).

Figure 14.9g shows how one peptide bond forms between the most recently attached amino acid and the next one brought to the ribosome. Here, you might wish to look once more at Section 3.5, which includes a step-by-step description of peptide bond formation during protein synthesis.

elongation

binding site for mRNA

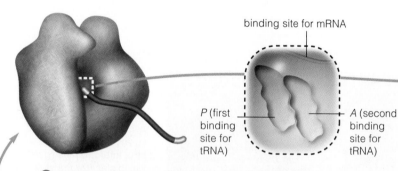

P (first binding site for tRNA)

A (second binding site for tRNA)

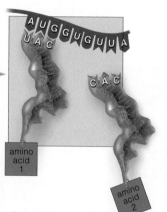

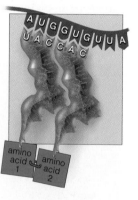

c Initiation ends when a large and small ribosomal subunit converge and bind together. In elongation, the second stage of translation, mRNA occupies a binding site at one end of a tunnel through the large subunit (Figure 14.8). tRNAs that deliver amino acids to the intact ribosome will occupy two other binding sites.

d The initiator tRNA binds to the ribosome. Its anticodon matches up with the mRNA START codon AUG, and it has the amino acid methionine attached to it. A second tRNA binds with the next codon (here, it is GUG).

e One of the rRNA molecules that make up the large ribosome catalyzes formation of a peptide bond between the amino acids (here, methionine and valine).

b *Initiation*, the first stage of translating mRNA, will start when an initiator tRNA binds to a small ribosomal subunit. The small subunit/tRNA complex will attach to the start of the mRNA, move along the transcript, and scan it for the START codon AUG.

initiation

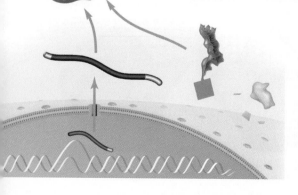

a A mature mRNA transcript leaves the nucleus through a pore in the nuclear envelope. It enters the cytoplasm, which has many free amino acids, tRNAs, and ribosomal subunits.

Figure 14.9 *Animated!* Stages of translation, the second step of protein synthesis. Here, we track a mature mRNA transcript that formed inside the nucleus of a eukaryotic cell. It passes through pores across the nuclear envelope and enters the cytoplasm, which contains pools of many free amino acids, tRNAs, and ribosomal subunits.

During *termination*, the last stage of translation, the mRNA's STOP codon enters the ribosome. No tRNA has a corresponding anticodon. Proteins called release factors bind to the ribosome. Binding triggers enzyme activity that detaches the mRNA *and* the polypeptide chain from the ribosome (Figure 14.9*i–k*).

In cells that are quickly using or secreting proteins, you often see many clusters of ribosomes (polysomes) on an mRNA transcript, all translating it at the same time. This is what happens in unfertilized eggs, which usually stockpile mRNA transcripts in the cytoplasm in preparation for the cell divisions that lie ahead.

Many newly formed polypeptide chains carry out their functions in the cytoplasm. Others have a special sequence of amino acids. The sequence is a shipping label that gets them into ribosome-studded, flattened sacs of rough ER (Section 4.6). In the organelles of the endomembrane system, the chains will take on final form before shipment to their ultimate destinations as structural or functional proteins.

> *Translation is initiated when a small ribosomal subunit and an initiator tRNA arrive at an mRNA transcript's START codon, and a large ribosomal subunit binds to them.*
>
> *tRNAs deliver amino acids to a ribosome in the order dictated by the linear sequence of mRNA codons. A polypeptide chain lengthens as peptide bonds form between the amino acids.*
>
> *Translation ends when a STOP codon triggers events that cause the polypeptide chain and the mRNA to detach from the ribosome.*

f The first tRNA is released, and the ribosome moves to the next codon position.

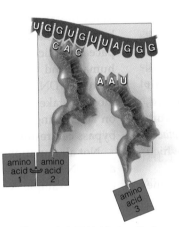

g A third tRNA binds with the next codon (here it is UUA). The ribosome catalyzes peptide bond formation between amino acids 2 and 3.

h Steps **f** and **g** are repeated as the ribosome moves along the mRNA transcript.

termination

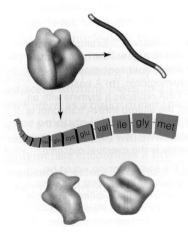

i A STOP codon moves into the area where the chain is being built. It is the signal to release the mRNA transcript from the ribosome.

j The new polypeptide chain is released from the ribosome. It is free to join the pool of proteins in the cytoplasm or to enter rough ER of the endomembrane system.

k The two ribosomal subunits now separate, also.

15.1 When Controls Come Into Play

LINKS TO
SECTIONS
6.4, 9.1, 14.1

*Ultimately, **gene expression** refers to controls over the kinds and amounts of proteins that are in a cell in any specified interval. Tremendous coordination goes into synthesizing, stockpiling, using, exporting, and degrading thousands of types of proteins.*

SOME CONTROL MECHANISMS

Regulatory elements interact with DNA, RNAs, new polypeptide chains, and final proteins. Different kinds respond to shifts in concentrations of substances or to outside signals, such as hormones. Many responses exert *negative* control; they slow or stop some activity. Others exert *positive* control; they enhance it.

For instance, **promoters** are short stretches of base sequences in DNA where regulatory proteins gather and control transcription of specific genes, often in response to a hormonal signal. **Enhancers** are binding sites where such proteins increase transcription rates.

Chemical modification also can exert control. Many methyl groups ($-CH_3$) are "painted" on parts of newly replicated DNA to block access to genes. Acetyl groups ($-CH_3CO^-$) are attached to DNA to make genes accessible. **Methylation** and **acetylation** are the names for the addition of such groups to DNA or any other molecule.

When, how, and to what extent any of these controls come into play depends on the type of cell, its functions, its chemical environment, and signals from the outside. Later chapters provide rich examples. For now, become familiar with the points at which control is exerted.

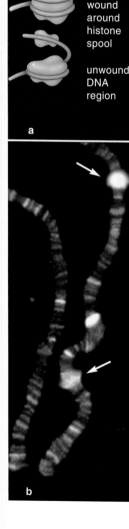

DNA wound around histone spool

unwound DNA region

a

b

Figure 15.2 Examples of gene control mechanisms. (**a**) Loosening a chromosome's DNA–histone units may expose genes for transcription. Attaching an acetyl group to a histone makes it loosen its grip on the DNA wound around it. Transcription enzymes attach and detach these groups.

(**b**) *Drosophila* polytene chromosomes. To sustain a rapid growth rate, *Drosophila* larvae eat continuously and use a lot of saliva. Giant chromosomes in their salivary glands form by repeated DNA replications. Each has hundreds or thousands of the same DNA molecule, aligned side by side.

An insect hormone, ecdysone, serves as a regulatory protein; it promotes gene transcription. In response to the hormonal signal, these chromosomes loosen and puff out in regions where genes are being transcribed. Puffs are largest and most diffuse where transcription is most intense (*arrows*).

POINTS OF CONTROL

Controls Before Transcription The *access* to genes is under control. Remember how histones and other proteins help keep eukaryotic chromosomes organized (Section 9.1)? Where a DNA molecule is wound up tightly, polymerases cannot access genes. Acetylation can make histones loosen their grip (Figure 15.2*a*). As another example, a maternal *or* paternal allele at any locus in a diploid cell may become methylated, which can block the gene's influence on a trait.

Controls also affect *how* a gene will be transcribed. For instance, some gene sequences can be rearranged or multiplied. In immature amphibian eggs and gland cells of certain insect larvae, the chromosomal DNA is copied repeatedly in interphase. The copying results in *polytene* chromosomes, which contain hundreds or thousands of side-by-side copies of genes. The repeats allow these cells to churn out copious amounts of the gene products necessary for survival (Figure 15.2*b*).

Control of Transcript Processing Many controls influence mRNA transcript processing. Remember, the pre-mRNA transcripts are modified in ways that affect whether, when, and how they are translated (Section 14.1). Consider what happens in different muscle cells. Exons of the gene for troponin, a contractile protein, are put together in different combinations in different muscle cells. As a result, each type of muscle cell gets mRNA transcripts that are unique in a small region. The structure and functioning of the troponin product vary in subtle ways among them.

Also, the nuclear envelope helps control when the mRNA transcript reaches a ribosome. The transcript cannot pass through a nuclear pore complex unless proteins become attached to it. A base sequence in the untranslated end of mRNA is like a zip code. Controls "read" the code and attach proteins to it. The bound proteins help move the transcript to the region where it is supposed to be translated or stored. Destinations are vital. Which mRNAs—and, in time, gene products —end up in different regions of an immature egg's cytoplasm are "maternal messages" on how to start to construct the body plan of a new embryo.

Unfertilized eggs that stockpile maternal messages keep them silent with the help of controls called Y-box proteins. When phosphorylated, Y-box proteins bind and help stabilize mRNA. When many of the proteins bind to a transcript, they block its translation. In other words, phosphorylation of Y-box proteins is a control mechanism in mRNA inactivation. You will read about such controls in later sections of the book.

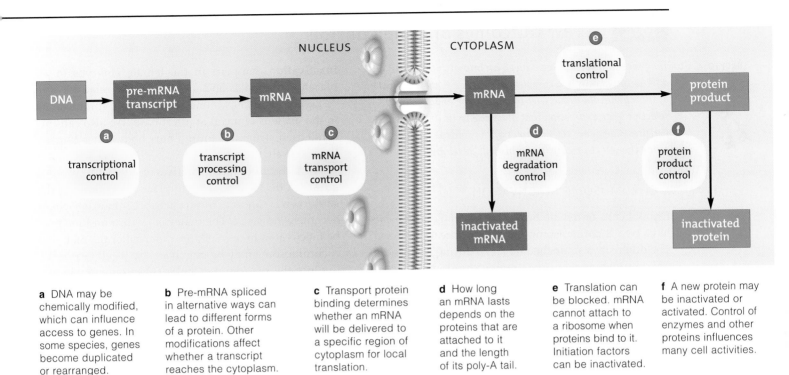

a DNA may be chemically modified, which can influence access to genes. In some species, genes become duplicated or rearranged.

b Pre-mRNA spliced in alternative ways can lead to different forms of a protein. Other modifications affect whether a transcript reaches the cytoplasm.

c Transport protein binding determines whether an mRNA will be delivered to a specific region of cytoplasm for local translation.

d How long an mRNA lasts depends on the proteins that are attached to it and the length of its poly-A tail.

e Translation can be blocked. mRNA cannot attach to a ribosome when proteins bind to it. Initiation factors can be inactivated.

f A new protein may be inactivated or activated. Control of enzymes and other proteins influences many cell activities.

Figure 15.3 *Animated!* Controls that influence whether, when, and how a gene in eukaryotic DNA will be expressed.

Control of Translation Many kinds of molecules function in coordinated ways during translation, and each is controlled independently. Some controls work on initiation factors and ribosome components. Others work through mRNA transcript stability. The longer a transcript lasts, the more times it can be translated. Enzymes start nibbling at the poly-A tail of a mature mRNA transcript within minutes of its appearance in the cytoplasm. How fast they digest it depends on the tail's length, its base sequences, and the proteins that have become attached to it (Section 14.1).

Controls After Translation Control is exerted over new enzymes and other proteins. For instance, Y-box proteins become activated only when enzymes attach a phosphate group to them. Other controls activate, inhibit, and stabilize diverse molecules that take part in protein synthesis. Allosteric control of tryptophan synthesis is a case in point (Section 6.4).

SAME GENES, DIFFERENT CELL LINEAGES

Later in the book, you will read about how complex organisms develop. For now, tentatively accept this premise: All cells of your body started out life with the same genes, because every one arose by mitotic cell divisions from the same fertilized egg. They all transcribe many of the same genes and are alike in most aspects of structure and housekeeping activities.

In other ways, however, *nearly all of your body cells became specialized in composition, structure, and function.*

This process—**cell differentiation**—is central to the development of all multicelled species. By selecting particular subsets of genes, specialized cells and their descendants give rise to different tissues and organs.

Here is an example: Cells generally transcribe the genes coding for enzymes of glycolysis all the time. But immature red blood cells alone transcribe genes for hemoglobin. Liver cells transcribe genes required to make enzymes that neutralize some toxins, but they are the only ones that do so. While your eyes formed, certain cells accessed genes necessary for synthesizing crystallin. No other cells in your body can activate the genes for this protein, which makes up the transparent fibers of the lens in each eye.

Figure 15.3 summarizes the main control points over gene expression in eukaryotic cells.

Gene expression is controlled by regulatory elements that interact with one another, with control elements built into the DNA, with RNA, and with newly synthesized proteins. Different forms of controls work before, during, and after transcription and translation.

Control also is exerted through chemical modifications that activate, inactivate, or restrict access to specific gene regions in DNA.

During development of all multicelled organisms, cells become different in composition, structure, and function as genes are activated and suppressed in selective ways.

15.2 A Few Outcomes of Gene Controls

LINKS TO
SECTIONS
12.1, 12.2. 12.5

The preceding section introduced an important idea. All differentiated cells in a complex, multicelled body use most of their genes the same way, but each type engages in selective gene expression that gives rise to its distinctive features. Consider two examples of the controls that guide the selections during embryonic development.

X CHROMOSOME INACTIVATION

Diploid cells of female humans and female calico cats have two X chromosomes. One is in threadlike form. The other stays scrunched up, even during interphase. This scrunching is a programmed shutdown of about 75 percent of the genes on *one* of two homologous X chromosomes. That shutdown, called **X chromosome**

inactivation, happens in the female embryos of all placental mammals and their marsupial relatives.

Figure 15.4*a* shows one condensed X chromosome in the nucleus of a cell at interphase. We also call this condensed structural form a Barr body (after Murray Barr, who first identified it).

An X chromosome is inactivated when XX embryos are still a tiny ball of cells. In placental mammals, the shutdown is random, in that *either* chromosome could become condensed. The maternal X chromosome may be inactivated in one cell; the paternal or the maternal X chromosome may be inactivated in a cell next to it.

Once the random molecular selection is made in a cell, all of that cell's descendants make the exact same selection as they go on dividing to form tissues. What is the outcome? *A fully developed female has patches of tissue where genes of the maternal X chromosome are being expressed and patches of tissue where genes of the paternal X chromosome are being expressed.* She is a "mosaic" for the expression of X-linked genes.

When alleles on two homologous X chromosomes are not identical, patches of tissues through the body often show variation. Mosaic tissues can be observed in women who are heterozygous for a rare mutant allele that causes an absence of sweat glands. Sweat glands form in some patches of skin only. Where sweat glands are absent, the mutant allele is on the active X chromosome. The mosaic effect is especially apparent in females affected by *anhidrotic ectodermal dysplasia* (Figure 15.4*b*). Abnormalities in the skin and structures derived from it, including teeth, hair, nails, and sweat glands, are signs of this heritable disorder.

A different mosaic tissue effect shows up in female calico cats, of the sort shown in Figure 15.5. These cats are heterozygous for a certain coat color allele on their X chromosomes.

According to the theory of **dosage compensation**, the shutdown is not an accident of evolution; it is a gene control mechanism. In mammals, recall, males are XY, which means that females have twice as many X chromosome genes (Section 12.5). Inactivating one of their two X chromosomes balances gene expression

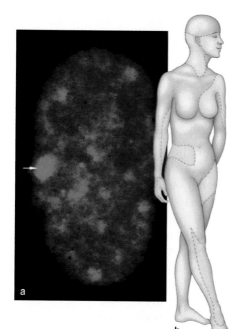

a

b

Figure 15.4 (**a**) In the somatic cell nucleus of a human female, a condensed X chromosome, also called a Barr body (*arrow*). The X chromosome in cells of human males is not condensed this way.

(**b**) A mosaic tissue effect that becomes apparent in women who are affected by anhidrotic ectodermal dysplasia. Some patches of skin have sweat glands, but other patches (color-coded *yellow*) have none.

Figure 15.5 *Animated!* Why is this female cat "calico"? In her body cells, one of her two X chromosomes has a dominant allele for the brownish-black pigment melanin. Expression of the allele on her other X chromosome codes for orange fur. When this cat was still an embryo, one X chromosome was inactivated at random in each cell that had formed by then. Patches of different colors reflect which allele was shut down in cells that formed a given tissue region. (White patches are an outcome of an interaction that involves a different gene, the product of which blocks melanin synthesis.)

petal

carpel

stamen

sepal

a Wild-type flower

b The abnormal flowers of four mutant plants

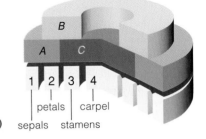

c sepals | petals | carpel stamens | 1 2 3 4 | B A C

Figure 15.6 *Animated!* Controls over formation of flowers, based on mutations in *Arabidopsis thaliana*. This plant's flowers have four sepals, four petals, six stamens, and two fused carpels. The pattern of expression of floral identity genes affects whorls of cells that differentiate near the tip of newly forming flowers. Whorl 1 becomes sepals; whorl 2, petals; whorl 3, stamens; and whorl 4, the carpel. The above model represents the relative locations of tissues in which each set of floral identity genes—*A*, *B*, or *C*—is expressed. Gene products "tell" cells in each whorl what to do.

between the sexes. The normal development of female embryos depends on this type of control.

How, in a single nucleus, does only one of two X chromosomes get shut down? Methylation of histones and the action of *XIST*, an X-linked gene, do the trick. The *XIST* product, a large RNA molecule, sticks like masking paint to chromosomal DNA. Although we do not know why, the *XIST* gene on only one of the two chromosomes is active. That chromosome and its genes get painted with RNA. The other one remains paint-free; its genes remain available for transcription —sort of. Be sure to read *Critical Thinking* question 9 on page 241. It puts a twist on this generalized picture of X chromosome inactivation.

GENE CONTROL OF FLOWER FORMATION

Plants, too, offer fine examples of gene controls. For instance, when some plant shoots put on new growth, young plant cells right behind the tips differentiate in ways that produce flowers. Whorls of the new tissues become sepals, petals, stamens, and carpels (Figure 15.6). Studies of mutations in the common wall cress plant, *Arabidopsis thaliana*, support an **ABC model** for how all of the specialized parts of a flower develop in a predictable pattern. Three sets of master genes—*A*, *B*, and *C*—guide the process. As you will see, they are like the genes that control how body parts of animal embryos form in predictable patterns.

The cells dividing at the tip of a floral shoot form whorls of tissue, one over the other, like onion layers. What will the cells in each whorl become? It depends on which genes of the *ABC* group are activated. In the outermost whorl, only *A* genes are switched on, and their products trigger events that cause sepals to form. Moving inward, cells in the next whorl express *A* and *B* genes; they give rise to petals. Farther in are cells that express *B* and *C* genes; they give rise to stamens (male floral structures). Cells of the innermost whorl express *C* genes only; they give rise to a fused carpel (a female floral structure).

Support for the model comes from mutations in genes of the *ABC* group (Figure 15.6b). Mutation in an *A* group gene alters the two outermost whorls. The flower that forms has stamens and carpels but no petals. Mutation in a *B* group gene affects the second and third whorls; sepals replace petals, and a carpel replaces stamens. Mutation in a *C* group gene alters the innermost whorls. The resulting flower is sterile (no stamens, no carpel) but has a profusion of petals.

The product of a different gene, *Leafy*, controls the activation of the sets of *ABC* genes. Certain mutations in *Leafy* keep flowers from forming on shoots where we normally expect to see them. And what switches on *Leafy*? Evidence points to a steroid hormone.

Dosage compensation in mammals is an example of gene control in eukaryotes. Most of the genes on one of the X chromosomes in females (XX) are inactivated so that early development proceeds the same as it does in males (XY).

As another example, selective expression of ABC genes controls how flowers develop.

15.3 There's a Fly in My Research

Patterns in the body plan emerge as an embryo develops, and they are both beautiful and fascinating. Researchers have correlated many of the patterns with expression of specific genes at particular times, in particular tissues. Tiny fruit flies yielded big clues to the connection.

For about a hundred years, *Drosophila melanogaster* has been the fly of choice for laboratory experiments. Why? It costs almost nothing to feed this fruit fly, which is only about 4.6 centimeters (1.8 inches) long and can live in bottles. Also, *D. melanogaster* reproduces fast and has a short life cycle, and disposing of dozens of spent bodies after an experiment is a snap. We now know how all of its 13,601 genes are distributed along the length of its four pairs of chromosomes.

Anatomical, cytological, biochemical, and genetic studies of *Drosophila* continue to reveal gene controls over development. In addition, they yield insights into evolutionary connections among groups of animals.

Discovery of Homeotic Genes

Like fruit flies, most eukaryotic species have **homeotic genes**, a class of master genes that contain information about mapping out the basic body plan. The genes code for regulatory proteins that include a "homeodomain," a sequence of about sixty amino acids. This sequence binds to control elements in promoters and enhancers.

Different homeotic genes are transcribed in specific parts of a developing embryo, so their products become concentrated in local tissue regions. Body parts form as the products interact with one another and with control elements to switch on other genes along the length of the body's main axis, according to an inherited plan.

Researchers discovered homeotic genes in mutant fruit flies that had body parts growing out of the wrong places. As an example, the *antennapedia* gene is supposed to be transcribed only in embryonic tissues that give rise to a thorax, complete with legs. This gene normally is not transcribed in cells of all other tissue regions. But Figure 15.7*a* shows what happened after a mutation altered some control over transcription and the gene was wrongly transcribed in the tissue destined to become a head.

Plants, too, have master genes. You already read about how they control floral development. Similarly, a different master gene helps leaf veins in corn plants form straight, parallel lines. When the gene mutates, the veins twist.

More than 100 homeotic genes have been identified in diverse eukaryotes—and the same mechanisms control their transcription. Many of the genes are functionally interchangeable among species as evolutionarily distant as yeasts and humans, so we can expect that they evolved in the most ancient eukaryotic cells. Their protein products often differ only in modest substitutions. In other words, one amino acid has replaced another, but its chemical properties are still similar.

Knockout Experiments

Drosophila researchers made more discoveries about how embryos develop. For instance, with **knockout experiments**, a wild-type gene is mutated in a way that prevents its transcription or translation. If genetically engineered knockout individuals differ in form or behavior from wild-type individuals, this may be a clue to the function of the missing gene. Such experiments have yielded insights into the functions of many hundreds of genes in different organisms.

Researchers tend to name the genes based on what happens in their absence. For instance, *eyeless* is a control gene expressed in fruit fly embryos. In its absence, no eyes form. *Dunce* is a regulatory gene required for learning and memory. *Wingless*, *wrinkled*, and *minibrain* genes are self-explanatory. *Tinman* is necessary for heart development. Among other things, *groucho* prevents overproduction of whisker bristles. Figure 15.7 shows a few of the mutants.

In other experiments, researchers add special promoters to a gene so that they can control its expression with an external cue, such as temperature. They delete genes from one part of the *Drosophila* genome and put them back someplace else. This molecular sleight of hand revealed that expression of the *eyeless* gene can induce an eye to form not only on the fruit fly head, but also on the wings and legs (Figure 15.7*c*).

Figure 15.7 (**a**) Experimental evidence of controls over where body parts develop. In *Drosophila* larvae, activation of genes in one group of cells normally results in antennae on the head. A mutation that affects *antennapedia* gene transcription puts legs on the head. This is one of the genes controlled by regulatory proteins with homeodomains. (**b**) Model for a homeodomain binding to a transcriptional control site in DNA. (**c**) More *Drosophila* mutations.

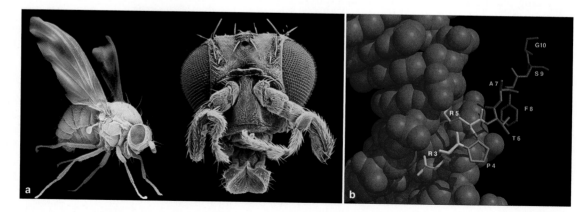

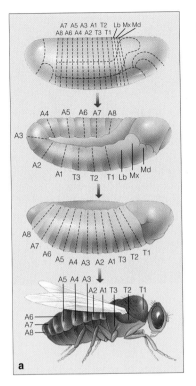

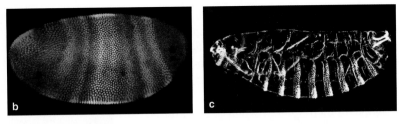

Figure 15.8 Genes and *Drosophila*'s segmented body plan. (**a**) Fate map for the surface of a *Drosophila* zygote. Such maps indicate where each differentiated cell type in the adult originated. The pattern starts with the polar distribution of maternal mRNA and proteins in the unfertilized egg. This polarity dictates the future body axis. A series of segments will develop along this axis. Genes specify whether legs, wings, eyes, or some other body parts will develop on a particular segment.

Briefly, here is how it happens: Maternal gene products prompt expression of gap genes. Different gap genes become activated in regions of the embryo with higher or lower concentrations of different maternal gene products. Gap gene products influence each other's expression as well. They form a primitive spatial map.

Depending on where they occur relative to concentrations of gap gene products, embryonic cells express different pair-rule genes. Products of pair-rule genes accumulate in seven transverse stripes that mark the onset of segmentation (**b**). They activate other genes, the products of which divide the body into units (**c**). These interactions influence the expression of homeotic genes, which collectively govern the structural and functional identity of each segment.

The *eyeless* gene even has counterparts in humans (the *PAX6* gene), mice (*Pax-6*), and squids (also *Pax-6*). Humans who have no functional *PAX6* genes have eyes with malformed irises. *PAX-6* inserted into any tissue of an eyeless mutant fly induces an eye to form wherever it is expressed. This kind of molecular evidence points to a shared ancestor among animals as evolutionarily distant as insects, cephalopods, and mammals.

Filling In Details of Body Plans Let's take stock. As an embryo develops, cells in different body regions become organized in different ways. Cells divide and differentiate. They migrate or stick to cells of the same type in tissues. They live or die after performing their function. These are genetically programmed events that fill in details of the body in orderly patterns, in keeping with the expression of master genes. Those genes are switched on in specific tissues, at specific stages of development. The products of those genes deal mainly with transcription. In effect, they form a three-dimensional map along the main body axis.

Depending on where undifferentiated cells are relative to the map, they are the start of specialized tissues and organs. That is how the sequential expression of master genes along the body axis gives rise to the body segments of fruit flies (Figure 15.8).

Pattern formation is the name for the emergence of embryonic tissues and organs in orderly patterns, at times and in places where we expect them to be. Section 43.5 offers a closer look at the controlled gene interactions that fill in details of the animal body plan.

c A few more *Drosophila* mutations that yielded clues to gene function *Left to right,* an eye that formed on a leg, yellow miniature, curly wings, vestigial wings, and a double thorax.

15.4 Prokaryotic Gene Control

LINK TO
SECTION
8.2

In prokaryotic cells, gene controls deal mainly with quickly slowing down and starting up transcription in response to short-term shifts in environmental conditions. A diversity of long-term controls is not required; none of these species slowly develops into a complex, multicelled form (Table 15.1).

When nutrients are plentiful and when other external conditions also favor growth and reproduction, all prokaryotic cells rapidly transcribe genes that specify all of the enzymes required for nutrient absorption and other growth-related tasks. Genes that are tapped most often occur one after the other as a set of genetic information in the DNA. They all can be transcribed together, which yields a single RNA strand.

NEGATIVE CONTROL OF THE LACTOSE OPERON

With this bit of background, consider an example of how one kind of prokaryote responds to the presence or absence of lactose. *Escherichia coli* lives in the gut of mammals, where it dines on nutrients traveling past. Milk typically nourishes mammalian infants. It does not contain glucose, the sugar of choice for *E. coli*. It does contain lactose, a different sugar.

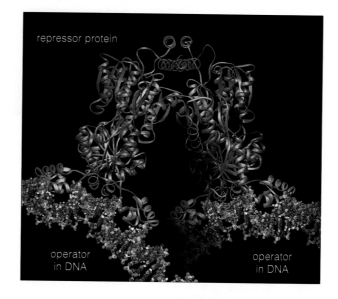

Figure 15.9 Model for the repressor protein of the lactose operon when it is bound to two operators in DNA.

| Table 15.1 | Prokaryotic Versus Eukaryotic Gene Control |

Prokaryotic Gene Control

1. Control mechanisms adjust enzyme-mediated reactions in response to short-term changes in nutrient availability and other environmental conditions.

2. Operons control the expression of more than one gene at a time.

3. Transcriptional controls are *reversibly* inhibited when conditions do not favor growth and reproduction.

4. Translation starts immediately; prokaryotic RNA transcripts have no introns, no processing controls.

Eukaryotic Gene Control

1. Some control mechanisms adjust enzyme-mediated reactions in response to short-term changing conditions.

2. In multicelled species, other controls activate sets of genes at different times, in different tissues. They induce generally *irreversible* events that are part of a long-term program of growth and development.

3. Diverse controls operate during gene transcription and translation, and on the gene products. mRNA transcripts are processed in the nucleus; controls govern the timing and rate of their translation in the cytoplasm.

After being weaned, infants of most species drink little (if any) milk. Even so, *E. coli* cells can still use lactose if and when it shows up in the gut. They can activate a set of three genes for lactose-metabolizing enzymes. In *E. coli* DNA, a promoter precedes all three genes, and two operators flank it. Each **operator** is a binding site for a type of regulatory protein known as repressor, which stops transcription (Figure 15.9).

Any arrangement in which a promoter and a set of operators control access to more than one prokaryotic gene is called an **operon**.

In the absence of lactose, a repressor molecule binds to the set of operators. Binding causes the DNA region that contains the promoter to twist into a loop, as in Figure 15.10. RNA polymerase, the workhorse that transcribes genes, is not able to bind to a looped-up promoter. The result is that operon genes are not used when they are not required.

When lactose *is* in the gut, *E. coli* converts some of it to allolactose. This sugar binds to the repressor and changes its molecular shape. The altered repressor cannot bind to operators. The looped DNA unwinds and RNA polymerase transcribes the genes, so lactose-degrading enzymes are produced when required.

POSITIVE CONTROL OF THE LACTOSE OPERON

E. coli cells pay far more attention to glucose than to lactose. Even when lactose is in the gut, the lactose operon is not used much—unless there is no glucose.

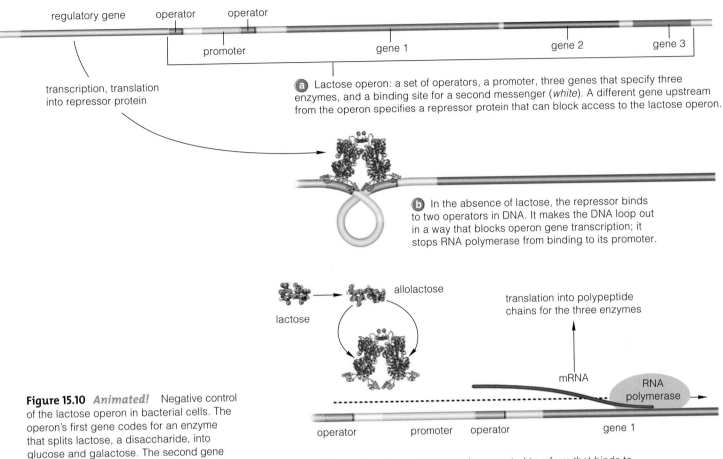

regulatory gene   operator   operator

promoter   gene 1   gene 2   gene 3

transcription, translation into repressor protein

a Lactose operon: a set of operators, a promoter, three genes that specify three enzymes, and a binding site for a second messenger (*white*). A different gene upstream from the operon specifies a repressor protein that can block access to the lactose operon.

b In the absence of lactose, the repressor binds to two operators in DNA. It makes the DNA loop out in a way that blocks operon gene transcription; it stops RNA polymerase from binding to its promoter.

allolactose

translation into polypeptide chains for the three enzymes

lactose

mRNA

RNA polymerase

Figure 15.10 *Animated!* Negative control of the lactose operon in bacterial cells. The operon's first gene codes for an enzyme that splits lactose, a disaccharide, into glucose and galactose. The second gene codes for an enzyme that helps transport lactose into cells. The third gene's product helps metabolize certain sugars.

operator   promoter   operator   gene 1

c When lactose is present, some is converted to a form that binds to the repressor and alters its shape. The altered repressor cannot bind to operators, so RNA polymerase is free to transcribe the operon genes.

When that happens, an activator called CAP (short for catabolite activator protein) exerts positive control over the lactose operon. It makes a promoter far more inviting to RNA polymerase. But CAP cannot issue the invitation until it has already become bound to a chemical messenger: cAMP (short for cyclic adenosine monophosphate). When the activator and cAMP join together as a complex to the promoter, they make it much easier for RNA polymerase to start transcribing genes. Such complexes are called transcription factors.

When glucose is plentiful, ATP forms by glycolysis (Section 8.2), but synthesis of an enzyme necessary to synthesize cAMP is blocked. The blocking ends when glucose is scarce and lactose becomes available. cAMP accumulates, the CAP–cAMP complexes form, and the lactose operon genes are transcribed fast. The gene products allow lactose to be converted to glucose, the preferred sugar of *E. coli*.

Unlike cells of *E. coli*, many of us develop *lactose intolerance*. Cells making up the lining of our small intestine make and then secrete lactase into the gut. As many people age, however, concentrations of this lactose-digesting enzyme decline. Lactose accumulates and is moved on to the large intestine, or colon. It promotes population explosions of resident bacteria. As the bacterial cells busily digest the lactose, a gaseous metabolic product accumulates, the gas distends the colon's wall and causes pain. The short fatty acid chains that form during bacterial metabolism cause diarrhea, which can be severe.

Prokaryotic cells do not require extensive controls over long-term development of complex bodies; they are small, fast reproducers. The main controls guide the transcription of enzyme-coding genes in response to short-term shifts in nutrient availability and other outside conditions.

Summary

Section 15.1 Gene expression within a cell changes in response to chemical conditions and signals from the outside. In complex multicelled species, it is subject to long-term controls over growth and development.

Control mechanisms govern whether, when, and how a gene is expressed. Hormones, activator proteins, and other regulatory elements interact with one another before, during, and after transcription and translation.

With negative control mechanisms, regulatory elements slow or stop a cell activity. With positive control mechanisms, they promote it.

Control is exerted before, during, and after gene transcription and translation. Transcription is a major control point for most eukaryotic genes because so many of the participating molecules can be controlled independently of the others.

Diverse controls guide embryonic development of multicelled eukaryotes. Each cell in an embryo inherits the same genes, but they start selectively activating and suppressing some in unique ways. The outcome of selective gene expression is called cell differentiation: cell lineages become unique in one or more aspects of composition, structure, and function. Those lineages are the start of specialized tissues and organs.

Biology⊘Now
Review the control points for gene expression with the animation on BiologyNow.

Section 15.2 Two examples are given of eukaryotic gene controls, one in mammals, the other in plants:

In the embryos of all female mammals, X chromosome inactivation is an outcome of the interactions between a product of the *XIST* gene and control elements in one of the two X chromosomes in cells. This control mechanism maintains a required balance of gene expression between the sexes while mammalian embryos are developing.

Studies of mutations in *Arabidopsis thaliana* support an ABC model for the formation of flowers. Three sets of master genes (*A*, *B*, and *C*) guide the differentiation of whorls of cells into sepals, petals, stamens, and carpels.

Biology⊘Now
Observe how eukaryotic gene controls influence development with the animation on BiologyNow.

Section 15.3 Many gene controls were identified through experiments with mutant forms of *Drosophila melanogaster*. Others were discovered by knockout experiments in which individual genes are deactivated before a new wild-type fly develops.

While embryos of this fruit fly and of most other eukaryotes develop, homeotic genes are activated in sequence. The protein products of this class of master genes become more or less concentrated along the body's main axis, which maps out the basic body plan. Cells differentiate according to their location along the map. Their descendants fill in the details of the body plan by forming specialized tissues and organs in patterns where we expect them to be.

Section 15.4 Prokaryotic cells do not have great structural complexity and do not undergo development. Most of the gene controls reversibly adjust transcription rates in response to environmental conditions, especially nutrient availability. Bacterial operons are examples of prokaryotic gene controls. The lactose operon controls three genes, the three products of which digest lactose. It has a promoter region in DNA (binding sites for RNA polymerase). Two operators flank it and are binding sites for a repressor protein that can block transcription.

Biology⊘Now
Explore the structure and function of the bacterial lactose operon with the animation on BiologyNow.

Self-Quiz   *Answers in Appendix II*

1. The expression of a given gene depends on the _____ .
   a. type and function of cell   c. environmental signals
   b. chemical conditions   d. all of the above

2. Control mechanisms adjust gene expression in response to changing _____ .
   a. nutrient availability   c. signals from other cells
   b. solute concentrations   d. all of the above

3. Regulatory elements interact with _____ .
   a. DNA   c. gene products
   b. RNA   d. all of the above

4. At _____ in DNA, regulatory proteins gather and control transcription of specific genes.
   a. promoters   c. operators
   b. enhancers   d. both a and b

5. Eukaryotic gene controls govern _____ .
   a. transcription   e. mRNA degradation
   b. RNA processing   f. gene products
   c. translation   g. a through e
   d. RNA transport   h. all of the above

6. Eukaryotic genes guide _____ .
   a. fast short-term activities   c. development
   b. overall growth   d. all of the above

7. Cell differentiation _____ .
   a. occurs in all complex multicelled organisms
   b. requires unique genes in different cells
   c. involves selective gene expression
   d. both a and c
   e. all of the above

8. During X chromosome inactivation _____ .
   a. many genes are shut down c. sweat glands form
   b. RNA paints chromosomes   d. both a and b

9. A cell with a Barr body is _____ .
   a. prokaryotic   c. from a female mammal
   b. from a male mammal   d. infected by Barr virus

10. Homeotic gene products _____ .
    a. are binding sites that flank a bacterial operon
    b. map out a developing embryo's body plan
    c. control X chromosome inactivation
    d. both a and c

11. Knockout experiments mutate _____ genes.
    a. bacterial   c. engineered
    b. wild-type   d. both a and c

12. A(n) _____ is a promoter and a set of operators that control access to two or more prokaryotic genes.
    a. lactose molecule
    b. operon
    c. dosage compensator
    d. both b and c

13. Match the terms with the most suitable description.
    _____ ABC model
    _____ *XIST* gene
    _____ operator
    _____ Barr body
    _____ process of cell differentiation
    _____ methylation

    a. a big RNA is its product
    b. binding site for repressor
    c. cells become specialized in composition, function, etc.
    d. inactivated X chromosome
    e. how flowers develop
    f. —CH₃ additions to DNA

Additional questions are available on Biology⊜Now™

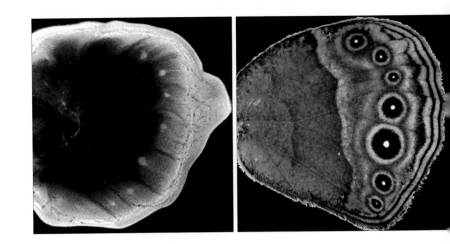

Figure 15.11 *Left*, seven spots in the embryonic wing of a moth larva identify the presence of a gene product that will induce the formation of seven "eyespots" in the wing of the adult (*right*).

Critical Thinking

1. Do all transcriptional controls operate in prokaryotic as well as eukaryotic cells? Why or why not?

2. If all cells in your body start out life with the same inherited information on how to build proteins, then what caused the differences between a red blood cell and a white one? Between a white blood cell and a nerve cell?

3. Unlike most rodents, guinea pigs are well developed at the time of birth. Within a few days, they can eat grass, vegetables, and other plant material.

    Suppose a breeder decides to separate baby guinea pigs from their mothers three weeks after they were born. He wants to raise the males and the females in different cages. However, he has trouble identifying the sex of young guinea pigs. Suggest how a quick look through a microscope can help him identify the females.

4. Calico cats are almost always female. A male calico cat is usually sterile. Find out why.

5. Reflect on the mutant *Arabidopsis thaliana* flowers in Figure 15.6. Small changes in the structure of control genes brought about those changes. Would you predict that such changes figured in the evolution of more than 295,000 kinds of plants, each with distinctive flowers?

    Also reflect on the *Drosophila melanogaster* mutants shown in Figure 15.7. Would you predict that homeotic gene mutations figured in the evolution of the more than 1.5 million known species of animals?

6. *Duchenne muscular dystrophy*, a genetic disorder, affects boys almost exclusively. Muscles begin to atrophy (waste away) in affected children, who typically die in their teens or early twenties (Section 12.7).

    Muscle biopsies of a few women who carry an allele that is associated with the disorder identified some body regions of atrophied muscle tissue. They also showed that muscles adjacent to a region of atrophy were normal or even larger and more chemically active, as if to compensate for the weakness of the adjoining region.

    Form a hypothesis about the genetic basis of Duchenne muscular dystrophy that includes an explanation of why the symptoms might appear in some body regions but not others.

7. Figure 15.11 shows seven "spots" that emerge in the wings of a developing moth larva. The spots identify where seven distinct eyespots will appear on the wings of adult moths. What is the name of the class of genes responsible for mapping out such details of the body plan of developing embryos, including insect larvae?

8. Geraldo isolated an *E. coli* strain in which a mutation has hampered the capacity of CAP to bind to a region of the lactose operon, as it would do normally. How will this mutation affect transcription of the lactose operon when the *E. coli* cells are exposed to the following conditions? Briefly state your answers:
    a. Lactose and glucose are both available.
    b. Lactose is available but glucose is not.
    c. Both lactose and glucose are absent.

9. About 300 million years ago, before mammals began their great adaptive radiation, their X and Y chromosomes were about the same in size. When paired, the two sex chromosomes typically synapsed and exchanged alleles along their length. The X chromosome now carries 1,141 genes. Over time, however, the Y chromosome lost most of itself and now contains only 255 genes. Its big claim to fame is ownership of the *SRY* gene, the master of sex determination.

    Think about the dosage compensation theory, sketched out in Section 15.2. According to this theory, X *chromosome inactivation* is nature's way of compensating for a double dose of X-linked genes in XX embryos, because there are not enough genes left on the puny Y chromosome to balance out their expression. And yet, about 15 percent of the genes on an inactivated X chromosome escape being painted to varying degrees—which means women make more copies of certain proteins than men do.

    Besides this, another 10 percent of the X-linked genes might or might not get painted in individual embryos—which means women differ significantly from one another in which X-linked genes are active.

    Now consider this: Human and chimpanzee genomes differ by 1.5 percent. Women differ from men by 1 percent! Go ahead and let your brain chew on that one.

    You may wish to start with this recent article: L. Carel and H. Willard, "X-Inactivation Profile Reveals Extensive Variability in X-Linked Gene Expression in Females," *Nature* 2005; 434(7031):400–404.

Golden Rice, or Frankenfood?

Not too long ago, the World Health Organization made a conservative estimate that 124 million children around the world show vitamin A deficiencies. Their skin, eyes, and mucous membranes are dry and vulnerable to infection. They do not grow and develop as they should, and they show signs of mental impairment. Each year at least a million die of malnutrition, and about 350,000 end up permanently blind.

Ingo Potrykus and Peter Beyer wanted to help. As they knew, beta-carotene is a yellow pigment in all plant leaves, and it also is a precursor for vitamin A. These geneticists borrowed three genes from garden daffodils (*Narcissus pseudonarcissus*) and a bacterium, and transferred them to rice plants. The plants transcribed the genes and did something they could not do before. They made beta-carotene not only in their leaves but also in their *seeds*—the grains of Golden Rice (Figure 16.1).

Why rice? Rice is the main food for 3 billion people in impoverished countries. There, the poor cannot afford leafy vegetables and other sources of beta-carotene. Getting beta-carotene into rice grains would be the least costly way to deliver the vitamin to those who need it the most, but doing so was beyond the scope of conventional breeding practices. Research continues, and the amount of beta-carotene in SGR1, a more recent version of Golden Rice, is twenty-three times higher than the prototype.

No one wants children to suffer or die. However, many people oppose the idea of genetically modified (GM) foods, including golden rice. Possibly they are unaware of the history of agrarian societies, because it is not as if our ancestors were twiddling their green thumbs. For thousands of years, their artificial selection practices coaxed new plants and new breeds of cattle, cats, dogs, and birds from wild ancestral stocks. Meatier turkeys, huge watermelons, big juicy corn kernels from puny hard ones—the list goes on (Figure 16.1).

And we are newcomers at this! During the 3.8 billion years before we even made our entrance, nature busily conducted uncountable numbers of genetic experiments by way of mutation, crossing over, and gene transfers between species. These processes introduced changes in the molecular messages of inheritance, and today we see their outcomes in the sweep of life's diversity.

Perhaps the unsettling thing about the more recent human-directed changes is that the pace has picked up, hugely. We are getting much better at tinkering with the genetics of many organisms. We do this for pure research and for useful, practical applications.

a   b

Figure 16.1 Where one genetic engineering success story started: (**a**) Researchers transferred genetic information from ordinary daffodils into rice plants, which then used it to stockpile beta-carotene in their seeds—rice grains. (**b**) Two successive generations of Golden Rice compared with grains from a regular rice plant at lower left. *Facing page*, an artificial selection success story—a big kernel from a modern strain of corn next to tiny kernels of an ancestral corn species discovered in a prehistoric cave in Mexico.

For instance, many crop plants, including corn, beets, and potatoes, have been modified. They are now widely planted. They are less temperamental about their living conditions than rice plants are, and they have not run rampant through ecosystems. After a decade-long study in the United Kingdom, researchers concluded that the new crop plants being monitored were doing no harm. Throughout Arizona, farmers grow cotton plants that are genetically engineered for pest resistance. The plantings have not put the environment at risk and might even be less disruptive compared to current agricultural practices. University of Arizona entomologist Bruce Tabashnik, who is monitoring cotton fields, notes that farmers have cut applications of chemical pesticides by 75 percent.

Take stock of how far you have come in this unit. You started with cell division mechanisms that allow parents to pass on DNA to new generations. You moved to the chromosomal and molecular basis of inheritance, then on to gene controls that guide life's continuity. The sequence parallels the history of genetics. And now, you have arrived at the point in time where geneticists hold molecular keys to the kingdom of inheritance. What they are unlocking is already having impact on life in the biosphere.

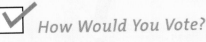

How Would You Vote?

Nutritional labeling is required on all packaged food in the United States, but genetically modified food products may be sold without labeling. Should food distributors be required to label all products made from genetically modified plants or livestock? See BiologyNow for details, then vote online.

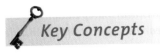

Key Concepts

MAKING RECOMBINANT DNA

Researchers routinely make recombinant DNA molecules. They use restriction enzymes to isolate, cut, and join gene regions from DNA of different species. They use plasmids and other vectors to insert the recombinant molecule into target cells.   Section 16.1

ISOLATING AND AMPLIFYING DNA FRAGMENTS

Researchers isolate and make many copies of genes that interest them. PCR is now the gene amplification method of choice. The genes are copied in amounts large enough for research and practical applications.   Section 16.2

DECIPHERING DNA FRAGMENTS

Sequencing methods reveal the linear order of bases in a sample of DNA. Automated methods complete the task with impressive speed.   Sections 16.3, 16.4

MAPPING AND ANALYZING WHOLE GENOMES

Genomics is concerned with mapping and sequencing of the genomes of humans and other species. Comparative genomics yields evidence of evolutionary relationships among groups of organisms.   Section 16.5

USING THE NEW TECHNOLOGIES

Genetic engineering results in transgenic organisms, which incorporate genes from another species. With gene therapy, a mutated or altered gene is isolated, modified, and copied. Copies are inserted back into the individual to cover the gene's function. The new technologies raise social, legal, ecological, and ethical questions.   Sections 16.6–16.10

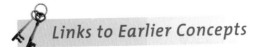

Links to Earlier Concepts

This chapter builds on earlier explanations of the molecular structure of DNA (Sections 3.7, 13.2), and DNA replication and DNA repair (13.4). You may wish to review quickly the nature of mRNA transcript processing (14.1) and controls over gene transcription (14.1). You will come across more uses for radioisotopes (2.2) and fluorescent light (6.6). You will be reminded of why it is useful to know about membrane proteins (5.2). You will see why the lactose operon is not necessarily of obscure interest (15.4).

16.5 The Rise of Genomics

LINKS TO
SECTIONS
3.7, 13.2, 14.5

The potential benefits of sequencing and analyzing the thousands of genes in the genome of selected organisms—say, the human genome—soon became apparent. Automated gene sequencing techniques were developed in response.

THE HUMAN GENOME PROJECT

By 1986, scientists were arguing about sequencing the 3 billion bases of the human genome. Many insisted that benefits for medicine and pure research would be incalculable. Others insisted that the mapping would divert funds from other work that was more urgent and had a better chance of success.

Automated sequencing had just been invented, as had PCR, the polymerase chain reaction. At the time, both techniques were cumbersome, expensive, and far from standardized, but many sensed their potential. Waiting for faster methods seemed the most efficient approach to sequencing the human genome—but who would decide when the technology was fast enough?

Several independent organizations launched their own versions of the Human Genome Project. Walter Gilbert started one company and declared he would sequence and patent the human genome. In 1988, the National Institutes of Health (NIH) annexed the entire Human Genome Project by hiring James Watson as its head and providing 200 million dollars per year to researchers. A public consortium formed between the NIH and institutions working on different versions of

the project. Watson set aside 3 percent of the funding for studies into ethical and social issues arising from the research. He then resigned in 1992 because of a disagreement with the NIH about patenting partial gene sequences. Francis Collins replaced him in 1993.

Amid ongoing squabbles over patent issues, Craig Venter started Celera Genomics (Figure 16.10). Venter cheekily declared that his new company would be the first to finish and patent the genome sequence. This prompted the public consortium to move its gene sequencing efforts into high gear.

Sequencing of the human genome was officially completed in 2003—fifty years after the discovery of the structure of DNA. About 99 percent of the coding regions in human DNA have been deciphered with a high degree of accuracy. A number of other genomes also have been fully sequenced.

What do we do with this vast amount of data? The next step is to investigate questions about precisely what each sequence means—what the genes do, what the control mechanisms are, and how they operate.

At this writing, 19,438 are confirmed as genes, and another 2,188 are probably genes. This does not mean that geneticists have learned what the genes encode.

Among the bizarre discoveries: Protein-encoding genes make up less than 2 percent of our genome. Millions of transposable elements repeated over and over make up more than half of it. There are almost as many *pseudogenes*—inactivated, nonfunctional copies of genes—as there are genes!

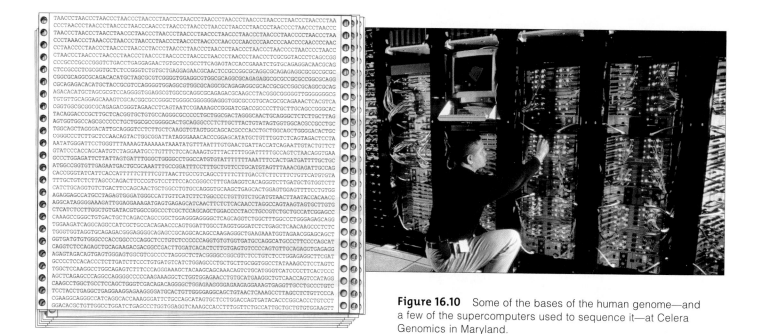

Figure 16.10 Some of the bases of the human genome—and a few of the supercomputers used to sequence it—at Celera Genomics in Maryland.

Figure 16.11 Complete yeast genome array on a DNA chip about 19 millimeters (3/4 inch) across. *Green* spots pinpoint genes that are active during fermentation. *Red* pinpoints the genes used in aerobic respiration, and *yellow*, the ones that are active in both pathways.

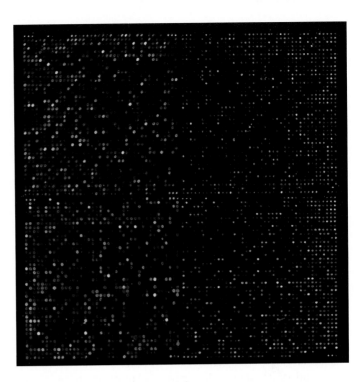

GENOMICS

Research into genomes of humans and other species has converged into a new research field—**genomics**. *Structural* genomics focuses on actual mapping and sequencing of the genomes of individuals. *Comparative* genomics sifts through the maps for similarities and differences that point to evolutionary connections.

Comparative genomics has practical applications as well as potential for research. The basic premise is that the genomes of all existing organisms are derived from common ancestors. For instance, pathogens share some conserved genes with human hosts even though they are only remotely related. Shared gene sequences, how they are organized, and where they differ might hold essential clues to where our immune defenses against pathogens are strongest or the most vulnerable.

Genomics has potential for **human gene therapy**— the transfer of one or more normal or modified genes into a person's body cells to correct a genetic defect or boost resistance to disease. However, even though the human genome is fully sequenced, it still is not easy to manipulate within the context of a living individual.

Today, experimenters use stripped-down viruses as vectors that inject genes into human cells. Some gene therapies deliver modified cells into a patient's tissue. In many cases, therapies make a patient's symptoms subside even when the modified cells are producing just a small amount of a required protein.

A caveat: No one can yet predict whether a virus-injected gene will be delivered to the right tissues and whether cellular mechanisms will maintain it.

DNA CHIPS

Analysis of genomes is now advancing at a stunning pace. Researchers pinpoint which genes are silent and which are being expressed with the use of **DNA chips**. These are microarrays of thousands of gene sequences representing a large subset of an entire genome—all stamped onto a glass plate that is about the size of a small business card.

A cDNA probe is built by using mRNA from, say, cells of a cancer patient. The free nucleotides used to synthesize the complementary strand of DNA have been labeled with a fluorescent pigment. Only genes that are expressed at the time the cells are harvested are making mRNA, so those genes alone make up the resulting probe population. The labeled probe is then incubated along with a chip made from genomic DNA. Wherever the probe binds with complementary base sequences on the chip, there will be a spot that glows under fluorescent light. Analysis of which spots on the chip are glowing reveals which of the thousands of genes inside the cells are active and which are not.

DNA chips are being used to compare different gene expression patterns between cells. Examples are yeasts grown in the presence and absence of oxygen, and different types of cells from the same multicelled individual. RNA from one set of cells is transformed into green fluorescent cDNA, and RNA from the other set into red fluorescent cDNA. The cDNAs are mixed and incubated with a genomic DNA chip. Green or red fluorescence indicates expression of genes in the different cell types. Yellow is a mixture of both red and green, and it indicates that both genes were being expressed at the same time in a cell (Figure 16.11).

In genomics, automated gene sequencing, the use of DNA chips, and other techniques let researchers rapidly evaluate and compare genome-spanning expression patterns.

16.6 Genetic Engineering

Genetic engineering is the deliberate modification of an individual's genome. Genes from another species may be transferred to an individual. Conversely, the individual may have its own genes isolated, modified and copied, and then receive copies of the modified genes.

Genetic engineering started with bacterial species, so consider them first. The kinds that take up plasmids are now widely used in basic research, agriculture, medicine, and industry. Plasmids, again, function as vectors for transferring fragments of foreign or modified DNA into an organism.

For instance, like you, bacterial cells have the metabolic machinery to make complex organic compounds. Genetically engineered types can be employed to transcribe genes that have been transferred to plasmids and synthesize desired proteins. Immense populations do this; they make useful amounts of medically valued proteins in huge stainless steel vats. *E. coli* cells were the first to transcribe and translate synthetic genes for human insulin. Their descendants were the first large-scale, cost-effective bacterial factory for proteins. In addition to insulin, vats of microbes churn out human somatotropin (growth hormone), hemoglobin, blood-clotting factors, interferon, and a variety of drugs and vaccines that we have come to depend upon.

Certain bacteria also hold potential for industry and for cleaning up environmental messes—that is, for *environmental remediation*. In nature, they break down organic wastes as part of their metabolic activities and help cycle nutrients through ecosystems. Modified types digest crude oil into less harmful compounds. When sprayed on oil spills, as from a shipwrecked supertanker, they can help mop up oil. Other species sponge up excess phosphates, heavy metals, and other pollutants, even radioactive wastes.

> *Genetic engineering refers to the directed alteration of an individual's genome. Microbes were the first targets.*
>
> *In some cases, DNA is transferred between individuals of different species, the outcome being a transgenic organism.*
>
> *In other cases, genes or gene regions from an individual are isolated, modified, then copied and inserted into the same individual.*

16.7 Designer Plants

Think back on those Golden Rice plants described in the chapter introduction. They are a prime example of genetic engineering that can produce valuable transgenic plants. There is some urgency surrounding much of this work, as you will now read.

As crop production expands to keep pace with human population growth, it puts unavoidable pressure on ecosystems everywhere. Irrigation leaves mineral and salt residues in soils. Tilled soil erodes, taking topsoil with it. Runoff clogs rivers, and fertilizer in it causes algae to grow so much that fish suffocate. Pesticides harm humans, other animals, and beneficial insects.

Pressured to produce more food at lower cost and with less damage to the environment, some farmers are turning to genetically engineered crop plants.

Cotton plants with a built-in insecticide gene kill only the insects that eat it, so farmers that grow them are not required to use as many pesticides. Certain transgenic tomato plants can grow, develop, and bear fruit in salty soils that would wither other plants. They also absorb and store excess salt in their leaves, thus purifying saline soil for future crops.

The cotton plants in Figure 16.12*a* were genetically engineered for resistance to a relatively short-lived herbicide. Spraying fields with this herbicide will kill all weeds—but not the engineered cotton plants. As you read in the chapter's introduction, the practice means that farmers can use reduced amounts of less toxic chemicals. They do not have to till the soil as much to control weeds, so river-clogging runoff can be reduced. As another example, Figure 16.12*b* shows transgenic aspen seedlings that grow well and do not make as much lignin. Lignin-deficient trees are better for making paper and other forest products.

Engineering plant cells starts with vectors that can carry genes into plant cells. *Agrobacterium tumefaciens* is a bacterial species that infects eudicots, including beans, peas, potatoes, and other major crops plants. Genes in its plasmids cause tumors to form on these plants; hence the name Ti plasmid (*Tumor-inducing*). Researchers use the Ti plasmid to transfer foreign or modified genes into plants.

Researchers excise the tumor-inducing genes, then insert a desired gene into the plasmid (Figure 16.13). Some plant cells cultured with the modified plasmid may take it up. Whole plants may be regenerated.

Modified *A. tumefaciens* bacteria deliver genes into monocots that also are food sources, including wheat, corn, and rice. Researchers can even transfer genes into plants by way of electric shocks, chemicals, and blasts of microscopic particles coated with DNA.

Figure 16.12 (**a**) *Left,* control cotton plant. *Right,* cotton plant genetically engineered for herbicide resistance. Both plants were sprayed with a weed killer that is widely applied in cotton fields.

(**b**) Control plant (*left*) and four genetically engineered aspen seedlings. Vincent Chiang and coworkers suppressed a control gene involved in a lignin biosynthetic pathway. The modified plants synthesized normal lignin, but not as much. Lignin synthesis dropped by as much as 45 percent—yet cellulose production increased 15 percent. Root, stem, and leaf growth were greatly enhanced. Plant structure did not suffer. Wood harvested from such trees might make it easier to manufacture paper and some clean-burning fuels, such as ethanol. Lignin, a tough polymer, strengthens secondary cell walls of plants. Before paper can be made from wood, the lignin must be chemically extracted.

a   b

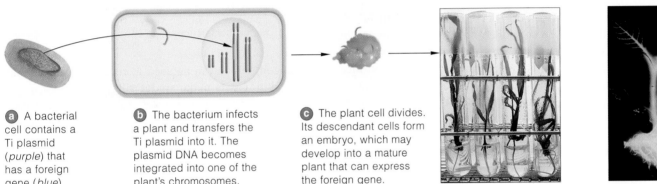

(**a**) A bacterial cell contains a Ti plasmid (*purple*) that has a foreign gene (*blue*).

(**b**) The bacterium infects a plant and transfers the Ti plasmid into it. The plasmid DNA becomes integrated into one of the plant's chromosomes.

(**c**) The plant cell divides. Its descendant cells form an embryo, which may develop into a mature plant that can express the foreign gene.

(**d**) Transgenic plants

(**e**) Example of a young plant with a fluorescent gene product.

Figure 16.13 *Animated!* (**a–d**) Ti plasmid transfer of an *Agrobacterium tumefaciens* gene to a plant cell. (**e**) A transgenic plant expressing a firefly gene for the enzyme luciferase.

Consider another compelling reason for modifying plant species: The food supply for most of the human population is extremely vulnerable. Farmers usually want to plant crops that give them the highest yields. Over time, genetically similar varieties have replaced the more diverse, older varieties. However, genetic uniformity makes food crops far more vulnerable to many pathogenic fungi, viruses, and bacteria.

That is why botanists comb the world for seeds of the older, diverse varieties of plants and of the wild ancestors of potatoes, corn, and other crop plants. They send their prizes—seeds with genes of a plant's lineage—to **seed banks**. These safe storage facilities are designed to preserve genetic diversity. They are now being tapped by genetic engineers as well as by traditional plant breeders.

Crop vulnerability is a huge problem. At one time, *Southern corn leaf blight* destroyed much of the United States corn crop. All of the plants carried the gene that conferred susceptibility to the fungal pathogen. Ever since that devastating epidemic, seed companies have been much more attentive to offering genetically diverse corn seeds. They tap seed banks, the treasure houses of plant genes.

Transgenic plants help farmers grow crops more efficiently and with less impact on the environment.

Genetic engineers as well as traditional plant breeders are tapping seed banks, which are safe storage facilities designed to preserve genetic diversity of plants.

16.8 Biotech Barnyards

LINKS TO
SECTIONS
5.2, 15.3

Laboratory mice were the first mammals to be genetically engineered. Today, featherless chickens, drug-producing goats, and transgenic pigs are part of the biotech barnyard.

TRANSGENIC ANIMALS

Traditional cross-breeding practices have produced unusual animals, including the featherless chicken in Figure 16.14. Now transgenic types are on the scene. The first ones arrived in 1982. Researchers isolated a gene for human somatotropin (growth hormone) and inserted it into a plasmid. They injected copies of the recombinant plasmids into fertilized mouse eggs that were later implanted into female mice. A third of the offspring of the surrogate mothers grew much larger than their littermates (Figure 16.15). The rat gene had become integrated into the host DNA and was being expressed in the transgenic mice.

Transgenic animals are used routinely for medical research. The functions of many gene products and how they can be controlled have been discovered by inactivating genes in "knockout mice" and analyzing the effect on phenotype (Section 15.3). Strains of mice, genetically modified mice to be susceptible to human diseases, help researchers study both the diseases and potential cures without experimenting on humans.

Genetically engineered animals also are sources of medically valued proteins. As a few examples, goats synthesize quantities of CFTR protein to treat cystic fibrosis and TPA protein to counter the bad effects of heart attacks. Rabbits make human interleukin-2, a protein that triggers divisions of immune cells called T lymphocytes. Cattle, too, may soon produce human

Figure 16.15 Evidence of a successful gene transfer. Two ten-week-old mouse littermates. *Left,* This one weighed 29 grams. *Right,* This one weighed 44 grams. It grew from a fertilized egg into which a gene for human somatotropin had been inserted.

collagen, which can be used to repair cartilage, bone, and skin. Goats make spider silk protein that might be used to make bullet-proof vests, medical supplies, and equipment for use in space. Different goats make human antithrombin, which is used to treat people with blood-clotting disorders (Figure 16.14*b*).

Genetic engineers have developed pigs that make environmentally friendlier manure. They have made freeze-resistant salmon, low-fat pigs, heftier sheep, and cows that are resistant to mad cow disease. Within a few years, they may give us allergen-free cats.

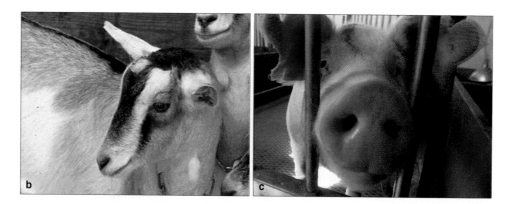

Figure 16.14 Genetically modified animals. (**a**) Featherless chicken developed by traditional cross-breeding methods in Israel. Such chickens survive in hot deserts where cooling systems are not an option. Chicken farmers in the United States have lost millions of feathered chickens in extremely hot weather. (**b**) Mira, a goat transgenic for human antithrombin III, an anticlotting factor. (**c**) Inquisitive transgenic pig at the Virginia Tech Swine Research facility.

Tinkering with the genetics of animals for the sake of human convenience does raise ethical questions. However, transgenic animal research may be viewed as an extension of thousands of years of acceptable barnyard breeding practices. Techniques have changed, but not the intent. Humans continue to have a vested interest in improving livestock.

KNOCKOUT CELLS AND ORGAN FACTORIES

Each year, about 75,000 people are on waiting lists for an organ transplant, but human donors are in short supply. There is talk of harvesting organs from pigs (Figure 16.14c), because pig organs function a lot like ours do. Transferring an organ from one species into another is called **xenotransplantation**.

The human immune system battles anything that it recognizes as "nonself." It rejects a pig organ at once, owing to a glycoprotein on the plasma membrane of cells that make up the blood vessels in pig organs. Antibodies circulating in human blood swiftly latch on to the sugar component and call for a response. In less than a few hours, blood inside the vessels coagulates massively and dooms the transplant. Drugs suppress this immune response, but a side effect is serious: the drugs make organ recipients vulnerable to infections.

Pig DNA contains two copies of *Ggta1*, the gene for an enzyme that catalyzes a key step in biosynthesis of alpha-1,3-galactose. This is the pig sugar that human antibodies recognize. Researchers have knocked out both copies of the *Ggta1* gene in transgenic piglets. Without the gene product, and the sugar, a pig tissue or organ may be less prone to rejection by the human immune system. Tissues and organs from such animals could help millions of people, including the ones with organs that have been severely damaged as a result of diabetes and Parkinson's disease.

Critics of xenotransplantation are concerned that, among other things, pig–human transplants would invite pig viruses to cross a species barrier and infect humans, perhaps catastrophically. Their concerns are not unfounded. In 1918, an influenza pandemic killed twenty million people worldwide. It originated with a swine flu virus—in pigs.

Genetic engineering started more than two decades ago. The kinds of animals being sought are beyond the scope of traditional breeding practices. Pigs engineered as donors for human organs are among the more startling cases.

Many years have passed since the first transfer of foreign DNA into a plasmid. That transfer ignited an ongoing debate about potential dangers of transgenic organisms entering the environment before rigorous testing.

In 1972, Paul Berg and his associates were the first to make recombinant DNA. Researchers knew that DNA was not toxic, but they could not predict what would happen every time they fused genetic material from different organisms into the recombinant molecules. Would they accidentally make superpathogens? Could they create a new form of life by the fusion of DNA from two normally harmless organisms? What if their creation escaped from the laboratory and transformed other organisms in the natural environment?

In a remarkably quick and responsible display of self-regulation, scientists reached a consensus on the safety guidelines for DNA research. Adopted at once by the NIH, their guidelines listed precautions for laboratory procedures. They covered the design and use of host organisms that could survive only under the narrow range of conditions inside the laboratory. Researchers stopped using DNA from pathogenic or toxic organisms for recombination experiments until proper containment facilities were developed.

As added precautions, "fail-safe" genes are now built into genetically engineered bacteria. They remain silent unless the bacteria escape and are exposed to environmental conditions—whereupon the genes get activated, with lethal results for the cell. Suppose that the package has a *hok* gene next to a promoter of the lactose operon (Section 15.4). Sugars are plentiful in the environment. If they were to activate the *hok* gene in a bacterial cell that escaped, the gene's product would destroy membrane function and the escapee.

Even so, does Murphy's law also apply to genetic engineering? As with any human endeavor, things can go wrong. After rabbits started taking over much of Australia, researchers were tinkering with a rabbit-killing virus in a containment laboratory on an island. Maybe the virus escaped in flying insects. However it happened, the virus is out and about, and killing lots of rabbits (Section 46.10). It is an example of why researchers are expected to expect the unexpected.

Rigorous safety guidelines for DNA research have been in place for decades in the United States. They have been adopted by the NIH, and researchers are expected to comply with their stringent standards.

16.10 Modified Humans?

We as a society continue to work our way through the ethical implications of applying the new DNA technologies. Even as we are weighing the risks and benefits, however, the manipulation of individual genomes has begun.

WHO GETS WELL?

Human gene therapy is often cited as one of the most compelling reasons for embracing the new research. We already have identified more than 15,500 genetic disorders. Many are rare in the population at large. Collectively, however, they show up in 3 to 5 percent of all newborns, and they cause 20 to 30 percent of all infant deaths every year. They account for about half of mentally impaired patients and nearly a fourth of all hospital admissions. They contribute to many age-related disorders that await all of us.

Rhys Evans, shown below, was born with a severe immune deficiency known as SCID-X1, which stems from mutations in gene *IL2RG*. Children affected by this disorder can live only in germ-free isolation tents, a "bubble," because they cannot fight infections.

In 1998, doctors withdrew stem cells from the bone marrow of eleven SCID-X1 boys. Stem cells, recall, are forerunners of other cell types, including white blood cells of the immune system. The doctors used a virus to insert nonmutated copies of *IL2RG* into each boy's stem cells, which they then infused back into his bone marrow. Months later, ten of the boys left isolation tents for good; gene therapy had successfully repaired their immune system. Since then, other gene therapy trials have freed many other SCID-X1 patients from life in a bubble. Rhys Evans is one of them.

In 2002, to the shock of researchers, two boys from the 1998 trial developed leukemia and one died. The researchers had anticipated that any cancer related to the therapy would be extremely rare. The very gene targeted to do the repair work—*IL2RG*—may be a problem, especially when combined with the viral vector that delivered the gene into stem cells. One other child who took part in a gene therapy experiment for SCID-X1 has developed leukemia. That it developed at all is evidence that our understanding of the human genome lags behind our ability to modify it.

WHO GETS ENHANCED?

When all is said and done, the idea of using human gene therapy to cure genetic disorders seems like a socially acceptable goal to most of us. Now see if your comfort level can move one step further. Would it also be acceptable to modify genes of some individual who falls within the normal range if he or she simply would like to minimize or enhance a particular trait?

We have already crossed the threshold of a brave new world. Researchers who are adept at transferring genes have already engineered strains of mice with enhanced memory and improved learning abilities. Perhaps their work is a beacon to those whose very lives have been turned upside down by Alzheimer's disease. Perhaps it draws others who are enchanted with the idea of simply getting more brain power.

The idea of selecting the most desired human traits is referred to as *eugenic engineering*. Yet who decides which forms of traits are most desirable? Realistically, cures for many severe but rare genetic disorders will not happen, because the payback for research is not financially attractive. Eugenics, however, might turn a profit. Just how much would potential parents pay to engineer tall or blue-eyed or fair-skinned children? Would it be okay to engineer "superhumans" with breathtaking strength or intelligence? How about an injection that would help you lose that extra weight and keep it off permanently? Where exactly is the line between interesting and abhorrent?

In a survey conducted in the United States, more than 40 percent of those interviewed said it would be fine to use gene therapy to make smarter and cuter babies. In one poll of British parents, 18 percent would be willing to use genetic enhancement to keep their child from being aggressive, and 10 percent would use it to keep a child from growing up to be homosexual.

Some argue that we must never alter the DNA of anything. The concern is that we just do not have the wisdom to bring about any genetic changes without causing irreparable damage to ourselves and nature.

One is reminded of our peculiar human tendency to leap before we look. And yet, something about the human experience gave us the capacity to imagine wings of our own making, a capacity that carried us to the frontiers of space. It gave one individual the dream of enhancing the rice plant genome to keep millions of children from going blind.

In this brave new world, two questions are before you: Should we be more cautious, because the risk takers may go too far? And what do we stand to lose if risks are not taken?

Be engaged; our understanding of the meaning of the human genome is changing even as you read this.

Summary

Section 16.1 Recombinant DNA technology uses restriction enzymes that can cut DNA into fragments. DNA ligases can splice the fragments into plasmids or some other cloning vector. Recombinant plasmids may be taken up by rapidly dividing cells, such as bacteria. When the host cells replicate, they make multiple, identical copies of the foreign DNA as well.

Bacteria cannot correctly express eukaryotic genes, which contain introns. Reverse transcriptase, a viral enzyme, can make a complementary DNA strand on mRNA. The hybrid molecule can then be converted to cDNA for cloning.

Biology Now
Explore the tools used to make recombinant DNA with the animation on BiologyNow.

Section 16.2 A gene library is a mixed collection of cells that have taken up cloned DNA. Researchers can isolate a gene of interest from a library by using a probe, a short stretch of DNA that can base-pair with the gene and that is traceable (it is labeled with a detectable tag, such as a radioisotope). Probes can help researchers locate and base-pair with one clone among millions. Such base pairing between nucleotide sequences from different sources is known as nucleic acid hybridization.

The polymerase chain reaction (PCR) is a technique for rapidly copying DNA fragments. A sample of a DNA template is mixed with nucleotides, primers, and a heat-resistant DNA polymerase. Each round of PCR proceeds through a series of temperature changes that amplifies the number of DNA molecules exponentially.

Biology Now
Learn how researchers isolate and copy genes with the interaction on BiologyNow.

Section 16.3 Automated DNA sequencing rapidly reveals the order of nucleotides in DNA fragments. As DNA polymerase copies a template DNA, progressively longer fragments stop growing as soon as one of four different fluorescent nucleotides becomes attached to them. Electrophoresis separates the labeled fragments into bands according to length. The order of the colored bands as they migrate through the gel reflects which fluorescent base was added to the end of each fragment, and so indicates the template DNA base sequence.

Biology Now
Investigate DNA sequencing with the animation on BiologyNow.

Section 16.4 Tandem repeats are multiple copies of a short DNA sequence that follow one another along a chromosome. The number and distribution of tandem repeats, unique in each person, can be revealed by gel electrophoresis; they form a DNA fingerprint.

Biology Now
Observe the process of DNA fingerprinting with the animation on BiologyNow.

Section 16.5 The entire human genome has been sequenced and is now being analyzed. Genomes of other organisms also have been fully sequenced.

The new field of genomics is concerned with the mapping and analysis of genomes. One branch, called comparative genomics, uses similarities and differences between DNA sequences of major groups of organisms to identify their evolutionary relationships.

DNA chips are microarrays used to compare patterns of gene expression within a genome.

Sections 16.6–16.8 Recombinant DNA technology and the mapping and analysis of genomes is the basis for genetic engineering. Genetic engineering is the directed modification of the genetic makeup of an organism, often to modify its phenotype. Researchers insert normal or modified genes from one organism into another of the same or different species. Gene therapies insert copies of modified genes into individuals to cover the functions of a mutant or altered gene.

Genetically engineered bacteria that contain plasmid vectors have diverse uses in basic research, medicine, agriculture, industry, and ecology. Transgenic crop plants help farmers use less toxic pesticides and produce food more efficiently. Genetic engineering of animals allows commercial production of human proteins, as well as research into genetic disorders.

Biology Now
See how the Ti plasmid is used to genetically engineer plants with the animation on BiologyNow.

Section 16.9 There is always a risk that genetically modified experimental organisms can escape from the laboratory. Typically, potentially dangerous types have fail-safe genes built into their genome that will destroy them when exposed to conditions that exist anywhere except in the laboratory. Rigorous tests for safety must precede the release of any modified organism into the environment.

Section 16.10 The goal of human gene therapy is to transfer normal or modified genes into body cells to correct genetic defects. As with any new technology, the benefits must be weighed against potential risks.

Self-Quiz   *Answers in Appendix II*

1. Researchers can cut DNA molecules at specific sites by using _____ .
   a. DNA polymerase   c. restriction enzymes
   b. DNA probes   d. reverse transcriptase

2. Fill in the blank: A _____ is a small circle of bacterial DNA that contains only a few genes and is separate from the bacterial chromosome.

3. By reverse transcription, _____ is assembled on a(n) _____ template.
   a. mRNA; DNA   c. DNA; ribosome
   b. cDNA; mRNA   d. protein; mRNA

4. PCR stands for _____ .
   a. polymerase chain reaction
   b. polyploid chromosome restrictions
   c. polygraphed criminal rating
   d. politically correct research

5. Automated DNA sequencing relies on _____ .
   a. supplies of standard and labeled nucleotides
   b. primers and DNA polymerases
   c. gel electrophoresis and a laser beam
   d. all of the above

6. By gel electrophoresis, fragments of DNA can be separated according to _____ .
   a. sequence   b. length   c. species

7. _____ can be used to insert genes into human cells.
   a. PCR   c. Xenotransplantation
   b. Modified viruses   d. DNA microarrays

8. For each species, all _____ in a haploid number of chromosomes is the _____ .
   a. genomes; phenotype   c. mRNA; start of cDNA
   b. DNA; genome   d. cDNA; start of mRNA

9. Match the terms with the most suitable description.
   ____ DNA fingerprint   a. selecting "desirable" traits
   ____ Ti plasmid   b. mutations, crossovers
   ____ nature's genetic   c. used in some gene transfers
        experiments   d. a person's unique collection
   ____ nucleic acid   of tandem repeats
        hybridization   e. base pairing of nucleotide
   ____ eugenic   sequences from different
        engineering   DNA or RNA source

Additional questions are available on **Biology ⓔ Now™**

Critical Thinking

1. Lunardi's Market put out a bin of tomatoes having vine-ripened redness, flavor, and texture. A sign identified them as genetically engineered produce. Most shoppers selected unmodified tomatoes in the adjacent bin even though those tomatoes were pale pink, mealy textured, and tasteless. Which tomatoes would you pick? Why?

2. Biotechnologists envision a new Green Revolution. As they see it, designer plants hold down food production costs, reduce dependence on pesticides and herbicides, enhance crop yields, offer improved flavor and nutritional value, and often produce plants with salt tolerance and drought tolerance. Fruits and vegetables can be designed for flavor, nutritional value, and extended shelf life.

Genetically engineered food crops are widespread in the United States. At least 45 percent of cotton crops, 38 percent of soybean crops, and 25 percent of corn crops have been modified to withstand weedkillers or make their own pesticides. For years, modified corn and soybeans have been used in tofu, cereals, soy sauce, vegetable oils, beer, and soft drinks. They are fed to farm animals.

In Europe especially, public resistance to modified food runs high. Besides arguing that modified foods might be toxic and have lower nutritional value, many people worry that designer plants might cross-pollinate wild plants and produce "superweeds." The chorus of critics in Europe has forced American farmers to keep genetically engineered crops separated from traditional crops. Traditional crops only are exported to Europe. Such separation is both costly and difficult.

Read up on scientific research related to this issue and form your own opinions. The alternatives are to be swayed either by media hype (the term Frankenfood, for instance) or by sometimes biased reports from groups (such as chemical manufacturers), which have their own agendas.

3. The sequencing of the human genome is completed, and knowledge about many genes is being used to detect genetic disorders. Many insurance companies will pay for their female subscribers to take advantage of genetic testing for breast cancer and are willing to allow them to keep the results confidential.

Explain how a health insurance company might benefit financially if it were to encourage its subscribers to take confidential tests for breast cancer susceptibility.

4. Scientists at Oregon Health Sciences University produced Tetra, the first primate clone. They also made the first transgenic primate by inserting a jellyfish gene into a fertilized egg of a rhesus monkey. (The gene codes for a bioluminescent protein that fluoresces green; refer to Section 6.6). The egg was implanted in a surrogate monkey's uterus, where it developed into a male.

The long-term goal of this gene transfer project is not to make glowing-green monkeys. It is the transfer of human genes into primates whose genomes are most like ours. Transgenic primates could yield insight into genetic disorders. That insight might lead to the development of cures for those who are affected and of vaccines for those who are at risk.

Something more controversial is at stake. Will the time come when foreign genes can be inserted into human embryos? Would it be ethical to transfer a chimpanzee or monkey gene into a human embryo to cure a genetic defect? To bestow immunity against a potentially fatal disease such as AIDS? Think about it.

III Principles of Evolution

Two male frigate birds (Fregata minor) in the Galápagos Islands, far from the coast of Ecuador. Each male inflates a gular sac, a balloon of red skin at his throat, in a display that may catch the eye of a female. The males lurk together in the bushes, sacs inflated, until a female flies by. Then they wag their head back and forth and call out to her. Like other structures that males use only in courtship, the gular sac is probably an outcome of sexual selection—one of the topics you will read about in this unit.

17 EVIDENCE OF EVOLUTION

Measuring Time

How do you measure time? Is your comfort level with the past limited to your own generation? Probably you can relate to a few centuries of human events. But geologic time? Comprehending the distant past requires a huge intellectual leap from the familiar to the unknown.

Perhaps the possibility of an asteroid slamming into our planet can help you make the leap. Asteroids are rocky, metallic bodies hurtling through space. They are a few meters to 1,000 kilometers across. When our solar system's planets were forming, their gravitational force swept up most of the asteroids. At least 6,000 asteroids, including the one shown in Figure 17.1, still orbit the sun in a belt between Mars and Jupiter. Millions more frequently zip past Earth. They are hard to spot; they do not emit light. We cannot identify most of them until after they have passed close by. Some have passed too close for comfort.

We have evidence that big asteroid impacts influenced the history of life. For instance, researchers found a thin layer of iridium around the world, and it dates precisely to a mass extinction that wiped out the last of the dinosaurs (Figures 17.1 and 17.2). Iridium is rare on Earth but not in asteroids. There are plenty of fossils of dinosaurs below this layer. Above it, there are none, anywhere.

It has only been about 100,000 years since the first modern humans (*Homo sapiens*) evolved. We have fossils of dozens of humanlike species that lived in Africa during the 5 million years before our own species even showed up. So why are we the only ones left?

Unlike today's large, globally dispersed populations of humans, those early species lived in small bands. What if most were casualties of the twenty asteroids that collided with Earth while they were alive? What if *our* ancestors were just plain lucky? About 2.3 million years ago, a huge object from space hit the ocean west of what is now Chile. If it had struck the rotating Earth just a few hours earlier, it would have hit southern Africa instead of the ocean. Our ancestors could have been incinerated.

Now that we know what to look for, we are seeing more and more craters in satellite images of Earth. One crater, in Iraq, is less than 4,000 years old. Energy released during that particular impact was equivalent to the detonation of hundreds of nuclear weapons.

Watch the video online!

Figure 17.1 *Left,* an asteroid nineteen kilometers (about twelve miles) long, hurtling through space. *Right,* part of the worldwide, iridium-rich layer of sediment (*black*) that dates to the Cretaceous–Tertiary (K–T) boundary. This vertical section through stacked layers of rock is evidence of an asteroid impact. The red pocketknife gives you an idea of its thickness.

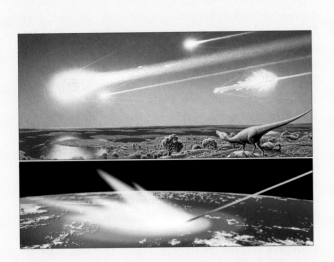

Figure 17.2 Artist's interpretation and computer-generated model for the last few minutes of the Cretaceous.

If we can figure out what an asteroid impact will do to us, then we can figure out how impacts affected life in the past. We *can* comprehend life long before our own, including its brushes with good and bad cosmic luck.

You are about to make an intellectual leap through time, to places that were not even known about a few centuries ago. We invite you to launch yourself from this premise: *Any aspect of the natural world, past as well as present, has one or more underlying causes.*

That premise is the foundation for scientific research into the history of life. It guides probes into physical and chemical aspects of Earth. It guides the studies of fossils and comparisons of species. It guides tests of hypotheses by way of experiments, models, and new technologies. This research represents a shift from experience to inference—from the known to what can only be surmised. And it has given us astonishing glimpses into the past.

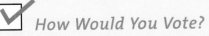

How Would You Vote?

A large asteroid could obliterate civilization and much of Earth's biodiversity. Should nations around the world contribute resources to locating and tracking asteroids? See BiologyNow for details, then vote online.

Key Concepts

EMERGENCE OF EVOLUTIONARY THOUGHT

As long ago as the 1400s, Western scientists started to learn about previously unknown species and how organisms are distributed around the world. They documented similarities and differences in the traits among living organisms and among fossil species that were being unearthed in layers of sedimentary rock. Sections 17.1, 17.2

A THEORY TAKES FORM

The emerging discoveries suggested that evolution, or changes in lines of descent, had occurred. Charles Darwin and Alfred Wallace independently developed a theory of natural selection to explain how the heritable traits that define each species might evolve. Section 17.3

EVIDENCE FROM FOSSILS

The fossil record will never be complete, so gaps are to be expected. Even so, it offers powerful evidence of change in lines of descent over great time spans. Sections 17.4, 17.5

EVIDENCE FROM BIOGEOGRAPHY

Evolutionary theories, reinforced by plate tectonics theory, help explain patterns in the distribution of species through the environment and through time. Section 17.6

EVIDENCE FROM COMPARATIVE MORPHOLOGY

The adults and embryos of different lineages often show similarities in one or more body parts that hint at descent from a common ancestor. Sections 17.7, 17.8

EVIDENCE FROM COMPARATIVE BIOCHEMISTRY

Today, researchers are using biochemical and molecular comparisons to illuminate the history of life and to clarify the evolutionary relationships among diverse species and lineages. Section 17.9

Links to Earlier Concepts

Section 1.4 sketched out key premises of the theory of natural selection. Here you will read about evidence that led to its formulation. As you read, remember that science does not deal with the supernatural, and it cannot answer subjective questions about nature (1.7).

You may wish to refresh your memory of radioisotopes (2.2), protein structure (3.5), mutation (1.4, 3.6, 14.4), nucleic acid hybridization (16.2), and automated gene sequencing (16.3). These topics are basic to understanding the biochemical and molecular comparisons that are clarifying evolutionary relationships among species.

17.1 Early Beliefs, Confounding Discoveries

LINK TO
SECTION
1.7

Prevailing beliefs can influence how we interpret clues to natural processes and their observable outcomes.

QUESTIONS FROM BIOGEOGRAPHY

Two thousand years ago the seeds of biological inquiry were taking hold in the West. Aristotle was foremost among the early naturalists. There were no books or instruments to guide him, and yet he was more than a collector of random observations. In his descriptions we see evidence that he was connecting observations in an attempt to explain the order of things. As others did, he saw nature as a continuum of organization, from lifeless matter through complex forms of plants and animals. By the fourteenth century, scholars had transformed his ideas into a rigid view of life. A Chain of Being was seen as extending from "lowest" forms to humans, and on to spiritual beings. Each kind of being, or **species**, was one separate link in the chain. All links had been designed and forged at the same time at one center of creation. They had not changed since. Once naturalists discovered and described all the links, the meaning of life would be revealed.

Then Europeans embarked on their globe-spanning explorations. They soon discovered that the world is a lot bigger than Europe. Tens of thousands of unique plants and animals from Asia, Africa, the New World, and the Pacific islands were brought back home to be carefully catalogued as one more link in the chain.

Later, Alfred Wallace and a few other naturalists moved beyond cataloguing species for its own sake. They started to identify *patterns* in where species live and how species might or might not be related. They were pioneers in **biogeography**: the study of patterns in the geographic distribution of individual species and entire communities. They were the first to think about the ecological and evolutionary forces in play.

Some patterns were intriguing. For example, many plants and animals are found only on islands in the middle of the ocean and other remote places. Many species that are strikingly similar live far apart, and vast expanses of open ocean or impassable mountain ranges keep them separated, in isolation.

Consider: Flightless, long-necked, long-legged birds are native to three continents (Figure 17.3a–c). Why are they so much alike? The plants in Figure 17.3d,e live on separate continents. Both have spines, tiny leaves, and short fleshy stems. Why are *they* so much alike?

Curiously, the flightless birds all live in the same kind of environment about the same distance from the equator. They sprint about in flat, open grasslands of dry climates, and they raise their long necks to keep an eye on predators in the distance. Both plant species also live the same distance from the equator in the

Figure 17.3 Species that resemble one another, strikingly so, even though they are native to distant geographic realms.

(**a**) South American rhea, (**b**) Australian emu, and (**c**) African ostrich. All three types of birds live in similar habitats. They are unlike most birds in several traits, most notably in their long, muscularized legs and their inability to get airborne.

(**d**) A spiny cactus native to the hot deserts of the American Southwest. (**e**) A spiny spurge native to southwestern Africa.

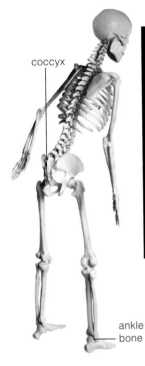

coccyx

ankle
bone

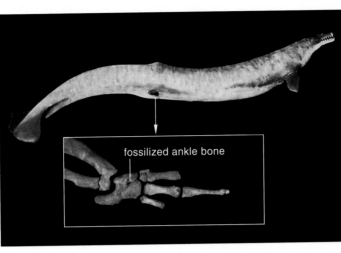

fossilized ankle bone

Figure 17.4 Body parts that have no apparent function. *Above,* reconstruction of an ancient whale (*Basilosaurus*), with a head as long as a sofa. This marine predator was fully aquatic, so it did not use its hindlimbs to support body weight as you do—yet it had ankle bones. We use our ankles, but not our coccyx bones.

Figure 17.5 Fossilized ammonites. These large marine predators lived hundreds of millions of years ago. Their shell is similar to the shell of a modern chambered nautilus.

same kind of environment—hot deserts—where water is seasonally scarce. Both species store water in their fleshy stems. Their stems have a notably thick, water-conserving cuticle and rows of sharp spines that deter thirsty, hungry herbivores.

If all birds and plants were created in one place, then how did such similar kinds end up in the same kind of environment in such distant, remote places?

QUESTIONS FROM COMPARATIVE MORPHOLOGY

The similarities and differences among species raised questions that gave rise to **comparative morphology**: the study of body plans and structures among groups of organisms. For instance, the bones in a human arm, whale flipper, and bat wing differ in size, shape, and function. As Section 17.7 explains, the different bones are located in similar body regions. They are made of the same kinds of tissues that are arranged in similar patterns. They also develop in much the same way in embryos. Naturalists who discovered the similarities wondered: Why do species that differ so much in some features look so much alike in other features?

By one hypothesis, body plans are so perfect there was no need to make a new design for each organism at the time of its creation. Yet if that were so, then why did some organisms have useless body parts? For instance, an ancient aquatic whale had ankle bones but it did not walk (Figure 17.4). Why the bones? Our coccyx is like some tailbones in many other mammals. We do not have a tail. Why do we have parts of one?

QUESTIONS ABOUT FOSSILS

About the same time, geologists were mapping layers of rock exposed by erosion or quarrying. As you will read later on, they added to the confusion when they found fossils in the same kinds of layers in different parts of the world. **Fossils** came to be recognized as the stone-hard evidence of earlier forms of life. Figures 17.4 and 17.5 have examples.

A puzzle: Many deep layers held fossils of simple marine life. Some layers above them contained fossils that were structurally similar but more intricate. In higher layers, fossils were like modern species. What did sequences in complexity among fossils of a given type mean? Were they evidence of lines of descent?

Taken as a whole, the findings from biogeography, comparative morphology, and geology did not fit with prevailing beliefs of the nineteenth century. Scholars floated novel hypotheses. If a simultaneous dispersal of all species from a center of creation was unlikely, *then perhaps species originated in more than one place.* If species had not been created in a perfect state—and fossil sequences and "useless" body parts implied they had not—*then perhaps species had become modified over time.* Awareness of evolution was in the wind.

Awareness of biological evolution emerged over centuries, through the cumulative observations of many naturalists, biogeographers, comparative anatomists, and geologists.

17.2 A Flurry of New Theories

LINK TO
SECTION
1.7

Nineteenth-century naturalists found themselves trying to reconcile the evidence of change with a traditional conceptual framework that simply did not allow for it.

SQUEEZING NEW EVIDENCE INTO OLD BELIEFS

A respected anatomist, Georges Cuvier, was among those trying to make sense of the growing evidence for change. For years he had compared fossils with living organisms. He was aware of the abrupt changes in the fossil record and was the first to recognize that they marked times of mass extinctions.

By Cuvier's hypothesis, a single time of creation had populated the world, which was an unchanging stage for the human drama. Monstrous earthquakes, floods, and other major catastrophes did happen, and many people died. Each time, survivors repopulated the world. By Cuvier's reckoning, there were no *new* species. Naturalists simply had not yet found all of the fossils that would date to the time of creation.

His hypothesis enjoyed support for a long time. It even became elevated to the rank of theory, one that later became known as **catastrophism**.

Still, many other scholars kept at the puzzle. For example, in Jean Baptiste Lamarck's view, offspring inherit traits that a parent *acquired in its lifetime.* By his hypothesis, environmental pressure and internal needs promote permanent changes in an individual's body form and functions, which offspring then inherit. By

this proposed process, life was created long ago in a simple state, and it gradually improved. The force for change was an intense drive toward perfection, up the Chain of Being. Lamarck thought that the force, centered in nerves, directed an unknown "fluida" to body parts needing change.

Try using his hypothesis to explain a giraffe's long neck. Suppose a short-necked, hungry ancestor of the modern giraffe kept stretching its neck in order to browse on leaves beyond the reach of other animals. Lengthier and lengthier stretches directed fluida into its neck and thus made the neck permanently longer. Offspring inherited a longer neck, and they stretched their necks, too. Generations that strained to reach ever loftier leaves led to the modern giraffe.

As Lamarck correctly inferred, the environment *is* a factor in changes in lines of descent. However, his hypothesis, and Cuvier's, has not been supported by experimental tests. Environmental factors can alter an individual's phenotype, as when a male builds large muscles through strength training. But any child of a muscle-bound parent will not be born muscle-bound. It can inherit genes, but not increased muscle mass.

VOYAGE OF THE *BEAGLE*

In 1831, in the midst of the confusion, Charles Darwin was twenty-two years old and wondering what to do with his life. Ever since he was eight, he had wanted

Figure 17.6 (**a**) Charles Darwin. (**b**) Replica of the *Beagle* sailing off a rugged coastline of South America. During one of his trips, Darwin ventured into the Andes. He discovered fossils of marine organisms in rock layers 3.6 kilometers above sea level.

(**c–e**) The Galápagos Islands are isolated in the ocean, far to the west of Ecuador. They arose by volcanic action on the seafloor about 5 million years ago. Winds and currents carried organisms to the once-lifeless islands. All of the native species are descended from those travelers. At far right, a blue-footed booby, one of many species Darwin observed during his voyage.

to hunt, fish, collect shells, or just watch insects and birds—anything but sit in school. Later, at his father's insistence, he did attempt to study medicine in college. The crude, painful procedures being used on patients in that era sickened him. His exasperated father urged him to become a clergyman, and so Darwin packed for Cambridge. His grades were good enough to earn a degree in theology. Yet he spent most of his time with faculty members who embraced natural history.

John Henslow, a botanist, perceived Darwin's real interests. He hastily arranged for Darwin to become ship's naturalist aboard the *Beagle*, which was about to leave on a five-year voyage around the world. The young man who had hated school and had no formal training quickly became an enthusiastic naturalist.

The *Beagle* sailed first to South America to finish work on mapping the coastline (Figure 17.6). During the Atlantic crossing, Darwin collected and studied marine life. He read Henslow's parting gift, the first volume of Charles Lyell's *Principles of Geology*. He saw diverse species in environments ranging from sandy shores of remote islands to high mountains. He also started circling the question of evolving life, which was now on the minds of many individuals.

Darwin started by mulling over a radical theory. As Lyell and other geologists were arguing, erosion and other gradual, natural processes of change had more impact on Earth history than rare catastrophes. Geologists for years had chipped away at sandstones,

limestones, and other rocks that form after sediments slowly accumulate in the beds of lakes, rivers, and seas. They took earthquakes and other less frequent events into account. As they knew, immense floods, more than a hundred big earthquakes, and twenty or so volcanic eruptions happen in a typical year, which means catastrophes are not that unusual.

The idea that gradual, repetitive change had shaped Earth became known as the **theory of uniformity**. It challenged the prevailing views of Earth's age.

The theory bothered scholars who firmly believed that Earth could be no more than 6,000 years old. They believed people had recorded all that had happened in those 6,000 years—and in all that time, no one had mentioned seeing a species evolve. Even so, by Lyell's calculations, it must have taken millions of years to sculpt the present landscape. *Was that not enough time for species to evolve in many diverse ways?* Later, Darwin thought so. But exactly *how* did they evolve? He would end up devoting the rest of his life to that burning question.

Prevailing beliefs can influence how we interpret clues to natural processes and their observable outcomes.

Darwin's observations during a global voyage helped him think about species in a novel way.

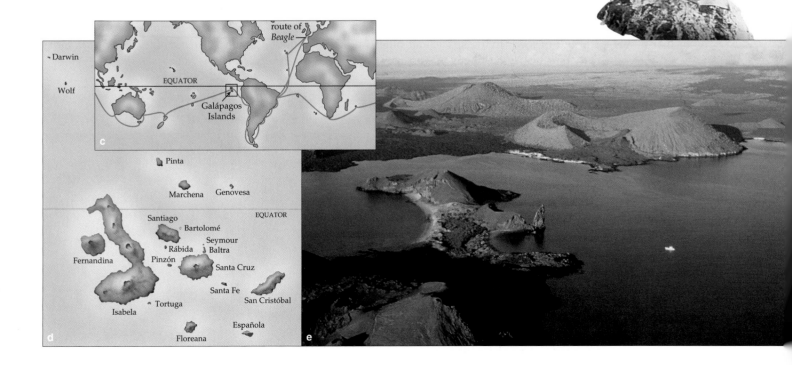

17.3 Darwin, Wallace, and Natural Selection

LINK TO
SECTION
1.4

Darwin's observations of thousands of species in different parts of the world helped him see how species might evolve.

OLD BONES AND ARMADILLOS

Darwin brought thousands of specimens with him to England. Although he had page after page of notes, he had been careless about recording where each species lived and what its habitat was like. He had left much of the "geography" out of biogeography. Colleagues helped him fill in some of the blanks, and in time he was able to explain how species might evolve.

Among the specimens were fossils of glyptodonts from Argentina. Of all animals, only living armadillos are like the now-extinct glyptodonts (Figure 17.7). Of all places on Earth, armadillos live only in the places where glyptodonts once lived.

If these animals had been created at the same time, lived in the same place, and were so alike in certain odd traits—such as body armor made of overlapping scales—then why is only one still with us? What if the glyptodonts were ancient relatives of armadillos? What if some traits of their common ancestor had changed in the line of descent that led to armadillos? *Descent with modification*—it did seem possible. What, then, could be the driving force for evolution?

A KEY INSIGHT—VARIATION IN TRAITS

While Darwin assessed his notes, an essay by Thomas Malthus, a clergyman and economist, made him reflect on a topic of social interest. Malthus had correlated population size with famine, disease, and war. He said that humans run out of food, living space, and other resources because they reproduce too much. The larger a population gets, the more individuals there are to reproduce. Individuals compete with one another for dwindling resources. Many starve, get sick, or engage in war or other forms of competition.

Darwin deduced that *any* population has a capacity to produce more individuals than the environment can support. Even one sea star can produce 2,500,000 eggs per year, but the seas do not fill with sea stars. (For one thing, predators eat many of the eggs and larvae.)

Darwin also reflected on species he had observed during his voyage. Individuals of those species were not alike in their details. They varied in size, color, and other traits. *It dawned on Darwin that variations in traits influence an individual's ability to secure resources and to survive and reproduce in the environment.*

He thought about the Galápagos Islands, separated from South America by 900 kilometers of open ocean. Nearly all of their finch species live nowhere else, yet they share traits with mainland species. Perhaps fierce storms had blown a few mainland birds out to sea. Perhaps prevailing winds and currents had dispersed them to the Galápagos. Perhaps those modern species were island-hopping descendants of the colonizers.

As he knew, different species live in diverse habitats near coasts, in dry lowlands, and in mountain forests. One strong-billed type is better than others at cracking open hard seeds (Figure 17.8). A drought that lasts for several years will make soft seeds harder to find. A strong-billed individual will survive and reproduce more than the others. Its bill size has a heritable basis, so its offspring will be favored, also. Conditions in the prevailing environment "select" individuals that have strong bills, which in time become more frequent in the population. *And a population is evolving when forms of heritable traits change over the generations.*

Figure 17.7 (a) From Texas, a modern armadillo, about a foot long excluding the tail. (b) A Pleistocene glyptodont, which was about as big as a Volkswagen Beetle, and now extinct. Glyptodonts shared unusual traits and a restricted distribution with the existing armadillos. Yet the two kinds of animals are widely separated in time. Their similarities were a clue that helped Darwin develop a theory of evolution by natural selection.

a

b

Figure 17.8 Three of thirteen finch species on the Galápagos Islands. (**a**) A big-billed seed cracker, *Geospiza magnirostris*. (**b**) *G. scandens* eats cactus fruit and insects in cactus flowers. (**c**) *Camarhynchus pallidus* uses cactus spines and twigs to probe for wood-boring insects. Differences in bill shape depend in large part on when and where the signaling molecule BMP4 is switched on in bird embryos. Mutations in regulatory elements may cause the bill variations. Compare the different bills of Hawaiian honeycreepers (Chapter 19).

NATURAL SELECTION DEFINED

Let's now put Darwin's observations and conclusions in the context of what we have learned from genetics and molecular biology:

1. *Observation:* Natural populations have an inherent reproductive capacity to increase in size over time.

2. *Observation:* No population can indefinitely grow in size, because its individuals will run out of food, living space, and other resources.

3. *Inference:* Sooner or later, individuals will end up competing for dwindling resources.

4. *Observation:* Individuals share a pool of heritable information about traits, encoded in genes.

5. *Observation:* Variations in traits start with alleles, slightly different molecular forms of genes that arise through mutations.

6. *Inferences:* Some forms of traits prove better than others at helping an individual compete for resources, survive, and reproduce. In time, alleles for adaptive forms become more frequent relative to other alleles in the population. They lead to increased **fitness**—an increase in adaptation to the environment as measured by the genetic contribution to future generations.

7. *Conclusions:* **Natural selection** is the outcome of differences in reproduction among individuals of a population that vary in shared traits. Environmental agents of selection act on the range of variation, and the population may evolve as a result.

Darwin kept on looking for patterns in his data and filling in gaps in his reasoning. He also wrote out his theory but let ten years pass without publishing it. He waited too long. Alfred Wallace sent him an essay he was working on that outlined the same theory! Today, Wallace is known as the father of biogeography. He did brilliant fieldwork in the Amazon River Basin, Malay Archipelago, and elsewhere (Figure 17.9). He had written earlier letters to Lyell and Darwin about patterns in the geographic distribution of species. He, too, had connected the dots.

Figure 17.9 Alfred Wallace. For one account of the Darwin–Wallace story, read David Quammen's *Song of the Dodo*.

In 1858, just weeks after Darwin received Wallace's essay, their similar theories were presented jointly at a scientific meeting. Wallace was still in the field and knew nothing about the meeting, which Darwin did not attend. The next year, Darwin published *On the Origin of Species*, which laid out detailed evidence in support of his theory.

You may have heard that Darwin's book fanned an intellectual firestorm, but most scholars were quick to accept the idea that diversity is a result of evolution. The theory of natural selection *was* fiercely debated. Decades passed before experimental evidence from a new field, genetics, led to its widespread acceptance.

> As Darwin and Wallace perceived, natural selection is the outcome of differences in survival and reproduction among traits. Natural selection can lead to increased adaptation to the environment, as measured by fitness—the relative genetic contribution to future generations.

17.4 Fossils—Evidence of Ancient Life

Turn now to fossil evidence of the connection between life's evolution and the evolution of Earth.

About 500 years ago, Leonardo da Vinci was puzzled by seashells entombed in the rocks of northern Italy's high mountains, hundreds of kilometers from the sea. How did they get there? By the prevailing belief, water from a stupendous and divinely invoked flood had surged up into the mountains, where it deposited the shells. But many shells were thin, fragile, and intact. If they had been swept across such great distances, then wouldn't they be battered to bits?

Leonardo also brooded about the rocks. They were stacked like cake layers. Some layers had shells, others had none. Then he remembered how large rivers swell with spring floodwaters and deposit silt in the sea. Did such depositions happen in ancient seasons? If so, then shells in the mountains could be evidence of layered communities of organisms that once lived in the seas!

By the 1700s, fossils were being accepted as remains and impressions of organisms that lived in the past. (*Fossil* comes from the Latin word for "something that was dug up.") People were still interpreting fossils through the prism of cultural beliefs, as when a Swiss naturalist unveiled the remains of a giant salamander and excitedly announced that they were the skeleton of a man who had drowned in the great flood.

By midcentury, naturalists were questioning these interpretations. Mining, quarrying, and excavations for canals were under way. Diggers were discovering similar rock layers and similar sequences of fossils in distant places, such as the cliffs on both sides of the English Channel. If those layers had been deposited over time, then a vertical series of fossils embedded in them might be a record of past life—*a fossil record.*

HOW DO FOSSILS FORM?

Most fossils are bones, teeth, shells, seeds, spores, and other hard parts (Figure 17.10). Even fossilized feces (coprolites) hold parts of organisms that were eaten. Indirectly, imprints of leaves, stems, tracks, burrows, and other *trace* fossils offer more evidence of past life.

Fossilization is a slow process that starts when an organism or traces of it become covered by sediments or volcanic ash. Water slowly infiltrates the remains, and metal ions and other inorganic compounds that are dissolved in it replace the minerals in bones and other hardened tissues. As sediments accumulate, they exert increasing pressure on the burial site. In time, the pressure and mineralization processes transform those remains into stony hardness.

Remains that become buried quickly are less likely to be obliterated by scavengers. Preservation is also favored when a burial site stays undisturbed. Usually, however, erosion and other geologic assaults deform, crush, break, or scatter the fossils. This is one reason fossils are relatively rare.

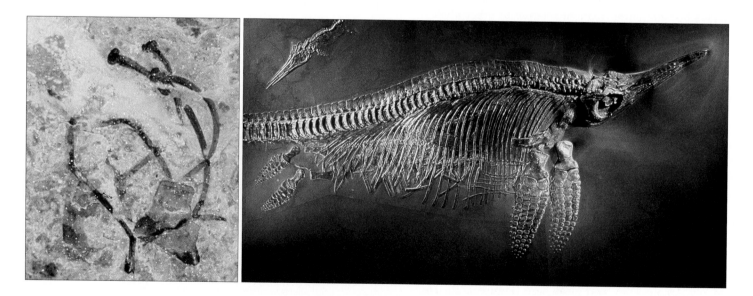

Figure 17.10 Two of the more than 250,000 species known from the fossil record. *Left*, fossilized parts of the oldest known land plant (*Cooksonia*). Its stems were a little taller than the length of a toothpick. *Right*, fossilized skeleton of an ichthyosaur. This marine reptile lived 200 million years ago.

Figure 17.11 A slice through time—Butterloch Canyon, Italy, once at the bottom of a sea. Its sedimentary rock layers slowly formed over hundreds of millions of years. Later, geologic forces lifted the stacked layers above sea level. Later still, the erosive force of river water carved the canyon walls and exposed the layers. Scientists Cindy Looy and Mark Sephton are climbing to reach the Permian–Triassic boundary layer, where they will look for fossilized fungal spores.

Other factors affect preservation. Organic materials cannot decompose in the absence of free oxygen, for instance. They may endure when sap, tar, ice, mud, or another air-excluding substance protects them. Insects in amber and frozen woolly mammoths are examples.

FOSSILS IN SEDIMENTARY ROCK LAYERS

Stratified (stacked) layers of sedimentary rock formed long ago from deposits of volcanic ash, silt, sand, and other materials. Sand and silt piled up after rivers transported them from land to the sea, as Leonardo suspected. Sandstones formed from sand, and shales from silt. Depositions were sometimes interrupted, in part because the sea level changed as ice ages began. Tremendous volumes of water froze in glaciers, rivers dried up, and the depositions ended in some regions. Later in time, when the climate warmed and glaciers melted, depositions resumed.

The formation of sedimentary rock layers is called **stratification**. The deepest layer was the first to form; those closest to the surface were the last. Most formed horizontally, as in Figure 17.11, because particles tend to settle in response to gravity. You may see tilted or ruptured layers, as along a road that was cut into a mountainside. Major crustal movements or upheavals disturbed them after they formed.

Most fossils are in sedimentary rock. Understand how rock layers form, and you know that fossils in them formed at specific times in the past. Specifically, *the older the rock layer, the older the fossils.*

INTERPRETING THE FOSSIL RECORD

We have fossils for more than 250,000 known species. Judging from the current range of biodiversity, there must have been many, many millions more. Yet the fossil record will never be complete. Why is this so?

The odds are against finding evidence of an extinct species. Why? At least one specimen had to be buried gently before it decomposed or something ate it. It had to escape erosion, flowing lava, and other forces of nature. The fossil had to end up where someone can find it. For instance, many have become exposed on canyon walls after a river or glacier slowly carved its way through sedimentary rock layers (Figure 17.11).

Also, most ancient species did not lend themselves to preservation. Unlike bony fishes and hard-shelled mollusks, for instance, the soft-bodied jellyfishes and worms do not show up as much in the fossil record. Probably they were just as common, or more so.

Also think about population density and body size. One plant population might release millions of spores in a single season. The earliest humans lived in small bands and raised few offspring. What are the odds of finding even one fossilized human bone compared to finding spores of plants that lived at the same time?

Finally, imagine one line of descent, a **lineage**, that vanished when its habitat on a remote volcanic island sank into the sea. Or imagine two lineages, one lasting only briefly and the other for billions of years. Which is more likely to be represented in the fossil record?

Fossils are physical evidence of organisms that lived in the remote past, a stone-hard historical record of life. In general, the oldest are in the deepest sedimentary rocks.

The fossil record is incomplete. Geologic events obliterated much of it. The record is slanted toward species that had hard parts, dense populations, and wide distribution, and that persisted a long time.

Even so, the fossil record is now substantial enough to help us reconstruct patterns and trends in the history of life.

17.5 Dating Pieces of the Puzzle

LINK TO
SECTION
2.2

How do we assign fossils to a place in time? In other words, how do we know how old fossils really are?

RADIOMETRIC DATING

At one time, people could assign only *relative* ages to their fossil treasures, not absolute ones. For instance, a fossilized mollusk embedded in a layer of rock was said to be younger than a fossil below it and older than a fossil above it, and so forth.

Things changed with **radiometric dating**. This is a way to measure proportions of a daughter isotope and the parent radioisotope of some element trapped in a rock since the time the rock formed. A radioisotope is a form of an element with an unstable nucleus (Section 2.2). Its atoms lose energy and subatomic particles—they decay—until they reach a more stable form.

We cannot predict the exact instant of one atom's decay, but a predictable number of a radioisotope's atoms decay in a characteristic time span. Like the ticking of a perfect clock, the rate of decay for each isotope is constant. Changes in pressure, temperature, or chemical state do not alter it. The time it takes for half of a quantity of a radioisotope's atoms to decay is its **half-life** (Figure 17.12*a*).

For instance, uranium 238's half-life is 4.5 billion years. It decays into thorium 234, which then decays into something else, and so on through intermediate isotopes to lead, the final, stable daughter element for this series. By measuring the uranium 238/lead ratio in the oldest known rocks, geologists figured out that Earth formed more than 4.6 billion years ago.

Radiometric dating has an error factor of less than 10 percent. Recent fossils still hold some carbon and can be dated on the basis of their carbon 14/carbon 12 ratio, as in Figure 17.12*b–d*. Older fossils are dated on the basis of isotope ratios in volcanic rocks or ashes buried with them in the same sedimentary layer.

PLACING FOSSILS IN GEOLOGIC TIME

Early geologists carefully counted backward through layers of sedimentary rock, then used their counts to construct a chronology of Earth history, or a **geologic time scale** (Figure 17.13). By comparing evidence from around the world, they found four abrupt transitions in fossil sequences and used them as boundaries for four great intervals. They named the first interval the Proterozoic, to indicate that it predates fossils of early animals. They named different intervals the Paleozoic,

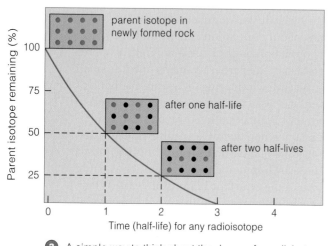

a A simple way to think about the decay of a radioisotope to a more stable form, as plotted against time.

Figure 17.12 *Animated!* (**a**) The decay of radioisotopes at a fixed rate to more stable forms. The half-life of each kind of radioisotope is the time it takes for 50 percent of a given sample to decay. After two half-lives, 75 percent of the sample has decayed, and so on.

(**b–d**) Radiometric dating of a fossil. Carbon 14 (^{14}C) forms in the atmosphere. There, it combines with free oxygen, the result being carbon dioxide. Along with far greater quantities of its more stable isotopes, trace amounts of carbon 14 enter food webs by way of photosynthesis. All organisms incorporate carbon into body tissues.

b Long ago, trace amounts of ^{14}C and a lot more ^{12}C were incorporated into the tissues of a living mollusk. Those atoms of carbon were part of organic compounds making up tissues of its prey. As long as that mollusk lived, the proportion of ^{14}C to ^{12}C in its tissues remained the same.

c When the mollusk died, it stopped gaining carbon. Over time, the proportion of ^{14}C to ^{12}C in its remains fell because of the radioactive decay of ^{14}C. Half of the ^{14}C had decayed in 5,370 years, half of what was left was gone after another 5,370 years, and so on.

d Fossil hunters find the fossil. They measure its ^{14}C/^{12}C ratio to determine half-life reductions since death. In this example, the ratio turns out to be one-eighth of the ^{14}C/^{12}C ratio in living organisms. Thus the mollusk lived about 16,000 years ago.

Eon	Era	Period	Epoch	Millions of Years Ago	Major Geologic and Biological Events That Occurred Millions of Years Ago (mya)
PHANEROZOIC	CENOZOIC	QUATERNARY	Recent	—0.01—	1.8 mya to present. Major glaciations. Modern humans evolve. The most recent *extinction crisis* is under way.
			Pleistocene	—1.8—	
		TERTIARY	Pliocene	—5.3—	65–1.8 mya. Major crustal movements, collisions, mountain building. Tropics, subtropics extend poleward. When climate cools, dry woodlands, grasslands emerge. *Adaptive radiations* of flowering plants, insects, birds, mammals.
			Miocene	—22.8—	
			Oligocene	—33.7—	
			Eocene	—55.5—	65 mya. Asteroid impact; *mass extinction* of all dinosaurs and many marine organisms.
			Paleocene	—65—	
	MESOZOIC	CRETACEOUS	Late	—99—	99–65 mya. Pangea breakup continues, inland seas form. Adaptive radiations of marine invertebrates, fishes, insects, and dinosaurs. Origin of angiosperms (flowering plants).
			Early	—145—	145–99 mya. Pangea starts to break up. Marine communities flourish. *Adaptive radiations* of dinosaurs.
		JURASSIC		—213—	145 mya. Asteroid impact? Mass extinction of many species in seas, some on land. Mammals, some dinosaurs survive.
		TRIASSIC		—248—	248–213 mya. *Adaptive radiations* of marine invertebrates, fishes, dinosaurs. Gymnosperms dominate land plants. Origin of mammals.
	PALEOZOIC	PERMIAN		—286—	248 mya. *Mass extinction*. Ninety percent of all known families lost. 286–248 mya. Supercontinent Pangea and world ocean form. On land, *adaptive radiations* of reptiles and gymnosperms.
		CARBONIFEROUS		—360—	360–286 mya. Recurring ice ages. On land, *adaptive radiations* of insects, amphibians. Spore-bearing plants dominate; cone-bearing gymnosperms present. Origin of reptiles.
		DEVONIAN		—410—	360 mya. *Mass extinction* of many marine invertebrates, most fishes. 410–360 mya. Major crustal movements. Ice ages. *Mass extinction* of many marine species. Vast swamps form. Origin of vascular plants. *Adaptive radiation* of fishes continues. Origin of amphibians.
		SILURIAN		—440—	440–410 mya. Major crustal movements. *Adaptive radiations* of marine invertebrates, early fishes.
		ORDOVICIAN		—505—	505–440 mya. All land masses near equator. Simple marine communities flourish until origin of animals with hard parts.
		CAMBRIAN		—544—	544–505 mya. Supercontinent breaks up. Ice age. *Mass extinction*.
PROTEROZOIC				—2,500—	2,500–544 mya. Oxygen accumulates in atmosphere. Origin of aerobic metabolism. Origin of eukaryotic cells. Divergences lead to eukaryotic cells, then protists, fungi, plants, animals.
ARCHEAN AND EARLIER					3,800–2,500 mya. Origin of photosynthetic prokaryotic cells. 4,600-3,800 mya. Origin of Earth's crust, first atmosphere, first seas. Chemical, molecular evolution leads to origin of life (from proto-cells to anaerobic prokaryotic cells).

Figure 17.13 *Animated!* Geologic time scale. *Red* boundaries mark times of the greatest mass extinctions. If these spans were to the same scale, the Archean and Proterozoic portions would extend downward, spill off the page, and spill across the room. Compare Figure 17.14.

Mesozoic, and lastly the "modern" era, the Cenozoic. Researchers now correlate the geologic time scale with **macroevolution**, or major patterns, trends, and rates of change among lineages. Also, they have subdivided the Proterozoic into finer intervals because it was far more immense than early researchers suspected. Life originated in one of those intervals, the Archean eon.

The geologic time scale now has absolute dates assigned to its boundaries, based on radiometric dating methods. The time scale has been correlated with macroevolution: major patterns, trends, and rates of change among lineages.

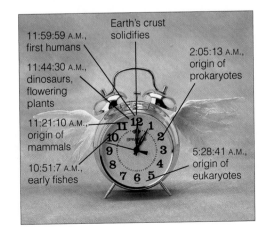

Figure 17.14 How time flies—a geologic time clock. Think of the spans as minutes on a clock that runs from midnight to noon. The recent epoch started after the very last 0.1 second before noon. And where does that put you?

17.6 Drifting Continents, Changing Seas

By clinking their hammers against the rocks, early geologists realized that "solid earth" does not stay put. It moves.

When geologists were first starting to map the vertical stacks of sedimentary rock, the theory of uniformity prevailed. They thought that mountain building and erosion had repeatedly altered Earth's surface in the same ways over time. Eventually, however, it became clear that those recurring geologic events were only part of the picture. Like life, the "unchanging" Earth had changed irreversibly.

AN OUTRAGEOUS HYPOTHESIS

For instance, the Atlantic coasts of South America and Africa seemed to "fit" like jigsaw puzzle pieces. Were all continents once part of a bigger one that had split into fragments and drifted apart? One model for the proposed supercontinent—**Pangea**—took into account the world distribution of fossils and existing species. It also took into account glacial deposits, which held clues to ancient climate zones.

Most scientists did not accept the continental drift hypothesis. Continents drifting on their own across

Earth's mantle seemed to be an outrageous idea, and they preferred to think that continents did not move.

Evidence kept piling up. For example, iron-rich rocks are molten when they form, and their bits of iron become oriented toward Earth's magnetic poles. They stay that way after the rocks harden. Yet in North and South America, the tiny iron compasses in rocks that formed 200 million years ago didn't point to the north and south poles. So scientists made a map to fit the north–south alignment of their compasses. Their map put North America and western Europe right next to each other, no Atlantic Ocean between them.

Later, deep-sea probes revealed that the seafloor is spreading away from the mid-oceanic ridges (Figure 17.15). Molten rock that is spewing from a ridge flows sideways in both directions, then it hardens into new crust. The formation of more crust forces older crust into trenches elsewhere in the seafloor. These ridges and trenches are the edges of enormous crustal plates, like pieces of a gargantuan cracked eggshell. They all move at almost imperceptibly slow rates. Over great time spans, land masses end up in different locations.

These findings put continental drift into a broader explanation of crustal movements, now known as the

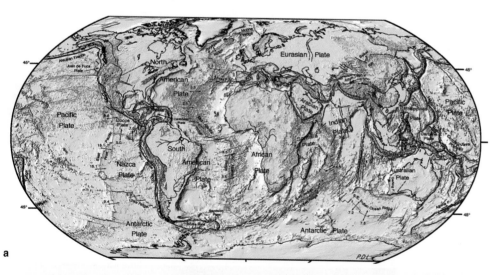

a

b island arc   oceanic crust   oceanic ridge   trench   continental crust

lithosphere (solid layer of mantle)   hot spot   athenosphere (plastic layer of mantle)   subducting plate

Figure 17.15 Forces of geologic change.

(**a**) Present configuration of Earth's crustal plates. These immense, rigid portions of the crust split, drift apart, and collide at almost imperceptible rates. In Appendix VIII, this map is greatly enlarged to show details.

(**b**) Huge plumes of molten material drive the movement. They well up from the interior, then spread laterally under the crust and rupture it at deep, mid-oceanic ridges. The molten material seeps out, cools, and slowly hardens into new seafloor, which displaces plates away from the ridges.

The advancing edge of one plate can plow under an adjacent plate and lift it up. The Cascades, Andes, and other great mountain ranges paralleling the coasts of continents formed this way. When 2004 drew to a close, the Indian Plate lurched violently under the Eurasian Plate and caused huge tsunamis. These earthquake-generated ocean waves traveled 600 miles per hour across the Indian Ocean and killed more than 240,000 people.

Besides these forces, superplumes ruptured the crust at what are now called "hot spots" in the mantle. The Hawaiian Archipelago has been forming this way. Continents also can rupture in their interior. Deep rifting and splitting are happening now in Missouri, at Lake Baikal in Russia, and in eastern Africa.

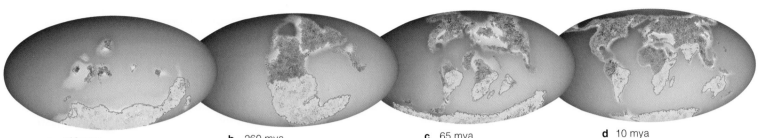

a 420 mya   **b** 260 mya   **c** 65 mya   **d** 10 mya

Figure 17.16 *Animated!* A series of reconstructions of drifting continents. (**a**) The early supercontinent Gondwana (*yellow*). (**b**) Later, all major land masses collided and formed the supercontinent Pangea. (**c**) Positions of fragments that drifted apart after Pangea split apart 65 million years ago, and (**d**) their positions 10 million years ago.

About 260 million years ago, seed ferns and other plants lived nowhere except on the area of Pangea that had once been Gondwana. So did mammal-like reptiles named therapsids. (**e**) Fossilized leaf of one of the seed ferns, *Glossopteris*. (**f**) *Lystrosaurus*, a therapsid about 1 meter (3 feet) long. This tusked herbivore fed on fibrous plants in dry floodplains.

plate tectonics theory. Researchers soon found ways to apply the new theory's predictive power.

For example, the same series of basalt formations, coal seams, and glacial deposits occurs in Africa, India, Australia, and South America. Each of these southern continents has fossils of the seed fern *Glossopteris* and of a therapsid, *Lystrosaurus* (Figure 17.16). This plant's seeds and the therapsid were too heavy to float across the ocean from one continent to the other. Researchers suspected that the organisms had evolved together on **Gondwana**, a supercontinent that preceded Pangea.

Like other land masses in the Southern Hemisphere, Antarctica formed after Gondwana broke up. Someone predicted that fossils of *Glossopteris* and *Lystrosaurus* would be discovered in a series of basalt formations, coal seams, and glacial deposits in Antarctica. In time, explorers did find the series and the fossils. Evidence supported the prediction and plate tectonics theory.

A BIG CONNECTION

Let's take stock. In the remote past, slow movements of Earth's crustal plates put immense land masses on collision courses. Over time, land masses converged and formed supercontinents, which later split at deep rifts and formed new ocean basins. Gondwana drifted south from the tropics, across the south pole, then north until it piled into other land masses. The result was Pangea, a supercontinent that extended from pole to pole with a single world ocean lapping against its coasts. All the while, erosive forces of water and wind resculpted the land surface. Asteroids and meteorites slammed into Earth's crust. The major impacts and their aftermath caused long-term changes in the global temperature, atmosphere, and regional climates.

Such changes on land and in the ocean and atmosphere influenced life's evolution. Imagine early life in shallow, warm waters along continents. Shorelines vanished as continents collided and wiped out many lineages. Yet, even as old habitats vanished, new ones opened up for survivors—and evolution took off in new directions.

Over the past 3.8 billion years, slow movements in Earth's crust as well as catastrophic events have changed the land, the atmosphere, and the ocean. Those changes have had profound effects on the evolution of life.

17.7 Divergences From a Shared Ancestor

To biologists, remember, evolution simply means heritable changes in lines of descent. Comparisons of the body form and structures of major groups of organisms yield clues to evolutionary trends.

Comparative morphology, again, is the study of body forms and structures of major groups of organisms, such as vertebrates and flowering plants. (The Greek *morpho–* means body form.) Sometimes comparisons reveal similarities in one or more body parts between groups, which may be evidence of a common ancestor. Such body parts are **homologous structures** (*homo–* means the same). Even when different groups use the parts for different functions, the genes for constructing those parts point to shared ancestry.

MORPHOLOGICAL DIVERGENCE

Populations of a species diverge genetically after gene flow ends between them (Chapter 18). In time, some morphological traits that help to define their species commonly diverge, also. Change from the body form of a common ancestor is a major macroevolutionary pattern called **morphological divergence**.

Even if the same body part of two related species became dramatically different, some other aspects of the species may remain alike. A careful look beyond unique modifications may reveal the shared heritage.

For instance, all vertebrates on land descended from the first amphibians. Divergences led to what we call reptiles, then to birds and mammals. We know about "stem reptiles" that probably were ancestral to these groups. Fossilized, five-toed limb bones tell us that the ancestral species crouched low to the ground (Figure 17.17a). Later, descendants diversified into many new habitats on land. A few descendants that had become adapted for walking on land even returned to the seas after environmental conditions changed.

A five-toed limb was evolutionary clay. It became molded into different kinds of limbs having different functions. In lineages that eventually led to penguins and porpoises, it became modified into flippers used in swimming. In the lineage leading to modern horses, it became modified into long, one-toed limbs suitable for running fast. Among moles, it became stubby and useful for burrowing into dirt. Among elephants, it became strong and pillarlike, suitable for supporting a great deal of weight.

The five-toed limb also became modified into the human arm and hand. Later on, a thumb evolved in opposition to the four fingers of the hand. It was the basis of stronger and more precise motions.

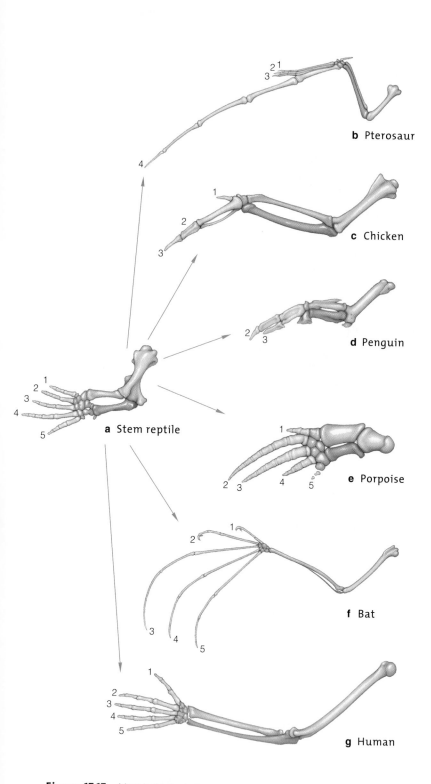

b Pterosaur

c Chicken

d Penguin

a Stem reptile

e Porpoise

f Bat

g Human

Figure 17.17 Morphological divergence among vertebrate forelimbs, starting with bones of a stem reptile (a cotylosaur). Similarities in the number and position of skeletal elements were preserved when diverse forms evolved. Some bones were lost over time (compare the numbers 1 through 5). The drawings are not to the same scale.

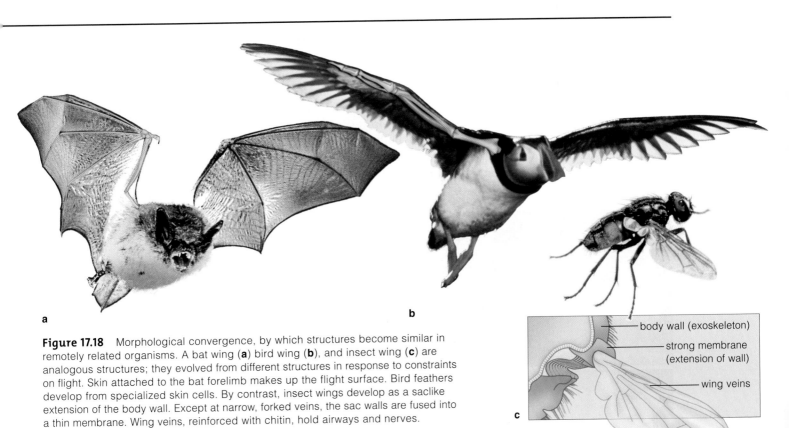

Figure 17.18 Morphological convergence, by which structures become similar in remotely related organisms. A bat wing (**a**) bird wing (**b**), and insect wing (**c**) are analogous structures; they evolved from different structures in response to constraints on flight. Skin attached to the bat forelimb makes up the flight surface. Bird feathers develop from specialized skin cells. By contrast, insect wings develop as a saclike extension of the body wall. Except at narrow, forked veins, the sac walls are fused into a thin membrane. Wing veins, reinforced with chitin, hold airways and nerves.

body wall (exoskeleton)
strong membrane (extension of wall)
wing veins

Even though forelimbs are not the same in size, shape, or function from one group of vertebrates to the next, they clearly are alike in the structure and positioning of their bony elements. They also are alike in the patterns of nerves, blood vessels, and muscles that develop inside them. In addition, comparisons of the early embryos of different vertebrates reveal strong resemblances in patterns of bone development. Such similarities point to a shared ancestor.

MORPHOLOGICAL CONVERGENCE

Body parts with similar form or function in different lineages are not *always* homologous. Sometimes they evolved independently in remote lineages. Parts that differed at first may have evolved in similar ways as organisms became subjected to similar environmental pressures. **Morphological convergence** refers to cases where dissimilar body parts evolved in similar ways in evolutionarily distant lineages.

For instance, you just read about the homologous forelimbs of birds and bats. Bones aside, are bird and bat wings homologous, too? No. The flight surface of birds evolved as a sweep of feathers, all derived from skin. The forelimb structurally supports it. The flight surface for bats is a thin membrane, an extension of the skin itself. The bat wing is attached to reinforcing bony elements inside the forelimb (Figure 17.18*a*,*b*).

The insect wing, too, resembles bird and bat wings in its function—flight. Is it homologous with them? No. This wing develops as an extension of an outer body wall reinforced with chitin. No underlying bony elements support it (Figure 17.18*c*).

The differences between bat, bird, and insect wings are evidence that each of these animal groups adapted independently to the same physical constraints that govern how a wing can function in the environment. The wings of all three are **analogous structures**. They are not modifications of comparable body parts in different lineages. They are three different responses of different body parts to similar challenges. The Greek *analogos* means similar to one another.

With morphological divergence, comparable body parts became modified in different ways in different lines of descent from a common ancestor.

Such divergences resulted in homologous structures. Even if these body parts differ in size, shape, or function, they have an underlying similarity because of shared ancestry.

With morphological convergence, dissimilar body parts became similar in lineages that are not closely related. Such body parts are analogous structures. They converged in form only as an outcome of similar pressures.

17.8 Changes in Patterns of Development

LINKS TO
SECTIONS
14.5, 15.2, 15.3

*Comparing the patterns of embryonic development
often yields evidence of evolutionary relationships.*

Multicelled embryos of plants and animals develop step-by-step from a fertilized egg, and there are built-in constraints on how the body plan develops. Most mutations and changes in chromosomes tend to be

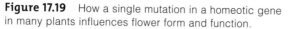

Figure 17.19 How a single mutation in a homeotic gene in many plants influences flower form and function.

(**a**) Normal flower of field mustard (*Brassica oleracea*). A mutation in the *Apetala1* gene causes a badly distorted flower (**b**) to form. (**c**) Wild-type flower of common wall cress (*Arabidopsis thaliana*). Mutation of the *Apetala1* gene in this plant causes flowers with no petals to form (**d**).

selected against, because most mutations disrupt the inherited developmental program. Even so, a mutation with neutral or beneficial effects can move a lineage past one of the constraints.

Master gene mutations can do this. Recall, from Sections 15.2 and 15.3, that homeotic genes guide the formation of tissues and organs in orderly patterns. A mutation in a homeotic gene can disrupt the patterns, sometimes drastically. Such disruptions typically lead to huge problems, but once in a while an altered body plan proves to be advantageous.

You already saw some examples of how homeotic genes guide when and how flowers form. To reinforce the point, here is another example: A single mutation in the homeotic gene known as *Apetala1* causes male floral reproductive structures (anthers) to form where petals are supposed to form in the flowers of field mustard, *Brassica oleracea* (Figure 17.19). At least in the laboratory, such abundantly anthered mutants are exceptionally fertile plants.

Another example: The embryos of some vertebrate lineages are alike in the early stages of development. Their tissues form in similar ways when cells divide, differentiate, and interact. The gut and heart, bones, skeletal muscles, and other parts start to grow and develop in orderly spatial patterns that are strikingly similar among these groups.

How, then, did adults of different groups get to be so different? We can expect that heritable changes in the onset, rate, or completion of developmental steps led to many of the differences. Some changes could have increased or decreased relative sizes of tissues and organs. Some changes could have ended growth during a juvenile stage; the adults of certain species still do have some juvenile features.

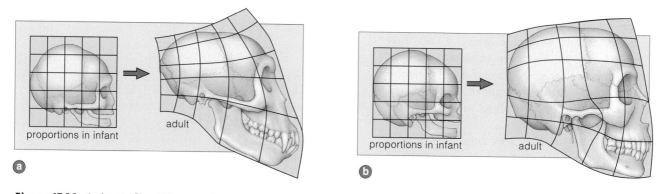

Figure 17.20 *Animated!* Differences between two primates, a possible outcome of mutations that changed the timing of steps in the body's development. The skulls are depicted as paintings on a rubber sheet divided into a grid. Stretching both sheets deforms the grid in a way that corresponds to differences in growth patterns between these primates. (**a**) Proportional changes in chimpanzee skull, and (**b**) human skull.

Figure 17.21 How many legs? Mutations in ancestral genes may help explain why animals differ in the number of legs and other appendages. *Dll* is a homeotic gene that initiates limb development, and other genes control its expression. Fluorescent *green* reveals *Dll* expression in (**a**) a velvet walking worm, and (**b**) a sea star; and (**c**) *blue* in a mouse embryo's foot. (**d**) Cambrian legs.

Modifications in genes that influence growth rates might have caused the major proportional differences between chimpanzee and human skull bones (Figure 17.20). For humans, the facial bones and skull bones around the brain increase in size at fairly consistent rates, from infant to adult. The rate of growth is faster for chimp facial bones, so the proportions of an infant skull and an adult skull differ significantly.

In addition, did transposons cause some variation among lineages? As Section 14.5 explains, these short DNA segments spontaneously and repeatedly slip into new places in genomes. Depending on where they end up, they can have powerful effects on gene expression.

Only primates carry the 300 base-pair transposons called *Alu* elements, and they have done so for at least 30 million years. *Alu* elements are noncoding, and yet they have sequences that resemble intron–exon splice signals. When inserted into coding regions of DNA, they promote duplications of themselves. *Alu* elements have had big effects on the expression of genes for estrogen, thyroid hormones, and other proteins that control growth and development. Were they pivotal in primate evolution? Possibly.

About 1 million *Alu* elements make up 10 percent of the human genome. The most recent shared ancestors of chimpanzees and humans diverged between 6 and 4 million years ago. More than 98 percent of human DNA is identical with chimpanzee DNA. Something in the remaining 2 percent accounts for the differences. Uniquely positioned *Alu* elements may be one factor.

As a final example, body appendages as different as crab legs, beetle legs, butterfly wings, sea star arms, fish fins, and mouse feet all start out as buds of tissue from the body surface. A bud will form wherever the *Dll* gene product is expressed. The product, a protein, is a signal for clusters of dividing embryonic cells to "stick out from the body" in an expected pattern, as in Figure 17.21. Normally, *Hox* genes help sculpt the body's details by suppressing expression of *Dll* where appendages are not supposed to form.

The *Dll* gene is expressed in similar ways across many phyla, which is strong evidence for its ancient origin. Indeed, in some Cambrian fossils, it looks like it was not suppressed at all (Figure 17.21*d*). Probably layers of gene controls evolved over time, resulting in the variable numbers and locations of appendages we observe today among all complex animals, including humans and other vertebrates.

Similarities and differences in patterns of development are often clues to shared ancestry, especially for the embryos of plants and animals.

Heritable changes that alter key steps in a developmental program may be enough to bring about major differences in the adult forms of related lineages.

Mutations in master genes and insertions of transposons into coding regions of DNA could have been enough to launch body plans in new evolutionary directions.

17.9 Clues in DNA, RNA, and Proteins

LINKS TO
SECTIONS
3.6, 14.4, 16.3, 16.4

All species are a mix of ancestral and novel traits, including biochemical ones. The kinds and numbers of traits they do or do not share are clues to relationships.

Each species has its own DNA base sequence, which encodes instructions for making RNAs and proteins (Sections 3.6 and 14.4). We can expect genes to mutate in any line of descent. The more recently two lineages have diverged, the less time each will have had to accumulate unique mutations. That is why RNA and proteins of closely related species are more similar than those of more distantly related ones.

Identifying biochemical similarities and differences among species is now rapid, thanks to the methods of automated gene sequencing (Section 16.3). Extensive sequence data for numerous genomes and proteins are compiled in internationally accessible databases. With such data, we know (for example) that 31 percent of the 6,000 genes of yeast cells have counterparts in our genome. So do 40 percent of the 19,023 roundworm genes and 50 percent of the fruit fly genes.

PROTEIN COMPARISONS

Similarities in amino acids can be used to decipher connections between species and to study why certain proteins are highly conserved. When two species have many proteins with similar or identical amino acid sequences, they probably are closely related. When sequences differ considerably, many mutations have been built in, which indicates that a long time passed since the two species shared a common ancestor.

A few essential genes have evolved very little; they are highly *conserved* across diverse species. One such gene encodes cytochrome *c*. Species that range from aerobic bacteria to humans must make this protein component of electron transfer chains. In humans, its primary structure consists of only 104 amino acids. Figure 17.22 shows the striking similarity between the entire amino acid sequences for cytochrome *c* from a yeast, a plant, and an animal. And think about this: The *entire* amino acid sequence of human cytochrome

c is identical with that of chimpanzee cytochrome *c*. It differs by merely 1 amino acid in rhesus monkeys, 18 in chickens, 19 in turtles, and 56 in yeasts. With this biochemical information in hand, would you predict that humans are more closely related to chimpanzees or to rhesus monkeys? Chickens or yeast?

NUCLEIC ACID COMPARISONS

Mutations that cause differences between species are dispersed through DNA's nucleotide sequences. Some regions of those sequences are unique to each lineage. Researchers use the regions to estimate evolutionary distances. They isolate and then compare DNA from the nuclei, mitochondria, and chloroplasts of different species (Figure 17.23). They also compare DNA regions that encode ribosomal RNA (rRNA).

Nucleic acid hybridization refers to base-pairing between DNA strands from different sources (Section 16.2). In the hybrid molecule, more hydrogen bonds form between matched bases than mismatched bases. Strands with more matches associate strongly with each other. The amount of heat required to separate two strands of a hybrid can be used as a comparative measure of their similarity. Why? It takes more heat to disrupt hybrid DNA of closely related species.

Evolutionary distances are still being measured by DNA–DNA hybridizations, although automated gene sequencing now gives faster, more quantifiable results. Remember DNA fingerprinting (Section 16.4)? DNA restriction fragments from different species can be compared after gel electrophoresis has separated them.

Mitochondrial DNA (mtDNA), which mutates fast, is used to compare individuals of eukaryotic species. In sexually reproducing species, it is inherited intact from one parent—usually the mother. Thus, changes between maternally related individuals probably were caused by mutations, not by genetic recombinations.

Computer programs quickly compare collections of DNA sequencing data. Comparative analyses either reinforce or invite modification of evolutionary trees based on morphological findings and the fossil record.

$^+$NH$_3$-gly asp val glu lys gly lys lys ile phe ile met lys cys ser gln cys his thr val glu lys gly gly lys his lys thr gly pro asn leu his gly leu phe gly arg lys thr gly gln ala pro gly tyr

$^+$NH$_3$-ala ser phe ser glu ala pro pro gly asn pro asp ala gly ala lys ile phe lys thr lys cys ala gln cys his thr val asp ala gly ala gly his lys gln gly pro asn leu his gly leu phe gly arg gln ser gly thr thr ala gly tyr

$^+$NH$_3$-thr glu phe lys ala gly ser ala lys lys gly ala thr leu phe lys thr arg cys leu gln cys his thr val glu lys gly gly pro his lys val gly pro asn leu his gly ile phe gly arg his ser gly gln ala glu gly tyr

Figure 17.22 *Animated!* Comparison of the primary structure of cytochrome *c* from a yeast (*top row*), wheat plant (*middle*), and primate (*bottom*). *Gold* highlights parts of the amino acid sequence that are identical in all three. The probability that such a pronounced molecular resemblance resulted by chance is extremely low. Cytochrome *c* is a vital component of electron transfer chains in cells. Its amino acid sequence has been highly conserved even in these three evolutionarily distant lineages.

RACCOON   RED PANDA   GIANT PANDA

SPECTACLED   SLOTH SUN   BLACK   POLAR BROWN
BEAR   BEAR   BEAR   BEAR   BEAR   BEAR

DIVERGENCE
15–20 million years ago

DIVERGENCE
approximately
40 million years ago

Figure 17.23 Example of biochemical comparisons that help construct and refine evolutionary trees. This tree for red pandas, giant pandas, and brown bears was confirmed using mitochondrial DNA sequence comparisons. Recently, researchers analyzed sequence data from three mitochondrial DNAs and one intron. They also studied more potential relatives. They found that red pandas may be more closely related to skunks, weasels, and otters than to raccoons.

However, gene transfers between species can slant the results. For example, after hybridization between two different species of plants, hybrid offspring may cross back to either parental species, thus transferring genes from one species into the other. Gene swapping is rampant among prokaryotic species.

MOLECULAR CLOCKS

Some researchers estimate the timing of divergence by comparing the numbers of neutral mutations in genes that have been highly conserved in different lineages. Because the mutations have little or no effect on the individual's survival or reproduction, we can expect that neutral mutations have accumulated in conserved genes at a fairly constant rate.

The accumulation of neutral mutations to the DNA of a lineage has been likened to the predictable ticks of a **molecular clock**. Turn the hands of such a clock back, so that the total number of ticks unwinds down through past geologic intervals. Where does the last tick stop? Theoretically, it stops close to the time when molecular, ecological, and geographic events put the lineage on its unique evolutionary road.

How are molecular clocks calibrated? The number of differences in DNA base sequences or amino acid sequences between species can be plotted against a series of branch points that researchers have inferred from the fossil record. Section 19.6 will show how this is done. Such diagrams may reflect relative times of divergences among species, phyla, and other groups.

Biochemical similarity is greatest among the most closely related species and smallest among the most remote.

ala asn lys asn lys gly ile ile trp gly glu asp thr leu met glu tyr leu glu asn pro lys lys tyr ile pro gly thr lys met ile phe val gly ile lys lys lys glu glu arg ala asp leu ile ala tyr leu lys lys ala thr asn glu-COO⁻

ala asn lys asn lys ala val glu trp glu glu asn thr leu tyr asp tyr leu leu asn pro lys lys tyr ile pro gly thr lys met val phe pro gly leu lys lys pro gln asp arg ala asp leu ile ala tyr leu lys lys ala thr ser ser-COO⁻

ala asn ile lys lys asn val leu trp asp glu asn asn met ser glu tyr leu thr asn pro lys lys tyr ile pro gly thr lys met ala phe gly gly leu lys lys glu lys asp arg asn asp leu ile thr tyr leu lys lys ala cys glu-COO⁻

Summary

Section 17.1 Awareness of evolution, or heritable changes in lines of descent, emerged long ago through biogeography, geology, and comparative morphology.

Section 17.2 Prevailing cultural belief systems influence our interpretation of natural events. In the nineteenth century, naturalists worked to reconcile traditional belief systems with a growing body of physical evidence in support of evolution.

Biology⊜Now
Read the InfoTrac article "Typecasting a Bit Part," Stephen J. Gould, The Sciences, *March 2000.*

Section 17.3 Charles Darwin and Alfred Wallace proposed a novel theory that natural selection can bring about evolution. Here are the theory's main premises:

Any population tends to grow in size until resources dwindle. Its individuals must compete more for them.

Individuals with forms of traits that make them more competitive tend to produce more offspring.

Over the generations, more competitive forms of traits that have a heritable basis increase in frequency in the population relative to less competitive forms.

Thus nature "selects" variations in traits that are more effective at helping individuals survive and reproduce; such traits are more adaptive in a given environment.

Biology⊜Now
Read the InfoTrac article "What Darwin's Finches Can Teach Us About the Evolutionary Origin and Regulation of Biodiversity," B. Rosemary Grant and Peter Grant, Bioscience, *March 2003.*

Section 17.4 Fossils are stone-hard evidence of life in the distant past. Many fossils are embedded in stacked layers of sedimentary rock. Generally, the oldest layers are near the bottom of the sequence and more recently deposited layers are on top. Although the fossil record is not complete, it reveals much about life in the past.

Biology⊜Now
Learn more about fossil formation and the geologic time scale with the animations on BiologyNow.

Section 17.5 Fossil sequences were the basis for the first geologic time scale, which used abrupt transitions in the fossil record as boundaries for different eras. Through radiometric dating of fossils, absolute dates have since been assigned to the scale. This dating method has a relatively small margin of error.

Biology⊜Now
Learn more about the half-life of a radioisotope's atoms with the animated interaction on BiologyNow.

Section 17.6 The global distribution of land masses and fossils, magnetic patterns in volcanic rocks, and evidence of seafloor spreading from mid-oceanic ridges support the plate tectonic theory. According to this theory, slow movements of Earth's crustal plates raft land masses to new positions. Such movements had profound impacts on the directions of life's evolution.

Biology⊜Now
Learn more about drifting continents with the interaction on BiologyNow.

Section 17.7 Comparative morphology reveals evidence of evolution. Homologous structures are one of the clues. These body parts recur in different lineages, but they became modified in different ways after the lineages diverged from a shared ancestor. These parts are not the same as analogous structures. Such body parts did not start out being alike; they became similar in independent lineages as a response to similar kinds of environmental pressures.

Section 17.8 Similarities in patterns and structures of embryonic development suggest common ancestry. Even minor genetic changes can alter the onset, rate, and completion time of developmental stages. They can have major impact on the adult form.

Biology⊜Now
Explore proportional changes in embryonic development with the animated interaction on BiologyNow.

Section 17.9 We are now clarifying evolutionary relationships through comparisons of DNA, RNA, and proteins between different species. The investigative methods include nucleic acid hybridization and, more recently, automated gene sequencing and DNA fingerprinting, as explained in Chapter 16.

Some researchers estimate the times of divergences from ancestral lineages by comparing the number of neutral mutations in highly conserved genes. Such mutations alter the base sequence in DNA, but flexibility built into the genetic code keeps the change from altering the amino acid sequence of the specified protein. They may accumulate in the DNA of a species at a constant rate, like ticks of a molecular clock.

Biology⊜Now
Learn more about amino acid comparisons with the interaction on BiologyNow.

Self-Quiz
Answers in Appendix II

1. Biogeographers deal with _____ .
   a. patterns in which continents drift
   b. patterns in the world distribution of species
   c. mainland and island biodiversity
   d. both b and c are correct
   e. all are correct

2. _____ have influenced the fossil record.
   a. Sedimentation and compaction
   b. Crustal plate movements
   c. Volcanic ash deposition
   d. a through c

3. Life originated in the _____ eon.
   a. Archean   c. Phanerozoic
   b. Proterozoic   d. Cambrian

4. Which of these supercontinents formed first: Pangea or Gondwana?

5. Through _____ , the same body parts became modified differently in different lines of descent from a common ancestor.
   a. morphological convergence
   b. morphological divergence
   c. ancestral analogy
   d. ancestral homology

6. Homologous structures among major groups of organisms may differ in _____ .
   a. size   c. function
   b. shape   d. all of the above

7. By altering steps in the program by which embryos develop, _____ may lead to major differences between adults of related lineages.
   a. automated gene sequencing   c. transposons
   b. homeotic gene mutations   d. b and c

8. Molecular clocks are based on comparisons of _____ mutations in _____ genes.
   a. beneficial; moderately conserved
   b. neutral; moderately conserved
   c. neutral; highly conserved
   d. lethal; highly conserved

9. Match the terms with the most suitable description.
   ____ stratification
   ____ fossils
   ____ homeotic genes
   ____ half-life
   ____ homologous structures
   ____ uniformity
   ____ analogous structures

   a. evidence of life in distant past
   b. theory of repetitive change only in Earth history
   c. body parts of similar size, shape, or function in different lineages with shared ancestor
   d. insect wing and bird wing
   e. time it takes to decay half of a quantity of a radioisotope's atoms into something else
   f. big role in development
   g. layers of sedimentary rock

Additional questions are available on **Biology** ⑤ **Now**™

Critical Thinking

1. At one time, all species were ranked in a great Chain of Being, from lowly forms to Man, then to spiritual beings. Even some modern scientists still call the traits of species "rudimentary" or "advanced." Does the theory of natural selection imply that all species become more complex over time? Why or why not?

2. At the end of your backbone is a coccyx, a few small, fused-together bones (Figure 17.4). Is the human coccyx a *vestigial* structure—all that is left of the tail of some distant vertebrate ancestors? Or is it the start of a newly evolving structure?

   Formulate a hypothesis, then design a way to test the predictions you make based on the hypothesis.

3. Think about the species living around you. From the evolutionary perspective, which ones are most successful in terms of sheer numbers, geographic distribution, and how long their lineage has endured on Earth?

4. Comparative biochemistry can help us estimate evolutionary relationship and approximate times for divergences from ancestral stocks. Base sequence

Figure 17.24 Reconstruction of *Rodhocetus* based on fossils discovered in Pakistan. This cetacean lived 47 million years ago, along the shores of the Tethys Sea. Its ankle bones are strong evidence of a close evolutionary link between early whales and hoofed land mammals. Compare Figure 17.4.

comparisons and amino acid comparisons yield good estimates. Reflect on the genetic code (Section 14.2), then suggest why it may be a useful measure of mutations, mutation rates, and biochemical relatedness.

5. For some time, evolutionists accepted that the ancestors of whales were four-legged animals that walked on land, then took up life in water about 55 million years ago. Fossils reveal gradual changes in skeletal features that made an aquatic life possible. But which four-legged mammals were the ancestors?

   The answer recently came from Philip Gingerich and Iyad Zalmout. While digging for fossils in Pakistan, they found remains of early aquatic whales. Intact, sheeplike ankle bones *and* archaic whale skull bones were in the same fossilized skeletons (Figures 17.4 and 17.24).

   Ankle bones of fossilized, early whales from Pakistan have the same form as the unique ankle bones of extinct and modern artiodactyls. Modern cetaceans no longer have even a remnant of an ankle bone. Here is evidence of an evolutionary link between certain aquatic mammals and a major group of mammals on land.

   The radiometrically dated fossils are real. Yet no one was around to witness this transitional time. Because we did not see ancient life evolving, do you think there can be absolute proof of evolution in the distant past? Is the circumstantial evidence of fossil morphology enough to convince you that the theory is not wrong?

18 MICROEVOLUTIONARY PROCESSES

Rise of the Super Rats

Slipping in and out of the pages of human history are rats—*Rattus*—the most notorious of mammalian pests. One kind of rat or another has distributed pathogens and parasites that cause bubonic plague, typhus, and other deadly infectious diseases (Figure 18.1). The death toll from fleas that bit infected rats and then bit people has exceeded the dying in all wars combined.

The rats themselves are far more successful. By one estimate, there is one rat for every person in urban and suburban centers of the United States. Besides spreading diseases, rats chew their way through walls and wires of homes and cities. In any given year, they cause economic losses approaching 19 billion dollars.

For years, people have been fighting back with traps, ratproof storage facilities, and poisons, including arsenic and cyanide. During the 1950s, they used baits laced with warfarin. This synthetic organic compound interferes with blood clotting. Rats ate the baits. They died within days after bleeding internally or losing blood through cuts or scrapes.

Warfarin was extremely effective. Compared to other rat poisons, it had a lot less impact on harmless species. It quickly became the rodenticide of choice.

In 1958, however, a Scottish researcher reported that warfarin did not work against some rats. Similar reports from other European countries followed. About twenty years later, 10 percent of the urban rats caught in the United States were warfarin resistant. *What happened?* To find out, researchers compared warfarin-resistant rat populations with still-vulnerable rats. They traced the difference to a gene on one of the rat chromosomes.

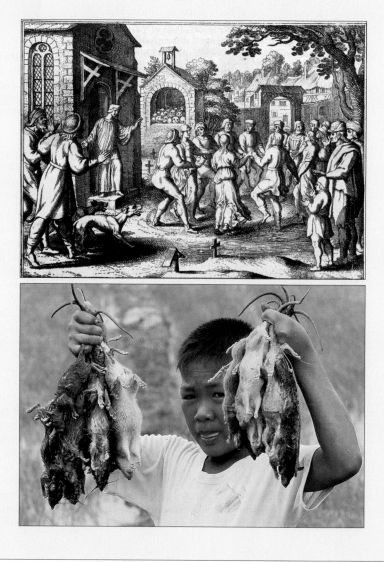

Figure 18.1 *Above*, medieval attempts to deal with a bubonic plague pandemic—the Black Death—that may have killed half the people in Europe alone. Not knowing that the disease agent hitches rides on rats, Europeans tried to protect themselves by praying and dancing until they dropped. Physicians wore bird masks, such as the mask shown on the facing page. They filled the "beak" with herbs that supposedly purified the air that plague victims had breathed. For the next 300 years, anyone accused of causing an outbreak of the plague, no matter how absurd the evidence, was burned alive.

Below, example of rats in this century. Rats infest 80,000 hectares of the rice fields in the Philippine Islands. They ruin more than 20 percent of the annual crops. Rice is the main food source for people in Southeast Asia.

Today we douse agricultural land and buildings with ever more potent rat poisons. By doing so, we have unwittingly promoted the rise of super rats. Three centuries from now, how will people be viewing *our* actions?

At that gene locus, a dominant allele was common among the warfarin-resistant rat populations but rare among the vulnerable ones. The dominant allele's product actually neutralizes warfarin's effect on blood clotting.

"What happened" was evolution by natural selection. As warfarin started to exert pressure on rat populations, the rat populations changed. The previously rare dominant allele suddenly proved to be adaptive. The lucky rats that inherited the allele survived and produced more offspring. The unlucky ones that inherited the recessive allele had no built-in defense, and they died. Over time, the dominant allele's frequency increased in all rat populations exposed to the poison.

Selection pressures can and often do change. When warfarin resistance increased in rat populations, people stopped using warfarin. Not surprisingly, the frequency of the dominant allele declined. Now the latest worry is the evolution of "super rats," which the newer and even more potent rodenticides cannot seem to kill.

The point is, when you hear someone question whether life evolves, remember this: With respect to life, **evolution** simply means heritable change is occurring in some line of descent. The actual mechanisms that can bring about such change are the focus of this chapter. Later chapters highlight how these mechanisms have contributed to the evolution of new species.

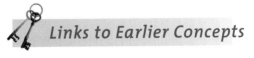
Watch the video online!

How Would You Vote?

Antibiotic-resistant strains of bacteria are becoming dangerously pervasive. Standard animal husbandry practice includes continually dosing healthy animals with antibiotics—the same antibiotics prescribed for people. Should this practice stop? See BiologyNow for details, then vote online.

Key Concepts

WHAT IS MICROEVOLUTION?

Individuals of all natural populations share a gene pool but differ in which alleles they inherit. As a result, they show variations in phenotypes.

An individual does not evolve. Rather, a *population* evolves, which means its shared pool of alleles is changing. Over the generations, any allele may increase in its frequency among individuals, or it may become rare or lost.

Microevolution refers to changes in allele frequencies as an outcome of mutation, natural selection, genetic drift, and gene flow.   Sections 18.1, 18.2

NATURAL SELECTION

Natural selection is the outcome of variation in heritable traits that influence which individuals of a population survive and reproduce in each generation. Selective agents operating in the environment can stabilize, disrupt, or cause directional shifts in the range of variation.   Sections 18.3–18.6

GENETIC DRIFT

Sometimes chance events bring about random changes in allele frequencies over time. The magnitude of this genetic drift is greatest in small populations, where it can lead to a loss of genetic diversity.   Section 18.7

GENE FLOW

Gene flow is the physical movement of alleles into and out of a population. It tends to oppose the effects of mutation, natural selection, and genetic drift; it keeps populations of a species similar to one another.   Section 18.8

ADAPTATION AND THE ENVIRONMENT

An evolutionary adaptation is a heritable aspect of form, function, behavior, or development that contributes to the fit between an individual and its environment. The challenge is to identify environmental conditions to which a given trait is presumably adapted.   Section 18.9

Links to Earlier Concepts

Before starting this chapter, review the premises of the theory of natural selection as outlined in Sections 1.4 and 17.3 as well as the definitions of basic terms in genetics (11.1).

You will be drawing upon your knowledge of mutation (14.5) and the chromosomal basis of inheritance (12.5 especially). We urge you to scan earlier sections on causes of continuous variation in populations (11.5) and on how the environment can modify gene expression (11.6).

18.1 Individuals Don't Evolve, Populations Do

LINKS TO
SECTIONS 11.4,
11.6, 11.7, 14.5, 17.9

As Charles Darwin and Alfred Wallace perceived long ago, individuals don't evolve; populations do. Each **population** *is a group of individuals of the same species in a specified area. To understand how it evolves, start with variation in the traits that characterize it.*

VARIATION IN POPULATIONS

The individuals of a population share certain features. Pigeons have two feathered wings, three toes forward, one toe back, and so on. These are *morphological* traits (*morpho–*, form). The individuals share *physiological* traits, including metabolic activities that help the body function in the environment. They respond the same way to certain basic stimuli, as when babies imitate adult facial expressions. These are *behavioral* traits.

However, the individuals of a population also show variation in the details of the shared traits. You know this just by thinking about the variations in the color and patterning of pigeon feathers or butterfly wings or snail shells. Figure 18.2 only hints at the range of variations in human skin color and distribution, color, texture, and amount of hair. Almost every trait of any species may vary, but variation can be dramatic among sexual reproducers.

For sexually reproducing species, at least, we may define the population as a group of individuals that are interbreeding, that are reproductively isolated from other species, and that produce fertile offspring. The offspring typically have two parents, and they have mixes of the parental forms of traits.

Many traits show *qualitative* differences; they have two or more distinct forms, or morphs. Remember the purple or white pea plant flowers that Gregor Mendel studied? The persistence of two forms of a trait in a population is a case of **dimorphism**. The persistence of three or more forms is **polymorphism**. In addition, for many traits, the individuals of a population show *quantitative* differences, a range of incrementally small variations in a specified trait (Section 11.7).

THE GENE POOL

Genes encode information about heritable traits. The individuals of a population inherit the same number and kind of genes (except for a pair of nonidentical sex chromosomes). Together, they and their offspring represent a **gene pool**—a pool of genetic resources.

For sexual reproducers, nearly all genes available in the shared pool have two or more slightly different molecular forms, or **alleles**. Any individual might or might not inherit identical alleles for any trait. This is the source of variations in *phenotype*, or differences in the details of shared traits. Whether you have black, brown, red, or blond hair depends upon which alleles you inherited from your two parents.

You read about the inheritance of alleles in earlier chapters. Here we summarize the key events involved:

Gene mutation

Crossing over at meiosis I (puts novel combinations of alleles in chromosomes)

Independent assortment at meiosis I (puts mixes of maternal and paternal chromosomes in gametes)

Fertilization (combines alleles from two parents)

Change in chromosome number or structure (loss, duplication, or repositioning of genes)

Only mutation creates new alleles. The other events shuffle existing alleles into different combinations, but what a shuffle! Each gamete gets one of many millions of possible combinations of maternal and paternal chromosomes that may or may not be identical at each locus. Unless you are an identical twin, it is extremely unlikely that another person with your precise genetic makeup has ever lived or ever will.

One other point about the nature of the gene pool: Offspring do not inherit phenotypes; they inherit *genes*. Section 11.6 describes how environmental conditions, too, bring about variation in the range of phenotypes, but the effects last no longer than the individual.

MUTATION REVISITED

Being the original source of new alleles, mutations are worth another look—this time in the context of their impact on populations. Usually, gene mutations that have beneficial or neutral effects are transmitted to a new generation. We cannot predict precisely when or in which individual a particular gene will mutate. We *can* predict rates of mutation, or the probability that a mutation will happen in a specified interval (Section 14.5). For instance, one estimated rate for mammalian genomes is 2.2^{-9} mutations per base pair per year.

Many mutations give rise to structural, functional, or behavioral alterations that reduce an individual's chances of surviving and reproducing. Even a single biochemical change may be devastating. For instance, skin, bones, tendons, lungs, blood vessels, and many other vertebrate organs incorporate collagen. Thus, when the collagen gene has mutated, drastic problems may ripple all through the body. Compare Section 11.4.

Any mutation that results in severe disruptions in phenotype usually causes death. It is a **lethal mutation**.

A **neutral mutation**, recall, alters the base sequence in DNA, but the change has no discernible effect on survival or reproduction (Section 17.9). It neither helps nor hurts the individual. For instance, if you carry a mutant gene that keeps your earlobes attached to the head instead of swinging freely, this in itself should not stop you from surviving and reproducing as well as anybody else. Therefore, natural selection does not affect the frequency of the trait in the population.

Every so often, a mutation proves useful. A mutant gene product that affects growth might make a corn plant grow larger or faster and thereby give it the best access to sunlight and nutrients. A neutral mutation might prove helpful if conditions in the environment change. Even if a mutant gene bestows only a slight advantage, natural selection or a chance event might favor its preservation in DNA and its transmission to the next generation.

Mutations are rare, so they usually have little or no immediate effect on a population's allele frequencies. But they have been slipping into genomes for billions of years. Cumulatively, they have served as reservoirs for change, for biodiversity that is staggering in its breadth. Think of it. The reason you don't look like a bacterium or an avocado or earthworm or even your neighbors down the street began with mutations that arose at different times, in different lines of descent.

STABILITY AND CHANGE IN ALLELE FREQUENCIES

Researchers typically track **allele frequencies,** or the relative abundances of alleles of a given gene among all individuals of a population. They can start from a theoretical reference point, **genetic equilibrium,** when a population is *not* evolving with respect to that locus.

Genetic equilibrium can only occur if five conditions are being met: There is no mutation, the population is infinitely large, the population is isolated from other populations of the same species, individuals mate at random, and all individuals survive and produce the same number of offspring.

If you are interested, the following section offers a closer look at the nature of genetic equilibrium—the point at which a population is not evolving.

As it happens, genetic equilibrium is exceedingly rare in nature. Why? Mutations are rare but inevitable, and they might throw a wild card in the game of who survives and reproduces. Also, three processes—called *natural selection, genetic drift,* and *gene flow*—can drive populations out of equilibrium. **Microevolution** refers to small-scale changes in allele frequencies that arise as an outcome of mutation, natural selection, genetic drift or gene flow, or some combination of these.

Figure 18.2 A sampling of the phenotypic variation in populations of humans and snails, the outcome of variations in frequencies of alleles.

We partly characterize a natural population or species by morphological, physiological, and behavioral traits, most of which are heritable.

At any gene locus, different alleles give rise to variations in individual phenotypes—to differences in the details of shared structural, functional, and behavioral traits.

The individuals of a population share a pool of genetic resources—that is, a pool of alleles.

Only mutation creates new alleles. Natural selection, genetic drift, and gene flow affect only the frequencies of various alleles at a given gene locus in the population.

Most populations are slowly evolving, which simply means that the frequencies of the alleles for a specified trait are changing from one generation to the next.

18.2 When Is A Population *Not* Evolving?

How do researchers know whether or not a population is evolving? They can start by tracking deviations from the baseline of genetic equilibrium.

The Hardy–Weinberg Formula Early in the twentieth century, Godfrey Hardy (a mathematician) and Wilhelm Weinberg (a physician) independently applied the rules of probability to sexually reproducing populations. Like the geneticists who came after them, they perceived that gene pools can remain stable only when five conditions are being met:

1. There is no mutation.
2. The population is infinitely large.
3. The population is isolated from all other populations of the species (no gene flow).
4. Mating is random.
5. All individuals survive and produce the same number of offspring.

In other words, allele frequencies for any gene in the shared pool will remain stable unless the population is evolving. Hardy and Weinberg developed a simple formula that can be used to track whether a population of any sexually reproducing species is slipping out of that state of genetic equilibrium.

Consider tracking a hypothetical pair of alleles that affect butterfly wing color. A protein pigment is specified by dominant allele *A*. If a butterfly inherits two *AA* alleles, it will have dark-blue wings. If it inherits two recessive alleles *aa*, it will have white wings. If it inherits one of each (*Aa*), the wings will be medium-blue (Figure 18.3).

At genetic equilibrium, the proportions of the wing-color genotypes are

$$p^2(AA) + 2pq(Aa) + q^2(aa) = 1.0$$

where *p* and *q* are the frequencies of alleles *A* and *a*. This is what became known as the *Hardy–Weinberg equilibrium equation*. It defines the frequency of a dominant and a recessive allele for a gene that controls a particular trait in a population.

The frequencies of *A* and *a* must add up to 1.0. To give a specific example, if *A* occupies half of all the loci for this gene in the population, then *a* must occupy the other half (0.5 + 0.5 = 1.0). If *A* occupies 90 percent of all the loci, then *a* must occupy 10 percent (0.9 + 0.1 = 1.0). No matter what the proportions,

$$p + q = 1.0$$

At meiosis, recall, paired alleles segregate and end up in different gametes. So the proportion of gametes having the *A* allele is *p*. The proportion having the *a* allele is *q*. The Punnett square on the next page reveals the genotypes possible in the next generation (*AA*, *Aa*, and *aa*).

490 *AA* butterflies
dark-blue wings

490 *AA* butterflies
dark-blue wings

490 *AA* butterflies
dark-blue wings

420 *Aa* butterflies
medium-blue wings

420 *Aa* butterflies
medium-blue wings

420 *Aa* butterflies
medium-blue wings

90 *aa* butterflies
white wings

90 *aa* butterflies
white wings

90 *aa* butterflies
white wings

Starting Population

Next Generation

Next Generation

Figure 18.3 *Animated!* How to determine whether a population is evolving. The frequencies of wing-color alleles among all individuals in this hypothetical population of morpho butterflies have not changed because all five assumptions upon which the Hardy–Weinberg rule is based are being met.

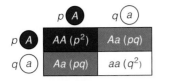

The frequencies add up to 1.0: $p^2 + 2pq + q^2 = 1.0$.

Suppose that the population has 1,000 individuals and that each one produces two gametes:

490 *AA* individuals make 980 *A* gametes
420 *Aa* individuals make 420 *A* and 420 *a* gametes
90 *aa* individuals make 180 *a* gametes

The frequency of alleles *A* and *a* among 2,000 gametes is

$$A = \frac{980 + 420}{2,000 \text{ alleles}} = \frac{1,400}{2,000} = 0.7 = p$$

$$a = \frac{180 + 420}{2,000 \text{ alleles}} = \frac{600}{2,000} = 0.3 = q$$

At fertilization, gametes combine at random and start a new generation. If the population size is still 1,000, you will find 490 *AA*, 420 *Aa*, and 90 *aa* individuals. Because the allele frequencies for dark-blue, medium-blue, and white wings are the same as they were in the original gametes, they will give rise to the same phenotypic frequencies that occurred in the preceding generation.

As long as the assumptions that Hardy and Weinberg identified continue to hold, the pattern will persist. If traits show up in different proportions from one generation to the next, however, then one or more of the five assumptions is not being met. The hunt can begin for one or more of the evolutionary forces driving the change.

Applying the Rule   So how does the Hardy–Weinberg formula work in the real world? For one thing, researchers use it to estimate the frequency of carriers of alleles that cause genetic traits and disorders.

For example, about 1 percent of people of Irish ancestry are affected by *hemochromatosis*. They absorb too much iron from their food. Symptoms of this autosomal recessive disorder include liver problems, fatigue, and arthritis. We can use the number to estimate the frequency of carriers of the recessive allele. If $p^2 = 0.01$, then *p* is 0.1, *q* is 0.9, and the carrier frequency ($2pq$) must be 0.18 among Irish populations. Such information is useful to doctors and public health professionals.

Another example: A deviation from the frequencies predicted by the Hardy–Weinberg formula suggests that a mutant allele for *BRCA2* may be lethal to female embryos. The allele also has been linked to breast cancer. For one study, researchers tracked the frequency of the mutant allele among newborn girls. There were fewer homozygotes than expected, based on the number of heterozygotes and the Hardy–Weinberg formula. By itself or in combination with other alleles, a pair of mutant *BRCA2* alleles may cause the spontaneous abortion of the early embryo.

18.3   Natural Selection Revisited

LINKS TO
SECTIONS
1.4, 17.3

Natural selection, again, is the outcome of differences in reproduction among individuals of a population that vary in their shared traits, some of which prove more adaptive than others under prevailing environmental conditions.

Natural selection may be the most influential process of microevolution. Its impact shows up at all levels of biological organization, which is the reason you were introduced to it early on, in Chapter 1. You also came across simple examples in other chapters, and Sections 17.2 and 17.3 offered you a glimpse of the history that preceded its discovery. Turn now to major categories of selection, as sketched out in Figure 18.4.

With *directional* selection, the range of variation for a trait shifts in a consistent direction; individuals at one end of the range of variation are selected against and those at the other end are favored. With *stabilizing* selection, the forms at one or both ends of the range are selected against. With *disruptive* selection, forms at one or both ends are favored and intermediate forms are selected against.

Diverse selection pressures acting on a population might favor forms at one end in the range of variation for a trait, or intermediate forms within that range, or extreme forms at both ends of the range.

a Extreme form at one end of the range of phenotypes favored

b Intermediate form of the range of phenotypes favored

c Extreme forms at both ends of the range of phenotypes favored

Figure 18.4   Overview of the outcomes of three modes of natural selection: (**a**) directional, (**b**) stabilizing, and (**c**) disruptive.

18.4 Directional Selection

LINKS TO
SECTIONS
1.4, 16.7

*With **directional selection**, allele frequencies shift in a consistent direction, so forms at one end of a phenotypic range become more common than midrange forms, as in Figure 18.5. Directional change in the environment or novel conditions can cause the shift.*

RESPONSES TO PREDATION

The Peppered Moth Populations of peppered moths (*Biston betularia*) offer us a classic case of directional selection. The moths feed and mate at night and rest motionless on trees during the day. Their behavior and coloration (mottled gray to nearly black) camouflage them from day-flying, moth-eating birds.

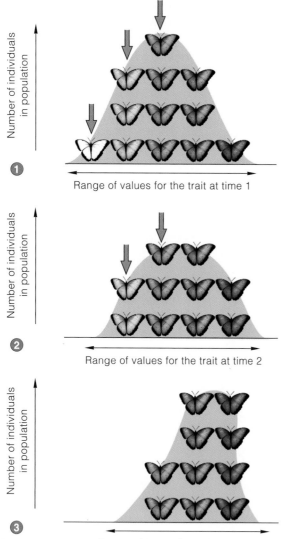

Figure 18.5 *Animated!* Directional selection. These bell-shaped curves signify a range of continuous variation in a butterfly wing-color trait. *Medium-blue* is between two phenotypic extremes—*white* and *dark purple. Orange* arrows signify which forms are being selected against over time.

In the 1850s, the industrial revolution started in England, and factory smoke altered conditions in much of the countryside. Before then, light moths were the most common form, and a dark form was rare. Also, light-gray speckled lichens had grown thickly on tree trunks. Light moths but not dark moths that rested on the lichens were camouflaged (Figure 18.6a).

Lichens are sensitive to air pollution. Between 1848 and 1898, soot and other pollutants started to kill the lichens and darken tree trunks. The dark moth form was better camouflaged (Figure 18.6b). Researchers hypothesized: If the original conditions favored light moths, then the *changed* conditions favored dark ones.

In the 1950s, H. B. Kettlewell used a *mark–release–recapture method* to test the possibility. He bred both moth forms in captivity and marked hundreds so that they could be easily identified after being released in the wild. He released them near highly industrialized areas around Birmingham and near an unpolluted part of Dorset. His team recaptured more dark moths in the polluted area and more light ones near Dorset:

	Near Birmingham (pollution high)	Near Dorset (pollution low)
Light-Gray Moths		
Released	64	393
Recaptured	16 (25%)	54 (13.7%)
Dark-Gray Moths		
Released	154	406
Recaptured	82 (53%)	19 (4.7%)

Observers also hid in blinds near moths that had been tethered to trees. They observed birds capturing more light moths around Birmingham and more dark ones around Dorset. Directional selection was in play.

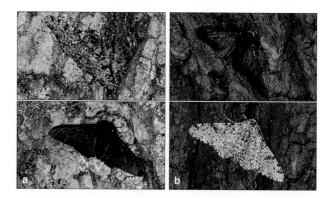

Figure 18.6 Natural selection of two forms of the same trait, body surface coloration, in two settings. (**a**) Light moths (*Biston betularia*) on a nonsooty tree trunk are hidden from predators. Dark ones stand out. (**b**) The dark color is more adaptive in places where soot darkens tree trunks.

Figure 18.7 Visible evidence of directional selection in a population of rock pocket mice relative to a neighboring population, as documented by Michael Nachman, Hoi Hoekstra, and Susan D'Agostino. **(a)** Lava basalt flow at the study site. The two color morphs of rock pocket mice, each posed on two different backgrounds: **(b)** tawny fur and **(c)** dark fur.

Pollution controls went into effect in 1952. Lichens made a comeback, and tree trunks became largely free from soot. Phenotypes shifted in the reverse direction. Where pollution has decreased, the frequency of dark moths has been decreasing as well.

Pocket Mice Directional selection is at work among rock pocket mice (*Chaetodipus intermedius*) of Arizona's Sonoran Desert. Of more than eighty genes known to affect coat color in mice, researchers found a gene that governs a difference between two populations of this mouse species (Figure 18.7).

Rock pocket mice are small mammals that spend the day in underground burrows and forage for seeds at night. Some live in tawny-colored outcroppings of granite. In this habitat, individuals with tawny fur are camouflaged from predators (Figure 18.7*b*).

A smaller population of pocket mice lives in the same region, but these mice scamper over dark basalt of ancient lava flows. They have dark coats, so they, too, are camouflaged from predators (Figure 18.7*c*).

We can expect that night-flying predatory birds are selective agents that affect fur color. For instance, owls have an easier time seeing mice with fur that does not match the rocks.

Michael Nachman used genetic data on laboratory mice to formulate a hypothesis on differences in coat color in the two wild populations of pocket mice. He predicted that a mutation of either the *Mclr* gene or *agouti* gene could cause the difference. He collected DNA from dark pocket mice at a lava flow and from light mice at adjacent granite outcroppings.

DNA analysis showed that the *Mclr* gene sequence for all dark mice differed by four nucleotides from that of their light-furred neighbors. In the population of dark mice, the allele frequencies had evolved in a consistent direction as a result of selection pressure, so dark fur became more common.

RESISTANCE TO PESTICIDES AND ANTIBIOTICS

Pesticides can cause directional selection, as they did for the super rats. Typically, a heritable aspect of body form, physiology, or behavior helps a few individuals survive the first pesticide doses. As the most resistant ones are favored, resistance becomes more common. About 450 species of pests are now resistant to one or more types of pesticides. Also, some pesticides kill off the natural predators. Freed from natural constraints, resistant populations flourish and inflict more damage. This result of directional selection is *pest resurgence*. Some genetically engineered crop plants resist pests. In time, they too may exert selection pressure.

Antibiotics also can result in directional selection. Certain microbes produce natural antibiotics that can kill bacterial competitors for nutrients. We use natural and synthetic antibiotics to fight pathogenic bacteria. Streptomycins, for example, inhibit protein synthesis in bacterial cells. The penicillins disrupt covalent bonds that hold a bacterial cell wall together.

Yet antibiotics have been overprescribed, often for simple infections that would clear up on their own. Genetic variation in bacterial gene pools allows some cells with certain genotypes to survive as others die. So overuse of antibiotics favors the resistant bacterial populations, which will be harder to eradicate in the millions of people who contract cholera, tuberculosis, and other bacterial diseases each year. Also, healthy farm animals are routinely dosed with antibiotics to prevent infection. Consider: In eggs that look slightly fluorescent green, tetracycline is showing through.

With directional selection, allele frequencies underlying a range of variation tend to shift in a consistent direction in response to some change in the environment.

18.5 Selection Against Or in Favor of Extreme Phenotypes

LINK TO SECTION 17.3

Consider now two more categories of natural selection. One works against phenotypes at the fringes of a range of variation; the other favors them.

STABILIZING SELECTION

With **stabilizing selection**, intermediate forms of a trait in a population are favored, and extreme forms are not. This mode of selection can counter mutation, genetic drift, and gene flow. It tends to preserve intermediate phenotypes in the population (Figure 18.8*a*).

As an example, prospects are not good for human babies who weigh far more or far less than average at birth. Also, pre-term instead of full-term pregnancies increase the risk, as Figure 18.9 indicates.

Newborns weighing less than 5.51 pounds or born before thirty-eight weeks of pregnancy are completed tend to develop high blood pressure, diabetes, and heart disease when they are adults. Researchers now suspect that the mother's blood concentration of a stress hormone, cortisol, is linked to low birth weight and illnesses that develop later in life.

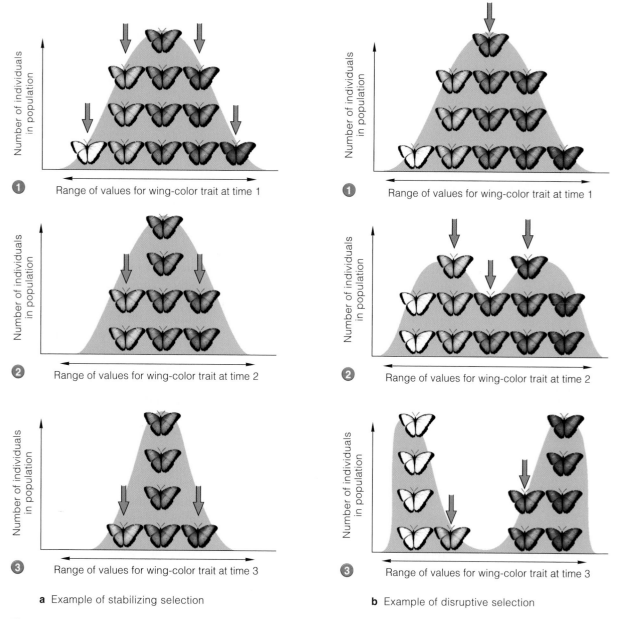

a Example of stabilizing selection

b Example of disruptive selection

Figure 18.8 *Animated!*   Selection against or in favor of extreme phenotypes, with a population of butterflies as the example. (**a**) stabilizing selection and (**b**) disruptive selection. The *orange* arrows show forms of the trait being selected against.

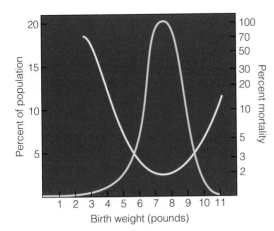

Figure 18.9 Weight distribution for 13,730 human newborns (*yellow* curve) correlated with death rate (*white* curve).

Figure 18.10 Adult sociable weaver (*Philetairus socius*), a native of the African savanna. These birds cooperate in constructing and using large communal nests in a region where trees and other good nesting sites are scarce.

lower bill 12 mm wide   lower bill 15 mm wide

Figure 18.11 Disruptive selection in African finch populations. Selection pressures favor birds with bills that are about 12 *or* 15 millimeters wide. The difference is correlated with competition for scarce food resources during the dry season.

Rita Covas and her colleagues gathered evidence of stabilizing selection on the body mass of juvenile and adult sociable weavers (*Philetairus socius*), as in Figure 18.10. Between 1993 and 2000, they captured, measured, tagged, released, and recaptured 70 to 100 percent of the birds living in communal nests during the breeding season. Their field studies supported a prediction that body mass is a trade-off between risks of starvation and predation. Intermediate-mass birds have the selective advantage. Foraging is not easy in this habitat, and lean birds do not store enough fat to avoid starvation. We can expect that fat ones are more attractive to predators and not as good at escaping.

DISRUPTIVE SELECTION

With **disruptive selection**, forms at both ends of the range of variation are favored and intermediate forms are selected against (Figure 18.8*b*).

Consider the black-bellied seedcracker (*Pyrenestes ostrinus*) of Cameroon. Females and males of these African finches have large or small bills—but no sizes in between (Figure 18.11). It is like everyone in Texas being four feet *or* six feet tall, with no one in between.

The pattern holds all through the geographic range. If unrelated to gender or geography, what causes it? If only two bill sizes persist, then disruptive selection may be eliminating birds with intermediate-size bills. Factors that affect feeding performance are the key. Cameroon's swamp forests flood in the wet season; lightning-sparked fires burn in the hot, dry season. Most plants are fire-resistant, grasslike sedges. One species produces hard seeds and the other, soft seeds.

Remember the bills of Galápagos finches (Section 17.3)? Here, also, the ability to crack hard seeds affects survival. All Cameroon seedcrackers prefer soft seeds,

but birds with large bills are better at cracking hard ones. In the dry season, the birds compete fiercely for scarce seeds. A scarcity of both types of seeds during recurring episodes of drought has a disruptive effect on bill size in the seedcracker population. Birds with intermediate sizes are being selected against, and now all bills are either 12 *or* 15 millimeters wide.

In these seedcrackers, bills of a particular size have a genetic basis. In experimental crosses between two birds with the two optimal bill sizes, all offspring had a bill of one size or the other, nothing in between.

With stabilizing selection, intermediate phenotypes are favored and extreme phenotypes at both ends of the range of variation are eliminated.

With disruptive selection, intermediate forms of traits are selected against and extreme forms in the range of variation are favored.

18.6 Maintaining Variation in a Population

LINKS TO
SECTIONS
3.6, 11.1

Natural selection theory helps explain diverse aspects of nature, including male–female differences and the relationship between sickle-cell anemia and malaria.

SEXUAL SELECTION

The individuals of many sexually reproducing species show a distinct male or female phenotype, or **sexual dimorphism** (*dimorphos,* having two forms). Often the males are larger and flashier than females. Courtship rituals and male aggression are common.

These adaptations and behaviors seem puzzling. All take energy and time away from an individual's survival activities. Why do they persist if they do not contribute directly to survival? The answer is **sexual selection**. By this mode of natural selection, winners are the ones that are better at attracting mates and successfully reproducing compared to others of the population. The most adaptive traits help individuals defeat same-sex rivals for mates or are the ones most attractive to the opposite sex.

By choosing mates, a male or female is a selective agent acting on its own species. For example, females of some species shop among a congregation of males, which vary in appearance and courtship behavior. The selected males and the females pass on their alleles to the next generation.

Flashy body parts and behaviors show up among species in which males provide little or no help with raising offspring. The female apparently chooses her partner on the basis of observable signs of health and vigor. Such traits may improve the odds of producing healthy, vigorous offspring (Figure 18.12).

You might be wondering whether we can correlate genes with specific forms of sexual behavior. The sexual deception practiced by an Australian orchid is a case in point. The flowers of *Chiloglottis trapeziformis* attract male wasps by secreting a substance that is identical with a sex pheromone—which female wasps release to attract male wasps. Flowers get pollinated as males attempt to copulate with them.

This orchid is stingy. It gives a male wasp nothing in return, not a single drop of nectar, even though it is the orchid's exclusive pollinator. The female wasps are wingless. They hatch in soil. When males do not lift and carry them to a food source, they starve to death.

When *C. trapeziformis* puts out blooms, male wasps waste precious time and metabolic energy trying to find females. Evolutionary biologist Florian Schiestl has proposed that selection pressure is afoot for wasps that can produce a new sex pheromone, one that the orchid cannot duplicate.

This interaction exploits male wasps, but Wittko Francke thinks it might put pressure on their brains to evolve. In an orchid patch, the average tiny-brained male wasp copulates blindly with whatever smells right. It will try to copulate even with the head of a pin that has a few micrograms of pheromone sprayed on it. However, a few wasps with a slightly less robotic brain might be able to identify the females by other cues, such as visual ones. Alternatively, both species could face extinction, another pattern in nature.

SICKLE-CELL ANEMIA—LESSER OF TWO EVILS?

With *balancing* selection, two or more alleles of a gene are being maintained at relatively high frequencies in the population. Their persistence is called **balanced polymorphism** (*polymorphos,* having many forms). The allele frequencies might shift slightly, but often they return to the same values over the long term. We may see this balance when conditions favor heterozygotes. In some way, their nonidentical alleles for a given trait

Figure 18.12 One male bird of paradise in a flashy courtship display. He caught the eye (and, interest) of the smaller, less colorful female. The males of this species compete fiercely perhaps, the sexual for females, which function as selective agents. (Why do you suppose the females are drab-colored?) *(Paradisaea raggiana)* is engaged in display.

grant them higher fitness compared to homozygotes, which, recall, have identical alleles for the trait.

Consider the environmental pressures that favor an Hb^A/Hb^S pairing in humans. The Hb^S allele codes for a mutant form of hemoglobin, an oxygen-transporting protein in blood. Homozygotes (Hb^S/Hb^S) develop the genetic disorder *sickle-cell anemia* (Section 3.6).

The Hb^S frequency is highest in both tropical and subtropical regions of Asia and Africa. Often, Hb^S/Hb^S homozygotes die in their early teens or early twenties. Yet, in these same regions, heterozygotes (Hb^A/Hb^S) make up nearly a third of the population! Why is this combination maintained at such high frequency?

The balancing act is most pronounced in areas that, historically, have had the highest incidence of *malaria* (Figure 18.13). Mosquitoes transmit the parasitic agent of malaria, *Plasmodium*, to human hosts. The parasite multiplies in the liver and then in red blood cells. The target cells rupture and release new parasites during severe, recurring bouts of infection (Section 22.7).

It turns out that Hb^A/Hb^S heterozygotes are more likely to survive malaria than people who make only normal hemoglobin. Several survival mechanisms are possible. In heterozygotes, the infected cells take on a sickle shape under normal conditions. The abnormal shape marks them as targets for the immune system, which destroys them, along with the parasites inside. In addition, heterozygotes have one functioning Hb^A allele. Although they are not completely healthy, they still produce enough normal hemoglobin to prevent sickle-cell anemia. That is why heterozygotes are more likely to survive long enough to reach reproductive age, compared to Hb^S/Hb^S homozygotes.

In short, the persistence of the "harmful" Hb^S allele may be a matter of relative evils. Malaria has been a selective force for thousands of years in tropical and subtropical areas of Asia, the Middle East, and Africa. Through that time span, natural selection has favored the Hb^A/Hb^S combination in all of the malaria-ridden regions, because heterozygotes show more resistance to the disease. In such environments, the combination has proved to have more survival value than either the Hb^S/Hb^S or the Hb^A/Hb^A combination.

With sexual selection, some version of a gender-related trait gives the individual an advantage in reproductive success. Sexual dimorphism is one outcome of sexual selection.

In a population showing balanced polymorphism, natural selection is maintaining two or more alleles at frequencies greater than 1 percent over the generations.

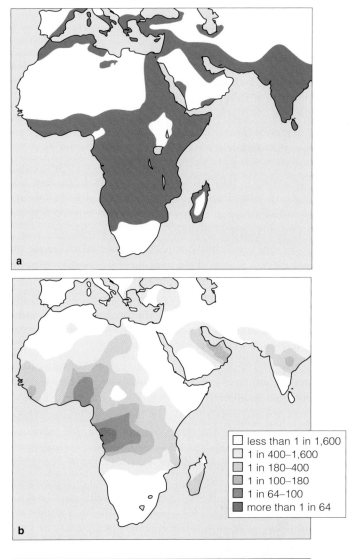

□ less than 1 in 1,600
□ 1 in 400–1,600
□ 1 in 180–400
■ 1 in 100–180
■ 1 in 64–100
■ more than 1 in 64

Figure 18.13 (**a**) Distribution of malaria cases reported in Africa, Asia, and the Middle East in the 1920s, before the start of programs to control mosquitoes, the vector for *Plasmodium*. (**b**) Distribution and frequency of people with the sickle-cell trait. Notice the close correlation between the maps. (**c**) Physician searching for *Plasmodium* larvae in Southeast Asia.

18.7 Genetic Drift—The Chance Changes

LINKS TO
SECTIONS
11.2, 12.10

Especially in small populations, random changes in allele frequencies can lead to a loss of genetic diversity.

Genetic drift is a random change in allele frequencies over time, brought about by chance alone. Researchers measure it in terms of probability rules. *Probability* is the chance that something will happen relative to the number of times it could happen (Section 11.2). We can measure an event's relative frequency as a fraction on a scale from zero to 1—or 0 to 100 percent of the time. For instance, if 10 million people enter a drawing for a month-long vacation in Hawaii, all expenses paid, each has an equal chance of winning: 1/10,000,000, or an exceedingly improbable 0.00001 percent.

By one probability rule, the expected outcome of some event is less likely to occur if the event happens only rarely. Each time you flip a coin, for example, there is a 50 percent chance it will turn up heads. With 10 flips, odds are high that the proportions of heads and tails will deviate greatly from 50:50. With 1,000 flips, large deviations from 50:50 are less likely.

We can apply the same rule to populations. Because population sizes are not infinite, there will be random changes in allele frequencies. These random changes tend to have minor impact on large populations. They greatly increase the odds that an allele will become more or less prevalent when populations are small.

Steven Rich and his coworkers used small and large populations of the flour beetle (*Tribolium castaneum*) to study genetic drift. They started with beetles that bred true for allele b^+ and other beetles that bred true for mutant allele b. (The superscript plus signifies a wild-type allele.) They hybridized individuals from both groups to get a population of F_1 heterozygotes

(b^+b), which they divided into sets of twelve. Different sets consisted of 10, 20, 50, and 100 randomly selected male and female beetles, and the subpopulation sizes were maintained for twenty generations.

Figure 18.14 shows two of the test results. Drift was greatest in the sets of 10 beetles and least in the sets of 100 beetles. Notice the loss of b^+ from one of the small populations (one graph line ends at 0 in Figure 18.14a). Only allele b remained. When all of the individuals of a population have become homozygous for one allele only at a locus, we say that **fixation** has occurred.

Thus, *random change in allele frequencies leads to the homozygous condition and a loss of genetic diversity over time.* This is genetic drift's outcome in all populations; it simply happens faster in small ones (Figure 18.14). Once alleles from the parent population have become fixed, their frequencies will not change again unless mutation or gene flow introduces new alleles.

BOTTLENECKS AND THE FOUNDER EFFECT

Genetic drift is pronounced when a few individuals rebuild a population or start a new one. This happens after a **bottleneck**, a drastic reduction in population size brought about by severe pressure. Suppose that contagious disease, habitat loss, or hunting nearly wipes out a population. Even if a moderate number of individuals survive a bottleneck, allele frequencies will have been altered at random.

In the 1890s, hunters killed all but twenty of a large population of northern elephant seals. Government restrictions allowed the population to recover to about 130,000 individuals. Each is homozygous for all of the genes analyzed so far.

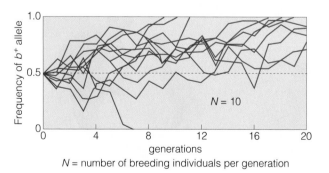

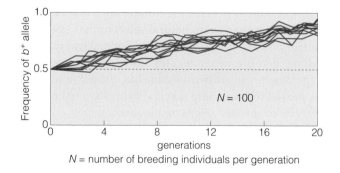

a The size of twelve populations of beetles was maintained at 10 breeding individuals per generation for twenty generations. Allele b^+ was lost and b became fixed in one population. Notice that alleles can be fixed or lost even in the absence of selection.

b The size of twelve populations was maintained at 100 individuals per generation for twenty generations. Allele b did not become fixed. Drift was far less in each generation than it was in the small populations tracked in (**a**).

Figure 18.14 *Animated!* Genetic drift's effect on allele frequencies in small and large populations. The starting frequency of mutant allele b^+ was 0.5.

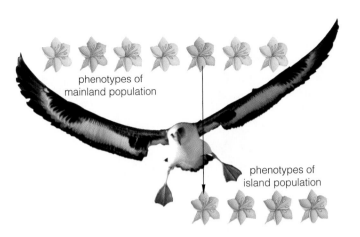

Figure 18.15 Founder effect. This wandering albatross carries seeds, stuck to its feathers, from the mainland to a remote island. By chance, most of the seeds carry an allele for orange flowers that are rare in the original population. Without further gene flow or selection for color, genetic drift will fix the allele on the island.

phenotypes of mainland population

phenotypes of island population

Unpredictable genetic shifts can occur after a few individuals establish a new population. This form of bottlenecking is a **founder effect**. Genetic diversity might be greatly reduced relative to the original gene pool, as when a lone seed founds a population on a remote island in the middle of the ocean (Figure 18.15).

INBRED POPULATIONS

Genetic drift is less pronounced in inbred populations. **Inbreeding** is nonrandom mating among very close relatives, which share many identical alleles. It leads to the homozygous condition. It also lowers fitness if harmful recessive alleles are increasing in frequency.

Most human societies forbid or discourage incest (inbreeding between parents and children or siblings). Inbreeding among other close relatives is common in geographically or culturally isolated small groups. The Old Order Amish in Pennsylvania are moderately inbred. One outcome is a rather high frequency of a recessive allele that causes *Ellis–van Creveld syndrome*. Affected individuals have extra fingers, toes, or both (Section 12.10). The allele might have been rare when a few founders entered Pennsylvania. Now, about 1 in 8 individuals of the community are heterozygous for the allele, and 1 in 200 are homozygous for it.

Genetic drift is the random change in allele frequencies over the generations, brought about by chance alone. The magnitude of its effect is greatest in small populations, such as one that endures a bottleneck.

18.8 Gene Flow

Individuals, and their alleles, move into and away from populations. The physical flow of alleles counters changes introduced by other microevolutionary processes.

Individuals of the same species don't always stay put. A population loses alleles when an individual leaves it for good, an event called *emigration*. The population gains alleles when individuals permanently move in, an event called *immigration*. In both cases, **gene flow** —the physical movement of alleles into and out of a population—occurs. This microevolutionary process counters mutation, natural selection, and genetic drift.

Later chapters will give historical examples of how gene flow has kept separated populations genetically similar. For now, simply consider the acorns that blue jays disperse when they gather nuts for the winter. Each fall, jays visit acorn-bearing oak trees repeatedly, then bury acorns in the soil of home territories that may be as much as a mile away (Figure 18.16). Alleles flowing in with the "immigrant acorns" help decrease genetic differences between stands of oak trees.

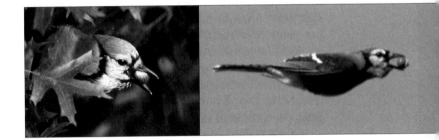

Figure 18.16 Blue jay, a mover of acorns that helps keep genes flowing between separate oak populations.

Or think of the millions of people from politically explosive, economically bankrupt countries who seek a more stable home. The scale of their emigrations is unprecedented, but the flow of genes is not. Human history is rich with cases of gene flow that minimized many of the genetic differences among geographically separate groups. Remember Genghis Khan? His genes flowed from China to Vienna (Section 12.10). Similarly, the armies of Alexander the Great brought alleles for green eyes from Greece all the way to India.

Gene flow is the physical movement of alleles into and out of a population, through immigration and emigration. It tends to counter the effects of mutation, natural selection, and genetic drift.

Summary

Section 18.1 Individuals of a population generally have the same number and kinds of genes for the same traits. Alleles are different molecular forms of a gene. Individuals who inherit different allele combinations vary in details of one or more traits. An allele at any locus may become more or less common relative to other kinds or may be lost.

Mutations are rare in individuals, but they have accumulated in natural populations of all lineages. Mutations are the original source of alleles, the raw material for evolution.

Microevolution refers to changes in allele frequencies of a population brought about by mutation, natural selection, genetic drift, and gene flow (Table 18.1).

Section 18.2 Genetic equilibrium is a state in which a population is not evolving. According to the Hardy–Weinberg equilibrium formula, this occurs only if there is no mutation, the population is infinitely large and isolated from all other populations of the species, there is no natural selection, mating is random, and all individuals survive and produce the same number of offspring. Deviations from this theoretical baseline indicate microevolution is in play.

Biology Now
Investigate gene frequencies and genetic equilibrium with the interaction on BiologyNow.

Section 18.3 Natural selection is the outcome of differences in reproduction among individuals of a population that show variations in their shared traits. Three major modes are directional, stabilizing, and disruptive selection. Selection pressures operating on the range of phenotypic variation shift or maintain allele frequencies in the population's gene pool.

Section 18.4 Directional selection shifts the range of phenotypic variation in a consistent direction. The individuals at one end of the range of variation are selected against and those at the other end are favored.

Biology Now
View the animation of directional selection on BiologyNow.

Read the InfoTrac article "AIDS in Africa Has Potential to Affect Human Evolution,"AIDS Weekly, June 2001.

Section 18.5 Stabilizing selection works against extremes in the range of phenotypic variation, and it favors intermediate forms. Disruptive selection favors forms at both extremes of the range; individuals in the intermediate range are selected against.

Biology Now
View the animation of disruptive and stabilizing selection on BiologyNow.

Read the InfoTrac article "Portraits of Evolution: Studies of Coloration in Hawaiian Spiders," Geoffrey S. Oxford, Rosemary G. Gillespie, Bioscience, July 2001.

Section 18.6 Sexual selection, by females or males, leads to forms of traits that favor reproductive success. Persistence in phenotypic differences between males and females (sexual dimorphism) is one outcome.

Selection may result in balanced polymorphism, with nonidentical alleles for a trait being maintained over time at relatively high frequencies.

Biology Now
Read the InfoTrac article "High-Risk Defenses," Gregory Cochran, Paul W. Ewald, Natural History, Feb. 1999.

Section 18.7 Genetic drift is a random change in a population's allele frequencies over time due to chance occurrences alone. It tends to lead to the homozygous condition and loss of genetic diversity.

The effect of genetic drift is most pronounced in very small populations, such as ones that have passed through a bottleneck or that arose from a small group of founders. Genetic drift has less effect on inbred populations, which are characterized by nonrandom mating of very close relatives.

Biology Now
Learn more about genetic drift with the interaction on BiologyNow.

Section 18.8 Gene flow moves alleles into or out of a population by immigration or emigration. The process helps keep populations of the same species genetically alike by countering the effects of mutation, natural selection, and genetic drift.

Section 18.9 Long-term, heritable adaptations are aspects of form, function, behavior, or development that improve the chance of surviving and reproducing, or at least did so under conditions that prevailed when genes for the trait first evolved.

Often it is not easy to correlate an adaptive trait with the particular environmental conditions to which it is assumed to be adapted.

Table 18.1	Summary Definitions for Microevolutionary Events
Mutation	A heritable change in DNA; original source of alleles in a population
Natural selection	Outcome of differences in reproduction among individuals of a population that show variation in their shared, heritable traits. Can shift the range of phenotypes in a consistent direction, disrupt it, or stabilize it
Genetic drift	Random changes in a population's allele frequencies through the generations as an outcome of chance alone
Gene flow	Individuals move their alleles into and out of a population by way of immigration and emigration; tends to counter the changes caused by mutation, natural selection, and genetic drift

Self-Quiz
Answers in Appendix II

1. Individuals don't evolve, _____ do.

2. Biologists define evolution as _____ .
   a. purposeful change in a lineage
   b. heritable change in a line of descent
   c. acquiring traits during the individual's lifetime
   d. both a and b

3. _____ is the original source of new alleles.
   a. Mutation
   b. Natural selection
   c. Genetic drift
   d. Gene flow
   e. All are original sources of new alleles

4. Natural selection may occur when there are _____ .
   a. differences in forms of traits
   b. differences in survival and reproduction among individuals that differ in one or more traits
   c. both a and b

5. Directional selection _____ .
   a. eliminates common forms of alleles
   b. shifts allele frequencies in a consistent direction
   c. favors intermediate forms of a trait
   d. works against adaptive traits

6. Disruptive selection _____ .
   a. eliminates uncommon forms of alleles
   b. shifts allele frequencies in one direction only
   c. doesn't favor intermediate forms of a trait
   d. both b and c

7. Sexual selection, especially competition between males for access to fertile females, frequently influences aspects of body form and leads to _____ .
   a. inbreeding
   b. genetic drift
   c. sexual dimorphism
   d. both b and c

8. The persistence of malaria and sickle-cell anemia in a population is a case of _____ .
   a. bottlenecking
   b. balanced polymorphism
   c. natural selection
   d. artificial selection
   e. both b and c

9. _____ tends to counter changes that occur in the allele frequencies among populations of a species.
   a. Genetic drift
   b. Gene flow
   c. Mutation
   d. Natural selection

10. Match the evolution concepts.
    _____ gene flow
    _____ natural selection
    _____ mutation
    _____ genetic drift

    a. source of new alleles
    b. changes in a population's allele frequencies due to chance alone
    c. allele frequencies change owing to immigration, emigration, or both
    d. outcome of differences in survival, reproduction among individuals of a population that vary in the details of shared traits

Additional questions are available on Biology Now™

Critical Thinking

1. Occasionally, a few of the families in a remote region of Kentucky produce *blue offspring*, a condition caused by an autosomal recessive disorder. Skin of affected individuals appears dark blue. Homozygous individuals do not have

Figure 18.20 Two designer dogs: the Great Dane (*legs, left*) and the chihuahua (*possibly fearful of being stepped on, right*).

the enzyme that maintains hemoglobin in its normal molecular form. Without it, a blue form of hemoglobin accumulates in blood and shows through the skin.

Formulate a hypothesis to explain the recurrence of the blue offspring trait among a cluster of families.

2. Martha is studying a population of tropical birds. The males have brightly colored tail feathers and the females don't. She suspects this difference is maintained by sexual selection. Design an experiment to test her hypothesis.

3. About 50,000 years ago, humans began domesticating wild dogs. By 14,000 years ago, they started to favor new varieties (breeds) by way of artificial selection. Individual dogs having desirable forms of traits were selected from each new litter and, later, encouraged to breed. Those with undesired forms of traits were passed over.

After favoring the pick of the litter for hundreds or thousands of generations, we ended up with sheep-herding border collies, badger-hunting dachshunds, bird-fetching retrievers, and sled-pulling huskies. And at some point we began to delight in the odd, extraordinary dog.

In practically no time at all, evolutionarily speaking, we picked our way through the pool of variant dog alleles and came up with such extreme breeds as Great Danes and chihuahuas (Figure 18.20).

Sometimes the canine designs have exceeded the limits of biological common sense. How long would a tiny, nearly hairless, nearly defenseless, finicky-eating chihuahua last in the wild? Not long. What about English bulldogs, bred for a stubby snout and compressed face? Breeders thought these traits would let the dogs get a better grip on the nose of a bull. (Why they wanted dogs to bite bulls is a story in itself.) So now the roof of the bulldog mouth is ridiculously wide and often flabby, so bulldogs have trouble breathing. Sometimes they get so short of breath they pass out.

Why do you suppose many people easily accept that artificial selection practices can produce startling diversity but will not accept that natural selection might do the same in the wild?

19 EVOLUTIONARY PATTERNS, RATES, AND TRENDS

Last of the Honeycreepers?

More than 5 million years ago, Kauai rose above the surface of the sea. It was the first of the big islands of the Hawaiian Archipelago. Several million years later, a few quite possibly terrified finches reached it after bobbing 4,000 kilometers (2,500 miles) across the open ocean. Were they unwilling pioneers, blown away from the mainland during a fierce storm? We may never know, but their chance geographic dispersal was the start of something big.

No predatory mammals had preceded the finches onto that isolated, volcanically born island. But tasty insects and plants that bore tender leaves, nectar, seeds, and fruits were already there. The finches thrived. Their descendants quickly radiated into habitats along the coasts, through dry lowland forests, and into rain forests of the highlands.

Between 1.8 million and 400,000 years ago, volcanic eruptions created the rest of the archipelago. Generation after generation, descendants of the first finches traveled on the winds to vacant habitats in the new islands. They foraged in many shrublands and forests, each with special food sources and nesting sites. Diverse agents of natural selection operated in each place, and differences in bill sizes and shapes, feather coloration and patterns, and territorial songs evolved. In this way, a spectacular family of birds, the Hawaiian honeycreepers, originated.

One existing Hawaiian honeycreeper has a bill that fits in the long, curving nectar tubes of *Lobelia* flowers (Figure 19.1). One probes tree bark with its sickle-shaped upper bill, then scoops out beetle larvae with its shovel-shaped lower bill. Other species use a thickened, strong, parrotlike bill to crush or pry open hard seed pods. The po'-ouli (Figure 19.2) is the only species that preferentially eats native tree snails.

Ironically, the very isolation that favored specialized adaptations to conditions in unique habitats made these birds vulnerable to extinction. When conditions changed, they had nowhere to go. They had no built-in defenses against predatory mammals and avian diseases of the mainland, against humans who coveted cloaks made of their eye-catching feathers, or against climate change.

Accompanying humans to the islands were brown tree snakes, rats, cats, and other voracious predators. People also imported chickens and other birds that happened to be infected with disease agents. Over time, people cleared more and more of the forests. Imported crop plants and plant-eating mammals became established. The Hawaiian honeycreeper habitats shrank. Today, with a long-term increase in global temperatures—global warming—the forests at higher elevations are not as cool as they once were. They have been infiltrated by mosquitoes, which

Figure 19.1 *Left*, the Hawaiian honeycreeper known informally as Iiwi (*Vestiaria coccinea*). It evolved in the Hawaiian Archipelago (*right*), far from the mainland. It is a descendant of a spectacular adaptive radiation—one of the patterns explained in this chapter.

Figure 19.2 Male po'ouli—rare, old, and missing one eye. Ecologists captured this small honeycreeper on the east slope of Haleakala, Hawaii, as part of a last-ditch effort to save the remaining population of three birds. This male was already suffering the effects of avian malaria, and it died in 2004. The perpetuation of its species—which was not even known about until 1973—now rests on the two remaining birds. The two have not been seen in many months.

thrive in warm climates. Mosquitoes happen to be vectors for pathogens that cause avian malaria and other diseases.

At one time there were approximately fifty species of Hawaiian honeycreepers. As many as twenty-four species colonized a single island. Half of the known species are extinct. The initial wave of extinction followed the arrival of the first Polynesians, and another ten species are now endangered. The remaining species are being studied in earnest, and efforts are under way to protect them. It may be a case of too little, too late; but time will tell.

How do we know so much about a group of birds on an island chain in the middle of the Pacific Ocean? Scientific theories and tools, particularly radiometric dating and automated gene sequencing, helped shine light on their rise and impending fall. The age of volcanic rocks on each island, as well as the DNA of different species, were among the clues. Such clues give us glimpses into **macroevolution** —the long-term patterns, rates, and trends in the origin and ultimate fate of Earth's many millions of species.

How Would You Vote?

Often, when a species is on the brink of extinction, some individuals are captured and brought to zoos for captive breeding programs. Some people object to this practice. They say keeping a species alive in a zoo is a distraction from more meaningful conservation efforts, and captive animals seldom are successfully restored to the wild. Do you support captive breeding of highly endangered species? See BiologyNow for details, then vote online.

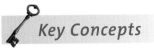

Key Concepts

HOW DO SPECIES ARISE?

All sexually reproducing species consist of one or more populations of individuals that interbreed under natural conditions, produce fertile offspring, and are reproductively isolated from other such populations.   Section 19.1

MODELS FOR SPECIATION

Speciation is a process that varies in details and duration among lineages. It starts when gene flow stops between populations of a species. Microevolutionary events occur independently in the reproductively isolated populations. The process ends when daughter species form.   Sections 19.2, 19.3

PATTERNS IN THE HISTORY OF LIFE

The timing, rate, and direction of speciation differ among branches of a lineage and between lineages. Adaptive radiations and extinction punctuate the history of life. Section 19.4

CLASSIFICATION SYSTEMS

Patterns in life's history are being identified and interpreted. Taxonomy identifies, names, and then classifies species. Systematics infers evolutionary relationships by analytical methods. Phylogenetic classification systems are efficient tools for retrieving information about the history of life. Section 19.5

PIECING TOGETHER FAMILY TREES

Biologists construct evolutionary tree diagrams that use derived traits to determine branch points. A current tree subsumes six traditionally defined kingdoms into three domains: Bacteria, Archaea, and Eukarya. It reveals how all species interconnect through shared ancestors, some remote, others recent.   Sections 19.6–19.9

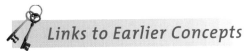

Links to Earlier Concepts

Before starting this chapter, be sure you understand how gene flow can help keep populations of the same species genetically similar by countering the impact of mutation, natural selection, and genetic drift (Sections 18.1, 18.8). You will be appying your knowledge of changes in chromosome structure and function (12.8 and 15.4). Quickly review how comparisons of morphology (17.7) and of genes and proteins of different lineages (17.9) yield clues to shared ancestry.

Also reflect on the major geologic forces. They have been a factor in the origin of many species, especially on island chains (17.2, 17.6). You also will be taking a closer look at the three-domain system of classification (1.3).

19.1 Reproductive Isolation, Maybe New Species

LINKS TO
SECTIONS
11.6, 18.1, 18.8

Speciation is a macroevolutionary process. It starts when a population becomes reproductively isolated from others of the species and ends when daughter species have formed.

WHAT IS A SPECIES?

Species is a Latin word that means "kind," as in "one kind of plant." This generic definition does not help much when we are trying to figure out whether, say, a population of plants in one place belongs to the same species as a population of plants somewhere else. You may see variations in traits between the populations and variation within them, because plants can inherit diverse combinations of alleles. Also, some individuals may grow in very different environments that cause changes in gene expression (Section 11.6 and Figure 19.3). In other words, we might not be able to identify a biological species on the basis of appearance alone.

Evolutionary biologist Ernst Mayr came up with a **biological species concept**: A species is one or more groups of individuals that interbreed, produce fertile offspring, and are reproductively isolated from other such groups. This definition is reasonable for species that reproduce sexually—which most species do. It does not apply to asexual reproducers, and it cannot be used to interpret the fossil record.

A more recent definition has wider applicability: A **species** is one or more populations of individuals that share at least one structural, functional, or behavioral trait—*the legacy of a common ancestor*—that sets them apart from other species. This definition is based on comparative morphology, biochemistry, and the fossil record. It applies to sexually or asexually reproducing species. Unlike the biological species concept, it does not directly address how a species attains and then maintains its separate identity. That clue, for sexual reproducers at least, is *reproductive isolation*—the end of gene exchanges between populations.

REPRODUCTIVE ISOLATING MECHANISMS

Gene flow, recall, is the movement of alleles into and out of a population (Section 18.8). Speciation begins when gene flow, or the potential for it, ends between natural populations. Once it stops, gene pools start to change and populations undergo **genetic divergence**, because mutation, natural selection, and genetic drift are free to operate independently in each one (Section 18.8). As you will see later, speciation may result from gradual genetic divergence. It also may be completed within a few generations, as commonly occurs among flowering plants.

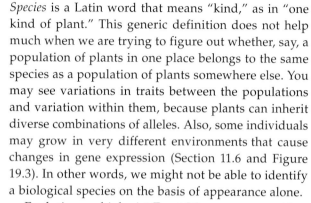

a

b

Figure 19.3 Morphological differences between plants of the same species (*Sagittaria sagittifolia*) growing (**a**) in water and (**b**) on land. The leaf shapes are responses to different environmental conditions, not to different genetic programs.

Figure 19.4 *Animated!* (**a**) Mechanical isolation. Few pollinating insects fit as well as wasps on a zebra flower. Petals form a landing platform below stamens.

(**b**) Temporal isolation. *Magicicada septendecim*, a periodical cicada that matures underground and emerges to reproduce every seventeen years. Its populations often overlap the habitats of a sibling species (*M. tredecim*), which reproduces every thirteen years. Adults live only a few weeks.

(**c**) Behavioral isolation. Courtship displays precede sex among many kinds of birds, including these albatrosses. Individuals recognize tactile, visual, and acoustical signals, such as a prancing dance followed by back arching, a skyward pointing bill and an exposed throat, and wing spreading.

a

b

c

Figure 19.5 *Animated!* When certain reproductive isolating mechanisms prevent interbreeding. There are barriers to (**a**) getting together, mating, or pollination, (**b**) successful fertilization, and (**c**) survival, fitness, or fertility of hybrid embryos or offspring.

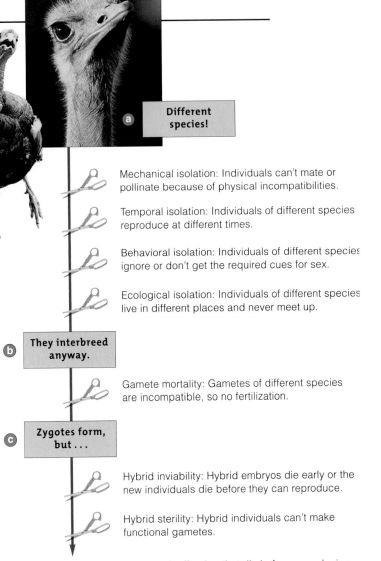

a Different species!

Mechanical isolation: Individuals can't mate or pollinate because of physical incompatibilities.

Temporal isolation: Individuals of different species reproduce at different times.

Behavioral isolation: Individuals of different species ignore or don't get the required cues for sex.

Ecological isolation: Individuals of different species live in different places and never meet up.

b They interbreed anyway.

Gamete mortality: Gametes of different species are incompatible, so no fertilization.

c Zygotes form, but . . .

Hybrid inviability: Hybrid embryos die early or the new individuals die before they can reproduce.

Hybrid sterility: Hybrid individuals can't make functional gametes.

No offspring, sterile offspring, or weak offspring that die before reproducing

Either way, **reproductive isolating mechanisms** evolve. All of these heritable aspects of body form, function, or behavior block interbreeding between populations. *Prezygotic* mechanisms, as in Figure 19.4, stop cross-pollination or cross-breeding, the formation of gametes, or fertilization. *Postzygotic* mechanisms kill hybrids or make them weak or infertile. Let us start off with the prezygotic isolating mechanisms listed in Figure 19.5a,b.

Mechanical isolation. The body parts of a species are not a physical match with those of a species that could otherwise serve as a mate or pollinator. Figure 19.4a shows the fit between a zebra plant and its preferred pollinator. Similarly, the pollen-bearing stamens of the flowers of one sage species extend above petals that act as a landing platform. Big-bodied pollinators get a dusting of pollen when they land and collect nectar. Pollen-gathering bees are not large enough to brush against these stamens. But the other sage species has its stamens poised above a bee-sized platform that is too small and fragile to hold big, heavy pollinators.

Temporal isolation. Diverging populations cannot interbreed when their timing of reproduction differs. Cicada species that differ in form and behavior often live in the same habitat in the eastern United States. They all mature underground and feed on juicy roots. Every 17 years, three species emerge and reproduce (Figure 19.4b). Each one has a *sibling* species of similar form and behavior. But siblings emerge on a 13-year cycle. This means that each species and its sibling do not get together except once every 221 years!

Behavioral isolation. Behavioral differences bar gene flow between related species. Before male and female birds copulate, they may engage in courtship displays (Figure 19.4c). A female bird is genetically prewired to recognize the singing, wing spreading, prancing, or head bobbing of a male of her species as an overture to sex. Females of different species usually do not.

Ecological isolation. Populations occupying different microenvironments may be ecologically isolated. Two manzanita species live in seasonally dry foothills of the Sierra Nevada, one at elevations between 600 and 1,850 meters, the other between 750 and 3,350 meters. They hybridize rarely, and only where the two ranges overlap. Water-conserving mechanisms operate in dry seasons. But one species is adapted to sites where water stress is not intense. The other lives in drier, exposed sites on rocky hillsides, so cross-pollination is unlikely.

Gamete mortality. Gametes of different species may have molecular incompatibilities. Example: If pollen lands on a plant of another species, it usually does not respond to the plant's molecular signals to germinate.

Postzygotic isolating mechanisms act in an embryo (Figure 19.5c). Unsuitable interactions among genes or gene products cause early death, sterility, or weak hybrids with low survival rates. Certain hybrids are sturdy but sterile. Mules, which are the offspring of a female horse and male donkey, are infertile hybrids.

A species is one or more populations of individuals having a unique common ancestor. Its individuals share a gene pool, produce fertile offspring, and remain reproductively isolated from individuals of other species.

Speciation is the process by which daughter species form from a population or subpopulation of a parent species. The process varies in its details and duration, but all modes of speciation are based on reproductive isolation.

19.2 The Main Model for Speciation

LINKS TO
SECTIONS
17.2, 17.6, 17.9

Three models for speciation differ in their basic premise of how populations become reproductively isolated.

START WITH GEOGRAPHIC ISOLATION

The genetic changes leading to a new species usually begin with *physical separation* between populations, so allopatry might be the most common speciation route. By a model for **allopatric speciation**, physical barriers stop gene flow among populations or subpopulations of a species. (*Allo–* means different; *patria* can be taken to mean the homeland.) In both groups, reproductive isolating mechanisms develop. In time, speciation is complete. Interbreeding is no longer possible even if daughter species come into contact with one another.

Whether a geographic barrier can block gene flow depends on an organism's means of travel (deliberate or accidental), how fast it can travel, and whether it is inclined to disperse. Populations of most species are some distance apart, and gene flow is intermittent. Barriers may arise abruptly and end the flow entirely. In the 1800s, a major earthquake buckled part of the Midwest and the Mississippi River changed course. It cut through the habitats of populations of insects that could not swim or fly. It ended the gene flow between those adjoining populations.

The fossil record suggests that geographic isolation generally happens slowly. For example, it happened after vast glaciers advanced into North America and Europe during the ice ages and cut off populations of plants and animals from one another. After glaciers retreated and the descendants of related populations met, some were no longer reproductively compatible. They were separate species. Genetic divergence was not as great between other separated populations, so descendants still interbred. In their case, reproductive isolation was incomplete; speciation did not follow.

Also, remember how Earth's crust is fractured into gigantic plates? Slow, colossal movements inevitably alter the configurations of land masses (Section 17.6). As Central America formed, part of an ancient ocean basin was uplifted, and it became a land bridge—now called the Isthmus of Panama. Some camelids crossed the bridge into South America. Geographic separation led to new species: llamas and vicunas (Figure 19.6).

THE INVITING ARCHIPELAGOS

An **archipelago** is an island chain some distance from a continent. Many chains are so close to the mainland that gene flow is more or less unimpeded, so there is little if any speciation. The Florida Keys are like this. As you read earlier, the Hawaiian Islands, Galápagos Islands, and other remote, isolated archipelagos favor adaptive radiations and speciation (Figures 17.3 and 19.1). The islands are only the tops of volcanoes that started building up on the seafloor. In time they broke the surface of the ocean. We can therefore assume that their fiery surfaces were initially barren, with no life.

In one view, winds or ocean currents carry a few individuals of some mainland species to such islands, as shown in Figure 19.7a. Descendants colonize other

Figure 19.6 Allopatric speciations. The earliest camelids, no bigger than a jackrabbit, evolved in the Eocene grasslands and deserts of North America. By the end of the Miocene, they included the now-extinct *Procamelus*. The fossil record and comparative studies indicate that this may have been the common ancestral stock for llamas (**a**), vicunas (**b**), and camels (**c**). One of the descendant lineages dispersed into Africa and Asia and evolved into modern camels. A different lineage, ancestral to the llamas and vicunas, dispersed into South America after gradual crustal movements formed a land bridge between the two continents.

Late Eocene paleomap, before a land bridge formed between North and South America. At that time, North America and Eurasia were still connected by a land bridge

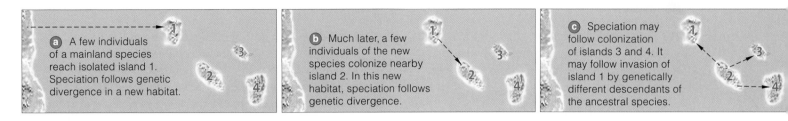

Akepa (*Loxops coccineus*)
Insects, spiders from buds twisted apart by bill; some nectar; high mountain rain forest

Akekee (*L. caeruleirostris*)
Insects, spiders, some nectar, high mountain rain forest

Nihoa finch (*Telespiza ultima*)
Insects, buds, seeds, flowers, seabird eggs; rocky or shrubby slopes

Palila (*Loxioides bailleui*)
Mamane seeds ripped from pods; buds, flowers, some berries, insects; high mountain dry forests

Maui parrotbill (*Pseudonestor xanthrophrys*)
Rips dry branches for insect larvae, pupae, caterpillars; mountain forest with open canopy, dense underbrush

Apapane (*Himatione sanguinea*)
Nectar, especially of ohi'a-lehua flowers; caterpillars and other insects; spiders; high mountain forests

Po'ouli (*Melamprosops phaeosoma*)
Tree snails, insects in understory; last known male died in 2004

Alauahio (*Paroreomyza montana*)
Bark or leaf insects; some nectar, high mountain rain forest

Kauai Amakihi (*Hemignathus kauaiensis*)
Bark-picker; insects, spiders; nectar; high mountain rain forest

Akiapolaau (*H. munroi*)
Probes, digs insects from big trees; high mountain rain forest

Akohekohe (*Palmeria dolei*)
Mostly nectar from flowering trees; some insects, pollen; high mountain rain forest

Iiwi (*Vestiaria coccinea*)
Mostly nectar (Ohia tree flowers, lobelias, some mints); some insects, high mountain rain forest

d The ancestor of Hawaiian honeycreepers might have resembled this housefinch (*Carpodacus*), based on morphological studies, and comparisons of chromosomal DNA and mitochondrial DNA sequences for proteins, such as cytochrome *b*.

Figure 19.7 *Animated!* (**a–c**) Allopatric speciation on an isolated archipelago. (**d**) Twelve of fifty-seven known species and subspecies of Hawaiian honeycreepers, with a sampling of their dietary and habitat preferences. Honeycreeper bills are adapted to diverse foods, such as insects, seeds, fruits, and nectar in floral cups.

islands that form in the chain. Habitats and selection pressures differ within and between these islands, so allopatric speciation proceeds by way of divergences. Later, new species may even invade islands that were colonized by their ancestors. Distances between islands in archipelagos are enough to favor divergence but not enough to stop the occasional colonizers.

The big island of Hawaii formed less than 1 million years ago. Its habitats range from old lava beds, rain forests, and grasslands to snow-capped volcanoes. The first birds to colonize it found a buffet of fruits, seeds, nectars, tasty insects, and few competitors for them. The near absence of competition spurred rapid speciations into vacant adaptive zones. Figure 19.7*d*

shows some of the Hawaiian honeycreepers described earlier. Like thousands of other species of animals and plants, they are unique to this island. As still another example of their potential for speciation, the Hawaiian Islands combined make up less than 2 percent of the world's land masses. Yet they are the original home of 40 percent of all species of fruit flies (*Drosophila*).

> By one allopatric speciation model, some type of physical barrier intervenes between populations or subpopulations of a species and prevents gene flow among them. Gene flow ends, and genetic divergences give rise to daughter species.

19.3 Other Speciation Models

LINKS TO
SECTIONS 4.10,
12.8, 15.4, 17.9

There is evidence that some species have arisen and are being maintained by less common mechanisms in which environmental barriers do not play a role.

ISOLATION WITHIN THE HOME RANGE

By the model for **sympatric speciation**, a species may form *within* the home range of an existing species, in the absence of a physical barrier. *Sym–* means together with, as in "together with others in the homeland."

Evidence From Cichlids in Africa In Cameroon, West Africa, many species of freshwater fishes called cichlids may have arisen by sympatric speciation. The fish live in lakes that formed in the collapsed cones of small volcanoes (Figure 19.8). The cichlids probably colonized the lakes before volcanic action severed the inflow from a nearby river system.

Figure 19.8 A small, isolated crater lake in Cameroon, West Africa, where different species of cichlids may have originated by way of sympatric speciation.

Figure 19.9 Love those polyploids! Among them are several cotton species (including the kind shown here), sugarcane, seedless watermelons, bananas, plums, sweet potatoes, coffee plants with 22, 44, 66, or 88 chromosomes, and marigolds, azaleas, and lilies.

Remember how gene sequences can be compared (Section 17.9)? Ulrich Schliewen looked at differences in nuclear DNA and mitochondrial DNA for eleven cichlid species in Barombi Mbo, one of the small crater lakes. He also compared the samples with DNA from cichlid species in nearby lakes and rivers. He found that cichlid species in Barombi Mbo are more closely related to one another than to neighboring species. He concluded that all of the Barombi Mbo cichlids are descended from the same ancestral species—and that speciation must have occurred *within* this lake.

What could have cause the divergences that led to speciation? The lake is only 2.5 kilometers across, so it is not likely that cichlid populations were separated from one another by any type of barrier. Also, physical and chemical conditions are uniform throughout the lake. Another point: Cichlids are good swimmers, so individuals of different species often meet up.

However, the Barombi Mbo species do show some *ecological* separation. Feeding preferences put species in different places. Some feed in open waters, others at the lake bottom. Yet they all breed close to the lake bottom, in sympatry. Was this small-scale ecological separation enough to promote sexual selection among potential mates? Possibly. Over time, it may have led to reproductive isolation, then speciation.

Polyploidy's Impact Reproductive isolation might happen within a few generations through **polyploidy**, in which individuals inherit three or more sets of the chromosomes characteristic of their species (Section 12.8). Either a somatic cell fails to divide mitotically after its DNA is duplicated, or nondisjunction occurs at meiosis and results in an unreduced chromosome number in gametes. Offspring usually cannot breed or mate successfully with the parent species, but they may be able to reproduce asexually.

*Auto*polyploids arise by a doubling of the parental chromosome number. This event arises spontaneously in nature but can be in artificially induced in plant breeding laboratories. Breeders expose dividing plant cells to colchicine which, recall, stops microtubular spindles from forming during mitosis (Section 4.10). Without the spindle, duplicated chromosomes do not separate, and cells with the unreduced chromosome number may function as gametes.

*Allo*polyploids originate through (1) spontaneous or induced hybridization between closely related species and (2) doubling of the chromosome number. Figure 19.9 is one example. As genome studies reveal, many stable allopolyploids originated long ago. The kinds produced in the laboratory may or may not prove to

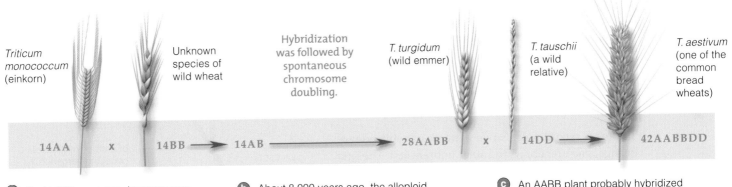

(a) By 11,000 years ago, humans were cultivating wild wheats. Einkorn has a diploid chromosome number of 14 (two sets of 7). It probably hybridized with another wild wheat species having the same number of chromosomes.

(b) About 8,000 years ago, the alloploid called wild emmer originated from an AB hybrid wheat plant in which the chromosome number doubled. Wild emmer is tetraploid, or AABB; it has two sets of 14 chromosomes.

(c) An AABB plant probably hybridized with *T. tauschii*, a wild relative of wheat. Its diploid chromosome number is 14 (two sets of 7 DD). Common bread wheats have a chromosome number of 42 (six sets of 7 AABBDD).

be stable and fertile. Attempts are more successful if the species are close relatives.

Plant speciation is rapid when polyploids produce fertile offspring by self-fertilizing or cross-fertilizing with an identical polyploid. The ancestor of common bread wheat apparently was a wild species, *Triticum monococcum*, which spontaneously hybridized about 11,000 years ago with another wild species (Figure 19.10). Much later in time, a spontaneous chromosome doubling gave rise to *T. turgidum*, an alloploid species with two sets of chromosomes (AABB). Later still, another hybridization resulted in *T. aestivum*, a bread wheat with a chromosome number of 42.

About 95 percent of fern species and 30–70 percent of flowering plants are polyploid species. So are a few conifers, mollusks, insects, and other arthropods, as well as fishes, amphibians, and reptiles. What about mammals? In 1999, A polyploid species of rat with a chromosome number of 102 was found in Argentina.

ISOLATION AT HYBRID ZONES

Parapatric speciation might proceed when different selection pressures operating across a broad region affect populations that are in contact along a common border. Hybrids that form in the contact zone are less fit than individuals on either side of it. Because the hybrids are being selected against, they appear in the hybrid zone only (Figure 19.11).

> By a sympatric speciation model, daughter species arise from a group of individuals within an existing population. Polyploid flowering plants probably formed this way.
>
> By a parapatric speciation model, populations maintaining contact along a common border evolve into distinct species.

Figure 19.10 *Animated!* Presumed sympatric speciation in wheat. Wheat grains 11,000 years old and diploid wild wheats have been found in the Near East, and chromosome analysis indicates that they hybridized. Later, in a self-fertilizing hybrid, homologous chromosomes failed to separate at meiosis, and it produced fertile polyploid offspring. A polyploid descendant hybridized with a wild species. We make bread from grains of their hybrid descendants.

Figure 19.11 Example of parapatric speciation on the island of Tasmania, directly south of eastern Australia. (**a**) Giant velvet worm, *Tasmanipatus barretti* and (**b**) blind velvet worm, *T. anophthalmus*.

(**c**) Both of these rare species of velvet walking worms live in adjoining regions of northeastern Tasmania. Their habitats overlap in a hybrid zone. Hybrid offspring are sterile, which may be the main reason these two species are maintaining separate identities in the absence of an obvious physical barrier between their habitats.

19.4 Patterns of Speciation and Extinction

LINK TO
SECTION
17.7

All species, past and present, are related by descent. They share genetic connections through lineages that extend back in time to the molecular origin of life.

BRANCHING AND UNBRANCHED EVOLUTION

The fossil record reveals two patterns of evolutionary change, one branching, the other unbranched. The first is known as **cladogenesis** (from *klados*, branch; and *genesis*, origin). In this pattern, a lineage splits when one or more of its populations become reproductively isolated and diverge genetically. It might be the main speciation pattern. It is the one introduced earlier, in Section 19.1.

In the second pattern, **anagenesis**, changes in allele frequencies and morphology accumulate in a single line of descent. (In this context, *ana*– means renewed.) Directional change is confined within that lineage, as gene flow continues among its populations. In time, allele frequencies and morphology shift so much that the new type differs significantly from the ancestral type, so it is classified as a separate species.

RATES OF CHANGE IN FAMILY TREES

Evolutionary trees summarize information on the relationships among groups. Figure 19.12 can start you thinking about how to construct these tree diagrams. Each branch represents one line of descent from a common ancestor. A *branch point* represents a time of genetic divergence.

When plotted against time, a branch that ends before the present (the treetop) signifies that the lineage is extinct. A dashed line signifies that we know something about the lineage but not exactly where it fits in the tree.

The **gradual model of speciation** holds that species originate by slight morphological changes over long time spans. The model fits with many fossil sequences. For example, sedimentary rock layers often hold vertical sequences of fossilized shells of foraminiferans, as in Figure 19.13. The sequence reflects gradual morphological change.

The **punctuation model of speciation** offers a different explanation for patterns of speciation. Most morphological changes are said to evolve

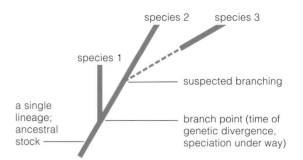

Figure 19.12 Some elements of evolutionary tree diagrams.

in a relatively brief geologic period, within the tens to hundreds of thousands of years when populations are starting to diverge. Directional selection, genetic drift, the founder effect, bottlenecks, or some combination of them favor rapid speciation. The daughter species recover fast from the adaptive wrenching, then they change very little over long periods.

The fossil record shows that stability prevailed for all but 1 percent of the history of most lineages, but it also reveals episodes of abrupt change. As it turns out, both models help explain speciation patterns. Changes have been gradual, abrupt, or both. Species originated at different times and have differed in how long they last. Some did not change much over millions of years; others were the start of adaptive radiations.

ADAPTIVE RADIATIONS

An **adaptive radiation** is a burst of divergences from a single lineage that leads to many new species. This is the pattern that gave rise to the family of Hawaiian honeycreepers. It requires **adaptive zones**, or a set of niches that come to be filled by a group of usually related species. Think of a *niche* as a way of life, such as "burrowing into seafloor sediments" or "catching winged insects in the air at night." Either the lineage enters a vacant adaptive zone or it competes with the resident species well enough to displace them.

You will read more about niches in Chapter 46, in the context of community structure. For now, be aware of two concepts. First, a species must have physical access to a niche when it opens up. Mammals were once distributed in the uniformly tropical regions of Pangea. That supercontinent broke up into huge land

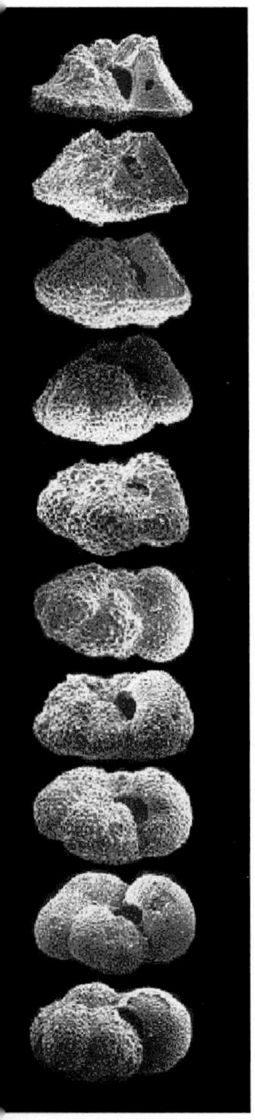

Figure 19.13 Fossilized foraminiferan shells from a vertical sequence of sedimentary rock layers. The first shell (*bottom*) is 64.5 million years old. The most recent (*top*) is 58 million years old. Analysis of shell patterns confirm that the evolutionary order matches the geological sequence.

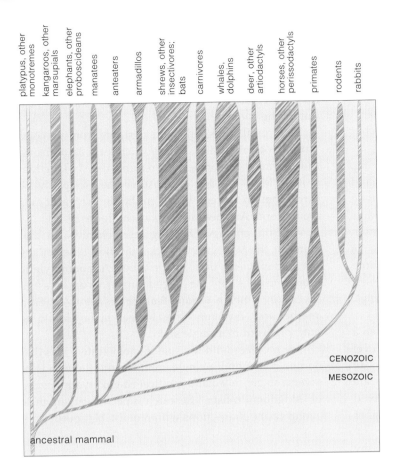

platypus, other monotremes
kangaroos, other marsupials
elephants, other proboscideans
manatees
anteaters
armadillos
shrews, other insectivores; bats
carnivores
whales, dolphins
deer, other artiodactyls
horses, other perissodactyls
primates
rodents
rabbits

CENOZOIC

MESOZOIC

ancestral mammal

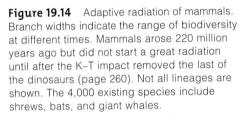

Figure 19.14 Adaptive radiation of mammals. Branch widths indicate the range of biodiversity at different times. Mammals arose 220 million years ago but did not start a great radiation until after the K–T impact removed the last of the dinosaurs (page 260). Not all lineages are shown. The 4,000 existing species include shrews, bats, and giant whales.

The photograph shows a fossil of *Eomaia scansoria* (Greek for ancient mother climber). About 125 million years ago, this insectivore crawled on low shrubs and branches. At this writing, it is the earliest placental mammal we know about.

masses, which drifted apart. Habitats and resources changed in different ways, in different places, and set the stage for independent radiations (Figure 19.14).

Second, a species may enter an adaptive zone by a **key innovation**: A chance modification in some body structure or function gives it the opportunity to exploit the environment more efficiently or in a novel way.

Once a species has entered an adaptive zone, genetic divergences can give rise to other species, which can fill a variety of niches within the zone. For example, when the forelimbs of certain vertebrate evolved into wings, novel niches opened up for the ancestors of modern birds and bats (Section 17.7).

EXTINCTIONS—THE END OF THE LINE

An **extinction** is the irrevocable loss of a species. By some estimates, more than 99 percent of all species that ever lived are extinct. The chapter introduction gave examples of typical causes, including imports of new predators and climate change.

In addition to ongoing, small-scale extinctions, the fossil record indicates there were at least twenty or more **mass extinctions**, or catastrophic losses of entire families or other major groups. They differed in size.

For example, 250 million years ago, 95 percent of all known species were abruptly lost. At other times, fewer groups were lost. Afterward, biodiversity slowly recovered as new species filled vacant adaptive zones.

Luck, again, had a lot to do with it. Many species were wiped out by global climate change. When one asteroid struck Earth and the last dinosaurs vanished, mammals were among the survivors that could radiate into vacated adaptive zones. Asteroids, imperceptibly drifting continents, climatic change—all contributed to past patterns of major extinctions and recoveries. In the next unit, you will have plenty of examples.

Lineages have changed gradually, abruptly, or both. Their member species originated at different times and have differed in how long they have persisted.

An adaptive radiation is the rapid origin of many species from a single lineage. It happens when an adaptive zone, a set of similar niches, opens up and the lineage has physical, evolutionary, and ecological access to it.

Repeated and often large extinctions happened in the past. After times of reduced biodiversity, new species originated and occupied new or vacated adaptive zones.

19.5 Organizing Information About Species

LINK TO
SECTION
1.3

So far, you have been thinking about what a species is, how it originates, and what has become of the many, many millions of them that originated. Turn now to what taxonomists do with the information.

SETS OF ORGANISMS—THE HIGHER TAXA

One field of biology, *taxonomy*, deals with identifying, naming, and classifying species. It goes hand in hand with *systematics*, or the study of relationships among organisms. Any organism that has been identified as representing a new species is assigned a unique two-part scientific name, the first part being the genus. As outlined in Chapter 1, species are grouped into more inclusive categories, such as families, orders, classes, phyla (or divisions, which is an equivalent ranking). Figure 19.15 has a few examples.

Each set of organisms in a given category is called a **taxon** (plural, taxa). The sets above the species level are known as the higher taxa, which are the units of classification systems. Most classification systems are now phylogenic, meaning that they reflect perceived evolutionary connections within and between higher taxa as well as patterns of evolutionary change.

A **six-kingdom classification system** promoted by Robert Whittaker prevailed for some time. It assigned all of the prokaryotic species to kingdoms Eubacteria and Archaea, and all single-celled eukaryotes (as well as many multicelled species) to kingdom Protista. It

bestowed separate kingdom status on animals, plants, and fungi. Figure 19.16 shows the six kingdoms of this system. Section 1.3 sketched out a few defining traits for their representatives, which are topics of the next unit of the book.

New fossil finds, and new insights from geology, morphological studies, and biochemical comparisons, caused many researchers to rethink the six-kingdom system. Most decided to subsume the groups into a **three-domain system**, in which the three highest taxa are Bacteria, Archaea, and Eukarya (Figure 19.16).

Why the change? Ongoing research revealed that many of the taxa in earlier classification schemes are not monophyletic, or a "single tribe." A **monophyletic group** includes only the descendants from an ancestral species in which a unique feature first evolved. Said another way, the branchings in each taxon should be outgrowths from a single stem.

One problem with the six-kingdom system was that no one could find a single stem for the thousands of diverse single-celled and multicelled eukaryotic species of "kingdom Protista." Research is now clarifying their evolutionary connections with remarkable speed.

A CLADISTIC APPROACH

Think of each set of species descended from just one ancestral species as a **clade** (from *klados*, a Greek word for branch or twig). A cladistic classification system

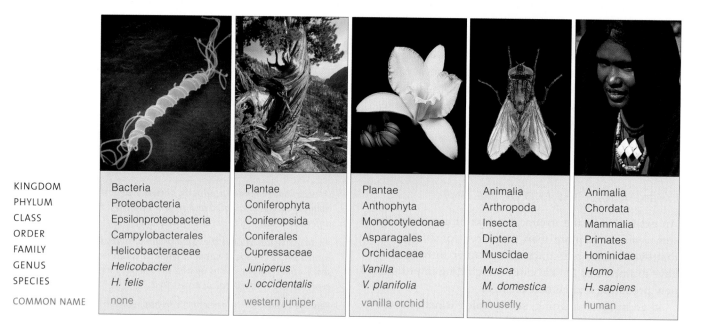

KINGDOM	Bacteria	Plantae	Plantae	Animalia	Animalia
PHYLUM	Proteobacteria	Coniferophyta	Anthophyta	Arthropoda	Chordata
CLASS	Epsilonproteobacteria	Coniferopsida	Monocotyledonae	Insecta	Mammalia
ORDER	Campylobacterales	Coniferales	Asparagales	Diptera	Primates
FAMILY	Helicobacteraceae	Cupressaceae	Orchidaceae	Muscidae	Hominidae
GENUS	*Helicobacter*	*Juniperus*	*Vanilla*	*Musca*	*Homo*
SPECIES	*H. felis*	*J. occidentalis*	*V. planifolia*	*M. domestica*	*H. sapiens*
COMMON NAME	none	western juniper	vanilla orchid	housefly	human

Figure 19.15 Taxonomic classification of five species. Each species has been assigned to ever more inclusive sets of organisms—in this case, from species to kingdom.

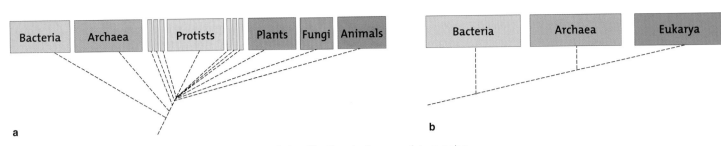

a

b

Figure 19.16 *Animated!* **(a)** Six-kingdom system of classification. In time, protists may be divided into more kingdoms. **(b)** The more recent three-domain system of classification. Protists, plants, fungi, and animals share features that unite them in domain Eukarya.

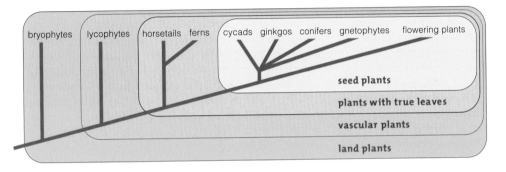

Figure 19.17 Evolutionary tree for land plants, with monophyletic groups nested as sets within sets. All but bryophytes have vascular tissues. All but bryophytes and lycophytes have true leaves. All but bryophytes, lycophytes, horsetails, and ferns produce seeds.

defines clades in terms of the history of divergences, or branch points in time. Only species that share traits derived from the last common ancestor are in the same clade. A **derived trait** is a novel feature that evolved in one species and is present only in its descendants. This emphasis means the descendants within a clade can differ—sometimes exuberantly so—in other traits.

Evolutionary tree diagrams called **cladograms** use the position of branch points from the last shared ancestor to convey inferred evolutionary relationships (phylogenies) among taxa. A cladogram is an estimate of "who came from whom." It has no time bar with absolute dates, so it cannot convey differences in rates of evolution among taxa. Even so, a cladistic approach has already reinforced part of the fossil record, and it is making us reevaluate interpretations of the past.

A tree of life only looks like a simple stick drawing. Many thousands of morphological and biochemical traits were analyzed during the attempts to make its evolutionary connections. For example, as you will see in the next unit of the book, detailed comparisons of genes and ribosomal RNAs have revealed sometimes surprising similarities and differences among groups. The evolutionary tree of life in Section 19.6 is based on such combined evidence.

The derived traits used to construct a cladogram also help us visualize different monophyletic groups as *sets within sets*. For example, Figure 19.17 shows a cladogram for major sets of land plants that have been nested into ever larger categories. We assume that the cycads, ginkgos, conifers, gnetophytes, and flowering plants form one set, because only they have a common ancestor that was the first seed-producing plant. The seed plants are nested in a larger set—plants with true leaves—that includes horsetails and ferns but excludes lycophytes and bryophytes. Only the bryophytes are not nested in the still-larger set called vascular plants; they do not have tubelike tissues that deliver water and solutes throughout the plant body.

The section to follow shows you how to construct a cladogram. It provides a closer look at the advantages and some of the pitfalls of a cladistic approach.

Taxonomists identify, name, and classify sets of organisms into ever more inclusive categories, the higher taxa.

Classification systems organize and simplify the retrieval of information about species. Phylogenetic systems attempt to reflect evolutionary relationships among species.

Reconstructing the evolutionary history of a given lineage is based on detailed understanding of the fossil record, morphology, life-styles, and habitats of its representatives, and on biochemical comparisons with other groups.

Recent evidence, especially from comparative biochemistry, favors the grouping of organisms into a three-domain system of classification—Archaea, Bacteria, and Eukarya (protists, plants, fungi, and animals).

19.6 How To Construct a Cladogram

LINK TO
SECTION
17.6

In case you would like to know how a cladogram can be constructed, here is a step-by-step approach.

Suppose you want to make a cladogram for vertebrates. You select an *ingroup* of organisms with traits that suggest they might be related—in this case, jaws and paired appendages. You focus on sharks, mammals, crocodiles, and birds because they also differ clearly in some morphological, physiological, and behavioral traits, or characters. Now you must select a different vertebrate that can be used as a reference point for estimating evolutionary distances within the ingroup.

To keep things simple, you check for the presence (+) or absence (–) of seven traits and tabulate them, as in Figure 19.18a. After scanning Chapter 26, you decide that lampreys are only distantly related to the other four vertebrates. For instance, although a tubular structure called a notochord forms in its embryos, as it does for all other vertebrates, only lampreys have no jaws or paired appendages, such as lateral fins and legs. They can be the *outgroup*, the one with the fewest derived traits when compared to the others.

Derived traits, recall, are evidence of morphological divergence and branching in an evolutionary tree. In Figure 19.18b, each zero (0) across the columns of traits for each vertebrate indicates an ancestral condition. Each numeral one (1) means the vertebrate shows the derived trait.

Now you look for derived traits that the selected groups do or do not share. For example, the crocodile, mammal, and bird share five derived traits, but the bird and the shark share only three. You can now make a simple cladogram, although systematists often use many traits of many taxa. Typically they use a computer to analyze data and find the pattern that is best supported by a lot of information.

Figure 19.18c–g shows how a cladogram develops as you keep adding information to it. Start with the presence or absence of jaws and paired appendages, two traits that all

groups except lampreys derived from a common ancestor. What about lungs? Like lampreys, sharks do not have them, so you have identified another branch point in vertebrate evolution. Past that branch point, only mammals have hair, only crocodiles and birds have some form of gizzard. Only birds and their immediate ancestors have feathers.

How do you "read" the final cladogram? Remember, it is an estimate of *relative* relatedness, which implies common ancestry. Birds are more closely related to crocodiles than they are to mammals. Crocodiles are not the ancestor of birds (they are modern organisms, too), but both share a more recent common ancestor than either does with mammals. Birds, crocodiles, and mammals are closer to one another evolutionarily than they are to the shark.

The higher up a branch point is on a family tree, the more derived traits are shared. The lower the position of the branch point between two groups in the diagram, the fewer traits are shared with other groups being investigated.

A few words of caution: Interpretations of evolutionary relationships are more reliable when many traits are used, and there must be strong evidence that shared traits are derived. This helps counter the impact of a bad choice, such as including a trait that is a result of morphological convergence rather than divergence (Section 17.6).

The choice of derived traits is essential. If you were to select body size, for instance, you might wrongly perceive an evolutionary connection between *Sauroposeidon* (a dinosaur that weighed 60 tons), blue whales (mammals that weigh 200,000 pounds), quaking aspen (one plant has 50,000 stems and weighs an estimated 13 million pounds), and a honey mushroom that has been growing for 2,400 years (its underground body extends through 2,200 acres).

Cladograms are only as good as the choices made for their construction—and good choices start with a broad, deep knowledge of life.

Figure 19.18 (a) Charting out a selection of traits among vertebrate groups that can be used for the construction of a simple cladogram.

(b) A trait's absence in an outgroup or ingroup indicates an ancestral state (here indicated by a zero). Its presence in the set of vertebrates selected as the ingroup is taken to mean it is a derived trait, indicated by a numeral one.

(c–g) Step-by-step construction of a cladogram, as explained in the text.

Taxon	Traits, or Characters						
	Notochord in Embryo	Jaws	Paired Appendages	Lungs	Hair	Gizzard	Feathers
Lamprey	+	–	–	–	–	–	–
Shark	+	+	+	–	–	–	–
Crocodile	+	+	+	+	–	+	–
Mammal	+	+	+	+	+	–	–
Bird	+	+	+	+	–	+	+

a

Taxon	Traits, or Characters						
	Notochord in Embryo	Jaws	Paired Appendages	Lungs	Hair	Gizzard	Feathers
Lamprey	1	0	0	0	0	0	0
Shark	1	1	1	0	0	0	0
Crocodile	1	1	1	1	0	1	0
Mammal	1	1	1	1	1	0	0
Bird	1	1	1	1	0	1	1

b

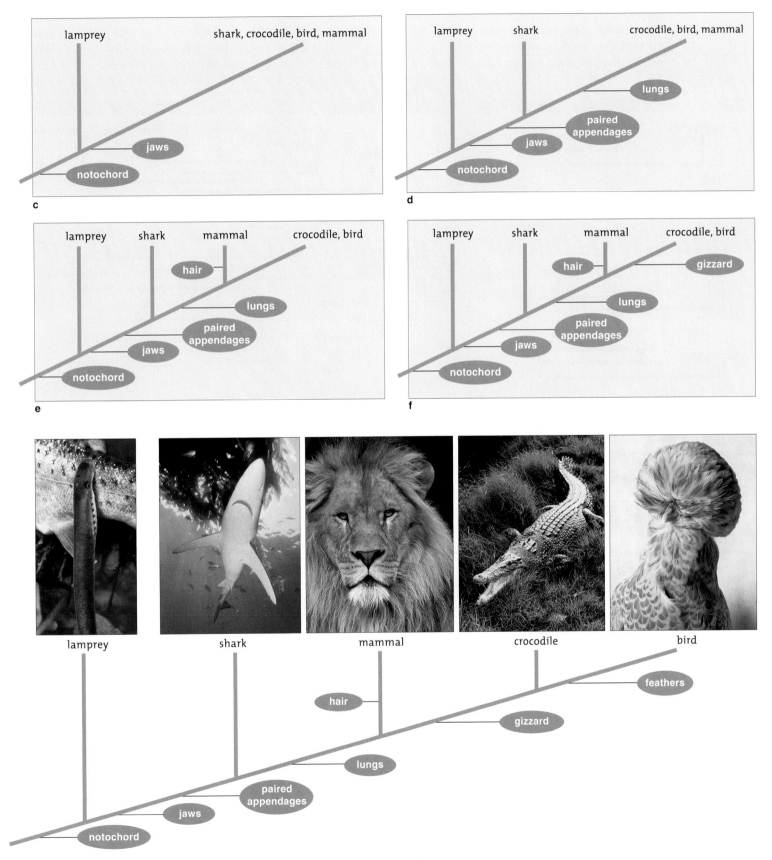

c

lamprey   shark, crocodile, bird, mammal

jaws

notochord

d

lamprey   shark   crocodile, bird, mammal

lungs

paired
appendages

jaws

notochord

e

lamprey   shark   mammal   crocodile, bird

hair

lungs

paired
appendages

jaws

notochord

f

lamprey   shark   mammal   crocodile, bird

hair   gizzard

lungs

paired
appendages

jaws

notochord

lamprey   shark   mammal   crocodile   bird

feathers

hair

gizzard

lungs

paired
appendages

jaws

notochord

g The completed diagram, livened up with photographs of representative species.

Chapter 19 Evolutionary Patterns, Rates, and Trends   313

19.7 Preview of Life's Evolutionary History

Figure 19.19, a tree of life, shows the macroevolutionary links among major groups of organisms, as described in the next unit. Each set of organisms (taxon) has living representatives. Each branch point represents the last common ancestor of the set above it. The small boxes within domains Archaea and Eukarya highlight taxa that are currently being recognized as the equivalent of kingdoms in earlier classification systems.

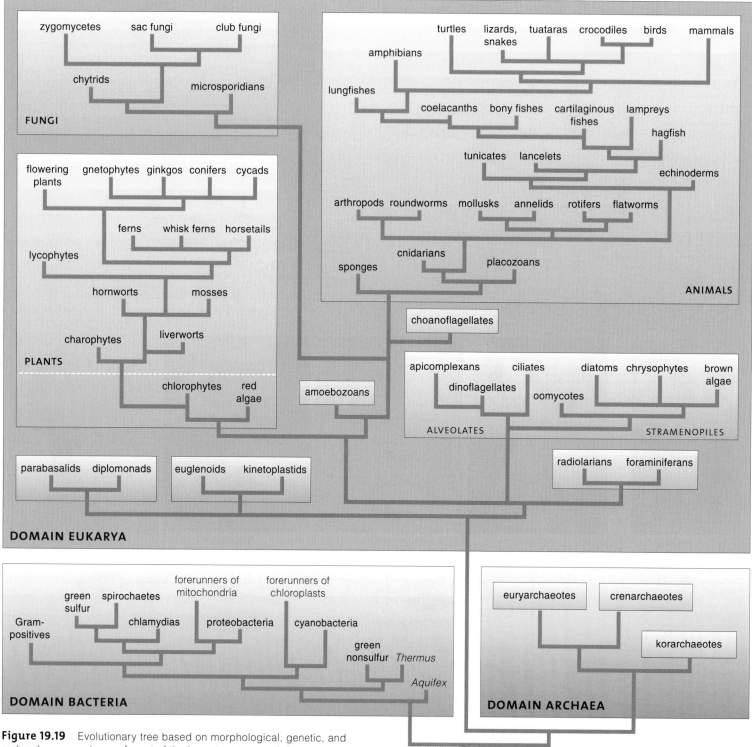

Figure 19.19 Evolutionary tree based on morphological, genetic, and molecular comparisons of most of the important groups. It is a work in progress, subject to refinements as more information comes in.

biochemical and molecular origin of life

19.8 Madeleine's Limbs

*So what does macroevolution have to do with
you and me? Everything.*

In August of 1994, about 900 million years after the
first animals appeared on Earth, Madeleine made *her*
entrance. Her grandmothers and aunts made a quick
count—arms, legs, ears, and eyes, two of each; fully
formed mouth and nose—just to be sure these were
present and accounted for. One grandmother, having
been too long in the company of biologists, had an
epiphany as she witnessed Madeleine's birth. In that
instant she sensed ancestral connections between the
distant past and, through this child, the future.

Madeleine's body plan did not emerge out of thin
air. Thirty-five thousand years ago, people just like us
were having children just like Madeleine. If we are
reading the fossil record correctly, then five million
years ago, the offspring of individuals on the road to
modern humans resembled her in some respects but
not others. Sixty million years ago, primate ancestors
of those individuals were giving birth precariously, up
in the trees. Two hundred fifty million years ago, the
mammalian ancestors of those primates were giving
birth—and so on back in time to the very first animals,
which had no limbs or eyes or noses at all.

We know little about the very first animals. Yet one
thing is clear. By the dawn of the Cambrian, they had
given rise to all major groups of invertebrates and to
Madeleine's backboned, jawed ancestors. We know this
from fossils. For example, one Cambrian community
flourished 530 million years ago in, on, and above the
dimly lit mud in a submerged basin between a steep
reef and the coast of an early continent. About 500 feet
below the surface, the water was oxygenated and clear
(Figure 19.20). Like a castle built from wet sand, their
home was unstable and an underwater avalanche buried
them. Over great time spans, compaction and chemical
change transformed those small, flattened animals into
fossils. By 1909, tectonic activity had moved the fossils
high into the eastern mountains of British Columbia,
and there a fossil hunter found them.

In the next unit, you will compare body plans of
diverse organisms. Such comparisons give insight into
evolutionary relatedness and help us construct family
trees. As you poke through the tree branches, make use
of the evolutionary perspective. At each branch point,
the processes of microevolution gave rise to workable
changes in body plans. Your collection of conserved
and modified traits, and Madeleine's, evolved earlier
in countless generations of vertebrates and, even before
them, in ancient invertebrate forms.

Our family tree is a record of conserved and derived traits.

Figure 19.20 *Left*, reconstruction of a few Cambrian animals known from fossils
of the Burgess Shale in British Columbia. *Right*, Madeleine.

Summary

Section 19.1 Populations of each species share at least one unique trait, a legacy of a common ancestor. In sexually reproducing species, individuals interbreed, produce fertile offspring under natural conditions, and are reproductively isolated from all other species.

If gene flow ends between populations, divergences may lead to new species. Mutation, natural selection, and genetic drift operate independently and may give rise to reproductive isolating mechanisms (Table 19.1). In some cases, reproductive isolation occurs in a few generations.

Prezygotic isolating mechanisms stop interbreeding. They include incompatibilities between reproductive parts or between gametes, differences in reproductive timing or behavior, and ecological restriction to different microenvironments in the same area. The postzygotic mechanisms lead to early death, sterility, or unfit hybrid offspring. They come into play after fertilization.

Biology⊛Now
Use the animation and interaction on BiologyNow to explore how species become reproductively isolated.
Read the Infotrac article "Tracking the Red-Eyed, Sluggish, and Ear-splitting," Tabitha M. Powledge, American Scientist, *July 2004.*

Section 19.2 By the allopatric speciation model, a geographic barrier cuts off gene flow between two or more populations. Genetic divergence and reproductive isolation are favored and may result in a new species.

Biology⊛Now
Learn more about speciation on an archipelago with the animation on BiologyNow.

Section 19.3 By a sympatric speciation model, populations in physical contact diverge from each other. Polyploid species of many plants and some animals have originated by chromosome doublings and hybridizations.

By a parapatric speciation model, different selection pressures across a broad region act on populations that are in contact along a common border. Unfit hybrids form in the contact zone, so populations on either side diverge independently from each other.

Biology⊛Now
Explore the effects of sympatric speciation in wheat with the animation on BiologyNow.

Section 19.4 Macroevolution refers to the timing, duration, and direction of speciation in the history of life. These features differ among lineages. The major speciation patterns are unbranched (evolution within a single lineage) or branching (divergences from ancestral stock). Most lineages remain stable for long periods, but abrupt episodes of change also have occurred.

Radiations occur in adaptive zones, a similar set of niches that come to be filled by a (usually) related group of species. A niche is a way of life, such as catching insects in the air at night. Species must have physical, evolutionary, and ecological access to these zones.

A key innovation is a chance modification in some body structure or function that lets an organism exploit the environment more efficiently or in a novel way.

Most species are now extinct. Mass extinctions, slow recoveries, and adaptive radiations are major patterns.

Section 19.5 Each species has a unique, two-part scientific name. Taxonomy deals with identifying, naming, and classifying species. Systematics deals with reconstructing life's evolutionary history (phylogeny). In classification systems, sets of organisms (taxa) are organized into ever more inclusive categories as a way to retrieve information about species.

A current three-domain classification system is based largely on phylogenetic evidence. It recognizes three domains: Bacteria, Archaea, and Eukarya. The Eukarya includes diverse lineages known informally as protists, as well as plants, fungi, and animals.

Biology⊛Now
Review biological classification systems with the animation on BiologyNow.

Section 19.6 A cladistic classification system recognizes monophyletic groups. Each group is a clade, a set of species that includes only descendants that display a derived trait, inherited from an ancestor in which that trait first evolved. It is the equivalent of all branches growing from the same point on a stem.

Biology⊛Now
Read the InfoTrac article "How Taxonomy Helps Us Make Sense Out of the Natural World," Sue Hubbell, Smithsonian, *May 1996.*

Sections 19.7, 19.8 Representing life's history as a tree with branchings from ancestral stems brings clarity to the view that all organisms are related by descent.

Table 19.1	Summary of Processes and Patterns of Evolution		

Microevolutionary Processes

Mutation	Original source of alleles	Stability or change in heritable traits that define populations, and the species, is the outcome of balances or imbalances among all of these processes. Population size and prevailing conditions in the environment influence the outcome.
Gene flow	Preserves species cohesion	
Genetic drift	Erodes species cohesion	
Natural selection	Preserves or erodes species cohesion, depending on environmental pressures	

Macroevolutionary Processes

Genetic persistence	The basis of the unity of life. The biochemical and molecular basis of inheritance extends from the origin of first cells through all subsequent lines of descent.
Genetic divergence	Basis of life's diversity, as brought about by adaptive shifts, branching, and radiations. Rates and times of change varied within and between lineages.
Genetic disconnect	End of the line for a species. Mass extinctions are catastrophic events in which major groups abruptly and simultaneously are lost.

Figure 19.21
Rama the cama
displaying his
unexpected
short temper.

Self-Quiz
Answers in Appendix II

1. _____ can isolate one population from others.
   a. Structural traits   c. Behavioral traits
   b. Functional traits   d. all of the above

2. Reproductive isolating mechanisms _____ .
   a. stop interbreeding   c. reinforce genetic divergence
   b. stop gene flow   d. all of the above

3. Most species originate by a (an) _____ route.
   a. allopatric   c. parapatric
   b. sympatric   d. parametric

4. In evolutionary trees, a branch point represents a
   _____ ; and a branch that ends represents _____ .
   a. single species; incomplete data on lineage
   b. single species; extinction
   c. time of divergence; extinction
   d. time of divergence; speciation complete

5. Fossil evidence supports the _____ model of
   evolutionary change.
   a. punctuation   b. gradual   c. both are correct

6. *Pinus banksiana, Pinus strobus,* and *Pinus radiata*
   are _____ .
   a. three families of pine trees
   b. three different names for the same organism
   c. three species grouped in the same genus
   d. both a and c

7. Individuals of a monophyletic group _____ .
   a. are all descended from an ancestral species
   b. demonstrate morphological convergence
   c. have a derived trait that first evolved in their
      last shared ancestor
   d. both a and c

8. A(n) _____ classification system reflects presumed
   evolutionary relationships.
   a. epigenetic   c. phylogenetic
   b. tectonic   d. both b and c

9. In modern classification systems, groupings of sets
   of taxa range from _____ to _____ .
   a. kingdom; genera and species
   b. kingdom; genera and domain
   c. genera; domain and kingdom
   d. species; kingdom and domain

10. Match these terms suitably.
    ____ phylogeny   a. now the most inclusive taxon
    ____ extinction   b. tree of branching lineages
    ____ domain   c. many lineages diverge from
    ____ derived trait   one in a new adaptive zone
    ____ cladogram   d. end of a species or lineage
    ____ adaptive   e. evolutionary history of species
         radiation   f. only in descendants of ancestor
                            in which it first evolved

Additional questions are available on Biology🌀Now™

Critical Thinking

1. You notice several duck species in the same lake habitat, with no physical barriers hampering the ducks' movements. All the females of the various species look quite similar to one another. But the males differ in the patterning and coloration of their feathers. Speculate on which forms of reproductive isolation may be keeping each species distinct. How does the appearance of the male ducks provide a clue to the answer?

2. *Rama the cama*, a llama-camel hybrid, was born in 1997 (Figure 19.21). Camels and llamas have a shared ancestor but have been separated for 30 million years. Veterinarians collected semen from a male camel that weighed close to 1,000 pounds, then used it to artificially inseminate a female llama one-sixth his weight. The idea was to breed an animal having a camel's strength and endurance and a llama's gentle disposition.

   Instead of being large, strong, and sweet, Rama is smaller than expected and has a camel's short temper. Rama resembles both parents, with a camel's long tail and short ears but no hump, and llama-like hooves rather than camel footpads. Now old enough to mate, he is too short to get together with a female camel and too heavy to mount a female llama. He has his eye on Kamilah, a female cama born in early 2002, but will have to wait several years for her to mature. The question is, will any offspring from such a match be fertile?

   What does Rama's story tell you about the genetic changes required for irreversible reproductive isolation in nature? Explain why a biologist might not view Rama as evidence that llamas and camels are the same species.

3. Speculate on what might have been a key innovation in human evolution. Describe how that innovation might be the basis of an adaptive radiation in environments of the distant future.

4. Shannon thinks there are too many major taxa and sees no reason to make a new one for something as tiny as archaeans. "Keep them with the other prokaryotes!" she says. Taxonomists would call her a "lumper." But Andrew is a "splitter." He sees no reason to withhold separate status from archaeans simply because they are part of a microscopic world that not many people know about. Which may be the most useful: more or fewer boundaries between groups? Explain your answer.

5. Richard Lenski uses bacterial populations in culture tubes to develop model systems for studying evolution. Bacteria produce several generations in a day. Researchers can store them in the deep freeze, then bring them back to active form, unaltered, to directly compare ancestors and their descendants. Are bacterial models relevant to any evolutionary studies of sexually reproducing organisms? Before you answer, read a short article by P. Raine and M. Travisano entitled "Adaptive Radiation in a Heterogeneous Environment" (*Nature*, July 2, 1998: 69–72).

20 LIFE'S ORIGIN AND EARLY EVOLUTION

Looking for Life in All the Odd Places

In the 1960s, microbiologist Thomas Brock was looking for signs of life in the hot springs and pools in Yellowstone National Park (Figure 20.1). He found a simple ecosystem of microscopically small cells, including *Thermus aquaticus*. This prokaryote uses simple carbon compounds dissolved in the water as its energy source. It is known as one of the thermophiles, or "heat lovers," for good reason. *T. aquaticus* withstands temperatures on the order of 80°C (176°F)!

Brock's work had two unexpected results. First, it put researchers on paths that led them to a great domain of life, the Archaea. Second, it led to a faster way to copy DNA and end up with useful amounts of it. *T. aquaticus* happens to make a heat-resistant enzyme, and it can catalyze the polymerase chain reaction—PCR. Synthetic forms of the enzyme helped trigger a revolution in biotechnology.

Bioprospecting became the new game in town. Many companies started to look closely at thermal pools and other extreme environments for species that might yield valuable products. They found forms of life adapted to extraordinary levels of temperature, acidity, alkalinity, salinity, and pressure.

To extreme thermophiles on the seafloor, Yellowstone's hot water would be too cool. They live in the superheated, mineral-rich water near hydrothermal vents. One kind even grows and reproduces at 121°C (249°F). Different species live in acidic springs, where the pH approaches zero, and in highly alkaline soda lakes. In Earth's polar regions, some types cling to life in salt ponds that never freeze and in glacial ice that never melts.

Extreme environments also support some eukaryotic species of ancient lineages. Populations of snow algae tint mountain glaciers red. Another red alga, *Cyanidium caldarium,* is a resident of acidic hot springs. Free-living photosynthetic cells called diatoms live in extremely salty lakes, where the hypertonicity would make cells of most organisms shrivel and die.

What could top that? Nanobes. Australian researchers found nanobes growing 3.8 kilometers (3 miles) below Earth's surface in truly hot rocks—170°C (338°F). Being one-tenth the size of most bacteria, nanobes cannot be observed without electron microscopes. Outwardly, they look something like the simplest fungi (Figure 20.2).

Nanobes are probably too small to be alive. They do not seem to be big enough to hold all of the metabolic machinery that now runs life processes. Even so, nanobes do contain DNA. And they appear to grow. Are they like proto-cells, which preceded the origin of the first living cells? Maybe.

Watch the video online!

Figure 20.1 From a thermal pool in Yellowstone National Park, cells of *Thermus aquaticus*, a prokaryotic species that is immensely admired by recombinant DNA researchers for its heat-resistant enzymes.

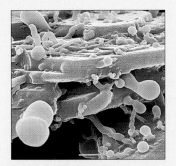

Figure 20.2 Nanobes, possibly like proto-cells. Australian researchers found them in hot rocks far beneath Earth's surface. They are only fifteen to twenty nanometers across; this image has been magnified 20,000 times. However, they do have DNA and other organic compounds enclosed within a membrane, and they grow.

What is the point of these examples? Simply this: *Life can take hold in almost any environment that has sources of carbon and energy.*

This chapter is your introduction to a sweeping slice through time, one that cuts back to Earth's formation and to life's chemical origins. The picture it paints sets the stage for the next unit, which will take you along lines of descent that led to the present range of biodiversity.

The picture is incomplete. Even so, evidence from many avenues of research points to a concept that can help us organize information about an immense journey: *Life is a magnificent continuation of the physical and chemical evolution of the universe, and of the planet Earth.*

 How Would You Vote?

Private companies make millions of dollars selling an enzyme first isolated from cells in Yellowstone National Park. Should the federal government let private companies bioprospect within the boundaries of national parks, as long as it shares in the profits from any discoveries? See BiologyNow for details, then vote online.

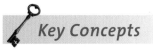 **Key Concepts**

ABIOTIC SYNTHESIS OF ORGANIC COMPOUNDS

The origin and early evolution of life correlate with the physical and chemical evolution of the universe, the stars, and Earth. The first step toward life was the spontaneous formation of complex organic compounds from simpler substances present on the early Earth. Section 20.1

ORIGIN AND EARLY EVOLUTION OF CELLS

Laboratory studies and computer simulations yield indirect evidence that self-assembly of membranes, combined with chemical and molecular evolution, gave rise to the structural and functional forerunners of cells.

The first cells were anaerobic prokaryotes. Some gave rise to bacteria, others to archaeans and to the ancestors of eukaryotic cells. Evolution of the noncyclic pathway of photosynthesis added oxygen to the atmosphere, which became a major selection pressure. Sections 20.2, 20.3

HOW THE FIRST EUKARYOTIC CELLS EVOLVED

Organelles help define eukaryotic cells. The nucleus and ER membranes may have evolved through infoldings of the plasma membrane. Mitochondria and chloroplasts may be descended from bacterial parasites or prey that took up permanent residence in host cells. Section 20.4

VISUAL PREVIEW OF THE HISTORY OF LIFE

A timeline for milestones in the history of life highlights the shared connections among all organisms. Section 20.5

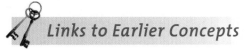 **Links to Earlier Concepts**

This chapter starts your survey of the sweep of biodiversity, as introduced in Section 1.3. This is where all of those details of cell metabolism, genetics, and evolutionary theory start to converge and help you make sense of life's fabulous journey. Now you can correlate prokaryotes (4.3) and eukaryotes (4.4) with a timeline of Earth history (17.5).

You will use your knowledge of how organic compounds are assembled (3.2), and of amino acids (3.5), membranes (5.1), enzymes (6.3), and the link between photosynthesis and aerobic respiration (Chapter 7). You may find yourself referring to the sections on DNA replication (13.3), RNAs and protein synthesis (14.1), and the genetic code (14.2). You will consider how the nucleus, ER, mitochondria, and chloroplasts (4.5–4.8) may have originated.

20.1 In the Beginning . . .

LINKS TO
SECTIONS
3.2, 3.5

Life originated when Earth was a thin-crusted inferno, so we may never find evidence of the first cells. Still, answers to three questions can yield clues to their origins. What were conditions like? Did cells emerge as a result of chemical and molecular evolution? Can experimental tests disprove that they did? Let's take a look.

Some clear evening, look up at the moon. *Five billion trillion times* the distance between it and you are the systems of stars, or galaxies, at the edge of the known universe. Light energy travels far faster than anything else, millions of meters a second, yet wavelengths of light that originated from faraway galaxies billions of years ago are just now reaching Earth. By all known measures, all near and distant galaxies in the space of the universe are moving away from one another. The entire universe, it seems, is expanding. One theory of how the colossal expansion started might account for every bit of matter in every living thing.

Think about how you can rewind a videotape on a VCR, then imagine "rewinding" the universe. As you do, the galaxies start moving closer together. After 12 to 15 billion years of rewinding, all galaxies, all matter and space are compressed into a hot, dense volume at one single point. You have arrived at time zero.

That incredibly hot, dense state lasted only for an instant. What happened next is called the **big bang**, the nearly instantaneous distribution of all matter and energy throughout the universe. Within minutes, the temperature dropped a billion degrees. Nuclear fusion reactions created most of the simplest elements, such as helium, which still are the most abundant kinds in the universe. Radio telescopes have detected a relic of the big bang—cooled, diluted background radiation left over from the beginning of time.

Over the next billion years, uncountable numbers of gaseous particles collided, and gravitational forces condensed them into the first stars. When stars were massive enough, nuclear reactions ignited inside them and gave off tremendous light and heat as the heavier elements formed. Stars have a life history, from birth to an often explosive death. In what might be called the original stardust memories, the heavier elements released from dying stars were swept up when new stars formed and helped form even heavier elements.

When explosions of dying stars ripped through our galaxy, they left behind a dense cloud of dust and gas that extended trillions of kilometers in space. As the cloud cooled, countless bits of matter gravitated toward one another. By 5 billion years ago, the shining star of our solar system—the sun—was born.

CONDITIONS ON THE EARLY EARTH

Figure 20.3 shows part of one of the vast clouds in the universe. It is mostly hydrogen gas, along with water, iron, silicates, hydrogen cyanide, ammonia, methane, formaldehyde, and other small inorganic and organic substances. Between 4.6 billion and 4.5 billion years ago, the cloud that became our solar system probably had a similar composition. Clumps of minerals and ice at the cloud's perimeter grew more massive. They became planets; one was the early Earth.

By four billion years ago, gases blanketed the first patches of Earth's thin, fiery crust (Figure 20.4). Most likely, this first atmosphere was a mixture of gaseous hydrogen, nitrogen, carbon monoxide, carbon dioxide. There was little free oxygen. How can we tell? When free oxygen is present, some binds to iron in rocks. However, geologists have discovered that such "rust" did not form until fairly recently in Earth's history.

Figure 20.3 Part of the Eagle nebula, a hotbed of star formation. Each pillar is wider than our solar system. New stars shine on the tips of gaseous streamers.

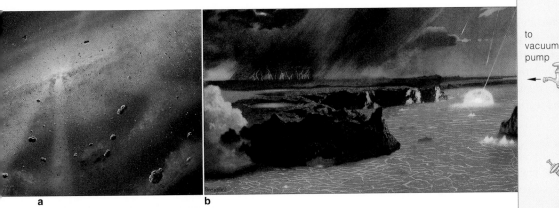

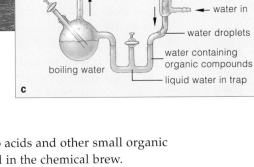

Figure 20.4 *Animated!* (**a**) What the cloud of dust, gases, rocks, and ice around the early sun might have looked like. (**b**) Less than 500,000 million years later, Earth was a thin-crusted inferno. (**c**) Sketch of the apparatus Stanley Miller used to test whether small organic compounds could form spontaneously in such a harsh environment.

The relatively low oxygen levels on the early Earth probably made the origin of life possible. Free oxygen is highly reactive. If it had been present, the organic compounds characteristic of life would not have been able to form and persist. Oxygen radicals would have attacked and destroyed compounds as they formed.

What about water? All of the water that fell on the molten surface would have evaporated at once. After the crust cooled and became solid, however, rainfall and runoff eroded mineral salts from rocks. Over many millions of years, salty water collected in crustal depressions and formed early seas. If liquid water had not accumulated, membranes could not have formed, because they take on their bilayer structure in water. No membrane, no cell, and no life.

ABIOTIC SYNTHESIS OF ORGANIC COMPOUNDS

Cells appeared less than 200 million years after the crust solidified, so complex carbohydrates and lipids, proteins, and nucleic acids must have formed by then. We know that meteorites, Mars, and Earth all formed at the same time, from the same cosmic cloud. Their rocks contain simple sugars, fatty acids, amino acids, and nucleotides, so we can expect that the precursors of biological molecules were on the early Earth, too.

Synthesizing organic compounds requires energy. On the early Earth, lightning, sunlight, or heat from hydrothermal vents might have fueled the reactions. Stanley Miller was the first to test the hypothesis that the simple compounds that now serve as the building blocks of life can form by chemical processes. He put water, methane, hydrogen, and ammonia in a reaction chamber. He kept circulating the mixture and zapping it with sparks to simulate lightning (Figure 20.4c). In less than a week, amino acids and other small organic compounds had formed in the chemical brew.

Recent geologic evidence suggests that Earth's early atmosphere was not quite like the Miller mixture. But in simulations that used other gases, different organic compounds formed—including certain types that can act as nucleotide precursors of nucleic acids.

By another hypothesis, simple organic compounds formed in outer space. Researchers detect amino acids in interstellar clouds and in some of the carbon-rich meteorites that have landed on Earth. One meteorite found in Australia contains eight amino acids that are identical with those in living organisms.

What about proteins, DNA, and the other *complex* organic compounds? Where could they form? In open water, hydrolysis reactions would have broken them apart as fast as they assembled. By one hypothesis, the clay of tidal flats bound and protected the newly forming polymers. Certain clays contain mineral ions that attract amino acids or nucleotides. Experiments show that once some of these molecules stick to clay, other molecules bond to them and form chains that resemble the proteins or nucleic acids in living cells.

Another hypothesis that is currently getting a lot of attention is this: The first biological molecules were synthesized near hydrothermal vents. Certainly the ancient seafloor was oxygen-poor. Experiments show that amino acids, at least, will condense into protein-like structures when heated in water.

Experiments provide indirect evidence that the complex organic molecules characteristic of life could have formed under conditions that probably prevailed on the early Earth.

20.2 How Did Cells Emerge?

LINKS TO
SECTIONS 5.1,
6.3, 13.4, 14.1, 14.3

Metabolism and reproduction are defining characteristics of life. In the first 600 million years or so of Earth history, enzymes, ATP, and other essential organic compounds assembled spontaneously. If they did so in the same places, their close association might have promoted the start of metabolic pathways and self-replicating systems.

ORIGIN OF AGENTS OF METABOLISM

Before cells appeared, chemical processes may have favored the formation of proteins and other complex organic compounds (Figure 20.5). However proteins originated, their molecular structure dictated their behavior. If some promoted reactions by acting like weak enzymes, they could interact with more amino acids and enzyme helpers, such as metal ions.

Visualize an early estuary, where seawater mixed with mineral-rich water that drained from the land. Beneath the sun's rays, organic molecules got stuck to clay in the mud (Figure 20.6a). At first, there were quantities of an amino acid; call it **D**. Molecules of **D** became incorporated into proteins—until **D** started to run out. Close by, however, was a weakly catalytic protein. This protein could speed the formation of **D** from a plentiful, simpler substance **C**.

By chance, clumps of organic molecules included the enzyme-like protein. Such clumps had an edge in the acquisition of starting materials. Suppose that the **C** molecules became scarce. The advantage tilted to molecular clumps that promoted the formation of **C** from simpler substances **B** and **A**. Suppose that **B** and **A** were carbon dioxide and water. The atmosphere and seas contain unlimited amounts of both. Thus, chemical selection favored a synthetic pathway:

$$A + B \longrightarrow C \longrightarrow D$$

Were some clumps better at absorbing and using energy? Think back on chlorophyll *a* (Section 7.1). A group of rings in this pigment absorbs light and gives up electrons. The same kinds of ring structures occur in electron transfer chains in all photosynthetic and aerobically respiring cells. They form spontaneously from formaldehyde (Figure 20.5)—one of the legacies of cosmic clouds. Were similar structures transferring electrons in early metabolic pathways? Probably.

The point is, long before cells emerged, a form of chemical competition was under way. Enzymes and other reactive organic compounds had the competitive edge in the acquisition of energy and materials.

ORIGIN OF THE FIRST PLASMA MEMBRANES

All living cells have an outer membrane that controls which substances enter and leave the cytoplasm in a given interval (Section 5.1). By a current hypothesis, proto-cells were transitional forms between simple organic compounds and the first living cells. These **proto-cells** were no more than membrane-bound sacs

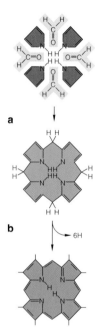

Figure 20.5 Hypothetical sequence of the chemical evolution of (**a**) an organic compound, formaldehyde, into (**c**) porphyrin.

Formaldehyde was present on the early Earth. Porphyrin is the light-absorbing and electron-donating part of chlorophyll molecules (**d**). It also is part of cytochrome, a protein component of the electron transfer chains in many metabolic pathways. It also is part of the heme of hemoglobin.

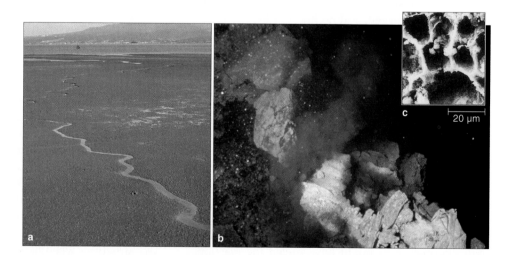

Figure 20.6 Where did the cells originate? Two likely candidates: (**a**) Clay templates in mud flats, and (**b**) iron sulfide-rich rocks at hydrothermal vents, which contain cell-sized chambers (**c**). Experiments show that such chambers are protected microenvironments in which membranes can form spontaneously. Iron sulfides projecting from the walls of such chambers catalyzed the synthesis of short peptide chains and other substances, as happens in metabolism. Many reactions in living cells use iron-sulfide cofactors. Are the cofactors a metallic legacy from a deep-sea ancestor? Perhaps.

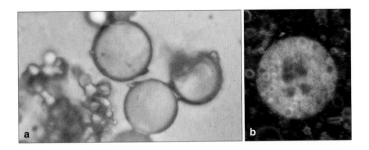

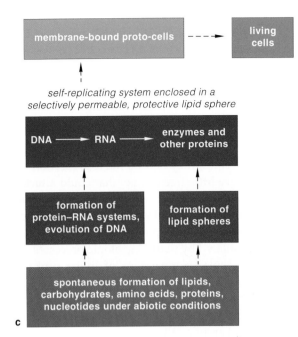

that contained systems of enzymes and other agents of metabolism, and that were self-replicating.

Experiments reveal that membrane sacs can form spontaneously. Under conditions that simulate ancient sunbaked tidal flats, amino acids do form chains that surround a volume of fluid (Figure 20.7a). Fatty acids and alcohols spontaneously form vesicles, especially when clays rich in minerals are present (Figure 20.7b).

Or did proto-cells form from organic compounds at hydrothermal vents? Cell-sized chambers occur in mineral-rich rocks at existing vents (Figure 20.6b,c). Were the chamber walls replication templates for RNA, proteins, DNA, and lipids? The molecules would have accumulated inside, favoring the chemical conditions required for the emergence of living cells.

Figure 20.7 Laboratory-grown proto-cells. (**a**) Selectively permeable sacs. Heated amino acids formed protein chains. When moistened, the chains assembled into a membrane. (**b**) A membrane of fatty acids and alcohols (*green*) enclosing RNA-coated clay (*red*). The mineral-rich clay catalyzes RNA polymerization and promotes the formation of a membrane sac. (**c**) Model for steps in the chemical processes that led to the first living, self-replicating, membrane-bound cells.

ORIGIN OF SELF-REPLICATING SYSTEMS

Life also is characterized by reproduction, which now starts with protein-building instructions in DNA. As you know from Section 14.3, it takes RNA, enzymes, and other molecules to translate DNA into proteins.

Coenzymes and metal ions assist most enzymes— and certain coenzymes are structurally identical with RNA subunits. When you mix and heat RNA subunits with very short chains of phosphate groups, they self-assemble into strands of RNA. Simple self-replicating systems of RNA, enzymes, and coenzymes have been made in laboratories. So we know RNA can serve as an information-storing template for making proteins.

Also, remember that one of the rRNA components of ribosomes catalyzes protein synthesis (Sections 14.3 and 14.4). The structure and function of ribosomes have been conserved over time; ribosomes of the most complex eukaryotes are extremely similar to those in prokaryotic cells of ancient lineages. rRNA's catalytic behavior probably evolved early in Earth history.

Did an **RNA world** *precede* the emergence of DNA? That is, were short RNA strands the first templates for protein synthesis? As you know, RNA and DNA are similar. Three of their four bases are identical. RNA's uracil differs from DNA's thymine by a single functional group. But DNA's *helically coiled, double-*

stranded structure is more stable than RNA, and it can store much more protein-building information in less space. There would have been selective advantage in functionally separating the storage of protein-building information (DNA) from protein synthesis (RNA).

Until we identify chemical ancestors of RNA and DNA, the history of life's origin will not be complete. But clues are coming in. For instance, researchers fed data about inorganic compounds and energy sources into a supercomputer. They programmed the computer to simulate random chemical reactions among organic compounds, which may well have happened untold billions of times in the distant past. Then they ran the program again and again.

The outcome of their experiment was always the same. *Simple precursors evolved. Then they spontaneously organized themselves into large, complex molecules. And they began to interact as complex systems.*

There are gaps in our knowledge of life's origin. But diverse laboratory experiments and computer simulations show that chemical processes can result in all organic molecules and structures that we think of as being characteristic of life.

20.3 The First Cells

LINKS TO
SECTIONS
4.3, 6.4, 7.8

The first cells apparently evolved during the Archaean, an eon that lasted from 3.8 billion to 2.5 billion years ago. Not long afterward, divergences gave rise to three great lineages that have persisted to the present.

THE GOLDEN AGE OF PROKARYOTES

Fossils indicate that the first cells were like existing prokaryotes; they had no nucleus (Section 4.3). There was very little free oxygen that could attack them. Its absence is a clue to their mode of nutrition. Anaerobic pathways would allow them to obtain energy from simple organic compounds and mineral ions that had accumulated by natural geologic processes in the seas.

Molecular comparisons of living prokaryotes tell us that some populations diverged not long after life originated. One lineage gave rise to the bacteria. The other gave rise to the shared ancestors of archaeans and eukaryotic cells.

Microscopically small fossils in 3.5-billion-year-old rocks give clues to what some of the first prokaryotes looked like (Figure 20.8a). Other fossils clearly show that chemoautotrophic forms had become established near deep-sea hydrothermal vents by 3.2 billion years ago. In some groups, pigments probably detected the type of weak infrared radiation (heat) that has been measured at hydrothermal vents. Pigments may have helped cells detect and avoid boiling water, as they do for some existing hydrothermal vent species.

Gene mutations arose independently in some of the prokaryotic populations. They led to modifications in radiation-sensitive pigments, electron transfer chains, and other bits of metabolic machinery that started a novel mode of nutrition. We call it the cyclic pathway of photosynthesis. Those bacterial populations were photoautotrophic; they had tapped into sunlight, an unlimited energy source (Section 7.8).

As they reproduced, those self-feeding populations of tiny cells grew on top of one another. They became flattened mats, infiltrated with calcium carbonate and other dissolved mineral ions, and fine sediments. In time, they were transformed into dome-shaped fossils known as **stromatolites**. Radiometric dating tells us that some are 3 billion years old (Figure 20.9).

When the Proterozoic dawned 2.7 billion years ago, stromatolites were abundant. By that time, a noncyclic pathway of photosynthesis had evolved in a bacterial lineage, the cyanobacteria. Cyanobacterial populations increased, and so did the pathway's waste product—free oxygen. At first, oxygen slowly accumulated in the surface waters of the seas, then in air. So now we return to events sketched out in Chapter 7.

An atmosphere enriched with free oxygen had two irreversible effects. First, *it stopped the further chemical origin of living cells*. Except in a few anaerobic habitats, complex organic compounds could no longer assemble spontaneously and stay intact; they could not escape attacks by oxygen radicals. Second, *aerobic respiration evolved and in time became the dominant energy-releasing pathway*. In many prokaryotic lineages, selection had favored this pathway, which neutralized oxygen by *using* it as an electron acceptor. Aerobic respiration was a key innovation that contributed to the rise of all complex, multicelled eukaryotes.

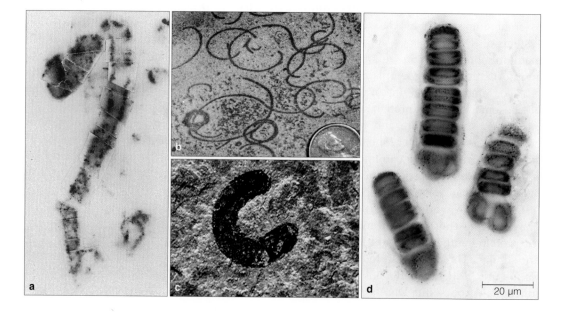

Figure 20.8 A sampling of early life. (**a**) A strand of what might be walled prokaryotic cells dates back 3.5 billion years. (**b**) One of the oldest known eukaryotic species, *Grypania spiralis*, which lived 2.1 billion years ago. Its fossilized colonies are large enough to see without a microscope. (**c**) Fossil of *Tawuia*, another early eukaryotic species that lived during the Proterozoic. (**d**) Fossils of a red alga, *Bangiomorpha pubescens*. This multicelled species lived 1.2 billion years ago, and it reproduced sexually.

20 μm

THE RISE OF EUKARYOTES

Eukaryotic cells also evolved during the Proterozoic. Traces of the kinds of lipids that existing eukaryotic cells produce have been isolated from rocks dated at 2.8 billion years old. But the first complete eukaryotic fossils are about 2.1 billion years old (Figure 20.8b,c). Those ancient species had organelles.

As you know, organelles are the defining features of eukaryotic cells. Where did they come from? The next section presents a few plausible hypotheses.

We still do not know how the earliest eukaryotes fit in evolutionary trees. The earliest known form we can assign to a modern group is the filamentous alga *Bangiomorpha pubescens*. This red alga, which lived 1.2 billion years ago, is the first multicelled eukaryotic species to be discovered. Its cells were differentiated. Some cells in its strandlike body served as anchoring structures. Others formed two types of sexual spores. Spore production certainly makes *B. pubescens* one of the earliest practitioners of sexual reproduction.

By 1.1 billion years ago the supercontinent Rodinia had formed. Stromatolites dotted its vast shorelines, but 300 million years later, they were in decline. Were the cyanobacteria a vast food source for predators and parasites? By then, protists, fungi, animals, and the algae that would later give rise to plants were sharing the shoreline with them. Also, 570 million years ago, when oxygen in the atmosphere approached modern levels, animals began their first adaptive radiations in the Cambrian seas. A coevolutionary arms race that continues to this day was off and running.

Figure 20.9 Some stromatolites. (**a**) A painting of how one shallow sea might have looked early in the Proterozoic.

(**b**) In Australia's Shark Bay are mounds that are 2,000 years old. They are structurally similar to stromatolites that formed 3 billion years ago.

(**c**) A cut stromatolite reveals many layers of fine sediments and mineral deposits. The cyanobacterial cells often were preserved as well.

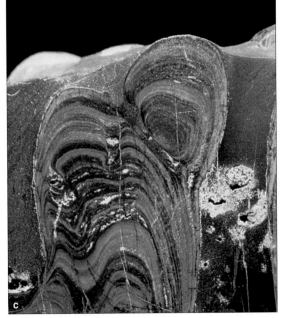

The first living cells evolved by 3.8 billion years ago, in the Archaean eon. All were prokaryotic, and they obtained energy by anaerobic pathways. Not long afterward, the ancestors of archaeans and eukaryotic cells diverged from the lineage that led to modern bacteria.

After the noncyclic pathway of photosynthesis evolved, free oxygen accumulated in the atmosphere and ended the further spontaneous chemical origin of life. The stage was set for the evolution of eukaryotic cells.

20.4 Where Did Organelles Come From?

LINKS TO
SECTIONS
4.3, 4.5–4.8, 14.2

Thanks to globe-hopping microfossil hunters, we have considerable evidence of early life, including the fossil treasures shown in Sections 4.3 and 20.3. Today, most descendant species contain a profusion of organelles. Where did the organelles come from?

ORIGIN OF THE NUCLEUS AND ER

Prokaryotic cells, recall, do not have an abundance of organelles. Some do have infoldings of their plasma membrane, which incorporates many enzymes and other components used in metabolic reactions (Figure 20.10a). Applying the theory of natural selection, we may hypothesize that infoldings originated among ancestors of eukaryotic cells. What advantages did the infoldings offer? They became channels that could concentrate nutrients, organic compounds, and other substances. Also, a membrane with a greater surface area could be a physical platform for more metabolic

machinery as well as transport proteins. Remember the surface-to-volume ratio?

The channels of endoplasmic reticulum (ER) may have evolved this way. They also may have protected the metabolic machinery from uninvited guests. From time to time, metabolically "hungry" foreign cells do enter the cytoplasm of existing prokaryotic cells.

Some infoldings might have extended around the DNA, the start of a nuclear envelope (Figure 20.10b). A nuclear envelope would have been favored because it helped protect the cell's hereditary material from foreign DNA. Bacteria and the simple eukaryotic cells called yeasts can transfer plasmids among themselves. Early eukaryotic cells with a nuclear envelope could copy and use their messages of inheritance, free from metabolic competition from a potentially disruptive hodgepodge of foreign DNA.

ORIGIN OF MITOCHONDRIA AND CHLOROPLASTS

Early in the history of life, cells became food for one another. Heterotrophs engulfed autotrophs and other heterotrophs. Intracellular parasites dined inside their hosts. In some cases, the engulfed meals or parasites struck an uneasy balance with the host cells. They were protected, they withdrew some nutrients from the cytoplasm, and—like their hosts—they continued to divide and reproduce. Over time, they evolved into mitochondria, chloroplasts, and some other organelles.

The novel partnerships are one premise of a theory of **endosymbiosis**, as championed by Lynn Margulis and others. (*Endo–* means within and *symbiosis* means living together.) The symbiont species lives out its life inside a host species, and the interaction benefits one or both of them.

By this theory, eukaryotic cells evolved after the noncyclic pathway of photosynthesis emerged and permanently changed the atmosphere. By 2.1 billion years ago, remember, certain prokaryotic cells had adapted to the concentration of free oxygen and were already engaged in aerobic respiration. The ancestors of eukaryotic cells preyed upon some aerobic bacteria and were parasitized by others (Figure 20.11a). At that time, endosymbiotic interactions began.

The host began to use ATP produced by its aerobic symbiont. The aerobe no longer had to spend energy on acquiring raw materials; the host did this work for it. DNA regions that specified proteins produced by both host and symbiont were free to mutate and lose their function in one partner or the other. In time, both types of cells became incapable of independent life.

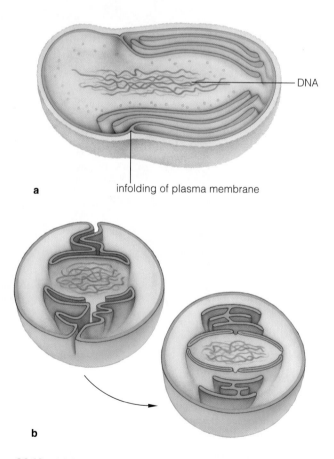

a   infolding of plasma membrane

DNA

b

Figure 20.10 (a) Sketch of a bacterial cell (*Nitrobacter*) that lives in soil. Cytoplasmic fluid bathes permanent infoldings of the plasma membrane. (b) Model for the origin of the nuclear envelope and the endoplasmic reticulum. In prokaryotic ancestors of eukaryotic cells, infoldings of the plasma membrane may have evolved into these organelles.

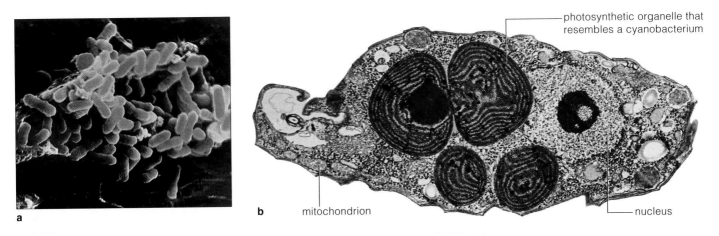

photosynthetic organelle that
resembles a cyanobacterium

a　　　**b**　　mitochondrion　　　　　　　　　　　　　　　nucleus

Figure 20.11　Clues to ancient endosymbiotic interactions. (**a**) What the ancestors of mitochondria may have looked like. The protist *Reclinomonas americana* has the structurally simplest mitochondria. The mitochondrial genes resemble genes of *Rickettsia prowazekii*, a parasitic bacterium that causes typhus. Like mitochondria, *R. prowazekii* divides only inside the cytoplasm of eukaryotic cells. Enzymes in the cytoplasm catalyze the partial breakdown of organic compounds—a task that is completed inside aerobically respiring mitochondria. (**b**) *Cyanophora paradoxa* is one of the flagellated protists called glaucophytes. Its mitochondria resemble aerobic bacteria in size and structure. Its photosynthetic structures resemble cyanobacteria—they even have a wall like that of cyanobacteria.

EVIDENCE OF ENDOSYMBIOSIS

Is such a theory far-fetched? A chance discovery in Jeon Kwang's laboratory suggests otherwise. In 1966, a rod-shaped bacterium had infected his culture of *Amoeba discoides*. Some infected cells died right away. Others grew more slowly, and they were smaller and vulnerable to starving to death. Kwang maintained the infected culture. Five years later, infected amoebas were harboring many bacterial cells, yet they were all thriving. Exposure to antibiotics killed the bacterial cells (but not the amoebas).

Infection-free cells were stripped of their nucleus and got a nucleus from an infected cell. They died. Yet more than 90 percent survived when a few bacteria were included with the transplant. As other studies showed, the infected amoebas had lost their ability to synthesize an essential enzyme. They depended on the bacterium to make it for them! Invading bacterial cells had become symbiotic with the amoebas.

When you think about it, mitochondria do resemble bacteria in size and structure. Each has its own DNA and divides independently of cell division. The inner membrane of a mitochondrion resembles a bacterial cell's plasma membrane. Its DNA has just a few genes (thirty-seven in human mitochondrial DNA). Also, a few of the codons are slightly different from those of the near-universal genetic code (Section 14.2).

We can predict that chloroplasts, too, originated by endosymbiosis. In one scenario, photosynthetic cells were engulfed by predatory aerobic bacteria, but they escaped digestion. They started to absorb nutrients in the host's cytoplasm and continued to function. They also released oxygen when they photosynthesized. By releasing oxygen inside the aerobically respiring hosts, they acted as agents favoring endosymbiosis.

In their metabolism and their overall nucleic acid sequence, existing chloroplasts resemble cyanobacteria. The chloroplast DNA replicates itself independently of cellular DNA. Chloroplasts and the cells in which they reside divide independently of each other.

Or consider the protists called glaucophytes. They have unique photosynthetic organelles that resemble cyanobacteria. These organelles even have their own cell wall (Figure 20.11*b*).

However they arose, the first eukaryotic cells had a nucleus, an endomembrane system, mitochondria and, in some lineages, chloroplasts. They were the world's first protists. They had efficient metabolic systems, and they evolved fast. In no time at all, evolutionarily speaking, some of their descendants evolved into the plants, fungi, and animals. The next section provides a time frame for these pivotal events.

> *A nucleus and other organelles are defining features of eukaryotic cells. The nucleus and ER may have evolved by infoldings of the plasma membrane. Mitochondria and chloroplasts may have evolved through endosymbiosis between heterotrophic host cells and their prey or parasites.*

20.5 Time Line for Life's Origin and Evolution

LINKS TO
SECTIONS
1.1, 17.5

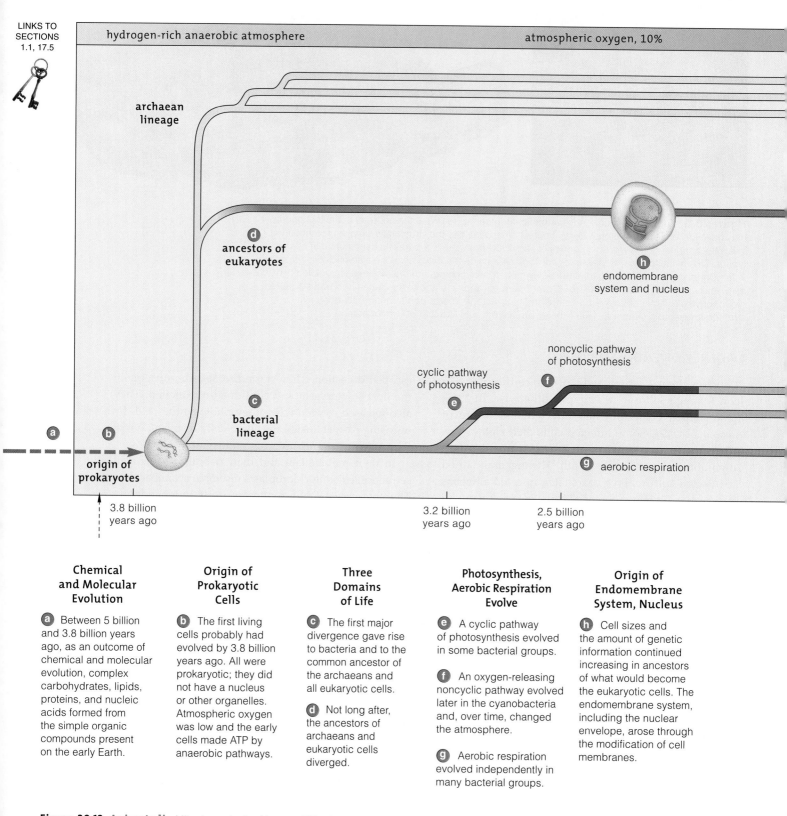

Chemical and Molecular Evolution

(a) Between 5 billion and 3.8 billion years ago, as an outcome of chemical and molecular evolution, complex carbohydrates, lipids, proteins, and nucleic acids formed from the simple organic compounds present on the early Earth.

Origin of Prokaryotic Cells

(b) The first living cells probably had evolved by 3.8 billion years ago. All were prokaryotic; they did not have a nucleus or other organelles. Atmospheric oxygen was low and the early cells made ATP by anaerobic pathways.

Three Domains of Life

(c) The first major divergence gave rise to bacteria and to the common ancestor of the archaeans and all eukaryotic cells.

(d) Not long after, the ancestors of archaeans and eukaryotic cells diverged.

Photosynthesis, Aerobic Respiration Evolve

(e) A cyclic pathway of photosynthesis evolved in some bacterial groups.

(f) An oxygen-releasing noncyclic pathway evolved later in the cyanobacteria and, over time, changed the atmosphere.

(g) Aerobic respiration evolved independently in many bacterial groups.

Origin of Endomembrane System, Nucleus

(h) Cell sizes and the amount of genetic information continued increasing in ancestors of what would become the eukaryotic cells. The endomembrane system, including the nuclear envelope, arose through the modification of cell membranes.

Figure 20.12 *Animated!* Milestones in the history of life. As you read the next unit on life's past and present diversity, refer to this visual overview. It can serve as a simple reminder of the evolutionary connections among all groups of organisms, from the structurally simple to the most complex.

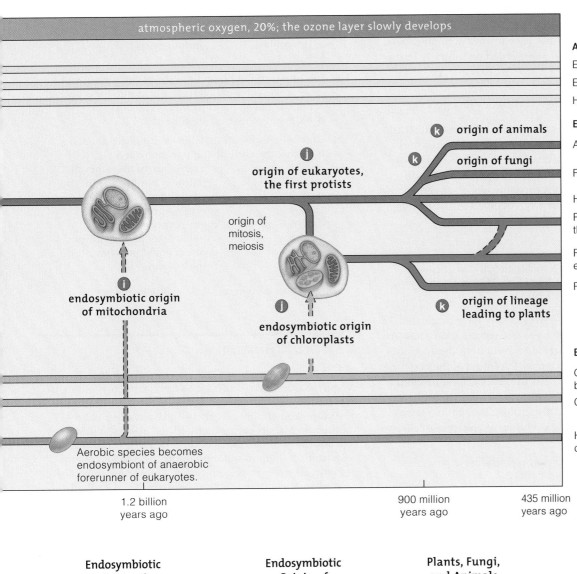

atmospheric oxygen, 20%; the ozone layer slowly develops

ARCHAEA

Extreme thermophiles

Extreme thermophiles and mesophiles

Halophiles and methanogens

EUKARYA

Animals

Fungi

Heterotrophic protists

Photosynthetic protists with chloroplasts that evolved from red and green algae

Red and green algae; their chloroplasts evolved from cyanobacterial symbionts

Plants

k origin of animals

k origin of fungi

j origin of eukaryotes, the first protists

origin of mitosis, meiosis

i endosymbiotic origin of mitochondria

j endosymbiotic origin of chloroplasts

k origin of lineage leading to plants

BACTERIA

Oxygen-releasing photosynthetic bacteria (cyanobacteria)

Other photosynthetic bacteria

Heterotrophic bacteria, including chemoheterotrophs

Aerobic species becomes endosymbiont of anaerobic forerunner of eukaryotes.

1.2 billion years ago

900 million years ago

435 million years ago

Endosymbiotic Origin of Mitochondria

i Before about 1.2 billion years ago, aerobic bacterial species and an anaerobic ancestor of eukaryotic cells entered into close symbiotic interaction. The endosymbiont evolved into the mitochondrion.

Endosymbiotic Origin of Chloroplasts

j Cyanobacteria entered into a close symbiotic interaction with early protists and evolved into chloroplasts. Later, photosynthetic protists would evolve into chloroplasts inside other protist hosts.

Plants, Fungi, and Animals Evolve

k By 900 million years ago, all major lineages—including fungi, animals, and the algae that would give rise to plants—had evolved along shorelines of the first supercontinent.

Lineages That Have Endured to the Present

l Today, organisms live in all regions of Earth's waters, crust, and atmosphere. They are related by descent and share certain traits. However, each lineage encountered different selective pressures, and each has evolved its own characteristic traits.

Summary

Section 20.1   Earth formed more than 4 billion years ago. Experimental tests, information on the formation of stars and planets, and other lines of research offer indirect evidence that the complex organic compounds characteristic of life could have formed spontaneously under the conditions that prevailed on the early Earth.

Biology⊘Now
See experiments on how organic compounds can form spontaneously with the animation on BiologyNow.

Section 20.2   The emergence of the first cells was preceded by chemical evolution that led to enzymes and other agents of metabolism, the self-assembly of membranes on environmental templates, and a self-replicating system. RNA probably was the template for protein synthesis before DNA evolved as an efficient way to store protein-building information.

Biology⊘Now
Read the InfoTrac article "Transitions from Nonliving to Living Matter," Steen Rasmussen et al., Science, February 2004. Also "First Cell," David Deamer, Discover, November 1995.

Section 20.3   The first cells may have originated 3.8 billion years ago. They were anaerobic prokaryotes. An early divergence separated bacteria from the ancestors of archaeans and eukaryotes. Evolution of the noncyclic pathway of photosynthesis in cyanobacteria resulted in an accumulation of free oxygen in the atmosphere, which favored aerobic respiration. This pathway was a key innovation in the evolution of eukaryotic cells.

Biology⊘Now
Explore levels of biological organization with the interaction on BiologyNow.

Section 20.4   The internal membranes of eukaryotic cells may have evolved through infoldings of the cell membrane. Mitochondria and chloroplasts most likely evolved by endosymbiosis, at the times indicated in the Section 20.5 visual summary.

Section 20.5   Key events in life's origin and early evolution can be correlated with the geologic time scale.

Biology⊘Now
Investigate the history of life with the animated interaction on BiologyNow.

Self-Quiz

Answers in Appendix II

1. An abundance of _____ in the atmosphere would have prevented the spontaneous (abiotic) assembly of organic compounds on the early Earth.
    a. hydrogen   b. methane   c. oxygen   d. nitrogen

2. The prevalence of iron-sulfide cofactors in living organisms may be evidence that life arose _____ .
    a. in outer space   c. near deep-sea vents
    b. on tidal flats   d. in the upper atmosphere

3. The evolution of _____ resulted in an increase in the levels of atmospheric oxygen.
    a. sexual reproduction
    b. aerobic respiration
    c. the noncyclic pathway of photosynthesis
    d. the cyclic pathway of photosynthesis

4. Mitochondria may have evolved from _____ .
    a. chloroplasts   c. early protists
    b. bacteria   d. archaeans

5. Infoldings of the plasma membrane into the cytoplasm of some prokaryotes may have evolved into the _____ .
    a. nuclear envelope c. primary cell wall
    b. ER membranes   d. both a and b

6. Chronologically arrange the evolutionary events, with 1 being the earliest and 6 the most recent.
    _____ 1   a. emergence of the noncyclic
    _____ 2   pathway of photosynthesis
    _____ 3   b. origin of mitochondria
    _____ 4   c. origin of proto-cells
    _____ 5   d. emergence of the cyclic
    _____ 6   pathway of photosynthesis
                 e. origin of chloroplasts
                 f. the big bang

Additional questions are available on **Biology⊘Now™**

Critical Thinking

1. Mars formed about 5 million years earlier than Earth, and it has a similar composition but is far richer in iron. It is farther from the sun and much chillier, with an average surface temperature of −63° C. Today, nearly all of the water on Mars is permanently frozen in soil. To some researchers, photographs of certain geological features indicate that liquid water might have flowed across the planet's surface during an earlier and warmer time. The Martian atmophere is now richer in carbon dioxide than Earth's, but very low in nitrogen and oxygen. Based on this information, would you rule out the possibility that life could have existed on Mars or that simple life forms could currently exist there? Explain your reasoning.

2. What if it were possible to create life in test tubes? That is the idea behind modeling and perhaps creating minimal organisms: living cells having the smallest set of genes required to survive and reproduce.
    Craig Venter and Claire Fraser found that *Mycoplasma genitalium*, a bacterium that has only 517 genes (and 2,209 transposons), is a good candidate for such an experiment. By disabling its genes one at a time, they discovered it may have only 265–350 essential protein-coding genes.
    What if those genes were synthesized one at a time and inserted into an engineered cell consisting only of a plasma membrane and cytoplasm? Would the cell come to life? The possibility that it might prompted Venter and Fraser to seek advice from a panel of bioethicists and theologians. No one on the panel objected to synthetic life research. They said that much good might come of it, provided scientists did not claim to have found "the secret of life." The December 10, 1999, issue of *Science* includes an essay from the panel and an article on *M. genitalium* research. Read both, then write down your thoughts about "creating" life in a test tube.

IV Evolution and Biodiversity

From the Green River formation near Lincoln, Wyoming, the stunning fossilized remains of a bird trapped in time. During the Eocene, some 50 million years ago, sediments that had been slowly deposited in layers at the bottom of a large inland lake became its tomb. In this same formation, fossils of palms, cattails, sycamore, and other plants tell us that the climate was warm and moist. Fossils from places all around the world yield major clues to life's early history.

21.1 Characteristics of Prokaryotic Cells

LINKS TO
SECTIONS
4.3, 5.5, 7.8, 8.5

Of all organisms, prokaryotic cells are the smallest and the most far-flung, abundant, and metabolically diverse. Deserts, hot springs, glaciers, the seafloor, and rocks 2,800 meters below Earth's surface are home to different kinds. Billions live in a handful of rich soil. The ones living in your gut and on your skin outnumber your body cells!

Prokaryotes are the most ancient lineages. Trace any line of descent back far enough and you find ancestral prokaryotes. From *Escherichia coli* to clams, elephants, and redwoods, all of life interconnects, regardless of evolutionary distance.

Recall, from Section 4.3, that "prokaryotic" means these cells originated before the kinds with a nucleus. Membrane-bound, internal sacs of any kind are rare; metabolic reactions occur at the plasma membrane or in the cytoplasm. But structural simplicity does not mean that prokaryotic cells are inferior to eukaryotic cells. Being tiny, fast reproducers, they survive very well without great internal complexity. Table 21.1 and Figure 21.2 present their basic characteristics.

Table 21.1 Characteristics of Prokaryotic Cells

1. No membrane-bound nucleus.
2. Generally a single chromosome (a circular DNA molecule); many species also contain plasmids.
3. Cell wall present in most species.
4. Reproduction mainly by prokaryotic fission.
5. Collectively, great metabolic diversity among species.

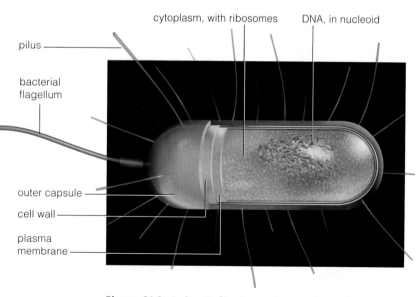

pilus

bacterial flagellum

cytoplasm, with ribosomes

DNA, in nucleoid

outer capsule

cell wall

plasma membrane

Figure 21.2 *Animated!* Generalized prokaryotic cell.

SIZES AND SHAPES

By now you have a sense of the microscopically small sizes of prokaryotic cells. Their width and length are between 0.5 and 1 micrometer, on the average. A few species are as large as 10 micrometers.

Three basic shapes are common among them. A spherical shape is a **coccus** (plural, cocci; from a word meaning berries). A rod shape is a **bacillus** (plural, bacilli, meaning small staffs). A cell body having one or more twists is called a **spirillum** (plural, spirilla):

coccus   bacillus   spirillum

Figure 21.3 shows examples of these shapes, but there are variations on the basic plan. Cocci may be oval or flattened. Bacilli may be long and thin or tapered like a cigar. Some spiral species are curled like a comma or a corkscrew (Figure 19.15). Extensions give some species a star shape, and one archaean looks a bit like a postage stamp. Also, when cells are dividing, the daughter cells often stick together in chains, sheets, or other aggregations, as in Figure 21.3*b*.

STRUCTURAL FEATURES

For most prokaryotic species, a **cell wall** encloses the plasma membrane (Figure 21.2). This somewhat rigid, permeable structure helps the cell maintain its shape and resist rupturing when the internal fluid pressure increases, as Section 5.5 explains. Unlike the cell wall of archaeans and some eukaryotic species, a bacterial wall consists of peptidoglycan, a unique compound in which peptide bonds crosslink many polysaccharide strands to one another.

When doctors must diagnose an infectious disease, they may use **Gram staining**. This procedure can help identify many bacterial species by their wall staining properties. The unknown species is exposed to purple dye, then iodine, then an alcohol wash, and finally a counterstain. The wall of *Gram-positive* species stays purple. The wall of *Gram-negative* species loses color at first, but the counterstain turns it pink (Figure 21.4).

A sticky mesh, or **glycocalyx**, often encloses the cell wall. It consists of polysaccharides, polypeptides, or both. When highly organized and attached firmly to the wall, it forms a capsule. When less organized and loosely attached, it forms a slime layer. The mesh helps the cell attach to teeth, mucous membranes of the intestinal or vaginal wall, rocks in streambeds, and other interesting surfaces. It helps some encapsulated types resist being engulfed by phagocytic, infection-fighting cells of a host organism.

Many species have one or more **bacterial flagella** (Figure 14.6*a*). Unlike eukaryotic flagella, these motile structures do not have microtubules, and they do not bend side to side. Instead, they rotate like a propeller. Also, **pili** (singular, pilus) are common. These are thin, filamentous proteins that project above the cell wall, as in Figure 21.3*c,d*. Some pili help the cell adhere to surfaces. A kind called a sex pilus helps one cell pull another cell next to it as a prelude to conjugation, an interaction explained in the next section.

METABOLIC DIVERSITY

Remember the Chapter 7 introduction? All organisms must acquire energy and carbon. Compared to other organisms, however, the prokaryotic cells collectively show the most diversity in how they get it.

Like plants, *photoautotrophic* prokaryotes are self-feeders that make their own food by photosynthesis. Cyanobacteria are like chloroplasts. They get electrons and hydrogen for the reactions from water molecules and release free oxygen as a product. Light-trapping pigments and electron transfer chains are embedded in their plasma membrane. Different photoautotrophs are strict anaerobes; free oxygen kills them. They get the electrons and hydrogen from gaseous hydrogen, hydrogen sulfide, and other inorganic compounds.

There also are self-feeding *chemoautotrophic* species. They obtain carbon from CO_2, and they obtain energy by oxidizing organic or inorganic compounds, such as iron and sulfur (Section 7.8).

Photoheterotrophic prokaryotes are not self-feeders. They tap the sun's energy—but they must get carbon from organic compounds. *Chemoheterotrophic* species are by far the most common types, and they are not self-feeders, either. The parasites get nutrients from a living host. The saprobes, like fungi, digest organic products or remains of organisms in the environment, then absorb the breakdown products.

Nearly all prokaryotic cells are microscopic in size. Unlike eukaryotic cells, they do not have great internal complexity, nor do they require it. Most have a cell wall that often is enclosed in a capsule or slime layer. The bacterial cell wall is uniquely made of peptidoglycan.

Surface specializations include bacterial flagella and pili.

Collectively, prokaryotic cells show the most diversity in acquiring energy and carbon building blocks.

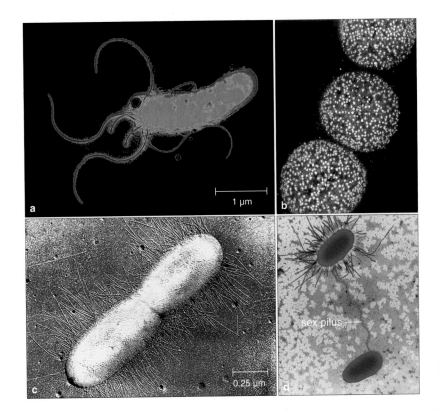

Figure 21.3 A few prokaryotic cell shapes and structures. (**a**) *Helicobacter pylori*, with a tuft of flagella. This pathogen can colonize the stomach lining. Untreated infections invite gastritis, peptic ulcers, and possibly stomach cancer. *H. pylori* can contaminate water and food, especially unpasteurized milk. A combination of antibiotics and an antacid kills it. (**b**) *Thiomargarita namibiensis*, a bacterial cell visible to the naked eye. A nitrate-filled vacuole occupies most of its cytoplasm. (**c**) An *Escherichia coli* cell, with its profusion of pili. This cell is dividing. (**d**) Sex pilus reeling in an *E. coli* cell (lower right) to the cell above, which will then transfer genetic material to it.

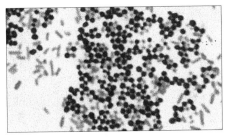

■ stain with purple dye
▣ stain with iodine
☐ wash with alcohol
▨ counterstain with safranin

Figure 21.4 Gram staining. Cocci and bacilli smeared on a slide were stained with a purple dye (such as crystal violet), washed, and stained with iodine. Staining turned all of the cell walls purple. The slide was then washed with alcohol, which removed the stain from the Gram-negative cells and made them colorless.

Next, the slide was counterstained (with safranin), washed, and dried. The Gram-positive cells (here, *Staphylococcus aureus*) stayed purple. However, the counterstain tinted the Gram-negative cells (*E. coli*) light pink.

21.2 Prokaryotic Growth and Reproduction

LINKS TO
SECTIONS
10.6, 16.1

Compared to eukaryotic cells (Section 10.6), prokaryotic cells divide by a more straightforward mechanism.

THE NATURE OF GROWTH

Prokaryotic cells grow by increasing their component parts between divisions. We measure **growth** in terms of increases in size for large, multicelled organisms, but doing so for a microscopically small cell would be a bit pointless. Instead, we measure the growth of any prokaryotic species in terms of increases in the number of cells in its populations. Under ideal conditions, a prokaryotic cell divides in two, then division of two cells results in four cells, four result in eight, and so on. Many types can divide every half hour; a few can do so every ten or twenty minutes. Such rates of increase can result in large population sizes in short order.

Some cells reproduce rarely, but their populations grow where no others can. This is the case for the few species clinging to life in Antarctic glaciers, the Negev Desert, and deep in Earth's crust. Even K–12, a strain of *E. coli* originally isolated from the human gut, has been cultivated for such a long time in the laboratory that it no longer can grow when reintroduced into its natural habitat. As an outcome of microevolutionary processes, it has become adapted to the conditions in its artificial laboratory environment.

PROKARYOTIC FISSION

A prokaryotic cell nearly doubles in size, then divides in two. Each daughter cell inherits a single **bacterial chromosome**—a circularized, double-stranded DNA molecule that has a few proteins associated with it. In some species, the daughter cell merely buds from the parent cell. Most often, however, a cell reproduces by a division mechanism called **prokaryotic fission**.

Prokaryotic fission starts when a cell replicates its DNA (Figure 21.5). The parent molecule and the copy are both anchored to the plasma membrane at adjacent sites. Meanwhile, the cell is synthesizing proteins and lipids, which become added to the plasma membrane between the two attachment sites. The additions make the membrane grow, which moves the molecules of DNA apart. New wall material is deposited onto the growing membrane, and growth continues on through the cell midsection. It cuts the cytoplasm in two, the result being two genetically equivalent daughter cells.

Especially in microbiology, you may hear someone refer to this division mechanism as *binary* fission, but such usage might be confusing. The same term applies to an asexual reproductive mode among flatworms and some other animals. It refers to growth by mitotic cell divisions, then division of the whole body into two parts of the same or different sizes.

a The bacterial chromosome is attached to the plasma membrane before DNA replication.

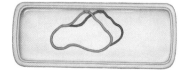

b Replication starts and proceeds in two directions from some point in the bacterial chromosome.

c The DNA copy is attached at a membrane site near the attachment site of the parent DNA molecule.

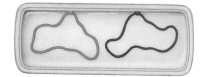

d The two DNA molecules are moved apart by membrane growth between the two attachment sites.

e New membrane and new wall material are added transversely, through the cell's midsection.

f The ongoing, orderly deposition of membrane and wall material at the midsection cuts the cell in two.

Figure 21.5 *Animated!* Prokaryotic fission. Only bacteria and archaeans reproduce by this cell division mechanism. *Right*, micrograph of the cytoplasmic division of *Bacillus cereus*. The arrow points to the deposition of new wall material and membrane at its midsection.

a A conjugation tube has already formed between a donor and a recipient cell. An enzyme has nicked the donor's plasmid.

b DNA replication starts on the nicked plasmid. The displaced DNA strand moves through the tube and enters the recipient cell.

c In the recipient cell, replication starts on the transferred DNA.

d The cells separate from each other; the plasmids circularize.

Figure 21.6 *Animated!* Conjugation between two prokaryotic cells. For clarity, the plasmid's size has been greatly increased and the chromosome is not shown.

CONJUGATION BETWEEN CELLS

Salmonella, *Streptococcus*, and *E. coli* and other species also inherit plasmids. As you know, each **plasmid** is a small, self-replicating circle of DNA with a few genes (Section 16.1). An F (Fertility) plasmid has genes that confer the means to engage in a form of **conjugation**. A donor cell transfers plasmid DNA to a recipient cell. Such transfers have been noted between *E. coli* and yeast cells in the laboratory. The F plasmid contains instructions for making a sex pilus (Figure 21.3d). Sex pili at a donor cell's surface can hook onto a recipient cell and pull it next to the donor. Shortly after the two cells make contact, a tiny conjugation tube develops between them. After this, plasmid DNA is transferred through the tube, in the manner shown in Figure 21.6.

Only bacteria and archaeans reproduce by a cell division mechanism, prokaryotic fission, that follows replication of DNA. Each daughter cell inherits one DNA molecule (one chromosome). Many species also transfer plasmid DNA.

21.3 Classifying Prokaryotes

How can we classify prokaryotes? Except for stromatolites, the most ancient groups are not well represented in the fossil record. Most are not represented at all.

LINKS TO SECTIONS 1.4, 16.1, 16.2, 19.5

Traditionally, prokaryotes have been classified mainly by **numerical taxonomy**. An unidentified prokaryotic cell is compared against a known group on the basis of shape, motility, wall staining attributes, nutritional requirements, metabolism, and other traits. The more traits that the cell shares with the known group, the closer is their inferred relatedness.

Automated gene sequencing and other methods of comparative biochemistry are clarifying phylogenies (Sections 16.2 and 16.3). The comparisons of ribosomal RNAs are especially revealing. As you will read soon, mutations that accumulated in the rRNAs of different prokaryotic lineages are being measured directly.

Biochemical analyses are uniting some groups that did not seem to be related on the basis of other tests. As Section 19.5 explains, they revealed the first genetic divergence that occurred shortly after life originated. One branching led to **Bacteria**. The other gave rise to **Archaea** and to the ancestors of eukaryotic cells:

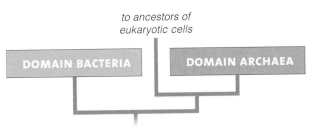

to ancestors of eukaryotic cells

DOMAIN BACTERIA   **DOMAIN ARCHAEA**

biochemical and molecular origin of life

By now, you probably have sensed that *species* is the basic unit in prokaryotic classification schemes. Even so, the definition that fits sexually reproducing species does not fit prokaryotes, which are not reproductively isolated populations of interbreeding individuals. A prokaryotic cell generally does its own thing. Also, many variations that do show up among species are determined by relatively few genes. If two cells under study show minor differences, one of them might be classified as a **strain**, not a new species. The next two sections take these considerations into account in their listings of major groups.

Prokaryotic cells are now being classified on the basis of biochemical comparisons as well as numerical taxonomy, or the total percentage of observable traits they have in common with a known prokaryotic group.

21.4 Domain Bacteria

LINKS TO
SECTIONS
7.8, 16.7 20.1–20.3

Figure 21.7 shows which of the many thousands of bacterial groups we sample throughout this book.

REPRESENTATIVE DIVERSITY

Bacteria originated when unstable crustal plates were colliding and forming the proto-continents. Volcanoes were spewing lava into superheated water and venting gases into a steaming atmosphere that big meteorites repeatedly pierced. For the next 3 billion years, cells did not change much, except in their metabolism. The first kinds may have resembled *Aquifex aeolicus*, one of the most extreme thermophilic species known (Figure 21.7). This one lives in water as hot as 96°C (204.8°F).

Cyanobacteria were ancient species (Section 20.3). Like their modern descendants, they released oxygen during photosynthesis. Collectively, these cells and the chloroplasts descended from them probably account for most of the free oxygen in the atmosphere. Today, cyanobacteria also cycle considerable carbon, oxygen, nitrogen, and other nutrients through aquatic habitats and soils. A few are symbionts with fungi, in lichens.

After mitotic cell division, daughter cyanobacterial cells often remain attached as mucus-sheathed chains that form slimy mats (Figure 21.8*a,b*). When nitrogen is scarce, some cells in chains of *Anabaena* develop into heterocysts, which convert nitrogen gas to ammonia. Ammonia dissolves at once into a form that can be used to synthesize nitrogen compounds. Heterocysts share the compounds with other cells in the chain and get carbohydrates in return. Cyanobacterial species are notable specialists in nitrogen fixation.

Proteobacteria are the most diverse monophyletic group of bacteria. They are all Gram-negative species; staining tints their walls pink. *Chromatium* species are anaerobic photosynthesizers. Their bacteriochlorophyll pigments are structurally similar to chlorophylls but have a slightly different absorption spectra.

Many proteobacteria are important nutrient cyclers. *T. namibiensis* helps connect sulfur and nitrogen cycles in the seas. It strips electrons from sulfur compounds as an energy source and also stores nitrate for use as a final electron acceptor. *Rhizobium* in the roots of peas and many other legumes fixes nitrogen. *Agrobacterium tumefaciens* causes tumors in plants but it also is a fine vector for genetic engineering (Section 16.7).

Make note of the genus names in Figure 21.7. You already met some of these pathogenic proteobacteria in earlier chapters and will encounter the rest later on.

Chlamydia is a group of intracellular parasites that cannot make ATP; they pilfer it from animal cells. One species causes a sexually transmitted disease you will read about in Section 44.8. **Spirochaetes** are free-living cells, parasites, or symbionts. All are motile. All look like stretched-out springs (Figure 21.8*c*). Symbionts, including *Pillotina* and others, in the termite gut digest cellulose in wood. *Borrelia burgdorferi* travels in ticks that bite many wild animals and humans. In humans, this spirochaete infection results in Lyme disease, a focus of Section 25.16.

Gram-positives are still being sorted out; they are not a monophyletic group. Staining

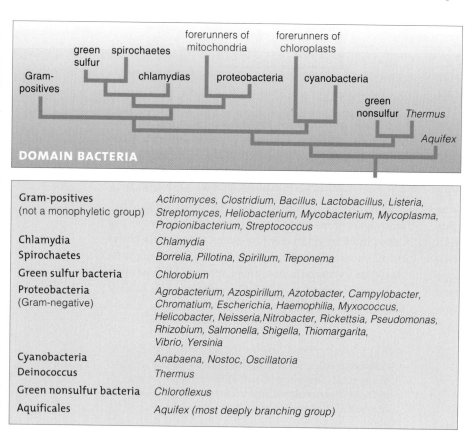

	forerunners of mitochondria	forerunners of chloroplasts
green sulfur	spirochaetes	
Gram-positives	chlamydias	proteobacteria	cyanobacteria
		green nonsulfur *Thermus*
		Aquifex

DOMAIN BACTERIA

Gram-positives (not a monophyletic group)	*Actinomyces, Clostridium, Bacillus, Lactobacillus, Listeria, Streptomyces, Heliobacterium, Mycobacterium, Mycoplasma, Propionibacterium, Streptococcus*
Chlamydia	*Chlamydia*
Spirochaetes	*Borrelia, Pillotina, Spirillum, Treponema*
Green sulfur bacteria	*Chlorobium*
Proteobacteria (Gram-negative)	*Agrobacterium, Azospirillum, Azotobacter, Campylobacter, Chromatium, Escherichia, Haemophilia, Myxococcus, Helicobacter, Neisseria, Nitrobacter, Rickettsia, Pseudomonas, Rhizobium, Salmonella, Shigella, Thiomargarita, Vibrio, Yersinia*
Cyanobacteria	*Anabaena, Nostoc, Oscillatoria*
Deinococcus	*Thermus*
Green nonsulfur bacteria	*Chloroflexus*
Aquificales	*Aquifex* (most deeply branching group)

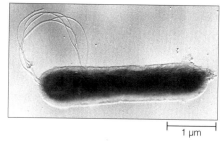

1 μm

Figure 21.7 Domain Bacteria. This evolutionary tree is by no means inclusive. It is meant only to convey some relationships among the prokaryotic groups mentioned in this book. *Above*, micrograph of *Aquifex aeolicus*, a descendant of what may be one of the earliest bacterial lineages.

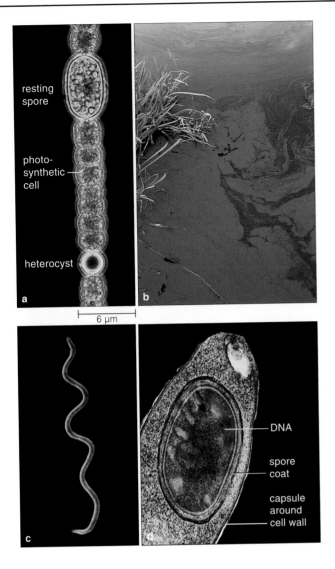

Figure 21.9 Magnetotactic bacterium. Inside the cytoplasm is a chain of magnetite particles that acts like a compass as the cell moves.

Figure 21.8 A few bacteria. (**a**) An *Anabaena* chain with a resting spore and a heterocyst. (**b**) Cyanobacterial chains forming a mat on a nutrient-rich pond surface. (**c**) *Borrelia burgdorferi*, a spirochete that causes Lyme disease in humans. (**d**) Endospore forming in *Clostridium tetani*.

resting spore

photo-synthetic cell

heterocyst

6 µm

DNA

spore coat

capsule around cell wall

a

b

c

d

tints their multilayered wall purple. Most species are chemoheterotrophs. We use the fermenting activities of *Lactobacillus* species to make many popular foods, including yogurt. *L. acidophilus* lowers the pH of skin and of intestinal and vaginal linings, which helps keep pathogenic bacteria and fungi from forming colonies.

The Gram-positive bacteria *Clostridium* and *Bacillus* make **endospores**. This resting structure encloses the bacterial chromosome and a bit of cytoplasm (Figure 21.8*d*). It can resist heat, irradiation, drying out, acids, disinfectants, and boiling water. It germinates when conditions favor growth. A bacterium emerges from it, and normal metabolic function resumes.

Endospores do damage inside the human body. In 2001, someone slipped *Bacillus anthracis* endospores into envelopes and mailed them. A few people opened the mail, breathed in endospores, and died of anthrax. *Clostridium tetani* endospores that slip inside the body through cuts or wounds can cause tetanus, a disease described in Section 37.9. *C. botulinum* endospores can taint canned food. Toxins synthesized by bacteria that emerge from germinating endospores cause botulism, a dangerous form of food poisoning.

REGARDING THE "SIMPLE" BACTERIA

Bacteria are small. Their insides are not elaborate. *But bacteria are not simple.* They sense and move toward areas where nutrients are more plentiful and where other conditions also favor growth. Aerobes move to oxygen; anaerobes move away from it. Photosynthetic types move toward light but away from light that is too intense. Many species avoid toxins or predators.

Magnetotactic bacteria migrate in the ocean, and they grow best in oxygen-poor seawater. They contain a compass—chains of magnetic particles that respond to Earth's magnetic field (Figure 21.9). For instance, in the Northern Hemisphere, geomagnetic north points downward at a slight angle from the equator toward the North Pole. Cells responding to the angle move farther down from the surface waters, where oxygen concentrations are not as great.

Some bacteria show a collective behavior, as when millions of free-living myxobacteria cells glide about as a "predatory" colony. They secrete enzymes that digest "prey," such as other bacteria. They move and change direction as a unit when following chemical gradients toward food. When food dwindles, many of the cells mass together, interact chemically, and form spore-bearing structures. Some cells differentiate and form a stalk; others form branching stalks or clusters of spores. Each spore holds a single, living cell that can germinate and give rise to a new colony.

Domain bacteria is the most ancient lineage of prokaryotic cells. Its groups include photoautotrophs, chemoautotrophs that cycle nutrients in habitats, and chemoheterotrophs. The chemoheterotrophs are most diverse; many species are dangerous parasites and pathogens.

21.5 Archaeans

LINKS TO
SECTIONS 4.3, 7.8,
8.5, 16.2, 19.5

Archaeans are prokaryotic, but in some respects they are as similar to eukaryotes as they are to bacteria. New species are turning up almost everywhere.

THE THIRD DOMAIN

It is easy to see why all prokaryotes were once placed in a single kingdom. As sketched out in Section 4.3, archaeans resemble bacteria in size and shape, and in not having a nucleus. They all have operons and other shared features. But archaean cell walls are different. Their membrane phospholipids are mirror images of those in bacterial membranes, and different enzymes make them. Also, the archaeans resemble eukaryotes in several ways. For instance, they, too, make histones. They make the same codon (methionine) to start gene transcription; bacteria make formylmethionine. Their RNA polymerases and some transcription factors are more eukaryotic than prokaryotic.

By the 1970s a molecular biologist, Carl Woese, was comparing the ribosomal RNAs of prokaryotes. The genes for rRNA are essential for protein synthesis, yet some of their base sequences have mutated without compromising rRNA's function. Many slight changes have accumulated in rRNAs of different lineages and can be measured directly. To Woese's great surprise, many lineages turned out to be somewhere *between* bacteria and eukaryotic cells in genetic distances! He proposed that the six-kingdom classification system be subsumed into three domains (Section 19.5).

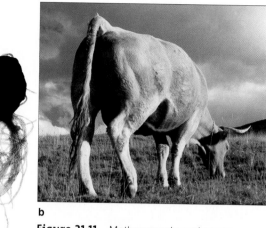

b

Figure 21.11 Methanogenic archaeans. (**a**) *Methanococcus jannaschii* lives near hydrothermal vents in 85°C water. (**b**) Three kinds of methanogenic archaeans have been isolated from the cattle gut.

a

Today there is wide acceptance that archaeans form a "third" domain, with lineages that may not have changed much since life originated (*archae–* means ancient). Based on ongoing comparisons, the domain already has been subdivided into three major groups. They are Euryarchaeota, Crenarchaeota, and the more recently discovered Korarchaeota (Figure 21.10).

HERE, THERE, EVERYWHERE

With respect to physiology, most of the archaeans are **methanogens** (methane makers), **extreme halophiles** (salt lovers), and **extreme thermophiles** (heat lovers). Methanogens and halophiles are euryarchaeotes. Most of the domain's thermophiles are crenarchaeotes.

The three informal designations can be confusing, because many bacteria also are methanogens, extreme halophiles, and extreme thermophiles. Even so, they will persist. Why? Taxonomists might be pleased with the formal names they bestowed upon the three sets of archaeans, but students would have an easier time learning names that are less of a mouthful.

You read about the methanogenic archaeans at the start of Chapter 3. Different kinds have been found in marshes, Antarctica, the ocean, and deep in the Earth. A few are symbionts in the gut of termites and some other animals (Figure 21.11). All are strict anaerobes; free oxygen kills them. When forming ATP, they strip electrons from hydrogen gas (H_2) or acetate. Carbon from CO_2 is the final electron acceptor, and methane (CH_4) forms as the product. Collectively, methanogens in both prokaryotic domains release about 2 billion tons of methane annually. They have major impact on the global cycling of carbon.

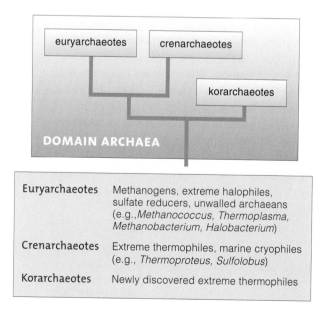

Euryarchaeotes	Methanogens, extreme halophiles, sulfate reducers, unwalled archaeans (e.g., *Methanococcus*, *Thermoplasma*, *Methanobacterium*, *Halobacterium*)
Crenarchaeotes	Extreme thermophiles, marine cryophiles (e.g., *Thermoproteus*, *Sulfolobus*)
Korarchaeotes	Newly discovered extreme thermophiles

Figure 21.10 The major groups of Domain Archaea. A few of the known representatives are listed.

Figure 21.12 Life in extreme environments. (**a**) In salty evaporation ponds in Utah's Great Salt Lake, extreme halophiles (certain archaeans and red algae) tint the water pink. (**b**) Yellowstone's exceedingly hot springs and pools are among the habitats of many extreme thermophiles. (**c**) The parasitic *Nanoarchaeum equitans* (smaller *blue* spheres) grows only when attached to *Ignicoccus* (larger spheres). Both marine archaeans were isolated from 100°C water near a hydrothermal vent.

Extreme halophilic archaeans live in the Dead Sea, the Great Salt Lake, saltwater evaporation ponds, and other highly salty habitats (Figure 21.12*a*). Most get ATP by aerobic reactions but they can harness light energy when free oxygen is scarce. Bacteriorhodopsin is a unique light-activated pigment embedded in their plasma membrane. When it absorbs sunlight energy, it changes shape and pumps protons (H+) out from the cell. H+ flows back into it, through a type of ATP synthase, and drives ATP formation.

Extreme thermophilic archaeans also live in sulfur-rich hot springs (Figure 21.12*b*). Like methanogens, they are strict anaerobes. Unlike them, they use sulfur as an electron acceptor or donor in their ATP-forming reactions. *Sulfolobus* cells grow in acidic hot springs.

Other extreme thermophiles are the start of food webs near hydrothermal vents, where temperatures exceed 110°C. These archaeans use hydrogen sulfide escaping from the vents as an electron source for ATP-forming reactions. Their existence is cited as evidence that life could have originated on the seafloor.

Researchers came across *Nanoarchaeum equitans* as they were exploring hydrothermal vents near Iceland. At 400 nanometers across, this euryarchaeote is the smallest known cell and the only parasitic archaean (Figure 21.12*c*). At this writing, its genome has been sequenced. It is the smallest genome yet found.

Seawater samples collected from the depths off the coast of Antarctica and California hold more archaeans (mostly crenarchaeotes) than the bacteria in seawater near the surface. Some archaeans are as-yet unnamed cryophiles; they are adapted to temperatures below 15°C (59°F). Mounds of methane hydrate on the ocean floor and sediments from the Great Lakes contain archaeans. So do agricultural fields, natural grasslands, northern coniferous forests, and the Siberian tundra. Today, biologists are finding new species of archaeans almost everywhere, including inside the human gut.

Woese now compares the discovery of Archaea to the discovery of a continent, which he and others are exploring and mapping. They have come to recognize that these prokaryotes are diverse and widespread. They also have discovered that archaeans and bacteria in the same habitats swap genes, and may do so often.

Like bacteria, archaeans are prokaryotic cells, but in some respects they resemble eukaryotes. Enough differences have been identified at the molecular level to grant archaeans equivalent ranking with bacteria and eukaryotes.

Archaeans were once thought to be confined to extreme environments. Many are now being discovered alongside bacteria in more hospitable habitats.

21.6 The Viruses

LINKS TO
SECTIONS
13.1, 13.2, 14.1

In ancient Rome, virus meant "poison" or "venomous secretion." In the late 1800s, this rather nasty word was bestowed on newly discovered pathogens, smaller than the bacteria being studied by Louis Pasteur and others. Many viruses deserve the name. They attack humans, cats, cattle, birds, insects, plants, fungi, protists, and bacteria. You name it, there are viruses that infect it.

DEFINING CHARACTERISTICS

Today, we define a **virus** as a noncellular infectious agent having two characteristics. First, a viral particle consists of a protein coat wrapped around a nucleic acid core—that is, its genetic material. Second, a virus cannot reproduce by itself. It can be reproduced only after its genetic material enters a host cell and directs that cell's biosynthetic machinery into making many copies of itself.

Different viruses contain DNA *or* RNA. Their coat consists of one or more types of protein monomers organized into a characteristic shape, such as a rod or polyhedron (with many sides), as in Figure 21.13. The coat protects the genetic material during the journey to a new host cell. It incorporates proteins that bind to specific receptors on host cells. Complex viruses have a sheath, tail fibers, and other structures attached to the coat. In some viruses, a bit of plasma membrane of an infected cell surrounds the coat when the virus particle buds off or when the membrane is ruptured. Glycoproteins spike above the membrane envelope.

The vertebrate immune system detects certain viral proteins. But genes for many viral proteins mutate at high frequencies, so viruses often elude the immune fighters. People susceptible to lung infections get new "flu shots" each year because the envelope spikes on influenza viruses keep changing.

EXAMPLES OF VIRUSES

Each kind of virus can multiply only in specific hosts. It cannot be studied easily except in cultures of living host cells. This is why much of our understanding of viruses comes from the **bacteriophages**, which infect bacterial cells. Unlike the cells of complex, multicelled species, bacterial hosts can be cultured easily and fast. Remember how bacteria and bacteriophages were used in early experimental studies of DNA (Section 13.1)? They are still used in genetic engineering.

Table 21.2 lists important groups of animal viruses. These viruses contain double- or single-stranded DNA or RNA, which gets replicated in various ways. Their sizes range from 18 nanometers (parvoviruses) to 350 nanometers (the brick-shaped poxviruses). Many cause diseases, including the common cold, certain cancers, warts, herpes, and influenza (Figure 21.14a,b). As you will see in Chapter 39, HIV is the trigger for AIDS. By destroying certain white blood cells, it weakens the immune system's ability to fight infections that may not otherwise be life threatening.

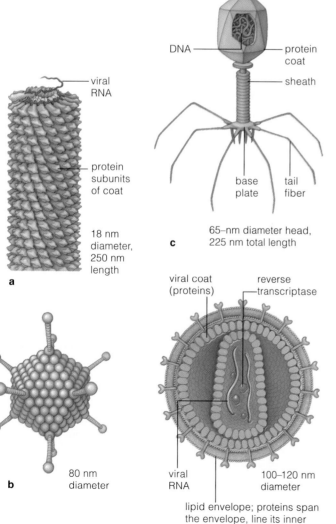

Figure 21.13 *Animated!* Structure of viruses. (**a**) *Helical* viruses, such as this tobacco mosaic virus, have a rod-shaped coat of protein subunits coiled helically around nucleic acid. (**b**) This adenovirus and other *polyhedral* viruses have a many-sided coat. (**c**) T-even bacteriophages and other *complex* viruses incorporate accessory parts attached to the coat. (**d**) Membrane encases the coat of the *enveloped* viruses. HIV, shown here, is an example.

Table 21.2	Classification of Some of the Major Animal Viruses

DNA Viruses	Some Diseases and Outcomes
Parvoviruses	Gastroenteritis; roseola (fever, rash) in small children; aggravation of symptoms of sickle-cell anemia
Adenoviruses	Respiratory infections (fever, cough, sore throat, rash), diarrhea in infants, conjunctivitis (inflamed, pebbly eye membranes); some cause tumors
Papovaviruses	Benign and malignant warts
Orthopoxviruses	Smallpox, cowpox, monkeypox
Herpesviruses:	
H. simplex type I	Oral herpes, cold sores
H. simplex type II	Genital herpes (Section 44.8)
Varicella–zoster	Chicken pox, shingles
Epstein–Barr	Infectious mononucleosis; cancers of skin, liver, cervix, pharynx; Burkitt's lymphoma (malignant tumor of jaw, face)
Cytomegalovirus	Hearing loss, mental impairment
Hepadnavirus	Hepatitis B (severe liver infection)

RNA Viruses	Some Diseases and Outcomes
Picornaviruses:	
Enteroviruses	Polio, hemorrhagic eye disease, hepatitis A (infectious hepatitis)
Rhinoviruses	Common cold
Hepatitis A virus	Inflammation of liver, kidneys, spleen
Togaviruses	Forms of encephalitis (inflammation in the brain), rubella
Flaviviruses	Yellow fever (fever, chills, jaundice), dengue (fever, severe muscle pain), St. Louis encephalitis
Coronaviruses	Upper respiratory infections, colds
Rhabdoviruses	Rabies, other animal diseases
Filoviruses	Hemorrhagic fevers, as by _Ebola_ virus (Section 21.8)
Paramyxoviruses	Measles, mumps, respiratory ailments
Orthomyxoviruses	Influenza
Bunyaviruses	
Bunyamwera virus	California encephalitis
Phlebovirus	Hemorrhagic fever, encephalitis
Hantavirus	Hemorrhagic fever, kidney failure
Arenaviruses	Hemorrhagic fevers
Retroviruses:	
HTLV-I, HTLV-II*	Adult T-cell leukemia
HIV	AIDS
Reoviruses	Respiratory and intestinal infections

* Human T-cell leukemia virus.

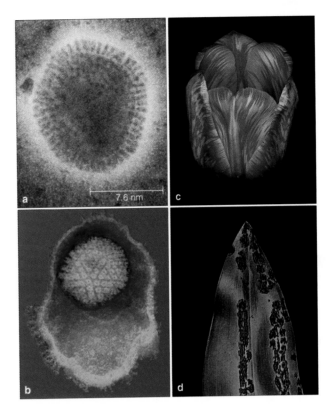

Figure 21.14 Examples of viruses and their effects. (**a**) Influenza virus. Spikes of glycoprotein project above its envelope. (**b**) One of the herpes viruses. Its envelope was pulled aside for this image. (**c**) Streaking in tulip petals. A harmless virus infected cells in the colorless parts and disrupted pigment formation. (**d**) An orchid leaf infected by a rhabdovirus.

Researchers who attempt to develop drugs against HIV and other diseases use HeLa cells and other cell lineages for initial experiments (Chapter 9). Later on, they must use laboratory animals, and then human volunteers, to test any potential drug for toxicity and effectiveness. Why? A functioning immune system is necessary to test responses to the drugs.

Plant viruses must breach plant cell walls to cause diseases. They typically hitch rides on the piercing or sucking devices of insects that feed on plant juices. Some RNA viruses infect tobacco plants (the tobacco mosaic virus), barley, potatoes, and other major crop plants. Certain DNA viruses infect such valued crops as cauliflower and corn. Figure 21.14c,d shows visible effects of two viral infections.

A virus is a noncellular infectious particle that consists of nucleic acid enclosed in a protein coat and sometimes an outer envelope. It cannot multiply without pirating the metabolic machinery of a specific type of host cell.

Nearly all organisms are targets of specific viruses.

21.7 Viral Multiplication Cycles

Even the simplest prokaryotes can replicate their own genetic material and reproduce. If we define life in terms of metabolism, protein synthesis mechanisms, and reproduction, then viruses are not alive. The catch is that they do contain DNA or RNA.

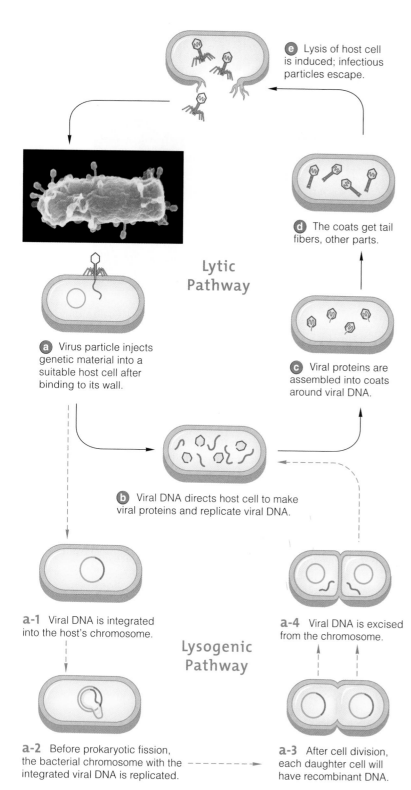

e Lysis of host cell is induced; infectious particles escape.

Lytic Pathway

d The coats get tail fibers, other parts.

a Virus particle injects genetic material into a suitable host cell after binding to its wall.

c Viral proteins are assembled into coats around viral DNA.

b Viral DNA directs host cell to make viral proteins and replicate viral DNA.

a-1 Viral DNA is integrated into the host's chromosome.

a-4 Viral DNA is excised from the chromosome.

Lysogenic Pathway

a-2 Before prokaryotic fission, the bacterial chromosome with the integrated viral DNA is replicated.

a-3 After cell division, each daughter cell will have recombinant DNA.

THE BASIC STEPS

Once different viruses infect host cells, they multiply in a variety of ways. Even so, nearly all multiplication cycles proceed through five steps, as outlined here:

1. *Attachment.* A virus particle attaches to a host cell by molecular groups that can chemically recognize and lock on to specific molecular groups of receptors at the cell surface.

2. *Penetration.* Either the virus particle or its genetic material alone crosses the plasma membrane of a host cell and enters the cytoplasm.

3. *Replication and then synthesis.* In an act of molecular piracy, the viral DNA or RNA directs the host cell's transcription and translation mechanisms into making many copies of viral nucleic acids and viral proteins, including enzymes.

4. *Assembly.* The viral nucleic acids and viral proteins become organized as new infectious particles.

5. *Release.* By one mechanism or another, the newly formed virus particles are released from the cell.

We can use the lytic and lysogenic pathways that are common among bacteriophage multiplication cycles to give you an idea of what happens in these steps.

Lytic Pathway In a **lytic pathway**, steps 1 through 4 proceed rapidly, and new particles are released when the host cell undergoes **lysis**. In this context, "lysis" means that the damage to a cell's plasma membrane, wall, or both is allowing the cytoplasm to dribble out. New virus particles escape while the ruptured cell is dying. Late into most lytic pathways, the host cell synthesizes a viral enzyme, and that enzyme's action triggers the cell's swift destruction (Figure 21.15).

Lysogenic Pathway During a **lysogenic pathway**, a latent period extends the duration of the cycle. In this case, the virus does not kill its host outright. Instead,

Figure 21.15 *Animated!* Generalized multiplication cycle for some bacteriophages. Virus particles may be produced and released by a lytic pathway. For certain viruses, the lytic pathway may expand to include a lysogenic pathway.

a viral enzyme cleaves the host's chromosomal DNA, then integrates the viral genes into its base sequence.

An infected cell may not divide for a while. When it does, it replicates its DNA, including all of the foreign genes in the recombinant molecule. As a result of one instance of genetic recombination, miniature time bombs are passed on to its descendants, and all of their descendants. At some point, however, a molecular signal or some other stimulus may reactivate the multiplication cycle, as shown in Figure 21.16.

Latency is part of the multiplication cycles of many kinds of viruses, not just bacteriophages. One case in point is Type I *Herpes simplex*, the cause of *cold sores* (fever blisters). Almost all humans harbor this virus. It remains latent in our facial tissues inside a ganglion. (A ganglion is a cluster of nerve cell bodies that each have one or more long, thin extensions called axons projecting from them.) Sunburn and other stress factors can reactivate the virus. When this happens, the virus particles move down to the axon endings and arrive near the surface of skin. There they infect epithelial cells and cause painful skin eruptions.

REVERSE TRANSCRIPTION OF VIRAL RNA

The multiplication cycle of RNA viruses has a twist to it. Inside a host cell's cytoplasm, their RNA functions as a template for synthesizing either DNA or mRNA. For example, HIV, a retrovirus, carts its own enzymes into cells. These assemble a DNA strand on viral RNA by reverse transcription (Sections 16.1 and 39.10).

Like other enveloped viruses, herpes viruses enter a host cell by their own version of endocytosis. Then new particles bud from the plasma membrane. Figure 21.16 shows how they accomplish this.

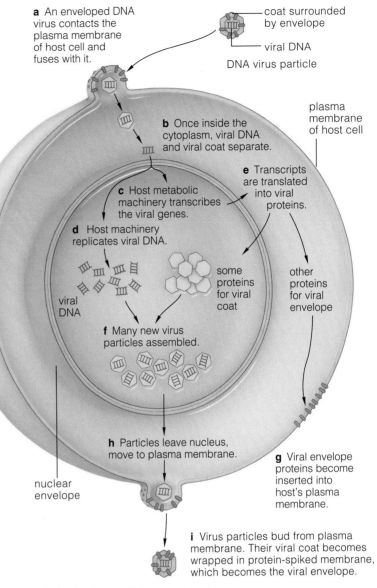

a An enveloped DNA virus contacts the plasma membrane of host cell and fuses with it.

coat surrounded by envelope

viral DNA

DNA virus particle

b Once inside the cytoplasm, viral DNA and viral coat separate.

plasma membrane of host cell

e Transcripts are translated into viral proteins.

c Host metabolic machinery transcribes the viral genes.

d Host machinery replicates viral DNA.

viral DNA

some proteins for viral coat

other proteins for viral envelope

f Many new virus particles assembled.

nuclear envelope

h Particles leave nucleus, move to plasma membrane.

g Viral envelope proteins become inserted into host's plasma membrane.

i Virus particles bud from plasma membrane. Their viral coat becomes wrapped in protein-spiked membrane, which becomes the viral envelope.

j The finished particle is equipped to infect a new potential host cell.

Figure 21.16 *Animated!* Multiplication cycle of one type of enveloped DNA virus, here infecting a generalized animal cell. Notice how part of the plasma membrane surrounds the budding viral particle. This is how some viruses get their envelope.

Viral multiplication cycles include five steps: Attachment to a suitable host cell, cell penetration, viral DNA or RNA replication and synthesis of viral proteins, assembly of new viral particles, and release from the infected cell.

Some bacteriophage replication cycles follow a rapid, lytic pathway and an extended, lysogenic pathway.

The replication cycles of RNA viruses involve the use of viral RNA as a template for synthesizing DNA or mRNA.

21.8 Evolution and Infectious Diseases

LINKS TO
SECTIONS
17.3, 18.4

Just by being human, you are a targeted host for a staggering variety of pathogenic bacteria, viruses, fungi, protozoans, and parasitic worms.

The Nature of Disease   "Disease" is something that just about everybody knows about but often has trouble explaining. Start with a few basic definitions. When a pathogen breaches the body's surface barriers and enters the internal environment, it may multiply inside cells or tissues. This is what **infection** means. **Disease** follows when the body's defenses cannot be mobilized quickly enough to keep a pathogen's activities from interfering with normal body functions. With *contagious* diseases, mucus, blood, or some other body fluid that harbors the pathogen must directly contact a new host.

In an *epidemic*, a disease spreads fast through part of a population in a limited time span, then subsides. In a *pandemic*, a disease breaks out in several countries at the same time. AIDS is pandemic. Sporadic diseases, such as whooping cough, occur irregularly and affect few people. Endemic diseases pop up more or less continually but do not spread far in large populations. Tuberculosis is like this. So is impetigo, a highly contagious bacterial infection that typically spreads no further than, say, a single day-care center.

Mycobacterium tuberculosis

AIDS is a pandemic that has no end in sight. A recent outbreak of *SARS* (severe acute respiratory syndrome) was a brief pandemic. It started in China, and travelers quickly carried it to countries around the world. Before government-ordered quarantines arrested its spread, thousands were sickened and hundreds died. A previously unknown coronavirus causes SARS. It has reappeared in China and may not subside completely.

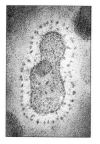

SARS virus

An Evolutionary Perspective   Consider disease in terms of a pathogen's prospects for survival. A pathogen stays around only for as long as it has access to outside sources of energy and raw materials. To a microscopic organism or virus, a human is a treasurehouse of both. With bountiful resources, a pathogen can multiply or replicate itself to amazing population sizes. Evolutionarily speaking, the ones that leave the most descendants win.

Two barriers prevent pathogens from evolving to a position of world dominance. First, any species that has a history of being attacked by a specific pathogen has coevolved with it and has built-in defenses against it, as by the vertebrate immune system. Second, if a pathogen kills too suddenly, it might disappear along with the individual host. This is one reason why most pathogens have less-than-fatal effects. After all, infected individuals who live longer spread more germs and contribute to a pathogen's reproductive or replicative success.

Usually, an individual dies only when it becomes host to overwhelming numbers of a pathogen, when it is a novel host with no coevolved defenses, or when a mutant pathogenic strain has emerged and has breached the current defenses.

Being equipped with an evolutionary perspective, you probably can perceive the connection on your own. The greater the population density of host individuals, the more kinds and greater frequencies of infectious diseases transmitted among them. This brings us to the bad news.

Emerging Pathogens   Thanks to planes, trains, and automobiles, people travel often and in droves around the world. Among their exotic destinations are virgin tropical forests and other remote regions where the human body had been an infrequent or nonexistent opportunity for pathogens. Strange and often dangerous pathogens are opportunistic about the two-legged packages of nutrients entering their habitats. Within just a few hours, they may infect travelers, who often carry pathogens back home.

Some of these emerging pathogens have been around for a long time and are only now taking advantage of the presence of novel human hosts. Others are newly mutated strains of existing pathogens.

Ebola virus

Consider *Ebola*, one of several viruses that cause a dangerous hemorrhagic fever. It might have coevolved with monkeys in the tropical forests of Africa. By 1976, it was infecting humans. It kills between 70 and 90 percent of the people it infects. No vaccine or treatment is available for the disease, which starts with high fever and flu-like aches. Within a few days, nausea, vomiting, and diarrhea begin. Blood vessels are destroyed. Blood seeps from the circulatory system into the surrounding tissues and out through all the body's orifices. The liver and kidneys may rapidly turn to mush. Patients often become deranged and then die of

Table 21.3	The Eight Deadliest Infectious Diseases		
Disease	Main Agents	Estimated New Cases per Year	Estimated Deaths per Year
Acute respiratory infections*	Bacteria, viruses	1 billion	4.7 million
Diarrheas**	Bacteria, viruses, protozoans	1.8 billion	3.1 million
Tuberculosis	Bacteria	9 million	3.1 million
Malaria	Sporozoans	110 million	2.5–2.7 million
AIDS	Virus (HIV)	5.6 million	2.6 million
Measles	Viruses	200 million	1 million
Hepatitis B	Virus	200 million	1 million
Tetanus	Bacteria	1 million	500,000

* Includes pneumonia, influenza, and whooping cough.
** Includes amoebic dysentery, cryptosporidiosis, and gastroenteritis.

Figure 21.17 (**a**) Charlene Singh, the first known vCJD case in the United States, being cared for by her mother. She probably contracted the disease when she was growing up in Britain, before moving to Florida. Mood swings, balance problems, and memory loss started in 2001. She became confined to bed and could not control her body functions or communicate. She died in 2004.

(**b**) Section through a brain damaged by BSE. The light-colored "holes" are areas where tissue was destroyed. Below this micrograph is a model for a normal prion. The vCJD-causing version misfolds into a different three-dimensional structure.

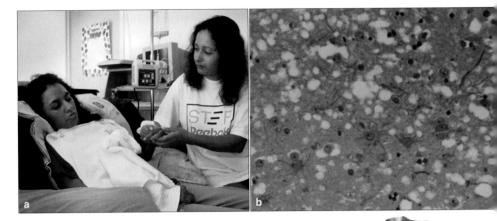

circulatory shock. Understandably, at the start of an *Ebola* outbreak, government agencies throughout the world are mobilized. Quarantines can limit the pathogen's spread.

Drug-Resistant Strains There is an old saying that when you attack nature, it will come back at you with a pitchfork. Antibiotic resistance, explained in Section 18.4, has contributed to an upsurge in infectious diseases. For example, preschoolers have an immune system that is still developing. Increasing numbers are enrolled in day-care centers. *Streptococcus pneumoniae* and other pathogens slip through the crowded populations of hosts, causing pneumonia, meningitis, and chronic middle-ear infections. Each year, *S. pneumoniae* kills 40,000 to 50,000 people of all ages, even in hospitals. The drug-resistant strains are becoming the rule, not the exception.

Most of our 6.4 billion selves live in crowded cities. In any given interval, as many as 50 million of us are on the move within and between countries in search of a better life. Is it any wonder that agents of cholera, tuberculosis, and other diseases are spreading globally (Table 21.3)?

Foodborne Diseases and Mad Cows Each year, according to estimates from Centers for Disease Control, contaminated food sickens as many as 80 million people in the United States. The most vulnerable are the very young, the very old, or those with a weakened immune system. *Campylobacter, Salmonella, Staphylococcus aureus,* and *Listeria* are frequent contaminants of dairy products, meat, and poultry. All can cause abdominal pain, nausea, and diarrhea. *Listeria* infection during pregnancy may cause miscarriage or premature birth.

0157:H7 is one of the pathogenic strains of *E. coli*, a normally harmless bacterial species that inhabits your intestinal tract. Tons of ground beef contaminated with 0157:H7 have been recalled from vendors. Vegetarians also are at risk, because 0157:H7 has been found in fruit juices, lettuce, green onions, and alfalfa sprouts.

Bacteria are not the only foodborne threats. Consider *BSE* (bovine spongiform encephalopathy). The first case in

the United States was reported in 2003. A single cow in Washington State developed this so-called *mad cow disease*. The government recalled meat from the herd and reassured consumers that none had entered the human food supply. What if it had?

Before 2000, thousands of cattle with BSE entered food chains, mainly in Great Britain. So far, that batch of meat has caused about 153 cases of a fatal brain disease, *vCJD* (variant Creutzfeldt–Jakob disease). Charlene Singh, an American, was one of the cases (Figure 21.17*a*).

A widely accepted hypothesis is that **prions** cause BSE, vCJD, and several other degenerative diseases in humans, cattle, and some wild animals, such as deer and elk.

Prions are certain protein particles that are present in normal tissues of the nervous system, where they are folded into a characteristic configuration (Figure 21.17*b*). Infectious prions are misfolded. In some unknown way, the altered protein shape induces the normal prions to become misfolded the same way.

Misfolded prions accumulate as massive deposits in the brain. Brain tissue samples taken from individuals who died from prion-induced diseases are riddled with holes. Collectively, the holes give brain tissue a spongelike (spongiform) appearance, as in Figure 21.17*b*.

The kinds of prions that cause BSE and vCJD are nearly identical. They might be mutated strains of the prions that cause *scrapie*, a degenerative disease in sheep. Some prions apparently were transferred from sheep to cattle by way of livestock feed that contained animal parts. Governments banned the use of such feed.

Prions are not alive, so they cannot be killed. It is not easy to destroy them by denaturation methods of boiling, baking, irradiation, and application of most disinfectants.

Should you worry about eating American beef? The USDA says no. If you are a beef lover and a skeptic, you might wish to avoid eating brains or meat cuts that contain spinal cord, such as oxtails. Ground beef and sausage may accidentally contain nervous tissue, so watch how it is being made or make your own. You may also want to look for meat from grass-fed cattle.

Summary

Section 21.1 Bacteria and archaeans are prokaryotic cells; they do not have a nucleus. Some have membrane infoldings and other structures in the cytoplasm. None has a profusion of organelles.

Three cell shapes are common: cocci (spheres), bacilli (rods), and spirilla (spirals). Nearly all bacteria have a cell wall with peptidoglycans that protects the plasma membrane and resists rupturing. Often a sticky mesh of polysaccharides (a glycocalyx) surrounds the wall as a capsule or slime layer. Some species have one or more bacterial flagella, or motile structures that rotate like a propeller. Many have pili: filamentous proteins that help cells adhere to a surface or that facilitate conjugation.

Collectively, prokaryotes show metabolic diversity, as in modes of acquiring energy and carbon. Cyanobacteria and other *photoautotrophs* use sunlight and carbon dioxide during photosynthesis. Some bacteria and archaeans are *photoheterotrophs*; they use sunlight energy but get carbon from organic compounds. *Chemoautotrophs*, such as the nitrifying bacteria, use carbon dioxide but make ATP by stripping electrons from organic or inorganic substances. Most prokaryotes are *chemoheterotrophs* that range from decomposers, pathogens, and parasites to saprobes.

Biology☰Now
Explore prokaryotic structure with the animation on BiologyNow.

Section 21.2 Only bacteria and archaeans divide by prokaryotic fission: replication of a single, circular bacterial chromosome and division of a parent cell into two genetically equivalent daughter cells. Many species have plasmids: small circles of DNA that are replicated independently. They can transfers plasmids to cells of the same or different species by bacterial conjugation.

Biology☰Now
Observe prokaryotic fission and conjugation with the animation on BiologyNow.

Section 21.3 Traditionally, any newly discovered prokaryote was classified by the total number of traits it shares with a known prokaryotic group. Biochemical comparisons are clarifying phylogenies. Groups are being assigned to domains Bacteria and Archaea.

Section 21.4 Bacteria are the most ancient lineages of prokaryotic cells, and they are metabolically diverse. Most are chemoheterotrophs, and many of these species are parasites and pathogens.

Some major groups are cyanobacteria, proteobacteria, chlamydia, spirochaetes, and Gram-positive bacteria. The proteobacteria are the most diverse monophyletic group.

Section 21.5 Archaeans are prokaryotic. They are like bacteria in size, shape, not having a nucleus, having operons, and other respects. They are unique in other respects, as in their cell wall composition. Archaeans also resemble eukaryotes—for example, in some aspects of gene transcription and translation. They are now classified in their own domain, separate from bacteria. These prokaryotes show far greater distribution and diversity than was previously suspected.

Comparative rRNA studies give evidence of three archaean groups: euryarchaeotes, crenarchaeotes, and korarchaeotes. Most archaean methanogens (methane makers) and halophiles (salt lovers) belong to the first group, and most of the archaean extreme thermophiles belong to the second group.

Archaeans and bacteria coexist in many hospitable as well as hostile habitats, and apparently they engage in gene transfers.

Section 21.6 All viruses are noncellular infectious particles that attack specific kinds of host species.

Each virus particle consists of a core of DNA or RNA and a protein coat that sometimes is enclosed in a lipid envelope. Many glycoproteins project like spikes from the envelopes. The coats of complex viruses also have sheaths, tail fibers, and other accessory structures.

A virus particle multiplies only after its genetic material enters a host cell and directs synthesis of the molecules necessary to produce new virus particles.

Biology☰Now
Compare viral forms with the animation on BiologyNow.

Section 21.7 Nearly all viral multiplication cycles have five steps: attachment to a suitable host cell, cell penetration, viral DNA or RNA replication followed by protein synthesis, assembly of new viral particles, and release. Two pathways are common in multiplication cycles of bacteriophages (bacteria-infecting viruses).

In a lytic pathway, multiplication is fast; new viral particles are released by lysis.

In a lysogenic pathway, the infection enters a latent period. A host cell is not killed outright, and the viral nucleic acid may undergo recombination with a host cell chromosome.

Multiplication cycles of viruses are diverse. Besides varying in their duration, the penetration and release of most enveloped types occur by endocytosis and budding. DNA viruses spend part of the cycle in the nucleus of a host cell. RNA viruses complete the replication cycle in the cytoplasm. The viral RNA is the template for mRNA synthesis and for protein synthesis.

Biology☰Now
Learn how viruses can multiply with the animation on BiologyNow.

Section 21.8 Infection is the invasion of a cell or multicelled body by a pathogen, which causes disease when its activities interfere with body functions. Hosts and pathogens coevolve. Antibiotics and antiviral drugs select for drug-resistant strains. Many bacteria cause food poisoning. Prions are infectious proteins that kill by destroying the brain and nervous tissue.

Biology☰Now
Read the InfoTrac article "Origins of HIV: The Interrelationship Between Nonhuman Primates and the Virus," Myrna Watanabe, Bioscience, September 2004.

Self-Quiz

Answers in Appendix II

1. Only _____ are prokaryotic.
   - a. archaeans
   - b. bacteria
   - c. prions
   - d. both a and b

2. Bacteria transfer plasmids by _____ .
   - a. prokaryotic fission
   - b. endospore formation
   - c. conjugation
   - d. the lytic pathway

3. The _____ are all oxygen-releasing photoautotrophs.
   - a. spirochetes
   - b. chlamydias
   - c. cyanobacteria
   - d. proteobacteria

4. The normally harmless *E. coli* in your gut are _____ .
   - a. spirochetes
   - b. chlamydias
   - c. cyanobacteria
   - d. proteobacteria

5. All _____ are intracellular parasites of vertebrates.
   - a. spirochetes
   - b. chlamydias
   - c. cyanobacteria
   - d. proteobacteria

6. Some Gram-positive bacteria (e.g., *Bacillus anthracis*) survive harsh conditions by forming a(n) _____ .
   - a. pilus
   - b. heterocyst
   - c. endospore
   - d. plasmid

7. Only _____ reproduce by prokaryotic fission.
   - a. viruses
   - b. archaeans
   - c. bacteria
   - d. b and c are correct

8. DNA or RNA may be the genetic material of _____ .
   - a. bacteria   b. a virus   c. archaeans   d. a prion

9. Is this statement false: All known viruses consist of genetic material, a protein coat, and an outer envelope.

10. Bacteriophages can multiply by _____ .
    - a. prokaryotic fission
    - b. the lytic pathway
    - c. the lysogenic pathway
    - d. both b and c

11. Match the terms with their most suitable description.
    - _____ archaean
    - _____ bacteria
    - _____ virus
    - _____ plasmid
    - _____ extreme halophile
    - _____ prion
    - _____ sex pilus

    - a. infectious protein
    - b. nonliving infectious particle; nucleic acid core, protein coat
    - c. prelude to conjugation
    - d. prokaryotes that most closely resemble eukaryotes
    - e. most common prokaryotic cells
    - f. small circle of bacterial DNA
    - g. salt lover

Additional questions are available on Biology ⒺNow™

Critical Thinking

1. Annual costs of treating known cases of food poisoning are between 5 billion and 22 billion dollars, and pathogens in kitchens may cause at least half of the cases. Electron microscopes show microbes on wood or plastic cutting boards, even stainless steel knives (Figure 21.18). Carlos Enriquez and his colleagues at the University of Arizona sampled 75 dishrags and 325 sponges in some homes. They found *Salmonella*, *E. coli*, *Pseudomonas*, and *Staphylococcus* colonies in most of them. Bacteria can live for as long as two weeks in a wet sponge (which can be sanitized in one dishwasher cycle). If your class has access to a good light microscope, each of you bag one of your kitchen sponges or dishrags and compare what might be living with you.

Figure 21.18 On a kitchen knife's blade, bits of food and *Pseudomonas*. The cell's pili attached to the blade. Think about it next time you use any unwashed kitchen utensil that contacted raw beef, poultry, or seafood.

2. One way to prevent bacterial food poisoning is by *food irradiation*, exposing food to high-energy rays or beams that kill pathogenic bacteria. The process also slows spoilage and prolongs shelf life. Some think this is a safe way to protect consumers. Others think the treatment could produce harmful chemicals. They say irradiation does not kill endospores that can cause botulism. In their view, the best way to prevent food poisoning is to tighten and enforce food safety standards.

Irradiated meat, poultry, and fruits are now available in many supermarkets. By law, they must be marked with the symbol shown at right. Would seeing this symbol on a package make you more or less likely to purchase the product? Explain your answer.

3. *Thermotoga maritima* was discovered in geothermically heated marine sediments and prefers water of about 80°C. Analysis of rRNA puts it in the bacterial group, but gene sequencing of the entire genome reveals a more complex picture. About 50 percent of *T. maritima* genes resemble bacterial genes, about 25 percent of its genes are unique, and about 25 percent resemble those of archaeans, the thermophile *Pyrococcus horikoshii* in particular. Suggest two possible explanations for the similarity between these prokaryotes of different domains.

4. Some picornavirus strains cause the highly contagious *foot and mouth disease*. Hosts include cattle, water buffalo, sheep, and pigs. Symptoms include extensive lesions and major tissue erosion, lameness, weight loss, milk loss, and often death. The viral strain causing a recent epidemic was first identified in Asia, then in Europe and elsewhere. It devastated farmers still recovering from a BSE outbreak. There is no vaccine; at this writing, more than a million animals have been destroyed to stop the spread of the disease. Research the impact of this virus on travel and the global economy. Hint: Start at http://www.cdc.gov/.

5. Do you think that the whole world would be better off without viruses? If so, consider this: Curtis Suttle at the University of British Columbia studies interactions among viruses, bacteria, and algae in ocean water. He selectively removed viruses from seawater samples—and the algae in the water stopped growing. Suttle found that they were dependent on nutrients released by dying bacteria, which lyse after viral infection. Make a list of some other ways in which viruses might be integrated into ecosystems.

22 "PROTISTS"—THE SIMPLEST EUKARYOTES

Tiny Critters, Big Impacts

Go for a swim just about anywhere in the natural world and you will be sharing the water with multitudes of diverse organisms traditionally called protists. Like their most ancient ancestors, almost all of these eukaryotic species are aquatic. Different kinds are photosynthesizers, predators, and decomposers in fresh water or the seas. Others live as parasites inside the cells or moist tissues of host species. Many are symbionts with plants, fungi, and animals.

Structurally, single-celled protists are the simplest of all eukaryotes. Although most are microscopically small, they often have big collective impact on our world.

For instance, the kinds called foraminiferans and coccolithophores drift in the surface waters of the world ocean. Most are tiny cells, with a shell or plates typically hardened with calcium carbonate. Figure 22.1 shows a sampling of them. Long ago, the calcium-rich remains of ancient species started to accumulate on the seafloor. Over great spans of time, the deposits were compressed into stone. Chalk and other limestones are the legacy of those uncountable numbers of single cells.

Dover, England, is renowned for its white chalk cliffs (Figure 22.1). The cliffs are a monument to individual coccolithophores that lived, died, and drifted down to the seafloor. Each year, they collectively formed deposits that

were only 0.5-millimeter thick. But they did so, silently, for *many millions* of years. Dover is not the only immensely stacked graveyard. For instance, Austin, Texas, was built on top of 400-meter-thick chalk deposits that formed in a similarly ancient sea.

In 3200 B.C. on Malta, foraminiferan-rich limestone deposits were quarried to build the Hagar Qim temple, now a World Heritage site (Figure 22.1). A few thousand years later, other limestones were used to construct the Egyptian pyramids and Sphinx. Look closely at the walls of the Empire State Building, the Pentagon, or Chicago's Tribune Tower, and you will find shells of foraminiferans.

Tiny protists with us today also make big statements. *Phytophthora infestans*, one of the water molds, infects valuable crop plants. As you will read in this chapter, it destroyed potato crops that were the main food staple in Europe. Millions of people starved to death or emigrated to other countries.

Today, related species threaten forests in the United States, Europe, and Australia. When conditions favored its growth, *P. ramorum* started an epidemic of sudden oak death in California. Tens of thousands of oaks have died. The pathogen has now jumped to madrone, redwoods, and other novel hosts. Cascading ecological changes will reduce sources of food and shelter for forest species.

Figure 22.1 What shells of foraminiferans (including *Spiroloculina* and *Peneroplis*) look like through a handheld lens. On average, the shells are about the size of the period that ends this sentence. Even so, foraminiferans have contributed to architectural marvels, such as those shown on the facing page: *Above*, Hagar Qim temple, Malta. *Below*, the white chalk cliffs of Dover, England, towering 91 meters (about 300 feet) above the English Channel.

Watch the video online!

Contrasting these two examples of protists can help you keep the global perspective in mind. Some of these simple eukaryotes do cause disease or damage, and we tend to assign "value" to them in terms of their direct impact on our lives. There is nothing wrong with battling the harmful species and admiring the beneficial ones. Even so, do not lose sight of how protists or any other kind of organism fit in nature's larger picture.

How Would You Vote?

The pathogen that causes sudden oak death has already infected twenty-six kinds of plants in California and Oregon. Some infected species are commonly sold as nursery stock. Should the States that are free of the pathogen be allowed to prohibit shipping of all plants from the States that are affected? See BiologyNow for details, then vote online.

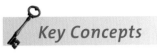

Key Concepts

SORTING OUT THE "PROTISTS"

For some time, single-celled eukaryotic organisms have been set aside in the catch-all kingdom Protista. Molecular comparisons and studies of ultrastructure are clarifying their evolutionary connections with one another and to multicelled plants, fungi, and animals.   Section 22.1

BASAL GROUPS

A group informally called flagellated protozoans includes two of the earliest lineages of eukaryotic cells. Another group of ancient origins includes the foraminiferans and radiolarians.   Sections 22.2–22.4

THE ALVEOLATES

Ciliated protozoans, dinoflagellates, and apicomplexans belong to a branch of the alveolates, which are unique in having a layer of tiny sacs under their plasma membrane. Sections 22.5–22.7

THE STRAMENOPILES

Chrysophytes, diatoms, and brown algae are groups of stramenopiles, most of which are photoautotrophs. The colorless, parasitic oomycotes are an offshoot of the stramenophile branch.   Sections 22.8, 22.9

THE CLOSEST RELATIVES OF LAND PLANTS

Evolutionarily, red algae and green algae are closer to the ancestors of plants than they are to any of the other "simple" eukaryotes. Molecular research indicates that green algae and plants are a monophyletic group. Sections 22.10, 22.11

DISTANT RELATIVES OF FUNGI AND ANIMALS

A diverse and formerly puzzling assortment of amoeboid cells are now assigned their own taxon, the amoebozoans. They are close to the lineages that gave rise to fungi and to animals.   Section 22.12

Links to Earlier Concepts

If we liken Section 19.7 to an evolutionary road map, then this chapter zeros in on villages of single-celled eukaryotes. Before you begin, reflect on the logic behind constructing cladograms (19.5, 19.6). You also may wish to review the molecular tools that are revealing much of the evidence for the new groupings (16.2, 16.5, 17.8–17.9).

You will encounter examples of endosymbiosis in the history of life (20.5). You will be building on your early introduction to reproductive phases of life cycles (10.5).

22.1 An Evolutionary Road Map

LINKS TO
SECTIONS
16.2, 16.5, 19.5

Traditionally, the mostly single-celled eukaryotes were assigned to kingdom Protista. Researchers sensed that they were not members of a monophyletic group, but only now do they have the tools to sort them out.

Chlorophytes (green algae)	Acetabularia, Chlamydomonas, Chlorella, Codium, Udotea, Ulva, Volvox, Scenedesmus
Red algae	Antithamnion, Porphyra
Amoebozoans	Amoeba, Dictyostelium, Physarum
Stramenopiles	
Chrysophytes	Emiliania, Mischococcus, Synura
Diatoms	Thalassiosira
Brown Algae	Laminaria, Macrocystis, Postelsia, Sargassum
Oomycotes	Phytophthora, Plasmopara, Saprolegnia
Alveolates	
Ciliates	Didinium, Paramecium, Stylonychia
Dinoflagellates	Gonyaulax, Gymnodinium, Karenia, Noctiluca
Apicomplexans	Plasmodium
Foraminiferans	Peneroplis, Spiroloculina
Radiolarians	Pterocorys, Stylosphaera
Kinetoplastids	Trypanosoma, Leishmania
Euglenoids	Euglena
Diplomonads	Giardia
Parabasalids	Trichomonas, Trichonympha

Of all existing species, **protists** are the most like the first eukaryotic cells. Unlike prokaryotes, protist cells have a nucleus. Most also have mitochondria, ER, and Golgi bodies. Their ribosomes are larger than those in bacteria. They have more than one chromosome, each consisting of DNA with many proteins attached. They have a cytoskeleton that includes microtubules. Many cells have chloroplasts. Also, unlike prokaryotes, they divide by way of mitosis, meiosis, or both.

Most protists are single cells, but there are colonial and multicelled species. Many are photoautotrophs, but others are predators, parasites, and decomposers. A number of them form spores.

Of course, these characteristics also show up among plants, fungi, and animals, so they scarcely define what a protist is. Until recently, protists were defined mainly in terms of what they are *not*, as in "not bacteria, not plants, not fungi, not animals." They were lumped into a kingdom that represented an evolutionary crossroads between prokaryotes and "higher" forms of life.

Comparative studies of these organisms have been accelerating as a result of gene sequencing and more advanced methods for analyzing fine structures. One outcome is that this former polyphyletic kingdom is being split into monophyletic groups—including the major groups shown in Figure 22.2. You may use this evolutionary tree as a road map through the chapter. Also shown is a list of representative genera that we use as examples throughout the book.

Most species known as protists are single celled, but nearly all lineages include multicelled forms. The diversity within and between lineages is astounding. Even so, seven distinct monophyletic groups have now been identified.

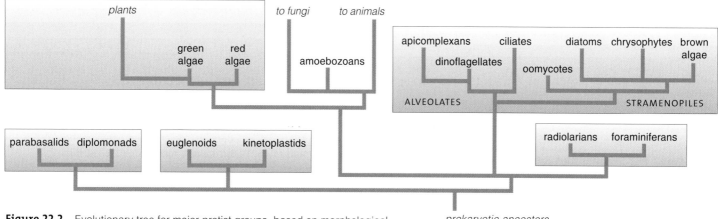

Figure 22.2 Evolutionary tree for major protist groups, based on morphological, molecular, and genetic comparisons. All groups shown have living representatives.

22.2 Parabasalids and Diplomonads

*Existing parabasalids and diplomonads represent two
of the earliest lineages of single-celled eukaryotes.*

"Protozoa" is a word with historical meaning (*proto–*,
first; and *–zoa*, animal). At one time, biologists were
looking for the ancestor of all animals among a loose
confederation of motile, single-celled, predatory, and
parasitic species. As we now know, most of the groups
are not close to the base of the animal family tree.

Among the most evolutionarily distant groups are
the **parabasalids** and **diplomonads** (Figure 22.2). Both
consist of heterotrophic flagellates. There are hundreds
of saclike or elongated species living in many oxygen-
poor or anaerobic habitats. As you might expect, they
have few mitochondria or none at all. Based on genetic
comparisons, the shared ancestor of these groups had
a rudimentary version of a mitochondrion or a well-
developed one that later became modified or lost.

All parabasalids have the equivalent of a backbone: bundled microtubules extending the length of
the cell. In *Trichonympha campanula,* hundreds to thousands of flagella are derived from the bundle. This
profusely flagellated cell is a cellulose-digesting symbiont in the gut of termites and wood roaches.

The parabasalid *Trichomonas vaginalis* is a parasite.
It latches on to the epithelial linings of the vagina or
male reproductive tract and sucks out nutrients for its
fermentation activities. It has four anterior flagella. A
thin extension of the plasma membrane billows around
another flagellum at the tailing end (Figure 22.3a).

Worldwide, *T. vaginalis* has infected more than 170
million people, primarily during sexual intercourse.
Reinfections among sexual partners are common, and
infected pregnant women can transmit the parasite to
a fetus during childbirth. Without prompt treatment,
infections damage the urinary and reproductive tracts.

Diplomonads have three flagella at their anterior
end and one at the trailing end. *Giardia lamblia* is an
intriguing species (Figure 22.3b). It has no lysosomes,
mitochondria, or Golgi bodies, and it does not form
a bipolar spindle at mitosis (Section 9.3). These clues
and others suggest that its lineage may have started

more than a billion years ago, when the ancestors of
eukaryotes were first diverging from prokaryotes.

Even so, *G. lamblia* is quite successful without great
internal complexity. It is a common intestinal parasite
of humans, cattle that forage on the open range, and
wild animals. Its sucking device leaves behind telltale
imprints on cells of the intestinal epithelium (Figure
22.3c). Although *G. lamblia* is anaerobic, it can survive
outside the host body, as cysts, in water that has been
contaminated with feces. Among microbes, a **cyst** is a
resting stage with a covering of cell secretions.

Ingesting cysts of *G. lamblia* may result in *giardiasis.*
Symptoms of this disease range from mild cramps to
severe diarrhea that may last for weeks. Giardiasis is
common among toddlers and day-care workers, and
people who drink water from streams. It is prevalent
in overcrowded regions where water quality is poor
owing to inadequate to nonexistent sewage treatment.

*Genetic comparisons, and the absence of well-developed
mitochondria and some other organelles, suggest that the
ancestors of parabasalids and diplomonads were among
the first eukaryotic cells to evolve.*

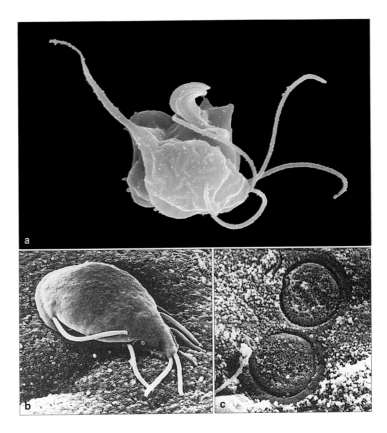

Figure 22.3 (**a**) Scanning electron micrograph of
the motile feeding stage of *Trichomonas vaginalis,*
This parasite causes the sexually transmitted disease
trichomoniasis. (**b**) Scanning electron micrographs of
Giardia lamblia and (**c**) examples of the imprints that
its sucking disk leaves on intestinal epithelium.

23.4 Seedless Vascular Plants

The first seedless vascular plants were branching and had no roots and leaves. A spectacular adaptive radiation led to diverse leafy and treelike forms in the Carboniferous.

Existing **lycophytes**, **horsetails**, and **ferns** had their beginnings in Devonian swamps. Like their ancestors, they differ in important ways from the bryophytes. Their sporophytes contain xylem and phloem and are not attached to gametophytes. Also, the sporophytes are the larger, longer lived phase of the life cycle.

Seedless vascular plants require a moist habitat to complete the sexual phase of their life cycle; a film of water must be present for flagellated sperm to reach the eggs. Like bryophytes, a few species live in dry habitats, but they reproduce sexually during pulses of seasonal rains. *Selaginella lepidophylla*, the resurrection plant, shrivels in the dry season but revives fast. You may see it growing in Mexico, New Mexico, and Texas.

WHISK FERNS

Whisk ferns are not ferns and are shaped like whisk brooms. They are native to New Zealand, Australia, Japan, and the southeastern United States. *Psilotum* is a unique vascular plant (Figure 23.11*a*). It probably evolved from an ancestral species that formed roots. However, its sporophytes have **rhizomes**: branching, short, mainly horizontal absorptive stems that grow underground. *Psilotum* also has reduced leaves—tiny outgrowths on photosynthetic branched stems.

LYCOPHYTES

Tree-sized lycophytes lived in Carboniferous swamp forests, but the 1,100 or so modern species are a great deal smaller. The most widespread are club mosses, which live in habitats from tundra to the tropics. Most club mosses have vascularized stems and roots, which grow from a branching, underground rhizome.

Microphylls—tiny leaves, each with an unbranched vein—are a defining trait of lycophytes. Chambers for spore production develop at the base of some of these leaves. In certain lycophytes, the leaves are organized around a central stem as a **strobilus** (plural, strobili). This general term refers to any conelike reproductive structure derived from modified leaves. Some of the *Lycopodium* species are informally called ground pines, partly because their appearance reminds some people of pine trees (Figure 23.11*b*). The sporophytes of many other plants, including horsetails and gymnosperms, also produce strobili during the life cycle.

HORSETAILS

Seedless vascular plants called sphenophytes grew as tall as trees in Carboniferous swamp forests. About thirty smaller species of genus *Equisetum* are with us today; you may know them as horsetails and scouring rushes (Figure 23.11*c–e*). They grow alongside many streams, railroad tracks, roads, and vacant lots.

The sporophytes of most horsetails have rhizomes, hollow stems, and scalelike leaves at stem nodes. A cylinder of xylem and phloem runs parallel with the stems. The stems have horizontal ribs reinforced with silica granules, so they feel like sandpaper. As they moved westward, American pioneers traveled light and used scouring rushes as disposable pot scrubbers.

Figure 23.11 Seedless vascular plants. (**a**) *Psilotum*, a whisk fern. (**b**) One of the lycophytes, *Lycopodium*. Horsetails: (**c**) Vegetative stem of *Equisetum arvense*. (**d**) Stems and (**e**) strobilus of scouring rushes (*E. hyemale*).

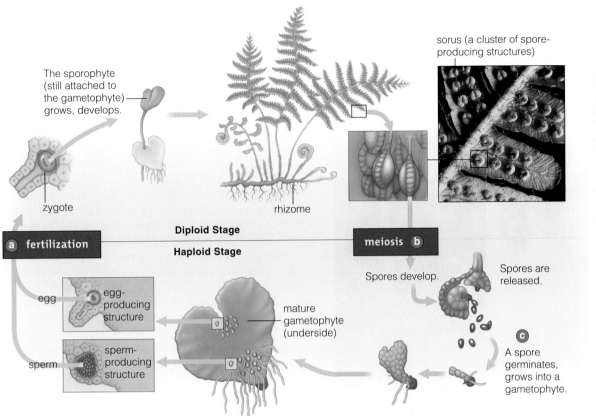

The sporophyte (still attached to the gametophyte) grows, develops.

zygote

sorus (a cluster of spore-producing structures)

Diploid Stage

Haploid Stage

rhizome

a fertilization

meiosis **b**

egg

egg-producing structure

sperm

sperm-producing structure

mature gametophyte (underside)

Spores develop.

Spores are released.

c

A spore germinates, grows into a gametophyte.

Figure 23.12 *Animated!* Life cycle of a chain fern (*Woodwardia*), one of the seedless vascular plants. The photograph below shows part of a forest of tree ferns (*Cyathea*) in Australia's Tarra-Bulga National Park. Tree fern forests like this once cloaked much of New Zealand. Most have now been cleared to make way for farming.

Figure 23.11 shows the photosynthetic vegetative stems and a fertile stem of two *Equisetum* species. The strobilus at a fertile stem tip releases haploid spores. Free-living gametophytes about one millimeter to one centimeter across grow from the germinating spores.

FERNS

With 12,000 or so species, the ferns are the largest and most diverse group of seedless vascular plants. All but 380 or so species are native to the tropics, but we find them in homes and gardens all over the world. They are highly diverse in size. Some floating ferns have leaves less than 1 centimeter wide. Certain tree ferns may grow 25 meters (82 feet) tall.

Young fern leaves, or fronds, develop in a pattern that looks a bit like the coil of a fiddlehead. They are uncoiled by maturity. Fronds of chain ferns and many other species are divided into leaflets (Figure 23.12).

Roots and leaves usually grow from vascularized rhizomes. Tropical tree ferns, such as those shown in Figure 23.12, are exceptions. So are species that live as epiphytes. *Epiphyte* refers to any aerial plant, which grows attached to tree trunks or branches.

Rust-colored patches form on the lower surface of most fern fronds. Each patch, a tiny cluster of spore-forming chambers, is a sorus (plural, sori). Usually, chamber walls are one cell thick. They pop open, and

haploid spores are catapulted through the air. After germination, each spore gives rise to a heart-shaped gametophyte a few centimeters across (Figure 23.12).

Large, independent sporophytes with xylem, phloem, and leaves dominate the life cycle of seedless vascular plants. Sperm reach eggs by swimming through films or droplets of water. In lycophytes and horsetails, spores form in protective clusters of modified leaves (strobili).

23.5 Ancient Carbon Treasures

LINKS TO
SECTIONS
4.9, 17.5, 19.5

Three hundred million years or so ago, in the middle of the Carboniferous, mild climates prevailed and swamp forests carpeted the wet lowlands of continents. The absence of pronounced seasonal swings in temperature favored plant growth through much of the year. Plants with lignin-reinforced tissues and well-developed root and shoot systems had the competitive edge.

In vast forests of the Carboniferous, massively stemmed lycophyte trees—the giant club mosses—topped out at almost forty meters (Figure 23.13). Being so high above the forest floor, dispersing spores was a cinch. Their strobili released as many as 8 billion microspores or hundreds of megaspores. Giant horsetails, including species of *Calamites*, were close to twenty meters tall. Aboveground stems often grew quickly from underground rhizomes that spread far, and they formed dense thickets.

As it happened, the sea level rose and fell *fifty times* in the Carboniferous. When the sea receded, steamy swamp forests flourished. When the sea moved back in, the trees became submerged and buried in sediments that protected them from decomposers. Over time, sedimentary layers accumulated over the remains, and the weight squeezed water out of the saturated, undecayed remains. Pressure continued, and the compaction of organic remains released heat energy. Pressure and heat transformed the compacted mass into great seams of **coal** (Figure 23.13).

With its high percentage of carbon, coal is rich in energy and one of our premier "fossil fuels." It took a staggering amount of photosynthesis, burial, and compaction to form each major seam of coal in the ground. It has taken us only a few centuries to deplete much of the world's known coal deposits. Often you will hear about annual production rates for coal or some other fossil fuel. How much do we really produce each year? We produce nothing. We simply *extract* it from the ground. Coal is a nonrenewable source of energy.

Lepidodendron,
a tree-sized
club moss

stem of a giant lycophyte
(*Lepidodendron*)

seed fern (*Medullosa*), one of the early seed-bearing plants

stem of a giant horsetail
(*Calamites*)

Figure 23.13 Reconstruction of a Carboniferous forest. Above it, a photograph of part of a seam of coal.

23.6 The Rise of Seed-Bearing Plants

In diversity, numbers, and distribution, seed producers became the most successful groups of the plant kingdom.

Let's take stock. Seed-bearing plants appeared in the late Devonian. The first were close relatives of ferns, yet seeds—in some cases as big as walnuts—formed on fernlike fronds (Figure 23.13). Those "seed ferns" flourished, but long before the last species vanished, cycads, conifers, and other gymnosperms had started their adaptive radiations. (What some folks call the Age of Dinosaurs, botanists call the Age of Cycads.) By 120 million years ago, the flowering plants were rapidly diversifying and in time assumed supremacy.

Think back on the factors that promoted the rise of seed plants. Structural modifications, such as a water-conserving cuticle, were one reason they endured in seasonally dry, often cold habitats. Just as important were their two kinds of spores.

Today, we observe how their **microspores** are the start of pollen grains: walled structures that protect male gametophytes inside. We see how pollen grains are released from a parent plant and journey on their own to eggs. They withstand cold and drought. They are tiny and easily dispersed through air or by insects and other animals. Regardless of the type of dispersing agent, **pollination** refers to the actual arrival of pollen on female reproductive parts of a seed plant.

In short, with the evolution of pollen grains, it no longer mattered whether water was available outside the plant body. Sperm could travel to eggs without it.

Unlike microspores, **megaspores** form in ovules that stay attached to the parent plant. An **ovule** starts as a tiny mass of sporophyte tissue. The megaspore

part of it gives rise to a female gametophyte that has an egg cell. Other parts form nutritious tissue and an outer, multilayered coat. Sperm fertilize the egg, and an embryo develops inside the ovule. *What we call a "seed" is a mature ovule.* Its coat protects the embryo when conditions force a parent sporophyte to enter dormancy. Its nutrient-rich tissue, the equivalent of a power bar, jump-starts the embryo's renewed growth as soon as conditions favor germination.

A footnote to this story: Seed plants recruit more than pollinators. They started recruiting us at least 500 million years ago, when *Homo erectus* stashed nuts and rose hips in caves and roasted seeds. By 11,000 years ago, humans were domesticating seed plants as reliable sources of food. We now recognize 3,000 or so species as edible and grow staggering numbers of 200 or so as crops (Figure 23.14). We also grow thousands more to grace homes and gardens. We grow them as sources of oils for perfumes and medicines and balms. Juices of *aloe vera* leaves soothe damaged skin; digitalin from foxgloves stabilizes blood circulation; alkaloids from periwinkle slow the growth of some cancer cells. People cultivate tobacco and marijuana for the mind-altering properties; they grow coca plants, the source of cocaine, for medicinal or suspicious reasons. In such ways, we promote the evolutionary success of many seed-bearing plants.

LINK TO SECTION 17.5

pine pollen grains

Gymnosperms and, later, flowering plants radiated into dry habitats with the help of structural modifications, pollen grains, and ovules that mature into seeds.

Figure 23.14 Edible treasures from flowering plants. (**a**) Fruits, which function in seed dispersal. (**b**) Mechanized harvesting of bread wheat, *Triticum*. (**c**) Indonesians picking shoots of tea plants (*Camellia sinensis*). Leaves of plants on hillsides in moist, cool regions have the best flavor. Only the terminal bud and two or three youngest leaves make the finest teas. (**d**) In Hawaii, a field of sugarcane, *Saccharum officinarum*. Sap extracted from its stems is boiled to make table sugar and syrups.

23.7 Gymnosperms—Plants With Naked Seeds

*With this bit of history behind us, we turn to a survey of some modern **gymnosperms**. Their seeds are perched, in exposed fashion, on a spore-producing structure. Gymnos means naked; sperma is taken to mean seed.*

CONIFERS

The 600 or so species of **conifers** are woody trees and shrubs, typically with needlelike or scalelike leaves. Conifers have female **cones**, or reproductive structures with many ovules wedged between clusters of papery or woody scales. Most shed some leaves all year long but remain leafy, or *evergreen*. A few shed all leaves in autumn; they are *deciduous*. The most abundant trees of the Northern Hemisphere (pines), the tallest ones (redwoods), and the oldest (bristlecone pine) are all conifers. Figures 23.15a and 23.16 show two species. Also in this group are dawn redwoods, firs, spruces, junipers, cypresses, larches, and podocarps.

LESSER KNOWN GYMNOSPERMS

About 130 species of **cycads** made it to the present. Pollen-bearing strobili and seed-bearing strobili form on separate plants (Figure 23.15b). Pollen moves from male to female plants on air currents or on pollinator beetles. Most cycads evolved in tropical or subtropical regions. In Guam, the Chamorro people grind cycad seeds into flour. The seeds contain a toxic amino acid that becomes concentrated as it moves through food chains (Section 47.3). These people have the world's highest rate of neurological disorders, such as ALS. Many cycads are widely sold as ornamental plants, but some species in the wild face extinction.

Ginkgos (Figure 23.15c–f) were diverse in dinosaur times. The maidenhair tree, *Ginkgo biloba*, is the only surviving species. Like a few other gymnosperms, it is deciduous. Ginkgos were widely planted in China a few thousand years ago. Their natural populations nearly vanished, maybe because firewood was scarce. Ginkgos have attractive fan-shaped leaves and resist insects, disease, and air pollutants. Once again these hardy trees are widely planted—male trees, at least. The thick, fleshy seeds of female trees are stinky.

Gnetophytes include tropical trees, leathery leafed vines, and desert shrubs. An herbal stimulant, now banned, was extracted from photosynthetic stems of *Ephedra* (Figure 23.15g). A single species of *Welwitschia* lives in deserts of Africa. The sporophyte has a deep taproot and a woody stem with strobili. Its two strap-shaped leaves may grow five meters long. A mature plant looks ragged because the leaves split lengthwise repeatedly during growth (Figure 23.15h).

A REPRESENTATIVE LIFE CYCLE

Before leaving the gymnosperms, take a quick look at the reproductive strategy of a conifer. For most plants,

Figure 23.15 Representative gymnosperms. (**a**) Bristlecone pine (*Pinus longaeva*) on a windswept mountain in the Sierra Nevada. (**b**) Male strobilus of *Dioon*, a cycad. *Ginkgo biloba*: (**c**) Fleshy seeds, (**d**) new leaves, (**e**) fossilized leaf, and (**f**) fall foliage. Two gnetophytes: (**g**) *Ephedra viridins* and (**h**) *Welwitschia mirabilis*, with strappy leaves and seed-bearing strobili.

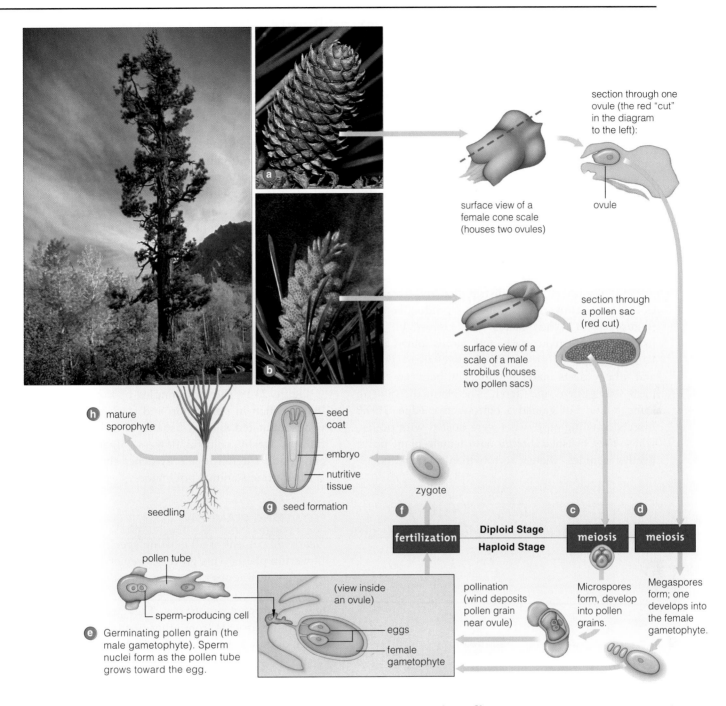

section through one
ovule (the red "cut"
in the diagram
to the left):

surface view of a
female cone scale
(houses two ovules)

ovule

section through
a pollen sac
(red cut)

surface view of a
scale of a male
strobilus (houses
two pollen sacs)

h mature
sporophyte

seed
coat

embryo

nutritive
tissue

zygote

seedling

g seed formation

pollen tube

f fertilization

Diploid Stage
Haploid Stage

c meiosis

d meiosis

sperm-producing cell

e Germinating pollen grain (the
male gametophyte). Sperm
nuclei form as the pollen tube
grows toward the egg.

(view inside
an ovule)

eggs

female
gametophyte

pollination
(wind deposits
pollen grain
near ovule)

Microspores
form, develop
into pollen
grains.

Megaspores
form; one
develops into
the female
gametophyte.

the time between pollination and fertilization can be measured in hours, but it is a year or more in conifers. Consider a mature pine tree. This sporophyte makes female cones with megaspore-containing ovules on tiers of woody scales, as in Figure 23.16. The ovules are exposed, not embedded in woody tissue. In a pine's male strobili (pollen-bearing cones), microspores form and become pollen grains. At a suitable time, millions of pollen grains drift away on air currents. Some land on ovules of the same sporophyte or a different one. After this pollination event, a pollen grain germinates, and part of it starts growing as a tubular structure. This is the sperm-bearing male gametophyte. It grows

Figure 23.16 *Animated!* Life cycle of a conifer, the ponderosa pine.

slowly through the ovule's tissues for about one year. After it penetrates the female gametophyte, its sperm reaches the egg, fertilization occurs, and a new pine tree zygote starts down a developmental road.

Existing conifers, ginkgos, cycads, and gnetophytes are gymnosperms. Like their early ancestors, they are adapted to seasonally dry climates. Depending on the group, seeds form on exposed surfaces of strobili or female cones.

23.8 Angiosperms—The Flowering Plants

LINK TO SECTION 17.6

*What does "angiosperm" mean? Sperma, recall, refers to seeds. Angio–, derived from a Greek word for vessel, refers to an enclosed, protected vessel or chamber called an **ovary**. Flowering plant ovules mature into seeds inside ovaries.*

COEVOLUTION WITH POLLINATORS

When other plant groups started to decline during the Mesozoic, flowering plants started their spectacular, ongoing adaptive radiation (Figure 23.17). Now there are at least 260,000 species in meadows and forests, in parched deserts, and on windswept mountaintops. A few even live in lakes and streams, salt marshes, and shallow marine habitats.

What accounts for their distribution and diversity? Consider the **flower**, a specialized reproductive shoot (Figure 23.18). When plants first evolved on land, insects that ate plant parts and spores joined them. Later, after pollen-producing plants evolved, beetles and other insects made a connection between "plant parts with pollen" and "food." The plants did give up some pollen but gained a reproductive edge. How? Insects crawling on flowers were dusted with pollen, which they brushed *directly* onto female plant parts. Beetles were less chancy vectors than air currents.

Later still, flowering plants coevolved in ways that became more attractive to insects, as with vivid colors, fragrances, and nectar. Instead of wasting energy on random searches for food, certain insects coevolved with particular flowering plant species. They became specialists at finding and collecting pollen or nectar from them—and making direct pollen deliveries. Not coincidentally, the adaptive radiation of flowering plants coincided with an adaptive radiation of insects.

Coevolution refers to two or more species jointly evolving because of their close ecological interactions. Heritable changes in one exert selection pressure on the other, which evolves also. That is how **pollinators** —agents that deliver pollen of one species to female parts of the same species—coevolved with the seed plants. As you will see in Section 31.2, they include bees and other insects, as well as bats and birds. By recruiting pollinators to assist in sexual reproduction, flowering plants have maintained their status as the dominant group for the past 100 million years.

Figure 23.19 shows examples of floral structures that function in pollination and seed formation. The wind-pollinated species usually form inconspicuous flowers. Flashy, colorful flowers, sugar-rich nectars, and heady fragrances are evidence that a particular species attracts animal pollinators.

FLOWERING PLANT DIVERSITY

Nearly 90 percent of all existing species of plants are flowering plants. This group is tremendously diverse even in size. Species range from aquatic duckweeds (1 millimeter long) to *Eucalyptus* trees 100 meters tall. A few are not photoautotrophs. They are parasites that withdraw nutrients from other plants or mycorrhizal fungi, or they are carnivorous plants.

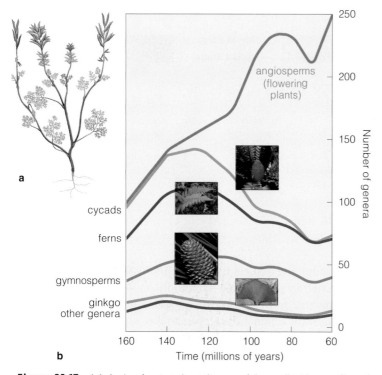

Figure 23.17 (**a**) *Archaefructus sinensis*, one of the earliest known flowering plants. Apparently it grew in shallow lakes. (**b**) Diversity of vascular plants in Mesozoic times. Conifers and other gymnosperms started to decline even before flowering plants started a major adaptive radiation.

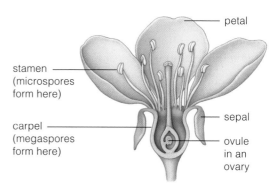

Figure 23.18 Structures of a typical modern flower. Like the earliest flowering plants, it has male and female parts (stamens and carpels). Unlike them, it also has petals and sepals.

Figure 23.19 (**a**) Representative flowers: reproductive structures with roles in pollination and seed formation. Their colors, patterns, and shapes attract pollinators. (**b**) This hummingbird pollinator has a long bill that fits the long, delicate nectar tube of the flower of a columbine (*Aquilegia*) plant.

(**c**) Sacred lotus (*Nelumbo nucifera*), an aquatic species. The radial pattern of petals is typical of ancient lineages. Some taxonomists classify it as one of the water lilies. Others group it with a basal eudicot group that includes sycamores. (**d**) More recent lineages, including these pansies (*Viola*), have a bilateral pattern, with roughly equivalent left and right parts. (**e**) Dwarf mistletoe (*Arceuthobium*) is nonphotosynthetic, and it has no pollinator-attracting petals. It parasitizes trees and stunts their growth. (**f**) Evolutionary tree diagram for flowering plants.

Flowering plants were once divided into only two groups based on the number of cotyledons, the tiny leaves that form on embryo sporophytes inside seeds. (Cotyledons are also called "seed leaves"). Plants with two cotyledons were named dicots; those with one were named monocots. It now appears that monocots branched off the more ancient dicot lineage.

As you can see from the evolutionary tree diagram in Figure 23.19*f*, we have identified three lineages that arose first. Water lilies, star anise, and *Amborella* are among their living descendants. Later on, divergences gave rise to the three dominant groups: **magnoliids**, **eudicots** (true dicots), and **monocots**. You will read about their characteristics in a later unit. For now, it is enough to get a sense of the range of their diversity.

Magnolias, avocado trees, nutmeg trees, and pepper plants are among the 9,200 kinds of magnoliids.

The 170,000 or so eudicot species include most of the herbaceous (nonwoody) plants, including daisies, lettuces, and cabbages. Most of the flowering shrubs and trees, such as roses, maples, oaks, elms, and fruit trees, as well as cacti are in this group.

Among the 80,000 named species of monocots are the orchids, palms, lilies, and grasses. Rye, sugarcane, rice, wheat, corn, barley, and some other grasses are our most important crop plants.

Angiosperms are the most successful plants. They alone produce flowers. These specialized reproductive structures have pollen-producing male parts and female parts with ovaries, protected chambers in which seeds develop.

Flowers coevolved with pollinators, which contribute greatly to the reproductive success of angiosperms.

Magnoliids, eudicots, and monocots are the main lineages of flowering plants. The largest group, eudicots, includes 170,000 named species.

25.6 Annelids—Segments Galore

LINK TO SECTION 17.8

A master gene that induces appendages to form on body segments is exuberantly expressed in annelids (Section 17.8).

ADVANTAGES OF SEGMENTATION

Annelids are 12,000 or so species of bilateral animals that include polychaetes, leeches, and oligochaetes such as earthworms (Figures 25.19 through 25.21). Of all existing animals, annelids have the most segments. Except in leeches, nearly all segments have clusters or pairs of chitin-reinforced bristles called chaetae or setae. Hence the names oligochaete and polychaete (*oligo–*, few; *poly–*, many). Bristles shoved into soil or sediments afford traction for crawling or burrowing.

Some annelids show the evolutionary potential of segmentation. Most earthworm segments are similar. The segments at both ends of a leech have a sucker. Polychaetes have an elaborate head and fleshy-lobed parapods ("closely resembling feet"). Existing species hint at the modifications in body plans that favored increases in size, more complex internal organs, and segments adapted for specialized tasks.

ANNELID ADAPTATIONS—A CASE STUDY

We can use an earthworm as a representative annelid. All earthworms are scavengers in moist habitats. Their flexible cuticle is permeable enough for gas exchange, so it cannot conserve body water. The segmented body has about 150 coelomic chambers, each with muscles, blood vessels, nerves, and other organs. A complete digestive system that has specialized regions extends through all the chambers, like a tube in a tube (Figure 25.21*a*). A muscular pharynx squeezes moist, detritus-rich soil down into the gut. **Detritus** means decaying particles of organic matter. An earthworm can eat the equivalent of its weight in detritus each day. As they burrow and feed, earthworms collectively aerate soil.

Like most annelids, earthworms also have a closed circulatory system with multiple hearts. Contractions of the heart and of muscularized blood vessels keep blood flowing in one direction. Smaller vessels service the gut, nerve cord, and body wall (Figure 25.21*b*).

An earthworm's head holds a fused pair of ganglia. Connected to this rudimentary "brain" are a pair of **nerve cords**: two lines of communication that extend down the length of the body. Signals from the brain travel along the cords and coordinate activities for the whole worm (Figure 25.21*c*). In each segment, both

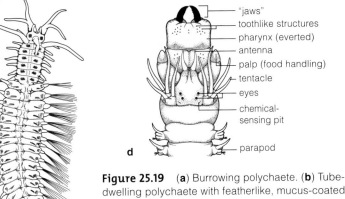

- "jaws"
- toothlike structures
- pharynx (everted)
- antenna
- palp (food handling)
- tentacle
- eyes
- chemical-sensing pit
- parapod

Figure 25.19 (**a**) Burrowing polychaete. (**b**) Tube-dwelling polychaete with featherlike, mucus-coated structures on its head. Polychaetes are common animals along coasts and by far the most diverse annelids, with stunning arrays of highly modified segments (**c**,**d**). Many types are predators or scavengers; others graze on algae.

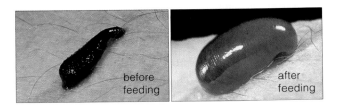

before feeding

after feeding

Figure 25.20 Leeches. Most are aquatic scavengers or predators with sharp jaws and a blood-sucking device. Shown here, *Hirudo medicinalis*, a freshwater species used for at least 2,000 years as a blood-letting tool to "cure" nosebleeds, obesity, and some other conditions. Leeches are still used. They draw off pooled blood after a doctor reattaches a severed ear, lip, or fingertip. A patient's body cannot do this on its own until blood circulation routes are reestablished.

the body. They now form before those on the left, and they pull organs along with them when they grow.

Was torsion—which put the anus above the mouth—a bad evolutionary experiment? Possibly, but lineages got around it. Often, cilia sweep out wastes. Sea slug and nudibranch larvae twist, but then they untwist.

HIDING OUT, OR NOT

Maybe it was their fleshy, soft bodies, so forgiving of chance evolutionary changes, that gave mollusks the potential to diversify so many ways. Consider a few examples. If you were small, soft of body, and *tasty*, a shell could discourage predators. For instance, when something disturbs it, a chiton contracts foot muscles that draw down its eight-piece shell, and its mantle is pressed like a suction cup against the substrate (Figure 25.24*b*). A shell protects scallops, oysters, mussels, and other bivalves. Its two hinged parts (valves) can snap shut. When a scallop claps the two together, it forces water out and propels itself backward, maybe enough to get away from a hungry sea star (Figure 25.25).

Finally, if you were small, soft of body, and *toxic* to a potential predator, a shell would be superfluous. Some nudibranchs secrete bad-tasting substances, such as sulfuric acid. Others graze on cnidarians and then incorporate ingested nematocysts into their tissues. The Spanish shawl nudibranchs flash red, nematocyst-studded respiratory organs while they mate (Figure 25.24*e*). Predators attracted to the colors get stung by nematocysts and learn to avoid the species.

> *Mollusks are bilateral, soft-bodied, coelomate animals, and only they form a mantle over the body mass.*

Figure 25.24 Mollusks. (**a**) Ventral view of an aquatic snail crawling on aquarium glass. (**b**) Chiton, with a shell divided into eight plates. (**c**) Land snail. (**d**) Octopus swimming. (**e**) Two Spanish shawl nudibranchs (*Flabellina iodinea*). Unlike other mollusks, which have males and females, nudibranchs are hermaphrodites; they exchange sperm and deposit chains of fertilized eggs into their partner. Their ancestors lost most of the mantle cavity and the respiratory organ. Most now have thin outgrowths that serve in gas exchange.

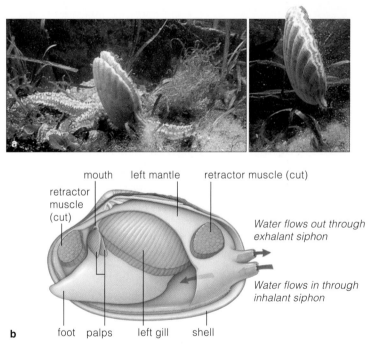

mouth   left mantle   retractor muscle (cut)

retractor muscle (cut)

Water flows out through exhalant siphon

Water flows in through inhalant siphon

b   foot palps   left gill   shell

Figure 25.25 Bivalves—clams, scallops, oysters, mussels, and other animals with a "two-valved shell." (**a**) A scallop escaping from a sea star by clapping its valves and producing a propulsive water jet. (**b**) Body plan of a clam, with half of its protective shell removed. In nearly all bivalves, gills collect food and exchange gases. As water moves through the mantle cavity, mucus on gills traps food. Cilia move the mucus and food to palps, where suitable bits are sorted out and driven to the mouth.

Some bivalves are 1 millimeter across. A few giant clams are more than 1 meter across and weigh 225 kilograms (close to 500 pounds). Humans have been eating one type of bivalve or another since prehistoric times.

25.9 On the Cephalopod Need for Speed

LINK TO
SECTION
17.5

Lively glimpses into the past emerge when evolutionary theory is enlisted to interpret the fossil record and the basis of existing species diversity.

Five hundred million years ago, the soft-bodied mollusks called **cephalopods** were supreme predators of the seas (Figure 25.26a). There were thousands of species with an elaborately chambered shell. All but one of their modern descendants have no shell at all or, at most, a shell that is extremely reduced in size (Figure 25.26b).

What happened? The loss or reduction of the shell correlates with the adaptive radiation of fishes during the Devonian, some 400 million years ago (Section 17.5). Among the fishes that hunted cephalopods or were competitors for the same prey, larger and swifter species emerged. In response, in what appears to have been a long-term race for speed and wits, most cephalopods lost the external shell and became active, streamlined and— for invertebrates—smart.

For cephalopods, jet propulsion became the name of the game. They now moved faster by forcing a water jet through a funnel-shaped siphon, out from beneath the mantle cavity. Modern species show how it works. Muscles in the mantle relax, which draws water into the cavity. They contract while the mantle's free edge closes over the head. The quick squeeze shoots water through the siphon. The brain controls the siphon's action, hence the direction of escape or pursuit. The increases in speed correlate with the evolution of complex eyes and far more efficient respiratory and circulatory systems. Cephalopods are the only mollusks with a closed circulatory system. Their heart pumps blood to two gills. Then two accessory (booster) hearts pump blood for oxygen uptake and carbon dioxide removal in metabolically active tissues, muscles especially.

Of all invertebrates, they became the fastest (squids), largest (giant squid), and smartest (octopuses). Of all mollusks, octopuses have the largest brain relative to body size, and they show the most complex behavior.

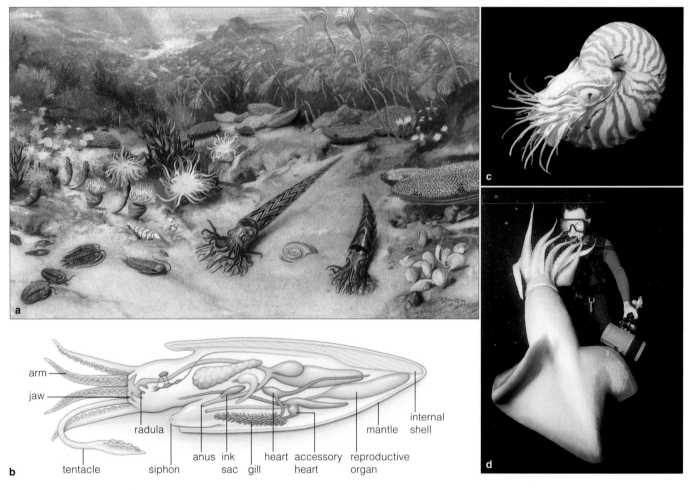

Figure 25.26 *Animated!* (**a**) Some animals of the vast, tropical seaways of the Ordovician: trilobites, the cephalopods called nautiloids, and crinoids that looked like stalked plants. (**b**) Generalized body plan of a cuttlefish, one of the cephalopods. (**c**) Chambered nautilus, a living descendant of Ordovician nautiloids. (**d**) Diver and squid (*Dosidicus*) inspecting each other.

25.10 Roundworms

Roundworms are among the most abundant living animals. Sediments in shallow water may hold a million per square meter; a cupful of topsoil teems with them.

The 22,000 or so kinds of roundworms, or nematodes, have a cylindrical body with bilateral features, tapered ends, a complete gut, and a false coelom packed mainly with reproductive organs (Figure 25.27). Roundworms shed their flexible cuticle during periodic molts. Many species are less than five millimeters long. Most are free-living decomposers that help cycle nutrients.

The free-living *Caenorhabditis elegans* is a cherished experimental organism among biologists. It has the same basic body plan and tissue types as far more complex animals, but it is transparent and tiny, with only 959 somatic cells. Researchers can monitor the developmental fate of each cell. Also, the generation time is short, and the genome is one-thirtieth the size of the human genome. Researchers also can enlist self-fertilizing hermaphroditic forms to get populations of offspring that are homozygous for desired alleles.

Many roundworms infect roots of crop plants; they are significant agricultural pests. Others are internal parasites of animals that include humans, dogs, and insects. *Ascaris lumbricoides*, a large roundworm, has now infected an estimated 1 billion people worldwide, mostly in Latin America and Asia (Figure 25.28a).

Pigs or game animals can carry *Trichinella spiralis*. Adults of this parasite lock on to the intestinal lining. Juveniles develop from eggs, travel the bloodstream to muscles, and form cysts (Figure 25.28b). The resulting disease, *trichinosis*, can be fatal. The encysted juveniles are hard to detect when inspecting fresh meat.

Repeated infections by the roundworm *Wuchereria bancrofti* can result in *elephantiasis*, a bad case of edema. Adult parasites in tissue fluid being sent back to the bloodstream get lodged inside lymph nodes (Section 38.10). There they obstruct the flow, so fluid backs up into tissue spaces. Increased fluid pressure makes the legs, feet, and other regions swell abnormally (Figure 25.28c). Mosquitoes serve as intermediate hosts.

The parasitic guinea worm makes thin, serpentlike ridges in human skin. For thousands of years, healers have extracted this "serpent" from infected people by very slowly winding it out around a stick. Pinworms (*Enterobius vermicularis*) are most commonly parasites of young children. At night, females migrate to the anal region and deposit eggs. When they thrash about, they make the infected person's skin itch.

Adult hookworms live inside the small intestine. Using toothlike devices or sharp ridges around their mouth, they cut into the intestinal wall, feed on blood and other tissues, and withdraw nutrients. Each day, adult females can release a thousand eggs, which exit in feces. When juveniles contact human skin, they cut their way inside, travel in blood to the lungs, and move up the windpipe. When the host swallows, they enter the stomach and then the small intestine, where they may mature and live for several years.

Roundworms are abundant bilateral animals with a false coelom and a complete digestive system. They include free-living decomposers and assorted parasites.

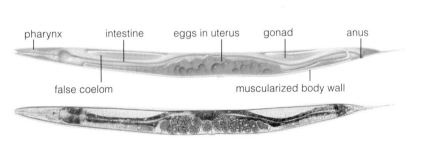

pharynx · intestine · eggs in uterus · gonad · anus

false coelom · muscularized body wall

Figure 25.27 *Animated!* Body plan and micrograph of the roundworm *Caenorhabditis elegans.*

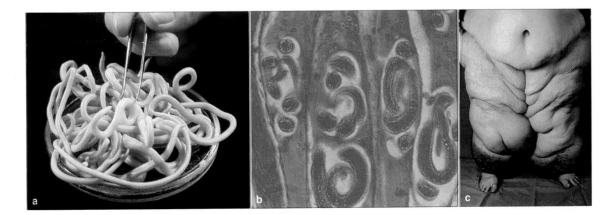

Figure 25.28 (**a**) Living roundworms (*Ascaris lumbricoides*). Infection by these intestinal parasites causes abdominal pain, vomiting, and appendicitis. (**b**) *Trichinella spiralis* juveniles in muscle tissue of a host animal. (**c**) One example of elephantiasis caused by the roundworm *Wuchereria bancrofti.*

25.11 Why Such Spectacular Arthropod Diversity?

LINKS TO
SECTIONS
8.1, 15.1

Evolutionarily speaking, "success" means having the most species in the most habitats, fending off competition and threats efficiently, exploiting the greatest amounts and kinds of food, and producing the greatest number of offspring. These features characterize the arthropods.

Arthropods are bilateral animals that have a hardened, jointed exoskeleton and specialized appendages. They have a complete gut, a greatly reduced coelom, and an *open* circulatory system, in which blood flows out of small vessels or hearts, then flows back after trickling through tissues. One group, the trilobites (Figure 25.8), became extinct long ago. Chelicerates and crustaceans, insects, and myriapods are groups that endured. The last group includes millipedes and centipedes.

We consider representatives in sections that follow. For now, start thinking about six important arthropod adaptations. They contributed to the stunning success of arthropods in general and insects in particular.

A hardened exoskeleton. Arthropods have a cuticle of chitin, proteins, and waxes, often stiffened by calcium carbonate deposits. It is a type of protective external skeleton, or **exoskeleton**. The arthropod exoskeleton may have evolved as a deterrent to predators. It took on added functions when some groups invaded land. It restricts water loss, and it helps support the weight of a body removed from the buoyancy of water.

Such exoskeletons do not restrict increases in size, because arthropods molt after each spurt in growth. **Molting** is the periodic shedding of a too-small body covering (or hair, or feathers, or horns) during the life cycle. It is under hormonal control. In arthropods, the hormones trigger the formation of a soft, new cuticle beneath the old one, which is then shed, as in Figure 25.29. Before the new cuticle hardens, the body mass increases by way of the rapid uptake of air or water and continuous mitotic cell divisions.

Jointed appendages. If any cuticle were uniformly hardened, it would prevent movements. The arthropod cuticle thins where it spans *joints*—regions where two body parts abut. The contraction of adjoining muscles makes the cuticle bend at joints, which shifts the abutting body parts relative to one another. A jointed exoskeleton was a key innovation.

It was like evolutionary clay that became molded into wings, antennae, legs, and other spectacularly diverse appendages. ("Arthropod" means jointed leg.)

Specialized and fused-together segments. Among the first arthropods, all of the body segments were more or less similar. However, in most lineages that made it to the present, some segments became fused together or modified in other specialized ways. As one example, compare the nearly identical segments of a centipede with the winged body of a butterfly.

Respiratory structures. Many freshwater and marine arthropods use a type of gill for gas exchange. Land-dwelling arthropods use a system of air-conducting tubes, called tracheas. Insect tracheas start at surface pores and branch into finer tubes that deliver oxygen into all internal tissues. Flight and other activities that use a great deal of ATP depend on the rapid uptake of oxygen in aerobically respiring tissues (Section 8.1).

Specialized sensory structures and organs. For many arthropods, diverse sensory structures contributed to their success. For example, each insect eye consists of many individual light-sensitive units that collectively sample the visual field in many directions.

Specialized stages of development. Many arthropods, especially insects, divide up the tasks of surviving and reproducing among different stages of development. For some species, the new individual is a juvenile, a miniaturized version of the adult that grows in size until sexually mature. Individuals of other species undergo **metamorphosis**. Between the embryonic and adult form, the body changes, often in drastic ways, as tissues get reorganized and remodeled. Hormones, recall, guide the developmental steps (Section 15.1).

Many immature stages in the life cycle specialize in eating and growing fast, whereas the adult specializes in reproduction and dispersal of the new generation. Think of caterpillars chewing leaves, then butterflies mating and depositing eggs. Differences in the stages of development are adaptations to specific conditions in the environment, including seasonal shifts in food sources, water, and availability of receptive mates.

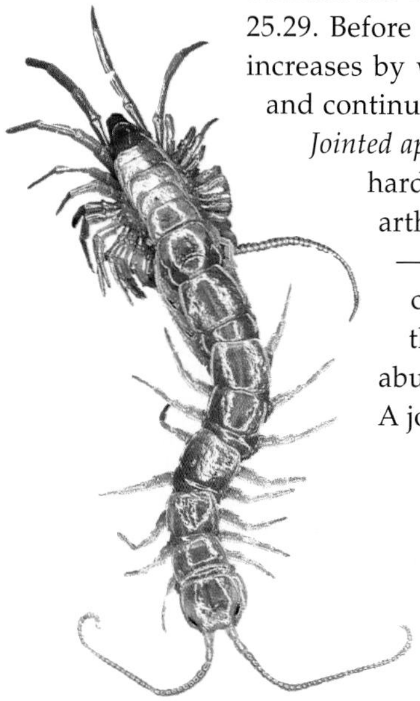

As a group, the arthropods are abundant and widespread, with diverse life-styles. Their success correlates with their hardened, jointed exoskeleton, with modified and often fused segments, and with highly specialized appendages, respiratory structures, and sensory structures.

In many species—insects especially—success also arises from a division of labor among different stages of the life cycle, such as larvae, juveniles, and adults.

Figure 25.29 Example of molting. This red-orange centipede is busily wriggling out of its old exoskeleton.

25.12 Spiders and Their Relatives

Chelicerates arose in shallow seas early in the Paleozoic. They are named for their first pair of feeding appendages (chelicerae). Horseshoe crabs are among the few living marine species. Of the familiar species on land—spiders, scorpions, ticks, and chigger mites—we might say this: Never have so many been loved by so few.

The body plan of horseshoe crabs (Figure 25.30*a*) has changed little since the Devonian. A horseshoe-shaped dorsal shield, or hardened carapace, protects the body from predators. The horseshoe crab's closest relatives on land are arachnids—spiders, scorpions, ticks, and chigger mites. Unlike arachnids, which have four pairs of legs, a horseshoe crab has five pairs. Its chelicerae move marine worms and scavenged food to its mouth.

All spiders and scorpions are aggressive predators (Figures 25.30*a* and 25.31*b*). Most spiders sting, bite, or stun prey before injecting venom into them. One spits glue and venom at prey. Another spins a thread with a ball of sticky material at the end, then uses one of its legs to swing the ball at insects passing by.

A spider's segments are fused into a forebody and hindbody. An open circulatory system extends through both regions. The heart pumps blood into tissues and takes it back through small openings in its wall. The respiratory organs are leaflike "book lungs" (Section 40.2). The paired, jointed forebody appendages include the legs, chelicerae that inflict wounds and discharge venom, and pedipalps that have mostly sensory roles (Figure 25.31). The spider's hindbody has one or more pairs of spinners, which put out silk threads for webs and egg cases. Most webs are netlike.

Figure 25.30 (**a**) Horseshoe crab (*Limulus*). Its long spine helps steer its body. (**b**) Fat-bodied scorpion of Australia.

Twenty-five or so kinds of spiders bite humans and other mammals when threatened or disturbed; Section 25.16 considers two of them. Unfortunately, they have given the whole group a bad name even though insects would overrun our world without spiders.

Mites, including the dust mites in Figure 25.31*c*, are among the smallest, most diverse, and widespread of all arachnids. Section 25.16 focuses on the direct and indirect impact of these parasites our lives.

> *Spiders, scorpions, and their relatives are predators or parasites. Their first pair of appendages, chelicerae, are structurally and functionally unique feeding structures.*

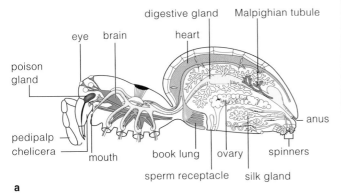

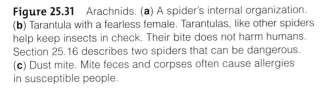

Figure 25.31 Arachnids. (**a**) A spider's internal organization. (**b**) Tarantula with a fearless female. Tarantulas, like other spiders help keep insects in check. Their bite does not harm humans. Section 25.16 describes two spiders that can be dangerous. (**c**) Dust mite. Mite feces and corpses often cause allergies in susceptible people.

25.13 A Look at the Crustaceans

LINK TO
SECTION
7.8

*The 35,000 or so species of **crustaceans** got their name because they have a hard yet flexible "crust," an external skeleton, but so do nearly all arthropods. Nearly all live in marine habitats, where they are so abundant that they are dubbed "insects of the seas."*

Shrimps, lobsters, crabs, barnacles, and pillbugs are familiar crustaceans. The lobsters and crabs are the giants of the group; most species are less than a few centimeters long. All have major roles in food webs, and humans harvest many edible types.

In one respect, the simplest crustaceans display an ancient feature: They have pairs of similar appendages along most of their body length. In different lineages, unspecialized appendages evolved into many diverse structures of the sort shown in Figure 25.32. To give two examples, the strong claws of lobsters and crabs are used to collect food, intimidate other animals, and sometimes dig burrows. The feathery appendages of barnacles comb bits of food from the water.

Many crustaceans have sixteen to twenty segments; some have more than sixty. Their head has pairs of antennae, and jawlike and food-handling appendages. The lobsters, shrimps, and crabs have five pairs of walking legs. A dorsal cuticle extends back from the head like a shield over the thoracic segments, which are fused together (Figure 25.32*a,d*).

Of all arthropods, only barnacles have a calcium-hardened shell that they secrete around themselves. The shell protects them from predators, drying winds, and strong currents or surf (Figure 25.32*b*). The adult barnacles spend their lives cemented to piers, rocks, and even other animals, including whales. Once they are attached, they cannot move on, so you would

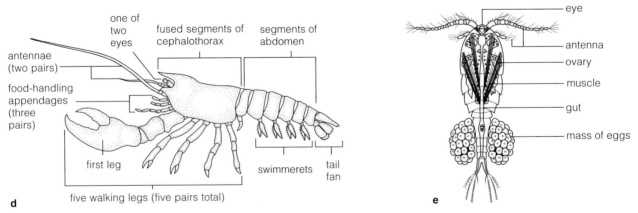

Figure 25.32 (a) Lobster in the Caribbean Sea, near Belize. (b) Goose barnacles. Adults are cemented to one spot. You might mistake them for mollusks until they open their hinged shell to filter-feed and you see jointed appendages—the hallmark of arthropods. (c) Copepod. Different kinds are free-living filter feeders, predators, and parasites. (d) Body plan of a lobster. The head has two pairs of antennae and pairs of food-handling appendages, including two jawlike mandibles. Lobsters, crayfish, crabs, shrimps, and their relatives have five pairs of walking legs. (e) Body plan of a copepod.

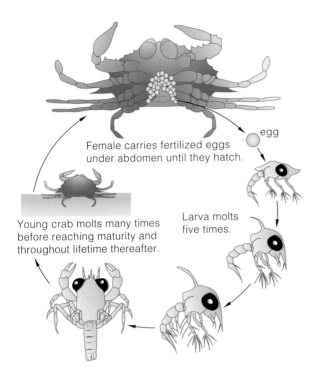

Female carries fertilized eggs under abdomen until they hatch.

egg

Young crab molts many times before reaching maturity and throughout lifetime thereafter.

Larva molts five times.

Figure 25.33 *Animated!* Life cycle of a crab. The larval and juvenile stages molt repeatedly and grow in size.

think that mating might be just a bit tricky for them. But barnacles tend to settle down and live in groups, and most species are hermaphrodites. One individual extends its penis, which can be several times its body length, out to neighboring barnacles.

Figure 25.32*c* shows a copepod. Copepods are less than two millimeters long and are the most numerous animals in aquatic habitats. They also are abundant on land. About 1,500 kinds parasitize invertebrates and fishes. However, the majority of copepods—8,000 species—eat phytoplankton, those aquatic "pastures" of photoautotrophs that you read about in Section 7.8. Some copepods prey on larvae or small invertebrates and fish eggs, which they grab with a pair of food-handling appendages. In turn, immense populations of copepods become food for different invertebrates, fishes, and baleen whales.

Like other arthropods, crustaceans repeatedly molt and shed the exoskeleton during the life cycle. Figure 25.33 shows larval stages of a crab as they periodically increase in size and molt their outgrown covering.

Crustaceans differ greatly in the number and kind of appendages. They grow in stages and periodically replace the hardened external skeleton by molting.

Millipedes and centipedes have a long, segmented body with many, many legs—a sure sign of that unfettered Dll *gene you read about in Section 17.8.*

LINK TO SECTION 17.8

Although millipedes do not have "a thousand" legs, as the name implies, most have about 100 pairs. One individual with an overexpressed *Dll* gene grew 752. Centipedes have between 15 and 177 pairs of legs, not a nicely rounded "one hundred."

As a millipede develops, its pairs of segments fuse, and each segment in its cylindrical body ends up with two pairs of legs (Figure 25.34*a*). Millipedes scavenge decaying plant material in soil and forest litter.

Centipedes have a flattened body, and all but two segments have a pair of walking legs. They are quick, aggressive predators with fangs and venom glands that subdue insects, earthworms, and snails. The one in Figure 25.34*b* hunts small lizards, toads, and frogs. A house centipede (*Scutigera*) often hides in buildings, where it hunts for cockroaches, flies, and other pests. Although helpful in this respect, most of us would rather do without its assistance.

Mild-mannered, scavenging millipedes and aggressive, predatory centipedes do not lend themselves to leg counts as they walk by.

Figure 25.34 (**a**) Millipede. (**b**) A Southeast Asian centipede.

25.15 A Look at Insect Diversity

LINKS TO
SECTIONS
6.4, 9.1, 14.1

Insects are the most diverse group of animals. There are more species of dragonflies alone than there are species of mammals. Many insects also produce staggering numbers of offspring. Ants and termites account for as much as a third of the biomass of all animals on land.

In most adult insects, the segmented body is divided into three distinct parts: a head, thorax, and abdomen. Their head has paired sensory antennae and paired mouthparts that are specialized for biting, chewing, or other functions (Figure 25.35). The thorax has three pairs of legs and, usually, two pairs of wings. Insects are the only winged invertebrates.

Insects have a complete digestive system divided into a foregut, a midgut where food is digested, and a hindgut, where water can be reabsorbed. A system of **malpighian tubules** disposes of wastes and maintains ion concentrations in body fluids. After proteins are digested, residues diffuse from blood into the tubes. There, enzymes convert them to crystals of uric acid, which are eliminated in feces. The tubules help land-dwelling insects get rid of potentially toxic wastes without losing precious water.

Larvae, nymphs, and pupae are immature stages of many insect life cycles (Figure 25.36). Like a human child, some insect nymphs have the form of an adult, although in miniature. Unlike children, however, these immature stages molt. In addition, most insect larvae undergo **metamorphosis**. By this process, tissues must be reorganized and remodeled, sometimes drastically, for the transition to the adult form (Figure 25.36*b,c*).

Figure 25.37 shows a few of the more than 800,000 species. If numbers and distribution are the measure, the most successful insects are small and reproduce often. Great numbers grow and reproduce on a plant that would be only an appetizer for another animal. By one estimate, if all the offspring of a single female fly survived and reproduced for six more generations, she would have more than 5 trillion descendants!

The most successful insects are winged. They move among food sources that are too widely scattered to be exploited by other kinds of animals. The capacity for flight contributed hugely to their success on land.

The very features that contribute to insect success also make them our most aggressive competitors for crops and forest products, such as paper or lumber.

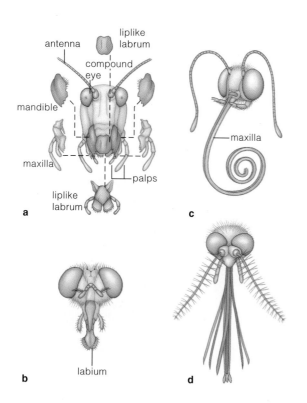

Figure 25.35 *Animated!* Examples of insect appendages. Head parts typical of (**a**) grasshoppers, which chew food; (**b**) flies, which sponge up nutrients; (**c**) butterflies, which siphon nectar; and (**d**) mosquitoes, which pierce hosts and suck up blood.

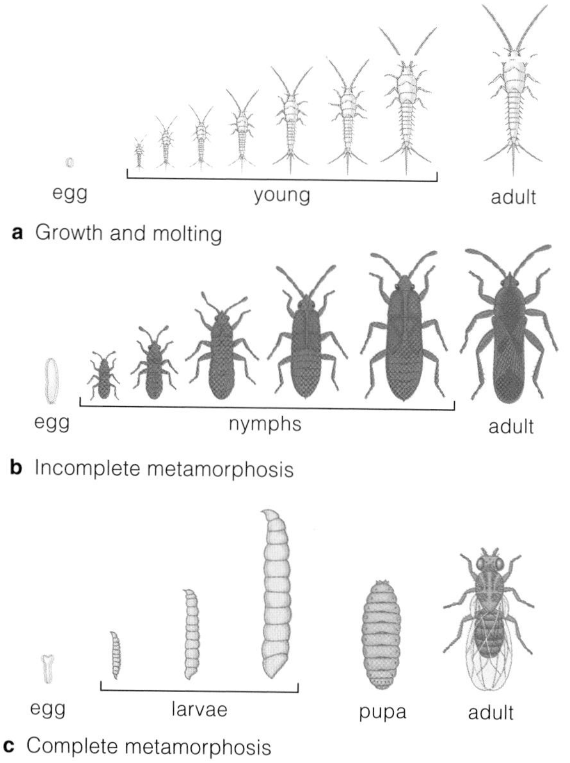

a Growth and molting

b Incomplete metamorphosis

c Complete metamorphosis

Figure 25.36 Insect development. (**a**) Young silverfish change mostly in size and proportion as they mature into adults. (**b**) True bugs show *incomplete* metamorphosis, or gradual change from the immature form until the last molt. (**c**) Fruit flies show *complete* metamorphosis. Larval tissues are destroyed and replaced before the adult emerges.

Some insects transmit pathogens that sicken us. We fight "bad" insects by spraying pesticides. Yet most are essential decomposers or pollinators of flowering plants—including many major crop plants. Also, the "good" insects are predators or parasites of the ones we would rather do without. Think about it.

Like other arthropods, insects have a hardened exoskeleton, jointed appendages; modified body segments, respiratory structures, and specialized sensory organs. Many kinds compartmentalize tasks among different stages in the life cycle. The most successful insects are winged.

In terms of distribution, number of species, population sizes, competitive adaptations, and exploitation of diverse foods, insects are the most successful animals.

Figure 25.37 Examples of insects. (**a**) Mediterranean fruit fly. Its larvae destroy citrus fruit and other crops. (**b**) Duck louse. It eats bits of feathers and skin. (**c**) European earwig, one of the common household pests. The long curved forceps at the tail end indicate that this individual is a male.

(**d**) Flea, with strong legs good for jumping onto and off of its animal hosts. (**e**) Stinkbugs, newly hatched. (**f**) In the center of this group of honeybee workers, one bee is dancing in a way that communicates the position of a food source to her hive mates, as explained in Section 49.4.

With more than 300,000 species of beetles, Coleoptera is the largest order in the animal kingdom. (**g**) Ladybird beetles swarming. (**h**) Staghorn beetle (*Cyclommatus*).

(**i**) Swallowtail butterfly and (**j**) luna moth. (**k**) An adult damselfly, which preys on flying insects. Its larvae are voracious predators of aquatic invertebrates and small fishes.

25.16 Unwelcome Arthropods

LINKS TO
SECTIONS
18.4, 22.7

*Very few of the 6.4 billion people on the planet encounter
arthropods that cause real pain. All things considered, we
collectively do more harm to more species than, say, the
spiders, most of which do good by eating uncountable
numbers of insect pests. Even so, the "good" ones have
a few "bad" relatives, and it doesn't hurt to know them
when we see them.*

Figure 25.38 (**a**)
Brown recluse, which
has a violin-shaped
mark on its forebody.
Its bite can be severe
to fatal. (**b**) Female
black widow, which
has a red, hourglass-
shaped marking on
the underside of
her shiny black
abdomen. Males
are smaller and
do not bite.

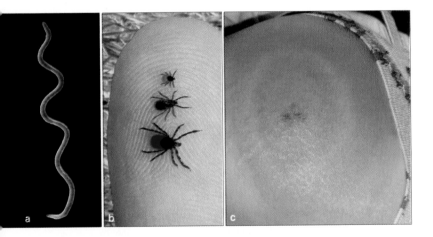

Figure 25.39 (**a**) *Borrelia burgdorferi*. This spirochete causes Lyme disease,
now the most prevalent tick-borne disease in the United States. (**b**) Deer ticks
(*Ixodes*) are the common vector for this bacterium. Bottom to top on this finger,
a female, male, and nymph. All stages can transmit the parasite. (**c**) Example
of a bull's-eye rash, the reaction in about 30 percent of those bitten.

Harmful Spiders   All spiders of the genus *Loxosceles*
are venomous to humans and other mammals, but we
seldom encounter most of them. One exception, the brown
recluse (*L. reclusa*), favors hiding in warm places. It turns
up in houses, especially near refrigerator motors and
clothes dryers, and in folds of clothing or linens in closets.
You can identify it by the violin-shaped marking on its
cephalothorax (Figure 25.38*a*). A bite ulcerates skin and
heals poorly. Some bitten people have required skin grafts,
and a few have died. *L. laeta*, a related species in Chile,
Argentina, and Peru, is similarly dangerous.

Every year in the United States, about five people
die after being bitten by the black widow (*Latrodectus*).
Besides causing pain, this spider's neurotoxin is paralytic;
medical attention must be prompt (Figure 25.38*b*).

Mighty Mites   Of all arachnids, mites are among the
smallest, most diverse, and most widely distributed. Of
an estimated 500,000 species, a few parasitic mites—ticks
especially—are harmful. Figure 25.39 shows a deer tick
(*Ixodes*), a vector for *Lyme disease*, which is most common
in northeastern and upper midwestern states. Symptoms
include fever, a stiff neck, headache, listlessness, muscle
and joint pain, and blurred vision. They may disappear and
then show up years later. Early diagnosis and treatment
are critical. In a few untreated people, infection can cause
many organ systems to function abnormally.

Different ticks spread the disease in the western and
southeastern states. Usually the tick feeds on white-tail
deer, mice, other mammals, and birds. *Borrelia burgdorferi*,
which causes Lyme disease, is transmitted to humans
mainly by the nymphal stages, probably because these are
no bigger than a pinhead and escape detection. Detection
is crucial, because ticks normally transmit the bacterium
after they have been feeding two days or more. Adult ticks
are easier to spot and more likely to be removed quickly.

Ticks do not fly or jump. They crawl onto grasses and
shrubs, then onto animals that brush past, then typically
into the scalp, groin, or other hairy, hidden body parts. Look
for them after walks in the wild. Different types transmit
pathogens that cause serious diseases, including Rocky
Mountain spotted fever, tularemia, babesiasis, scrub
typhus, and encephalitis. A few can kill directly. Their toxin
interferes with nerves and invites progressive paralysis,
from the lower extremities upward. The tick must be
removed before the bitten person stops breathing.

Ending on a less scary but irritating note, think of the
house dust mites. They don't kill you, but the allergies
they provoke in people can be irritating to the extreme.

Scorpion Country   Oversized, prey-seizing pincers, a
venom-dispensing stinger at the tip of a narrowed, jointed
abdomen—what's not to love about scorpions? Actually,
scorpions seldom dispense venom unless their prey—other
small arthropods, the occasional small lizard or mouse—
puts up a fight. In the wild they hunt at night and rest

Figure 25.40 *Centruroides sculpturatus*, the bark scorpion, which is about five centimeters (two inches) long. In the past century, it has killed more people in Arizona than all poisonous snakes combined. Antivenin is available for scorpion stings.

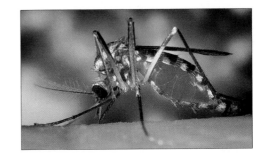

Figure 25.41
(**a**) Adult western corn rootworm (*Diabrotica virgifera*).

(**b**) Larva on a corn root. (**c**) A row of damaged corn plants.

Figure 25.42
Mosquito (*Aedes triseriatus*) busily siphoning blood from a human. This species is one of the vectors for the West Nile virus.

during the day in burrows or under logs, stones, and bark. They can hide without eating and still live for a year.

We find scorpions in cold regions, including Canada and the southern Andes and southern Alps. Some live in moist forests. Most live in hot, dry regions, where they are active all year. The ones that can hurt or kill people are not the fiercest looking, and there are not very many of them. *Centruroides sculpturatus* is the most feared species in the United States (Figure 25.40). Its close relatives in Mexico cause fatalities, especially among small children.

When in scorpion country, it is not a good idea to walk about barefoot at night or to put on shoes without first turning them upside down and shaking them vigorously.

Beetles Everywhere   First in America's Midwest and now in the Balkans, *Diabrotica virgifera* has been making the rounds in cornfields as a representative of one of our major competitors for food. Figure 25.41 shows what it looks like and the damage it causes. Its larvae feed on corn roots until the damaged plants keel over. Each year, they cause about a billion dollars' worth of crop losses.

The astronomical losses compel farmers to spread about 30 million pounds of pesticides annually through cornfields —about half the total for all row crops in the United States. Many rootworms now show pesticide resistance (Section 18.4). Inventive researchers targeted adults, which are fond of a bitter plant juice called cucurbitacin. The juice evolved as a natural pesticide in cucumbers, melons, squashes, and other plants. Rootworms opportunistically use the chemical as a defense against bird predators. (Birds learn to avoid rootworms after a taste trial that ends in vomiting.)

Robert Schroder and others concocted a bait. In one test, they laced watermelon juice with red dye 28, a photoactive chemical. Adult rootworms gorged themselves, turned deep red, and after five minutes of sun exposure, dropped dead. The sun's rays triggered oxidation reactions that destroyed the insect tissues. Beneficial insects avoid the dyed juice. In this ongoing contest, chalk one up for the researchers.

The Mosquito Menace   Which animal is, indirectly, one of most dangerous threats to human health? The tiny mosquito. Give them credit: Mosquitoes *are* integral parts of many communities. Adults fit into the food webs that sustain birds; larvae fit into the food webs that sustain fishes. But mosquitoes also are vectors for pathogens.

Each year, mosquito-borne diseases kill about 3 million people around the world. In addition to malaria and West Nile virus, many species of mosquitoes act as intermediate hosts for pathogens that cause Rift Valley Fever, dengue fever, and yellow fever, and different forms of encephalitis. Mosquitoes also transmit the parasitic roundworms that cause elephantiasis (Section 25.10).

Only female mosquitoes are bloodsuckers, and most will target any warm-blooded animal that they happen to find. The males feed on plant nectar and juices. After a female gets a good blood meal, she produces a batch of eggs, which must be laid in standing water. Repellents can help prevent bites in the first place, but eliminating breeding sites is the best way to control mosquito populations.

Here is a final thought: All mosquitoes thrive in warm, humid climates. With climates now undergoing a long-term warming trend, mosquitos are likely to become active during more of the year, in ever expanding habitats.

25.17 The Puzzling Echinoderms

Reflect, for a moment, on Figure 25.43. At a major branch point in time, bilateral animals gave rise to protostomes, including the roundworms, arthropods, annelids, and mollusks you just read about. Genetic divergences put deuterostomes on a separate evolutionary journey, with echinoderms and chordates being major travelers.

The name **echinoderm** means spiny-skinned; it refers to a kind of body covering that has interlocking spines and plates stiffened with calcium carbonate. All 6,000 of the known species of echinoderms live in the seas. They include the sea stars, brittle stars, sea urchins, sea cucumbers, feather stars, sand dollars, sea biscuits, sea pens, and sea lilies or, more formally, the crinoids. Figures 25.43 and 25.44 show representatives.

Most adults are radial with a few bilateral features, and most of them have bilateral larvae. Did the group originate from a bilateral ancestor? DNA sequencing studies now indicate that they probably did.

Almost all adult echinoderms are bottom dwellers. Some sea lilies live attached to the seafloor by a stalk, although most echinoderms creep about. They have a decentralized nervous system, not a concentration of integrative nerve cells in a brain. Their system allows them to detect and respond to food, danger, and mates in any direction. For instance, any sea star arm that touches a tasty mussel can become the leader for the rest of the body, which then moves toward the prey.

Echinoderms have tube feet. These are fluid-filled, muscular structures, sometimes with suckerlike disks at the tips (Figure 25.44*a,g*). Focus on sea stars, which use tube feet in walking, burrowing, clinging to rocks, and gripping a clam or snail about to become a meal.

Tube feet are part of a **water–vascular system** that is unique to the echinoderms. In sea stars, that system includes a main canal in each arm. Short side canals extend from them and deliver water to the tube feet (Figure 25.44*f*). Inside each tube foot is a fluid-filled, muscular structure shaped like the rubber bulb on a medicine dropper. When the bulb contracts, it forces fluid into the foot and causes it to lengthen.

Hundreds of tube feet change shape continually as muscle contractions redistribute fluid all through the water–vascular system. Each little tube foot lets go of the substrate, swings forward, reattaches, then swings back and lets go before swinging forward once again. The motions are so coordinated that sea stars can glide smoothly instead of lurching along.

Most sea stars are active predators with a feeding apparatus on their ventral surface, as in Figure 25.44*g*. Some swallow prey whole. Others push part of their stomach outside of their mouth and around prey, then start digesting the meal even before swallowing it. Sea stars expel coarse, undigested residues through their mouth, not their small anus.

In sea urchins, the calcium carbonate plates form a stiff, rounded covering from which spines protrude (Figure 25.44*b*). Sea urchin roe (eggs) are ingredients in some sushi. Each year, Japan imports 20 million pounds or so of sea urchins that were captured off the California coast. Harvests are largely unregulated and are probably destroying the most prized species.

Some sea cucumber species, also favorite foods, are declining. These are the echinoderms with the softest body; their plates have been reduced to microscopic spicules (Figure 25.44*d*). They crawl on the seafloor, where they trap prey and bits of organic debris with sticky tentacles. When a predator starts to attack it, a sea cucumber shoots its internal organs out through its anal region as a distraction. If the sea cucumber makes a getaway, it can regenerate its missing parts.

With that jarring image in mind, we now leave the closest ancestors of *chordates*—the branch of the animal family tree that is the topic of the next chapter.

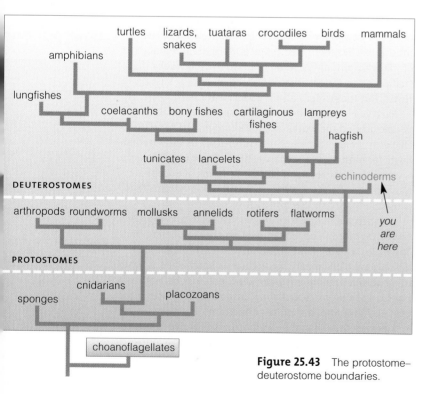

Figure 25.43 The protostome–deuterostome boundaries.

With their odd combination of radial and bilateral traits, echinoderms make a case for this point: There are exceptions to the major trends in animal evolution.

Figure 25.44 *Animated!* Some echinoderms. (**a**) Sea star. (**b**) Underwater "forest" of sea urchins, which can move about on spines and a few tube feet. (**c**) Brittle star. Its slender arms (rays) make rapid, snakelike movements. (**d**) Sea cucumber, with rows of tube feet along its body.

We can use the body plan of a sea star to show the degree of compartmentalization and integration that evolved in this group of invertebrates. (**e**) Major components of the central body and the radial arms, with a close-up of its little tube feet. (**f**) Organization of the water–vascular system. In combination with many tube feet, it is the basis of locomotion. (**g**) A sea star's toothy feeding apparatus.

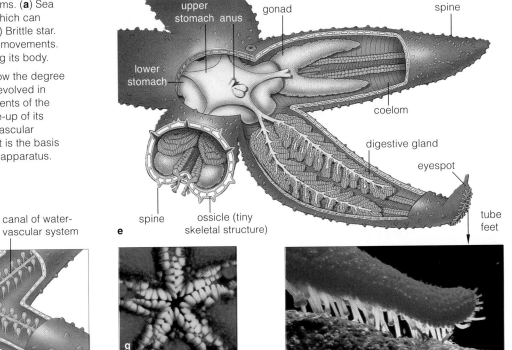

Summary

Section 25.1 Animals apparently arose from colonial flagellates. All species are multicelled heterotrophs. Their embryos develop through a series of stages, at least one of which is motile during the life cycle. Two or three primary tissue layers (ectoderm, endoderm, mesoderm) form in embryos. Cells of those layers rise to epithelial tissues, connective tissues, and extracellular matrixes.

The animal lineage reflects a trend toward increases in size, compartmentalization of tasks (a division of labor among cells, tissues, and organs), and integration.

Section 25.2 Except for placozoans and sponges, animals have a radial or bilateral body plan, and they have a saclike or a tubular gut.

A tubular gut has a mouth and anus. Its formation in embryos is one of many defining traits for two major lineages of bilateral animals. In protostomes, the mouth forms first. In deuterostomes, the anus forms first.

Protostomes and deuterostomes differ in whether they have a coelom (a main body cavity between their gut and body wall) and in how that cavity develops. In some lineages the coelom was lost or reduced; such animals are called acoelomates and pseudocoelomates.

Bilateral animals differ in the extent to which they became cephalized (how concentrated sensory and nerve cells became at the head end) and the extent to which the body became segmented, if at all. As a group, they show a stunning range of specializations along the anterior–posterior body axis.

Biology ⒺNow
Compare the different types of animal body plans with the animation on BiologyNow.

Section 25.3 Sponges are simple animals, with no body symmetry, tissues, or organs. Different cell types carry out specialized tasks (e.g., flagellated collar cells capture food; they line a skeletal framework made of proteins and spicules in a gelatinous matrix). The soft placozoan body has no symmetry, but it has two tissues analogous to ectoderm and endoderm.

Biology ⒺNow
Discover the body plan of a sponge with the animation on BiologyNow.

Section 25.4 Cnidarians, such as jellyfishes, sea anemones, corals, and hydroids, are radial animals. They have an outer epidermis, inner gastrodermis with gland cells, and a jellylike substance in between that has a few scattered cells. All are carnivores. They alone make nematocysts that function in prey capture and defense.

Biology ⒺNow
Explore cnidarian body plans and life cycles with the animation on BiologyNow.

Section 25.5 Flatworms are among the simplest protostomes. They are bilateral, with a saclike gut. The mostly free-living turbellarians and the parasitic tapeworms and flukes are flatworms.

Biology ⒺNow
Learn about planarian organ systems and a tapeworm life cycle with the animation on BiologyNow.

Section 25.6 Annelids are segmented worms called polychaetes (marine worms), leeches, and oligochaetes (e.g., earthworms). Most move by a hydrostatic skeleton (coelomic chambers), anchors (e.g., setae), and waves of muscle contractions down the length of the body.

Biology ⒺNow
Investigate the body plan of an earthworm with the animation on BiologyNow.

Section 25.7 Rotifers are bilateral, cephalized animals with ciliated lobes at their head end and a false coelom packed with organs. They are relatives of flatworms.

Biology ⒺNow
Watch the video of rotifers on BiologyNow.

Sections 25.8, 25.9 With 100,000 named species, the coelomate, bilateral mollusks are one of the largest animal groups. Only mollusks have a mantle, a sheetlike part of the body mass that is draped back on itself. All mollusks are soft-bodied; many are shelled or have a reduced shell. Examples are gastropods (such as snails) and cephalopods. Gastropods undergo torsion during their development, so that the anus ends up above the mouth. The fastest (squids), largest (giant squids), and smartest (octopuses) invertebrates are cephalopods.

Biology ⒺNow
Compare the molluscan body plans with the animation on BiologyNow.

Section 25.10 Roundworms (nematodes) have a cylindrical body with bilateral features, a flexible cuticle, a complete gut, and a false coelom. At least 22,000 kinds are free-living decomposers or parasites.

Biology ⒺNow
Learn about the roundworm body plan with the animation on BiologyNow.

Sections 25.11–25.16 There are more than a million named species of bilateral, coelomate animals called arthropods. Major groups are chelicerates, crustaceans, centipedes, millipedes, and insects.

The success of arthropods as a group is attributable to a hardened exoskeleton, jointed appendages, specialized and fused segments, efficient respiratory and sensory structures, and division of labor among developmental stages adapted to specific environmental conditions.

Chelicerates include horseshoe crabs and arachnids (spiders, scorpions, ticks, and mites). They are predators, parasites, or scavengers. The mostly marine crustaceans include crabs, lobsters, barnacles, and copepods, all with specialized appendages and diverse life-styles. Mild-mannered millipedes and aggressive centipedes have a highly segmented, many-legged body.

Insects are the most successful of all animal groups, and the only winged invertebrates. They are our major competitors for food.

Figure 25.45 (a) A soft, branching coral (*Telesto*). (b) Shell of a chambered nautilus, cutaway view.

Figure 25.46
You just flipped it over on a beach. What is it?

Biology🅢Now
Explore the crab life cycle with the animation on BiologyNow.
Look at insect specializations with the animation on BiologyNow.

Section 25.17 Echinoderms, such as sea stars, are invertebrates of the deuterostome lineage. They have an exoskeleton of spines, spicules, or plates of calcium carbonate. Adults are radial, but bilateral ancestry is evident in their larval stages and other features.

Biology🅢Now
Look into the body plan of a sea star and watch tube feet in action on BiologyNow.

Self-Quiz
Answers in Appendix II

1. All animals _____ .
   a. are motile for at least some stage in the life cycle
   b. consist of tissues arranged as organs
   c. can reproduce asexually as well as sexually
   d. both a and b

2. A coelom is a _____ .
   a. lined body cavity   c. sensory organ
   b. resting stage   d. type of bristle

3. Cnidarians alone produce _____ .
   a. nematocysts   c. a hydrostatic skeleton
   b. a mantle   d. malpighian tubules

4. Flukes are most closely related to _____ .
   a. planarians   c. arthropods
   b. roundworms   d. echinoderms

5. Which group has the greatest number of species?
   a. crustaceans   c. mollusks
   b. insects   d. roundworms

6. Earthworm nephridia perform a function most similar to the _____ .
   a. gemmules of sponges   c. flame cells of planarians
   b. chelicerae of spiders   d. tube feet of echinoderms

7. The _____ are coelomate and radial as adults.
   a. cnidarians   c. roundworms
   b. echinoderms   d. both a and c

8. Match the organisms with their appropriate descriptions.
   _____ choanoflagellate   a. complete gut, false coelom
   _____ sponges   b. simplest organ systems
   _____ cnidarians   c. no tissues, no organs
   _____ flatworms   d. jointed exoskeleton
   _____ roundworms   e. mantle over body mass
   _____ annelids   f. segmented worms
   _____ arthropods   g. tube feet, spiny skin
   _____ mollusks   h. nematocyst producers
   _____ echinoderms   i. sister taxon of animals

Additional questions are available on Biology🅢Now™

Critical Thinking

1. You are diving in the calm, warm waters between a large tropical reef that formed around an island. You come across the branching soft coral shown in Figure 25.45a. You observe that it grows only on the side of the reef's spine facing the island, not the side exposed to the open ocean. Develop a hypothesis to explain its one-sided distribution. How you would go about testing the hypothesis?

2. A nautiloid shell develops as a symmetrical series of ever larger chambers (Figure 25.45b). Do some research and then write a brief report that correlates the functions of the chambers with the nautiloid life-style.

3. The animal groups called flatworms, roundworms, and annelids all include species that are parasites of mammals, whereas the groups called sponges, cnidarians, mollusks, and echinoderms do not. Propose an explanation for the difference.

4. In 2000, only 10 percent of the lobster population in Long Island Sound survived after a massive die-off. Many lobstermen think the deaths followed heavier spraying of the pesticides used to control mosquitoes that are vectors for West Nile virus. Speculate on why a chemical designed to target mosquitoes might also harm lobsters.

5. You are walking along a beach and see what looks like a shield with a spine sticking out from under it. You flip it over and see five pairs of legs (Figure 25.46). Is this animal in the same group as lobsters and crabs? Or in the same group as spiders and ticks? How do you know?

Interpreting and Misinterpreting the Past

In Charles Darwin's time, major groups of organisms were already identified. A big obstacle to accepting his theory of evolution by natural selection was the seeming lack of transitional forms. If new species evolve from older ones, then where were the "missing links" in the fossil record—forms with intermediate traits that bridge major groups?

Ironically, workmen at a limestone quarry in Germany had already unearthed one. The pigeon-sized fossil looked like a small meat-eating dinosaur (*Dromaeosaurus*). It had three long clawed fingers on each forelimb, a long bony tail, and short, spiky teeth. Later, diggers found another one. Later

still, someone noticed the feathers. If those were birds, why did they have teeth and a bony tail? If dinosaurs, what were they doing with *feathers*? In time the specimen was named *Archaeopteryx*, meaning ancient winged one (Figure 26.1).

Between 1860 and 1988, six *Archaeopteryx* specimens and a fossilized feather were found. Anti-evolutionists tried to dismiss them as forgeries. Someone, they said, pressed modern bird bones and feathers against wet plaster and the imprints only looked like fossils. Microscopic examination confirmed the fossils are real. Further confirmation of their age came from the remains of obviously ancient jellyfishes, worms, and other species in the same limestone layers.

Radiometric dating shows that *Archaeopteryx* lived 150 million years ago. Why are its remains so well preserved? This vertebrate lived in a tropical forest near a large, warm, stagnant lagoon, and reefs kept out oxygenated water from the sea. It was not an inviting place for scavenging animals that might have eaten the remains of *Archaeopteryx* and other organisms that fell from the sky or drifted offshore.

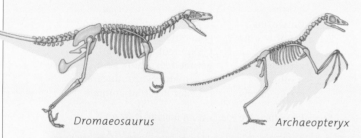

Dromaeosaurus   *Archaeopteryx*

Figure 26.1 Placing *Archaeopteryx* in time. This painting is based on fossils of plants and animals that lived in a tropical forest during the Jurassic. In the center foreground, gliding *Archaeopteryx*. In the background, left to right, *Stegosaurus* and *Apatosaurus* (herbivores), *Saurophaganax* ("king of reptile eaters"), and *Camptosaurus* (a beaked herbivore). At far right, a climbing mammal. Tiny-brained *Apatosaurus* grew as long as 27 meters (90 feet) and weighed 33 to 38 tons; it was one of the largest land animals that ever lived. On the facing page, one of the *Archaeopteryx* fossils.

Watch the video online!

With each storm surge, fine sediments driven over the reefs gently buried carcasses on the bottom of the lagoon. Over time, soft, muddy sediments were slowly compacted and hardened. They became a limestone tomb for more than 600 species, including *Archaeopteryx*.

No one was around to witness such transitions in the history of life. But fossils are real, just as the morphology, biochemistry, and molecular makeup of living organisms are real. Radiometric dating assigns fossils to their place in time. *Biological evolution is not "just a theory."* Remember, a scientific theory differs from speculation because its predictive power has been tested in nature many times, in many different ways. If a theory stands, there is a high probability that it is not wrong.

Evolutionists argue all the time among themselves. They argue over how to interpret the evidence and which mechanisms and events can explain life's history. At the same time, they do not ignore evidence—which is there for us to gather and interpret. Here is one account of vertebrate evolution, including our own origins.

 How Would You Vote?

Private collectors find and protect fossils, but a private market for rare vertebrate fossils raises the cost to museums and encourages theft from protected fossil beds. Should private collecting of vertebrate fossils be banned? See BiologyNow for details, then vote online.

 Key Concepts

CHARACTERISTICS OF CHORDATES

Four features develop in chordate embryos and set them apart from other animals: a supporting rod (notochord), a dorsal nerve cord, a pharynx with gill slits in the wall, and a tail extending past the anus. Section 26.1

TRENDS AMONG VERTEBRATES

In some early vertebrate lineages, a backbone replaced the notochord as a partner of muscles used in locomotion. Jaws evolved, sparking the evolution of novel sensory organs and brain expansions. On land, lungs replaced gills, and more efficient blood circulation enhanced gas exchange. Fleshy fins with skeletal supports evolved into limbs of amphibians, reptiles, birds, and mammals. Section 26.2

TRANSITION FROM WATER TO LAND

Vertebrates evolved in the seas, and the greatest diversity still resides in lineages of cartilaginous and bony fishes. Mutations in master genes that control body plans were pivotal in the rise of aquatic tetrapods and their move onto dry land. Sections 26.3–26.11

THE AMNIOTES

Amniotes—known informally as the reptiles, birds, and mammals—are vertebrate lineages that radiated into nearly all habitats on land. Sections 26.6–26.9

EARLY HUMANS AND THEIR ANCESTORS

Primates that were ancestral to the human lineage became physically and behaviorally adapted to long-term changes in climate, geography, and resource availability. Humans have dispersed throughout the world through behavioral flexibility and the force of culture. Sections 26.12–26.15

 Links to Earlier Concepts

Take a moment to assess how far you have come on life's evolutionary road by reviewing the family tree in Section 19.7. Refer to the geologic time scale (17.5) to see where you are heading. Remember an earlier point, that highly active life-styles are not possible without aerobic respiration (8.1)? Here you will see how tetrapods hit the jackpot.

You will come across more examples of how small changes in master genes changed the course of evolution (15.3, 17.8). You will draw on your understanding of speciation, adaptive radiation, and extinctions (19.4). You will revisit the dinosaurs (Chapter 17) and biochemical and molecular tools (17.9).

26.1 The Chordate Heritage

LINKS TO
SECTIONS
19.7, 25.12

As you consider vertebrates and their chordate heritage, remember this: Each kind of organism is a mosaic of traits—many conserved from remote ancestors, and others unique to its branch on the family tree.

CHORDATE CHARACTERISTICS

The preceding chapter ended with echinoderms, one of the earliest lineages of deuterostomes. Dominating this branch of the animal family tree are the **chordates** (Figure 26.2). These coelomate, bilateral animals are characterized in part by four features. *First*, a rod of stiffened tissue—not bone or cartilage—helps support the body. It is a notochord. *Second*, a nerve cord runs parallel with the notochord and gut; its anterior end develops into a brain. Unlike invertebrate nerve cords, this one is dorsal. *Third*, the wall of their pharynx, a muscular tube, has **gill slits**—openings that function in feeding, respiration, or both. *Fourth*, a tail extends past the anus. These features develop in all chordate embryos, but they do not always persist in adults.

Like the sea stars and other echinoderms, about 2,100 species of chordates are "invertebrates," which have no internal backbone. Far more—about 48,000—are **vertebrates**. These animals contain a backbone of cartilage or bone. They also have a cranium, which is a chamber of cartilage or bone that encloses the brain. Appendix I has an expanded classification system for vertebrates. Unit VI explores their body plans and functions. Here we start with their closest relatives.

INVERTEBRATE CHORDATES

Urochordates Tunicates are among the bag-shaped animals called **urochordates**, which are characterized in part by larvae that have a firm, flexible notochord extending through a tail. Adults secrete a gelatinous or leathery "tunic" around a pharynx with gill slits (Figure 26.3). Most are only a few centimeters long. They range from intertidal zones to the deep ocean.

The adults are **filter feeders**. They filter food from seawater flowing through a siphon, past gill slits in a pharynx, then out through another siphon. They spurt out water when something irritates them; hence their familiar name, sea squirts. Their pharynx also acts as a respiratory organ. It takes up dissolved oxygen, which diffuses into blood vessels. Carbon dioxide wastes of aerobic respiration diffuse into water flowing out.

An intriguing point is that the sea squirt larvae are bilateral swimmers. Their notochord is like a torsion bar. It interacts with bands of muscles located under the epidermis. When muscles of one or the other side

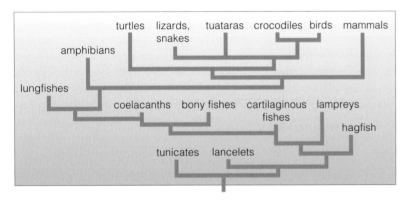

Figure 26.2 Family tree for the chordates which, together with echinoderms and some other lesser known groups, belong to the deuterostome lineage.

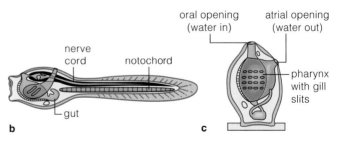

Figure 26.3 (**a**) Adult sea squirts (*Rhopalaea crassa*). (**b**) Sea squirt larva, which swims briefly and then undergoes drastic tissue remodeling and reorganization. Its metamorphosis starts when its head attaches to a substrate. Then its tail, notochord, and most of the nervous system are resorbed (recycled and used to form new tissues). Many gill slits form in the pharynx wall. Organs rotate until openings through which water flows into and out from the pharynx are pointing away from the substrate, as in (**c**).

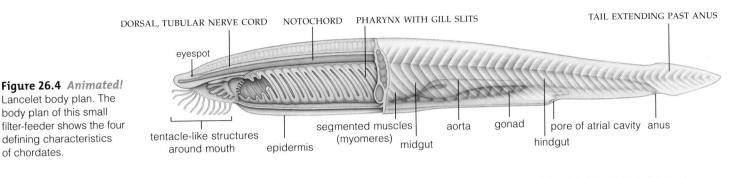

Figure 26.4 *Animated!* Lancelet body plan. The body plan of this small filter-feeder shows the four defining characteristics of chordates.

DORSAL, TUBULAR NERVE CORD   NOTOCHORD   PHARYNX WITH GILL SLITS   TAIL EXTENDING PAST ANUS

eyespot

tentacle-like structures around mouth   epidermis   segmented muscles (myomeres)   midgut   aorta   gonad   hindgut   pore of atrial cavity   anus

of the tail contract, the notochord bends, then springs back when muscles relax. Strong, side-to-side motion propels the larva forward. In most fishes, muscles and a backbone bring about the same kind of motion.

Cephalochordates At one end of their nerve cord, the fish-shaped **cephalochordates**, or lancelets, have a simple brain in a head that develops as it does in the vertebrates (*cephalo–*, head). Nerve cells (neurons) in the brain control reflex responses to light and other stimuli. The body is three to seven centimeters long and tapers at both ends. It clearly displays chordate features (Figure 26.4).

Also like vertebrates, the lancelets have segmented muscles. Contractile units in muscle cells run parallel with the body's long axis. Contractile force directed against the notochord allows side-to-side swimming motions. Even so, lancelets spend most of the time burrowing into shallow sediments and filter-feeding. Coordinated beating of cilia that line numerous gill slits move water into their pharynx, where food gets caught in mucus before moving through the gut.

A gill-slitted pharynx, a simple brain, a nerve cord, and bands of muscles that parallel the body's long axis—as you will see, these features turned out to have amazing evolutionary possibilities for vertebrates.

A WORD OF CAUTION

We are describing chordate features that will help give you insight into the evolution of vertebrates, but don't assume tunicates and lancelets are missing links between vertebrates and the earliest chordates. These groups do share some traits that preceded the origin of vertebrates, but each also has unique traits that put it on a separate branch of the animal family tree.

Similarly, hagfishes are jawless fishes that resemble lancelets in some respects and vertebrates in others, but certain traits set them apart. A few cartilage plates do help protect their brain, but their endoskeleton is not hardened with mineral deposits as it is in vertebrates.

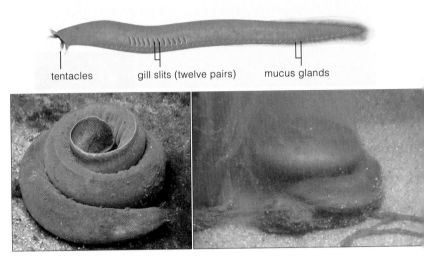

tentacles   gill slits (twelve pairs)   mucus glands

Figure 26.5 Hagfish body plan. The two photographs show a hagfish before and after it had coated its body with slimy mucous secretions.

The hagfish is less than a meter long and nearly blind, but it easily finds food with sensory tentacles (Figure 26.5). Its rasping, tonguelike structure draws small invertebrates and tissues of a dead or dying fish into its mouth. Hagfishes are burrowers, not highy active animals. But low metabolic rates work for them. They can get along without food for more than half a year.

If a hagfish is threatened, it secretes up to a gallon of sticky mucus. Fishermen who snag hagfishes view the behavior as disgusting. Even so, sliming potential predators has been a fine defense for this lineage of otherwise vulnerable, soft-bodied chordates.

> *All chordate embryos have a notochord, a tubular dorsal nerve cord, a pharynx with gill slits in its wall, and a tail that extends past the anus. These traits are legacies from a shared ancestor, an early invertebrate chordate.*
>
> *Existing filter-feeding lancelets and tunicates are among the groups closest to the most ancient chordate lineages.*
>
> *Hagfishes are at the next level in chordate complexity. Some cartilage protects a portion of their brain.*

26.2 Evolutionary Trends Among the Vertebrates

LINKS TO
SECTIONS
17.4, 17.5

The first vertebrates—jawless fishes—arose in Cambrian times. Before the Cambrian ended, an adaptive radiation had given rise to all modern groups of vertebrates.

EARLY CRANIATES

All fishes, amphibians, reptiles, birds, and mammals alive today are **craniates**, with a cartilaginous or bony chamber enclosing their brain. The first craniates had evolved by at least 530 million years ago, and they were active swimmers. Like the lancelets, they had a

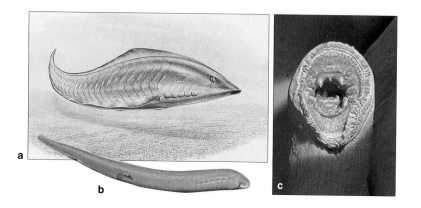

Figure 26.6 (**a**) Painting of an early craniate (*Myllokunmingia*). (**b**) Larva of a lamprey, an existing jawless fish. Most lampreys are parasitic. They attach to prey with a suckerlike oral disk and rasp at it with horny mouthparts (**c**). In the 1800s, sea lampreys probably entered the Hudson River, then canals built for commerce. By 1946, they were established in all of the Great Lakes of North America. Trout, salmon, and other natives have no defenses against them.

notochord and segmented muscles. Unlike lancelets, they had fins. Adults resembled the larvae of modern jawless fishes called lampreys (Figure 26.6).

Jawless fishes originated from craniate ancestors 500 million years ago. They included **ostracoderms**, which had armorlike plates of bony tissue and dentin, the same hard tissue in your teeth. Armor may have worked against pincers of the giant sea scorpions that hunted on the seafloor. It did not work against **jaws**— hinged, bony feeding structures. Jawed craniates had evolved, and when they began an adaptive radiation, most jawless fishes disappeared.

Placoderms were among the earliest fishes having a mineral-hardened backbone and jaws. Armor plates protected their head, but not much else. Their jaws were expansions of the first in a series of hard parts that structurally supported the gill slits (Figure 26.7). In certain species, the jaws had sharp bony plates.

KEY INNOVATIONS

The appearance of jawed fishes profoundly changed the course of animal evolution. Unlike their nearly brainless, filter-feeding relatives, jawed fishes had cells that could form **bone tissue** with mineral-hardened secretions. They formed a linear series of cartilaginous or bony segments—**vertebrae** (singular, vertebra)— which replaced the notochord. That vertebral column was part of an inner skeleton, and it was more flexible and less clunky than armor plates. The hard segments

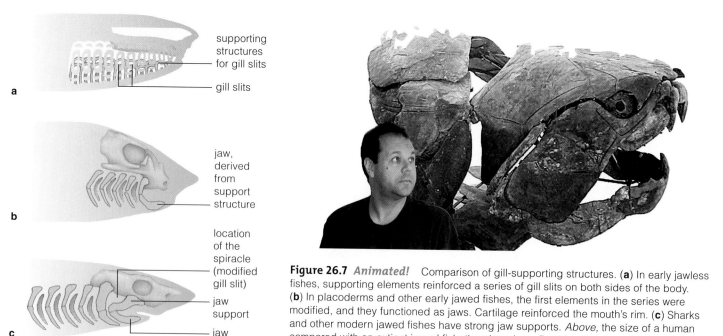

Figure 26.7 *Animated!* Comparison of gill-supporting structures. (**a**) In early jawless fishes, supporting elements reinforced a series of gill slits on both sides of the body. (**b**) In placoderms and other early jawed fishes, the first elements in the series were modified, and they functioned as jaws. Cartilage reinforced the mouth's rim. (**c**) Sharks and other modern jawed fishes have strong jaw supports. *Above*, the size of a human compared with an extinct jawed fish, the placoderm *Dunkleosteus*.

were paired with muscle segments, which permitted more flexibility and more forceful contractions. *The vertebral column was a starting point for the evolution of fast-moving, agile predators and prey in the seas.*

Jaws sparked another trend. Early chordate feeding modes were limited to filtering, sucking, and rasping food. When placoderms started biting and ripping off chunks of prey, better defenses evolved, as did ways to overcome them. For instance, fishes with better eyes for detecting predators or bigger brains that could plan pursuit or escape now had a competitive edge. *A trend toward more complex sensory organs and nervous systems started among ancient lineages of fishes, and much later in time, it continued among vertebrates on land.*

Another trend started when pairs of fins evolved. **Fins** are appendages that help propel, stabilize, and guide a body through water. They go by these names:

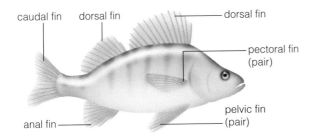

caudal fin   dorsal fin   dorsal fin

pectoral fin (pair)

anal fin

pelvic fin (pair)

Some Devonion fishes had fleshy pelvic and anal fins with internal skeletal supports. *Paired, fleshy fins were a starting point for all legs, arms, and wings that evolved among amphibians, reptiles, birds, and mammals.*

A shift in respiration set another trend in motion. In lancelets, oxygen and carbon dioxide diffuse across skin. In most early vertebrates, paired gills evolved. **Gills** are respiratory organs having moist, thin folds serviced by blood vessels. They have a large surface area that exchanges gases between the environment and the body. Oxygen diffuses *from* water inside the mouth into blood vessels within each gill, *and* carbon dioxide diffuses *into* the surrounding water.

Gills became more efficient in larger, more active fishes. But gills do not function out of water; their thin surfaces stick together unless flowing water moistens them. In fishes ancestral to land vertebrates, two tiny outpouchings developed on the gut wall and evolved into **lungs**: internally moistened sacs for gas exchange. Lungs supplemented and then replaced gills during the invasion of land. In a related trend, modifications to the heart enhanced the uptake of oxygen and removal of carbon dioxide. *The earliest vertebrates on land relied more on a pair of lungs than on gills. Increased efficiency of circulatory systems accompanied the evolution of lungs.*

In the Carboniferous, the heavy-plated placoderms and other jawed fishes died out as faster, more agile, brainier predators replaced them. By adaptation, key innovations, and plain luck, the descendants of some jawless and jawed groups made it to the present.

MAJOR VERTEBRATE GROUPS

Figure 26.8 shows how different vertebrate groups are related to one another. Besides lampreys, the major groups are known as cartilaginous fishes, bony fishes, amphibians, reptiles, birds, and mammals. The rest of this chapter considers their characteristics, using the evolutionary trends as the conceptual framework.

Vertebrates arose in the Cambrian. All of the major modern groups evolved during an adaptive radiation of jawed fishes.

Jaws, a vertebral column, paired fins, and lungs were pivotal developments in the evolution of lineages that gave rise to amphibians, reptiles, birds, and mammals.

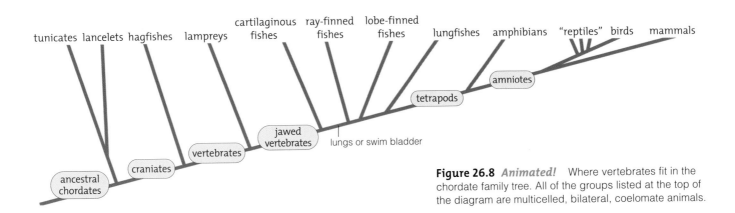

tunicates lancelets hagfishes lampreys cartilaginous fishes ray-finned fishes lobe-finned fishes lungfishes amphibians "reptiles" birds mammals

amniotes

tetrapods

jawed vertebrates

lungs or swim bladder

vertebrates

craniates

ancestral chordates

Figure 26.8 *Animated!* Where vertebrates fit in the chordate family tree. All of the groups listed at the top of the diagram are multicelled, bilateral, coelomate animals.

26.3 Jawed Fishes and the Rise of Tetrapods

Unless you study underwater life, you may not know that the number and diversity of fishes exceed those of all other vertebrate groups combined.

The body form and behavior of fishes offer clues to the challenges of life in water. Water is 800 times more dense than air and resists movements through it. One response to this physical constraint is a streamlined body that reduces friction during chases. Sharks and other predators of the open ocean have such a body, with strong muscles for forward propulsion (Figure 26.9a). Species that do not make high-speed runs tend to have small-finned, flattened bodies that easily slip through crevices, as in reefs. Fish scales also assist in motion. Collectively, these small and usually thin bony plates protect the body without weighing it down.

A trout suspended in shallow water is an example of adaptation to the density of water. Like many other fishes, it maintains neutral buoyancy by adjusting the volume of its **swim bladder**, a flotation device that exchanges gases with blood. It adjusts the volume of its swim bladder as it gulps air at the water's surface.

CARTILAGINOUS FISHES

Cartilaginous fishes (Chondrichthyes) include about 850 species of skates, sharks, and chimaeras (Figure 26.9). These marine predators have prominent fins, a skeleton of cartilage, and five to seven gill slits on both sides. At fifteen meters in length, some sharks are among the largest living vertebrates. They all shed and replace their teeth on an ongoing basis. The teeth are modified scales having a dentin core and an outer enamel layer. Many strong-jawed sharks grab and rip chunks off prey. They have been ocean predators for hundreds of millions of years, but the rare attacks on humans give the whole group a bad reputation.

Most skates and rays have flattened teeth that crush shelled prey on the seafloor. Manta rays are up to six meters across. Stingrays have a venom gland in the tail; other rays have electric organs that can stun prey with as much as 200 volts of electricity. Thirty or so species of chimaeras prefer eating small mollusks. All have a venom gland in front of the dorsal fin.

Figure 26.9 Cartilaginous fishes. (**a**) Galápagos sharks. (**b**) Manta ray. Two projections on its head unfurl and waft plankton to a mouth between them. Notice the gill slits on both fishes. (**c**) Chimaera, or ratfish.

Representative ray-finned fishes. This is the largest and most diverse group of bony fishes. (**d**) Sea horse, which anchors itself to substrates with its tail. (**e**) Coral grouper, one of the many species of bony fishes that fishermen are overharvesting. (**f**) Long-nose gar.

"BONY FISHES"

Jawed fishes having a bony endoskeleton evolved 400 million years ago. The **ray-finned fishes** have flexible fin supports derived from skin and thin scales. With more than 21,000 species in freshwater and marine habitats, they are the most diverse vertebrates of all. Paddlefish, sturgeons, and gars are in one group. The group called *teleosts* encompasses about 95 percent of the world's fish species, and half of them inhabit reefs and shallow seas. Teleosts include herrings, true eels, moray eels, puffers, killifishes, scorpionfishes, and the perciforms. Perches, salmon, mullets, groupers, basses, parrotfishes, blennies, cichlids, tuna, mackerels, and the barracudas are representative perciforms. So are sea horses, which are poor swimmers but good hiders.

You can see a few examples of ray-finned fishes in Figure 26.9 and in Sections 1.1, 27.5, 45.6, and 46.5.

Coelacanths (*Latimeria*) are the only surviving group of **lobe-finned fishes**. We know of two populations that may be separate species. The ventral fins of these fishes are fleshy extensions of the body, with skeletal support elements inside (Figure 26.10*a*). Like the ray-finned fishes, coelacanths have gills.

Lungfishes (Figure 26.10*b*) have gills *and* one or two small, modified outpouchings of the gut wall. These sacs help take in oxygen and remove carbon dioxide. Lungfishes must surface and gulp air; they drown if held under water. When streams dry out seasonally, lungfishes encase themselves in slimy secretions and mud. They stay inactive until the rainy season starts.

Do such fishes share a common ancestor with four-legged walkers—**tetrapods**? Probably. Like tetrapods, a lungfish has a separate blood circuit to the lungs. Its skullbones are arranged the same way. It has the same tooth enamel. Also, the fins of both lobe-finned fishes and lungfishes are about the same as tetrapod limbs in structure, position, and sizes. A lungfish even uses its limbs to move forward on underwater substrates.

Limb joints, digits, and other traits of fossils imply that walking originated in water, not on land. By the late Devonian, some swimming and crawling species were using limbs and digits (Figure 26.10*c–e*).

Ray-finned fishes are now the most diverse and abundant vertebrates. The lobe-finned fishes have fleshy ventral fins reinforced with skeletal parts. The lungfishes have simple lunglike sacs that supplement respiration by gills.

Walking probably started under water during the Devonian, among the aquatic forerunners of tetrapods on land.

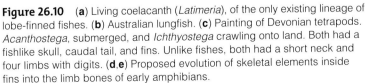

inside lobed fins, bony or cartilaginous structures undergoing modification

limb bones of an early tetrapod

Figure 26.10 (**a**) Living coelacanth (*Latimeria*), of the only existing lineage of lobe-finned fishes. (**b**) Australian lungfish. (**c**) Painting of Devonian tetrapods. *Acanthostega*, submerged, and *Ichthyostega* crawling onto land. Both had a fishlike skull, caudal tail, and fins. Unlike fishes, both had a short neck and four limbs with digits. (**d,e**) Proposed evolution of skeletal elements inside fins into the limb bones of early amphibians.

26.8 Portfolio of Existing "Reptiles"

The name reptile *is derived from the Latin* repto, *which means to creep. Some existing forms do creep. Others race, lumber, or swim about. These diverse lineages include all living amniotes that are not birds and not mammals.*

GENERAL CHARACTERISTICS

More than 8,160 species of turtles, lizards, tuataras, snakes, and crocodilians live in a variety of habitats on land and in water. All are cold-blooded. Both sexes have a cloaca, an opening that functions in excretion and reproduction. The females are fertilized internally. Even aquatic species typically lay their eggs on land. Figure 26.16 shows one of the distinctive body plans.

MAJOR GROUPS

Turtles About 305 existing species of turtles live in a shell attached to their skeleton. If threatened, turtles can pull their head and limbs inside (Figure 26.17a,b). Their body plan has been around since Triassic times, although the shell's size became reduced in lineages of sea turtles and other highly mobile types.

Instead of teeth, turtles have horny plates that are used to grip and chew food. Strong jaws and a fierce disposition help fend off predators. But turtle eggs are vulnerable to many predators on land. Add hunting and destruction of nesting sites to the mix, and most sea turtles are now poised at the brink of extinction.

Lizards With 4,710 species, the lizards are the most diverse reptiles. The smallest, *Sphaerodactylus ariasae*, can fit on a dime (*left*). The largest, *Varanus komodoensis* (Komodo dragon) is a monitor lizard that may grow 3 meters (10 feet long). This ambush predator lurches out of bushes at deer, wild boar, an occasional human, and other prey. Like most predatory lizards, it snags its prey with sharp, peglike teeth. Chameleons (page

557) use accurate flicks of a sticky tongue, which can be longer than their body. Iguanas, including marine iguanas of the Galápagos, are herbivores.

Being small themselves, most lizards are prey for other animals. Some try to outrun predators; others try to startle them by flaring their throat fan (Figure 26.17c,d). Many will give up their tail when a predator grabs it. The detached tail wriggles for a bit and may distract a predator from its fleeing owner.

Tuataras Two species of tuataras survive on small islands near New Zealand (Figure 26.17e). They are descendants of a group that had its heyday in the Triassic. They are reptiles, yet they resemble modern amphibians in their brain and locomotion. Like some lizards, the tuataras have a third, middle "eye" under the skin, with retina, a rudimentary lens, and links to the brain. It may detect changes in daylength and light intensity and affect hormonal control of reproduction.

Snakes In the Cretaceous, short-legged, long-bodied lizards gave rise to elongated, limbless snakes. Some snakes retain bony remnants of ancestral hindlimbs. Most move their body in S-shaped waves, much like salamanders do (Section 42.7). All 2,955 living species are carnivores that have flexible skull bones and jaws. Many swallow prey wider than they are. Rattlesnakes and other fanged species bite and subdue prey with venom (Figure 26.17f). Snakes in general tend not to attack humans, but venomous types bite about 7,000 people annually in the United States and kill a few of them. About 40 percent of those bitten reported that they were handling the snake or otherwise disturbing it. Worldwide, the annual death toll from snake bites is estimated at 30,000–40,000.

All snakes are vulnerable during their life cycle, because birds and other predators relish snake eggs. Female snakes store sperm and lay several clutches

Figure 26.16
Animated! Body plan of a crocodile which, like its lifestyle, has remained much the same for nearly 200 million years. The group's future is not rosy. Human habitats now encroach on many crocodilian habitats. Alligator belly skin is in high demand for wallets, handbags, and shoes.

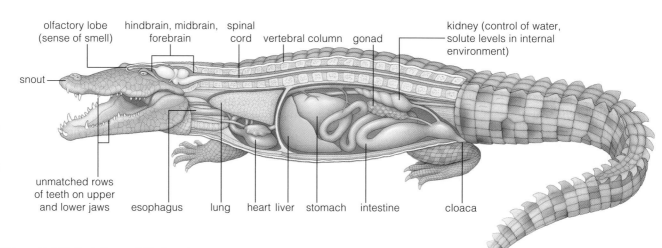

olfactory lobe (sense of smell)   hindbrain, midbrain, forebrain   spinal cord   vertebral column   gonad   kidney (control of water, solute levels in internal environment)

snout

unmatched rows of teeth on upper and lower jaws   esophagus   lung   heart   liver   stomach   intestine   cloaca

hard shell — vertebral column

b

Figure 26.17 (**a**) Galápagos tortoise. (**b**) Turtle shell and skeleton. (**c**) Lizard fleeing and (**d**) lizard confronting. (**e**) Tuatara (*Sphenodon*). (**f**) Rattlesnake. (**g**) Spectacled caiman, a crocodilian. Its upper and lower rows of teeth do not match up, as mammalian teeth do.

of fertilized eggs at intervals after they mate, which improves the odds that at least some will hatch.

Crocodilians Almost a dozen species of crocodiles, alligators, and caimans are the closest living relatives of birds. All are predators in or near water. They have powerful jaws, a long snout, and sharp teeth (Figure 26.17*g*). They all drag prey into water, tear it apart as they spin around, and gulp down chunks. The feared southern Asia crocodile and Nile crocodile can weigh as much as 0.25 to 1 ton (500 to 2,000 pounds).

Like other reptiles and birds, the crocodilians can adjust body temperature by altering certain behaviors and adjusting metabolic rates (Section 42.7). However, they are the only reptiles that have a four-chambered heart, as mammals and birds do. Also like most birds, crocodilians engage in parental behavior, as when they guard their nests and assist the hatchlings.

venom gland

hollow fang

Turtles and tuataras have body plans that have changed very little since the Triassic. Lizards and snakes are the most diverse reptilians. Crocodilians have the most complex brain and heart structure. They are the closest relatives of birds.

26.9 Birds—The Feathered Ones

A capacity for flight evolved in pterosaurs, which are now extinct, and in insects, birds, and bats. But only birds, like a few extinct dinosaurs that preceded them, make feathers—lightweight structures used in flight, as body insulation, and often in courtship displays (Figures 26.18 and 26.19).

Birds are warm-blooded, feathered vertebrates. Like the dinosaurs and crocodilians, their closest relatives, they have scales on their legs and some of the same internal organs, including a cloaca that functions in excretion and reproduction. Birds, too, make amniote

Figure 26.18 *Confuciusornis sanctus*, which lived about the same time as *Archaeopteryx*. It is the earliest known bird with a bill. Fossils show how the clawed limbs of its four-legged ancestor were modified into wings. Each wing had three digits, one a reduced claw with flight feathers.

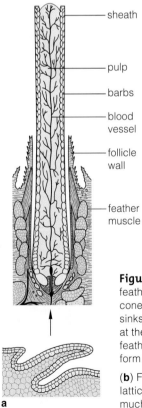

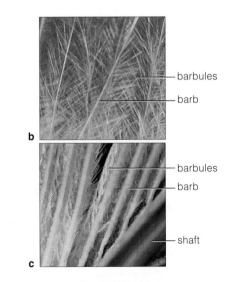

Figure 26.19 Feathers. (**a**) in chick embryos, a typical feather develops from actively dividing epidermal cells. A cone-shaped bud starts growing out from the skin. Its base sinks, and the space around it becomes a follicle. Horny cells at the cone's surface differentiate and form a sheath, then a feather forms inside it. Blood vessels in the underlying dermis form pulp, which nourishes the growing feather.

(**b**) Flight feathers. Their hollow central shaft and interlocked lattice of barbs and barbules impart strength without adding much weight. (**c**) Innermost down feathers are insulative.

eggs that are fertilized internally in females (Figure 26.20). Internal and external modifications to the basic body plan led to the capacity for flight.

Birds originated when Mesozoic reptiles began an adaptive radiation. They diverged from a lineage of small theropod dinosaurs, a type of carnivore that ran about on two legs, and their feathers evolved from highly modified reptilian scales. *Archaeopteryx* was in or near that lineage; *Confuciusornis sanctus* definitely was an early bird (Figures 26.1 and 26.18).

Unlike reptiles, birds have built-in mechanisms that adjust body temperature. They gain most of their heat not from outside sources but from metabolism. When it gets cold outside, they fluff feathers as insulation. When it gets too hot, elastic air sacs connected to the lungs speed the outflow of warm air, which can help dissipate excess metabolic heat (Sections 40.3 and 42.7).

The evolution of bird flight involved far more than feathers. It involved modification of the entire body, from internal bone structure to highly efficient modes of respiration and circulation. The high metabolic rates that sustain flight depend on a strong flow of oxygen through the body. The elastic air sacs greatly enhance oxygen uptake. Also, like mammals, birds have a four-chambered heart that pumps oxygen-rich blood to the lungs and to the rest of the body in separate circuits, as it probably did in reptilian ancestors of birds.

Flight also demands an airstream, low weight, and a powerful downstroke that can provide *lift*, or a force at right angles to an airstream. Each wing, a modified forelimb, consists of feathers and lightweight bones attached to powerful muscles. Its bones do not weigh much, owing to profuse air cavities in the bone tissue. The flight muscles attach to an enlarged breastbone (sternum) and to the upper limb bones attached to it. When the muscles contract, they produce a powerful downstroke (Figures 26.21 and 26.22).

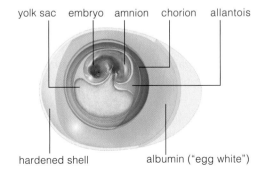

Figure 26.20 *Animated!* One type of amniote egg.

A wing's long flight feathers work like airfoils on planes. The bird normally folds these feathers slightly on each upstroke, which lessens the wing surface and presents less resistance to air. On the downstroke, the bird spreads out the feathers, which increases the area pushing against the air (Figure 26.21a).

There are about 28 orders of birds and 9,000 named species that vary in size, proportions, coloration, and capacity for flight. The tiniest, the bee hummingbird, weighs 1.6 grams (0.6 ounce). The ostrich, the largest living bird, can weigh 150 kilograms (330 pounds). It cannot fly, but powerful leg muscles help it sprint. Warblers, parrots, and some other perching species that show complex social behaviors often have vivid feather colors and patterns. Bird song and other forms of behavior are among the topics of Chapter 49.

We see one of the most dramatic forms of behavior among birds that migrate with the changing seasons. **Migration** is a recurring pattern of movement from one region to another in response to environmental rhythms. Seasonal change in daylength is one factor that influences the internal timing mechanisms that we call "biological clocks." It causes physiological and behavioral changes that induce migratory birds to make round trips between breeding grounds and wintering grounds. Arctic terns migrate the farthest. They spend summers in arctic regions and winters in antarctic regions. One monitored bird took off from an island near Great Britain and landed three months later in Melbourne, Australia. It racked up more than 22,000 kilometers (14,000 miles) on that one journey.

Of all animals, birds alone produce feathers, which they use in flight, heat conservation, and socially significant communication displays. A bird's feathers, lightweight bones, and highly efficient respiratory and circulatory systems sustain flight.

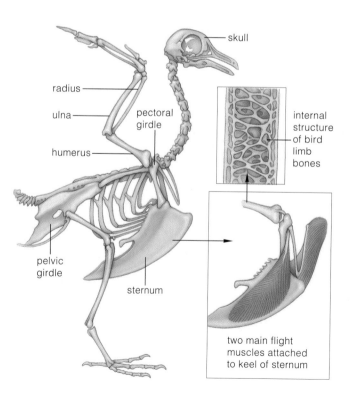

Figure 26.21 Bird flight. (**a**) Of all living animals, only birds and bats fly by *flapping* wings. A bird wing is a system of lightweight bones and feathers. Section 19.4 looks at the evolution of wing bones. (**b**) Laysan albatross. Its wingspan is more than two meters, yet this seabird weighs less than 10 kilograms (22 pounds). It is so at home in the air, it even sleeps while riding the winds. One bird monitored by researchers flew 24,843 miles in ninety days as it searched for food and then brought it back to its nestlings. That is the equivalent of a trip around the globe.

skull

radius

ulna

pectoral girdle

humerus

internal structure of bird limb bones

pelvic girdle

sternum

two main flight muscles attached to keel of sternum

Figure 26.22 *Animated!* Flight muscles and the keeled breastbone (sternum).

26.10 The Rise of Mammals

LINKS TO
SECTIONS 17.6, 17.8,
CHAPTER 3
INTRODUCTION

Remember the premise of cause and effect? Christine Janis put it this way: In a world dominated by the dinosaurs, mammals survived by being small, efficient at exploiting food, and flexible in behavior. They originated before the dinosaurs and outlasted them, largely by being artful dodgers and because of chance, catastrophic events.

Which traits set **mammals** apart from other animals? These are the only animals that make hair, although species differ in the amount, type, and distribution. Female mammals alone secrete milk for their young. This nutrient-rich fluid forms in mammary glands that have ducts to the body surface. Jaws and teeth also help set mammals apart. Most have four types of upper and lower teeth that match up (Figure 26.23).

Reptiles have a hinged two-part jaw, and all teeth are one type only and do not match up. Most mammals care for their offspring during an extended period of dependency. The young typically have the capacity to learn and repeat behaviors with survival value. Some species show notable *behavioral flexibility*—a capacity to expand on basic activities with novel behaviors.

Mammals are the only synapsids that made it to the present. Synapsids, again, were the first amniotes to branch from stem reptiles in the late Carboniferous. They became dominant on land, but many were lost in a mass extinction that ended the Permian (Chapter 3).

The therapsid branch endured. During the Triassic, the four limbs of certain species became positioned under the body instead of at its sides. The change made it easier to walk erect, but a trunk higher from the ground was not as stable. Not coincidentally, the volume of the cerebellum—a brain region that controls balance and spatial positions—expanded at that time.

Modern mammals evolved early in the Jurassic. By 130 million years ago, there were three lineages: the **monotremes** (egg-laying mammals), the **marsupials** (pouched kinds), and **eutherians** (placental kinds). In the Paleocene, with no more pressure from dinosaurs, a great mammalian radiation began. Climates were mild; forests or woodlands extended into polar areas. Carnivores, omnivores, and herbivores evolved. Their teeth could nip, shred, or grind food (Figure 26.24).

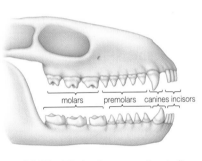

b

Figure 26.23 Distinctly mammalian traits. (**a**) A human baby, already with a mop of hair, being nourished by milk from mammary glands. (**b**) Four types of mammalian teeth; upper and lower rows match up.

Figure 26.24 A few of the early Paleocene mammals that lived in sequoia forests in what is now Wyoming. On the ground, raccoonlike *Chriacus* faces a tree-climbing rodent (*Ptilodus*). Higher up is *Peradectes*, a marsupial.

Figure 26.25 (**a**) *Indricotherium*, the "giraffe rhinoceros." At 15 tons and 5.5 meters high at the shoulder, it was the largest mammal we know about. (**b**) Saber-tooth cat (*Smilodon*) in a dry woodland of the Pleistocene. *Smilodon* fossils have been recovered from the pitch pools in Rancho LaBrea, California.

a About 150 million years ago, during the Jurassic, the first monotremes and marsupials evolved and migrated throughout the supercontinent Pangea.

b Between 130 and 85 million years ago, in the Cretaceous, placental mammals emerged and started to spread. Monotremes and marsupials of the southern supercontinent evolved in isolation from placental mammals.

c About 20 million years ago, in the Miocene, placental mammals expanded in range and diversity. On Antarctica, mammals vanished. Marsupials and early placental mammals displaced monotremes in South America.

d About 5 million years ago, in the Pliocene, advanced placental mammals invaded South America. They drove most marsupials and the early placental species to extinction.

Figure 26.26 *Animated!* (**a–d**) Adaptive radiations of mammals. *Right,* one consequence—a case of convergent evolution. (**e**) Australia's spiny anteater, one of two living monotremes. (**f**) Africa's aardvark and (**g**) South America's giant anteater. Compare the specialized ant-snuffling snouts.

Many modern species, such as rodents, primates, bats, even-toed hoofed mammals (hippos, camels, and other artiodactyls), and odd-toed mammals such as the tiny, earliest horse (*Hyracotherium*) arose in the early Eocene. Later on, parts of Pangea that would become Australia and Antarctica drifted apart, rafting some monotremes and marsupials away from the placental mammals radiating through other land masses. Then a passage opened between the Arctic and the North Atlantic oceans, and deep circulation patterns shifted. An Antarctic ice cap formed, and northern lands grew cooler and drier. Woodlands and vast open grasslands replaced tropical forests, and plant growth patterns changed. Most early mammals became extinct.

From the Oligocene on through the Pliocene, many grazing and browsing animals evolved, including gazelle-like camels, the "giraffe rhinoceros," and the saber-tooth cat (Figure 26.25). On the land mass that became South America, monotremes were replaced by marsupials and early placental mammals. Using a land bridge from North America, the highly evolved placental mammals radiated southward and rapidly replaced many of their previously isolated relatives (Figure 26.26). Only the opossums and a few other species invaded land masses in the other direction.

What gave placental mammals a competitive edge? They could use higher metabolic rates to control body temperature, and they had a better way to nourish embryos. Later chapters explain these traits. For now, note that the adaptive radiation of highly specialized placental mammals came about at the expense of less competitive relatives.

The climate cooled again in the Miocene as great mountain ranges arose and altered air circulation and rainfall patterns. Grasslands formed, and new kinds of herbivores emerged. Beginning in the Pliocene, ice ages started alternating with interglacials, or warmer periods. Species richness declined in deserts, prairies, steppes, and pampas, which are still with us today. As you will see, early humans faced these challenges.

One more point: Many lineages originally evolved on separate continents but became adapted to similar habitats. In time, they came to resemble one another in form and function. For example, Australia's spiny anteater, South America's giant anteater, and Africa's aardvark all have similar snouts adapted to preying on ants (Figure 26.26*e–g*). Such species are examples of *convergent* evolution, as described in Section 17.4.

> Mammals are the only animals with hair, and only female mammals produce milk from mammary glands.
>
> The three major groups are the egg-laying monotremes, the pouched marsupials, and the placental mammals. Placental mammals are now the dominant group in most regions.

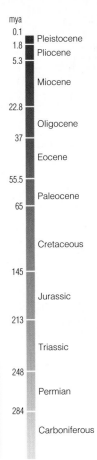

mya
0.1 Pleistocene
1.8
5.3 Pliocene
Miocene
22.8
Oligocene
37
Eocene
55.5
Paleocene
65
Cretaceous
145
Jurassic
213
Triassic
248
Permian
284
Carboniferous

26.11 Portfolio of Existing Mammals

Reflect now on the current range of mammalian diversity—the monotremes, marsupials, and placental mammals.

The only living monotremes are two species of spiny anteaters (Figure 26.26) and the duck-billed platypus (Figure 26.27a). One spiny anteater lives in Australia, and the other in Papua New Guinea. These burrowing mammals eat ants. Like porcupines, they bristle with protective spines (modified hairs). As is the case for platypuses, females lay eggs. Unlike platypuses, they do not dig out nests. They incubate a single egg, and their hatchling suckles and completes its development in a skin pouch that forms temporarily (through muscle contractions) on the mother's ventral surface.

Most of the 260 existing species of marsupials are native to Australia and nearby islands; a few live in the Americas (Figure 26.27b,c). Tiny, blind, hairless newborns suckle and finish developing in a permanent pouch on the mother's ventral surface. The Tasmanian devil, the largest carnivorous marsupial, bares fangs as a threat display (Figure 26.27c). It makes hair-raising screeches, coughs, and snarls but is only a scavenger, famous for its rowdy communal feeds at carcasses.

Placental mammals are named after the **placenta**, a spongy organ made of maternal and fetal membranes (Figure 26.28a). A placenta forms in a chamber (uterus) that protects an embryo as it develops. It gets oxygen and nutrients to the embryo and removes its metabolic wastes (Section 44.11). In general, such embryos have a developmental advantage over their closest relatives. They can grow faster than marsupial embryos do in a pouch. Also, many species are fully formed, thus less vulnerable to predators, at birth. Appendix I lists the major groups; Figure 26.28 shows representatives.

The body form, function, behavior, and ecology of mammals will occupy our attention later in the book. In the next chapter, we will take a look at mammals that are now threatened by a current mass extinction. Here, we leave the three existing lineages by inviting you to reflect on Table 26.1. It highlights classic cases of convergent evolution that occurred among them.

Most existing monotremes and marsupials are native to the southern continents and islands. Placental mammals show far greater distribution and diversity.

Figure 26.27 (**a**) Platypus, a monotreme. Two marsupials: (**b**) Adult koala (*Phascolarctos cinereus*). Its ancestors evolved millions of years ago, when the climate turned drier and drought-resistant plants evolved. Koalas eat only eucalyptus leaves. European settlers in Australia cleared eucalyptus forests for farmland. Millions of slow-moving koalas were shot for their pelts. In the 1930s they became a protected species, but their numbers declined because nothing protected the eucalyptus trees. This still is the case in much of their home range. (**c**) Young Tasmanian devil.

Table 26.1	Convergences Among Groups of Mammals	
Life-Style	Home	Mammalian Family
Aquatic invertebrate eater	North America Central America Australia	Water shrew (Soricidae) Water mouse (Cricetidae) Platypus (Ornithorhynchidae)
Carnivore on land	North America Australia	Wolf (Canidae) Tasmanian wolf (Thylacinidae)
Anteater on land	South America Africa Australia	Giant anteater (Myrmecophagidae) Aardvark (Orycteropodidae) Spiny anteater (Tachyglossidae)
Ground-dwelling leaf, tuber eater	North America South America Eurasia	Pocket gopher (Geomyidae) Tuco-tuco (Ctenomyidae) Mole rat (Spalacidae)
Tree-dwelling leaf eater	South America Africa Madagascar Australia	Howler monkey (Cebidae) Colobus monkey (Cercopithecidae) Woolly lemur (Indriidae) Koala (Phascolarctidae)
Tree-dwelling nut, seed eater	Southeast Asia Africa Australia	Flying squirrel (Sciuridae) Flying squirrel (Anomaluridae) Flying squirrel (Phalangeridae)

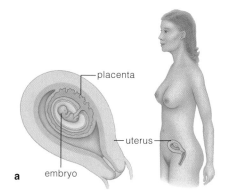

Figure 26.28 Placental mammals. (**a**) Location of the placenta in a pregnant human female. (**b**) Blue whale and (**c**) the size of its skeleton relative to a human. At 200 tons, an adult is the biggest living animal. (**d**) A manatee lives in water and eats seaweed. Early sailors imagined it was a mermaid. (**e**) A camel traverses hot deserts while (**f**) harp seals swim in frigid waters after prey and sunbathe on ice. (**g**) Flying squirrel, really just a glider. The only flying mammals are bats (**h**); this one is a Kitti's hog-nosed bat. (**i**) Red fox hiding in blue spruce. Thick, insulative fur protects it from winter cold.

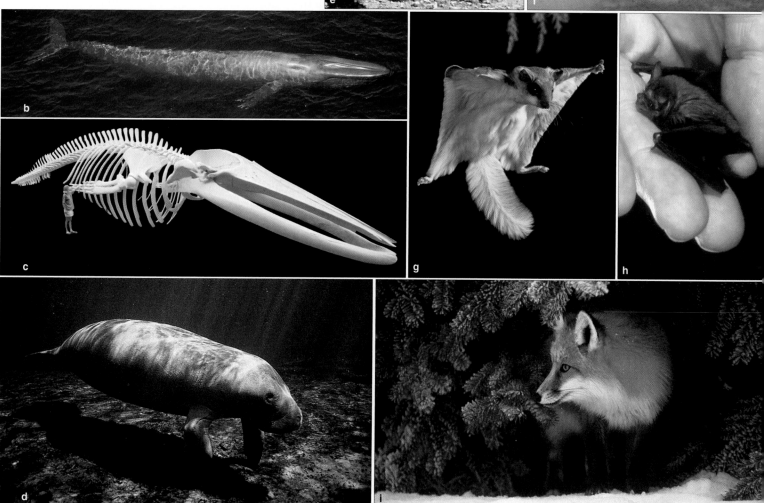

26.12 Trends in Primate Evolution

LINK TO
SECTION
17.4

So far, you have traveled 570 million years through time, from tiny flattened invertebrates to craniates, then on to the jawless and jawed fishes with a backbone. You encountered some early tetrapods, which were walking on four limbs that had evolved from fleshy lobed fins. You moved on to amniotes—reptiles, birds, and mammals. Now you are about to travel primate roads that led to modern humans.

a Hole at back of skull; the backbone is habitually parallel with ground or a plant stem

b Hole close to center of base of skull; the backbone is habitually perpendicular to ground

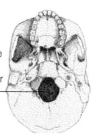

Figure 26.29 Hole in the head as a diagnostic tool—the foramen magnum.

WHAT IS A PRIMATE?

The **primate** lineage originated from squirrel-sized mammals of the Mesozoic. By 60 million years ago, its earliest forms—*prosimians*—had evolved. These tree-dwelling (arboreal) species radiated through northern forests. Before the Oligocene, one branch gave rise to *monkeys* and *apes* that would, from the Miocene on, displace prosimians as the dominant primates. Today, prosimians endure only in isolated places, including Madagascar, or where they can forage at night while monkeys and apes are sleeping.

Gibbon, siamang, orangutan, gorilla, chimpanzee, bonobo—these primates are classified as the apes. All monkeys, apes, and *humans* are further classified as **anthropoids**. In biochemistry and body form, apes and humans are close relatives. They, and their closest extinct ancestors, are the only **hominoids**. All human-like and human species, past and present, are known as the **hominids**. They will be our focus here.

KEY EVOLUTIONARY TRENDS

Most primates live in the trees of forests, woodlands, or savannas (grasslands with a scattering of trees). Few features set them apart from other mammals, and each lineage has its own defining traits. The five trends that define the lineage that led to modern humans were set in motion among the first tree-dwelling lineages.

First, a reliance on the sense of smell became less important than daytime vision. *Second*, skeletal changes promoted **bipedalism**—upright walking—which freed hands for new tasks. *Third*, altered bones and muscles made hands more versatile. *Fourth*, teeth became less specialized. *Fifth*, the evolution of the brain, behavior, and culture became interlocked. **Culture** is the sum of a social group's behavioral patterns, passed on through the generations by learning and symbolic behavior. In brief, *"uniquely" human traits emerged by modification of traits that had evolved earlier, in ancestral forms.*

Enhanced Daytime Vision Early primates had an eye on each side of the head, which was held parallel to the ground. Later on, many species had eyes facing forward, which was better for sampling shapes and movement in three-dimensional space. Visual systems became increasingly responsive to variations in light intensity (dim to bright) and to colors.

Upright Walking How a primate walks depends on the length, shape, and skeletal position of its bones, especially the backbone, shoulder blades, and pelvic girdle. At the base of its skull bones is an opening, the **foramen magnum**, where the spinal cord can connect with the brain (Figure 26.29). If this hole is at the back of the skull, you can be sure that the animal moved on all fours. If it is almost centered at the skull's base, the animal held its backbone perpendicular to the ground.

Tarsiers and monkeys are four-legged walkers, with front and back limbs of about the same length; gorillas walk on two legs and knuckles of their long arms. But their foramen magnum is not at the back of the skull, as it was in their ancestors. When sitting, they can hold their head erect. In humans, the foramen magnum is close to center at the base of the skull. Of all primates, humans have the shortest, most flexible backbone, one curved into an S shape. These are structural traits that make bipedalism possible (Figures 26.29*b* and 26.30).

Power Grip and Precision Grip So how did we get our versatile hands? Early mammals spread their toes apart to support their weight as they walked or ran on four legs. In ancient tree-dwelling primates, the bones of each hand changed, so fingers could curl around objects (*prehensile* movements), and the thumb could touch all the fingertips (*opposable* movements). In time, the hands were freed from load-bearing functions and became modified in ways that permitted powerful or precision gripping of objects:

power grip

precision grip

mya
0.1
1.8
5.3

22.8

37

55.5

65

Pleistocene
Pliocene

Miocene

Oligocene

Eocene

Paleocene

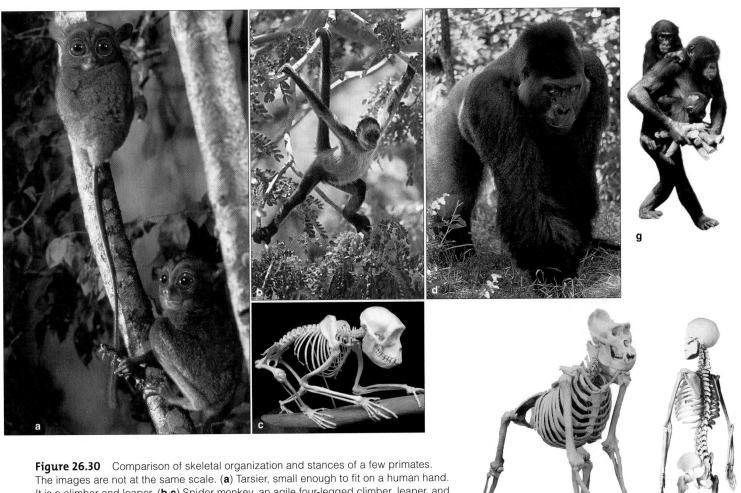

Figure 26.30 Comparison of skeletal organization and stances of a few primates. The images are not at the same scale. (**a**) Tarsier, small enough to fit on a human hand. It is a climber and leaper. (**b,c**) Spider monkey, an agile four-legged climber, leaper, and runner. It has long, thin limbs relative to its torso, and its long, thin digits are good at gripping, not at supporting the body's weight. (**d,e**) Gorillas use their forelimbs to climb and support body weight when walking on knuckles and two legs. (**f**) A human is a two-legged walker. (**g**) In captivity, a bonobo is upright 2 percent of the time on land, and 90 percent of the time when it is wading after edible fishes and aquatic invertebrates.

Having the capacity for versatile hand positions gave the ancestors of humans the capacity to make and use tools. Refined prehensile and opposable movements led to the development of technologies and cultures.

Teeth for All Occasions As you will see shortly, modifications to the jaws and teeth in certain groups accompanied a shift from eating insects, to fruits and leaves, to a mixed diet. Rectangular jaws and lengthy canines evolved in monkeys and apes. Among early hominids, a bow-shaped jaw evolved. So did teeth that were smaller and all about the same length.

Brains, Culture, and Behavior Over time, shifts in reproductive and social behavior accompanied a shift to life in the trees. In many lineages, parents put more

effort in fewer offspring, and maternal care became intense. The more the young had to learn to survive on their own, the longer they depended on parents.

Culture is the sum of a human group's patterns of social behavior. It endures largely because of a system of recognized signs and symbols—language. It evolved in hominids as selection pressures acted on the brain regions dealing with sensory inputs. The brain became larger and more complex, but only after mutations led to increases in the volume of its bony chamber.

Complex, forward-directed vision; bipedalism, refined hand movements, generalized teeth, and interlocked elaboration of brain regions, behavior, and culture were key adaptations on the road from arboreal primates to modern humans.

26.13 From Early Primates to Hominids

LINKS TO
SECTIONS
17.4, 19.4

Fossils from central, eastern, and southern Africa show that the Miocene through the Pliocene was a "bushy" time of hominid evolution, meaning many forms were rapidly evolving. We still do not know how they are related.

ORIGINS AND EARLY DIVERGENCES

Primate-like mammals emerged between 85 million and 65 million years ago, in tropical Paleocene forests. Like small tree shrews (Figure 26.31), they had a long snout and a good sense of smell, suitable for snuffling food or predators. They clambered about in stems and branches, although not with speed or grace. They, too, had huge appetites. They foraged at night for seeds, buds, insects, and eggs under trees.

Prosimians evolved by the Eocene. These primates climbed higher and their snout was shorter. They had a larger brain, better daytime vision, and better ways to grasp objects. *How did these traits evolve?*

Consider the trees. Trees provided food and safety from ground-dwelling predators, but they also were zones of uncompromising selection. Picture an Eocene morning, leaves swaying in the breeze, colorful fruit, and predatory birds. An odor-sensitive, lengthy snout would not have been of much use, because breezes disperse odors. But skeletal changes and a brain that could assess motion, depth, shape, and color must have been favored. For instance, eye sockets were now facing front, not on the skull's sides, which is better for depth perception. For climbers and leapers, a brain that could estimate body weight, distance, wind speed,

and suitable destinations would have been favored; adjustments had to be quick for a body in motion.

By 36 million years ago, anthropoids had evolved. These tree-dwellers were on or close to the lineage that led to monkeys and apes. One kind had forward-directed eyes, a flattened face, and an upper jaw that sported shovel-shaped front teeth. It probably used the digits on its forelimbs to grasp food. Some lived in swamp forests infested with predatory reptiles. Was that a reason why it became imperative to think fast, grip strongly, and avoid the ground? Possibly.

THE FIRST APELIKE AND HUMANLIKE FORMS

Between 23 and 18 million years ago, in the Miocene, apes evolved. They were the first of the *hominoids*, and they spread through Africa, Asia, and Europe when the climate was changing (Figure 26.32). Africa became cooler and drier, with more seasonal change. Tropical forests, with their fruits, leaves, and insects, gave way to open woodlands and, later, to savannas. Food had become drier, harder, and more difficult to find.

Hominoids that had evolved in lush forests could move into new adaptive zones or die out. Most died, but not the common ancestor of modern apes and humans. Did upright stances and nimble fingers help their descendants survive? Did both traits evolve on land or in water? (Captive bonobos stay upright when wading into lakes and estuaries in pursuit of fishes and invertebrates.) Another puzzle: Increased cranial volume could not have occurred without mutation in

Figure 26.31 (a) Tree shrew (*Tupaia*), a squirrel-size primate of tropical forests in Southeast Asia. Skull shape and teeth of early primates: (b) This Paleocene primate, *Plesiadapis*, had rodentlike teeth. (c) Monkey-sized *Aegyptopithecus*, an Oligocene anthropoid, predates the split to Old World apes and monkeys. (d) Apelike Miocene dryopith. Some dryopiths were as big as chimpanzees.

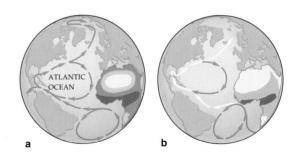

Figure 26.32 Model for a long-term shift in global climate. (a) Before the Isthmus of Panama formed, the salinity of surface ocean currents was similar around the world. Ocean circulation kept Arctic waters from getting too cold. (b) After the isthmus formed, circulation patterns changed. The North Atlantic surface currents grew saltier and heavier, and sank before reaching the Arctic. Water got colder in polar regions. An Arctic ice cap formed; climates became cooler and drier.

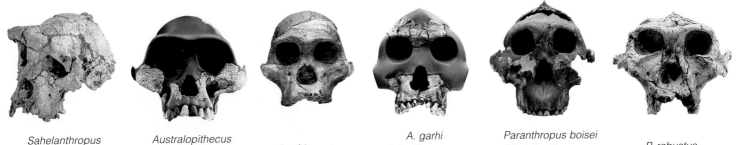

Sahelanthropus tchadensis
7–6 million years

Australopithecus afarensis
3.6–2.9 million years

A. africanus
3.2–2.3 million years

A. garhi
2.5 million years
(first tool user?)

Paranthropus boisei
2.3–1.4 million years
(huge molars)

P. robustus
1.9–1.5 million years

Figure 26.33 *Animated!* Fossils of African hominids that lived between 7 and 1.5 million years ago.

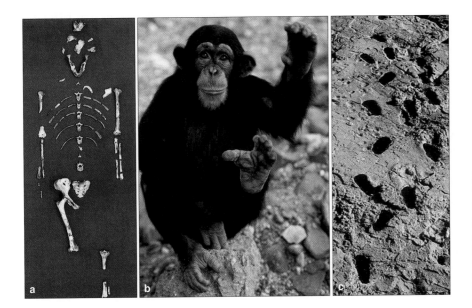

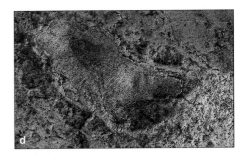

Figure 26.34 (**a**) Fossilized bones of a female dubbed "Lucy" (*Australopithecus afarensis*), a two-legged wader or walker that lived 3.2 million years ago.

Unlike chimpanzees and other apes, the early hominids did not have a splayed-out big toe (**b**). So how do we know? (**c,d**) At Laetoli in Tanzania, Mary Leakey found footprints made in soft, damp, volcanic ash 3.7 million years ago. The arch, big toe, and heel marks of these footprints are signs of bipedal hominids.

the genes that control metabolism and bone growth (Section 17.9). Concurrently, the diet would have had to provide enough fuel for a larger, more active brain. By one hypothesis, a switch to eating fishes, so rich in fatty acids, fueled the expansion in brain volume.

Sahelanthropus tchadensis was an ape or a hominid that lived in Central Africa about 6 or 7 million years ago. Its brain was chimpanzee-sized (Figure 26.33). Even so, like hominids that evolved later on, it had a flattened face, a prominent brow ridge, and smaller canines. It lived near the shore of a lake.

Australopithecus and *Paranthropus* were two groups, known informally as **australopiths** ("southern apes"). *A. anamensis*, *A. afarensis*, and *A. africanus* had slight builds (Figure 26.34a). *P. boisei* and *P. robustus* were muscular and heavily built. Like apes, all had a large face relative to brain size, and their jaws protruded. Given their thick-enameled molars, they had to be eating harder food. Also, they walked upright, either

straight forward or with a side-to-side motion, like a skater or someone wading through water.

However upright walking arose, we know that the hominids were already bipedal by 3.7 million years ago, because they left footprints. At that time, two *A. afarensis* individuals scuttled across a layer of newly deposited volcanic ash. A light rain was falling, and it transformed the ash into fast-drying cement—which preserved the footprints. Compare Figure 26.34*d*.

> By the Eocene, prosimians—the first primates—had evolved from mammals that resembled modern tree shrews. The first apelike forms (hominoids) evolved in the Oligocene. One branch gave rise to apes in the Miocene.
>
> Australopiths and certain hominids that preceded them walked upright. Long-term climate change drove them out of the trees in pursuit of food, either by wading in lakes and estuaries or by walking about in open woodlands.

26.14 Emergence of Early Humans

Judging from the fossil record, the earliest members of the human lineage emerged about 2.5 million years ago, in the great East African Rift Valley.

What do the fossilized fragments of early hominids tell us about human origins? The record is still too sketchy for us to interpret how all the diverse forms were related, let alone which might have been our ancestors. Besides, exactly which traits should we use to define **humans**—members of the genus *Homo*?

Well, what about brains? Our brain is the basis of unsurpassed analytical skills, verbal skills, complex social behavior, and technological innovations. Our brain sets us apart from apes that have a far smaller cranial volume (Figure 26.35). Yet this feature alone cannot tell us when a lineage of hominids made the evolutionary leap to becoming human. Their brain size fell within the range of apes. They did use simple tools, but so do chimps and some birds. We have no clue to their social behavior.

We are left to speculate on the evidence of physical traits among diverse fossils. They include a skeleton that permitted bipedalism, nimble fingers, a smaller face, a larger cranium, and smaller, thickly enameled teeth. These traits emerged late in the Miocene and are features of what might be the first humans—*Homo habilis*. The name means "handy man" (Figure 26.36).

Most of the early known forms of *Homo* evolved in Olduvai Gorge, Lake Turkana, and other formations

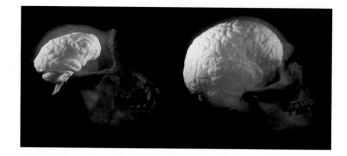

Figure 26.35 Comparison of brain size for a chimpanzee (*left*) and a modern human (*right*).

of the East African Rift Valley (Figure 26.37). This long valley had started to form earlier in the Miocene, when crustal plates were colliding. Later in time, woodlands flourished on its western slope. However, mountains blocked most of the seasonal rainfall to the west, and there, savannas formed.

Fossil teeth indicate that these early humans ate hard-shelled nuts, dry seeds, soft fruits and leaves, and insects—all seasonal foods. *H. habilis* had to think ahead, and plan when to gather and store foods that could last through the cool, dry seasons ahead.

H. habilis shared its habitat with carnivores such as saber-tooth cats, which could impale prey and tear flesh but could not crush open the marrow bones. Carcasses with meat shreds clinging to bones offered nutrients in nutrient-stingy places. Judging from its

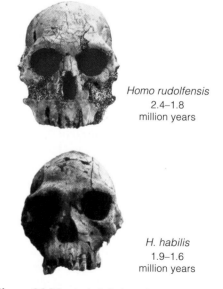

Homo rudolfensis
2.4–1.8
million years

H. habilis
1.9–1.6
million years

Figure 26.36 *Left:* Painting of a band of *Homo habilis* in an East African woodland. Two australopiths are shown in the distance. *Above:* Two fossils of early humans.

Figure 26.37 Olduvai Gorge in Africa, and a sampling of stone tools. *Left to right:* crude chopper, more refined chopper, hand ax, and cleaver.

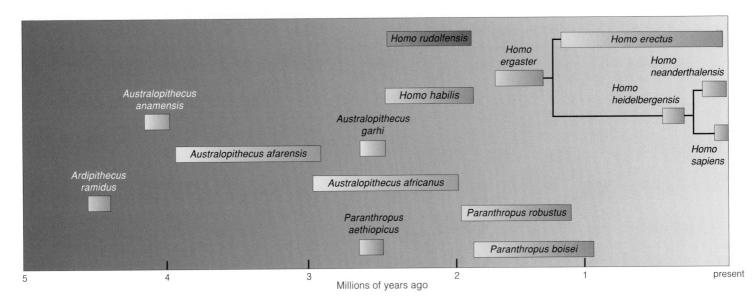

Figure 26.38 Timeline for the appearance and extinction of major hominid species, currently placed in four genera. Graph lines at upper right depict one view of the evolutionary relationships among the most recent species. The number of species, which fossils should be assigned to each group, and how species are related are all matters of heated debate among paleontologists.

array of teeth, *H. habilis* apparently was not one of the full-time carnivores. However, it might have enriched its diet opportunistically, by scavenging carcasses.

Which was the first toolmaker? The earliest stone tools we know about, from a site in Ethiopia, date to about 2.5 million years ago. Pieces of volcanic rock, chipped to create a sharp edge, were found alongside bones of animals that appear to have been butchered with similar tools. Fossils of *A. garhi* that were found nearby date to the same period. Did *A. garhi* make the tools? Perhaps.

The layers of Tanzania's Olduvai Gorge document later refinements in toolmaking skills (Figure 26.37). The deepest of these sedimentary rock layers date to about 1.8 million years ago and hold crudely chipped pebbles similar to the earliest ones in Ethiopia. They

also hold the fossil remains of australopiths and of *H. habilis*. We don't know which group created the tools. More recently deposited layers hold somewhat more complex tools. But all innovations in tool form were limited. The same types appear in layers that formed slowly, over a period of about 1 million years.

Before turning the page, take a look at Figure 26.38 to get a sense of where we have been on the primate evolutionary road, and where we go from here.

Once again, biological and geologic evolution converged. Crustal movements formed the great East African Rift Valley, which became cloaked in woodlands and forests that promoted the evolution of the first humans.

26.15 Emergence of Modern Humans

LINKS TO
SECTIONS
17.8, 19.4

Ancestors of modern humans remained in Africa until Homo erectus *evolved. Then cultural evolution took off.*

EARLY BIG-TIME WALKERS

Homo erectus is a species related to modern humans (Figures 26.39 and 26.40). Its name means "upright man." Its forerunners walked upright, but some *H. erectus* populations did the name justice. They walked out of Africa, turned left into Europe, and also right, all the way to China. Middle East and Southeast Asia fossils date from 1.8 million to 1.6 million years ago. So *H. erectus* had to adapt to brutally cold climates of that period, and to ice sheets that advanced several times into northern Europe and North America.

Whatever pressures triggered the far-flung travels, this was a time of physical changes, as in skull size and leg length. It was a time of cultural lift-off for the human lineage. *H. erectus* had a larger brain and was a more creative toolmaker. Its social organization and

communication skills must have been well developed. How else can we explain its dispersal? From southern Africa to England, populations used the same kinds of hand axes and other tools. They met environmental challenges by building fires and using fur for clothing, starting with an early Pleistocene ice age.

One fossil from Ethiopia shows that *Homo sapiens* had evolved by about 160,000 years ago. Compared to *H. erectus,* that species had smaller teeth, facial bones, and jawbones. Many individuals had a novel feature —a chin. They had a higher, rounder skull, a bigger brain, and possibly complex language.

A different group, the massively built and large-brained Neandertals, lived in Europe and the Near East from 200,000 to 30,000 years ago (Figure 26.39). Their extinction coincided with the arrival of modern humans in the same regions. Did they interbreed or war with the new arrivals? We cannot say. Neandertal mitochondrial DNA does have unique sequences that are not found in modern gene pools. By all molecular evidence, they have no modern descendants.

WHERE DID MODERN HUMANS ORIGINATE?

Was Africa the cradle for us all? At this writing, no one has discovered fossils of *Homo* that are older than 2 million years *except* in Africa. Between 2 million and 500,000 years ago, some *H. erectus* groups left Africa in waves. Some still lived in Southeast Asia between 53,000 and 37,000 years ago.

But when did *H. sapiens* originate? *Here we find one case of how the same body of evidence can be interpreted in different ways.* The *multiregional* model and the *African emergence* model explain the distribution of early and modern humans differently. Even so, they both rely on measurements of genetic distance between modern populations. Biochemical and immunological studies correlate *H. sapiens* fossils to specific times and places in the past (Figure 26.41). They also reveal that the greatest genetic distance is between the populations native to Africa and all other populations. The next greatest distance separates populations of Southeast Asia and Australia from others (Figure 26.42).

According to the **multiregional model**, groups of *H. erectus* in different regions faced different selection pressures, and they gave rise to subpopulations (races) of *H. sapiens.* In 2003, fossils of early humans that date to 18,000 years ago turned up on an Indonesian island (Flores). They were about as tall as australopiths. Like *H. erectus,* they had a small brain relative to body size and heavy brow ridges. But *H. erectus* vanished 200,000 years ago. The species, named *H. floresiensis,* may have

Figure 26.39
Fossils of early modern human forms.

H. erectus
2 million–53,000? years

H. neandertalensis
200,000–30,000 years

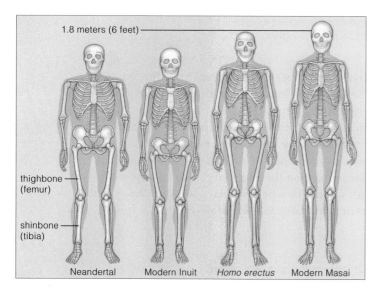

1.8 meters (6 feet)

thighbone (femur)

shinbone (tibia)

Neandertal   Modern Inuit   *Homo erectus*   Modern Masai

Figure 26.40 Body build correlated with climate. Humans adapted to cold climates have a heat-conserving body: stockier and with shorter legs, compared to humans adapted to hot climates.

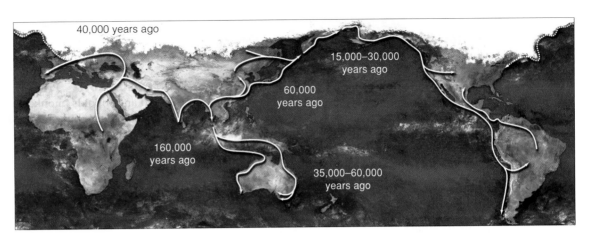

H. sapiens
fossil from Ethiopia,
160,000 years old

Figure 26.41 Estimated times when early *H. sapiens* colonized different parts of the world, based on radiometric dating of fossils. White lines are presumed dispersal routes. Molecular and immunological data imply that only a few populations left Africa. Genetically, all modern populations are highly similar.

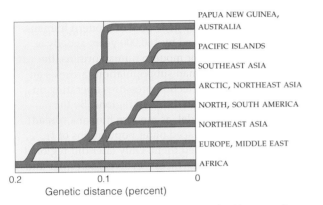

Figure 26.42 One proposed family tree for *Homo sapiens* populations native to different regions. The tree is based on analysis of many genes (such as those for mitochondrial DNA and the ABO blood group) and immunological comparisons.

Figure 26.43 Cave paintings at Lascaux, France.

descended from *Homo erectus*. As we know from many studies in biogeography, reduction in the body size of big organisms actually is a common adaptation to life on small islands, which have limited resources.

The **African emergence model** does not contradict fossil evidence that *H. erectus* migrated out of Africa and evolved further in different regions. But it has *H. sapiens* arising in sub-Saharan Africa, then dispersing into other regions (Figure 26.41). As modern humans spread out of Africa, they replaced archaic *H. erectus* populations that had preceded them. Only later did regional phenotypic differences—races—evolve.

Fossil evidence and considerable DNA sequence data from mitochondria, X and Y chromosomes, and autosomes support the African emergence model. So does evidence from archaeology. The oldest *H. sapiens* fossils are from Africa. In Zaire, intricately wrought barbed-bone tools indicate that populations in Africa were as skilled at making tools as *Homo* populations in Europe. Gene patterns from forty-three ethnic Asian

groups suggest modern humans moved from Central Asia, along India's coast, then on into Southeast Asia and southern China. They dispersed north and west into China and Siberia, then into the Americas.

From 40,000 years ago to today, culture became an immense part of human evolution, and we must leave that story. A point to remember: Humans dispersed rapidly through the world by devising *cultural* means to deal with the environment, and their art reflects a hunting heritage (Figure 26.43). Today, hunters and gatherers persist in parts of the world; other groups moved on from "stone-age" technology to the age of "high tech." The coexistence of these groups attests to the deep behavioral plasticity of the human species.

Cultural evolution has outpaced the biological evolution of the only remaining human species, H. sapiens. Today, humans everywhere rely on cultural innovation to adapt rapidly to a broad range of environmental challenges.

27.1 On Mass Extinctions and Slow Recoveries

Era	Period	MASS EXTINCTION UNDER WAY
CENOZOIC	QUATERNARY — 1.8 mya — TERTIARY	With high population growth rates and cultural practices (e.g., agriculture, deforestation), humans become major agents of extinction.
		MASS EXTINCTION
MESOZOIC	—— 65 —— CRETACEOUS —— 145 —— JURASSIC —— 213 —— TRIASSIC	Slow recovery after Permian extinction, then adaptive radiations of some marine groups and plants and animals on land. Asteroid impact at K–T boundary, 85% of all species disappear from land and seas.
		MASS EXTINCTION
PALEOZOIC	—— 248 —— PERMIAN —— 286 —— CARBONIFEROUS	Pangea forms; land area exceeds ocean surface area for first time. Asteroid impact? Major glaciation, colossal lava outpourings, 90%–95% of all species lost.
		MASS EXTINCTION
	—— 360 —— DEVONIAN —— 410 —— SILURIAN	More than 70% of marine groups lost. Reef builders, trilobites, jawless fishes, and placoderms severely affected. Meteorite impact, sea level decline, global cooling?
		MASS EXTINCTION
	—— 440 —— ORDOVICIAN —— 505 —— CAMBRIAN	Second most devastating extinction in seas; nearly 100 families of marine invertebrates lost.
		MASS EXTINCTION
	—— 544 —— (precambrian)	Massive glaciation; 79% of all species lost, including most marine microorganisms.

a

Based on many lines of evidence gathered over the past few centuries, an estimated 99 percent of all species that have ever lived are extinct. Even so, the full range of biodiversity is greater now than it has been at any time in the past.

Reflect on what you have read in this unit. For at least the first 2 billion years after the origin of life, single-celled prokaryotes dominated the evolutionary stage. They did so until Cambrian times, when free oxygen started to approach its current concentration in the atmosphere. That became a trigger for rapid genetic divergences and intricate species interactions that led, ultimately, to millions of eukaryotic species, the kinds we call protists, plants, fungi, and animals.

You saw how mass extinctions slashed biodiversity on land and in the seas. Each major episode promoted evolutionary changes and radiations into the vacated adaptive zones. Each time, however, recovery to the same level of biodiversity has been exceedingly slow, requiring 20 to 100 million years. Figure 27.2a reviews the pattern of five major extinctions and recoveries.

The pattern is just a composite of what happened to the major taxa. Lineages, remember, differ in their time of origin, the extent to which the member species diverged, and how long they endured. If you consider ongoing survival and reproduction to be measures of success, then each became a loser or a winner when environmental conditions changed in drastic or novel ways. To appreciate this point, reflect on Figure 27.2b. It shows the changing ranges of biodiversity of some

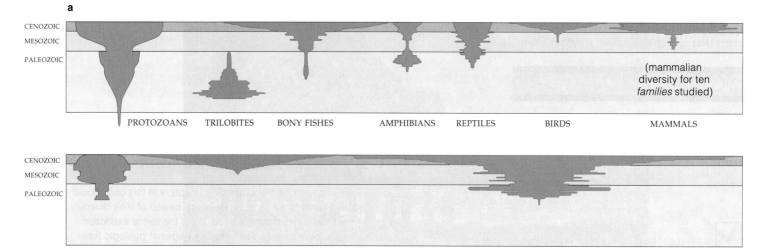

b

Figure 27.2 *Animated!* (**a**) Review of the five greatest mass extinctions and subsequent slow recoveries in the past. More mass extinctions occurred than are shown here. Compare Figure 17.13, in Section 17.5. (**b**) Within the framework of this generalized graph, patterns of extinction and recoveries differed significantly for different major taxa, as these selected examples indicate.

Figure 27.3 Two extinct and one threatened species. (**a**) Charles Knight's magnificent painting of *Tylosaurus*, one of the mosasaurs. This powerful marine lizard flourished in shallow, nearshore waters of the Cretaceous seaways. Indonesia's Komodo dragon, which grows 10 feet (3 meters) long, is a living relative. We think the Komodo dragon is a giant, yet imagine meeting up with a mosasaur. Some were 40 feet (12 meters) long.

(**b**) The dodo (*Raphus cucullatus*), a large, flightless bird that evolved on Mauritius, then vanished more than 300 years ago. Certain trees (*Calvaria major*) also evolved on that island, as did tortoises. Did the tree coevolve with dodos, tortoises, or both? Either may have fed on the tree's large fruits. For example, the large dodo gizzard could have partially digested the thick-walled coat of seeds in the fruits, just enough to help seeds germinate after leaving the gut. Either way, after the dodo became extinct, the seeds stopped germinating. Only thirteen trees were still standing by the mid-1970s. By some estimates they are more than 300 years old. Each year they produce seeds, but these apparently cannot break out of the seed coats without help. Botanists now employ turkeys or gem polishers as substitute grinders.

b

representative lineages. By combining many histories, we get the overall pattern of Figure 27.2*a*.

What causes mass extinctions? The answer is not always obvious. The K–T asteroid impact did deliver the coup de grace for numerous lineages, including dinosaurs and mosasaurs, yet many insect lineages sailed right on through it (Figure 27.3*a*). Besides, many dinosaur species had already vanished during the 10 million years that preceded that particular impact.

Maybe tectonic change was a contributing factor. Pangea started breaking up long before the impact, and deep, cold currents from the south had moved into warmer, equatorial seas. Many marine groups perished. Major changes in ocean circulation affected the atmosphere, the climate, and through them, life on land. The fossil and geologic records show that the K–T asteroid did abruptly eliminate some lineages. But they also show that circumstances were bringing about mass extinctions long before the asteroid hit.

Now extend the premise that mass extinctions may have complex causes to individual species. Long ago Dutch sailors clubbed to death every last dodo. This flightless bird lived only on Mauritius. When there were no more dodos, a species of tree native to the island stopped reproducing. It was not until the 1970s that a hypothesis emerged: If the tree had depended intimately on a coevolved species that became extinct, then the tree would be vulnerable to extinction, also. Was its partner the dodo? Maybe (Figure 27.3*b*).

Biodiversity is greater than it has ever been in the past. Its current range is largely an outcome of an overall pattern of mass extinctions and slow recoveries in the history of life. Within the pattern, lineages differ in which member species persisted or became extinct.

The loss of individual species, as well as mass extinctions, may have obvious or complicated causes.

27.2 The Expanding Human Habitats

LINKS TO
SECTIONS
7.1, 23.11

Like other great cities of the world, San Francisco has a vibrant, culturally diverse heritage. It is an exciting place to live, and many of its inhabitants marvel over the beauty of the bay, the hills, the forests around their homes. Now look at the maps in Figure 27.4. The human population is becoming more and more dense in this region, at the expense of natural habitats. When, if ever, will it stop? Conversely, who has the right to keep more people out?

Sometimes change is abrupt, and right before our eyes, but most of the time it happens almost imperceptibly in places far removed from home. For instance, take another look at the world map on page 1 of this book. Think of those points of light as markers where human habitats have replaced natural habitats.

To gain perspective on the impact we are having, think about something we take for granted—the air around us. The composition of Earth's atmosphere is a result of geologic and metabolic events—most notably, photosynthesis—that began billions of years ago. The first humans evolved by 2.4 million years ago. Like us, they breathed oxygen from an atmosphere of ancient origins. Like us, they were protected from the sun's ultraviolet radiation by the ozone layer (Section 7.1). Population sizes were not much to speak of, and their effect on biodiversity was trivial.

About 11,000 years ago, however, agriculture began in earnest. As you will read in Section 45.7, it was the start of increases in population size. Briefly, starting a few centuries ago, industrial revolutions, mechanized agriculture, then advances in medicine and hygiene improved the human condition—and population size skyrocketed in a blip of evolutionary time.

Today, it takes the extraction of huge amounts of energy and other resources from natural habitats to keep places like San Francisco running. At the same time, people create huge amounts of wastes that have to be disposed of somewhere—and guess where that is.

The wastes become **pollutants**. In biology, the term refers to substances that accumulate to harmful levels in habitats because resident species have no adaptive mechanisms for neutralizing or disposing of them.

In a few developed countries, population growth has more or less stabilized. Resource consumption per individual has dropped a bit. But those individuals are already using a lot of resources and generating most of the pollution. In the developing countries in Central America, Africa, and elsewhere, populations, resource consumption, and levels of pollution are growing fast. Half of the individuals of those populations are already struggling to survive on less than three dollars' worth of resources per day (Sections 49.7 through 49.9).

1900   1940   1954   1962

It will take many decades, even centuries, to reverse some trends that are in motion. Scattered attempts will not be enough. It may well be that individuals of all nations will unite to reverse the trends only when they perceive that the dangers of not doing so outweigh personal benefits of ignoring them.

Does this seem pessimistic? Think of the exhaust fumes released to the air each time you travel in a car or bus. Think of oil refineries, food-processing plants, and paper mills that supply you with goods and also release chemical wastes into the nation's waterways. Think of the billions of impoverished individuals all around the world who are struggling simply to stay alive. Who changes behavior first?

As an individual, you might choose to cherish or brood about or ignore any aspect of nature. Whatever your choice, the bottom line is this: Interactions among species extend through the biosphere, and it may be high time to start mending some broken connections.

> *Human population growth has been skyrocketing since the mid-eighteenth century. Humans now have the population size, the technology, and the cultural inclination to use resources and alter the environment at astonishing rates.*

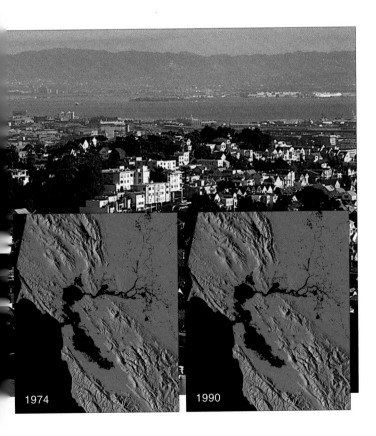

1974   1990

Figure 27.4 Tracking human population growth in one small part of the world, based on historical data and satellite imaging. *Red* denotes areas of dense human settlement in and around the San Francisco Bay area and Sacramento since 1900.

Can one individual make a difference? Rachel Carson never stopped to ask; she just went out and did so.

In 1951 Rachel Carson (Figure 27.5), an oceanographer and marine biologist, wrote a book about human impact on the ocean. *The Sea Around Us* was translated into thirty-two languages.

Seven years later, officials happened to spray mosquito-controlling DDT near the home and private bird sanctuary of one of her friends, who grew agitated over the subsequent agonizing deaths of a number of birds. That friend begged Carson to find someone who might study the effects of pesticides on birds and other wildlife.

As Carson soon discovered, independent, critical research on the subject was almost nonexistent. She conducted an extensive survey of the literature, then methodically developed information about the harmful effects of the widespread use of pesticides.

In 1962, she published her findings in book form. *Silent Spring*, the book's title, alluded to the silencing of "robins, catbirds, doves, jays, wrens, and scores of other bird voices" after exposure to pesticides.

Many scientists, politicians, and policy makers read *Silent Spring*, and the public embraced it. Manufacturers of chemicals viewed Carson as a threat to their booming pesticide sales, and they swiftly mounted a campaign to discredit her. Critics and industry scientists claimed that the book was full of inaccuracies, made selective and biased use of research findings, and failed to balance the picture with an account of the benefits of pesticides. Some even claimed that, as a woman, Carson simply was incapable of understanding such a highly technical subject. Others charged that she was an hysterical woman and a radical nature lover trying to scare the public in order to sell books.

During those intense personal attacks, Carson knew she had terminal cancer. Yet she strongly defended her research and successfully countered her critics. She died in 1964—eighteen months after publication of *Silent Spring*—without knowing that her efforts would be the impetus for what is now known as the environmental movement in the United States. The new field of conservation biology is one of the outgrowths of that movement.

Figure 27.5 Rachel Carson, who helped awaken public interest in human impact on nature. She died without knowing she helped start the environmental movement.

27.4 Threats to Biodiversity

LINKS TO CHAPTER 17 INTRODUCTION, SECTIONS 23.10, 26.10

No biodiversity-shattering asteroids have struck Earth for 65 million years. Yet the sixth major extinction event is under way. There is evidence that humans are driving many species to extinction.

ON THE NEWLY ENDANGERED SPECIES

An **endangered species** is any endemic species that is extremely vulnerable to extinction. *Endemic* means it originated in one region and is found nowhere else.

Some lineages are now more vulnerable than others (Figure 27.2). The mammalian lineage outlasted the dinosaurs, but humans have been hunting many kinds out of existence over the past 2 million years. Only 4,500 or so species have endured, and hundreds are endangered. We may be saying goodbye to elephants, rhinos, otters, most whales, tigers, and the red wolf, giant panda, gorilla, chimpanzee, and snow leopard.

The list cuts across major taxons. Who cares about the impending loss of the whooping crane, Philippine eagle, leatherback and loggerhead and Ridley turtles, polar bear, great white shark, coelocanth, the common sturgeon? Who cares about hundreds of invertebrate species that vanish each year? Does it matter? It will take more than one chapter to convey why it might. For now, know that communities of life can fall apart like a house of cards.

Example: James Estes and others at UC Santa Cruz traced a drastic decline in sea otters off the Aleutian Islands in Alaska to greater predation by killer whales.

Figure 27.6 Humans as competitor species—an example of dominance in the competition for living space in an arid habitat.

Killer whales prefer to eat Steller sea lions and harbor seals. Populations of these endangered species have declined recently, possibly through overharvesting of ocean perch and herring—which seals prefer to eat. Fisheries managers did not believe ocean perch and herring were overfished, but they did not know what was going on in other marine communities, including kelp forest habitats (Section 22.8). When killer whales caused sea otters to decline, populations of *sea urchins* —the preferred prey of sea otters—skyrocketed. The sea urchins, which eat kelp, effectively clear-cut kelp forests along the Alaskan coast. The loss may impact populations of fishes that use kelp forests as habitats, and on bald eagles and other coastal birds that prey on the fishes.

For now, start thinking about some practices that have had the most impact over the past forty years: conversion of natural habitats, species introductions, overharvesting, and illegal wildlife trading.

CAUSES AND EFFECTS

Habitat Loss and Fragmentation Edward Wilson has defined **habitat loss** as the physical reduction in suitable places to live, as well as the loss of suitable habitat as a result of pollution. The losses are putting huge pressure on more than 90 percent of the endemic species that now face extinction.

Biodiversity is greatest in the tropics, where lands are most threatened by deforestation (Section 23.10). Other examples: 98 percent of the tallgrass prairies, 50 percent of the wetlands, and as much as 95 percent of the United States old-growth forests have disappeared. Even arid lands are not exempt (Figure 27.6).

Habitats often become fragmented, or chopped into patches that each have a separate periphery. Species at any habitat's periphery are more exposed to wind, fire, temperature changes, predators, and pathogens. Also, fragmented patches often are not large enough to support the population sizes required for successful breeding. Patches may be too small to have sufficient food or other resources to sustain a species.

Together with Robert MacArthur, Wilson devised an **equilibrium model of island biogeography**. By this model, a 50 percent loss of a habitat will drive about 10 percent of its endemic species to extinction. A 90 percent loss will drive about 50 percent of the species in a habitat to extinction. The model is being used to estimate the number of current and future extinctions in "islands of shrinking habitat." Such islands need not be true islands; many are patches of natural habitat surrounded by degraded areas or by

Figure 27.7 A minke whale kill. Despite a worldwide ban, minke whales are still being harvested, even within the boundaries of marine sanctuaries.

encroaching suburban developments. Many national parks, tropical forests, reserves, and lakes are like this.

Some species help warn us of changes in habitats and impending loss of diversity. Birds are one of the key **indicator species**. Different types live in all major land regions and climate zones, they react quickly to changes, and they are fairly easy to track. Indicator species are used as a standard in identifying similar communities. Malcolm Ramsay at the University of Saskatchewan has been tracking populations of polar bears as an indicator species to assess the impact of global warming on marine mammals in the Arctic. The potential for significant increases in temperature over the next few decades is high, and the capacity of mammals to adapt rapidly to the resulting changes in their habitat may be very low.

Species Introductions Introductions of species are another threat to endemic ones. Each habitat has only so many resources, and all of its occupants compete for a share. Introduced species often escape predation and other factors that keep population sizes in check; endemic predators have not coevolved with them. For such reasons, they have been a key factor in almost 70 percent of cases where endemic species are being driven to extinction (Sections 46.9 and 46.10).

Overharvesting Species of commercial value, such as whales, are being overharvested. People kill whales for food, lubricating oil, fuel, cosmetics, fertilizer, even pet food, yet substitutes are available for all whale-derived products. A few nations ignore a moratorium

Figure 27.8 Confiscated tiger skins, leopard skins, rhinoceros horns, and elephant tusks. An illegal wildlife trade threatens the continued existence of more than 600 species, yet all but 10 percent of this illegal trade escapes detection. A few of the black market prices: Live, mature saguaro cactus ($5,000–15,000), bighorn sheep, head only ($10,000–60,000), polar bear ($6,000), grizzly ($5,000), bald eagle ($2,500), peregrine falcon ($10,000), live chimpanzee ($50,000), live mountain gorilla ($150,000), Bengal tiger hide ($100,000), and rhinoceros horn ($28,600 per kilogram).

on hunting large whales. Some whalers continue to poach within the boundaries of marine sanctuaries set aside for population recoveries (Figure 27.7).

Illegal Wildlife Trading In a sad commentary on human nature, the more rare a wild animal becomes, the more its value soars in the black market. Figure 27.8 gives an idea of the money that changes hands in illegal wildlife trading.

Human populations have contributed to accelerated rates of extinction as outcomes of hunting, habitat losses, species introductions, and overharvesting.

27.5 The Once and Future Reefs

LINKS TO
SECTIONS
1.1, 25.4

What does it mean when someone says that each species is part of a web of interactions? Here is a case in point.

Coral reefs are wave-resistant formations consisting of accumulated remains of marine organisms. They develop mainly in clear, warm waters between latitudes 25° north and 25° south (Figures 27.9 and 27.10). Hard coral parts have formed each reef's spine. Red algae such as *Corallina* contributed mineral-hardened cell walls to it. Secretions from other organisms helped cement things together.

The massive, pocketed spine is home to living corals and many other species. Australia's Great Barrier Reef supports 500 coral species, 3,000 fish species, 1,000 kinds of mollusks, and 40 kinds of sea snakes. Figure 27.10e hints at the wealth of warning colors, tentacles, and stealthy behavior—all signs of fierce competition for resources among species packed together in limited space.

Dinoflagellates, which are photosynthetic, often live as symbionts in tissues of reef-building corals (Section 25.4). The dinoflagellates find protection in the tissues. They provide the coral polyp with oxygen and recycle its mineral wastes. When stressed, coral polyps expel the symbionts. When stressed for more than a few months, they die; only bleached hard parts remain (Figure 27.11).

Abnormal, widespread bleaching in the Caribbean and tropical Pacific began in the 1980s. So did increases in sea surface temperature, which may be a key stress factor. Is the damage one outcome of global warming (Chapter 47)? If so, as marine biologists Lucy Bunkley–Williams and Ernest Williams suggest, the future looks grim for reefs. They may be devastated within three decades.

Also, people can directly destroy reefs, as by raw sewage discharges into nearshore waters of populated islands. Massive oil spills, as occurred in the Persian Gulf conflict

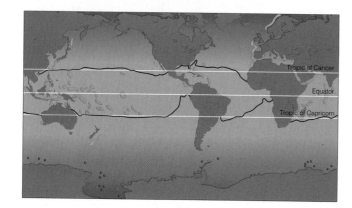

Figure 27.9 Distribution map for coral reefs (*orange*) and coral banks (*green*). Nearly all reef-building corals occur in regions with warm sea surface temperatures, here enclosed in dark lines. Beyond latitudes 25° north and south, solitary and colonial corals (*red*) form coral banks in temperate seas and in cold seas above the continental shelves.

Figure 27.11 Coral bleaching in 1998 at Australia's Great Barrier Reef, which parallels Queensland's coast for 2,500 kilometers. This is the largest example of biological architecture, but it is not one continuous reef. It is a string of thousands of reefs, some of which are 150 kilometers (95 miles) across.

Figure 27.10 Three types of coral reef formations. *Fringing* reefs form near land when rainfall and runoff are light, as on the downwind side of the most recently formed volcanic islands. Many reefs in the Hawaiian Islands and in Moorea (**a**) are like this. *Barrier* reefs form parallel with the shore of continents and volcanic islands, as in Bora Bora (**b**). Behind them are calm lagoons a few meters to sixty meters deep. Ring-shaped *atolls* are coral reefs and coral debris. They fully or partly enclose a shallow lagoon that a channel often links to the open ocean (**c**). Biodiversity is not great in shallow water, which can get too hot for corals to survive. (**d**) Coralline alga, one of the reef builders. (**e**) The facing page has a sampling of biodiversity from different coral reefs.

(Chapter 1), have calamitous effects. So do commercial dredging operations and mining for coral rock.

Where commercial fishermen from Japan, Indonesia, and Kenya move in, reef life is vanishing. No simple nets for these fellows; they drop dynamite in the water. Fish hiding in the coral are blasted out and float dead to the surface. Also, sodium cyanide squirted into the water stuns fish, which float to the surface. Some of the survivors end up as tropical fishes in pet stores. Endangered species are transported to exotic restaurants, where they are killed and served up as exorbitantly priced status symbols. On small, native-owned islands, fishing rights are traded for paltry sums. The fishermen destroy the reefs, which then no longer sustain the small human populations that have depended on them for survival.

Reef biodiversity is in danger around the world, from Australia and Southeast Asia to the Hawaiian Islands, Galápagos Islands, Gulf of Panama, Florida, and Kenya. To give a final example of this, the biodiversity on the coral reef off Florida's Key Largo has been reduced by 33 percent since 1970.

CORAL REEF IN THE RED SEA

LIONFISH

MORAY EEL

CHAMBERED NAUTILUS

SEA ANEMONE

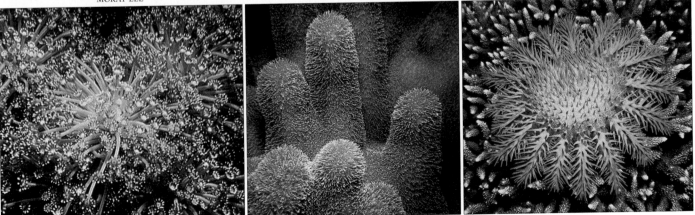

DAISY CORAL

PILLAR CORAL

CROWN-OF-THORNS SEA STAR

e

Chapter 27 Biodiversity in Perspective   471

27.6 Sustaining Biodiversity *and* Human Populations

LINKS TO
SECTIONS
21.5, 23.11

Increasingly, efforts to save endangered species are focusing on identifying and protecting the most threatened communities having the most diversity.

CONSERVATION BIOLOGY

Awareness of the current extinction rates gave rise to **conservation biology**. The goals for this field of pure and applied research are to (1) systematically survey the full range of biodiversity, (2) make sense of its evolutionary and ecological origins, and (3) identify ways to maintain and use it in ways that may benefit human populations. *The plan is to conserve and utilize, in sustainable ways, as much biodiversity as possible.*

Figure 27.12 is a chart of the approximate number of named species in some major groups. Our current taxonomic knowledge is fairly complete for numerous groups, including the flowering plants, conifers, and fishes, birds, mammals, and the other vertebrates. We tend to know more about large organisms, especially land dwellers, and about the showy ones, including birds and butterflies. Yet we have a lot of evidence that astonishing numbers of fungi, protists, bacteria, and archaeans are still out there, waiting to be identified. For example, biologists recently surveyed a hot spring in Yellowstone National Park. They found more kinds of archaeans than all those previously identified.

IDENTIFYING AREAS AT RISK

As noted earlier, the vast majority of existing species have not even been described. We do not know where most of them live. Conservation biologists are now focusing on identifying the **hot spots**, the habitats of a large number of species that are found nowhere else and that are in greatest danger of extinction.

At the first survey level, workers target a limited area, such as an isolated valley. A complete survey is impractical, so they make an inventory of flowering plants, birds, mammals, fishes, butterflies, and other indicator species in the habitat. At the next level up, broader areas that are major or multiple hot spots are explored systematically. The widely separated forests of Mexico are one example of a key multiple hot spot. Research stations have been set up at many different latitudes and at different elevations across the region.

At the highest survey level, hot spot inventories are combined with existing knowledge of biodiversity in ecoregions. An **ecoregion** is a broad land or ocean region defined by climate, geography, and producer species. Figure 27.13, a map of regions identified as essential reservoirs of biodiversity, includes 142 land, 53 freshwater, and 43 marine habitats. Of these, 25 are high-priority targets for conservation.

ECONOMIC FACTORS AND SUSTAINABLE DEVELOPMENT

Biological Wealth Every nation enjoys three forms of wealth—material, cultural, and biological wealth. Until recently, few countries comprehended the value of biological wealth—biodiversity—which can be the source of food, medicine, and other products. Much of the world's biological wealth has already been lost.

Again, by 2050, the size of the human population may reach 8.9 billion, with most increases occurring in the developing countries. Each person must have raw materials, energy, and living space. How many species will be crowded out? Ironically, the countries with the greatest monetary wealth also have the least biological wealth and use the most natural resources. Countries with the least monetary wealth and the most biological wealth have the fastest growing populations.

People who are locked in poverty often must choose between saving themselves or an endangered species. They hunt animals and dig up plants in "protected" parks and reserves. They try to raise crops and herds on marginal land. Conservation biologists attempt to identify ways for such people to earn a decent living from their biological wealth by using the biodiversity in threatened habitats in ways that sustain it.

Monteverde Cloud Forest There are some success stories. For instance, a nonprofit, Costa Rican research group owns the Monteverde Cloud Forest Reserve (Figure 27.14). It is one of the world's most popular

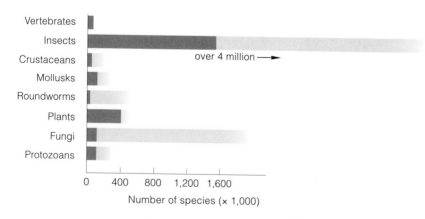

Figure 27.12 Current species diversity for a few major taxa. *Red* signifies the number of named species; *gold* signifies the estimated number of species yet to be discovered and named.

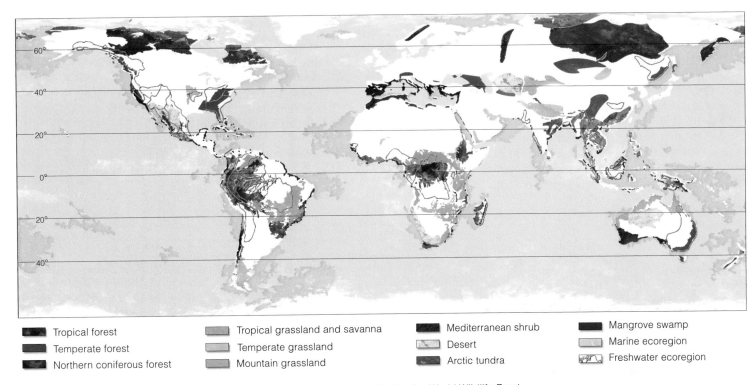

Tropical forest	Tropical grassland and savanna	Mediterranean shrub	Mangrove swamp
Temperate forest	Temperate grassland	Desert	Marine ecoregion
Northern coniferous forest	Mountain grassland	Arctic tundra	Freshwater ecoregion

Figure 27.13 Map of the most vulnerable regions of land and seas, compiled by the World Wildlife Fund.

ecotourism spots. Years ago, a graduate student from the University of California, Davis was doing research on birds in a part of Costa Rica where deforestation was under way. He got the idea to buy some land and set it aside as a bird sanctuary. His vision inspired individuals and conservation groups. One Swedish schoolboy thought up the Children's Rainforest. All over the world, children raised funds that eventually helped purchase 14,200 hectares for a sanctuary.

We find endangered species among the reserve's named plants and animals, including more than 100 mammalian species, 400 bird species, and 120 species of amphibians and reptiles. It is one of the few habitats left for the jaguar, ocelot, puma, and their relatives.

The conservation efforts have a ripple effect in the region. Ecotourism offers employment opportunities. Farmers are accepting the idea of sustainable harvests. Habitat for Humanity has built homes for low-income families. A nonprofit school inside the reserve helps educate the area's children.

Protection cannot stop at the reserve's boundaries. Extensive deforestation is causing exposed soils in the lowlands to heat up, which is modifying the way that clouds form in the region. In the absence of moisture from clouds, the forest becomes threatened.

Not all cloud forests are protected, and they are high on the list of endangered ecoregions. For a long time, people did not want to live in such perpetually damp places, but population pressures are driving

Figure 27.14 A view of the Monteverde Cloud Forest. Although such forests generally have less rainfall than tropical rain forests, they are more humid and have more cloud cover.

them into the mountains. They are clearing land for subsistance farming. The cleared land erodes, which drives the farmers higher into the forests.

Conservation biology entails a systematic survey of the full range of biodiversity, analysis of its evolutionary and ecological origins, and identification of methods to maintain and use biodiversity for the benefit of humans.

Preserving biodiversity requires identifying and protecting regions that support the highest levels of biodiversity.

Sustainable development involves utilizing biodiversity in ways that benefit humans without shredding the web of interactions that sustains a natural community.

27.7 Thinking Outside the Box

LINKS TO
SECTIONS
1.1, 23.10

We each have a big, complex brain. We can use it to find solutions to big, complex problems. Examples:

Strip Logging Tropical forests yield wood for local economies and exotic woods that are prized in developed countries. How to sustain them? Gary Hartshorn was the first to propose *strip logging* in sloped forests where there are a number of streams. As explained in Figure 27.15, logging can be sustained even while maintaining the maximum biodiversity possible.

Cattle and Riparian Zones Developed countries also benefit from sustaining biological wealth, as in a *riparian zone*: a narrow corridor of vegetation along a stream or a river. Plants are a line of defense against flood damage by sponging up water during spring runoffs and summer

storms. Shade cast by a canopy of taller shrubs and trees in a riparian zone helps conserve water during droughts. A riparian zone provides wildlife with food, shelter, and shade, particularly in arid and semiarid regions. In the western United States, 67 to 75 percent of the endemic species spend all or part of their life cycle in riparian zones. Among them are 136 kinds of songbirds, some of which will nest only in the plants of riparian zones.

Compared with wild ungulates, cattle drink a lot more water, so they tend to congregate at rivers and streams. There, they trample and feed until grasses and herbaceous shrubs are gone. It takes only a few head of cattle to destroy a riparian zone. All but 10 percent of the riparian vegetation of Arizona and New Mexico is already gone, mainly into the stomachs of grazing cattle.

In some places, however, cattle are now being restricted from riparian zones and given water away from streams. Livestock is being rotated and given supplemental feed in different grazing areas. Figure 27.16 shows one outcome.

Peace and Cooperation Nations in the Middle East are unfortunately renowned for being locked in long-term conflicts with one another. Chapter 1 referred to some of the disastrous consequences for the environment as well as for their human populations. *And yet.* In 1994, Jordan and Israel, in cooperation with the United States, founded the Red Sea Marine Peace Park in the Gulf of Aqaba. Some enlightened individuals from all three nations argued persuasively that spectacular coral reefs in the northern waters of the Gulf should be preserved (Figure 27.17).

The take-home lesson? There is much good in the world, and much to do. Be part of it.

uncut forest

cut 1 year ago

dirt road

cut 3–5 years ago

cut 6–10 years ago

uncut forest

stream in watershed

Figure 27.15 *Above,* one example of squandering biological wealth—clearing large swaths of tropical forests, with only short-term benefits.

Below, an example of sustaining biological wealth for the future—strip logging. The practice may protect biodiversity while it permits logging on tropical slopes. A narrow corridor paralleling the land's contours is cleared. A new roadbed is made at the top to haul away logs. After a few years, saplings grow in the cleared corridor. Another corridor is cleared above the roadbed. Nutrients leached from exposed soil trickle into the first corridor. There they are taken up by saplings, which benefit from all the nutrient input by growing faster. Later, a third corridor is cut above the second one—and so on in a profitable cycle of logging, which the habitat sustains over time.

Figure 27.16 Riparian zone restoration. This example comes from Arizona's San Pedro River, shown before and after restoration efforts.

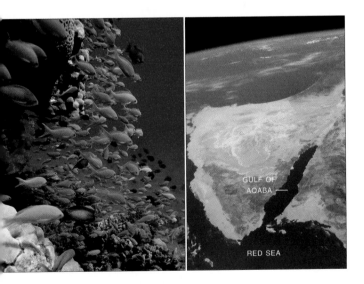

Figure 27.17 Gulf of Aqaba, first introduced in Section 1.1. Here, the Red Sea Marine Peace Park is being sustained through international cooperative efforts.

Summary

Section 27.1 Global biodiversity is greater now than ever, but the current rate of species losses is high enough to suggest that an extinction crisis is under way. After mass extinctions, it has taken 20 million to 100 million years for biodiversity to recover to the previous level.

Taxa have different histories involving, for instance, mass extinctions and adaptive radiations. Extinctions of individual species, as well as mass extinctions, may have obvious or hidden, complicated causes.

Biology⊘Now
Compare the patterns of extinction and adaptive radiations with the interaction on BiologyNow.

Sections 27.2, 27.3 The human population is growing rapidly and is expected to reach 8.9 billion by 2050. It is a key factor in the sixth major extinction event. It is encroaching more on natural habitats, using their resources, and giving back pollutants.

Sections 27.4, 27.5 Starting long ago, hunting, habitat losses and fragmentation, introduction of species into novel habitats, overharvesting, and illegal wildlife trading have threatened endemic species. An endemic species originated in and is limited to one geographic region. An endangered species is defined as an endemic species extremely vulnerable to extinction.

An example is the human impact on coral reefs, as when commercial fishermen dynamite them to harvest fish. Humans also may be affecting them indirectly, as by contributing to global warming (with concurrent rises in sea surface temperatures and sea levels).

Habitat loss refers to physical reduction or chemical pollution of suitable places for species to live. Habitat fragmentation is carving a habitat into isolated patches. It puts species at risk, as by splitting up populations to sizes that cannot promote successful breeding.

Island biogeography models help predict the number of current and future extinctions. A habitat island is a habitat in a sea of possibly destructive human activities, such as logging. Generally, destruction of 50 percent of an island habitat (or habitat island) will drive one-tenth of its species to extinction. Destruction of 90 percent will drive one-half to extinction.

Indicator species are birds and other easily tracked species that can provide warning of changes in habitats and impending widespread loss of biodiversity.

Sections 27.6, 27.7 Conservation biology is a field of pure and applied research. Its goal is to conserve and use biodiversity in sustainable ways through:

a. A systematic survey at three ever more inclusive levels of the full range of biodiversity.

b. Analysis of biodiversity's origins in evolutionary and ecological terms

c. Identification of methods that might maintain and use biodiversity for the benefit of the human population, which may otherwise destroy it.

Figure 27.18 One of the realities of the modern world. Would you prefer to be buried in a casket in the ground or mixed with concrete as part of a long-lasting foundation for an artificial reef?

The systematic survey of conservation biology is now proceeding at three levels:

a. Local hot spots (for example, an isolated valley) are identified and indicator species inventoried. Hot spots are habitats that have many species in great danger of extinction because of human activities.

b. Major hot spots or multiple ones are inventoried. Then research stations are set up across broader areas to gather data by latitude and elevation.

c. Data gathered at the first two levels are combined with data on ecoregions; these are the most vulnerable broad regions of land and seas throughout the world. Data gathered for the regional maps will be refined to give researchers an increasingly detailed picture of global biodiversity.

Protecting biodiversity depends on finding ways for people to make a living from it without destroying it. To counter growing economic demands, biodiversity's future economic value must be determined. Methods by which local economies can tap that biodiversity in sustainable ways are being developed. A tremendous amount of work remains to be done.

Biology⊛Now

Read the following InfoTrac articles:

"Can Cows and Conservation Mix?" Mari Jensen, Bioscience, *February 2001.*

"On the Wings of Hope," Don Boroughs, International Wildlife, *July–August 2000.*

Self-Quiz

Answers in Appendix II

1. Following mass extinctions, recovery to the same level of biodiversity has taken many _____ of years.
   a. hundreds   b. millions   c. billions

2. Humans are contributing to the _____ major extinction event in world history.
   a. first   c. third   e. fifth
   b. second   d. fourth   f. sixth

3. _____ are factors in the current extinction event.
   a. habitat loss and habitat fragmentation
   b. overharvesting and illegal wildlife trading
   c. species introductions
   d. all of the above

4. An endangered species is _____ and is highly vulnerable to extinction.
   a. endemic to a region
   b. any endemic or introduced species
   c. a plant or an animal only
   d. all of the above

5. The goal(s) of conservation biology is (are) to _____ .
   a. conduct a three-level, systematic survey of all biodiversity
   b. analyze biodiversity's evolutionary and ecological origins
   c. identify ways to maintain and use biodiversity for people
   d. all of the above

6. Nations have _____ wealth.
   a. material   d. all of the above
   b. cultural   e. both a and b
   c. biological   f. both b and c

7. Strip logging _____ .
   a. can sustain forests   c. destroys wild habitats
   b. is profitable   d. both a and b

Additional questions are available on **Biology⊛Now™**

Critical Thinking

1. Visit or study a riparian zone near where you live. Imagine visiting it ten years from now. Given its location, what kinds of changes do you predict for it?

2. Mentally transport yourself to a tropical rain forest of South America. Imagine you don't have a job. There are no jobs available, not even in overcrowded cities some distance away. You have no money or contacts to get you anywhere else. Yet you are the sole supporter of a large family. Today a stranger approaches you. He tells you he will pay good money if you can capture alive a certain brilliantly feathered parrot in the forest. You know the parrot is seldom seen, but you have an idea of where it lives. What will you do?

3. Here's an alternative to formal burial on land. Ashes of cremated people are mixed with pH-neutral concrete and used to build *artificial reefs*. Each perforated concrete ball weighs 400 to 4,000 pounds and is designed to last 500 years. So far, 100,000 of these reefs are sustaining marine life in 1,500 locations (Figure 27.18). Proponents say that the burial costs less, does not waste land, and is better for the environment. Your thoughts?

4. Ecologist Robin Tyser once told us that students might find the current biodiversity crisis too overwhelming to contemplate. But people can make a difference. Many helped reestablish bald eagles in the continental United States. Others have reestablished wolves in northern Wisconsin. Daniel Janzen is working to recreate a dry forest ecosystem in Costa Rica. Do a library search or computer research and report on one or two success stories you find inspiring.

Principles of Anatomy and Physiology

Two bottlenose dolphins leaping from the sea, to which they are most exquisitely adapted, into air—where they cannot survive for long. This transient bridging of two very different environments invites you to ask: Are there fundamental constraints on how any organism is put together and how it functions, regardless of where it lives? Can we identify patterns in the diverse responses to recurring challenges?

Too Hot To Handle

When was the last time the word **homeostasis** slipped into one of your conversations? Years ago? Or maybe never? So why should it? Homeostasis is a state in which the body's internal environment is being kept within a range that its cells can tolerate. In humans, that happens when activities of trillions of living cells of many organ systems are being integrated. To understand "homeostasis," you have to know the details of how complex, multicelled organisms are put together and how their parts function.

Sometimes terrible things can happen to the body in the absence of that understanding. In the summer of 2001 Korey Stringer, a football player for the Minnesota Vikings, collapsed after morning practice (Figure 28.1). On that day, his team had been working out in full uniform on a field where temperatures were in the high 90s. High humidity put the heat index above 100°F.

Stringer's internal body temperature had soared to 108.8°F, and his blood pressure was too low even to record.

He was rushed to the hospital, where doctors immersed him in an icewater bath, then wrapped him in cold, wet towels. It was too late.

Stringer's blood clotting mechanism shut down and he started to bleed internally. His kidneys faltered and he was placed on dialysis. He stopped breathing on his own and was attached to a respirator. However, his heart gave out. Less than twenty-four hours after football practice had started, Stringer was pronounced dead. He was twenty-seven years old.

All organisms function best within a limited range of internal operating conditions. For humans, "best" is when the body's internal temperature remains between about 97°F and 100°F. Past 104°F, metabolism is in an uproar, and blood is transporting a great deal of metabolic heat. Controls over blood transport divert heat from the brain and other internal organs to the skin. Then skin transfers heat to the outside environment—as long as it is not too hot outside. Profuse sweating can dissipate more heat, but not on hot, humid days.

These normal cooling mechanisms fail above 105°F. The body cannot sweat as much, and its temperature starts to climb rapidly. The heart beats faster and the individual becomes confused or faints. These are signs of *heat stroke*. The high internal temperature that causes it can affect the enzymes and other proteins that keep us alive. When heat stroke is not countered fast enough, brain damage or death is the expected outcome.

We use this sobering example as our passport to the world of anatomy and physiology. *Anatomy* is the study

Figure 28.1 Temperature control mechanisms that have worked for millions of years. Water helps dissipate excess body heat when it can move out through pores of glands at the skin surface (*facing page*). Also, blood circulating through the body reaches fine blood capillaries in skin. When the skin's tissues are not as warm as the blood circulating past, heat is transferred into them, then into the air—if the air is cooler still. In Korey Stringer's case, cooling could not happen fast enough.

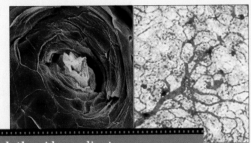

of body form—that is, its morphology. *Physiology* is the study of patterns and processes by which an individual survives and reproduces in the environment. It deals with how the body's parts are put to use and how metabolism and behaviors are adjusted when conditions change. It also deals with how, and to what extent, physiological processes can be controlled.

On a personal level, this information can help you monitor what is going on in your own body. More broadly, it also can help you identify commonalities among animals and plants. Regardless of the species, *the structure of any given body part almost always has something to do with a present or past function.* Most aspects of form and function are long-standing adaptations that evolved as responses to environmental challenges.

That said, nothing in the evolutionary history of the human species suggests that the organ systems making up our body are fine-tuned to handle intense football practice on hot, humid days.

 How Would You Vote?

Heat sickness can affect a person's ability to think clearly. Should a coach who does not stop team practice or a game when the heat index soars be held responsible if a player dies from heat stroke? See BiologyNow for details, then vote online.

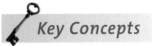 Key Concepts

MANY LEVELS OF STRUCTURE AND FUNCTION

Anatomy is the study of body form at successive levels of structural organization, from molecules on up through organ systems. Physiology is the study of how the body functions in response to the environment. The structure of most body parts correlates with current or past functions, and it emerges during stages of growth and development. Section 28.1

SIMILARITIES BETWEEN ANIMALS AND PLANTS

Animals and plants are alike in their requirements for gas exchange, internal transport, maintaining the volume and composition of their internal environment, and cell-to-cell communication. They show variations in their response to resources and threats. Section 28.2

HOMEOSTASIS

Extracellular fluid bathes all living cells in the multicelled body. Cells, tissues, organs, and organ systems contribute to maintaining this internal environment within a range that individual cells can tolerate. This concept yields insight into the functions and interactions of body parts.

Homeostasis is the name for stable operating conditions in the internal environment. Negative and positive feedback mechanisms are among the controls that work to maintain these conditions. Sections 28.3, 28.4

CELL COMMUNICATION IN MULTICELLED BODIES

Cells of tissues and organs communicate with one another by secreting hormones and other signaling molecules into extracellular fluid, and by selectively responding to signals from other cells. Section 28.5

 Links to Earlier Concepts

Look back on the road map through the levels of biological organization in Section 1.1. You are about to see how new properties of life emerge through interactions among the tissues, organs, and organ systems of plants and animals.

You will draw on your understanding of the constraints on multicelled body plans—the surface-to-volume ratio (4.1, 25.1), diffusion and demands for gas exchange (5.3), and internal transport (5.4, 23.2, 25.2). You will find examples of how signaling molecules help control growth, development, day-to-day activities, and reproduction (5.2, 22.12). You also will consider specific cases of evolutionary adaptation (8.9).

28.1 Levels of Structural Organization

LINKS TO
SECTIONS
1.1, 4.9, 23.2, 25.1

We introduce important concepts in this chapter. They have broad application across the next two units of the book. Become familiar with them; they can deepen your sense of how plants and animals function under environmental conditions, both favorable and stressful.

FROM CELLS TO MULTICELLED ORGANISMS

Most plants and animals have cells, tissues, organs, and organ systems that split up the task of survival. A separate cell lineage gives rise to body parts that will function in reproduction. Said another way, the plant or animal body shows a division of labor.

A **tissue** is a community of cells and intercellular substances that are interacting in one or more tasks. For example, wood and bone are tissues that function in structural support. Each **organ** is a structural unit of at least two tissues, organized in certain proportions and patterns, that carries out one or more common tasks. A leaf adapted for photosynthesis and an eye responsive to light in the surroundings are examples. An **organ system** has two or more organs interacting physically, chemically, or both in the performance of one or more tasks. The organs of photosynthesis and reproduction of a flowering plant are like this (Figure 28.2). So is an animal's digestive system, which takes in food, breaks it up into bits of nutrients, absorbs the bits, and expels the unabsorbed leftovers.

GROWTH VERSUS DEVELOPMENT

A plant or animal becomes structurally organized as it grows and develops. For any multicelled species, **growth** refers to an increase in the number, size, and volume of cells. **Development** is a series of stages in which specialized tissues, organs, and organ systems form. That is why we measure growth in *quantitative* terms and development in *qualitative* terms.

STRUCTURAL ORGANIZATION HAS A HISTORY

The structural organization of each tissue, organ, and organ system has an evolutionary history. Remember how plants invaded land? Section 23.2 already gave you an idea of how their structure relates to function. The pioneers encountered an abundance of sunlight and carbon dioxide for photosynthesis, which meant more oxygen for aerobic respiration and a foundation for increases in size. However, in leaving the aquatic cradle, they faced a new challenge—how to keep from drying out in air.

Think about that challenge when micrographs show the internal structure of plant roots, stems, and leaves (Figure 28.2). Pipelines conduct streams of water from soil to leaves. Stomata, the small gaps across a leaf's epidermis, open and close in ways that conserve water (Section 30.4). Collectively, lignin-reinforced cell walls support the upright growth of stems. Remember that same challenge when you come across examples of

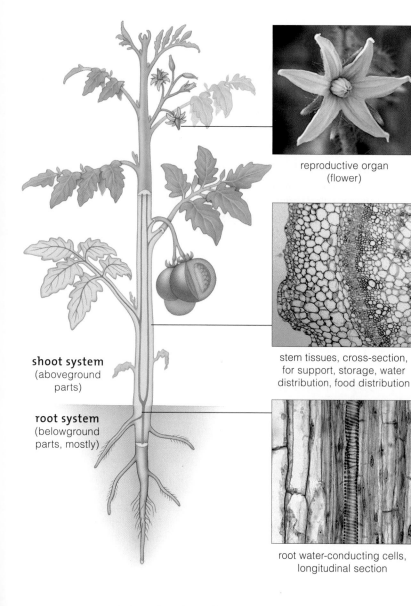

reproductive organ
(flower)

stem tissues, cross-section,
for support, storage, water
distribution, food distribution

shoot system
(aboveground
parts)

root system
(belowground
parts, mostly)

root water-conducting cells,
longitudinal section

Figure 28.2 *Animated!* Morphology of a tomato plant (*Solanum lycopersicon*). Different cell types make up vascular tissues that conduct water, dissolved mineral ions, and organic compounds. These tissues thread through others that make up most of the plant body. Another tissue covers all surfaces exposed to the surroundings.

Figure 28.3 Some of the structures that function in human respiration. Their cells carry out specialized tasks. Airways to a pair of organs called lungs are lined with ciliated cells that whisk away bacteria and other airborne particles that might cause infections. Inside the lungs are tubes (blood capillaries), a fluid connective tissue we call blood, and thin air sacs (epithelial tissue). All of these components indirectly or directly facilitate the flow of oxygen into the internal environment and the flow of carbon dioxide out of it.

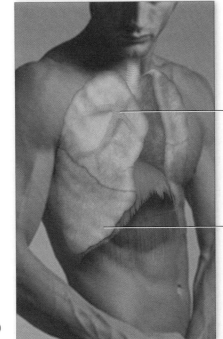

organs (lungs), part of an organ system (the respiratory tract) of a whole organism

ciliated cells and mucus-secreting cells that line respiratory airways

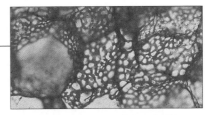

lung tissue (tiny air sacs) laced with blood capillaries—one-cell-thick tubular structures that hold blood, which is a fluid connective tissue

root systems. Any patch of soil with plenty of water and dissolved mineral ions stimulates root growth in its direction; hence the branching roots.

Similarly, respiratory systems of land animals are adaptations to life in air. Gases only move into and out of their body by diffusing across a moist surface. This is not a problem for aquatic organisms, but moist surfaces dry out in air. The animals on dry land have moist sacs for gas exchange *inside* their body (Figure 28.3), which brings us to the internal environment.

THE BODY'S INTERNAL ENVIRONMENT

Plant and animal cells must be bathed in a fluid that delivers nutrients and carries away metabolic wastes. In this they are no different from free-living single cells. But each plant or animal has thousands to many trillions of living cells. All must draw nutrients from the fluid bathing them and release wastes into it.

Body fluid *not* inside cells—extracellular fluid—is an **internal environment**. Changes in its composition and volume affect cell activities. The type and number of ions are vital; they must be kept at concentrations compatible with metabolism. It makes no difference whether the plant or animal is simple or complex. *It requires a stable fluid environment for all of its living cells.* This concept is central to understanding how plants and animals work.

START THINKING "HOMEOSTASIS"

The next two units describe how a plant or an animal carries out these basic functions: The body works to maintain favorable conditions for all of its living cells. It acquires water, nutrients, and other raw materials, distributes them, and disposes of wastes. It passively and actively defends itself against attack. It has the capacity to reproduce. Parts of it nourish and protect gametes, and, in most species, the embryonic stages of the next generation.

Each living cell engages in metabolic activities that will favor its own survival. Collectively, however, the activities of cells in tissues, organs, and organ systems sustain the body as a whole. Their interactions keep the operating conditions of the internal environment within tolerable limits—*the state we call homeostasis.*

The structural organization of plants and animals emerges during stages of growth and development.

Cells, tissues, and organs require a favorable internal environment, and they collectively maintain it. All body fluids not contained in cells make up that environment.

Acquiring materials and distributing them to cells, getting rid of wastes, protecting cells and tissues, reproducing, and often nurturing offspring are basic body functions.

28.2 Recurring Challenges to Survival

LINKS TO
SECTIONS
4.1, 5.3, 5.4

Plants and animals have such diverse body plans that we sometimes forget how much they have in common.

GAS EXCHANGE IN LARGE BODIES

How often do you think of similarities between, say, Tina Turner and a tulip (Figure 28.4)? Connections are there. Cells in both would die if oxygen and carbon dioxide stopped diffusing across their body surface. Tina, a heterotroph, supplies her aerobically respiring cells with oxygen and disposes of the carbon dioxide products. The plant, an autotroph, exchanges oxygen for carbon dioxide, but it keeps some for its own cells, which also engage in aerobic respiration.

All multicelled species respond, structurally and functionally, to this same challenge. *They must quickly move gaseous molecules to and from individual cells.*

Remember **diffusion**? When they are concentrated in one place, ions or molecules of any substance tend to move to a place where they are not as concentrated. Animals and plants keep gases diffusing in directions most suitable for metabolism and cell survival. How? That question will lead you to stomata at leaf surfaces (Chapter 30) and to the circulatory and respiratory systems of animals (Chapters 38 and 40).

INTERNAL TRANSPORT IN LARGE BODIES

Metabolic reactions happen fast. If reactants take too long to diffuse through the body or to and from the surface, systems shut down. This is one reason cells and multicelled species have the sizes and shapes that they do. As they grow, their volume increases in three dimensions—in length, width, and depth—but their surface area increases in two dimensions only. That is the essence of the **surface-to-volume ratio** (Section 4.1). If the body were to develop into a densely solid mass, it would not have enough surface area for fast, efficient exchanges with the environment.

When a body or some body part is thin, as it is for the flatworm and lily pads in Figure 28.5, substances easily diffuse between individual cells and the outer environment. In massive bodies, individual cells that are far from an exchange point with the environment depend on systems of rapid internal transport.

Most plants and animals have vascular tissues, or systems of tubes through which substances move to and from all living cells. In land plants, xylem moves soil water and mineral ions. Phloem moves sucrose to cells from leaves. Each leaf vein has long strands of xylem and phloem (Figure 28.5c).

In big animals, the vascular tissues extend from a surface facing the environment to living cells inside. Each time Tina belts out a song, she is moving oxygen into her lungs and carbon dioxide out of them. Blood vessels thread through lung tissue where the gases are exchanged. Large vessels transport oxygen and branch into small capillaries in all tissues—where interstitial fluid and cells exchange gases (Figure 28.5d).

In both plants and animals, the vascular system also transports diverse substances, such as nutrients, water, and hormones. In animals, it moves infection-fighting white blood cells and chemical weapons about. As you will see, phloem distributes defensive chemicals that are produced in response to infection or wounds.

MAINTAINING THE WATER–SOLUTE BALANCE

Plants and animals continually gain and lose water and solutes, and produce metabolic wastes. Given all the inputs and outputs, how does the volume and composition of their internal environment stay within a tolerable range? Plants and animals differ hugely in this respect. Yet we still can find common responses by zooming down to the level of molecular motions.

Substances tend to follow concentration gradients when moving into and out of the body or from one body compartment into another. At such interfaces, the individual cells of sheetlike tissues passively and actively transport substances. **Active transport**, recall, pumps substances against the direction in which their concentration gradient would take them (Section 5.4).

Active transport mechanisms in roots help control which solutes can move into the plant. In leaves, they help control water loss and gas exchange by closing

Figure 28.4 What do these organisms have in common besides their good looks?

Figure 28.5 (**a**) Flatworm gliding through water, (**b**) waterlily leaves floating on water, (**c**) veins in a decaying dicot leaf, (**d**) human veins and capillaries. Which constraint has influenced these body plans and structures?

and opening stomata at different times. For animals, such mechanisms occur in kidneys and other organs. As you will see again and again in the next two units, *active and passive transport help maintain the internal environment and metabolism itself by adjusting the kinds, amounts, and directional movements of substances.*

CELL-TO-CELL COMMUNICATION

Plants and animals show another major similarity in their structure and function. A number of their cells release signaling molecules that coordinate and also control events inside the body as a whole. Different signaling mechanisms guide how the plant or animal body grows, develops, and maintains itself. Section 28.5 has a good example of this.

ON VARIATIONS IN RESOURCES AND THREATS

A **habitat** is the place where individuals of a species normally live. Each has different resources and poses a unique set of challenges. What are its physical and chemical characteristics? Is water plentiful, with the right kinds and amounts of solutes? Is the habitat rich or poor in nutrients? Is it sunlit, shady, or dark? Is it warm or cool, hot or icy, windy or calm? How much does the outside temperature vary from day to night? How much do conditions vary with the seasons?

And what about biotic (living) components of the habitat? Which producers, predators, prey, pathogens, or parasites live there? Is competition for resources and reproductive partners fierce? Such ever changing variables promote diversity in form and function.

Even with all the diversity, we may still see similar responses to similar challenges. Sharp cactus spines or porcupine quills deter most animals that might eat a cactus or porcupine (Figure 28.6). Modified epidermal cells give rise to both spines and quills. Deliveries from vascular tissues and other body parts kept those cells alive. In return, by forming spines or quills and other

Figure 28.6 Protecting body tissues from predation: (**a**) Cactus spines. (**b**) Quills of a porcupine (*Erethizon dorsatum*).

defensive structures at the body's surface, epidermal cells help protect the vascular tissues and other body parts against outside threats.

Plant and animal cells function in ways that help ensure survival of the body as a whole. At the same time, tissues and organs that make up the body function in ways that allow the continued survival of individual living cells.

The connection between each cell and the body as a whole is evident in the requirements for—and contributions to—gas exchange, nutrition, internal transport, stability in the internal environment, and defense.

28.3 Homeostasis in Animals

LINKS TO
SECTIONS
1.2, 6.4

In preparation for your trek through the next two units, focus again on what homeostasis means to survival.

Like other adult humans, your body consists of more than 65 trillion living cells. Each must draw nutrients from and dump wastes into the same fifteen liters of fluid—less than four gallons. Fluid not inside cells is *extracellular* fluid. Much is **interstitial fluid**, meaning it fills the spaces between cells and tissues. The rest is **plasma**, the fluid portion of blood. Interstitial fluid exchanges substances with cells and with blood.

Homeostasis, again, is a state in which the internal environment is being maintained within a range that cells can tolerate. Sensory receptors, integrators, and effectors are in charge of it. Collectively, they detect, process, and respond to information on *how things are* compared to preset points of *how they should be*.

Sensory receptors are cells or cell parts that detect stimuli, which are specific forms of energy. A kiss, for example, is a form of mechanical energy that changes pressure on the lips. Receptors in lip tissues translate a kiss into signals that reach the brain. The brain is an **integrator**, a central command post that receives and processes information about stimuli. It issues signals to **effectors**—muscles, glands, or both—that carry out suitable responses to the stimulation (Figure 28.7).

NEGATIVE FEEDBACK

Feedback mechanisms help control what goes on in cells (Section 6.4). They also are major homeostatic controls over how the multicelled body functions. By **negative feedback mechanisms**, some activity changes a specific condition in the internal environment, and if the condition changes past a certain point, a response reverses the change.

Think of a furnace with a thermostat. A thermostat senses the surrounding air temperature relative to a preset point on a thermometer built into the furnace's control system. When the temperature falls below the

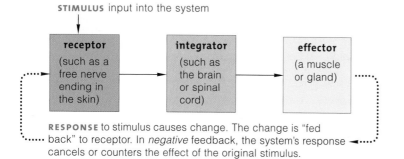

STIMULUS input into the system

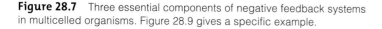

RESPONSE to stimulus causes change. The change is "fed back" to receptor. In *negative* feedback, the system's response cancels or counters the effect of the original stimulus.

Figure 28.7 Three essential components of negative feedback systems in multicelled organisms. Figure 28.9 gives a specific example.

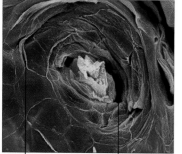

dead, flattened skin cells   sweat gland pore

Figure 28.8 *Animated!* Major homeostatic controls over a human body's internal temperature. *Solid* arrows signify the main control pathways. *Dotted* arrows signify the feedback loop.

The scanning electron micrograph shows a sweat gland pore at the skin surface. Such glands are among the effectors for this control pathway.

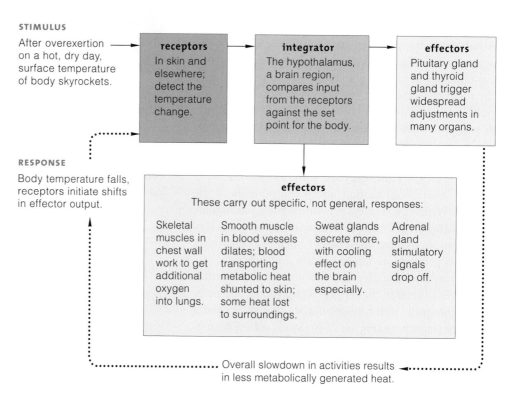

STIMULUS

After overexertion on a hot, dry day, surface temperature of body skyrockets.

receptors
In skin and elsewhere; detect the temperature change.

integrator
The hypothalamus, a brain region, compares input from the receptors against the set point for the body.

effectors
Pituitary gland and thyroid gland trigger widespread adjustments in many organs.

RESPONSE
Body temperature falls, receptors initiate shifts in effector output.

effectors
These carry out specific, not general, responses:

Skeletal muscles in chest wall work to get additional oxygen into lungs.

Smooth muscle in blood vessels dilates; blood transporting metabolic heat shunted to skin; some heat lost to surroundings.

Sweat glands secrete more, with cooling effect on the brain especially.

Adrenal gland stimulatory signals drop off.

Overall slowdown in activities results in less metabolically generated heat.

preset point, it signals a switching mechanism, which turns on the furnace. When the air gets hot enough to match the preset level, the thermostat senses that match and signals the switch to shut off the furnace.

Similarly, feedback mechanisms help keep the internal body temperature of humans and many other mammals near 98.6°F (37°C) even during hot or cold weather. When someone runs on a hot summer day, the body becomes hot. Receptors trigger changes that slow down the entire body *and* its cells (Figure 28.8). Normally, such controls counter overheating. They curb activities that naturally generate metabolic heat, and give up excess heat to the surrounding air.

As another example, think about how we breathe to take in oxygen and get rid of carbon dioxide. Most of us live at low elevations, where we can get enough oxygen. The brain is responsive to receptors that are sensitive to how much carbon dioxide is dissolved in blood. It compares receptor signals against a set point for what the carbon dioxide level is supposed to be. It calls for adjustments in breathing and other activities that can shift the concentration of carbon dioxide in blood so that it is more in line with the set point.

Above 3,300 meters (10,000 feet), oxygen is scarce (Figure 28.9). Its level in blood declines. Receptors sensitive to the oxygen level in blood signal the brain, which responds by making us hyperventilate (breathe faster and deeper than usual). If the level remains too low, shortness of breath, heart palpitations, headaches, nausea, and vomiting may follow. These are warnings that cells are screaming for oxygen.

POSITIVE FEEDBACK

In certain situations, **positive feedback mechanisms** operate. These controls initiate a chain of events that *intensify* change from an original condition, and after a limited time, the intensification reverses the change. Positive feedback mechanisms are usually associated with instability in a system.

For instance, rapid flows of sodium ions across the plasma membrane of neurons help the body send and receive signals quickly. A suitable signal opens a few sodium channels across a neuron's membrane. Ions flow into the neuron, and the increased concentration inside makes more channels open, then more, until sodium ions flood inside. Such ion flows start the "messages" that travel along a neuron's surface.

As another example, during labor, the fetus exerts pressure on the wall of the chamber enclosing it, the uterus. Pressure induces the production and secretion of a hormone, oxytocin, that makes muscle cells in the

Figure 28.9 A climber near the summit of Mount Everest, where the normal act of breathing is not enough to sustain the human body.

wall contract. Contractions exert pressure on the fetus, which puts pressure on the wall to expand, and so on until the fetus is expelled from the mother's body.

We have been introducing a pattern of detecting, evaluating, and responding to the flow of information about the internal and external environments. During this activity, organ systems work together. Soon, you will be asking these questions about their operation:

1. Which physical or chemical aspects of the internal environment are organ systems working to maintain?

2. How are organ systems kept informed of change?

3. By what means do they process the information?

4. What mechanisms are set in motion in response?

As you will read in Unit Six, organ systems of nearly all animals are under neural and endocrine control.

The internal environment is all the fluid not inside the body's cells. It consists of interstitial fluid and plasma, the fluid portion of blood.

Homeostatic controls, such as negative feedback and positive feedback mechanisms, help maintain physical and chemical aspects of the body's internal environment within ranges that its individual cells can tolerate.

28.4 Does Homeostasis Occur in Plants?

LINKS TO
SECTIONS
23.7, 24.6

Plants differ from animals in important respects. Even so, they, too, have controls over their internal environment.

Direct comparisons between plants and animals are not always possible. In all young plants, new tissues arise at the tips of actively growing roots and shoots. In animal embryos, tissues form all through the body. Plants do not have a centralized integrator that serves as the equivalent of an animal brain. They do have some decentralized mechanisms that can protect the internal environment and work to keep the body as a whole functioning. Later chapters explain how, but two simple examples here will make the point.

WALLING OFF THREATS

Unlike people, trees consist mostly of dead and dying cells. Also unlike people, trees cannot run away from attacks. When a pathogen infiltrates their tissues, trees cannot unleash infection-fighting phagocytic cells in response, because they have none. However, plants do have **system acquired resistance** to infections and to injured tissues. This mechanism starts with signaling molecules that cells release in an affected tissue. These particular signals induce the synthesis of defensive compounds. Signals also diffuse to undamaged tissues farther away. They induce cells to produce and release compounds that will protect tissues from attack for days or months to come. Synthetic versions of organic compounds with roles in systemic defenses are now being developed and used to boost disease resistance in crops and ornamental plants.

Also, most trees protect the internal environment by walling off wounds, unleashing phenols and other toxic compounds, and often secreting resins. A heavy flow of gooey compounds saturates and protects bark and wood at an attack site. It also can seep into soil around roots. Some toxins are so potent that they also kill cells of the tree itself. As a result, compartments form around injured, infected, or poisoned sites, and new tissues grow right over them. This plant response to attack is called **compartmentalization**.

Drill holes into a tree species that makes a strong compartmentalization response and it quickly walls off that wound (Figure 28.10). In a species that makes a moderate response, decomposers that cause wood to decay expand farther, into tissues more distant from the holes. Drill holes into a weak compartmentalizer, and decomposers will cause massive decay.

Even strong compartmentalizers live only so long. After they form too many walls, they shut off the flow of water and solutes to their living cells. You may ask, What about a bristlecone pine? One was almost 5,000 years old. Such trees live in habitats that are too harsh or remote to favor the survival of most pathogens, so they are not attacked much. They spend most of the year in snow and the rest of it growing slowly under intense radiation from the sun (Section 23.7).

SAND, WIND, AND THE YELLOW BUSH LUPINE

Anybody who has tiptoed barefoot across sand near the coast on a hot, dry day has a tangible clue to why few plants grow in it. One exception is the yellow bush lupine, *Lupinus arboreus* (Figure 28.11).

L. arboreus is native to warm, dry areas of Central and Southern California. It is a hardy colonizer of soil exposed by fires or abandoned after being cleared for agriculture. Like all other legumes, this species has nitrogen-fixing symbionts in its young roots (Section 24.6). The interaction gives it a competitive edge in

strong

moderate

weak

Figure 28.10 *Animated!* You think it's easy being a tree? Drilling patterns for an experiment to test compartmentalization responses. From top to bottom, the decay patterns (*green*) in stems of three species of trees that make strong, moderate, or weak compartmentalizing responses.

Figure 28.11 Yellow bush lupine, *Lupinus arboreus*, in a sandy shore habitat. On hot, windy days, its leaflets fold up longitudinally along the crease that runs down their center. This helps minimize evaporative water loss.

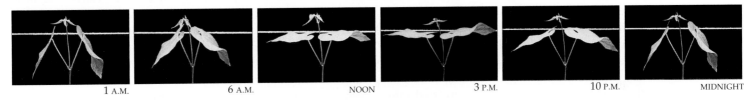

| 1 A.M. | 6 A.M. | NOON | 3 P.M. | 10 P.M. | MIDNIGHT |

Figure 28.12 *Animated!* Observational test of rhythmic leaf movements by a young bean plant (*Phaseolus*). Physiologist Frank Salisbury kept the plant in darkness for twenty-four hours. Its leaves kept on folding and unfolding independently of sunrise (6 A.M.) and sunset (6 P.M.).

nitrogen-deficient soil. In the early 1900s, this species was planted in northern California to stabilize coastal dunes. Unfortunately, it is too successful in its new habitat. It outgrows and displaces native plants.

One big environmental challenge near the beach is the lack of fresh water. Leaves of a yellow bush lupine are structurally adapted for water conservation. Each leaf has a surprisingly thin cuticle, but a dense array of fine epidermal hairs projects above it, particularly on lower leaf surfaces. Collectively, all of the hairs trap moisture escaping from stomata. The trapped, moist air slows evaporation and helps maintain water levels inside the leaf at levels that best favor metabolism.

These leaves make homeostatic responses to the environment. They fold along their length, like two parts of a clam shell, and resist the moisture-sucking force of the wind. Each folded, hairy leaf is better at stopping moisture loss from stomata (Figure 28.11).

Leaf folding by *L. arboreus* is a controlled response to changing conditions. When winds are strong and the potential for water loss is greatest, the leaves fold tightly. The least-folded leaves are close to the plant's center or on the side most sheltered from the wind. Folding is a response to heat as well as to wind. When air temperature is highest during the day, leaves fold at an angle that helps reflect the sun's rays from their surface. The response minimizes heat absorption.

ABOUT RHYTHMIC LEAF FOLDING

In case you think leaf folding couldn't possibly be a coordinated response, take a look at the bean plant in Figure 28.12. Like some other plants, it holds its leaves horizontally during the day but folds them closer to a stem at night. Keep the plant in full sun or darkness for a few days and it will continue to move its leaves in and out of the "sleep" position, independently of sunrise and sunset. The response might help reduce heat loss at night, when air cools, and so maintain the plant's internal temperature within tolerable limits.

Rhythmic leaf movements are just one example of a **circadian rhythm**, a biological activity repeated in cycles that each last for close to twenty-four hours. Circadian means "about a day." As you will see in a later chapter, a pigment molecule called phytochrome may be part of a control mechanism over leaf folding.

Control mechanisms that help maintain homeostasis are at work in plants, although they are not governed from central command posts as they are in most animals.

System acquired resistance, compartmentalization, and rhythmic leaf movements in response to environmental challenges are examples of these mechanisms.

28.5 How Cells Receive and Respond to Signals

LINKS TO
SECTIONS 4.9,
5.2, 9.4, 22.12

*Signal reception, transduction, and response—this is
a fancy way of saying cells chatter among themselves
in ways that bring about changes in their activities.*

Reflect on Section 4.9, the overview of how adjoining cells communicate as by plasmodesmata in plants and gap junctions in animals. Also think back on how free-living *Dictyostelium* cells issue signals that induce them to converge and make a spore-bearing structure. They do so in response to dwindling supplies of food, an environmental cue for change (Section 22.12).

In big multicelled organisms, one cell type signals others in response to cues from both the internal and external environment. Local and long-distance signals can trigger local and regional changes in metabolism, gene expression, growth, and development.

Molecular mechanisms by which cells "talk" to one another evolved early in the history of life. They often have three parts. *First*, a specific receptor is activated, as by reversibly binding a signaling molecule. *Second*, the signal is transduced—it is converted into a form that can operate inside the cell. *Third*, the cell makes a functional response to the signal (Figure 28.13a).

Most receptors are membrane proteins of the sort shown in Section 5.2. An activated receptor starts the signal transduction. It may activate an enzyme that in turn activates many molecules of a different enzyme, which activates many molecules of another kind, and so on. These chains of cascading reactions inside the cell greatly amplify the original signal.

In the next two units, you will come across diverse cases of signal reception, transduction, and response. For now, consider this example to get a sense of the kinds of events that the signals set in motion.

The first cell of a new multicelled individual holds marching orders that guide its descendants through growth, development, reproduction, and often death. As part of that program, many cells heed calls to self-destruct when their function is over. **Apoptosis** is the process of programmed cell death. It starts with signals that unleash proteases and other digestive enzymes, which each body cell produces and stockpiles (Figure 28.13b). Proteases chop up structural proteins, such as the building blocks of cytoskeletal elements and of proteins that structurally organize DNA. Nucleases, enzymes that snip nucleic acids, are also stockpiled.

A cell undergoing apoptosis shrinks away from its neighbors in the tissue. Its surface bubbles in and out. Its chromosomes bunch up near the nuclear envelope. The nucleus and then the cell break apart. Phagocytic cells that patrol and protect tissues engulf dying cells and their remnants. Lysosomes inside the phagocytes digest engulfed bits, which are recycled.

Many cells committed suicide as your hands were developing (Section 9.4). Each hand starts forming as a paddlelike structure. Normally, apoptosis in linear rows of cells divides the paddle into fingers within a

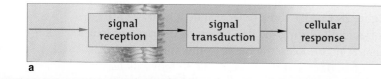

a

Signal to die docks at receptor.

Signal leads to activation of protein-destroying enzymes.

b

Figure 28.13 (**a**) Generalized signal transduction pathway. A signaling molecule docks at a membrane receptor. The signal activates enzymes or other cytoplasmic components that cause changes in metabolism, gene expression, or membrane properties. (**b**) Artist's depiction of a suicidal cell. Normally, body cells self-destruct when they finish their functions or become infected or cancerous, which could threaten the body as a whole.

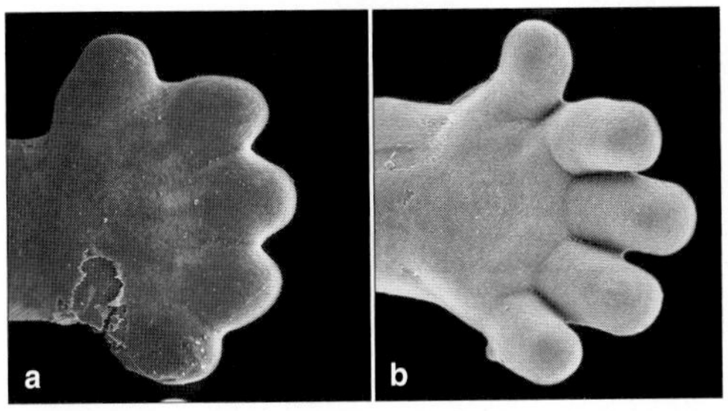

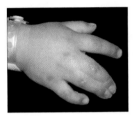

Figure 28.14 *Animated!* Formation of human fingers. (**a**) Forty-eight days after fertilization, tissue webs connect embryonic digits. (**b**) Three days later, after apoptosis by cells making up the tissue webs, the digits are separated.

Figure 28.15 Digits that remained attached when embryonic cells did not commit suicide on cue.

few days (Figure 28.14). When the cells do not die on cue, the paddle does not split properly (Figure 28.15).

The timing of programmed cell death is usually predictable. For instance, keratinocytes are cells that last about three weeks. Collectively, the dead ones form protective layers; you can see some of them in Figure 33.14. Keratinocytes can be induced to die ahead of schedule. So can other cells that are supposed to last a lifetime. All it takes is receptors for molecular signals that can call up the enzymes of death.

Control genes suppress or trigger programmed cell death. One gene, *bcl-2*, helps keep normal body cells from dying before their time. In certain cancers, this gene has mutated and cells do not respond to signals to die. Apoptosis is no longer a control option.

What about plants? The controlled death of xylem cells creates walled pipelines for water. Also, when a tissue is being attacked, signals may trigger the death of nearby cells in a pattern that walls off the threat.

Plant and animal cells communicate with one another by secreting signaling molecules into extracellular fluid and selectively responding to signals from other cells.

Communication involves receiving signals, transducing them, and inducing change in a target cell's activity.

Signal transduction requires membrane receptors and other membrane proteins. It often involves a cascade of reactions that amplify the initial signal.

Summary

Section 28.1 Anatomy is the study of the multicelled body's form. Physiology is the study of how that body functions in the environment. The structure of most body parts correlates with current or past functions.

A plant or animal's structural organization emerges during growth and development. Each living cell carries out many metabolic functions that keep it alive. At the same time, cells are organized in tissues, organs, and often organ systems. These structural units function in coordinated ways in the performance of specific tasks.

Biology⊘Now
Investigate the structural organization of a tomato plant with the interaction on BiologyNow.

Section 28.2 Plants and animals have responded in similar ways to environmental challenges, such as the constraint imposed by the surface-to-volume ratio.

Plants and animals exchange gases with the outside environment, transport substances to and from cells, and maintain water and solute concentrations of their internal environment at tolerable levels. They all have mechanisms of integrating and controlling body parts in ways that favor survival of the whole organism. They also have mechanisms for responding to signals from other cells and to signals or cues from the outside.

Section 28.3 In all complex multicelled organisms, homeostasis is a state in which the body's internal environment is being maintained within a range that individual cells can tolerate.

In animal cells, sensory receptors, integrators, and effectors interact in ways that maintain these tolerable conditions, often by way of feedback loops.

With negative feedback mechanisms, a change in a certain aspect of the internal environment triggers a response that brings about the reversal of the change. For example, one mechanism corrects deviations from a set point for the body's inner (core) temperature. Another mechanism corrects deviations from the set point for the concentration of carbon dioxide in blood.

With positive feedback mechanisms, a change in the internal environment leads to a response that intensifies the condition that caused it. This type of mechanism causes the expulsion of the fetus during childbirth.

Biology⊘Now
Observe the effects of negative feedback on temperature control in humans with the animation on BiologyNow.

Section 28.4 Plants have decentralized mechanisms of homeostasis, such as systemic resistance, that maintain the internal environment and help ensure their survival.

Many species of trees also wall off infected or injured tissues by secreting resins and toxins. This defensive response is called compartmentalization. Some plants respond to changes in environmental conditions by folding their leaves closer to the stem. Rhythmic leaf

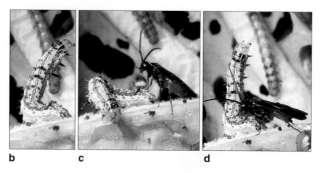

Figure 28.16 (**a**) Consuelo De Moraes studies plant stress responses. (**b,c**) A caterpillar chewing on tobacco causes the plant to secrete chemicals that attract a wasp. (**d**) The wasp grabs the caterpillar and lays an egg inside it.

folding is a circadian rhythm that apparently helps a plant maintain its internal temperature.

Biology⊜Now
Learn about plant defense mechanisms with the animation on BiologyNow.

Section 28.5 Cell-to-cell communication involves signal reception, signal transduction, and a response by a target cell. Many signals are transduced by membrane proteins that trigger reactions in the cell. Reactions may alter the activity of genes or enzymes that take part in cellular events. An example is a signal that unleashes the protein-cleaving enzymes of apoptosis; a target cell self-destructs.

Biology⊜Now
See the formation of a human hand with the animation on BiologyNow.

Self-Quiz
Answers in Appendix II

1. An increase in the number, size, and volume of plant cells or animal cells is called _____ .
   a. growth   c. differentiation
   b. development   d. all of the above

2. The internal environment consists of _____ .
   a. all body fluids   c. all body fluids outside cells
   b. all fluids in cells   d. interstitial fluid

3. _____ influences the concentrations of water and solutes in the internal environment.
   a. Diffusion   c. Passive transport
   b. Active transport   d. all are correct

4. Cell communication typically involves signal _____ .
   a. reception   c. response
   b. transduction   d. all are correct

5. Match the terms with their most suitable description.
   ____ circadian rhythm a. programmed cell death
   ____ homeostasis   b. 24-hour or so cyclic activity
   ____ apoptosis   c. stable internal environment
   ____ negative   d. an activity changes some
        feedback   condition, then the change
                         triggers its own reversal

Additional questions are available on **Biology⊜Now**™

Critical Thinking

1. Many plants protect themselves with thorns or toxins or nasty-tasting chemicals that deter plant-eating animals. Some get help from wasps.

Consuelo De Moraes, currently at Pennsylvania State University, studies interactions among plants, caterpillars, and parasitoid wasps. *Parasitoids* are a special class of parasites; their larvae eat a host from the inside out.

When a caterpillar chews on a tobacco plant leaf, it secretes a lot of saliva. Some chemicals in its saliva are an external signal that triggers a chemical response from leaf cells. The cells release certain molecules that diffuse through the air. Parasitoid wasps follow the concentration gradient to the stressed leaves. They attack a caterpillar, and each wasp deposits one egg inside it. The wasp eggs grow, develop, and become caterpillar-munching larvae (Figure 28.16).

As De Moraes discovered, plant responses are highly specific. Leaf cells release different chemicals in response to different caterpillar species. Each chemical attracts only the wasps that parasitize the particular kind of caterpillar that triggers the chemical's release.

Are the plants "calling for help"? Not likely. Give a possible explanation for this plant–wasp interaction in terms of cause, effect, and natural selection theory.

2. The Arabian oryx (*Oryx leucoryx*), an endangered antelope, evolved in the harsh deserts of the Middle East. Most of the year there is no free water, and temperatures routinely reach 117°F (47°C). The most common tree in the region is the umbrella thorn tree (*Acacia tortilis*). List common challenges that the oryx and the acacia face. Also research and report on the morphological, physiological, and behavioral responses of both organisms.

3. In the summer of 2003, a record heat wave lasted for weeks and caused the death of more than 5,000 people in France. The elderly and very young were at greatest risk. High humidity increases the chance of heat-related illness because it reduces the rate of evaporative cooling. So does tight clothing that does not "breathe," like the uniform Korey Stringer wore during his last practice. Other risk factors are obesity, poor circulation, dehydration, and alcohol intake. Using Figure 28.8 as a reference, briefly suggest how each factor may overwhelm homeostatic controls over the body's internal temperature.

V How Plants Work

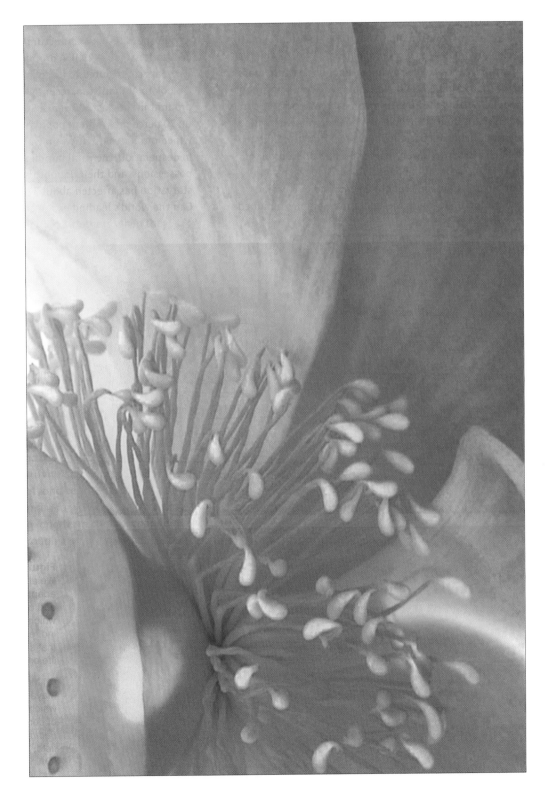

*The sacred lotus,
Nelumbo nucifera,
busily doing what
its ancestors did for
well over 100 million
years—flowering
spectacularly during
the reproductive
phase of its life cycle.*

29.3 Primary Structure of Shoots

Ground, vascular, and dermal tissues of monocot and dicot stems become organized in characteristic patterns during growth.

BEHIND THE APICAL MERISTEM

The structural organization of a new flowering plant has become mapped out by the time it is an embryo sporophyte inside a seed coat. As you will read later, a tiny primary root and shoot have already formed as part of the embryo. Both are poised to resume growth and development as soon as the seed germinates.

Terminal buds are a shoot's main zone of primary growth. Just beneath a terminal bud's surface, cells of shoot apical meristem divide continuously during the growing season. Some of its descendants divide and differentiate into specialized tissues. Each descendant cell lineage divides in orderly directions, at different rates, and its individual cells go on to differentiate in size, shape, and function (Figure 29.10).

Just below the terminal bud, along the sides of the apical meristem, tiny bulges of tissue develop. Each is the start of a leaf. As the stem continues to lengthen, new leaves form and mature in orderly tiers, one after another. Each stem region where one or more leaves have formed is a node, and the region between two successive nodes is an internode (Figure 29.2).

Lateral buds, or *axillary* buds, are dormant shoots of mostly meristematic tissue. Each forms in a leaf axil, which is just above the petiole's upper surface where the leaf attaches to the stem. Often, protective bud scales, or modified leaves, encase it. Different axillary buds are the start of side branches, leaves, or flowers. As you will read in Section 32.2, a hormone secreted by a terminal bud can keep lateral buds dormant.

INSIDE THE STEM

In most flowering plants, the cells of primary xylem and phloem form inside the same cylindrical sheath

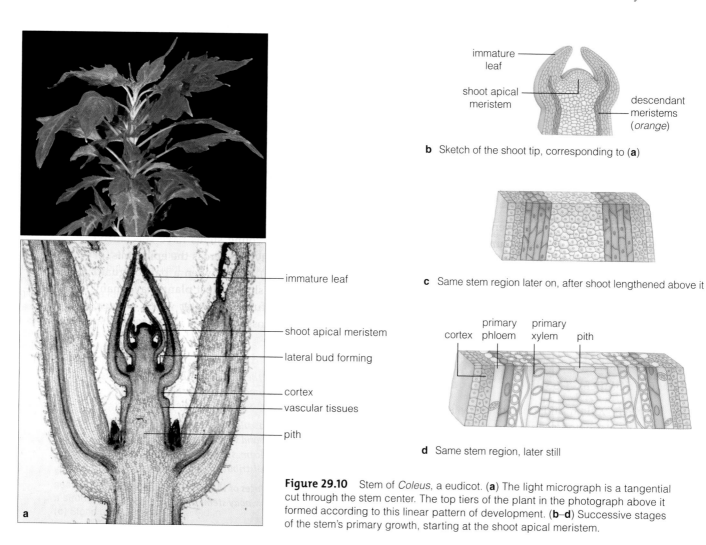

b Sketch of the shoot tip, corresponding to (**a**)

immature leaf

shoot apical meristem

descendant meristems (*orange*)

c Same stem region later on, after shoot lengthened above it

d Same stem region, later still

immature leaf

shoot apical meristem

lateral bud forming

cortex

vascular tissues

pith

cortex   primary phloem   primary xylem   pith

Figure 29.10 Stem of *Coleus*, a eudicot. (**a**) The light micrograph is a tangential cut through the stem center. The top tiers of the plant in the photograph above it formed according to this linear pattern of development. (**b–d**) Successive stages of the stem's primary growth, starting at the shoot apical meristem.

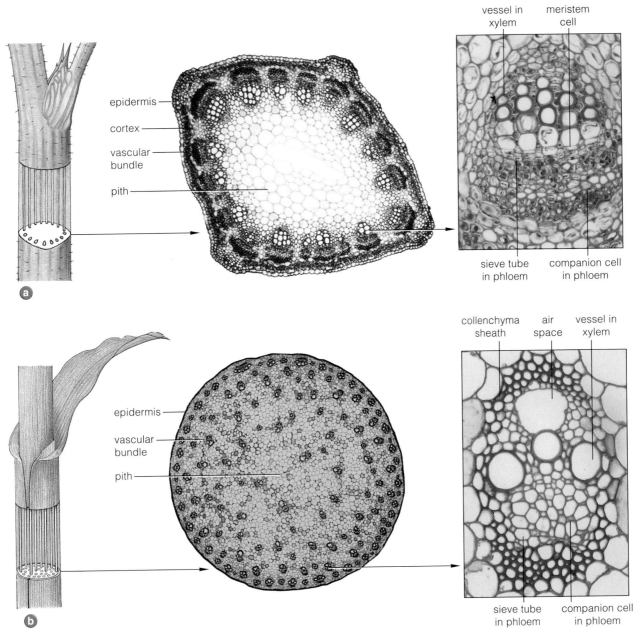

Figure 29.11 *Animated!* Internal stem organization. (**a**) Alfalfa (*Medicago*), a eudicot.
(**b**) Corn (*Zea mays*), a monocot.

of cells, as distinctively long, **vascular bundles**. These multistranded cords thread lengthwise through every shoot's ground tissue system.

Vascular bundles form in two genetically dictated patterns. In most eudicots, they develop in a ringlike array that runs parallel with the shoot's long axis. The ring divides parenchyma of the ground tissue into a cortex and pith (Figure 29.11*a*). The stem *cortex* is the portion between the ring of vascular bundles and the epidermis. The *pith* is the part inside the ring.

Most monocot and some magnoliid stems have a different arrangement. Their long vascular bundles do not form a ring; rather, they are distributed all through the ground tissue (Figure 29.11*b*). The next chapter explains how these vascular tissues take up, conduct, and give up water and solutes throughout the plant.

Ground, vascular, and dermal tissues of monocot and eudicot stems have distinct organizational patterns. A stem's primary growth starts at apical meristems in terminal and lateral buds.

The meristematic activity gives rise to the primary plant body, which develops a distinctive internal structure, as in the pattern in which its vascular bundles are arranged.

29.4 A Closer Look at Leaves

LINKS TO
SECTIONS
4.9, 7.7, 23.2

*Each **leaf** is a metabolic factory with many photosynthetic cells. Even so, leaves vary greatly in size, shape, surface specializations, and internal structure.*

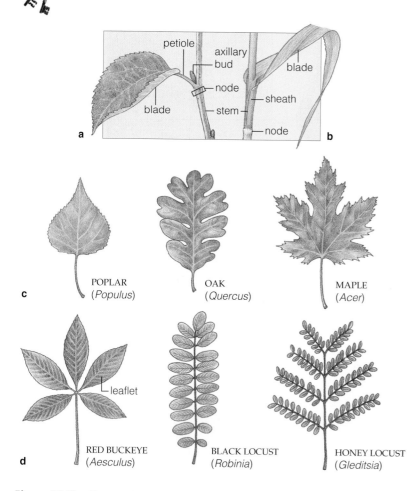

Figure 29.12 Common leaf forms of (**a**) eudicots and (**b**) monocots. Examples of (**c**) simple leaves and (**d**) compound leaves.

Figure 29.13 Example of leaf specializations. Leaves of the cobra lily (*Darlingtonia californica*) form a "pitcher." The pitcher becomes partially filled with plant secretions that hold digestive enzymes. It also gives off chemical odors that some insects find irresistible. Insects lured in often cannot find the way back out; light that is shining through the patterned dome of the pitcher's leaves confuses them. They just wander around and down, adhering to downward-pointing leaf hairs—which are slickened with wax above the potent vat.

LEAF SIMILARITIES AND DIFFERENCES

Figures 29.12 and 29.13 hint at some of the structural and functional variations among leaves, which also differ in size. A duckweed leaf is 1 millimeter (0.04 inch) across; leaves of one palm (*Attalea*) are 12 meters across. Leaves are shaped like needles, blades, spikes, cups, tubes, and feathers. They differ in color, odor, and edibility; many form toxins. Leaves of birches and other *deciduous* species wither and drop from stems as winter nears. Leaves of camellias and other *evergreen* plants also drop but not all at the same time.

A typical leaf has a flat blade and a petiole, or stalk, attached to the stem (Figure 29.12*a*). *Simple* leaves are undivided, although many are lobed. *Compound* leaves have blades divided as leaflets. Most monocots, such as ryegrass and corn, have flat blades, the base of which encircles and sheathes the stem.

Leaf shapes and orientations function to intercept sunlight and to exchange gases. Most leaves are thin, with a high surface-to-volume ratio, and most reorient themselves during the day so that they stay perpendicular to the sun's rays. Leaves often project from the same stem in a pattern that keeps one another in sunlight, as the four-leaf clover at right is doing. In hot, arid places, however, the leaves of many desert plants orient themselves parallel with the sun's rays and so reduce heat absorption. The thick leaves and stems of a cactus and some other plants also help store water.

LEAF FINE STRUCTURE

A leaf's fine structure, too, is adapted to intercept the sun's rays and enhance gas exchange. Many leaves also have surface specializations.

Leaf Epidermis Epidermis covers every leaf surface exposed to the air. This surface tissue may be smooth, sticky, or slimy, with "hairs," scales, spikes, hooks, glands, and other specializations. A cuticle covers the sheetlike, compact array of epidermal cells; it restricts water loss (Figures 29.9 and 29.14). Most leaves have far more stomata on the lower surface. In arid or cold habitats, stomata and thickly coated epidermal hairs often occur in depressions in the leaf surface. Both leaf adaptations help conserve water.

Mesophyll—Photosynthetic Ground Tissue Each leaf holds **mesophyll**, a photosynthetic parenchyma in which the individual cells are exposed to air spaces

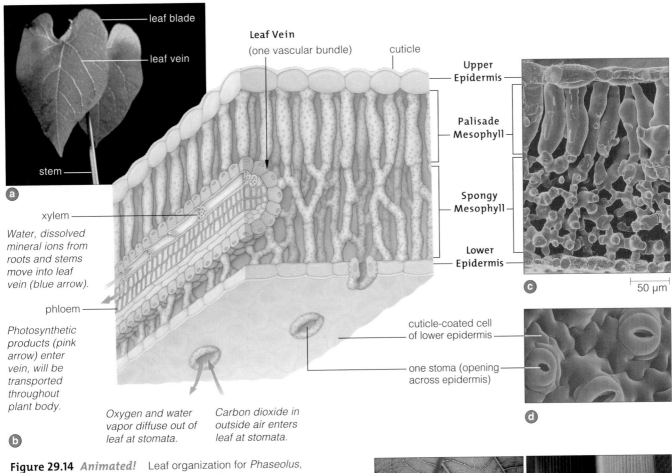

leaf blade

leaf vein

stem

(a)

Leaf Vein
(one vascular bundle)

cuticle

Upper Epidermis

Palisade Mesophyll

Spongy Mesophyll

Lower Epidermis

(c)

50 μm

xylem

Water, dissolved mineral ions from roots and stems move into leaf vein (blue arrow).

phloem

Photosynthetic products (pink arrow) enter vein, will be transported throughout plant body.

(b)

Oxygen and water vapor diffuse out of leaf at stomata.

Carbon dioxide in outside air enters leaf at stomata.

cuticle-coated cell of lower epidermis

one stoma (opening across epidermis)

(d)

Figure 29.14 *Animated!* Leaf organization for *Phaseolus*, a bean plant. (**a**) Foliage leaves. (**b–d**) Leaf fine structure.

(Section 7.7 and Figure 29.14). Carbon dioxide reaches these cells by diffusing into the leaf through stomata; oxygen wastes of photosynthesis diffuse out the other way. Plasmodesmata functionally connect these cells. This type of junction aligns across the walls of two adjoining cells and allows substances to flow rapidly into and out of their cytoplasm (Section 4.9).

Leaves oriented perpendicular to the sun have two mesophyll regions. Attached to the upper epidermis is *palisade* mesophyll. These columnar parenchymal cells have more chloroplasts and photosynthetic potential than cells of the *spongy* mesophyll layer below (Figure 29.14). In grass blades and other monocot leaves, the mesophyll is not divided into two layers. Such leaves grow vertically and intercept light from all directions.

Veins—The Leaf's Vascular Bundles Leaf **veins** are vascular bundles, usually strengthened with fibers. Their continuous strands of xylem rapidly move water and dissolved nutrients to all mesophyll cells, and the continuous strands of phloem carry photosynthetic products—especially sugars—away from them. In most eudicots, veins branch lacily into a number of minor

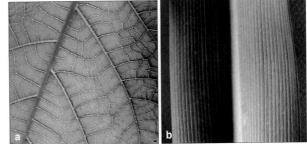

Figure 29.15 Typical vein patterns in flowering plants. (**a**) A netlike array of veins is common among eudicots. A stiffened midrib often runs from the petiole to the leaf tip, and ever smaller veins branch from it. (**b**) A strongly parallel orientation of veins is typical of monocot leaves. Like our blood-transporting veins, leaf veins are conduits for dissolved substances. Like an umbrella's ribs, stiffened veins also help maintain leaf shape.

veins embedded in the mesophyll. In most monocots, the veins are more or less similar in length and run parallel with the leaf's long axis (Figure 29.15).

A leaf's shape, orientation, and structure typically function in sunlight interception, gas exchange, and distribution of water and solutes to and from its living cells. Its epidermis encloses photosynthetic parenchyma (mesophyll) and veins.

29.5 Primary Structure of Roots

Roots function mainly in providing plants with a large surface area for absorbing water and essential mineral ions dissolved in it.

Unless tree roots start to buckle a sidewalk or choke off a sewer line, most of us do not pay much attention to flowering plant root systems. As we are walking above them, roots are busily mining the soil for water and minerals, and most grow no deeper than 2 to 5 meters. In hot deserts, where free water is scarce, one hardy mesquite shrub sent its roots down 53.4 meters (175 feet) near a stream bed. Some cacti have shallow roots radiating outward for 15 meters.

Someone once measured the roots of a young rye plant that had been growing for four months in 6 liters of soil water. If the surface area of that root system were laid out as one sheet, it would occupy more than 600 square meters, or close to 6,500 square feet!

INTERNAL STRUCTURE OF ROOTS

A root's structural organization, recall, is laid out in a seed. When a seed germinates, a primary root is the first structure to poke through the seed coat. In nearly all eudicot seedlings, the primary root thickens.

Look at the Figure 29.16*a* root tip. Some cellular descendants of root apical meristem give rise to a root cap, a dome-shaped mass of cells that helps protect the soft, young root as it grows through soil. Other cells give rise to three primary meristems—which divide, enlarge, elongate, and differentiate into the dermal, ground, and vascular tissue systems.

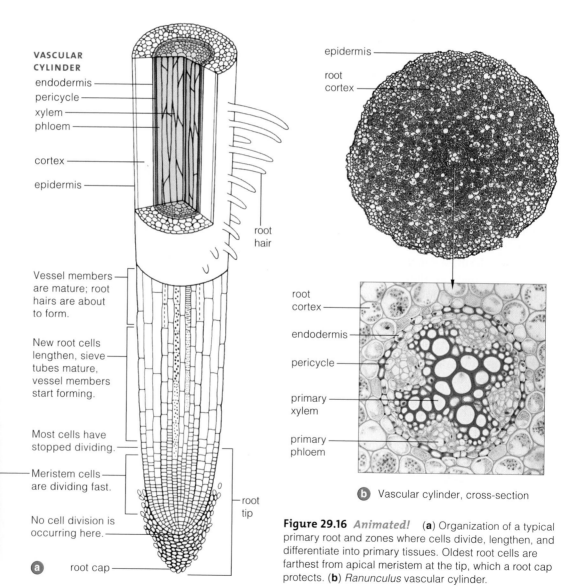

VASCULAR CYLINDER
endodermis
pericycle
xylem
phloem

cortex

epidermis

root hair

Vessel members are mature; root hairs are about to form.

New root cells lengthen, sieve tubes mature, vessel members start forming.

Most cells have stopped dividing.

Meristem cells are dividing fast.

No cell division is occurring here.

a root cap

root tip

epidermis
root cortex

root cortex
endodermis
pericycle
primary xylem
primary phloem

b Vascular cylinder, cross-section

Figure 29.16 *Animated!* (**a**) Organization of a typical primary root and zones where cells divide, lengthen, and differentiate into primary tissues. Oldest root cells are farthest from apical meristem at the tip, which a root cap protects. (**b**) *Ranunculus* vascular cylinder.

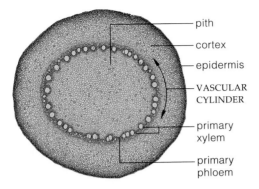

Figure 29.17 Root of corn (*Z. mays*), transverse section. Its vascular cylinder divides the ground tissue into two zones—cortex and pith.

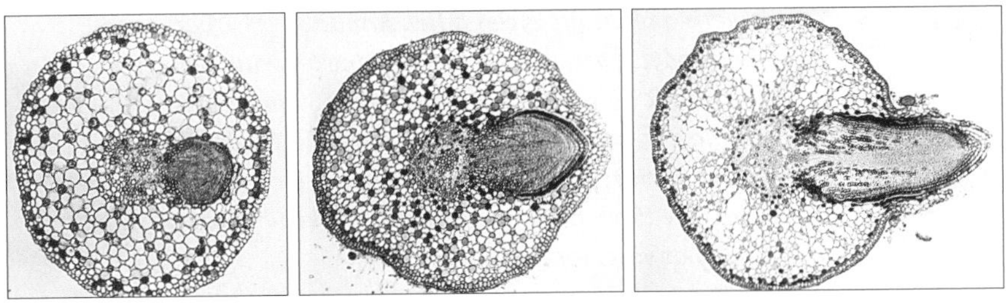

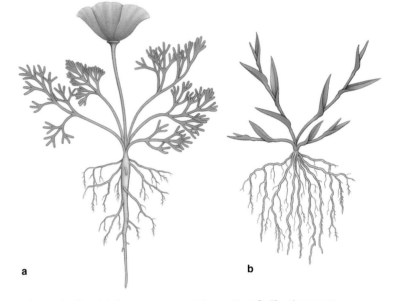

Figure 29.18 Lateral root formation from the pericycle, a cylindrical sheet of cells, one cell thick, inside the endodermis. Transverse sections.

Root epidermis is the plant's absorptive interface with soil. Some epidermal cells send out extensions called **root hairs**. Collectively, root hairs enormously increase the surface area available for taking up water and dissolved oxygen and mineral ions. We consider the nutritional role of root hairs in the next chapter. For now, simply note that an abundance of spaces between the cells of the ground tissue system allows oxygen to diffuse easily through all roots. Like other living cells in the plant, root cells make ATP by aerobic respiration and require oxygen for the reactions.

Root apical meristem also gives rise to the ground tissue system and to the **vascular cylinder**. A vascular cylinder consists of primary xylem and phloem inside a pericycle: one or more layers of parenchyma cells (Figure 29.16*b*). In corn and some other monocots, the vascular cylinder divides the root ground tissue into cortex and pith (Figure 29.17).

Although pericycle cells are differentiated, some retain the capacity to divide. They divide repeatedly in a direction perpendicular to the root axis. Masses of cells erupt through the cortex and epidermis as the start of new, lateral roots (Figure 29.18).

As you will see in the next chapter, water entering a root moves from cell to cell until it reaches the **endodermis**—a layer of cells around the pericycle. Abutting endodermal cells are waterproofed, so water must pass *through* the cytoplasm of these cells to reach the vascular cylinder. Transport proteins built into the plasma membrane exert some control over the uptake of water and dissolved substances.

TAPROOT AND FIBROUS ROOT SYSTEMS

Root primary growth results in one of two kinds of root systems. In eudicots, a **taproot system** consists of a primary root and its lateral branchings. Dandelions,

Figure 29.19 (**a**) Taproot system of the eudicot California poppy. (**b**) Fibrous root system of a grass plant, a monocot.

carrots, oak trees, and poppies are among the plants having this system (Figure 29.19*a*). By comparison, the primary root of most monocots, such as grasses, is short-lived. Adventitious roots arise from the stem in its place, and then lateral roots branch from them. (*Adventitious* refers to any plant structure that forms at an unusual place, relative to most species.) Lateral roots are more or less similar in diameter and length. Such roots form a **fibrous root system** (Figure 29.19*b*).

Roots provide a plant with a tremendous surface area for absorbing water and solutes. Inside each is a vascular cylinder, with long strands of primary xylem and phloem.

Taproot systems consist of a primary root and lateral branchings. Fibrous root systems consist of similar roots that replace the primary root.

29.6 Accumulated Secondary Growth—The Woody Plants

LINKS TO
SECTIONS
23.2, 28.2

Flowering plant life cycles differ. **Annuals**, *or herbaceous (nonwoody) types, survive for only one growing season.* **Biennials** *form roots, stems, and leaves in one season, then flower, make seeds, and die the next.* **Perennials** *grow and make seeds year after year. Their roots and stems, like those of some biennials, thicken by way of activity at vascular cambium.*

WHAT HAPPENS AT VASCULAR CAMBIUM?

Some monocots, many eudicots, the magnoliids, and most gymnosperms add secondary growth in two or more growing seasons. They are *woody* plants. Early in life, their stems and roots are like those of nonwoody plants. Differences emerge as their lateral meristems become active. These are the meristems that produce

secondary xylem and phloem. When roots and stems add secondary growth season after season, periderm eventually replaces the epidermis.

Figure 29.20 shows the growth pattern at vascular cambium. Secondary xylem forms on the *inner* face of this meristem. Secondary phloem forms on its *outer* face. The inner core of xylem thickens and displaces meristematic cells toward the surface of the stem. The displaced cells maintain the ring of vascular cambium by dividing sideways, in a widening circle.

In spring, as primary growth resumes at the stem's buds, secondary growth is added *inside* it. When fully formed, vascular cambium in a stem is like a cylinder, one or a few cells thick. Some cells (fusiform initials) give rise to secondary xylem and phloem that extend *lengthwise* through the stem. Other cells (ray initials)

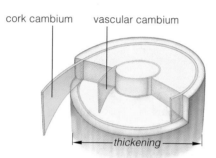

cork cambium   vascular cambium

thickening

ⓐ Two lateral meristems in older stems and roots of woody plants produce secondary growth, or increases in diameter. *Vascular cambium* gives rise to secondary vascular tissues. *Cork cambium* gives rise to periderm.

ⓑ Pattern of activity at vascular cambium. Reading left to right, ongoing cell divisions enlarge the inner core of secondary xylem and displace vascular cambium toward the stem or root surface.

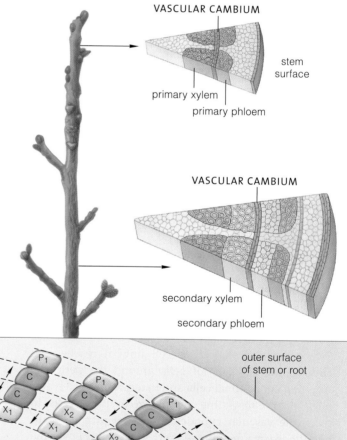

VASCULAR CAMBIUM

stem surface

primary xylem

primary phloem

VASCULAR CAMBIUM

secondary xylem

secondary phloem

Figure 29.20
Animated!
(**a**) Two meristems that become active during secondary growth. (**b**) Twig from a walnut tree in winter, after leaves dropped. In spring, *primary* growth resumes at terminal and lateral buds. *Secondary* growth resumes at vascular cambium. (**c**) The overall pattern of growth at vascular cambium.

ⓒ

division   *division*

One of the cells of vascular cambium at the start of secondary growth.

One of the two daughter cells differentiates into a xylem cell (coded *blue*), and the other remains meristematic.

One of the two daughter cells differentiates into a phloem cell (coded *pink*), and the other remains meristematic.

The same pattern of cell division and differentiation into xylem and phloem cells continues through the growing season.

outer surface of stem or root

form *horizontal* rays of parenchyma, in a pattern that is a bit like the spokes of a bike wheel. Through these secondary vascular tissues, water and solutes travel up, down, and sideways through an ever enlarging woody stem (Figure 29.21).

In different perennials, vascular cambium has been reactivated each growing season for tens, hundreds, or thousands of years. Some individual trees become giants. At last measure, the massive trunk of a coast redwood (a gymnosperm) had grown more than 110 meters (360 feet) above the forest floor. Its secondary growth might weigh close to 120 tons. As another example, the tree having the largest girth is one of the chestnuts (*Castanea*). It is growing in Sicily. To walk completely around the base of it, you would have to pace off 58 meters, or 190 feet.

DON'T FORGET THE WOODY ROOTS

We have been focusing on how stems can thicken year after year through secondary growth. Bear in mind, secondary xylem and phloem also form at vascular cambium in the plant's roots. Figure 29.22 represents the overall pattern of secondary growth at vascular cambium in the root of a typical woody plant.

WOODY AND NONWOODY PLANTS COMPARED

Woody stems and roots have selective advantages. Like other organisms, plants compete for resources. Plants with taller stems or broader canopies that defy the pull of gravity intercept more energy streaming in from the sun. By tapping a greater supply of energy for photosynthesis, they have the metabolic means to form large root and shoot systems. With larger root and shoot systems, they can be more competitive than their neighbors in acquiring resources. Other factors being equal, they ultimately will be more successful, in reproductive terms, in particular habitats.

In woody plants, secondary vascular tissues form at a ring of vascular cambium inside older stems and roots. Wood is an accumulation of secondary xylem especially.

With their sturdier tissues, woody plants defy gravity and grow taller and broader. Where competition for sunlight is intense, the ones that intercept the most sunlight win.

When other factors are equal, plants that can secure more energy than their neighbors to drive photosynthesis have advantages in terms of metabolic capabilities, growth, and reproductive success.

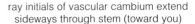
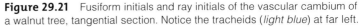
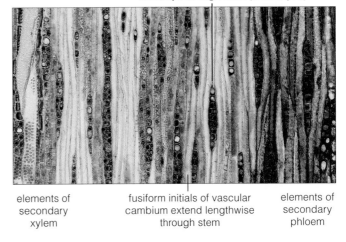

Figure 29.21 Fusiform initials and ray initials of the vascular cambium of a walnut tree, tangential section. Notice the tracheids (*light blue*) at far left.

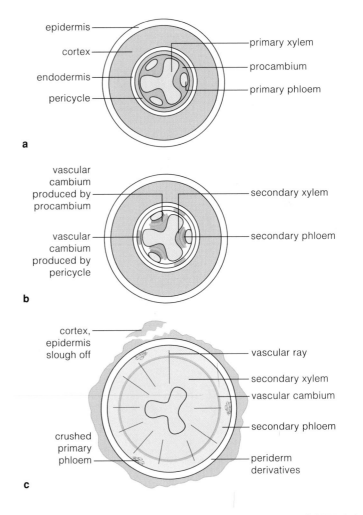

Figure 29.22 Secondary growth in one type of woody root. (**a**) This is how tissues are organized as primary growth ends. (**b,c**) A thin cylinder of vascular cambium forms and gives rise to secondary xylem and phloem. Cell divisions are parallel with the vascular cambium. The cortex ruptures as the root thickens.

29.7 A Closer Look At Wood and Bark

LINK TO SECTION 28.4

A core of secondary xylem—wood—makes up 90 percent of the mass of some trees. Secondary phloem, a narrow zone outside the vascular cambium, has thin-walled, living parenchyma cells and sieve tubes, often interspersed between bands of thick-walled reinforcing fibers. The only functioning sieve tubes lie within a centimeter or so of the vascular cambium. The rest died, but they help protect the living cells beneath them.

FORMATION OF BARK

As the seasons pass, a tree ages and the inner core of xylem continues its outward expansion. The resulting pressure is directed toward the stem or root surface.

Figure 29.23 Thick, fire-resistant bark of a coast redwood. This gymnosperm is a champion of secondary growth.

Eventually it ruptures the cortex and the outer part of secondary phloem. When that happens, parenchyma cells in this region start dividing, and they give rise to the cork cambium. Where rupturing causes the cortex and epidermis to split away, ongoing cell divisions at the cork cambium give rise to the **periderm**. This new dermal tissue is composed of parenchyma and cork, as well as the cork cambium that produces it.

Collectively, the periderm and secondary phloem constitute **bark**. In other words, *bark consists of living cells and dead tissues on the outside of vascular cambium* (Figures 29.23 and 29.24).

The **cork** component of bark has densely packed rows of cells, each with a wall thickened by a fatty substance called suberin. Only the innermost cells of this tissue are alive, because they alone have access to nourishment from xylem and phloem. With its many suberized layers, cork can protect, insulate, and also waterproof the stem or root surface. Cork also forms over wounded tissues. When leaves are about to drop from the plant, cork forms at the place where petioles attach to stems.

Like all living plant cells, the cells in woody stems and roots require oxygen for aerobic respiration and give off carbon dioxide wastes. So how do these gases get across the suberized, corky surface of bark? They can cross through lenticels, which are localized areas where the packing of cork cells is loosened up a bit. Those dark spots you might have noticed on a wine bottle's cork are all that is left of lenticels.

HEARTWOOD AND SAPWOOD

Wood's appearance and function change as a stem or root ages. The core becomes **heartwood**, a dry tissue that no longer transports water and solutes but helps the tree defy gravity. Metabolic wastes, such as resins, tannins, gums, and oils, collect in heartwood. In time, they fill and clog the oldest xylem pipelines. They often darken the heartwood, which becomes stronger, more aromatic, and prized by furniture builders.

In the early 1900s, when lumbermen were active in California's groves of old-growth redwoods, someone cut a tunnel through heartwood of a few big trees, the better to drive an automobile through them.

Sapwood is all of the secondary growth in between the vascular cambium and heartwood (Figure 29.24a). Unlike heartwood, sapwood is wet, usually pale, and not as strong, as in maple trees. Each spring, New Englanders insert tubes into sugar maple sapwood. Sap, a sugar-rich fluid in the secondary xylem, drips through the tubes, into buckets positioned below.

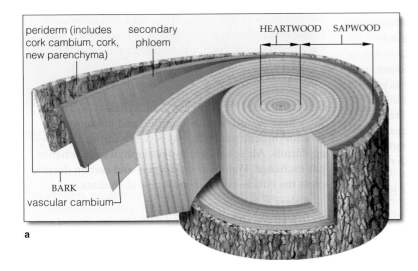

periderm (includes cork cambium, cork, new parenchyma)

secondary phloem

HEARTWOOD SAPWOOD

BARK

vascular cambium

a

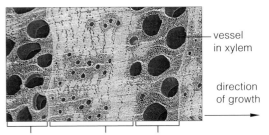

vessel in xylem

direction of growth

b early wood late wood early wood

Figure 29.24 *Animated!* (**a**) One type of woody stem. (**b**) Early and late wood cut from red oak. Late wood is evidence that a tree did not waste energy making large-diameter xylem cells for water uptake during a dry summer or drought.

EARLY WOOD, LATE WOOD, AND TREE RINGS

Vascular cambium becomes inactive in cool winters or long dry spells. *Early* wood, with large-diameter, thin-walled cells, starts forming with the first rains of the growing season. *Late* wood forms in dry summers and has small-diameter, thick-walled xylem cells. The two form one annual **growth ring**. A transverse cut from a trunk reveals alternating bands, because early and late wood reflect light differently. The differences are growth rings, which also are known informally as "tree rings" (Figures 29.24 and 29.25).

Seasonal change is predictable in temperate zones, and trees growing there usually add one growth ring per year. In deserts, thunderstorms rumble through at different times of year, and trees respond by adding more than one ring of early wood in the same season. In the tropics, seasonal change is almost nonexistent, so growth rings are not a feature of tropical trees.

Oak, hickory, and other eudicot trees that evolved in temperate and tropical zones are **hardwoods**, with vessels, tracheids, and fibers in their xylem. Pines, redwoods, and the other conifers are **softwood** trees; their xylem has tracheids and rays of parenchyma, but no vessels or fibers. Lacking fibers, the trees are weaker and less dense than hardwoods (Figure 29.25).

LIMITS TO SECONDARY GROWTH

Some species, such as redwoods and bristlecone pines, add secondary growth over centuries. Most die far sooner from old age and environmental assaults. One response, compartmentalization, counters threats but eventually shuts off the flow of essential water and solutes through the vascular system (Section 28.4).

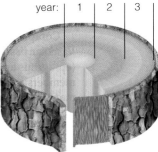

year: | 1 | 2 | 3

Figure 29.25 Tree rings, each corresponding to one growing season. In its first season, a stem puts on primary and some secondary growth; in later years, it adds secondary growth. Growth layers of (**a**) pine, (**b**) oak, and (**c**) elm. Pine, a softwood, is lightweight, resists warping, and grows faster than hardwoods. It is commercially farmed as a source of relatively inexpensive lumber. Oak and elm are strong, durable hardwoods. The elm made this series between 1911 and 1950. Differences in growth layer widths correspond to shifts in climate, including water availability. Count the rings and you have clues to a tree's age and to climates and life in the past.

Bark consists of all living and nonliving tissues outside the vascular cambium, both secondary phloem and periderm. Periderm consists of cork (the outermost covering of woody stems and roots), cork cambium, and new parenchyma.

Wood may be classified by its location and functions (as in heartwood versus sapwood) and by the type of plant (many dicots produce hardwood, and conifers produce softwood).

Leafy Clean-Up Crews

From World War I until the 1970s, the United States Army used a weapons testing and disposal site at Aberdeen Proving Grounds in Maryland (Figure 30.1). When chemical weapons and explosives became obsolete, workers burned them in open pits at the site, along with assorted plastics and other wastes.

Toxic levels of lead, arsenic, mercury, and other metals contaminated the soil and the water at Aberdeen Proving Grounds. Dozens of harmful organic compounds, such as trichloroethylene (TCE) seeped into groundwater. TCE is used as a solvent for cleaning metals. This colorless liquid can adversely affect the nervous system, lungs, and the liver in ways that can lead to coma and death. Today, the contaminated groundwater is gradually seeping toward nearby marshes and the Chesapeake Bay. Parts of the bay are now so polluted from other sources that they are a dead zone, where no marine life survives.

The Army, in concert with the Environmental Protection Agency, is now repairing the site, which is known as J-Field. Workers cannot dig up and cart off the contaminated soil; there is too much of it. As an alternative, they have planted hybrid poplars (*Populus trichocarpa x deltoides*). The trees are taking up TCE and other organic solvents and thereby cleansing the groundwater (Figure 30.1).

What is going on at Aberdeen Proving Grounds is an example of *phytoremediation*, the use of plant species to cleanse blighted regions by taking up and concentrating or degrading contaminants.

After poplar roots take up dissolved contaminants as well as mineral ions, living cells in their tissues degrade some of them. Gaseous forms of other compounds that are still somewhat harmful are released into the surrounding air. Such airborne contaminants are the lesser of two evils. For instance, TCE persists for a long time in groundwater. It breaks down faster in polluted air.

Elsewhere, different species of plants are taking up contaminants. In some cases, they are degrading targeted compounds or releasing them into the air, as in J-Field. In other cases, microbial symbionts of plants degrade them, and plant roots take up the breakdown products.

Some kinds of plants used in phytoremediation store contaminants in their tissues. Workers then remove the plants from the site for safer, convenient disposal.

The best plants for phytoremediation take up many contaminants, grow fast, and grow big. Not many plants can tolerate toxic substances, but genetically engineered ones might expand the range of choices. For instance, the alpine pennycress (*Thlaspi caerulescens*) shown in Figure 30.1c absorbs ions of zinc, cadmium, and other potentially toxic minerals dissolved in soil water. Unlike most plants, its living cells store zinc and cadmium out of the way, in their central vacuole. Genetic researchers are attempting

Watch the video online!

Figure 30.1 (a) J-Field, once a weapons testing and disposal site. (b) Today, hybrid poplars are helping to remove substances that contaminate the field's soil and groundwater. (c) Pennycress, which can take up toxic metals and survive.

to transfer a gene that confers the toxin-storing capacity of pennycress to other plants.

We use phytoremediation as your introduction to *plant physiology*, the study of how plants function in their environment. Many of the adaptations by which the toxin-busters cleanse the environment are the same ones they use to absorb and distribute water and solutes through the plant body during normal growth and development.

When considering the nature of these adaptations, it helps to remember a key point. Rarely in nature do plants have unlimited supplies of the resources they require to nourish themselves. Of every 1 million molecules of air, for example, only 350 are carbon dioxide. Most of the soils of natural habitats are frequently dry. Nowhere except in overfertilized gardens does soil water hold lavish amounts of dissolved minerals. In short, *plant structure and function are, in large part, responses to low concentrations of vital environmental resources.*

How Would You Vote?

Phytoremediation using genetically engineered plants can increase the efficiency with which a contaminated site is cleaned up. Do you support planting genetically engineered plants for such projects? See BiologyNow for details, then vote online.

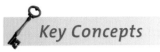

Key Concepts

UPTAKE OF NUTRIENT-LADEN WATER

Many aspects of a vascular plant's structure and function are adaptive responses to low concentrations of water, mineral ions, and other environmental resources.

A plant's root system takes up water from soil and mines the soil for nutrients. For many land plants, mycorrhizae and bacterial symbionts assist in the uptake. Soils in different habitats are in different stages of development, and they affect water and nutrient availability.   Sections 30.1, 30.2

WATER MOVEMENT THROUGH PLANTS

Xylem distributes absorbed water and solutes throughout the plant. Transpiration is the evaporation of water from plant parts exposed to dry air. The evaporative water loss creates a continuous negative tension in xylem that pulls unbroken columns of water from roots to leaves. The water molecules in xylem are hydrogen-bonded to one another, and they replace the ones lost.   Section 30.3

WATER LOSS VERSUS GAS EXCHANGE

A cuticle and stomata help plants conserve water, a scarce resource in most land habitats. Although closed stomata stop water loss, they also stop gas exchange. Certain plant adaptations represent trade-offs between the requirements for water conservation and photosynthesis.   Section 30.4

SUGAR DISTRIBUTION THROUGH PLANTS

Translocation, an energy-requiring process, distributes sucrose and other organic compounds from photosynthetic cells in leaves to all other living cells in the plant. Organic compounds are actively loaded into conducting cells of phloem and then unloaded in actively growing regions or storage regions.   Section 30.5

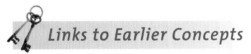

Links to Earlier Concepts

This chapter moves you from photosynthesis in individual cells (Sections 7.3, 7.7) to functional adaptations that sustain photosynthesis. You caught glimpses of these adaptations in ecological (7.8) and evolutionary (18.9, 23.2) contexts. Here you will consider actual mechanisms of the acquisition and distribution of resources in the plant body.

You will draw on your knowledge of ionization and hydrogen bonding (2.4), water's cohesive properties (2.5), membrane transport mechanisms (5.4), and osmosis and turgor (5.5). You will use your knowledge of vascular tissues (29.2), primary stems and roots (29.3. 29.5), and the fine structure of leaves (29.4). You will once again glimpse interactions between plants and their fungal symbionts (24.6), this time in the context of plant nutrition.

30.1 Plant Nutrients and Availability in Soil

LINK TO SECTION 2.4

*We've mentioned nutrients in passing. But exactly what are they? A **nutrient** is any element that is essential in an organism's life because no other element can indirectly or directly fulfill its metabolic role.*

THE REQUIRED NUTRIENTS

Sixteen elements are essential for plant growth. They become ionized in soil water, as Section 2.4 explains. Examples are ions of calcium (Ca^{++}) and potassium (K^+). Nine elements are *macro*nutrients, required in amounts above 0.5 percent of the plant's dry weight.

Table 30.1	Plant Nutrients and Symptoms of Deficiencies
Carbon Hydrogen Oxygen	No symptoms; all three macronutrients are available in abundance from water and carbon dioxide
Nitrogen	Stunted growth; young leaves turn yellow and die (these are symptoms of chlorosis)
Potassium	Reduced growth; curled, mottled, or spotted older leaves; burned leaf edges; weakened plant
Calcium	Terminal buds wither; deformed leaves; stunted roots
Magnesium	Chlorosis; drooped leaves
Phosphorus	Purplish veins; stunted growth; fewer seeds, fruits
Sulfur	Light-green or yellowed leaves; reduced growth
Chlorine	Wilting; chlorosis; some leaves die
Iron	Chlorosis; yellow, green striping in leaves of grasses
Boron	Terminal buds, lateral branches die; leaves thicken, curl, become brittle
Manganese	Dark veins, but leaves whiten and fall off
Zinc	Chlorosis; mottled or bronzed leaves; abnormal roots
Copper	Chlorosis; dead spots in leaves; stunted growth
Molybdenum	Pale green, rolled or cupped leaves

Table 30.2	What to Ask When Home Gardening Hits a Wall

1. What are the symptoms? (e.g., brown, yellow, curled, wilted, chewed leaves)
2. What is the species? Is part of one plant, a whole plant, or many plants affected?
3. Is the planting soil loose or compact? Were amendments added? Are fertilizers used, and how often?
4. Is watering by hand, hose, sprinklers, drip system? When and how often?
5. Is the plant indoors? Outdoors, in full sun or partial or full shade? In wind?
6. Dig gently to expose a few small feeder roots. Are they black and mushy (overwatering), brown and dry (not enough water), or white with a crisp "snap"?
7. Do you see insects, or insect droppings, webs, cast skins, or slime?
8. Some unique symptoms of infections rather than nutrient deficiencies:

   Viral: Leaves or petals stunted, with mottling, colored rings, distorted shapes.

   Bacterial: Tissues have a soaked, slimy texture, often a rotting smell.

   Fungal: Leaves with dry texture, discolored spots with distinct margins, usually with concentric rings (usually tan at the center, then brown, then light yellow at edge of infection).

That is the weight after all water has been removed from an organism. At least seven more elements are *micro*nutrients. They make up traces—typically a few parts per million—of the dry weight. A deficiency in any nutrient causes problems (Tables 30.1 and 30.2).

PROPERTIES OF SOIL

Soil consists of mineral particles mixed with variable amounts of decomposing organic material, or **humus**. The minerals form by the weathering of hard rocks. Humus forms from dead organisms and organic litter: fallen leaves, feces, and so on. Water and air occupy spaces between the particles and organic bits.

Soils differ in their proportions and compaction of mineral particles. The three main sizes of particles are sand, silt, and clay. The biggest sand grains are 0.05 to 2 millimeters across. You can see individual grains by dribbling beach sand through your fingers. Rub silt between your fingers and you cannot see individual particles; they are only 0.002 to 0.05 millimeter across. Clay particles are the finest of all.

How suitable is a given soil for plant growth? Is it gummy when wet because it does not have enough air spaces? Does it form hard clods when dry? The answer depends partly on its proportions of sand, silt, and clay. The more clay, the finer the soil's texture.

Each clay particle consists of thin, stacked layers of aluminosilicates with negatively charged ions at their surfaces. As water trickles through soil, clay attracts dissolved, positively charged mineral ions as well as water molecules, both of which cling reversibly to it. The charged surfaces are the reason clay is so good at latching on to many nutrients for plant growth.

Most plants do not grow well in soil with too much clay. Without enough sand and silt, clay particles pack so tightly that they exclude air. Root cells cannot get oxygen for aerobic respiration. Also, water does not soak into heavy clay soils; it tends to flow along the surface, as runoff. Runoff ends up in streams, taking soil and dissolved minerals with it. Soils having the best oxygen and water penetration are **loams**, which have roughly equal proportions of sand, silt, and clay.

Humus, too, promotes plant growth. Its negatively charged organic acids attract mineral ions of opposite charge. Also, humus swells and shrinks as it quickly absorbs and releases water. Such rapid changes aerate soil by opening up spaces that air can penetrate.

In general, soils that have 10 to 20 percent humus are best for plants. The worst soils have less than 10 percent humus or more than 90 percent humus. Bogs and swamps have notably poor soils.

Soils develop slowly, over thousands of years, and they are in different stages of development in different places. They generally form in layers, or horizons, that are distinct in color and other properties (Figure 30.2). The layers help us profile the soil in a particular place. For instance, the A horizon is **topsoil**. Topsoil is the layer most essential for plant growth, and it is deeper in some places than in others. Section 48.5 shows soil profiles for some of the world's major land regions.

LEACHING AND EROSION

Leaching is the downward percolation of water, and small quantities of dissolved nutrients, through soil. Leaching is fastest in sandy soils, which are not as good as clay at binding nutrients. During spring rains and the ensuing runoff, leaching occurs more in forests than in grasslands. Why? Grass plants grow fast, and actively growing plants absorb more water.

Soil erosion is a loss of soil under the force of wind and water. Strong winds, fast-moving water, and poor vegetation cover cause the greatest losses. For example, erosion from croplands puts about 25 billion metric tons of topsoil into the Mississippi River, then the Gulf of Mexico each year. Figure 30.3 has other examples.

Either way, the nutrient losses affect plant growth and all organisms that depend on plants for survival.

Nutrients are essential elements. No other element can substitute for their direct or indirect roles in the metabolic activities that sustain growth and survival. Plants require nine macronutrients and at least seven micronutrients.

Soils consist mainly of particles ranging from large-grained sand to silt and fine-grained clay. Clay especially promotes plant growth by attracting and reversibly binding many water molecules and dissolved mineral ions.

Soil contains humus, a reservoir of organic material rich in organic acids and in different stages of decay. Most plants grow best in soils having equal proportions of sand, silt, and clay, as well as 10 to 20 percent humus.

O HORIZON
Fallen leaves and other organic material littering the surface of mineral soil

A HORIZON
Topsoil, with decomposed organic material; variably deep (only a few centimeters in deserts, elsewhere extending as far as thirty centimeters below the soil surface)

B HORIZON
Compared with A horizon, larger soil particles, not much organic material, more minerals; extends thirty to sixty centimeters below soil surface

C HORIZON
No organic material, but partially weathered fragments and grains of rock from which soil forms; extends to underlying bedrock

BEDROCK

Figure 30.2 Example of soil horizons. These developed in a habitat in Africa.

Figure 30.3 (a) Erosion. In Chiapas State, Mexico, forests that once sponged up water from soil were cut down. Out-of-control runoff from rains cut deeper and wider gullies and carried away topsoil. (b) Why the Great Plains of North America became known as the Dust Bowl. Drought and strong winds prevail in this region. In the 1850s, native prairies were plowed under for farms. In time, a third of the once-deep topsoil and half the nutrients were blown away. By the 1930s, monstrous clouds of dust were forming. Not surprisingly, they ushered in erosion control practices.

30.2 How Do Roots Absorb Water and Mineral Ions?

LINKS TO
SECTIONS
5.4, 24.6, 29.5

Mining soil for water and minerals clinging to clay particles takes energy. Where a soil's composition and texture change, new roots form, replace old ones, and infiltrate different regions. The roots are not "exploring" soil. Rather, gradients are simply stimulating their growth toward patches of soil with higher concentrations of water and minerals.

SPECIALIZED ABSORPTIVE STRUCTURES

A mature corn plant absorbs as much as three liters of water per day. Plants in general use a lot of water and dissolved mineral ions. Mycorrhizae, root nodules, and root hairs of the plant itself enhance the uptake.

a Longitudinal section through infected plant cells and bacteria inside them. The mass of dividing cells is becoming a root nodule.

b Fully formed root nodule of a soybean plant

Mycorrhizae As Section 24.6 explains, a **mycorrhiza** (plural, mycorrhizae) is a form of mutualism between a young root and a fungus, in which both species gain benefits. The fungal hyphae grow as a velvety covering around the root or penetrate its cells. Collectively, the hyphae have a far larger surface area and can absorb scarce minerals from a larger volume of soil than the root can do on its own. The root's cells give up some sugars and nitrogen-rich compounds to the fungus, which gives up some minerals to the plant.

Root Nodules Certain bacteria in soil are mutualists with clover, peas, and other legumes which, like other plants, require nitrogen for growth. Gaseous nitrogen (N≡N, or N_2) is plentiful in air, but plants do not have enzymes that can break its three covalent bonds. The bacteria have enzymes that split the bonds, after which the atoms become rearranged as ammonia. The metabolic conversion of gaseous nitrogen to ammonia is called **nitrogen fixation**. Ammonia gets converted to forms that plants can absorb. As Section 47.11 explains, nitrogen fixation is a vital stage of the nitrogen cycle.

Nitrogen-fixing bacteria infect roots, then become symbionts in localized swellings called **root nodules** (Figure 30.4). The bacteria pilfer some photosynthetic products. The plants absorb some of the nitrogen that bacterial cells assimilated from the atmosphere.

Root Hairs As most plants put on primary growth, their root system may develop billions of **root hairs** (Figure 30.5). Collectively, these thin extensions of root epidermal cells enormously increase the surface area

Figure 30.4 Nutrient uptake at root nodules of legumes that are mutualists with nitrogen-fixing bacteria (*Rhizobium* and *Bradyrhizobium*). **(a)** When infected by the bacteria, root hair cells form a thread of cellulose deposits. Bacteria use the thread as a highway to invade plant cells in the root cortex.

(b) Infected plant cells and bacterial cells inside them divide repeatedly, forming a swollen mass that becomes a root nodule. Bacteria start fixing nitrogen when membranes of plant cells surround them. The plant takes up some of the nitrogen; the bacteria take up some photosynthetic compounds.

(c) Soybean plants growing in nitrogen-poor soil show the effect of root nodules on growth. Only the plants in the rows at *right* were inoculated with *Rhizobium* bacteria and formed nodules.

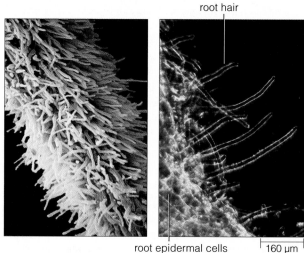

root hair

root epidermal cells   160 μm

Figure 30.5 Two views of a profusion of root hairs. These extensions of a young root's epidermal cells specialize in absorbing water and dissolved ions.

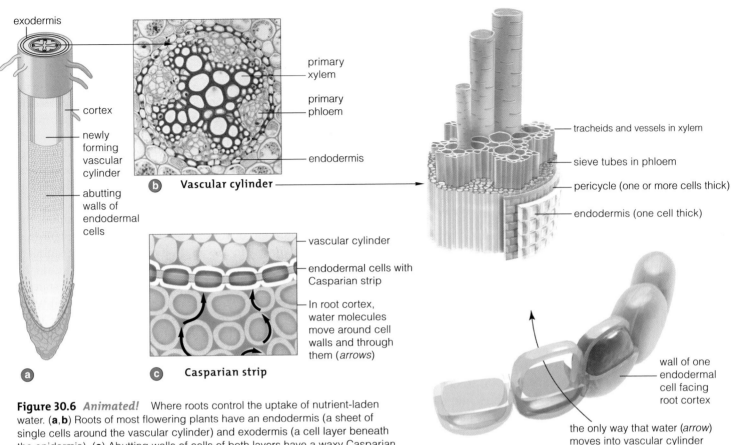

exodermis

cortex

newly forming vascular cylinder

abutting walls of endodermal cells

a

primary xylem

primary phloem

endodermis

b Vascular cylinder

tracheids and vessels in xylem

sieve tubes in phloem

pericycle (one or more cells thick)

endodermis (one cell thick)

vascular cylinder

endodermal cells with Casparian strip

In root cortex, water molecules move around cell walls and through them (*arrows*)

c Casparian strip

wall of one endodermal cell facing root cortex

the only way that water (*arrow*) moves into vascular cylinder

d Waxy, water-impervious Casparian strip (*gold*) in abutting walls of endodermal cells that control water and nutrient uptake

Figure 30.6 *Animated!* Where roots control the uptake of nutrient-laden water. (**a**,**b**) Roots of most flowering plants have an endodermis (a sheet of single cells around the vascular cylinder) and exodermis (a cell layer beneath the epidermis). (**c**) Abutting walls of cells of both layers have a waxy Casparian strip that keeps water from slipping past cells. Water must move through the cytoplasm of endodermal cells. (**d**) Transport proteins in the plasma membrane of the cells selectively control water and nutrient uptake.

available for absorption. These are fragile structures that do not become roots. They grow no more than a few millimeters through soil and die after a few days. New ones form just behind the root tip (Section 29.5).

HOW ROOTS CONTROL WATER UPTAKE

Section 29.5 introduced you to the tissue organization of typical roots. Turn now to how roots carry out their absorptive function. Water molecules in soil are only weakly bound to clay particles, so they readily move across root epidermis and continue on to a column of vascular tissue. This is the root's **vascular cylinder**. A cylindrical sheet of endodermal cells is all that lies between the root cortex and the cylinder's xylem and phloem—the pipelines to the rest of the plant.

Wherever endodermal cells abut, we find a band of waxy deposits. This **Casparian strip** is a barrier to the unrestricted flow of water and solutes into the vascular cylinder. It forces water and solutes to cross only at unwaxed wall regions and through the cell. They move across the plasma membrane facing the cortex, then

through cytoplasm, then across the plasma membrane facing the vascular cylinder (Figure 30.6).

Like all living cells, endodermal cells have a lot of transport proteins embedded in the plasma membrane. The proteins let some solutes but not others cross the membrane. *The transport proteins of endodermal cells are control points where a plant adjusts the quantity and types of solutes absorbed from soil water.*

Roots of many plants also have an **exodermis**, a cell layer just beneath their surface (Figure 30.6*a*). Walls of exodermal cells commonly have a Casparian strip that functions like the one next to the root vascular cylinder.

Root hairs, root nodules, and mycorrhizae greatly enhance a plant's uptake of water and dissolved nutrients.

Roots control the type and amount of solutes that can enter their vascular cylinder. A waxy strip seals abutting cell walls in two sheetlike layers, an endodermis and exodermis. Membrane transport proteins of these cells selectively control water and nutrient uptake.

30.3 How Does Water Move Through Plants?

LINKS TO
SECTIONS
2.4, 2.5, 29.2–29.5

By now, you have a sense of how the distribution of water and dissolved mineral ions to all living cells is central to plant growth and functioning. Turn now to a model that explains how water moves throughout the plant.

TRANSPIRATION DEFINED

Plants hold on to just a fraction of the water absorbed for growth and metabolism. Most water is lost, mainly through stomata. The evaporation of water molecules from leaves, stems, and other plant parts is a process called **transpiration**.

COHESION–TENSION THEORY

Start with a basic question: How does water move from soil, into roots, and all the way up into leaves? What gets individual water molecules to the top of plants, including redwoods and other trees that may be more than 100 meters tall?

Inside a vascular plant body, water moves through a complex tissue called xylem. Section 29.2 introduced **tracheids** and **vessel members**, the water-conducting cells of xylem. Figure 30.7 gives a closer look at their structure. These cells are dead at maturity; only lignin-impregnated walls are left behind. What this means is that xylem's conducting cells cannot be expending energy to pull water "uphill."

Some time ago, the botanist Henry Dixon came up with an explanation of how water is transported in plants. By his **cohesion–tension theory**, water inside xylem is pulled upward by air's drying power, which creates a continuous negative pressure called tension. The tension extends all the way from leaves to roots. Figure 30.8 illustrates Dixon's theory. As you review this figure, consider the following points:

First, air's drying power causes transpiration: the evaporation of water from all parts of the plant that are exposed to the air, but most notably at stomata. Transpiration puts water molecules that are inside the waterproof conducting tubes of xylem into a state of tension. The tension extends from veins inside leaves, down through the stems, and on into young roots where water is being absorbed.

Second, continuous, fluid columns of water show *cohesion*, which means that they resist breaking into droplets as they are being pulled up under tension. Remember how water shows cohesion (Section 2.5)? The collective strength of hydrogen bonds among the water molecules imparts cohesion in xylem.

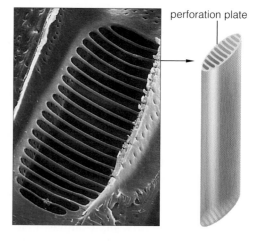

Tracheids have tapered, unperforated end walls. Pits in adjoining tracheid walls match up.

Three adjoining members of a vessel. Thick, finely perforated walls of these dead cells connect as long vessels, another type of water-conducting tube in xylem.

Perforation plate at the end wall of one type of vessel member. Perforated ends allow water to flow unimpeded.

Figure 30.7 A few types of tracheids and vessel members from xylem. Interconnected, pitted walls of cells that died at maturity form these water-conducting tubes. The pectin-coated pits may help control water distribution to specific regions. When hydrated, the pectins swell and close off water flow. During droughts, they shrink, and water moves freely through open pits toward leaves.

mesophyll (photosynthetic cells)   vein   upper epidermis

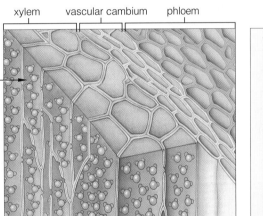

stoma

The driving force of evaporation in air

a Transpiration is the evaporation of water molecules from aboveground plant parts, especially at stomata. The process puts the water in xylem in a state of tension that extends from roots to leaves.

xylem   vascular cambium   phloem

Cohesion in root, stem, leaf xylem plus water uptake in growth regions

b The collective strength of hydrogen bonds among water molecules, which are confined within the narrow water-conducting tubes in xylem, imparts cohesion to water. Hence the narrow columns of water in xylem can resist rupturing under the continuous tension.

vascular cylinder endodermis cortex water molecule root hair cell

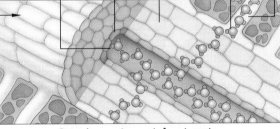

Ongoing water uptake at roots

c For as long as water molecules continue to escape by transpiration, that tension will drive the uptake of replacements from soil water.

Figure 30.8 *Animated!* Key points of the cohesion–tension theory of water transport in vascular plants.

Third, for as long as individual molecules of water escape from a plant, the continuous tension inside the xylem permits more molecules to be pulled upward from the roots, and therefore to replace them.

Hydrogen bonds are strong enough to hold water molecules together inside the water-conducting tubes of xylem. However, the bonds are not strong enough to stop the water molecules from breaking away from one another during transpiration and then escaping from leaves, through stomata.

Air's drying power causes transpiration: the evaporation of water from plant parts, especially through stomata.

By a cohesion–tension theory, transpiration puts water in xylem in a state of tension from leaf veins down to roots where water is being absorbed.

As transpiration pulls continuous, fluid columns of water upward, collective strength of hydrogen bonds between water molecules resists rupturing under the tension.

30.4 How Do Stems and Leaves Conserve Water?

LINKS TO
SECTIONS 5.5, 7.7,
7.8, 18.9, 23.2

At least 90 percent of the water transported from roots to a leaf evaporates right out. Only about 2 percent gets used in photosynthesis, membrane functions, and other activities, but that amount must be maintained.

Think of a young plant cell. As it grows, water diffuses in and exerts turgor pressure on its soft primary wall (Section 5.5). The wall—and the cell—expands until turgor pressure is enough to counter osmotic pressure, or the tendency of water to follow its concentration gradient and diffuse into the cell. When soil dries or gets too salty, the balance can tilt badly (Section 18.9 and Figure 30.9). However, plants are not entirely at the mercy of their surroundings. They have a cuticle, and they have stomata.

THE WATER-CONSERVING CUTICLE

Even mildly water-stressed plants would wilt and die without a cuticle (Figure 30.10). Epidermal cells secrete this translucent, water-impermeable layer, which coats cell walls exposed to air. Cuticle is made of waxes,

pectin, and fibers of cellulose embedded in cutin, an insoluble lipid polymer. A cuticle does not stop rays of light from reaching photosynthetic tissues. It does restrict water loss. It also restricts the *inward* diffusion of carbon dioxide necessary for photosynthesis, and *outward* diffusion of oxygen formed as a by-product.

CONTROLLED WATER LOSS AT STOMATA

Photosynthesis in plants requires carbon dioxide and releases free oxygen (Section 7.7). Both gases diffuse across cuticle-covered epidermis at small, collapsible openings called stomata (singular, stoma). Water also moves out, but when soil holds enough water, roots can replace the losses. Stomata of most plants close at night; CAM plants are exceptions (Section 7.7). Water is conserved, and carbon dioxide collects in leaves as cells make ATP by way of aerobic respiration.

A pair of specialized parenchyma cells define each stoma; we call them guard cells (Figure 30.11). When the pair swell with incoming water, they bend slightly and move apart. The gap between them is the stoma. When the pair lose water, the loss in turgor lets their cell walls collapse against each other, so the gap closes.

Whether stomata are open or closed is influenced by several cues, such as carbon dioxide's concentration inside a leaf, incoming light, and temperature. For example, photosynthesis starts after the sun comes up. As the morning progresses, carbon dioxide levels fall in all photosynthetic cells, including guard cells. The decrease stimulates active transport of potassium ions into guard cells. Blue light, another cue, activates kinases in each guard cell's plasma membrane. These

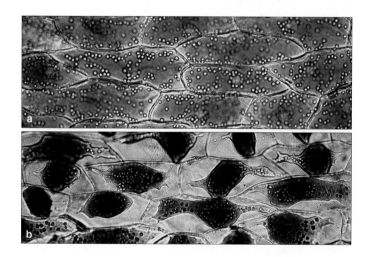

Figure 30.9 Plant wilting. When a plant with soft green leaves is growing well, the soil water is dilute (hypotonic) compared to fluids in its living cells. Water moves osmotically into its cells, and internal fluid pressure builds up against the cell walls. Water also is squeezed out when this turgor becomes great enough to counter the attractive force of cytoplasmic fluid. (Cytoplasm usually has more solutes compared to soil water.)

In soft, erect plant parts, as much water is moving into cells as is moving out. The constant pressure keeps cells plump. When the soil dries or gets too salty, water's concentration gradient reverses and the cells lose water. Osmotically induced shrinkage of cytoplasm in the young cells results in wilting.

(**a**) Cells from an iris petal, plump with water. Their cytoplasm and central vacuole extend to the cell wall. (**b**) Cells from a wilted iris petal. The loss of turgor pressure resulted in plasmolysis; their cytoplasm and central vacuole shrank, and the plasma membrane moved away from the wall.

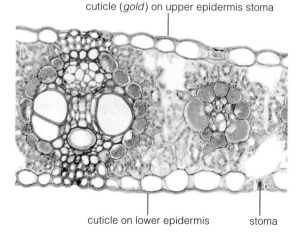

cuticle (*gold*) on upper epidermis stoma

cuticle on lower epidermis   stoma

Figure 30.10 Upper and lower cuticle on a basswood leaf.

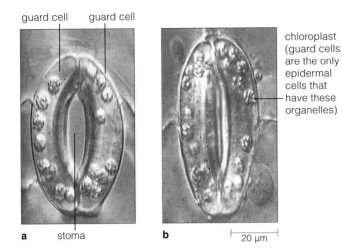

guard cell   guard cell

chloroplast (guard cells are the only epidermal cells that have these organelles)

a   stoma   **b**   20 μm

Figure 30.11 Stomata in action. Whether a stoma is open or closed at any given time depends on the shape of two guard cells that define this small gap across a cuticle-covered leaf epidermis. (**a**) This stoma is open. High turgor pressure in the guard cells caused them to bulge outward, which opened a gap between the paired cells. (**b**) This stoma is closed. Water diffused out of the guard cells, which caused them to collapse against each other and close the gap between them.

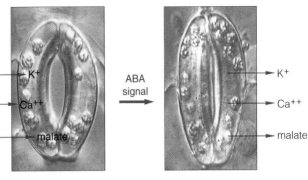

K+   Ca++   malate

ABA signal

K+   Ca++   malate

a Stoma is open; water has moved in.

b Stoma is closed; water has moved out.

Figure 30.12 Hormonal control of stomatal closure. (**a**) When a stoma is open, high solute concentrations in the cytoplasm of both guard cells have raised the turgor pressure, which keeps both cells plump. (**b**) In a water-stressed plant, the hormone abscisic acid binds to receptors on the guard cell plasma membrane. It activates a signal transduction pathway that lowers the solute concentrations in the cells, which lowers the turgor pressure and causes the stoma to close.

enzymes cause potassium ions to flow into the guard cells. Water follows the ions, and the stoma opens.

A water-stressed plant closes stomata in response to the hormone abscisic acid (ABA). Remember the Section 28.5 preview of signal reception, transduction, and response? ABA is a signal. It binds to receptors on a guard cell's plasma membrane, which causes gated channels across the membrane to open. Calcium ions flow into the cells. They cause other channels to open, so potassium ions and other substances flow out from the cytoplasm. When many ions leave, water follows its gradient and moves out of the guard cells. As the water exits, the stoma closes. It opens as potassium ions move back into the guard cells (Figure 30.12).

As this section makes clear, plant survival depends on stomatal function. Think about it when you are out and about on smog-shrouded days (Figure 30.13).

Water-dependent events in plants are severely disrupted when water loss exceeds uptake at roots for extended periods. Wilting is one observable outcome.

Transpiration and gas exchange occur mainly at stomata. These numerous small openings span all plant epidermal surfaces exposed to air, including the waxy cuticle.

Plants open and close stomata at different times to control water loss, carbon dioxide uptake, and oxygen disposal, all of which affect rates of photosynthesis and plant growth.

Figure 30.13 (**a**) Smog in Central Europe. (**b**) Stomata at a holly leaf surface. (**c**) A holly plant growing in industrialized regions becomes covered with gritty airborne pollutants that clog stomata and prevent sunlight from reaching photosynthetic cells in the leaf.

30.5 How Do Organic Compounds Move Through Plants?

LINKS TO
SECTIONS
5.4, 5.5, 7.7, 29.2

Xylem distributes water and minerals through plants. The vascular tissue called phloem distributes organic products of photosynthesis.

CONDUCTING TUBES IN PHLOEM

Phloem is a vascular tissue having organized arrays of conducting tubes, fibers, and strands of parenchyma cells. Unlike xylem, it has **sieve tubes** through which organic compounds rapidly flow. *Living* cells form the long tubes, the cells of which are positioned side by side and end to end. Their abutting end walls, called

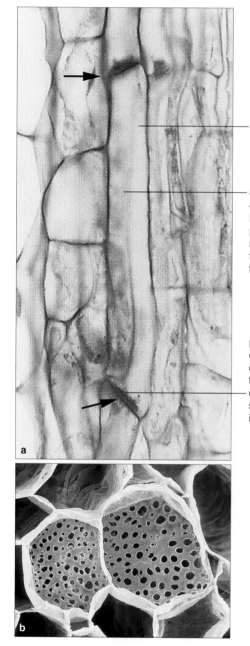

one of a series of living cells that abut, end to end, and form a sieve tube

companion cell (in the background, pressed right against the sieve tube)

perforated end plate of sieve tube cell, of the sort shown in (**b**)

Figure 30.14 (**a**) Part of a sieve tube inside phloem. Arrows point to perforated ends of individual tube members. (**b**) Scanning electron micrograph of the sieve plate on the end of two side-by-side sieve tube members.

sieve plates, are porous (Figure 30.14*a,b*). **Companion cells** are pressed against the tubes. These cells help load organic compounds into neighboring sieve tubes by active transport mechanisms.

Some organic products of photosynthesis are used in leaf cells that make them. The rest move to roots, stems, buds, flowers, and fruits (Section 7.7). Starch is the main carbohydrate storage form. Starch molecules are too big for transport across the plasma membrane of cells and too insoluble for transport. Cells convert them to sucrose, which is more easily transportable.

Experiments with insects show that sucrose is the main carbohydrate transported in phloem. Aphids were anesthetized by exposing them to high levels of carbon dioxide while they were feeding on the juices inside phloem's conducting tubes (Figure 30.15). Then researchers detached the body from the mouthparts, which they left attached to the plant. They collected and analyzed the exuded fluid. For most of the plants studied, sucrose was the most abundant carbohydrate in the fluid that was being forced out of the tubes.

TRANSLOCATION

Translocation is the formal name for the process that moves sucrose and other organic compounds through phloem. High fluid pressure drives the movement (Section 5.5). The pressure in phloem's conducting tubes is often five times higher than the air pressure inside an automobile tire.

Phloem translocates photosynthetic products along declining pressure and solute concentration gradients. The *source* of the flow is any region of the plant where organic compounds are being loaded into sieve tubes. Common sources are mesophylls—the photosynthetic

Figure 30.15 Honeydew exuding from an aphid after this insect's mouthparts penetrated a sieve tube. High pressure in phloem forced this droplet of sugary fluid out through the terminal opening of the aphid gut.

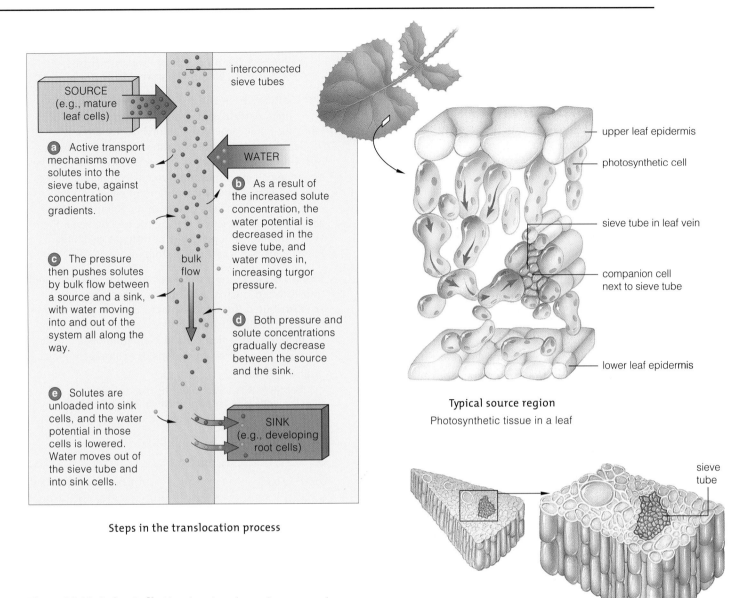

a Active transport mechanisms move solutes into the sieve tube, against concentration gradients.

b As a result of the increased solute concentration, the water potential is decreased in the sieve tube, and water moves in, increasing turgor pressure.

c The pressure then pushes solutes by bulk flow between a source and a sink, with water moving into and out of the system all along the way.

d Both pressure and solute concentrations gradually decrease between the source and the sink.

e Solutes are unloaded into sink cells, and the water potential in those cells is lowered. Water moves out of the sieve tube and into sink cells.

SOURCE (e.g., mature leaf cells)

interconnected sieve tubes

WATER

bulk flow

SINK (e.g., developing root cells)

Steps in the translocation process

upper leaf epidermis

photosynthetic cell

sieve tube in leaf vein

companion cell next to sieve tube

lower leaf epidermis

Typical source region
Photosynthetic tissue in a leaf

sieve tube

Typical sink region
Actively growing cells in a young root

Figure 30.16 *Animated!* Translocation of organic compounds. Review Section 7.7 to get an idea of how translocation relates to photosynthesis in vascular plants.

tissues in leaves. The flow ends at a *sink*, which is any plant region where products are being used or stored. For instance, while flowers and fruits are forming, on the plant, they are sink regions.

Why do organic compounds flow from a source to a sink? According to the **pressure flow theory**, internal pressure builds up at the source end of the sieve tube system and *pushes* the solute-rich solution on toward any sink, where solutes are being removed.

Use Figure 30.16 to track what happens to sucrose as it moves from the photosynthetic cells into small leaf veins. By energy-requiring reactions, companion cells in veins load sucrose into sieve tube members. When the sucrose concentration increases in the tubes, water also moves into them, by osmosis. The rising

fluid volume exerts more pressure on the wall of sieve tubes. With sufficient turgor pressure, sucrose-laden fluid inside the tubes is forced out of the leaf, into the stem, and on toward the sink.

Plants store carbohydrates as starch but distribute them in the form of sucrose and other small, water-soluble units.

Translocation is the distribution of organic compounds to different plant regions. It depends on concentration and pressure gradients in the sieve tube system of phloem.

Gradients last as long as companion cells load compounds into sieve tubes at sources, such as mature leaves, and as long as compounds are unloaded at sinks, such as roots.

Summary

Section 30.1   Plant nutrition requires water, mineral ions, and carbon dioxide. Mineral-laden water and the products of photosynthesis are distributed throughout the plant (Figure 30.17). Nutrients are essential elements; no other element can perform their metabolic functions.

Plants obtain nine macronutrients (such as carbon, oxygen, hydrogen, nitrogen, and phosphorus) and at least seven micronutrients (such as iron) from air, water, and soil. The properties of a given soil greatly affect the accessibility of water, oxygen, and nutrients to plants.

Section 30.2   Roots absorb water and nutrients that often are scarce in soil. Root hairs greatly increase their absorptive surface. Fungi are mutualists with young roots in mycorrhizae. Certain bacteria in root nodules are mutualists with plants. In both cases, the symbionts give up some dissolved mineral ions to the plant, which gives up some products of photosynthesis in return.

Roots exert some control over absorption. A waxy Casparian strip seals abutting walls of endodermal (and exodermal) cells that form one-cell thick cylinder inside roots. Water and dissolved mineral ions cannot reach the vascular cylinder for distribution through the plant without moving through the cytoplasm of endodermal cells. Their inward diffusion is controlled to a large extent at active transport proteins in cell membranes.

Biology⑧Now
See how vascular plant roots control nutrient uptake with the animation on BiologyNow.

Section 30.3   Plants distribute nutrient-laden water through tracheids and vessel members of the vascular tissue called xylem. Both cell types are dead at maturity, but their interconnected walls form narrow pipelines.

Transpiration is the evaporation of water from plant parts, mainly at stomata, into air. A cohesion–tension theory explains it as a force that pulls water upward through xylem by causing continuous negative pressure (tension) from leaves to roots. Water molecules escape from leaves, but more are pulled into the leaf under tension. Collectively, hydrogen bonds among water molecules resist rupturing; they impart cohesion, so water is pulled upward as continuous fluid columns.

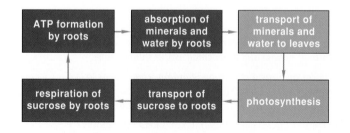

Figure 30.17   Summary of interdependent processes that sustain plant growth. All living cells in plants require at least sixteen nutrients. They all produce ATP, which drives metabolic activities.

Biology⑧Now
Learn about water transport in vascular plants with the animation on BiologyNow.
Read the InfoTrac article "How Plants Get High," Adam Summers, Natural History, March 2005.

Section 30.4   A cuticle is a waxy, waterproof cover on all plant parts in contact with the surroundings. It helps the plant conserve water on hot, dry days.

Gas exchange and transpiration occur at stomata, openings across the cuticle-covered epidermis of leaves and many stems. In response to environmental cues, ions and water move into and out of paired guard cells, which alternately plump up or collapse against each other, opening and closing the gap between them. When stomata are open, they permit gas exchange. Plants conserve water when stomata are closed.

Section 30.5   By an energy-requiring process called translocation, sucrose and other organic compounds are distributed throughout plants. Sucrose produced by photosynthesis is loaded into vessels of phloem with the help of companion cells. It is then unloaded at the plant's actively growing regions or at storage regions.

Biology⑧Now
Observe how vascular plants distribute organic compounds with the animation on BiologyNow.

Self-Quiz   *Answers in Appendix II*

1. Carbon, hydrogen, and oxygen are examples of _____ for plants.
   a. macronutrients   d. essential elements
   b. micronutrients   e. both a and d
   c. trace elements

2. A _____ strip between abutting endodermal cell walls forces water and solutes to move through these cells rather than around them.
   a. cutin   c. Casparian   b. lignin   d. cellulose

3. The nutrition of some plants depends on a root–fungus association known as a _____ .
   a. root nodule   c. root hair
   b. mycorrhiza   d. root hypha

4. Water evaporation from plant parts is called _____ .
   a. translocation   c. transpiration
   b. expiration   d. tension

5. Water transport from roots to leaves occurs by _____ .
   a. pressure flow
   b. differences in source and sink solute concentrations
   c. the pumping force of xylem vessels
   d. cohesion–tension among water molecules

6. In daytime, most plants lose _____ and take up _____ .
   a. water; carbon dioxide   c. oxygen; water
   b water; oxygen   d. carbon dioxide; water

7. At night, most plants conserve _____ , and _____ accumulates.
   a. carbon dioxide; oxygen c. oxygen; water
   b. water; oxygen   d. water; carbon dioxide

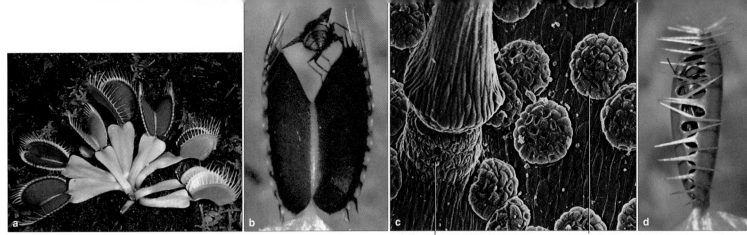

base of epidermal hairlike trigger   epidermal gland

Figure 30.18 (**a**) Venus flytrap (*Dionaea muscipula*), a carnivorous plant. (**b**) A fly stuck in sugary goo on a lobed leaf. (**c**) It brushes against hairlike triggers that activate the leaf, which snaps shut in half a second. (**d**) While a trap is open, mesophyll cells below the epidermis are compressed. Spring the trap and turgor pressure abruptly decompresses the cells. Whoosh!

8. Match the concepts of plant nutrition and transport.

_____	stomata	a. evaporation from plant parts
_____	nutrient	b. harvesting soil nutrients
_____	sink	c. balance water loss with carbon dioxide requirements
_____	root system	
_____	hydrogen bonds	d. cohesion in water transport
		e. sugars unloaded from sieve tubes
_____	transpiration	f. organic compounds distributed through the plant body
_____	translocation	
		g. element with roles in metabolism that no other element can fulfill

Additional questions are available on **Biology❸Now**™

Critical Thinking

1. Home gardeners, like farmers, must be sure that their plants have access to nitrogen from either nitrogen-fixing bacteria or fertilizer. Insufficient nitrogen stunts plant growth; leaves yellow and then die. Which biological molecules incorporate nitrogen? How would nitrogen deficiency affect biosynthesis and cause these symptoms?

2. You just returned home from a three-day vacation. Your severely wilted plants tell you they weren't watered before you left. Being aware of the cohesion–tension theory of water transport, explain what happened to them.

3. When moving a plant from one location to another, it helps to include some native soil around the roots. Explain why, in terms of mycorrhizae and root hairs.

 4. In the sketch at *left*, label the stoma. Now think about Cody, who discovered a way to keep all of a plant's stomata open at all times. He also figured out how to keep those of another plant closed all the time. Both plants died. Explain why.

5. Allen is studying the rate of transpiration from tomato plant leaves. He notices that several environmental factors, including wind and relative humidity, affect the rate. Explain how they might do so.

6. The Venus flytrap (*Dionaea muscipula*) is a flowering plant native to bogs of North and South Carolina. Its two-lobed, spine-fringed leaves open and close like a steel trap (Figure 30.18a–d). Like all other plants, it cannot grow well without nitrogen and other nutrients, which are scarce in bogs. Plenty of insects fly in from places around the bogs.

Epidermal glands on its leaf surfaces secrete sticky sugars that attract insects. As insects land, they brush against hairlike structures—triggers for the trap. When an insect touches two hairs at the same time or the same hair twice in rapid succession, the two lobes of the leaf snap shut. Digestive juices pour out from leaf cells, pool around the insect, and dissolve the prey. This plant makes its own nutrient-rich water, which it proceeds to absorb!

The Venus flytrap is only one of several species of *carnivorous plants*. Their mode of nutrient acquisition is a form of extracellular digestion and absorption. Not all carnivorous plants actively spring traps. Some have fluid-filled traps into which prey slip, slide, or fall and then simply drown. Given the variety and numbers of insects and other animals that attack plants, you can just imagine how endearing the carnivorous plants are to botanists. With their plucky modes of nutrition, these plants also can give you glimpses into the diversity of adaptations by which plants function in their environment.

All carnivorous plants evolved in habitats where nitrogen and other nutrients are hard to come by. They take hold even in shallow freshwater lakes and streams, which have only dilute concentrations of dissolved minerals. Insects and other small invertebrates are fair game, and so is the occasional tiny amphibian.

As a class activity, divide into research groups, each focusing on one genera of the carnivorous plants listed below. Gather data on the number of known species, their distribution, abiotic and biotic conditions in their habitats, the type and numbers of prey captured in a given interval, and mechanisms by which they capture prey. Note whether one or more species in the genera are threatened or endangered. If so, note possible causes. Present an oral or written report of your findings.

Byblis   Rainbow plant
Cephalotus   Western Australian pitcher plant
Darlingtonia   Cobra lily
Dionaea   Venus flytrap
Drosera   Sundews
Drosophyllum   Dewy pine
Nepenthes   Monkey cup (Tropical pitcher plants)
Pinguicula   Butterworts
Sarrecenia   North American pitcher plants
Utricularia   Bladderworts

Imperiled Sexual Partners

Imagine a world with no chocolates, no chocolate ice cream, no cocoa, no brownies. Too far-fetched? Chocolate comes from seeds of *Theobroma cacao* (Figure 31.1). This tree grows in shady tropical rain forests, but it also grows well in sunny plantations that have been carved out of rain forest habitats. There is just one problem. Trees in sunny plantations do not form many seeds.

As plantation owners found out, *T. cacao* has exclusive pollinators: midges, a type of fly that can fit on a pinhead. Midges carry *T. cacao* pollen from flower to flower. They live and breed only in the damp, deep shade of tropical rain forests. No forests, no midges, and no new chocolate.

Some growers started planting cacao trees in the shade of rain forests. Besides keeping trees supplied with midges, they are preserving otherwise threatened habitats for endangered birds, orchids, and other rain forest species.

Dwindling numbers of pollinators are a global problem. An estimated 75 percent of all crop plants set more fruit and complete the life cycle with the help of pollinators—especially bees, butterflies, flies, moths, and beetles.

Honeybees (*Apis mellifera*) pollinate most crop plants in the United States. Beekeepers easily pack up and transport artificial hives with thousands of bees from field to field. However, diseases, parasitic mites, pesticides, and other threats are wiping out the honeybees.

Certain species of bees are required pollinators for one or a few crop plants. For alfalfa plants, the alfalfa leaf-cutting bee (*Megachile rotundata*) is the pollinator of choice. For blueberry plants, it is the southeastern blueberry bee (*Habropoda laboriosa*). The blueberry bee is dwindling in numbers as an outcome of habitat losses, pesticides, and (whoops) accidentally imported ants that eat its larvae.

Seeds house the new generation. More than 60 percent of all flowering plant species are not making as many of them. In the United States, about 570 species are listed as endangered and another 140 as threatened. Some of North

Figure 31.1 Flowers, fruits, and seeds of *Theobroma cacao*, a plant pollinated only by midges. Each fruit holds as many as forty seeds—the "cocoa beans." We process the seeds into cocoa butter and essences of chocolate, which are mildly addictive. The average American, including the chocolate ice-cream lovers shown on the facing page, buys 8 to 10 pounds of chocolate annually. The average Swiss citizen craves a whopping 22 pounds. With that kind of demand, *T. cacao* trees are in luck; many humans are intensely interested in promoting their survival and reproduction.

Watch the video online!

America's most beautiful wildflowers, including the prairie white-fringed orchids, are skittering toward extinction. Pollinator shortages are a factor, and their impact ripples through communities. Seeds and fruits are food for an astounding number of animal species, including songbirds.

It will not be easy to protect the pollinators. Many face small but cumulative assaults in different habitats. Think of the migratory pollinators, such as monarch butterflies, rufous hummingbirds, and the lesser long-nosed bat. They move seasonally between breeding grounds in the United States and wintering grounds in Mexico. When the fruits and seeds that sustain them dwindle in either habitat—or anywhere along their migratory routes—so do their populations.

For the vast majority of flowering plants, reproductive success depends on animal pollinators that coevolved with them. That same dependence now puts many at risk.

 How Would You Vote?

Microencapsulated pesticides are easy to apply and are effective for long periods. But they are about the size of pollen grains and are a tempting but toxic threat to certain pollinators. Should we restrict their use? See BiologyNow for details, then vote online.

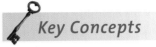 **Key Concepts**

STRUCTURE AND FUNCTION OF FLOWERS

Sexual reproduction is the dominant reproductive mode of flowering plant life cycles. It involves the formation of spores and gametes in flowers, which are specialized reproductive shoots. The vast majority of flowering plant species complete the life cycle with the help of pollinators. Sections 31.1, 31.2

LIFE CYCLE OF A EUDICOT

Haploid microspores form in a flower's male reproductive parts and develop into pollen grains, the sperm-bearing male gametophytes of seed plants.

Haploid megaspores develop in ovules, which form on the inner ovary wall of a flower's female reproductive parts. A mature female gametophyte forms from a megaspore. One of its cells is the egg.   Section 31.3

STRUCTURE AND FUNCTION OF SEEDS AND FRUITS

After sperm fertilize eggs, ovules mature into seeds. Each seed consists of an embryo sporophyte and tissues that function in its nutrition and protection.

As seeds develop, tissues of the ovary and sometimes adjacent structures mature into fruits, which function in seed dispersal. Air currents, water currents, and animals function as dispersal agents.   Sections 31.4–31.6

ASEXUAL REPRODUCTIVE MODES

Many species of flowering plants also reproduce asexually by mechanisms of vegetative growth, parthenogenesis, and tissue culture propagation.   Section 31.7

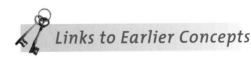 **Links to Earlier Concepts**

This chapter returns to the reproductive mode of flowering plants, which form pollen and seeds (Sections 23.2, 23.6). You may wish to reflect briefly on the evolutionary history of flowering plants and their pollinators (23.8). You will be contrasting a eudicot life cycle with that of a monocot (23.9). You will be using your knowledge of plant tissues and the meristems that give rise to them (29.1, 29.2).

31.1 Reproductive Structures of Flowering Plants

LINKS TO
SECTIONS
29.1, 29.3

Sexual reproduction is the main reproductive mode of flowering plant life cycles. It involves the formation of spores and gametes, both of which originate inside specialized reproductive shoots called flowers.

Each **flower** is a specialized reproductive shoot that develops on a sporophyte. A **sporophyte**, remember, is a spore-producing body that grows by mitotic cell divisions from a fertilized egg. Two kinds of haploid spores form inside flowers, and these spores are the start of **gametophytes**—structures that produce male or female gametes (Figure 31.2).

Sperm develop in male gametophytes, and eggs develop in female gametophytes. Fusion of a haploid sperm with a haploid egg at fertilization results in a diploid zygote that grows into a new sporophyte.

FLORAL STRUCTURE AND FUNCTION

A flower, recall, develops from a lateral bud (Section 29.3). Four whorls of modified leaves form on the end of the new floral shoot, the receptacle. The outermost whorl differentiates into a ring of sepals, called a calyx.

The whorl just inside forms a ring of nonfertile petals, the flower's corolla (Figure 31.3).

Like other plant parts, sepals and petals consist of dermal, ground, and vascular tissues. Epidermal cells of many petals make fragrant oils. Cells in the ground tissue of many petals make pigments, such as yellow to red-orange carotenoids and red-to-blue anthocyanins. Some contain small, light-refracting crystals that make the petals shimmer. As Section 31.2 explains, the colors and patterning, arrangement, and often fragrance of petals function in attracting and exploiting pollinators.

Inside the corolla, the other two whorls of modified leaves form fertile parts: stamens and carpels. **Stamens** are male floral reproductive structures, often fused to the calyx at their base. Each is an anther, usually borne on a thin filament (Figure 31.3*a*). A typical anther has two to six pairs of pollen sacs. Walled, haploid spores form in the sacs and give rise to **pollen grains**, which are structures that contain male gametophytes.

Carpels are female floral reproductive parts. Take a look at the sketches below. Many flowers have one carpel; others have two or more carpels fully or partly fused together. Carpels may be positioned higher or lower than the other three whorls; in some cases they are even embedded in receptacle tissues. Their sticky or hairlike upper portion, a stigma, can gently trap pollen grains. The stigma often is borne on a slender stalk called the style (Figure 31.3*a*).

Each carpel's lower portion, the **ovary**, is a chamber in which one or more ovules form. An **ovule**, recall, is a female reproductive structure. It consists of a female gametophyte (with egg cell), nutrient-rich tissue, and a jacket of cell layers. Sperm fertilize eggs inside ovaries. After this, ovules mature into seeds; their jacket becomes a seed coat (Section 23.6). In other words, each **seed** is a mature ovule.

carpels differ

Perfect flowers have stamens and carpels. But some species produce *imperfect* flowers, with stamens *or* carpels. These "male" and "female" flowers form on the same plant or on a different plant of the species. A cucumber or corn plant has male and female flowers. Male flowers form on some asparagus plants and the female flowers form on others. Kiwi plants also do this.

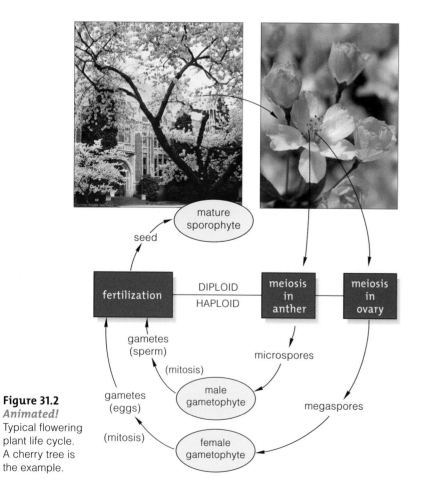
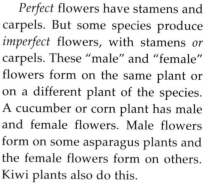

Figure 31.2
Animated!
Typical flowering plant life cycle. A cherry tree is the example.

mature sporophyte

seed

fertilization — DIPLOID / HAPLOID — meiosis in anther — meiosis in ovary

gametes (sperm)

(mitosis)

microspores

gametes (eggs)

male gametophyte

megaspores

(mitosis)

female gametophyte

ovary positions differ

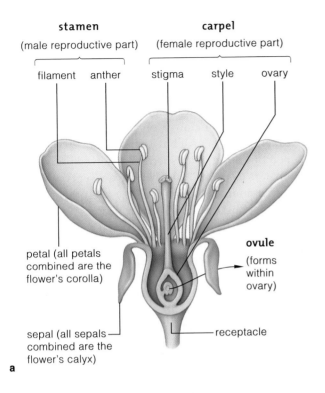

stamen
(male reproductive part)

carpel
(female reproductive part)

filament   anther   stigma   style   ovary

petal (all petals combined are the flower's corolla)

ovule
(forms within ovary)

sepal (all sepals combined are the flower's calyx)

receptacle

a

corolla (all petals combined)

calyx (all sepals combined)

receptacle

b

Figure 31.3 *Animated!* (**a**) Structure of a cherry blossom (*Prunus*). Like many flowers, it has a single carpel and stamens. (**b**) Prized corolla of one of those roses (*Rosa*) you read about in Chapter 11.

REGARDING THE POLLEN ALLERGIES

Pollen grains often have a sticky or spiked, sculpted wall that sticks to hairs, fur, or feathers of a pollinator. The wall structure also helps pollen grains adhere to and germinate on a receptive stigma. But some walls have proteins that cause problems for many people.

At different times of year, trees, woody shrubs, and herbaceous plants release pollen. Many pollen grains land in a nose or a throat instead of on flowers. The proteins of a pollen grain wall are normally harmless, but they can bring about *seasonal allergic rhinitis*, or "hay fever," in hypersensitive individuals. Many millions of people are genetically predisposed to overreact to certain kinds of pollen. Infections, emotional stress, and temperature changes also trigger the abnormal response to pollen, which causes a profusely runny nose, reddened and itchy eyelids, and sneezing.

Hot, dry conditions spur pollen release, and pollen counts are highest in the morning. Plants having the least showy flowers cause the most problems. They typically make staggering numbers of tiny, light pollen grains that drift on air currents. In North America, ragweed pollen is the biggest offender (Figure 31.4). Pollen from sagebrush, bermuda grass, and Kentucky bluegrass are not far behind. Those lightweight kinds get around! Ragweed pollen has been found drifting 3 kilometers above Earth and more than 640 kilometers out to sea.

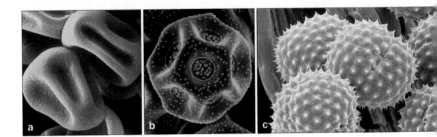

Figure 31.4 Pollen grains of (**a**) grass, (**b**) chickweed, and (**c**) ragweed plants. Pollen grains of most families of plants differ in size, wall details, and the number of wall pores.

Seasonal allergic rhinitis is a big thing. You can get daily pollen counts for your very own city from the National Allergy Bureau (www.aaaai.org/nab/).

Life cycles of flowering plants alternate between spore-producing stages (sporophytes) and gamete-producing stages (gametophytes). Flowers are reproductive shoots that form on the sporophytes. They typically have sepals and petals as well as reproductive parts.

Male gametophytes form in stamens and are contained in pollen grains. Female gametophytes with egg cells form as part of ovules, which develop in the ovaries of carpels.

An ovule consists of a jacket of cell layers around nutritive tissues and the female gametophyte. Once the egg inside is fertilized, an ovule matures into a seed.

31.2 A Closer Look at Flowers and Their Pollinators

LINK TO
SECTION
23.8

When they are busy at sexual reproduction, flowering plants put out floral invitations to third parties—pollinators that help get sperm to eggs. Long before humans thought of it, plants were using colors and perfumes to improve the odds for sexual success.

THE CUES AND BEHAVIORAL REINFORCERS

Pollination vectors are any agents that deliver pollen grains to structures that house female gametophytes. Winds are a fine pollination vector for a single species that grows densely in grasslands. Winds easily shake clouds of tiny, dry pollen grains away from long, thin stamens and drop them onto big, sticky stigmas that are typical of these species. But dozens to hundreds of species may be crowded together in part of a tropical rain forest or deciduous forest, where the wind could deposit a lot of pollen on the wrong flowers. There is selective advantage in having insects, birds, or other animal pollinators visit one species among many, get dusted with pollen, then continue searching for one particular kind of flower on plants that may be far apart. No surprise: All but 10 percent of 295,000 or so named flowering plant species coevolved with animal pollinators (Section 23.8). They have visual or olfactory cues that animals recognize. They keep them coming back by giving up some of their nectar and pollen as food, which functions as a behavioral reinforcer.

Visual Cues Flowers of specific sizes, shapes, colors, and patterns attract specific kinds of pollinators. Bee-pollinated flowers, for instance, have yellow, blue, or purple petals, often with pigments that absorb and reflect ultraviolet light. The pigments are distributed in patterns that bees recognize as visual guides to the nectar (Figure 31.5a). We cannot see the floral patterns without special camera filters; our eyes do not have sensory receptors that respond to ultraviolet light.

Red and yellow flowers, especially those with both colors in strong patterns, attract birds and butterflies (Figure 31.5b). Birds have acute daytime vision but a poor sense of smell.

Olfactory Cues Some pollinators do not have acute vision, but special sensory receptors help them follow the concentration gradient of a volatile chemical to its source. *Manduca sexta*, a hawkmoth, has about 55,000 receptors. The night-flying nectar-sipping bats home in on flowers with intense fruity or musky odors (Figure

Figure 31.5 (**a**) Bee-attracting pattern of a gold-petaled marsh marigold, which we see by shining ultraviolet light on it. (**b**) Bahama woodstar sipping nectar from a hibiscus blossom. Like other hummingbirds, it sips nectar in midflight. Its long, narrow bill coevolved with long, narrow floral tubes.

Figure 31.6 Flowers of the giant saguaro (*Carnegia gigantea*). Birds and insects sip nectar from these large, white flowers by day, and bats sip by night. Nearly all nectar-feeding bats prefer sipping solutions of hexose (a monosaccharide), not the typical sucrose-rich nectar.

31.6). Strong odors that smell like fermenting fruit and dung beckon the beetles and flies.

Reinforcements Nectar and pollen are tasty rewards that encourage pollinators to revisit specific flowering species. **Nectar** is a sucrose-rich fluid that is secreted from a tissue (the nectary) connected to phloem. It is dilute enough for pollinators to sip easily. Nectar is the only food for most butterflies and the fuel of choice for hummingbird flight. Nectar that honeybees collect condenses to honey back at the hive and helps feed bees through the winter. Pollen is a richer food source and, unlike nectar, offers a lot of vitamins and mineral ions. Pollen-laden stamens of most flowers are positioned so that they are certain to brush against a pollinator body or lob pollen onto it.

Floral sizes and shapes do not promote visits from animals that are not good at delivering pollen for the plant. They fit best with coevolved partners. Nectar-filled floral tubes or spurs are the same length as their pollinator's feeding siphon (proboscis). Flowers such as daisies actually are many tiny flowers, each with a tiny nectar tube of no interest to, say, finches or bats. Flowers that have tall, thin stems and flimsy landing platforms exclude heavy beetles, which fly about and land without much precision and delicacy.

CASE STUDY: LONG FLORAL TUBES AND THE HAWKMOTHS

While Charles Darwin was exploring Madagascar, he discovered what came to be known as the Christmas Star—the orchid *Angraecum sesquipedale*. This orchid has large, star-shaped flowers with a nectar spur that can grow more than 30.5 centimeters (12 inches) long, and nectar fills only the bottom 3.8 centimeters. The orchid releases a strong fragrance only at night. Some naturalists mocked Darwin after he predicted that its pollinator would be a night-flying hawkmoth with a proboscis as long as the floral spur.

Fifty years after Darwin passed away, someone found *Xanthopan morgani praedicta* in the same forest habitat as the orchid. Figure 31.7 shows the uncoiled proboscis of this hawkmoth. Like other hawkmoths around the world, it has strong wings and can hover before a flower, without landing. This hawkmoth, too, uncoils its long proboscis into a floral spur and then pumps nectar upward and into its esophagus.

Hawkmoth-pollinated flowers in general release a strong, fragrant attractant just before the sun sets or at night. They do not have a landing platform that can support heavy, nonhovering pollinators. Each flower

Figure 31.7 *Xanthopan morgani praedicta*, the hawkmoth with a proboscis long enough to reach nectar at the base of the equally long floral spur of the orchid *Angraecum sesquipedale*.

of *A. sesquipedale* produces more than four grams of nectar. It is a big energy source for *X. morgani*—a big, energy-demanding moth with a six-inch wingspan.

Hawkmoth-pollinated flowers are not the favorites of butterflies and other kinds of daytime pollinators; they have no vivid pigments. The nocturnal perfumes hold no allure for birds, and their floral tubes are too skinny to be of use to night-flying bats.

The mutualistic partnership between a hawkmoth and a plant has advantages flowing both ways. A big disadvantage is that, when outside factors make one mutualistic partner vulnerable, the other might not have optional partners available (Figure 31.8).

The vast majority of flowering plants attract animal pollinators with visual cues, olfactory cues, or both, and offer rewards—mainly nectar and pollen—to them. The sizes and shapes of flowers correlate with the body sizes, shapes, and behavior of preferred pollinators.

Figure 31.8 Flower of the eastern prairie fringed orchid (*Platanthera leucophaea*), which is pollinated by several species of hawkmoths that sip nectar while hovering before it. This flowering plant grows in moist tallgrass prairie and in peat moss bogs. At one time, it was common, but it has become extremely rare because of habitat losses, casual plucking, and overly enthusiastic collecting.

31.4 From Zygotes to Seeds and Fruits

Inside a flowering plant ovule, a new zygote embarks on mitotic cell divisions and becomes an embryo sporophyte. Outside the ovule, the ovary wall and sometimes other tissues mature into what we call a fruit.

ovule positions differ in ovaries

If you were to make a lateral cut through an ovary of different flowering plants, as in the examples shown to the left, you would see that some are divided into more than one chamber, with more than one ovule attached to the ovary wall. Part of the wall may even become a fleshy tissue mass in the center of the ovary, as in the third example shown. Make the same kind of cut through mature fruits, and you get a sense of the variation in seed positions.

THE EMBRYO SPOROPHYTE

We also see variation in how the embryo sporophytes develop in flowering plant ovules. For now, focus on shepherd's purse (*Capsella*). Like all eudicot embryos, it has two **cotyledons**, or seed leaves, which develop from two lobes of meristematic tissue (Figure 31.10). Embryos of all monocots have one. *Capsella* embryos absorb nutrients from endosperm and store them in cotyledons. By contrast, most monocot embryos will not tap into nutritive tissue until after germination.

SEED AND FRUIT FORMATION

Until embryo sporophytes are fully formed, parent plants transfer nutrients to an ovule's tissues. Food accumulates in endosperm or in the cotyledons. Over time, the ovule wall pulls away from the ovary wall. A seed coat forms from the thickened and hardened integuments. The embryo, food reserves, and the coat have become a seed—that is, a mature ovule.

Only flowering plants form seeds in ovaries, and only they make fruits. Botanists categorize fruits by origin, composition, or appearance. *Simple* fruits, such as cherries and apples, originate in a single or fused carpel. Strawberries, raspberries, and other *aggregate* fruits originate in several unfused carpels and become a cluster of several fruits. *Multiple* fruits such as figs and pineapples start out as a cluster of individually pollinated flowers that grow together and fuse into a single body. Figure 31.11 shows examples.

Fruits also can be categorized on the basis of which tissues make up the fruit. *True* fruits are the ovarian wall and its contents. In *accessory* fruits, other floral parts, such as the receptacle, expand right along with the ovary. Watermelons and apples are like this.

To categorize a fruit based on appearance, the first step is to describe it as dry or juicy (fleshy). Dry fruits are dehiscent or indehiscent. If *dehiscent*, the fruit wall splits along definite seams to release the seeds inside. *Capsella* pods and pea pods are like this.

A dry fruit is *indehiscent* if the wall does not split open; seeds are dispersed inside intact fruits. Acorns and grains (such as corn) are dry indehiscent fruits, as are fruits of sunflowers, maples, and—suprisingly—strawberries. Strawberry fruits are not juicy and not berries. A strawberry's red flesh is an accessory to the dry indehiscent fruits on its surface (Figure 31.11g).

Three major categories of fleshy fruits are drupes, berries, and pomes. *Drupes* have a pit—a stone-hard jacket around one seed (sometimes more), and fleshy fruit that encloses the pit. Cherries, peaches, apricots, almonds, and olives are examples of drupes.

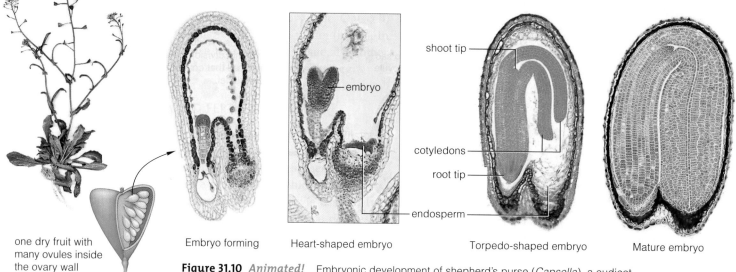

fruit

one dry fruit with many ovules inside the ovary wall

Embryo forming

Heart-shaped embryo

embryo

shoot tip

cotyledons

root tip

endosperm

Torpedo-shaped embryo

Mature embryo

Figure 31.10 *Animated!* Embryonic development of shepherd's purse (*Capsella*), a eudicot.

wall of one dry fruit | expanded receptacle | embryo sporophyte with two cotyledons

remnants of sepals, petals
ovary tissue
seed
enlarged receptacle

d

Figure 31.11 From flowers to fruits. (**a–d**) Fruit formation on an apple tree (*Malus*). After eggs are fertilized, petals fall from the flowers. (**e–h**) Strawberry (*Fragaria*). Most of the flesh arises from the receptacle. Many dry fruits are positioned at the surface of the mature receptacle; each has a hard fruit wall around the embryo sporophyte. (**i**) Pineapple (*Ananas*), a multiple fruit. Native Americans cultivated pineapples, which also were their hospitality symbol. They used the juice as a base for an alcoholic beverage. Christopher Columbus thought the pineapple looked like a pinecone, hence the name.

A *berry* has one to many seeds, no pit, and fleshy fruit. Grapes and tomatoes are in this category. So are lemons, oranges, grapefruits, and other citrus fruits; a pithy layer encloses the berries, and an oily, leathery rind encloses the pith. Each "section" of a citrus fruit started out as one ovary of a fused carpel. Pumpkins, watermelons, and cucumbers are still another type of berry in which a hard rind of accessory tissues forms over the somewhat slippery true fruit.

A *pome* has seeds in a somewhat elastic core tissue and fleshy accessory tissues that enclose its core. Two familiar pomes are apples and pears (Figure 31.11*a*).

And so, if you were to categorize, say, an apple, you could say that it originates from one flower, so it is a simple fruit. A fleshy receptacle expands around five carpels, so an apple also is an accessory fruit. The carpels form an elastic core in fleshy accessory tissue, so the apple also is a pome (Table 31.1).

A seed is a mature ovule, which consists of an embryo sporophyte, food reserves, and a protective coat. A mature ovary, with or without accessory tissues, is a fruit.

We can categorize a fruit in terms of how it originated, its composition, and whether it is dry or fleshy.

Table 31.1 Three Ways To Classify Fruits

How did the fruit originate?

1. Simple fruit	One flower, single or fused carpels
2. Aggregate fruit	One flower, several unfused carpels; becomes cluster of several fruits
3. Multiple fruit	Cluster of individually pollinated flowers that grow and fuse together

What is the fruit's tissue composition?

1. True fruit	Only ovarian wall and its contents
2. Accessory fruit	Ovary as well as other floral parts, such as the receptacle

Is the fruit dry or fleshy?

1. Dry:		
	a. Dehiscent	Dry fruit wall splits on definite seam to release seeds
	b. Indehiscent	Seeds dispersed from the parent plant inside intact, dry fruit wall
2. Fleshy:		
	a. Drupe	Fleshy fruit around hard pit with one (usually) seed inside
	b. Berry	Fleshy fruit, no pit, one to many seeds
		Pepo: Hard rind on ovary wall
		Hesperidium: Leathery rind on ovary wall
	c. Pome	Fleshy accessory tissues, seeds in elastic core

31.5 Seed Dispersal—The Function of Fruits

LINKS TO
SECTIONS
17.4, 23.11

The fruits you read about in the preceding section have one overriding function: seed dispersal. Some fruits can disperse seeds all by themselves. Others have wings or parachutes, or explode open, or hitch rides on or inside the gut of animals.

Tulip trees (*Liriodendron*), American elms (*Ulmus*), and maples (*Acer*) have winged fruits. Figure 31.12*a* shows some maple fruits. Part of a maple fruit extends out, like a pair of thin, lightweight wings. It breaks in half, drops from a tree, and spins sideways as it interacts with air currents. The spinning motion moves it far enough away that the embryo sporophytes inside the seeds will not have to compete with the parent plant for water, mineral ions, and sunlight.

Dandelions, thistles, cattails, and milkweed have tiny, lightweight fruits equipped with an outward-pluming "parachute." Brisk winds have been known to transport some as far as ten kilometers from a tree. Orchid seeds do not even require parachutes. They are as fine as dust particles and simply drift through the air. *Impatiens capensis* sends off seeds with a bang. Its fruit capsule pops open explosively, propelling the small seeds inside away from the parent plant.

Fleshy fruits or nuts travel to new locations on or in animals, including diverse birds, mammals, and insects (Figure 31.12*b*). The seed coat surrounding the embryo sporophytes inside the fleshy tissues survive being assaulted by digestive enzymes in the animal gut. Besides digesting the fruit's flesh, the enzymes digest some of the seed coat. A hard seed coat that has been partially digested is easier to break through after the seed is expelled from the animal body, inside feces, and later germinates. Germination is one of the topics of the next chapter.

Other fruits have hooks, spines, hairs, and sticky surfaces that can adhere to feathers, feet, or fur. Fruits of cocklebur, bur clover, and bedstraw are like this (Figure 31.12*c*).

Some water-dispersed fruits have heavy wax coats. Others, including sedge fruits, have sacs of air that help them float. With its thick, water-repelling layers, a coconut palm fruit is adapted to travel the open ocean. If it does not wash onto a beach soon enough, saltwater may seep into the layers and kill the embryo. Coconuts can drift hundreds of kilometers before then.

Finally, remember Figure 31.1? Humans are grand agents of dispersal. In the past, explorers transported seeds all over the world, although most places now control imports. Cacao, oranges, corn, and other plants were domesticated and encouraged to reproduce in new habitats, in far greater numbers than they did on their own. And in evolutionary terms, reproductive success is what life is all about.

> *Fruits are structurally adapted to disperse seeds through the air or drifting on water currents. Many kinds are dispersed by animals.*

wing   seed (in carpel)

Figure 31.12 Examples of fruit dispersal. (**a**) Simple, dry fruits of maple (*Acer*) use winds to lift their "wings" and spin the seeds inside away from the parent plant. (**b**) Mountain ash berries, a simple fleshy accessory fruit, attracts cedar waxwings. (**c**) Cocklebur, a prickly accessory fruit, sticks to the fur of animals that brush past.

31.6 Why So Many Seeds and So Few Fruits?

In the Sonoran Desert of Arizona (Figure 31.13a), the giant saguaro produces as many as a hundred large, showy flowers at the tips of tall, spiny arms (Figure 31.6). Each flower blooms for only twenty-four hours. Insects visit it by day, and bats by night. The animals benefit by getting nectar and pollen. The giant saguaro benefits because the animals deliver their dusting of pollen grains to other giant saguaros. But therein lies a puzzle.

After giant saguaro flowers are pollinated, courtesy of insects or bats, they begin the task of producing seeds and fruits. Petals wilt, then they wither as the many egg cells inside the ovary become fertilized. Each egg-containing ovule expands and matures into a seed. During the same interval, the ovary wall matures into a fruit (Figure 31.13). Weeks later, the plum-sized fruit wall splits open, which exposes the dark seeds and bright-red tissues inside.

White-winged doves feast on the ripe fruits. When they fly away, each might carry hundreds of seeds inside its gut. A few seeds may elude the gut's digestive enzymes and later end up on the ground. There, each seed has a slim chance of growing up and becoming a giant saguaro.

One of the puzzling aspects of this annual event is the frequency with which saguaro flowers fail to give rise to fruit (Figure 31.13b, for example). Why would a cactus invest so much energy constructing a hundred flowers if only thirty or so of them will set fruit? Compare one of these "inefficient" plants with a species that produces a hundred flowers, all of which develop into fruit. Which do you think will leave more descendants?

Common sense might lead you to conclude, "The more fruits, the better." Yet saguaros—like many other plants—commonly produce far more flowers than they do mature fruits. They seem to be producing flowers excessively and giving up chances to make seeds and leave descendants. Doing so is not a reproductive adaptation that you would expect, based on Darwinian evolutionary theory.

Well, *suppose that some of the plants simply did not receive enough pollen to have all their egg cells fertilized.* After all, unfertilized flowers cannot produce seeds and fruits. This hypothesis has been experimentally tested for some plant species. Researchers carefully brushed pollen on every single flower. In many instances, the plants still failed to produce fruit for every flower!

Alternatively, *suppose a presumed "excess" of flowers is formed strictly to produce pollen for export to other plants.* Pollen grains are small. Energetically speaking, they are inexpensive to produce compared to large, calorie-rich, seed-containing fruits. For a fairly small investment, a plant might reap a big reward in offspring that carry its genes.

The idea that some kinds of plants actually set aside flowers exclusively for pollen export is only now being tested for saguaros. Perhaps you might like to design and carry out experiments that will help clear up the mystery of the "excess" flowers of saguaros and similar plants.

Figure 31.13 (**a**) Giant saguaro growing in Arizona's Sonoran Desert. (**b**) Two fruits. The one above set, and the one below it failed to set. (**c**) A red, maturing fruit.

31.7 Asexual Reproduction of Flowering Plants

LINKS TO
SECTIONS
9.2, 10.1

Sexual reproduction is the predominant reproductive mode among flowering plants. But many species also reproduce asexually, which permits rapid production of genetically identical offspring. We put this alternative to use when we propagate plants that interest us.

ASEXUAL REPRODUCTION IN NATURE

Many flowering plant species reproduce asexually by modes of **vegetative growth**, as listed in Table 31.2. In essence, the new roots and shoots grow right out from extensions or fragments of parent plants. Such asexual reproduction involves mitotic cell divisions and cell differentiation. Remember the introduction to Chapter 10? This means that a new generation of offspring is genetically identical to the parent; it is a clone.

A "forest" of quaking aspen (*Populus tremuloides*) is actually an example of vegetative reproduction. Figure 31.14 shows the shoot systems of an individual plant. That parent plant's root system keeps on giving rise to new shoots. One aspen forest in Colorado stretches across hundreds of acres and includes 47,000 shoots.

Barring rare mutations, trees at the north end of this clone are genetically identical with those at the south end. Water travels from roots near a lake all the way to shoot systems living in much drier soil. Dissolved ions travel in the opposite direction.

As long as environmental conditions favor growth and regeneration, such clones are about as close as an organism can get to being immortal. No one knows how old the aspen clones are. The oldest known clone might be a ring of creosote bushes (*Larrea divaricata*) growing in the Mojave Desert. It has been around for the past 11,700 years.

The reproductive possibilities are amazing. Watch strawberry plants send out horizontal, aboveground stems (runners), then watch the new roots and shoots develop at every other node. Most navel oranges we eat are produced by the descendants of one particular California tree that reproduced by **parthenogenesis**. By this asexual reproductive mode, embryos develop from unfertilized eggs or somatic cells in ovules.

Parthenogenesis is sometimes stimulated when a pollen grain contacts a stigma but a pollen tube does not grow through the style. In some cases, hormones released by the stigma or pollen grains diffuse to the unfertilized egg and stimulate it to develop into an embryo. That embryo is $2n$ as a result of fusion of the products of mitotic cell division in the egg. A $2n$ cell outside the gametophyte also may be stimulated to develop into an embryo. In all cases, all of the plants that are produced by parthenogenesis have the same genome.

Table 31.2   Asexual Reproductive Modes of Flowering Plants

Mechanism	Examples	Characteristics
Vegetative Reproduction on Modified Stems		
1. Runner	Strawberry	New plants arise at nodes along aboveground horizontal stems.
2. Rhizome	Bermuda grass	New plants arise at nodes of underground horizontal stems.
3. Corm	Gladiolus	New plants arise from axillary buds on short, carbohydrate-storing, underground stems.
4. Tuber	Potato	New shoots arise from axillary buds (tubers are enlarged tips of slender underground rhizomes).
5. Bulb	Onion, lily	New bulbs arise from axillary buds on short underground stems.
Parthenogenesis		
	Orange, rose	Embryo develops without nuclear or cellular fusion (for example, from an unfertilized haploid egg or by developing adventitiously, from tissue surrounding the embryo sac).
Vegetative Propagation		
	Jade plant, African violet	New plant develops from tissue or structure (a leaf, for instance) that drops from the parent plant or gets separated from it.
Tissue Culture Propagation		
	Orchid, lily, wheat, rice, corn, tulip	New plant induced to arise from a parent plant cell that is not irreversibly differentiated.

INDUCED PROPAGATION

Most houseplants, woody ornamentals, and orchard trees are clones. People typically propagate them from cuttings or fragments of shoot systems. For example, with suitable encouragement, a severed African violet leaf may form a callus from which adventitious roots develop. A callus is one type of meristem. Remember, a meristem is a localized region of young plant cells that retain the potential for mitotic cell division.

A twig or bud from one plant may be grafted onto a different variety of some closely related species. As one example, many vintners in France now graft their prized but disease-prone grapevines onto disease-resistant root stock imported from America.

Frederick Steward was one of the pioneers in the field of **tissue culture propagation**. He placed small bits of phloem from the roots of carrot plants (*Daucus carota*) in flasks along with a liquid growth medium. The liquid included sucrose, minerals, and vitamins.

Figure 31.14 Quaking aspen (*Populus tremuloides*), a plant that reproduces asexually by runners, which give rise to a stand of trees that is an impressive clone.

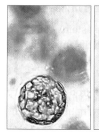

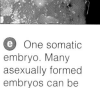

a Protoplast. This is a "naked" living plant cell without a wall, which digestive enzymes have stripped away.

b Protoplast regenerates the missing cell wall, and then mitotic cell divisions begin.

c Repeated divisions result in a clump of living cells. All of these cells are not yet differentiated.

d Cell divisions continue and result in a callus, a mass of undifferentiated cells that can be induced to give rise to a plant embryo.

e One somatic embryo. Many asexually formed embryos can be produced at the same time in tanks (bioreactors).

f Young plant that grew from a somatic embryo. Clones of thousands of plants are produced by such micropropagation methods.

Figure 31.15 One method of tissue culture propagation.

It also contained coconut milk, which Steward knew was rich in a then-unidentified chemical that induces cell divisions. As the flasks rotated, individual cells torn away from the tissue bits divided and formed multicelled clumps. Some clumps gave rise to new plants (Figure 31.15). These experiments were among the first to show that cells of specialized plant tissues still retain the genetic instructions that are required to produce a whole individual.

Researchers now use shoot tips and other parts of individual plants for tissue culture propagations. The techniques often prove useful when an advantageous mutant arises. For instance, a mutant may show better resistance to a disease that can cripple other plants of the same species. Tissue culture propagation is used to produce hundreds or thousands of identical plants from a single promising mutant. These techniques are now being used to improve food crops, such as corn, wheat, soybeans, and rice. They also are being used commercially to increase production of hybrid lilies, orchids, and other prized ornamentals.

Flowering plants reproduce asexually by diverse processes. Clonal propagation in nature has resulted in some of the world's largest individual plants, and the oldest.

Plants can also be propagated asexually by culturing bits of plant tissue in the laboratory. Asexual propagation yields genetically uniform copies of the plants of interest.

32.2 Plant Hormones and Other Signaling Molecules

LINKS TO
SECTIONS
15.1, 15.2, 28.5

As a plant grows and develops, its diverse cells increase in number, size, and volume—and specialized tissues form. These events require chemical communication among different cell types that secrete and respond to hormones and growth regulators.

MAJOR TYPES OF PLANT HORMONES

Table 32.1 lists five major classes of plant hormones—gibberellins, auxins, cytokinins, ethylene, and abscisic acid. These hormones interact in ways that stimulate or inhibit growth and development.

Gibberellins Let us first look at those **gibberellins** you read about in the chapter introduction. They are acidic compounds synthesized in seeds and young shoot tissues. As seeds germinate, gibberellins induce primary roots and shoots to grow. Their action causes the release of stored nutrients, which fuel cell divisions and elongation —especially between stem nodes, as in Figure 32.5. They also induce biennials and long-day plants to

Figure 32.5 Foolish cabbages! At left, in front of the ladder, are two untreated cabbages that were the controls. At right, three cabbage plants treated with gibberellins.

flower (Section 32.5). They are present in all flowering plants, mosses, gymnosperms, ferns, and some fungi.

Remember Mendel's dwarf pea plants (Section 11.3)? As in other dwarf varieties of plants, the short stems result from a mutation that affects gibberellin's action.

Auxins Cells at apical meristems of roots and shoots produce **auxins**. So do developing leaves and seeds. In monocots, an auxin is synthesized in the coleoptile, a protective sheath that surrounds the early seedling. As Section 32.3 explains, auxins induce cell walls to soften, so young cells can expand. Gradients of auxin affect which genes are transcribed in different tissues. They cause leaves to grow in certain patterns, stems to bend toward a light source, and roots to grow down through soil. Auxins induce vascular tissue formation and the division of vascular cambium cells. Maturing seeds produce auxins that encourage fruit formation.

Auxins also have inhibitory effects. Auxin diffusing down from a lengthening shoot tip stops lateral buds from growing. This inhibitory control is called *apical dominance*. When gardeners pinch off shoot tips, lateral buds along stems can grow and plants can get bushier.

Auxins help to prevent **abscission**—the dropping of leaves, flowers, or fruits. IAA (indoleacetic acid) is the most pervasive auxin in nature. Synthetic auxins have important commercial applications (Table 32.2).

Cytokinins Unlike other plant hormones, **cytokinins** are also synthesized by animal cells. They signal cells to start dividing rapidly. Root cells produce most of a plant's cytokinins, and xylem moves them into shoots. Cytokinins induce divisions in apical meristems and in maturing fruits. They also can release lateral buds

Table 32.1 Major Classes of Plant Hormones and Their Main Effects

Hormone	Source and Mode of Transport	Stimulatory or Inhibitory Effects
Gibberellins	Young tissues of shoots, seeds, possibly roots. May travel in xylem and phloem	Make stems lengthen greatly (stimulates cell division, elongation). Help seeds germinate; help induce flowering in some plants
Auxins	Apical meristems of shoots, coleoptiles, seeds. Polar transport through parenchyma cells of shoots toward base of root	Lengthen shoots, coleoptiles. Promote vascular cambium activity, vascular tissue differentiation. Inhibit abscission. Fruit formation. Block lateral bud formation (apical dominance)
Cytokinins	Mainly the root tip. Travels from roots to shoots inside xylem	Stimulate cell division in roots and shoots, leaf expansion. Inhibit leaf aging. Applications release buds from apical dominance
Ethylene (a gas)	Most tissues undergoing ripening, aging, or stress. Diffusion in all directions	Affects orientation of tissues by promoting or inhibiting cell growth; fruit ripening; promotes abscission; aging and death (senescence)
Abscisic acid (ABA)	Root cells in response to water stress. Travels from roots to leaves inside xylem	Stimulates stomatal closure. Induces transport of photosynthetic products from leaves to seeds. Stimulates embryo formation in seeds. May induce and maintain dormancy in some species

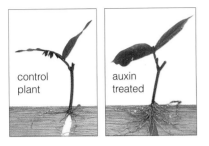

control plant

auxin treated

Figure 32.6 Plant cutting dipped into a rooting powder with 0.8 percent synthetic auxin. It developed more roots than the untreated control.

from apical dominance, and they can stop leaves from aging prematurely.

Ethylene The only gaseous hormone, **ethylene**, can promote or inhibit cell growth so that tissues expand in the most suitable directions. Ethylene also induces fruit ripening. Its concentration is highest in apples, bananas, avocados, and other fruits that dramatically step up aerobic respiration as they mature. Ethylene concentrations also are high when a plant is stressed, as happens in autumn or near the end of the life cycle. At such times, ethylene induces abscission of leaves and fruits, and often the death of the whole plant.

Abscisic Acid Growth and reproduction depend on water and dissolved minerals. When a plant is water stressed, root cells produce more **abscisic acid** (ABA), which xylem moves swiftly to leaves. ABA is part of a stress response that causes stomata to close so that the plant minimizes water loss (Section 30.4). Also, when the growing season ends, ABA's influence overrides growth-promoting effects of gibberellins, auxins, and cytokinins. ABA causes photosynthetic products to be diverted from leaves to seeds, where it promotes the synthesis of storage forms of proteins and also helps the embryo mature. It may affect dormancy in some plants. This growth-inhibiting hormone actually was misnamed; it has little to do with abscission.

OTHER SIGNALING MOLECULES

As we now know, other signaling molecules have roles in plant growth and development. *Brassinosteroids* are like animal steroid hormones in their structure. They help promote cell division and elongation; stems stay short in their absence. *Jasmonates*, derived from fatty acids, help other hormones control seed germination, root growth, and tissue defense. *FT protein* is part of a signaling pathway that induces flower formation.

Signaling molecules also function in plant defenses. *Salicylic acid* is a phenol structurally similar to aspirin.

Together with *nitric oxide*, it induces synthesis of gene products that help plants survive pathogenic attacks. *Systemin* forms when insects feed on plant tissues. This peptide activates transcription of genes for substances that cripple an insect's capacity to digest proteins.

COMMERCIAL USES

Table 32.2 lists natural and synthetic plant hormones of commercial interest. An auxin in a rooting powder makes cuttings of a desired plant develop roots faster (Figure 32.6). Applications of ethylene make orchard fruits ripen all together, and quickly. Many fruits are harvested "green" to minimize bruising when shipped, and they are doused with ethylene after they arrive at distribution centers. Gibberellin applications promote increases in fruit sizes. Growers spray synthetic auxins on unpollinated flowers to get seedless fruits. Another synthetic auxin, dichlorophenoxyacetate (2,4–D), finds wide use as an herbicide. It so accelerates the growth of eudicot weeds that cells cannot take up or distribute enough water, nutrients, and photosynthetic products to keep up. The weeds grow themselves to death.

Interactions among hormones and other kinds of signaling molecules govern the normal growth, development, daily functioning, and reproduction of plants.

Plant hormones are signaling molecules secreted by specific cells that alter activities in target cells. At specific times, they promote or arrest growth by stimulating or inhibiting cell division, elongation, differentiation, and other events.

The five main classes are gibberellins, auxins, cytokinins, abscisic acid, and ethylene. Brassinosteroids and other growth regulators have been discovered.

32.3 Mechanisms of Plant Hormone Action

LINKS TO
SECTIONS 2.6,
3.2, 3.3, 5.3, 28.5

How do plant hormones exert their effects? Here are two examples, one from a germinating seed, and the other from a lengthening seedling.

SIGNAL TRANSDUCTION

Plants, too, have pathways of cell communication, as introduced in Chapters 15 and 28. First, one cell type secretes a hormone or another signaling molecule that binds with a receptor on a target cell. Then the signal is transduced to a form that may influence a metabolic pathway, gene expression, or membrane properties. Often the receptor's shape changes, which activates enzymes or other cytoplasmic components. Reactions begin, and they bring about a cellular response to the initial signal. Figure 32.7 is one way to think about this pathway of cell-to-cell communication.

HORMONE ACTION IN GERMINATION

Figure 32.8 shows a barley seed. During germination, the imbibed water moves through cells of the embryo, which release gibberellin. Molecules of this hormone act in the endosperm's aleurone, a layer of cells that stores protein. (The brown part of brown rice is an example of the aleurone.) The hormonal signal induces transcription of the gene for amylase, an enzyme that can hydrolyze starch molecules (Sections 3.2 and 3.3).

Meanwhile, incoming water is activating protein-digesting enzymes in the aleurone. They break down proteins to free amino acids, some of which are taken up as building blocks in the synthesis of amylase.

The amylase is released into the starch-rich part of endosperm, where its action makes sugar monomers available. The transportable sugars, as well as amino acids, fuel the young plant's rapid growth.

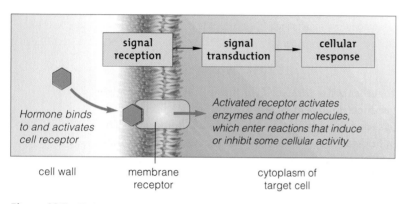

signal reception → signal transduction → cellular response

Hormone binds to and activates cell receptor

Activated receptor activates enzymes and other molecules, which enter reactions that induce or inhibit some cellular activity

cell wall   membrane receptor   cytoplasm of target cell

Figure 32.7 Pathway of cell communication in plants.

a Imbibed water stimulates cells of embryo to release gibberellin, which water moves to cells of aleurone. Water also activates protein-digesting enzymes.

b In cells of the aleurone layer, this hormone triggers transcription and translation of a gene for amylase, an enzyme that digests starch into a transportable sugar.

c Amylase moves into endosperm's starch-rich cells. Sugar monomers released from starch fuel aerobic respiration.

d The ATP from aerobic respiration provides the energy for growth of the primary root and shoot.

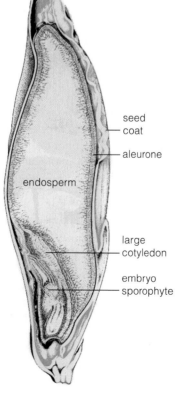

seed coat

aleurone

endosperm

large cotyledon

embryo sporophyte

Figure 32.8 Action of gibberellin in barley seed germination.

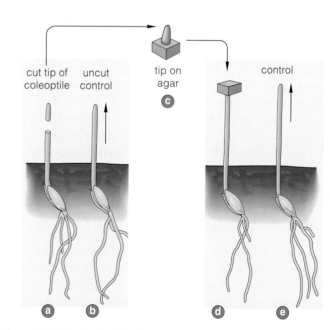

cut tip of coleoptile   uncut control

tip on agar

c

control

a   **b**   **d**   **e**

Figure 32.9 *Animated!* Experiment to test whether auxin from a coleoptile tip induces seedling elongation. (**a**) A seedling with its tip cut off will not lengthen as much as an uncut control seedling (**b**). (**c**) Put a tiny block of agar under a cut tip for several hours so auxin can move into it. (**d**) Place the agar on top of a different de-tipped coleoptile. Cells below it will lengthen about as fast as the control (**e**).

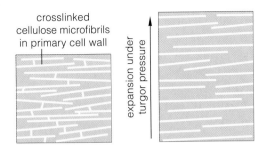

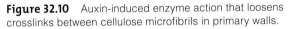

Figure 32.10 Auxin-induced enzyme action that loosens crosslinks between cellulose microfibrils in primary walls.

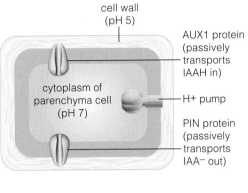

POLAR TRANSPORT OF AUXIN

Auxin concentration gradients start forming during early cell divisions of the embryo sporophyte. Cells exposed to higher concentrations transcribe different genes than those exposed to lower concentrations. The regionally different gene products help form plant parts, such as leaves, in expected patterns. Auxin also helps young cells elongate. It causes crosslinks among cellulose microfibrils in cell walls to loosen, so shoots and roots can lengthen (Figures 32.9 through 32.11).

The auxin concentration is highest at its source, the apical meristem in a shoot (or coleoptile). From there, auxin is transported down, toward the shoot's base. This polar transport takes place in parenchyma cells, at membrane transporters that actively move ionized auxin into and out of their cytoplasm (Figure 32.11). Auxin gives up hydrogen in each cell, which alters the cytoplasmic pH. Membrane pumps actively transport the H⁺ outside, which lowers the pH of the moist cell wall. Enzymes in the wall become active at a low pH. They cleave crosslinks between the microfibrils, which support the wall. Meanwhile, water is diffusing into the cell, so turgor pressure builds up against the wall. Because the microfibrils are now free to move apart, the wall is free to expand, and so the cell lengthens.

The pH changes also activate transcription factors. Within twenty minutes after auxin exposure, proteins that help a cell assume its new shape are synthesized.

Plant hormones exert their effects by pathways of signal reception, signal transduction, and a cellular response.

The nutrients in endosperm become available for growth when the embryo sporophyte releases gibberellin.

Auxin is actively transported down from shoot tips. By inducing pH changes, it causes cellulose microfibrils to loosen in cell walls, so the walls (and cells) can expand during growth. It also triggers regional gene transcription.

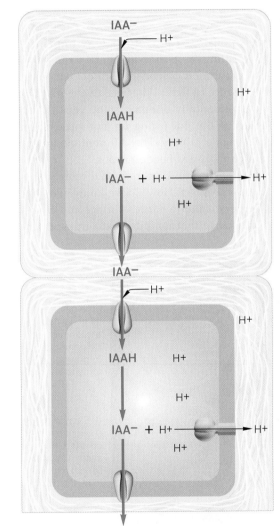

a As auxin diffuses through cell wall, the low pH makes H⁺ bind to it. The resulting non-ionized form is IAAH.

b AUX1 transports IAAH into cytoplasm.

c The higher cytoplasmic pH makes auxin give up H and revert to ionized form.

d H⁺ pumped out of cell, pH in wall decreases and prods enzymes to cleave cellulose crosslinks.

e PIN protein passively transports auxin out.

f Steps **a** through **e** are repeated in each adjoining parenchyma cell. The auxin transport shows polarity, from its source in a shoot tip and leaves, downward toward the base of the stem.

Auxin moves down to roots through parenchyma cells of the vascular cylinder. A different transport process takes over in the root tip. Transport proteins in the parenchyma cells of the root epidermis and cortex move auxin up toward the root–shoot junction.

Figure 32.11 Polar transport of auxin (IAA) in primary shoots and roots. The uppermost sketch simply identifies the components involved. Parenchyma cells transport auxin from apical meristem downward, to the shoot–root junction. The resulting changes in hydrogen ion concentration gradients across the plasma membrane cause an enzyme-induced loosening of cellulose crosslinks in the cell wall. The wall, and the cell, can expand during growth, as turgor pressure builds up in the cytoplasm.

A different mechanism moves auxin molecules upward from the root tip to the shoot–root junction.

32.4 Adjusting the Direction and Rates of Growth

LINKS TO
SECTIONS
7.1, 29.5

Young plant roots and shoots adjust their direction of growth by turning toward or away from an environmental stimulus. Such responses are tropisms (after the Greek trope, for turning). Roots and shoots also alter their growth patterns in response to mechanical stress.

RESPONSES TO GRAVITY

Figure 32.12*a* shows the normal direction of growth of a corn seedling. Its primary root—the first to break through the seed coat—curves down, and the coleoptile and primary shoot curve up. Turn the seedling upside down, and its primary root curves down and the shoot curves up, as in Figure 32.12*b*. Any growth response to Earth's gravitational force is a form of **gravitropism.**

Turn the seedling on its side in a dark room and it still makes a gravitropic response. Its shoot curves up even in the absence of light. What makes the plant do this? In that horizontally oriented stem, cells do not elongate much on the side facing up, but those on the lower side elongate rapidly. The different elongation rates result in an upward-bended shoot.

Auxin, and auxin transporters in cell membranes, function in gravitropism. How do we know? Position seedlings perpendicular to Earth's surface and expose them to a substance that inhibits auxin transporters.

Unlike untreated control seedlings, the seedlings will not bend (Figure 32.12*c,d*).

Gravity-sensing mechanisms of many organisms are based on **statoliths.** Plant statoliths are modified plastids made denser by clusters of starch grains. In response to gravity, statoliths in specialized root cap cells settle until they rest in the lowest cytoplasmic region (Figure 32.13). When a root is reoriented with respect to gravity, the shifting statoliths (and possibly other organelles) may redistribute auxin in root cells. Such a change initiates the gravitropic response.

RESPONSES TO LIGHT

When light streams in from one direction only, a stem or leaf adjusts its rate and direction of growth so that it grows toward the light source. This response is a form of **phototropism.** In plants, it orients photosynthetic cells in ways that maximize light interception.

Charles Darwin was the first to make note of the directional bending of a coleoptile in response to light. Later, Fritz Went proposed that a growth-promoting substance, which he called auxin, moves down from the coleoptile tip and causes the bending. Because the auxin moves down faster through cells on the side less exposed to light, it makes them elongate faster than

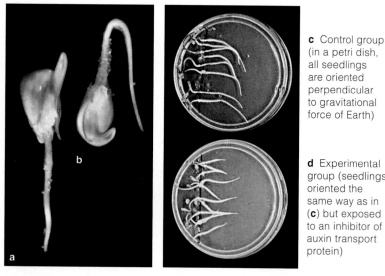

c Control group (in a petri dish, all seedlings are oriented perpendicular to gravitational force of Earth)

d Experimental group (seedlings oriented the same way as in (**c**) but exposed to an inhibitor of auxin transport protein)

Figure 32.12 Gravitropic responses of a corn primary root and shoot growing in a normal orientation (**a**), and turned upside down (**b**).

(**c**) Experimental test of whether gravitropism requires auxin transport proteins in root cell membranes. In primary roots turned on their side, auxin is laterally transported to the down-facing side, where it stops cells from lengthening. Cells of the upward-facing side continue to lengthen, so the roots bend downward. (**d**) In an experimental group exposed to auxin transport inhibitors, downward bending stopped. Mutations in genes that encode auxin transport proteins also stop the bending.

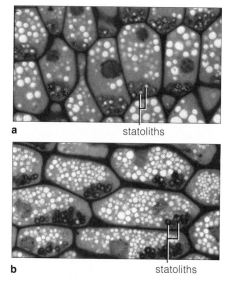

a statoliths

b statoliths

Figure 32.13 *Animated!* Gravity and statolith distribution. (**a**) Normal orientation of gravity-sensing cells in a corn root cap. Statoliths in the cells settle downward. (**b**) Turn the root sideways. Ten minutes later, the statoliths have settled to the new "bottom" of the cells. A gravity-sensing mechanism using statoliths may induce auxin redistribution in root tips. A difference in auxin concentrations makes cells on the top side of a horizontal root lengthen faster than cells on the bottom. Differences in elongation rates make the root tip curve downward.

Rays of sunlight strike one side of a coleoptile.

b The coleoptile bends after the auxin diffuses from tip to cells on shaded side.

a

cells on the illuminated side. The difference in growth rates bends the coleoptile, as Figure 32.14a,b indicates. Position seedlings of a sun-loving plant in a dark room next to a window through which sunlight is streaming. They, too, bend toward the light (Figure 32.14c).

In plants, wavelengths of blue light stimulate the strongest phototropic responses. *Flavoproteins* absorb them. These nonphotosynthetic, yellow pigments are light-sensitive membrane receptors. They transduce energy of sunlight into signals that can stimulate the redistribution of auxin to the shaded side of a shoot or coleoptile. Auxin is moved by the polar transport mechanism explained earlier, in Figure 32.11.

Figure 32.14 *Animated!* Phototropism. (**a,b**) Hormone-mediated differences in cell elongation rates along its length will induce a coleoptile to bend toward light. (**c**) Flowering shamrock (*Oxalis*) seedlings responding to light.

RESPONSES TO CONTACT

In **thigmotropism**, auxin and ethylene induce plants to adjust the direction of growth if they contact objects. Soft, flimsy vines and tendrils (new, modified leaves or stems) cannot grow upright unless they make such a response. When cells at the shoot tip touch a stable object, cells on the contact side stop elongating and the cells on the other side keep growing. Their unequal rates of growth make the vine or tendril curl around the object, often more than once (Figure 32.15). Then cells on both sides resume growth at the same rate.

Figure 32.15 Passion flower (*Passiflora*) tendril twisting thigmotropically.

RESPONSES TO MECHANICAL STRESS

Mechanical stress, as inflicted by prevailing winds and grazing animals, inhibits stem lengthening. Trees high up in windswept mountains are stubbier than sheltered trees of the same species at lower elevations (Section 23.7). Similarly, plants grown outdoors commonly have shorter stems than the same kinds of plants grown in a greenhouse. Briefly shake a plant every day and you will inhibit its overall growth (Figure 32.16).

Plants adjust the direction and rate of growth in response to environmental stimuli, as when hormones induce a shoot to bend by making some cells lengthen faster than others. Plants also adjust growth patterns by changing gene expression in response to mechanical stimulation.

Figure 32.16 Effect of mechanical stress on tomato plants. (**a**) This plant, the control, grew in a greenhouse. (**b**) Each day for twenty-eight days, this plant was mechanically shaken for thirty seconds. (**c**) This one had two shakings each day.

32.5 Seasonal Shifts in Growth

LINKS TO
SECTIONS
7.1, 15.2, 28.3

*Plant growth responses correlate with environmental cues,
such as seasonal shifts in night length and temperature.*

A *circadian* cycle is one that is completed in a period
of about twenty-four hours. **Photoperiodism** refers to
a biological response to alternations in the length of
darkness relative to daylight during a circadian cycle.
For example, the number of hours a plant spends in
darkness and in daylight shifts with the seasons. The
shift is more pronounced the farther away you move
from Earth's equator, as Section 48.1 explains. The life
cycles of different plant species show adaptations to
shifts in the hours of daylight (Figure 32.17).

Like other organisms, plants have **biological clocks**:
internal mechanisms that preset the time for recurring
shifts in daily tasks or seasonal patterns of growth,
development, and reproduction. Some clocks in plants
use **phytochrome**. This blue-green pigment functions
as a receptor for red and far-red light. The red light at
sunrise causes phytochrome to shift from its inactive
form (Pr) to its active form (Pfr). The far red light at
sunset, and darkness, shift phytochrome to inactive
form (Figure 32.18). The longer the nights, the longer
the interval when phytochrome is inactivated.

Phytochrome's active form—Pfr—can induce gene
transcription. The gene products are diverse signaling
molecules that bring about seed germination, shoot
elongation and branching, leaf expansion, and flower,
fruit, and seed formation, then dormancy.

*Photoperiodic responses involve phytochrome. The activated
form of this photoreceptor triggers secretion of one or more
hormones that can induce and inhibit plant growth.*

32.6 When To Flower?

*Hormones activate the genes for flower formation.
But what activates the hormone-secreting cells?*

RESPONSES TO HOURS OF DARKNESS

Different species are known as short-day, long-day,
and day-neutral plants. The names are misleading. The
cue is length of *darkness*. "Short-day" plants flower in
early spring or fall, when nights are longer than some
critical value; "long-day" plants flower in summer,
when nights are shorter than a critical value (Figure
32.19a,b). Day-neutral plants flower whenever they
are mature enough to do so.

Phytochrome is the trigger for flowering. In certain
experiments, a pulse of red light interrupted the dark
phase of the circadian cycle for a long-day plant and a
short-day plant (Figure 32.20a,b). The light activated
phytochrome, which made the plants act as if nights
were short. In other experiments with the same kinds
of plant, a pulse of far-red light after a pulse of red
light inactivated phytochrome molecules. Both plants
acted as if nights were long, and again they responded
in predictable ways (Figure 32.20c).

Detection of a photoperiod occurs in leaves, where
hormones inhibit or induce a shift from leaf growth to
flower formation. Signals from leaves may travel in
phloem to floral buds. In one experiment, a short-day
plant called cocklebur was stripped of all but one leaf,
which was kept in darkness for 8–1/2 hours a day to
simulate long nights. The plant flowered. Later on, the
leaf was grafted onto a cocklebur that had *not* been
exposed to long dark periods. The plant flowered.

Apparently, flower-inducing and flower-inhibiting
substances participate in flowering. Intriguingly, the
gibberellins can induce flowering in some species.

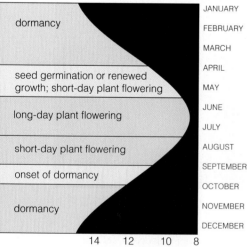

Figure 32.17 Plant growth
and development correlated
with seasonal changes in
hours of darkness. The
data reflect photoperiodic
responses of diverse plants
in northern temperate zones,
where the temperature and
rainfall shift over the year.

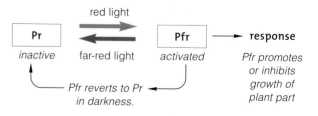

Figure 32.18 *Animated!* Conversion of phytochrome,
a light-sensitive receptor, from inactive to active form: Pr to
Pfr. When activated, this light-sensitive receptor induces gene
transcription. The signal induces cells to take up free calcium
ions (Ca^{++}) or it induces organelles to release them. The
response starts as the ions combine with calcium-binding
proteins. At sunset, at night, or in the shade, far-red light
predominates, and so the signal transduction pathway is
reversed. At such times, Pfr reverts to Pr.

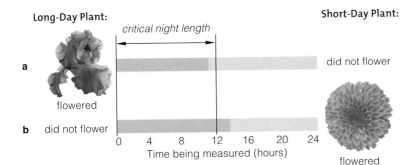

Long-Day Plant:

critical night length

a

flowered

b did not flower

Time being measured (hours)
0 4 8 12 16 20 24

Short-Day Plant:

did not flower

flowered

Figure 32.19 *Animated!* Experiments showing that plants flower in response to night length. Each horizontal bar represents 24 hours. *Yellow* signifies daylight; *blue-gray* signifies night. (**a**) This long-day plant, an iris, flowered only when hours of darkness were *less* than the value that is critical for flowering of its species. (**b**) This short-day plant, a chrysanthemum, flowered only when hours of darkness were *more* than a critical value for its species.

REVISITING THE MASTER GENES

In Sections 15.2 and 17.8, you read about the genetic basis for flowering. Briefly, three groups of master genes, designated *A*, *B*, and *C*, control the formation of floral structures from the whorls of a floral shoot.

For some time, no one knew what switches on the master genes. Indirect evidence pointed to some kind of signal, which was tentatively named "florigen." We now know that, in response to photoperiods or other environmental cues, leaf cells transcribe a flowering gene. The mRNA transcripts travel in phloem to as-yet undifferentiated floral buds, where they are translated into *FT* protein. This signaling molecule, along with a transcription factor, turns on master genes that cause cellular descendants of an as-yet uncommitted bud of meristematic tissue to develop into a flower.

VERNALIZATION

Unless certain biennials and perennials are exposed to low winter temperatures, flowers will not form on their stems in spring (Figure 32.21). Low-temperature stimulation of flowering is known as **vernalization** (from *vernalis*, meaning "to make springlike"). In one experiment, stored seeds of winter rye (*Secale cereale*) were exposed to near-freezing temperatures in late spring (not winter) and then planted. Plants grown from the seeds still flowered in summer.

a

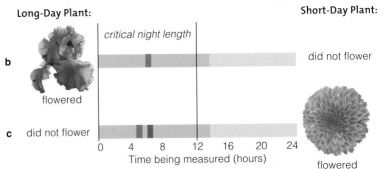

Long-Day Plant:

critical night length

b

flowered

c did not flower

Time being measured (hours)
0 4 8 12 16 20 24

Short-Day Plant:

did not flower

flowered

Figure 32.20 (**a**) Flowering responses of chrysanthemum, a short-day plant, to (*left*) long-day exposure and (*right*) short-day exposure. Two experiments that pointed to phytochrome as a trigger for flowering in long-day and short-day plants. (**b**) When an intense red flash interrupts a long night, both plants respond as if it were a short night (long day). The long-day plant flowers; the short-day plant does not. (**c**) A short pulse of far-red light after the red flash cancels the disruptive effect of the red flash by inactivating phytochrome. The long-day plant does not flower, but the short-day plant does.

The main environmental cue for flowering is night length, which varies seasonally. Genes influence the time of flowering in different species.

Photoperiodic responses to red and far-red light involve phytochrome. The activated form of this photoreceptor, Pfr, may trigger secretion of one or more hormones that induce and inhibit flowering at different times of year.

Low winter temperatures stimulate the flowering of many plant species in spring.

Figure 32.21 Local effect of cold temperature on dormant buds of a lilac (*Syringa*) plant. For this experiment, a single branch was positioned to protrude from a greenhouse through a cold winter. The rest of the plant was kept inside and exposed only to warm temperatures. Only buds exposed to the low outside temperatures resumed growth and flowered in springtime.

32.7 Entering and Breaking Dormancy

LINKS TO
SECTIONS
4.9, 29.2, 30.4

In midsummer, days almost imperceptibly begin to grow shorter, and many herbaceous and woody perennials of temperate zones start to shut down growth. They do so even when temperatures are mild, the sky is bright, and water is plentiful. They prepare for **dormancy,** *a period of arrested growth that will not end until the arrival of special environmental cues.*

ABSCISSION AND SENESCENCE

As leaves and fruits grow in early summer, their cells make auxin, which moves into stems. Together with cytokinins and gibberellins, the auxin helps maintain growth. By midsummer, the hours of daylight start to decrease. Plants start to divert nutrients from leaves, stems, and roots to flowers, fruits, and seeds. As the growing season comes to an end, deciduous species move nutrients to storage sites in twigs, stems, and roots before the leaves die and drop. The dropping of flowers, leaves, fruits, and other plant parts is called **abscission.** Parenchyma cells form an abscission layer as well as a protective, suberized layer at the base of a petiole or some other structure about to drop from the plant. Figure 32.22 shows a suberized layer.

Abscission occurs in response to decreasing hours of daylight, even to drought, injuries, and nutrient deficiencies. Auxin production declines in leaves and fruits. A different signal stimulates cells of abscission zones to make ethylene. Cells enlarge, deposit suberin in their walls, and make enzymes that digest cellulose and pectin in the middle lamella, the cementing layer between plant cell walls (Section 4.9). Cells separate from their neighbors as they enlarge and as enzymes digest the walls. Eventually leaves or other structures above the abscission zones drop away.

What happens when you interrupt the diversion of nutrients into flowers, seeds, and fruits? Remove all flowers or seed pods from a plant, and its leaves and stems stay green much longer (Figure 32.23).

Senescence refers to the phase from full maturity until the eventual death of plant parts or the whole plant. Aging and death are part of plant life cycles.

BUD DORMANCY

By midsummer, buds of many trees do not lengthen even when the meristems are active. Their embryonic shoots, complete with nodes and internodes as well as rudimentary leaves, are housed in bud scales that will insulate them and keep them from drying out.

Short days, long, cold nights, and dry soil that is deficient in nitrogen are strong cues for dormancy. Researchers tested this by interrupting the long dark period of some Douglas firs with a short period of red light. The plants responded as if nights were shorter and days longer. They continued to grow taller (Figure

Figure 32.22 In autumn, the changing colors of leaves of a horse chestnut tree (*Aesculus hippocastanum*). The leaf scar at *right* is all that remains of an abscission zone. Before the leaf detached from the tree, a tissue formed in a horseshoe-shaped zone (hence the tree's name). The tissue helps prevent water loss and infection after plant parts drop away.

control (pods   experimental plant
not removed)   (pods removed)

Figure 32.23 The observable results from an experiment in which seed pods were removed from a soybean plant. Removal delayed its senescence.

32.8 Plants Move

FOCUS ON THE
ENVIRONMENT

LINKS TO
SECTIONS
28.4, 30.4

Figure 32.24 Experimental test of the effect of night length on Douglas firs. The fir at left was exposed to 12-hour light/12-hour dark cycles for one year. Its buds stayed dormant; "daylength" was too short. The fir at right was exposed to 20-hour light/4-hour dark cycles. It continued growing. The middle fir was exposed to 12-hour light/11-hour dark cycles. But 1 hour of light interrupted the middle of the dark period and prevented bud dormancy. The interruption caused Pfr to form at a sensitive time in the normal day–night cycle.

In this chapter, we have focused primarily on plant growth responses to seasonal changes in environmental cues. Bear in mind, plants also respond to daily changes in ways that indirectly affect growth. Some of these responses may be considered homeostatic, because they help maintain internal operating conditions for cells. Remember those folding leaves in Section 28.4? Here is one more example for you to consider.

Many plants track the course of the sun as it appears to arc overhead during the day. On hot, dry days, the plants move their soft petioles so that the flat surfaces of leaves (and flowers) are less perpendicular to the sun. A petiole, remember, is the stalk that attaches a leaf to a stem. In this orientation, heat absorption is minimized. At other times, the plants continually reposition the flat surfaces of leaves and flowers to intercept the sun's rays. Those sunflowers shown in Figure 32.25 make this response. So do lupine, cotton, and soybean plants.

Any photoperiodic response to the changing angle of the sun through the day is called **solar tracking**. Unlike the photo*tropic* responses explained in Section 32.4, such a response does not involve asymmetric growth (whereby cells on the shaded side of a shoot elongate faster than those in the light). Rather, petiole cells undergo reversible changes. The mechanisms involved are similar to those associated with the opening and closing of stomata; they cause reversible changes in turgor pressure in the cells. Collectively, the changes bend and straighten the petiole.

32.24). In this experiment, conversion of Pr to Pfr by red light during the dark period prevented dormancy. In nature, buds might enter dormancy because less Pfr can form when daylength shortens in late summer.

The requirement for multiple cues for dormancy has adaptive value. If temperature were the only cue, warm autumn weather might make plants flower and seeds germinate—and winter frost would kill them.

Dormancy-breaking mechanisms operate between late fall and spring. Temperatures often grow milder, and rains and nutrients become available. Breaking dormancy requires gibberellins and abscisic acid, and often prior exposure to low temperatures at specific times of year. Generally, trees in the southern United States require less cold exposure than trees growing in the northern states and Canada.

With environmental cues and hormone interactions, many plants seasonally enter and break dormancy.

Figure 32.25 From the American Midwest, a field of sunflowers (*Helianthus*) that are busily demonstrating solar tracking.

32.9 Regarding the World's Most Nutritious Plant

In this unit, you explored the nature of plant structure, function, and behavior. Consider, finally, an example of how individuals put knowledge of plants to use in ways that have tremendous impact on human lives.

Alejandro Bonifacio was raised in poverty in rural Bolivia. As a child he spoke Aymaran, a language that predates the Incas. He learned Quechua, the language of the Incas, then learned Spanish before college. He earned a bachelor's degree in agriculture and became a plant breeder for Bolivia's equivalent of the United States Department of Agriculture.

His research interest is *Chenopodium quinoa*, a plant that originated in the Andes. Quinoa is a leafy eudicot, a distant relative of spinach and beets. Its seeds are not a cereal grain, but they are nutritional treasures, and for many thousands of years the seeds have been a staple of Latin American diets.

Quinoa seeds are 16 percent protein, on average. By comparison, wheat seeds are about 12 percent protein and rice seeds, about 8 percent. However, the proteins in both wheat and rice are deficient in the amino acid lysine. The array of amino acids in quinoa is as good as that in milk. The seeds also contain more iron than most cereal grains and are a notable source of calcium, phosphorus, and many B vitamins. In addition, quinoa plants are highly resistant to drought, frost, and saline soils. It is the only food crop that can grow in the arid salt deserts that prevail in much of Bolivia.

Far to the north, even before Alejandro received his college scholarship, Daniel Fairbanks had accepted a position as a botanist at Brigham Young University (BYU). He, too, became interested in quinoa because of its potential to feed millions in Bolivia and Peru, the most impoverished countries of Latin America. There, the majority of families eke out a living as subsistence farmers, and *kwashiorkor* is common.

Kwashiorkor is a form of malnutrition that results from protein-deficient diets. Fatigue, drowsiness, and irritability are its early symptoms. Continued protein deficiency adversely affects body growth. Edema sets in, and muscle mass decreases. The immune system starts to weaken. The abdomen of a malnourished person often protrudes. Epidermal conditions, such as vitiligo and thinning hair, are common. Severe cases result in mental and physical problems that can lead to coma and death. Kwashiorkor is most common in impoverished countries, but as many as 50 percent of the elderly in nursing homes also may be affected.

In 1991, Alejandro and Daniel met at a conference on Andean crops and became friends. Later, the World Bank awarded Alejandro a fellowship to study in the United States, and Dan became his advisor at BYU. Alejandro earned his PhD and also learned a fourth language—English.

The two are now codirectors of an international program of quinoa research, funded by the McKnight Foundation. They have taken a holistic approach to improving quinoa production for farmers who are now locked in poverty. As they look for new ways to conserve, improve, and use genetic diversity, they also research the economic impact of new quinoa strains, agricultural technologies, nutrition, hygiene, health, and community programs. Today, more than twenty scientists in four countries are a part of the program.

Thousands of Bolivian families are now producing more food, thanks to the new quinoa strains. Children who would otherwise have died from kwashiorkor are now attending school.

In a recent letter, Dan told us that he learned more from Alejandro than Alejandro learned from him. He appended a photograph of his colleague in a research field, standing next to one of his new quinoa varieties, so that we can put a face with the name.

> *Plants enrich our world in uncountable ways. Researchers around the world are taking part in wide-ranging efforts to develop new varieties that can improve human life.*

Figure 32.26 Alejandro Bonifacio in a field of hybrid quinoa plants.

Summary

Section 32.1   The seeds dispersed from a flowering plant enter a period of dormancy. The body plan of the embryo sporophyte is already mapped out.

A seed germinates by way of imbibition. It absorbs water, resumes growth, and breaks through its seed coat. The seedling increases in volume and mass. Its tissues and organs grow and develop.

From the time of germination until the end of its life cycle, a plant's growth and development require gene products—enzymes, and other proteins with diverse roles in cell structure and function. Hormones and other signaling molecules affect how and when genes are transcribed and when gene products function.

Plant hormones, which have many and sometimes overlapping roles, bring about predictable patterns of growth and development by stimulating or inhibiting gene expression. They are synthesized in response to environmental cues, such as seasonal changes in night length, temperature, and oxygen and water in soil.

Biology ⊘ Now
Compare the growth and development of a monocot and a eudicot with the animation on BiologyNow.

Section 32.2   Like other signaling molecules, plant hormones are secreted by one cell type and dock at receptors on target cells. At specific times in the life cycle, they promote or arrest growth by stimulating or inhibiting cell division, elongation, differentiation, and other events. Five main classes are gibberellins, auxins, cytokinins, abscisic acid, and ethylene. Other growth regulators include brassinosteroids, which promote cell divisions and cell elongations, and stimulators and inhibitors of flowering, as yet unidentified.

Gibberellins promote stem elongation, help seeds and buds break dormancy in spring, and stimulate the flowering process in some species.

Auxins promote the elongation of shoots and coleoptiles. They also function in phototropism and gravitropism, which are both adjustments in the direction of elongation during primary growth.

Cytokinins stimulate cell division, promote leaf expansion, and retard leaf aging.

Ethylene promotes fruit ripening and abscission (the dropping of leaves, fruits, and other plant parts).

Abscisic acid promotes bud and seed dormancy, and it limits water loss by promoting stomatal closure.

Section 32.3   Plant hormones act in pathways of signal reception, signal transduction, and response.

One pathway starts in the seed. Gibberellin released from cells of the embryo sporophytes cause target cells to release endosperm's nutrients, which thus become available as fuel for rapid growth of the seedling.

Another pathway involves the polar transport of auxin from the tip to the base of shoots. Concentration gradients of auxin form and trigger transcription of

different genes in different tissue regions. They also promote cell elongation. The polar transport brings about shifts in pH between the cytoplasm and wall of target cells. When pH decreases in the wall, crosslinks between cellulose microfibrils in the wall break apart. Because the young cell is expanding fast under turgor pressure, the softened wall is free to expand also.

Biology ⊘ Now
Observe the effect of auxin on plant growth with the animation on BiologyNow.

Section 32.4   Hormones control patterns of growth and development, but they also trigger adjustments in the direction and rate of growth in response to environmental stimuli.

In gravitropism, roots grow downward and stems grow upright in response to Earth's gravity. The polar transport of auxin causes this pattern. When a primary root is turned on its side, unequal rates of transport cause cells to elongate faster on the top side, so the root bends down. Statoliths (clusters of particles in cells) are gravity-sensing mechanisms in some plants.

In phototropism, stems and leaves adjust rates and directions of growth toward or away from light. A flavoprotein is the receptor for blue wavelengths that trigger the strongest response.

With thigmotropism, plants adjust their direction of growth in response to contact with solid objects.

Plants respond to mechanical stress, as when strong winds inhibit stem elongation and plant growth.

Biology ⊘ Now
Investigate plant tropisms with the animation on BiologyNow.

Section 32.5   A circadian cycle is some event that is completed in about a twenty-four-hour period. Plants have biological clocks that measure the cycle. That is, internal mechanisms set the time for a shift in specific cell activities that underlie daily tasks as well as seasonal patterns of growth, development, and reproduction.

In plants, photoperiodism is a biological response to the length of darkness relative to daylength in the circadian cycle. Phytochrome, a blue-green pigment, functions as a receptor for red and far-red light. It is like an alarm button for an internal timing mechanism, or biological clock. It promotes or inhibits germination, stem elongation, leaf expansion, stem branching, and flower, fruit, and seed formation.

Section 32.6   Short-day plants flower mainly in spring or fall, when there are more hours of darkness relative to daylight in the circadian cycle. Long-day plants flower in summer, when there are fewer hours of darkness relative to daylight hours. Day-neutral plants flower whenever they are mature enough to do so.

Biology ⊘ Now
Learn how plants respond to daylength (night length) with the animation on BiologyNow.

Section 32.7 Dormancy is a period of arrested growth that does not end until the arrival of specific dormancy-breaking environmental cues. Declining Pfr levels may trigger dormancy. Typically it involves the dropping of flowers, leaves, fruits, and other plant parts, a process known as abscission.

Breaking dormancy might involve exposure to certain temperatures and hormonal action, including gibberellins and abscisic acid.

Section 32.8 In response to environmental cues, many plants make daily movements that affect growth only indirectly. Solar tracking, a reversible repositioning of leaves and flowers relative to the angle of the sun overhead during the day, is an example.

Section 32.9 Knowledge of plant structure and function can improve the human condition.

Self-Quiz
Answers in Appendix II

1. Seed germination is over when the _____ .
   a. embryo sporophyte absorbs water
   b. embryo sporophyte resumes growth
   c. primary root pokes out of the seed coat
   d. cotyledons unfurl

2. Which of the following statements is false?
   a. Auxins and gibberellins promote stem elongation.
   b. Cytokinins promote cell division but retard leaf aging.
   c. Abscisic acid promotes water loss and dormancy.
   d. Ethylene promotes fruit ripening and abscission.

3. Plant hormones _____ .
   a. interact with one another
   b. are influenced by environmental cues
   c. are active in plant embryos within seeds
   d. are active in adult plants
   e. all of the above

4. Plant growth depends on _____ .
   a. cell division   c. hormones
   b. cell enlargement   d. all of the above

5. _____ are the strongest stimulus for phototropism.
   a. Red wavelengths   c. Green wavelengths
   b. Far-red wavelengths d. Blue wavelengths

6. Light of _____ wavelengths makes phytochrome switch from inactive to active form; light of _____ wavelengths has the opposite effect.
   a. red; far-red   c. far-red; red
   b. red; blue   d. far-red; blue

7. The flowering process is a _____ response.
   a. phototropic   c. photoperiodic
   b. gravitropic   d. thigmotropic

8. Leaves and fruits drop from a plant when there is a decrease in _____ in their tissues and an increase in _____ at abscission zones.
   a. auxin; ethylene   c. phytochrome; gibberellin
   b. ethylene; auxin   d. gibberellin; abscisic acid

9. Match the plant reproduction and development terms.
   ____ vernalization   a. water moves into seeds
   ____ polar transport   b. unequal growth following
   ____ imbibition   contact with solid objects
   ____ thigmotropism   c. lateral bud formation inhibited
   ____ apical   d. low-temperature stimulation
        dominance   of the flowering process
                           e. auxin from shoot tip toward
                              base of shoot

Additional questions are available on **Biology** 🌀 **Now**™

Critical Thinking

1. Given what you know about the growth of plants (Chapter 29), would you expect hormones to influence primary growth only? What about secondary growth in, say, a coast redwood or a hundred-year-old oak tree?

2. Plant growth depends on photosynthesis, which depends on inputs of light energy from the sun. How, then, can seedlings that were germinated in a dark room grow taller than different seedlings that germinated in the sun?

3. Belgian scientists isolated a mutated gene in common wall cress (*Arabidopsis thaliana*) that produces excess amounts of auxin molecules. Predict what some of the resulting phenotypic traits might be.

4. Cattle typically are given somatotropin, an animal hormone that makes them grow bigger (the added weight means greater profits). There is a major concern that such hormones may have unforeseen side effects on beef-eating humans. Would you think plant hormones applied to crop plants can affect humans also? Why or why not?

5. Like tulips and many other plants, a spinach plant flowers in spring under the mediation of gibberellins. It grows best when night length is about ten hours. Figure 32.27 shows the results from an experiment involving its flowering response. Which plant was grown under short-day conditions? Which was grown under long-day conditions?

Figure 32.27 Experiment involving flowering responses of spinach (*Spinacia oleracea*), a long-day plant.

VI How Animals Work

How many and what kinds of body parts does it take to function as a lizard in a tropical forest? Make a list of what comes to mind as you start reading Unit VI, then see how resplendent the list can become at the unit's end.

33 ANIMAL TISSUES AND ORGAN SYSTEMS

Open or Close the Stem Cell Factories?

Hundreds of millions of years ago, in the forerunners of animals, cells started to interact as part of functional units called tissues. Today, animal tissues arise from the tiny ball of cells that form by way of divisions of a fertilized egg. The cells in that ball are not yet committed to being a particular kind of cell; they are undifferentiated (Figure 33.1). However, they are the body's first **stem cells**. Cell lineages descended from them will differentiate and give rise to all of the body's specialized tissues and organs.

In the laboratory, embryonic stem cells divide again and again. Their descendants often can be coaxed to develop into the specialized cells of blood, muscle, nerve, and other tissues. Why coax them to do so? Stem cells might one day help mend tissue damage resulting from injury or diseases, including Parkinson's disease, leukemia, and heart attacks.

Actor Christopher Reeve, known for his motion picture role as Superman, stoked public awareness of the potential for stem cell research. In a 1995 showjumping competition, Reeve was thrown from his horse and landed on his head. The fall paralyzed him; he could not even breathe on his own. Medical advances kept him alive and gave him hope. He became an advocate for the disabled and a champion of human embryonic stem cell research (Figure 33.2).

Where did the laboratory stocks of embryonic stem cell lineages come from? They were established from cells removed from aborted embryos, cells that formed by *in vitro* fertilization but were not used, and cells in the blood of discarded umbilical cords. Such cells have already shown potential for repaired spinal cord injuries. In laboratory rats, dividing and differentiating cells replaced damaged tissues, and the rats regained some use of their legs.

Embryonic stem cell research is advancing fast. In 2005, for instance, Korean researchers established eleven stem cell lines that are an exact genetic match to donor patients affected by diseases. As Section 13.5 explains, however, this research has opponents as well as proponents. Do *adult* stem cells hold the same potential? Adults retain patches of stem cells in bone marrow, fat, and other tissues. Duke University researchers experimenting with stem cells from liposuctioned fat have already prodded descendant cells to become bone, cartilage, or nerve cells.

Before such lineages can be used for treating diseases and injuries, researchers must learn how to control the steps from stem cells to normal—and stable—differentiated cells. In this respect, adult stem cells are not as versatile as the embryonic ones. They form fewer kinds of tissues. Also, the lineages now being maintained in culture for long periods of time tend to lose the capacity to differentiate.

Stem cell research is a fitting introduction to this unit, which deals with animal anatomy (how the body is put together) and physiology (how it works in the environment). These cells invite you to reflect on who we are, where we came from, and where medical research may be taking us.

In this chapter, you will start with the four basic types of animal tissues—epithelial, connective, muscle, and nervous tissues. A **tissue**, remember, is a community of cells and intercellular substances that carry out one or more tasks, such as muscle tissue contraction. An **organ** is a structural unit of two or more tissues organized in proportions and patterns necessary to carry out specific tasks. Your heart is an organ that has certain proportions and arrangements of all four types of tissues. In **organ systems**, two or more

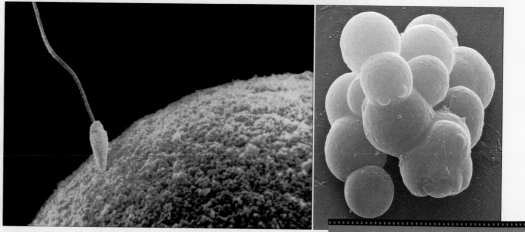

Figure 33.1 *Left,* moment before the union of two cells— sperm and egg. *Right,* after fertilization, a tiny ball of stem cells, not yet differentiated with a capacity to give rise to all of the human body's specialized cells, tissues, and organ systems.

Watch the video online!

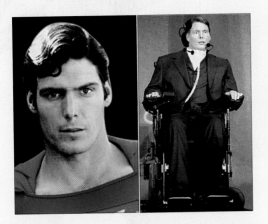

Figure 33.2 Actor Christopher Reeve before and after a spinal cord injury left him paralyzed. He died in 2004.

organs and other body components interact physically, chemically, or both in a common task, as when a beating heart forces blood through interconnected vessels.

Throughout this unit, you will come across examples of a concept outlined in Chapter 28. Cells, tissues, and organs interact smoothly when the body's internal environment is being maintained within a range that individual cells can tolerate. That state, recall, is called **homeostasis**. In most kinds of animals, blood and interstitial fluid make up the internal environment. It is worth a reminder: Whether you are considering the body of a flatworm or salmon, a bird or human, you will see that it must do the following:

1. Function in ways that ensure homeostasis.

2. Acquire and distribute raw materials to individual cells and dispose of wastes.

3. Protect tissues against injury or attack.

4. Reproduce and, in many species, nourish and protect offspring through early growth and development.

 How Would You Vote?

Many researchers believe that embryonic stem cell studies would greatly benefit medical science. Others object to the use of any cells from human embryos. Should researchers be allowed to start embryonic stem cell lines from human embryos that were frozen but were never used for in vitro fertilization? See BiologyNow for details, then vote online.

 *Key Concepts*

BASIC TYPES OF ANIMAL TISSUES

Epithelial, connective, muscle, and nervous tissues are the basic categories of tissues in nearly all animals.

Epithelia line the body surface and its internal cavities and tubes. They have protective and secretory functions.

Connective tissues bind, support, strengthen, protect, and insulate other tissues. They include soft connective tissues, cartilage, bone, blood, and adipose tissue.

Muscle tissues help move the body and its parts. The three kinds are skeletal, cardiac, and smooth muscle tissue.

Nervous tissue provides local and long-distance lines of communication among cells. It consists of neurons and neuroglia. Sections 33.1–33.4

INTRODUCING ANIMAL ORGAN SYSTEMS

Vertebrate organ systems compartmentalize the tasks of survival and reproduction for the body as a whole; they show a division of labor. Different systems arise from ectoderm, mesoderm, and endoderm, the primary tissue layers that form in the early embryo. Section 33.5

CASE STUDY: AN INTEGUMENTARY SYSTEM

Human skin is an example of an organ system. It includes epithelial layers, connective tissue, adipose tissue, glands, blood vessels, and sensory receptors. It helps protect the body from injury and some pathogens, conserve water and control internal temperature, excrete certain wastes, and detect some external stimuli. Section 33.6

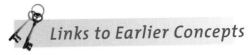 *Links to Earlier Concepts*

With this unit, you have now arrived at the tissue and organ system levels of biological organization for animals (Section 1.1). This chapter expands on the nature of multicelled body plans (26.1, 28.1–28.3). It builds on your knowledge of the origin and evolution of animal tissues (25.1, 25.2). You may wish to refresh your understanding of cell junctions (4.9), membrane transport proteins (5.2, 5.4), aerobic respiration (8.1), and energy conversion pathways (8.6).

33.1 Epithelial Tissue

LINKS TO
SECTIONS 4.9,
5.2, 5.4, 25.1

Recall, from Section 25.1, that ectoderm is the first layer of cells to form in the embryos of nearly all animals. Epithelia arises from this primary tissue layer.

GENERAL CHARACTERISTICS

Epithelium (plural, epithelia) is a sheetlike tissue of cells that are close together, with little extracellular material between them. One free surface is exposed to the outside environment or to some body fluid (Figure 33.3). At the opposite surface, epithelial cell secretions form a basement membrane that incorporates many adhesion proteins, including integrins and cadherins. The proteins anchor epithelium to other tissues.

Most epithelial cells have a squamous (flattened), cuboidal, or columnar shape, as in Figure 33.3. Certain types are highly specialized for absorbing substances. Others are highly specialized for secreting products; they release them at their free surface. Secretion is not the same as excretion, which refers to a concentration and removal of substances of no use to the body.

In *simple* epithelium, cells form a layer that is only one cell thick. In *stratified* epithelium, the cells form two or more layers. The outer layer of your own skin is mostly stratified squamous epithelium.

GLANDULAR EPITHELIUM

Gland cells occur only in epithelia. Such cells secrete products, unrelated to their own metabolism, that are

free surface
of epithelium

a

simple
epithelium

basement
membrane

b

underlying connective tissue

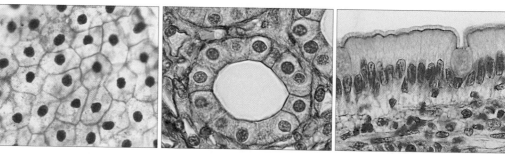

Figure 33.3 Characteristics of epithelium, an animal tissue with a free surface exposed to a body fluid or the outside environment.

(**a**) Section near the surface of stratified squamous epithelium. It consists of multiple layers and cells that become flattened near the free surface. (**b**) In all epithelia, the opposite surface rests on a basement membrane that anchors it to an underlying tissue.

(**c**) Light micrographs of three simple epithelia, showing the three most common shapes of cells in this tissue.

c

TYPE: Simple squamous
DESCRIPTION: Friction-reducing slick, single layer of flattened cells
COMMON LOCATIONS: Lining of blood and lymph vessels, heart; air sacs of lungs; peritoneum
FUNCTION: Diffusion, filtration, secretion of lubricants

TYPE: Simple cuboidal
DESCRIPTION: Single layer of squarish cells
COMMON LOCATIONS: Ducts, secretory part of small glands; retina; kidney tubules; ovaries, testes; bronchioles
FUNCTION: Secretion, absorption

TYPE: Simple columnar
DESCRIPTION: Single layer of tall cells; free surface may have cilia, mucus-secreting glandular cells, microvilli
COMMON LOCATIONS: Glands, ducts; gut; parts of uterus; small bronchi
FUNCTION: Secretion, absorption; ciliated types move substances

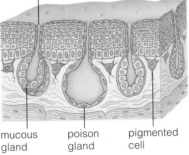

pore that opens at skin surface

mucous gland   poison gland   pigmented cell

Figure 33.4 Frog glandular epithelium. This frog (*Dendrobates*) makes one of the most lethal glandular secretions known. Natives of one Colombian tribe use its exocrine gland secretion to poison tips of blowgun darts. Pigment-rich epithelial cells impart color to the skin. The skin of all poisonous frogs has distinctive colors and patterns that evolved as a clear warning signal. In essence, it says to predators, "Don't even think about it."

to be used elsewhere. In most animals, gland cells are concentrated in **glands**, which are saclike, secretory organs that open to the free epithelial surface.

Exocrine glands secrete many substances, such as oils, mucus, saliva, tears, milk, digestive enzymes, and earwax. They have ducts or tubes that open onto the free epithelial surface (Figures 33.3*c* and 33.4).

Endocrine glands have no ducts; they secrete their products, hormones, directly into interstitial fluid. The hormone molecules typically diffuse into neighboring blood capillaries, and the circulatory system transports them to target cells in tissues that are typically some distance away (Chapter 36).

CELL JUNCTIONS

In epithelia, as in most tissues, cell junctions connect adjoining cells. **Adhering junctions** function like spot welds and lock adjoining cells together (Figure 33.5*a*). They are profuse in skin and other tissues subjected to ongoing abrasion. Collectively, **tight junctions** stop most substances from leaking across a tissue. Rows of proteins fuse each cell to its neighbors and form tight seals (Figure 33.5*b*). Dissolved substances must pass *through* cells to get to the opposite surface. Membrane transport proteins are selective about which ions and molecules can enter and leave the cells (Section 5.4).

Think about the stomach's acidic fluid. If it were to leak across the stomach's epithelial lining, it would digest proteins of your body instead of the proteins you eat. That is what happened in people with peptic ulcers. The stomach lining was breached, typically by a bacterial infection, and strong acid leaked into the abdominal cavity (Section 21.1).

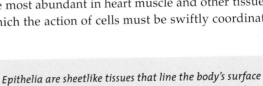

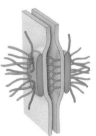

a

Adhering junction

Adjoining cells are welded together at a mass of proteins, which is anchored under the plasma membrane by tufts of intermediate filaments of the cytoskeleton.

b

Tight junction

Strands (rows of proteins) running parallel with the free surface of the tissue; they block leaking between adjoining cells.

c

Gap junction

Cylindrical arrays of proteins span the plasma membrane of adjoining cells. They pair up as open channels for signals between cells.

Figure 33.5 *Animated!* Examples of animal cell junctions.

Gap junctions permit ions and small molecules to pass freely from the cytoplasm of one cell to another (Figure 33.4*c*). These open communication channels are most abundant in heart muscle and other tissues in which the action of cells must be swiftly coordinated.

Epithelia are sheetlike tissues that line the body's surface and its cavities, ducts, and tubes. Epithelia have one free surface exposed to the outside environment or a body fluid. Glands are secretory organs derived from epithelium. Cell junctions structurally and functionally link adjoining cells.

33.2 Connective Tissues

LINKS TO
SECTIONS
3.4, 8.6, 26.3

Connective tissues have "connecting" roles in the body. They structurally or functionally support, bind, separate, and in one case insulate other tissues. They are the body's most abundant and widely distributed tissues.

Connective tissues consist of cells scattered within an extracellular matrix of their own secretions. In all but one connective tissue (blood), fibroblasts are the main type of cell. They make and secrete structural fibers of collagen and elastin into the matrix. Some tissues are classified as soft (loose and dense connective tissues). Others are specialized (cartilage, bone tissue, adipose tissue, and blood). Each kind has characteristic types, proportions, and arrangements of components. White blood cells patrol all of them.

SOFT CONNECTIVE TISSUES

Loose and dense connective tissues actually have the same components but in different proportions. In **loose connective tissue**, fibroblasts and fibers are dispersed widely through the matrix. Figure 33.6*a* is an example. This tissue, the most common type in the vertebrate body, helps hold organs and epithelia in place.

In **fibrous, irregular connective tissue**, the matrix is packed with many fibroblasts and collagen fibers that are positioned every which way (Figure 33.6*b*). Dense, irregular connective tissue is a component of skin. It supports intestinal muscles and also forms protective capsules around organs that do not stretch much.

Fibrous, regular connective tissue has orderly rows of fibroblasts between parallel, tightly packed bundles of fibers (Figure 33.6*c*). This organization helps keep the tissue from being torn apart when placed under mechanical stress. Tendons and ligaments are mainly dense, regular connective tissue. All tendons connect skeletal muscle to bones; ligaments attach one bone to another. In ligaments, elastic fibers in the tissue matrix facilitate movements around joints.

SPECIALIZED CONNECTIVE TISSUES

Cartilage is a tissue of fine collagen fibers packed in a rubbery, compression-resistant matrix. Specialized cells secrete the rubbery material, chondrin, which in time imprisons them (Figure 33.6*d*). Sharks, recall, have a skeleton of cartilage. Your own skeleton started out as cartilage, but bone tissue replaced most of it. Cartilage supports the outer ear, nose, and throat. It protects and cushions joints between limb bones, and between bones of the vertebral column. Unlike unspecialized connective tissues, it has no blood vessels; substances move to and from cells by diffusion. Also unlike those tissues, its cells do not divide often in adults.

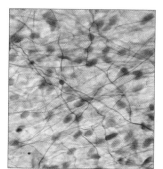

collagenous fiber
fibroblast
elastic fiber
a

TYPE: Loose connective tissue
DESCRIPTION: Fibers, fibroblasts, other cells loosely arranged in extensive ground substance
COMMON LOCATIONS: Beneath skin and most epithelia
FUNCTION: Elasticity, diffusion

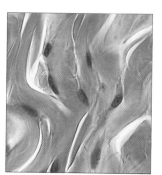

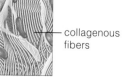
collagenous fibers
b

TYPE: Fibrous, irregular connective tissue
DESCRIPTION: Collagen fibers, fibroblasts occupy most of the ground substance
COMMON LOCATIONS: In skin and in capsules around some organs
FUNCTION: Structural support

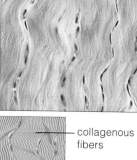

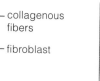
collagenous fibers
fibroblast
c

TYPE: Fibrous, regular connective tissue
DESCRIPTION: Collagen fibers bundled in parallel, long rows of fibroblasts, little ground substance
COMMON LOCATIONS: Tendons, ligaments
FUNCTION: Strength, elasticity

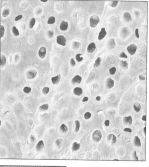

ground substance with very fine collegen fibers
cartilage cell (chondrocyte)
d

TYPE: Cartilage
DESCRIPTION: Chondrocytes inside pliable, solid ground substance
COMMON LOCATIONS: Nose, ends of long bones, airways, skeleton of cartilaginous fish, vertebrate embryo
FUNCTION: Support, flexion, low-friction surface for joint movements

Figure 33.6 Characteristics of connective tissue.

Bone tissue is a hardened connective tissue with living cells imprisoned in their mineralized secretions (Figure 33.6e). It is the main tissue of bones, the organs that interact with muscles to move the body and that support and protect soft internal organs. Figure 33.7 shows a limb bone, which has load-bearing functions. As explained in Sections 37.3 and 37.4, some bones are sites of blood cell formation.

Adipose tissue is an energy reservoir. Fat droplets form in many cells as excess carbohydrates and lipids are converted to fats (Section 3.4). However, cells of adipose tissue get so swollen with stored fat that their nucleus and a few fibroblast nuclei are flattened and pushed to one side (Figure 33.6f). This tissue has little extracellular matrix but many fine blood vessels that rapidly move fats to and from individual cells. Fat deposits under the skin form an insulating layer and cushion certain body parts. They accumulate around some organs, such as the kidneys and heart.

Blood is considered a connective tissue because its cellular components arise from stem cells in bone, a connective tissue. Blood cells are suspended in plasma, a fluid extracellular matrix that functions in transport and heat transfer. Plasma is mostly water with diverse proteins, gases, ions, sugars, and other substances dissolved in it. Red blood cells, white blood cells, and platelets tumble through it (Figure 33.8). Red blood cells get oxygen to metabolically active tissues and get rid of carbon dioxide wastes. White blood cells defend and repair tissues. Platelets function in blood clotting.

> Connective tissues support, protect, organize, or insulate other tissues. They consist of cells within an extracellular matrix. Except for blood, each contains fibroblasts.
>
> The matrix of soft connective tissues contains characteristic proportions and arrangements of fibroblasts and fibers.
>
> Cartilage, bone, adipose tissue, and blood are specialized connective tissues. Cartilage and bone are both structural materials. Adipose tissue is a reservoir of stored energy. Blood, a fluid connective tissue, has transport functions.

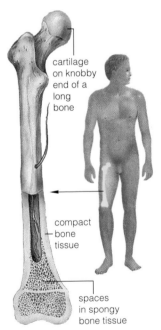

cartilage on knobby end of a long bone

compact bone tissue

spaces in spongy bone tissue

Figure 33.7 Cartilage and bone tissue. Spongy bone tissue has needlelike hard parts with spaces between. Compact bone tissue is more dense. Bone, a load-bearing tissue, resists being compressed and gives giraffes and other big animals selective advantages. Big animals can ignore most predators and roam farther for food and water. They gain or lose heat more slowly than smaller animals do; they have a lower surface-to-volume ratio.

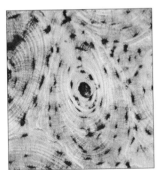

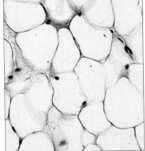

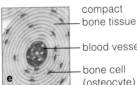

compact bone tissue
blood vessel
bone cell (osteocyte)

e

TYPE: Bone tissue

DESCRIPTION: Collagen fibers, osteocytes occupying extensive calcium-hardened ground substance

LOCATION: Bones of all vertebrate skeletons

FUNCTION: Movement, support, protection

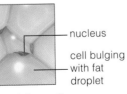

nucleus
cell bulging with fat droplet

f

TYPE: Adipose tissue

DESCRIPTION: Large, tightly packed fat cells occupying most of ground substance

COMMON LOCATIONS: Under skin, around the heart and kidneys

FUNCTION: Energy storage, insulation, padding

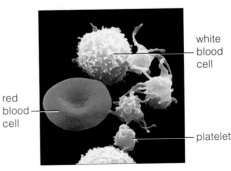

red blood cell

white blood cell

platelet

Figure 33.8 Cellular components of human blood. Many diverse proteins, nutrients, oxygen and carbon dioxide, and other substances also are dissolved in plasma, blood's straw-colored fluid portion.

33.3 Muscle Tissues

LINKS TO
SECTIONS
3.4, 4.10, 8.1

Vertebrates have three types of muscle tissue: skeletal, cardiac, and smooth muscle tissues. Each type has unique properties that reflect its functions.

In muscle tissues, cells *contract*, or forcefully shorten in response to stimulation, then they relax and passively lengthen. These tissues consist of many cells arranged in parallel with one another, in tight or loose arrays. Coordinated contractions of layers or rings of muscles move the whole body or its component parts. We find smooth and striated muscle cells among invertebrates, but focus here on the kinds found in vertebrates.

SKELETAL MUSCLE TISSUE

Skeletal muscle tissue, the functional partner of bone (or cartilage), helps move and maintain the positions of the body and its parts. Skeletal muscle tissue has parallel arrays of long, cylindrical *muscle fibers* (Figure 33.9*a*). The fibers are not single cells. While embryos are developing, groups of cells fuse together and form each fiber, which ends up with multiple nuclei. Inside the fiber are myofibrils—long strands with row after row of contractile units. These rows are so regular that skeletal muscle has a striated, or striped, appearance.

Each unit, a sarcomere, is contractile. It has parallel, interacting arrays of the contractile proteins actin and myosin (Section 4.10). The structure and function of skeletal muscle is the focus of Section 37.6.

Skeletal muscle tissue makes up 40 percent or so of the weight of an average human. Reflexes activate it, but we also make it contract simply by thinking about it, as is happening in Figure 33.9*a*. That is why skeletal muscles are commonly called "voluntary" muscles."

CARDIAC MUSCLE TISSUE

Cardiac muscle tissue occurs only in the heart wall (Figure 33.9*b*). Like skeletal muscle tissue, it contains sarcomeres and looks striated. Unlike skeletal muscle tissue, it consists of single, branching cells that have a nucleus. Cardiac muscle cells abut at their ends, where adhering junctions help keep them from being ripped apart during forceful contractions. Signals to contract

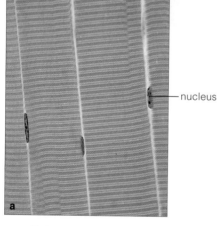

— nucleus

TYPE: Skeletal muscle

DESCRIPTION: Bundles of cylindrical, long, striated muscle fibers and many mitochondria; often reflex-activated but can be consciously controlled

LOCATIONS: Partner of skeletal bones, against which it exerts great force

FUNCTION: Locomotion, posture; head, limb movements

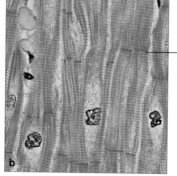

— nucleus

— adjoining ends of abutting cells

TYPE: Cardiac muscle

DESCRIPTION: Cylindrical, unevenly striated muscle fibers that abut at their ends; signal flow through gap junctions make them contract rapidly as a unit

LOCATIONS: Heart wall

FUNCTION: Pump blood forcefully through circulatory system

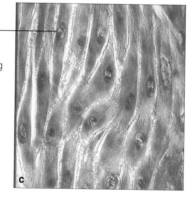

nucleus —

TYPE: Smooth muscle

DESCRIPTION: Contractile cells tapered at both ends; not striated

LOCATIONS: Wall of arteries, sphincters, stomach, intestines, urinary bladder, many other soft internal organs

FUNCTION: Controlled constriction; motility (as in gut); arterial blood flow

Figure 33.9 (**a**) Skeletal muscles, the functional partners of bones, move the vertebrate body. Row after row of contractile units give skeletal muscle cells a striated appearance. (**b**) Striated cells of cardiac muscle tissue. Adhering junctions are profuse in the horizontal bands; they hold abutting cells together. (**c**) Smooth muscle tissue, showing tapered cells and no striations.

pass so swiftly from cell to cell at gap junctions that all of the cells in cardiac muscle tissue contract as a unit. This tissue has more mitochondria than we find in skeletal or smooth muscle tissue, because it takes a continuous supply of ATP from aerobic respiration to keep the heart beating nonstop. Cardiac muscle tissue does not store as much glycogen, so glycolysis cannot do much when oxygen is scarce. If something ends the flow of oxygen to them, cardiac muscle cells will falter or die. That is what happens during a heart attack.

Like smooth muscle tissue, cardiac muscle tissue is said to be "involuntary" muscle; we usually cannot make its cells contract just by thinking about it.

SMOOTH MUSCLE TISSUE

We find layers of **smooth muscle tissue** in the wall of many soft internal organs, including the stomach, bladder, and uterus. It has single, unbranching cells, tapered at both ends, with one centrally positioned nucleus. Contractile units are not arranged in orderly repeating fashion, as they are in skeletal and cardiac muscle tissue, so smooth muscle tissue is *not* striated (Figure 33.9c). Even so, cells of this tissue do contain actin and myosin filaments, which are anchored to the plasma membrane by intermediate filaments.

Smooth muscle tissue contracts more slowly than skeletal muscle, but its contractions can be sustained much longer. Contractions drive many internal events, as when they propel material through the gut, shrink the diameter of arteries, and close sphincters.

> *Muscle tissue, which functions in movement, contracts in response to stimulation.*
>
> *Skeletal muscle is the functional partner of bones. Cardiac muscle is present only in the heart wall. Smooth muscle tissue is present in many soft internal organs.*

33.4 Nervous Tissue

Of all animal tissues, the one with the communication lines made of neurons exerts the most control over how the body senses and responds to changing conditions.

Nervous tissue is composed of neurons and a variety of cells, collectively called neuroglia, that structurally and functionally support them. **Neurons** are a kind of excitable cell that makes up the communication lines in most nervous systems. Figure 33.10 offers a look at one type, a motor neuron.

All cells respond to stimulation, but the neuron is highly excitable in a specific way. When it is suitably stimulated, it propagates a message along its plasma membrane, all the way to some output zone, without altering it. There, the message triggers the release of signaling molecules called neurotransmitters. These signals diffuse to another cell that is almost but not quite touching the neuron that sent them.

Your nervous system contains more than 100 billion neurons, and half of its volume consists of neuroglial cells that keep neurons positioned and functioning as they should. Sensory neurons detect specific stimuli, such as light and pressure. Neurons in your brain and spinal cord are called interneurons. They receive and integrate sensory information, store the bits that hold meaning, and coordinate the body's short-term and long-term responses to stimuli. Motor neurons relay commands from the brain and spinal cord to muscle cells, as in Figure 33.11, and to glands. Such signals stimulate or inhibit activity in target cells, which is a topic of later chapters.

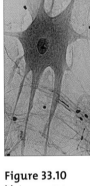

Figure 33.10 Motor neuron, which relays signals from the nervous system to muscle cells.

> *Neurons are the basic units of communication in nervous tissue. Different kinds detect specific stimuli, integrate information, and issue or relay commands to other tissues.*
>
> *Nervous tissue also contains neuroglia. Diverse cells in this category structurally and functionally support the neurons.*

Figure 33.11 One good example of the coordinated interaction between muscle tissue and nervous tissue. Interneurons in the brain of this lizard, a chameleon, calculate the distance and the direction of a tasty fly. In response to this stimulus, signals from the interneurons flow along certain motor neurons and reach muscle fibers inside the lizard's long, coiled-up tongue. The tongue uncoils swiftly in the precise direction of the fly.

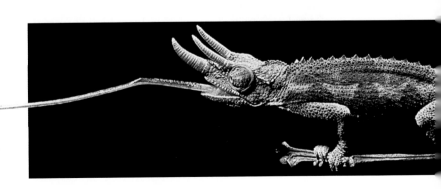

33.5 Overview of Major Organ Systems

LINKS TO
SECTIONS 1.1, 10.5,
25.1, 28.1–28.3

*Organ systems perform compartmentalized functions,
such as gas exchange, blood circulation, and locomotion.*

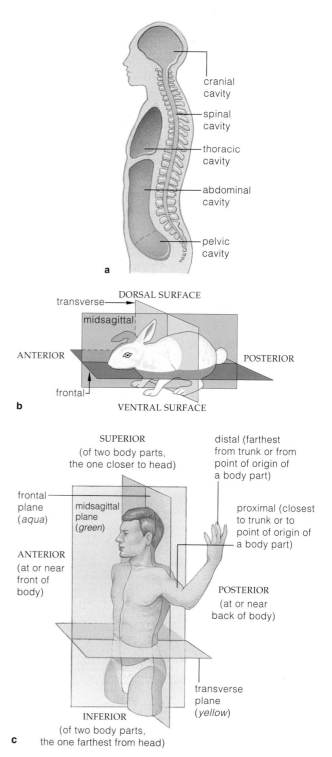

TISSUE AND ORGAN FORMATION

Figure 33.12 has an overview of eleven organ systems of a typical vertebrate, an adult human. It lists some terms used when describing the positions of organs. It also shows major body cavities in which a number of important organs are located.

The amazing thing is that the internal environment stays within tolerable limits even though all of these organ systems are putting various amounts and kinds of substances into it and withdrawing substances from it while they grow, develop, and maintain themselves. Talk about clockwork!

The clock starts ticking when germ cells, a type of immature reproductive cell, give rise to a sperm and an egg. (All other body cells, remember, are somatic.) At fertilization, a zygote forms. Mitotic cell divisions turn it into a ball of cells that arrange themselves into three *primary tissue layers*, which are forerunners of all adult tissues and organs. **Ectoderm**, the outer layer, is the first to form. It will give rise to epidermis and the nervous system. **Mesoderm**, the middle layer, is the start of muscles, bones, and most of the circulatory, urinary, and reproductive systems. **Endoderm** is the inner primary tissue layer. It is the start of the lining of the digestive tract and organs derived from it.

REGARDING THE DIVISION OF LABOR

Remember, new properties emerge at higher levels of biological organization (Section 1.1). Collectively, the organ systems of a multicelled body show a **division of labor**—a compartmentalization of functions—that help the body survive in ways that no one tissue can offer. Organ systems divide up the tasks of securing, processing, and distributing materials, and expelling wastes, protecting the body, integrating its activities, and reproducing. In Chapter 25, you glimpsed how this emergent property—the division of labor—turned out to be a key innovation when large-bodied animals evolved. In the rest of this unit, you will come across organ systems that reflect its extraordinary potential.

Animal organ systems compartmentalize many specialized tasks which, taken together, contribute to the survival and reproduction of the whole body.

Vertebrate organ systems arise from three primary tissue layers: ectoderm, mesoderm, and endoderm.

Figure 33.12 *Animated!* (**a**) Major cavities in the human body. (**b,c**) Directional terms and planes of symmetry for the body. Most vertebrates move with the main body axis parallel to Earth's surface. For them, *dorsal* refers to their back (upper surface) and *ventral* to the opposite, or lower surface. In humans, who are upright, *anterior* refers to the front of a standing person; it corresponds to ventral. *Posterior*, the back, is equivalent to dorsal in the rabbit. (**d**) Human organ systems and their functions.

Integumentary System

Protects body from injury, dehydration, and some pathogens; controls its temperature; excretes certain wastes; receives some external stimuli.

Nervous System

Detects external and internal stimuli; controls and coordinates the responses to stimuli; integrates all organ system activities.

Muscular System

Moves body and its internal parts; maintains posture; generates heat by increases in metabolic activity.

Skeletal System

Supports and protects body parts; provides muscle attachment sites; produces red blood cells; stores calcium, phosphorus.

Circulatory System

Rapidly transports many materials to and from cells; helps stabilize internal pH and temperature.

Endocrine System

Hormonally controls body functioning; works with nervous system to integrate short-term and long-term activities.

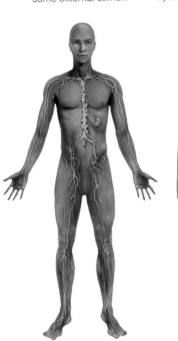

Lymphatic System

Collects and returns some tissue fluid to the bloodstream; defends the body against infection and tissue damage.

Respiratory System

Rapidly delivers oxygen to the tissue fluid that bathes all living cells; removes carbon dioxide wastes of cells; helps regulate pH.

Digestive System

Ingests food and water; mechanically, chemically breaks down food and absorbs small molecules into internal environment; eliminates food residues.

Urinary System

Maintains the volume and composition of internal environment; excretes excess fluid and blood-borne wastes.

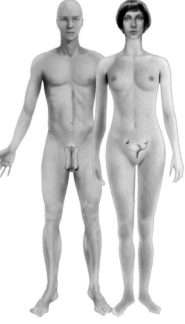

Reproductive System

Female: Produces eggs; after fertilization, affords a protected, nutritive environment for the development of new individuals. *Male:* Produces and transfers sperm to the female. Hormones of both systems also influence other organ systems.

d

33.6 Vertebrate Skin—Example of an Organ System

LINKS TO
SECTIONS
9.5, 26.11, 26.12

Go back in time to the Cambrian seas, when jawed fishes were first evolving. The body covering—integument—of some species had protective, heavy armor plates that hindered speed and precision movements. Among some bony fishes, a more flexible integument evolved into vertebrate skin.

Of all vertebrate organ systems, the outer body covering called **skin** has the largest surface area. It consists of two layers—an underlying dermis and outer dermis (Figures 33.13 and 33.14). Skin stretches, conserves water, and fixes small cuts or burns. It helps make vitamin D and dissipates excess metabolic heat. Its sensory receptors help the brain assess the outside world. White blood cells patrolling it help defend the body from external threats.

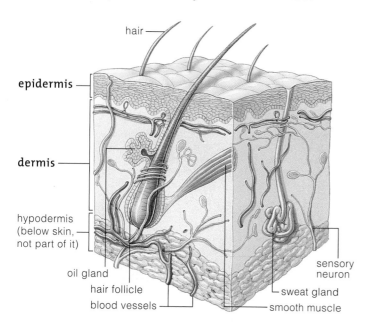

Figure 33.13 *Animated!* Structure of skin that includes hair, oil glands, and sweat glands. Skin components differ in different body regions.

hair
epidermis
dermis
hypodermis (below skin, not part of it)
oil gland
hair follicle
blood vessels
sensory neuron
sweat gland
smooth muscle

The Dermis The **dermis** is primarily a dense connective tissue with many fibers of stretch-resistant elastin and supportive collagen. Blood vessels, lymph vessels, and sensory receptors thread through it. The dermis rests on a hypodermis, which is not part of skin. The hypodermis contains loose connective tissue and adipose tissue that insulates or cushions some body parts (Figure 33.13).

Human skin has many exocrine glands, including about 2.5 million sweat glands, in the dermis. Sweat glands help dissipate heat. Their fluid secretions are 99 percent water in which salts, traces of ammonia, vitamin C, and other substances are dissolved. Secretions from some sweat glands increase during stress, pain, and sexual foreplay, and prior to menstruation. Except on the palms and soles, the dermis contains oil glands (sebaceous glands). The secretions lubricate and soften hair and skin, and they kill many surface bacteria. When bacteria do infect oil gland ducts, they can cause *acne*, an inflammation of skin.

The Epidermis Epidermis is a stratified squamous epithelium with an abundance of adhering junctions and no extracellular matrix. Ongoing mitotic cell divisions in the deepest epidermal layers push previously formed cells toward the skin's surface. Wear and tear from the surface, together with the pressure exerted by the perpetually growing cell mass, flatten and kill epidermal cells before they reach the surface. Dead ones are continually rubbed off or flake away. The main types of epidermal cells are keratinocytes, melanocytes, and dendritic cells.

When some vertebrates invaded the land, two kinds of skin cells—keratinocytes and melanocytes—proved useful. The *keratinocytes* secrete keratin, a tough, water-resistant protein that makes skin waterproof and more durable. In mammals, dead, flattened keratinocytes make up most of the flexible structures called hairs (Figure 33.15). Figure 33.16 shows how hair can be curled or straightened.

An average human scalp has about 100,000 hairs, but genes, nutrition, and hormones affect hair growth. Protein deficiency thins hair; amino acids are required for keratin

Figure 33.14
Section through human skin.

outermost epidermal layer (all dead cells)

keratinized cells being flattened

rapidly dividing cells of epidermis

dermis

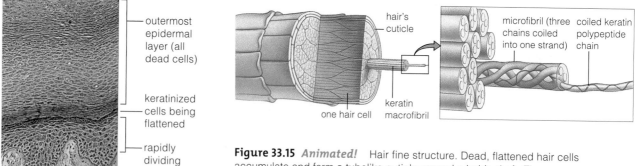

hair's cuticle
microfibril (three chains coiled into one strand)
coiled keratin polypeptide chain
one hair cell
keratin macrofibril

Figure 33.15 *Animated!* Hair fine structure. Dead, flattened hair cells accumulate and form a tubelike cuticle around a hair's shaft. They are derived from modified skin cells that synthesize polypeptide chains of the protein keratin. Disulfide bridges link three chains into thin fibers, which become bundled into larger fibers. The fibers almost fill the cells, which in time die off. Figure 33.16 shows how people play with the disulfide bridges.

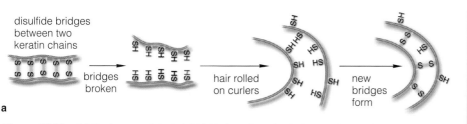

Figure 33.16 (**a**) Curly or straight hair? Hair's long keratin chains are bonded together by disulfide bridges, which break when exposed to chemicals. Ironing (flattening out) hair or rolling it around curlers holds the unbonded chains in new positions. Exposure to a different chemical makes new disulfide bridges form between different sulfur-bearing amino acids. The displaced bonding locks chains in new positions. That is how many women, including actress Nicole Kidman (**b**,**c**), straighten their naturally curly hair.

synthesis. High fever, emotional stress, and too much vitamin A in the diet also cause hair to thin.

Melanocytes in the epidermis produce the brownish-black pigment melanin, then give up these pigments to the keratinocytes. Melanin is one of the body's barriers to harmful ultraviolet (UV) radiation in the sun's rays.

Variations in skin color arise from differences in the distribution and activity of melanocytes. In pale skin, little melanin forms. Such skin often appears pink because the red color of the iron in hemoglobin shows through thin-walled blood vessels and epidermis. An orange pigment, carotene, also contributes to skin color.

Langerhans cells migrate through the epidermis. These phagocytic cells engulf bacteria or viruses and notify the immune system of the threat (Section 39.4). Ultraviolet (UV) radiation damages Langerhans cells. When that happens, skin becomes much more vulnerable to viral outbreaks, such as the cold sores caused by *Herpes* virus.

Lab-grown epidermis is used to protect tissues and aid wound healing in some patients. One company makes it by growing cells of foreskins that were discarded earlier from circumcised male infants. The resulting product is missing some cell types, such as melanoctyes, and it has no glands.

On Suntans and Shoe-Leather Skin
Sunlight can burn unprotected light skin, sometimes severely. UV light stimulates the melanocytes in skin to make melanin, which gives skin the "tan" that many light-skinned people covet (Figure 33.17). Melanin production accelerates slowly, then peaks about ten days after tanning starts.

At first, dark-skinned people are better protected. But prolonged or repeated UV exposure damages collagen and causes elastin fibers to clump. Chronically tanned skin gets less resilient and starts to look like shoe leather. UV also attacks DNA, which invites skin cancer (Section 9.5).

As we age, epidermal cells divide less often. Skin thins and becomes less elastic as collagen and elastin fibers become sparse. Glandular secretions that kept it soft and moist dwindle. Wrinkles deepen. Many people needlessly accelerate the aging process by indulging in tanning or smoking, which shrinks the skin's blood supply.

The Vitamin Connection
UV light also stimulates melanocytes to synthesize a precursor of vitamin D called cholecalciferol. "Vitamin D" is a generic name for steroid-like compounds that help the body absorb calcium ions from food. At the same time, the UV exposure causes the breakdown of folate, a B vitamin. Folate deficiency can lead to many problems. For one thing, the nervous system of an embryo cannot develop normally without it.

Variations in skin color among human populations may be adaptations to differences in sunlight exposure. Our early ancestors evolved in Africa, where the sun's rays are intense (Sections 26.14 and 26.15). Their melanin-rich skin protected folate but still produced enough vitamin D. After some human populations migrated to colder regions, their descendants spent much of the time indoors or bundled up outside. We can hypothesize that embryo-protecting dark skin became disadvantageous and that paler skin, which does not block as much UV light, proved advantageous.

Researchers tested the hypothesis by comparing annual UV levels with the skin color of people who were native to more than fifty countries. The data support the hypothesis that latitudinal decreases in the annual levels of UV light correlate with latitudinal increases in lighter skin.

Figure 33.17 Demonstration of how to encourage the formation of shoe-leather skin.

Summary

Section 33.1 A tissue is an aggregation of cells and intercellular substances that interact in performing one or more common tasks. Epithelial tissues cover the body surface and line internal cavities and tubes. Epithelium has one free surface exposed to extracellular fluid or the environment. Gland cells and glands are derived from epithelium. Endocrine glands are ductless and secrete hormones. Exocrine glands secrete other products, such as sweat or milk, through ducts to body surfaces.

Animal tissues have a variety of cell-to-cell junctions. Adhering junctions cement neighboring cells together. Tight junctions prevent substances from leaking across a tissue. Gap junctions are open channels that connect the cytoplasm of abutting cells. They permit the rapid transfer of ions and small molecules between cells.

Biology⊗Now
Compare the structure and function of the main types of animal cell junctions with the animation on BiologyNow.

Section 33.2 Connective tissues structurally and functionally "connect" other animal tissues. Different types bind, organize, support, strengthen, protect, and insulate other tissues. All contain cells scattered within an extracellular matrix of their own secretions. Except in blood, the main cell type is the fibroblast. It makes and secretes structural fibers of collagen and elastin into the extracellular matrix.

Loose connective tissue and dense connective tissue have the same components but differ in the proportions. They are classified as soft connective tissues. Cartilage, bone tissue, adipose tissue, and blood are classified as specialized connective tissues.

Section 33.3 Muscle tissues contract (shorten), then passively lengthen. They help move the body and its component parts. The three types are skeletal muscle, cardiac muscle, and smooth muscle tissue. Only skeletal muscle and cardiac muscle tissues are striated. Only skeletal muscle is under voluntary control.

Section 33.4 Neurons in nervous tissue make up communication lines through the body. Different kinds detect, integrate, and assess stimuli about internal and external conditions, and deliver commands to muscles and glands that carry out responses. Nervous tissue also contains diverse cells collectively called neuroglia, which protect and support the neurons.

Section 33.5 An organ system consists of two or more organs that interact chemically, physically, or both in tasks that help keep individual cells as well as the whole body functioning. Most vertebrate organ systems contribute to homeostasis; they help maintain tolerable conditions in the internal environment that benefit individual cells and the body as a whole.

All tissues and organs of an adult animal arise from three primary tissue layers that form in early embryos: ectoderm, mesoderm, and endoderm. Ectoderm, the outer primary tissue layer, forms first and gives rise to epidermis and part of the nervous system. The other two give rise to the rest of the internal tissues and organs.

Collectively, the organ systems of a multicelled body show a division of labor—a compartmentalization of functions—that help the body survive in ways that no one tissue can offer.

Biology⊗Now
Investigate the function of vertebrate organ systems and learn about terms used to describe their locations with the animation on BiologyNow.

Section 33.6 An organ system called skin functions in protection, temperature control, detection of shifts in external conditions, vitamin D production, and defense.

Biology⊗Now
Explore the structure of skin and hair with the animation on BiologyNow.

Self-Quiz   *Answers in Appendix II*

1. The four light micrographs at lower left show four types of animal tissues. Identify each type and write out a brief description of its defining features.

2. _____ tissues are sheetlike with one free surface.
   a. Epithelial   c. Nervous
   b. Connective   d. Muscle

3. _____ function in cell-to-cell communication.
   a. Tight junctions   c. Gap junctions
   b. Adhering junctions d. all of the above

4. In most animals, glands are located in _____ tissue.
   a. epithelial   c. muscle
   b. connective   d. nervous

5. Most _____ have many collagen and elastin fibers.
   a. epithelial tissues c. muscle tissues
   b. connective tissues d. nervous tissues

6. _____ is mostly plasma.
   a. Adipose tissue   c. Cartilage
   b. Blood   d. Bone

7. Your body converts excess carbohydrates and proteins to fats. _____ specializes in storing the fats.
   a. Epithelial tissue   c. Adipose tissue
   b. Dense connective tissue d. both b and c

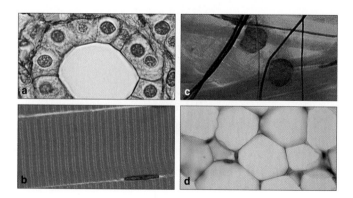

Figure 33.18 Assaults on the integument: skin piercings.

Figure 33.19 In the Kalahari Desert, gray meerkats (*Suricata suricatta*). Each morning, they come out of their burrows and face the sun's warming rays.

8. In your body, cells of _____ can shorten (contract).
  a. epithelial tissue   c. muscle tissue
  b. connective tissue   d. nervous tissue

9. Only _____ muscle tissue has a striated appearance.
  a. skeletal   c. cardiac
  b. smooth   d. a and c

10. _____ detects and integrates information about changes and controls responses to those changes.
  a. Epithelial tissue   c. Muscle tissue
  b. Connective tissue   d. Nervous tissue

11. Match the terms with the most suitable description.
  ____ exocrine gland   a. strong, pliable; like rubber
  ____ endocrine gland   b. secretion through duct
  ____ cartilage   c. outermost primary tissue
  ____ ectoderm   d. contracts, not striated
  ____ smooth muscle   e. cements cells together
  ____ blood   f. fluid connective tissue
  ____ adhering   g. ductless secretion
       junction

Additional questions are available on **Biology ⒺNow™**

Critical Thinking

1. The nose, lips, tongue, navel, nipples, and genitals are often targets for *body piercing:* cutting holes into the body so jewelry can be threaded through them (Figure 33.18). *Tattooing*, or using permanent dyes to make patterns in skin, is another fad. Besides being painful, both skin invasions can invite bacterial infections, chronic viral hepatitis, AIDS, and other diseases if the piercers and tattooers reuse unsterilized needles, dye, razors, gloves, swabs, and trays. Months may pass before any problems develop, so the cause-and-effect connection isn't always obvious. If despite the risks you think tissue invasions are okay, how can you be sure the equipment used is sterile?

2. Adipose tissue and blood are often said to be atypical connective tissues. Compared to other types of connective tissues, which features are *not* typical?

3. Many people oppose the use of animals for testing the safety of cosmetics. They say alternative test methods are available, such as the use of *lab-grown tissues* in some cases. Given what you learned in this chapter, speculate on the advantages and disadvantages of tests that use specific lab-grown tissues as opposed to living animals.

4. After a cold night in Africa's Kalahari Desert, animals small enough to fit inside a coat pocket emerge stiffly from burrows. These "meerkats" are a type of mongoose. They stand on tiny hind legs and face east, exposing their chilled bodies to the warm rays of the morning sun (Figure 33.19). Once meerkats warm up, they fan out and search for food. Into the meerkat gut go insects and the occasional lizard. These are pummeled, dissolved, and digested into glucose and other nutritious bits small enough to move across the gut wall, into blood, and on to the body's cells.

Name as many tissues as you can that might have roles in (1) keeping the meerkat body warm, (2) moving the body, as during foraging and heart-thumping flights from predators, and (3) digestion and absorption of nutrients, and elimination of the residues.

5. *Porphyria* is a name for a set of rare genetic disorders. Affected people lack one of the enzymes in the metabolic pathway that forms heme, the iron-containing group of hemoglobin. As a result, intermediates of heme synthesis (porphyrins) accumulate in the body. In some forms of the disorder, porphyrins pile up in the bones, skin, and teeth. When porphyrins in the skin are exposed to sunlight, they absorb energy and release energized electrons. Electrons careeening around the cell can break bonds and cause free radicals to form. The most notable result is the formation of lesions and scars on skin (Figure 33.20). In the most extreme cases, gums and lips can recede, which makes some front teeth—the canines—look more fanglike.

Affected individuals must avoid sunlight, and garlic can exacerbate their symptoms. By one hypothesis, people who were affected by the most extreme forms of porphyria may have been the source for vampire stories. Would you consider this hypothesis plausible? What other kinds of historical data might support or disprove it?

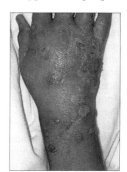

Figure 33.20 Skin lesions of an individual affected by a severe form of porphyria.

34.1 Evolution of Nervous Systems

LINKS TO
SECTIONS
25.1–25.7, 26.2

All animals above the sponge level of organization have a nervous system, a means of sensing and responding to information about conditions inside and outside the body.

You probably know that your own nervous system has communication lines made of three classes of neurons. *Sensory* neurons detect information about stimuli, such as light. *Inter*neurons accept the sensory input, process it, and signal other neurons. *Motor* neurons relay new signals to effectors—muscles and glands—that carry out responses (Figure 34.2). Other cells—*neuroglia*—structurally and metabolically support the neurons. Your system is far more complex than the one in, say, a jellyfish. Nevertheless, even a jellyfish offers clues to how your own system originated.

REGARDING THE NERVE NET

Animals first evolved in the seas, and the ones with the simplest nervous systems still live in water. They are cnidarians, such as sea anemones, jellyfishes, and other groups having a *radial* body plan (Sections 25.2 and 25.4). Cnidarians have two epithelial tissues—an epidermis and a gastrodermis. An asymmetrical mesh of neurons, a **nerve net**, controls simple movements of both. Sensory neurons of the epidermis signal motor neurons that extend all through the two tissues. The motor neurons activate epithelial cells that have long contractile extensions, which are organized as sheets and rings in the body wall (Figure 34.3*a* and Section 37.1). By making them contract, the nerve net changes the diameter of the mouth or body or bends tentacles.

The cnidarians will never dazzle you with speed or acrobatics. Even so, their nerve net lets them capture food that randomly drifts into them and then move it into the gut. The radial system is equally responsive to tidbits arriving from any direction.

ON THE IMPORTANCE OF HAVING A HEAD

Bilateral animals, remember, have a top-to-bottom and front-to-back body plan, which master genes map out as the embryos develop (Section 25.2). These genes evolved before radial animals did, but their expression is mostly suppressed in cnidarians. Cnidarian nerve nets still control a bilateral array of muscles that ring the opening to the gut. In some groups, the nerve net also controls the motions of bilateral larvae.

Today, flatworms are the simplest animals with a bilateral nervous system. They have **nerves**, or long-distance cables. Branching nerves join two nerve cords in a ladderlike array (Figure 34.3*b*). In some species, nerve cords expand to form ganglia in their head end. A **ganglion** (plural, ganglia) is a cluster of nerve cell

Figure 34.2 The line of communication.

stimulus
↓
receptors
sensory neurons
↓
integrators
interneurons
↓
motor neurons
↓
effectors
muscles,
glands
↓
response

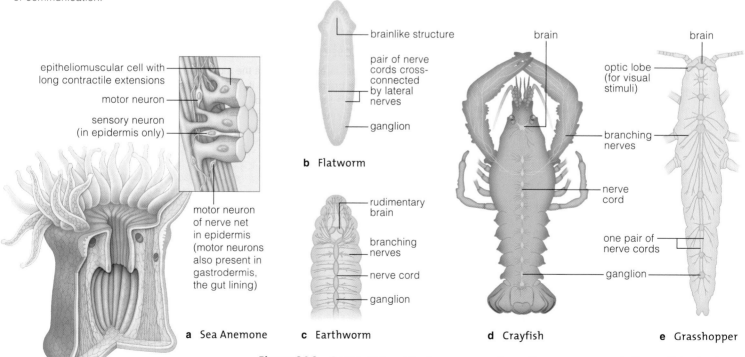

epitheliomuscular cell with long contractile extensions

motor neuron

sensory neuron (in epidermis only)

motor neuron of nerve net in epidermis (motor neurons also present in gastrodermis, the gut lining)

a Sea Anemone

brainlike structure

pair of nerve cords cross-connected by lateral nerves

ganglion

b Flatworm

rudimentary brain

branching nerves

nerve cord

ganglion

c Earthworm

brain

nerve cord

d Crayfish

brain

optic lobe (for visual stimuli)

branching nerves

one pair of nerve cords

ganglion

e Grasshopper

Figure 34.3 Gallery of invertebrate nervous systems. Sea anemones and other radial animals have a nerve net. The other representatives shown have a cephalized, bilateral nervous system.

bodies that function as a local integrating center. In flatworms, ganglia in the head end integrate signals from paired sensory organs, such as eyespots. Other ganglia in the body exert local control over nerves.

However it occurred, **cephalization**—the formation of a head—evolved in concert with bilateral nervous systems in nearly all animal groups. We see evidence of the bilateral heritage in the rudimentary brain and paired ganglia of most invertebrates (Figure 34.3c–e). We see it in paired sensory structures, brain centers, nerves, and skeletal muscles of all vertebrates.

So which lineages of bilateral animals are now the brainiest? Generally, when an animal has a big brain relative to its body size, it shows notable behavioral complexity. Cephalopods, again, are the brainiest of all invertebrates. They have a complex brain and sensory organs, and they precisely maneuver their streamlined body and its individual tentacles. Precision comes in handy when cephalopods hunt prey (Section 25.9).

What about vertebrates? Over evolutionary time, nervous tissue thickened at the anterior end of their dorsal nerve cord. In most lineages, brains got bigger because bigger brains gave individuals a competitive edge in assessing and responding to food and danger. Also, diverse stimuli greeted the first vertebrates that invaded land. In this new setting, agents of selection favored expansions of sensory structures, motor skills, and brain centers that could coordinate, process, and direct the body's responses to novel stimuli.

THE VERTEBRATE NERVOUS SYSTEM

The nervous system of vertebrates has two functional divisions (Figure 34.4). The brain and spinal cord are the *central* nervous system. Nerves extending through the rest of the body make up most of the *peripheral* nervous system (Figure 34.5). Their sensory fibers are *afferent*; they deliver signals into the central system. Motor fibers are *efferent*; they carry signals out of it.

Radial animals have a nerve net. Nearly all other animals have a bilateral, cephalized nervous system. All nervous systems evolved as an outcome of mutations in genes that control how the basic body plan develops in embryos.

In vertebrates, the nervous system has two functional divisions. The brain and spinal cord are the central nervous system. Threading through the rest of the body are spinal and cranial nerves, the major communication lines of the peripheral nervous system.

Inside the nerves, sensory fibers carry signals into the brain and spinal cord. Motor fibers carry signals away from them.

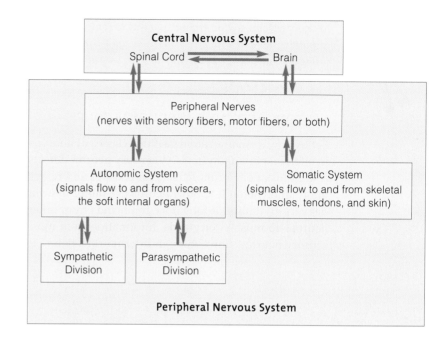

Figure 34.4 Functional divisions of vertebrate nervous systems. The spinal cord and brain are its central portion. The peripheral nervous system includes spinal nerves, cranial nerves, and their branchings, which extend through the rest of the body. They carry signals to and from the spinal cord and brain.

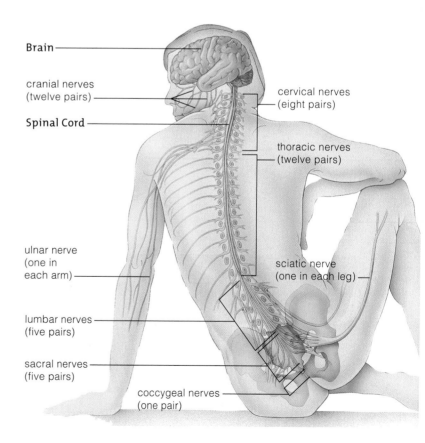

Figure 34.5 Some of the major nerves of the human nervous system.

34.2 Neurons—The Great Communicators

LINKS TO
SECTIONS
5.2, 5.4, 5.5, 6.1

Nervous systems, again, evolved as a way to sense and respond faster and with precision to changing conditions inside and outside the body. They do so by way of many communication lines, the units of which are neurons.

NEURONS AND THEIR FUNCTIONAL ZONES

Let us now formally define the three classes of neurons in bilateral nervous systems. A **sensory neuron** detects a stimulus at one or more receptor endings and relays information about it to other neurons. A **motor neuron** delivers excitatory or inhibitory commands from other neurons to muscles or glands. Information from most sensory neurons flows through **interneurons** before it gets to motor neurons. Interneurons receive, process, and often store sensory information, and they interact to integrate most of the responses to it. Your brain and spinal cord contain hundreds of billions of them.

A neuron cell body has a nucleus and one or more fibers, or slender cytoplasmic extensions. Two classes of fibers—**dendrites** and **axons**—differ in number and length among neurons (Figure 34.6a–c). Most often, the cell body and dendrites are *input* zones, where signals arrive and cause an electrical disturbance across the plasma membrane. A large disturbance may spread to the *trigger* zone, an adjoining patch of the membrane where information about a stimulus becomes encoded in action potentials. As you will see, action potentials usually propagate themselves along an axon, which is a neuron's *conducting* zone. Most axons have branched endings that are *output* zones. Here, action potentials become transduced into signals that can be sent on to neighboring cells (Figure 34.6d).

MEMBRANE GRADIENTS AND POTENTIALS

When a neuron is "at rest," or not being stimulated, mechanisms are maintaining a slight electric gradient across the plasma membrane. The cytoplasmic fluid near the membrane has a slight negative charge with respect to the fluid outside. As in a car battery, these separated charges have potential energy, which can be measured in millivolts (thousandths of a volt). The steady voltage difference across the neuron's plasma membrane is called the **resting membrane potential**. In many cases, it is about −70 millivolts.

All living cells maintain an electric gradient across their plasma membrane, but only in neurons, muscle cells, and a few other *excitable* cells does that electric gradient briefly reverse itself in an abrupt response to stimulation. We call the reversal an **action potential**. It

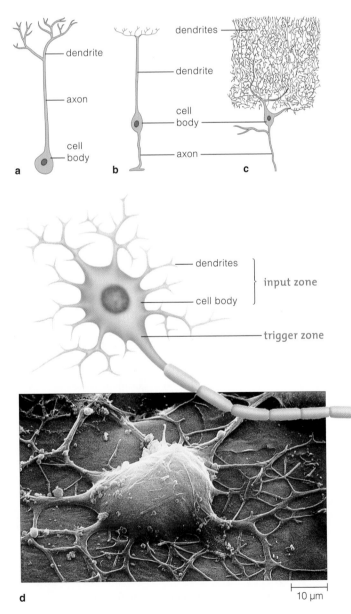

a — dendrite, axon, cell body

b — dendrites, dendrite, cell body, axon

c

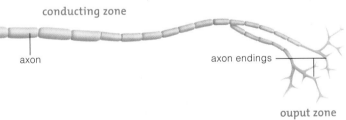

d — dendrites, cell body } input zone; trigger zone; conducting zone; axon; axon endings; ouput zone

Figure 34.6 Examples of neurons that differ in the number of their cytoplasmic extensions. (**a**,**b**) Some types have one axon only or one axon and one dendrite. Many sensory neurons are like this. (**c**) Others have one axon and a profusion of dendrites; many kinds in the mammalian brain are like this. (**d**) This scanning electron micrograph and sketch show functional zones of a motor neuron.

10 μm

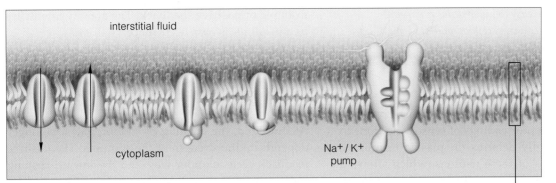

interstitial fluid

cytoplasm

Na+ / K+ pump

Figure 34.7 *Animated!*
How ions cross the plasma membrane of a neuron. They are selectively allowed to cross at protein channels and pumps that span the membrane.

Passive transporters with open channels let ions continually leak across the membrane.

Other passive transporters have voltage-sensitive gated channels that open and shut. They assist diffusion of Na+ and K+ across the membrane as they follow concentration gradients.

Active transporters pump Na+ and K+ across the membrane, against their concentration gradients. They counter ion leaks and restore resting membrane conditions.

lipid bilayer of neural membrane

sets in motion a series of fleeting reversals that travels from a neuron's trigger zone to its output zone.

Before tracking an action potential, take a moment to become familiar with certain membrane properties on which it is based. Diverse transport proteins work with and against ion concentration gradients across a neuron's plasma membrane. Specifically, potassium ions (K+), sodium ions (Na+), and certain other ions cannot cross the lipid bilayer on their own. Transport proteins, of the sort represented in Figure 34.7, must passively or actively help them across.

As you know from Section 5.2, transporters have an interior channel that opens to both surfaces of the membrane. Some transporters are like open channels for certain ions, which leak through them all the time. Other transporters have a molecular gate at one end of the channel. The gates are tightly closed when the neuron is at rest. They open during action potentials.

When a neuron is at rest, there are about 15 sodium ions in the fluid just inside the plasma membrane for every 150 outside. There are 150 potassium ions inside for every 5 outside. We can show the ion concentration gradients across the membrane in this fashion, where larger letters represent the higher concentration:

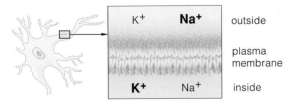

K+   **Na+**   outside

plasma membrane

K+   Na+   inside

Without these ion gradients, an action potential cannot arise. When a membrane is maintaining the gradients, it is largely impermeable to sodium, because the gates

on Na+ channels are shut. Some potassium is leaking out, through many open K+ channels. The leaks make the cytoplasmic fluid slightly more negative, so some K+ is attracted back inside. K+ shows no more net movement when outward-directed diffusion balances the inward pull of electric charge.

Even so, a tiny fraction of the K+ that leaked out is still outside, and a tiny fraction of Na+ is leaking in through a few channels that are not *quite* shut. Do the leaks mean that the ion gradients will disappear? No. **Sodium–potassium pumps** maintain the ion gradients. They also restore the gradients after they are reversed during an action potential. These active transporters span the membrane (Section 5.4). When activated by a phosphate-group transfer from ATP, each pumps two K+ ions into the cell and three Na+ out of it. Both ions are pumped *against* their concentration gradient.

With this bit of background on the gradients across the neural membrane, we are ready to look at how an action potential can arise at the trigger zone and then propagate itself, undiminished, to an output zone.

Sensory neurons, interneurons, and motor neurons make up the communication lines of nervous systems.

In a neuron at rest, transport proteins are maintaining ion concentration gradients by assisting or restricting the diffusion of ions across the lipid bilayer of the plasma membrane. The steady voltage difference associated with the ion gradients is the resting membrane potential.

An action potential is an abrupt, fleeting reversal in the voltage difference across the plasma membrane of any excitable cell. Sodium–potassium pumps maintain and restore the ion gradients, which are required for a reversal.

34.3 | A Look at Action Potentials

LINKS TO
SECTIONS
5.4, 5.5, 28.3

Action potentials are easy to understand when you remember that ions can cross cell membranes only through the interior of transport proteins.

APPROACHING THRESHOLD

Tap your wrist gently, and the mechanical pressure stimulates the receptor endings of sensory neurons in your skin. It slightly deforms the plasma membrane at the input zones of these neurons, which allows a few ions to slip across and shift the voltage difference just a bit. The light pressure has resulted in a local, *graded* potential. "Graded" means that disturbances to an input zone vary in magnitude, because some kinds of stimuli are more intense or last longer than others. Graded potentials do not spread far from the point of stimulation. It takes certain kinds of ion channels to spread farther, and the input zones of neurons simply do not have them.

When a stimulus *is* intense or long-lasting, graded signals spread from an input zone into an adjoining trigger zone. This membrane patch is richly endowed with voltage-sensitive gated channels for sodium ions. When the voltage difference across the membrane is disturbed by a certain amount—a *threshold* level—the gates open and trigger an action potential.

The opened gates allow positively charged sodium ions to flow through the interior of transport proteins and into the neuron (Figure 34.8). That influx makes the cytoplasmic side of the membrane less negative, which causes more gates to open and more sodium to enter. This ever increasing, inward flow of ions is one example of **positive feedback**. The activity intensifies as a result of its own occurrence:

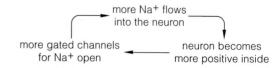

At threshold, the opening of sodium gates no longer depends on the strength of the stimulus. The positive feedback cycle is under way, and the inward-rushing sodium itself is enough to open more gates.

AN ALL-OR-NOTHING SPIKE

You can make a recording of an action potential by putting one electrode in an axon and another outside, and connecting both to a recording device. Figure 34.9 shows what a recording looks like before, during, and after an action potential. At threshold, the change in membrane potential for a given neuron always spikes with the same intensity, as an *all-or-nothing* event.

Each spike lasts for just a millisecond or so. Where the charge is reversed, gates on sodium channels shut and cut off the Na⁺ inflow. Also, halfway through the reversal, gated *potassium* channels open, so K⁺ flows out. The outflow restores the voltage difference at the membrane patch, but not the particular ion gradients that are required for another action potential. Again, sodium–potassium pumps adjust the distribution of Na⁺ and K⁺ across the plasma membrane after each action potential ends. These active transporters pump sodium out and potassium in (Figure 34.8c).

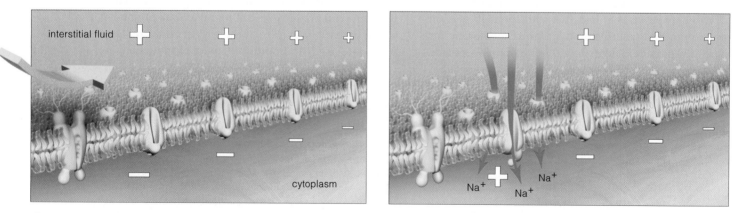

a Membrane at rest; inside of neuron negative with respect to the outside. An electrical disturbance (*yellow* arrow) spreads from an input zone to an adjacent trigger region of the membrane, which has a great number of gated sodium channels.

b A strong disturbance initiates an action potential. Sodium gates open. The sodium inflow decreases the negativity inside the neuron. The change causes more gates to open, and so on until threshold is reached and the voltage difference across the membrane reverses.

Figure 34.8 *Animated!* Propagation of an action potential along the axon of a motor neuron.

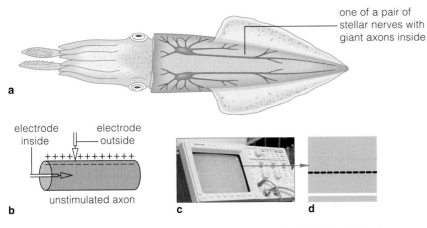

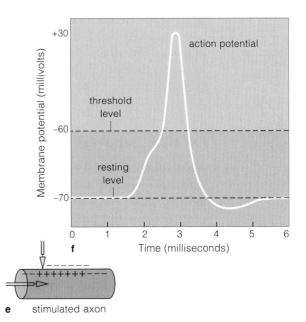

one of a pair of stellar nerves with giant axons inside

electrode inside   electrode outside

+++++V+++++++++

unstimulated axon

b

c

d

Figure 34.9 Action potential recordings. (**a**) A squid (*Loligo*) yielded the first recordings. (**b**) Researchers put electrodes inside and outside one of the squid's "giant" axons and connected them to a monitoring device called an oscilloscope (**c**). (**d**) The resting membrane potential showed up as a horizontal beam of light (the *white* line) across the oscilloscope's screen. (**e,f**) Strong stimulation of the axon deflected the beam. The resulting waveform in (**f**) is typical of action potentials.

+++++++---

e   stimulated axon

f

DIRECTION OF PROPAGATION

An action potential causes gated channels to open in an adjoining membrane patch, which causes channels to open in the *next* patch, and so on down the axon. This positive feedback event is self-propagating; it does not weaken with distance. In addition, it always moves *away* from the trigger zone. Why? As an action ends, gated sodium channels are inactivated until the sodium–potassium pumps are done restoring the ion gradients across the plasma membrane. Meanwhile, down the line, the adjoining patch of membrane has an abundance of sodium channels with closed gates that are ready to be opened.

In neurons, an action potential starts after the membrane potential reaches a threshold level. The voltage difference across the membrane reverses as gated sodium channels open in an ever accelerating way. The disturbance causes self-propagating reversals at each consecutive patch of membrane along an axon, with no loss in magnitude.

An action potential ends when potassium flows out of the neuron and restores the voltage difference. Gated sodium channels are inactivated until sodium–potassium pumps restore the ion gradients across the patch of membrane, which are required for the next action potential.

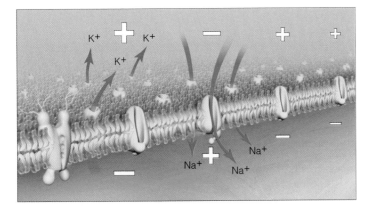

c With the reversal, sodium gates shut and potassium gates open (*red* arrows). Potassium follows its gradient out of the neuron. Voltage is restored. The disturbance triggers an action potential at the adjacent site, and so on, away from the point of stimulation.

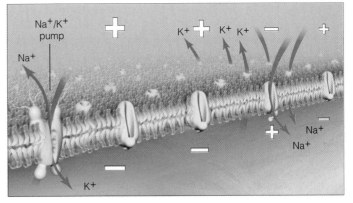

d Following each action potential, the inside of the plasma membrane becomes negative once again. However, the sodium and potassium concentration gradients are not yet fully restored. Active transport at sodium–potassium pumps restores them.

34.4 How Neurons Send Messages to Other Cells

LINKS TO
SECTIONS
5.4–5.6

So far, you have tracked the energy of a stimulus, from its transduction into the electrochemical energy of an action potential to its propagation down to the output zone of a neuron. What happens next?

CHEMICAL SYNAPSES

A thin cleft separates the output zone of one neuron from a neighboring neuron, gland cell, or muscle cell, as in Figure 34.10a,b. At this zone, the electrochemical energy of an action potential is transduced to the form of chemical signal that can diffuse across the cleft and activate or inhibit the target cell. Hence the name for this type of functional bridge between a neuron and some other cell: **chemical synapse**. *Synapse* is derived from a Greek word meaning to fasten together.

Figure 34.10 shows an axon ending of a *pre*synaptic neuron. Inside are many synaptic vesicles filled with **neurotransmitter**, a type of signaling molecule that is synthesized in neurons only. The plasma membrane has many gated channels for calcium ions. In between action potentials, there are more calcium ions outside than inside, and the gates stay shut tightly. An action potential, however, makes the gates open.

As calcium ions flow in, the synaptic vesicles move through the cytoplasm in the axon ending and fuse with the membrane. These exocytic vesicles lose their identity when they merge with the plasma membrane, and the neurotransmitter molecules are released into the synaptic cleft (Section 5.6).

Diffusion alone quickly moves neurotransmitters to receptors on the *post*synaptic cell membrane. In some cells, the receptors are part of proteins with gated ion channels. When neurotransmitter binds to them, the gates open, so ions diffuse in (Figure 34.10c). Different receptors indirectly induce change in the membrane's permeability. When neurotransmitter binds to them, it induces enzymes to enter reactions that open up ion channels elsewhere in the membrane.

A postsynaptic cell's response depends on the type and number of neurotransmitter molecules, receptors, and gated ion channels. Often a neurotransmitter has an *excitatory* effect, in that it can drive the membrane toward the threshold of an action potential. In other cases, it has an *inhibitory* effect; it pulls the membrane away from threshold.

To give one example, a **neuromuscular junction** is a type of chemical synapse between a motor neuron

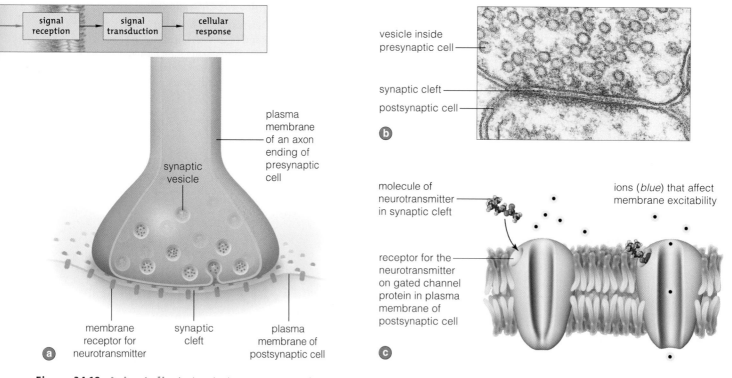

Figure 34.10 *Animated!* A chemical synapse, part of a signaling pathway in the nervous system. (**a**,**b**) An action potential arrriving at an axon ending of a presynaptic cell causes the release of neurotransmitter. (**c**) These signaling molecules diffuse across a cleft to a postsynaptic cell membrane. When they bind to receptors on gated membrane proteins, the gates open, and the signal is transduced into a flow of ions into the cell. That flow, a form of electrochemical energy, may activate or inhibit the target cell's activity.

and a skeletal muscle fiber. The motor neuron releases **acetylcholine**, or ACh, which diffuses across the cleft and binds to membrane receptors on muscle fibers (Figure 34.11). ACh has an excitatory effect on these cells and might stimulate muscle contraction (Section 37.7). However, as you will see later, ACh released by the axons of different nerves binds to cardiac muscle cells and inhibits contraction. Also, in the brain, ACh inhibits activity of cells that have roles in memory.

SYNAPTIC INTEGRATION

Between 1,000 and 10,000 communication lines reach a typical interneuron in your brain, which is only one of at least 100 billion others that may interact at 100 trillion synapses! At any time, many excitatory and inhibitory signals are helping to maintain the resting membrane potential of a given neuron or are driving it closer to or farther away from threshold.

Signals that cross a chemical synapse trigger two kinds of graded potentials in the postsynaptic cell. An **EPSP** (short for excitatory postsynaptic potential) has a *depolarizing* effect; it can drive a membrane closer to threshold. An **IPSP** (inhibitory postsynaptic potential) can have a *hyperpolarizing* effect. Depending upon the state of the target cell, it can help drive the membrane farther away from threshold or help maintain it at the resting level.

With **synaptic integration**, a postsynaptic neuron sums all signals that are arriving at its input zone on more than one communication line. Summation can have several outcomes. Two or more incoming signals might be dampened, suppressed, reinforced, or sent on to other cells. Figure 34.12 has an example of how this integrative process works. It is a composite of the recordings of an EPSP, an IPSP, and their summation.

Neurons also integrate signals that arrive one after another. For example, this happens when an ongoing stimulus triggers a series of action potentials in the presynaptic cell and stimulates it to bombard a target cell with waves of neurotransmitter molecules.

> *Neurotransmitters are signaling molecules secreted into a synaptic cleft from a neuron's output zone. They may have excitatory or inhibitory effects on a postsynaptic cell.*
>
> *Synaptic integration is the summation of all excitatory and inhibitory signals that are arriving at a postsynaptic cell's input zone. By this process, the information flowing through different parts of the nervous system can be reinforced or downplayed, sent onward or suppressed.*

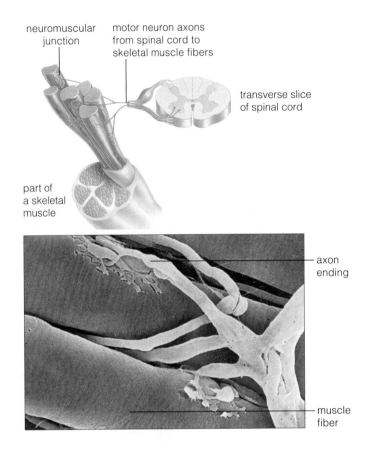

Figure 34.11 A common chemical synapse: a neuromuscular junction. Micrographs reveal many of these junctions between the axon endings of motor neurons and the muscle fibers in skeletal muscle.

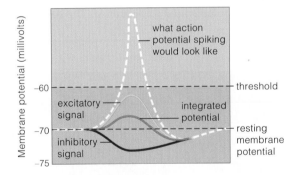

Figure 34.12 Synaptic integration. Typically, many excitatory and inhibitory signals arrive at the input zone of a postsynaptic neuron at the same time.

The *yellow* waveform in this composite graph shows how one postsynaptic cell responded to the arrival of an excitatory signal of a certain magnitude. This signal was not strong enough to drive the membrane to threshold. The *purple* waveform shows how it responded to an inhibitory signal. What if both signals arrived at the same time? The inhibitory signal would lower the excitatory signal's effect, so the *red* waveform would result. Summation of the two signals used in this example would not lead to an action potential (the *white* line).

34.5 A Smorgasbord of Signals

LINKS TO
SECTION
6.4

What do neurotransmitters have to do with how we respond to a kiss, the perfect pair of shoes, a win by the home team? Everything. They are part of communication pathways that govern every sensation and every move we make.

NEUROTRANSMITTER DIVERSITY

In the early 1920s, Austrian scientist Otto Loewi was working to find out what controls the heart's beating. He surgically removed a frog heart—with a nerve still attached—and put it in saline solution. Loewi already knew that the heart would still beat on its own for a while. He stimulated the nerve and noticed that the heartbeat slowed a bit. Perhaps the stimulated nerve released a chemical signal. To test this hypothesis, he put two frog hearts into a saline-filled chamber and stimulated the nerve connected to one of them. Both hearts started to beat more slowly. As expected, the nerve had released a chemical that not only affected the attached heart, it also diffused through the liquid and slowed the beating of the second heart!

Loewi had discovered one of the responses to ACh, the neurotransmitter you read about in the preceding section. ACh acts on skeletal muscle, smooth muscle, the heart, a variety of glands, and the brain. When it binds to receptors in skeletal muscle, it can stimulate contraction. Binding causes sodium ions to enter the muscle fibers, which drives the membrane potential closer to threshold. When ACh binds to a different type of receptor in heart muscle, it slows contraction. It initiates a cascade of reactions in which different enzymes are activated in sequence. The result? Gated channels for potassium ions open, and the membrane potential is pulled away from threshold.

ACh is only one of many neurotransmitters. Other major types are **norepinephrine**, **epinephrine** (also called adrenalin), and **dopamine**. These three are made from the amino acid tyrosine. Both norepinephrine and epinephrine prime the body to respond to stress.

Dopamine affects fine motor control and pleasure-seeking behaviors. Destruction of dopamine-secreting neurons in one brain region causes *Parkinson's disease* (Figure 34.13). Usually, a minor tremor in the hands is the earliest symptom. The sense of balance is affected, so walking may become difficult. Heritable mutations increase the risk. Exposure to certain chemicals in the environment also may be a contributing factor.

Signaling pathways that involve dopamine are part of the learning process in vertebrates and a number of invertebrates. In humans, changes in dopamine levels influence mood, motivation, and attention spans. Some dopamine-secreting neurons fire off action potentials spontaneously. Injecting heroin into a vein accelerates the firing rate. The abnormal stimulation amplifies the sense of pleasure in addictive ways (Section 34.13).

Serotonin is a small neurotransmitter derived from the amino acid tryptophan. As you read earlier in the chapter, serotonin affects mood and memory. Ecstasy interferes with its functions. Low serotonin levels are related to depression. Prozac (fluoxetine) and similar antidepressants alter serotonin levels.

GABA (for gamma amino butyric acid) is derived from glutamate. In the brain, it is the major inhibitor of neurotransmitter release by other neurons.

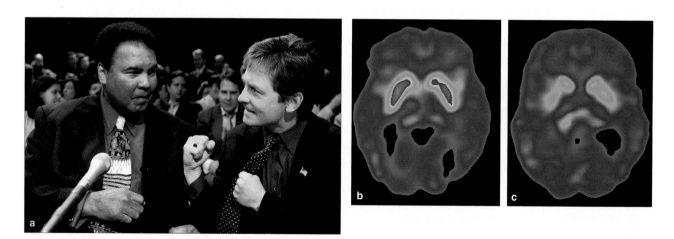

Figure 34.13 Battling Parkinson's disease. (**a**) At this writing, this neurological disorder has affected former heavyweight champion Muhammad Ali, actor Michael J. Fox, and about 500,000 other people in the United States. PET scans of an unaffected individual used as a control and (**b**) of an affected person (**c**). *Red* and *yellow* reveal the levels of metabolic activity by dopamine-secreting neurons.

CLEANING UP THE CLEFT

After they work, neurotransmitter molecules must be promptly removed from synaptic clefts so that new signals can be sent. Some diffuse away. Membrane transport proteins actively pump others back into the presynaptic cell or neuroglial cells. Enzymes secreted into the cleft break down specific molecules, as when acetylcholinesterase cleaves molecules of ACh.

When neurotransmitter accumulates inside the cleft, it disrupts the signaling pathways. That is how *sarin* and other nerve gases exert their effects. After sarin is inhaled, it binds to acetylcholinesterase and blocks its active site—so ACh cannot be degraded (Section 6.4). The ACh buildup causes paralysis of skeletal muscles, confusion, slurred speech, and headaches. High levels of sarin kill by interfering with signaling pathways that control breathing. In 1995, a few members of a cult released sarin gas into a crowded Tokyo subway. Eleven people died; about 5,000 required treatment.

Ecstasy, remember, slows serotonin uptake. Prozac and similar drugs do the same thing. We return to the topic of these and other psychoactive drugs later in this chapter.

THE NEUROPEPTIDES

Some neurons also produce neuropeptides, which are larger than neurotransmitter molecules. These act as **neuromodulators**; they magnify or reduce the effects of neurotransmitter on neurons that are either close to the secreting cell or some distance away.

One neuromodulator, **substance P**, enhances pain perception. **Enkephalins** and **endorphins** are natural painkillers and resemble morphine in their structure. Both inhibit the release of substance P, and both are secreted in response to strenuous activity or injuries. Endorphins also are released when people laugh, reach orgasm, or get a soothing acupuncture treatment or a calming massage. Besides suppressing pain, both of these neuropeptides reportedly can elevate mood and enhance the function of certain immune cells.

ACh, norepinephrine, epinephrine, dopamine, serotonin, and GABA are small neurotransmitters with diverse effects. All are derived from amino acids.

Once neurotransmitters work, they must be removed from the synaptic cleft to keep signaling pathways open.

Endorphins, enkephalins, and other neuromodulators magnify or reduce the effects of neurotransmitters.

34.6 Let's Hear It for Neuroglia!

Like celebrities, neurons get all the attention. Even so, they would quickly fall to pieces without their supporting staff. Similarly, we now know that neuroglia are more than bit players; neurons cannot act at all without them.

Neuroglial cells outnumber neurons in a human brain by about 10 to 1. Many are the framework that holds neurons in place; *glia* means glue in Latin. When the nervous system is developing, new neurons migrate to their final positions along highways of neuroglial fibers, which extend outward from the forming brain.

Oligodendrocytes in the brain make myelin, a fatty substance in the insulating sheaths of long-distance axons. Schwann cells do the same in peripheral nerves.

The brain's most abundant cells are star-shaped astrocytes (Figure 34.14). Astrocytes have diverse functions. They control local concentrations of ions and neurotransmitters. They play a role in immune defenses, make lactate that fuels active neurons, and make nerve growth factor.

A **growth factor** is a type of signaling molecule secreted from one cell that targets receptors on another cell, which responds by dividing or differentiating.

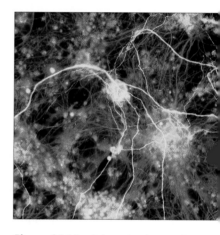

Figure 34.14 Astrocytes (*orange*) and a neuron (*yellow*) in brain tissue. The cells in this light micrograph were stained by immunofluorescence. This procedure attaches fluorescent dye molecules to antibodies designed to target specific molecules on a cell.

Once most neurons mature, they no longer enter mitotic cell divisions. However, when they are exposed to nerve growth factor, they make more synaptic connections with neighboring cells. As you will see, this growth response also is the basis of new connections that are required in the storage of memories.

Microglia, another cell type, hang out in the brain in resting form. When a tissue is injured, they become active, motile cells. They prowl the brain and engulf dead or dying tissue. They also issue chemical signals that summon immune cells to the threatened tissue.

Most cells of the vertebrate nervous system are neuroglia. These diverse cells help organize neurons as the nervous system is developing. They structurally support and nourish neurons, and promote synapse formation and immunity.

Astrocytes, oligodendrocytes, and microglia are major components of the brain. Myelin-producing Schwann cells sheathe many axons of peripheral nerves.

34.7 Nerves and Reflex Arcs

Through synaptic integration, messages arriving at a neuron may be reinforced and sent on to its neighbors. In which direction will a given message travel? That depends on how neurons are organized in the body.

BLOCKS AND CABLES OF NEURONS

Remember, information about stimuli generally flows from sensory receptors to interneurons, then to motor neurons, then to effector cells. However, many signals also loop about in amazing ways.

Consider the billions of interneurons in your brain. They take part in circuits that integrate messages and organize responses. In the *diverging* circuits, dendrites and axons of neurons extend out from one block and communicate with other blocks. In *converging* circuits, signals from many neurons zero in on just a few. In certain circuits, neurons synapse back on themselves, repeating signals like gossip that just won't go away. As one example, such *reverberating* circuits make your eye muscles twitch rhythmically while you sleep.

Information also flows rapidly through **nerves**, the long-distance cables between body regions. A nerve contains dendrites of sensory neurons, axons of motor neurons, or both, bundled inside connective tissues. Figure 34.15*a* is an example of nerve structure.

The neuroglial cells called Schwann cells wrap like jelly rolls around axons of most peripheral nerves, one after another, as a **myelin sheath**. A sheath functions as an electronic insulator; it speeds the propagation of action potentials. How? Ions cannot cross the neural membrane at sheathed regions. The ion disturbances associated with an action potential must spread down the axon's cytoplasm until they reach a tiny exposed gap (node) between two Schwann cells (Figure 34.15*b*). The membrane at each node is loaded with gated Na^+ channels. When the gates open, the voltage difference reverses abruptly. By jumping from node to node, the signal can be moved as fast as 120 meters per second down long axons. By contrast, the maximum speed in unmyelinated axons is about 10 meters per second.

In *multiple sclerosis*, or MS, certain white blood cells wrongly identify a type of protein in myelin sheaths as foreign and destroy it. This autoimmune response leads to inflammation of axons in the brain and spinal cord and the death of oligodendrocytes. Some people are genetically predisposed to develop the disorder, but viral infection may set it in motion. Either way, information flow is disrupted. Dizziness, numbness, muscle weakness, fatigue, visual problems, and other symptoms follow. About 500,000 people in the United States are now affected by MS.

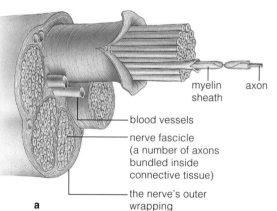

myelin sheath · axon

blood vessels

nerve fascicle (a number of axons bundled inside connective tissue)

the nerve's outer wrapping

a

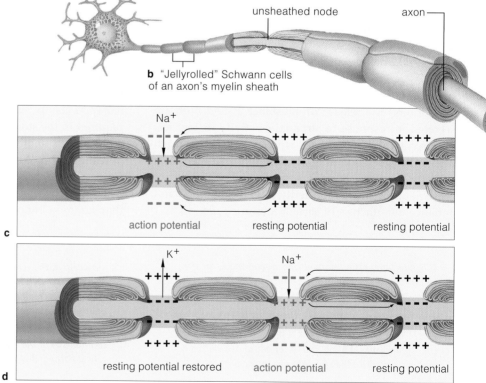

unsheathed node · axon

b "Jellyrolled" Schwann cells of an axon's myelin sheath

Na^+

action potential · resting potential · resting potential

c

K^+ · Na^+

resting potential restored · action potential · resting potential

d

Figure 34.15 *Animated!* (**a**) Structure of one type of nerve. (**b–d**) In axons with myelin sheaths, ions flow across the neural membrane at nodes, or tiny unsheathed gaps where many gated channels for sodium ions are exposed. When a disturbance caused by an action potential reaches each node, sodium gates open and start a new action potential. The disturbance spreads swiftly to the next node, where it triggers a new action potential, and so on down the line to the output zone.

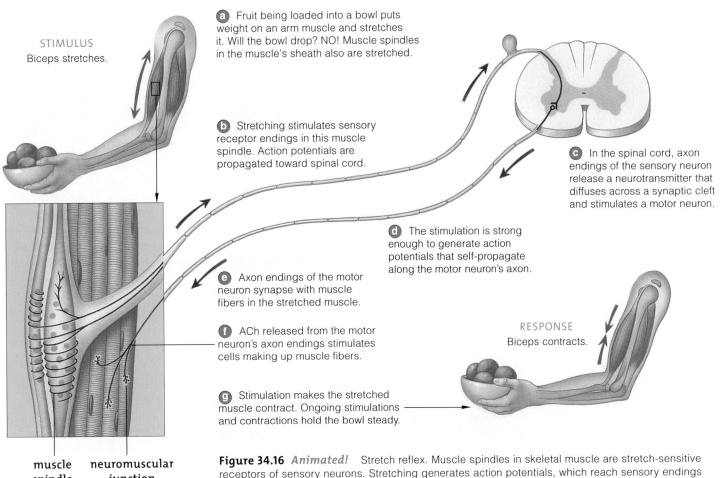

a Fruit being loaded into a bowl puts weight on an arm muscle and stretches it. Will the bowl drop? NO! Muscle spindles in the muscle's sheath also are stretched.

STIMULUS
Biceps stretches.

b Stretching stimulates sensory receptor endings in this muscle spindle. Action potentials are propagated toward spinal cord.

c In the spinal cord, axon endings of the sensory neuron release a neurotransmitter that diffuses across a synaptic cleft and stimulates a motor neuron.

d The stimulation is strong enough to generate action potentials that self-propagate along the motor neuron's axon.

e Axon endings of the motor neuron synapse with muscle fibers in the stretched muscle.

RESPONSE
Biceps contracts.

f ACh released from the motor neuron's axon endings stimulates cells making up muscle fibers.

g Stimulation makes the stretched muscle contract. Ongoing stimulations and contractions hold the bowl steady.

muscle spindle   neuromuscular junction

Figure 34.16 *Animated!*   Stretch reflex. Muscle spindles in skeletal muscle are stretch-sensitive receptors of sensory neurons. Stretching generates action potentials, which reach sensory endings in the spinal cord. These synapse with a motor neuron. It carries signals calling for contraction, from the spinal cord back to the stretched muscle. The muscle contracts, steadying the arm.

REFLEX ARCS

Reflexes are the most ancient paths of information flow. A **reflex** is a movement or some other response to a stimulus that happens automatically, no thought required. In the simplest reflex arcs, sensory neurons synapse directly on motor neurons. In more complex reflexes, sensory neurons interact with one or more interneurons, which then stimulate or suppress all motor neurons required for a coordinated response.

The *stretch reflex* is a simple reflex arc that causes a muscle to contract after gravity or some other force stretches it. Suppose you hold a bowl while someone puts peaches into it. The load makes your hand drop a bit, which stretches an arm muscle called a biceps. As a biceps is stretched, the receptor endings of some sensory organs within it are stretched as well. These organs are **muscle spindles**. Their endings, enclosed in a sheath that runs parallel with the muscle, are the input zones of certain sensory neurons (Figure 34.16).

The rate of signal transmission along the axons of these sensory receptors depends on the extent to which the muscle is stretched. In the spinal cord, axons of muscle spindles synapse with motor neurons—the axons of which lead directly back to the muscle. The action potentials reach the axon endings of the motor neurons, where they trigger the release of ACh. This neurotransmitter causes the biceps to contract, which helps steady the arm against the added load.

The knee-jerk reflex is a type of stretch reflex. A tap just below the knee shortens the thigh muscle, signals flow to the spinal cord, and the leg jerks in response.

In complex animals, neurons are organized in blocks and in cables, some of which arc back to the point of stimulation.

Nerves are long-distance cables between body regions. Most long axons of their sensory neurons, motor neurons, or both have a myelin sheath, which speeds signal propagation.

Reflex arcs, in which sensory neurons synapse directly on motor neurons, are the simplest paths of information flow.

34.8 What Are the Major Expressways?

Now you are ready to consider the peripheral nervous system and the spinal cord. The two interconnect as the body's main expressways for information flow.

PERIPHERAL NERVOUS SYSTEM

Somatic and Autonomic Systems   In humans, the peripheral nervous system includes thirty-one pairs of *spinal* nerves, which connect with the spinal cord. It also has twelve pairs of *cranial* nerves, which connect directly with the brain. Most cranial nerves, and all spinal nerves, contain bundles of sensory and motor fibers inside a tough outer wrapping.

Nerves of the peripheral system are classified by function. The sensory part of **somatic nerves** relays information from receptors in the skin, tendons, and skeletal muscles to the central nervous system. Their motor axons deliver commands from the brain and spinal cord to skeletal muscles. The **autonomic nerves** relay information to and from the viscera. *Viscera* refers to the soft internal organs, such as cardiac muscles, smooth muscles, and glands.

Sympathetic and Parasympathetic Divisions   The nerves of the autonomic system fall in two categories: sympathetic and parasympathetic. Both service most organs and work antagonistically, meaning the signals from one type oppose signals from the other (Figure 34.17). **Sympathetic neurons** are most active in times of stress, excitement, and danger. Their axon endings release norepinephrine. **Parasympathetic neurons** are most active in times of relaxation. The release of ACh from their axon endings promotes daily housekeeping tasks, such as digestion and urine production.

What happens when something startles or scares you? The parasympathetic input decreases and gives

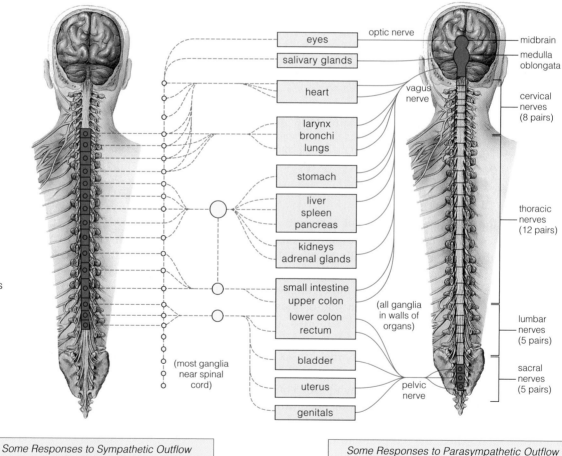

Figure 34.17 *Animated!*
(**a**) Major sympathetic and (**b**) parasympathetic neurons of the autonomic system. These nerves are paired; the body's right and left halves have one of each. The ganglia (clusters of nerve cell bodies) are local control centers. Their axons are bundled together inside nerves.

Some Responses to Sympathetic Outflow
Heart rate increases
Pupils of eyes dilate (widen, let in more light)
Glandular secretions decrease in airways to lungs
Salivary gland secretions thicken
Stomach and intestinal movements slow down
Sphincters (rings of muscle) contract

a Sympathetic outflow from the spinal cord

Some Responses to Parasympathetic Outflow
Heart rate decreases
Pupils of eyes constrict (keep more light out)
Glandular secretions increase in airways to lungs
Salivary gland secretions become dilute
Stomach and intestinal movements increase
Sphincters (rings of muscle) relax

b Parasympathetic outflow from the spinal cord and brain

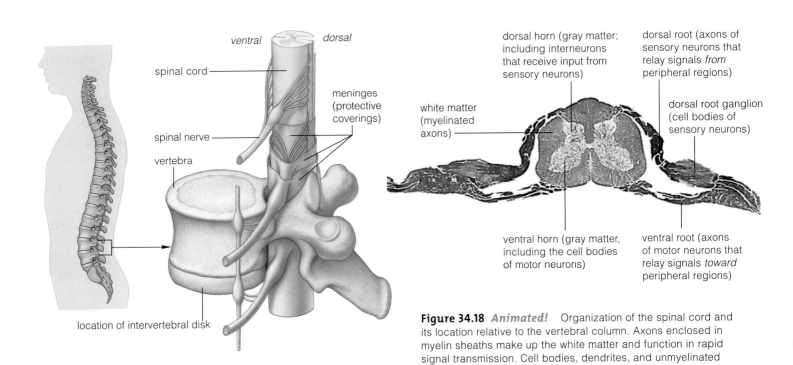

ventral   dorsal

spinal cord

spinal nerve

vertebra

meninges (protective coverings)

location of intervertebral disk

dorsal horn (gray matter; including interneurons that receive input from sensory neurons)

dorsal root (axons of sensory neurons that relay signals *from* peripheral regions)

white matter (myelinated axons)

dorsal root ganglion (cell bodies of sensory neurons)

ventral horn (gray matter, including the cell bodies of motor neurons)

ventral root (axons of motor neurons that relay signals *toward* peripheral regions)

Figure 34.18 *Animated!* Organization of the spinal cord and its location relative to the vertebral column. Axons enclosed in myelin sheaths make up the white matter and function in rapid signal transmission. Cell bodies, dendrites, and unmyelinated axons of neurons, along with neuroglia, make up the gray matter.

way to sympathetic signals. These signals raise your heart rate and blood pressure, make you sweat more and breathe faster, and make adrenal glands secrete epinephrine. By their action, the signals help put you in a state of intense arousal, primed to fight or make a fast getaway. Hence the term *fight–flight response.*

An organ often receives opposing sympathetic and parasympathetic signals. Consider the smooth muscle cells in the gut wall. Even as sympathetic neurons are releasing norepinephrine at synapses with the smooth muscle cells, parasympathetic neurons are releasing ACh at other synapses with the same cells. One signal tells the gut to slow its contraction; the other calls for increased activity. Synaptic integration finely adjusts the muscle's actual response.

SPINAL CORD

Inside the spinal cord and deep in the brain is *white matter,* or specialized tracts of myelin-sheathed axons. The rest of the nerve tissue is *gray matter*—neuroglia and cell bodies, dendrites, and (mostly) unmyelinated axons of motor neurons and interneurons.

The **spinal cord** connects the peripheral nervous system with the brain and controls some reflexes. The reflexes affect basic tasks, such as bladder emptying and limb movements; remember the stretch reflex?

The spinal cord threads through ligaments and the bones of the vertebral column, which protect it. It also

is protected by three coverings, called the meninges, that enclose the brain as well. In *meningitis,* a viral or bacterial infection has inflamed the coverings. Severe headaches, fever, a stiff neck, and nausea are among the symptoms that follow.

Sensory neurons, which are afferent, connect to the spinal cord at one of a pair of structures called dorsal roots. Motor neurons, which are efferent, connect to the spinal cord at ventral roots. A ganglion is visible as a bulge in each dorsal root (Figure 34.18).

In amphibians, spinal reflexes play a greater role in motor activity. Between the frog brain and spinal cord are circuits that make bent legs straighten. Cut these circuits near the brain and the legs become paralyzed —but only for about a minute. Reflex pathways in the spinal cord set the frog hopping again. Humans and other primates depend more on brain centers, so they show little or no recovery from a similar injury.

Nerves of the peripheral nervous system connect the brain and spinal cord with the rest of the body.

The somatic division of the peripheral nervous system deals with skeletal muscle movements. Its autonomic division deals with smooth muscle, cardiac muscle, and glands.

The spinal cord is a vital expressway for signals between the brain and the peripheral nerves.

34.9 The Vertebrate Brain

LINKS TO SECTIONS 25.1, 26.2, 26.13

The spinal cord is continuous with the brain, the body's master control center. The brain receives, integrates, stores, retrieves, and issues information. It coordinates responses to sensory input. As is the case for the spinal cord, bones and membranes (meninges) enclose and protect the brain.

THE BRAIN'S SUBDIVISIONS

A hollow, tubular nerve cord forms in every chordate embryo. In vertebrates, it develops into a spinal cord and brain. Genes that control segmented body plans divide the brain into specialized regions: the forebrain, midbrain, and hindbrain (Figure 34.19). A **brain stem**, the most ancient nervous tissue, persists in all three regions and is continuous with the spinal cord.

The hindbrain's **medulla oblongata** houses reflex centers for respiration, circulation, and other essential tasks. It integrates motor responses and governs some reflexes, such as coughing. It also affects sleep. The **cerebellum** uses inputs from the eyes, ears, muscle spindles, and forebrain regions to help control motor

skills and posture. Axons from its two halves reach the **pons** (meaning bridge). The pons is like a traffic officer; it controls signal flow between the cerebellum and integrating centers in the forebrain.

Fishes and amphibians have the most pronounced midbrain, which sorts out most of their sensory input and initiates motor responses. In all vertebrates, the midbrain has centers for visual input. Especially when primates evolved, the forebrain took over the task of integrating most visual stimuli (Section 26.13).

Vertebrates first evolved in water, where chemical odors diffusing from predators, prey, and mates were vital cues. They relied heavily on olfactory lobes and paired outgrowths from the brain stem that integrated olfactory input and responses to it. Especially among land vertebrates, the outgrowths expanded into two halves of the **cerebrum**, the two cerebral hemispheres.

The **thalamus** became a forebrain center for sorting out sensory input and relaying it to the cerebrum. The **hypothalamus** ("under the thalamus") evolved into the main center for homeostatic control of the internal environment. It assesses and regulates all behaviors related to internal organ activities, such as thirst, sex, and hunger. It also governs related emotions, such as sweating with passion and vomiting from fear.

PROTECTION AT THE BLOOD–BRAIN BARRIER

The neural tube's lumen (the space inside it) persists in adult vertebrates as a system of cavities and canals filled with a clear *cerebrospinal fluid* (Figure 34.20). The fluid forms inside the brain ventricles, but it seeps out and bathes the tissues of the brain and spinal cord. It cushions them against potentially jarring movements.

A **blood–brain barrier** protects the spinal cord and brain from harmful substances. It exerts some control over which solutes enter cerebrospinal fluid. No other

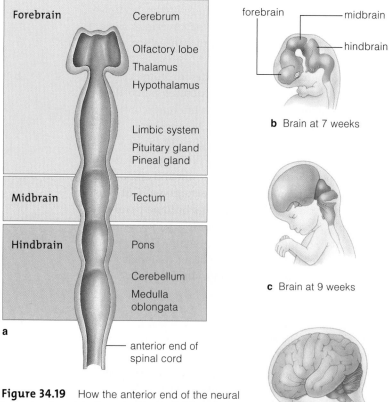

Forebrain	Cerebrum
	Olfactory lobe
	Thalamus
	Hypothalamus
	Limbic system
	Pituitary gland Pineal gland
Midbrain	Tectum
Hindbrain	Pons
	Cerebellum
	Medulla oblongata

a

anterior end of spinal cord

Figure 34.19 How the anterior end of the neural tube develops into three brain subdivisions in a human embryo. (**a**) This sketch flattens out the tube to list the dominant brain regions that develop along its length. The tube actually curves forward in the embryo, as in (**b–d**).

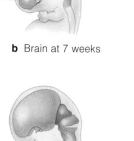

forebrain — midbrain — hindbrain

b Brain at 7 weeks

c Brain at 9 weeks

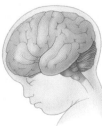

d Brain at birth

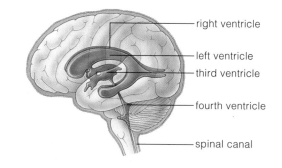

right ventricle
left ventricle
third ventricle
fourth ventricle
spinal canal

Figure 34.20 Cerebrospinal fluid (*blue*). This extracellular fluid is produced in the brain's four interconnected ventricles (cavities) and in the spinal cord's central canal.

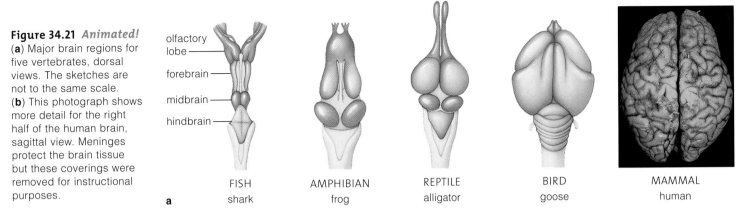

Figure 34.21 *Animated!* (a) Major brain regions for five vertebrates, dorsal views. The sketches are not to the same scale. (b) This photograph shows more detail for the right half of the human brain, sagittal view. Meninges protect the brain tissue but these coverings were removed for instructional purposes.

olfactory lobe
forebrain
midbrain
hindbrain

a

FISH
shark

AMPHIBIAN
frog

REPTILE
alligator

BIRD
goose

MAMMAL
human

part of extracellular fluid has solute concentrations maintained within such narrow limits. Even changes brought on by eating and exertion are limited. Why? Hormones and other chemicals in the blood can alter neural function. Also, changes in ion concentrations can alter the threshold for action potentials.

The barrier works at the wall of blood capillaries that service the brain. In most parts of the brain, tight junctions form a seal between the abutting cells of the capillary wall, so water-soluble substances must pass *through* the cells to reach the brain. Transport proteins in the plasma membrane of these cells allow glucose, other vital nutrients, and some ions to cross. They bar many toxins and wastes, including urea. The barrier does not keep out small, fat-soluble molecules, such as oxygen, carbon dioxide, alcohol, caffeine, nicotine, and mercury vapor. Inflammation or traumatic blows can damage it and compromise neural function.

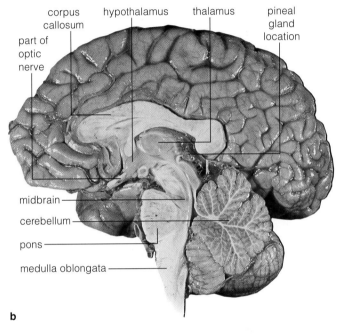

corpus callosum
hypothalamus
thalamus
pineal gland location
part of optic nerve
midbrain
cerebellum
pons
medulla oblongata

b

SOME ARE BRAINIER THAN OTHERS

On average, the human brain weighs 1,300 grams, or 3 pounds. Again, it has about 100 billion interneurons, and neuroglia makes up more than half of its volume. Compared to the brains of other vertebrates, including those in Figure 34.21, the human midbrain is smaller. The hindbrain's cerebellum is larger. It is about the size of a fist and has more interneurons than all other brain regions combined. As in other vertebrates, it deals with the sense of balance and coordination but took on other functions as humans evolved. It affects learning of motor and mental skills, such as language.

What about the forebrain? A deep fissure divides its cerebrum into two halves, the cerebral hemispheres (Figure 34.21). The next section takes a look at their thin outer layers, the cerebral cortex. Each half deals mainly with input from the opposite side of the body. For instance, signals about pressure on the right arm travel to the left hemisphere. Activities of both halves

are coordinated by signals that flow both ways across a thick band of nerve tracts, the corpus callosum.

The pineal gland is located near the hypothalamus. The hypothalamus receives signals about light sources from the retina and relays them to this light-sensitive endocrine gland. Chapter 36 describes the functions of these forebrain regions and explores connections between the nervous and endocrine systems.

The vertebrate brain develops from a hollow neural tube, the lumen of which persists in adults as a system of cavities and canals filled with cerebrospinal fluid. The fluid cushions nervous tissue from sudden, jarring movements.

Nervous tissue is subdivided into a hindbrain, forebrain, and midbrain. The brain stem is the most ancient tissue. The forebrain has the most complex integrating centers.

34.10 The Human Cerebrum

LINKS TO SECTIONS 5.3, 26.2, 26.4, 26.10

Our "humanness" starts in the outer layer of gray matter of our cerebral cortex, which governs conscious behavior. The cerebral cortex processes and coordinates responses to sensory input. It interacts with the limbic system, which governs emotions and contributes to memory.

FUNCTIONAL AREAS OF THE CORTEX

Each half of the cerebrum, or cerebral hemisphere, is divided into four lobes: frontal, temporal, occipital, and parietal. At the **cerebral cortex**—the gray matter at the surface of each lobe—distinct areas receive and

process different signals, but they still interact. *Motor* areas influence voluntary motor activity. *Sensory* areas assist in our perceptions of what a specific sensation means. Diverse *association* areas integrate information that brings about conscious actions.

The two hemispheres overlap in function, but there are some specialized differences. For example, the left hemisphere's cortex is more concerned with analytical skills, mathematics, and speech. The cortex of the right hemisphere interprets music, judges spatial relations, and assesses visual inputs.

Motor Areas The body is spatially mapped out in the primary motor cortex of each frontal lobe—which controls and coordinates the movements of skeletal muscles on the opposite side of the body. Much of the motor cortex is devoted to finger, thumb, and tongue muscles. Figure 34.22 hints at the control required for voluntary hand movements and verbal expression.

The premotor cortex of each frontal lobe governs learned patterns of motor skills. Dribble a basketball, play a piano, use a keyboard—repetitive movements are evidence that simultaneous and sequential actions of different muscle groups are being coordinated.

Broca's area helps translate thoughts into speech by controlling tongue, throat, and lip muscles. It gives us our capacity to create complex sentences. In most individuals, Broca's area is in the frontal cortex of the left hemisphere (Figure 34.23*a*). Damage to it prevents normal speech, although an affected individual is still able to understand language.

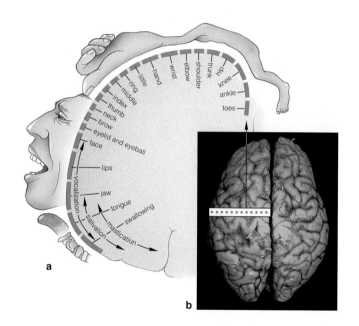

Figure 34.22 (**a**) A slice of the primary motor cortex, through the region indicated in (**b**). The sizes of body parts draped over the artful slice are distorted to show which ones get the most precise control.

Sensory Areas The primary somatosensory cortex is located at the front of the parietal lobe. Like the motor cortex, it is organized as a map that corresponds to the

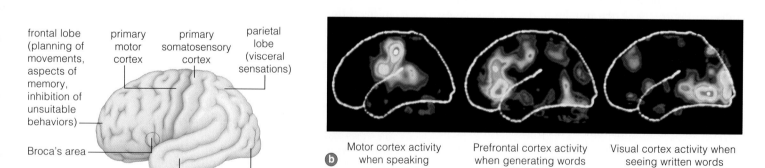

frontal lobe (planning of movements, aspects of memory, inhibition of unsuitable behaviors)

Broca's area

temporal lobe (hearing, advanced visual processing)

primary motor cortex

primary somatosensory cortex

parietal lobe (visceral sensations)

occipital lobe (vision)

Motor cortex activity when speaking

Prefrontal cortex activity when generating words

Visual cortex activity when seeing written words

Figure 34.23 (**a**) Primary receiving and integrating centers of the human cerebral cortex. Association areas coordinate and process sensory input from diverse receptors. (**b**) Three PET scans identifying which areas were active when an individual performed three different tasks.

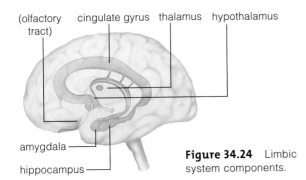

(olfactory tract) cingulate gyrus thalamus hypothalamus

amygdala

hippocampus

Figure 34.24 Limbic system components.

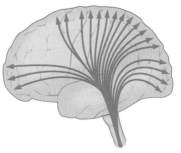

Figure 34.25 Communication pathways making up the reticular formation. This evolutionarily ancient and diffuse network of neurons extends from the spinal cord into the highest integrative areas of the cerebral cortex. Parts of the reticular formation can release serotonin and help control states of consciousness.

body's parts. It is a receiving center for sensory input from skin and joints (Section 35.2). Also in the cortex of the parietal lobe, another sensory area deals with taste perception. At the back of the occipital lobe, the primary visual cortex receives signals from the eyes. Perceptions of sound and odor arise in sensory areas of each temporal lobe.

Association Areas Association areas are scattered through the cortex, but not in the primary motor and sensory areas. Each integrates diverse inputs (Figure 34.23a). For instance, a visual association area around the primary visual cortex compares what we see with visual memories. The most recently evolved area, the prefrontal cortex, is the foundation of our personality and intellect, of abstract thought, judgment, planning, and concern for others.

CONNECTIONS WITH THE LIMBIC SYSTEM

The **limbic system** encircles the upper brain stem. It governs emotions, assists in memory, and correlates organ activities with self-gratifying behavior, such as eating and sex. That is why the limbic system is called our emotional–visceral brain. It can put a heart on fire with passion and a stomach on fire with indigestion. These and other "gut reactions" can be overridden by signals from the prefrontal cortex.

The system includes the hypothalamus, part of the thalamus, and the cingulate gyrus, hippocampus, and amygdala (Figure 34.24). The cingulate (belt-shaped) gyrus is a fold in brain tissue right above the corpus callosum. It affects motivation, and it is more active in extroverts and risk takers than in the introverted or cautious. The hypothalamus correlates emotions with visceral activities. The almond-shaped amygdala is necessary for emotional stability and for interpreting social cues. It is responsive to fright and anxiety.

The limbic system is evolutionarily related to the olfactory lobes. Olfactory input causes signals to flow to the hippocampus, amygdala, and hypothalamus as well as the olfactory cortex. That is one reason why you feel warm and fuzzy when you recall the scent of someone special. Signals about taste also travel to the limbic system and can call up emotional responses.

THE RETICULAR FORMATION

An ancient network of interneurons extends from the upper spinal cord, through the brain stem, and into the cerebral cortex. This **reticular formation** is a low-level path to motor centers in the medulla oblongata and spinal cord. It affects many parts of the nervous system, including the cerebral cortex (Figure 34.25).

Part of the reticular formation promotes chemical changes that affect states of consciousness, such as sleeping or waking. One of its sleep centers produces serotonin. High serotonin levels cause drowsiness and sleep. Substances released from another brain center counter the effect and bring about wakefulness.

Electroencephalograms, or *EEGs*, are recordings of the summed electrical activity of the brain's neurons. They are used to study states of consciousness. EEGs from electrodes placed on the scalp show up as wave forms. The pattern for a person who is meditating is an alpha rhythm. During the transition to sleep, wave forms become larger, more widely spaced, and more erratic. People who are aroused from slow-wave sleep usually say that they were not dreaming; often they were mulling over recent, ordinary events. Slow-wave sleep is punctuated by a pattern of REM sleep, with rapid eye movements, irregular breathing, increased heartbeat, twitching fingers, and often vivid dreams.

The cerebral cortex, each hemisphere's outermost layer of gray matter, contains motor, sensory, and association areas that interact to govern conscious behavior. It also interacts with the limbic system, which affects emotions and contributes to memory.

34.11 Sperry's Split-Brain Experiments

The two cerebral hemispheres are connected by a thick band of axons called the corpus callosum. Neurobiologist Roger Sperry discovered the importance of this connection and revealed the dual nature of human consciousness.

As mentioned in the preceding section, the two cerebral hemispheres look alike but differ a bit in their functions. The differences first became apparent in the mid-1800s, through studies of people who had injuries to particular brain regions. For instance, damage to Broca's area in the left frontal cortex interfered with the ability to vocalize words. Injury to Wernike's area in the left temporal lobe did not interfere with the capacity to say words, but the affected person could not put words into sentences.

Fast-forward to the 1960s. Evidence of the importance of the left hemisphere continued to flow in, and doctors were wondering what role, if any, the right hemisphere plays in the advanced functions of typical right-handed people. Roger Sperry and his coworkers decided to find out.

Sperry was interested in "split-brain" patients—people who had undergone surgery to sever the corpus callosum. At the time, this was an experimental way to treat severe *epilepsy*. Epileptic seizures are like electrical storms in the brain. Surgeons severed a patient's corpus callosum to stop the flow of disturbed electrical signals from one hemisphere to the other. After a brief recovery period, patients were able to lead what seemed to be normal lives, with fewer seizures.

But were those patients really normal? The surgery had stopped the flow of information across 200 million or so axons in the corpus callosum. Surely *something* had to be different. Something was.

Sperry devised elegant experiments to examine the split-brain experience by presenting the two halves of affected patients with two different parts of a visual stimulus. At the time, researchers already knew that the visual connections to and from one hemisphere are mainly concerned with the opposite half of the visual field, as in Figure 34.26. Sperry projected a word—say, COWBOY—onto a screen so that COW fell in the left half of the visual field, and BOY fell in the right (Figure 34.27).

The subjects of this experiment reported *seeing* the word BOY. The left hemisphere, which controls language, recognized the word. However, when asked to *write* the word with the left hand—which was hidden from view—the subject wrote COW. The right hemisphere "knew" the other half of the word (COW) and had directed the left hand's motor response. But it could not tell the left hemisphere what was going on because of the severed corpus callosum. The subject knew a word was being written but could not say what it was!

"The surgery," Sperry reported, "left these people with two separate minds—two spheres of consciousness." Sperry concluded that both hemispheres contribute to normal perception by sharing information that shapes the experience we call consciousness.

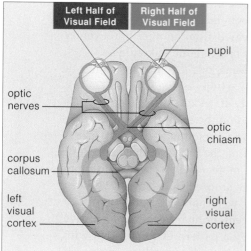

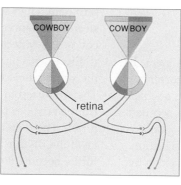
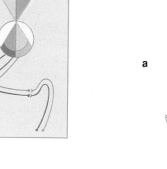
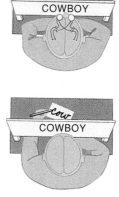

Figure 34.26 *Animated!* **(a)** Pathway by which sensory input about visual stimuli reaches the visual cortex of the human brain.

(b) Each eye gathers visual information at the retina, a thin layer of densely packed photoreceptors at the back of the eyeball (Section 35.7). Light from the *left* half of the visual field strikes receptors on the right side of both retinas. Parts of two optic nerves carry signals from the receptors to the right cerebral hemisphere. Light from the *right* half of the visual field strikes receptors on the left side of both retinas. Parts of the optic nerves carry signals from them to the left hemisphere.

Figure 34.27 One example of the response of a split-brain patient to visual stimuli. As described in the text, this type of experiment demonstrated the importance of the corpus callosum in coordinating activities between the two cerebral hemispheres.

34.12 Storing and Retrieving Memories

*What do we know about **memory**—the brain's capacity to retrieve information about past sensory experiences?*

Embedded in your brain are memory banks. Without them, learning and behavioral adjustments based on experiences would be impossible. The cerebral cortex is bombarded with sensory information, but only a tiny fraction become memories, which form in stages.

Short-term memory lasts just a few seconds or hours. This stage holds a few bits of information—a set of numbers, the words of a sentence, and so on. In *long-term memory*, a seemingly unlimited quantity of larger bits is stored more or less permanently (Figure 34.28).

Different forms of input are stored and called up by different mechanisms. Retention is greatest for *skill* memories. Once you learn how to drive a car, dribble a basketball, or play a violin, you don't forget how—even if you rarely do so again. The skill memories are created when you consciously repeat an activity over and over. As the skill is being learned, the prefrontal cortex signals motor areas of the cortex. Signals flow to the sensory cortex, the cerebellum, and the corpus striatum—a part of the basal ganglia (Figure 34.29a). Once the skill is mastered, the corpus striatum is able to call for appropriate movements, which frees you from having to consciously think about how to make the movements on a second-by-second basis.

Declarative memory allows you to remember how a lemon smells, that a quarter is worth more than a dime, and where you had breakfast yesterday. It starts with signals from the sensory cortex to the amygdala, which acts as the memory gatekeeper. The amygdala connects to the hippocampus, which functions as an association center (Figure 34.29b). Signals must loop repeatedly through the hippocampus and cortex, basal ganglia, and thalamus for a memory to be retained.

Emotional states influence memory retention. For instance, epinephrine released in stressful times can assist the shuffling of short-term memories into long-term storage. This makes evolutionary sense, because an animal that recalls what happened when it dealt with a threat or some other stress may be more likely to survive if the threat recurs. This mechanism makes it difficult for us to forget traumatic events. Chapter 36 returns to the effects of long-term stress on memory.

Amnesia is the loss of declarative memory. It often happens when the hippocampus, amygdala, or both have been damaged. *Alzheimer's disease* usually starts late in life and involves changes in the hippocampus and cerebral cortex. Commonly, the affected person is able to recall long-known facts, including a childhood address, but has trouble remembering recent events.

> *Memory, the storage and retrieval of sensory information, arises from circuits between the cerebral cortex and the limbic system, thalamus, and hypothalamus.*

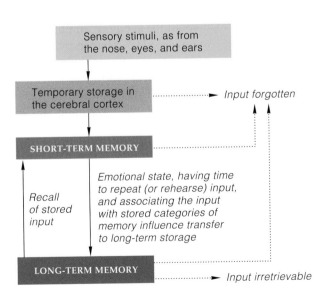

Figure 34.28 Stages of memory processing, starting with temporary storage of sensory inputs in the cerebral cortex.

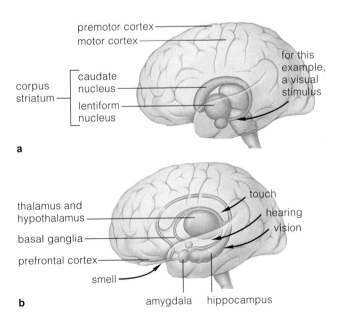

Figure 34.29 Simple diagrams of possible circuits involved in (**a**) skill memory and (**b**) declarative memory.

34.13 Drugs, The Brain, and Behavior

LINKS TO
CHAPTER 6
INTRODUCTION

Psychoactive drugs are substances that affect the function or action of neurotransmitters. Some induce the release of a neurotransmitter. Others prevent its breakdown or uptake. Still others block receptors on the membrane of a postsynaptic cell where it would bind.

ROOTS OF ADDICTION

People take some legal or illegal pyschoactive drugs to mediate illness or stress. They take others to fan the pleasure associated with sex and other self-gratifying behaviors. Even when the body functions well without them, a user may continue to take drugs for real or imagined relief. The body often develops a tolerance of such drugs; it takes larger or more frequent doses to get the same effect. Habituation and tolerance lead into **drug addiction**, a form of chemical dependence in which a drug has rewired the brain and assumed an "essential" biochemical role in the body. Table 34.1 lists warning signs of drug addiction. Three or more may be cause for concern.

Addicts abruptly deprived of their drugs undergo biochemical upheaval, which causes physical pain and mental anguish. However, continued addiction is far more harmful. Besides rewiring the brain, drugs that are inhaled can damage airways and lungs. Sharing needles may invite AIDS, hepatitis B, and hepatitis C, and damage blood vessels. The liver and kidneys are strained when they detoxify and eliminate any drug.

Addictive drugs stimulate the release of dopamine, the neurotransmitter directly involved in the sense of pleasure. For instance, most people think smoking is just a habit. Actually, when nicotine binds to neurons in the brain, it causes a dopamine spike. All addicts specifically crave the dopamine spike.

Table 34.1 Warning Signs of Drug Addiction

1. Tolerance—it takes increasing amounts of the drug to get the same effect.
2. Habituation—it takes continued drug use over time to maintain the self-perception of functioning normally.
3. Inability to stop or curtail drug use, even if the desire to do so persists.
4. Concealment—not wanting others to know of the drug use.
5. Extreme or dangerous actions to get and use a drug, as by stealing, by asking more than one doctor for prescriptions, or by jeopardizing employment by using drugs at work.
6. Deterioration of professional and personal relationships.
7. Anger and defensiveness if someone suggests there may be a problem.
8. Drug use preferred over previous customary activities.

EFFECTS OF PSYCHOACTIVE DRUGS

Stimulants These drugs make you alert, then they depress you. An example is the *caffeine* in coffee, tea, chocolate, and many soft drinks. Low doses acting in the cerebral cortex increase nervousness and suppress fine motor coordination. *Nicotine* in tobacco products is a stimulant that mimics ACh. It affects a variety of sensory receptors. Chapter 40 looks at the impacts of nicotine addiction on health.

Millions are *cocaine* abusers. This stimulant creates feelings of pleasure by blocking uptake of dopamine, norepinephrine, and other signals. Postsynaptic cells are not released from stimulation. Blood pressure and sexual appetite rise. In time, the signaling molecules are cleared away, but the body can't replace them fast enough. The sense of pleasure is permanently lost as hypersensitized postsynaptic cells demand and cannot get stimulation. Figure 34.30 shows a long-term effect.

Abusers inhale granular cocaine or burn crack and inhale the smoke. This extremely addictive drug has staggering social and economic costs. Also, there is no antidote for overdoses, which have caused seizures, respiratory failure, and heart failure.

Amphetamines induce the oversecretion of dopamine and norepinephrine. Addicts smoke, snort, inject, or ingest various forms. Each form can trigger euphoria, sexual arousal, heart pounding, agitation, dry mouth, tremors, and often paranoia. Amphetamines kill the appetite; some people became addicted while taking them to lose weight. Dopamine and norepinephrine synthesis decline over time because the brain depends more and more on the artificial stimulation.

You already read about the synthetic amphetamine MDMA (*Ecstasy*, Adam, or XTC). Methamphetamine hydrochloride, also known as *crystal meth*, is another widely abused form. It is easy to produce, and its use is epidemic, especially in rural areas. During 2004, law enforcement personnel shut down 1,472 meth labs in the state of Iowa alone. The labs themselves can be a danger to public health. They release toxic fumes and have a tendency to explode.

Depressants, Hypnotics Depressants sedate, and hypnotics induce sleep (not a hypnotic trance). Their effects depend on the dose and on physiological and emotional states. They invite calm, drowsiness, sleep, coma, and even death. Low doses impact inhibitory synapses the most, so users get excited or euphoric at first. Larger doses suppress excitatory synapses and lead to depression. Both drugs amplify each other, as when alcohol and barbiturates heighten depression.

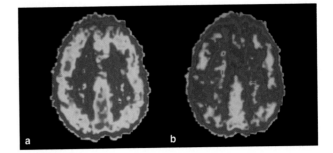

Figure 34.30 Want to be one of the brain-dead? (**a**) This is a PET scan of normal brain activity. (**b**) This PET scan reveals cocaine's long-term effect. *Red*, regions of greatest activity, then *yellow*, *green*, and *blue* for the least activity.

Alcohol, or ethyl alcohol, has wide-ranging effects on the nervous system. Like nicotine and cocaine, it can cross the blood–brain barrier. Among its effects is a rise in the hypothalamic production of endorphins. It binds to membrane proteins, including receptors for GABA, dopamine, serotonin, and norepinephrine. Binding alters receptor shapes and activities. Even one drink skews judgment and motor skills. Liver failure results from chronic alcohol abuse (Chapter 6).

Analgesics When severe stress leads to pain, your brain makes endorphins and enkephalins. These are natural *analgesics*, or pain killers. Narcotic analgesics include *morphine*, *codeine*, and *heroin*, as well as newer prescription drugs, such as *fentanyl* and *oxycodone*. In addition to pain relief, the drugs cause euphoria and lead to severe addiction. Withdrawal symptoms may include chills, fever, anxiety, vomiting, and diarrhea.

Hallucinogens These drugs alter the user's sensory perception. *LSD* (*lysergic acid diethylamide*) mimics the effects of serotonin. It is not addictive, but distorted perceptions can be deadly. For instance, some users leaped out of buildings, believing that they could fly.

Ketamine and *PCP* (phencyclidine), two chemically related drugs, were developed as anesthetics. Ketamine is still used as one. Both give users an out-of-the-body experience and numb the extremities by slowing the clearing of dopamine, norepinephrine, and serotonin. Both can trigger seizures, hyperthermia, and kidney failure. PCP is also known as angel dust. It can induce schizophrenia-like psychosis and agitation which, in some individuals, lasts for more than a week.

The hallucinogen *marijuana* is made from parts of the plant *Cannabis*. In low doses it is like a depressant. It slows but does not stop motor activity. It relaxes the body and may elicit mild euphoria, but it may cause acute panic attacks. Marijuana often disrupts short-term memory. As alcohol does, it affects how well we do complicated tasks, such as driving a car. Smoking marijuana, like smoking tobacco, invites respiratory ailments and cancers of the mouth, throat, and lungs. Frequent marijuana use lowers testosterone levels in blood and sperm counts. It results in abnormal sperm.

Researchers are not sure how this drug exerts effects on sperm, but they agree that men who wish to be fathers should avoid using it.

BRAIN DEVELOPMENT AND RISK BEHAVIOR

At one time the brain was said to be fully developed, with all 100 billion neurons hardwired, before the end of puberty. However, all of the necessary connections are not completed until the early twenties. Like arms, legs, and other body parts, different parts of the brain develop on different schedules.

For instance, the prefrontal cortex is one of the last brain regions to finish developing. This coordinating center, recall, keeps tabs on other regions, including the limbic system. It mediates decision making, sifts through and makes sense of ambiguous signals, and reinforces or dampens raw emotions that arise in the limbic system. The limbic system develops faster than the prefrontal cortex. In effect, a teenager is like a car with a revved-up engine and no brakes.

Without a finished prefrontal cortex to act as traffic cop, teenagers are more open than adults to invading the brain's dopamine-releasing pleasure center. Novel behaviors, especially those with an element of risk or danger, stimulate that center. Sneak out late for a rock concert? Sure. Snort cocaine? Why not?

If you are not yet in your twenties, don't jump to the conclusion that you do not have to nurture your brain (as by reading) or make choices (as in behaving responsibly instead of impulsively). How you decide to exercise your brain helps shape the forming neural circuits, which in turn will profoundly influence your behavior and success later in life.

> Here is the connection: It took hundreds of millions of years to put together all of the intricate wiring and signaling mechanisms in the human brain. Those mechanisms can unravel in one individual's lifetime.

Summary

Section 34.1 Radial animals have simple nerve nets. Most animals have a bilateral nervous system with a brain at their head end. Vertebrate nervous systems are functionally divided into central and peripheral regions. The central nervous system consists of a brain and spinal cord. The peripheral nervous system's paired nerves connect the brain and spinal cord to the rest of the body.

Section 34.2 Neurons are excitable cells. In most bilateral animals, sensory neurons relay signals into a spinal cord and brain. Interneurons in both receive, process, and integrate information. Motor neurons carry signals to effectors (muscles and glands). Neurons have input, conducting, and output zones for signals.

Biology Now
Study the structure and membrane properties of neurons with the animation on BiologyNow.

Section 34.3 In a resting neuron, transport proteins are maintaining ion concentration gradients by assisting or restricting the diffusion of ions across the lipid bilayer of the plasma membrane. The steady voltage difference associated with the ion gradients is the resting membrane potential. In any excitable cell, an action potential is an abrupt, fleeting reversal in the voltage difference across the plasma membrane. The action of sodium–potassium pumps maintains and restores the ion gradients, and so keeps the membrane ready for another action potential.

Biology Now
View an action potential step by step with the animation on BiologyNow.

Section 34.4 Neurons signal other neurons, muscle cells, or gland cells at chemical synapses. Arrival of an action potential at a presynaptic cell's output zone triggers the release of neurotransmitter molecules. These diffuse across a thin cleft and bind to postsynaptic cell receptors. All signals arriving in the same interval are summed by a process of synaptic integration.

Biology Now
See what occurs at a synapse between a motor neuron and a muscle cell with the animation on BiologyNow.

Sections 34.5, 34.6 Many neurotransmitters bind to a variety of receptor proteins on different cells, with different effects. Neuromodulators can mediate the effects. Vertebrate nervous systems have many diverse neuroglial cells that support, protect, and assist neurons.

Section 34.7 Nerves are long-distance cables, with bundles of fibers that carry signals through the body. Myelin sheaths enclose most of their axons and speed signal conduction rates. Reflexes are simple, automatic responses to stimulation. In the simplest reflex arcs, a sensory neuron synapses directly on a motor neuron.

Biology Now
Observe what happens during a stretch reflex with the animation on BiologyNow.

Section 34.8 The peripheral nervous system's somatic nerves act on skeletal muscles; its autonomic nerves act on soft internal organs. Its sympathetic and parasympathetic divisions often work in opposition to control the same organs. Sympathetic signals dominate during danger or heightened awareness. In less stressful times, parasympathetic signals dominate. The spinal cord links the peripheral nervous system to the brain.

Biology Now
Explore the structure of the spinal cord and compare the effects of sympathetic and parasympathetic stimulation with the animation on BiologyNow.

Section 34.9 In vertebrate embryos, a neural tube develops into a spinal cord and brain. The brain stem is the most evolutionarily ancient neural tissue. It governs many reflex centers for breathing and other vital tasks.

Cells making up blood capillaries in the brain form a blood–brain barrier that stops many harmful substances from reaching brain cells. The cerebral cortex, the most recently evolved neural tissue, governs most complex

Table 34.2 Summary of the Central Nervous System*

FOREBRAIN	Cerebrum	Localizes, processes sensory inputs; initiates, controls skeletal muscle activity. Governs memory, emotions, abstract thought in the most complex vertebrates
	Olfactory lobe	Relays sensory input from nose to olfactory centers of cerebrum
	Thalamus	Has relay stations for conducting sensory signals to and from cerebral cortex; has role in memory
	Hypothalamus	With pituitary gland, a homeostatic control center; adjusts volume, composition, temperature of internal environment. Governs organ-related behaviors (e.g., sex, thirst, hunger) and expression of emotions
	Limbic system	Governs emotions; has roles in memory
	Pituitary gland (Chapter 36)	With hypothalamus, provides endocrine control of metabolism, growth, development
	Pineal gland (Chapter 36)	Helps control some circadian rhythms; also has role in mammalian reproductive physiology
MIDBRAIN	Roof of midbrain (tectum)	In fishes and amphibians, its centers coordinate sensory input (as from optic lobes), motor responses. In mammals, its reflex centers swiftly relay sensory input to forebrain
HINDBRAIN	Pons	Tracts bridge cerebrum and cerebellum; other tracts connect spinal cord with forebrain. With the medulla oblongata, controls rate and depth of respiration
	Cerebellum	Coordinates motor activity for moving limbs and maintaining posture, and for spatial orientation
	Medulla oblongata	Its tracts relay signals between spinal cord and pons; its reflex centers help control heart rate, adjustments in blood vessel diameter, respiratory rate, vomiting, coughing, other vital functions
SPINAL CORD		Makes reflex connections for limb movements. Its tracts connect brain, peripheral nervous system

*The reticular formation extends from the spinal cord to the cerebral cortex.

Figure 34.31 Sea hare (*Aplysia*). This brainless mollusk is a boon for studies into the neural basis of behavior in a simple organism.

functions. Table 34.2 lists the central nervous system's components and functions.

Biology⊗Now
Review the structure and function of human brain regions with the interaction on BiologyNow.

Sections 34.10, 34.11 The cerebral cortex has motor, sensory, and association areas. It is the seat of conscious behavior. Its centers also interact with the limbic system to control emotion and memory. The reticular formation connects lower brain centers with the cerebral cortex.

Section 34.12 Memory forms in multiple stages by different pathways. Emotional states affect the transfer of memories from short-term to long-term storage.

Biology⊗Now
Read the InfoTrac Article, "Alzheimer's—Searching for a Cure," Linda Bren, FDA Consumer, July–August 2003.

Section 34.13 Psychoactive drugs act at synapses, where they mimic neurotransmitters or disrupt their release or uptake. Habitual use of psychoactive drugs can rewire the brain and cause addiction.

Self-Quiz
Answers in Appendix II

1. _____ relay messages from the brain and spinal cord to muscles and glands.
   a. Motor neurons   c. Interneurons
   b. Sensory neurons   d. Neuroglia

2. When a neuron is at rest _____ .
   a. it is at threshold potential
   b. gated sodium channels are open
   c. the sodium–potassium pump is operating
   d. both a and c

3. Action potentials occur when _____ .
   a. a neuron receives adequate stimulation
   b. sodium gates open in an ever accelerating way
   c. sodium–potassium pumps kick into action
   d. both a and b

4. Neurotransmitters are released by _____ .
   a. axon endings   c. dendrites
   b. the cell body   d. the myelin sheath

5. The most abundant cells in the brain are _____ .
   a. Schwann cells   c. astrocytes
   b. microglia   d. neurons

6. Skeletal muscles are controlled by _____ .
   a. sympathetic signals   c. somatic nerves
   b. parasympathetic signals d. both a and b

7. When you sit quietly on the couch reading, output from the _____ system prevails.
   a. sympathetic   c. both
   b. parasympathetic   d. neither

8. Skeletal muscles contract in response to _____ .
   a. ACh   c. serotonin
   b. dopamine   d. all of the above

9. The cerebrum is part of the _____ .
   a. forebrain   c. hindbrain
   b. midbrain   d. brain stem

10. Match each item with its description.
    _____ muscle spindle   a. start of brain, spinal cord
    _____ neurotransmitter   b. connects the hemispheres
    _____ limbic system   c. protects brain and spinal
    _____ corpus callosum   cord from some toxins
    _____ cerebral cortex   d. type of signaling molecule
    _____ neural tube   e. neurons' support team
    _____ neuroglia   f. stretch-sensitive receptor
    _____ gray matter   g. roles in emotion, memory
    _____ blood–brain   h. most complex integration
          barrier   i. unmyelinated axons and
                                 cell bodies

Additional questions are available on Biology⊗Now™

Critical Thinking

1. The sea hare *Aplysia californica* is a marine mollusk about the length of your hand (Figure 34.31). Its nervous system only has 20,000 or so neurons, some with large cell bodies that make it easy for scientists to place electrodes or inject dyes. The limited behavior includes responses to food and to touch. Researchers found that dopamine affects its feeding behavior. This neurotransmitter affects human reward-seeking behavior, including addiction to drugs. How might studies of dopamine's effects on sea hares help shed light on treating drug addiction?

2. In human newborns, especially premature ones, the blood–brain barrier is not yet fully developed. Why is this one reason to pay careful attention to their early diet?

3. In humans, the axons of some motor neurons extend from the base of the spinal cord to the big toe, a distance of more than a meter. In a giraffe the longest axons are several meters long. What are some of the functional challenges involved in the development and maintenance of such lengthy cellular extensions?

4. When Jennifer was six years old, a man lost control of his car and hit a tree in front of her house. She ran over to him and screamed when she saw blood from a wound dripping on a bush of red roses. Thirty-five years later, someone gave her a bottle of *Tea Rose* perfume. When she sniffed it, she became frightened and extremely anxious. A few minutes later she also had a vivid recollection of the accident. Explain this incident in terms of what you learned about memory, olfaction, and the limbic system.

5. Eric typically drinks one cup of coffee nearly every hour, all day long. By midafternoon, he has a lot of trouble concentrating on his studies, and he becomes tired and more than a little clumsy. Drinking another cup of coffee does not make him more alert. Explain how the caffeine in coffee might produce such symptoms.

A Whale of a Dilemma

Imagine yourself in the sensory world of a whale. You are 200 meters below the ocean surface, where little sunlight penetrates, so you cannot see well as you move through the water. Many fishes detect the wavelike motion of your body with their lateral line system, which responds to differences in water pressure. Like other fishes, they also use chemicals dissolved in the water as navigational cues. However, a whale has no lateral line, it has few chemical receptors, and its olfactory lobes are greatly reduced.

So how does a whale sense where it is going? It uses sounds—acoustical cues. Water is an ideal medium for transmitting sound waves, which move five times faster in water than in air. Also, compared to mammals on land, whales have far more neurons that collect and integrate auditory information.

Unlike humans, whales do not have external ear flaps that collect sound waves. Some species do not even have a canal that leads from the body surface to each middle ear. Other species have ear canals packed with wax. How, then, do whales hear? Their jaws pick up vibrations traveling through water. The vibrations are transmitted from the jaws, through a layer of fat, to a pair of middle ears.

Whales use sound to communicate, locate food, and map out the contours of their surroundings. A killer whale or some other species of toothed whale uses *echolocation*. It emits high-pitched sounds and listens as echoes bounce off objects, including prey. Its ears are especially sensitive to high-frequency sounds. Baleen whales, including the humpback whale, make very low-pitched sounds that can travel across an entire ocean basin. Their ears are attuned to sounds in this range.

The ocean is becoming a lot noisier, and the superb acoustical adaptations of whales put them at risk. In 2001, for instance, a number of whales beached themselves near an area where the United States Navy was testing a sonar system (Figure 35.1). This sonar system uses echoes of low-frequency sounds to detect a new generation of missile-launching submarines that can run in near-silence. Humans cannot hear the intense sounds. Whales can.

Autopsies showed that the beached whales had blood in their ears and in acoustic fat. Apparently the intense sounds—as high as 235 decibels—made the whales race to the surface in fear. As they fled, the rapid change in pressure damaged their internal tissues.

Figure 35.1 A few children drawn to one of the whales that beached itself during military testing of a new sonar system. Of sixteen stranded whales, six died on the beach. Volunteers pushed the others out to sea, and their fate is unknown.

Public outcry halted the deployment of the new sonar system. Testing continues, however, because the threat of stealth attacks against the United States is real.

Besides, noise pollution from commercial shipping may be a more pervasive problem for whales. Each day, huge tankers generate low-frequency sounds that frighten whales or drown out acoustical cues. Realistically, the global shipping of oil and other resources that nations require is not going to stop. If research shows that whales are at risk, will those same nations be willing to design and deploy new tankers that minimize the damage?

With this chapter we turn to sensory systems. With these organ systems, animals receive signals from inside and outside the body, then decode them in ways that give rise to awareness of sounds, sights, odors, and other sensations. Most systems contain sensory neurons, nerve pathways, and specialized brain regions. Depending on the type and the numbers of sensory receptors, animals sample the environment in different ways, and they differ in their perception of it. How to meet pressing human needs while respecting sensory diversity can be something of a dilemma.

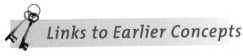

Watch the video online!

How Would You Vote?

Shipping and other human activities generate an underwater ruckus. To what extent should we limit these activities to protect whales against potential harm? Would you support banning activities that exceeded a certain noise level from United States territorial waters? If so, how would you get other nations to do the same? See BiologyNow for details, then vote online.

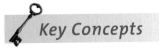

Key Concepts

HOW SENSORY PATHWAYS WORK

Sensory systems are front doors of the nervous system. Each has sensory receptors, nerve pathways to the brain, and brain regions that receive and process sensory input. A stimulus is a form of energy that activates a specific type of sensory receptor. Information becomes encoded in the number and frequency of action potentials sent to the brain along particular nerve pathways.   Section 35.1

SOMATIC SENSES

Touch, pressure, pain, temperature, and muscle sense are somatic sensations, which start at mechanoreceptors in skin, muscles, and the wall of internal organs.   Section 35.2

CHEMICAL SENSES

Olfaction (smell) and taste require chemoreceptors, which bind molecules of specific substances that have become dissolved in the fluid bathing them.   Section 35.3

BALANCE AND HEARING

The sense of balance starts at mechanoreceptors, which detect gravity, velocity, acceleration, and other forces that affect the position and motion of the body or its parts.

In vertebrates, the sense of hearing involves structures that collect, amplify, and sort out pressure variations caused by sound waves. The variations trigger action potentials in mechanoreceptors called hair cells.   Sections 35.4, 35.5

VISION

All organisms have light-sensitive pigments, but vision requires eyes with a dense array of photoreceptors and image formation in the brain. Cephalopod and vertebrate eyes are like cameras. They collect and process data on the distance, shape, position, brightness, and movement of visual stimuli. A sensory pathway starts at the retina and ends in the visual cortex.   Sections 35.6–35.9

Links to Earlier Concepts

We did not expect you to memorize Chapter 34, but you will find yourself drawing repeatedly upon its content. We provide ample cross-references to specific sections. You may wish to reflect on the properties of light (7.1) and how lenses bend light (4.2). You will come across specific cases of signal transduction pathways (28.5). You will reflect again on the evolution of fast-moving cephalopods (25.9) and on the challenges facing the first vertebrates on land (26.1, 26.2).

35.1 Overview of Sensory Pathways

LINKS TO
SECTIONS
34.2, 34.7

Sensory systems are portions of a nervous system. Each typically consists of cell communication pathways from sensory neurons and nerves to specific brain regions.

Reflect on Section 28.5. In all *sensory* communication pathways, energy from a stimulus activates receptors, which transduce it to a form that travels to the brain and may trigger a sensation or perception in response:

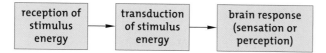

A **stimulus** (plural, stimuli) is a form of energy that activates receptor endings of a sensory neuron (Section 34.2). That energy is transduced to the electrochemical energy of action potentials—the messages by which brain cells continually monitor and respond to the presence or absence of stimuli. The brain's responses are *sensations*, or conscious awareness of a stimulus. A sensation is not the same as *perception*, which is an understanding of what a sensation means.

The **mechanoreceptors** detect forms of mechanical energy: changes in pressure, position, or acceleration. Figure 35.2 shows an example. **Pain receptors** detect damage to tissues. **Thermoreceptors** are sensitive to heat or cold. **Chemoreceptors** detect chemical energy of substances dissolved in the fluid bathing them. **Osmoreceptors** detect changes in the solute levels of

some body fluid. **Photoreceptors** detect differences in the energy of visible and ultraviolet light (Table 35.1).

Regardless of the differences, all sensory receptors transduce stimulus energy into action potentials. But action potentials are not like an ambulance siren; they do not vary in amplitude. How, then, does the brain assess each stimulus? It assesses *which* nerve pathways are carrying action potentials, the *frequency* of action potentials traveling on each axon in the pathway, and the *number* of axons recruited by the stimulus.

First, an animal's brain is prewired, or genetically programmed, to interpret action potentials in certain

Table 35.1	Major Categories of Sensory Receptors	
Category	Examples	Stimulus
Mechanoreceptors		
Touch, pressure	Certain free nerve endings and Pacinian corpuscles in skin	Mechanical pressure against body surface
Baroreceptor	Arterial baroreceptors (Figure 38.11)	Pressure changes in the fluid that bathes them
Stretch	Muscle spindle in skeletal muscle	Stretching
Auditory	Hair cells in organ inside ear	Vibrations (sound or ultrasound waves)
Balance	Hair cells in organ inside ear	Movement of some body fluid in confined space
Pain Receptors (Nociceptors)*	Certain free nerve endings	Tissue damage (e.g., distortions, burns)
Thermoreceptors	Certain free nerve endings	Change in temperature (heating, cooling)
Chemoreceptors		
Internal chemical sense	Carotid bodies in blood vessel wall	Substances (O_2, CO_2, etc.) dissolved in extracellular fluid
Taste	Taste receptors of tongue	Substances dissolved in saliva, etc.
Smell	Olfactory receptors of nose	Odors in air, water
Osmoreceptors	Hypothalamic osmoreceptors (Section 42.4)	Change in solute concentration (water volume) of the fluid that bathes them
Photoreceptors		
Visual	Rods, cones of eye	Wavelengths of light

* Extremely intense stimulation of any sensory receptor also may be perceived as pain.

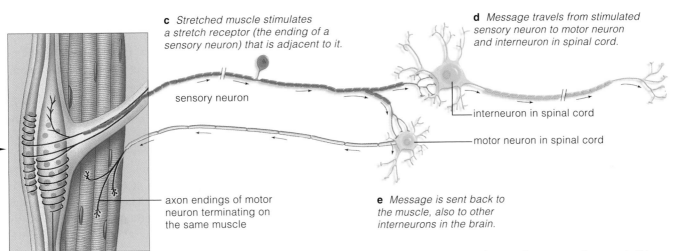

c *Stretched muscle stimulates a stretch receptor (the ending of a sensory neuron) that is adjacent to it.*

sensory neuron

d *Message travels from stimulated sensory neuron to motor neuron and interneuron in spinal cord.*

interneuron in spinal cord

motor neuron in spinal cord

axon endings of motor neuron terminating on the same muscle

e *Message is sent back to the muscle, also to other interneurons in the brain.*

b muscle spindle

Figure 35.2 Example of a how we make use of sensory pathways—in this case, from receptors in many muscles throughout the body to the spinal cord and brain.

ways. That is why you can "see stars" after an eye gets poked, even in a dark room. Many photoreceptors in the eye were mechanically disturbed and sent signals down one of two optic nerves to the brain, which will interpret any signals from an optic nerve as "light."

Second, a strong signal makes receptors fire action potentials more often and longer. The same receptor might detect the sound of a throaty whisper or a wild screech. The brain interprets such differences through variations in the frequency of signals (Figure 35.3).

Third, a strong stimulus recruits far more sensory receptors, compared to a weak stimulus. Gently tap a spot of skin on one of your arms and you will activate a few receptors. Press harder on the same spot and you will activate more receptors in a larger area. A big disturbance translates into action potentials in many sensory axons, and the brain interprets the combined activity as an increase in stimulus intensity.

In certain cases, the frequency of action potentials declines or ends even when the stimulus continues at constant strength. For instance, after you put on a sock, your awareness of the pressure the sock is exerting on your skin ceases. Such a diminishing response to an ongoing stimulus is called **sensory adaptation**.

Some mechanoreceptors adapt fast to a sustained stimulus and only signal its onset and removal. Other receptors adapt more slowly to a stimulus or not at all; they continue to keep the brain informed about it.

The gymnast in Figure 35.2 is holding his position in response to signals from his skin, skeletal muscles, joints, tendons, and ligaments. For example, how fast and how far a muscle stretches depends on activation of stretch receptors in muscle spindles (Section 34.7). By responding to changes in the length of muscles, the brain helps him maintain his balance and posture.

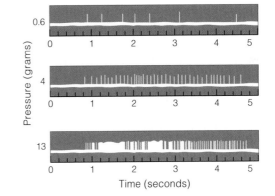

Figure 35.3 *Animated!* Recordings of action potentials from a pressure receptor with endings in a human hand. They correspond to variations in stimulus strength. A thin rod was pressed against skin with the amount of pressure indicated to the left of each diagram. Vertical bars above each thick horizontal line record individual action potentials. Increases in their frequency correspond to increases in the stimulus strength.

In sections to follow, you will consider the *somatic* sensations that arise from sensory receptors located in tissues throughout the body. You also will consider *special senses*, such as taste, smell, balance, hearing, and vision. These senses depend on receptors that are restricted to certain body parts, such as ears and eyes.

A sensory system has sensory receptors for specific stimuli, nerve pathways that conduct information from receptors to the brain, and brain regions that receive the information.

The brain assesses a stimulus according to which nerve pathway is signaling it, the frequency of action potentials from the pathway, and the number of axons that the stimulus recruited.

35.2 Somatic Sensations

LINKS TO
SECTIONS
34.5, 34.7, 34.8, 34.10

Somatic sensations are responses to receptors near the body surface, in skeletal muscles, and in the wall of internal organs. These receptors are most highly developed in birds and mammals. Fishes and amphibians have only a few.

Somatic sensations arise in the cerebrum's outer layer of gray matter, the cerebral cortex. In its sensory areas, interneurons are organized like a map corresponding to the body surface (Figure 35.4). The largest regions of the map correspond to body parts having the most sensory acuity, such as fingers, thumbs, and lips. Less sensitive regions are represented by fewer neurons.

RECEPTORS NEAR THE BODY SURFACE

Mammals discern sensations of touch, pressure, cold, warmth, and pain near the body surface. Regions with the greatest number of sensory receptors, such as the fingertips and tip of the tongue, are most sensitive to stimulation. Less sensitive regions, such as the back of the hand, do not have nearly as many.

Free nerve endings are the unmyelinated or thinly myelinated, branched endings of sensory neurons in skin and internal tissues. These nerve endings are the simplest kinds of mechanoreceptors, thermoreceptors, and pain receptors. All adapt slowly to stimulation. One subpopulation gives rise to a sense of prickling pain, as when you jab a finger with a pin. Another contributes to itching or warming sensations that are responses to chemicals, such as histamine. Two kinds

are most sensitive to temperatures that are higher and lower than the body's normal core temperature. One mechanoreceptive type coils around hair follicles and detects the motion of hair (Figure 35.5).

Encapsulated receptors near the body surface also detect somatic sensations. Four types are called the Meissner's corpuscle, bulb of Krause, Ruffini endings, and Pacinian corpuscle (Figure 35.5). The Meissner's corpuscle is notably abundant in the fingertips, lips, eyelids, nipples, and genitals. It can adapt very slowly to low-frequency vibrations. The bulb of Krause, one of the thermoreceptors, is activated at 20°C (68°F) or lower. Below about 10°C, it contributes to painful freezing sensations. Slowly adapting Ruffini endings detect steady touch and pressure, and temperatures above 45°C (113°F). The Pacinian corpuscle is sensitive to fine textures. It occurs in the dermis, near freely movable joints, and in some internal organs. Its outer capsule is an onionlike layering of plasma membrane with fluid-filled spaces in between (Figure 35.5). The layering is most sensitive to rapid pressure changes brought on by touch and vibrations.

MUSCLE SENSE

You read about stretch receptors in muscle spindle fibers in Section 34.7. These receptors increase their firing rate as a muscle stretches. Along with receptors in the skin and near joints, they signal the brain about the positions of the body's limbs.

PAIN

Pain is the perception of a tissue injury. *Somatic* pain is a response to signals from pain receptors in skin, skeletal muscles, joints, and tendons. *Visceral* pain is associated with internal organs. It a response to high chemical stimulation, muscle spasms, muscle fatigue, excessive distension of the gut, inadequate blood flow to organs, and other abnormal conditions.

Superficial somatic pain arises at or near the skin surface. It is sharp or prickling and often does not last long. Deep somatic pain arises deeper in the skin or in

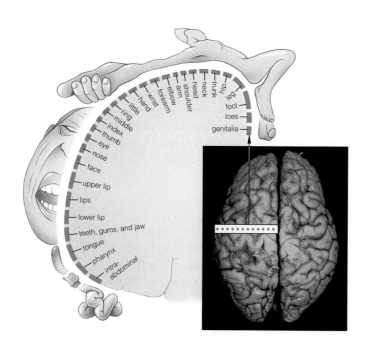

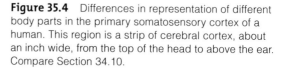

Figure 35.4 Differences in representation of different body parts in the primary somatosensory cortex of a human. This region is a strip of cerebral cortex, about an inch wide, from the top of the head to above the ear. Compare Section 34.10.

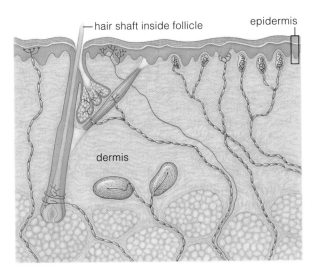

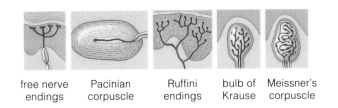

free nerve endings | Pacinian corpuscle | Ruffini endings | bulb of Krause | Meissner's corpuscle

Figure 35.5 *Animated!* A sampling of receptors in human skin.

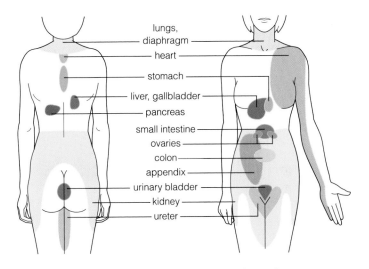

Figure 35.6 Referred pain. Receptors in some internal organs detect painful stimuli. Instead of localizing the pain at the organs, the brain projects the sensation to the skin areas indicated.

muscles or joints. It is more diffuse and lasts longer. This is also the case for visceral pain. We characterize both as burning or gnawing pain, or a dull ache.

When cells are injured, they release chemicals that activate neighboring pain receptors. The most potent, bradykinins, open the floodgates for an outpouring of histamine, prostaglandins, and other participants in inflammatory responses (Section 39.3). They also may bind with and activate pain receptor endings.

Signals from pain receptors enter the spinal cord and cause the release of a neuromodulator, substance P, which activates neurons that can signal the sensory cortex. There, neurons assess the intensity and type of pain. Different signals rouse the body and mediate emotional responses. They bring about the release of natural opiates—endorphins and enkephalins—that lower pain perception (Section 34.5).

Emotional states, cultural factors, and possibly age influence pain tolerance (older people tend to handle it better). Intense or long-lasting pain also may lead to *hyperalgesia*, a condition in which pain is amplified. For example, pain gets worse when someone with a bad sunburn steps into a hot shower.

REFERRED PAIN

Pain perception depends on the brain's capacity to identify the affected tissue and project a sensation back to it. Get smacked in the face with a snowball and you "feel" the contact on facial skin. But sometimes the brain projects pain from internal organs to skin. Such *referred pain* is the perception of visceral sensations as somatic sensations. For instance, a heart attack often is wrongly perceived as pain radiating from the chest wall, from the neck, and along the left shoulder and

arm (Figure 35.6). How can the brain make such a bad mistake? We find the answer in the construction of the nervous system. Sensory inputs from skin and certain internal organs enter the same segments of the spinal cord (Section 34.8). The skin encounters more painful stimuli than internal organs do, so its signals travel more often along the somatic pathway to the brain. The brain may interpret most sensory input as arriving from skin—the more frequent source.

Referred pain is not the same as the *phantom pain* reported by amputees. They often sense the presence of a missing body part as if it were still there. In some undetermined way, severed sensory nerves continue to respond to the amputation. The brain projects the pain back to the missing part, past the healed region.

Sensory receptors near the body surface and in internal organs detect touch, pressure, temperature, pain, motion, and positional changes of body parts.

35.3 Sampling the Chemical World

LINKS TO
SECTIONS
2.6, 34.10

In the remainder of this chapter, we will sample the special senses, which have receptors in specific organs. They include smell, taste, balance, hearing, and vision.

The sensory pathways of olfaction (smell) and taste start at chemoreceptors, which become activated by binding molecules of a substance that is dissolved in the fluid bathing them. In both sensory pathways, a stimulus can trigger signals that travel along nerve expressways through the thalamus. The signals end up in the cerebral cortex. There, perceptions of certain stimuli take shape and undergo fine-tuning. Sensory information also reaches the limbic system, which acts to integrate it with emotional states and with stored memories (Section 34.10).

Olfactory receptors fire off signals when they are exposed to water-soluble or volatile (easily vaporized) chemicals. The lining of your nose has about 5 million olfactory receptors. That of a bloodhound nose has more than 200 million. Receptor axons lead into one of two olfactory bulbs. In these small brain structures, axons synapse with clusters of cells that sort out the components of a scent. From there, information flows along an olfactory tract to the cerebrum, where it is further processed (Figure 35.7).

Many animals use olfactory cues to navigate, find food, and communicate socially, as with pheromones. As later chapters explain, **pheromones** are signaling molecules secreted by one individual that change the social behavior of other individuals of its species. As one example, olfactory receptors on the antennae of a male silk moth help him find a pheromone-secreting female that may be more than a kilometer upwind.

In reptiles and most mammals (but not primates), a cluster of sensory cells forms a *vomeronasal organ* that detects pheromones. Humans have a reduced version of this organ about 12.7 millimeters (1/2 inch) inside the nose. It may or may not be functional, but recent studies show that human females and males manage to respond to certain pheromones (Chapter 49).

Different animals have **taste receptors** on antennae, legs, or tentacles, or inside the mouth. On the surface of your mouth, throat, and upper part of the tongue especially, chemoreceptors are located in about 10,000 sensory organs called taste buds (Figure 35.8). Fluids in the mouth reach the receptors by entering a pore in each taste bud.

You perceive many tastes, but all are a combination of five main sensations: sweet (elicited by glucose and the other simple sugars), sour (acids), salty (NaCl or other salts), bitter (plant toxins, including alkaloids), and *umami* (elicited by amino acids such as glutamate, which has a savory taste typical of aged cheese and meat). With its strong savory taste and no texture or odor of its own, MSG (monosodium glutamate) has become a common flavor enhancer.

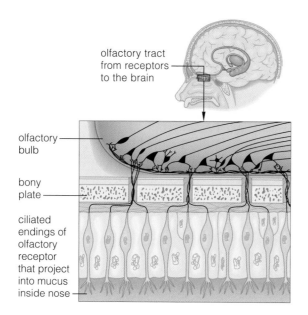

Figure 35.7 Pathway from sensory endings of olfactory receptors in the human nose to the cerebral cortex and the limbic system. Receptor axons pass through holes in a bony plate between the lining of the nasal cavities and the brain.

olfactory tract from receptors to the brain

olfactory bulb

bony plate

ciliated endings of olfactory receptor that project into mucus inside nose

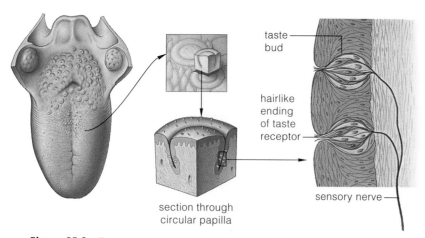

taste bud

hairlike ending of taste receptor

sensory nerve

section through circular papilla

Figure 35.8 Taste receptors in the human tongue. The structures called circular papillae ring epithelial tissue that contains taste buds. A human tongue has approximately 10,000 of these sensory organs, each of which has as many as 150 chemoreceptors.

The senses of smell and taste start at chemoreceptors. Both involve sensory pathways that lead to processing regions in the cerebral cortex and in the limbic system.

35.4 Keeping the Body Balanced

All animals assess and respond to displacements from equilibrium, when the body is balanced in relation to gravity, acceleration, and other forces that may affect its position and movement. Even brainless jellyfishes right themselves when turned upside down.

The first organs of equilibrium evolved in fishes. You have a pair of them, one inside each ear. A **vestibular apparatus** in each ear consists of two sacs—the utricle and the saccule—along with three semicircular canals (Figure 35.9). The sacs and canals interconnect into a continuous, fluid-filled system. In this system, many mechanoreceptors are stimulated whenever you move about or rotate your head.

Inside the bulging base of a semicircular canal is an organ of *dynamic* equilibrium. It has a gelatinous mass into which hair cells, a type of mechanoreceptor, project. Any rotation of the head displaces fluid in the canals. Hydrostatic pressure exerted by the moving fluid shifts the gelatinous mass, which in turn bends the hair cells. The bending causes signals to flow from sensory neurons to the vestibular nerve, which carries signals about motion to the brain.

Inside each utricle and saccule is an organ of *static* equilibrium. These organs send messages to the brain about how the head is oriented relative to the ground. In each organ, a thick membrane rests on hair cells that project upward from the floor of the sac. That membrane contains a mass of crystals, which weigh it down. When your head is held upright, the weighted membrane presses down on hair cells, which bends them slightly. This sends a steady stream of action potentials from the sensory neurons along the vestibular nerve to the brain. If your posture changes or if you speed up or slow down movement in one direction, the position of the membrane above these hair cells will shift. Depending upon the position, a hair cell will now have more or less weight on it, so signaling from sensory cells will step up or slow down. To interpret the body's posture and movement, the brain constantly compares and integrates the signals from sensory cells in all organs of static and dynamic equilibrium in the inner ears on both sides of the head.

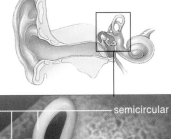

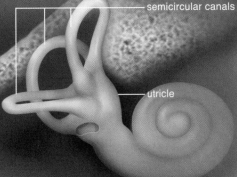

Figure 35.9 The vestibular apparatus inside the human ear. Organs of dynamic and static equilibrium inside this fluid-filled organ contribute to the sense of balance.

At the same time, the brain evaluates input from receptors in the skin, joints, and tendons. Integration of all the information allows your brain to control eye muscles and keep the visual field in focus, even as your position shifts or your head rotates. Integration also helps maintain awareness of the body's position and motion in space, as demonstrated by champion figure skater Sarah Hughes.

A stroke, an inner ear infection, or loose particles in the semicircular canals cause *vertigo*, a sensation that the world is moving or spinning around. Vertigo also arises from conflicting sensory inputs, as when you stand on the top viewing deck of a skyscraper and look down. The vestibular apparatus reports that you are motionless even as your eyes report that your body is floating in space far above the ground.

Mismatched signals also give rise to the dizziness and nausea of *motion sickness*. Passengers in a vehicle moving fast on a hilly, curvy road experience changes in acceleration and direction that scream "motion" to each vestibular apparatus. At the same time, signals from the eyes about objects inside of the vehicle tell the brain that the body is standing still. Being the driver minimizes motion sickness because a driver is forced to focus on the scenery rushing past, so visual signals are consistent with vestibular signals.

Organs of equilibrium help keep the body balanced in relation to gravity, velocity, acceleration, and other forces that influence its position and movement.

35.5 Collecting, Amplifying, and Sorting Out Sounds

LINKS TO
SECTIONS
26.1, 26.4, 28.5

Many arthropods and most vertebrates perceive sounds; they have a sense of **hearing**. *Land vertebrates have ears that capture and sort out sound waves traveling in air.*

PROPERTIES OF SOUND

All sounds are forms of mechanical energy. They arise when a vibrating object causes pressure variations in air, water, or some other medium. We can represent pressure variations as wave forms. The *amplitude* of a sound corresponds to loudness (intensity), which we measure in decibels. Every ten decibels is a tenfold increase above the faintest sounds that humans hear. A sound's *frequency* is the number of wave cycles per second. Each cycle extends from the peak (or trough) of one wave to the peak (or trough) of the next in line. The more cycles per second, the higher the frequency and the perceived pitch (Figure 35.10).

Unlike a tuning fork's pure tone, most sounds are combinations of waves of different frequencies. Their timbre, or quality, varies. Differences in timbre make your voice nasal or deep.

THE VERTEBRATE EAR

Water readily transfers vibrations to body tissues. As you might expect, fishes do not require elaborate ears to detect them. However, sounds spread out in air. When some vertebrates invaded land, the capacity to collect and then amplify vibrations became important (Sections 26.1, 26.2). Some structures evolved in ways that trap sound waves moving through air and that also amplify and make sense of them.

Focus on the structures of the paired human ears (Figure 35.11*a*). The *outer* ear is adapted for gathering sounds from air, as it is for most mammals. Its pinna is a much-folded flap of cartilage, sheathed in skin, that projects from the side of the head. An auditory canal leads from the pinna into the middle ear.

The *middle* ear amplifies and transmits air waves to the inner ear. It has an eardrum, which originated in early reptiles as a shallow depression on each side of

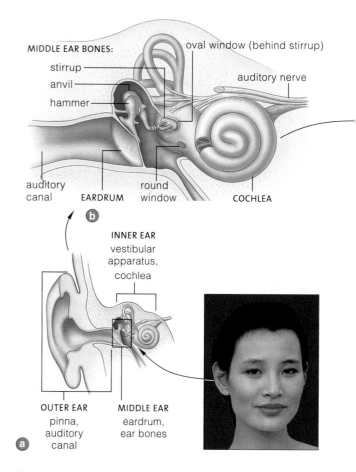

Figure 35.11 *Animated!* Components of the human ear.

the head. The "drum," a thin membrane, vibrates fast in response to pressure waves. Behind the drum is an air-filled cavity and small bones: the hammer, anvil, and stirrup (Figure 35.11*b*). By interacting, the bones transmit the force of sound waves from the eardrum to a smaller surface: the oval window. The "window" is an elastic membrane in front of the inner ear.

The *inner* ear, remember, has a vestibular apparatus (Section 35.4). It also has a **cochlea**, which in humans is a pea-sized, fluid-filled structure that resembles a coiled snail shell (the Greek *koklias* means snail). The transduction of waves of sound into action potentials takes place inside the cochlea (Figure 35.11*c*).

When sound waves make the stirrup vibrate, this middle earbone pushes against the oval window. It transmits pressure waves to the fluid in two of three cochlear ducts (scala vestibuli and scala tympani). The waves end at another membrane, the round window, which bows inward and outward in response.

The third cochlear duct sorts out pressure waves. Its wall, a *basilar* membrane, is stiff and narrow near the oval window, then it broadens and becomes more

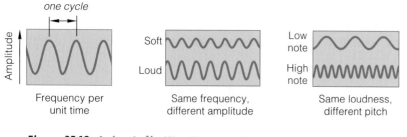

Figure 35.10 *Animated!* Wavelike properties of sound.

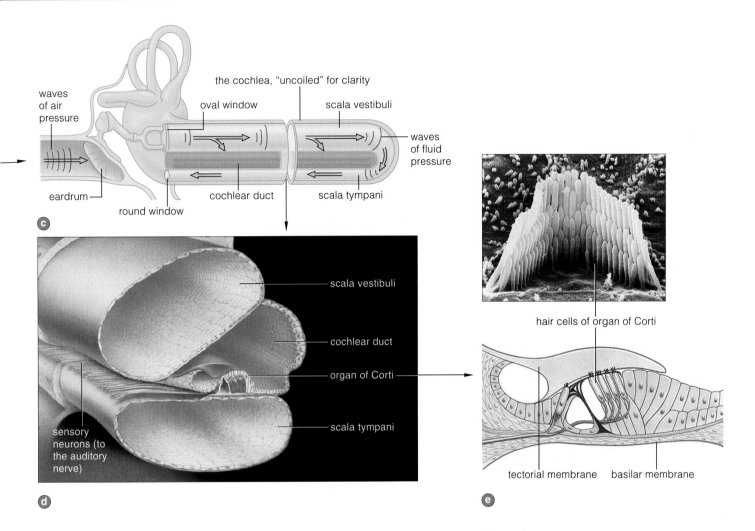

The cochlea, "uncoiled" for clarity — oval window — scala vestibuli — waves of fluid pressure — waves of air pressure — eardrum — round window — cochlear duct — scala tympani

c

scala vestibuli — cochlear duct — organ of Corti — scala tympani — sensory neurons (to the auditory nerve)

d

hair cells of organ of Corti — tectorial membrane — basilar membrane

e

flexible deeper in the coil. Because of the differences, frequency variations have different effects along the length of the basilar membrane. High-pitched sounds make the stiff, narrow part of the cochlear duct vibrate; low-pitched sounds make the flexible part vibrate.

Attached to one surface of the basilar membrane is an acoustical organ. This organ of Corti has arrays of **hair cells**, a type of acoustical receptor having a tuft of modified cilia at one end. The cilia project into a wall extension—a *tectorial* membrane—draped over them (Figure 35.11d,e). The cilia bend when pressure moves the basilar membrane. The mechanical energy is transduced into action potentials that reach the brain by an auditory nerve. Hearing loss occurs when hair cells are damaged, as by chronic exposure to intense sounds. Figure 35.12 is an example of the damage.

Ears of land vertebrates collect, amplify, and sort out sound waves. In the inner ear, sound waves produce fluid pressure variations and trigger action potentials in hair cells.

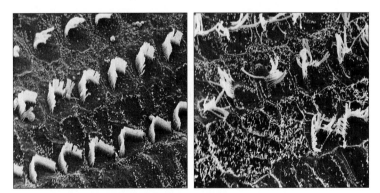

Figure 35.12 Results of an experiment on the effect of intense sound on the inner ear. *Left:* From a guinea pig ear, two rows of hair cells that normally project into the tectorial membrane in the organ of Corti. *Right:* Hair cells inside the same organ after twenty-four hours of exposure to noise levels comparable to extremely loud music.

To give you a sense of how sound intensity is measured, a ticking watch measures 10 decibels. Normal conversation is about 60 decibels, a food blender operating at high speed is about 90 decibels, and an amplified rock concert is about 120 decibels. The perceived loudness of a sound depends not only on its intensity but also on its frequency.

35.6 Do You See What I See?

LINKS TO
SECTIONS
4.2, 7.2, 25.9, 34.10

Shine a light on a single-celled amoeba and it abruptly stops moving. Sunflowers track the sun as it arcs overhead. Most organisms are sensitive to light, but few "see" as you do.

REQUIREMENTS FOR VISION

At the minimum, a sense of **vision** requires eyes and image perception in brain centers that can interpret patterns of visual stimulation. Images are assembled from information on the shapes, brightness, positions, and movements of visual stimuli. **Eyes** are sensory organs that contain a tissue of many densely packed photoreceptors. Regardless of variation in the details, all **photoreceptors** have this in common: *They contain pigment molecules that can absorb photon energy, which can be converted to excitation energy in sensory neurons.*

There are two types of photoreceptors, but both are derived from epidermal cells that have one cilium. In *ciliary* photoreceptors, the plasma membrane around the cilium develops into the photosensitive surface. Cnidarians, some flatworms, and all vertebrates have this type. By contrast, in *rhabdomeric* photoreceptors, the photosensitive surface develops from microvilli around the cilium. This type occurs in most flatworms and mollusks, annelids, arthropods, and echinoderms.

A SAMPLING OF INVERTEBRATE EYES

In earthworms and some other eyeless invertebrates, photoreceptors are dispersed through the integument. The simplest kind of eye is an ocellus (plural, ocelli). Photoreceptors project into these pigmented spots or shallow cups of the integument (Figure 35.13*a*). The pigments partially shade the photoreceptors, which helps the animal determine the direction of the light source. Such animals use light as a cue to orient the body, detect a predator's shadow, or adjust biological clocks, but that is about it.

Vision arose among swift predatory fishes that had to discriminate among prey and other objects in their rapidly shifting visual field. A **visual field** is the part of the outside world that the eye sees (Section 34.11). At the least, photoreceptors must sample the intensity of light from different parts of the visual field, which the brain interprets as contrasting details of a visual stimulus. However, fine visual detail requires a lot of photoreceptors, and the eyes of most invertebrates are too small to have enough of them. The pigment cup of a planarian has about 200 photoreceptors, compared to 70,000 per square millimeter in an octopus eye.

Also, an eye **lens** helps image formation. This type of transparent structure bends all light rays from a given point in the visual field so that they converge onto photoreceptors (Sections 4.2 and 7.2). Abalones have a round lens, but it does not bend incoming rays to the same points, so blurred images form (Figure 35.13*b*). Images sharpen a bit when photoreceptors are some distance away from the lens, as in the snail eye (Figure 35.13*c*). They sharpen more with a **cornea**, a transparent cover that directs light rays onto the lens, as happens in a snail or conch eye (Figure 35.13*d,e*).

Invertebrates that evolved on land are far better at discriminating among objects in the visual field. For example, consider the simple and more complex eyes of arthropods. A *simple* arthropod eye has one lens for all photoreceptors. Spiders have these (Section 25.12). Crustaceans and insects have **compound eyes**, with many closely packed rhabdomeric units, as in Figure 35.14*a*. Some compound eyes have thousands of units, of a sort called ommatidia (singular, ommatidium). Inside each unit is a photoreceptor with rhabdomeric

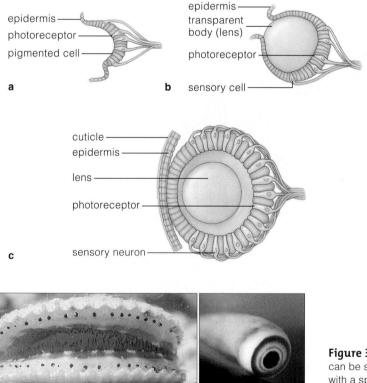

epidermis ——
photoreceptor ——
pigmented cell ——

a

epidermis ——
transparent
body (lens) ——
photoreceptor ——
sensory cell ——

b

cuticle ——
epidermis ——
lens ——
photoreceptor ——
sensory neuron ——

c

d

e

Figure 35.13 Invertebrate eyes. There are far more photoreceptors than can be shown in the simple diagrams. (**a**) Limpet ocellus. (**b**) Abalone eye, with a spherical, transparent lens. (**c**) Eye of a land snail. (**d**) Ocelli at the mantle margin of a scallop. (**e**) Well-developed eye of a conch, peering into the waters of the Great Barrier Reef along the east coast of Australia.

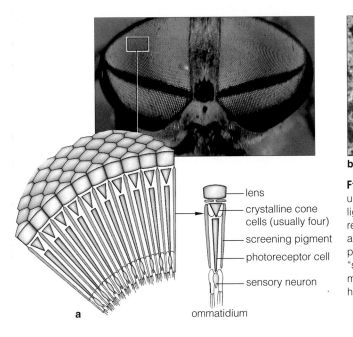

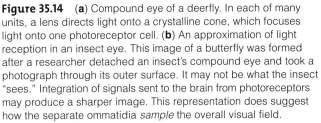

b

Figure 35.14 (**a**) Compound eye of a deerfly. In each of many units, a lens directs light onto a crystalline cone, which focuses light onto one photoreceptor cell. (**b**) An approximation of light reception in an insect eye. This image of a butterfly was formed after a researcher detached an insect's compound eye and took a photograph through its outer surface. It may not be what the insect "sees." Integration of signals sent to the brain from photoreceptors may produce a sharper image. This representation does suggest how the separate ommatidia *sample* the overall visual field.

microvilli. A visual pigment, rhodopsin, is embedded in the membrane of these photosensitive structures.

By the mosaic theory of image formation, each unit in an arthropod's compound eye samples a small part of the visual field. The brain then constructs images based on signals about differences in light intensities among units, with each unit contributing a small bit to the formation of a visual mosaic (Figure 35.14*b*).

The cephalopods called octopuses and squids have the most complex invertebrate eyes (Section 25.9). Like vertebrates, they have **camera eyes**, so named because the eyeball is structured like a camera. Light enters the interior, a dark chamber, through the pupil, an opening in a ring of contractile tissue called the iris (equivalent to the camera diaphragm). Behind the pupil, a lens focuses light on a **retina**, a tissue with many photoreceptors (equivalent to a light-sensitive film inside a camera). In this region, axons of sensory nerves converge to form an optic tract (Figure 35.15). One tract from each eye extends into the brain.

The structural similarities between cephalopod and vertebrate eyes might be an outcome of convergent evolution. Genes that affected the development of the nervous system in both groups were appropriated for the task of eye formation. This may have happened more than once in different animal lineages.

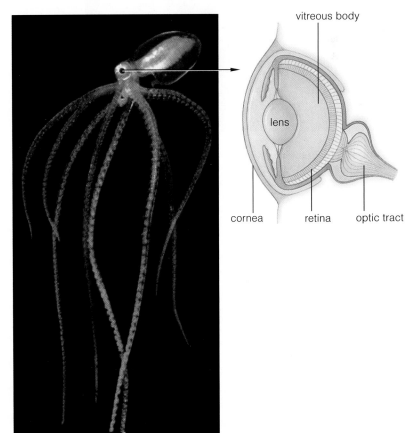

Vision requires eyes with a dense array of photoreceptors and image formation in the brain. Brain centers receive and process patterns of information about brightness, shapes, positions, and movement in the visual field.

Figure 35.15 Octopus eye. The pupil constricts into slits and flares open in response to stimuli, changing light conditions, or the presence of a potential mate or enemy. Experiments show that cephalopods form distinct images. They sense the size, shape, and vertical and horizontal orientation and possibly the color of objects. They are nearsighted. Even so, they can snag tiny prey three meters away in the water.

35.7 Structure and Function of Vertebrate Eyes

LINKS TO
SECTIONS
4.2, 7.1

Like cephalopods, vertebrates have a pair of camera eyes. In most species, the eye has a three-layered wall, and one layer has a tissue of densely packed photoreceptors. The wall forms a chamber. A jellylike substance fills the chamber, and a lens lets in light from the outside.

Figure 35.16 Here's looking at you! In owls and some other birds of prey, photoreceptors are concentrated more on top of the inner eyeball, not back. Such birds look down more than up when they fly and scan the ground for a meal. When they are on the ground, they cannot see something overhead very easily unless they turn their head almost upside-down.

COMPONENTS OF THE EYE

In a vertebrate eye, the transduction of light energy starts at the retina, a tissue that has densely packed photoreceptors. The retina usually is at the back of the eyeball, but in birds of prey, such as hawks and owls, it is closer to the roof of the eyeball (Figure 35.16). Remember, vision requires more than arrays of photoreceptors. It also requires image formation in brain centers that receive and interpret patterns of stimulation from different parts of the eye. Images are based on information about the shapes, brightness, positions, and movements of visual stimuli. Figure 35.17 has a cutaway view of the human eye and a list of its component parts. Like most vertebrate eyes, a human eye has a three-layer wall. The retina rests on the eyeball's inner layer.

The middle layer has a choroid, ciliary body, iris, and pupil. The choroid is a tissue rich in pigments and blood vessels. It absorbs light that photoreceptors do not and prevents it from scattering inside the eye. Suspended behind the cornea is a doughnut-shaped, pigmented iris (the Latin *iris* refers to the rings of a rainbow). The dark "hole" in the center of an iris is the pupil, the entrance for light. When bright light strikes an eye, circular muscles embedded in the iris contract, so the pupil diameter shrinks. In dim light, radial muscles in the iris contract, so the pupil dilates.

The eyeball's outermost layer consists of the sclera and cornea. The sclera, the dense, fibrous "white" of an eye, protects most of the eyeball. The cornea, made of transparent collagen fibers, covers the rest of it.

Wall of eyeball	(three layers)
Sensory Tunic (inner layer)	*Retina.* Absorbs, transduces light energy
	Fovea. Increases visual acuity
Vascular Tunic (middle layer)	*Choroid.* Blood vessels nutritionally support wall cells; pigments prevent light scattering
	Ciliary body. Its muscles control lens; shape; its fine fibers hold lens upright
	Iris. Adjusting iris controls incoming light
	Pupil. Serves as entrance for light
	Start of optic nerve. Carries signals to brain
Fibrous Tunic (outer layer)	*Sclera.* Protects eyeball
	Cornea. Focuses light

Interior of eyeball	
Lens	Focuses light on photoreceptors
Aqueous humor	Transmits light, maintains pressure
Vitreous body	Transmits light, supports lens and eyeball

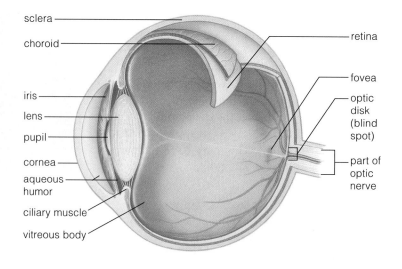

Figure 35.17 *Animated!* Structure and components of the human eye.

Light enters the eyeball's chamber only through a lens that is constructed of layers of transparent fibers of proteins. A clear fluid, the aqueous humor, bathes the lens. A jellylike substance called the vitreous body fills the eyeball chamber behind the lens.

Eyes that have a large pupil and iris are clues to animal life-styles. Species that move about actively at night or in dimly lit habitats are less likely to stumble, bump into objects, or fall off cliffs when they intercept as much of the available light as they can. The greater the amount of incoming light, the better is the visual discrimination. Large pupils let in more light. Large irises can be dilated more to let in far more light than small ones, which is a useful option for animals that are active at night as well as during the day.

When light rays converge at the back of the eye, they stimulate the retina in a distinct pattern. Because the cornea and lens both have a curved surface, all of the rays from some point in the visual field hit them at different angles, so their trajectories change. Rays of light bend at the boundaries between substances of different densities, and bending sends them off in new directions (Sections 4.2 and 7.1). Because of their newly angled trajectories, rays that converge at the back of the eye stimulate the retina in a pattern that is upside-down and reversed left to right, relative to the original light source. Figure 35.18a is a simple sketch of this outcome.

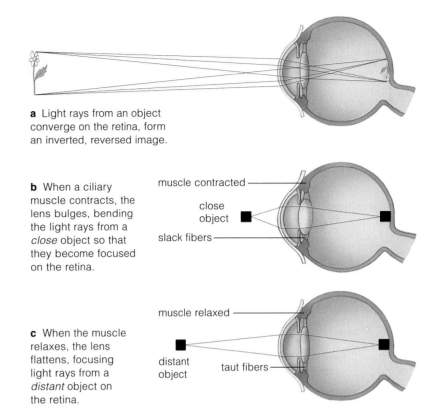

a Light rays from an object converge on the retina, form an inverted, reversed image.

b When a ciliary muscle contracts, the lens bulges, bending the light rays from a *close* object so that they become focused on the retina.

muscle contracted

close object

slack fibers

c When the muscle relaxes, the lens flattens, focusing light rays from a *distant* object on the retina.

muscle relaxed

distant object

taut fibers

Figure 35.18 (**a**) Pattern of retinal stimulation in a human eye. A curved, transparent cornea in front of the pupil changes the trajectories of light rays as they enter the eye. (**b,c**) Two focusing mechanisms use a ciliary muscle that encircles the lens and attaches to it.

VISUAL ACCOMMODATION

Light rays from sources at varying distances from the eye strike the cornea at different angles. They could become focused at different distances in back of it and impair image formation. However, by mechanisms of **visual accommodation**, the lens position or its shape are adjusted in ways that focus all incoming rays onto the retina. Without these normal adjustments, rays of light from distant objects would become improperly focused in front of the retina, and the rays from close objects would be focused behind it.

In fish and reptilian eyes, muscles move the lens forward or back, like a camera's focusing apparatus. Extending the distance between the lens and retina moves the focal point forward; shrinking it moves the focal point back. The lens shape is adjusted in birds and mammals. A ciliary muscle encircles the lens and is anchored to it by fiberlike ligaments (Figure 35.17). When this muscle relaxes, the lens flattens and moves the focal point farther back (Figure 35.18b). When the muscle contracts, the lens bulges and moves the focal point forward, as in Figure 35.18c.

In some cases, the lens position or shape cannot be adjusted enough to make the focal point match up precisely with the retina. For instance, in some people, the eyeball is not shaped quite right, and the position of the lens is either too close or too far away from the retina. When this is the case, accommodation alone cannot bring about an exact match. Eyeglasses often correct both visual disorders, which you may know as nearsightedness and farsightedness. Section 35.9 takes a look at both visual disorders as well as others.

For most vertebrate eyes, the eyeball has three layers, and its interior has a lens, aqueous humor, and vitreous body. Adjustments in the positioning or shape of the lens focus incoming visual stimuli onto the retina inside.

A protective sclera and a light-focusing cornea make up the outer layer. The middle layer has a vascularized, pigmented choroid and other parts that admit and control incoming light. Photoreception occurs at the retina of the inner layer.

35.8 A Closer Look at the Retina

LINKS TO
SECTIONS
7.1, 28.5, 34.9–34.11

Each eye is an outpost of the brain, busily collecting and analyzing information about the distance, shape, position, brightness, and movement of visual stimuli. Its sensory pathway starts at the retina and ends in the brain.

THE RETINA'S RODS AND CONES

In between the retina and the choroid is a pigmented epithelium. Anchored to the epithelium are densely packed arrays of **rod cells** and **cone cells**, which are two categories of ciliary photoreceptors (Figure 35.19). Rod cells and cone cells transduce photon energy into action potentials, the messages by which brain cells monitor the visual field.

In most mammals, rod cells are the most abundant outside the fovea, a small circular area near the center of the retina. Rod cells detect very dim light and are the basis for coarse perception of movement across a visual field.

Part of the rod cell membrane is folded into several hundred disks. Each disk has about 10^8 molecules of the visual pigment rhodopsin. The membrane folding and the high density of pigments enormously increases the odds of intercepting photons, the packets of light energy you read about in Section 7.1. When photons of blue-to-green light stimulate it, rhodopsin changes shape, and that is the start of signal transduction. The change triggers a cascade of reactions that affect ion distributions across the rod cell's plasma membrane. The outcome is a reduction in the release of a certain

neurotransmitter that suppresses activity in neurons that are adjacent to the rods.

Cone cells, which detect bright light, are the basis of sharp vision and color perception. Our sense of color and daytime vision starts when red, green, and blue cone cells, each with a different kind of visual pigment, absorb photons. Photon energy is transduced as it is in rod cells. The fovea has the greatest density of cone cells, and it is the basis of the greatest visual acuity—the most precise discrimination between any two points in the visual field.

A SIGNAL TRANSDUCTION PATHWAY

Layers of distinct types of sensory neurons lie above the rods and cones between the pigmented epithelium and lens on the opposite side of the eyeball (Figure 35.20). These neurons receive, process, and start to integrate signals that arise from the transduction of photon energy. Their activity is organized as receptive fields, or tiny circles where a signal may stimulate or inhibit neural activity.

At the first level of synaptic integration, input from about 125 million rods and cones converges on the retinal neurons known as bipolar cells (Figure 35.20). Humans have eleven types of these bipolar neurons—ten for cones and one for rods. They sort out objects that are lighter than darker ones in the visual field's background. Information also flows laterally among amacrine cells and horizontal cells.

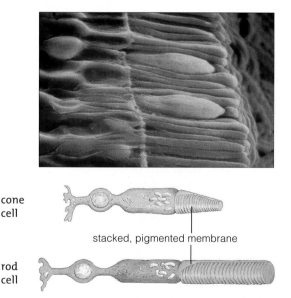

cone cell

stacked, pigmented membrane

rod cell

Figure 35.19 Scanning electron micrograph and sketches of rod cell and cone cell structure.

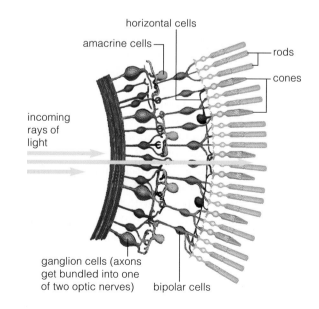

horizontal cells

amacrine cells

rods

cones

incoming rays of light

ganglion cells (axons get bundled into one of two optic nerves)

bipolar cells

Figure 35.20 *Animated!* General organization of photoreceptors and sensory neurons in the retina. The array is about 0.5 millimeter thick.

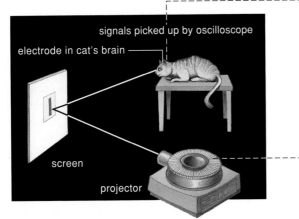

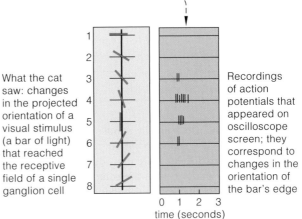

signals picked up by oscilloscope

electrode in cat's brain

screen

projector

What the cat saw: changes in the projected orientation of a visual stimulus (a bar of light) that reached the receptive field of a single ganglion cell

Recordings of action potentials that appeared on oscilloscope screen; they correspond to changes in the orientation of the bar's edge

time (seconds)

Figure 35.21 *Animated!* Example of experiments into the nature of receptive fields for visual stimuli. David Hubel and Torsten Wiesel implanted an electrode in an anesthetized cat's brain. They placed the cat in front of a small screen upon which different patterns of light were projected—here, a hard-edged bar. Light or shadow falling on part of the screen excited or inhibited signals sent to a single neuron in the visual cortex. Tilting the bar at different angles produced changes in the neuron's activity. A vertical bar image produced the strongest signal (*numbered 5 in the sketch*). When the bar image tilted slightly, signals were less frequent. When it tilted past a certain angle, signals stopped.

The messages converge on 1 million ganglion cells. These are the output neurons; their axons are the start of an optic nerve that carries action potentials to the brain. Different ganglion cells respond to rapid shifts in light intensity, a spot of one color, motion, and so on in their tiny receptive field. Figure 35.21 highlights one example. Thus, *before a transduced signal leaves the retina, neurons start integrating and processing it.*

Humans have two optic nerves, one from the retina in each eye (Figure 35.22). Each optic nerve delivers signals concerning a stimulus from the *left* visual field to the right cerebral hemisphere. Each optic nerve also delivers signals from the *right* visual field to the left hemisphere, as Section 34.11 explains.

Optic nerve axons end in a layered brain region named the lateral geniculate nucleus (Figure 35.23). Each layer has a map corresponding to the receptive fields and deals with one kind of visual stimulus, such as form, movement, depth, color, and texture. After early processing, signals rapidly and simultaneously reach different parts of the visual cortex. There, final integration organizes the incoming action potentials and produces visual sensations.

Organization of visual signals begins at receptive fields in the retina. Signals are further processed in a layered brain region and finally integrated in the visual cortex.

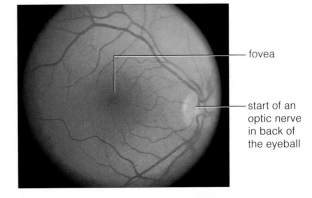

fovea

start of an optic nerve in back of the eyeball

Figure 35.22 Location of the fovea and start of the optic nerve, which rings the entrance for blood vessels that service the eye.

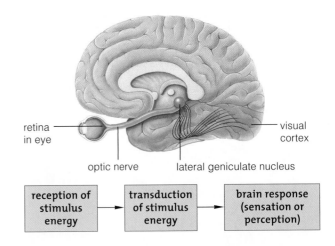

retina in eye

visual cortex

optic nerve

lateral geniculate nucleus

| reception of stimulus energy | transduction of stimulus energy | brain response (sensation or perception) |

Figure 35.23 Sensory pathway from the retina into centers of the brain. Photon energy is transduced into the electrical energy of action potentials in the retina. Action potentials travel along an optic nerve from each eye into a layered brain region, the lateral geniculate nucleus. Each layer processes messages about one kind of visual signal, then relays information to the visual cortex. Synaptic integration in the cortex results in a sense of vision.

35.9 Visual Disorders and Diseases

LINK TO
SECTION
12.7

Given their essential part in monitoring the outside world, a lot of attention is paid to eye problems that result from injuries, inherited abnormalities, diseases, and advancing age. Each year, many millions of people deal with these problems.

Color Blindness   Occasionally, some or all of the cone cells that selectively respond to light of red, green, or blue wavelengths fail to develop. Rare individuals who have only one of three kinds of cones are completely color-blind. They perceive the world in shades of gray.

You read about *red–green color blindness* in Section 12.7. This X-linked recessive trait shows up most often in males. Some or all of the cone cells that respond to light of red or green wavelengths are missing from the retina. Individuals who are affected by this condition typically have trouble distinguishing red from green in dim light. Some cannot distinguish them even in bright light.

Focusing Problems   Heritable alterations in the shape of the eyeball affect the focusing of light. *Astigmatism*, for example, results from unevenly curved corneas, which cannot properly bend all of the incoming light rays to the same focal point.

Nearsightedness, or *myopia*, often occurs when the horizontal axis of the eyeball is longer than the vertical axis. It also occurs when the ciliary muscle responsible for adjusting the lens contracts too strongly. The outcome is that images of distant objects get focused in front of the retina instead of on it (Figure 35.24*a*).

Farsightedness, or *hyperopia*, is the opposite of myopia. The eyeball's vertical axis is longer than the horizontal axis, or the lens is "lazy." Either way, light rays from close objects get focused behind the retina (Figure 35.24*b*).

Even a normal lens loses some of its natural flexibility as a person grows older. That is why people over forty years old often start wearing eyeglasses. Most focusing problems can be corrected with glasses or surgery. Today, lasers are used to reshape the cornea. In 2002, the most common procedure, LASIK, was performed on 1.8 million Americans. When all goes well, laser surgery eliminates the need for glasses during most activities, although the elderly may still need reading glasses. Results vary. Chronic eye irritation is common, and the rare complications can be serious.

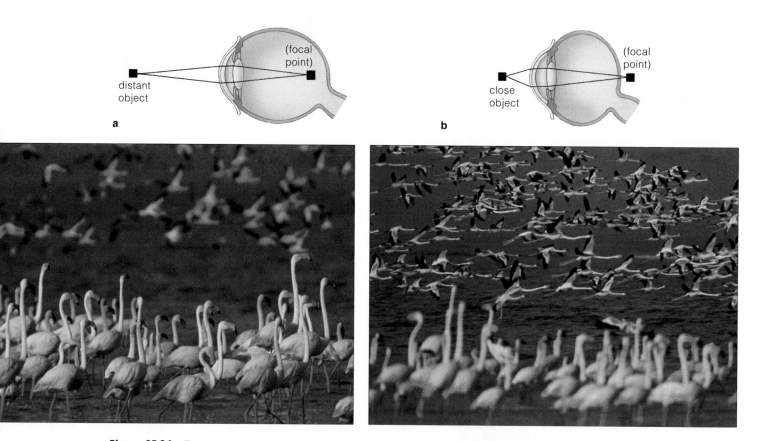

Figure 35.24 Focusing problems. In nearsightedness (**a**) light rays from distant objects converge in front of the retina. In farsightedness (**b**) light rays from close objects have not converged when they arrive at the retina.

Age-Related Disorders The macula is a retinal region with a high density of photoreceptors. It fully surrounds the fovea (Figure 35.22). Without it, we cannot see enough detail to read, drive, even to recognize faces. *Macular degeneration* is one of the most common causes of blindness among those who are age fifty-five or older. In the United States, about 13 million people have age-related macular degeneration (AMD). Destruction of photoreceptor cells in the macula and fovea clouds the center of the visual field more than the periphery. The causes of AMD are not fully known, although there is a genetic component. Smoking, obesity, and high blood pressure increase the risk.

AMD cannot be cured, but its progression can be slowed with eye drops, vitamins, and laser therapy. An experimental treatment filters cholesterol and certain proteins from an affected person's blood. So far, it has helped about 30 percent of those treated.

Glaucoma results when too much aqueous humor builds up inside the eyeball. The excess damages blood vessels and ganglion cells, and it can interfere with peripheral vision and visual processing. Glaucoma can be treated with medication and surgery. Although we often associate chronic glaucoma with advanced age, conditions that give rise to the disorder actually start to develop in middle age. When doctors detect the fluid pressure that tends to build up in the eye early enough, they can prescribe drugs or perform surgery before the damage becomes severe.

A *cataract*, a gradual clouding of the lens, alters the amount of light that reaches the retina and where light is focused. Possibly cataracts form when the transparent proteins that make up the eye's lens undergo structural changes. When the lens becomes fully opaque, vision is impossible. The clouded lens can be replaced with an artificial implant. Each year, millions of people undergo cataract surgery. It is a problem associated with aging, although an injury or diabetes also may cause it.

Eye Diseases The eyes are vulnerable to infectious diseases. For instance, a fungus that causes the lung disease *histoplasmosis* sometimes invades the eyes as well. Partial or total loss of vision may follow. *Herpes simplex*, a virus that causes skin sores, also can infect the cornea and result in its ulceration. You read about these diseases in Chapter 21.

Trachoma is an extremely contagious disease that has blinded millions, mostly in North Africa and the Middle East. The infectious agent is a bacterium that also is responsible for the sexually transmitted disease chlamydia (Section 44.8). The eyeball and the lining of the eyelids (the conjunctiva) become damaged.

These damaged tissues are entry points for other kinds of pathogenic bacteria that may cause secondary infections. In time, the cornea may become so scarred that blindness follows.

Summary

Section 35.1 Sensory pathways are examples of cell communication, from reception and transduction of a signal to a suitable response. In this case, different classes of sensory receptors transduce the energy of a specific stimulus, such as pressure, into electrochemical energy of action potentials. The main classes are called mechanoreceptors, pain receptors, thermoreceptors, chemoreceptors, osmoreceptors, and photoreceptors.

The brain evaluates action potentials from sensory receptors based on which nerve delivers them, their frequency, and the number of axons firing in a given interval. Continued stimulation of a receptor may lead to a diminished response (sensory adaptation).

Receptors for somatic sensations, such as touch and warmth, are not localized in a special organ or tissue. Receptors for special senses—taste, smell, hearing, balance, and vision—are in special sensory organs.

Biology⊘Now
See how the intensity of a stimulus affects action potential frequency with the animation on BiologyNow.

Section 35.2 Somatic sensations start at free nerve endings, encapsulated receptors, and stretch receptors near the body surface, in the wall of internal organs, and in skeletal muscles. Signals arrive in sensory areas of the cerebral cortex, where interneurons are organized like maps for individual parts of the body surface.

Biology⊘Now
Learn about the sensory receptors in human skin with the animation on BiologyNow.

Section 35.3 The senses of taste and smell involve pathways from chemoreceptors to processing regions in the cerebral cortex and limbic system. Taste receptors are concentrated in taste buds in the tongue and mouth. Olfactory receptors line the nasal passages. Pheromones are chemical signals secreted by many animals that also have specialized organs or structures to detect them.

Section 35.4 The vestibular apparatus is a fluid-filled organ of equilibrium in the vertebrate inner ear. It detects gravity, acceleration, and other forces that affect the body's position and movement.

Section 35.5 Sound is a form of mechanical energy. A vibrating object generates pressure waves, which can vary in amplitude and frequency. Humans have a pair of ears with three regions: the outer, middle, and inner ear. The outer ear collects sound waves. The middle ear amplifies sound waves and transmits them to the inner ear, which includes the vestibular apparatus and the cochlea: a coiled, fluid-filled structure with three ducts.

Pressure waves traveling through the fluid inside the cochlea bend mechanoreceptors called hair cells, which are embedded in one of the cochlear membranes. Sounds get sorted out according to their amplitude. In this case, mechanical energy is transduced to action potentials that are relayed along auditory nerves to the brain.

Biology (∂) Now
Explore the structure and function of the human ear with the animation on BiologyNow.

Sections 35.6, 35.7 All organisms are sensitive to light, but vision requires eyes and brain centers that can process visual information. An eye is a sensory organ that contains a dense array of photoreceptors.

Like squids and octopuses, humans have camera eyes, each with an iris that adjusts incoming light and a lens that focuses light on the retina in the back of the eyeball chamber. The retina has densely packed photoreceptors.

Section 35.8 Rods detect dim light, and cones detect bright light and colors. These photoreceptors interact with other retinal cells to process visual information even before sending it on to the brain. Two optic nerves carry signals that eventually reach the cerebral cortex.

Biology (∂) Now
Investigate the structure and function of the human eye with the animation on BiologyNow.

Section 35.9 Abnormalities in eye shape, in the lens, and with retinal cells often impair vision.

Biology (∂) Now
Learn about the organization of the retina and how visual stimuli are processed with the animation on BiologyNow.

Self-Quiz
Answers in Appendix II

1. A stimulus is a specific form of energy in the outside environment that is detected by _____ .
    a. a sensory receptor   c. the brain
    b. nerves   d. all of the above

2. _____ is defined as a decrease in the response to an ongoing stimulus.
    a. Perception   c. Sensory adaptation
    b. Visual accommodation   d. Somatic sensation

3. Which is a somatic sensation?
    a. taste   c. touch   e. a through c
    b. smell   d. hearing   f. all of the above

4. Chemoreceptors play a role in the sense of _____ .
    a. taste   c. touch   e. both a and b
    b. smell   d. hearing   f. all of the above

5. In the _____ neurons are arranged like maps that correspond to different parts of the body surface.
    a. cerebral cortex   c. basilar membrane
    b. retina   d. all of the above

6. Mechanoreceptors in the _____ send signals to the brain about the body's position relative to gravity.
    a. eye   b. ear   c. tongue   d. nose

7. The middle ear functions in _____ .
    a. detecting shifts in body position
    b. amplifying and transmitting sound waves
    c. sorting sound waves out by frequency
    d. both b and c

8. Label the components of the human eye, as indicated in the following diagram:

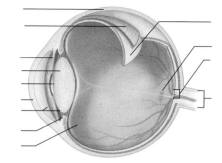

9. Match each structure with its description.
    ____ rod cell   a. sensitive to vibrations
    ____ cochlea   b. function in balance
    ____ lens   c. detects color
    ____ hair cell   d. detects dim light
    ____ cone cell   e. contains chemoreceptors
    ____ taste bud   f. focuses rays of light
    ____ vestibular   g. sorts out sound waves
         apparatus   h. helps brain assess heat,
    ____ free nerve ending   pressure, pain

Additional questions are available on Biology (∂) Now™

Critical Thinking

1. Laura loves to eat broccoli and brussels sprouts. Lionel cannot stand them. Everyone has the same five kinds of taste receptors, so what is going on? Is Lionel just being difficult? Perhaps not. The number and distribution of receptors that respond to bitter substances vary among individuals of a population—and studies now indicate that some of this variation is heritable.

    People who have the greatest number of receptors for bitter substances find many fruits and vegetables highly unpalatable. These *supertasters* make up about 25 percent of the general population. They tend to be slimmer than average but are more likely to develop colon polyps and colon cancer. How might Lionel's highly sensitive taste buds put him at increased risk for colon cancer?

2. Above and below a python's mouth are rows of pits that contain thermoreceptors (Figure 35.25a). They detect body heat, or infrared energy, of nearby prey. Name the type of prey organisms the receptors can detect. Which kinds of otherwise edible animals would they miss so that the snake would slither on by?

3. Think about the bats. Nearly all of these mammalian species sleep during the day and spread their webbed wings at dusk. Different kinds take to the air in search of nectar, fruit, frogs, or insects. Many sensory receptors in their eyes, nose, ears, mouth, and skin are not all that different from yours. Others are very different. They help even tiny-eyed, nearly blind species navigate and capture flying insects swiftly in the dark.

    As one of these bats flies, it emits a steady stream of about ten clicking sounds per second. The bat shown in Figure 35.25b is emitting these clicks. You cannot hear a bat's clicks. They are ultrasounds, of an intensity beyond the

range of sound waves that receptors in human ears can detect. When a bat hears a pattern of distant echoes from, say, an airborne mosquito, it increases the rate of ultrasonic clicks to as many as 200 per second. That is faster than a machine gun can fire bullets. In the few milliseconds of silence between the clicks, sensory receptors in the bat's pair of inner ears detect the echoes. The bat brain swiftly constructs a "map" of the sounds, which the bat follows during its maneuvers through the night world.

Which part of the cochlear duct inside the bat ear do the ultrasounds stimulate?

4. The vestibular apparatus of all modern whales and dolphins is much smaller, relative to their body size, than it is in other mammals (Figure 35.26). Whales, remember, are descended from land-dwelling mammals (page 281). Fossils of early whales show that the vestibular apparatus became modified as their ancestors made the transition back to a life in the seas.

Think about how the sense of balance in an aquatic animal might differ from that of a land mammal that walks on four legs. Speculate on why natural selection might have favored a rapid reduction in the size of the vestibular apparatus during the transition from a terrestrial to an aquatic life-style.

5. Are organs of dynamic equilibrium, static equilibrium, or both activated during a heart-thumping roller coaster ride, as in Figure 35.27?

6. In humans, photoreceptors are most concentrated at the very back of the eyeball. In birds of prey, such as owls and hawks, the greatest density of photoreceptors is in a region closer to the eyeball's roof. When these birds are on the ground, they cannot see objects even slightly above them unless they turn their head almost upside down. What is the adaptive advantage of this peculiar type of retinal organization?

7. The strength of Earth's magnetic field and its angle relative to the surface vary with latitude. Diverse species sense these differences and use them as cues for assessing their location and direction of movement. Behavioral experiments have shown that sea turtles, salamanders, and spiny lobsters use information from Earth's magnetic field during their migrations. Whales and some burrowing rodents also seem to have a magnetic sense. Evidence about humans is contradictory. Is it likely that humans have such a sense? Suggest an experiment that might support or disprove the possibility.

Figure 35.25 Examples of sensory receptors. (**a**) Thermoreceptors in pits above and below a python's mouth detect body heat, or infrared energy, of nearby prey. (**b**) Some bats listen to echoes of their own high-frequency sounds. Their brain constructs a sound map from the echoes that bounce back from prey and other objects in the surroundings. Such maps help this night-flying bat capture insects in midair without the help of eyes.

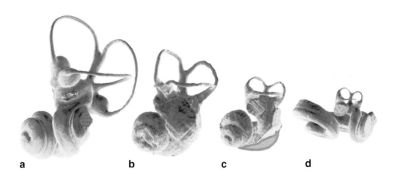

Figure 35.26 Inner ear structures from (**a**) a bushbaby, a notably agile land mammal, (**b**) a forerunner of whales (*Ichthyolestes*) that lived on land 50 million years ago, (**c**) a marine whale (*Indocetus*) that lived 45 million years ago, and (**d**) a dolphin. These computer reconstructions are adjusted to account for differences in body size.

Figure 35.27 One way to test whether your organs of equilibrium are working.

Hormones in the Balance

Atrazine has been widely used as an herbicide for more than forty years. Each year, people in the United States apply about 76 million pounds of it, mostly to fields of corn and other monocot crops. They also apply it to lawns at homes, and golf courses. Where does it go from there? Into soil and water. Atrazine molecules break down in less than a year, but they still turn up in groundwater, ponds, wells, and rain. Atrazine contamination is common in soils and water of the American Midwest.

Is this bad? University of California biologist Tyrone Hayes thinks so (Figure 36.1). Atrazine, he suspects, may be one of the synthetic organic compounds with unexpected side effects. The compounds mimic, block, or boost the action of natural hormones involved in reproduction and development. They are *endocrine disruptors*.

Hayes looked at atrazine's effects on African clawed frogs (*Xenopus laevis*) and leopard frogs (*Rana pipiens*) in his laboratory. After he exposed male tadpoles to the herbicide, some became hermaphrodites, with male *and* female reproductive organs. Atrazine concentrations as low as 0.1 part per billion could cause such changes.

Did atrazine have similar effects in the wild? To find out, Hayes collected native leopard frogs from ponds and ditches across the Midwest. In every one of the atrazine-contaminated ponds sampled, male frogs had abnormal sex organs. In the pond with the most atrazine, 92 percent of the males had feminized sex organs.

Researchers funded by herbicide industries dispute the findings, but other studies confirmed atrazine can cause or amplify frog deformities. The Environmental Protection Agency found the data intriguing. Among other things, this agency regulates the application of chemicals in agriculture. It has called for further study of atrazine's potentially harmful effects on amphibians. It has been encouraging farmers to do more to minimize the runoff of atrazine-laden water from fields.

Meanwhile, other hormone-disrupting chemicals are infiltrating aquatic habitats. For instance, estrogens from birth control pills are excreted from the body in urine and cannot be removed by standard wastewater treatments. When estrogen-tainted water enters streams or rivers, it can cause male fish to develop female traits. So can estrogen-mimicking pollutants from industries.

In some pesticide-contaminated lakes in Florida, male alligators have low testosterone levels and an abnormally small penis. In New York, salamanders developed bizarre skin coloration patterns after being exposed to herbicide-rich runoff from a new golf course in the neighborhood.

Figure 36.1 Benefits and costs of herbicide applications. *Left*, Atrazine can keep cornfields nearly weed-free; no need for constant tilling that causes soil erosion. Tyrone Hayes (*right*) suspects this chemical may be scrambling amphibian hormonal signals.

not a good
time to be
a frog

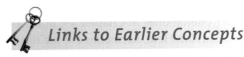

Watch the video online!

Are amphibians, fishes, and alligators too removed from your interests? The endocrine disruptors may be meddling with humans, too. Between 1938 and 1990, sperm counts of males in Western countries declined by about 40 percent. Would you be surprised to learn that men in agricultural regions may have the lowest counts? As University of Missouri researchers discovered, semen samples from men in Columbia, Missouri, contained only about half as many motile sperm as the samples from men in New York City and Los Angeles.

What is killing off sperm? Estrogen-like chemicals are the main suspects. Estrogen is known as a female sex hormone, yet both males and females produce it, and both have receptors for it. In males, sperm cannot mature unless estrogen docks at the receptors. However, many other chemicals, including kepone and DDT, can latch on to estrogen receptors. Both insecticides are now banned in the United States.

This chapter focuses on hormones—their sources, targets, and interactions. All vertebrates have similar hormone-secreting glands and systems. If the details start to seem remote, remember endocrine disruptors. What you learn here can help you evaluate research that may influence your health and life-style.

 How Would You Vote?

Crop yields that sustain the human population currently depend on agricultural pesticides, some of which may disrupt hormone function in frogs and other untargeted species. Should chemicals that may cause problems remain in use while researchers investigate them? See BiologyNow for details, then vote online.

 Key Concepts

THE ENDOCRINE SYSTEM

Hormones and other signaling molecules are components of pathways that control metabolism, growth, development, and reproduction. The same hormone sources make up the endocrine system of nearly all vertebrates. Section 36.1

SIGNALING MECHANISMS

A hormone signals target cells, which are any cells that have receptors for it. Receptor activation leads to transduction of the signal and the cellular response. Section 36.2

A MASTER INTEGRATING CENTER

In vertebrates, the hypothalamus and pituitary gland are connected structurally and functionally. Together, they coordinate activities of many other glands throughout the body. Section 36.3

OTHER HORMONE SOURCES

Negative feedback loops to the hypothalamus and pituitary control the secretion of many glands. Local changes in the internal environment control the secretion of other glands. Sections 36.4–36.10

INVERTEBRATE HORMONES

Hormones control molting and other events in the life cycles of invertebrates. Vertebrate hormones and their receptors started to evolve in invertebrate lineages. Section 36.11

THE HORMONAL SYMPHONY

Mammalian organ systems are a focus of many chapters in this unit. This section pulls together most of their tissue-specific responses to hormones, which can be traced in part to variations in receptors. Section 36.12

Links to Earlier Concepts

This chapter builds on your understanding of membrane proteins (Sections 5.2, 5.4), gene controls (15.1), feedback loops (28.3), and cell-to-cell signaling (28.5). You will draw on your understanding of organic metabolism (8.6) and brain function (34.9–34.12). You will be looking more closely at the nature of molting (25.11, 25.13, 25.15).

An earlier section introduced glandular epithelium (33.1). All hormone-secreting cells are embedded in this kind of tissue, which may form a gland or a sheetlike lining for some organs.

36.3 The Hypothalamus and Pituitary Gland

LINKS TO
SECTIONS
33.1, 34.8

In all vertebrates, the hypothalamus interacts with the pituitary gland as a major integrating center. Together, they monitor and control the activities of other organs, a number of which secrete hormones of their own.

The **hypothalamus,** a forebrain region, includes some neurons that secrete hormones, not neurotransmitters. A tissue stalk connects the base of the hypothalamus to a **pituitary gland**. In humans, this lobed gland is about the size of a pea. Axons of hormone-secreting neurons extend down through the stalk and into the pituitary gland's *posterior* lobe, which stores and secretes their hormones. The *anterior* lobe makes its own hormones, but hypothalamic signals control their release. The pituitary gland of many vertebrates (not humans) has an *intermediate* lobe between the other two. One of its secretions governs reversible changes in skin and fur color. Together, the hypothalamus and pituitary gland function as a master control center that integrates the activities of many glands and the nervous system.

POSTERIOR PITUITARY FUNCTION

Some hypothalamic neurons secrete ADH (antidiuretic hormone) and oxytocin into the posterior lobe of the pituitary (Figure 36.4). The hormones diffuse into an adjacent capillary bed, then travel the bloodstream. In the kidneys, ADH targets cells that adjust how much water is excreted in urine. Oxytocin makes the uterus contract during childbirth and makes milk move into secretory ducts of mammary glands during lactation.

ANTERIOR PITUITARY FUNCTION

Other hypothalamic hormones act on anterior pituitary cells. Most are **releasers**; they induce secretions from target cells. Others are **inhibitors**, which slow target cell secretions. Releasers or inhibitors diffusing out of the axon endings of hypothalamic neurons are picked up by a capillary bed in the stalk above the pituitary (Figure 36.5). They flow down into a second capillary bed in the anterior pituitary, and there they diffuse out into the tissues. They control secretions from six types of anterior lobe hormones:

ACTH	Adrenocorticotropin
TSH	Thyrotropin
FSH	Follicle-stimulating hormone
LH	Luteinizing hormone
PRL	Prolactin
STH (or GH)	Somatotropin (growth hormone)

What are the targets of these hypothalamic releasers and inhibitors? As you will see, ACTH stimulates the release of cortisol from a pair of adrenal glands. TSH stimulates the release of thyroid hormones from the thyroid gland. FSH and LH affect gamete formation in the reproductive organs and other aspects of sexual reproduction. STH, or growth hormone, has targets in most tissues. It triggers secretions from liver cells that promote growth of bone and soft tissues in the young; it influences metabolism in adults. Prolactin initiates and maintains milk production in mammary glands after other hormones prime the tissues for activity.

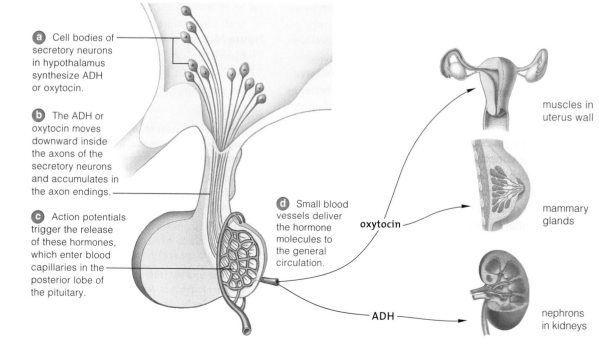

Figure 36.4 *Animated!* Functional links between the hypothalamus and the posterior lobe of the pituitary gland. Targets of posterior lobe secretions are shown.

ⓐ Cell bodies of secretory neurons in hypothalamus synthesize ADH or oxytocin.

ⓑ The ADH or oxytocin moves downward inside the axons of the secretory neurons and accumulates in the axon endings.

ⓒ Action potentials trigger the release of these hormones, which enter blood capillaries in the posterior lobe of the pituitary.

ⓓ Small blood vessels deliver the hormone molecules to the general circulation.

oxytocin

muscles in uterus wall

mammary glands

ADH

nephrons in kidneys

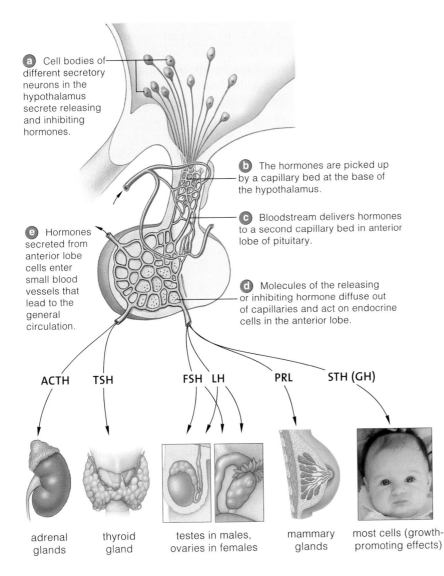

a Cell bodies of different secretory neurons in the hypothalamus secrete releasing and inhibiting hormones.

b The hormones are picked up by a capillary bed at the base of the hypothalamus.

c Bloodstream delivers hormones to a second capillary bed in anterior lobe of pituitary.

e Hormones secreted from anterior lobe cells enter small blood vessels that lead to the general circulation.

d Molecules of the releasing or inhibiting hormone diffuse out of capillaries and act on endocrine cells in the anterior lobe.

ACTH   TSH   FSH   LH   PRL   STH (GH)

adrenal glands   thyroid gland   testes in males, ovaries in females   mammary glands   most cells (growth-promoting effects)

Figure 36.5 *Animated!* Functional links between the hypothalamus and the pituitary gland's anterior lobe. Also shown are the main targets of anterior lobe secretions. The photograph above shows one case of abnormal output of a posterior pituitary hormone. This male is twelve years old. He is affected by pituitary gigantism caused by overproduction of STH. At six feet, five inches tall, he towers over his mother.

ABNORMAL PITUITARY OUTPUTS

The vertebrate body does not churn out hormones in vast amounts. Roger Guillemin and Andrew Schally had to purify seven tons of hypothalamic tissue to get a single milligram of TSH, the first known releaser. Even so, tiny amounts of hormones have big impacts. When control of their secretion fails, the body's form, function, or both become altered.

Overproduction of STH during childhood leads to *pituitary gigantism*. Affected adults have normal body form, but are much larger, as in Figure 36.5. Low STH secretion during childhood leads to *pituitary dwarfism*. Affected adults have the form of an average person but are smaller. High STH secretion during adulthood produces *acromegaly*. Long bones cannot lengthen, but the hands and feet become enlarged and swollen. Too much cartilage and bone form, which distorts facial features. Skin thickens; internal organs may enlarge.

Another example: A severe blow to the head can damage the posterior pituitary, and slow or stop ADH secretion. This is one cause of *diabetes insipidus*. In this condition, the body loses too much water in the urine. The resulting dehydration can be life threatening.

The functional links between the nervous and endocrine systems are most evident in the close interaction between the hypothalamus and pituitary gland.

The hypothalamus produces ADH and oxytocin. Both hormones are stored in and secreted from the pituitary's posterior lobe. Other hypothalamic hormones stimulate or inhibit secretion of anterior pituitary hormones.

The anterior pituitary makes and secretes ACTH, TSH, FSH, LH, PRL, and STH. These hormones trigger secretion of hormones from different glands and have diverse effects.

36.4 Thyroid and Parathyroid Glands

LINKS TO
SECTIONS
6.4, 28.3

Amines produced by the thyroid have roles in development and metabolism. A feedback loop to the pituitary and hypothalamus controls their secretion. Parathyroid glands make a hormone that controls calcium levels in the blood.

FEEDBACK CONTROL OF THYROID FUNCTION

A human **thyroid gland** looks vaguely like a butterfly resting at the base of the neck, in front of the trachea (Figure 36.6). It secretes two amines, triiodothyronine and thyroxine, but we can refer to both as "thyroid hormone." Thyroid hormone has targets throughout the body, and no vertebrate can develop properly in

its absence. In birds and mammals, it is an important part of controls over metabolic rates.

Negative feedback loops to the anterior pituitary and hypothalamus control thyroid hormone secretion. In these loops, an increase in a hormone's level above a set point inhibits secretion of that hormone. Figure 36.7 shows what happens when the level of thyroid hormone in blood declines. Sensing the decline, the hypothalamus secretes TRH into the anterior lobe of the pituitary. This releaser makes the pituitary step up its secretion of thyroid-stimulating hormone (TSH). The TSH induces the thyroid gland to secrete thyroid hormone, and the hormone's blood level rises back to the set point. At that point, the secretions of TRH and TSH from the master control center slow down.

The importance of feedback control of hormones is brought into sharp focus by cases where controls fail. Consider that synthesis of thyroid hormone requires iodine, which we obtain from food. Low levels of this nutrient lead to an enlarged thyroid, or *simple goiter* (Figure 36.8a). Thyroid hormone cannot be produced without iodine, and its blood level falls below the set point. The anterior pituitary responds by secreting TSH, which stimulates the thyroid gland. However, with no iodine, the thyroid gland cannot synthesize the hormone and shut down TSH secretion. Feedback to the anterior pituitary keeps calling for TSH, which in time causes the thyroid gland to enlarge.

With *hypothyroidism*, the thyroid hormone level in blood remains too low. This condition stunts growth during childhood and delays the maturation of sexual organs. Hypothyroidism arising from iodine deficiency is prevalent in mothers-to-be around the world. It is the major cause of preventable mental impairment. Although the use of iodized salt can prevent dietary hypothyroidism, it is not easy to get in some remote areas of less developed countries.

Graves' disorder and other *toxic goiters* result from *hyperthyroidism*—an excess of thyroid hormone in the blood. Symptoms typically include anxiety, tremors, trouble sleeping, heat intolerance, protruding eyes, and erratic heartbeat. Some cases arise when the body mistakenly makes antibodies against thyroid tissues. Others result after inflammation or tumor formation in the thyroid gland.

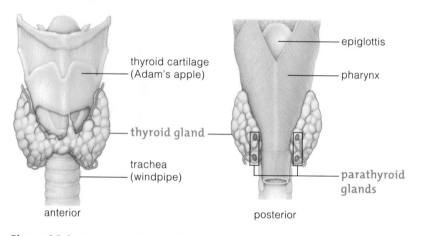

Figure 36.6 Location of human thyroid and parathyroid glands.

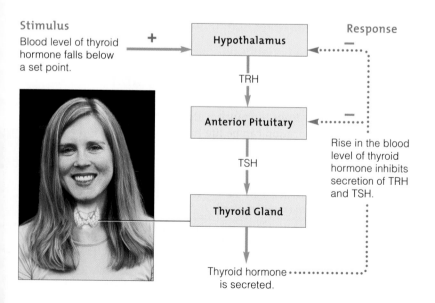

Figure 36.7 Negative feedback loop to the hypothalamus and anterior lobe of the pituitary gland that governs thyroid hormone secretion.

PARATHYROID GLANDS AND CALCIUM LEVELS

Four **parathyroid glands** are located on the thyroid's posterior surface, as Figure 36.6 shows. They release parathyroid hormone (PTH) in response to a decline in the level of calcium in blood. Calcium ions, recall,

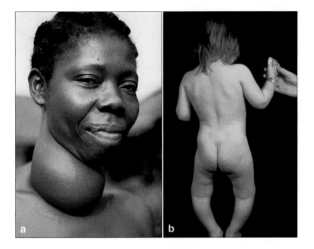

Figure 36.8 (a) A case of goiter caused by a diet low in iodine, a micronutrient. (b) A child affected by rickets shows the characteristic bowed legs.

have roles in blood clotting, enzyme activity, muscle contraction, and other processes.

PTH targets certain bone cells and kidney cells. In bones, it induces osteoclasts to secrete bone-digesting enzymes. Section 37.4 explains how it works. For now, simply note that calcium and other minerals released from bone enter interstitial fluid, then blood. In the kidneys, PTH stimulates tubule cells to reabsorb more calcium. It also stimulates the secretion of enzymes that activate vitamin D_3, which stimulates cells in the intestinal lining to absorb calcium from food.

Young children with vitamin D deficiencies have bone problems. They absorb too little calcium, so they cannot form enough new bone tissue. A low calcium level also triggers an increase in PTH secretion, which leads to the breakdown of existing bone. The resulting disorder is called *rickets*. Bowed legs and deformities in pelvic bones are common symptoms (Figure 36.8*b*).

Calcitonin is another peptide hormone secreted by the thyroid, and it opposes the effect of PTH in bones. In humans, calcitonin has a neglible effect on day-to-day calcium regulation. When the thyroid gland is surgically removed, calcium levels are not affected.

Negative feedback loops from many glands lead to the hypothalamus and anterior pituitary gland. Secretion of thyroid hormone is regulated by such feedback.

Thyroid hormone acts during development and influences metabolic rates. Its synthesis requires iodine.

Parathyroid glands secrete PTH, the main regulator of calcium levels in blood.

36.5 Twisted Tadpoles

This chapter's introduction focused on hormone disruptors that may cause frog deformities. Studies of frog thyroid function also indicate that all is not well in the water.

A tadpole is an aquatic larval stage in the life cycle of frogs and toads. It undergoes a big remodeling in body form, or *metamorphosis*, when it makes the transition to an adult. For instance, it sprouts legs, lungs replace its gills, and tissues of its tail are resorbed (Section 43.2). A surge in thyroid hormone triggers the changes. When its thyroid tissue is experimentally removed, a tadpole keeps on growing, but the adult form never emerges.

Some water pollutants may be the chemical equivalent of thyroid removal. For one study, investigators exposed embryos of African clawed frogs (*X. laevis*) to water drawn from lakes in Minnesota and Vermont. Half of the water samples came from lakes where deformity rates were low. The other half came from "hot spots," where the water has as many as twenty kinds of dissolved pesticides and where deformity rates are high.

The embryos that were raised in hot-spot water often developed into tadpoles that had a bent spine and other abnormalities, as in Figure 36.9. Some tadpoles never did metamorphose into adults. By contrast, control embryos raised in water from other lakes developed normally.

To find out if something in the water was interfering with thyroid hormone, the researchers dissolved some of the hormone in hot-spot water. Embryos raised in this mix developed into tadpoles that had fewer deformities or none at all.

Frogs are highly sensitive to disturbances in thyroid function, and thyroid disruptions are easy to detect. That is why toxicologists use laboratory frogs to test whether chemicals are thyroid disruptors. They also use them to track how the disruptive chemicals exert their effects.

Among the chemicals under study are perchlorates, which are widely used in explosives, propellants, and batteries. Perchlorates can interfere with the metabolism of iodine. As little as 5 parts per billion may stop a frog's forelimbs from developing.

Figure 36.9 Evidence that pollutants disrupt thyroid function and interfere with metamorphosis. The uppermost *Xenopus laevis* tadpole in this photographic series was raised in water from a lake with few deformed frogs. Tadpoles below it developed in water taken from three "hot spot" lakes with increasingly higher concentrations of dissolved chemical compounds. As later tests showed, supplemental thyroid hormone can lessen or eliminate hot-spot deformities.

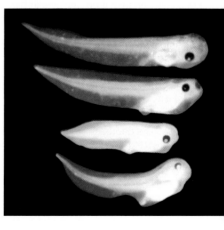

36.6 Pancreatic Hormones

Many hormones work antagonistically, in that the action of one opposes the action of another. Insulin and glucagon act in this way. Controls over when these two pancreatic hormones are secreted maintain the glucose level in blood.

The pancreas lies in the abdominal cavity, in back of the stomach. This organ has exocrine and endocrine functions. Its exocrine cells secrete digestive enzymes into a duct to the small intestine. Its endocrine cells are grouped in clusters called pancreatic islets. Each islet contains three types of hormone-secreting cells.

Alpha cells secrete the peptide hormone glucagon. In between meals, cells throughout the body take up glucose from the blood. When the glucose blood level falls below a set point, alpha cells secrete glucagon, and the liver is the main target. Section 36.2 explains how the binding of glucagon activates many enzymes that break down glycogen to its glucose monomers. Thus, *glucagon raises the level of glucose in blood.*

Beta cells, the most abundant cells in the pancreatic islets, secrete insulin—the only hormone that causes target cells to take up and store glucose. After a meal, high levels of glucose in blood stimulate beta cells to release insulin. The targets are mainly liver, fat, and skeletal muscle cells. Insulin particularly stimulates muscle and fat cells to take up glucose. In all target cells, it activates enzymes that function in protein and fat synthesis, and it inhibits the enzymes that catalyze protein and fat breakdown. As a result of its actions, *insulin lowers the level of glucose in the blood.*

Delta cells secrete somatostatin. This hormone helps control digestion and nutrient absorption. Also, it can inhibit the secretion of insulin and glucagon.

In sum, discontinuous eating patterns and shifts in cell activities alter the blood glucose level. Pancreatic secretions of glucagon and insulin work in opposition to maintain the level within a homeostatic range that can keep the body's cells functioning (Figure 36.10).

Glucagon triggers the breakdown of glycogen; it raises the blood level of glucose. Insulin helps cells take up and store more glucose; it lowers the blood level of glucose.

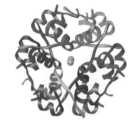

Figure 36.10 *Animated!* *Above,* model for insulin. *Right,* how cells that secrete insulin and glucagon respond to change in the level of glucose circulating in blood. These two hormones work antagonistically to return the glucose level to its homeostatic range.

(**a**) *After* a meal, glucose enters blood faster than cells can take it up. Its level in blood increases. In the pancreas, the increase (**b**) stops alpha cells from secreting glucagon and (**c**) stimulates beta cells to secrete insulin. In response to insulin (**d**), muscle and adipose cells take up and store glucose, and liver cells synthesize more glycogen. The result? (**e**) Insulin *lowers* the glucose blood level.

(**f**) *Between* meals, the glucose level in blood declines. The decrease (**g**) stimulates alpha cells to secrete glucagon and (**h**) stops beta cells from secreting insulin. (**i**) In the liver, glucagon causes cells to convert glycogen back to glucose, which enters the blood. The outcome? (**j**) Glucagon *raises* the blood level of glucose.

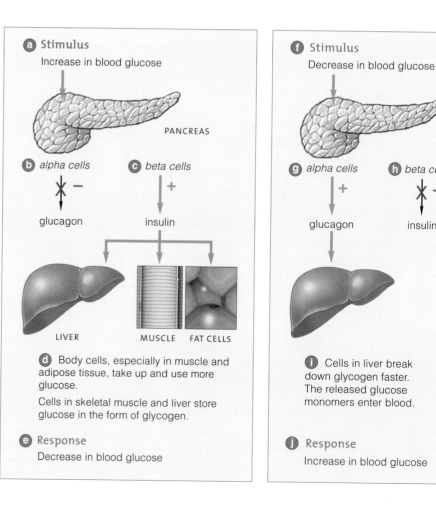

a Stimulus
Increase in blood glucose

PANCREAS

b *alpha cells* **c** *beta cells*

glucagon — insulin +

LIVER   MUSCLE   FAT CELLS

d Body cells, especially in muscle and adipose tissue, take up and use more glucose.

Cells in skeletal muscle and liver store glucose in the form of glycogen.

e Response
Decrease in blood glucose

f Stimulus
Decrease in blood glucose

g *alpha cells* **h** *beta cells*

glucagon + insulin —

i Cells in liver break down glycogen faster. The released glucose monomers enter blood.

j Response
Increase in blood glucose

LINK TO
SECTION
8.6

The cells of all tissues burn glucose as fuel. Glucose is the preferred energy source for brain cells and the only source for red blood cells. The importance of glucose homeostasis becomes clear when we study people who cannot produce enough insulin or whose target cells cannot respond to it.

Diabetes mellitus is a metabolic disorder in which cells have trouble taking up glucose. The sugar accumulates in blood and in urine. Complications develop in tissues throughout the body (Table 36.2). Affected people become abnormally thirsty and urinate too much, which disrupts the water–solute balance in the internal environment. Over time, the kidneys become damaged, and they may fail completely.

Remember Section 8.6? In the absence of a steady supply of glucose, cells turn to fats and proteins as energy sources. Weight loss is one outcome. Another is ketone accumulation in the blood and urine. Ketones are normal acidic products of fat breakdown, but when too many build up, the result is *ketoacidosis*. The altered acidity and solute levels can interfere with normal brain function. Extreme cases may lead to death. Uncontrolled diabetes also damages blood vessels and nerves, especially in the extremities. Diabetics account for more than 60 percent of the lower limb amputations in populations at large.

Diabetes has two main forms. *Type 1 diabetes* develops after the body mounts an autoimmune response against its insulin-secreting beta cells. Certain white blood cells mistakenly identify these cells as being foreign (nonself) and destroy them. Environmental factors complicate a genetic predisposition to the disorder. Type 1 diabetes accounts for only 5 to 10 percent of all reported cases, but it is the most dangerous in the short term.

Symptoms usually start to appear in childhood and adolescence, which is why this metabolic disorder is also known as juvenile-onset diabetes. All affected individuals require insulin injections, and they must monitor their blood sugar carefully (Figure 36.11).

Type 2 diabetes is by far the most common form of the disorder. Insulin levels are normal or even high. However, target cells do not respond to the hormone as they should, and blood sugar levels remain elevated. The symptoms typically start to develop when insulin production declines in middle age. Genetics is a factor, but obesity increases the risk.

Diet, exercise, and oral medications can control most cases of type 2 diabetes. Even so, if glucose levels are not lowered, pancreatic beta cells will be stimulated nonstop. Eventually, they will falter, and so will insulin production. When that happens, a type 2 diabetic may require insulin injections.

Worldwide, rates of type 2 diabetes are soaring. By one estimate, more than 150 million people are now affected. Western diets and sedentary life-styles are contributing factors. The prevention of diabetes and its complications is acknowledged to be among the most pressing public heath priorities around the world.

Finally, in *hypoglycemia*, blood glucose level falls so low that normal body function is disrupted. Rare tumors that secrete insulin can cause it. But most cases occur after an insulin-dependent diabetic miscalculates and injects a bit too much insulin to balance food intake. The result is *insulin shock*. The brain stalls as its fuel supply dwindles. Symptoms of the disorder commonly include dizziness, confusion, and difficulty speaking. Insulin-shock can be life threatening, but an injection of glucagon quickly reverses the condition.

Table 36.2	Some Complications of Diabetes
Eyes	Changes in lens shape and vision; damage to blood vessels in retina; blindness
Skin	Increased susceptibility to bacterial and fungal infections; patches of discoloration; thickening of skin on the back of hands
Digestive system	Gum disease; delayed stomach emptying that causes heartburn, nausea, vomiting
Kidneys	Increased risk of kidney disease and failure
Heart and blood vessels	Increased risk of heart attack, stroke, high blood pressure, and atherosclerosis
Hands and feet	Impaired sensations of pain; formation of calluses, foot ulcers; possible amputation of a foot or leg because of necrotic tissue that formed owing to poor circulation

Figure 36.11 A diabetic checks his blood glucose by placing a blood sample into a glucometer. Compared with Caucasians, Hispanics and African Americans are about 1.5 times more likely to be diabetic. Native Americans and Asians are at even greater risk. Proper diet helps control blood sugar, even in type 1 diabetics.

36.8 The Adrenal Glands

LINKS TO
SECTIONS
8.6, 34.7, 34.11

The preceding section gave an example of how the action of one hormone may counter the action of another. Signals from the nervous system also can intervene to stimulate or dampen the secretion of a hormone, hence its effects.

THE ADRENAL CORTEX

Humans have a pair of adrenal glands, one on top of each kidney (Figure 36.12). Some cells of the **adrenal cortex**, the gland's outer layer, secrete cortisol. Two negative feedback loops to the hypothalamus and to the anterior lobe of the pituitary keep the blood level of cortisol from rising or falling too far.

Figure 36.12 shows what happens when the level of cortisol in blood decreases below the body's set point. The decline triggers secretion of CRH (corticotropin releasing hormone) from the hypothalamus. The CRH prods cells in the anterior pituitary to secrete ACTH (adrenocorticotropin). In turn, ACTH stimulates cells of the adrenal cortex to secrete cortisol.

Secretion continues until the blood level of cortisol rises above the set point. Then, the hypothalamus and anterior pituitary issue signals that inhibit secretion of CRH and ACTH, and so cortisol secretion slows.

Cortisol helps maintain the blood level of glucose when food is not being absorbed from the gut. Most notably, it induces liver cells to break down their store of glycogen, and it suppresses the uptake of glucose by other cells. By doing this, cortisol indirectly helps keep glucose continually available for brain activity. Cortisol also induces adipose cells to degrade fats and skeletal muscles to degrade proteins. The breakdown products—fatty acids and amino acids—enter blood, and they serve as alternative energy sources for cells other than those of the brain (Section 8.6).

With injury, illness, or anxiety, the nervous system overrides the feedback loop, so the level of cortisol in blood can soar. In the short term, this is an adaptive response. It helps the body get enough glucose to the brain when food supplies are likely to be low. Cortisol also helps keep inflammatory responses in check.

As the next section explains, *long-term* stress causes problems. A chronically high level of cortisol disrupts production and secretion of other hormones. It also suppresses the immune system.

LOCAL FEEDBACK AND THE ADRENAL MEDULLA

The **adrenal medulla**, the inner region of the adrenal gland, also houses neurons. They secrete epinephrine and norepinephrine, which are neurotransmitters in some contexts and hormones in others. Signals reach them by sympathetic neurons from the hypothalamus. Suppose the signals trigger norepinephrine secretion. Molecules of this hormone accumulate in the synaptic cleft between target cells in the adrenal medulla and axon endings from the nerve. When there is too much norepinephrine, a local negative feedback mechanism kicks in. Norepinephrine binds to receptors for it on the axon endings and inhibits its further release.

Excitement or stress triggers the *fight-flight response* (Section 34.8). Epinephrine and norepinephrine make the heart beat faster, which increases blood flow rates, and fat and carbohydrate metabolism. They widen or narrow the diameter of arterioles in different regions. More of the total blood volume—and oxygen—reaches heart and muscle cells and sustains their demands for ATP production through aerobic respiration.

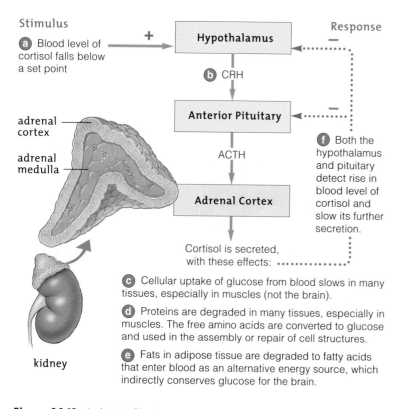

Stimulus

a Blood level of cortisol falls below a set point

Hypothalamus

Response

b CRH

Anterior Pituitary

ACTH

Adrenal Cortex

f Both the hypothalamus and pituitary detect rise in blood level of cortisol and slow its further secretion.

adrenal cortex

adrenal medulla

Cortisol is secreted, with these effects:

c Cellular uptake of glucose from blood slows in many tissues, especially in muscles (not the brain).

d Proteins are degraded in many tissues, especially in muscles. The free amino acids are converted to glucose and used in the assembly or repair of cell structures.

e Fats in adipose tissue are degraded to fatty acids that enter blood as an alternative energy source, which indirectly conserves glucose for the brain.

kidney

Figure 36.12 *Animated!* Structure of the human adrenal gland. An adrenal gland rests on top of each kidney. The diagram shows a negative feedback loop that governs cortisol secretion.

Negative feedback loops to the hypothalamus and pituitary control the secretion of cortisol from the adrenal cortex, but signals from the nervous system can override the loop.

The nervous system, together with a local negative feedback mechanism, governs secretions from the adrenal medulla.

36.9 Hormones, Stress, and the Brain

We can expect that the stress response to an immediate threat has adaptive value, because the genetic basis for its neural and hormonal components has persisted through time. Yet here again we must ask: Adaptation to what? What happens when stress provocations do not let up?

Each summer, a troop of olive baboons (*Papio anubis*) on East Africa's Serengeti plains gets visitors. For more than twenty years, neurobiologist Robert Sapolsky and Kenyan researchers have been recording how the animals interact and how their social position affects health. Their approach of combining physiological measurements with long-term observations yielded insights into the effects of chronic social stress.

Remember, when the body is stressed, commands from the nervous system trigger secretion of cortisol, epinephrine, and norepinephrine. As these secretions find their targets, they deal with the immediate threat by diverting resources from the long-term tasks of growing and surviving. This *stress response* is highly adaptive for short bursts of activity, as when it allows an animal to flee and escape from a predator.

Sometimes stress does not end. The baboons live in big troops with a clearly defined dominance hierarchy. Those on top get first access to food, grooming, and sexual partners. Those at the bottom must relinquish food, grooming, and sex to a higher ranking baboon or face attack (Figure 36.13). No surprise: Low-ranking baboons have elevated cortisol levels.

Physiological responses to chronic stress interfere with growth, the immune system, sexual function, and cardiovascular function. In monkeys, chronically high cortisol levels also can damage or destroy cells in the hippocampus, a brain structure that is central to memory and learning.

We see the impact of long-term elevated cortisol levels in humans affected by *Cushing's syndrome*, or hypercortisolism. This rare metabolic disorder might be triggered by an adrenal gland tumor, oversecretion of ACTH by the anterior pituitary, or ongoing use of the drug cortisone. Doctors often prescribe cortisone to relieve chronic pain, inflammation, or other health problems. The body converts it to cortisol.

The symptoms of hypercortisolism include a puffy, rounded "moon face" and fat deposition, especially around the torso. Blood pressure and the blood level of glucose become abnormally high. White blood cell counts are low; patients are more prone to infections. Thinning skin, losses in bone density, and weakened muscles are common. Wounds may be slow to heal. Women's menstrual cycles are erratic or nonexistent. Men often become impotent.

Figure 36.13 A dominant baboon (*right*) raising the stress level of a less dominant member of its troop.

As you might expect from Sapolsky's research, an impaired memory is one of the frequent complications of Cushing's syndrome. Magnetic resonance imaging (MRI) studies show that the hippocampus becomes reduced in many cases. Patients who have the highest cortisol levels also have the greatest reduction in the volume of the hippocampus, and the worst memory.

We also know that combat soldiers as well as air traffic controllers have elevated cortisol levels. Studies of other stressed individuals indicate that repetitive stress may destroy the body's capacity to mediate the stress response. The response mechanism just seems to burn out. Cortisol output no longer falls after a threat ends, but neither can it rise to meet a new threat.

Something else to think about: People who are low in a socioeconomic hierarchy tend to have more health problems—obesity, hypertension, and diabetes—than those who are higher up. The differences persist even after researchers factor out the obvious causes, such as variations in diet and access to health care. Are high cortisol levels and low status the missing link between poverty and poor health? Maybe.

Social connections seem to moderate the effects of stress. For humans, the impact of chronic stress can be mediated by friends, family, and counselors.

The stress response evolved as a physiological means to deal with short-term stress. When stress continues, the response can have damaging effects on the body.

What is the take-home lesson? Long-term adaptation to certain conditions in the environment may not be so adaptive in other contexts.

36.10 Hormones and Reproductive Behavior

LINKS TO
SECTIONS
25.11, 34.11, 35.7

An individual's growth, development, and reproduction start with genes and hormones, and so does behavior. Environmental factors often influence the timing and rates of secretion for hormones that help orchestrate these aspects of the life cycle.

THE GONADS

Gonadal hormones govern reproductive function by way of homeostatic feedback loops. The gonads are primary reproductive organs. In vertebrates, gonads in males are testes (singular, testis). In females, they are called ovaries. These organs produce gametes and also secrete sex hormones—estrogens, progesterone, and androgens such as testosterone.

Puberty is a postembryonic stage of development when the reproductive organs and structures mature. During puberty, a female mammal's ovaries step up estrogen production, which causes breasts and other female secondary sexual traits to develop. Estrogens and progesterone control egg formation and ready the uterus for pregnancy. In males, a rise in testosterone output triggers the onset of sperm formation and the development of secondary sexual traits.

Females, too, make a small amount of testosterone, which influences *libido*, or a desire for sex. Removal of ovaries, as during cancer treatment, decreases female libido. A testosterone patch is currently being tested in women who have undergone such treatment.

Remember, sperm cannot mature without the small amount of estrogen that male mammals produce. We will return to the topic of sex hormones and controls over their secretion in Chapter 44.

THE PINEAL GLAND

Deep within the vertebrate brain is a small **pineal gland** (Figure 36.2). It secretes melatonin, an amine hormone that is part of an internal timing mechanism, or **biological clock**. Melatonin secretion decreases in response to light. Remember, signals from the retina flow through optic nerves to the brain in response to light (Section 35.7). The amount of light varies from day to night, and it varies with the seasons. As you might expect, so does melatonin production.

Variations in melatonin level affect the functions of gonads in many species. Think of a male songbird in winter, when nights are long. Elevated blood levels of melatonin indirectly suppress his sexual activity. In the spring, there are fewer hours of darkness, and so melatonin levels decline in blood. With the lifting of hormonal brakes, the male's gonads secrete hormones that promote singing and other behaviors associated with territoriality and courtship (Figure 36.14).

Does melatonin affect human gonads? Possibly. A drop in the blood level of melatonin correlates with the onset of puberty and might trigger it. Also, a few disorders that increase melatonin production delay puberty. Others that cause abnormally low melatonin levels hasten it. Documented decreases in the age at puberty might be caused in part by an increase in the evening hours children spend in front of a television screen or computer monitor, both of which emit light.

Melatonin also influences neurons that can lower body temperature and make us drowsy in low light. Just after sunrise, less melatonin is secreted, so body temperature rises and we wake up. A biological clock responsive to light regulates a cycle of sleeping and arousal. Travelers are often advised to spend time in the sun to reset this clock and minimize jet lag.

What about people who sit at their computers late into the night and have trouble falling asleep? Staring at a bright monitor may lower melatonin levels and disrupt normal sleep cycles.

In winter, *seasonal affective disorder* (SAD) hits some people, who become exceedingly depressed, binge on carbohydrates, and crave sleep. These "winter blues" might develop when a biological clock is out of sync with the seasonal decline in daylight hours. Clinically administered melatonin can make symptoms worse. Exposure to artificial light, which shuts down pineal activity, can cause dramatic improvement.

Figure 36.14 Male white-throated sparrow belting out a song that began, indirectly, with an environmentally induced decrease in melatonin secretion from the pineal gland.

Gonads make hormones with roles in sexual reproduction and in the development of secondary sexual traits. Testes produce androgens, including testosterone. Ovaries produce progesterone and estrogens.

Melatonin from the pineal gland may affect the onset of puberty. It controls a biological clock, and its production is inhibited by time spent in a well-lit environment.

36.11 Comparative Look at a Few Invertebrates

This chapter focused on vertebrates, but most animals produce hormones of one sort or another. How did this diverse array of signaling molecules arise?

LINKS TO SECTIONS
25.9, 25.10, 25.12

EVOLUTION OF RECEPTOR DIVERSITY

How did vertebrates get so many diverse hormones and hormone receptors? Molecular evidence points to gene duplications and subsequent divergences by way of mutations. Genetic analysis reveals the invertebrate beginnings of some hormone receptors. For example, sea anemones have receptors at the plasma membrane that are structurally similar to vertebrate receptors for TSH, LH, FSH, and other signaling molecules. Also, the genes that encode them have similar nucleotide sequences in both kinds of animals, and they have the same number and type of introns in similar regions. Presumably, the gene for this receptor protein arose millions of years ago in a common ancestor.

CONTROL OF MOLTING

Do not assume that because invertebrate hormones evolved first, they are somehow primitive in structure and function. To get a sense of what they do, think back on **molting**, the periodic shedding of a too-small body covering (or hair, feathers, or horns) during the life cycle. In arthropods, remember, a soft new cuticle forms beneath an old one, which is then shed (Section 25.11). Before the new cuticle hardens, the body mass increases by the rapid uptake of air or water and by continuous mitotic cell divisions. Details vary among groups, but in all cases, molting is largely under the control of **ecdysone**, a steroid hormone.

An arthropod molting gland produces and stores ecdysone and releases it for distribution through the body at molting time. Hormone-secreting neurons in the brain seem to control its release. They respond to a combination of internal signals and environmental cues, including light and temperature.

Figure 36.15 is an example of the control steps in crabs and other crustaceans. Right before and during an episode of molting, coordinated interactions among ecdysone and other hormones bring about structural and physiological changes. The interactions make the old cuticle separate from the epidermis and muscles. They induce changes that dissolve inner layers of the cuticle and recycle the remnants. These interactions trigger changes in metabolism and in the composition of the internal environment. They promote rapid cell divisions, secretions, and pigment formation that help make a new cuticle. At the same time, the hormonal

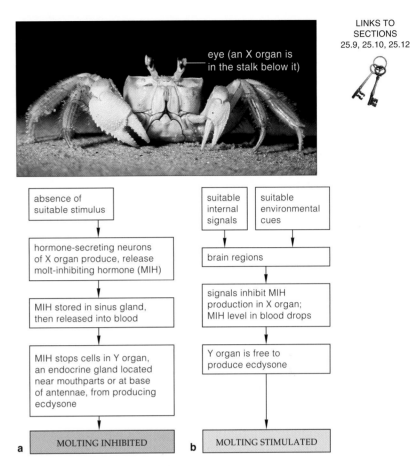

eye (an X organ is in the stalk below it)

Figure 36.15 Steps in hormonal control of molting in crabs and other crustaceans. Ecdysone is a hormone that stimulates molting. (**a**) In the absence of environmental cues for molting, an X organ in each crab eye stalk produces a hormone that inhibits ecdysone synthesis. (**b**) Right before and during molts, signals from the brain turn off cell activities in the X organ, and so ecdysone is produced and secreted.

interactions govern heart rate, muscle action, changes in body coloration, and other processes.

The steps differ a bit in insects, which do not have a molt-inhibiting hormone. Rather, stimulation of the insect brain sets in motion a cascade of signals that trigger the production of molt-inducing ecdysone.

Chemicals that mimic or interfere with ecdysone and juvenile hormone are often used as insecticides. Also, nematodes that parasitize plants are controlled with the use of ecdysone inhibitors.

Gene duplications and divergences have produced a diverse array of animal hormones and receptors.

Ecdysone, a steroid hormone, controls the molting process in nematodes and arthropods, such as insects. Ecdysone secretion is influenced by environmental cues.

36.12 The Hormone Connection

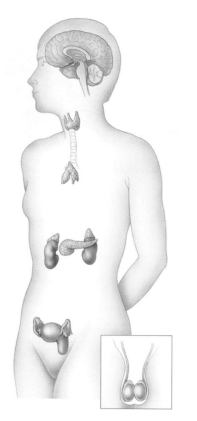

If we liken hormones to violins, harps, and other musical instruments, then the whole body is attuned to a sweeping symphony, and the hypothalamus and pituitary gland are its premier conductors.

Take a moment to reflect on Tables 36.3 and 36.4. This chapter introduced many of the hormones listed. Some other hormones are included for reference purposes, and still others are topics of later chapters.

The point is not to memorize this information but rather to get a sense of what vertebrate hormones do. No tissue or organ is beyond their reach. The cells in most tissues have receptors for different hormones that may compete for or reinforce a cellular response. For instance, every skeletal muscle cell has receptors for glucagon, insulin, cortisol, epinephrine, estrogen, testosterone, somatotropin, somatostatin, and thyroid hormone, as well as others. What happens at a given moment in each muscle cell depends in part on blood levels of these hormones, on the interactions among them, and on the effect of other signaling molecules coursing through the body.

Remember, hormones often bind to receptors in many different tissues. Receptors may vary in their

Table 36.3 Primary Actions of Hypothalamic and Pituitary Hormones

Pituitary Lobe	Secretions	Designation	Main Targets	Primary Actions
Posterior Nervous tissue (extension of hypothalamus)	Antidiuretic hormone (or vasopressin)	ADH	Kidneys	Induces water conservation as required during control of extracellular fluid volume and solute concentrations
	Oxytocin	OCT	Mammary glands	Induces milk movement into secretory ducts
			Uterus	Induces uterine contractions during childbirth
Anterior Glandular tissue, mostly	Adrenocorticotropin	ACTH	Adrenal cortex	Stimulates release of cortisol, an adrenal steroid hormone
	Thyrotropin	TSH	Thyroid gland	Stimulates release of thyroid hormones
	Follicle-stimulating hormone	FSH	Ovaries, testes	In females, stimulates estrogen secretion, egg maturation; in males, helps stimulate sperm formation
	Luteinizing hormone	LH	Ovaries, testes	In females, stimulates progesterone secretion, ovulation, corpus luteum formation; in males, stimulates testosterone secretion, sperm release
	Prolactin	PRL	Mammary glands	Stimulates and sustains milk production
	Somatotropin (or growth hormone)	STH (GH)	Most cells	Promotes growth in young; induces protein synthesis, cell division; roles in glucose, protein metabolism in adults
Intermediate* Glandular tissue, mostly	Melanocyte-stimulating hormone	MSH	Pigmented cells in skin and other integuments	Induces color changes in response to external stimuli; affects some behaviors

* An intermediate pituitary lobe is present in most vertebrates, but not in humans. MSH is associated with the human anterior lobe.

structure, so binding may call up different responses. For example, in Chapter 42, you will see how ADH (antidiuretic hormone) secreted by the posterior lobe of the pituitary acts on kidney cells in ways that help maintain homeostasis in the internal environment. ADH is sometimes called vasopressin, because it also binds to receptors in the wall of blood vessels and causes their diameter to narrow. In many mammals (not humans), it helps control blood pressure. ADH even acts on brain cells in certain mammals to alter sexual and social behavior, as Section 49.1 explains. Such diversity in the responses to a single hormone is an outcome of variations in ADH receptors. After the hormone binds to it, each kind of receptor summons a different cellular response.

> *During any interval, vertebrate cells are exposed to and are selectively responding to a complex mix of hormones, the secretions of which are under endocrine and neural control.*

Table 36.4 Sources and Primary Actions of Additional Vertebrate Hormones

Source	Examples of Secretion(s)	Main Targets	Primary Actions
Adrenal cortex	Glucocorticoids (including cortisol)	Most cells	Promote breakdown of glycogen, fats, and proteins as energy sources; thus help raise blood level of glucose
	Mineralocorticoids (including aldosterone)	Kidney	Promote sodium reabsorption (sodium conservation); help control the body's salt–water balance
Adrenal medulla	Epinephrine (adrenaline)	Liver, muscle, adipose tissue	Raises blood level of sugar, fatty acids; increases heart rate and force of contraction
	Norepinephrine	Smooth muscle of blood vessels	Promotes constriction or dilation of certain blood vessels; thus helps control the flow of blood volume to different body regions
Thyroid	Triiodothyronine, thyroxine	Most cells	Regulate metabolism; have roles in growth, development
	Calcitonin	Bone	Lowers calcium level in blood
Parathyroids	Parathyroid hormone	Bone, kidney	Elevates calcium level in blood
Gonads			
Testes (in males)	Androgens (including testosterone)	General	Required in sperm formation, development of genitals, maintenance of sexual traits, growth, and development
Ovaries (in females)	Estrogens	General	Required for egg maturation and release; preparation of uterine lining for pregnancy and its maintenance in pregnancy; genital development; maintenance of sexual traits; growth, development
	Progesterone	Uterus, breasts	Prepares, maintains uterine lining for pregnancy; stimulates development of breast tissues
Pancreatic Islets	Insulin	Liver, muscle, adipose tissue	Promotes cell uptake of glucose; thus lowers glucose level in blood
	Glucagon	Liver	Promotes glycogen breakdown; raises glucose level in blood
	Somatostatin	Insulin-secreting cells	Inhibits digestion of nutrients, hence their absorption from gut
Thymus	Thymopoietin, thymosin	T lymphocytes	Poorly understood regulatory effect on T lymphocytes
Pineal	Melatonin	Gonads (indirectly)	Influences daily biorhythms, seasonal sexual activity
Stomach, Small Intestine	Gastrin, secretin, etc.	Stomach, pancreas, gallbladder	Stimulate activities of stomach, pancreas, liver, gallbladder; required for food digestion, absorption
Liver	Somatomedins	Most cells	Stimulate cell growth and development
Kidneys	Erythropoietin	Bone marrow	Stimulates red blood cell production
	Angiotensin*	Adrenal cortex, arterioles	Helps control secretion of aldosterone (hence sodium reabsorption, and blood pressure)
	1,25-hydroxyvitamin D_6* (calcitriol)	Bone, gut	Enhances calcium reabsorption from bone and calcium absorption from gut
Heart	Atrial natriuretic hormone	Kidney, blood vessels	Increases sodium excretion; lowers blood pressure

* Kidneys produce *enzymes* that modify precursors of this substance, which enters blood as an activated hormone.

Summary

Section 36.1 Hormones, neurotransmitters, local signaling molecules, and pheromones are signaling molecules. They are chemical secretions from one cell type that adjust the behavior of other, target cells. Any cell is a target if it has receptors for a signaling molecule at its plasma membrane or in the cytoplasm or nucleus.

All vertebrates have an organ system of endocrine glands and cells. In most cases, the hormonal secretions travel through the bloodstream to nonadjacent targets.

Section 36.2 Steroid hormones are lipid soluble and derived from cholesterol. Some kinds enter a target cell and interact directly with DNA. Others bind to the cell's plasma membrane and alter the membrane properties. Amines also enter and act inside their target cells.

The peptide and protein hormones bind to plasma membrane receptors. Binding may lead to the formation of a second messenger, such as cAMP, that relays the signal into the cytoplasm. There, the transduced signal causes a cascade of enzyme activations.

Biology⊗Now
Compare the mechanisms of steroid and protein hormone action by viewing the animation on BiologyNow.

Section 36.3 The hypothalamus, a forebrain region, is structurally and functionally linked with the pituitary gland. They are a major center for homeostatic control.

Some hypothalamic neurons make ADH or oxytocin, two hormones that the posterior pituitary gland secretes. ADH acts in kidneys. Oxytocin acts on the uterus and milk ducts. Other hypothalamic neurons make six releasers and inhibitors that have anterior pituitary targets. These releaser and inhibitor hormones control the secretion of ACTH, TSH, FSH, LH, PRL, and STH.

ACTH acts on the adrenal cortex, TSH on the thyroid, FSH and LH on male and female gonads, and PRL on mammary glands and the uterus. STH (somatotropin, or growth hormone) has growth-promoting effects on cells in tissues throughout the body.

Biology⊗Now
Use the animation on BiologyNow to study how the hypothalamus and pituitary interact.

Section 36.4 A negative feedback loop to the anterior pituitary gland and hypothalamus governs thyroid hormone secretion. Iodine deficiency as well as exposure to certain chemicals in the environment may disrupt the function of the thyroid gland.

The parathyroid glands are the main regulators of calcium levels in the blood. They release parathyroid hormone (PTH) in response to low calcium levels. PTH acts on bone cells and kidney cells in ways that raise calcium levels in the blood.

Section 36.5 Thyroid hormone plays a role in amphibian metamorphosis. Chemical pollutants that mimic or interfere with thyroid hormone action may be contributing to a rise in amphibian deformities.

Sections 36.6, 36.7 Two pancreatic hormones, insulin and glucagon, are central to organic metabolism. They are secreted in response to shifts in the blood level of glucose. Pancreatic beta cells secrete insulin when the glucose level is high. It stimulates glucose uptake by muscle and liver cells, which lowers the blood level. Pancreatic alpha cells secrete glucagon, which stimulates the release of glucose, when the blood level is too low.

Biology⊗Now
Use the animation on BiologyNow to see how the actions of insulin and glucagon regulate blood sugar.

Sections 36.8, 36.9 Cortisol secretion by the adrenal gland is governed by a negative feedback loop to the anterior pituitary gland and hypothalamus. In times of stress, the central nervous system can override the feedback controls so that cortisol levels rise.

Negative feedback governs secretion of epinephrine and norepinephrine by the adrenal medulla.

Biology⊗Now
Watch the animation on BiologyNow to see how cortisol levels are maintained by negative feedback.

Section 36.10 Environmental cues influence the secretion of some hormones, such as the sex hormones: estrogens, progesterone, and androgens (including testosterone). Sex hormones control gamete formation and the development of secondary sexual traits.

Melatonin secretion by the vertebrate pineal gland is part of a biological clock, a type of internal timing mechanism. In humans, it affects the onset of puberty and the daily sleep/wake cycle. Increased daylight hours, as in spring, suppresses melatonin production.

Section 36.11 Some vertebrate hormone receptors may have evolved from similar receptor proteins in invertebrates. In arthropods and roundworms, molting is controlled by ecdysone, a steroid hormone that is secreted in response to environmental cues.

Section 36.12 The big picture is this: All cells in the vertebrate body are bathed in an array of hormones and selectively respond to them according to the types of receptors in different tissues, competing or enhancing hormone interactions, and other factors.

Self-Quiz
Answers in Appendix II

1. _____ are signaling molecules released from one type of cell that can alter target cell activities.
    a. Hormones   d. Local signaling molecules
    b. Neurotransmitters   e. both a and b
    c. Pheromones   f. a through d

2. ADH and oxytocin are hormones produced in the hypothalamus but distributed from the _____ .
    a. anterior lobe of pituitary   c. pancreas
    b. posterior lobe of pituitary   d. pineal gland

3. Overproduction of _____ causes gigantism.
    a. somatotropin   c. insulin
    b. ADH   d. melatonin

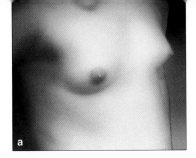

Figure 36.16 Two examples of hormone effects in humans. (**a**) At *twenty-three months*, this Puerto Rican girl already shows breast development. Industrial chemicals that mimic estrogen may be a factor. (**b**) Researcher Hiralal Maheshwari, with two men who have a heritable form of dwarfism.

Figure 36.17 A blue crab that has just molted its old shell. For twelve hours or so, it will be a soft-shelled crab, considered a delicacy by many seafood lovers.

4. Which do not stimulate hormone secretions?
   a. neural signals   d. environmental cues
   b. local chemical changes e. All of the above can
   c. hormonal signals   stimulate secretion.

5. _____ lowers blood sugar levels; _____ raises it.
   a. Glucagon; insulin   c. Gastrin; insulin
   b. Insulin; glucagon   d. Gastrin; glucagon

6. The pituitary detects a rising hormone concentration in blood and inhibits the gland secreting the hormone. This is a _____ feedback loop.
   a. positive   c. long-term
   b. negative   d. b and c

7. The _____ has endocrine and exocrine functions.
   a. hypothalamus   c. pineal gland
   b. pancreas   d. parathyroid gland

8. Match the hormone source listed at left with the most suitable description at right.
   ____ adrenal medulla   a. affected by daylength
   ____ thyroid gland   b. potent local effects
   ____ parathyroid   c. raises blood calcium
        glands   level
   ____ pancreatic islets   d. epinephrine source
   ____ pineal gland   e. insulin, glucagon
   ____ prostaglandin   f. hormones require iodide

Additional questions are available on Biology ⑤ Now™

Critical Thinking

1. In the late 1990s, Marcia Herman-Giddens asked this question of pediatricians throughout the United States: At what age are female patients developing secondary sexual traits? The compiled answers showed that girls are developing breasts six months to one year earlier than they did in earlier decades.

Herman-Giddens's data were criticized because the girls were not a random sample; they had appointments to see pediatricians. Nevertheless, many researchers suspect that her findings reflect a real trend. What might be the cause of this *precocious breast development*?

Evidence that chemicals may play a role comes from the island of Puerto Rico, which has the world's highest incidence of premature breast development. A recent study found that most affected girls started to develop breasts

when they were between six and twenty-four months old (Figure 36.16*a*). Sixty-eight percent had high blood levels of phthalates—chemicals used in the manufacture of some plastics and pesticides. By comparison, none of the normal girls that researchers examined had high phthalate levels.

What type of information might be used to test the hypothesis that exposure to phthalates may be contributing to early breast development among young girls in the mainland United States?

2. Many abnormally small but normally proportioned people live in a remote village in Pakistan. On average, the men are about 130 centimeters (a little over 4 feet) tall, and the women about 115 centimeters (3 feet, 6 inches) tall. Northwestern University researchers interviewed the people. They found out that all were part of an extended family. The abnormality, a type of *dwarfism*, is one case of autosomal recessive inheritance (Figure 36.16*b*).

Biochemical and genetic analysis show that affected individuals have a low level of somatotropin but a normal gene that codes for this growth hormone. The abnormality starts with a mutant gene that specifies a receptor for one of the hypothalamic releasing hormones. That receptor is on cells in the anterior lobe of the pituitary gland. Explain how a mutant receptor could cause the abnormality.

3. The blue crab (*Callinectes sapidus*) is found in estuaries along the Gulf and Atlantic coasts of the United States. Like other arthropods, these crabs molt as they grow (Figure 36.17). After molting, it takes about twelve hours for the new shell to harden. During this interval, the crab is vulnerable to natural predators and to human crabbers who can sell it as an expensive "soft-shelled" crab. Blue crab populations have declined in some areas. Chemical pollutants that bind to and block ecdysone receptors may be a factor. How might a chemical that affects ecdysone action interfere with the crab's life cycle?

4. Tuberculosis and other infectious diseases can damage the adrenal glands, slowing or halting cortisol secretion. The result is *Addison's disease*. In developed countries, this hormonal disorder most commonly arises as a result of autoimmune attacks on the adrenal gland. President John F. Kennedy showed this form of the disorder. Emotional or physical stress can push an affected person into a crisis state in which many organ systems spin out of control. Explain why a lack of cortisol would have this effect.

Pumping Up Muscles

Want to be more muscular and stronger, with a lot more endurance? Just use our pills or powders and be like the guy in Figure 37.1. That is the message in advertisements for many dietary supplements that target body builders and other athletes. The supplements are easily purchased from health food stores and through the Internet. They have not been classified as drugs by the Food and Drug Administration (FDA), so testing for their effectiveness and long-term side effects has been negligible. Independent monitoring of quality control over their commercial production ranges from little to none.

Consider androstenedione, or "andro." Sales of this drug got a big boost in 1998, after Mark McGwire admitted that he used it during his successful attempt to break Major League Baseball's home-run record. Androstenedione forms naturally in the body as an intermediate in the synthesis of testosterone. Testosterone *does* have tissue-building, anabolic effects that are well documented. Andro is said to raise the blood level of testosterone, which in turn boosts the rate of protein synthesis in muscles.

Does andro work? Probably not. In controlled studies, males of an experimental group used an andro supplement. They did not gain any more muscle mass or strength than males of a control group who were given a placebo. At most,

the andro supplement raised the testosterone level in blood for a few hours.

Andro also forms as an intermediate during the synthesis of estrogen. This sex hormone has feminizing effects. Its known side effects on males include shrunken testicles, the development of female-like breasts, and hair loss. The users in both sexes risk liver damage, acne, and a lower blood level of "good" cholesterol (HDL). Females commonly develop masculinized patterns of hair growth and speech, and they can expect menstrual cycle disruptions.

In early 2004, the FDA issued an advisory that andro supplements have serious side effects. Companies were ordered to stop distributing the drug immediately.

Creatine is another supplement that is being touted as a performance enhancer. Creatine is only a short chain of amino acids. The body normally makes some creatine and takes in more from food. When muscles are called upon to contract hard and fast, they use phosphorylated creatine as an instant energy source.

Unlike andro, creatine supplements might work. In several controlled studies, they improved performance during brief, high-intensity exercise. Clinical trials are under way to determine whether creatine may benefit individuals affected by muscular dystrophy and other

Watch the video online!

Figure 37.1 Overabundance of contractile tissue, of the sort shown on the facing page.

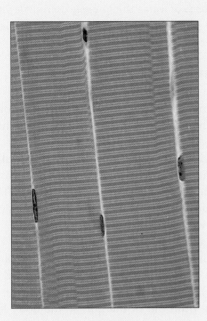

muscle disorders. Nevertheless, excessive creatine intake does put a strain on the kidneys, and it is too soon to know whether it has long-term side effects. No regulatory agency checks to see how much creatine is actually present in any commercial product.

With this chapter, we turn to the point of having muscles in the first place. To most of us, the directional movement of the body or portions of it is one of the hallmarks of nearly all animals. Regardless of the species, movement requires contractile cells and some enclosed fluid or skeletal element against which the contractile force can be applied. The structure of skeletal and muscular systems, and how they work, has an evolutionary history—one that can help you evaluate how far both systems can be pushed in the pursuit of enhanced performance.

How Would You Vote?

Dietary supplements are largely unregulated. Should they be placed under the jurisdiction of the Food and Drug Administration, which could subject them to more stringent testing for effectiveness and safety? See BiologyNow for details, then vote online.

Key Concepts

HOW ANIMALS MOVE

Animals apply contractile force against a hydrostatic skeleton, exoskeleton, or endoskeleton. Muscle cells evolved from epithelial cells in which contractile filaments became organized in functional arrays.   Section 37.1

DIVERSE INVERTEBRATE SKELETONS

In hydrostatic skeletons, a volume of fluid confined in a body chamber accepts the force of contraction. Muscles and often hydraulic pressure move exoskeletons. Endoskeletons are internal skeletons partnered with muscles.   Section 37.2

VERTEBRATE SKELETONS

Bones are collagen-rich, mineralized organs that function in movement, protection and support of soft organs, and mineral storage. Blood cells form in some bones. Cartilage or ligaments connect bones at joints. Tendons attach skeletal muscles to bones.   Sections 37.3, 37.4

THE MUSCLE—BONE PARTNERSHIP

Skeletal muscles are bundles of muscle fibers that interact with bones and with one another. Some cause movements by working as pairs or groups. Others oppose or reverse the action of a partner muscle.   Section 37.5

HOW SKELETAL MUSCLE CONTRACTS

Inside a skeletal muscle, many myofibrils are transversely divided into sarcomeres, the basic unit of contraction. ATP energy forces parallel arrays of actin and myosin filaments in each sarcomere to interact. The interactions shorten the sarcomeres, which collectively accounts for contraction. Sections 37.6–37.8

VARIATIONS IN MUSCLE FUNCTION

Cross-bridges in sarcomeres collectively exert tension. A whole muscle shortens only when this mechanical force exceeds other, opposing forces. Exercise enhances the properties of whole muscles, and aging diminishes them. Sections 37.9, 37.10

Links to Earlier Concepts

In this chapter you will return to the contractile proteins actin and myosin (Sections 4.10, 4.11, 5.4). You will look once more at some skeletons of invertebrates (17.7, 25.4, 25.17) and vertebrates (26.2, 26.11–26.13), and at the fine structure of bone tissue and muscle tissues (33.2, 33.3). You will draw on your knowledge of active transport (5.4), ATP function (6.2), and pathways of organic metabolism (8.1, 8.6). You also will take a closer look at hormonal control of the body's calcium levels (36.4) and at how ACh triggers contraction (34.4).

37.1 An Evolutionary Heritage

37.2 Invertebrate Skeletons

LINKS TO
SECTIONS 4.9,
17.7, 25.6, 25.17, 34.1

All animals make directional movements in response to stimuli. Even a mussel that lives out its adult life glued to a rock moves some of its structures. Animal larvae use cilia or flagella as motile structures. Larvae also change direction through interactions between their contractile cells and skeletal elements.

Where did animals get the capacity to move and bend the body in different directions? It started in epithelia of ancient invertebrates. Remember actin and myosin, the contractile proteins that are part of cytoskeletons (Section 4.9)? They became organized in longitudinal, circular, and diagonal arrays at the base of epithelial cells that formed the body's outer tissue layer. Being anchored to the rest of the body, the cells could change its shape by contracting and relaxing. They also could bend the body one way or another by contracting on one side at a time. Such "epitheliomuscular" cells are still present in cnidarians. Section 34.1 has a sketch of a few that are part of the body wall.

In other lineages, epitheliomuscular cells sank a bit below the free epithelial surface. Cells like this still occur in sweat glands, mammary glands, and the eye's iris. In many animals, however, these contractile cells sank completely into connective tissue. They evolved into muscle cells and became bundled in connective tissue inside muscles. Muscles are structural units that function exclusively in contraction.

Regardless of how they are organized in animals, contractile cells must interact with parts of a skeleton. A skeleton is a structural framework that functions in maintaining body shape, supporting and protecting cells, and accepting the force of contraction that can bring about movements. Three types are common.

With a **hydrostatic skeleton**, muscle cells apply the force of contraction against a body fluid and thereby redistribute it within a confined space. Try squeezing the middle of a long, skinny, water-filled balloon and you can see that fluid can offer considerable resistance to compression. With an **exoskeleton**, rigid or flexible structures at the body surface accept the applied force of contraction. An insect cuticle is one example. With an **endoskeleton**, *internal* body parts, such as bones, receive the applied force of muscle contraction.

> *The capacity of animals to move and bend in specific directions started with epithelial cells in which organized arrays of actin and myosin filaments evolved. Contractile cells of different animal groups are at different levels of organization. They direct their contractile force against a hydrostatic skeleton, exoskeleton, or endoskeleton.*

Different kinds of skeletons show up among the 1.3 million or so named species of invertebrates. Some species apply contractile force to a skeleton that is no more than a confined body of water. Others apply contractile force to an exoskeleton or endoskeleton, either of which may be rubbery or rigid.

HYDROSTATIC SKELETONS

Sea anemones, worms, and many other soft-bodied animals have a hydrostatic skeleton and a tubular or cylindrical body. Many contractile cells are oriented side by side, longitudinally, in the body wall. Others are oriented like rings around the body cavity. Stiff fibers often form a mesh in the body wall. They help prevent uncontrollable bulges when contraction makes fluid move inside their gut cavity.

Think about a sea anemone (Figure 37.2). When the longitudinal cells in its body wall contract and radial ones relax (lengthen), its body collapses and is squat. When the longitudinal cells relax, radial ones contract and force fluid out of the gut, so the body lengthens into an upright feeding position. A nerve net controls the directional movements (Section 34.1).

An earthworm, recall, has a hydrostatic skeleton. It applies contractile force against fluid-filled coelomic chambers of its highly segmented body. Muscles in

resting position, which typically is assumed at low tide when currents are not delivering food — feeding position

Figure 37.2 Directional movement of a sea anemone, which has a hydrostatic skeleton. Epitheliomuscular cells in this cnidarian's body wall run longitudinally to the main body axis and radially around the gut. *Left*, radial epitheliomuscular cells are relaxed; longitudinal ones are contracted. *Right*, the radial cells have contracted and longitudinal ones are relaxed. The body extends upward, to its feeding position.

Fluid confined by the body wall accepts the contractile force.

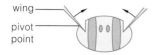

wing
pivot point

Dorsal-ventral muscles contract; exoskeleton pops out, wings move up.

Exoskeleton pops back, muscles relax, wings move down.

Figure 37.3 *Animated!* Motion of a fly wing. An insect cuticle forms a pliable hinge across gaps between body segments. Contracting muscles change the shape of the thorax, causing wings attached to it to move up and down.

Figure 37.4 Jumping spiders. A large muscle in their cephalothorax contracts and forces blood into the hind legs. The surge of high fluid pressure extends the legs outward. As you might predict, these spiders have enormous eyes relative to their body size, which comes in handy for looking before they leap.

each segment's body wall contract and relax one after another as many anchoring bristles (setae) alternately plunge into soil and pull away from it. Section 25.6 shows how the contractions are coordinated to bring about forward motion.

EXOSKELETONS

Lobsters, spiders, insects, and other arthropods have a hinged exoskeleton with attachment sites for sets of muscles that move hard parts like levers. Consider a winged insect's cuticle. It thins where it extends over gaps between body segments on both sides of each wing. Being pliable at certain gaps, the cuticle acts as a hinge when contracting muscles alter the angle of the wing's attachment site (Figure 37.3 and Section 17.7). Small contractions can bring about large movements.

Arthropods move by a combination of muscles and hydraulic pressure. *Hydraulic* refers to fluid pressure within a tube. For instance, all spiders have an open circulatory system; their heart pumps blood directly into body tissues. Muscles attached to an exoskeleton contract and pull the spider legs inward, but there are no opposing muscles to push them out again. Instead, a large muscle of the spider cephalothorax contracts, which causes blood to surge into the hind legs (Figure 37.4). It is a bit like squeezing a rubber glove partially filled with water to make the limp fingers become erect. The hydraulic pressure helps a jumping spider leap twenty-five times its own length.

DO ANY INVERTEBRATES HAVE ENDOSKELETONS?

Glasslike spicules enclosed in the two cellular linings of a sponge's body are sort of like an endoskeleton, not that sponges do much with it. Contractile cells of some species close pores in the body wall in response to stimuli. Some freshwater sponges can expel water, like a slow sneeze, by coordinated contractions.

Echinoderms have an endoskeleton located within their dermis. It consists of arrays of structural elements called ossicles, which are made of tiny calcite crystals. Ossicles are shaped like spines, rods, and plates, and they form a honeycombed framework that is strong yet light in weight. Epidermis covers the ossicles that project above the body surface and give the group its name. Echinoderm means spiny-skinned.

All echinoderms move by a combination of muscles and hydraulic pressure. Their water-vascular system includes fluid-filled bulbs and sucker-bottomed tube feet (Section 25.17). Contracting muscles squeeze the bulbs and force water into the feet, which extend and attach to substrates or prey. Each tube foot retracts as its longitudinal muscles contract and thereby squeeze fluid back into the bulb above it (Figure 37.5).

tube foot   ossicle

Figure 37.5 Ossicles of an echinoderm's endoskeleton

Cnidarians, worms, and many other invertebrates have a hydrostatic skeleton. Arthropods use muscles and hydraulic pressure to move their exoskeleton. Echinoderms have a honeycombed endoskeleton within their dermis.

37.3 Evolution of Vertebrate Skeletons

LINKS TO
SECTIONS
26.2, 26.11–26.13

From notochord to vertebral column, from supports for gills to jaws, from structural elements inside lobed fins to limbs—Section 26.2 introduced these evolutionary trends, which occurred among vertebrates that invaded land. Many other modifications helped those lineages make the transition to life in a new medium—air.

Your skeleton, like those of other vertebrates on land, holds evidence of a time when body weight became deprived of water's buoyancy. Remember, the pelvic and pectoral girdles of a fish function as a stable base for moving fins that propel, guide, and stabilize the body in water. Among early four-legged vertebrates, those girdles transferred the weight of the main body mass to limbs, and they developed more surface area to which muscles became attached. The limbs became repositioned closer to the body mass and helped hold the body above the ground, so that forward thrusting motions became easier. A cage of hard bones became connected to the backbone. It helped keep the heart, lungs, and other soft organs from collapsing under the weight of a body no longer supported by water.

Figure 37.6 shows the cartilaginous skeleton of a fish and the kinds of bony endoskeletons that evolved among the amphibians and, later, among reptiles and mammals on land. Many thousands of variations on the original skeletal plan show up in the fossil record.

For example, compare Figure 37.6 with 37.7, and you see some of the same structural elements in the human skeleton, which has 206 bones. First, notice its pectoral girdle (at the shoulders), pelvic girdle (at the hips), and the paired arms, hands, legs, and feet. This is the *appendicular* portion—a legacy from an ancient tetrapod. Notice the shoulder blades, which are easy to dislocate. Also notice the long, thin collarbones, which are the bones broken most often. These vulnerabilities of the human skeleton are one outcome of an ancient aquatic heritage, of being a "fish out of water."

What about the *axial* portion of the human skeleton? The jaws and other skull bones, twelve pairs of ribs, a breastbone, and twenty-six vertebrae are the legacies of early craniates and jawed vertebrates. The **vertebrae** (singular, vertebra), or bony segments of a backbone, extend from the base of the skull to the pelvic girdle.

What does the vertebral column do now? Its bony parts offer attachment sites for paired muscles and a protective canal for the spinal cord. The column itself transmits your torso's weight to the lower limbs of a two-legged walker (Sections 26.11 through 26.13).

Our hominid ancestors started walking on two legs at least 4 million years ago. Over time, the backbone had to curve into an S shape to keep the body's main axis in vertical alignment. It helps that its bones are

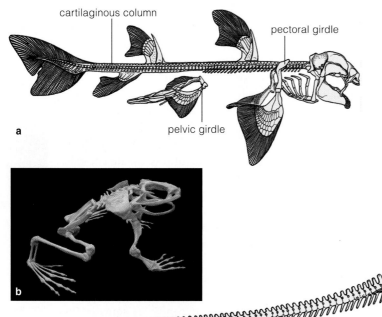

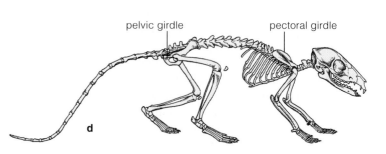

Figure 37.6 Comparison of skeletons of (**a**) a shark, (**b**) an amphibian, (**c**) a generalized early reptile, and (**d**) a generalized mammal. The sketches are not to the same scale. The skeleton in (**b**) belonged to the largest modern frog species, the goliath frog (*Conraua goliath*). The earliest known frog evolved during the late Jurassic, about 190 million years ago. Fossils indicate that the skeleton of this amphibian has not changed much since. Compare it with a bird skeleton (Section 26.9), whale skeleton (26.11), and primate skeletons (26.12–26.14).

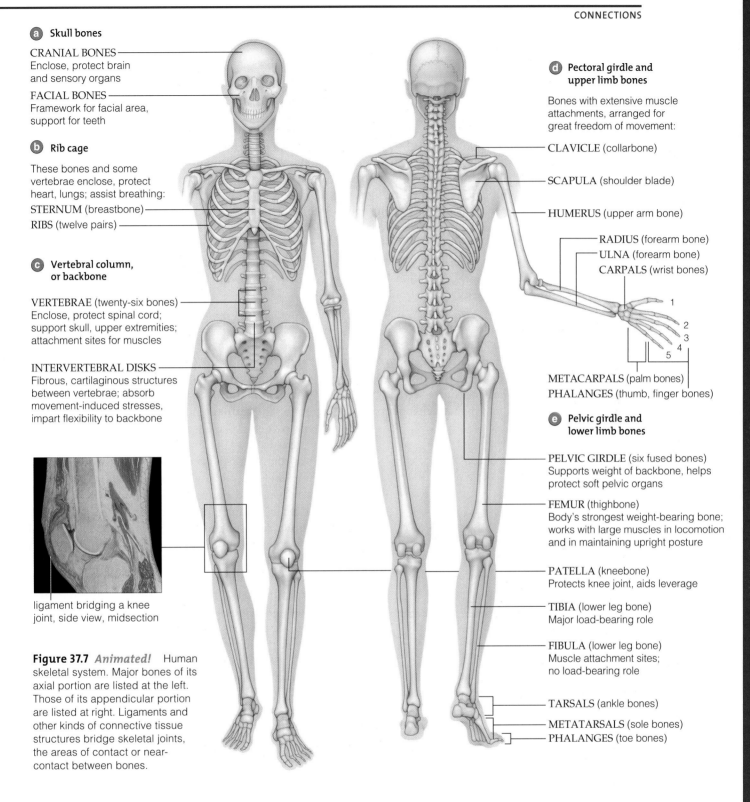

a Skull bones

CRANIAL BONES
Enclose, protect brain
and sensory organs

FACIAL BONES
Framework for facial area,
support for teeth

b Rib cage

These bones and some
vertebrae enclose, protect
heart, lungs; assist breathing:

STERNUM (breastbone)
RIBS (twelve pairs)

c Vertebral column,
or backbone

VERTEBRAE (twenty-six bones)
Enclose, protect spinal cord;
support skull, upper extremities;
attachment sites for muscles

INTERVERTEBRAL DISKS
Fibrous, cartilaginous structures
between vertebrae; absorb
movement-induced stresses,
impart flexibility to backbone

ligament bridging a knee
joint, side view, midsection

Figure 37.7 *Animated!* Human
skeletal system. Major bones of its
axial portion are listed at the left.
Those of its appendicular portion
are listed at right. Ligaments and
other kinds of connective tissue
structures bridge skeletal joints,
the areas of contact or near-
contact between bones.

d Pectoral girdle and
upper limb bones

Bones with extensive muscle
attachments, arranged for
great freedom of movement:

CLAVICLE (collarbone)

SCAPULA (shoulder blade)

HUMERUS (upper arm bone)

RADIUS (forearm bone)
ULNA (forearm bone)
CARPALS (wrist bones)

1
2
3
4
5

METACARPALS (palm bones)
PHALANGES (thumb, finger bones)

e Pelvic girdle and
lower limb bones

PELVIC GIRDLE (six fused bones)
Supports weight of backbone, helps
protect soft pelvic organs

FEMUR (thighbone)
Body's strongest weight-bearing bone;
works with large muscles in locomotion
and in maintaining upright posture

PATELLA (kneebone)
Protects knee joint, aids leverage

TIBIA (lower leg bone)
Major load-bearing role

FIBULA (lower leg bone)
Muscle attachment sites;
no load-bearing role

TARSALS (ankle bones)

METATARSALS (sole bones)
PHALANGES (toe bones)

separated from one another by **intervertebral disks**—
cartilaginous shock absorbers and flex points. But the
bones and disks are stacked against gravity. A rapid
shock can force a disk to slip out of place or rupture—
and this *herniated* disk can cause chronic back pain.

No longer aquatic, no longer tetrapods, the most
recent representatives of the human species must live

with the costs as well as benefits of getting around in
the world with the two legs of an upright walker.

*Skeletons of land vertebrates reflect early adaptations to
life in water and later modifications to a life deprived of
water's buoyancy.*

37.4 Zooming In on Bones and Joints

LINKS TO
SECTIONS
33.2, 36.4

Bones of a vertebrate skeleton are organs with diverse roles. They function in movement, in protection of soft internal organs, and as reservoirs for mineral ions. In certain bones, stem cells give rise to the body's blood cells (Table 37.1).

BONE STRUCTURE AND FUNCTION

Human bones range in size from middle ear bones smaller than lentils to clublike thighbones or femurs (Figure 37.7). Bone tissue, recall, consists of bone cells and collagen fibers in a calcium-hardened, organic matrix (Section 33.2). It has three types of bone cells. *Osteoblasts* are the bone-forming cells; they secrete the components of the matrix. Huge populations of them are present on the outer surfaces and internal cavities of the bones of adults. *Osteocytes* are osteoblasts that became imprisoned in small chambers after secreting matrix material around themselves. They are the most common bone cells in bony tissue of adults. *Osteoclasts* are bone cells that break down bone tissue by secreting acids and enzymes into the hardened matrix.

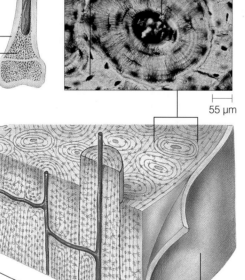

nutrient canal

location of
yellow marrow

compact
bone tissue

spongy
bone
tissue

a

space occupied
by living bone cell

blood
vessel

|—————|
55 µm

spongy
bone
tissue

compact
bone tissue

blood vessel

outer layer
of dense
connective tissue

b

Figure 37.8 *Animated!* (**a**) Structure of a human femur, or thighbone, and (**b**) a section through its spongy and compact bone tissues.

Figure 37.8 shows the two types of bone tissue in a femur. *Compact* bone tissue resists mechanical shock. This tissue's matrix is laid down as dense concentric rings around tiny canals for nerves and blood vessels. Osteocytes reside in narrow clefts between the rings. *Spongy* bone tissue is present in the femur's shaft and knobby ends. It is strong but does not weigh much; its hardened matrix is pocketed with open spaces.

Red marrow, the major site of blood cell formation, fills the spaces in spongy bone. The central cavity of the femur and most mature bones of adults is filled with **yellow marrow**. This marrow is mostly fat, but it can be converted to blood cell-producing red marrow in times of severe blood loss.

BONE FORMATION AND REMODELING

The first skeleton to form in all vertebrate embryos is made of cartilage. Adult cartilaginous fishes retain it. In all other vertebrates, the cartilage is just the model. Osteoblasts infiltrate the model and transform it into bone. The cartilage in bone shaft breaks down, and a marrow cavity opens up inside (Figure 37.9).

In healthy young adults, the total bone mass does not change much even though osteocytes and mineral ions are being removed and replaced all the time. The removals and deposits help maintain required blood levels of calcium and phosphorus while keeping bones strong. In an ongoing process called **bone remodeling**, osteoblasts help form new bone tissue, which makes up for bone tissue that osteoclasts are breaking down. The osteoclast action releases mineral ions that enter interstitial fluid and the bloodstream, which transports them to metabolic reaction sites all through the body.

Calcium ions are the most prevalent mineral ions stored in and released from bones. Neural function, muscle contraction, and many other activities would end without them. The blood calcium level is a tightly

Table 37.1 Functions of Bone
1. *Movement.* Bones interact with skeletal muscle and change or maintain the position of the body and its parts.
2. *Support.* Bones support and anchor muscles.
3. *Protection.* Many bones are organized as hard compartments that enclose and protect soft internal organs.
4. *Mineral storage.* Bones are a reservoir for calcium and phosphorus ions. Deposits and withdrawals of these minerals help maintain essential ion concentrations in body fluids.
5. *Blood cell formation.* Only certain bones contain regions where blood cells form.

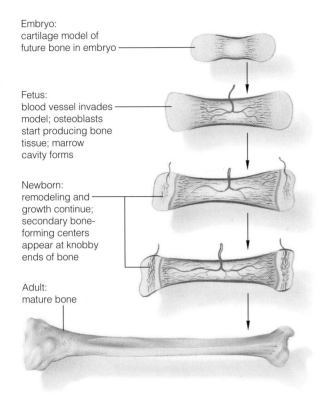

Embryo:
cartilage model of
future bone in embryo

Fetus:
blood vessel invades
model; osteoblasts
start producing bone
tissue; marrow
cavity forms

Newborn:
remodeling and
growth continue;
secondary bone-
forming centers
appear at knobby
ends of bone

Adult:
mature bone

Figure 37.9 Long bone formation, starting with osteoblast activity in a cartilage model formed earlier in the embryo. The bone-forming cells are active first in the shaft region, then at the knobby ends. In time, cartilage is left only at the ends.

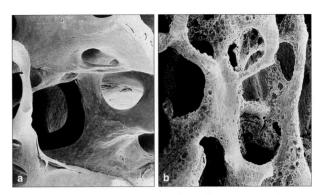

Figure 37.10
(**a**) Normal bone tissue. (**b**) Bone affected by osteoporosis.

controlled aspect of metabolism. The bones and teeth store all but about 1 percent of the body's calcium.

Remember, negative feedback loops help regulate calcium release and uptake (Section 36.4). When there is too much calcium in the blood, the thyroid gland secretes calcitonin. By inhibiting osteoclast action, this hormone slows calcium release into blood. When there is too little calcium in blood, the parathyroid glands release parathyroid hormone, or PTH. This hormone stimulates calcium release from bones and calcium reabsorption from both kidneys. It enhances osteoclast action and also activates vitamin D, which stimulates calcium absorption from the gut lumen.

Until humans are about twenty-four years old, the osteoblasts are secreting more matrix than osteoclasts can break down, and so the bone mass increases. Bones become denser and stronger. Later in life, osteoblast activity declines, and bones gradually weaken.

Significant loss in bone density is called *osteoporosis* (Figure 37.10). Deficient calcium or vitamin D intake, parathyroid problems, and physical inactivity add to the risk. Declining sex hormone levels in menopause, smoking, excessive alcohol intake, and prolonged use of steroids also can slow bone deposition.

WHERE BONES MEET—SKELETAL JOINTS

Connective tissue bridges **joints**, the areas of contact or near-contact between bones. **Ligaments** are straps of dense connective tissue at many joints, such as the knees. They attach one bone to another and let both move freely. The breastbone, vertebrae, and ribs move only a bit. Their joints of cartilage are pliable enough to cushion abutting bones and help absorb shocks.

For instance, knee joints let you swing, bend, and turn the long bones below them while absorbing the mechanical force of your weight. They also are easily injured. Abruptly twist a knee joint too far and you *strain* it. Tear its ligaments or tendons and you *sprain* it. Moving the wrong way might dislocate the attached bones. If a blow to the knee during football or another collision sport severs its ligaments, the ligaments must be surgically put together within ten days. Phagocytic white blood cells patrol fluid that bathes knee joints, where they clean up debris from daily wear and tear. Their action can turn torn ligaments to mush.

Joint inflammation and degenerative disorders are collectively called arthritis. In *osteoarthritis*, cartilage at freely movable joints wears away. Joints in fingers, knees, hips, and the backbone are affected most. With *rheumatoid arthritis*, joint membranes become inflamed. They thicken, cartilage degenerates, and bone deposits accumulate as a result of an autoimmune response. A bacterial or viral infection might be the trigger for the disorder, but genetics and smoking probably increase the risk. Rheumatoid arthritis can develop at any age, although symptoms usually appear before age fifty.

Bones are collagen-rich, mineralized organs that function in movement, support, protection, storage of calcium and other minerals, and blood cell formation.

Balancing ongoing calcium deposits with the withdrawals maintains bone mass and calcium concentrations in blood.

37.5 Skeletal–Muscular Systems

LINKS TO
SECTIONS
17.4, 23.11, 33.3

Skeletal muscles are the functional partners of bones. They contract (shorten) in response to stimulation, then passively return to their resting position (lengthen).

Skeletal muscle cells are not your typical cells. Before they can differentiate and mature in embryos, groups of them fuse together into one multinucleated **muscle fiber**. Bundles of muscle fibers are sheathed in dense connective tissue, which extends past them. The entire array is a skeletal muscle. The extension, a cordlike or straplike **tendon**, attaches it to bone.

Most of these attachment sites are like a gearshift. *They act as a lever system, in which a rigid rod is attached to a fixed point and moves about it.* Muscles connect to bones (rigid rods) near a joint (fixed point). When they contract, they transmit force that makes bones move.

The skeletal muscles also interact with one another. Some work in pairs or groups in ways that promote a movement. Others work in opposition; the action of one opposes or reverses the action of another. As an example, extend your right arm forward, then place your left hand over the biceps in your upper right arm and slowly bend the elbow, as in Figure 37.11. Feel the biceps contract? Even when a biceps contracts just a bit, it causes a large motion of the bone connected to it. This is the case for most leverlike arrangements, as the frog in Figure 37.12 obligingly illustrates.

Bear in mind, only *skeletal* muscle is the functional partner of bone. As you read in Section 33.3, smooth muscle is mainly a component of soft internal organs,

such as the stomach. Cardiac muscle forms only in the heart wall. Later chapters will consider the structure and function of smooth muscle and cardiac muscle.

The human body has close to 700 skeletal muscles, some near the surface, others deep in the body wall (Figure 37.13). The trunk has muscles of the thorax, backbone, abdominal wall, and pelvic cavity. Other muscles attach to the upper and lower limb bones. Later chapters describe how skeletal muscles assist in respiration and circulation. For now, turn to some of the mechanisms that bring about their contraction.

> Cordlike or straplike tendons of dense connective tissue attach skeletal muscles to bones. Only skeletal muscles transmit contractile force to bones, and many groups function in opposition to one another.

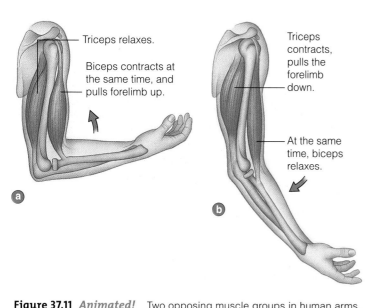

c The first muscle group in the frog's upper hindlimb contracts again and draws the leg back toward the body.

b An opposing muscle group attached to the limb forcefully contracts and pulls the limb back. The contractile force, applied against the ground, propels the frog forward.

a A frog muscle attached to each upper hindlimb contracts and pulls the leg slightly forward relative to the body's main axis.

Triceps relaxes.

Biceps contracts at the same time, and pulls forelimb up.

Triceps contracts, pulls the forelimb down.

At the same time, biceps relaxes.

(a)

(b)

Figure 37.11 *Animated!* Two opposing muscle groups in human arms. (**a**) When the triceps relaxes and its opposing partner (biceps) contracts, the elbow joint flexes and the forearm is pulled upward. (**b**) When the triceps contracts and the biceps relaxes, the forearm is extended down.

Figure 37.12 A frog demonstrating how a small change in the length of a contracting muscle can cause big movements.

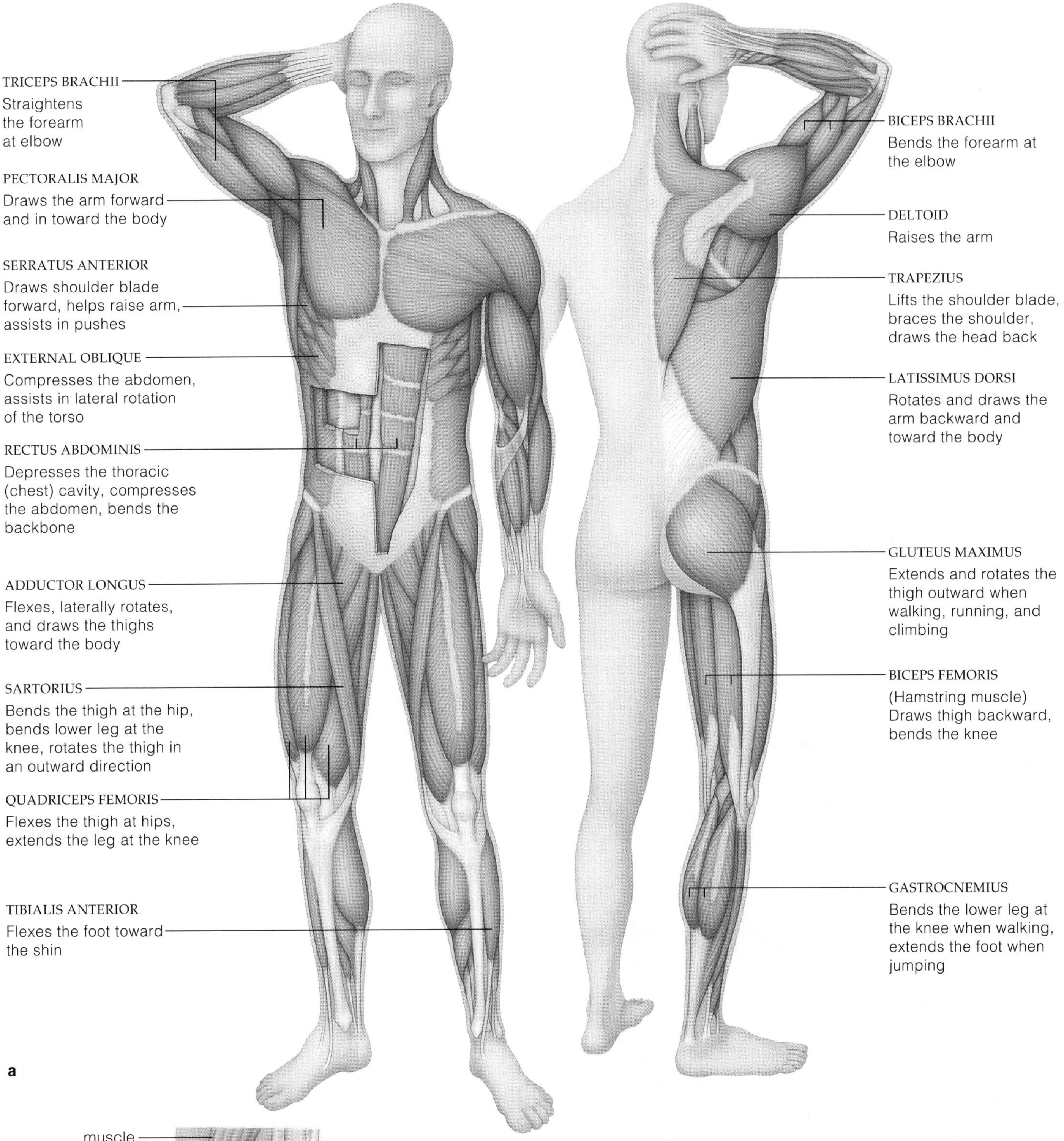

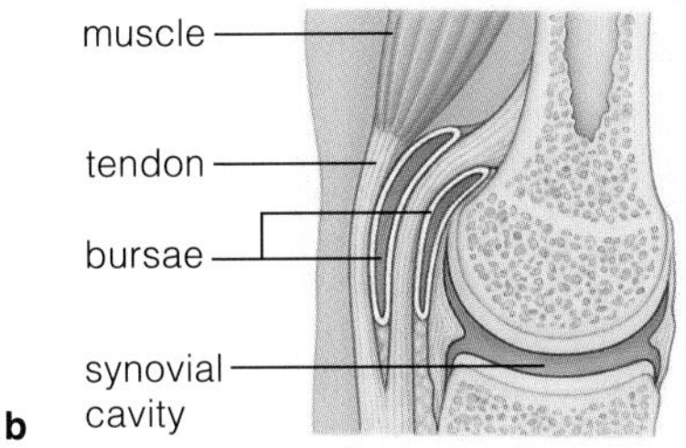

TRICEPS BRACHII
Straightens
the forearm
at elbow

PECTORALIS MAJOR
Draws the arm forward
and in toward the body

SERRATUS ANTERIOR
Draws shoulder blade
forward, helps raise arm,
assists in pushes

EXTERNAL OBLIQUE
Compresses the abdomen,
assists in lateral rotation
of the torso

RECTUS ABDOMINIS
Depresses the thoracic
(chest) cavity, compresses
the abdomen, bends the
backbone

ADDUCTOR LONGUS
Flexes, laterally rotates,
and draws the thighs
toward the body

SARTORIUS
Bends the thigh at the hip,
bends lower leg at the
knee, rotates the thigh in
an outward direction

QUADRICEPS FEMORIS
Flexes the thigh at hips,
extends the leg at the knee

TIBIALIS ANTERIOR
Flexes the foot toward
the shin

a

BICEPS BRACHII
Bends the forearm at
the elbow

DELTOID
Raises the arm

TRAPEZIUS
Lifts the shoulder blade,
braces the shoulder,
draws the head back

LATISSIMUS DORSI
Rotates and draws the
arm backward and
toward the body

GLUTEUS MAXIMUS
Extends and rotates the
thigh outward when
walking, running, and
climbing

BICEPS FEMORIS
(Hamstring muscle)
Draws thigh backward,
bends the knee

GASTROCNEMIUS
Bends the lower leg at
the knee when walking,
extends the foot when
jumping

muscle

tendon

bursae

synovial
cavity

b

Figure 37.13 *Animated!* (**a**) Major skeletal muscles of the human
skeletal–muscular system. Not all skeletal muscles are shown, but these
are the ones that are most familiar to body builders. (**b**) A typical tendon.
Bursae form between tendons and bones (or some other structure). Each
bursa is a flattened sac filled with synovial fluid. It helps reduce friction
between body parts during movements.

37.6 How Does Skeletal Muscle Contract?

LINKS TO
SECTIONS 4.10,
6.2, 33.3

Bones of a dancer or any other human in motion move in some direction when skeletal muscles attached to them shorten. A muscle shortens when its muscle fibers, and individual contractile units inside the fibers, shorten.

FINE STRUCTURE OF SKELETAL MUSCLE

A skeletal muscle's function arises from its internal organization. Long, slender muscle fibers run parallel with the muscle's long axis. The fibers are packed with **myofibrils**, each a bundle of contractile filaments that run from one end of the fiber to the other. Staining the myofibrils for microscopy reveals repeats of light-to-dark crossbands along their entire length. The bands give the muscle fiber a striated, or striped, appearance (Figures 37.1 and 37.14).

The banding corresponds to units of contraction, or **sarcomeres**, that are repeated one after another along the length of the myofibril. Each end of a sarcomere is anchored to its neighbor at a Z band, a dense mesh of cytoskeletal elements (Figure 37.14c).

Parallel arrays of thin filaments extend from both Z bands toward the sarcomere center but stop short of it. The thin filaments are mainly repeating units of the globular protein **actin** (Section 4.10 and Figure 37.14d).

An array of thicker filaments starts at the center of the sarcomere. It runs parallel with the thin filaments but does not extend all the way to the Z bands (Figure 37.15a). The thick filaments consist of **myosin**, a motor protein with a clublike head (Section 4.10 and Figure 37.14e). Each myosin head is positioned no more than a few nanometers away from a thin actin filament.

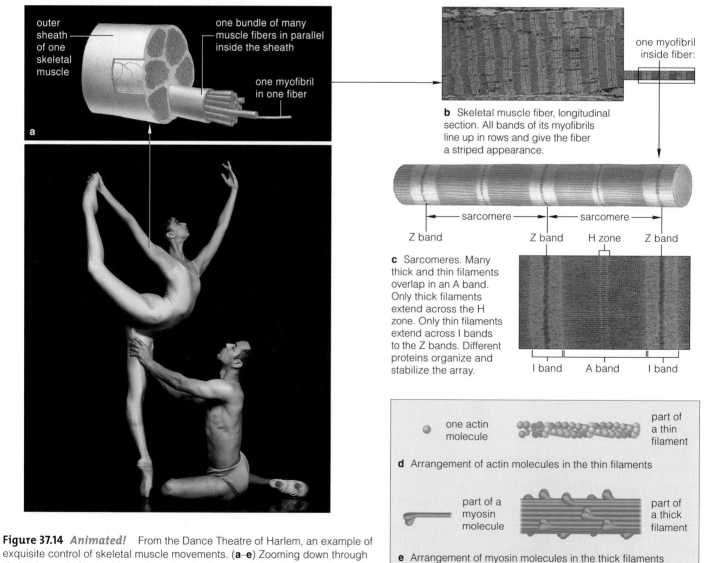

a

outer sheath of one skeletal muscle

one bundle of many muscle fibers in parallel inside the sheath

one myofibril in one fiber

one myofibril inside fiber:

b Skeletal muscle fiber, longitudinal section. All bands of its myofibrils line up in rows and give the fiber a striped appearance.

sarcomere — sarcomere

Z band — Z band — H zone — Z band

c Sarcomeres. Many thick and thin filaments overlap in an A band. Only thick filaments extend across the H zone. Only thin filaments extend across I bands to the Z bands. Different proteins organize and stabilize the array.

I band — A band — I band

one actin molecule — part of a thin filament

d Arrangement of actin molecules in the thin filaments

part of a myosin molecule — part of a thick filament

e Arrangement of myosin molecules in the thick filaments

Figure 37.14 *Animated!* From the Dance Theatre of Harlem, an example of exquisite control of skeletal muscle movements. (**a–e**) Zooming down through skeletal muscle from a biceps to molecules having contractile properties.

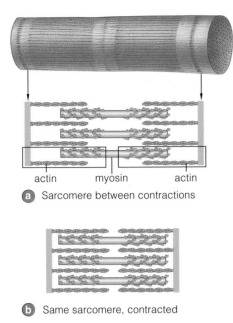

a Sarcomere between contractions

actin   myosin   actin

b Same sarcomere, contracted

Figure 37.15 *Animated!* A sliding-filament model for the contraction of a sarcomere in a myofibril of a skeletal muscle fiber. (**a**,**b**) Highly organized, overlapping arrays of actin and myosin filaments interact to reduce a sarcomere's width. (**c–g**) For clarity, we show the action of just two of many myosin heads. Each myosin head binds repeatedly and then slides an actin filament to the sarcomere's center. Their collective action makes the sarcomere shorten (contract).

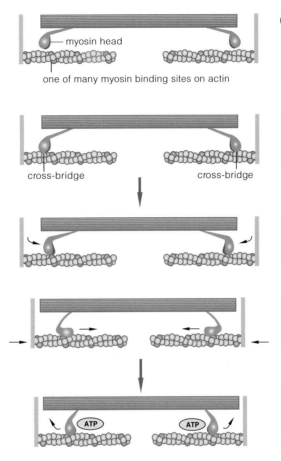

— myosin head

one of many myosin binding sites on actin

c A myosin filament in a muscle at rest. All the myosin heads were energized earlier by the binding of ATP, which they hydrolyzed to ADP and inorganic phosphate.

cross-bridge   cross-bridge

d Release of calcium from a cellular storage system allows formation of cross-bridges; myosin binds to sites on actin filaments.

e Binding makes each myosin head tilt toward the center of the sarcomere and slide the bound actin filaments along with it.

f ADP and phosphate are released as the myosin heads drag the actin filaments inward, pulling the Z lines closer together.

ATP   ATP

g New ATP binds to the myosin heads and they detach from actin. Hydrolysis of the ATP will return the myosin heads to their original orientation, ready to act again.

Muscle fibers, myofibrils, thin filaments, and thick filaments have the same orientation; they run parallel with the muscle's long axis. What is the point? *The repetitive orientation focuses the force of contraction, so that all sarcomeres in all fibers of a muscle work together to pull a bone in the same direction.*

SLIDING-FILAMENT MODEL FOR CONTRACTION

How do sarcomeres, and a skeletal or cardiac muscle, shorten? By a **sliding-filament model**, myosin heads move actin filaments toward the sarcomere's center by short, repetitive, ATP-driven power strokes. The myosin filaments stay in place, but the actin filaments slide past them. Both Z bands, which are connected to the actin filaments, are pulled inward with them, and that shortens the sarcomere (Figure 37.15b).

Each myosin head latches on to one binding site after another along an actin filament. Part of the head is enzymatic. It catalyzes a phosphate-group transfer from ATP, which is the energy that drives contraction. As you will read in the next section, myosin forms a

cross-bridge to actin when the local concentration of calcium ions rises and a binding site for the myosin's head is exposed. Once the head binds, it tilts toward the sarcomere's center, and the actin slides along with it. When another ATP boost breaks the grip on actin, the myosin head reverts to its resting position (Figure 37.15c–e). Depending on the calcium and ATP levels, the myosin head may attach to the next binding site on the actin filament, tilt in another stroke, and so on. It takes hundreds of myosin heads, all performing a rapid series of short strokes along the actin filaments, to bring about a single contraction of a sarcomere.

Sarcomeres are the basic units of contraction in skeletal (and cardiac) muscle. The parallel orientation of all of the muscle components focuses contractile force on a bone that is to be moved.

By energy-driven interactions between myosin and actin filaments, the many sarcomeres of a muscle cell shorten and collectively bring about the muscle's contraction.

37.7 From Signal to Response: A Closer Look at Contraction

LINKS TO
SECTIONS 4.6,
5.4, 34.2, 34.4

Section 34.4 introduced you to the idea that acetylcholine (ACh) can stimulate muscle contraction. Here is a closer look at what goes on after the nervous system commands a motor neuron to release this neurotransmitter into the synaptic cleft of a neuromuscular junction.

All cells at rest show a voltage difference across their plasma membrane, in that interstitial fluid has a slight positive charge relative to the cytoplasm next to the membrane. Only in muscle fibers, neurons, and some other *excitable* cells can the voltage difference reverse abruptly and briefly in response to stimulation. Such a reversal is an **action potential**.

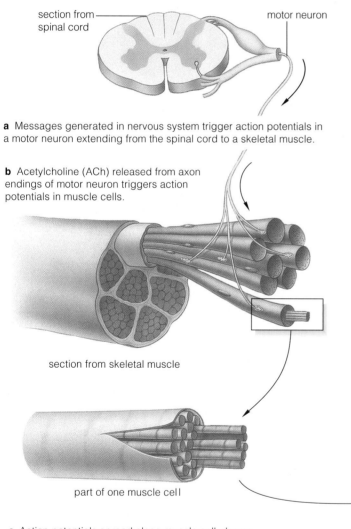

a Messages generated in nervous system trigger action potentials in a motor neuron extending from the spinal cord to a skeletal muscle.

b Acetylcholine (ACh) released from axon endings of motor neuron triggers action potentials in muscle cells.

c Action potentials spread along muscle cell plasma membrane and reach the sarcoplasmic reticulum.

Figure 37.16 Pathway by which the nervous system stimulates or inhibits skeletal muscle contraction. The plasma membrane of each muscle cell encloses myofibrils. Tubular extensions of it, the T tubules, have anchor points for actin filaments at the Z bands of all sarcomeres inside.

As Section 34.2 explains, an action potential arises as certain ions flow across the plasma membrane in an ever accelerating way. The disturbance self-propagates from the point of stimulation without diminishing.

Suppose signals from the nervous system strongly spread rapidly from the stimulation site, then along T tubules. The small tubes are extensions of the plasma membrane. Actin filaments in sarcomeres are attached to them, and so is a system of membranous chambers. That system, the **sarcoplasmic reticulum**, wraps lacily around the myofibrils. It takes up, stores, and releases calcium ions in controlled ways (Figure 37.16).

The arrival of action potentials causes calcium ions to flow out of the chambers. The released ions diffuse into the myofibrils and reach actin filaments. The actin binding sites for myosin heads are blocked in resting muscle fibers, but calcium ions clear them.

Figure 37.17 shows a cross-bridge binding site that is blocked by proteins. The proteins, tropomyosin and troponin, are positioned in or near grooves at the actin filament surface. When the calcium level is low, they are joined so tightly that the tropomyosin is forced out of the groove. It moves every so slightly, but that is enough to block the cross-bridge binding site.

When signals from the nervous system trigger an inflow of calcium into the sarcomere, enough calcium ions bind with troponin to cause its shape to change. The troponin now has a different molecular grip on the tropomyosin filament, which is free to slip back into the groove. The binding site is now exposed.

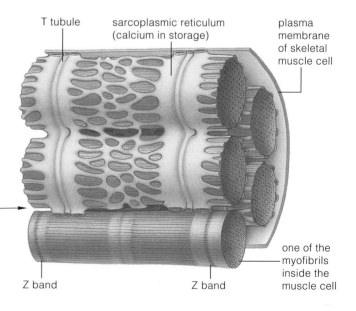

d Action potentials trigger release of calcium ions from sarcoplasmic reticulum threading among myofibrils.

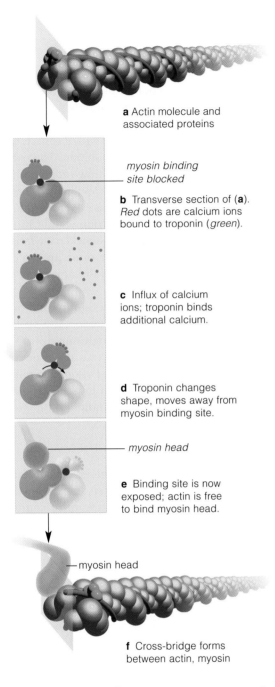

a Actin molecule and associated proteins

myosin binding site blocked

b Transverse section of (**a**). Red dots are calcium ions bound to troponin (*green*).

c Influx of calcium ions; troponin binds additional calcium.

d Troponin changes shape, moves away from myosin binding site.

myosin head

e Binding site is now exposed; actin is free to bind myosin head.

myosin head

f Cross-bridge forms between actin, myosin

Figure 37.17 *Animated!* The interactions among actin, tropomyosin, and troponins in a skeletal muscle cell.

When contraction ends, membrane proteins actively transport the calcium ions back into the sarcoplasmic reticulum. The muscle fiber is ready for another signal.

Certain commands from the nervous system initiate action potentials in muscle cells. These action potentials are signals for cross-bridge formation, hence for contraction.

37.8 Energy for Contraction

LINKS TO SECTIONS 6.2, 8.5, 8.6, 36.6

When a muscle fiber at rest is called upon to contract, the demand for phosphate donations from ATP increases 20 to 100 times. Where does all the energy come from?

A muscle fiber has only a small supply of ATP when it starts contracting. It can produce more very quickly by transferring one phosphate group from **creatine phosphate** to ADP (Figure 37.18). It has about six times as much creatine phosphate as ATP, but the supply only fuels 15 seconds or so of contraction. After the call to contract ends, the supply of creatine phosphate is restored; ATP donates phosphate to creatine.

During prolonged but moderate exercise, aerobic respiration typically provides most of the energy for contraction. In the first five to ten minutes of activity, a muscle fiber converts stored glycogen to glucose, the starting substrate. Glucose and fatty acid arriving from the bloodstream sustain the next half hour or so of activity. Fatty acids become the main energy source for further contraction (Section 8.6).

Not all fuel is burned in aerobic respiration. Even in a resting muscle, some pyruvate from glycolysis is converted to lactate by a fermentation route. Lactate production rises with exercise. Remember, not much ATP forms by this anaerobic pathway. But lactate is a transportable form of energy. Skeletal muscle fibers can give it up, and the bloodstream can deliver it to other muscle fibers or tissues. There, it can be used as energy or stored as glycogen (Sections 8.5 and 36.6).

Muscle contraction requires high concentrations of ATP. Muscle fibers make ATP by dephosphorylation of creatine phosphate, by aerobic respiration, and by glycolysis.

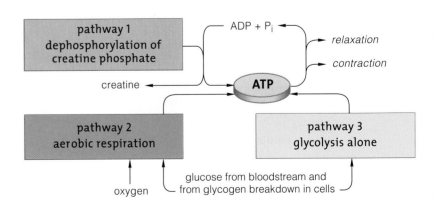

Figure 37.18 *Animated!* Three metabolic pathways by which ATP forms inside muscles in response to the demands of physical exercise.

37.9 Properties of Whole Muscles

LINKS TO
SECTIONS
12.7, 34.2, 34.4

Muscle contractions vary in duration and intensity. Exercise can strengthen or fatigue muscles. Age and certain genetic disorders can weaken them.

TYPES OF CONTRACTIONS

A motor neuron has many axon endings that synapse on different fibers in each muscle (Sections 34.2 and 34.4). A motor neuron and all of the muscle fibers that are functionally connected to it constitute one **motor unit**. When a motor neuron is briefly stimulated, all fibers in the motor unit contract. A fleeting contractile force is generated that lasts just a few milliseconds. That contraction is a **muscle twitch** (Figure 37.19a).

Applying a new stimulus before a response ends makes a muscle twitch again. Repeatedly stimulating a motor unit during a short interval makes all of the twitches run together, and it results in the sustained contraction called **tetanus**. This contraction generates three or four times the force of a single twitch. Figure 37.19c shows a recording of tetanic contraction.

Muscle tension is the mechanical force exerted by a muscle on an object. The number of fibers recruited into action influences it. Opposing this force is a load, either the weight of an object or gravity's pull on the muscle. Only when muscle tension exceeds opposing forces does a stimulated muscle shorten. *Isotonically* contracting muscles shorten and move a load (Figure 37.20a). *Isometrically* contracting muscles do develop tension, as when you attempt to lift something that is too heavy, but they cannot shorten (Figure 37.20b).

WHAT IS MUSCLE FATIGUE?

When ongoing, strong stimulation keeps a muscle in a state of tetanic contraction, muscle fatigue follows. **Muscle fatigue** is a decrease in a muscle's capacity to generate force, a decline in tension.

After a few minutes of rest, a fatigued muscle will contract again in response to stimulation. The extent of recovery depends largely upon how long and how often the muscle was stimulated previously. Muscles trained by a pattern of brief, intense exercise fatigue and recover fast. This happens during weight lifting. Muscles used in prolonged, moderate exercise fatigue slowly. They take longer to recover, often up to a day. Exactly what causes muscle fatigue is unknown, but glycogen depletion is one factor.

A *muscle cramp* is an abrupt involuntary, and often painful contraction that resists release. Any skeletal muscle can cramp, but calf and thigh muscles cramp most often. Motion usually aggravates the cramping; gentle stretching, massage, and heat often may relieve it. To avoid muscle cramps, stretch muscles regularly and avoid overexertion and dehydration.

WHAT ARE MUSCULAR DYSTROPHIES?

Muscular dystrophies are a class of genetic disorders in which muscles progressively weaken and degenerate. Duchenne muscular dystrophy is the most common form among children. Myotonic muscular dystrophy is the most common form among adults.

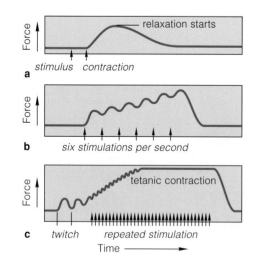

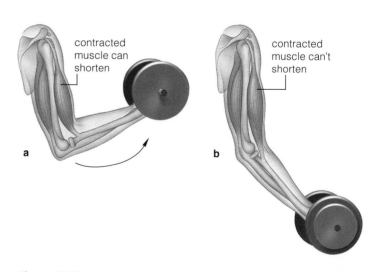

Figure 37.19 *Animated!* Three recordings of twitches in muscles exposed to artificial stimulation of different frequencies. (**a**) A single twitch. (**b**) Summation of twitches after six stimulations per second. (**c**) Tetanic contraction after twenty stimulations per second.

Figure 37.20 (**a**) Isotonic contraction. The load is less than a muscle's peak capacity to contract. The muscle can contract, shorten, and lift the load. (**b**) Isometric contraction. The load exceeds a muscle's peak capacity, so the muscle contracts but cannot shorten.

A single mutant gene on the X chromosome causes Duchenne muscular dystrophy. The gene encodes one of a group of proteins found in extensions of a muscle fiber's plasma membrane. The protein group attaches the plasma membrane to actin filaments at Z bands of sarcomeres. Unless actin binds firmly to proteins at the bands, sarcomeres cannot contract. Like all other X-linked disorders (Section 12.7), muscular dystrophy primarily affects males, at a frequency of about 1 in 3,500. Generally, affected individuals are confined to a wheelchair by their teens. They can expect to survive only into their twenties. There is no cure.

MUSCLES, EXERCISE, AND AGING

No matter how much you exercise, you will not make any more muscle fibers. Existing ones just get bigger, more active, and more resistant to fatigue. Think of *aerobic exercise*—not intense, but long in duration. It increases the number of mitochondria in muscles and the number of blood capillaries servicing them. These physiological changes improve endurance.

Strength training (brief but intense exercise, such as weight lifting) causes muscle fibers to thicken. Also, it stimulates the synthesis of certain enzymes necessary for glycolysis. Bulging muscles do form, but they do not have much endurance. They fatigue fast.

As people age, the number and size of their muscle fibers decline. Tendons that attach muscles to bone stiffen and are more likely to tear. Older people may exercise intensely for long periods, but their muscle mass no longer can increase as much. Even so, aerobic exercise does improve blood circulation, and modest strength training can slow the loss of muscle tissue.

As a beneficial side effect, aerobic exercise among the middle-aged and elderly can lift major depression as well as drugs do. It may improve memory and the capacity to plan and carry out complex tasks. Also, exercise increases blood flow and helps nerves make new connections. In short, exercise is good for more than muscles. It is good for the brain.

We conclude this chapter with a look at two bacterial infections that interfere with how muscles work.

LINKS TO SECTIONS 21.4, 34.4, 34.5

In Section 21.4, you read about *Clostridium botulinum*. This anaerobic soil bacterium can cause disease in humans. Food that has been stored in unsterilized cans or jars may contain its endospores. When the spores germinate, they start making an odorless botulinum toxin.

Now think about what happens in a person who ate toxin-tainted food. The poison enters motor neurons and stops the release of acetylcholine (ACh). Muscles cannot contract without this signal from the nervous system. The muscles become more and more flaccid and paralyzed. These are symptoms of *botulism*, a type of food poisoning. Affected people must be treated with antitoxin as quickly as possible, especially to safeguard their cardiac muscle as well as skeletal muscles with roles in breathing.

A related bacterium, *C. tetani*, lives in the gut of horses, cattle, and other grazing animals, and many people. Its endospores can survive for years in soil—manure-rich soils especially—as long as sunlight and oxygen do not reach them. The endospores resist strong disinfectants, heat, and boiling water. When they enter the body through a deep puncture or cut, they can germinate in necrotic (dead) tissues. Bacterial cells do not spread away from anaerobic, dead tissues. However, they make a toxin that blood or nerves deliver to the spinal cord and brain.

In the spinal cord, the toxin blocks the release of GABA and glycine from certain neurons. These neurotransmitters exert inhibitory control over motor neurons. In the absence of the controls, nothing can put the brakes on signals to contract, and so symptoms of the disease *tetanus* begin.

After four to ten days, the overstimulated muscles stiffen and cannot be released from contraction; they go into spasm. Prolonged, spastic paralysis follows. The fists and the jaws may remain clenched; lockjaw is a common name for the disease. The backbone may become locked in an abnormally arching curve. When respiratory and cardiac muscles become paralyzed, death nearly always follows.

Vaccines were not available for soldiers of early wars, when *C. tetani* lurked in dead cavalry horses and manure on battlefields (Figure 37.21). Vaccines have all but eradicated tetanus in the United States. Worldwide, the annual death toll is over 200,000, mostly due to unsanitary childbirths.

Cross-bridges in sarcomeres collectively exert a mechanical force called muscle tension. Muscles shorten only if this tension exceeds other opposing forces.

Muscle tension is related to the number of fibers that have been stimulated. A signal from a motor neuron stimulates all the fibers of the same motor unit.

Exercise, aging, and some genetic disorders influence the capacity of muscles to generate tension.

Figure 37.21 Painting of a casualty of a contaminated battle wound as he lay dying of tetanus in a military hospital.

Summary

Sections 37.1, 37.2   Various kinds of contractile cells occur in different animal groups. They all evolved from epithelial cells that contained filaments of contractile proteins—actin and myosin. Some evolved into muscle cells when they detached from epithelium and sank into connective tissue below it. In vertebrate muscles, muscle fibers are bundled in connective tissue sheaths.

Muscles cannot act alone. The force of contraction must be applied to skeletal elements to move the body or parts of it. Nearly all animals have a hydrostatic skeleton, an exoskeleton, or an endoskeleton.

In a hydrostatic skeleton, fluid confined inside a body chamber accepts the force of contraction. Sea anemones and earthworms have this type of skeleton.

Muscles and often hydraulic pressure interact with exoskeletons. Arthropods have this type of skeleton.

An endoskeleton is inside the body. In echinoderms, it consists of honeycombed ossicles inside the dermis. In vertebrates, the endoskeleton consists of cartilage or of bones and restricted regions of cartilage.

Biology Now
Learn how a fly moves its wings with the animation on BiologyNow.

Section 37.3   Humans, like other land vertebrates, have an endoskeleton with skull bones, a pelvic girdle, a pectoral girdle, and paired limbs. The move of ancestral vertebrates from water onto land involved modifications to fins, which evolved into the four limbs of amphibious tetrapods. Other modifications occurred in response to the challenges imposed on a body deprived of water's buoyancy. Benefits as well as costs were incurred when humans became upright, two-legged walkers.

Biology Now
Explore the vertebrate skeleton with the animation on BiologyNow.

Section 37.4   Bones are collagen-rich, mineralized organs. They function in mineral storage, movement, and protection and support of soft organs. Blood cells form in bones having red marrow. Ongoing mineral deposits and removals help maintain blood levels of mineral ions and also adjust bone strength. Parathyroid hormones induce bone breakdown and calcium release. Ligaments and cartilage hold bones together at joints.

Biology Now
Look inside a human femur with the animation on BiologyNow.

Section 37.5   Before they differentiate and mature in embryos, groups of muscle cells fuse together into one multinucleated muscle fiber. A skeletal muscle consists of bundles of muscle fibers sheathed in dense connective tissue that extends past them. The extensions, tendons, attach skeletal muscles to bones.

When skeletal muscles contract, they transmit force that moves bones. Many muscles work as opposing pairs. The internal organization of skeletal and cardiac muscle promotes strong, directional contraction.

Biology Now
Learn about the location and action of human skeletal muscles with the animation on BiologyNow.

Sections 37.6, 37.7   A skeletal muscle fiber runs parallel with the long axis of a muscle and extends to both ends of it. This fiber contains many myofibrils, or transversely banded filaments. Each myofibril is divided repeatedly along its length into sarcomeres, the basic unit of contraction in skeletal and cardiac muscle.

Each sarcomere has two arrays of actin filaments that are anchored to Z bands, or dense arrays of cytoskeletal elements that define the width of the sarcomere. Each sarcomere has an array of myosin filaments at its center that overlap with the actin filaments. All of the filaments run parallel with the muscle's long axis.

Cumulatively, the repetitive orientations focus the force of contraction, so that all sarcomeres in a muscle work together to pull a bone in the same direction.

Motor neurons relay signals from the nervous system to muscle fibers. When they release ACh, they call for action potentials in the sarcoplasmic reticulum. This is a membrane system of calcium-storing chambers around the fibers. Calcium floods out, and binding sites on actin filaments become exposed to the myosin heads close by. The sites are blocked in muscles at rest.

By a sliding-filament model, the myosin heads move actin filaments toward the sarcomere's center by short, repetitive, ATP-driven power strokes. Myosin filaments stay in place, but actin filaments slide past them. The Z bands, being connected to the actin filaments, are pulled inward with them, which shortens the sarcomere.

Biology Now
Investigate the molecular structure and function of human skeletal muscles with the animation on BiologyNow.
See how the nervous system controls skeletal muscle contraction with the animation on BiologyNow.

Section 37.8   Muscle fibers can get ATP energy for contraction by dephosphorylation of creatine phosphate, aerobic respiration, and glycolysis.

Biology Now
Compare different sources of energy for muscle contraction with the animation on BiologyNow.

Section 37.9   Muscle tension is a mechanical force caused by cross-bridge formation. A muscle shortens when muscle tension exceeds an opposing load. A motor neuron and all muscle fibers that form junctions with its endings are a motor unit. Repeated stimulation of a motor unit results in tetanus, or sustained contraction.

Biology Now
Observe what happens when a muscle is stimulated with the animation on BiologyNow.

Section 37.10   Bacterial toxins can interfere with nervous signals that stimulate muscle contraction.

1. Bones are _____ .
   a. mineral reservoirs
   b. skeletal muscle's partners
   c. sites where blood cells form (some bones only)
   d. all of the above

2. Bones move when _____ muscles contract.
   a. cardiac
   b. skeletal
   c. smooth
   d. all of the above

3. The _____ is the basic unit of contraction.
   a. osteoblast
   b. sarcomere
   c. muscle fiber
   d. myosin filament

4. In sarcomeres, phosphate-group transfers from ATP activate _____ .
   a. actin   b. myosin   c. both   d. neither

5. A sarcomere shortens when _____ .
   a. thick filaments shorten
   b. thin filaments shorten
   c. both thick and thin filaments shorten
   d. none of the above

6. ATP for muscle contraction can be formed by _____ .
   a. aerobic respiration
   b. glycolysis
   c. creatine phosphate breakdown
   d. all of the above

7. Match the words with their defining feature.
   ____ osteoblast   a. actin's partner
   ____ muscle twitch   b. all in the hands
   ____ muscle tension c. blood cell production
   ____ joint   d. decline in tension
   ____ myosin   e. bone-forming cell
   ____ red marrow   f. motor unit response
   ____ metacarpals   g. force exerted by cross-bridges
   ____ myofibrils   h. area of contact between bones
   ____ muscle fatigue i. muscle fiber's threadlike parts

Additional questions are available on Biology ⑤ Now™

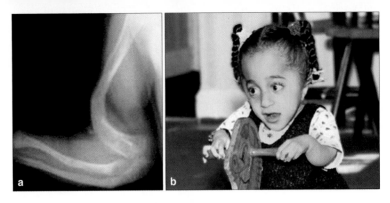

Figure 37.22 (**a**) X-ray of an arm bone deformed by *osteogenesis imperfecta* (OI). (**b**) Tiffany is affected by OI. She was born with multiple fractures in her arms and legs. By age six, she had undergone surgery to correct more than 200 bone fractures and to place steel rods in her legs. Every three months she receives intravenous infusions of an experimental drug that may help strengthen her bones.

Figure 37.23 Polar huskies warming up for an Arctic crossing.

Critical Thinking

1. The genetic disorder *osteogenesis imperfecta* (OI) arises from a mutant form of collagen. As bones develop, this protein forms a scaffold for the deposition of mineralized bone tissue. The scaffold forms improperly in children with OI, who end up with fragile, brittle, easily broken bones (Figure 37.22). OI is rare; most doctors never come across it. Parents are often accused of child abuse when they take an OI child with multiple bone breaks to an emergency room. Prepare a handout about OI that could be used to alert emergency room staffs to the condition. You may wish to start with resources on the Osteogenesis Imperfecta Foundation web site at www.oif.org.

2. Lydia is training for a marathon. While training, she plans to take creatine supplements because she heard that they give athletes extra energy. She asks you whether you think this is a good idea. What is your response?

3. Compared to most people, long-distance runners have far more mitochondria in skeletal muscles. In sprinters, skeletal muscle fibers have more of the enzymes required for glycolysis but not as many mitochondria. Suggest why.

4. In adults, the nuclei of muscle fibers are stuck in G1 of the cell cycle and cannot undergo division. Lots of adults increase the size of muscles by lifting weights. What accounts for the increased bulk of the muscles?

5. Botulinum toxin can kill you if you eat it, but some people pay hundreds of dollars to have it injected. These *botox injections* prevent facial muscles from contracting and pulling on the skin in ways that cause wrinkling. Most commonly, botox is injected into muscles between the eyebrows. It makes the eyelids droop for weeks in 3 percent of the patients. Explain why.

6. In 1989, Will Steger and his dogsled team walked on ice for seven months, enduring temperatures of –113°F and a seven-week blizzard. They crossed Antarctica, all 6,023 kilometers (3,741 miles) of it. Steger called his polar huskies the members of the team that worked hardest and pulled all the weight (Figure 37.23). Husky leg bones are sturdy yet lightweight. The forelegs move freely, thanks to a deep but not-too-broad rib cage. By contrast, human legs are adapted for long-distance walking, not long-distance, load-pulling motion. Compared to human legs, what kind of muscle fibers and muscle mass would you expect to find in a husky's hind legs?

And Then My Heart Stood Still

In the nineteenth century, physiologist Augustus Waller put the paws of his pet bulldog Jimmie in bowls of salty water, which wires connected to a simple recording device. Salt water is a good conductor of electricity. It picked up the faint electric signals from Jimmie's beating heart, and the recording device scratched out the world's first graph of a heartbeat—an *electrocardiogram* (Figure 38.1).

Now fast-forward to the present. A graph of your heart's normal activity would look much the same as Jimmie's. The pattern emerged a few weeks after you started growing from a fertilized egg and developing into an embryo. Early on, some of the embryonic cells differentiated into cardiac muscle cells and started to contract on their own. The first patch of cells to do so set the pace for all of the others, and they have acted as the natural pacemaker ever since. If all goes well, that patch of cardiac muscle cells will continue to contract and make your heart beat steadily, about seventy times per minute, until the day you die.

However, with *sudden cardiac arrest*, the heart abruptly stops beating in many people who are not expecting it, in a lot more places than hospitals. In the United States alone, the cases of sudden cardiac arrest exceed 250,000 per year. Tammy Higgins became one of those cases before she was

thirty years old. Just after leaving a church service, Tammy abruptly collapsed and stopped breathing.

Her husband could not find her pulse. With his training as a lifeguard, he knew the flow of oxygen-rich blood to her brain had stopped and had to resume fast. He started cardiopulmonary resuscitation, or *CPR*, a method of mouth-to-mouth respiration and repetitive compressions of the chest. He kept her alive until an ambulance arrived.

The emergency crew used a portable defibrillator. This device delivers an electric shock to the chest, which often can jump-start the heart into beating again (Figure 38.1). In the hospital, Tammy found out that her heartbeat is abnormal. She also found out that she was pregnant.

Surgeons implanted a tiny defibrillator in her chest wall. It constantly monitors her heart and shocks it when it stops beating. The implant saved her during pregnancy and once more during the three years that followed.

Tammy was lucky; her brain escaped damage and she did not die. Her husband and a few bystanders knew how to perform CPR. The chance of surviving sudden cardiac arrest increases by 50 percent when CPR starts within four to six minutes. For every minute that passes without defibrillation, the chance of survival decreases by about 7 to 10 percent.

a   BUCKET   BUCKET   b   JIMMIE   DR. WALLER

Figure 38.1 A bit of history in the making. (**a**) Jimmie the bulldog taking part in a painless experiment. (**b**) The physiologist Augustus Waller and his beloved pet bulldog sharing a moment in Waller's study after the experiment, which yielded a recording of a heartbeat (**c**). The P, QRS complex, and T designate three waves of electrical activity caused by the spread of action potentials across cardiac muscle, then recovery of the normal resting potential. *Facing page:* (**d**) Defibrillator paddles. With luck, they can shock a heart into beating again after cardiac arrest.

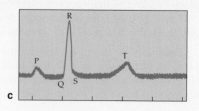

c

AEDs (automated external defibrillators) are about the size of a laptop computer. They have been distributed to many senior centers and shopping malls, hotels, and other public places. An AED checks for a heartbeat, and it shocks the heart if required. Its electronic voice commands trained bystanders to carry out appropriate steps. The American Heart Association, the American Red Cross, community groups, and adult education programs offer CPR and AED training to the public. In short, we have come a long way from Jimmie in checking out the heart and in helping the hearts that falter.

With this bit of history, we turn to the structure and function of the circulatory system. In humans, this organ system has a durable pump—a heart. The heart beats incessantly and generates the pressure required to drive blood through two elaborate loops of blood vessels, which lead back to the heart. As you will see, this system has structural and functional links with the lymphatic system.

 How Would You Vote?

CPR can make the difference between life and death after a cardiac arrest or a heart attack. Should public high schools in your state require all students to take a course in CPR? Is such a course worth diverting time and resources from the basic curriculum? See BiologyNow for details, then vote online.

 Key Concepts

OVERVIEW OF CIRCULATORY SYSTEMS

Animals have an open or a closed circulatory system that helps transport substances to and from cells.   Section 38.1

BLOOD COMPOSITION AND FUNCTION

Vertebrate blood is a fluid connective tissue. It consists of red blood cells, white blood cells, platelets, and diverse substances dissolved in plasma, the transport medium. Red blood cells function in gas exchange. White blood cells and platelets defend tissues. Diseases and abnormalities impair blood cell function.   Sections 38.2–38.4

THE HUMAN HEART AND TWO FLOW CIRCUITS

The human heart has four chambers. Blood flows into its two atria and is pumped out of its two ventricles, into separate circuits of blood vessels. One circuit extends through all body regions, the other through lung tissue only. Both circuits loop back to the heart.   Sections 38.5, 38.6

BLOOD VESSEL STRUCTURE AND FUNCTION

The heart pumps blood rhythmically, on its own. Blood pressure is highest in the ventricles, drops as it flows through arteries, capillaries, and veins, and is lowest in the heart's atria. Different blood vessels function in rapid transport, distribution of flow volume, and gas exchange. Sections 38.7, 38.8

WHEN THE SYSTEM BREAKS DOWN

Ruptured or clogged blood vessels or abnormal heart rhythms cause problems. Some problems have a genetic basis; most are related to age or life-styles.   Section 38.9

LINKS WITH THE LYMPHATIC SYSTEM

A lymph vascular system returns excess tissue fluid to blood. Lymphoid organs cleanse blood of infectious agents and other threats to health.   Section 38.10

 Links to Earlier Concepts

This chapter expands on earlier introductions to circulatory systems (Sections 25.6, 25.9, 26.2, 28.2), cardiac muscle (33.3), and muscle contraction (37.6, 37.7). You will draw on your knowledge of hemoglobin (3.6), diffusion and bulk flow (5.3), gas exchange (28.3), action potentials (34.3, 34.4), and cell junctions (33.1). You will revisit ABO blood typing (11.4). You will be invited to reflect on blood cell disorders (3.6, 18.6) and on connections between cardiovascular function, lipoproteins (3.4), and cholesterol (3.4).

38.1 The Nature of Blood Circulation

LINKS TO
SECTIONS 5.3,
25.6, 26.2, 28.1

Imagine an earthquake closing off your neighborhood's highway. Grocery trucks can't enter; waste-disposal trucks can't leave. Your food supplies dwindle and garbage piles up. Cells would be similarly stressed if something were to disrupt the flow along the body's highways.

FROM STRUCTURE TO FUNCTION

The **circulatory system** moves substances to and from cellular neighborhoods. **Blood**, its transport medium, typically flows inside tubular vessels under pressure generated by a muscular pump, a **heart**. Blood makes exchanges with fluid in tissue spaces between cells—or **interstitial fluid**—which exchanges substances with cells. Thus blood and interstitial fluid are functionally inseparable as an *internal* environment for body cells. Many organ systems interact to keep the composition, volume, and temperature of the environment within tolerable ranges for cell activities (Section 28.1).

Circulatory systems are open or closed. In the *open* systems of most mollusks and all arthropods, blood flows out of muscular vessels or one or more hearts, mingles with fluid in body tissues, then flows back in (Figure 38.2a,b). You and many other animals have a *closed* circulatory system. It confines blood in a heart and blood vessels that vary in thickness and diameter. One example is the earthworm's circulatory system,

part of which is shown in Figure 38.2c,d. We include diagrams of a few other closed systems in Figure 38.3.

Reflect on the overall "design" of a closed system. The total *volume* of blood is pumped continuously out of the heart, through vessels, then back to the heart. It travels fastest in large-diameter transport vessels. It slows in beds of small-diameter vessels called blood capillaries, where the blood and interstitial fluid have enough time to exchange substances by way of simple diffusion (Section 5.3 and Figure 38.3d).

EVOLUTION OF VERTEBRATE CIRCULATION

All vertebrates have a closed circulatory system, but fishes, amphibians, birds, and mammals differ in the pumps and plumbing. These differences evolved over hundreds of millions of years after some vertebrates left the water for land. The earliest vertebrates, recall, had gills. Like all respiratory structures, gills have a thin, moist surface, which oxygen and carbon dioxide diffuse across. In time, *internally moistened* sacs called lungs evolved and supported the move to dry land. So did modifications that helped blood move faster in a loop between the heart and lungs (Section 26.2).

In fishes, blood still flows in a single circuit (Figure 38.3a). The contractile force of a two-chambered heart drives blood through two capillary beds. Blood flows

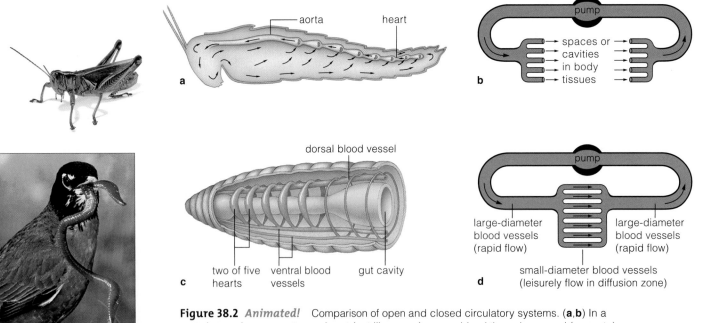

Figure 38.2 *Animated!* Comparison of open and closed circulatory systems. **(a,b)** In a grasshopper's open system, a heart (not like yours) pumps blood through a vessel (an aorta). From there, blood moves into tissue spaces. It mingles with fluid bathing cells, then reenters the heart at openings in the heart wall. **(c,d)** The closed system of an earthworm confines blood inside pairs of muscular hearts near the head end and inside many blood vessels.

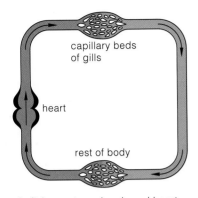
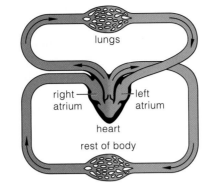
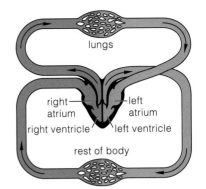

a In fishes, a two-chambered heart (atrium, ventricle) pumps blood in one circuit. Blood picks up oxygen in gills, delivers it to rest of body. Oxygen-poor blood flows back to heart.

b In amphibians, a heart pumps blood through two partially separate circuits. Blood flows to lungs, picks up oxygen, returns to heart. But it mixes with oxygen-poor blood still in the heart, flows to rest of body, returns to heart.

c In birds and mammals, the heart is fully partitioned into two halves. Blood circulates in two circuits: from the heart's right half to lungs and back, then from the heart's left half to oxygen-requiring tissues and back.

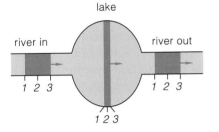

d Why have capillary beds? Picture two fast rivers flowing into and out from a lake. The flow *rate* is the same in all three places, with an identical volume of water moving from point *1* to point *3* in the same interval. However, flow *velocity* decreases in the lake. Why? The volume spreads out through a larger cross-sectional area and moves forward a shorter distance in the specified interval.

Figure 38.3 *Animated!* Comparison of flow circuits in the closed circulatory systems of fishes, amphibians, birds, and mammals.

through gill capillary beds into a large vessel, through capillary beds in body tissues and organs, and back to the heart. The blood is not under much fluid pressure when it leaves the gill capillaries, so it moves slowly on the way back to the heart.

Blood circulation picked up a bit when the heart of amphibians became partitioned into three chambers. Two atria emptied into a single ventricle. Oxygenated blood and oxygen-poor blood flowed in two circuits but mixed a bit in the ventricle (Figure 38.3*b*).

In birds and mammals, the heart became divided into right and left halves, each with two chambers. It pumps blood in two separate circuits (Figure 38.3*c*). In the **pulmonary circuit**, oxygen-poor, carbon dioxide-rich blood flows from the *right* half of the heart to the lungs. There, blood picks up oxygen, gives up carbon dioxide, and flows into the left half of the heart.

In the main, **systemic circuit**, the heart's *left* half pumps oxygenated blood to all tissues where oxygen is used and carbon dioxide forms. After giving up the oxygen and picking up carbon dioxide, blood flows into the heart's right half, then back to the lungs. A double circuit is a fast, efficient mode of exchanging gases. It supports the high levels of activity typical of vertebrates whose ancestors evolved on land.

LINKS WITH THE LYMPHATIC SYSTEM

The heart's pumping puts pressure on blood flowing through the circulatory system. Partly because of this, some water, nutrients, and a few proteins dissolved in blood are forced from capillaries into interstitial fluid. Some fluid and reclaimable solutes move back into capillaries. Any excess interstitial fluid drains into the circulatory system by way of lymph vessels, which are components of a **lymphatic system**. Other components help protect the internal environment (Section 38.10).

In most animals, a circulatory system moves substances to and from interstitial fluid that fills tissue spaces between metabolically active cells. A typical closed system confines blood inside a heart and blood vessels.

In vertebrates, blood flows fastest in large-diameter vessels that connect the heart with capillary beds, where blood slows enough for efficient exchanges with interstitial fluid.

In fishes and amphibians, oxygen-poor and oxygenated blood mix a bit. In birds and mammals, blood flows in two separate circuits, through a heart divided into two side-by-side pumps. Gases are exchanged more efficiently in the double circuit, which supports higher levels of activity.

38.2 Characteristics of Blood

LINKS TO
SECTIONS
3.6, 33.2, 37.4

Tumbling along in the fluid portion of blood are cells and substances that do more than move gases and nutrients to and from cellular neighborhoods. Many defend the body and help maintain internal operating conditions.

FUNCTIONS OF BLOOD

Blood, a fluid connective tissue, transports oxygen, nutrients, and other solutes to cells. It carries away metabolic wastes and secretions, including hormones. Blood helps stabilize internal pH and is a highway for cells and proteins that defend and repair tissues. In birds and mammals, it helps keep body temperature within tolerable limits by moving excess heat to skin, where it can be dissipated into the surroundings.

BLOOD COMPOSITION AND VOLUME

The volume of blood depends on body size and on the concentrations of water and solutes. In average-sized humans, it is about four or five quarts (6 to 8 percent of the total body weight). In all vertebrates, blood is a viscous fluid, thicker than water and slower flowing. Its components are plasma, red and white blood cells, and platelets. The cells and platelets arise from stem cells in bone marrow. Any **stem cell** is unspecialized, and it retains a capacity for mitotic cell division. Some portion of its daughter cells divide and differentiate into specialized types.

Plasma About 50 to 60 percent of the blood's total volume is plasma (Figure 38.4). **Plasma** is 90 percent water. Besides being the transport medium for blood cells and platelets, it acts as a solvent for hundreds of different plasma proteins, other molecules, and ions. Some of the proteins transport lipids and fat-soluble vitamins through the body. Others function in blood clotting. Other solutes include glucose, lipids, amino acids, vitamins, hormones, as well as oxygen, carbon dioxide, and gaseous nitrogen.

Red Blood Cells Erythrocytes, the **red blood cells**, are biconcave disks about 8 micrometers in diameter and 2 micrometers thick (Figure 38.5). They transport oxygen from lungs to aerobically respiring cells and carry carbon dioxide wastes from them. When oxygen first diffuses into blood, it binds to hemoglobin in red

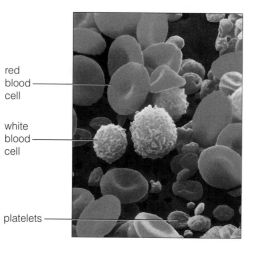

red blood cell

white blood cell

platelets

Figure 38.4 Typical components of human blood. The scanning electron micrograph above shows some cellular components.

The sketch of a test tube shows what happens when you keep a blood sample from clotting. The sample separates into the straw-colored plasma, which floats on a reddish-colored cellular portion. Blood makes up 6 to 8 percent of total body weight.

Components	Relative Amounts	Functions
Plasma Portion (50%–60% of total volume):		
1. Water	91%–92% of plasma volume	Solvent
2. Plasma proteins (albumin, globulins, fibrinogen, etc.)	7%–8%	Defense, clotting, lipid transport, roles in extracellular fluid volume, etc.
3. Ions, sugars, lipids, amino acids, hormones, vitamins, dissolved gases	1%–2%	Roles in extracellular fluid volume, pH, etc.
Cellular Portion (40%–50% of total volume):		
1. Red blood cells	4,800,000–5,400,000 per microliter	Oxygen, carbon dioxide transport
2. White blood cells:		
Neutrophils	3,000–6,750	Phagocytosis during inflammation
Lymphocytes	1,000–2,700	Immune responses
Monocytes (macrophages)	150–720	Phagocytosis in all defense responses
Eosinophils	100–360	Defense against parasitic worms
Basophils	25–90	Secrete substances for inflammatory response and for fat removal from blood
3. Platelets	250,000–300,000	Roles in clotting

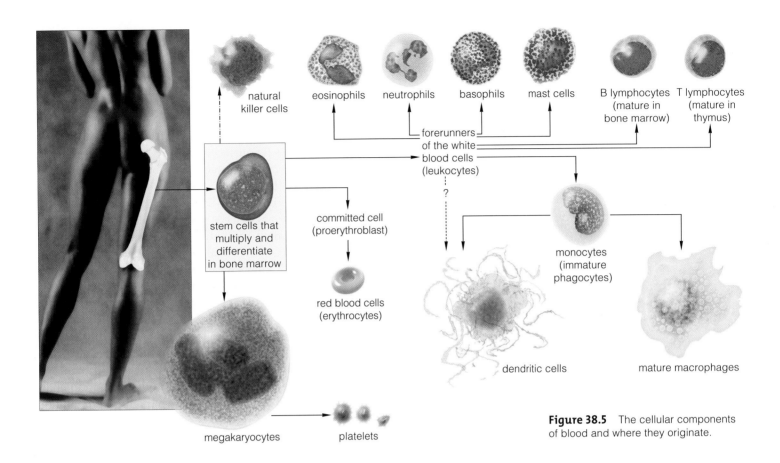

natural killer cells

eosinophils

neutrophils

basophils

mast cells

B lymphocytes (mature in bone marrow)

T lymphocytes (mature in thymus)

forerunners of the white blood cells (leukocytes)

stem cells that multiply and differentiate in bone marrow

committed cell (proerythroblast)

red blood cells (erythrocytes)

monocytes (immature phagocytes)

dendritic cells

mature macrophages

megakaryocytes

platelets

Figure 38.5 The cellular components of blood and where they originate.

blood cells. You learned about this protein in Section 3.6. Stored hemoglobin fills about 98 percent of the interior of red blood cells. It makes both the cells and oxygenated blood appear bright red. Oxygen-poor blood is dark red, and it appears blue through blood vessel walls near the body surface.

Mature red blood cells no longer have a nucleus, nor do they require it. They have enough hemoglobin, enzymes, and other proteins to last about 120 days. Phagocytes engulf the oldest red blood cells and those already dead. In individuals who are in good health, ongoing replacements keep the cell count fairly stable.

A **cell count** is a measure of the quantity of cells of a specified type in 1 cubic millimeter of blood. The number of red blood cells per cubic millimeter averages 5.4 million in human males and 4.8 million in females.

White Blood Cells A variety of leukocytes, or **white blood cells**, functions in housekeeping and defense. Some patrol tissues and engulf damaged or dead cells and anything else chemically recognized as "nonself." Some sound the alarm that tissues are under siege by specific viruses, bacteria, or other threats. Many form highly organized masses inside lymph nodes and the spleen, which are organs of the lymphatic system.

White blood cells differ in size, nuclear shape, and staining traits (Figure 38.5). Their numbers fluctuate with levels of activity and state of health. We consider these cells in the next chapter, but a few examples can give you a sense of what they do. The neutrophils and basophils are phagocytes with roles in inflammation. Macrophages and dendritic cells are phagocytes that can stimulate immune responses to particular threats. Two categories of lymphocytes, the B cells and T cells, are specialized for immune responses. Natural killer cells directly kill body cells that do not have normal self-markers or that have been tagged for destruction.

Platelets Megakaryocytes originate from some stem cells. They later shed membrane-wrapped cytoplasmic fragments of themselves as **platelets**. A platelet lasts five to nine days. Hundreds of thousands are always circulating in blood. When activated, platelets release substances that initiate blood clotting.

Blood, a fluid connective tissue, functions as a transport medium and solvent. Diverse proteins, gases, sugars, and other substances are dissolved in it. Its cellular components include red blood cells, white blood cells, and platelets.

38.3 Blood Disorders

LINKS TO
SECTIONS 3.6,
5.2, 11.4, 12.1, 18.6

The body continually replaces blood cells for good reason. Besides aging and dying off regularly, blood cells may become hosts to diverse pathogens that complete the life cycle inside them. In addition, sometimes blood cells malfunction as a result of gene mutations.

Red Blood Cell Disorders Too few red blood cells or deformed ones result in the disorders collectively called **anemias**. Abnormally low oxygen levels in blood cannot support metabolism. Shortness of breath, fatigue, and chills follow. *Hemorrhagic* anemias follow sudden blood loss, as from a bad wound; chronic anemias result from low red blood cell production or a slight but persistent blood loss, as happens from a bleeding ulcer.

Bacteria and protozoans that replicate in blood cells cause some *hemolytic* anemias; they kill infected cells as they escape, by lysis. Insufficient iron in the diet causes *iron deficiency* anemia; red blood cells cannot make enough normal hemoglobin without iron. *Sickle-cell anemia* arises from a mutation that modifies hemoglobin (Section 3.6). In *thalassemias*, mutations disrupt or stop synthesis of globin chains of hemoglobin. Too few red blood cells form. Those that do form are thin and fragile.

Polycythemias, or having far too many red blood cells, makes blood more viscous and elevates blood pressure. So does *blood doping*. Some athletes withdraw and store their red blood cells and reinject them a few days before competing in strenuous events. The withdrawal triggers red blood cell formation; the body attempts to replace the "lost" cells. When withdrawn cells are put back in the body, the cell count shoots up. The idea is to increase oxygen-carrying capacity and endurance. Other athletes elevate red cell counts artificially by injecting hormones that stimulate red cell production.

White Blood Cell Disorders An Epstein–Barr virus is the agent of *infectious mononucleosis*, a highly contagious disease that causes too many monocytes and lymphocytes to form. Most people recover after a few weeks of fatigue, muscle aches, low-grade fever, and a chronic sore throat.

Recovery is dicey for *leukemias*. These cancers originate in bone marrow and interfere with the formation of white blood cells (Figure 38.6). Either chemotherapy or radiation therapy kills the cancer cells, but side effects can be severe. Remissions, or symptom-free periods in chronic illness, may last months or years. New gene therapies put some leukemias into remission, but they are still experimental.

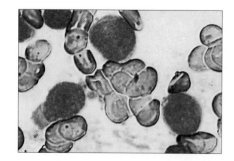

Figure 38.6 Blood sample typical of chronic myelogenous leukemia. Abnormal, immature white blood cells are starting to crowd out normal cells.

38.4 Blood Typing

Blood from donors can be transfused into patients who are affected by severe blood loss or a blood disorder. Such blood transfusions cannot be hit-or-miss. Red blood cells of a potential donor and recipient must bear the same kinds of recognition proteins at their surface.

Each kind of cell bears "self" markers, or recognition proteins at its surface that give it a unique identity (Section 5.2). If donated red blood cells have markers that are not also on the recipient's red blood cells, the recipient's immune system will attack them.

In a normal defense response called **agglutination**, proteins called antibodies bind foreign cells and make them form clumps that attract phagocytes. However, Figure 38.7 shows what happens when the blood from incompatible donors and recipients intermingle. Free antibodies circulating in the recipient's plasma bind to the nonself markers on the introduced cells and make them clump. There are so many foreign cells that the clumps clog small blood vessels and damage tissues. Without treatment, death may follow. The same thing can happen during a pregnancy if a mother and child differ in some red blood cell markers. The mother's immune system will attack the child's cells.

Blood typing—the analysis of the surface markers on red blood cells—can help prevent such problems.

ABO BLOOD TYPING

Molecular variations in one kind of self marker on red blood cells are analyzed with **ABO blood typing**. The genetic basis of such variation is described in Section 11.4. People with one form of the marker have type A blood; those with a different form have type B blood. If they have both forms of the marker on their red blood cells, their blood is type AB. If they have neither form of the marker, their blood is type O.

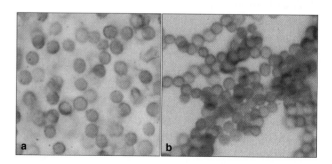

Figure 38.7 Light micrographs showing (**a**) an absence of agglutination in a mixture of two different yet compatible blood types and (**b**) agglutination in a mixture of incompatible types.

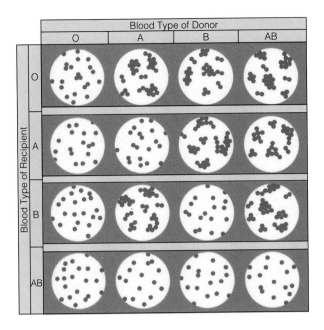

Figure 38.8 *Animated!* Responses in blood types A, B, AB, and O when mixed with samples of the same and different types.

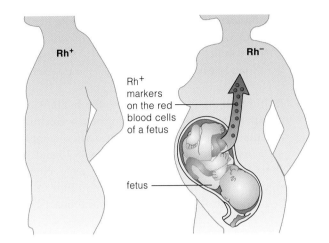

a A forthcoming child of an Rh⁻ woman and Rh⁺ man inherits the gene for the Rh⁺ marker. During childbirth, some of its cells bearing the marker may leak into the maternal bloodstream.

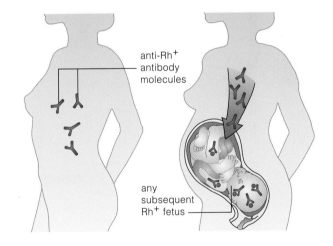

b The foreign marker stimulates antibody formation. If the same woman becomes pregnant again and if her second fetus (or any other) inherits the gene for the marker, her circulating anti-Rh⁺ antibodies will act against it.

Figure 38.9 *Animated!* Maternal production of antibodies in response to Rh⁺ markers on red blood cells of her fetus.

Consider Figure 38.8. If you are blood type A, your immune system will recognize blood cells with type B markers as foreign. If type B, it will react against cells with type A markers. If you are blood type AB, your immune system is familiar with both markers, and it will not react against either type, so you can receive blood transfusions from anyone. If you are blood type O, it will recognize both A and B markers as foreign. You can accept blood only from other people who are type O, but you can donate blood to anyone.

Rh BLOOD TYPING

Rh blood typing is based on the presence or absence of the Rh marker (first identified in blood of *Rh*esus monkeys). If you are type Rh⁺, your blood cells bear this marker. If you are type Rh⁻, they do not. People normally do not have antibodies against Rh markers, because their body has never been exposed to them. However, an Rh⁻ recipient of transfused Rh⁺ blood will make antibodies that will remain in the blood.

If an Rh⁻ woman is impregnated by an Rh⁺ man, there is a chance the fetus will be Rh⁺. Some fetal red blood cells usually leak into a woman's blood during childbirth, and her body makes antibodies against Rh (Figure 38.9). If she becomes pregnant again, some of her Rh antibodies may enter the blood of her fetus. If the fetus is Rh⁺, then her antibodies will cause its red blood cells to swell, rupture, and release hemoglobin.

Erythroblastosis fetalis is the name for any hemolytic disorder in a fetus caused by blood incompatibilities. Some newborns do not display any symptoms; others die not long after birth. Whether it be an ABO or Rh mismatch, diagnosis before birth can help. Fetal blood can be slowly replaced by transfusions of blood from a compatible donor.

Markers on red cell surfaces are the basis for ABO and Rh blood typing. Incompatible blood types cause problems in transfusions and in some pregnancies.

38.5 | Human Cardiovascular System

LINKS TO
SECTIONS
28.1, 28.5, 33.5

"Cardiovascular" comes from the Greek kardia *(for heart) and Latin* vasculum *(vessel). In the human cardiovascular system, a muscular heart pumps blood to the lungs and the tissues by way of two completely separate circuits.*

In humans, as in all mammals, the heart is a double pump that drives blood through two cardiovascular circuits. Arteries, arterioles, capillaries, venules, and veins make up each circuit (Figures 38.10 and 38.11). A short loop, the pulmonary circuit, oxygenates blood. Again, it leads from the heart's right half to capillary beds in both lungs, then back to the heart's left half.

The systemic circuit is a longer loop. The heart's left half pumps oxygenated blood into the main artery, the **aorta**. That blood gives up oxygen in all tissues, then oxygen-poor blood flows back to the heart's right half.

Most blood flows through only one capillary bed. Some flows through two. Blood picks up glucose and other substances at capillaries in the intestines, then flows to and through a capillary bed inside the liver. This organ has vital roles in metabolizing and storing nutrients. It also neutralizes many toxins.

As Figure 38.12 shows, the cardiovascular system rapidly distributes oxygen, nutrients, and many other substances that enter the body through the digestive and respiratory systems. It moves carbon dioxide and other metabolic wastes to the respiratory and urinary systems for disposal. Interactions among these organ systems help keep operating conditions of the internal environment within tolerable ranges. That is the state we call homeostasis (Sections 28.1 and 28.3).

> In the human cardiovascular system's pulmonary circuit, oxygen-poor blood flows from the heart's right half, through both lungs, then back to the heart. It takes up oxygen and gives up carbon dioxide in the lungs.
>
> In the systemic circuit, oxygenated blood flows from the heart's left half and aorta to capillary beds of all tissue regions. There it gives up oxygen and takes up carbon dioxide, then moves to the heart's right half.

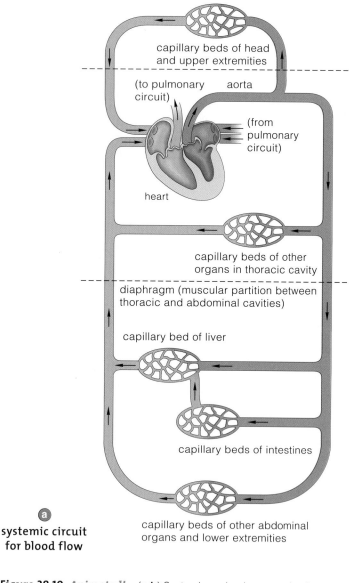

a
**systemic circuit
for blood flow**

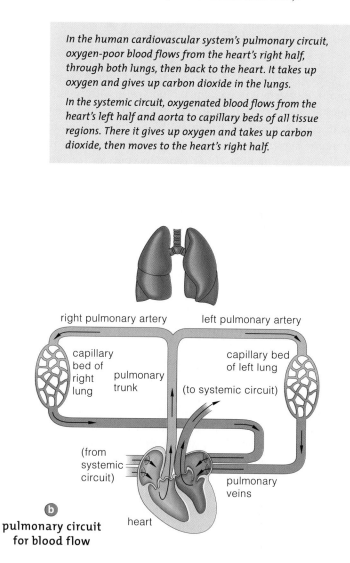

b
**pulmonary circuit
for blood flow**

Figure 38.10 *Animated!* (a,b) Systemic and pulmonary circuits for blood flow through the human cardiovascular system. Blood vessels carrying oxygenated blood are color-coded *red*. Those carrying oxygen-poor blood are color-coded *blue*.

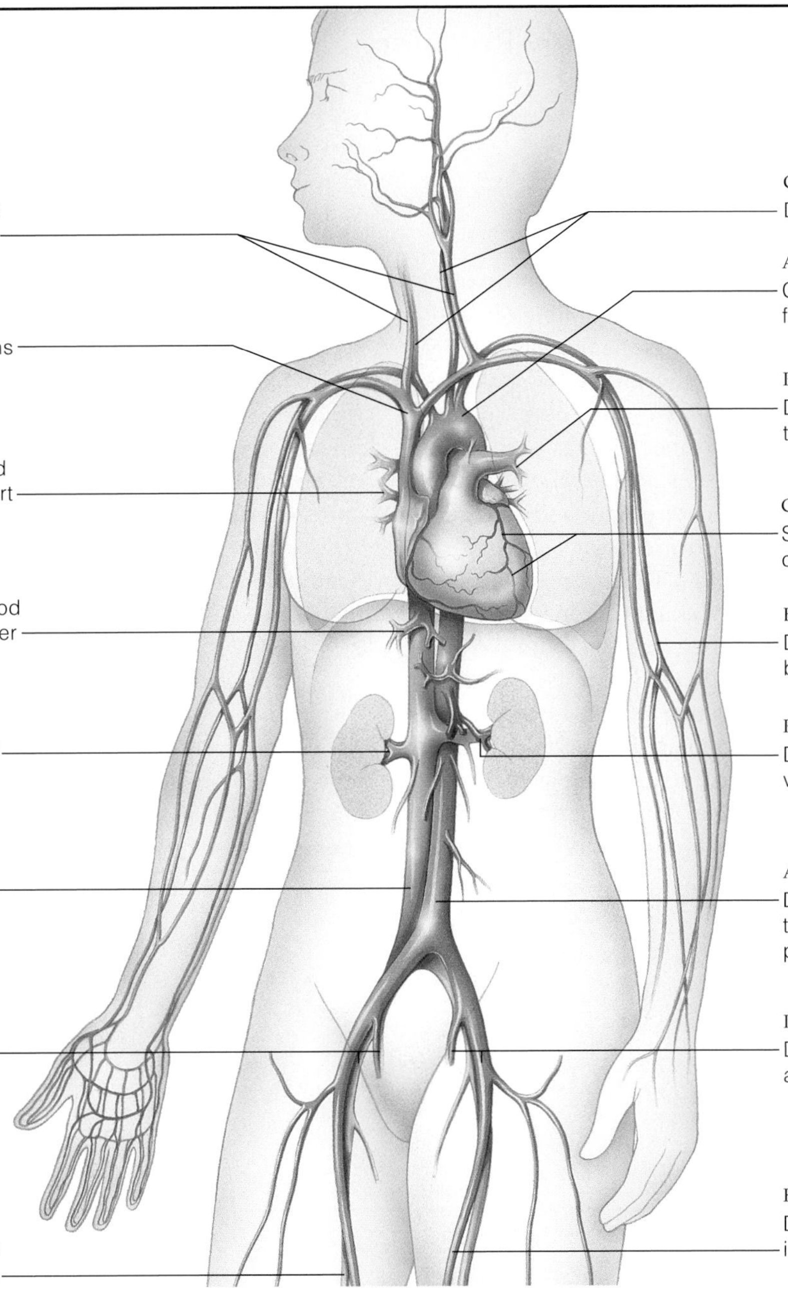

JUGULAR VEINS
Receive blood from brain and from tissues of head
SUPERIOR VENA CAVA
Receives blood from veins of upper body
PULMONARY VEINS
Deliver oxygenated blood from the lungs to the heart
HEPATIC VEIN
Delivers nutrient-rich blood from small intestine to liver for processing
RENAL VEIN
Carries processed blood away from kidneys
INFERIOR VENA CAVA
Receives blood from all veins below diaphragm
ILIAC VEINS
Carry blood away from the pelvic organs and lower abdominal wall
FEMORAL VEIN
Carries blood away from the thigh and inner knee
CAROTID ARTERIES
Deliver blood to neck, head, brain
ASCENDING AORTA
Carries oxygenated blood away from heart; the largest artery
PULMONARY ARTERIES
Deliver oxygen-poor blood from the heart to the lungs
CORONARY ARTERIES
Service the incessantly active cardiac muscle cells of heart
BRACHIAL ARTERY
Delivers blood to upper extremities; blood pressure measured here
RENAL ARTERY
Delivers blood to kidneys, where its volume, composition are adjusted
ABDOMINAL AORTA
Delivers blood to arteries leading to the digestive tract, kidneys, pelvic organs, lower extremities
ILIAC ARTERIES
Deliver blood to pelvic organs and lower abdominal wall
FEMORAL ARTERY
Delivers blood to the thigh and inner knee

Figure 38.11 ***Animated!*** Major blood vessels of the human cardiovascular system. This art is greatly simplified for clarity. For example, each arm has a complete set of the arteries and veins listed, and so does each leg. Humans, remember, have bilateral ancestry (Section 25.2).

Carotid bodies are located at the first branch point of the carotid arteries, and aortic bodies are in the aorta, where it arches above the heart. Both kinds of sensory receptors monitor chemical changes in blood. Baroreceptors in the same locations monitor blood pressure. In response to the receptor signals, the brain adjusts the heart's output and flow resistance in arterioles and veins.

food, water intake
oxygen intake
DIGESTIVE SYSTEM
RESPIRATORY SYSTEM
elimination of carbon dioxide
nutrients, water, salts
oxygen
carbon dioxide
CIRCULATORY SYSTEM
URINARY SYSTEM
water, solutes
elimination of food residues
rapid transport to and from all living cells
elimination of excess water, salts, wastes

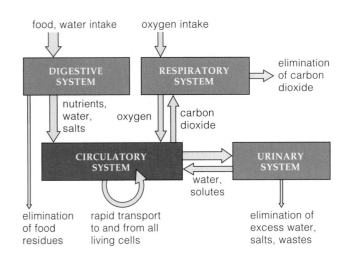

Figure 38.12 Functional links between the circulatory system and other organ systems that maintain the internal environment.

38.6 The Heart Is a Lonely Pumper

LINKS TO
SECTIONS 33.3,
34.3, 37.6–37.9

The human heart is a durable pump that spontaneously beats 2.5 billion times in a seventy-year life span.

HEART STRUCTURE

The human heart's durability arises from its structure. Outermost is the *pericardium*, a tough, double-layered, connective tissue sac that anchors the heart to nearby structures (Figure 38.13). In between the two layers is a fluid that lubricates the heart during its perpetual wringing motions. The sac's inner layer is part of the heart wall. Most of the wall is *myocardium*, or cardiac muscle cells tethered to fibers of elastin and collagen. The densely crisscrossed fibers are like a skeleton that accepts the force of contraction. The inner wall has an epithelial lining called *endothelium*. Coronary arteries branch off the aorta and move nutrients and oxygen to a capillary bed dedicated to cardiac muscle cells.

Each half of the heart has two chambers: an atrium (plural, atria) that receives blood, and a ventricle that pumps it out. Between each atrium and ventricle is an atrioventricular (AV) valve. Between each ventricle and the artery that leads away from it is a semilunar valve. Both are *one-way* valves. Fluid pressure forces them open and shut in an alternating way that helps keep the blood moving in the forward direction only.

Heart muscle alternately relaxes and contracts in a recurring **cardiac cycle**. All four chambers go through *diastole* (expansion) and *systole* (contraction). First, the relaxed atria expand with blood, and fluid pressure forces the AV valves to open. Blood now flows into the relaxed ventricles, which expand as the atria contract (Figure 38.14). The ventricles contract, and the fluid pressure in them rises so sharply above the pressure in the great arteries that both semilunar valves open, and so blood flows out. The ventricles relax while the atria are already filling for a new cycle. In short, atrial contraction only helps fill the ventricles. *Contraction of the ventricles is the driving force for blood circulation.*

HOW DOES CARDIAC MUSCLE CONTRACT?

Reflect on Sections 37.6 through 37.9, which explain how skeletal muscle contracts. **Cardiac muscle**, found only in the heart, contracts by the same ATP-driven sliding filament mechanism. Its cells, too, look striated

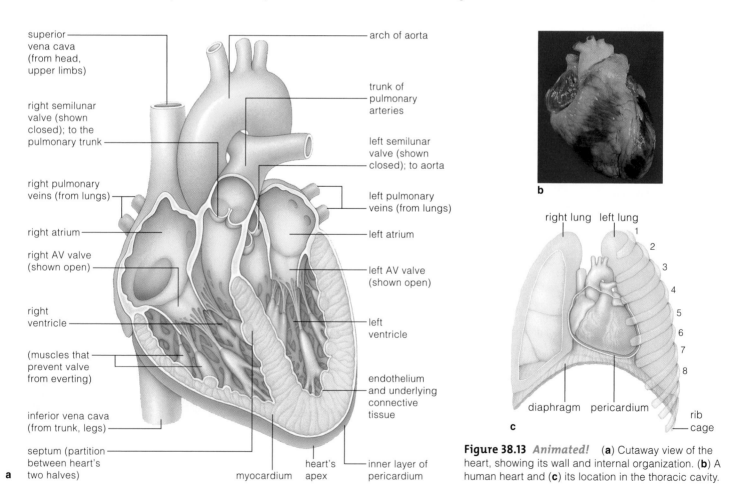

superior vena cava (from head, upper limbs)

right semilunar valve (shown closed); to the pulmonary trunk

right pulmonary veins (from lungs)

right atrium

right AV valve (shown open)

right ventricle

(muscles that prevent valve from everting)

inferior vena cava (from trunk, legs)

septum (partition between heart's two halves)

a

arch of aorta

trunk of pulmonary arteries

left semilunar valve (shown closed); to aorta

left pulmonary veins (from lungs)

left atrium

left AV valve (shown open)

left ventricle

endothelium and underlying connective tissue

myocardium

heart's apex

inner layer of pericardium

b

right lung left lung

1
2
3
4
5
6
7
8

diaphragm pericardium

rib cage

c

Figure 38.13 *Animated!* (**a**) Cutaway view of the heart, showing its wall and internal organization. (**b**) A human heart and (**c**) its location in the thoracic cavity.

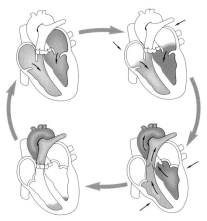

a Atria fill. Fluid pressure opens AV valves. Blood flows into ventricles.

b Atria contract. Fluid pressure in ventricles rises sharply.

d Ventricles relax as atria start filling and start another cycle.

c Ventricles contract, pump blood into aorta and pulmonary artery.

Figure 38.14 *Animated!* Cardiac cycle. We hear it as a "lub-dup" near the chest wall. At each "lub," AV valves are closing as ventricles are contracting. At each "dup," semilunar valves are closing as ventricles are relaxing.

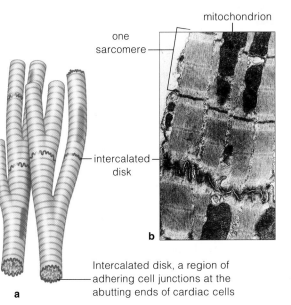
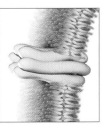

mitochondrion

one sarcomere

intercalated disk

b

a

Intercalated disk, a region of adhering cell junctions at the abutting ends of cardiac cells

c Gap junction. These communication junctions occur on the sides of cardiac muscle cells, where mechanical stress is not as great as at the ends. They connect the cytoplasm of adjoining cells. We removed one cell's plasma membrane for clarity.

Figure 38.15 (**a**) Branching cardiac muscle cells. (**b**) Intercalated disk, where a profusion of adhering junctions in the plasma membrane keeps abutting ends of cells from getting ripped apart by the heart's wringing motions. (**c**) On the sides of these cells, vast arrays of gap junctions across the plasma membrane are a pathway of low electrical resistance. Excitation spreads swiftly from cell to cell.

because of repeated contractile units (sarcomeres) that form a transverse banding pattern along their length. Unlike skeletal or smooth muscle, cardiac muscle has branching, nucleated cells that abut at their ends, and it has far more mitochondria. On the sides of cardiac muscle cells, action potentials pass right on through gap junctions, so that waves of excitation wash swiftly over the entire heart (Section 33.3 and Figure 38.15).

Where do the signals come from? In cardiac muscle, some specialized cells do not contract. They are part of the **cardiac conduction system**, which initiates and distributes signals that tell regular cardiac muscle cells to contract. As Figure 38.16 shows, the system consists of a sinoatrial (SA) node and an atrioventricular (AV) node, functionally linked by junctional fibers. These fibers are bundles of long, thin cardiac muscle cells.

The SA node, a clump of noncontracting cells in the right atrium's wall, is the **cardiac pacemaker**. Its cells have specialized membrane channels that let them fire action potentials again and again, seventy or so times a minute. The rhythmic firing rate starts in embryos and is on autopilot until death. Each firing triggers one cardiac cycle. First, a signal spreads through the atria and makes them contract. At the same time, the signal activates junctional fibers, which conduct it to the AV node. This clump of cells is the only electric bridge to the ventricles, which are insulated everywhere else by connective tissue. The time it takes for each signal to cross the bridge is enough to keep the ventricles from contracting before they fill. From this node, the signal

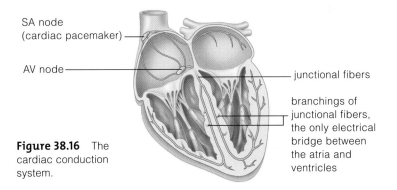

SA node (cardiac pacemaker)

AV node

junctional fibers

branchings of junctional fibers, the only electrical bridge between the atria and ventricles

Figure 38.16 The cardiac conduction system.

flows along a bundle of fibers. The fibers branch inside the septum, a partition between the heart's halves. The two branchings extend down to the heart apex and up the ventricle walls. Starting at the apex, cardiac muscle responds by contracting in a twisting motion, which ejects blood into the aorta and pulmonary arteries.

The nervous system can only adjust the rate and strength of contractions set by the natural pacemaker, which can keep the heart beating even when an injury severs all nerves to the heart. Those defibrillators you read about earlier work by resetting the SA node.

The four-chambered heart is partitioned into two halves, each with an atrium and a ventricle. Contraction of the ventricles pumps blood out of the heart, into arteries.

The SA node is the cardiac pacemaker. Its spontaneous, rhythmic signals make the branching, abutting cardiac muscle cells contract almost as if they were a single unit.

38.7 Pressure, Transport, and Flow Distribution

LINKS TO
SECTIONS
33.3, 35.1

During any specified interval, the pressure driving blood through your body depends on how fast and how strongly the heart is beating. It depends also on the total resistance to flow through the vascular system.

Figure 38.17 compares the structure of blood vessels. **Arteries** are rapid-transport vessels for blood pumped out of the heart's ventricles. They grade into **arterioles**, smaller vessels where controls over the distribution of blood flow operate. These grade into **capillaries**, the small blood vessels that form diffusion zones. **Venules** are small vessels located between capillaries and veins. **Veins** are large vessels that transport blood back to the heart, and they also are blood volume reservoirs.

Blood pressure is fluid pressure imparted to blood by ventricular contractions. It is highest in contracting ventricles, still high at the start of arteries, and lowest in the relaxed atria (Figure 38.18). In the pulmonary or systemic circuit, the difference in pressure between any two points affects the flow rate. Heartbeats establish high pressure at the start of either circuit. Blood rubs against the vessel walls, and friction impedes its flow. Resistance to flow depends mainly on the blood vessel diameter. Simply put, resistance rises as tubes narrow. A twofold decrease in the radius of a blood vessel will increase the resistance to flow sixteenfold.

RAPID TRANSPORT IN ARTERIES

With their large diameter and low resistance to flow, arteries are fast, efficient transporters of oxygenated blood. They also are pressure reservoirs that smooth out pulsations in pressure that are generated by each cardiac cycle. An artery's thick, muscular, elastic wall bulges from the large volume of blood forced into it by ventricular contraction. As the elastic wall recoils, it forces oxygen-rich blood farther through the circuit between contractions.

DISTRIBUTING BLOOD FLOW

No matter what you are doing, all of the blood from the heart's right half is flowing to your lungs. The flow distribution from the heart's left half to one organ or another along the systemic circuit can vary, but when you are just lounging around, it is probably close to the values shown in Figure 38.19.

What happens when you go for a run? A greater volume of blood is diverted to your skeletal muscles relative to the volume being distributed to your skin,

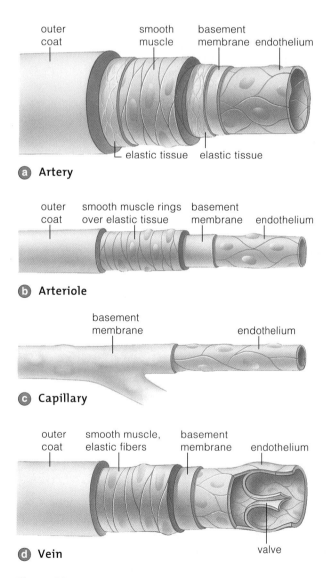

a Artery

b Arteriole

c Capillary

d Vein

Figure 38.17 Structural comparison of human blood vessels. Capillaries grade into venules (not shown), which have a similar function. These drawings are not drawn to the same scale.

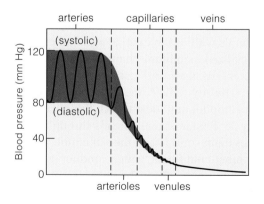

Figure 38.18 Plot of the decline in fluid pressure for a volume of blood that flowed through the systemic circuit.

kidneys, and gut. Like traffic cops, arterioles guide the flow according to instructions from a command post. The nervous and endocrine systems send signals that act on rings of smooth muscle cells in arteriole walls (Figure 38.17b). Specific signals make these cells relax, causing **vasodilation**; the diameter of the blood vessel enlarges. Different signals make them contract, causing **vasoconstriction**; the diameter shrinks. Dilate all arterioles in one tissue, and more blood flows to it. Constrict them, and less blood flows to the tissue.

Arterioles also respond to metabolic activities that shift the concentrations of substances in a tissue. Such local chemical changes are "selfish" in that they invite or divert blood flow to meet a tissue's own metabolic needs. For instance, as you run, your skeletal muscle cells use up oxygen, and the levels of carbon dioxide, hydrogen and potassium ions, and other solutes rise. The increase in ion levels causes arterioles in skeletal muscle tissues to dilate. More blood flowing through the region delivers more raw materials and carts off cell products and metabolic wastes. When the skeletal muscles relax, their demand for oxygen declines. Now the oxygen level rises and the arterioles constrict.

CONTROLLING BLOOD PRESSURE

We generally measure blood pressure at the brachial artery in an upper arm (Figure 38.20). In each cardiac cycle, *systolic* (peak) pressure is exerted by contracting ventricles against the arterial wall. *Diastolic* pressure, the lowest arterial blood pressure of the cardiac cycle, is reached when the ventricles are relaxed. An average resting blood pressure is recorded as 120/80, which is the systolic pressure/diastolic pressure.

Blood pressure depends on the total blood volume, how much blood is being pumped out of the heart (cardiac output), and arteriole resistance. *Baroreceptors* in the wall of some arteries, such as the carotids, signal a control center in the brain when fluid pressure rises or falls (Section 35.1). In response, the brain calls for changes in cardiac output and arteriole diameter. This reflex response is a key short-term control over blood pressure. Long-term controls, exerted in the kidneys, adjust the total volume and composition of the blood.

> *The rate and strength of heartbeats and resistance to flow through blood vessels dictates blood pressure. Pressure is greatest in contracting ventricles and at the start of arteries.*
>
> *Flow distribution is controlled mainly by adjustments in the diameter of arterioles in different tissues.*

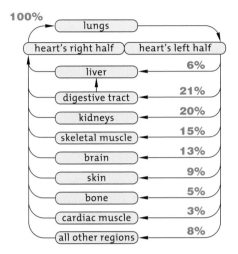

100%
lungs
heart's right half — heart's left half
liver — 6%
digestive tract — 21%
kidneys — 20%
skeletal muscle — 15%
brain — 13%
skin — 9%
bone — 5%
cardiac muscle — 3%
all other regions — 8%

Figure 38.19 Distribution of the heart's output in people napping. How much flows through a given region is adjusted through selective vasodilation and vasoconstriction at many arterioles all along the systemic circuit.

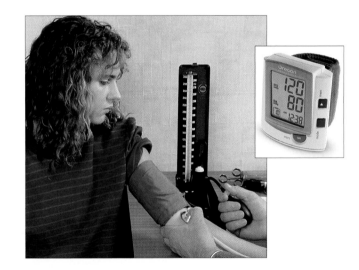

Figure 38.20 *Animated!* Measuring blood pressure. First, a hollow inflatable cuff attached to a pressure gauge is wrapped around the upper arm. A stethoscope is placed over the brachial artery, just below the cuff.

The cuff is inflated with air to a pressure above the highest pressure of the cardiac cycle, when ventricles contract. Above this pressure, you will not hear sounds through the stethoscope, because no blood is flowing through the vessel.

Air in the cuff is slowly released until the stethoscope picks up soft tapping sounds. The sounds are caused by blood spurting into the artery under the force of the strongest ventricular contraction. When sounds start, the gauge's value is typically about 120 mm Hg. That measured amount of pressure would force mercury (Hg) to move up 120 millimeters in a narrow glass column of standardized size.

More air is released from the cuff. Eventually the sounds stop. Blood is now flowing continuously, even when the ventricles are the most relaxed. The pressure when the sounds stop is the lowest during a cardiac cycle, usually about 80 mm Hg.

Monitors, such as the one shown in the inset, are now available that automatically record the systolic/diastolic blood pressure.

38.8 Diffusion at Capillaries, Then Back to the Heart

LINKS TO
SECTIONS 5.3, 5.5,
25.10, 28.3, 34.9

Every capillary bed is a diffusion zone for exchanges between blood and interstitial fluid. Living cells in any body tissue die quickly when deprived of the exchanges. Brain cells start dying off within four minutes.

CAPILLARY FUNCTION

Figure 38.21 shows some of the 10 billion to 40 billion capillaries that service the human body. Collectively, they offer a tremendous surface area for exchanging gases. At least one is next to living cells in nearly all tissues. The proximity is critical. Diffusion distributes molecules and ions slowly, and not very far.

Red blood cells are eight micrometers across. They have to squeeze single file through the capillaries. The squeeze puts oxygen-transporting red blood cells and solutes in plasma in direct contact with the exchange area—the capillary wall—or only a very short distance away from it.

A capillary is a sheet of endothelial cells organized as a cylinder, one cell thick, wrapped in a basement membrane. So how do oxygen and carbon dioxide get from blood, across the endothelial cells, and into the interstitial fluid? Like many other small, lipid-soluble substances, they diffuse through the lipid bilayer of each cell's plasma membrane, through its cytoplasm, then on through the bilayer on the other side. Certain proteins cross by way of endocytosis and exocytosis.

The capillaries in most body regions have very thin clefts between endothelial cells. Ions and small, water-soluble molecules squeeze through these clefts. So do white blood cells that are circulating in blood, as the next chapter explains. In capillaries in the brain, tight junctions join endothelial cells together, and clefts are nonexistent. Substances cannot "leak" in between the endothelial cells; they have to pass through them. The junctions help maintain the vital blood–brain barrier, as explained in Section 34.9.

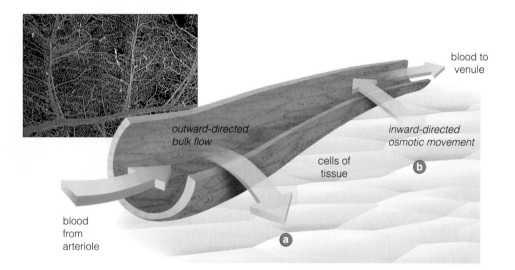

Figure 38.21 Bulk flow in a capillary bed. Fluid crosses a capillary wall by way of ultrafiltration and reabsorption. (**a**) At the capillary's arteriole end, a difference between blood pressure and interstitial fluid pressure forces out some plasma but few plasma proteins. Bulk flow moves plasma through clefts between endothelial cells of the capillary wall. Ultrafiltration is the bulk flow of fluid *out* of a capillary.

(**b**) Reabsorption is the osmotic movement of some interstitial fluid *into* the capillary. It happens when the water concentration between interstitial fluid and the plasma differs. Plasma, with its dissolved proteins, has a greater solute concentration and therefore a lower water concentration. Reabsorption near the end of a capillary bed tends to balance ultrafiltration at the start of it. Normally there is only a small *net* filtration of fluid, which the lymphatic system returns to the blood.

ARTERIOLE END OF CAPILLARY BED	
Outward-Directed Pressure:	
Hydrostatic pressure of blood in capillary:	35 mm Hg
Osmosis due to interstitial proteins:	28 mm Hg
Inward-Directed Pressure:	
Hydrostatic pressure of interstitial fluid:	0
Osmosis due to plasma proteins:	3 mm Hg
Net Ultrafiltration Pressure:	
(35 − 0) − (28 − 3) = 10 mm Hg	
ULTRAFILTRATION FAVORED	

VENULE END OF CAPILLARY BED	
Outward-Directed Pressure:	
Hydrostatic pressure of blood in capillary:	15 mm Hg
Osmosis due to interstitial proteins:	28 mm Hg
Inward-Directed Pressure:	
Hydrostatic pressure of interstitial fluid:	0
Osmosis due to plasma proteins:	3 mm Hg
Net Reabsorption Pressure:	
(15 − 0) − (28 − 3) = −10 mm Hg	
REABSORPTION FAVORED	

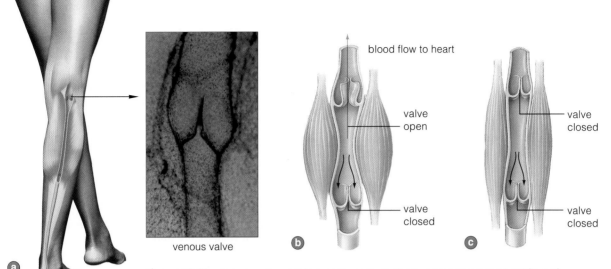

Figure 38.22 Vein function. (**a**) Valves in medium-sized veins prevent the backflow of blood. Neighboring skeletal muscles helps raise fluid pressure inside a vein. (**b**) These muscles bulge into the vein when they contract, which increases the pressure that keeps blood flowing forward. (**c**) When muscles relax, venous valves shut and stop backflow.

Diffusion is not the only process acting at capillary beds. **Bulk flow** is the movement of water and solutes in response to a difference in fluid pressure between regions. It does not affect diffusion much, but it helps maintain the distribution of fluid between blood and interstitial fluid. Figure 38.21 indicates how the water and solutes in blood and interstitial fluid influence the direction of bulk flow. At the start of a capillary bed, a small amount of protein-free plasma is pushed out in bulk through clefts in the capillary wall. This process is called **ultrafiltration**. Farther on, tissue fluid moves back into capillaries through the clefts in the capillary wall. This process is called **capillary reabsorption**.

Normally, there is a tiny *net* outward flow from a capillary bed, which the lymphatic system returns to blood. High blood pressure increases the amount of fluid leaving capillaries. Too much fluid in interstitial spaces is called *edema*. Commonly, gravity causes the fluid to pool in foot and ankle tissues. Edema often follows physical exercise, because arterioles in many tissues have dilated. It also can result from obstructed veins or heart failure. It is dramatic in *elephantiasis*, when roundworm infection slows the return of fluid to lymphatic vessels (Section 25.10).

VENOUS PRESSURE

Blood moving out of capillary beds enters venules, or "little veins," which merge into large-diameter veins. Functionally, venules are a bit like capillaries. Certain solutes diffuse across their thin walls.

Veins are large-diameter, low-resistance transport tubes that carry blood toward the heart. Within many veins, especially in the leg, valves prevent backflow (Figure 38.22). When gravity causes blood in the vein to reverse direction, the valves are pushed shut.

The vein wall can bulge quite a bit under pressure, far more so than an arterial wall. Thus, veins can act as reservoirs for variable volumes of blood. About 60 percent of the total blood volume is in the veins.

The vein wall contains some smooth muscle. When blood must be circulated faster, as during exercise, this smooth muscle contracts. It stiffens the wall, so that the vein cannot bulge out as much. As pressure in the lumen rises, more blood is driven toward the heart. Also, when limbs move, contracting skeletal muscles bulge against neighboring veins. This helps raise the venous pressure that moves blood toward the heart (Figure 38.22). Rapid breathing also contributes to an increase in venous pressure. Inhaling pushes air down on internal organs, which alters the pressure gradient between the heart and veins.

Sometimes venous valves lose their elasticity. The veins become enlarged and bulge near the surface of skin. Such *varicose veins* are common in legs. Around the anus, such bulges are called *hemorrhoids*. Exercise and maintaining an ideal weight reduce the risk.

Capillary beds are diffusion zones for exchanges between blood and interstitial fluid. Here, bulk flow contributes to the fluid balance between blood and interstitial fluid.

Venules overlap somewhat with capillaries in function. Veins are highly distensible blood volume reservoirs where the flow volume back to the heart can be adjusted.

38.9 Cardiovascular Disorders

LINKS TO
SECTIONS
3.5, 36.7

A stroke results from a blood clot or ruptured blood vessel in the brain. A heart attack occurs when the small blood vessels that service the heart become clogged. High blood pressure and atherosclerosis increase the risks, many of which can be minimized by changes in life-styles.

Good Clot, Bad Clot A process called **hemostasis** can stop blood loss from small blood vessels that have become ruptured or cut, and it can build a framework for repairs. As Figure 38.23 shows, in the first phase of hemostasis, smooth muscle in a damaged wall goes into spasm. This involuntary, abnormal contraction lasts for about thirty minutes and may stop the blood loss. In the second phase, platelets clump together briefly at the damaged site. They release substances that prolong the spasm and attract more platelets. In the third phase, plasma proteins help convert blood to a gel. Then fibrinogens (rodlike plasma proteins) form. They form long, insoluble threads that stick to exposed collagen fibers at the damaged site. Together,

they form a net that traps blood cells and platelets (Figure 38.23). The entire mass, a *blood clot*, retracts as a compact mass that seals the breach in the blood vessel wall.

Clot formation is essential for repairing blood vessels. However, clots cause problems when they completely block blood flow through a vessel, as you will now see.

Atherosclerosis With *arteriosclerosis*, arteries thicken and lose elasticity. With **atherosclerosis**, the condition worsens as lipids build up in the arterial wall and narrow the lumen. You have probably heard that cholesterol plays a role in this "hardening of the arteries." To be sure, the human body requires cholesterol to make cell membranes, myelin sheaths, bile salts, and steroid hormones. The liver makes enough cholesterol for cell structure and function, but more cholesterol is absorbed from food in the gut. People differ in how the excess is handled.

Most of the cholesterol dissolved in blood is bound to protein carriers. The complexes are known as *low-density lipoproteins*, or LDLs, and most cells can take them up. A lesser amount is bound up in *high-density lipoproteins*, or HDLs. Cells in the liver metabolize HDL. The metabolic wastes are eliminated by way of bile, which the liver secretes into the gut lumen for disposal (Section 41.4).

When the LDL levels in blood go up, so does the risk of atherosclerosis. The first sign of trouble is a build-up of lipids in an artery's endothelium (Figure 38.24). As lipids accumulate, smooth muscle proliferates in the arterial wall, which becomes inflamed. Fibrous connective tissue forms over the entire mass. The mass is an atherosclerotic plaque, and it makes the arterial wall bulge inward and narrow the lumen.

A hardened plaque can rupture an artery wall, thereby triggering clot formation. When the clot stays put, we call it a *thrombus*. When it becomes dislodged and travels in blood, we call it an *embolus*, and it can clog small vessels.

Coronary arteries are vulnerable. If they narrow by 25 percent, they invite *angina pectoris* (chest pain) or a heart attack. During an attack, some of the cardiac muscle cells die from lack of oxygen. A first attack is fatal about 30 percent of the time. Survivors often have a badly weakened heart.

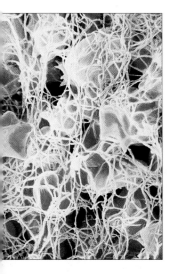

Stimulus
Blood vessel damage

Phase 1 response
Half-hour vascular spasm constricts vessel at site of damage, slows blood loss.

Phase 2 response
Platelets aggregate and stick together within fifteen seconds, thus plugging the site.

Phase 3 response
Clot formation starts after thirty seconds:

1. Enzymes activate factor X; prothrombin forms.
2. Prothrombin converts an enzyme precursor to thrombin.
3. Thrombin converts fibrinogen, a plasma protein, to insoluble protein threads (fibrin).
4. Fibrin forms a net that entangles blood cells and platelets; the entire mass is a blood clot.

Figure 38.23 Hemostasis. The photomicrograph shows a fibrous protein net that helps blood clots form.

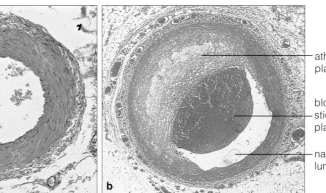

wall of artery, cross-section

unobstructed lumen of normal artery

atherosclerotic plaque

blood clot sticking to plaque

narrowed lumen

Figure 38.24 Sections from (**a**) a normal artery and (**b**) an artery with a lumen narrowed by an atherosclerotic plaque. A clot clogged this one.

With *coronary bypass surgery*, doctors stitch a section of a blood vessel from elsewhere in the body to the aorta and to the coronary artery below a clogged region (Figure 38.25). With *laser angioplasty*, laser beams vaporize the plaques. With *balloon angioplasty*, doctors inflate a small balloon in a blocked artery to flatten the plaques.

Hypertension—A Silent Killer

Hypertension refers to chronically high blood pressure even during periods of rest. Blood pressure stays above 140/90, often for unknown reasons. This chronic condition is known as a silent killer, because symptoms do not always show up. Heredity may be a factor; the disorder tends to run in families. Diet and lack of regular exercise are factors. In some people, high salt intake raises blood pressure and makes the heart pump harder. The heart may enlarge and fail to pump efficiently. High blood pressure also may contribute to atherosclerosis, which interferes with the delivery of oxygen to the brain, heart, and other vital organs. Of 23 million hypertensive Americans, most do not seek treatment. About 180,000 die each year.

Rhythms and Arrhythmias

As you read in Section 38.6, the SA node controls the rhythmic beating of the heart. Electrocardiograms, or ECGs, record the electrical activity of a beating heart (Figures 38.1 and 38.26a).

ECGs can reveal *arrhythmias*, which are abnormal heart rhythms (Figure 38.26b–d). Arrhythmias are not always dangerous. For instance, endurance athletes commonly experience *bradycardia*, a below-average resting cardiac rate. In response to ongoing exercise, their nervous system adjusts their cardiac pacemaker's rate of firing downward. Intense exercise or stress often results in 100 or more heartbeats per minute, a condition called *tachycardia*.

In *atrial fibrillation*, the atria do not contract normally. They quiver, which increases the risk of blood clots and stroke. *Ventricular fibrillation* is the most dangerous type of arrhythmia. It caused the collapse of Tammy Higgins, as described at the start of this chapter. Ventricles flutter and their pumping action falters or halts. The individual loses consciousness and faces death. A shock from a defibrillator might be able to restore the heart's normal rhythm.

Risk Factors

Cardiovascular disorders are the leading cause of death in the United States. Each year, about 40 million people experience cardiovascular disorders, and each year about 1 million die. Nine factors top the list of risks, and tobacco smoking tops them all (Section 40.8). Other factors include a genetic predisposition to heart attacks, hypertension, a high blood level of cholesterol, obesity, and diabetes mellitus, a condition described in Section 36.7. Advancing age also is a risk factor; the older you get, the greater the risk. Physical inactivity increases the risk. Regular exercise can lower the risk, even when an activity is not particularly strenuous. Gender, too, is a factor; until about age fifty, males are at greater risk.

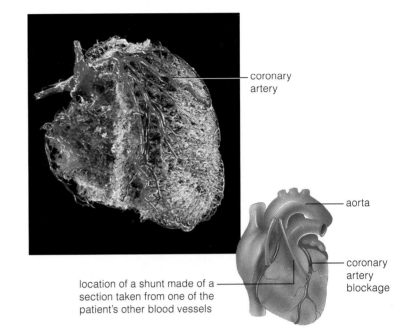

coronary artery

aorta

coronary artery blockage

location of a shunt made of a section taken from one of the patient's other blood vessels

Figure 38.25 Coronary arteries and other blood vessels of the heart. Resins were injected into them, then the rest of the cardiac tissues were dissolved, leaving this accurate, three-dimensional corrosion cast. The sketch shows two coronary bypasses (artificially colored *green*), which extend from the aorta past two clogged parts of the coronary arteries.

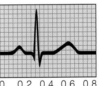

one normal heartbeat

0   0.2 0.4 0.6 0.8

a   time (seconds)

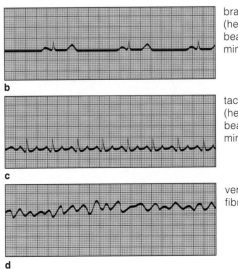

Figure 38.26 **(a)** ECG of one normal beat of the human heart. **(b–d)** Three arrhythmia recordings.

b   bradycardia (here, 46 beats per minute)

c   tachycardia (here, 136 beats per minute)

d   ventricular fibrillation

38.10 Interactions With the Lymphatic System

LINK TO SECTION 36.12

We conclude this chapter with a brief look at how the lymphatic system interacts with blood circulation. Think of this section as a bridge to the next chapter, on immunity, because the lymphatic system also helps defend the body against injury and attack.

LYMPH VASCULAR SYSTEM

A portion of the lymphatic system, called the **lymph vascular system,** consists of many tubes that collect and deliver water and solutes from interstitial fluid to ducts of the circulatory system. Its main components are lymph capillaries and vessels (Figure 38.27). Tissue fluid that moves into these vessels is called **lymph.**

The lymph vascular system serves three functions. First, its vessels are drainage channels for water and plasma proteins that have leaked out from the blood at capillary beds and must be delivered back to the circulatory system. Second, the system takes up fats that the body has absorbed from the small intestine and delivers them to the general circulation (Section 41.4). Third, it delivers pathogens, foreign cells, and cellular debris from tissues to the lymph vascular system's disposal centers, the lymph nodes.

TONSILS
Defense against bacteria and other foreign agents

RIGHT LYMPHATIC DUCT
Drains right upper portion of the body

THYMUS GLAND
Site where certain white blood cells acquire means to chemically recognize specific foreign invaders

THORACIC DUCT
Drains most of the body

SPLEEN
Major site of antibody production; disposal site for old red blood cells and foreign debris; site of red blood cell formation in the embryo

SOME OF THE LYMPH VESSELS
Return excess interstitial fluid and reclaimable solutes to the blood

SOME OF THE LYMPH NODES
Filter bacteria and many other agents of disease from lymph

BONE MARROW
Marrow in some bones is production site for infection-fighting blood cells (as well as red blood cells and platelets)

a

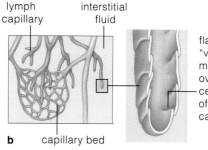

b capillary bed

lymph capillary   interstitial fluid

flaplike "valve" made of overlapping cells at tip of lymph capillary

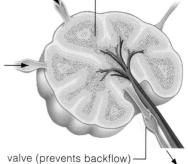

lymph trickles past organized arrays of lymphocytes

c valve (prevents backflow)

Figure 38.27 *Animated!* (**a**) Components of the human lymphatic system and their functions. Not shown are patches of lymphoid tissue in the small intestine and in the appendix. (**b**) Diagram of lymph capillaries at the start of a drainage network, the lymph vascular system. (**c**) Cutaway view of a lymph node. Its inner compartments are packed with organized arrays of infection-fighting white blood cells.

The lymph vascular system starts at all capillary beds. There, excess fluid enters the lymph capillaries. These capillaries have no obvious entrance; water and solutes move into clefts between cells. As you can see from Figure 38.27b, the endothelial cells overlap, and they form flaplike valves. Lymph capillaries merge into larger diameter lymph vessels. These vessels have some smooth muscle in the wall and valves that stop backflow. The lymph vessels converge into collecting ducts, which drain into veins in the lower neck.

LYMPHOID ORGANS AND TISSUES

The other portion of the lymphatic system has roles in the body's defense responses to injury and attack. We call its components *lymphoid* organs and tissues. They include the lymph nodes, spleen, and thymus, as well as the tonsils, and patches of tissue in the wall of the small intestine and appendix.

Lymph nodes are strategically located at intervals along lymph vessels (Figure 38.27c). Before entering blood, the lymph trickles through at least one node for filtration. Masses of lymphocytes take up stations in the nodes after they form in bone marrow. When they contact cellular debris or an invader, they divide rapidly and form large armies that destroy it.

The **spleen** is the largest lymphoid organ; it's about the size of a fist in an average adult. It functions as a site of red blood cell formation only in embryos. After childbirth, it filters pathogens and used-up red cells and platelets from blood vessels that branch through it. Phagocytic white cells inside the spleen engulf and digest defunct cells and alert the body to invaders. Lymphocytes in the spleen make antibodies. Even so, people survive without it. When the spleen has been damaged by trauma and must be removed, a greater risk of infection is the only bad consequence.

In the **thymus gland**, T lymphocytes differentiate in ways that allow them to recognize and respond to particular pathogens. The thymus gland also makes the hormones that influence these actions. It is central to immunity, the focus of the next chapter.

The lymph vascular part of the lymphatic system consists of many vessels that start in capillary beds. They return water and solutes from tissue fluid to blood, deliver absorbed fats to the bloodstream, and deliver pathogens to lymph nodes.

The lymphatic system also includes lymph nodes and other lymphoid organs that have specific roles in defending the body against tissue damage and infectious diseases.

Summary

Section 38.1 A circulatory system moves substances to and from interstitial fluid faster than simple diffusion could take them. This fluid fills tissue spaces between cells and exchanges substances with them.

Some invertebrates have an open circulatory system, in which blood spends part of the time mingling with tissue fluids. Vertebrates have a closed circulatory system, with blood confined inside a heart and blood vessels. They differ in whether blood flows through one or two circuits of blood vessels and in the number of chambers in their heart. As lungs evolved in the early vertebrates on land, the circulatory system underwent modifications that made gas exchange more efficient. Fluid leaking from blood vessels enters the lymphatic system, which filters it and then returns it to the circulatory system.

Biology ⋐Now
Compare animal circulatory systems with the animation on BiologyNow.

Section 38.2 Blood, a fluid connective tissue, consists of plasma, blood cells, and platelets. Plasma is mostly water in which diverse ions and molecules are dissolved. Red blood cells, or erythrocytes, contain the hemoglobin that functions in the rapid transport of oxygen and, to a lesser extent, carbon dioxide. Different white blood cells, or leukocytes, function in day-to-day tissue maintenance and repair as well as defense of threatened tissues. Platelets release substances that initiate blood clotting. All blood cells and platelets arise from stem cells in bone marrow.

Section 38.3 In a blood disorder, an individual has too many, too few, or abnormal red or white blood cells.

Section 38.4 Among the recognition proteins on the surface of red blood cells are self markers that identify an individual's blood types. ABO blood typing helps match the blood of donors and recipients to avoid blood transfusion problems. Rh blood typing and suitable treatment assure the survival of fetuses of prospective parents who have incompatible Rh blood types.

Biology ⋐Now
Learn about blood types with the animation on BiologyNow.

Section 38.5 The human heart is a durable pump, the contraction of which forces blood through two separate circuits that lead back to the heart.

In the pulmonary circuit, oxygen-poor blood from the heart's right half flows to the lungs, picks up oxygen, then flows to the heart's left half.

In the systemic circuit, the oxygen-rich blood flows from the heart's left half to all body tissues, then oxygen-poor blood flows to the heart's right half.

Most blood flows through one capillary system. In one route, it flows through intestinal capillaries, then liver capillaries. The liver metabolizes or stores nutrients and neutralizes a number of bloodborne toxins.

Biology⊘Now

Explore the human cardiovascular system with the animation on BiologyNow.

Section 38.6 The human heart is a thick-walled, double pump that beats perpetually. Each half has two chambers: an atrium and a ventricle.

In one cardiac cycle, all of the chambers undergo rhythmic expansion (diastole) and contraction (systole). When a cycle starts, each atria expands as blood fills it and opens a valve to a ventricle. Ventricles are already filling when the atria contract. When ventricles contract, they force blood into the aorta and pulmonary arteries.

The cardiac conduction system is the basis of the heart's beating. It consists of an SA node in the right atrium wall that is functionally linked by bundles of conducting fibers to an AV node.

The SA node is the cardiac pacemaker. Here, action potentials are spontaneously generated and set the pace for contraction. Waves of excitation wash over atria, down fibers in the heart's septum, then up the walls of the ventricles. The nervous system adjusts the rate and strength of contractions but does not initiate them.

Biology⊘Now

Learn about the structure and function of the human heart with the animation on BiologyNow.

Section 38.7 The blood pressure is highest in the contracting ventricles. It falls as blood flows through arteries, arterioles, capillaries, venules, and veins of the systemic or pulmonary circuit. It is lowest in relaxed atria. The flow rate depends on the strength and rate of heartbeat and on resistance to flow in different blood vessels. Adjustments in the diameter of arterioles in different parts of the body can redistribute flow to the tissues requiring the most metabolic support during a given interval.

Biology⊘Now

See how blood pressure is measured with the animation on BiologyNow.

Section 38.8 Capillary beds are zones of diffusion between blood and interstitial fluid. Ultrafiltration pushes a small amount of fluid out of capillaries. Fluid moves back in by capillary reabsorption. Normally, both processes are almost balanced, with just a small net outward flow of fluid from a capillary bed.

Venules overlap capillaries in function. Veins are rapid-transport vessels and a blood volume reservoir for adjusting the flow volume back to the heart.

Section 38.9 Hemostasis, a process that stops blood flow from small vessels after injury, results in clot formation. Blood clotting is beneficial except in some cardiovascular disorders. The most common disorders are atherosclerosis, hypertension (chronic high blood pressure), heart attacks, strokes, and certain arrythmias. Regular exercise, maintaining a normal body weight, and not smoking lower the risks for most people.

Section 38.10 The lymphatic system structurally and functionally supports the circulatory system. The vascular portion of the lymphatic system includes lymph vessels and lymph capillaries. It takes up excess water and plasma proteins from interstitial fluid, as well as absorbed fats, and transports them to blood. It transports bloodborne pathogens and foreign material to lymph nodes and the spleen.

The lymphoid organs and tissues of the lymphatic system are sites of maturation for some white blood cells. Some are battlegrounds where organized arrays of these cells screen lymph and battle disease-causing agents.

Biology⊘Now

Learn about the human lymphatic system with the animation on BiologyNow.

Self-Quiz   *Answers in Appendix II*

1. Cells directly exchange substances with _____ .
   - a. blood vessels
   - b. lymph vessels
   - c. interstitial fluid
   - d. both a and b

2. All vertebrates have _____ .
   - a. an open circulatory system
   - b. a closed circulatory system
   - c. a four-chambered heart
   - d. both b and c

3. Which are not found in the blood?
   - a. plasma
   - b. blood cells and platelets
   - c. gases and dissolved substances
   - d. All of the above are found in blood.

4. A person who has type O blood _____ .
   - a. can receive a transfusion of blood of any type
   - b. can donate blood to a person of any blood type
   - c. can donate blood only to a person of type O
   - d. cannot be a blood donor
   - e. both a and b

5. In the blood, most oxygen is transported _____ .
   - a. in red blood cells
   - b. in white blood cells
   - c. bound to hemoglobin
   - d. both a and c

6. Blood flows directly from the left atrium to _____ .
   - a. the aorta
   - b. the left ventricle
   - c. the right atrium
   - d. the pulmonary arteries

7. Contraction of _____ drives the flow of blood through the aorta and pulmonary arteries.
   - a. the atria
   - b. arterioles
   - c. the ventricles
   - d. skeletal muscle

8. Blood pressure is highest in the _____ and lowest in the _____ .
   - a. arteries; veins
   - b. arterioles; venules
   - c. veins; arteries
   - d. capillaries; arterioles

9. At rest, the largest volume of blood is in the _____ .
   - a. arteries
   - b. capillaries
   - c. veins
   - d. arterioles

10. Which is not a function of the lymphatic system?
   - a. filters out pathogens
   - b. returns fluid to the circulatory system
   - c. helps certain white blood cells mature
   - d. distributes oxygen to the tissues

11. Match the components with their functions.
_____ capillary bed
_____ lymph node
_____ blood
_____ ventricle
_____ SA node
_____ veins
_____ aorta

a. filters out pathogens
b. cardiac pacemaker
c. main blood volume reservoir
d. largest artery
e. fluid connective tissue
f. zone of diffusion
g. contractions drive blood circulation

12. Label the components of the heart diagram shown to the right:

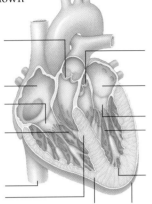

13. Identify these blood vessels and list their functions:

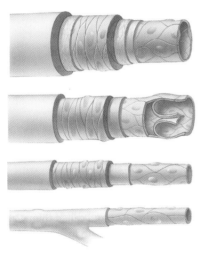

Additional questions are available on Biology **ⓔ**Now™

Critical Thinking

1. The highly publicized deaths of a few airline travelers led to warnings about *economy-class syndrome*. The idea is that sitting motionless for long periods on flights allows blood to pool and clots to form in legs. Low oxygen levels in airline cabins may increase clotting. If a clot gets large enough to block blood flow or breaks free and is carried to the lungs or the brain, the outcome can be deadly.

There might be a time lag between clot formation and health problems, so a connection to air travel might easily have been overlooked. Studies are now under way to

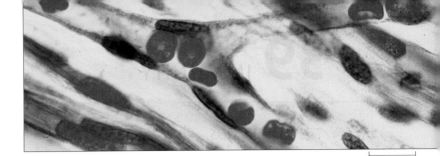

10 μm

Figure 38.28 Light micrograph of a branching blood vessel.

determine whether economy-class travel represents a significant risk. Given what you know about blood flow in the veins, explain why periodically getting up and moving around in the plane's cabin during a long flight may lower the risk of clot formation.

2. Consider the micrograph in Figure 38.28. It shows red blood cells moving through a blood vessel. What type of vessel is this? Explain how you came to this conclusion.

3. Some membrane proteins of *Streptococcus pyrogenes* are similar to those of cells in connective tissues throughout a human body. When this bacterium causes throat infections, weapons called antibodies go to work against the invader. However, they also go to work against connective tissues of the heart, joints, and elsewhere. Chronic inflammation over the course of a few years or even decades leads to *rheumatic heart disease*. The heart valves become damaged or deformed. Explain how this disease affects the heart's function and what problems might arise as a consequence.

4. Mitochondria occupy about 40 percent of the volume of human cardiac muscle but only 12 percent of the volume of skeletal muscle. Explain why there is such a difference.

5. Like other insects, the fruit fly (*Drosophila melanogaster*) has an open circulatory system. The transport medium is hemolymph, not blood. Contraction of one of the main vessels of the system drives a slow flow of hemolymph through the tiny fly body.

In 1993, researchers found that normal development of the fly "heart" requires the presence of a gene they named *tinman*. The gene's name is a reference to the character in *The Wizard of Oz* who had no heart. If the *tinman* gene is mutated, no heart forms in the embryonic fly, which dies. Genes having a similar base sequence have been found in zebrafish, the African clawed toad, chickens, mice, and humans. In all of these evolutionarily distant species, mutant versions result in abnormal heart development. In humans, mutations cause some genetic defects in the partition dividing the heart chambers and in the cardiac conduction system. Does it surprise you that a single gene would have such a major effect on the development of hearts or heartlike organs in so many different organisms? Explain your answer.

6. Miranda is nineteen, a college student, and a dedicated runner. About a month ago, a friend convinced her to switch to a *strict vegan diet*. She avoids all foods and other products that contain materials from animals, including eggs and milk. As she began training for a race, she found herself feeling unusually fatigued. Her friends commented that she looked pale. When she visited her doctor, she was told that the iron stores in her body were being depleted. Explain how iron deficiency would affect her blood and could cause her symptoms.

39.4 Tailoring Responses to Specific Antigens

LINKS TO
SECTIONS
4.6, 38.10

Sometimes surface barriers, fever, and inflammation are not enough to end a threat, and an infection becomes established. Long-term, specific defenses now begin.

FEATURES OF ADAPTIVE IMMUNITY

Given life's diversity, the number of different antigens is essentially infinite. No living system can recognize all of them, but vertebrate adaptive immunity comes close. Lymphocytes, phagocytes, and their signaling molecules interact to bring about the four defining characteristics of vertebrate adaptive immunity: self/nonself recognition, specificity, diversity, and memory.

Self versus nonself recognition starts at the particular molecular configurations that give every kind of cell or virus its unique identity. The plasma membrane of your own cells bears self-recognition proteins named human leukocyte antigens (HLA). These proteins also are known as **MHC markers**, after the group of genes that encode them (*Major Histocompatibility Complex*). Your T cells have **TCRs**, a class of antigen receptors at

their surface. T cells normally do not target a body cell that has unadorned MHC markers but will act against it if those markers have antigen bits attached.

Specificity means that a new B cell or T cell makes receptors for one—and only one—kind of antigen.

Diversity refers to the collection of antigen receptors on all B and T cells in the body. There are potentially billions of different antigen receptors, so an individual has the potential to counter billions of different threats.

Memory refers to the immune system's capacity to "remember" antigen that it vanquished. The first time lymphocytes recognize an antigen, it takes a few days for their populations to form. When the same antigen shows up again, the system makes a faster, heightened response. That is why you usually do not get as sick when the same disease agent infects you again.

FIRST STEP—THE ANTIGEN ALERT

Recognition of a specific antigen is only a first step in adaptive immunity. It stimulates repeated mitotic cell divisions that result in large populations of B and T cells, all primed to recognize the same antigen.

T cell receptors recognize an antigen only after an antigen-presenting cell finds it first. Macrophages, B cells, and dendritic cells do the presenting. First, they engulf anything bearing antigen, which vesicles move into the cytoplasm. The vesicles fuse with lysosomes, and lysosomal enzymes digest that antigen into bits (Section 4.6 and Figure 39.10). Some fragments bind to MHC markers. Then the antigen–MHC complexes are shuttled off to the plasma membrane, and there they are displayed. *When a cell's MHC markers become paired with antigen fragments, it becomes a call to arms.*

Odds are that at least one T cell has a receptor that can bind to this antigen–MHC complex. If it does, it becomes activated and secretes cytokines that induce divisions of B or T cells sensitive to the same antigen. Huge clonal populations of B and T cells form. Most of the cells are *effector* cells—differentiated lymphocytes that act immediately against the antigen. *Memory* cells are long-lived B and T cells that develop during the first exposure to an antigen. They are set aside for any future encounters with the same antigen.

TWO ARMS OF ADAPTIVE IMMUNITY

Like a boxer's one-two punch, adaptive immunity has two separate arms. Figure 39.11 is an overview of how the two work together to eliminate diverse threats.

Not all threats present themselves in the same way. For example, bacteria, fungi, or toxins circulating in

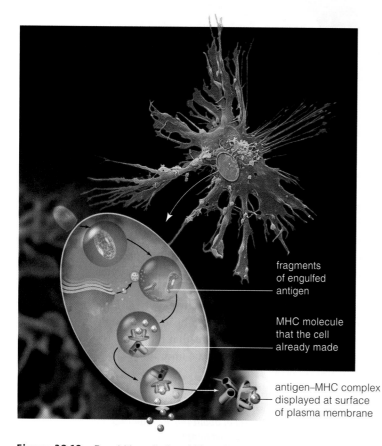

fragments
of engulfed
antigen

MHC molecule
that the cell
already made

antigen–MHC complex
displayed at surface
of plasma membrane

Figure 39.10 Dendritic cell. Dendritic cells patrol blood, internal organs, and skin. They ingest, process, and display antigen bound to MHC markers. Their primary function is to present antigen to T cells. Langerhans cells of skin, described in Section 33.6, are a type of dendritic cell.

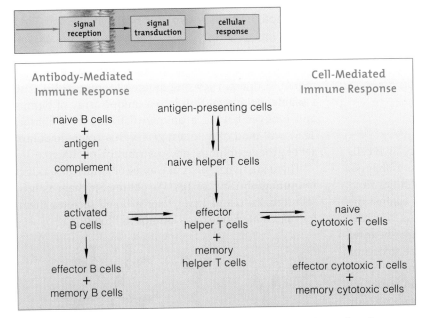

| signal reception | signal transduction | cellular response |

Antibody-Mediated Immune Response

naive B cells
+
antigen
+
complement

↓

activated B cells

↓

effector B cells
+
memory B cells

Cell-Mediated Immune Response

antigen-presenting cells

↑↓

naive helper T cells

↓

effector helper T cells
+
memory helper T cells

naive cytotoxic T cells

↓

effector cytotoxic T cells
+
memory cytotoxic cells

Figure 39.11 Overview of key interactions between antibody-mediated and cell-mediated responses—the two arms of adaptive immunity. A "naive" cell simply is one that has not made contact with its specific antigen.

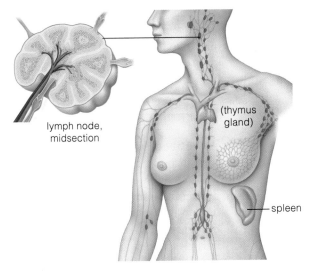

lymph node, midsection

(thymus gland)

spleen

Figure 39.12 Revisiting the big battlegrounds in adaptive immunity. Lymph nodes positioned along lymph vascular highways hold macrophages, dendritic cells, B cells and, deep in the node, T cells. The spleen filters free antigen from blood. Section 38.10 explains the structure and function of all of the lymphatic system's components.

blood or in interstitial fluid are intercepted quickly by phagocytes and B cells. These cells interact during an **antibody-mediated immune response**. As you will see, B cells execute most of this response, although certain types of T cells also support it.

However, some pathogens evade B cells. They hide and often reproduce inside body cells while draining the life out of them. These are *intracellular* pathogens, and they are vulnerable only for the brief time when they are slipping out of one cell and infecting others. You already know about several intracellular viruses, bacteria, fungi, and apicomplexans, as introduced in earlier chapters. You know about the tumor cells and other genetically altered body cells that pose different kinds of threats. All are targets of the **cell-mediated immune response**, which does not require antibodies. This response starts after antigen becomes positioned at the surface of infected or altered body cells—where phagocytes and cytotoxic T cells detect it.

INTERCEPTING AND CLEARING OUT ANTIGEN

Think back on Sections 33.6 and 38.10. After engulfing antigen, dendritic cells and macrophages enter a lymph node. Once inside, both kinds of phagocytes alert the T cells to the threat (Figure 39.12). Each day, about 25 billion lymphocytes pass through each lymph node.

Also, free antigen in interstitial fluid enters lymph vessels, which deliver it to lymph nodes. Inside the

nodes, it trickles past arrays of B cells, macrophages, and dendritic cells—which bind, process, and present it to T cells. Because lymph flows on to the bloodstream, there is a chance that free antigen will circulate to all body tissues. However, lymph nodes trap most of it. Antigen that manages to slip through is filtered out as blood flows through the spleen.

During infection, antigen-presenting T cells become trapped briefly in lymph nodes. The swollen "lumps" you may have noticed under your jaw and elsewhere when you are sick are signs of lymphocyte activity.

Immune responses subside once antigen is cleared away. The tide of battle turns as effector cells and their secretions kill most antigen-bearing agents. With less antigen present, fewer immune fighters are recruited. Complement proteins assist in the cleanup. They clear antibody–antigen complexes from blood and deposit them in the liver and spleen for disposal.

Adaptive immunity has four important characteristics: self/nonself recognition, specificity, diversity, and memory.

Antibody-mediated responses and cell-mediated responses are two interacting arms of adaptive immunity. The first targets antigen detected in blood or interstitial fluid. The second directly kills infected or altered body cells.

Lymph nodes house arrays of immune cells that recognize and trap most antigen-bearing particles.

39.5 Antibodies and Other Antigen Receptors

LINKS TO
SECTIONS
3.5, 14.1

Your body is continually exposed to a mind-boggling array of antigens. However, it continually produces B and T cells that collectively bear an equally mind-boggling array of receptors that are capable of recognizing billions of different antigens! How does the receptor diversity arise?

B CELL WEAPONS—THE ANTIBODIES

All **antibodies** are proteins synthesized only by B cells that encounter and bind antigen. Many are Y shaped. Most circulate in blood and enter the interstitial fluid during inflammation. Each acts specifically against the one antigen that prompted its synthesis. As you saw in Figure 39.8, some antibodies set up antigen-bearing particles for phagocytosis; others neutralize toxins or destroy bacteria by binding directly to them.

Think back on the four levels of protein structure (Section 3.5). Each antibody molecule consists of four polypeptide chains—two identical "light" ones and two identical "heavy" ones. Each chain has a *constant* region, an unvarying domain that forms the molecule's backbone. One end of each chain also has a *variable* region, a domain for one antigen (Figure 39.13). The variable regions fold up into a unique array of bumps and grooves having a unique distribution of charge. They will bind only to antigen with a complementary set of grooves, bumps, and distribution of charge.

There are five structural classes of antibodies, called **immunoglobulins**, or **Igs**. We abbreviate them as IgG, IgD, IgE, IgA, and IgM. After a B cell secretes them, they circulate alone or as clumps in blood:

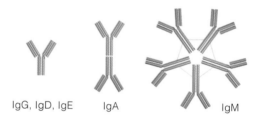

IgG, IgD, IgE   IgA   IgM

IgG makes up 80 percent of all immunoglobulins in blood. It induces complement cascades and neutralizes toxins. It also crosses the placenta, a blood-engorged organ that forms from maternal and embryonic tissues in pregnancy. IgG protects the fetus with its mother's acquired immunities. It also is secreted into early milk.

IgA is the main immunoglobulin in exocrine gland secretions, which include tears, saliva, and milk. It is a chemical weapon in the mucus that coats the linings of the respiratory, digestive, and reproductive tracts. Invading bacteria and viruses cannot bind to a body cell if IgA binds to them first. The IgA (and IgM) form larger structures that clump antigen-bearing threats together for more efficient removal from the body.

IgE induces inflammation after pathogen invasions. Constant regions of its heavy chains become anchored

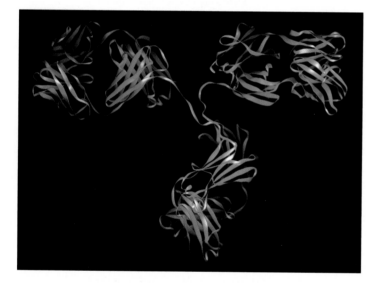

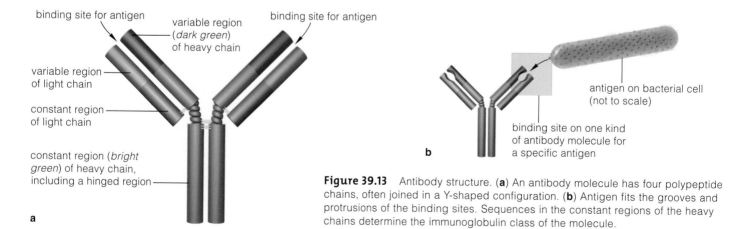

binding site for antigen

variable region
(*dark green*)
of heavy chain

binding site for antigen

variable region
of light chain

constant region
of light chain

constant region (*bright green*) of heavy chain, including a hinged region

a

antigen on bacterial cell
(not to scale)

binding site on one kind
of antibody molecule for
a specific antigen

b

Figure 39.13 Antibody structure. (**a**) An antibody molecule has four polypeptide chains, often joined in a Y-shaped configuration. (**b**) Antigen fits the grooves and protrusions of the binding sites. Sequences in the constant regions of the heavy chains determine the immunoglobulin class of the molecule.

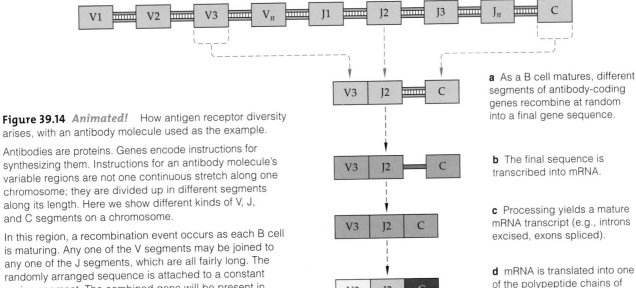

Figure 39.14 *Animated!* How antigen receptor diversity arises, with an antibody molecule used as the example.

Antibodies are proteins. Genes encode instructions for synthesizing them. Instructions for an antibody molecule's variable regions are not one continuous stretch along one chromosome; they are divided up in different segments along its length. Here we show different kinds of V, J, and C segments on a chromosome.

In this region, a recombination event occurs as each B cell is maturing. Any one of the V segments may be joined to any one of the J segments, which are all fairly long. The randomly arranged sequence is attached to a constant region segment. The combined gene will be present in all of the B cell's descendants.

a As a B cell matures, different segments of antibody-coding genes recombine at random into a final gene sequence.

b The final sequence is transcribed into mRNA.

c Processing yields a mature mRNA transcript (e.g., introns excised, exons spliced).

d mRNA is translated into one of the polypeptide chains of an antibody molecule.

at the surface of mast cells, basophils, monocytes, or dendritic cells. Antigen bound to any variable region makes these cells release histamines and cytokines. IgE is a factor in allergic reactions and in HIV infection.

IgM is the first to be secreted in a primary response and the first made by newborns. The surface of each new B cell is covered with hundreds of thousands of IgM or IgD antibodies, each of which recognizes the same antigen. These antibodies are *B cell receptors*, the surface immunoglobulins that function as the B cell's antigen receptors.

THE MAKING OF ANTIGEN RECEPTORS

A gene that encodes an antigen receptor is a unique combination of segments from a large precursor gene. Different parts of the sequence are located in different places in a B or T cell's DNA. *Which* parts are selected for transcription is random. The gene is snipped and spliced in different ways before transcription (Section 14.1 and Figure 39.14). *These random recombinations are the source of diversity in B and T cell receptors.*

Before a new B cell leaves bone marrow, it already is synthesizing unique antigen receptors. The constant region of each one is positioned in the lipid bilayer of the B cell's plasma membrane; the two variable arms project above it. In time the B cell bristles with more than 100,000 antigen receptors. It is now a "naive" B cell, meaning it has not yet met its antigen.

What about the T cells? They, too, form inside bone marrow, but they do not mature until they take a tour

through the thymus gland. After exposure to thymic hormones, the T cells get receptors for MHC markers. They also get TCRs—their unique antigen receptors—by gene splicing events similar to the ones that occur in B cells. But these recombinations are random. Some TCRs end up recognizing MHC markers rather than antigen. Most will not recognize MHC markers at all.

So how does an individual end up with a functional set of T cells—one that does not attack its own body? Thymus cells produce small peptides that are derived from a variety of the body's proteins. The peptides get attached to MHC markers, where they act as built-in quality controls to weed out the "bad" TCRs.

Any passing T cell that binds too tightly to one of the complexes has TCRs that recognize a self peptide. T cells that do not bind at all cannot recognize MHC markers. Both types of cells die. Therefore, by the time naive T cells leave the thymus to begin their journey through the circulatory system, their surface bristles with functional TCRs. In most circumstances, none of them will perceive the body's own cells as foreign.

Antibodies are antigen-recognition proteins made only by B cells. They can be secreted or membrane-bound.

Each individual has the potential to make B and T cell receptors that recognize billions of different antigens.

By random recombinations of segments from receptor-encoding regions of DNA, each B and T cell gets a gene sequence for one unique antigen receptor.

39.6 The Antibody-Mediated Response

LINKS TO
SECTIONS
5.6, 21.1

Think of B cells as assassins. Each has a genetic assignment to liquidate one specific target—an extracellular pathogen or toxin. Antibodies are their molecular bullets.

Suppose that you accidentally nick your finger. Being opportunists, *Staphylococcus aureus* cells hanging out on your skin invade your internal environment. But complement in interstitial fluid quickly latches on to carbohydrates in their bacterial cell wall and activates cascading reactions. Complement coats the bacteria. Within an hour, bacterial cells tumbling along in lymph vessels reach a lymph node in your elbow. There they are paraded past an army of naive B cells.

As it happens, a B cell bears immunoglobulins that tightly bind to a peptidoglycan in the bacterium's cell wall. Also, it bears complement receptors that bind to complement that coats the invading cell. These two events provoke the B cell to begin receptor-mediated endocytosis (Section 5.6). The bacterium enters the B cell, which is no longer naive; it is now activated.

Meanwhile, undaunted by complement and tissue inflammation, more *S. aureus* cells have been secreting chemotactic factors into interstitial fluid around the cut. These secretions attract phagocytes. A dendritic cell engulfs a few bacteria, then migrates to that same lymph node in your elbow. By now, it has digested the bacterial cells and displays antigen fragments bound to MHC markers on its surface.

Each hour, about 500 different naive T cells travel through the lymph node, inspecting resident dendritic cells. In this case, one of those T cells has TCRs that tightly bind to *S. aureus* antigen–MHC complexes on the dendritic cell. For the next twenty-four hours, the two cells interact. Transcription factors are activated in the T cell. After this, the two cells disengage, and and the T cell returns to the circulatory system.

Applying a theory of *clonal selection*, the *S. aureus* antigen "chose" that T cell because it bears a receptor that can bind to it. After the T cell is activated in this way, many descendants form by mitotic cell divisions. They are a clone—one lineage of genetically identical cells. This population differentiates into *helper T cells*, all with identical TCRs specific for *S. aureus* antigen.

Return to that B cell in the lymph node. By now, fragments of *S. aureus* are bound to MHC markers and

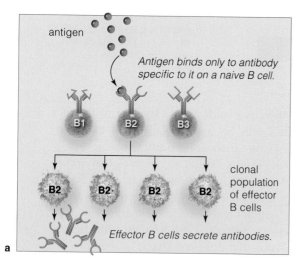

a

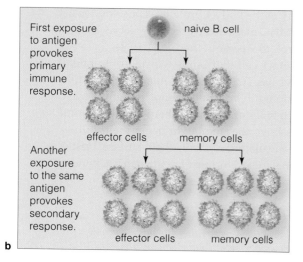

b

Figure 39.15 *Animated!* (**a**) Clonal selection of one B cell. Only B cells with receptors that bind to antigen divide and differentiate. (**b**) First exposure to antigen generates a primary immune response in which effector cells fight the infection. Memory cells also form in a primary response but are set aside, sometimes for decades. When activated, they start a secondary response (Figure 39.16).

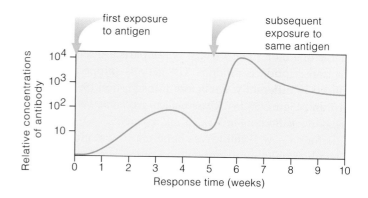

Figure 39.16 Comparison of one secondary immune response with the primary response that preceded it. A later encounter with the same antigen provoked memory cells to divide and differentiate rapidly, which became the basis of a faster, more intense secondary response. The new effector B cells produce mainly IgG.

a A dendritic cell ingests and digests a bacterium in interstitial fluid, and then migrates to a lymph node. There, it presents antigen–MHC complexes. Binding to this antigen-presenting cell induces naive helper T cells to proliferate and differentiate into effector helper T cells and memory helper T cells.

b Each naive B cell bristles with more than 100,000 identical receptors, all specific for the same antigen. This naive B cell binds to its specific antigen on the surface of a bacterium in a lymph node. After also binding complement, the bacterium enters the B cell by receptor-mediated endocytosis and is digested. Bacterial fragments bind to MHC molecules, and the complexes are displayed at the B cell's surface.

c TCRs of an effector helper T cell bind to the antigen–MHC complexes on the B cell. Binding makes the helper T cell secrete cytokines. These signals induce the B cell to divide, giving rise to a huge population of clones with identical antigen receptors. The cells differentiate into effector B cells and memory B cells.

d Effector B cells begin making and secreting huge numbers of IgA, IgG, or IgE, all of which recognize the same antigen as the original B cell receptor. The new antibodies circulate throughout the body, and bind to any remaining bacteria.

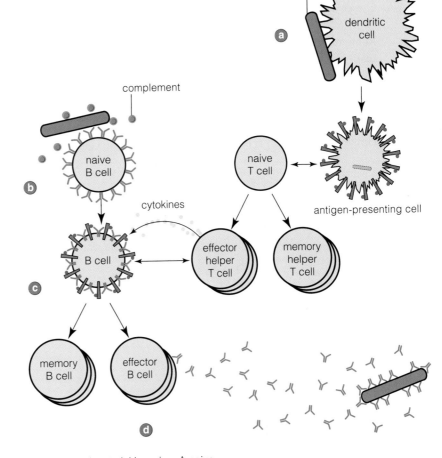

Figure 39.17 *Animated!* Antibody-mediated immune response to a bacterial invasion. A naive B cell is activated after encountering antigen, complement, and a helper T cell, or when microbial pattern receptors on its surface bind to a pathogen directly. It then proliferates and differentiates into huge populations of effector and memory cells, each with the same antigen specificity.

displayed at the B cell's surface. A new helper T cell recognizes the *S. aureus* antigen–MHC complexes on the B cell. Like long-lost friends, the two cells latch on to each other and exchange costimulatory signals. The helper T cell secretes several interleukins—cytokines that signal the B cell to divide and differentiate.

When the cells disengage, the B cell divides, and its clonal descendants form a huge population, all with identical receptors (Figure 39.15*a*). They differentiate into effector and memory cells, as shown in Figures 39.15*b* and 39.16).

The effector cells work immediately in the *primary* immune response, or against this initial exposure to antigen. Instead of making membrane-bound *IgM* as B cell receptors, they switch antibody classes. They start making and secreting *IgG*, *IgA*, or *IgE* instead. Each of the secreted antibody molecules has the same antigen specificity as the original B cell receptor.

Great numbers of antibody molecules specific for *S. aureus* are now circulating through the body. They bind any bacterial cells remaining in the bloodstream and interstitial fluid. In this way, they prevent them from attaching to body tissues and also tag them for disposal by NK cells and complement (Figure 39.17). In addition, they neutralize toxic antigens.

In an antibody-mediated immune response, a naive B cell becomes activated after it binds to antigen, to complement, and to an activated helper T cell.

The activated B cell divides. A huge clonal population forms, all with the same antigen receptor. These cells differentiate into effector and memory cell populations.

Effector B cells secrete antibodies that bind to any antigen remaining in the body, tagging it for destruction.

39.7 The Cell-Mediated Response

LINK TO
SECTION
28.5

If B cells are like assassins, cytotoxic T cells are like ninjas that specialize in cell-to-cell combat. They also touch-kill altered body cells that evade antibody-mediated responses.

Remember, the main targets of an antibody-mediated response are extracellular pathogens and toxins freely circulating in blood and interstitial fluid. Viruses and certain bacteria, fungi, and protists can slip into body cells. As long as they hide in a host cell, an antibody-mediated response cannot be initiated against them. However, during an acute inflammatory response, cell-mediated defenses against these insidious threats get under way in interstitial fluid. Typically, the plasma membrane of an infected body cell displays antigen—peptides of an intracellular pathogen or self proteins that were altered by cancerous transformation.

Dendritic cells recognize, engulf, and digest these antigens as bits of diseased or abnormal cells or their remains. Afterward, the dendritic cells travel to lymph nodes, where antigen–MHC complexes on their surface are presented to two different populations of naive T cells (Figure 39.18).

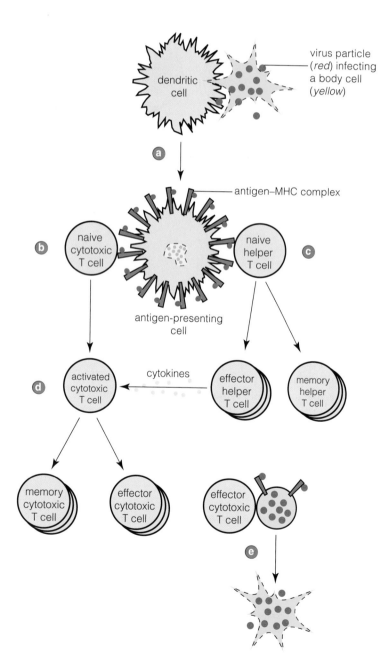

virus particle (*red*) infecting a body cell (*yellow*)

dendritic cell

a

antigen–MHC complex

naive cytotoxic T cell   naive helper T cell

b   **c**

antigen-presenting cell

activated cytotoxic T cell   cytokines   effector helper T cell   memory helper T cell

d

memory cytotoxic T cell   effector cytotoxic T cell   effector cytotoxic T cell

e

a A dendritic cell patrolling interstitial fluid encounters and engulfs the remains of a virus-infected cell. The digested antigen fragments bind to MHC molecules, and the complexes are displayed at the cell's surface. The dendritic cell migrates to a lymph node or to the spleen.

b In the lymph node or spleen, TCR receptors on a naive cytotoxic T cell bind to the complexes on the surface of the antigen-presenting dendritic cell. The interaction activates the cytotoxic T cell.

c In the lymph node, receptors of a naive helper T cell bind to antigen–MHC complexes on the dendritic cell. The interaction activates the helper T cell, which then divides to form a large clonal population. The daughter cells differentiate into effector and memory helper T cells.

d Cytokines secreted by effector helper T cells stimulate the activated cytotoxic T cell to divide as it returns to the circulatory system. Its clonal descendants differentiate into large populations of effector and memory cytotoxic T cells.

e One of the new circulating cytotoxic T cells encounters antigen–MHC complexes on the surface of a cell infected with the same virus. It touch-kills the cell by injecting it with perforin and proteases that induce apoptosis.

Figure 39.18 *Animated!* Primary cell–mediated immune response to a viral infection. A naive cytotoxic T cell binds to an antigen-presenting dendritic cell and is exposed to cytokines that helper T cells have secreted. It is thereby stimulated to divide. Its clonal descendants will differentiate into patrolling cytotoxic T cells that will destroy any body cell displaying the original antigen.

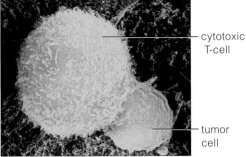

Figure 39.19 Cytotoxic T cell caught in the act of touch-killing a tumor cell.

Both types of naive T cells are activated when their receptors bind to the antigen–MHC complexes on the dendritic cells. Clonal descendants of one population differentiate into effector helper T cells, which secrete interleukins and other cytokines. These signals induce the other type of T cell to divide and differentiate into cytotoxic T cells (Figure 39.18c).

Cytotoxic T cells circulate through the body in blood and interstitial fluid. They bind to any cell bearing the original antigen that is complexed with MHC markers. Then they inject it with perforin and proteases. These toxins poke holes into the cell and induce it to die by apoptosis (Section 28.5 and Figure 39.19). Cytotoxic T cells are the body's primary weapons against infected body cells and tumors. However, they also cause the rejection of tissue and organ transplants.

Cytokines secreted by some helper T cells enhance macrophage action. They stimulate these phagocytes to secrete more inflammatory mediators and toxins that help kill tumor cells and larger parasites.

Helper T cell cytokines also stimulate cell divisions of NK cells. These "natural killers" attack cells that are tagged for destruction by antibodies. They also detect stress markers on infected and cancerous body cells. A cell that has normal MHC markers is not killed. But a cell with MHC markers that have been altered, as by an infection or malignant transformation, will die. NK cells and macrophages are crucial in killing such cells; neither depends on the presence of MHC markers.

In a primary cell-mediated response, antigen-presenting dendritic cells activate helper T cells.

Helper T cells secrete cytokines that induce formation of cytotoxic T cells, proliferation of NK cells, and enhance the activity of macrophages. These three cell types destroy infected or altered body cells.

39.8 Immunotherapies

LINKS TO
SECTION 9.5
CHAPTER 14

New immunotherapies are being designed to enhance the body's defenses against many cancers and infections.

As explained in Section 9.5, neoplasms in skin, bone, and other tissues sometimes turn malignant when viral attack, irradiation, or chemicals alter genes. Immune responses to altered surface proteins may not kill the transformed cells. When they are not killed or surgically removed, such cells can be lethal. *Immunotherapy* bolsters the defenses by manipulating the body's own immune mechanisms.

Interleukins that stimulate B and T cell divisions are being used to treat some cancers. Tumor necrosis factors kill or slow the proliferation of some tumor cells. Antiviral interferons are used to treat cancer and chronic viral infections. Hematopoietic growth factors are signaling molecules that stimulate the production of blood cells. They encourage the body to replace normal cells that are killed as a side effect of traditional cancer therapies—radiation and chemotherapy.

Cancer cells overexpress genes for surface proteins. Such proteins can be targeted by *monoclonal antibodies*, antibodies that are produced from a population of B cell clones. They are produced by exposing animals to an antigen, then fusing their extracted B cells to tumor cells that grow in vitro. These cultured hybrid cells make large amounts of antibody against the antigen (Figure 39.20).

Some kinds of metastatic breast cancer cells have an overabundance of HER2 proteins at their surface. The drug Herceptin (trastuzumab) is a monoclonal antibody that binds to the proteins and invites the attention of NK cells. However, a few HER2 proteins also occur on some normal body cells, which may be attacked as well.

Monoclonal antibodies can be bound to poisons to make *immunotoxins*. Remember ricin? It stops protein synthesis by disabling ribosomes (Chapter 14). Genetic engineers have attached ricin A chains to monoclonal antibodies specific for antigens on certain cancer cells. When these immunotoxins bind to antigen on a cancer cell, ricin enters the cell and blocks translation. Other researchers have bound ricin to monoclonal antibodies that can target body cells infected by HIV. So far, these immunotherapies are only in experimental stages.

Also on the horizon are *cancer vaccines*, which may prime the body to recognize and fight cancer cells. In some ongoing clinical trials, patients' cancer cells are removed and treated with a substance that makes them appear foreign. The treated cells are reinjected into a patient, the idea being that they will sound the alarm for immune responses against cancer. Other therapies include in vitro expansion and stimulation of a patient's lymphocytes. Dendritic cells are especially promising in this work.

Figure 39.20 Monoclonal antibody-producing cells stored in liquid nitrogen.

39.9 Defenses Enhanced or Compromised

Sometimes immune responses are not strong enough or are misdirected or compromised. Here are examples of what we can and cannot do about it.

IMMUNIZATION

Immunization refers to processes that may promote immunity. In *active* immunization, a preparation that contains antigen—a **vaccine**—is administered orally or injected into the body (Figure 39.21). A first injection elicits a primary immune response. A second one, a booster, elicits a secondary response. Additional effector and memory cells form, for lasting protection.

Many vaccines are made from weakened or killed pathogens, or inactivated bacterial toxins. Others consist of harmless viruses that have genes from other pathogens inserted into their DNA or RNA. After vaccination, body cells will synthesize the antigen encoded by those genes, and immunity will be established.

Recommended Vaccines	Recommended Ages
Hepatitis B	Birth–2 months
Hepatitis B booster 1	1–4 months
Hepatitis B booster 2	6–18 months
DTP Diphtheria, tetanus; and pertussis (whooping cough)	2, 4, and 6 months
DTP booster 1	15–18 months
DTP booster 2	4–6 years
DT	11–12 years
HiB (*Haemophilus influenzae*)	2, 4, and 6 months
HiB booster	12–15 months
Polio	2 and 4 months
Polio booster 1	6–18 months
Polio booster 2	4–6 years
MMR (Measles, Mumps, Rubella)	12–15 months
MMR booster	4–6 years
Pneumococcal	2, 4, and 6 months
Pneumococcal booster 1	12–15 months
Pneumococcal booster 2	2–18 years
Varicella	12–18 months
Hepatitis A series (in some areas)	2–12 years
Influenza	Yearly, 1–18 years

Figure 39.21 Centers for Disease Control and Prevention (CDC) recommended childhood immunization schedule as of 2005. Pediatricians routinely immunize infants and children. Low-cost or no-cost vaccinations are available at many community clinics and health departments.

Passive immunization helps if hepatitis B, tetanus, rabies, and a few other diseases have already started. It is based on injections of antibody purified from the blood of a person who already fought the disease. The antibodies do not activate the body's immune system, so memory cells do not form. Protection ends when the body disposes of the injected antibody molecules.

Vaccines occasionally can fail or have side effects. There are a few reports of vaccines that have caused immunological problems. There is also concern that the mercury preservative in some vaccines harms the nervous system. The preservatives have been removed from most vaccines. However, if you are considering a vaccination, it may be a good idea to first discuss the risks and benefits of the procedure with a doctor.

ALLERGIES

In millions of people, exposure to harmless proteins stimulates immune responses. Any substance that is ordinarily harmless yet provokes such responses is an **allergen**. Hypersensitivity to them is called an **allergy** (Figure 39.22). Drugs, foods, pollen, dust mites, fungal spores, and venom from bees, wasps, and some other insects are among the most common allergens.

Some people are genetically predisposed to having allergies. Infections, emotional stress, and changes in air temperature can trigger reactions. Exposure to an allergen stimulates the immune system to make IgE, which binds to mast cells. In a secondary response, antigen attaches to the bound IgE. These mast cells respond by secreting histamine and cytokines, which initiate an inflammatory response. Vasodilation in the respiratory tract causes copious amounts of mucus to be secreted. Airways constrict. Stuffed-up sinuses,

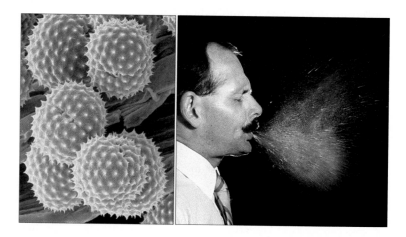

Figure 39.22 One of the effects of ragweed pollen, an allergen, on sensitive people.

Figure 39.23 A case of severe combined immunodeficiency (SCID). Cindy Cutshwall was born with a deficient immune system. She carries a mutated gene for adenosine deaminase (ADA). Without this enzyme, her cells could not break down adenosine, so a reaction product that is toxic to white blood cells accumulated in her body. High fevers, severe ear and lung infections, diarrhea, and an inability to gain weight were outcomes.

In 1991, when Cindy was nine years old, she and her parents consented to one of the first human gene therapies. Genetic engineers spliced the normal ADA gene into the genetic material of a harmless virus. The modified virus was the vector that delivered copies of the normal gene into her bone marrow cells. Some cells incorporated the gene in their DNA and started making the missing enzyme.

Now in her early twenties, Cindy is doing well. She still requires weekly injections to supplement her ADA production. Other than that, she is able to live a normal life. She is a strong advocate of gene therapy.

laborious breathing, sneezing, and a drippy nose are typical symptoms of *asthma* and *hay fever*.

Antihistamines, or anti-inflammatory drugs, often relieve symptoms of allergies. Also, in desensitization programs, skin tests are used to identify the allergens. Increasing doses of allergens are slowly administered. Each time, the body produces more IgG, which freely circulates as an allergen neutralizer.

Anaphylactic shock is a life-threatening response to an allergen. For instance, someone allergic to wasp or bee venom can die within minutes of a sting. Airways constrict and cut off air flow. Fluids leak from grossly permeable capillaries, and blood pressure plummets. The circulatory shock may quickly end in death.

AUTOIMMUNE DISORDERS

Sometimes lymphocytes and antibody molecules fail to discriminate between self and nonself. When that happens, they mount an **autoimmune response**, or a misdirected attack against one's own tissues.

For example, *rheumatoid arthritis* is an autoimmune disease in which auto-antibodies collect in the joints and often the heart, blood vessels, and lungs. T cell responses and complement cascades result in chronic, painful, and eventually immobilizing inflammation.

In some cases, auto-antibodies bind to hormone receptors. *Graves' disease*, or thyrotoxicosis, is like this. Auto-antibodies bind to stimulatory receptors on the thyroid gland, causing it to produce excess thyroid hormone. This quickens the body's overall metabolic rate. Feedback loops that normally balance hormone production do not include antibodies. So binding of antibodies continues, excessive amounts of thyroid hormones are synthesized, and metabolism spins out of control. Common symptoms are a rapid, irregular heartbeat; uncontrollable weight loss; sleeplessness; pronounced mood swings; and bulging eyes.

A common neurological disorder, *multiple sclerosis*, results if autoreactive T cells attack the myelin sheaths of axons and enter cerebrospinal fluid. The symptoms include paralysis and blindness. At least eight genes heighten susceptibility, but activation of white blood cells against viral infections may trigger the disorder.

Immune responses tend to be stronger in women than in men, and autoimmunity is far more frequent in women. We know estrogen receptors take part in controls over gene expression throughout the body. T cells bear them. By one hypothesis, estrogens enhance T cell activation in autoimmune diseases and thereby cause amplified B cell–T cell interactions.

DEFICIENT IMMUNE RESPONSES

Loss of immune function can have a lethal outcome. *Primary* deficiencies, present at birth, are outcomes of mutant genes or abnormal developmental steps. The severe combined immunodeficiencies (SCIDs) are like this. The genetic disorder called adenosine deaminase deficiency (ADA) is one case (Figure 39.23). *Secondary* immune deficiencies are losses of immune function after exposure to some outside agent, such as a virus. Severe immune deficiencies make individuals more vulnerable to infections by opportunistic agents that are otherwise harmless to those in good health.

AIDS (Acquired ImmunoDeficiency Syndrome) is the most common secondary immune deficiency. The next section describes its cause and its effects.

Immunization programs are designed to boost immunity to specific diseases.

Some heritable disorders, developmental abnormalities, and attacks by viruses and other outside agents result in misdirected, compromised, or nonexistent immunity.

39.10 AIDS Revisited—Immunity Lost

LINKS TO
SECTIONS
21.6–21.8

Worldwide, HIV infection rates continue to skyrocket. An effective vaccine still eludes us. For now, the best protection is avoiding unsafe behavior.

AIDS is a constellation of disorders that develop after infection by HIV (*Human Immunodeficiency Virus*). The virus cripples the immune system and makes the body highly susceptible to infections and rare forms of cancer. Worldwide, HIV has infected an estimated 34 million to 46 million individuals (Table 39.4).

There is no way to rid the body of known forms of the virus, HIV–I and HIV–II. *There is no cure for those already infected.* At first, an infected person appears to be in good health, maybe fighting "a bout of the flu." But symptoms emerge that foreshadow AIDS: fever, many enlarged lymph nodes, fatigue, chronic weight loss, and drenching night sweats. Then opportunistic infections strike, such as yeast infections of the mouth, esophagus, vagina, and a form of pneumonia caused by *Pneumocystis carinii*. Painless colored lesions often erupt, especially on legs and feet (Figure 39.24). The lesions are visible evidence of *Kaposi's sarcoma*, a cancer common among AIDS patients.

HIV INFECTION—A TITANIC STRUGGLE BEGINS

HIV is a retrovirus with a lipid envelope. Remember, this type of envelope is a bit of plasma membrane acquired as a virus particle buds from an infected cell (Section 21.6). Proteins spike out from the envelope, span it, and line its inner surface. Just beneath the envelope, more viral proteins enclose two strands of RNA and copies of reverse transcriptase. Figure 39.25 shows this structural arrangement.

HIV infects primarily macrophages, dendritic cells, and helper T cells. When it enters the body, dendritic cells in skin or in the gut or respiratory tract lining engulf it. These cells enter lymph nodes and present HIV antigen to naive T cells. Any T cells that bind to the antigen give rise to populations of helper T cells that interact with naive B cells and naive cytotoxic T cells. Virus-neutralizing IgG antibodies and cytotoxic T cells now form that can kill HIV-infected cells.

It is a typical adaptive response, and it clears the body of most of the HIV. However, during this first response, HIV infects just a few helper T cells inside a few lymph nodes. For years or decades, the immune system functions normally. The antiviral antibody and cytotoxic T cells keep HIV levels relatively low.

HIV particles multiplying in lymph nodes keep on shedding envelope proteins, which reach the blood. In response, helper T cells in uninfected lymph nodes secrete interleukin–4 (IL–4). This cytokine stimulates B cells to stop making the virus-neutralizing IgG and start making IgE instead.

Again, IgE anchors itself to mast cells, basophils, monocytes, and dendritic cells. Also, when it binds to antigen, IgE makes these cells release histamines and inflammatory substances. It normally protects mucus membranes from worms and arthropods, because its release induces lymphocytes and plasma proteins to enter and defend infested tissues.

But glycoproteins shed from HIV envelopes bind to IgE. Binding causes the anchoring cells to release inflammatory weapons, including IL–4. This prompts more B cells to make IgE instead of IgG. When less IgG is available to fight HIV, more IgE binds to free viral proteins, which causes more B cells to make IgE instead of IgG, and so on. IL–4 also inhibits cytokine secretion from helper T cells, so that cytotoxic T cell production falls as IL–4 levels rise. This vicious cycle makes adaptive immunity mechanisms less and less effective at fighting the HIV infection.

The number of virus particles rises. Up to 2 billion helper T cells become infected. Up to 1 billion virus particles are built every day. Every two days, half of the virus particles are destroyed and half of the helper T cells lost in battle are replaced. HIV reservoirs and masses of infected T cells collect in the lymph nodes.

Eventually the battle tilts as the body makes fewer replacement helper T cells. It may take a decade or more, but the erosion of helper T cell counts destroys the body's capacity for adaptive immunity.

Other viruses make far more particles on any given day, but the immune system usually wins. Some may even persist over a lifetime, but the immune system

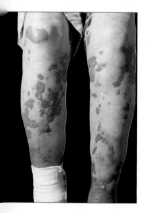

Figure 39.24 The lesions that are a sign of Kaposi's sarcoma.

Table 39.4	Global Cases of HIV and AIDS*	
Region	AIDS Cases	New HIV Cases
Sub-Saharan Africa	25,400,000	3,100,000
South/Southeast Asia	7,100,000	890,000
Latin America	1,700,000	240,000
Central Asia/East Europe	1,400,000	210,000
East Asia	1,100,000	290,000
North America	1,000,000	44,000
Western/Central Europe	610,000	21,000
Middle East/North Africa	540,000	92,000
Caribbean Islands	440,000	53,000
Australia/New Zealand	35,000	5,000

*Global estimates as of December 2004, *www.unaids.org*

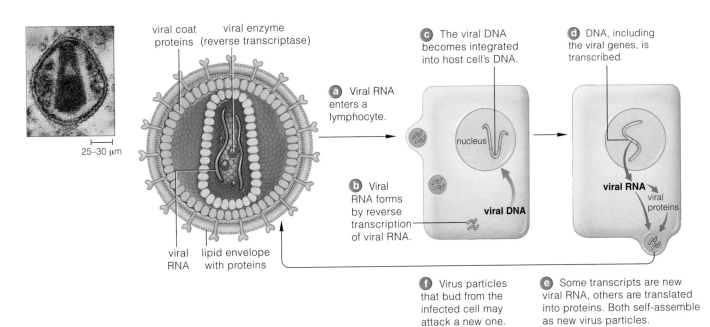

Figure 39.25 *Animated!* Replication cycle of HIV, the retrovirus that causes a disease, AIDS, which is incurable at this writing.

keeps them at bay. HIV wrecks the system. Secondary infections and tumors kill the patient.

HOW IS HIV TRANSMITTED?

Most often, HIV is transmitted by having unprotected sex with an infected partner. HIV virus in semen and vaginal secretions enters a partner through epithelial linings of the penis, vagina, rectum, and (rarely) the mouth. The risk of transmission increases when the linings are damaged by rough sex, anal intercourse, or other sexually transmitted diseases.

Infected mothers can transmit HIV to babies during vaginal birth and by breast-feeding. HIV also occurs in traces of infected blood in syringes that are shared by intravenous drug abusers and by patients in hospitals of developing countries. Transmission through food, air, water, casual contact, or insect bites is ineffective.

WHAT ABOUT DRUGS AND VACCINES?

Current drugs cannot cure HIV, but some can slow its progression by interfering with processes unique to viral replication. AZT and other nucleoside phosphate analogs disrupt reverse transcription; they replace the normal nucleotides in HIV cDNA. Protease inhibitors can be used to interrupt viral transcript processing.

The cost of current treatments is as much as 15,000 dollars a year. That amount puts treatments beyond the reach of most infected people in the developing countries, where the AIDS pandemic is now raging. At present, the best option for halting the spread of the virus appears to be the development of a safe, effective, low-cost vaccine.

A vaccine may not be on the immediate horizon. Traditional vaccine strategies do not work with HIV because the virus replicates in the immune system. Also, it mutates very quickly. Neutralizing antibody exerts a strong selective pressure on the virus. Given the immense number of viral replications in infected people, HIV genes have a high mutation rate. Vaccine or no vaccine, antibodies to the mutated products just cannot be produced fast enough to keep pace.

Even so, persistent researchers are using several strategies to develop an HIV vaccine. An immediate, strong immune response to a primary HIV exposure might clear the virus before it has a chance to infect helper T cells. The best option for a vaccine might be an engineered virus with enough HIV components to provoke the desired immune response, but not so many components that they would pose a different threat. Several of these recombinant vaccines are in clinical trials, including one being tested in India.

Learning about the structure and replication of viruses has practical application for millions of people worldwide. Ongoing research to understand and intervene in the HIV replication cycle is a case in point.

Summary

Section 39.1 Vertebrates fend off pathogens with physical and chemical barriers at body surfaces. They also are protected by innate responses to tissue damage and invasion. Adaptive responses of white blood cells counter specific threats (Table 39.5).

Section 39.2 Skin and linings of the body's tubes and cavities are physical barriers to infection. Chemical barriers include glandular secretions such as mucus as well as lysozyme in saliva, tears, and gastric fluid. Microbes that normally colonize these external surfaces usually have neutral or beneficial effects. Some cause disease when they breach the surface barriers.

Section 39.3 Innate immunity uses microbial pattern receptors and complement in fast, off-the-shelf responses to a fixed set of nonself cues. Macrophages and dendritic cells engulf and digest anything other than healthy tissue. The complement system helps coordinate defenses in both innate and adaptive immune responses.

Inflammation is a fast response to tissue damage, toxins, or invasion. It starts when mast cells release histamine, which increases blood flow to the tissue and makes blood capillaries leaky to phagocytes and plasma proteins. Fever fights infection by increasing the metabolic rate while slowing microbial growth.

Biology❂Now
View animation of inflammation and complement action on BiologyNow.

Section 39.4 Adaptive immunity is characterized by self/nonself recognition, sensitivity to specific threats, diversity, and memory. B and T cells are central to it.

Antigen is any molecule or particle that certain white blood cells recognize as foreign (nonself) and that elicits an immune response. Such responses start with the recognition of antigen, which stimulates repeated cell divisions that form huge populations of effector and memory B cells and T cells. Every cell of the clonal populations is sensitized to the same antigen.

Macrophages, dendritic cells, and B cells function as antigen-presenting cells. They engulf and digest antigen, then display its fragments complexed to MHC markers at their surface.

Section 39.5 Antibodies are proteins, made only by B cells, that bind specific antigens and tag them for death or neutralize them. Many have a Y shape.

B and T cells form in bone marrow. As each cell develops, random gene recombinations give each one a unique gene sequence for its antigen receptor.

T cells mature in the thymus gland. Thymus cells weed out T cells with receptors that bind too tightly to a self peptide or do not bind to MHC. Only T cells that distinguish between self and nonself survive.

Biology❂Now
Learn how antibody diversity is generated with the animation on BiologyNow.

Section 39.6 B cells carry out antibody-mediated responses, assisted by T cells and cytokines. Antigen presented by a dendritic cell prompts a naive T cell to divide. Descendant cells divide and differentiate into large populations of effector helper T cells and memory T cells, all of which recognize the same antigen.

Next, a naive B cell recognizes and binds to antigen and complement. It internalizes the antigen. Digested bits of antigen become complexed with MHC markers and are displayed on the B cell surface. Binding to an effector helper T cell induces this antigen-presenting B cell to divide. Its cellular descendants form huge clonal populations of antibody-secreting and memory B cells.

After a primary response to antigen, memory cells persist. A later encounter with the same antigen will provoke a faster, more intense secondary response.

Biology❂Now
Observe an antibody-mediated response with the animation on BiologyNow.

Section 39.7 T cells make cell-mediated responses to fight intracellular pathogens and altered body cells. Helper T cells secrete cytokines, which provoke naive T cells to divide and give rise to clonal populations of cytotoxic T cells and memory T cells.

Cytotoxic cells touch-kill their targets by injecting them with perforin and enzymes that cause the cell to die by apoptosis. Cytokines secreted by helper T cells cause natural killer (NK) cells to divide, and also enhance macrophage activity.

Biology❂Now
Observe a cell-mediated response with the animation on BiologyNow.

Table 39.5	Summary of White Blood Cells and Their Functions
Macrophage	Phagocyte. Presents antigen to helper T cells; secretes variety of cytokines during innate and adaptive immune responses.
Neutrophil	Most abundant, fast-acting phagocyte. Takes part in inflammation; most effective against bacteria.
Eosinophil	Secretory. Some of its enzymes punch holes in parasitic worms.
Basophil	Secretory cell that circulates in blood. Secretes histamine, other substances that cause inflammation.
Mast cell	Secretory cell in tissues. Secretes histamine, other substances that produce inflammation; contributes to allergies.
Dendritic cell	Circulating phagocyte. Presents antigen to naive T cells.
Lymphocytes:	*Take part in most immune responses. After antigen recognition, clonal populations of effector and memory cells form.*
B cell	Recognizes antigens via surface immunoglobulin receptors. It is the only cell that produces antibodies.
Helper T cell	Coordinates all immune responses with signaling molecules (cytokines); activates naive B cells and T cells.
Cytotoxic T cell	Recognizes antigen–MHC complexes by way of receptors (TCRs); touch-kills infected, cancerous, or foreign cells.
Natural Killer (NK) cell	Cytotoxic; kills stressed body cells lacking MHC markers; also kills antibody-tagged cells.

Section 39.8 Immunotherapies enhance defenses against persistent infection or cancer with the help of cytokines, monoclonal antibodies, or cancer vaccines.

Section 39.9 Immunization, as by administration of vaccines, provokes immunity to specific diseases. Allergens are normally harmless substances that induce immune responses; allergies are hypersensitivities to an allergen. In autoimmunity, the body's own cells are mistakenly recognized as foreign. Immune deficiency is a weakened or nonexistent capacity to mount an immune response.

Section 39.10 HIV, a retrovirus, causes AIDS. It destroys the immune system mainly by infecting helper T cells. At present, AIDS cannot be cured.

Biology③Now
See how HIV invades and replicates inside a cell with the animation on BiologyNow.

Figure 39.26 Two piglets developed from the same cell lineage at the University of Missouri. The yellow hooves and snout of the piglet at left are evidence that it is transgenic.

Self-Quiz
Answers in Appendix II

1. _____ are the first line of defense against threats.
   a. Skin, mucous membranes
   b. Tears, saliva, gastric fluid
   c. Urine flow
   d. Resident bacteria
   e. a through c
   f. all of the above

2. Complement proteins _____ .
   a. form pore complexes
   b. enhance resident bacteria
   c. promote inflammation
   d. neutralize toxins
   e. both a and c
   f. both c and d

3. _____ trigger immune responses.
   a. Interleukins
   b. Lysozymes
   c. Immunoglobulins
   d. Antigens
   e. Histamines
   f. all of the above

4. Recognition of specific threats is based on _____ .
   a. antigen receptor diversity
   b. recombinations of receptor genes
   c. liver proliferation
   d. both a and b
   e. all of the above

5. Antibody-mediated responses work against _____ .
   a. intracellular pathogens
   b. extracellular pathogens
   c. extracellular toxins
   d. both a and c
   e. both b and c
   f. all of the above

6. Immunoglobulin(s) _____ promotes antimicrobial activity in mucus-coated surfaces of some organ systems.
   a. IgA b. IgE c. IgG d. IgM e. IgD

7. _____ are targets of cytotoxic T cells.
   a. Extracellular virus particles in blood
   b. Virus-infected body cells or tumor cells
   c. Parasitic flukes in the liver
   d. Bacterial cells in pus
   e. Pollen grains in nasal mucus

8. Match the immunity concepts.
   ____ inflammation
   ____ antibody secretion
   ____ fast-acting phagocyte
   ____ immunological memory
   ____ autoimmunity

   a. neutrophil
   b. effector B cell
   c. nonspecific response
   d. immune response against own body
   e. secondary responses

Additional questions are available on Biology③Now™

Critical Thinking

1. As described in the chapter opener, Edward Jenner was lucky. He performed a potentially harmful experiment on a boy who managed to survive it. What would happen if a would-be Jenner tried to do the same thing today?

2. New research links some of our resident microbes with seemingly unrelated major illnesses. Periodontitis itself is not life-threatening, for instance, but it is a fairly good predictor for heart attacks. How do you suppose bacteria that cause gum disease may also be harmful to the heart?

3. Researchers have long attempted to develop ways to get the immune system to accept foreign tissue as "self." Remember Section 16.8? Pigs are considered the most likely animal candidates as organ donors for humans. As a first step, some researchers are developing transgenic pigs—animals that received and are expressing foreign genes (Figure 39.26). Which type of human genes might be inserted to stop the human immune system from attacking a pig-to-human transplant? Explain your reasoning.

4. As Section 39.4 explains, generation of tremendously diverse B cell and T cell antigen receptors is the basis of pathogen-specific immune responses. The genetic roots of this adaptive immunity are deep in the chordate family tree. Remember the lancelets (Section 26.1)? Researchers have found that these invertebrate chordates have genes that resemble the V-region genes of vertebrates. However, the ones in lancelets cannot undergo rearrangements, as they do in vertebrates. Lancelets can make nonspecific defense responses, but not immune responses. Why?

5. Elena developed *chicken pox* when she was in first grade. Later in life, when her children developed chicken pox, she remained healthy even though she was exposed to countless virus particles daily. Explain why.

6. Before each flu season, you get a flu shot, an influenza vaccination. This year, you come down with "the flu" anyway. What do you suppose happened? There are at least three explanations.

7. Vaccines hold promise for preventing or treating some cancers. *Therapeutic* vaccines, remember, are designed to help the body fight an existing cancer. They activate the immune system against antigens specific to tumor cells. In typical clinical trials, some patients receive a known cancer treatment and others get the new vaccine. If you were affected by a life-threatening cancer, would you take part in a clinical trial of a new vaccine?

Up in Smoke

Each day 3,000 or so teenagers join the ranks of habitual smokers in the United States. Most are not even fifteen years old. The first time they light up, they cough and choke on irritants in the smoke. They typically become dizzy and nauseated, and develop headaches.

Sound like fun? Hardly. Why, then, do they ignore the signals of the threat to the body and work so hard to be a smoker? Mainly to fit in. To many adolescents, the misguided perception of social benefits overwhelms the seemingly remote threat to health (Figure 40.1).

Changes that can make the threat a reality start at once. Normally, ciliated epithelial cells along the airways to the lungs sweep out airborne pathogens and pollutants, but smoke from just a single cigarette immobilizes them for hours. Smoke kills the white blood cells that patrol and defend the respiratory system's tissues. Gunk starts to clump up in the airways of young smokers. Pathogens become entrenched in the gunk, as biofilms. More colds, more asthma attacks, and bronchitis are on the way.

Nicotine in each smoke-filled inhalation quickly reaches the circulatory system and the brain. This highly addictive stimulant constricts blood vessels, which makes the heart work harder and raises blood pressure. It raises the blood level of "bad" cholesterol (LDL) and lowers that of "good" cholesterol (HDL). It makes blood stickier—more sluggish—and invites the formation of clots. Clogged arteries, heart attacks, and strokes are among the physiological costs of a social pressure that leads to addiction.

Which smokers have not heard that they are inviting painful, deadly lung cancers? Maybe they haven't heard that carcinogens in cigarette smoke can induce cancers in organs all through their body. For instance, we now know that females who start smoking when they are teenagers are about 70 percent more likely to develop breast cancer than those who never took up the habit.

The tissue damage inflicted by cigarette smoke is not confined to smokers. Families, coworkers, and friends often get unfiltered doses of carcinogens in tobacco smoke. As urine samples reveal, those carcinogens also end up in the tissues of nonsmokers who live with smokers. Each year in the United States, lung cancers arising from secondhand smoke cause an estimated 3,000 deaths. Children exposed to secondhand smoke face more than cancer down the road. They are more prone to develop chronic middle ear infections, asthma, and other respiratory ailments.

In the United States, smoking is being banned from airline cabins and airports, restaurants, theaters, and other enclosed spaces. Cigarette sales to minors are prohibited.

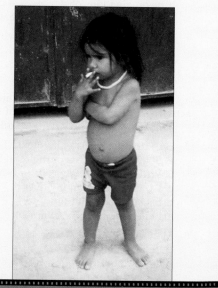

Watch the video online!

Figure 40.1 An addiction that has become pandemic. *Left*, learning to smoke is easy, compared with trying to quit. In one survey, two-thirds of female smokers who were sixteen to twenty-four wanted to stop smoking entirely. Of those who tried to quit, only about 3 percent remained nonsmokers for an entire year.

Right, a child in Mexico City, already proficient at smoking cigarettes. This behavior ultimately will endanger her capacity to breathe.

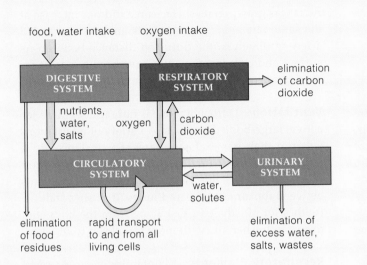

food, water intake   oxygen intake

Figure 40.2 Interactions between the respiratory system and other organ systems that are central to homeostasis in humans and other vertebrates.

Attempts to restrict advertising near elementary schools are ongoing. However, are tobacco companies now viewing children and women of developing countries as untapped markets? Mark Palmer, former ambassador to Hungary, thinks so. He argues that selling tobacco is the worst single thing the United States does to the rest of the world.

This chapter samples a few **respiratory systems**, which function in the exchange of gases between the internal and external environments. Together with other organ systems, a respiratory system contributes to homeostasis—that is, to maintaining internal operating conditions for the body's living cells (Figure 40.2). If you or someone you know has joined the culture of smoking, you might use the chapter as a guide to the short-term and long-term impact of smoking on health. For a more graphic preview, find out what goes on every day with smokers in hospital emergency rooms and intensive care units. No glamour there. It is not cool, and it is not pretty.

How Would You Vote?

Tobacco is a worldwide threat to health and a profitable product for American companies. As tobacco use by its citizens declines, should the United States encourage international efforts to reduce tobacco use around the globe? See BiologyNow for details, then vote online.

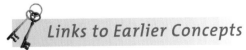

Key Concepts

THE WHOLE POINT OF GAS EXCHANGE

Aerobic respiration uses free oxygen, and its carbon dioxide wastes must be quickly removed before the pH of the internal environment fluctuates dangerously. Respiration is the sum of processes that move oxygen from the outside environment to metabolically active tissues, and that move carbon dioxide from the tissues to the outside. Section 40.1

GAS EXCHANGE IN INVERTEBRATES

In small aquatic invertebrates, the integument functions as the respiratory surface. In large invertebrates on land, gases are exchanged across a moist, internal respiratory surface or at fluid-filled tips of branching tubes, which start at the body surface and end next to cells. Section 40.2

GAS EXCHANGE IN VERTEBRATES

Gills, the skin, and paired lungs function in gas exchange in different animals. Breathing ventilates the lungs. Blood is a transport medium to and from the interstitial fluid that bathes metabolically active cells. It picks up oxygen and gives up carbon dioxide at the respiratory surface.

Gas exchange is most efficient when the rate of air flow matches the rate of blood flow at the respiratory surface. Respiratory centers adjust the rate and depth of breathing when demands for oxygen change. Sections 40.3–40.7

RESPIRATORY DISORDERS

Abnormally high concentrations of airborne particles, as in cigarette smoke, put extra workloads on the respiratory system and cause respiratory problems. Section 40.8

GAS EXCHANGE IN EXTREME ENVIRONMENTS

At high altitudes, the human body makes short-term and long-term adjustments to the thinner air. Built-in respiratory mechanisms and specialized behaviors allow sea turtles and diving marine mammals to remain under water, at great depths, for long periods. Section 40.9

Links to Earlier Concepts

In this chapter, you will put your knowledge of concentration and pressure gradients, diffusion, and the surface-to-volume ratio to use (Sections 4.1, 5.3, 25.1, 28.2). You will see how the properties of water affect respiration (2.5). You will look at organ systems that evolved in support of aerobic respiration as well as homeostasis (8.1, 26.1, 26.4, 28.1, 34.1). You will consider sensory receptors and brain centers that control breathing (34.9, 35.1). Hemoglobin! Red blood cells! Now you will see precisely how they function (3.6, 28.5, 34.9).

40.1 The Nature of Respiration

LINKS TO
SECTIONS 3.6, 4.1,
5.3, 8.1, 28.2, 34.2

Aerobic respiration is the only metabolic pathway that produces enough ATP to sustain the active life-styles of animals, especially large-bodied ones (Section 8.1). It requires free oxygen, and its carbon dioxide wastes must be eliminated as fast as they form, otherwise the pH of the internal environment would fluctuate dangerously.

THE BASIS OF GAS EXCHANGE

Gaseous oxygen (O_2) is abundant in the atmosphere and is dissolved in the water of most aquatic habitats. **Respiration** is the sum of physiological processes that move O_2 from the surroundings to all metabolically active tissues in the animal body and carbon dioxide (CO_2) from tissues to the outside. It uses the tendency of O_2 and CO_2 to diffuse down their concentration gradients—or, as we say for gases, pressure gradients—between the external and internal environments.

At sea level, air contains about 78 percent nitrogen, 21 percent oxygen, 0.04 percent carbon dioxide, and 0.04 percent other gases. Each gas is contributing to a total atmospheric pressure that measures 760 mm Hg with a mercury barometer, as in Figure 40.3. Oxygen's partial pressure (its contribution to total atmospheric pressure) is 760 × 21/100, or close to 160 mm Hg. For carbon dioxide, it is about 0.3 mm Hg.

Animals have a **respiratory surface**, an epithelium or another layer thin enough for gases to cross easily on their way to and from the internal environment. The surface must be kept moist at all times, because gases diffuse across only when dissolved in fluid.

How many gas molecules can cross the surface in a given interval? By **Fick's law**, the larger the surface area and the steeper the partial pressure gradient, the faster diffusion will proceed. As you will see, other factors influence diffusion rates.

FACTORS INFLUENCING GAS EXCHANGE

Surface-to-Volume Ratio As you know, the surface-to-volume ratio puts constraints on gas exchange (Section 4.1). That is why animals that have no respiratory organs are small, cylindrical, or flattened. Gases can diffuse directly across the body's surface layer. Remember the flatworm in Section 28.2? As it was growing, its surface area did not increase at the same rate as its volume. If its girth had exceeded a few millimeters, the diffusion distance between its surface and cells inside would have been too great for efficient gas exchange. The worm would have died.

Ventilation Diffusion alone is not rapid enough for active, large-bodied animals. A variety of adaptations enhance their gas exchange rates. For instance, dogs and humans breathe in and out, which ventilates their lungs. Breathing helps maintain the pressure gradients between the atmosphere and lungs. It pushes stale air (with a lot of exhaled CO_2) away from the body and draws in fresher air (with more O_2) closer to it.

Respiratory Pigments Maintaining steep pressure gradients across the respiratory surface is easier with **respiratory pigments**. These proteins incorporate one or more metal ions that bind oxygen atoms in oxygen-rich tissues and give them up in oxygen-poor tissues. Remember vertebrate red blood cells? They are packed with the iron-containing hemoglobin (Sections 3.6 and 34.2). Mollusks, annelids, and crustaceans also produce hemoglobin. The annelid hemoglobin molecule is six times bigger than yours; it has as many as 200 globin chains that each bind one oxygen atom. Hemerythrin (with iron groups) and hemocyanin (copper groups) reversibly bind oxygen in different invertebrates.

Myoglobin is structurally similar to a hemoglobin chain but serves in oxygen storage, not transport. It is abundant in some muscle cells. Once oxygen binds to myoglobin, it cannot diffuse out, so this pigment helps maintain the O_2 pressure gradient between blood and the cells. As you will see, myoglobin releases oxygen for aerobic respiration during muscle contraction.

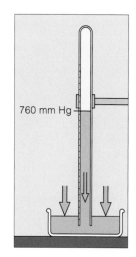

Figure 40.3 How a mercury barometer can measure the atmospheric pressure. A narrow column of mercury (Hg) rises or falls in a tube, depending on air pressure outside. At sea level, it rises to 760 millimeters (29.91 inches) from the tube's base. Altitude affects the level of atmospheric pressure. On top of Mount Everest, air pressure is about one-third of what it is at sea level.

760 mm Hg

> Respiration is the sum of processes that supply animal cells with oxygen for aerobic respiration and remove carbon dioxide, the metabolic pathway's wastes, from the body.
>
> Oxygen follows its partial pressure gradient from the atmosphere to a respiratory surface: a thin, moist boundary to the internal environment. Carbon dioxide follows its partial pressure gradient in the opposite direction.
>
> Gas exchange requires steep partial pressure gradients between the external and internal environments. Ventilation helps keep the gradients steep. So do hemoglobin and other respiratory pigments that reversibly bind oxygen.

40.2 Invertebrate Respiration

The body wall is the respiratory surface of the smallest and thinnest invertebrates. Larger kinds have internal respiratory surfaces or tubes with ever finer branchings that deliver oxygen right next to cells in all tissues.

LINKS TO SECTIONS 25.8, 25.12, 28.2

Small-bodied invertebrates of aquatic or continually moist habitats have the simplest forms of respiration. Gases just diffuse across the body surface covering— the integument (Figure 40.4a). This respiration mode is called **integumentary exchange**.

Many aquatic invertebrates have thin-walled, moist respiratory organs called **gills**, although these are not the same in structure as fish gills. Extensively folded gill walls increase the respiratory surface area and gas exchange rates between body fluids and the outside. Figure 40.4b shows the much-folded gills of a sea hare (*Aplysia*) that supplement integumentary exchange.

Spiders and other arthropods in dry habitats have a small, thick, or hardened integument that conserves water but cannot exchange gases. They use an *internal* respiratory surface. For instance, in most spiders, the surface forms book lungs: it folds back on itself into thin sheets of vascularized tissue (Section 25.12).

Insects, millipedes, centipedes, and certain spiders have a **tracheal system** of branching tubes inside the body. Figure 40.5 shows one insect's tracheal system. The tubes start at tiny openings (spiracles) across the integument. Each tube branches, then branches again. Fluid fills the tips of the finest branchings, which end right next to cells in all body tissues. There, gases are exchanged with interstitial fluid.

Larger, active insects on land have tracheal systems with water-conserving adaptations. For instance, fruit fly spiracles have muscled valves. They open during oxygen-demanding flight, and when the fly rests, they partly close and conserve body water. A grasshopper forcibly ventilates thin, pliable tracheal tubes. When its abdominal muscles contract, internal organs press into the tubes and force out air. When they relax, the organs spring back, the tracheal lumen expands, and air is sucked in. Air sacs attach to the largest tubes. They allow more air to be drawn in and expelled with each round of muscle contraction and relaxation.

Some small invertebrates of aquatic or moist habitats use integumentary exchange, with gases diffusing across the body wall. Others use a type of gill, with gases diffusing across a thin, folded internal respiratory surface. Most insects and spiders use tracheal respiration, with gases exchanged at fluid-filled tips of open, branching tubes that extend from the surface to internal tissues.

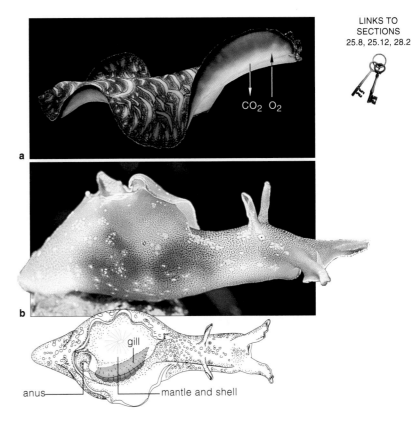

a

b

gill

anus —————— mantle and shell

Figure 40.4 Two aquatic invertebrates. (**a**) A flatworm, small enough to get along without an oxygen-transporting circulatory system. Dissolved oxygen in its habitat reaches individual cells simply by diffusing across the body surface. (**b**) Gill of a sea hare (*Aplysia*), one of the gastropods. Compare Sections 25.8 and 28.2.

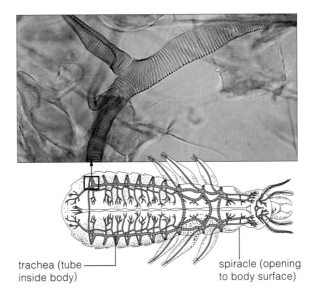

trachea (tube inside body)

spiracle (opening to body surface)

Figure 40.5 Generalized insect tracheal system. The photograph shows the chitin rings that reinforce many branching tubes in such respiratory systems. In grasshoppers, abdominal muscles and air sacs enhance ventilation of the tracheal system.

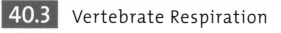

40.3 | Vertebrate Respiration

LINKS TO
SECTIONS
26.4, 35.5

Different vertebrates use gills, their skin, paired lungs, or a combination of skin and lungs as respiratory organs.

GILLS OF FISHES AND AMPHIBIANS

Gills are organs that have a thin, moist, vascularized respiratory surface, and they vary among vertebrates. In certain fish larvae and a few amphibians, *external* gills project into the water. Adult fishes have a pair of *internal* gills with overlapping rows of filaments. Slits between rows extend from the back of the mouth to the body surface (Figure 40.6a–c).

Fish respiration starts as water flows through the mouth and past gill filaments. A filament has rows of lamellae, or platelike folds, each with a capillary bed. Collectively, the capillaries greatly expand the surface area available for gas exchange between the water and blood. Oxygen-poor blood flows away from the main body mass inside an *efferent* vessel, through capillary beds, and into an *afferent* vessel that carries oxygen-enriched blood back to the main body mass.

The blood in capillaries flows in a direction that is counter to the direction of water flowing past (Figure 40.6d,e). Compared to the water, blood holds less O_2. So O_2 diffuses out of water and into blood at many points along each capillary bed. A flow of two fluids in opposing directions is called **countercurrent flow**. It lets a fish extract about 80 to 90 percent of the O_2

dissolved in water flowing past—more than it would get from a one-way flow, at less energy cost.

EVOLUTION OF PAIRED LUNGS

Some fishes and all amphibians, birds, and mammals have a pair of **lungs**. A lung is a saclike respiratory organ located in a body cavity, but airways connect it to the surface. Lungs evolved as tiny outpouchings of the gut wall. By the close of the Devonian, this key innovation was increasing the surface area available for gas exchange. Aquatic tetrapods were making the transition to land, and they no longer could depend on gills (Section 26.4). Gill filaments get stuck to one another unless water surrounds and moistens them.

Most amphibian larvae have gills, and adults have lungs. Some adult salamanders have neither lungs nor gills. Their skin functions in integumentary exchange. Frogs and toads use their small lungs for O_2 uptake, but CO_2 diffuses outward across their skin.

A frog does not breathe like you do. Air flows into its nostrils and is gulped into lungs. Stale air is forced from the lungs when muscles contract (Figure 40.7). Like you, the frog uses air to make sounds. It forces air back and forth between lungs and two pouches on the floor of its mouth. Air flows through a glottis, an opening between a pair of membrane folds at the start of an airway (larynx), and makes the folds vibrate. By

water flows into mouth

FISH GILL

Water flows over gills, then out.

a

mouth open

lid closed

b

mouth closed

lid open

c

Figure 40.6 Animated! (**a–c**) Fish gill ventilation. A fish forces water past paired gills when it opens its mouth and closes the lid (operculum) over each gill. Water flows out as the mouth closes and the lids open.

(**d**) Gas exchange in fish gills. Each gill filament is a vascularized respiratory surface folded into a series of platelike lamellae. Each lamella has a capillary bed, so gills have a tremendous surface area for exchanging gases between water and blood. The beds connect one vessel that transports oxygen-poor blood from the main body mass to a vessel that transports blood back into it.

(**e**) Capillaries move blood from the efferent vessel into the afferent one, counter to the direction of water flow. Countercurrent flow favors movement of oxygen (down its partial pressure gradient) from water into the blood.

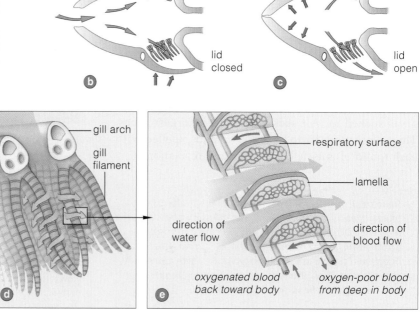

gill arch

gill filament

d

respiratory surface

lamella

direction of water flow

direction of blood flow

oxygenated blood back toward body

oxygen-poor blood from deep in body

e

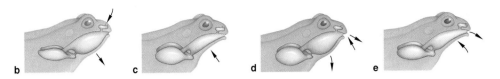

a

b   c   d   e

Figure 40.7 *Animated!* (**a**) How frogs breathe. (**b**) The frog lowers the floor of its mouth and inhales air through nostrils. (**c**) It closes the nostrils, opens the glottis, and elevates the mouth's floor. This *forces* air into the lungs. (**d**) Rhythmic ventilation assists in gas exchange. (**e**) Air is forced out when muscles in the body wall above the lungs contract and lungs elastically recoil.

amphibian
salamander;
like fishes
and early
amphibians

amphibian
frog; only
adults are
adapted to
dry habitats

reptile
lizard;
adapted
to dry
habitats

mammal
human;
adapted to
dry habitats

Figure 40.8 Lung structure of three kinds of vertebrates, suggestive of an evolutionary trend from simple sacs for gas exchange to larger, more complex respiratory surfaces.

controlling the frequency and magnitude of vibrations, the frog makes croaking sounds (Section 35.5).

Paired lungs are the dominant respiratory organs in modern reptiles, birds, and mammals (Figure 40.8). Breathing moves air by bulk flow into and out of the lungs. Blood capillaries thread lacily around the lung respiratory surfaces. Steep gradients draw oxygen and carbon dioxide swiftly across the surfaces, which are extensive. Oxygen enters the capillaries and circulates through the body. In tissues where its level is low, this gas diffuses into interstitial fluid and then into cells. Carbon dioxide diffuses out of cells, into interstitial fluid, into blood, and then is expelled from the lungs.

Birds alone have a system of air sacs that force air continuously *through* a pair of lungs, as the example in Figure 40.9 shows. With this one exception in mind, we turn next to the respiratory system of humans. Its operating principles apply to most vertebrates.

A countercurrent flow mechanism in fish gills boosts oxygen uptake from water. Gills do not work in dry land habitats, where internal air sacs—lungs—are more efficient.

Amphibians force oxygen into small lungs, but carbon dioxide diffuses out across the skin. Reptiles, birds, and mammals use paired lungs as their respiratory organs.

Birds have a unique ventilating system. Many air sacs force air continuously through tubes that thread through the vascularized tissue of a pair of small, inelastic lungs.

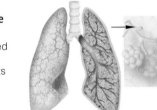

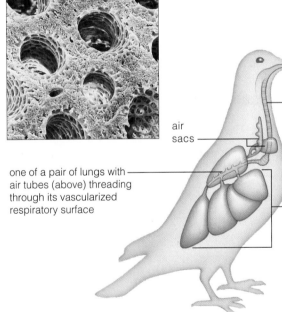

trachea

air sacs

one of a pair of lungs with air tubes (above) threading through its vascularized respiratory surface

air sacs

Figure 40.9 *Animated!* Respiratory system of a typical bird, a pigeon. Many air sacs are attached to two small, inelastic lungs. The bird inhales and draws air into the sacs through tubes, open at both ends, that thread through vascularized lung tissue. This tissue is the respiratory surface.

The bird exhales, forcing air from the sacs, through the tubes, and out of the trachea. Air is not only drawn into bird lungs. It is drawn continuously through them and across the respiratory surface. The unique ventilating system supports the high metabolic rates that birds require for flight and other energy-demanding activities.

40.4 Human Respiratory System

It will take at least 300 million breaths to get you to age seventy-five. The system that brings about each breath also functions in speech, in the sense of smell, and in homeostatic control of the internal environment.

THE SYSTEM'S MANY FUNCTIONS

Figure 40.10 shows the human respiratory system and lists its functions. Two paired lungs have about 600 million **alveoli** (singular, alveolus), or air sacs. Rates of air inflow and outflow to the alveoli are adjusted in ways that match metabolic demands for gas exchange.

The respiratory system functions in more than gas exchange. It assists in speech and the sense of smell. It helps blood in veins return to the heart and helps the body dispose of excess heat and water. Controls over breathing also contribute to maintaining the internal environment's acid–base balance. Parts of the system intercept many potentially dangerous airborne agents and substances before they can reach the lungs. Many bloodborne substances can be removed or neutralized before they can enter the internal environment.

The respiratory system ends at alveoli, and there it exchanges gases with blood vessels called *pulmonary*

NASAL CAVITY
Chamber in which air is moistened, warmed, and filtered, and in which sounds resonate

PHARYNX (THROAT)
Airway connecting nasal cavity and mouth with larynx; enhances sounds; also connects with esophagus

EPIGLOTTIS
Closes off larynx during swallowing

LARYNX (VOICE BOX)
Airway where sound is produced; closed off during swallowing

TRACHEA (WINDPIPE)
Airway connecting larynx with two bronchi that lead into the lungs

LUNG (ONE OF A PAIR)
Lobed, elastic organ of breathing; enhances gas exchange between internal environment and outside air

BRONCHIAL TREE
Increasingly branched airways starting with two bronchi and ending at air sacs (alveoli) of lung tissue

ORAL CAVITY (MOUTH)
Supplemental airway when breathing is labored

PLEURAL MEMBRANE
Double-layer membrane that separates lungs from other organs; the narrow, fluid-filled space between its two layers has roles in breathing

INTERCOSTAL MUSCLES
At rib cage, skeletal muscles with roles in breathing. There are two sets of intercostal muscles (external and internal)

DIAPHRAGM
Muscle sheet between the chest cavity and abdominal cavity with roles in breathing

a Human respiratory system

Figure 40.10 *Animated!* (**a**) Human respiratory system: its components and their functions. Muscles, including the diaphragm, and certain bones of the axial skeleton have secondary roles in respiration. (**b,c**) Location of alveoli relative to bronchioles and to lung (pulmonary) capillaries.

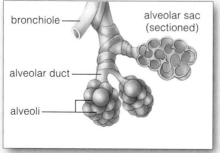

bronchiole

alveolar sac (sectioned)

alveolar duct

alveoli

b

alveolar sac

pulmonary capillary

c

capillaries (Latin *pulmo*, meaning lung). At this point, respiration requires the participation of another organ system. The circulatory system swiftly transports O_2 to all metabolically active tissues and CO_2 away from them (Section 38.1). Later, you will get a sense of how the systems interact in supporting aerobic respiration. You also will see how both contribute to homeostasis in the internal environment by swiftly removing CO_2 wastes from this metabolic pathway.

FROM AIRWAYS INTO THE LUNGS

Take a deep breath. Now look at Figure 40.10 to get an idea of where the air will travel in your respiratory system. Unless you are out of breath and panting, air has entered two nasal cavities, not your mouth. In the nose, mucus warms and moistens the air. Hairs and cilia of the nasal lining filter dust and particles from it, and sensory receptors detect odors. Air flows into the **pharynx**, or throat. This is the entrance to the **larynx**, a large airway with two paired folds of mucus-covered membrane projecting into it. The lower pair are vocal cords (Figure 40.11). Remember how a frog vocalizes? When you breathe, you force air past a **glottis**, a gap between the vocal cords. The air flow makes the cords vibrate. Humans make different sounds by controlling the vibrations (Section 35.5).

While the larynx muscles contract and relax, elastic ligaments in the vocal cords tighten or slacken to alter how much the folds are stretched. The nervous system coordinates the narrowing and widening of the glottis. When its signals cause increased tension in the larynx muscles, for example, the gap between the vocal cords narrows and so high-pitched sounds emerge. The lips, teeth, tongue, and the soft roof above the tongue can be enlisted to modify the basic patterns of sounds.

When infected or irritated, vocal cords swell, which interferes with their capacity to vibrate. Severe swelling can cause *laryngitis*, or a hoarse throat.

At the entrance to the larynx is an **epiglottis**. When this tissue flap points up, air moves into the **trachea**, or windpipe. When you swallow, the epiglottis flops over and points downward. In this position, it covers the entrance to the larynx, so food and fluids can enter a different tube. That tube, the esophagus, is not one of the airways. It connects the pharynx to the stomach.

The trachea branches into two airways: one to each lung. Each airway is a **bronchus** (plural, bronchi). Its epithelial lining has many ciliated and mucus-secreting cells that work against infection. Bacteria and airborne particles stick to the mucus; cilia sweep mucus to the mouth, where it can be expelled.

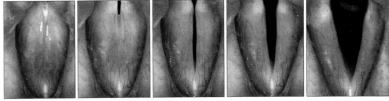

glottis closed

glottis open

vocal cords
glottis (closed)
epiglottis
tongue's base

Figure 40.11 Human vocal cords, where the sounds used in speech originate. Signals from the nervous system coordinate contractions in skeletal muscles that change the width of a gap between them, called the glottis. This series of five photographs shows how the width changes during a normal breath.

The paired human lungs are elastic, cone-shaped organs inside the thoracic cavity, one located on each side of the heart. The rib cage protects them. Below the lungs, a broad sheet of smooth muscle—the **diaphragm**—partitions the coelom into a thoracic and an abdominal cavity (Figure 40.10). A moistened, thin, saclike pleural membrane lines the outer lung surface and the inner thoracic cavity wall. The wall and lungs press the sac's two surfaces together.

A film of lubricating fluid cuts friction between the pleural membrane surfaces. In *pleurisy*, a respiratory ailment, the membrane gets inflamed and swollen. Its two surfaces rub each other, so breathing is painful.

Inside each lung, air moves through finer and finer branchings of a "bronchial tree." These branchings are **bronchioles**. The finest kinds, *respiratory* bronchioles, end in cup-shaped alveoli. Most alveoli are clustered as pouches, or alveolar sacs. There are great numbers of sacs, which collectively afford a tremendous surface area for gas exchange with blood. If all alveolar sacs in your body could be stretched out in one layer, they would cover the surface of a racquetball court!

The human respiratory system moves oxygen from the atmosphere into a pair of lungs, and carbon dioxide from the lungs into the atmosphere.

At the lungs, the circulatory system takes over the task of gas exchange with interstitial fluid throughout the body.

The respiratory system also functions in moving venous blood to the heart, vocalizing, adjusting the acid–base balance, and defending the body from harmful airborne agents or substances.

41.10 Weighty Questions, Tantalizing Answers

LINKS TO
SECTIONS
17.4, 23.11

For many people, weight is a touchy subject. Part of the problem is that the standard of what constitutes an "ideal weight" varies from culture to culture. Here we consider weight as it relates to health. Thinner people really do live longer, on average.

More than 108 million Americans—a whopping 60 percent of the United States population—are overweight, and about 300,000 or so die each year as an outcome of preventable, weight-related conditions. As weight increases, so does the risk of type 2 diabetes, hypertension, heart disease, breast cancer, colon cancer, gallstones, and many other ailments.

What Is the "Right" Body Weight?   Figure 41.14 shows one of the widely accepted weight guidelines for women and men. A different guideline is the *body mass index* (BMI), a measurement designed to help assess the health risk associated with weight gains. You can calculate your body mass index with this formula:

$$BMI = \frac{weight\ (pounds) \times 703}{height\ (inches)^2}$$

Generally, individuals with a BMI of 25 to 29.9 are considered to be overweight. A score of 30 or more indicates **obesity**. We define this term as an overabundance of fat in adipose tissue that may lead to severe health problems. How body fat is distributed also helps predict risks. Fat that becomes stored above the belt, as in "beer bellies," is associated with an increased likelihood of heart problems.

Dieting alone cannot lower an individual's BMI value. When you eat less, the body will slow its metabolic rate to conserve energy. So how do you function normally over the long term while maintaining acceptable weight? You must balance caloric intake with energy output. For most people, this means eating only the recommended portions of low-calorie, nutritious foods and exercising regularly.

Bear in mind, energy stored in food is expressed as kilocalories or Calories (with a big C). A kilocalorie is 1,000 calories, which are units of heat energy.

To calculate how many kilocalories you should take in daily in order to maintain a preferred weight, multiply that weight (in pounds) by 10 if you are not active physically, by 15 if you are moderately active, and by 20 if you are highly active. Subtract one of the following amounts from the multiplication result:

Age:	Subtract:
25–34	0
35–44	100
45–54	200
55–64	300
Over 65	400

Suppose you are 25 years old, highly active, and weigh 120 pounds. This means you will require 120 × 20 = 2,400 kilocalories daily to maintain weight. If you want to gain weight you will require more; to lose, you will require less. But this is only a rough estimate. Other factors, including height, must be considered. A person who is 5 feet, 2 inches tall and is active does not require as much energy as an active 6-footer whose body weight is the same.

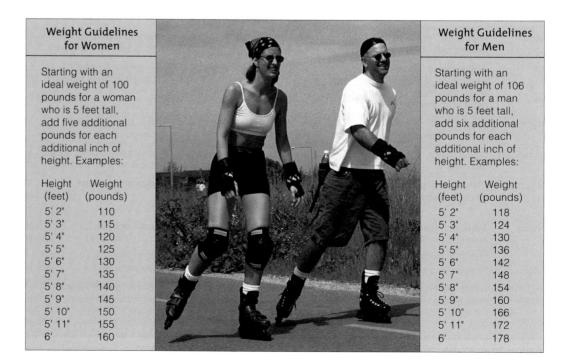

Figure 41.14 How to estimate "ideal" weights for adults. Values shown are consistent with a long-term Harvard study into the link between excess weight and risk of cardiovascular disorders. The "ideal" varies. It is influenced by specific factors such as having a small, medium, or large skeletal frame; bones are heavy.

Weight Guidelines for Women

Starting with an ideal weight of 100 pounds for a woman who is 5 feet tall, add five additional pounds for each additional inch of height. Examples:

Height (feet)	Weight (pounds)
5' 2"	110
5' 3"	115
5' 4"	120
5' 5"	125
5' 6"	130
5' 7"	135
5' 8"	140
5' 9"	145
5' 10"	150
5' 11"	155
6'	160

Weight Guidelines for Men

Starting with an ideal weight of 106 pounds for a man who is 5 feet tall, add six additional pounds for each additional inch of height. Examples:

Height (feet)	Weight (pounds)
5' 2"	118
5' 3"	124
5' 4"	130
5' 5"	136
5' 6"	142
5' 7"	148
5' 8"	154
5' 9"	160
5' 10"	166
5' 11"	172
6'	178

A Question of Portion Sizes

Have you noticed that portions of restaurant food are getting more and more "super-sized"? What was enough to feed two people in 1977 is now served on a plate for one. Correlated with the bigger portions are bigger waistlines and more increases in body weight. *How much* we eat affects weight gain as much as what we eat and how much we exercise.

Compared with the past, fast food now makes up a big part of the daily caloric increase. In one study, researchers discovered that their subjects were packing on an extra 49 kilocalories a day from soft drinks; 68 from french fries; 93 from potato chips, popcorn, and other salty snacks; 97 from hamburgers; and 133 from Mexican food.

Do these numbers seem small? An extra 10 kilocalories per day translates into a gain of 1 pound per year.

The FDA guidelines no longer say "servings" of food but rather specify amounts. Too many people have a distorted view of portion sizes—for instance, a serving of fruit is the size of a tennis ball, not a platter of bananas.

Genes, Hormones, and Obesity

Many experiments, including those highlighted in Figure 41.15, revealed that genes have a lot to do with body weight. In 1990, Claude Bouchard reported on a study of experimental overeating by twelve pairs of male twins. All were lean young men in their early twenties. For 100 days they lived in a dormitory, were told not to exercise, and followed a diet that delivered 6,000 more kilocalories a week than usual.

All the men gained weight. But some gained three times as much as others. Those who were twins tended to gain similar amounts of weight. Genetic differences apparently affected the response to overfeeding. For a different test, Bouchard started with sets of obese twins and put them on a limited calorie diet. The men varied in their responses, with twins losing similar amounts.

As the chapter introduction indicated, we are learning more about genes that contribute to obesity. The latest candidate for appetite-affecting hormones is PYY3-36. Glandular cells in the lining of the stomach and small intestine release it after a meal. It acts in the brain to suppress appetite. Rats injected with the hormone eat less and lose weight. Human volunteers who were given an intravenous dose of PYY3-36 before a buffet meal ate less than a control group that got a saline dose.

In such ways, researchers are attempting to identify all steps of the ancient pathways that control weight. They also plan to identify the contributions of those steps to other aspects of the body's physiology.

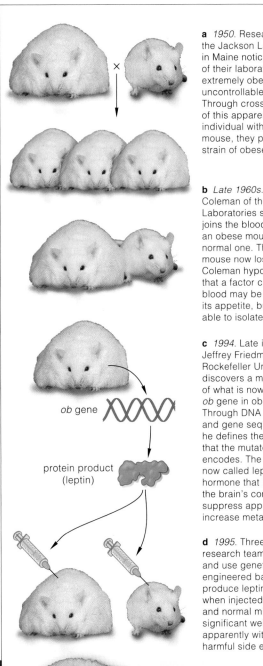

a *1950.* Researchers at the Jackson Laboratories in Maine notice that one of their laboratory mice is extremely obese, with an uncontrollable appetite. Through cross-breeding of this apparent mutant individual with a normal mouse, they produce a strain of obese mice.

b *Late 1960s.* Douglas Coleman of the Jackson Laboratories surgically joins the bloodstreams of an obese mouse and a normal one. The obese mouse now loses weight. Coleman hypothesizes that a factor circulating in blood may be influencing its appetite, but he is not able to isolate it.

c *1994.* Late in the year, Jeffrey Friedman of Rockefeller University discovers a mutated form of what is now called the *ob* gene in obese mice. Through DNA cloning and gene sequencing, he defines the protein that the mutated gene encodes. The protein, now called leptin, is a hormone that influences the brain's commands to suppress appetite and increase metabolic rates.

d *1995.* Three different research teams develop and use genetically engineered bacteria to produce leptin, which, when injected in obese and normal mice, triggers significant weight loss, apparently without harmful side effects.

ob gene

protein product (leptin)

Figure 41.15 Chronology of research developments that identified leptin as a heritable factor that affects body weight.

Summary

Section 41.1   A digestive system mechanically and chemically breaks down food into molecules small enough to be absorbed into the internal environment, and it eliminates unabsorbed residues. Its interactions with other organ systems promote homeostasis. The saclike systems are incomplete, with ony one opening. Most animals have a complete digestive system: a tube with two openings (mouth and anus) and specialized areas between them.

Biology⊛Now
Compare vertebrate digestive systems with the animation on BiologyNow.

Section 41.2   Table 41.4 summarizes the components and accessory organs of the human digestive system, as well as their functions.

Biology⊛Now
Tour the human digestive system with the animation on BiologyNow.

Sections 41.3, 41.4   Carbohydrate digestion starts in the mouth. Chewing mixes food with enzyme-rich saliva. Protein digestion starts in the stomach. This muscular sac has a glandular lining that secretes gastric fluid. Most digestion is completed in the small intestine.

Ducts leading from the pancreas and gallbladder empty into the small intestine. The pancreas secretes digestive enzymes. Bile, which assists in fat digestion, is made in the liver and stored in the gallbladder.

Local controls as well as the nervous and endocrine systems respond to the volume and composition of food in the gut. They cause changes in muscle activity and in secretion rates for hormones and enzymes.

Biology⊛Now
Explore levels of biological organization with the interaction on BiologyNow.

Section 41.5   The highly folded epithelial lining of the small intestine (intestinal mucosa) has absorptive structures called villi. Many of its cells secrete mucus, hormones, and lysozymes. Many others (brush border cells) have membrane proteins that cotransport sodium ions (Na^+) and simple sugars or amino acids from the intestinal lumen into the villi. A blood vessel in each villus takes up absorbed sugars and amino acids.

The resulting Na^+ concentration gradient sets up an osmotic gradient that draws 80 percent of the water out of the lumen's contents and into the villi.

Monoglycerides and fatty acids diffuse across the lipid bilayer of brush border cells. They recombine in the cells to form triglycerides, which are then secreted into interstitial fluid. From there, they enter lymph vessels that deliver them to blood.

Biology⊛Now
Learn about the structure of the small intestine and how it absorbs nutrients with the animation on BiologyNow.

Section 41.6   The large intestine (colon) absorbs water and mineral ions. It also compacts undigested residues and water into feces, which are stored in the rectum, the final part of the digestive tract.

Section 41.7   Small organic compounds absorbed from the gut lumen are stored, used in biosynthesis or as energy sources, or excreted by other organ systems.

Section 41.8   Dietary guidelines are in flux. There is general agreement that high intake of saturated fats and refined carbohydrates should be avoided.

Section 41.9   Vitamins are organic compounds and minerals are inorganic compounds with essential roles in normal metabolism. Usually a balanced diet provides adequate amounts.

Section 41.10   Obesity raises the risk of health problems and shortens life expectancy. To maintain body weight, energy (caloric) intake must balance energy (caloric) output. Heritable factors make it difficult for some people to maintain a weight that promotes overall health. Leptin, ghrelin, and other hormones are now known to affect appetite centers and metabolism.

Biology⊛Now
Calculate your body mass index with the interaction on BiologyNow.

Table 41.4	Summary of the Human Digestive System
Mouth (oral cavity)	Start of digestive system, where food is chewed, and moistened; polysaccharide digestion begins
Pharynx	Entrance to digestive and respiratory tract tubes
Esophagus	Muscular tube, moistened by saliva, that moves food from pharynx to stomach
Stomach	Stretchable sac where food mixes with gastric fluid and protein digestion starts; stores food taken in faster than can be processed; its fluid kills many microbes
Small intestine	Receives secretions from liver, gallbladder, pancreas; digests most nutrients; delivers unabsorbed material to colon
Colon (large intestine)	Concentrates and stores undigested matter (by absorbing mineral ions and water)
Rectum	Distension triggers expulsion of feces
Anus	Terminal opening of digestive system

Accessory Organs:

Salivary glands	Glands that secrete saliva, a fluid with polysaccharide-digesting enzymes, buffers, and mucus
Liver	Secretes bile; roles in carbohydrate, fat, and protein metabolism
Gallbladder	Stores and concentrates bile from the liver
Pancreas	Secretes enzymes that digest all major food molecules; buffers against HCl secretions from stomach lining

Figure 41.16 One success story: In 2000, after recovering from anorexia, Dutch cyclist Leontien Zijlaard won three Olympic gold medals. Four years earlier, she was too malnourished and weak to compete.

Figure 41.17 Exceptionally Big Mac of the snake world.

Self-Quiz

Answers in Appendix II

1. A digestive system functions in _____ .
   a. secreting enzymes
   b. absorbing compounds
   c. eliminating wastes
   d. all of the above

2. Protein digestion begins in the _____ .
   a. mouth
   b. stomach
   c. small intestine
   d. colon

3. Most nutrients are absorbed in the _____ .
   a. mouth
   b. stomach
   c. small intestine
   d. colon

4. Bile has roles in _____ digestion and absorption.
   a. carbohydrate
   b. fat
   c. protein
   d. amino acid

5. Monosaccharides and amino acids are both absorbed from the gut _____ .
   a. at membrane proteins
   b. at lymph vessels
   c. as fat droplets
   d. both b and c

6. The largest number of bacteria thrive in the _____ .
   a. stomach
   b. small intestine
   c. large intestine

7. _____ are inorganic substances with metabolic roles that no other substance can fulfill.
   a. Fats
   b. Minerals
   c. Proteins
   d. Vitamins
   e. Simple sugars
   f. both b and d

8. Match each organ with a digestive function.
   ____ gallbladder
   ____ colon
   ____ liver
   ____ small intestine
   ____ stomach
   ____ pancreas
   a. makes bile
   b. compacts undigested residues
   c. secretes most digestive enzymes
   d. absorbs most nutrients
   e. secretes gastric fluid
   f. stores, secretes bile

Additional questions are available on **Biology ⊜ Now**™

Critical Thinking

1. *Anorexia nervosa* is an eating disorder in which people, most often young women, starve themselves. The name means "nervous loss of appetite," but the affected people are usually obsessed with food and continually hungry. Like people who undergo stomach stapling to lose weight, anorexics are vulnerable to loss of bone tissue and brittle bones. Anorexia nervosa has complex causes, but genes play a role. Australian researchers looked at a gene for a protein that transports norepinephrine, a neurotransmitter, across cell membranes. Individuals with one mutant allele were twice as likely to be anorexic. The discovery may lead to new treatments. Even now, many recovered anorexics enjoy normal lives (Figure 41.16).

Another extreme eating disorder is *bulimia*, an out-of-control "oxlike appetite." A bulimic on an hour-long eating binge may take in 50,000 kilocalories' worth of food, then vomit or use laxatives to get rid of it. To some, the binge–purge routine is an "easy" way to lose weight. Others may not even like to eat but purge themselves to relieve anger and frustration. The binge–purge routines range from once a month to a few times a day. Purgings damage the gut. Chronic vomiting brings up gastric fluid that erodes teeth to stubs. In severe cases, the stomach can rupture and the heart and kidneys fail.

Are severe eating disorders rare? In which age bracket do most anorexic or bulimic females and males fall? How many die each year from these disorders? Research both conditions, and write up your findings to present in class.

2. How can a python swallow something that is wider than it is? Consider Figure 41.17, then do some research and write up a brief description of your findings.

3. A glassful of whole milk contains lactose (a sugar), proteins, butterfat, vitamins, and minerals. Explain what will happen to each component in your digestive tract.

4. A *peptic ulcer* is a hole in the lining of the stomach or duodenum (Figure 41.5a). Acid leaking from it can cause pain and may be life threatening if it erodes a blood vessel. Most ulcers form after an infection by the bactrium *Helicobacter pylori* (Figure 21.3). Ulcers are treated with a course of antibiotics that can extend for as long as two weeks. Side effects are common and include cramps and diarrhea. Women who are treated have double the normal risk of vaginal infections. Explain why a lengthy course of antibiotics would cause such symptoms.

5. Biologist Dianne Anderson presented the hypothetical nutrition label shown at right to her students at San Diego State University. Can you name the organism that would offer the kinds and proportions of ingredients listed?

Nutrition Facts		
Serving size: 1/2 cup (112g)		
Servings per container: About 68		
Amount per serving		
Calories		201
Calories from fat		121
Percent Daily Value		
Total Fat 13g		17%
Saturated fat 13g		78%
Total Carbohydrate 2g		
Dietary fiber 0g		0%
Sugars 0 g		0%
Other carbohydrates 2g		
Protein 18g		40%
Ingredients: _____?_____ , saltwater		

Truth in a Test Tube

Light or dark? Clear or cloudy? A lot or a little? Asking about and examining urine is an ancient art (Figure 42.1). As early as 1500 B.C., physicians in India recorded that some people were producing copious amounts of sweet-tasting urine that attracted insects. In time, the disorder was named *diabetes mellitus*, which loosely translates as "passing honey-sweet water." Doctors still diagnose it by testing for sugar in urine, although they have replaced the taste test with chemical analysis.

Today physicians routinely check the sugar level, pH, protein content, and concentration of urine. Acidic urine can signal metabolic problems. Alkaline urine can signal a bacterial infection. Too much protein dissolved in urine might mean that the kidneys are not working properly. Excretion of too many solutes may be a sign of dehydration or problems with hormones that control kidney function. Also, specialized urine tests can detect chemicals produced by cancers of the kidney, bladder, and prostate gland.

Do-it-yourself urine tests have now become popular. If a woman is hoping to become pregnant, she can use one test to keep track of the amount of LH —luteinizing hormone—in her urine. About midway through a menstrual cycle, LH triggers ovulation, the release of an egg from an ovary. Another over-the-counter urine test can reveal whether or not she is pregnant. Older women can test for declining hormone levels in urine, which is a sign that they are entering menopause.

Not everyone is in a hurry to have their urine tested. Olympic athletes can be stripped of their medals when mandatory urine tests reveal they use prohibited drugs. Major League Baseball players agreed to urine tests only after repeated allegations that certain star players had taken prohibited steroids. The National Collegiate Athletic Association (NCAA) tests urine samples from about 3,300 student athletes per year for any performance-enhancing substances as well as "street drugs."

If you use marijuana, cocaine, Ecstasy, or other kinds of drugs, urine tells the tale. After the active ingredient of marijuana enters blood, the liver converts it to another

Figure 42.1 *This page*, a seventeenth-century physician and a nurse examining a urine specimen. Urine's consistency, color, odor, and—at least in the past—taste afford clues to health problems. Urine forms inside kidneys, and it provides clues to abnormal changes in the volume and composition of blood and interstitial fluid. *Facing page*, testing for the presence of drugs in urine samples.

compound. Kidneys filter the blood and they add this telltale compound to the newly forming urine. It can take up to ten days for all of it to be metabolized and removed from the body. Until then, urine tests can detect it.

That urine is such a remarkable indicator of health, hormonal status, and drug use is a tribute to the urinary system. Each day, a pair of fist-sized kidneys filter all of the blood in an adult human body, and they do so more than thirty times. When all goes well, kidneys get rid of excess water and excess or harmful solutes, including many metabolites, toxins, hormones, and drugs.

So far in this unit, you have considered several organ systems that work to keep cells supplied with oxygen, nutrients, water, and other substances. Turn now to the kinds that maintain the composition, volume, and even the temperature of the internal environment.

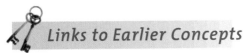
Watch the video online!

How Would You Vote?

Many companies use urine testing to screen for drug and alcohol use among prospective employees. Some people say this is an invasion of privacy. Do you think employers should be allowed to require a person to undergo urine testing before being hired? See BiologyNow for details, then vote online.

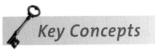
Key Concepts

MAINTAINING THE EXTRACELLULAR FLUID

Animals continually produce metabolic wastes, and they gain and lose water and solutes. Yet the overall composition and volume of a body's extracellular fluid must be kept within a range that individual cells can tolerate. In humans, as in other vertebrates, a urinary system interacts with several organ systems in this task. Section 42.1

THE HUMAN URINARY SYSTEM

The human urinary system consists of two kidneys, two ureters, a bladder, and a urethra. Inside a kidney, millions of nephrons filter water and solutes from the blood on an ongoing basis. Most of this filtrate is returned to the blood. Water and solutes not returned to the blood by way of the kidneys leave the body as urine. Section 42.2

WHAT KIDNEYS DO

Urine forms by processes of filtration, reabsorption, and secretion. Hormonal and behavioral responses to shifts in the internal environment continually adjust its concentration. The hormones ADH and aldosterone, as well as a thirst mechanism, influence whether urine becomes concentrated or dilute during any given interval. Sections 42.3–42.6

ADJUSTING THE CORE TEMPERATURE

The temperature at the core of the animal body reflects a balance between heat produced through metabolism, heat absorbed from the environment, and heat lost to the environment. Body temperature is maintained within a favorable range with controls over metabolic activity and adaptations in body form and behavior. Sections 42.7, 42.8

Links to Earlier Concepts

Vertebrates, recall, originated in water, and some moved onto land (Section 26.2). This chapter explores mechanisms that made the move possible, including protein-mediated transport and osmosis (5.4, 5.5). You will be tapping your understanding of pH and buffer systems (2.6), and of the circulatory system (38.5), especially capillary function (38.8). This chapter tracks waste products of organic metabolism (41.6) and returns to the effects of pituitary hormones and adrenal glands on the internal environment (36.3, 36.8).

Thermal homeostasis will be easier to understand if you review water's temperature-stabilizing effects (2.5), the nature of energy flow (6.1), and mitochondrial function (8.4).

42.1 Gains and Losses in Water and Solutes

LINKS TO
SECTIONS
26.2, 41.6

*About 425 million years ago, some lineages of animals
that evolved in salty water moved into freshwater habitats
and onto dry land. They were able to do so partly because
they brought along salty fluid inside their body tissues,
as an internal environment for individual cells.*

In most animals, recall, **interstitial fluid** fills tissue
spaces between cells, and a circulatory system moves
blood to and from tissues. Together, interstitial fluid
and blood are **extracellular fluid** that functions as an
internal environment for body cells. It does not matter
whether the animal lives on land or in the water. The
composition and volume of its internal environment
must be maintained within rather narrow ranges that
the body's individual cells can tolerate—a state called
homeostasis. Interactions among organ systems bring
about this fluid homeostasis (Figure 42.2).

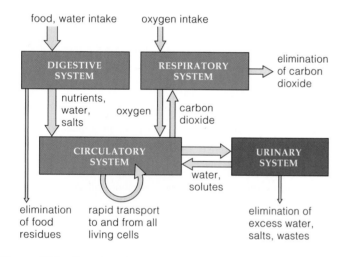

Figure 42.2 Overview of the functional links between the digestive,
respiratory, circulatory, and urinary systems. Guided by the nervous
and endocrine system, interactions among these systems contribute
to maintaining homeostasis in the internal environment.

CHALLENGES IN WATER

Osmosis, recall, is the movement of water across some
semipermeable membrane in response to differences
in solute concentrations between regions (Chapter 5).
In most marine invertebrates, such as crabs, interstitial
fluid is like seawater in its concentrations of solutes, so
little water is gained or lost by way of osmosis.

In vertebrates, body fluids are about one-third as
salty as seawater. Vertebrates have a **urinary system**,
a collection of interacting organs that counter shifts in
the composition and volume of extracellular fluid. All
have paired **kidneys** that filter blood, form urine, and
help maintain the body's water–solute balance. **Urine**,
a fluid excreted from the body, contains nitrogenous
wastes and excess amounts of other solutes and water.

Freshwater fishes and amphibians gain water and
lose solutes continually (Figure 42.3a). Neither kind of
animal can drink water. Water moves into the internal
environment by way of osmosis—by diffusing across
thin respiratory surfaces and skin. Excess water leaves
as dilute urine. The solute losses are balanced out by
the intake of more solutes with meals and the active
pumping of sodium into the cells of gills and skin at
membrane transport proteins.

Compared to seawater, marine bony fishes contain
less salt and continually lose water by osmosis. They
gulp in seawater, and gill cells pump out any excess
solutes (Figure 42.3b). Some of the dissolved wastes
are excreted in a tiny volume of concentrated urine.

CHALLENGES ON LAND

Pioneers on land and their descendants faced intense
sunlight, drying winds, more pronounced swings in
temperature, water of variable salt content, and often
no water at all. They kept their internal environment

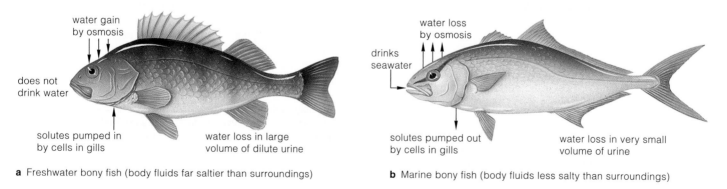

a Freshwater bony fish (body fluids far saltier than surroundings)

b Marine bony fish (body fluids less salty than surroundings)

Figure 42.3 Water–solute balance in a freshwater fish (**a**) and saltwater fish (**b**).

from spinning out of control. How? *How did the land-dwelling descendants of marine animals maintain internal operating conditions and so keep their cells functioning?*

Think about how animals *gain water*. Water enters the gut in food and drink, and then is absorbed into the internal environment. It also is a product of many metabolic reactions. The volume of water that enters the gut during a given interval can be monitored and adjusted by the thirst mechanism, which we explain later in the chapter.

Unlike fishes, vertebrates on land do not *lose water* by way of osmosis. They lose it in controlled ways, mainly by **urinary excretion**. With this process, excess water and solutes exit the body as urine. Some water evaporates from respiratory surfaces, especially those in mammalian lungs. Also, mammals are the only animals that lose some water by way of sweating. In addition, very little water is still present in food residues that leave the body in feces.

Now think about some processes that *add solutes* to extracellular fluid. Absorption of nutrients from the gut, secretions from cells, and the release of carbon dioxide wastes from aerobically respiring cells adds solutes. Air intake in the lungs adds oxygen to it.

Mammals *lose solutes* mainly in sweat, respiration, and urinary excretion. All exhale carbon dioxide, the most abundant waste material. Mammalian urine has ions and wastes that formed through metabolism. For example, toxic ammonia is formed as a by-product of protein metabolism. In the mammalian liver, ammonia is converted to **urea**, which is then excreted in urine. The distinctive yellowish color of urine comes from breakdown products of hemoglobin. In humans, other solutes that leave the body in urine are food additives, drugs, and many other synthetic chemicals.

Figure 42.4 compares the balancing act for humans and kangaroo rats. The daily gains and losses balance out in both, mainly by way of a urinary system.

In all animals in good health, daily gains in water and solutes balance the daily losses.

In all vertebrates, a urinary system counters unwanted shifts in the volume and composition of extracellular fluid. Paired kidneys filter blood, form urine, and help maintain solute levels within tolerable limits.

	Kangaroo Rat	Human
Daily water gain (milliliters):		
by ingesting solids	6.0	850
by ingesting liquids	0	1,400
by metabolism	54.0	350
	60.0	2,600
Daily water loss (millilters):		
in urine	13.5	1,500
in feces	2.6	200
by evaporation	43.9	900
	60.0	2,600

Figure 42.4 How a kangaroo rat gains and loses water, compared with humans. As in other animals, losses normally balance any gains in excess of the volume required to maintain the internal environment.

In a New Mexico desert, free water is scarce, except during a brief rainy season. A kangaroo rat waits out the heat of the day in a burrow, then forages at night for dry seeds and bits of plants. This tiny mammal hops rapidly and far, searching for seeds and fleeing from predators.

Hopping uses ATP energy and water, which seeds replace. Metabolic reactions release usable energy and water from carbohydrates and other compounds in seeds. Each day, "metabolic water" makes up a whopping 90 percent of a kangaroo rat's total water intake. Metabolic water is only about 12 percent of a human's total intake.

A kangaroo rat conserves and recycles water when it rests in its cool burrow. It moistens and warms air that it inhales. When it exhales, water condenses in its cooler nose, and some diffuses back into the body. When seeds are emptied from the kangaroo rat's cheek pouches, they soak up water dripping from the nose. The kangaroo rat reclaims water when it eats dripped-on seeds.

A kangaroo rat has no sweat glands and its feces hold only half the water that human feces do, relative to body size. It loses water when urinating, but its two specialized kidneys do not let it lose much. A kangaroo rat's kidneys excrete urine that is as much as three to five times more concentrated than yours.

42.2 Structure of the Urinary System

LINKS TO
SECTIONS
38.5, 38.6

In the urinary system of humans and other mammals, kidneys filter water, mineral ions, organic wastes, and other substances from the blood flowing through them. They adjust the composition of this filtrate and return all but about 1 percent or so to the blood. The unreclaimed water, solutes, and wastes become urine.

COMPONENTS OF THE SYSTEM

Again, a urinary system rids the body of metabolic wastes and adjusts water and solute concentrations in extracellular fluid. Drink too little water, gulp down salty potato chips, lose too much sodium in sweat, and the kidneys adjust how much water and salt the body excretes or saves. The human urinary system consists of two kidneys, two ureters, a urinary bladder, and a urethra. The kidneys are a pair of bean-shaped organs about the size of a typical adult fist. They are located between the peritoneal lining and abdominal cavity wall, where they flank the backbone (Figure 42.5a,b).

A thin but strong layer of dense connective tissue encapsulates each kidney (Figure 42.5c). Beneath this capsule is an outer tissue zone, the *kidney cortex*. The cortex is continuous with the *kidney medulla*, an inner zone with pyramid-shaped lobes of tissue. The lobes appear striated because each has a profusion of long tubes that extend down to a chamber, the renal pelvis.

Blood reaches each kidney by way of a renal artery, then flows through arterioles, capillaries, and venules in the cortex and medulla. The venules connect with a renal vein, which leads out of the kidney.

Urine forms in the kidneys, then enters one of two ureters that connect with a muscular sac: the urinary bladder. Stretch receptors inside the bladder wall are stimulated when urine expands and fills the sac. In a reflex response, smooth muscle in the wall contracts and forces urine into the urethra, a muscular tube that opens on to the body surface. An adult male's urethra is about 0.2 meter (8 inches) long and extends to the surface of the tip of the penis. The urethra of an adult female is less than 0.05 meter (2 inches) long.

After age two or three, urination can be voluntarily controlled by neural signals that act on a sphincter of skeletal muscle at the start of the urethra.

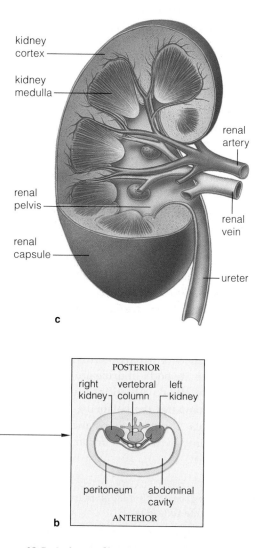

kidney (one of a pair)

Constantly filters water and all solutes except proteins from blood; reclaims water and solutes as the body requires and excretes the remainder, as urine

ureter (one of a pair)

Channel for urine flow from a kidney to the urinary bladder

urinary bladder

Stretchable container for temporarily storing urine

urethra

Channel for urine flow between the urinary bladder and body surface

Figure 42.5 *Animated!* (**a**) Human urinary system and its functions. (**b**) Its paired kidneys are located in between the abdominal cavity's wall and its lining, the peritoneum. (**c**) Structural organization of a human kidney.

Figure 42.6 *Animated!* **(a)** Diagram of a nephron. Through their interactions with two sets of blood capillaries, nephrons are the kidney's functional units.

(b) The arterioles and blood capillaries associated with each nephron. Large gaps in the cell wall of the glomerular capillaries make the capillaries a hundred times more permeable than any others in the body. Only a thin basement membrane separates each capillary wall from cells of the inner layer of Bowman's capsule, which cloak the wall. Cells of this inner layer have long extensions that interdigitate with one another, like interlaced fingers. Fluid flows through the narrow slits between them.

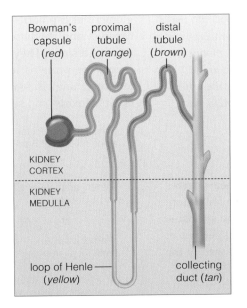

a Bowman's capsule and tubular regions of one nephron, cutaway view.

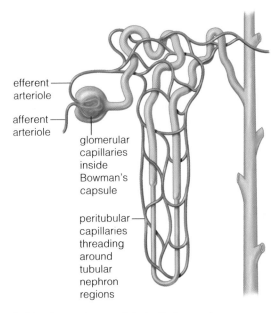

b Blood vessels associated with the nephron.

THE NEPHRON

Each kidney contains more than a million **nephrons**, microscopically small functional units that are only one cell thick all along their length. A nephron starts in the kidney cortex, where its wall balloons outward like a cup (Figure 42.6a). The nephron continues in the cortex as a twisting, convoluted tube. It straightens as it descends into the kidney medulla, then ascends into the cortex. There, another convoluted region connects with a collecting duct. As many as eight nephrons may empty into the same duct, which extends through the kidney medulla and opens onto the renal pelvis.

Now build on this simple picture of the nephron's structure. At its cup-shaped entrance, called *Bowman's capsule*, the thin nephron wall doubles back on itself and encloses the walls of highly porous blood vessels clustered inside. These are the **glomerular capillaries** (Figure 42.6b). Bowman's capsule and the glomerular capillaries interact as a blood-filtering unit. The unit is known as a renal corpuscle.

Bowman's capsule collects fluid that is forced out, under pressure, from the glomerular capillaries. This fluid enters the *proximal* convoluted tubule. "Proximal" simply means it is the region closest to the start of the nephron. The next tubular region, the *loop of Henle*, plunges into the medulla, makes a hairpin turn, and ascends out of it. The *distal* tubule is the convoluted part most distant from the entrance to the nephron.

Inside the kidneys, the renal artery branches into *afferent* arterioles, one for each nephron, that delivers blood to glomerular capillaries. The capillaries rejoin and form an *efferent* arteriole, which carries away the blood that did not get filtered into Bowman's capsule. This arteriole quickly branches to form **peritubular capillaries** that thread all around the nephron (*peri–*, around). The peritubular capillaries rejoin as venules, which join the renal vein leading out of the kidney.

Urine forms continually by three processes that exchange water and solutes between all of the nephrons, glomerular capillaries, and peritubular capillaries. The processes are called glomerular filtration, tubular reabsorption, and tubular secretion. Remember, each human kidney has more than 1 million nephrons. Each minute, nephrons of both kidneys filter close to 125 milliliters of fluid from blood flowing past, which amounts to 180 liters (about 47.5 gallons) per day. This means that, for an average-sized adult, the two kidneys filter the entire volume of blood plasma 65 times each day!

The human urinary system has two kidneys, two ureters, and a urinary bladder. The kidneys filter water and solutes from blood. The body reclaims most of the filtrate. The rest flows as urine through ureters into a bladder that stores it. Urine is excreted to the body surface through a urethra.

Millions of nephrons, and two sets of blood capillaries that interact with each one, are the functional units of human kidneys. They form urine by three processes: glomerular filtration, tubular reabsorption, and tubular secretion.

42.3 Urine Formation

LINKS TO
SECTIONS 5.3–5.6,
38.8, 40.3, 41.5

Blood pressure drives water and solutes from blood into the nephron. Transport proteins and variations in permeability along the nephron's tubular parts affect which components of the filtrate return to blood or leave in urine. Figure 42.7 introduces the text's step-by-step account of how the urine forms by glomerular filtration, tubular reabsorption, and tubular secretion.

GLOMERULAR FILTRATION

Blood pressure generated by the beating heart drives **glomerular filtration**, the first step of urine formation. In the glomerular capillaries inside Bowman's capsule of each nephron, the pressure forces out 20 percent of the volume of plasma. Collectively, the cell wall of the glomerular capillaries, the inner cell wall of Bowman's capsule, and the basement membrane between them are like a sieve. They do not let blood cells, platelets, or plasma proteins leave blood. They nonselectively let everything else—water, ions, glucose, amino acids, diverse toxins, and so forth—become the filtrate:

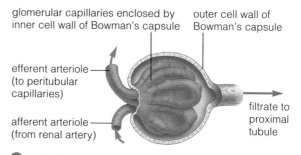

a **Filtration.** Water and solutes forced out across glomerular capillary wall collect in Bowman's capsule, which drains into the proximal tubule.

The volume of blood that kidneys must handle at any time is adjusted at afferent arterioles that deliver blood to the nephron. When intense exercise or stress makes blood pressure rise along the systemic circuit, these arterioles vasodilate in response to signals from sympathetic neurons. When the body rests and blood pressure decreases, the arterioles vasoconstrict. If too much fluid enters the body, then a reflex pathway will override the sympathetic signals. These blood vessels will vasodilate, and excess fluid will leave in urine.

The heart pumps out five liters or so of blood each minute. Between 20 and 25 percent goes right to the kidneys, which cleanse the blood even while helping to maintain homeostasis in the internal environment.

TUBULAR REABSORPTION

Only a fraction of the water and solutes making up the filtrate actually leaves the body. Instead, the nephron—especially the proximal tubule—gives back most of the filtrate, in the amounts required to maintain the volume and composition of the internal environment. By **tubular reabsorption**, a variety of substances leak or get pumped out of the nephron, diffuse through interstitial fluid, and then enter a peritubular capillary (Figure 42.7c). The process returns close to 99 percent of the filtrate's water, 100 percent of the glucose and amino acids, all but about 0.5 percent of the sodium ions (Na^+), and 50 percent of the urea to blood.

Cotransporters span the plasma membrane of cells making up the nephron's tubular wall. They resemble those cotransporters of cells in the lining of the small intestine (Section 41.5). Example: Passive transporters

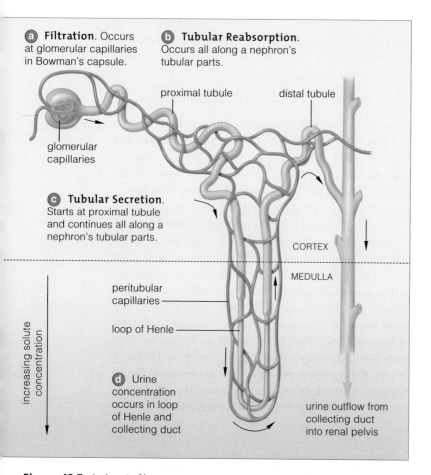

Figure 42.7 *Animated!* A nephron and associated blood capillaries. Solutes entering and leaving the nephron's tubular parts create a solute concentration gradient in interstitial fluid. The highest concentration is deep in the medulla.

(a) Glomerular filtration nonselectively moves water, ions, and solutes from blood into Bowman's capsule. **(b)** Most of the filtrate leaks or is transported out of the nephron's tubular parts into interstitial fluid, then is selectively reabsorbed into blood. **(c)** Secretion moves other solutes from blood into interstitial fluid, then into tubular parts of the nephron. **(d)** Urine becomes most concentrated as water moves by osmosis out of the loop of Henle and the collecting duct.

help Na$^+$ and glucose diffuse out of the filtrate, into tubule cells. Sodium–potassium pumps on the other free surface of the tubule cells actively transport Na$^+$ into interstitial fluid. Their pumping action sets up an electrochemical gradient. That gradient attracts more Na$^+$, Cl$^-$, and other negatively charged ions out of the filtrate, across the tubule wall, and on into interstitial fluid. One result is a solute concentration gradient, so water now moves into the cytoplasm of tubule cells and then into interstitial fluid by way of osmosis:

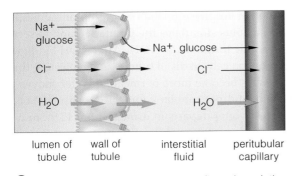

| lumen of tubule | wall of tubule | interstitial fluid | peritubular capillary |

b **Tubular Reabsorption**. As filtrate flows through the proximal tubule, ions and some nutrients are actively and passively transported outward, into interstitial fluid. Water follows, by osmosis. Cells making up peritubular capillaries transport them into blood. Water again follows by osmosis.

The tubule wall and the capillary wall each have a certain number of transporters that can bind a specific substance, such as glucose or amino acids. When all of the transporters get saturated, some percentage of the substance is not reabsorbed and is excreted in urine.

TUBULAR SECRETION

Metabolism generates many ions, mainly H$^+$ and K$^+$. Together with urea and other wastes, these ions end up in blood. With **tubular secretion**, transporters in the wall of peritubular capillaries move the ions into interstitial fluid. Then transporters in the tubular wall of nephrons move them from interstitial fluid into the filtrate, so that they may now be excreted in urine:

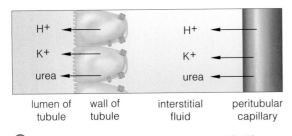

| lumen of tubule | wall of tubule | interstitial fluid | peritubular capillary |

c **Tubular Secretion**. Transporters move H$^+$, K$^+$, urea, and wastes out of peritubular capillaries. Transporters in tubular nephron regions move them into the filtrate.

As you will read shortly, secretion of H$^+$ is essential to maintaining the body's acid–base balance.

CONCENTRATING THE URINE

Sip soda all day and your urine will be dilute; sleep eight hours and it will be concentrated. However, even the most dilute urine still has far more solutes than plasma or most of the body's interstitial fluid. It gets that way by a concentrating mechanism in kidneys.

Inside the kidney cortex, the filtrate is isotonic with interstitial fluid, so there is no net movement of water from one to the other. Interstitial fluid is hypertonic in the kidney medulla. It draws water out of the filtrate, by osmosis, along the loop of Henle's descending limb. The filtrate becomes most concentrated and attracts the most water near the loop's hairpin turn (Figure 42.7d). It actually becomes more and more concentrated just before the turn until it is as salty as interstitial fluid.

However, the loop of Henle's wall *after* the turn is impermeable to water. Transporters in the ascending limb actively pump out Na$^+$ and Cl$^-$, which makes interstitial fluid in the medulla even saltier—which draws out even more water before the hairpin turn:

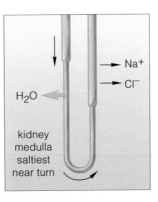

d Solutes in interstitial fluid are the most concentrated deep in the kidney medulla. The gradient draws more and more water from the *descending* limb of the loop of Henle by osmosis. But Na$^+$, Cl$^-$, urea, and other solutes cannot cross the descending limb wall, so the filtrate becomes more concentrated as it approaches the loop of Henle's hairpin turn.

Transporters in the thick-walled *ascending* limb actively pump Na$^+$ out of the filtrate. Cl$^-$ follows passively. Interstitial fluid gets saltier and attracts *more* water out of the descending limb and the collecting duct.

Remember countercurrent flow in fish gills? Here we see another countercurrent mechanism: Transporters are altering the composition of filtrate that is flowing in opposite directions inside the loop of Henle.

During glomerular filtration, blood pressure generated by heartbeats drives water and solutes into nephrons.

With tubular reabsorption, membrane transporters set up a Na$^+$ gradient that draws water, other ions, and selected solutes out of the filtrate and into peritubular capillaries.

By tubular secretion, transporters move urea, H$^+$, and K$^+$ from the capillaries into the nephron for excretion.

Mechanisms built into the loop of Henle concentrate urine.

42.4 Behavioral and Hormonal Adjustments

LINKS TO
SECTIONS 35.1,
36.3, 36.7, 38.7

Changes in what you eat and drink, the temperature and humidity, and levels in activity affect water and solute gains and losses. Hormonal and behavioral mechanisms adjust the sense of thirst, and the concentration of urine, in response.

A THIRST MECHANISM

When you do not drink enough on a hot day, you get thirsty. Why? The solute concentration in blood has risen, which slows saliva secretion into the mouth. A drier mouth stimulates nerve endings that signal the **thirst center**, which is a region of the hypothalamus. The center also receives signals from osmoreceptors that detect the rise in the blood level of sodium inside

the brain (Section 35.1). In response, the thirst center notifies other centers in the cerebral cortex, which in turn compel you to search for and drink fluid.

While thirst mechanisms are calling for the uptake of water, hormonal controls act to conserve the water already inside the body. As explained next, ADH and aldosterone act on cells in the walls of distal tubules and collecting ducts (Section 36.3 and Figure 42.8).

EFFECT OF ADH

As another response to the osmoreceptor signals, the hypothalamus stimulates the pituitary gland to secrete antidiuretic hormone or **ADH**. After ADH binds with receptors on cells of distal tubules and collecting ducts, both tubes become more permeable. Water moves out of them, and peritubular capillaries reabsorb more of it, so less water departs in urine (Figure 42.9). In time, the volume of extracellular fluid rises and its solute levels decline, so ADH secretion slows.

Nausea and vomiting also trigger ADH secretion. So does a big drop in blood pressure; baroreceptors in the wall of some arteries can detect it (Section 38.7).

ADH works through **aquaporins** in target cells to step up water reabsorption. These passive transporters selectively allow water to diffuse across the plasma membrane. When ADH binds to membrane receptors on the cells of nephron and collecting duct walls, it stimulates cytoplasmic vesicles to move to the plasma

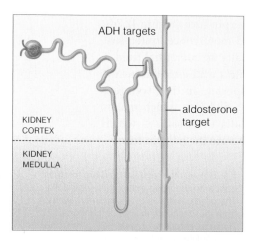

Figure 42.8 Sites of hormone action. ADH targets cells of distal tubules and parts of collecting ducts in the kidney cortex. Aldosterone targets distal tubule cells. Both hormones affect reabsorption. Directly or indirectly, they help the body conserve more water.

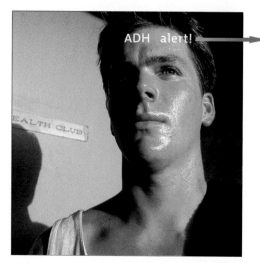

Stimulus

a Water loss lowers the blood volume. Sensory receptors in the hypothalamus detect a big deviation from the set point.

b The hypothalamus stimulates the pituitary gland to step up its secretion of ADH.

c ADH circulates in blood, reaches nephrons in the kidneys. By acting on cells of distal tubules and collecting ducts, it makes the tube walls more permeable to water.

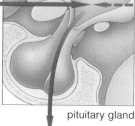

hypothalamus

pituitary gland

Response

f Sensory receptors in hypothalamus detect the increase in blood volume. Signals calling for ADH secretion slow down.

e The blood volume rises.

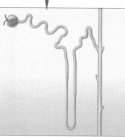

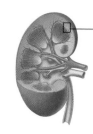

d More water is reabsorbed by peritubular capillaries around the nephrons, so less water is lost in urine.

Figure 42.9 Feedback control of ADH secretion, one of the negative feedback loops from kidneys to the brain that helps adjust the volume of extracellular fluid. Nephrons in the kidneys reabsorb more water when we do not take in enough water or lose too much, as by profuse sweating.

membrane. The vesicles contain aquaporin subunits. As they reach and fuse with the plasma membrane, the subunits self-assemble into pores for water across tube walls. There they facilitate the rapid diffusion of water out of the filtrate, back into interstitial fluid.

EFFECT OF ALDOSTERONE

Any decrease in the volume of extracellular fluid also activates cells in the efferent arteriole that brings blood to the nephron. These cells release renin, an enzyme that sets in motion a chain of reactions. The outcome is that the adrenal gland above each kidney secretes **aldosterone**. This hormone stimulates production of sodium–potassium pumps and their insertion into the plasma membrane of cells in the collecting duct wall. As they pump more sodium out of the collecting ducts, water follows, by osmosis, into interstitial fluid. At the same time, they assist the diffusion of potassium into the ducts, and so urine becomes more concentrated.

ABNORMALITIES IN HORMONAL CONTROL

A metabolic disorder known as *diabetes insipidus* arises when the pituitary gland secretes too little ADH, when ADH receptors do not respond to ADH, or when the aquaporin proteins are modified or missing. The main symptom is the formation of large volumes of highly dilute urine (Section 36.7). Typically, insatiable thirst accompanies copious urine production.

Certain cancers, infections, and drugs, including some antidepressants, stimulate ADH oversecretion. With an excess of ADH, the kidneys retain too much water. With more water, solute concentrations in the interstitial fluid decrease. Water begins to move by osmosis from the interstitial fluid into body cells, which are now relatively hyperosmotic. At the same time, solutes move out of body cells into this fluid. This is bad news, especially for brain cells, which are highly sensitive to solute concentrations. Unless solute balances are restored, the outcome can be deadly.

Adrenal gland tumors may cause oversecretion of aldosterone, or *hyperaldosteronism*. This in turn causes fluid retention and high blood pressure. Kidney, liver, or heart problems also can raise the aldosterone level, sometimes with similar consequences.

> ADH promotes water conservation at distal tubules and collecting ducts. Aldosterone promotes reabsorption of sodium, which indirectly increases water retention.

42.5 Acid–Base Balance

LINK TO SECTION 2.6

Besides maintaining the volume and composition of extracellular fluid, kidneys help keep it from getting too acidic or too basic (alkaline).

Hydrogen ions continuously enter extracellular fluid following amino acid breakdown, protein breakdown, lactate fermentation, and other metabolic processes. Even so, certain controls keep the H^+ concentration in extracellular fluid within a tight range—a state called an **acid–base balance**. Buffer systems, respiration, and urinary excretion contribute to the balancing act.

A **buffer system**, recall, involves substances that reversibly bind and release H^+ or OH^- ions, thereby minimizing pH changes as acidic or basic molecules enter or leave a solution (Section 2.6).

In humans, extracellular pH is maintained between 7.35 and 7.45. Any excess acids dissociate into H^+ and other fragments, and pH falls. The effect is minimized when excess hydrogen ions react with buffers, most notably the bicarbonate–carbonic acid buffer system:

$$H^+ + HCO_3^- \;\rightleftharpoons\; H_2CO_3 \;\rightleftharpoons\; CO_2 + H_2O$$
$$\text{BICARBONATE} \qquad \text{CARBONIC ACID}$$

When the pH of blood shifts, adjustments in the rate and magnitude of breathing offset that change. When the blood pH decreases, breathing quickens and CO_2 is expelled faster than it forms. As you can tell from the equation above, less CO_2 means less carbonic acid can form, so the pH rises. Slower breathing lets CO_2 accumulate, so more carbonic acid can form.

Control of bicarbonate reabsorption and secretion of H^+ can adjust the pH inside kidneys. Reabsorbed bicarbonate moves into peritubular capillaries, where it can buffer excess acid. H^+ secreted into tubule cells combines with phosphate or ammonia ions. In either form, it can be disposed of in the urine.

The importance of this balancing act becomes clear when *metabolic acidosis* develops. This condition arises if kidneys cannot excrete enough of the H^+ released during metabolism. It can be life threatening.

> The kidneys, buffering systems, and the respiratory system interact to neutralize acids. They tightly control the acid–base balance of extracellular fluid, so that it is neither too acidic nor too basic (alkaline) for cell functions.
>
> By reversible reactions, a bicarbonate–carbon dioxide buffer system neutralizes excess H^+. Shifts in the rate and depth of breathing affect this buffer system, hence the pH of blood.
>
> The kidneys also can shift the pH of blood when they adjust bicarbonate reabsorption and H^+ secretion.

42.6 When Kidneys Break Down

LINKS TO
SECTIONS 16.8,
36.7, 38.9, 41.8

At this point, you probably have figured out that good health depends on nephron function. Whether by illness or accident, when nephrons of both kidneys are damaged and no longer performing their regulatory and excretory functions, renal failure follows. It can be irreversible.

Causes of Renal Failure In the United States, most kidney problems are outcomes of *diabetes mellitus*, and *high blood pressure* accounts for many more cases. As you read in Sections 36.7 and 38.9, these chronic disorders can damage blood vessels, including capillaries that interact with nephrons. Some people are genetically predisposed to infections or conditions that damage kidneys. Kidneys also fail after lead, arsenic, pesticides, or other toxins enter the body. On rare occasions, repeated high doses of aspirin and some other drugs damage them beyond repair.

High-protein diets force the kidneys to work overtime to dispose of nitrogen-rich breakdown products (Section 41.8). Such diets also increase the risk for *kidney stones*. These hardened deposits form when uric acid, calcium, and other wastes settle out of urine and collect in the renal pelvis.

Most kidney stones are washed away in urine, but one can become lodged in a ureter or the urethra and cause severe pain. Any stone that blocks urine flow invites infections and permanent kidney damage.

We usually measure kidney function in terms of the rate of filtration through the glomerular capillaries. Renal failure occurs when the filtration rate falls by half, regardless of whether it is caused by low blood flow to the kidneys or by damaged tubules or blood vessels. Renal failure can be fatal. Wastes build up in the blood and interstitial fluid. The pH rises. The resulting changes in the concentrations of other ions, most notably Na^+ and K^+, interfere with metabolism.

Kidney Dialysis At one time or another, about 13 million people in the United States have experienced renal failure. A kidney dialysis machine is used to restore proper solute balances. Like kidneys, it selectively adjusts solutes in blood. "Dialysis" simply refers to exchanges of solutes across any artificial membrane between two different solutions.

With *hemodialysis*, the machine is connected to a vein or an artery. It pumps blood through semipermeable tubes submerged in a warm solution of salts, glucose, and other substances. As blood flows through the tubes, the wastes dissolved in it diffuse out, so that solute concentrations are returned to normal levels. Cleansed, solute-balanced blood is allowed to flow back into the patient's body.

With *peritoneal dialysis*, a fluid of a specific composition is pumped into a patient's abdominal cavity. It exchanges solutes with extracellular fluid, and then it is drained out. In this case the peritoneum, the lining of the abdominal cavity, functions as the membrane for dialysis.

Kidney dialysis is able to keep a person alive through an episode of temporary kidney failure. When the kidney damage is permanent, dialysis must be continued for the rest of a person's life or until a kidney becomes available for transplant surgery (Figure 42.10).

Regarding Kidney Transplants Each year in the United States, about 12,000 people are the recipients of *kidney transplants*. More than 40,000 others remain on the waiting list because there is a shortage of suitable donors. Most transplanted kidneys are made available from people who agreed to donate them after their death. Increasingly, more are becoming available from living donors, most often from a relative, spouse, or friend. A transplant from a living donor stands a better chance of success than one from a deceased person. A single kidney is adequate to maintain good health, so risks to a living donor are mainly related to the surgery—unless the donor's remaining kidney fails.

The benefits of organs from living donors, a lack of donated organs, and high dialysis costs have led some to suggest that people should be allowed to sell a kidney. Critics argue that it is unethical to tempt people to risk their health for money. Section 16.8 describes another potential alternative—*xenotransplantation*. Some day, genetically modified pigs may become organ factories.

Figure 42.10 What kidney trouble? Karole Hurtley has attitude as well as energy. At thirteen, she was the national champion for her age group and karate belt level. She also lives with renal failure.

Karole gets regular checkups from a home dialysis nurse for Denver's Children's Hospital. Her dad jokes that Karole takes so many pills that she rattles when she walks and is pricked "a gazillion times a year" for blood tests.

Eight or nine times a night, while Karole sleeps, a peritoneal dialysis machine beside her bed pumps a special solution into her torso and filters out waste products that her kidneys no longer can remove. She still embraces an active life-style. Inspirational attitude? You bet.

42.7 Heat Gains and Losses

We turn now to another major aspect of homeostasis. How does the body maintain the core of its internal environment within a tolerable temperature range when the surroundings become too hot or cold?

HOW THE CORE TEMPERATURE CAN CHANGE

The core temperature of an animal body rises when heat from the surroundings or metabolism builds up. When warm, a body loses heat to cool surroundings. The core temperature stabilizes when the rate of heat loss balances the rate of heat gain and production. The heat content of any complex animal depends on a balancing act between gains and losses:

$$\begin{array}{c} \text{change in} \\ \text{body heat} \end{array} = \begin{array}{c} \text{heat} \\ \text{produced} \end{array} + \begin{array}{c} \text{heat} \\ \text{gained} \end{array} - \begin{array}{c} \text{heat} \\ \text{lost} \end{array}$$

Heat is gained and lost through exchanges at the body surfaces. Radiation, conduction, convection, and evaporation are processes that drive these exchanges.

Thermal radiation is the emission of heat from any object in the form of radiant energy. Radiant energy from the sun can heat animals. Also, a metabolically active animal itself produces and radiates heat.

By **conduction**, heat is transferred between objects in direct contact with each other. An animal loses heat when it rests on objects cooler than it is. If it contacts objects that are warmer, the animal will gain heat.

By **convection**, moving air or water transfers heat. Conduction plays a part; heat moves down a thermal gradient between the body and air or water next to it. Mass transfer also plays a part. The heated air is less dense, and so it moves away from the body.

By **evaporation**, a liquid converts to gaseous form and heat is lost in the process. Evaporation from body surfaces cools the body; water molecules carry energy away with them. Evaporative heat loss rises with dry air and breeze. High humidity and still air slow it.

ENDOTHERM? ECTOTHERM? HETEROTHERM?

Fishes, amphibians, and reptiles are warmed mostly by heat gained from the environment rather than by metabolically generated heat. These animals are called **ectotherms**, which means "heated from outside." They can only adjust behaviorally to rising or falling outside temperatures. Most species have low metabolic rates and not much insulation. A rattlesnake (Figure 42.11*a*) is one example. When its body is cold, it basks in the sun. When hot, the snake moves into shade.

Most birds and mammals are **endotherms**, which means "heat from within." They have relatively high

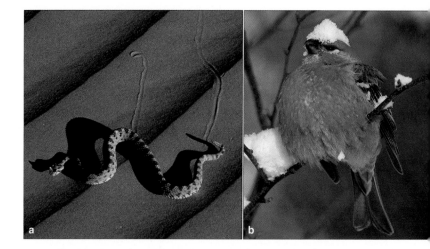

Figure 42.11 (**a**) Sidewinder, an ectotherm. (**b**) Pine grosbeak, an endotherm, using fluffed feathers as insulation against winter cold.

metabolic rates. Compared to, say, a foraging lizard of the same weight, a foraging mouse uses thirty times *more* energy. Metabolic heat helps endotherms remain active across a wider range of temperatures. Fur, fat, or feathers function as insulating layers and minimize heat transfers (Figure 42.11*b*).

LINKS TO SECTIONS 2.5, 6.1

Some birds and mammals are **heterotherms**. They can maintain a fairly constant core temperature some of the time but let it shift at other times. For example, hummingbirds have very high metabolic rates when foraging for nectar during the day. At night, metabolic activity decreases so much that the bird's body may become almost as cool as the surroundings.

Warm climates favor the ectotherms, which do not have to spend as much energy as endotherms do on maintaining the core temperature. In tropical regions, reptiles have the advantage, because they can spend more energy on reproduction and other tasks. There they exceed mammals in numbers and diversity. In all cool or cold regions, however, most vertebrates tend to be endotherms. As one example, you will not find reptiles running around in the arctic.

The core temperature of the internal environment is being maintained when heat losses balance heat gains. Ectotherms gain heat mainly from the outside, endotherms generate most from within. Heterotherms do both at different times.

Animals gain and lose heat by way of thermal radiation, conduction, convection, and evaporation. Different kinds make distinct morphological, physiological, and behavioral adjustments to changes in environmental temperatures.

42.8 Temperature Regulation in Mammals

LINKS TO
SECTIONS 2.5, 8.4,
28.3, 35.2, 39.3

Section 28.3 introduced the negative feedback controls that govern increases in human body temperature. Now compare the controls that generally come into play when mammals are subjected to heat stress and cold stress.

The hypothalamus, again, contains centers that help control the core temperature of the mammalian body. These hypothalamic centers receive input from many thermoreceptors inside skin and from others deep in the body (Figure 35.2). When the temperature deviates from a set point, the centers integrate the responses of skeletal muscles, smooth muscle of arterioles in the skin, and sweat glands. Negative feedback loops back to the hypothalamus inhibit the responses when the core temperature returns to the set point. Dromedary camels can recalibrate the set point so that it is *higher* during the hottest time of day (Figure 42.12).

RESPONSES TO HEAT STRESS

When any mammal becomes too hot, the temperature control centers in the hypothalamus issue commands for **peripheral vasodilation**: the diameter of the blood vessels in skin increases. More blood flows to the skin and delivers more metabolic heat that can be given up to the surroundings (Table 42.1).

Another response to heat stress, **evaporative heat loss**, occurs at moist respiratory surfaces and across skin. Animals that sweat lose some water this way. For instance, humans and some other mammals have sweat glands that release water and solutes through pores at the skin's surface. Remember Section 33.6? An average-sized adult human has 2–1/2 million or more sweat glands. For every liter of sweat produced, about 600 kilocalories of heat energy leave the body by way of evaporative heat loss.

Sweat dripping from skin dissipates little heat. The body cools greatly when sweat evaporates. On humid days, the air's high water content slows evaporation rates, so sweating is less effective at cooling the body. When you exercise strenuously, sweating helps offset heat production by skeletal muscles.

With the exception of marine species, nearly all mammals (and only mammals) can sweat. Many rely more on behavioral responses, such as licking fur or panting. "Panting" refers to shallow, rapid breathing. It assists evaporative water loss from the respiratory tract, nasal cavity, mouth, and tongue (Figure 40.22).

Sometimes peripheral blood flow and evaporative heat loss cannot counter heat stress, and *hyperthermia* follows. Core temperature rises above normal. As you read earlier in Chapter 28, a human body temperature above 105°F (41.5°C) can be dangerous.

What about a fever? Recall, from Section 39.3, that **fever** is not itself an illness. It is one of the responses to infection. When activated by a threat, macrophages release signaling molecules that stimulate part of the brain to secrete prostaglandins. These local signaling molecules cause the hypothalamus to allow the core temperature to rise a bit above the normal set point. The rise makes the body less hospitable for invaders and calls up more immune responses. Generally, the hypothalamus does not let the core temperature rise above 105°F. When a fever exceeds that point or when it lasts more than a few days, the condition causing it is life threatening and medical evaluation is essential.

RESPONSES TO COLD STRESS

Selectively distributing blood flow, fluffing hair, and shivering are all mammalian responses to cold stress. Birds make similar responses but maintain a higher

Figure 42.12 Short-term adaptation to desert heat stress. Dromedary camels are heterotherms; they let the core temperature rise by as much as 14 degrees during the hottest hours of the day, then let it fall at night. A hypothalamic mechanism raises the internal thermostat, so to speak, and doing so has no adverse effects.

Table 42.1	Mammalian Responses to Core Temperature Shifts	
Stimulus	Main Responses	Outcome
Heat stress	Widespread vasodilation in skin; behavioral adjustments; in some species, sweating, panting	Dissipation of heat from body
	Decreased muscle action	Heat production decreases
Cold stress	Widespread vasoconstriction in skin; behavioral adjustments (e.g., minimizing surface parts exposed)	Conservation of body heat
	Increased muscle action; shivering; nonshivering heat production	Heat production increases

Figure 42.13 Two responses to cold stress.

(**a**) Polar bears (*Ursus maritimus*, "bear of the sea"). A polar bear stays active even during severe arctic winters. It does not get too chilled after swimming because the coarse, hollow guard hairs of its coat shed water quickly. Thick, soft underhair traps heat. Brown adipose tissue, about 11.5 centimeters (4–1/2 inches) thick, insulates and helps generate metabolic heat.

(**b**) In 1912, the *Titanic* collided with an immense iceberg on her maiden voyage. It took about 2–1/2 hours for the *Titanic* to sink and rescue ships arrived in less than two hours. Even so, 1,517 bodies were recovered from the calm sea. All of the dead had on life jackets; none had drowned. Hypothermia killed them.

a

b

Table 42.2	Mammalian Physiological Responses to Increases in Cold Stress
Core Body Temperature	Physiological Responses
36°–34°C (about 95°F)	Shivering response; rise in respiration, metabolic heat output. Peripheral vasoconstriction, more blood deeper in body. Dizziness, nausea set in.
33°–32°C (about 91°F)	Shivering response ends. Metabolic heat output declines.
31°–30°C (about 86°F)	Capacity for voluntary motion is lost. Eye and tendon reflexes inhibited. Consciousness is lost. Cardiac muscle action becomes irregular.
26°–24°C (about 77°F)	Ventricular fibrillation sets in (Section 38.9). Death follows.

body temperature. Peripheral thermoreceptors detect cool outside temperatures and alert the hypothalamus, which signals smooth muscle in arterioles that service the skin to contract. Such a response to cold stress is called **peripheral vasoconstriction**. When arterioles in skin constrict, less metabolic heat reaches the body surfaces. When your fingers or toes are chilled, all but 1 percent of the blood that would usually flow to skin is diverted to other regions of the body.

Also, muscle contractions make hairs (or feathers) "stand up." This **pilomotor response** creates a layer of still air next to skin. It helps reduce convective and radiative heat loss. Forms of behavior also reduce heat loss from exposed surfaces, as when polar bear cubs cuddle against their mother (Figure 42.13*a*).

With prolonged cold exposure, the hypothalamus commands skeletal muscles to contract ten to twenty times each second. Although this **shivering response** increases heat production, it has a high energy cost.

Long-term or severe cold exposure also leads to a hormonal response—an increase in thyroid activity that raises the rate of metabolism. This **nonshivering heat production** mainly targets cells of *brown* adipose tissue, which has a large number of mitochondria. In this tissue, the main function of aerobic respiration is to generate heat; producing ATP is secondary.

Animals that hibernate or are active in cold regions, as well as the young of many others, contain brown adipose tissue. In human infants, this tissue accounts for about 5 percent of body weight. Unless exposure to cold is ongoing, the tissue shrinks as we age. Some Japanese and Korean commercial divers who harvest shellfish in frigid waters retain brown adipose tissue. Male Finlanders who work outside through the year also have a bit of it.

Failure to protect against cold leads to *hypothermia*, a condition in which core temperature plummets. In humans, a decline to 95°F changes brain function and causes confusion. Severe hypothermia can spiral into coma and death (Figure 42.13*b* and Table 42.2).

Temperature shifts are detected by thermoreceptors that send signals to an integrating center in the hypothalamus.

This center serves as the body's thermostat and calls for adjustments that maintain core temperature.

Mammals counter cold stress by vasoconstriction in skin, behavioral adjustments, increased muscle activity, and shivering and nonshivering heat production.

Mammals counter heat stress by widespread peripheral vasodilation in skin and evaporative water loss.

Summary

Section 42.1 Blood and interstitial fluid make up the extracellular fluid. Maintaining the volume and solute concentration of extracellular fluid is a vital aspect of homeostasis. All organisms must balance solute and fluid gains with solute and fluid losses. In vertebrates, that balancing act is the main function of the kidneys and other components of a urinary system.

Mammals gain water by absorption from the gut and metabolism. They lose it by urinary excretion, evaporation from body surfaces, and in exhalations and in feces. They gain solutes by absorption from the gut, secretion, respiration, and metabolism. Solutes are lost in urine and feces, and by respiration and sweating.

Section 42.2 The human urinary system consists of a pair of kidneys, a pair of ureters, a urinary bladder, and the urethra. The renal artery delivers blood into kidneys, where it branches into efferent arterioles.

Tubular structures called nephrons, and certain blood capillaries that interact with them, are the functional units that cleanse the blood and form urine.

A nephron starts in the kidney cortex at Bowman's capsule. It continues as a proximal convoluted tubule, a loop of Henle that descends into and ascends out of the kidney medulla, and a distal convoluted tubule that empties into a collecting duct. All collecting ducts drain into the renal pelvis.

Bowman's capsule and the set of highly permeable glomerular capillaries within it are a blood-filtering unit, called a renal corpuscle. Most of the filtrate that enters Bowman's capsule is reabsorbed along the nephron's tubular regions and is returned to blood by way of peritubular capillaries that thread around the nephron. The portion of the filtrate that is not returned is excreted from the urethra, as urine.

Biology⑧Now
Explore the anatomy of the human urinary system and kidneys with the animation on BiologyNow.

Section 42.3 Urine forms in nephrons by three processes: filtration, reabsorption, and secretion.

Filtration. Blood pressure drives water and small solutes (not proteins) out of the highly leaky glomerular capillaries and into Bowman's capsule, the entrance to tubular parts of the nephron.

Reabsorption. Water and solutes to be conserved move out of the tubular parts, then into peritubular capillaries. The tubules retain a small volume of water and solutes.

Secretion Urea, H^+, and some other solutes move out of peritubular capillaries and into the nephron for excretion, in urine.

Biology⑧Now
Learn about the processes that form urine with the animation on BiologyNow.

Section 42.4 Sensory receptors that detect high solute concentrations in the internal environment or a low blood volume signal a hypothalamic thirst center.

Signals from the hypothalamus stimulate water-seeking behavior and call for secretion of the hormone ADH. ADH and aldosterone are two hormones that affect reabsorption in the kidneys.

ADH is secreted by the pituitary gland and promotes water reabsorption across the walls of the proximal tubule and collecting ducts, so urine becomes more concentrated. Aldosterone is secreted by the adrenal cortex and promotes sodium reabsorption into the distal tubule. Because water follows sodium, the action of this hormone indirectly causes water to be conserved.

Section 42.5 The urinary system, buffer systems, and the respiratory system all contribute to maintaining the body's acid–base balance. The urinary system can excrete H^+ in the urine and reabsorb bicarbonate.

Section 42.6 When kidneys fail, frequent dialysis or a kidney transplant is required to sustain life.

Biology⑧Now
Read the InfoTrac article "The Kidney Swap: Adventures in Saving Lives," Denise Grady and Anahad O'Connor, The New York Times, *October 2004.*

Section 42.7 Animals produce metabolic heat. To maintain a core temperature, heat gains (by metabolism and from the surrounding environment) must balance heat losses to the environment.

For ectotherms, the core temperature depends more on heat exchange with the environment than on metabolic heat. These animals can regulate their core temperature mainly by modifications in their behavior.

For endotherms (most birds and mammals), a high metabolic rate is the primary source of heat. Their core temperature is governed largely by controls over the production and loss of metabolic heat.

For heterotherms, the core temperature is tightly controlled some of the time and allowed to fluctuate with environmental temperatures at other times.

Section 42.8 In mammals, the hypothalamus is the main integrating center for the control of temperature. It receives signals from thermoreceptors and stimulates responses in smooth muscle in arterioles, sweat glands, and other effectors. The core temperature is maintained by behavioral, metabolic, and physiological responses.

Self-Quiz   *Answers in Appendix II*

1. A freshwater fish gains most of its water by _____ .
   a. drinking   c. osmosis
   b. eating food   d. transport across the gills

2. Bowman's capsule, the start of the tubular part of a nephron, is located in the _____ .
   a. kidney cortex   c. renal pelvis
   b. kidney medulla   d. renal artery

3. Fluid filtered into Bowman's capsule flows directly into the _____ .
   a. renal artery   c. distal tubule
   b. proximal tubule   d. loop of Henle

4. Label the structures in the following diagrams and name their functions:

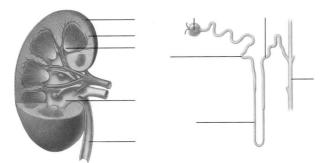

5. Water and small solutes from blood enter nephrons during _____ .
   a. filtration
   b. tubular reabsorption
   c. tubular secretion
   d. both a and c

6. Kidneys return most of the water and small solutes back to blood by way of _____ .
   a. filtration
   b. tubular reabsorption
   c. tubular secretion
   d. both a and b

7. ADH binds to receptors on distal tubules and collecting ducts making them _____ permeable to _____ .
   a. more; water
   b. less; water
   c. more; sodium
   d. less; sodium

8. Increased sodium reabsorption at distal tubules _____ .
   a. will make the urine more concentrated
   b. will make the urine more dilute
   c. is caused by insertion of aquaporins
   d. both a and c

9. Match each structure with a function.
   ____ ureter
   ____ Bowman's capsule
   ____ urethra
   ____ collecting duct
   ____ pituitary gland
   a. start of nephron
   b. delivers urine to body surface
   c. carries urine from kidney to bladder
   d. secretes ADH
   e. target of aldosterone

10. The main control center for maintaining the temperature of the mammalian body is in the _____ .
    a. anterior pituitary
    b. kidney cortex
    c. adrenal cortex
    d. hypothalamus

11. Negative feedback loops that maintain a mammal's core temperature involve _____ .
    a. the hypothalamus
    b. receptors in skin
    c. receptors in deep tissue
    d. all of the above

12. Is this statement true or false: Thirst behavior is initiated and controlled exclusively by a thirst center in the hypothalamus.

13. Match each term with the most suitable description.
    ____ endotherm
    ____ ectotherm
    ____ convection
    ____ conduction
    ____ thermal radiation
    a. environment dictates core temperature
    b. metabolism dictates core temperature
    c. heat transfer between objects that are in direct contact
    d. water, air current transfers heat
    e. emission of radiant energy

Additional questions are available on **Biology ⊗ Now™**

Critical Thinking

1. Protein breakdown produces toxic ammonia. In all mammals, ammonia is converted to urea, which can only be eliminated when dissolved in water in urine. Birds and reptiles convert it to uric acid, a solid that can be excreted when mixed with a small amount of water. It takes twenty to thirty times more water to excrete 1 gram of urea than it does to excrete 1 gram of uric acid. Formulate a hypothesis regarding how this difference might relate to the species richness of birds, mammals, and reptiles in different environments. Devise a way to test your hypothesis.

2. The kangaroo rat kidney efficiently excretes a tiny volume of urine (Section 42.1). Compared to a human nephron, its nephrons have a loop of Henle that is proportionally much longer. Explain how this helps the kangaroo rat conserve water.

3. Drinking too much water can be a bad thing (Figure 42.14). As marathoners or other endurance athletes sweat heavily and drink lots of water, their sodium levels drop. The resulting *water intoxication* can be fatal. Why is the sodium balance so important?

4. Two-thirds of the water and solutes that the body reclaims by tubular reabsorption are reclaimed in the proximal tubule region of nephrons. Proximal tubule cells have great numbers of mitochondria and demand a great deal of oxygen. Explain why.

5. Licorice (*Glycyrrhiza*) is used as a remedy in Chinese traditional medicine and also is a flavoring for candy. When licorice is eaten, one of its components triggers the formation of a compound that mimics aldosterone and binds to receptors for it. Based on this information, explain why people who have high blood pressure are advised to avoid eating much licorice.

6. Animals that live in cold habitats continually lose heat to their environment. Ectotherms are few and endotherms often show morphological adaptations to cold. Compared to closely related species that live in warmer areas, cold dwellers tend to have smaller appendages. For example, relative to its body size, the arctic hare has ears that are far shorter than those of hares that live in more temperate zones. Also, animals adapted to cool climates tend to be larger than related species that evolved in warmer places. For example, the largest bear species is the polar bear and the largest penguin is Antarctica's emperor penguin.

Think about heat transfers between animals and their habitat, then explain why smaller appendages and larger overall body size are advantageous in very cold climates.

Figure 42.14
Excessive water intake during an endurance event—not a good idea.

Sex and the Mammalian Heritage

Sex and romance! Reinforcement of the contrived linkage between the two starts early and often in Western cultures. The idea of sex is used to sell everything from underwear and flowers to automobiles and erectile dysfunction drugs —and it trivializes the mammalian reproductive heritage.

Victoria's Secret aside, the breasts of a female mammal function to nourish offspring that are too vulnerable to survive on their own. Any mammalian mother that bonds with offspring is committing herself to protect it through an extended time of dependency and learning. Extended care helps the new individual survive until it is old enough to survive on its own. We can expect that such behavior is a result of natural selection, for it increases the odds of reproductive success in the next generation.

We see this behavior among all mammals. Less than four months after mating, a lioness typically gives birth to no more than a few small, blind cubs (Figure 43.1). She hides them in marshes or rock outcroppings. She nurses them for six or seven months before leading them to nearby kills. The cubs will remain dependent on her for at least sixteen months. Even with protection, 80 percent will die before then, primarily as a result of starvation.

Males alone have a thick, showy, protective mane, and they are much larger and twice as heavy as females. This pronounced sexual dimorphism is one outcome of severe reproductive competition. Males often form coalitions to battle for the chance to monopolize the females. Once they succeed, they have about two years to perpetuate their genes before younger, stronger lions challenge them.

The intense competition may explain the infanticidal behavior among male lions that take over a pride. The first thing they do is kill all cubs they can catch. The lionesses fight fiercely to protect the cubs. When they fail, they often become hyperactive sexually without becoming pregnant. Is the delay a form of natural birth control that gives the strongest males time to turn back more takeover attempts and assume dominance? Possibly. Four or five months after the initial takeover, all the females in the pride ovulate, get pregnant, and bear cubs together. When one females hunts, others even nurse her cubs.

Figure 43.1 Glimpses into maternal care. This complex form of behavior has become most highly developed among mammals. The offspring of these sexually reproducing species require an extended period of development, dependency, and learning.

In such ways, synchronization improves the chances of reproductive success for the males as well as females.

The point is this: In nature, the main function of sex is not recreational but rather the perpetuation of one's genes. And so, with this example, we turn to one of life's great dramas—the reproduction and development of complex animals in the image of their parents. *How does a single fertilized egg of a lion or frog, a bird or human, become transformed into all of the specialized cells and structures of the adult form?* Some answers will start to emerge through this chapter's survey of basic principles that guide animal life cycles, from the time of reproduction, through embryonic and postnatal development, and on to aging and eventual death. More answers will emerge in the next chapter, which offers a case study of human reproduction and development.

 ## How Would You Vote?

Sanitation and medical advances have greatly extended the average human life span, especially in developed countries. Some researchers are now looking for ways to extend the human life span even further. Do you think research into life extension should be supported by federal research funding? See BiologyNow for details, then vote online.

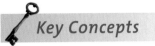 ## Key Concepts

COSTS AND BENEFITS OF SEXUAL REPRODUCTION

Biologically, separation into sexes is far more costly than asexual reproduction. But sexual reproduction expands the capacity for fast, adaptive responses to abiotic and biotic conditions. That capacity may be present in the range of variation among offspring, so that at least some have a better chance to survive and reproduce.   Section 43.1

SIX STAGES IN ANIMAL LIFE CYCLES

Animal life cycles typically proceed through six stages of reproduction and development: gamete formation, fertilization, cleavage, gastrulation, formation of organs, and growth and tissue specialization.   Section 43.2

FORMATION OF THE EARLY EMBRYO

Different cell lineages in an embryo set out on different developmental roads. Their fate is partly sealed by cleavage, when daughter cells receive different instructions that were localized in different parts of the fertilized egg's cytoplasm. Then cells in the embryo start signaling and responding to one another. Cell differentiation and morphogenesis are outcomes of the interactions.   Sections 43.3, 43.4

FILLING IN DETAILS OF THE BODY PLAN

The cytoplasmic localization and then inductive interactions among classes of master genes map the basic body plan. Gene products specify where and how body parts develop. Like beacons, they help cells assess their position in the embryo and how they will differentiate.   Section 43.5

AGING IN THE LIFE CYCLES

All species of multicelled animals that show extensive cell differentiation undergo aging. Cell structure and function gradually decline, which leads to the decline of tissues, organs, and eventually the body.   Sections 43.6, 43.7

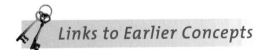 ## Links to Earlier Concepts

This chapter builds on the Chapter 10 introduction, which invited speculation on the costs and benefits of having separate sexes. You will be drawing on your knowledge of mitosis (Section 9.3), cleavage (9.4), and meiosis, gamete formation, and fertilization (10.3, 10.5). You will be revisiting the nature of cell differentiation and the role of master genes in laying out the basic body plan (13.4, 15.1–15.3, 17.8). You will take a closer look at the primary tissue layers that give rise to all tissues and organs of adult animals (25.1, 33.5).

43.1 Reflections on Sexual Reproduction

LINK TO
CHAPTER 10
INTRODUCTION

Sexual reproduction dominates the life cycle of most animals, even those that also can reproduce asexually. We therefore can expect that the benefits of sexual reproduction outweigh the costs. What are they?

SEXUAL VERSUS ASEXUAL REPRODUCTION

In earlier chapters, we considered the genetic basis of **sexual reproduction**. Again, meiosis and the formation of gametes typically occur in two prospective parents. At fertilization, a gamete from one parent fuses with a gamete from the other and forms the first cell of the new individual—the zygote. We looked at **asexual reproduction**, whereby a single organism—just one parent—produces offspring. We turn now to examples of the structural, behavioral, and ecological aspects of these two modes of animal reproduction.

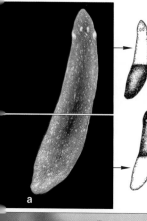

Think about a fragment torn away from a sponge body. It may well grow, by mitotic cell divisions, into a new sponge. Or think of a flatworm that splits spontaneously in two. If its body constricts at its midsection, the part below grips a substrate and starts a tug-of-war with the front. It splits off a few hours later. Then both parts go their separate ways and each grows what is missing, thus becoming a whole flatworm (Figure 43.2a).

Mutation aside, by *asexual* reproduction, one parent has all of its genes represented in the next generation; its offspring are all genetically identical. Phenotypic uniformity from one generation to the next helps when gene-encoded traits are adapted to fairly consistent abiotic and biotic conditions. In such circumstances, drastic variations in an adaptive gene package could be disastrous.

Most animals live where opportunities, resources, and danger are variable. They reproduce sexually, and their offspring inherit mixes of maternal and paternal alleles. Remember the Chapter 10 introduction? The capacity for rapid, adaptive responses to abiotic and biotic conditions is typically present in the expressed range of variation, so at least some offspring have a better chance to survive and reproduce.

COSTS OF SEXUAL REPRODUCTION

Separation into sexes is costly. Energy and resources must be allocated to forming and nurturing gametes. Often, reproductive structures that can help deliver or accept sperm must be built. A potential mate might have to be courted. The timing of gamete formation and mating must be synchronized between the sexes.

Reflect on *reproductive timing*. How do the sperm in one individual mature at the same time that eggs are maturing in a different individual? Timing requires energy outlays to construct, maintain, and use neural and hormonal control mechanisms in both parents. It requires responsiveness to environmental cues, such as daylength, that signal the best time to start making gametes and to produce offspring.

For example, moose become sexually active only in late summer and early fall. This timing is adaptive; it means that the offspring will be born in spring, when weather is milder and food more plentiful.

Think about what it takes to find and recognize a likely mate. Many animals invest energy to produce sex attractants called pheromones or make receptors for them. They invest in visual signals such as richly colored and patterned feathers. Many males attract mates and fend off rivals, as with bonding rituals or claws, horns, or a larger body mass that may make a difference in territorial defense (Figure 43.2b–d).

Figure 43.2 (a) Example of fast, easy, asexual reproduction. One of the flatworms (*Dugesia*) can reproduce asexually by spontaneous fission. If it divides into two pieces, each piece replaces what is missing, and each will be genetically identical to the original flatworm. All of the parent worm's genes are represented in both.

Biological costs associated with sexual reproduction. Reflect on the energy and raw materials directed into producing (**b**) sable antelope horns, (**c**) colorful feathers, and (**d**) the body mass of male northern elephant seals. The bulls are fighting for access to the far smaller female, lower right.

Figure 43.3 A look at where some invertebrate and vertebrate embryos develop, how they are nourished, and how (if at all) parents protect them.

Snails (**a**) and spiders (**b**) release eggs from which the young hatch later on. Most snails abandon their laid eggs. Spider eggs develop in a silk egg sac that the female anchors or carries around with her. Females often die soon after they make the sac. Some species guard the sac, then cart spiderlings about for a few days while they feed them.

(**c**) Ruby-throated hummingbirds and all other birds lay fertilized eggs with big yolk reserves. The eggs develop and hatch outside the mother. Unlike snails, one or both parent birds expend energy feeding and caring for the young.

(**d**) Embryos of most sharks, most lizards, and some snakes develop in their mother, receive nourishment continuously from yolk reserves, and are born live. Shown here, live birth of a lemon shark.

Embryos of most mammals draw nutrients from maternal tissues and are born live. (**e**) In kangaroos and other marsupials, embryos are born "unfinished." They finish their embryonic development inside a pouch on the mother's ventral surface. (**f**) Juvenile stages (joeys) continue to be nourished from mammary glands inside the pouch. A human female (**g**) retains a fertilized egg in her uterus. Her own tissues nourish the developing individual until birth.

Producing enough offspring so that at least some survive is costly (Figure 43.3). Many invertebrates, the bony fishes, and frogs release sperm, eggs, or both into the environment. If each adult were to make only one sperm or egg each season, chances would not be good for fertilization. These animals invest energy in making many gametes, often thousands of them.

As another example, nearly all animals on land use internal fertilization, or the union of sperm and egg *within* the female body. They invest metabolic energy to construct elaborate reproductive organs, such as a penis and a uterus. A penis deposits sperm inside the female, and a uterus is a chamber in which an embryo develops inside certain mammalian females.

Finally, animals set aside energy in forms that can *nourish the developing individual* until it has developed enough to feed itself. Nearly all animal eggs contain **yolk**. This thick fluid has an abundance of proteins and lipids that nourish embryonic stages.

The eggs of some species have much more yolk than others. Sea urchins make enormous numbers of tiny eggs with little yolk. Each fertilized egg develops into a freely moving, self-feeding larva in less than a

day. Very few escape predators. For sea urchins, then, reproductive success means allocating small amounts of energy and resources to making each egg.

Birds put a lot of energy and resources into making eggs with a lot of yolk, which has to nourish the bird embryo through an extended time inside an eggshell that forms after fertilization. *Your* mother placed huge strains on herself to nourish you through nine months of development from a nearly yolkless, fertilized egg. Physical exchanges with her bloodstream supported your embryonic development (Figure 43.3g).

Animals show great diversity in reproduction and development, as these examples suggest. However, as you will see in sections to follow, some basic patterns are widespread throughout the animal kingdom.

Separation into male and female sexes requires special reproductive cells and structures, neural and hormonal control mechanisms, and forms of behavior.

A selective advantage—variation in traits among offspring —offsets biological costs related to separation into sexes.

43.2 Stages of Reproduction and Development

LINKS TO
SECTIONS
10.5, 25.1

For animals more complex than sponges, the life cycle has six developmental stages, from gamete formation through growth and tissue specialization.

Figure 43.4 is an overview of the six stages of animal reproduction and development. In *gamete formation*, the first stage, eggs or sperm develop inside parental reproductive tissues or organs, as outlined in Section 10.5. At *fertilization*, the first cell of a new individual—the zygote—forms when a sperm penetrates a mature egg and their nuclei fuse. *Cleavage* repeatedly cuts the fertilized egg by mitotic cell divisions. The number of cells grows, but the egg's original volume does not. This third stage is over when a ball of cells, a **blastula**, has formed. A blastula's cells are called **blastomeres**. They enclose a fluid-filled cavity, the blastocoel.

The blastula enters *gastrulation*. In this fourth stage, cells do not divide; they reorganize themselves into a **gastrula**. As you read in Sections 25.1 and 33.5, this early embryonic form has two or three primary tissue layers, or germ layers. Its cellular descendants give rise to all of the tissues and organs of the adult animal.

Ectoderm, the outermost primary tissue layer, forms first in all animal embryos. It is the forerunner of cell lineages that give rise to the nervous system and the outer part of the integument. Innermost is **endoderm**, the start of the gut's inner lining and organs derived from it. In most animal embryos, **mesoderm** forms in between the outer and inner primary tissue layers. It is the forerunner of muscles, most of the skeleton, the circulatory, reproductive, and excretory systems, and connective tissues of the gut and integument. For instance, in vertebrates, some mesoderm gives rise to **somites**: a longitudinal series of paired segments that are the source of most bones, skeletal muscles of the trunk and head, and most of the overlying dermis. Once again, mesoderm evolved hundreds of millions of years ago (Section 25.1). It was a key innovation in the evolution of nearly all large, complex animals.

After the primary tissue layers form, cells start to signal one another and interact in ways that result in distinct subpopulations of cells. These cells become specialized in composition, structure, and function. By orderly processes of *organ formation*, they form tissues and then organs in expected patterns.

Growth and tissue specialization is the sixth stage of animal development. The tissues and organs continue to grow, and they slowly take on their final sizes, shapes, proportions, and functions. This stage continues into adulthood.

Figure 43.5 has examples of the stages for one kind of vertebrate, the leopard frog (*Rana pipiens*). Take a moment to study this figure, for it reinforces an important principle: *Body structures that emerge during one developmental stage are a foundation for the stage that will come after it.* In the sections to follow, you will come across evidence that reinforces this principle.

a Eggs form and mature in female reproductive organs. Sperm form and mature in male reproductive organs.

b A sperm penetrates an egg. Their nuclei fuse. A zygote has formed.

c Mitotic cell divisions form a ball of cells, a blastula. Each cell gets regionally different parts of the egg cytoplasm.

d A gastrula, an early embryo that has primary tissue layers, forms by cell divisions, cell migrations, and rearrangements.

e Details of the body plan fill in as different cell types interact and form tissues and organs in predictable patterns.

f Organs grow in size, take on mature form, and gradually assume specialized functions.

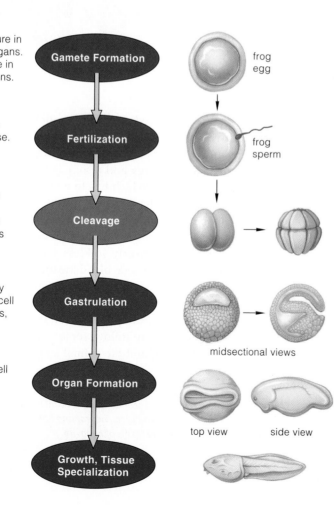

frog egg

frog sperm

midsectional views

top view   side view

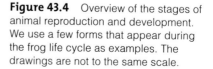

Most animal life cycles proceed through gamete formation, fertilization, cleavage, gastrulation, organ formation, and then growth and tissue specialization. Each stage builds on the stage that preceded it.

Figure 43.4 Overview of the stages of animal reproduction and development. We use a few forms that appear during the frog life cycle as examples. The drawings are not to the same scale.

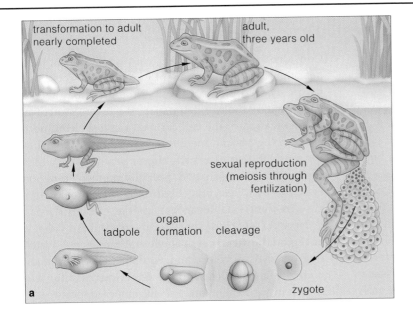

Figure 43.5 *Animated!* Reproduction and development in the life cycle of the leopard frog, *Rana pipiens*.

(a) We zoom in on the life cycle as a female releases her eggs into the surrounding water and a male releases sperm over the eggs. A frog zygote forms at fertilization. About one hour after fertilization, a surface feature called the gray crescent appears on this type of embryo. It establishes the frog's head-to-tail axis. Gastrulation will start here.

(b–e) Division planes of the first three cuts of cleavage and the blastula, formed by the end of cleavage. Gastrulation starts with this ball of cells, which contains a fluid-filled cavity (blastocoel).

(f–i) Starting at gastrulation, part of the ectoderm folds inward at a site called the dorsal lip. Three primary tissue layers form by cell migrations and rearrangements. A primitive gut cavity opens up. A neural tube, notochord, and other organs form from the primary tissue layers.

(j–l) The embryo becomes a tadpole, which develops into an adult.

Carving up the egg cytoplasm during cleavage:

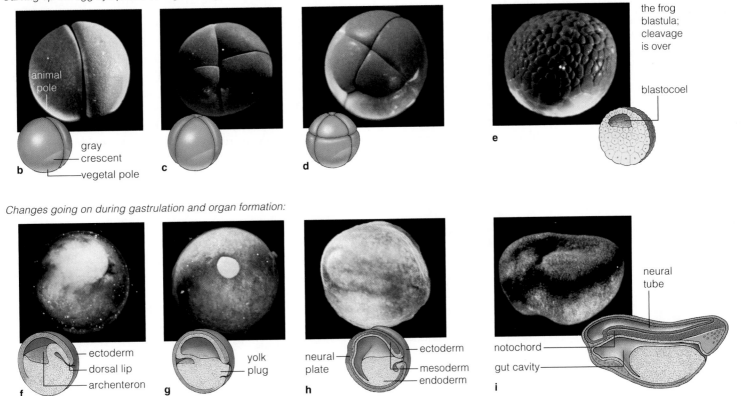

Changes going on during gastrulation and organ formation:

Changes in body form during growth and tissue specialization:

j Tadpole, a swimming larva with segmented muscles and a notochord extending into a tail.

k Metamorphosis to adult form under way. Limbs growing, tail tissues being resorbed.

l Sexually mature, four-legged adult leopard frog.

43.3 Early Marching Orders

LINKS TO
SECTIONS 4.10,
9.4, 10.5, 25.2

Why don't you have an arm attached to your nose or toes growing from your navel? The patterning of body parts starts with messages in immature, unfertilized eggs.

INFORMATION IN THE EGG

A **sperm**, recall, consists of paternal DNA and a bit of equipment that helps it reach and penetrate an egg. An **oocyte**, or immature egg, is much larger and more complex than the sperm (Section 10.5). As the oocyte is maturing, different numbers and kinds of enzymes, mRNA transcripts, and other factors are stockpiled in different parts of the cytoplasm. They are "maternal messages" for the forthcoming embryo. The messages are commonly used after fertilization, during the early rounds of DNA replication and mitotic cell divisions.

Also, remember how tubulin subunits assemble into microtubules (Section 4.10)? They become localized in specific parts of the egg cytoplasm—which establishes when microtubular spindles will form, and at which angles, for early cell divisions that follow fertilization. The amount of yolk affects how those divisions will cut up and parcel out cytoplasm, and its messages, to cells.

Some messages are "read" as early as fertilization. Most animal eggs show polarity, which establishes a front-to-back body axis for the embryo (Section 25.2).

Consider the frog oocyte. A lot of yolk is concentrated near the *vegetal* pole; pigment granules are stockpiled by the *animal* pole, or closest to the nucleus. When a sperm enters the egg cytoplasm, it triggers structural reorganization of the cell cortex—a cytoskeletal mesh beneath the plasma membrane (Section 4.10). Part of the mesh shifts toward the site of penetration, which exposes a crescent-shaped, partially pigmented region near the cell midsection (Figure 43.6). This region, the **gray crescent**, establishes the anterior–posterior axis.

How do we know that a gray crescent is evidence of regional differences in maternal messages? Watch frog embryos develop, and you see that gastrulation normally starts here each time. Experiments of the sort shown in Figure 43.6 offer additional evidence.

CLEAVAGE—THE START OF MULTICELLULARITY

Once an oocyte is fertilized, a zygote enters cleavage. By this process, recall, a ring of microfilaments just under the plasma membrane contracts and pinches the cell in two (Section 9.4). The zygote's cytoplasm does not grow in size during cleavage; the repeated cuts divide its volume into ever smaller blastomeres.

Simply by virtue of where the cuts are made, the blastomeres receive different maternal messages. This outcome of cleavage, called **cytoplasmic localization**, helps seal the developmental fate of each cell lineage. Cells of one lineage alone might inherit the cytoplasm with a protein that activates, say, a gene coding for a certain hormone. In most animal zygotes, those genes are silent through early cleavage; stockpiled maternal proteins and mRNAs control the cuts. In mammals,

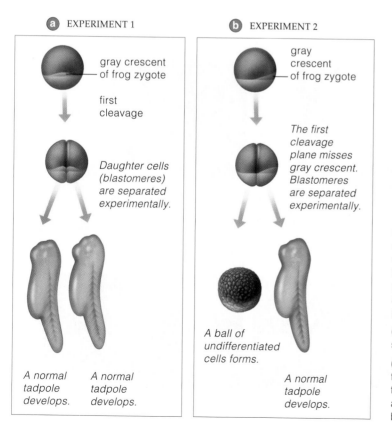

a EXPERIMENT 1

gray crescent of frog zygote

first cleavage

Daughter cells (blastomeres) are separated experimentally.

A normal tadpole develops.

A normal tadpole develops.

b EXPERIMENT 2

gray crescent of frog zygote

The first cleavage plane misses gray crescent. Blastomeres are separated experimentally.

A ball of undifferentiated cells forms.

A normal tadpole develops.

sperm as it penetrates a frog egg

pigmented egg cortex

yolk-rich cytoplasm

gray crescent

Figure 43.6 *Animated!* Two experiments that showed the effect of cytoplasmic localization on the fate of a frog embryo. The frog egg cortex has granules of dark pigment concentrated near one pole. At fertilization, part of the granule-containing cortex shifts toward the point of sperm entry and exposes lighter colored, yolky cytoplasm, as a crescent-shaped gray area. Normally, the first cleavage puts part of the gray crescent in both of the first two blastomeres.

(a) In one experiment, the first two blastomeres were physically separated from each other. Each still gave rise to a whole tadpole.

(b) In another experiment, a fertilized egg was manipulated so the cut through the first cleavage plane missed the gray crescent. Only one of the first two blastomeres got the gray crescent. It alone developed into a normal tadpole. Deprived of maternal messages in the cytoplasm beneath the gray crescent, the other cell could not develop normally.

a Sea urchin egg, with little yolk. Radial cleavage complete and blastomeres similar in size.

b Frog egg, with moderate, localized concentration of yolk. Radial cleavage complete but blastomeres unequal in size.

c *Drosophila* egg with an abundance of centrally located yolk. Superficial cleavage, blastomeres form only from the cytoplasm's periphery.

d At right, a highly yolky fertilized egg of a zebrafish. Incomplete cleavage; the blastomeres form in a disk-shaped area on top of yolk.

Figure 43.7 Examples of complete and incomplete cleavage patterns from different groups of animals.

certain genes must be activated first; cleavage cannot be completed without their protein products.

Each species has a characteristic cleavage pattern. Differences start with the first cut, which determines whether the first two cells will be equal or unequal in size and the types and proportions of messages they will receive. The pattern depends in part on whether and how much yolk forms, and where. It depends also on the way the spindle forms, which is a heritable trait.

Are the Cuts Complete or Incomplete?

When little yolk is present, the first cut divides all the cytoplasm. An abundance of yolk impedes the cut, so cleavage is incomplete. Sea urchin eggs do not have enough yolk to impede complete cuts, and blastomeres are similar in size (Figure 43.7*a*). Cleavage is complete for frogs and other amphibians, but concentrated yolk slows the cuts near the vegetal pole. Blastomeres form faster near the animal pole and are smaller than those carved from the yolky vegetal pole (Figure 43.7*b*). Cleavage is complete in the nearly yolkless eggs of mammals.

Drosophila and other insect eggs have a great deal of yolk concentrated in their center. Cuts are confined to the cytoplasm's periphery; cleavage is superficial (Figure 43.7*c*). Eggs of reptiles, birds, and most fishes are so yolky that cuts are exceedingly slow or blocked entirely, *except* in a small, disk-shaped region that has the least amount of yolk (Figure 43.7*d*).

How Are the Cuts Oriented?

In Section 25.2, you read about a major divergence in animal evolution. The split into protostomes and deuterostomes arose by modifications in patterns of development. For instance, cleavage is *spiral* in the mollusks, annelids, arthropods, and other protostomes. The first bipolar spindles that form in the egg cytoplasm are angled with respect to the anterior–posterior axis, so all daughter blastomeres are tilted left to right (Figure 43.8*a*). Cleavage is *radial* in deuterostomes, which include the echinoderms and chordates. In these organisms, the first bipolar spindles that form are parallel with the main axis of the embryo, and later ones are perpendicular to it (Figure 43.8*b*).

There are many variations on these basic patterns. Mammals have the slowest cleavage rate of all; twelve to twenty-four hours pass between each cut. The first cut runs parallel with the anterior–posterior axis. Of the two blastomeres that form, one also is cut parallel, but the other gets cut sideways; the spindle orientation rotates by 90 degrees. Similar rotation occurs in some of the descendent cells. This is one type of a *rotational* cleavage pattern. Also, mammalian blastomeres do not all divide at the same time, so the blastula that forms may consist of an odd number of cells.

The next chapter considers how human embryos develop. For now, it is enough to know that the first cuts result in eight blastomeres with spaces between. Tight junctions hold the loose collection together. More cuts result in a hollow ball of cells. These outer cells secrete fluid that fills the ball's cavity; others huddle in a mass against the cavity wall. This type of blastula forms in all mammals, and it is called a **blastocyst**. The inner cell mass gives rise to the embryo proper.

a Early protostome embryo. Its four cells are undergoing spiral cleavage, *oblique to* the anterior–posterior axis:

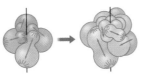

b Early deuterostome embryo. Its four cells are undergoing radial cleavage, *parallel with* and *perpendicular to* the anterior–posterior axis:

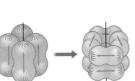

Figure 43.8 Spiral and radial cleavage patterns.

Unfertilized eggs contain maternal messages: localized regions with differences in the type and proportions of enzymes, mRNAs, tubulins, yolk, and other factors.

Cleavage divides a zygote into blastomeres. Simply by virtue of the cuts, each ends up with different maternal messages. Cleavage patterns differ among the major animal groups.

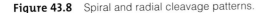

43.4 How Do Specialized Tissues and Organs Form?

LINKS TO
SECTIONS 13.4, 15.1,
22.12, 25.1. 28.5

Nearly all animals have a gut that digests nutrients for absorption. They have surface parts that protect organs inside and detect what is going on outside. Most have organs in between that function in structural support, motion, and circulation. This three-layered body plan emerges after cleavage, as gastrulation gets under way.

During gastrulation, the embryonic cells migrate and rearrange themselves into three primary tissue layers of the gastrula: ectoderm, endoderm, and mesoderm. Figure 43.9 shows an example. The first cavity to form in protostome gastrulas becomes a mouth, but it will become an anus in deuterostomes (Section 25.2). Also, the anterior–posterior body axis forms in gastrulas. In vertebrates, this main axis precedes the formation of a **neural tube**, the forerunner of the brain and spinal cord (Figures 43.10 and 43.11). These specializations arise through cell differentiation and morphogenesis.

CELL DIFFERENTIATION

All cells of a normal embryo have the same number and kinds of genes, having descended from the same zygote. They all activate the genes for products that assure their survival, such as histones and glucose-metabolizing enzymes. But from gastrulation onward,

selective gene expression occurs: Some cell lineages express different groups of genes than others do. This is the start of **cell differentiation**. By this process, cell lineages become specialized in composition, structure, and function (Section 15.1).

As an example, when your eye lenses formed, only one type of cell could activate the genes for crystallin proteins. Long, transparent crystallin fibers formed in the cells and forced them to lengthen and flatten. These differentiated cells impart unique optical properties to each lens. They are only 1 of 200 differentiated cell types in the human body.

As many experiments show, nearly all cells become differentiated with no loss of genetic information. For instance, John Gurdon stripped unfertilized frog eggs

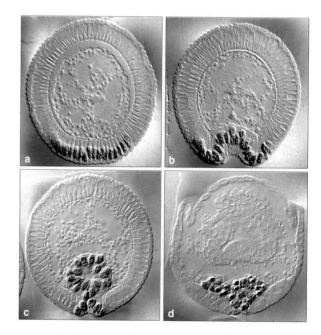

Figure 43.9 Gastrulation in a fruit fly (*Drosophila*), cross-section. After cleavage, the blastula is transformed into a gastrula. Some cells (stained *gold*) migrate inward through an opening that forms at the surface of the ball of cells. Fruit flies are protostomes; this opening will become a mouth.

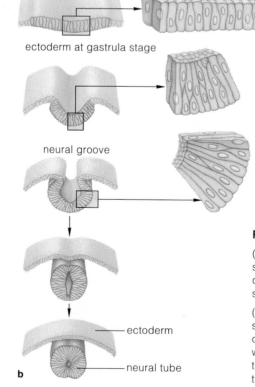

ectoderm at gastrula stage

neural groove

ectoderm

neural tube

Figure 43.10 *Animated!* Examples of what goes on during morphogenesis.

(**a**) Cell migration. The art shows the same embryonic neuron (*orange*) at successive times during its "climb" along cells that have already formed in developing brain tissue. The neuron is responding to chemical cues on the surface of a glial cell (*yellow*), which guide it to its destination in the embryo.

(**b**) Neural tube formation. By the end of gastrulation, ectoderm is a uniform sheet of cells. Along the axis of the future tube, microtubules in ectodermal cells lengthen. Rings of microfilaments constrict in some of these cells, which become wedge-shaped. The part of the ectodermal sheet where they are located folds back on itself, over the wedge-shaped cells, which then disengage from it as a separate tube.

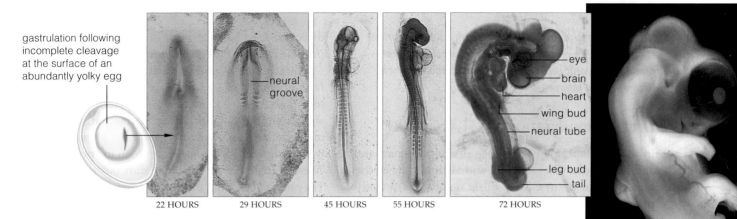

gastrulation following incomplete cleavage at the surface of an abundantly yolky egg

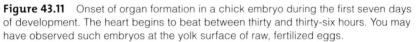

neural groove

eye
brain
heart
wing bud
neural tube
leg bud
tail

22 HOURS   29 HOURS   45 HOURS   55 HOURS   72 HOURS

168 HOURS (SEVEN DAYS OLD)

Figure 43.11 Onset of organ formation in a chick embryo during the first seven days of development. The heart begins to beat between thirty and thirty-six hours. You may have observed such embryos at the yolk surface of raw, fertilized eggs.

of their nucleus. He ruptured the plasma membrane of intestinal cells from tadpoles of the same species. He left the nucleus and much of the cytoplasm of the fully differentiated cells intact and inserted them into the enucleated eggs. Some eggs developed into a frog. Intestinal cells still had the same number and kinds of genes as the zygote; its nucleus still had all the genes required to make all cell types that make up a frog.

Each cell in an early human embryo also retains the capacity to give rise to a whole individual. That is how identical twins arise (page 769). That is the basis of adult cloning methods, as Section 13.4 explains.

MORPHOGENESIS

Tissues and organs of specific proportions, sizes, and shapes form by a program of orderly changes called **morphogenesis**. During this process, cells of different lineages divide, grow, disperse, and change in size. Tissues lengthen or widen and fold over. And some cells die in controlled ways at prescribed locations.

Think about active cell migration. *Cells send out and use pseudopods that move them along prescribed routes.* When they reach their destination, they connect with cells already there. Embryonic neurons migrate this way in a developing nervous system (Figure 43.10a).

How do these cells know where to move and when to stop? They respond to adhesive cues and chemical gradients. The cell migrations are coordinated by the synthesis, release, deposition, and removal of specific chemicals in the extracellular matrix.

The neuron shown in Figure 43.10a is responding to gradients in the "stickiness" of its surroundings. Adhesion proteins on a patch of its plasma membrane stuck to proteins on the surface of a glial cell. Now, cytoskeletal elements in the neuron lengthen from the attachment site and push the cytoplasm forward. The neuron will migrate until its surface adhesion proteins reach the spot in the embryo that is stickiest to them. The capacity for adhesion arose early in the evolution of multicelled animals (Sections 22.12 and 25.1).

Also, *whole sheets of cells expand and fold inward and outward as their cells change in shape.* Within these cells, microtubules grow longer, and rings of microfilament constrict. The controlled assembly and disassembly of these cytoskeletal components cause the changes.

Figures 43.10b and 43.11 hint at what happens after three primary tissues form in vertebrate embryos. At the embryo's midline, some ectodermal cells elongate and form the neural plate, the start of nervous tissue. The plate sinks inward as cells lengthen and become wedge shaped. At the edges of the resulting groove in the surface, flaps of tissue fold over and meet at the midline, forming the neural tube.

Finally, *programmed cell death helps sculpt body parts.* By this process, called **apoptosis**, signaling molecules from some cells activate tools of self-destruction that are stockpiled in other, target cells. Remember how a human hand forms from a paddle-shaped body part? Programmed cell suicide was at work (Section 28.5).

In cell differentiation, some cells selectively use certain genes that other cells do not use. Selective gene expression is the basis of cell differentiation. It results in cell lineages with characteristic structures, products, and functions.

Morphogenesis is a program of orderly changes in the size, shape, and proportions of developing tissues and organs.

Morphogenesis involves cell division, active cell migration, tissue growth and foldings, changes in cell size and shape, and programmed cell death, or apoptosis.

43.6 Why Do We Age and Die?

LINKS TO
PAGES 105, 206,
SECTIONS
12.4, 12.8, 33.6

As the years pass, all multicelled species undergo aging; tissues become harder to maintain and repair. Each species has a maximum life span—122 years for humans, 20 years for dogs, 12 weeks for butterflies, 35 days for fruit flies, and so on. The verifiably oldest human lived 122 years. We can expect that genes influence aging processes.

PROGRAMMED LIFE SPAN HYPOTHESIS

Do biological clocks influence aging? If so, an animal body might be analogous to a clock shop, with each type of cell, tissue, and organ ticking away at its own genetically set pace. Many years ago, Paul Moorhead and Leonard Hayflick tested this hypothesis. The two researchers cultured human embryonic cells—which divided about fifty times before dying out.

Hayflick also took cultured cells that were part of the way through the series of divisions and froze them for a few years. After he thawed the cells and placed them in a culture medium, they completed an *in vitro* cycle of fifty doublings and died on schedule.

No cell in a human body divides more than eighty or ninety times. You may well wonder: If an internal clock ticks off their life span, then how can *cancer* cells go on dividing? The answer provides us with a clue to why *normal* cells can't beat the clock.

Cells, remember, duplicate all chromosomes before they divide. Chromosomes have **telomeres**, or caps of DNA and proteins. Telomeres keep the chromosome ends from unraveling. Each time the nucleus divides, enzymes nibble off a bit of each telomere. When only a nub remains, cells stop dividing and die.

Cancer cells and germ cells are exceptions; both make telomerase, an enzyme that makes telomeres lengthen. Expose cells in culture to telomerase, and they go on dividing well beyond the normal life span.

CUMULATIVE ASSAULTS HYPOTHESIS

Another hypothesis: Over the long term, aging is the outcome of cumulative damage at the molecular and cellular levels. Environmental assaults and failure of DNA repair mechanisms are at work here.

For example, the rogue molecular fragments called free radicals attack all biological molecules, including DNA. This includes DNA of mitochondria, the power plants of eukaryotic cells. Structural changes in DNA can skew the synthesis of enzymes and other proteins necessary for normal life processes.

Free radicals are implicated in many age-related problems, such as cataracts, Alzheimer's disease, and atherosclerosis. In one recent experiment, researchers extended the average life span of some roundworms by 50 percent simply by providing them with synthetic antioxidant enzymes.

Other studies suggest that extending the life span may come at a cost. Researchers were able to double the life span of roundworms by knocking out a single gene. However, the mutant worms were sterile.

What about DNA replication and repair problems? *Werner's syndrome*, an aging disorder, is linked with a mutation in a gene that specifies one of the enzymes that unwind nucleotide strands. The mutated helicase probably does not compromise replication of DNA; affected people do not die right away. They start aging fast in their thirties and die before age fifty. But the nonmutated gene may be vital for repairs. People with Werner's syndrome accumulate mutations at high rates. Sooner or later, the damage interferes with cell division. Like skin cells of the elderly, skin cells of Werner's patients just do not divide many times.

It may be that both hypotheses have merit. Aging may be an outcome of many interconnected processes in which genes, hormones, environmental assaults, and a decline in DNA repair mechanisms come into play. Consider how living cells of all tissues depend upon exchanges of materials with extracellular fluid. Also consider how collagen is a structural component of many connective tissues. If something shuts down or mutates collagen-encoding genes, then missing or altered gene products may disrupt flow of oxygen, nutrients, hormones, and so forth to and from living cells through every connective tissue. Repercussions from such a mutation would ripple through the body.

Similarly, if mutations cause altered self markers on the body's cells, do T cells of the immune system perceive them as foreign and attack? If autoimmune responses were to become more frequent over time, they would promote greater vulnerability to disease and stress associated with old age.

In evolutionary terms, reproductive success means living long enough to produce and raise offspring. Humans reach sexual maturity in fifteen years *and* help children reach adulthood. We don't *need* to live longer than we do. But we among all animals have the capacity to think about it, and most of us decide that we like life better than the alternative. Eventually, however, we may all learn to accept the inevitability of our mortality with wisdom and grace.

Aging may be an outcome of time running out on internal biological clocks as well as cumulative, irreversible damage at the molecular and cellular levels.

43.7 Death in the Open

As a leading cancer specialist, Lewis Thomas reflected with compassion on the fear of dying. Before Thomas himself died of cancer, he gave us this gift of insight.

Everything in the world dies, but we only know about it as a kind of abstraction. If you stand in a meadow, at the edge of a hillside, and look around carefully, almost everything you catch sight of is in the process of dying, and most things will be dead long before you are. If it were not for the constant renewal and replacement going on before your eyes, the whole place would turn to stone and sand under your feet....

There are said to be a billion billion insects on the Earth at any moment, most of them with short life expectancies by our standards. Someone estimated that there are 25 million assorted insects hanging in the air over every temperate square mile, in a column extending upward for thousands of feet, drifting through the layers of atmosphere like plankton. They are dying steadily, some by being eaten, some just dropping in their tracks, tons of them around the Earth, disintegrating as they die, invisibly.

Who ever sees dead birds, in anything like the huge numbers stipulated by the certainty of the death of all birds? A dead bird is an incongruity, more startling than an unexpected live bird, sure evidence to the human mind that something has gone wrong. Birds do their dying off somewhere, behind things, under things, never on the wing.

Animals seem to have an instinct for performing death alone, hidden. Even the most conspicuous find ways to conceal themselves in time. If an elephant missteps and dies in an open place, the herd will not leave him there; the others will pick him up and carry the body from place to place, finally putting it down in some inexplicably suitable location. When elephants encounter the skeleton of an elephant in the open, they methodically take up the bones and distribute them, in a ponderous ceremony, over neighboring acres.

It is a natural marvel. All of the life on Earth dies all of the time, in the same volume as the new life that dazzles us each morning, each spring. All we see of this is the odd stump, the fly struggling on the porch floor of the summer house in October, the fragment on the highway. I have lived all my life with an embarrassment of squirrels in my backyard, they are all over the place, all year long, and I have never seen, anywhere, a dead squirrel.

I suppose it is just as well. If Earth were otherwise, and all the dying were done in the open, with the dead there to be looked at, we would never have it out of our minds. We can forget about it much of the time, or think of it as an accident to be avoided somehow. But it does make the process of dying seem more exceptional than it really is, and harder to engage in at the times when we must ourselves engage.

In our way, we conform as best we can to the rest of nature. The obituary pages tell us of the news that we are dying away, while birth announcements in finer print, off at the side of the page, inform us of our replacements, but we get no grasp from this of the enormity of the scale. There are now billions of us on the Earth, and all must be dead, on a schedule, within this lifetime. The vast mortality, involving something over 50 million each year, takes place in relative secrecy. We can only really know of the deaths in our households, among our friends. These, detached in our minds from all the rest, we take to be unnatural events, anomalies, outrages. We speak of our own dead in low voices; struck down, we say, as though visible death can occur only for cause, by disease or violence, avoidably. We send off for flowers, grieve, make ceremonies, scatter bones, unaware of the rest of the billions on the same schedule. All of that immense mass of flesh and bone and consciousness will disappear by absorption into the Earth, without recognition by the transient survivors.

Less than half a century from now, our replacements will have more than doubled in numbers. It is hard to see how we can continue to keep the secret, with such multitudes doing the dying. We will have to give up the notion that death is a catastrophe, or detestable, or avoidable, or even strange. We will need to learn more about the cycling of life in the rest of the system, and about our connection in the process. Everything that comes alive seems to be in trade for everything that dies, cell for cell. There might be some comfort in the recognition of synchrony, in the information that we all go down together, in the best of company.

— LEWIS THOMAS, 1973

Summary

Section 43.1 Compared with asexual reproduction, reproducing sexually takes more time and energy, in terms of the structures and behaviors required. Also, the genetic representation in the next generation of each parent is not as great. However, it might give the offspring a greater capacity to make faster adaptive responses to changing biotic and abiotic conditions.

Section 43.2 Most animal life cycles have six stages of embryonic development, and each must be completed successfully before the next begins:

Gamete formation. Oocytes (immature eggs) and sperm form in reproductive organs.

Fertilization. A sperm penetrates an egg cytoplasm; the sperm and egg nuclei fuse and form a zygote. This fertilized egg is the first cell of a new individual.

Cleavage. Mitotic cell divisions cut the egg cytoplasm into ever smaller blastomeres with no increase in volume.

Gastrulation. Two or three primary tissue layers (germ layers) form: ectoderm, endoderm, and often mesoderm. All tissues of the adult body develop from these layers.

Onset of organ formation. Different organs start developing by a tightly orchestrated program of cell differentiation and morphogenesis.

Growth and tissue specialization. Organs enlarge and acquire specialized chemical and physical properties.

Biology⌀Now

View the development of a frog with the animation on BiologyNow.

Section 43.3 As eggs mature, different types and proportions of enzymes, mRNAs, tubulins, and other maternal messages are stockpiled in different parts of the cytoplasm. After fertilization, cleavage puts these maternal messages in different daughter cells.

Cleavage patterns differ among animal groups. The cuts are complete in eggs with a sparse to moderate amount of yolk and may produce blastomeres of equal or unequal size. Cuts are incomplete in highly yolky eggs, and blastomeres form from all or part of the periphery of the cytoplasm only. How the cuts are oriented depends on the mitotic spindle's orientation, which is a heritable trait. Different maternal messages end up in different blastomeres, an outcome called cytoplasmic localization. Cleavage ends with the formation of a blastula, a tiny ball of cells with a fluid-filled cavity, the blastocoel.

Biology⌀Now

See a demonstration of the effects of cytoplasmic localization with the animation on BiologyNow.

Section 43.4 In cell differentiation, a cell selectively uses certain genes and synthesizes proteins not found in other cell types. The outcomes are subpopulations of specialized lineages of cells that differ from one another in their structure, biochemistry, and functioning.

Morphogenesis starts at gastrulation. By this program of orderly changes, tissues and organs of specific sizes,

shapes, and proportions emerge. It involves cell divisions, migrations, enlargements, and programmed cell death, as well as the lengthening, widening, and folding of tissues.

Biology⌀Now

Learn about neural tube formation with the animation on BiologyNow.

Section 43.5 By embryonic induction, embryonic cell lineages become committed to developing a certain way when they are exposed to signals from adjacent tissues. The signals are the basis of pattern formation, a sculpting of specialized tissues and organs from clumps of cells in the proper places in the embryo, in the proper order. Signals act on cells that form together, as well as on cells that make contact during morphogenesis.

Products of master genes map out the basic body plan. Different products are short-range and long-range beacons, as when they form gradients from signaling centers that help cells assess their position in the embryo and how they will contribute to the formation of tissues and organs. Master genes are similar and in some cases identical among all major animal groups.

Biology⌀Now

See experimental evidence of embryonic induction with the animation on BiologyNow.

Sections 43.6, 43.7 Aging may be partly a result of time running out of internal biological clocks, which are genetically preset. Aging also may be partly an outcome of cumulative assaults on DNA and other biological molecules during the life cycle.

Self-Quiz
Answers in Appendix II

1. Compared to an asexual reproducer, a male or female animal that reproduces sexually _____ .
   a. has spent more, biologically
   b. has fewer of its genes in the next generation
   c. gets variation in traits among offspring
   d. all of the above

2. A cell formed during cleavage is a _____ .
   a. blastula   b. morula   c. blastomere   d. gastrula

3. Three primary tissue layers form during _____ .
   a. gametogenesis   c. gastrulation
   b. implantation   d. pattern formation

4. _____ distributes different maternal messages to different blastomeres.
   a. Gametogenesis   c. Morphogenesis
   b. Cleavage   d. Pattern formation

5. Primary tissue layers first appear _____ .
   a. in the egg cortex   c. in the gastrula
   b. during cleavage   d. in primary organs

6. During development, the formation of subpopulations of different cell types is the outcome of _____ .
   a. selective gene expression   c. metamorphosis
   b. cell differentiation   d. a and b

7. Homeotic genes map out the _____ .
   a. cleavage planes   c. basic body plan
   b. primary tissue layer d. all of the above

8. _____ take part in pattern formation.
   a. Master genes   c. Regulatory proteins
   b. Morphogens   d. all of the above

9. The developmental fate of an embryonic cell lineage changes upon exposure to gene products from an adjacent tissue. This is a case of _____ .
   a. cleavage   c. cytoplasmic localization
   b. embryonic induction d. apoptosis

10. The cleavage pattern for *Drosophila* is _____ .
   a. superficial   c. radial
   b. superfluous   d. rotational

11. Match each term with the most suitable description.
   ____ gamete formation   a. blastomeres form
   ____ fertilization   b. cellular rearrangements form primary tissues
   ____ cleavage
   ____ gastrulation   c. eggs and sperm form
   ____ cell differentiation d. sperm and egg nuclei fuse
   ____ morphogenesis   e. tissues, organs of specific sizes and shapes emerge
                            f. in most species, result of selectively using genes

Additional questions are available on Biology 🕏 Now™

Critical Thinking

1. The zebrafish (*Danio rerio*) is special to developmental biologists. This small freshwater fish is easily maintained in tanks. A female produces hundreds of eggs, which can develop and hatch in three days. The transparent embryos let researchers directly observe developmental events (Figure 43.15). Single cells can be injected with dye to see how they change position, or they can be killed or injected with genes to observe the outcome. In evolutionary terms, *D. rerio* is only remotely related to humans. Explain why researchers expect the early development of this fish to yield useful information about human development.

2. The photograph at left shows an early embryonic stage of a sea urchin (*Lytechinus*). Is it a blastula? Or a gastrula? How do you know?

3. Before an amphibian egg enters cleavage, you divide it so that the gray crescent is parceled out to only one of the two blastomeres that form first. You separate the two blastomeres. Only the one with the gray crescent gives rise to an embryo with an anterior–posterior axis, notochord, nerve cord, and dorsal muscles. The other blastomere gives rise to a shapeless mass of immature gut cells and blood cells. Would you expect cytoplasmic localization or embryonic induction to exert more influence in bringing about these results? Explain your answer.

4. In normal *Drosophila* larvae, a specific cluster of cells is the embryonic source of antennae on the head. When a larva has a certain mutant gene, it develops legs instead of antennae on its head, as in Section 15.4. Would you say that the mutated gene is first expressed during cleavage or during organ formation?

5. Four hundred million years ago, six-legged insect lineages arose from crustaceans that had far more jointed legs. Based on what you know about the constraints on

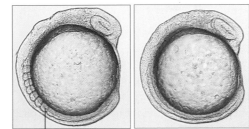

Figure 43.15 (**a**) Adult zebrafish. (**b**) Normal embryo. Somites, a series of segments that give rise to bone and muscle, are clearly visible. (**c**) Mutant embryo that could not form somites. Changes in developmental steps show up clearly in these transparent embryos.

b somite   **c**

drastic changes in morphology, formulate a hypothesis to explain why such a drastic reduction has persisted.

6. Once in a while in humans, the first two blastomeres, the inner cell mass, or the blastocyst splits in two. *Identical* twins, which have identical genes and are the same sex, may be the result. But identical offspring are the norm for the nine-banded armadillo (*Dasypus novemcinctus*), the state mammal of Texas and a common sight in the southeastern United States. Four offspring make up a normal litter. They formed from the same fertilized egg. After the second cleavage, the four blastomeres separated and each developed into an armadillo. Explain why this embryonic event reduces the genetic variability within an armadillo litter. In evolutionary terms, what might be some disadvantages of this event?

7. Sometimes mRNA transcripts cannot hang around in an embryonic cell after they have been used. For instance, some may have translated for an early developmental step that is over. How can the cell get rid of mRNA that is no longer needed and might cause problems in upcoming stages? In many eukaryotes, including roundworms, insects, fishes, and mammals, cells prevent transcription of unwanted mRNAs by *RNA interference*.

The process begins with the transcription of a gene that encodes a small interfering RNA (siRNA). The product is a long RNA sequence with a region complementary to the unwanted mRNA. A group of enzymes and other proteins attaches to a mature siRNA transcript. When the unwanted mRNA binds with the transcript, the proteins chop it up and destroy it.

How important are the siRNAs? One way to find out is to observe what happens if cells cannot make them. For example, you can inactivate Dicer, an enzyme that is required to produce mature siRNA. The results of such experiments suggest that siRNAS are essential for normal vertebrate development. When Dicer production was prevented in zebrafish, development was abnormal and stopped short of maturity. Attempts to breed Dicer-free mice have been unsuccessful. Homozygous mutants do not survive to birth.

Genes encode siRNAs. We know that genes can mutate. So it is reasonable to assume that some mutations will produce defective siRNAs. Think about how particular genes are turned on during different developmental stages. What types of genes would you expect to be targeted by siRNAs during development? What might happen if the products of these genes remained available in cells longer than they should be?

Mind-Boggling Births

In December of 1998, Nkem Chukwu of Texas gave birth ahead of schedule to six girls and two boys—the first octuplets to survive premature birth (Figure 44.1).

The combined weight of the eight newborns was a bit more than 10 pounds (4.5 kilograms). Odera, the smallest, weighed less than 1 pound (20 grams). She died of heart and lung failure six days later. The others had to spend three months in the hospital before going home. Two required abdominal surgery.

Chukwu was having trouble getting pregnant, so she asked for hormone injections. The hormones caused many eggs in her ovaries to mature and be released at the same time. Chukwu chose not to reduce the number of embryos that became fertilized. The first newborn, thirteen weeks premature, was delivered naturally. Chukwu's doctor used drugs to stop labor, then surgically delivered the rest.

Over the past two decades, the incidence of multiple births has increased by almost 60 percent. The incidence of *higher order multiple births*—triplets or more—has quadrupled. What is going on?

A woman's fertility peaks in her mid-twenties. By age thirty-nine, the likelihood of natural conception declines by about half. Yet the number of first-time mothers who are more than forty years old doubled in the past decade —and many were assisted by fertility drugs, in vitro fertilization, and other reproductive interventions.

Fertility drugs are driving the increase in higher order multiple births, and they worry many doctors. Carrying more than one embryo increases the risk of miscarriage, premature delivery, or delivery by cesarean section. The newborn weight is lower and mortality rates are higher, compared to normal births. A woman carrying quintuplets runs twice the risk of miscarriage. The newborns are 90 percent more likely to develop postdelivery complications. The parents also face far more physical, emotional, and financial challenges.

The preceding chapter sketched out some principles that govern reproduction and development of animals in general. This chapter applies the principles to humans. Besides the core sections, it has many optional reference sections on topics that will directly and indirectly have a profound impact on your future. What you read may help you work through health-related problems and ethical issues concerning how we individually and collectively deal with human fertility.

So start with a premise we offered in the preceding chapter's introduction: *In nature, the function of sex is not recreational, it is the perpetuation of one's genes.* As in other mammals, human genes become packaged in gametes that start forming in a pair of gonads, or primary reproductive organs. We call them testes in males and ovaries in females. As you will see, these

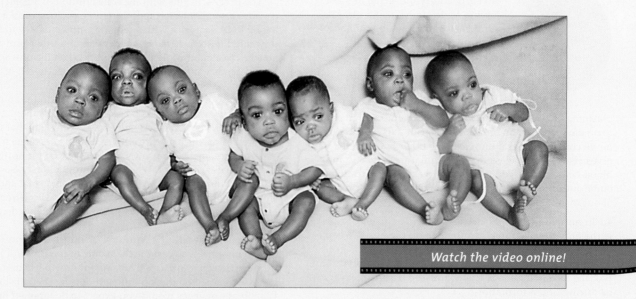

Watch the video online!

Figure 44.1 Testimony to the potency of fertility drugs—seven survivors of a set of octuplets. Besides manipulating so many other aspects of nature, humans are now manipulating their own reproduction. *Facing page*, a human embryo at an early stage of development.

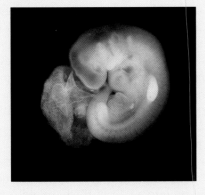

organs also secrete diverse sex hormones, which control reproductive functions as well as the development of the traits we associate with maleness and femaleness.

Remember also that early human embryos have neither male nor female traits (Section 12.5). Seven weeks after fertilization, however, a pair of ovaries start to develop in the embryos that did not inherit a Y chromosome, which carries the master gene for sex determination. Testes develop only in XY embryos. Ovaries and testes are fully formed at the time of birth. It takes about a decade for them to grow to their full size and become reproductively functional. And that is where our case study picks up.

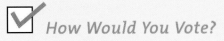

How Would You Vote?

Fertility drugs induce multiple ovulations at the same time and increase the likelihood of high-risk multiple pregnancies. Should we restrict the use of such drugs to conditions that limit the number of embryos formed? See BiologyNow for details, then vote online.

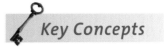

Key Concepts

HUMAN REPRODUCTIVE SYSTEMS

The primary reproductive organs, or gonads, are sperm-producing testes in human males and oocyte-producing ovaries in females. In response to the hypothalamus and the pituitary gland, gonads release sex hormones that guide reproductive functions and the development of secondary sexual traits. Sections 44.1–44.3

THE MENSTRUAL CYCLE

From puberty onward, human females are fertile on a cyclic basis. Each month during the reproductive years, an egg is released from an ovary, and the lining of the uterus is primed for possible pregnancy. Sections 44.4, 44.5

REGARDING SEX AND PREGNANCY

Sexual intercourse leads to pregnancy, which human interventions attempt to deflect or promote. Pathogens opportunistically exploit human sexual behavior as a means of transmission to new hosts. Sections 44.6–44.8

HUMAN EMBRYONIC DEVELOPMENT

Embryonic development starts with gamete formation and proceeds through fertilization, followed by implantation of the blastocyst in the uterine lining. The embryo develops by way of cleavage, gastrulation, organ formation, and growth and tissue specialization. It forms vital connections with the mother by way of a placenta. Sections 44.9–44.13

FROM BIRTH ONWARD

Each generation starts with the birth of a new individual and extends through the reproductive years to a time of gradual aging and death. The motivation to engage in sex during the life cycle invites reflection on the gift of human life and on its consequences when multiplied by the many billions of humans now on Earth. Sections 44.14, 44.15

Links to Earlier Concepts

This chapter builds on principles of animal reproduction and development (Chapter 43). It draws on your understanding of nuclear and cytoplasmic divisions (Sections 9.3, 9.4, 10.3) and gamete formation (10.5). You may wish to review how reproductive organs form in embryos (12.5). You will revisit master genes (15.2, 17.8) and consider some effects of diet (41.8), psychoactive drug use (34.13), and smoking (40.8). You will reconsider some endocrine controls (36.3, 36.8). You will be invited to reflect on sexual behavior in medical terms (21.4, 22.2, 39.10) and from an evolutionary perspective (20.4 and the Chapter 10 introduction).

44.1 Reproductive System of Human Males

LINKS TO
SECTIONS
2.7, 12.5, 33.10

A human male's genes become packaged in sperm. These male gametes form in a pair of gonads that also secrete sex hormones.

WHEN MALE GONADS FORM AND BECOME ACTIVE

Again, male gonads are called **testes** (singular, testis). Testes are the primary components of a reproductive system that also includes accessory organs, glands, and ducts (Table 44.1).

Earlier, Figure 12.8 showed how the testes start to form on the wall of an XY embryo's abdominal cavity. Before birth, they descend into the scrotum, which is a pouch of loose skin suspended below the pelvic girdle (Figure 44.2). Muscle contractions can draw the pouch closer to the main body mass, and muscle relaxation can lower it. The reflexive adjustments help keep the internal temperature suitable for sperm formation.

Packed within each testis are many small, highly coiled tubes called seminiferous tubules. Section 44.2 explains how sperm cells form and start maturing in these tubules in hormone-guided ways.

Sperm production and the emergence of secondary sexual traits start at **puberty**. This stage of postnatal development usually begins in boys between ages twelve and sixteen. Signs that it is under way include enlarging testes, growth spurts, a deepening voice, and the growth of more hair on the face, chest, armpits, and around the base of the scrotum and penis. Also, the amount and distribution of body fat and skeletal muscles undergo modification. Such secondary sexual traits do not play a direct role in reproduction.

STRUCTURE AND FUNCTION OF THE MALE REPRODUCTIVE SYSTEM

Mammalian sperm travel from testes through a series of ducts that lead into the urethra (Figure 44.3). They are not quite mature when they enter a pair of coiled, long ducts. Each duct is an epididymis, and secretions from glandular cells in its wall trigger the events that put the finishing touches on the maturing sperm cells. The last region of an epididymis stores mature sperm. During a male's reproductive years, about 100 million sperm mature every day. Unused ones are resorbed or excreted in urine.

In a sexually aroused male, smooth muscle in the walls of his reproductive organs contracts and propels mature sperm into a pair of thick-walled ducts called the vasa deferentia (singular, vas deferens). The sperm are propelled onward, through a pair of ejaculatory ducts, then into and finally out from the urethra. This last tubular duct threads vertically through the penis, the male sex organ, and opens at its tip. The urethra, remember, also functions in urinary excretion.

Sperm traveling to the urethra mix with glandular secretions and form semen, a thickened fluid that gets expelled from the penis during sexual activity. Paired seminal vesicles secrete fructose. Sperm use this sugar as an energy source. Prostaglandins secreted from the prostate gland and, to a lesser extent, seminal vesicles enter the mix. These signaling molecules may induce muscle contractions in the female reproductive tract, which may help sperm reach an egg. They also might help the sperm slip past the female's immune system, which is on guard against anything foreign.

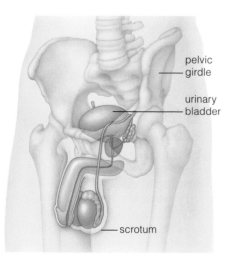

Figure 44.2 Position of the human male reproductive system relative to the pelvic girdle and urinary bladder.

pelvic girdle

urinary bladder

scrotum

Table 44.1	Organs and Accessory Components of the Human Male Reproductive System

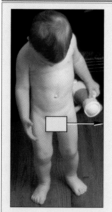

Reproductive Organs

Testis (2)	Sperm, sex hormone production
Epididymis (2)	Sperm maturation site and subsequent storage
Vas deferens (2)	Rapid transport of sperm
Ejaculatory duct (2)	Conduction of sperm to penis
Penis	Organ of sexual intercourse

Accessory Glands

Seminal vesicle (2)	Secretion of large part of semen
Prostate gland	Secretion of part of semen
Bulbourethral gland (2)	Production of mucus that functions in lubrication

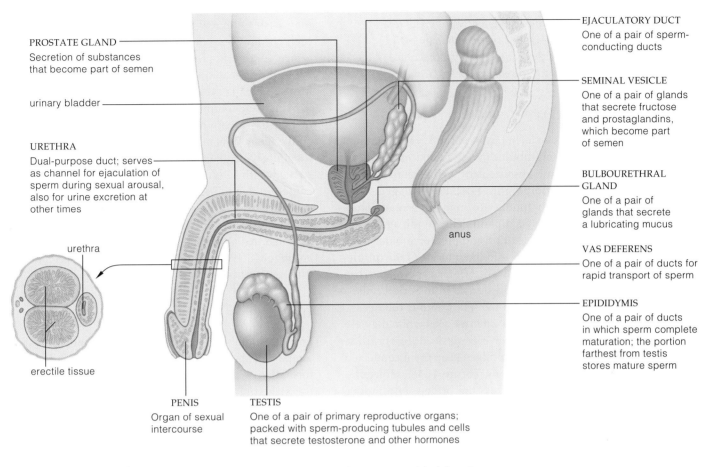

PROSTATE GLAND
Secretion of substances that become part of semen

urinary bladder

URETHRA
Dual-purpose duct; serves as channel for ejaculation of sperm during sexual arousal, also for urine excretion at other times

urethra

erectile tissue

PENIS
Organ of sexual intercourse

TESTIS
One of a pair of primary reproductive organs; packed with sperm-producing tubules and cells that secrete testosterone and other hormones

EJACULATORY DUCT
One of a pair of sperm-conducting ducts

SEMINAL VESICLE
One of a pair of glands that secrete fructose and prostaglandins, which become part of semen

BULBOURETHRAL GLAND
One of a pair of glands that secrete a lubricating mucus

VAS DEFERENS
One of a pair of ducts for rapid transport of sperm

EPIDIDYMIS
One of a pair of ducts in which sperm complete maturation; the portion farthest from testis stores mature sperm

anus

Figure 44.3 *Animated!* Components of the human male reproductive system and their functions.

Prostate gland secretions may also help buffer the acidic conditions in the female reproductive tract. The pH of vaginal fluid is about 3.5–4.0, but sperm swim more efficiently when the pH is 6.0. Finally, a pair of bulbourethral glands secrete mucus-rich fluid into the urethra during sexual arousal.

CANCERS OF THE PROSTATE AND TESTES

Prostate cancer is a leading cause of death among men, surpassed only by lung cancers. In 2004, more than 220,000 males were diagnosed in the United States, and about 30,000 died. In the same year, there were close to 9,000 cases of *testicular cancer*. In early stages, both cancers are painless. They might spread silently into lymph nodes of the abdomen, chest, neck, then lungs. If the cancer metastasizes, prospects are not good.

It is often said that if you examine the prostate of all the men who died of something else, you will find many who had prostate cancer and never knew it. But this might trivialize the risk. Some prostate cancers grow slowly, but other kinds now being detected in younger men often grow rapidly. Also bear in mind, heredity is a factor in prostate cancer. Other factors are advancing age, high-fat diets, exposure to toxic metals, tobacco smoking, and sedentary life-styles.

Doctors may detect prostate cancer by blood tests for increases in prostate-specific antigen (PSA) and by physical examinations. Adult males should examine their testes monthly—after a warm shower or bath—to check for enlargement, hardening, or new lumps. Treatment of testicular cancer has a high success rate; it is the easiest cancer to cure, provided it is caught before it spreads to other parts of the body.

Human males have a pair of testes. These gonads (primary reproductive organs) produce sperm. They also produce and secrete testosterone and other sex hormones.

In a duct leading out from each testis (an epidymis), sperm mature and are stored. Semen is a mix of mature sperm and noncellular secretions from a few accessory glands. During sexual arousal, it is propelled through a series of ducts. It is ejaculated from the last duct, the urethra.

44.2 Sperm Formation

LINKS TO
SECTIONS
9.1, 10.5, 36.3

A type of germ cell called a spermatogonium (plural, spermatogonia) gives rise to sperm in human males. Signaling pathways that connect the hypothalamus, pituitary gland, and testes control sperm formation.

FROM GERM CELLS TO MATURE SPERM

A testis is not even as big as a golfball, yet it contains two or three coiled tubules that would stretch out 125 meters, end to end. Hundreds of wedge-shaped lobes partition the interior (Figure 44.4). Pressed against the inner wall of each tubule are many spermatogonia. These undifferentiated diploid cells divide again and again, and their newest descendants force older ones away from the wall and into the interior of the tubule. The displaced, older cells are primary spermatocytes. They receive nutrients and signals from Sertoli cells, a type of supporting cell in the seminiferous tubules.

Primary spermatocytes enter meiosis while they are being displaced—but their cytoplasm does not quite divide. Thin cytoplasmic bridges keep them connected during the nuclear divisions (Figure 44.4c). Molecules and ions diffuse freely across the bridges and induce all cells of each generation to mature at the same time.

At the end of meiosis I, each cell that has formed is a secondary spermatocyte (Figure 44.4c). It is haploid, with chromosomes that are still duplicated. (Here you

may wish to review Section 10.5.) Sister chromatids of each chromosome move apart during meiosis II, after which spermatids form. These haploid daughter cells have unduplicated chromosomes. While they mature into sperm, the cytoplasmic bridges are broken.

A mature sperm is a flagellated cell. Its flagellum, or tail, has a core of microtubules, and the midpiece above it contains mitochondria that supply energy for the tail's whiplike motions (Figure 44.4d). The head is packed with DNA and has an acrosome, an enzyme-containing cap. The enzymes help a sperm penetrate an oocyte by partly digesting away its outer layer.

Sperm formation takes about 100 days, from start to finish. An adult male normally produces sperm on an ongoing basis, so that many millions of cells are in different stages of development on any given day.

THE SIGNALING PATHWAYS

Four hormones—testosterone, LH, FSH, and GnRH—are part of the signaling pathways that control sperm formation. **Testosterone**, a steroid hormone, governs the structure and function of the male reproductive tract as well as the formation of sperm. It induces the development of male secondary sexual traits. It also promotes sexual and aggressive behavior. Leydig cells inside the testes secrete testosterone.

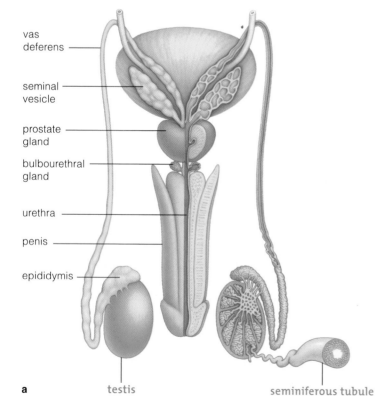

vas
deferens

seminal
vesicle

prostate
gland

bulbourethral
gland

urethra

penis

epididymis

a

testis   seminiferous tubule

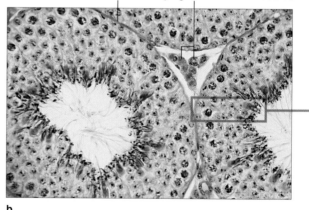

wall of seminiferous tubule   Leydig cells between tubules

b

Figure 44.4 *Animated!* (**a**) Male reproductive tract, posterior view.

Sperm formation. (**b**) Light micrograph of cells in three adjacent seminiferous tubules, cross-section. Leydig cells occupy tissue spaces between tubules. (**c**) How sperm form, beginning with a diploid germ cell. (**d**) Structure of a mature sperm, the male gamete.

As you read in Section 36.3, the anterior lobe of the pituitary gland makes and secretes **LH** and **FSH**. (The names are abbreviations for *Luteinizing Hormone* and *Follicle-Stimulating Hormone*, which actually refer to their effects in ovaries, which were discovered first. It later became clear that males have them, too.)

The hypothalamus controls the secretion of all three hormones (Figure 44.5). It secretes **GnRH** when the blood levels of testosterone and other factors are low. This releasing hormone makes the anterior lobe step up its LH and FSH secretions. In turn, LH stimulates Leydig cells to secrete testosterone, which helps sperm form and mature. At puberty, the FSH binds to Sertoli cells and jump-starts sperm formation.

Figure 44.5 also shows how feedback loops to the hypothalamus can slow down testosterone secretion and sperm formation. An elevated testosterone level in blood slows the release of GnRH. When the sperm count is high, Sertoli cells release inhibin. This protein hormone induces the hypothalamus and the pituitary to decrease GnRH and FSH secretion.

Sperm formation depends on the hormones LH, FSH, and testosterone. Negative feedback loops from the testes to the hypothalamus and pituitary gland control their secretion.

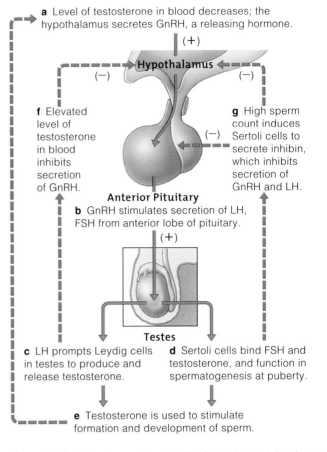

a Level of testosterone in blood decreases; the hypothalamus secretes GnRH, a releasing hormone.

f Elevated level of testosterone in blood inhibits secretion of GnRH.

g High sperm count induces Sertoli cells to secrete inhibin, which inhibits secretion of GnRH and LH.

Anterior Pituitary

b GnRH stimulates secretion of LH, FSH from anterior lobe of pituitary.

Testes

c LH prompts Leydig cells in testes to produce and release testosterone.

d Sertoli cells bind FSH and testosterone, and function in spermatogenesis at puberty.

e Testosterone is used to stimulate formation and development of sperm.

Figure 44.5 Signaling pathways in the formation and development of sperm. Negative feedback loops extend from the paired testes to the hypothalamus and the anterior lobe of the pituitary gland.

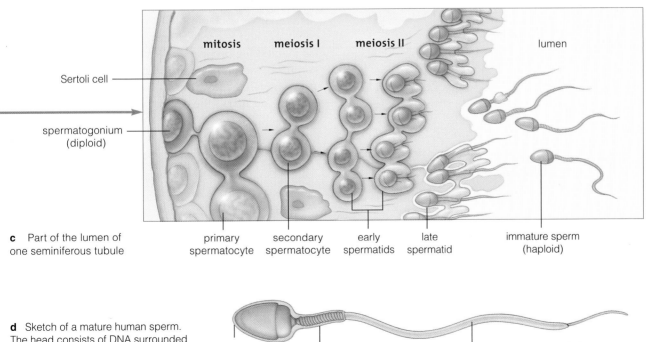

mitosis   meiosis I   meiosis II   lumen

Sertoli cell

spermatogonium (diploid)

c Part of the lumen of one seminiferous tubule

primary spermatocyte

secondary spermatocyte

early spermatids

late spermatid

immature sperm (haploid)

d Sketch of a mature human sperm. The head consists of DNA surrounded by the acrosome.

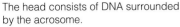

head   midpiece with mitochondria   tail, with its core of microtubules

44.3 Reproductive System of Human Females

LINKS TO
SECTIONS
12.9, 36.3, 43.3

The reproductive system of human females functions in the production of gametes and sex hormones. It also has a chamber in which a new, developing individual is protected and nourished until birth.

COMPONENTS OF THE SYSTEM

Figures 44.6 and 44.7 show the reproductive system of a human female, and Table 44.2 lists their functions. The gonads are a pair of **ovaries** that produce oocytes (immature eggs) and secrete sex hormones. An ovary releases oocytes on a cyclic basis to one of a pair of oviducts, also called Fallopian tubes. Oviducts lead to the **uterus**, a hollow, pear-shaped organ. Fertilization usually occurs and a blastocyst forms in an oviduct (Section 43.3). But the blastocyst tumbles on into the uterus, where embryonic development is completed.

A thick layer of smooth muscle, the myometrium, makes up most of the uterine wall. The uterine lining, or **endometrium**, consists of connective tissues, blood vessels, and glands. The cervix, or the narrowed-down part of the uterus, connects with a muscular tube, the vagina. Mucus-secreting epithelium lines the vagina, which extends from the cervix to the body's surface. The vagina receives sperm from the male and serves as part of the birth canal.

At the body's surface are genital organs of sexual stimulation. Outermost are the labia majora, a pair of skin folds padded with adipose tissue. They enclose the labia minora, a pair of smaller folds having many blood vessels but no adipose tissue. These folds partly enclose the clitoris, a female sex organ derived from the same embryonic tissue as the male penis. Like the penis, the clitoris also has an abundance of sensory receptors and is very sensitive to sexual stimulation. The urethra opens at the body surface about midway between the vaginal opening and the clitoris.

OVERVIEW OF THE MENSTRUAL CYCLE

Females of most mammalian species follow an *estrous* cycle, meaning they are fertile and "in heat" (sexually receptive to males) at only certain times in the cycle. Females of humans and some other primates follow a **menstrual cycle**. They are fertile intermittently, on a cyclic basis. Their fertile periods are not synchronized with sexual receptivity. Even though they can become pregnant at only certain times in the cycle, they may be receptive to sex at any time.

The next section explains this cycle, but here is a brief overview (Table 44.3). An oocyte matures inside an ovary and is released from it, and the endometrium is primed for pregnancy. When fertilization does not occur, blood and bits of the uterine lining flow from the vagina. The flow means "there's no embryo at this time," and marks the start of a new cycle.

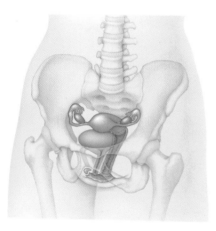

Figure 44.6 Location of the human female reproductive system relative to the pelvic girdle and urinary bladder.

Table 44.2	Organs of the Human Female Reproductive Tract
Ovaries	Oocyte production and maturation, sex hormone production
Oviducts	Ducts for conducting oocyte from ovary to uterus; fertilization normally occurs here
Uterus	Chamber in which new individual develops
Cervix	Secretion of mucus that enhances sperm movement into uterus and (after fertilization) reduces embryo's risk of bacterial infection
Vagina	Organ of sexual intercourse; birth canal

Table 44.3	Events of a Menstrual Cycle Lasting Twenty-Eight Days	
Phase	Events	Days of Cycle
Follicular phase	Menstruation; endometrium breaks down	1–5
	Follicle matures in ovary; endometrium rebuilds	6–13
Ovulation	Oocyte released from ovary	14
Luteal phase	Corpus luteum forms, secretes progesterone; the endometrium thickens and develops	15–28

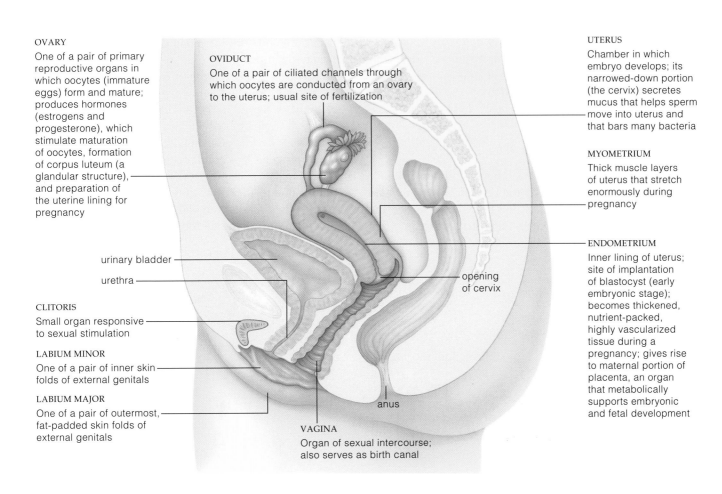

OVARY

One of a pair of primary reproductive organs in which oocytes (immature eggs) form and mature; produces hormones (estrogens and progesterone), which stimulate maturation of oocytes, formation of corpus luteum (a glandular structure), and preparation of the uterine lining for pregnancy

OVIDUCT

One of a pair of ciliated channels through which oocytes are conducted from an ovary to the uterus; usual site of fertilization

UTERUS

Chamber in which embryo develops; its narrowed-down portion (the cervix) secretes mucus that helps sperm move into uterus and that bars many bacteria

MYOMETRIUM

Thick muscle layers of uterus that stretch enormously during pregnancy

ENDOMETRIUM

Inner lining of uterus; site of implantation of blastocyst (early embryonic stage); becomes thickened, nutrient-packed, highly vascularized tissue during a pregnancy; gives rise to maternal portion of placenta, an organ that metabolically supports embryonic and fetal development

urinary bladder

urethra

CLITORIS

Small organ responsive to sexual stimulation

LABIUM MINOR

One of a pair of inner skin folds of external genitals

LABIUM MAJOR

One of a pair of outermost, fat-padded skin folds of external genitals

opening of cervix

anus

VAGINA

Organ of sexual intercourse; also serves as birth canal

Figure 44.7 *Animated!* Components of the human female reproductive system and their functions.

In the cycle's *follicular* phase, menstruation occurs, the uterine lining breaks down and starts rebuilding, and an oocyte starts maturing. At **ovulation**, which is the second phase, the ovary releases an oocyte. In the *luteal* phase, a glandular structure, the **corpus luteum**, forms in an ovary. Its hormonal secretions cause the endometrium to thicken in preparation for pregnancy.

Females now enter puberty between ages ten and sixteen. Pubic hair forms, fat is deposited in breasts and around the hips, and menstrual cycles start. Each cycle lasts for twenty-eight days, on average, but it is longer or shorter for some females. Cycles usually continue until the late forties or early fifties, when sex hormone secretions start to dwindle. The decline in secretions correlates with the onset of *menopause,* the twilight of a female's reproductive capacity.

The oocytes released late in a woman's life are more vulnerable to abnormal changes in the structure and number of chromosomes when meiosis resumes inside them. Down syndrome, explained in Section 12.9, is a prime example of the potential risks.

Ten million women in the United States may suffer *endometriosis,* a disorder caused by endometrial tissue that grew outside of the uterus. Hormones still act on cells in the mislocated tissue, so menstruation, sex, and urination are painful. Scar tissue forms on ovaries or oviducts and may cause infertility. Menstrual flow that backs up through oviducts and spills into the pelvic cavity may cause the disorder. Or it may be that a few embryonic cells ended up in the wrong tissue before birth and were stimulated to grow at puberty.

Ovaries, the primary reproductive organ of human females, produce immature eggs and sex hormones. Endometrium lines the uterus, a chamber in which embryos develop.

Estrogens and progesterone guide the cyclic growth and release of oocytes from the ovary as well as the breakdown and rebuilding of the endometrium, which depends on whether pregnancy occurs. These are the key events of menstrual cycles, which start at puberty.

44.4 Preparations for Pregnancy

LINKS TO
SECTIONS
12.5, 36.3

*Even before a human female is born, germ cells in her two ovaries enter meiosis, but the nuclear division process hits a wall in prophase I. Each ovarian cell that is arrested in prophase I is a **primary oocyte**. The only way it will finish meiosis is if a sperm fertilizes it much later in time.*

FROM PRIMARY TO SECONDARY OOCYTES

Every normal baby girl has about 2 million primary oocytes in her ovaries. By the time she is seven years old, she has about 300,000; her body has resorbed the rest. Beginning with her first menstrual cycle, meiosis resumes in one primary oocyte at a time. The primary oocyte, together with a layer of cells surrounds and nourishes it, is a primary follicle. Hormones stimulate the primary follicle to grow in size (Figure 44.8).

Eight to ten hours before being released from the ovary, the primary oocyte completes meiosis I, but its cytoplasm divides unevenly. One of the two haploid daughter cells, a **secondary oocyte**, receives most of the cytoplasm. The other cell is the first of three **polar bodies** that form by way of meiosis. It will degenerate later in the cycle. About 400 to 500 secondary oocytes will form during the reproductive years.

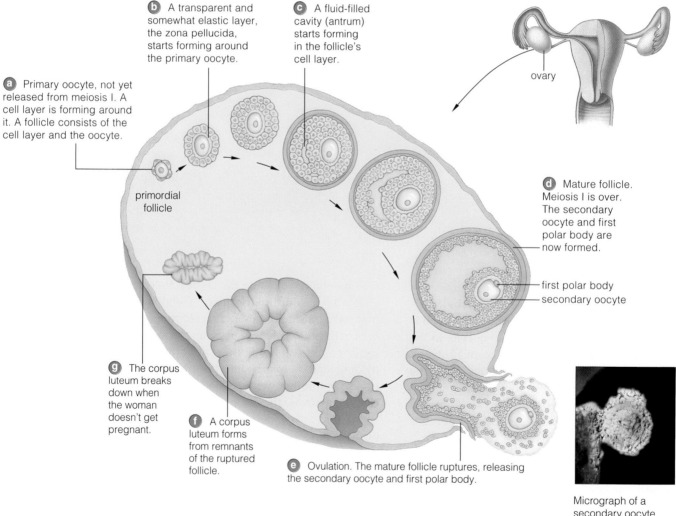

b A transparent and somewhat elastic layer, the zona pellucida, starts forming around the primary oocyte.

c A fluid-filled cavity (antrum) starts forming in the follicle's cell layer.

a Primary oocyte, not yet released from meiosis I. A cell layer is forming around it. A follicle consists of the cell layer and the oocyte.

primordial follicle

ovary

d Mature follicle. Meiosis I is over. The secondary oocyte and first polar body are now formed.

first polar body
secondary oocyte

g The corpus luteum breaks down when the woman doesn't get pregnant.

f A corpus luteum forms from remnants of the ruptured follicle.

e Ovulation. The mature follicle ruptures, releasing the secondary oocyte and first polar body.

Micrograph of a secondary oocyte escaping from the surface of an ovary

Figure 44.8 *Animated!* Cyclic events in a human ovary, cross-section. The follicle does not "move around" as in this diagram, which simply shows the *sequence* of events. All of these structures form in the same place during one menstrual cycle. In the cycle's first phase, a follicle grows and matures. At ovulation, the second phase, the mature follicle ruptures and releases a secondary oocyte. In the third phase, a corpus luteum forms from the follicle's remnants.

THE SIGNALING PATHWAYS

Four kinds of hormones—FSH, LH, estrogens, and progesterones—take part in signaling pathways that control the menstrual cycle. These pathways involve feedback loops from the ovaries to the hypothalamus and the anterior pituitary gland (Figure 44.9).

Estrogens are sex hormones. Remember, they help reproductive organs form in female embryos and they maintain secondary sexual traits (Section 12.5). Also, together with **progesterone**, they stimulate oocytes to mature and prime the endometrium for pregnancy.

Just as it does in males, the hypothalamus secretes GnRH, a releasing hormone that makes the anterior pituitary step up LH and FSH secretions. In females, however, these hormones stimulate the growth of the primary oocyte and the formation of more and more cells around it. Glycoprotein molecules are deposited beneath these cells. They form a noncellular layer that is called the zona pellucida (Figure 44.8b).

FSH and LH collect in fluid in the follicle and prod its cells to secrete estrogens, so the levels of estrogens in blood increase. About halfway through the cycle, the pituitary responds to the increases. It secretes LH in a brief pulse. The follicle swells in response, and its wall weakens and ruptures. Fluid—and the secondary oocyte—are released. *The midcycle surge of LH triggers ovulation, the release of a secondary oocyte from the ovary.*

Estrogens released early in the cycle also stimulate growth of the endometrium and its glands, which sets the stage for pregnancy. Before the midcycle surge of LH, the follicle cells were busy secreting progesterone and estrogens. Blood vessels grew fast in the thickened endometrium. At ovulation, estrogens prompted cells of the cervical canal to release a thin, clear mucus—an ideal medium for sperm to swim through.

The midcycle LH surge that brings about ovulation also stimulates cells in the ruptured follicle to form a corpus luteum. Progesterone and estrogens secreted by this structure cause the endometrium to thicken, in preparation for pregnancy.

WHAT IF NO PREGNANCY?

All the while, the hypothalamus has been preventing other follicles in the ovary from maturing, because it has been slowing FSH secretion. When a blastocyst does not burrow into the endometrium, however, the corpus luteum lasts no more than twelve days or so. During the last days of the menstrual cycle, it secretes a prostaglandin. This local signaling molecule induces cells of the corpus luteum to self-destruct.

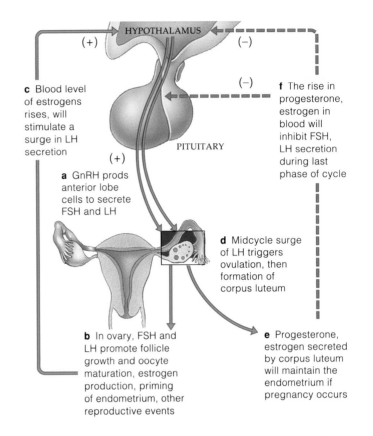

c Blood level of estrogens rises, will stimulate a surge in LH secretion

a GnRH prods anterior lobe cells to secrete FSH and LH

f The rise in progesterone, estrogen in blood will inhibit FSH, LH secretion during last phase of cycle

d Midcycle surge of LH triggers ovulation, then formation of corpus luteum

b In ovary, FSH and LH promote follicle growth and oocyte maturation, estrogen production, priming of endometrium, other reproductive events

e Progesterone, estrogen secreted by corpus luteum will maintain the endometrium if pregnancy occurs

Figure 44.9 Signaling pathways that control the menstrual cycle. A positive feedback loop from an ovary to the hypothalamus (*green*) triggers ovulation. After a secondary oocyte escapes, a negative feedback loop from the ovary to the hypothalamus and pituitary gland (*red*) inhibit hormone secretions and keep another follicle from maturing until the cycle is over.

Without the corpus luteum, levels of progesterone and estrogen decline fast, and the endometrium starts to break down. Deprived of oxygen and nutrients, its blood vessels constrict and tissues die. Blood escapes as weakened capillaries rupture. Blood and sloughed endometrial tissues form a menstrual flow that lasts three to six days. Afterward, rising levels of estrogens invite the repair and growth of the endometrium.

During a menstrual cycle, FSH and LH stimulate growth of an ovarian follicle. The primary oocyte undergoes the first meiotic cell division in the oocyte. The outcome is a secondary oocyte and the first polar body.

A midcycle surge of LH triggers ovulation—the release of the secondary oocyte and the polar body from the ovary.

Feedback loops to the hypothalamus and pituitary from the ovaries and, later, the corpus luteum control cyclic changes in the function of the ovary and in the structure of the uterine lining.

44.5 Visual Summary of the Menstrual Cycle

By now, you may have come to the conclusion that the human menstrual cycle is not a simple tune on a biological banjo. It's more like a full-blown hormonal symphony! Before continuing with your reading, take a moment to review Figure 44.10. It correlates cyclic changes in the ovary and the uterus with changing concentrations of hormones that trigger the events of each menstrual cycle.

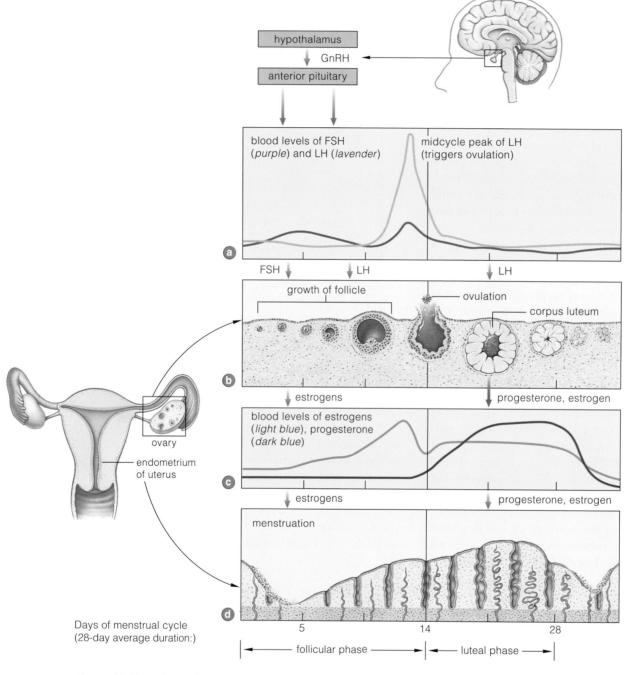

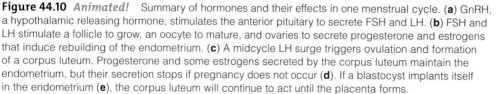

Figure 44.10 *Animated!* Summary of hormones and their effects in one menstrual cycle. (**a**) GnRH, a hypothalamic releasing hormone, stimulates the anterior pituitary to secrete FSH and LH. (**b**) FSH and LH stimulate a follicle to grow, an oocyte to mature, and ovaries to secrete progesterone and estrogens that induce rebuilding of the endometrium. (**c**) A midcycle LH surge triggers ovulation and formation of a corpus luteum. Progesterone and some estrogens secreted by the corpus luteum maintain the endometrium, but their secretion stops if pregnancy does not occur (**d**). If a blastocyst implants itself in the endometrium (**e**), the corpus luteum will continue to act until the placenta forms.

44.6 Pregnancy Happens

When a female and male engage in sexual intercourse, or coitus, the hormonal fog of the moment may obscure what can happen if a secondary oocyte is in an oviduct.

SEXUAL INTERCOURSE

The male sex act begins with an erection, whereby the penis stiffens and lengthens. The sex act culminates in ejaculation, the forceful expulsion of semen from the penis. As Figure 44.3 shows, the penis contains long cylinders of spongy tissue. The penis of an unaroused male stays limp because the large blood vessels that supply the spongy tissue are vasoconstricted. When a male becomes aroused, the vessels vasodilate; blood flow into the penis exceeds the amount of blood that is flowing out. Spongy tissue inside becomes engorged with blood. The penis stiffens and lengthens, which facilitates insertion into a female's vaginal canal.

Repetitive pelvic thrusting mechanically stimulates friction-activated receptors that abound at the tip of the penis. It also stimulates the female's clitoris and vaginal wall. In the male, the response is involuntary muscle contractions that force sperm-laden semen into the urethra, from which it is ejaculated.

Emotional intensity, hard breathing, strong heart pounding, and the contraction of skeletal muscles in general accompany a rhythmic throbbing of the pelvic muscles. During *orgasm*, the end of the sex act, strong sensations of physical release, warmth, and relaxation dominate. Similar sensations typify female orgasm.

You may have heard that a female will not become pregnant as long as she does not reach orgasm. Don't believe it.

FERTILIZATION

On average, an ejaculation can put 150 million to 350 million sperm in the vagina. Fertilization may occur if they arrive a few days before or after ovulation or any time in between. Less than thirty minutes after sperm arrive, contractions move them deep into the female's reproductive tract. A few hundred actually reach the upper portion of the oviduct, where eggs usually are fertilized (Figure 44.11).

Many sperm bind to the oocyte's zona pellucida. Binding triggers the release of enzymes from the cap over each sperm's head. Collectively, these digestive enzymes make a passage through the zona pellucida. Usually only one sperm enters the secondary oocyte. Only its nucleus and centrioles do not degenerate.

Remember, meiosis II was arrested in the primary oocyte. Now, upon sperm penetration, the secondary

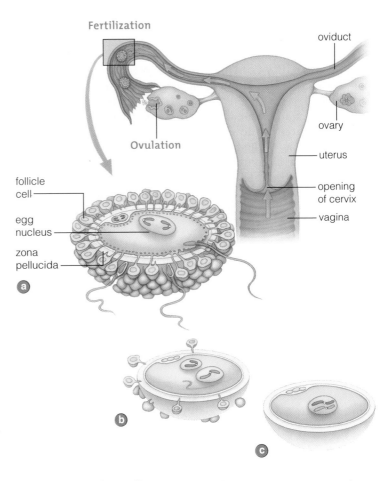

Figure 44.11 *Animated!* Fertilization. (**a**) Many human sperm travel rapidly from the vagina to an oviduct (*blue* arrows), where they surround a secondary oocyte. Digestive enzymes released from the cap of each sperm clear a path through the zona pellucida.

(**b**) A single sperm penetrates the secondary oocyte, which releases substances that make the zona pellucida impenetrable to the other sperm. Penetration also stimulates meiosis II of the oocyte's nucleus. (**c**) The sperm's tail degenerates; its nucleus enlarges and fuses with the nucleus of its target. Fertilization is over; a zygote has formed.

oocyte and the first polar body both complete meiosis II and cytoplasmic division. Now there is one mature egg—an **ovum** (plural, ova)—and three polar bodies. The egg nucleus fuses with the sperm nucleus (Figure 44.11*b*). Collectively, the chromosomes of both nuclei restore the diploid number for a brand new zygote.

> *The intense physiological events that accompany coitus put sperm on a collision course with an egg.*
>
> *Fertilization is over when a sperm nucleus and egg nucleus fuse. The diploid zygote that forms is the start of a new individual.*

44.7 Preventing or Seeking Pregnancy

LINKS TO
SECTIONS
1.7, 34.13

What options are available to those who decide to postpone, forgo, or seek pregnancy?

Fertility Control Options Let us start with the most effective way to avoid pregnancy—complete *abstinence*, or no sex at all. It takes great self-discipline to override the neural and hormonal sense of urgency associated with sex. That sexual drive, or libido, arises through interactions among the limbic system, hypothalamus, and other brain centers. It starts when the sex hormone floodgates open at puberty and peaks during the teens. The difficulty for teenagers is that the limbic system develops faster than the prefrontal cortex, through which self-discipline is exerted (Section 34.13).

Different versions of the *rhythm method* are forms of abstinence; a female simply avoids sex in her fertile period. She calculates when she is fertile by recording how long her menstrual cycles last, by taking her core temperature each morning, or both. The method is inexpensive; it really costs nothing after you buy a thermometer. It does not require fittings or periodic medical checkups. Its practitioners do run a risk of pregnancy (Figure 44.12). Miscalculations are frequent. Sperm deposited in the vagina a few days before ovulation may live long enough to meet up with an egg.

Withdrawal, or removing the penis from the vagina before ejaculation, dates at least to biblical times. It requires great willpower and still may fail, because fluid released from the penis before ejaculation can contain some sperm.

Douching, or chemically rinsing the vagina right after intercourse, is too chancy. Sperm travel out of reach of the douche ninety seconds after ejaculation. Frequent douching also irritates the reproductive tract.

Controlling fertility by surgical intervention is less chancy. Males who do not want children may opt for a *vasectomy,* a procedure that requires a local anesthetic. A doctor makes a small incision in the scrotum, then cuts and ties off each vas deferens. Sperm no longer can move out of the testes and become part of semen.

A vasectomy can be reversed surgically. However, so far, only about 60 percent of those who have reversed the surgery have been able to father a child.

Tubal ligation nearly always guarantees permanent infertility. The oviducts are cauterized or cut and tied. This procedure is now more common than vasectomy. Surgical reversal is about 70 percent successful.

Less drastic fertility control methods are based on physical and chemical barriers that stop sperm from reaching an egg. *Spermicidal foam* and *spermicidal jelly* poison sperm. An applicator is used to insert either one into the vagina before sex. These products are not always reliable, but using them with a diaphragm or a condom makes them more effective.

A *diaphragm* is a flexible, dome-shaped device. It is inserted into the vagina and positioned so that it covers

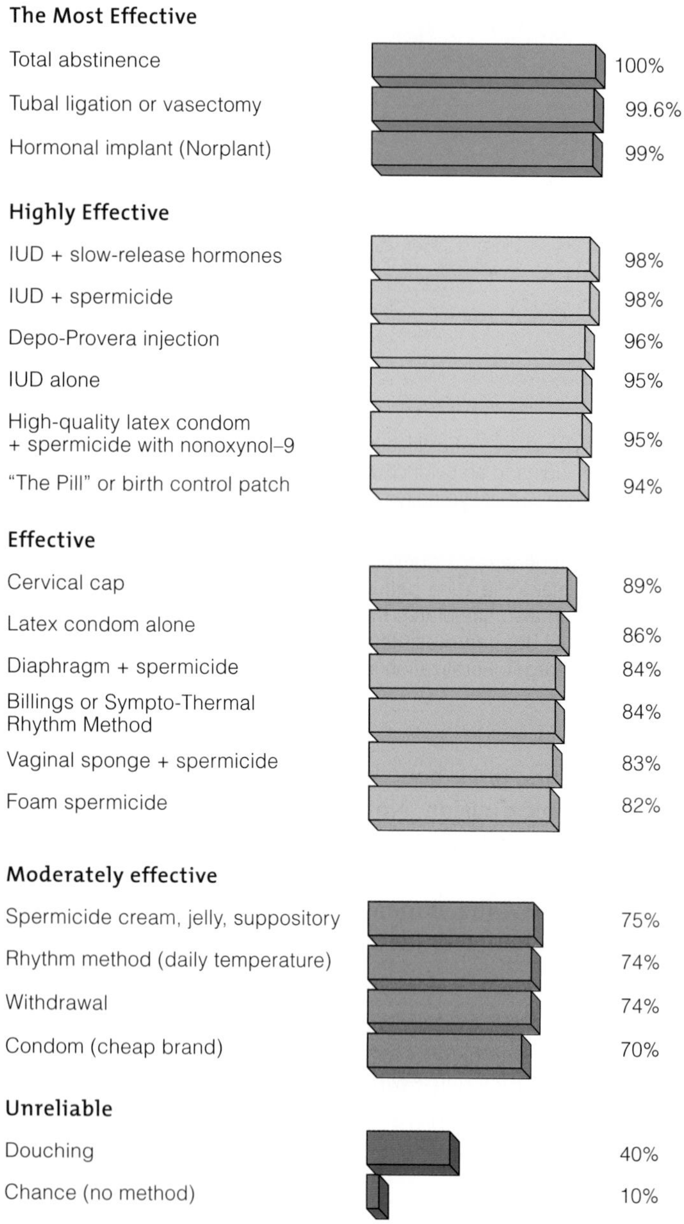

The Most Effective

Total abstinence	100%
Tubal ligation or vasectomy	99.6%
Hormonal implant (Norplant)	99%

Highly Effective

IUD + slow-release hormones	98%
IUD + spermicide	98%
Depo-Provera injection	96%
IUD alone	95%
High-quality latex condom + spermicide with nonoxynol–9	95%
"The Pill" or birth control patch	94%

Effective

Cervical cap	89%
Latex condom alone	86%
Diaphragm + spermicide	84%
Billings or Sympto-Thermal Rhythm Method	84%
Vaginal sponge + spermicide	83%
Foam spermicide	82%

Moderately effective

Spermicide cream, jelly, suppository	75%
Rhythm method (daily temperature)	74%
Withdrawal	74%
Condom (cheap brand)	70%

Unreliable

Douching	40%
Chance (no method)	10%

Figure 44.12 Comparison of the effectiveness of some methods of contraception. These percentages also indicate the number of unplanned pregnancies per 100 couples who use only that method of birth control for a year. For example, "94% effectiveness" for oral contraceptives means that 6 of every 100 females will still become pregnant, on average.

the cervix before intercourse. As Figure 44.12 indicates, a diaphragm is relatively effective when first fitted by a doctor, used in conjunction with a spermicidal foam or jelly, inserted correctly each time, and left in place for a prescribed length of time.

Condoms are thin, tight-fitting sheaths worn over the penis during intercourse. Good brands may be as much as 95 percent effective when used with a spermicide. Only condoms made of latex afford protection against sexually transmitted diseases. However, even the best ones can tear or leak, at which time they become useless.

A *birth control pill* delivers synthetic estrogens and progesterone-like hormones that block maturation of oocytes and ovulation. "The Pill" reduces menstrual cramps but can cause nausea, headaches, and weight gain. With at least 50 million users, it is the most common fertility control method in the United States. When used correctly, it is at least 94 percent effective. Its use lowers the risk of ovarian cancer but increases the risk of breast, cervical, and liver cancers.

A *birth control patch* is a small, flat adhesive patch applied to skin once a week for three weeks per month. The fourth week, when the menstrual period starts, is patch-free. The patch delivers the same hormones as an oral contraceptive and blocks ovulation the same way. Like birth control pills, it is not for everyone. Some women, especially smokers, develop life-threatening blood clots and other serious cardiovascular disorders.

Progestin injections or implants block ovulation. One Depo-Provera injection is effective for three months. Norplant works for five years. Both methods are quite effective, but they may cause sporadic, heavy bleeding. Also, removing Norplant rods can be tricky.

Some women use *morning-after pills* after a condom tears, or after unprotected consensual sex or rape. One brand, Previn, is a set of two pills with hormones that suppress ovulation and block corpus luteum secretions. Morning-after pills work best when taken early but are somewhat effective up to five days after intercourse.

Regarding Abortion

The methods just outlined are interventions in *fertility*, although they commonly are referred to as methods of birth control. They are meant to stop pregnancy from happening in the first place. By contrast, *abortion* is a deliberate intervention after pregnancy is under way. The word actually refers to the spontaneous as well as an induced dislodging and removal of an embryo or fetus from the uterus.

From a clinical standpoint, induced abortion usually is a rapid, relatively painless procedure that is free of complications when performed in the first three months after fertilization. The drug mifepristone (RU 486) and a prostaglandin can induce an abortion during the first nine weeks of pregnancy. Both substances bind to and block progesterone receptors in the uterus. The uterine lining, hence pregnancy, cannot be maintained.

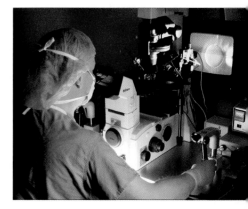

Figure 44.13 A glimpse into in vitro fertilization (IVF). A micromanipulator is used to insert a human sperm into an oocyte. The doctor is guided with the help of an image displayed on a video screen of a monitor that is attached to a microscope.

For both medical and moral reasons, the majority of people in the United States view sexually responsible behavior as being preferable to an abortion. Aborting a late-term fetus is highly controversial unless the mother's life is threatened.

Bear in mind, this textbook cannot offer you the "right" answer to a question about the morality of abortion or some other option, for reasons given in Section 1.7. It can only offer a serious explanation of how a new individual develops to help you objectively assess the biological basis of human life. Your choice of how to answer a question of what is "right" will be just that—your choice.

In Vitro Fertilization

Approximately 15 percent of all couples in the United States cannot have children because of sterility or infertility. In some of the cases, hormonal imbalances stop the female from ovulating. In other cases, the male's sperm count is so low that fertilization is next to impossible.

When a couple can make normal sperm and oocytes, they sometimes seek out *in vitro fertilization*. This medical intervention promotes conception outside the body. (In vitro literally means "in glass" petri dishes or test tubes.) First the female receives injections of a hormone that stimulates oocytes into maturing. Before an oocyte can be released from an ovary, it is withdrawn and placed in a solution that simulates fluid inside the female's oviducts. Then a doctor attempts to inject a sperm into it (Figure 44.13).

When the attempt is successful, cleavage produces a tiny cluster of cells within a few days. The cluster is then transferred to the female's uterus for development.

At this writing, each attempt at in vitro fertilization costs an average of 12,000 to 17,000 dollars. When infertility (not sterility) is the problem, attempts usually fail.

Each liveborn "test-tube" baby costs health care systems between 60,000 and 100,000 dollars. A childless couple may believe no cost is too great. But many people question the cost to society in an era of increased population growth and shrinking medical coverage. Another concern is the fate of the embryonic cell clusters that are created for IVF but never used. Court battles are being waged over this issue.

44.8 Sexually Transmitted Diseases

LINKS TO
SECTIONS
21.4, 22.2, 39.10

Unprotected sex exposes you to potential infection by any pathogens that your partner unknowingly may have picked up from a previous sexual partner.

CONSEQUENCES OF INFECTION

Each year, pathogens that cause **sexually transmitted diseases**, or STDs, infect about 15 million people in the United States (Table 44.4). Two-thirds of those infected are under age twenty-five. One-quarter are teenagers. Over 65 million Americans now live with an incurable STD. Treating STDs and secondary complications costs a staggering 8.4 billion dollars in an average year.

The social consequences are enormous. Females are more easily infected than males, and they develop more complications. For instance, pelvic inflammatory disease (PID), a secondary outcome of some bacterial STDs, affects about 1 million females annually. It scars the reproductive tract and can cause infertility, tubal pregnancies, and chronic pain (Figure 44.14a). Some fetuses acquire STDs before or during birth, then abort on their own or develop abnormally. Females commonly transmit the bacterial agent of chlamydia to newborns (Figure 44.14b). Type II *Herpes* virus kills 50 percent of the fetuses it infects and causes neural defects in 25 percent of the survivors.

Table 44.4	Estimated New STD Cases Per Year *	
STD	U.S. Cases	Global Cases
HPV infection	5,500,000	20,000,000
Trichomoniasis	5,000,000	174,000,000
Chlamydia	3,000,000	92,000,000
Genital herpes	1,000,000	20,000,000
Gonorrhea	650,000	62,000,000
Syphilis	70,000	12,000,000
AIDS	40,000	4,900,000

* Global data on HPV and genital herpes were last compiled in 1997.

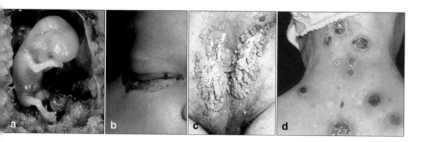

Figure 44.14 A few downsides of unsafe sex. (**a**) Tubal pregnancy. Scarring from STDs makes an embryo implant itself in an oviduct, not the uterus. Untreated tubal pregnancies can rupture an oviduct and cause bleeding, infection, and death. (**b**) One sign of a chlamydial infection transferred from a mother to her infant. (**c**) Genital warts. (**d**) Chancres typical of secondary syphilis.

MAJOR AGENTS OF STDS

HPV   Infection by human papillomaviruses (HPV) is the most widespread and fastest growing STD in the United States. At least 20 million are already infected. Of about 100 HPV strains, a few cause *genital warts*. These bumpy growths form on the vagina, cervix, and external genitals, and around the anus (Figure 44.14c). In males, they form on the penis and scrotum. Two strains, HPV 16 and HPV 18, cause *cervical cancer*. Sexually active females should have an annual pap smear to check for any cervical changes.

Trichomoniasis   *Trichomonas vaginalis*, a flagellated protozoan, causes the disease *trichomoniasis* (Section 22.2). Symptoms include vaginal soreness, itching, and a yellowish discharge. Infected males are usually symptom-free. Untreated infections damage the urinary tract, cause infertility, and invite HIV infection. A single dose of an antiprotozoal drug quickly cures an infection. To prevent reinfection, both sexual partners must be treated.

Chlamydia   *Chlamydial infection* is primarily a young person's disease. Forty percent of those infected are between ages fifteen and nineteen; 1 in 10 sexually active teenage girls is infected. *Chlamydia trachomatis* causes the disease (Section 21.4). Antibiotics can quickly kill this bacterium. Most infected females are undiagnosed; they have no symptoms. Between 10 and 40 percent of those who are untreated will develop PID. In about 50 percent of infected males, symptoms include abnormal discharges from the penis and painful urination. Untreated males risk inflammation of the epididymes and infertility.

Genital Herpes   About 45 million Americans have genital herpes, caused by type II *Herpes simplex* virus. Transmission to new hosts requires direct contact with active *Herpes* viruses or with sores that contain them. Mucous membranes of the mouth and genitals are vulnerable. Early symptoms are often mild or absent. Painful, small blisters may form on the vulva, cervix, urethra, or anal tissues of infected females. Blisters form on the penis and anal tissues of infected males. Within three weeks, the virus enters latency. Sores crust over and heal, but viral particles are hidden in the body.

The virus is reactivated sporadically, which causes painful sores at or near the original site of infection. Sexual intercourse, menstruation, emotional stress, or other infections trigger flare-ups. One antiviral drug, Acyclovir, decreases healing time and often the pain.

Gonorrhea   The STD *gonorrhea* is caused by *Neisseria gonorrhoeae* (Figure 44.15b). This bacterium commonly crosses mucous membranes of the urethra, cervix, or anal canal during sexual intercourse. An infected female may notice a slight vaginal discharge or burning sensation while urinating. If the bacterium enters her oviducts, it

Figure 44.15 Light micrographs of bacteria that cause (**a**) chlamydia, (**b**) gonorrhea, and (**c**) syphilis.

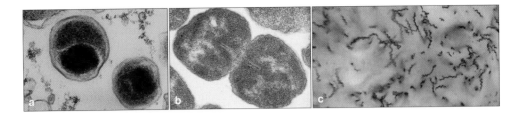

may cause cramps, fever, vomiting, and scarring that can result in sterility. Less than a week after a male is infected, yellow pus oozes from the penis. Urination becomes more frequent and may also be painful.

Prompt treatment with antibiotics quickly cures this disease, yet it still is rampant. Many females ignore early symptoms. Also, people wrongly believe infection confers immunity. Someone can contract gonorrhea over and over again, probably because there are at least sixteen strains of *N. gonorrhoeae*. Also, use of oral contraceptives invites infection by altering vaginal pH. Populations of resident bacteria decline, so *N. gonorrhoeae* is free to move in.

Syphilis The spirochete *Treponema pallidum* causes *syphilis*, a dangerous STD (Section 21.4 and Figure 44.15c). Having sex with an infected partner puts this bacterium on the surface of genitals or into the cervix, vagina, or oral cavity. *T. pallidum* also slips into the body through tiny cuts. One to eight weeks later, treponemes are twisting about in a flattened, painless chancre, or localized ulcer.

This chancre is a sign of the primary stage of syphilis. It usually heals, but treponemes multiply in the spinal cord, brain, eyes, bones, joints, and mucous membranes. In an infectious secondary stage, a skin rash develops and more chancres form (Figure 44.14d). In about 25 percent of the cases, immune responses succeed and symptoms subside. Another 25 percent are symptom-free. In the remainder, lesions and scars appear in the skin and liver, bones, and other organs. Few treponemes form in this tertiary stage, but a host's immune system is hypersensitive to them. Chronic immune reactions may damage the brain and spinal cord and cause paralysis.

Possibly because the symptoms are so alarming, more people seek early treatment for syphilis than they do for gonorrhea. Later stages require prolonged treatment.

AIDS As you read earlier in Section 39.10, an infection by HIV, the human immunodeficiency virus, leads to *AIDS* —Acquired *I*mmune *D*eficiency Syndrome. The immune system almost always loses the battle with HIV; this is an incurable STD. There may be no outward symptoms at first. Five to ten years later, a set of chronic disorders develops. The immune system weakens, which opens the door to opportunistic infectious agents. Normally harmless bacteria already living in and on the body are the first to take advantage of lowered resistance. Then dangerous pathogens take their toll. Eventually they overwhelm the compromised immune system.

Most often, HIV spreads by way of anal, vaginal, and oral intercourse and intravenous drug use. Virus particles in blood, semen, urine, or vaginal secretions enter a new host through cuts and abrasions in the epithelial lining of the penis, vagina, rectum, or oral cavity.

Free or low-cost, confidential testing for HIV exposure is available at public health facilities. It takes a few weeks to six months or more before the body forms detectable amounts of antibodies in response to the first exposure. Anyone who tests positive for HIV can spread the virus.

Public education may help slow the spread of HIV (Figure 44.16). Most health care workers advocate safe sex, although there is confusion over what "safe" means. The use of high-quality latex condoms together with a nonoxynol–9 spermicide helps prevent viral transmission. However, as mentioned in the preceding section, this practice still carries a slight risk. Open-mouth kissing with an HIV-positive individual carries a risk. Caressing is not risky if there are no lesions or cuts where HIV-laden body fluids can enter the body. Skin lesions caused by any other sexually transmitted disease are vulnerable points of entry for the virus.

New, costly drug therapies are prolonging some lives. In the late 1990s, the rate of infection started to climb again, possibly because of a misperception that AIDS is no longer a deadly threat. But AIDS kills, and the viral agent keeps on mutating. How long today's drugs can keep a lid on deaths is anybody's guess.

Figure 44.16 NBA legend Magic Johnson, one of the torch bearers of the 2002 Winter Olympics. He was diagnosed as HIV positive in 1991. He contracted the virus through heterosexual sex, and credits his survival to AIDS drugs and informed medical care. He continues to campaign to educate others about AIDS.

44.9 Formation of the Early Embryo

LINKS TO
SECTIONS
26.6, 26.9, 43.3

Nine months or so after the time of fertilization, a female gives birth. Besides taking longer to develop inside its mother, the new individual will require more intense care for a much longer time compared to other primates.

Pregnancy lasts an average of thirty-eight weeks from the time of fertilization. It takes about one week for a blastocyst to form. All major organs form during the *embryonic* period—the third to the end of the eighth week of pregnancy. When this period ends, the new individual is called a **fetus**. It has distinctly human features. In the *fetal* period, from the start of the ninth week until birth, organs grow and become specialized. We often refer to the first three months of pregnancy as the first trimester. The second trimester extends from the start of the fourth month to the end of the sixth, and the third trimester ends at birth.

CLEAVAGE AND IMPLANTATION

Three to four days after fertilization, the zygote has already started cleavage. Some of its genes are being expressed, because the early cuts need their products (Section 14.3). At the eight-cell stage, the cells huddle into a ball. A human blastocyst forms by the fifth day. It consists of a trophoblast (an outer layer of cells), a cavity filled with their secretions (a blastocoel), and an inner cell mass (Figure 44.17*d*). One or two days later, **implantation** is under way. The blastocyst adheres to the uterine lining. It invades the mother's tissues and forms connections that will metabolically support the pregnancy. By now, the inner cell mass has become two flattened layers of cells in the shape of a disk. This embryonic disk will give rise to the embryo proper.

EXTRAEMBRYONIC MEMBRANES

As implantation continues, membranes start to form outside the embryo. First, a fluid-filled *amniotic* cavity opens up between the embryonic disk and part of the blastocyst surface (Figure 44.17*f*). Many cells migrate around the wall of the cavity and form the **amnion**, a membrane that will enclose the embryo. Fluid in the cavity will function as a buoyant cradle in which an embryo can grow, move freely, and be protected from abrupt temperature changes and mechanical impacts.

As the amnion forms, other cells migrate around the inner wall of the blastocyst, forming a lining that

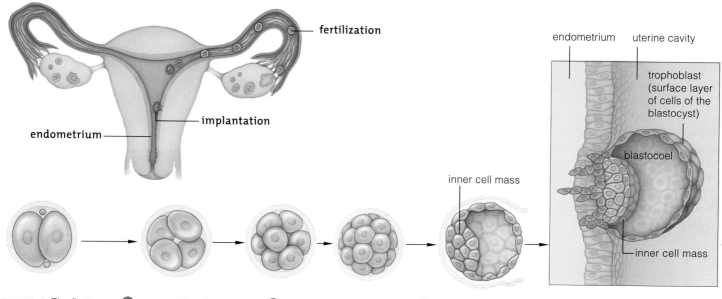

a **DAYS 1–2**. The first cleavage furrow extends between the two polar bodies. Later cuts are angled, so cells become asymmetrically arranged. Until the eight-cell stage forms, they are loosely organized, with space between them.

b **DAY 3**. After the third cleavage, cells abruptly huddle into a compacted ball, which tight junctions among the outer cells stabilize. Gap junctions formed along the interior cells enhance intercellular communication.

c **DAY 4**. By 96 hours there is a ball of sixteen to thirty-two cells shaped like a mulberry. It is a morula (after *morum*, Latin for mulberry). Cells of the surface layer will function in implantation and will give rise to a membrane, the chorion.

d **DAY 5**. A blastocoel (fluid-filled cavity) forms in the morula as a result of surface cell secretions. By the thirty-two-cell stage, differentiation is occurring in an inner cell mass that will give rise to the embryo proper. This embryonic stage is the blastocyst.

e **DAYS 6–7**. Some of the blastocyst's surface cells attach themselves to the endometrium and start to burrow into it. Implantation has started.

actual size

becomes a **yolk sac**. This extraembryonic membrane speaks of the evolution of land vertebrates. For most animals that produce shelled eggs, the yolk sac holds nutritive yolk. In humans, one portion of the yolk sac becomes a site of blood cell formation. Another will give rise to germ cells, the forerunners of gametes.

Before a blastocyst is fully implanted, spaces open in maternal tissues and become filled with blood that seeps in from ruptured capillaries. In the blastocyst, a new cavity opens up around the amnion and yolk sac. The lining of the cavity becomes the **chorion**, a third membrane that balloons like fingers of many rubber gloves into maternal tissues. It will become part of the spongy, blood-engorged tissue called the placenta.

After the blastocyst is implanted, an outpouching of the yolk sac will become the fourth extraembryonic membrane—the **allantois**. The allantois has different roles in different groups. Among reptiles, birds, and some mammals, it serves in respiration and in storing metabolic wastes. In humans, the urinary bladder and blood vessels for a placenta form from it (Table 44.5).

The blastocyst itself stops menstruation. Cells of the blastocyst secrete a hormone called human chorionic gonadotropin, or HCG. This hormone stimulates the corpus luteum to continue secreting progesterone and estrogens (Figure 44.10). It does so until the placenta

takes over secretion of HCG about eleven weeks later. By the start of the third week, HCG may be detected in samplings of the mother's blood or urine. At-home *pregnancy tests* have a treated "dipstick" that changes color when urine contains this pregnancy hormone.

Table 44.5	Human Extraembryonic Membranes
Amnion	Encloses, protects embryo in a fluid-filled, buoyant cavity
Yolk sac	Becomes site of red blood cell formation; germ cell source
Chorion	Lines amnion and yolk sac, becomes part of placenta
Allantois	Source of urinary bladder and blood vessels for placenta

Cleavage of the human zygote produces a cluster of cells that develops into the blastocyst that implants itself in the endometrium six or seven days after fertilization.

Projections from the blastocyst's surface invade maternal tissues, and connections start to form that in time will metabolically support the developing embryo.

Some parts of the blastocyst give rise to an amnion, yolk sac, chorion, and allantois. These extraembryonic membranes serve different functions. Together they are vital for the structural and functional development of the embryo.

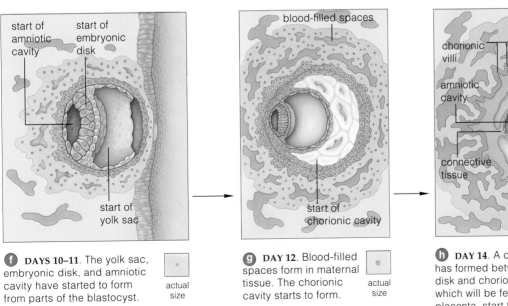

f **DAYS 10–11.** The yolk sac, embryonic disk, and amniotic cavity have started to form from parts of the blastocyst.   actual size

g **DAY 12.** Blood-filled spaces form in maternal tissue. The chorionic cavity starts to form.   actual size

h **DAY 14.** A connecting stalk has formed between the embryonic disk and chorion. Chorionic villi, which will be features of a placenta, start to form.   actual size

Figure 44.17 *Animated!* From fertilization through implantation. A blastocyst forms, and its inner cell mass will give rise to a disk-shaped early embryo. Three extraembryonic membranes (the amnion, chorion, and yolk sac) start forming. A fourth membrane (allantois) forms after the blastocyst is implanted.

44.10 Emergence of the Vertebrate Body Plan

LINKS TO
SECTIONS 15.1,
15.3, 17.8, 43.2, 43.4

By the time a female misses a first menstrual period after fertilization, cleavage is over. Gastrulation is under way.

Gastrulation, recall, is the stage when cell divisions, migrations, and rearrangements give rise to primary tissue layers (Section 43.4). The inner cell mass has been developing much as it did in reptilian ancestors of mammals. But the blastomeres are now arranged as a flattened, two-layered embryonic disk. That disk is reminiscent of the one that forms in the yolky eggs of living reptiles and birds (Figure 44.18a).

By now, the embryonic disk is surrounded by the amnion and chorion—except where a stalk joins it to the chorion wall. The yolk sac lining has formed from one of the two layers. The other layer now starts to become the embryo proper. A depression appears on the disk and its sides begin to thicken. Figure 44.18a shows this *primitive streak*. The next day, it lengthens and thickens more. Its appearance marks the onset of gastrulation. It defines the anterior–posterior axis and, in time, the bilateral symmetry of the embryo.

Endoderm and mesoderm form from cells that are migrating inward along this axis. Pattern formation starts. Embryonic inductions and interactions among classes of master genes map out the basic body plan, as they do in all vertebrates. Tissues and organs form in orderly steps according to predictable patterns.

For example, by the eighteenth day, the embryonic disk has two folds that will merge into a neural tube, the start of the spinal cord and brain (Figure 44.18b). Some mesoderm folds into a tube that develops into a notochord. The vertebrate notochord is no more than a structural model; bony segments form on it. In *spina bifida*, the neural tube, and one or more vertebrae do not form properly, and the spinal cord, its coverings, or both may protrude from the vertebral column.

Toward the end of the third week, multiple paired segments form from some mesoderm. These **somites** are embryonic sources of most bones, skeletal muscles of the head and trunk, and the dermis overlying these body parts (Section 43.2). Pharyngeal arches start to form; they will contribute to the pharynx, larynx, and the face, neck, mouth, and nose (Figure 44.18c). Small spaces open up in certain parts of the mesoderm. In time, they will interconnect as a coelomic cavity.

> *The basic vertebrate body plan emerges early in the development of the new individual.*
>
> *A primitive streak, neural tube, somites, and pharyngeal arches form during the embryonic period of all vertebrates. Formation of the primitive streak establishes the body's anterior–posterior axis and its bilateral symmetry.*

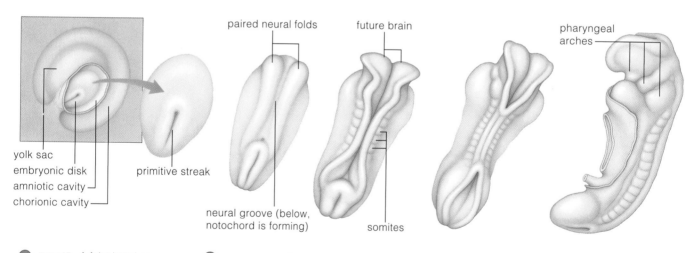

a **DAY 15.** A faint band appears around a depression along the axis of the embryonic disk. This is the primitive streak, and it marks the onset of gastrulation in vertebrate embryos.

b **DAYS 18–23.** Organs start to form through cell divisions, cell migrations, tissue folding, and other events of morphogenesis. Neural folds will merge to form the neural tube. Somites (bumps of mesoderm) appear near the embryo's dorsal surface. They will give rise to most of the skeleton's axial portion, skeletal muscles, and much of the dermis.

c **DAYS 24–25.** By now, some embryonic cells have given rise to pharyngeal arches. These will contribute to the formation of the face, neck, mouth, nasal cavities, larynx, and pharynx.

Figure 44.18 Hallmarks of the embryonic period of humans and other vertebrates. A primitive streak and then a notochord form. Neural folds, somites, and pharyngeal arches form later. (**a,b**) Dorsal views of the embryo's back. (**c**) Side view.

44.11 Why Is the Placenta So Important?

Even before the embryonic period starts, the uterus has been interacting with extraembryonic membranes in ways that will sustain the embryo's rapid growth.

By the third week, tiny fingerlike projections from the chorion have grown into the maternal blood that has pooled in the endometrial spaces. These projections are chorionic villi. They enhance the rate of exchange of substances between the mother and the embryo. The villi are functional components of the placenta.

The **placenta** is a blood-engorged organ composed of the uterine lining and extraembryonic membranes. At full term, it will make up about one-fourth of the inner surface of the uterus (Figure 44.19).

A placenta is the body's way of sustaining the new individual while allowing its blood vessels to develop separately from the mother's blood vessels. Oxygen and vital nutrients diffuse out of the maternal blood vessels, across the placenta's blood-filled spaces, then into embryonic blood vessels. The vessels converge in an umbilical cord, the lifeline between the placenta and the new individual. Carbon dioxide and other wastes diffuse in the other direction. The mother's lungs and kidneys dispose of the wastes (Section 40.5).

After the third month, the placenta itself takes over the task of maintaining the uterine lining. It starts to secrete the sex hormones progesterone and estrogens.

LINK TO SECTION 40.5

> The placenta is a blood-engorged organ of endometrial and extraembryonic membranes. It allows the new individual to take up oxygen and nutrients from the mother and give up wastes to her. It does so while allowing embryonic blood vessels to develop separately from the mother's.

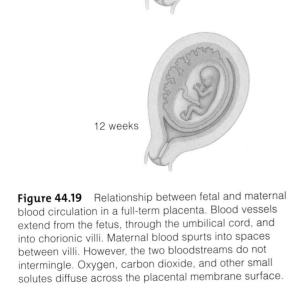

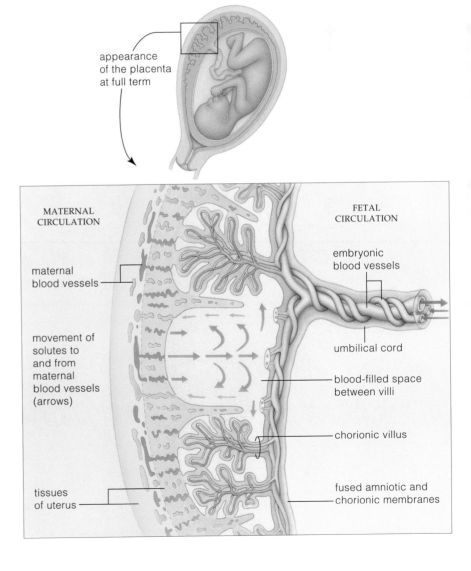

appearance of the placenta at full term

MATERNAL CIRCULATION

FETAL CIRCULATION

maternal blood vessels

movement of solutes to and from maternal blood vessels (arrows)

embryonic blood vessels

umbilical cord

blood-filled space between villi

chorionic villus

tissues of uterus

fused amniotic and chorionic membranes

4 weeks

8 weeks

12 weeks

Figure 44.19 Relationship between fetal and maternal blood circulation in a full-term placenta. Blood vessels extend from the fetus, through the umbilical cord, and into chorionic villi. Maternal blood spurts into spaces between villi. However, the two bloodstreams do not intermingle. Oxygen, carbon dioxide, and other small solutes diffuse across the placental membrane surface.

44.12 Emergence of Distinctly Human Features

LINKS TO
SECTIONS
9.4, 12.5, 28.5

Early on, a human embryo—with its gill arches and long tail—has a distinctly vertebrate appearance. The tail soon disappears, and by the beginning of the fetal period, the developing individual has distinctly human features.

When the fourth week ends, the embryo is 500 times its starting size. Weeks five and six are the boundary between the embryonic and fetal periods. Now growth slows as details of organs fill in. Limbs form; toes and

fingers are sculpted from paddles. The umbilical cord develops, and so does an intricate circulatory system. Growth of the all-important head now surpasses that of all other regions (Figure 44.20). Reproductive organs start forming, as explained in Section 12.5. At the end of the eighth week, the individual is no longer just "a vertebrate." Its features define it as a human fetus.

In the second trimester, as developing nerves and muscles connect up, reflexive movements begin. Legs

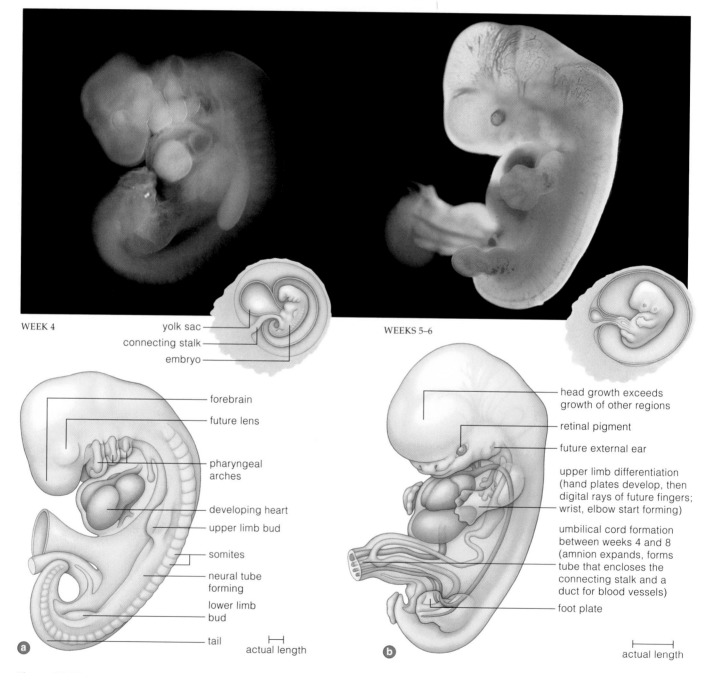

WEEK 4

yolk sac
connecting stalk
embryo

WEEKS 5–6

a
forebrain
future lens
pharyngeal arches
developing heart
upper limb bud
somites
neural tube forming
lower limb bud
tail
actual length

b
head growth exceeds growth of other regions
retinal pigment
future external ear
upper limb differentiation (hand plates develop, then digital rays of future fingers; wrist, elbow start forming)
umbilical cord formation between weeks 4 and 8 (amnion expands, forms tube that encloses the connecting stalk and a duct for blood vessels)
foot plate
actual length

Figure 44.20 Human embryo at successive stages of development.

kick, arms wave about, and fingers grasp. The fetus frowns, squints, puckers its lips, sucks, and hiccups. When the fetus is five months old, its heartbeat can be heard clearly through a stethoscope positioned on the mother's abdomen. The mother can sense movements of fetal arms and legs.

By now, soft, fetal hair (the lanugo) covers the skin; most will be shed before birth. A thick, cheesy coating protects the wrinkled, reddish skin from abrasion. In the sixth month, delicate eyelids and eyelashes form. Eyes open during the seventh month, the start of the final trimester. By this time all portions of the brain have formed and have begun to function.

In the fetal period, the primary tissues that formed in the early embryo become sculpted in ways that transform this vertebrate embryo into one with distinctly human features.

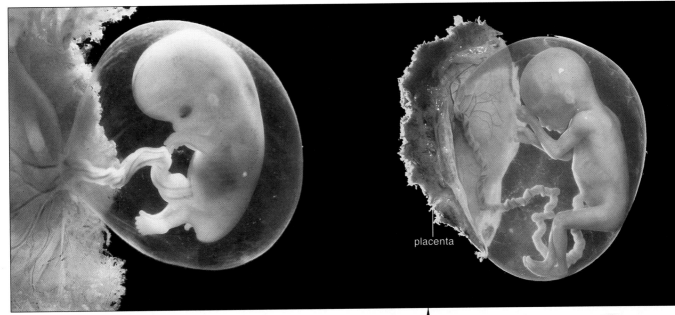

placenta

WEEK 8

final week of embryonic period; embryo looks distinctly human compared to other vertebrate embryos

upper and lower limbs well formed; fingers and then toes have separated

primordial tissues of all internal, external structures now developed

tail has become stubby

c |⊢——— actual length ———⊣|

WEEK 16 ⌐—
Length: 16 centimeters
           (6.4 inches)
Weight: 200 grams
           (7 ounces)

WEEK 29
Length: 27.5 centimeters
           (11 inches)
Weight: 1,300 grams
           (46 ounces)

WEEK 38 (full term) ——————→
Length: 50 centimeters
           (20 inches)
Weight: 3,400 grams
           (7.5 pounds)

During fetal period, length measurement extends from crown to heel (for embryos, it is the longest measurable dimension, as from crown to rump).

d

44.13 Mother as Provider, Protector, Potential Threat

LINKS TO
SECTIONS 18.5,
34.13, 40.5, 41.8

Each pregnant female is committing much of her body's resources to the growth and development of a brand-new individual. From fertilization until birth, her future child is at the mercy of her diet, health habits, and life-style.

NUTRITIONAL CONSIDERATIONS

When a mother-to-be eats a well-balanced diet, her embryo gets all the proteins, carbohydrates, and lipids it requires for growth and development (Section 41.8). However, her own body's demands for vitamins and minerals increase as the placenta preferentially absorbs them for the fetus from her blood. Medically supervised increases in her uptake of B-complex vitamins before and during early pregnancy reduce the embryo's risk of severe neural tube defects. Folate (folic acid) is especially important in this regard.

Dietary deficiencies adversely affect many developing organs. For example, the brain expands most in the weeks just before and after birth. Poor nutrition during this span may impair intelligence and other functions later in life.

A pregnant female must eat enough to gain twenty to twenty-five pounds, on average. If she does not, her newborn may be seriously underweight, at greater risk of postdelivery complications and, in time, impaired brain function.

INFECTIOUS DISEASES

Remember, IgG antibodies in a pregnant female's blood cross the placenta and protect the embryo or fetus from all but the most serious bacterial infections (Section 39.5). Some viral diseases are dangerous in the first six weeks after fertilization, a crucial time of organ formation.

Suppose the female contracts *rubella* (German measles) in this critical period. There is a 50 percent chance that some organs will not form properly. For instance, if she is infected while embryonic ears are forming, her newborn may be deaf (Figure 44.21). If she is infected at any time from the fourth month of pregnancy onward, this particular disease will have no notable effect. A female may avoid the risk entirely by getting vaccinated against the virus before pregnancy.

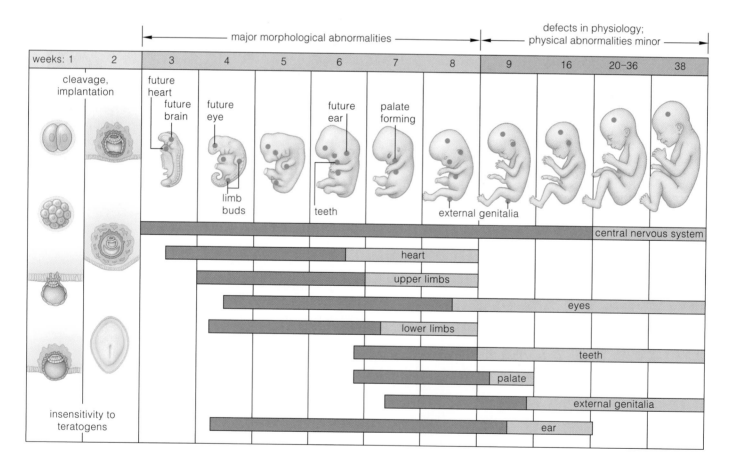

Figure 44.21 Teratogen sensitivity. Teratogens are drugs, infectious agents, and environmental factors that invite embryonic or fetal deformities, usually after organs form. They adversely affect growth, tissue remodeling, and tissue resorption. *Dark blue* signifies the highly sensitive period; *light blue* signifies periods of less severe sensitivity to teratogens. For example, the upper limbs are most sensitive to damage during weeks 4 through 6, and somewhat sensitive during weeks 7 and 8.

Figure 44.22 An infant with fetal alcohol syndrome—FAS. The obvious symptoms are low and prominently positioned ears, improperly formed cheekbones, and an abnormally wide, smooth upper lip. Growth-related complications and abnormalities of the nervous system can be expected.

FAS cases are grossly underdiagnosed, misdiagnosed, and underreported. Part of the problem is that medical schools do not always train doctors to recognize FAS symptoms. Patients of child-bearing age should be screened for alcohol use, just as they are for glucose levels and other indicators of health.

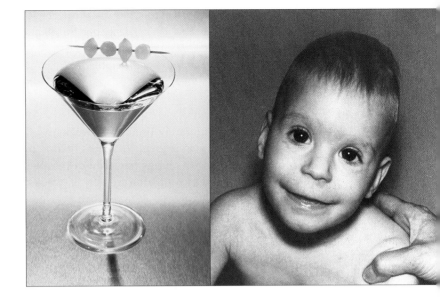

ALCOHOL, TOBACCO, AND OTHER DRUGS

Alcohol Remember the introduction to Chapter 6? Alcohol passes freely across cell membranes. It also passes freely across the placenta. When a pregnant female drinks, her developing embryo or fetus will quickly absorb alcohol. Excessive intake invites *fetal alcohol syndrome* (FAS).

Symptoms of this disorder include reduced brain size, mental impairment, facial deformities, a small head, slow growth, possible heart problems, and poor coordination (Figure 44.22). In some parts of the United States, the average incidence of FAS is as high as 1.5 cases per 1,000 live births. The damage is permanent; children affected by FAS never do catch up, physically or mentally.

Even moderate drinking during pregnancy may have negative effects. One study indicated that even a single episode of high alcohol intake can induce apoptosis in neurons of the developing brain. Increasingly, doctors are urging total abstinence from alcohol during pregnancy.

Tobacco Smoking or exposure to secondhand smoke increases the risk of miscarriage and adversely affects fetal growth and development. Remember, carbon monoxide in smoke outcompetes oxygen for binding sites on hemoglobin (Section 40.5), so the embryo or fetus cannot get enough oxygen. Nicotine levels in amniotic fluid actually can be higher than those in the mother's blood.

Smoking any tobacco regularly during pregnancy results in underweight newborns. This happens even when the female's weight, nutrition, and other key variables match those of pregnant nonsmokers. Tobacco smoke adversely affects nutrition as well. In one study, pregnant females who smoked lowered the blood concentration of vitamin C for themselves *and* for their fetuses, even when intake of that vitamin matched the intake of a control group.

The effects may be long term. Researchers tracked a group of children born in the same week for seven years. More children of smokers died of postdelivery complications. Those that survived were smaller, had twice as many heart defects. By age seven, they were nearly half a year behind children of nonsmokers in their "reading age."

Cocaine A pregnant female who uses cocaine of any kind increases the likelihood of miscarriage and premature delivery. As Section 34.13 explains, cocaine is a psychoactive drug, and it disrupts the development of the fetal nervous system. A child of a cocaine addict is likely to be abnormally small and irritable early in life. Some studies indicate that prenatal exposure to cocaine has long-term negative effects on intelligence and behavior.

Prescription Drugs Pregnant women should not take any drugs except under medical supervision. To underscore this point, the tranquilizer *thalidomide* was routinely prescribed in Europe. Infants of some of the women who used it during the first trimester had severely deformed arms and legs, or none at all. This drug has been withdrawn from the market. But other tranquilizers, sedatives, and barbiturates are still being prescribed, and they may cause some less severe damage. Certain *anti-acne drugs* might invite facial and cranial deformities.

Depression during pregnancy is not uncommon and may put the fetus at risk. A mother who does not feel like eating or otherwise taking care of herself can compromise fetal development. To avoid such problems, some pregnant women are treated with antidepressants. So far, research suggests that these drugs do not increase the miscarriage rate, slow fetal growth, or increase the incidence of birth defects. However, the infants of women who used the drugs right up to the time of delivery apparently can show some withdrawal symptoms.

A final note: Teenagers are much less likely than older women to receive timely prenatal care and are more likely to smoke during pregnancy. Because of these and other factors, the babies born to teenagers are more likely to be premature. They are at greater risk of serious and long-term illness, of delays in postnatal developments, and of dying in the first year of life (Section 18.5).

44.14 Birth and Postnatal Development

LINKS TO
SECTIONS 26.10,
35.9, 36.8, 43.6, 43.7

Human growth and development continue as the newborn embarks on a course of extended dependency and learning. As with all mammals, its early survival depends on nutritious milk, typically provided by the mother.

GIVING BIRTH

A fetus born too prematurely (before 22 weeks) will not survive. The risk also is great for births before 28 weeks, mainly because the lungs have not developed enough. The risk starts to drop after this. By 36 weeks, the survival rate is 95 percent. A fetus born between 36 and 38 weeks still has some trouble breathing and maintaining a core temperature even with the best of medical care. On average, the most favorable birthing time is 38 weeks after fertilization.

The properties of the cervix change as a fetus nears full term. Until now, the firm wall of the cervix has helped keep the fetus from prematurely slipping out of the uterus. Its connective tissue weakens in the last weeks of pregnancy. The crosslinks between collagen fibers loosen, the cervix becomes thinner, softer, and more flexible. These structural changes will allow it to stretch enough to permit the expulsion of the fetus.

The birth process is known as labor. Typically, the amnion ruptures right before birth, so amniotic fluid drains from the vagina. The cervical canal dilates and the fetus can move out of the uterus, then through the vagina and into the outside world (Figure 44.23).

Remember the hormone **oxytocin**? It causes smooth muscle inside the uterine wall to contract during labor. When the fetus was nearing full term, it "dropped," or shifted downward, so its head touched the cervix. Receptors in the cervix sensed the mechanical pressure and signaled the hypothalamus, which induced the posterior lobe of the pituitary to secrete oxytocin.

Binding of oxytocin to smooth muscle now causes an increase in the strength of contractions, which puts more mechanical pressure on the cervix. More of the hormone is secreted. And so, in a **positive feedback cycle**, the stretching causes oxytocin secretion, which causes more stretching, and so on until the fetus is expelled and there is no more mechanical pressure on the cervix. Intravenous injections of synthetic oxytocin can be given to induce or increase contractions.

The strong contractions also detach and expel the placenta from the uterus, as the afterbirth. They help stop bleeding at the site where the placenta attached to the wall of the uterus. They constrict blood vessels at the ruptured attachment site, so the umbilical cord can be cut and tied off. A few days after shriveling up, the cord's stump has become the navel.

Corticotropin-releasing hormone (CRH) affects the timing of labor, and it might contribute to *postpartum depression*. The hypothalamus makes this hormone in all humans, but the placenta also produces it. During pregnancy, its blood level may increase three times. CRH stimulates the adrenal cortex to secrete **cortisol** (Sections 36.8 and 36.9). This hormone might help the mother-to-be cope with the extraordinary physical and emotional strains of pregnancy and labor.

During pregnancy, a high blood level of cortisol can suppress CRH production by the hypothalamus. After the placenta is expelled, the CRH level briefly plummets. The decline commonly triggers a short-term depression that may continue until the hypothalamus resumes its normal production of CRH.

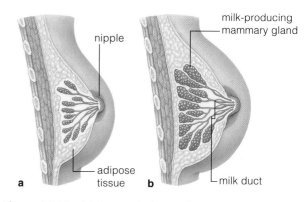

Figure 44.24 (**a**) Breast of a human female who is not pregnant. (**b**) Breast of a lactating female.

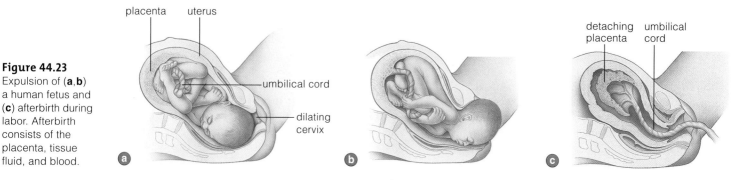

Figure 44.23
Expulsion of (**a,b**) a human fetus and (**c**) afterbirth during labor. Afterbirth consists of the placenta, tissue fluid, and blood.

NOURISHING THE NEWBORN

Once the lifeline to the mother is severed, a newborn enters the extended time of dependency and learning that is typical of all primates (Section 26.10 and page 461). Its early survival requires an ongoing supply of milk or its nutritional equivalent. **Lactation**, or milk production, occurs in mammary glands in a mother's breasts (Figure 44.24). Before pregnancy, breast tissue is largely adipose tissue and a system of undeveloped ducts. Their size depends on how much fat they hold, not on milk-producing capacity. During pregnancy, estrogens and progesterone stimulate development of a glandular system for milk production.

For the first few days after birth, mammary glands produce a fluid rich in proteins and lactose. **Prolactin**, a hormone that calls for synthesis of enzymes used in milk production, is secreted by the mother's anterior pituitary (Section 36.3). When the newborn suckles, the pituitary releases oxytocin, which triggers muscle contractions that force fluid into milk ducts. It also causes uterine contractions that help shrink this birth chamber back to its pre-pregnancy size.

Besides being nutrient-rich, human breast milk has immunoglobulins that enhance resistance to infection. Some other components stimulate growth of bacterial symbionts in the infant gut. However, alcohol, drugs, mercury, and other toxins in a mother's body also can be secreted in milk. As during pregnancy, a nursing mother should tailor her life-style and diet.

POSTNATAL DEVELOPMENT

As is the case for many species, humans change in size and proportion until they reach sexual maturity. Figure 44.25 shows a few of the proportional changes during the life cycle. Table 44.6 defines the prenatal ("before birth") and postnatal ("after birth") stages.

Postnatal growth is most rapid between the years thirteen and nineteen. Not until adulthood are bones fully mature. Long after the adults have put the new generation on a path toward sexual reproduction and development, they gradually age and then pass on in a natural turn of events in the life cycle (Sections 43.6 and 43.7). Selection has favored the perpetuation of genes, not morphological immortality.

The human life cycle flows naturally from the time of birth, growth, and development, to production of the individual's own offspring, and on through aging to the time of death.

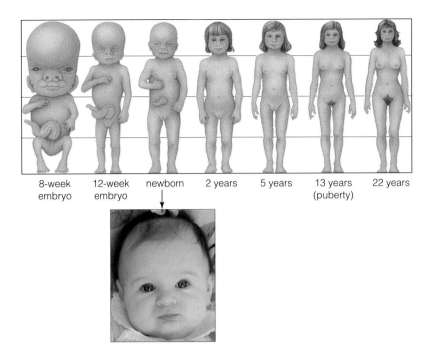

8-week embryo 12-week embryo newborn 2 years 5 years 13 years (puberty) 22 years

Figure 44.25 Observable, proportional changes in the human body during prenatal and postnatal periods of development. Changes in overall physical appearance are slow but noticeable until the teenage years. For example, compared to an embryo, the legs of teenagers are longer and the trunk shorter, so the head is proportionally smaller. Correlate these drawings with the stages in Table 44.6.

Table 44.6	Stages of Human Development
Prenatal period	
Zygote	Single cell resulting from fusion of sperm nucleus and egg nucleus at fertilization.
Morula	Solid ball of cells produced by cleavages.
Blastocyst (blastula)	Ball of cells with surface layer, fluid-filled cavity, and inner cell mass.
Embryo	All developmental stages from two weeks after fertilization until end of eighth week.
Fetus	All developmental stages from ninth week to birth (about 38 weeks after fertilization).
Postnatal period	
Newborn	Individual during the first two weeks after birth.
Infant	Individual from two weeks to about fifteen months after birth.
Child	Individual from infancy to about ten or twelve years.
Pubescent	Individual at puberty; secondary sexual traits develop; girls between 10 and 15 years, boys between 12 and 16 years.
Adolescent	Individual from puberty until about 3 or 4 years later; physical, mental, emotional maturation.
Adult	Early adulthood (between 18 and 25 years); bone formation and growth finished. Changes proceed slowly after this.
Old age	Aging processes result in expected tissue deterioration.

44.15 On Human Fertility

FOCUS ON BIOETHICS

LINK TO SECTION 20.4

The motivation to engage in sex during the life cycle has been evolving for hundreds of millions of years. A few centuries of moral arguments for self-control have not suppressed it. How will we reconcile our biological past with the need for a stabilized cultural present?

In this chapter, you tracked the transformation of a human zygote into an adult. That transformation raises profound questions. *When, precisely, does development begin?* As you have seen, major developmental events unfold even before fertilization. *When does life begin?* During her lifetime, a female may produce as many as 500 eggs, all of which are alive. With one ejaculation, a male may release a quarter of a billion sperm, all of which are alive. Before one sperm and one egg merge by chance and establish the genetic makeup of a new individual, they are as much alive as any other form of life.

It is scarcely tenable, then, to say that life begins at fertilization. *Life began more than 3.8 billion years ago—and each gamete, each zygote, and each sexually mature individual is but a fleeting stage in the continuation of that beginning.*

This greater perspective on life cannot diminish the meaning of conception. It is no small thing to entrust a new individual with the gift of life, wrapped in the unique evolutionary threads of our species and handed down through an immense sweep of time.

Yet how can we reconcile the marvel of individual birth with the astounding birth rate for our species? About 14,800 newborns are entering the world every hour. By the time you go to bed tonight, there may be 356,000 more—about as many as there are in Cincinnati. In less than four months, there may 36,800,000 more—about as many as there are now in the entire state of California.

Human population growth is outstripping resources. Many millions already face the horrors of starvation. Living where we do, few of us know what it means to give birth to a child, to give it the gift of life, and have no food to keep it alive.

And how can we reconcile the marvel of birth with the reality of unwanted pregnancies? In the United States, about half of all pregnancies are unintended, and half of those—750,000 to 1 million annually—end in abortion. Complex biological and social factors contribute to sexual behavior. Many teenagers have reported that they were not mature enough or willing to accept responsibility for sexual behavior that has unintended consequences. Far more adult women around the world say they simply have no resources to raise a child, or one more child. In 2004 alone, close to 46 million chose abortions.

Whether and how fertility should be controlled is a volatile issue. We return to this issue in the next chapter, in the context of principles that govern the growth and stability of all populations.

Summary

Sections 44.1–44.3 The human reproduction system consists of a pair of primary reproductive organs, or gonads, and accessory organs and ducts. Gonads produce gametes, the packages that perpetuate one's genes. They also produce sex hormones that orchestrate reproductive function and the development of gender-specific secondary sexual traits.

The male gonads are testes. Leading away from each is an epididymis, vas deferens, and ejaculatory duct in which sperm successively finish maturing, are stored, and then are rapidly transported to the urethra that opens onto the surface of the tip of the penis.

Sperm are mixed with secretions from two seminal vesicles, a bulbourethral gland, and a prostate gland to form semen, a thick fluid that is expelled from the penis during sexual activity.

Feedback loops from the testes to the hypothalamus and pituitary gland govern secretion of GnRH, LH, FSH, and testosterone, which control sperm formation and male reproductive function. Testosterone, which Leydig cells in testes secrete, promotes development of male secondary sexual traits as well.

Each sperm is a flagellated cell. Its head is packed with DNA and has an enzyme-filled cap. Sertoli cells in the testes nourish them as they mature.

The female gonads are ovaries. Other reproductive organs are a pair of oviducts (Fallopian tubes) that form a channel to the uterus, a muscular chamber in which embryos develop. The cervix, a narrowed neck of the uterus, opens to the vagina, the organ of sexual intercourse as well as the birth canal.

Biology Now

Learn about the reproductive system of the human male with the animation on BiologyNow.
See how sperm form with the animation on BiologyNow.
Learn about the reproductive system of the human female with the animation on BiologyNow.

Sections 44.4, 44.5 A menstrual cycle is a recurring cycle of fertility during the reproductive years. A human female cycle is monthly. Feedback loops from the ovaries to the hypothalamus and anterior lobe of the pituitary gland control the cycle's three phases:

Follicular phase. The hypothalamus secretes GnRH, which stimulates the anterior pituitary to secrete FSH and LH. These sex hormones act on a follicle: a primary oocyte, arrested in meiosis I, and a cell layer around it. The follicle matures and secretes estrogens that stimulate the endometrium, the lining of the uterine chamber, to thicken in preparation for pregnancy. The oocyte finishes meiosis I. Cytoplasmic division follows and results in a large secondary oocyte and one polar body, which quickly degenerates.

Ovulatory phase. A midcycle surge of LH triggers the release of the secondary oocyte from the ovary. This event is called ovulation.

Luteal phase. After ovulation, a glandular structure, the corpus luteum, forms from remnants of the follicle. It secretes progesterone and some estrogen that prime the endometrium for fertilization. If fertilization occurs, the corpus luteum will be maintained until a placenta forms. If fertilization does not occur, the corpus luteum will degenerate. The endometrium will break down and get sloughed off with blood, and a new cycle will begin.

Biology ⓔNow
Observe the cyclic changes in an ovary with the animation on BiologyNow.
Learn about the effects of hormones on the menstrual cycle with the animation on BiologyNow.

Section 44.6 As a sperm penetrates it, a secondary oocyte is stimulated to complete meiosis II; cytoplasmic division results in one mature egg (ovum) and two polar bodies. Fertilization is complete when the egg nucleus fuses with the sperm nucleus to form a zygote.

Biology ⓔNow
See what happens during fertilization with the animation on BiologyNow.

Sections 44.7, 44.8 Humans prevent pregnancy by abstinence, surgery, physical or chemical barriers, or manipulations of female sex hormones. Unsafe sex and other behaviors promote the spread of pathogens that cause sexually transmitted diseases, or STDs.

Biology ⓔNow
Read the InfoTrac article "Genital Herpes: A Hidden Epidemic," Linda Bren, FDA Consumer, March–April 2002.

Section 44.9 After fertilization, cleavage begins and transforms the zygote into a blastocyst, which implants itself in the endometrium. HCG, a hormone the blastocyst secretes, stimulates the corpus luteum to keep on maintaining the endometrium. Gastrulation starts with the formation of ectoderm, endoderm, and mesoderm. Four extraembryonic membranes also form:

The *amnion* becomes a fluid-filled sac around the embryo, which it protects from drying out, mechanical shock, and abrupt temperature changes.

In most shelled eggs, a *yolk sac* stores nutritive yolk. In humans, part of the sac becomes a major site of blood formation. Some of its cells give rise to germ cells that later give rise to sperm or eggs.

The *chorion*, a protective membrane, encloses the embryo and the other extraembryonic membranes. It becomes a major component of the placenta.

In humans, blood vessels for the placenta arise from the *allantois*, as does the urinary bladder.

Biology ⓔNow
Observe early human development with the animation on BiologyNow.

Section 44.10 The anterior–posterior body axis forms in the gastrula. The gastrula's neural disk gives rise to the neural tube, the forerunner of the brain and spinal cord. Somites are paired bumps of mesoderm that give rise to skeletal muscles, bones, and the overlying dermis.

Sections 44.11–44.13 A blood-engorged organ, the placenta, gradually forms from endometrial tissue and extraembryonic membranes. It permits embryonic blood vessels to develop independently of the mother's but also allows oxygen, nutrients, and wastes to diffuse between them. To some extent, the placenta serves as a protective barrier for the fetus. But it cannot protect the fetus from harmful effects of the mother's nutritional deficiencies, infections, intake of prescription drugs, illegal drugs, alcohol, and cigarette smoke.

The embryo has distinctly human features by the end of the eighth week of pregnancy and thereafter is called a fetus. A mother's health, nutrition, and life-style affect fetal growth and development.

Section 44.14 During labor, uterine contractions expel the fetus and afterbirth. Hormones trigger labor, the maturation of mammary glands, and milk flow. Most development and growth are over by adulthood.

Self-Quiz
Answers in Appendix II

1. Label all of the parts of the human male reproductive system and check to see that you can state their functions:

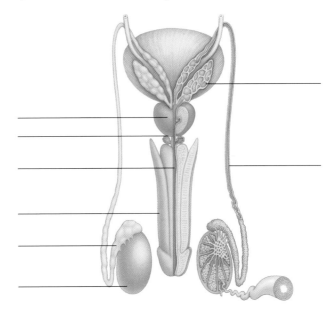

2. Label all of the parts of the human female reproductive system and check to see that you can state their functions:

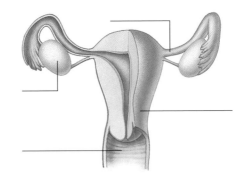

Figure 44.26 Identical twins Sabra and Nina, who started life as the same zygote. The first two blastomeres that formed at cleavage, the inner cell mass, or some other early stage split and gave rise to two genetically identical individuals.

3. Meiotic divisions of _____ produce mature sperm.
   a. Leydig cells   c. both a and b
   b. Sertoli cells   d. neither a nor b

4. During a menstrual cycle, a midcycle surge of _____ triggers ovulation.
   a. estrogens b. progesterone c. LH d. FSH

5. The corpus luteum secretes _____ .
   a. LH   b. FSH   c. progesterone   d. prolactin

6. Which of the following is (are) caused by bacteria?
   a. chlamydia   d. trichomoniasis
   b. gonorrhea   e. a and b
   c. genital warts   f. c and d

7. A _____ implants in the lining of the uterus.
   a. zygote   b. gastrula   c. blastocyst d. fetus

8. Which of the following puts human developmental stages in the correct order?
   a. zygote, blastocyst, embryo, fetus
   b. zygote, embryo, blastocyst, fetus
   c. zygote, embryo, fetus, blastocyst
   d. blastocyst, zygote, embryo, fetus

9. A human ovum is a(an) _____ .
   a. immature ooocyte   c. secondary oocyte
   b. primary oocyte   d. tertiary oocyte

10. The _____ , a fluid-filled sac, surrounds and protects an embryo and keeps it from drying out.
    a. yolk sac b. allantois c. amnion   d. chorion

11. At full term, a placenta _____ .
    a. is composed of extraembryonic membranes alone
    b. directly connects maternal and fetal blood vessels
    c. keeps maternal and fetal blood vessels separated

12. (A) _____ form(s) in all vertebrate embryos.
    a. neural tube   c. pharyngeal arches e. a through c
    b. somites   d. primitive streak   f. a through d

13. Distinctly human features emerge in the embryo by the end of the _____ week after fertilization.
    a. second   c. fourth   e. eighth
    b. third   d. fifth   f. sixteenth

14. Match each term with the most suitable description.
    ____ testis   a. maternal and fetal tissues
    ____ cervix   b. stores mature sperm
    ____ placenta   c. produces testosterone
    ____ vagina   d. produces estrogen and
    ____ ovary   progesterone
    ____ oviduct   e. usual site of fertilization
    ____ epididymis   f. lining of uterus
    ____ endometrium   g. birth canal
                         h. entrance to uterus

Additional questions are available on Biology Now™

Critical Thinking

1. A male erection occurs when blood flows into the penis faster than it flows out. Signals from the nervous system cause nerve endings in the penis to release nitric oxide, a neurotransmitter. When nitric oxide binds to receptors on postsynaptic cells in the organ's spongy tissue, it activates an enzyme that catalyzes production of cGMP (short for cyclic guanine monophosphate). The cGMP causes blood vessels to vasodilate, which lets more blood flow in. When all else is working smoothly, an erection occurs.

Erectile dysfunction is being treated with drugs such as sildenafil, sold under the brand name Viagra, which target PDE–5 (short for phosphodiesterase–5). This liver enzyme converts cGMP to inactive form. Sketch out a simple flow chart that shows the normal action of nitric oxide, cGMP, and PDE–5 and identify where sildenafil acts.

2. Drugs that inhibit signals of sympathetic neurons may be prescribed for males who have high blood pressure. How might such drugs interfere with sexual behavior?

3. On rare occasions, Leydig cells form tumors in young boys, and the testes secrete as much as 1,000 times the normal amount of testosterone. The boys grow up much shorter than would otherwise be expected. Explain why, and also speculate on what other symptoms may develop.

4. *Identical twins* form when an embryo splits, most often between the third and eighth day following fertilization. Worldwide, the birth rate of identical twins is about 4 in every 1,000, with no significant difference among ethnic groups. Because identical twins arise from the same fertilized egg, they have the same genotype and look much the same (Figure 44.26).

Fraternal twins, which are nonidentical genetically, arise when two oocytes mature, are released, and become fertilized at the same time. Such twins run in families. The incidence varies among ethnic groups and is highest among blacks and lowest in Asians. The variation might be an outcome of differences in gene products that affect the FSH level in blood.

Explain how a high FSH level in blood would increase the likelihood of fraternal twins. In addition, explain why variation in FSH levels would not affect the incidence of identical twins.

5. By UNICEF estimates, each year 110,000 people are born with abnormalities as a result of rubella infections. Major symptoms of *congenital rubella syndrome,* or CRS, are deafness, blindness, mental impairment, and heart problems. A fetus is at risk if a nonvaccinated female is infected during the first trimester, but not later. Review the developmental events that unfold during pregnancy and explain why this is the case.

6. Given what you have read about where a human embryo develops, explain why a tubal pregnancy, as shown in Figure 44.14*a*, must be surgically terminated.

7. In the United States, teenage pregnancies as well as STD infections are rampant. Suppose the office of the Surgeon General requests your participation in a task force that will recommend practices that might reduce the incidence of teenage pregnancies and STD infections. What practices might have the most success? Would they provoke enthusiasm or resistance among teenagers in your community? Among adults? Explain why.

VII Principles of Ecology

Two organisms—a fox in the shadows cast by a snow-dusted spruce tree. What are the consequences of their interactions with each other, with other kinds of organisms, and with their environment? By the end of this last unit, you might find worlds within worlds in such photographs.

45 POPULATION ECOLOGY

The Numbers Game

In 1944 Allied forces invaded Normandy, which marked the beginning of the end of Hitler's "Fortress Europe." On the other side of the world, the United States Coast Guard stationed nineteen men on a remote island in the Bering Sea. The enlisted men set up long-range navigational aids for ships and aircraft. They barged in twenty-nine reindeer (*Rangifer tarandus*) as an emergency food source.

Remember, reindeer preferentially eat lichens (Chapter 24). Thick mats of lichens carpeted St. Matthew, an island where winds howl across tundra and tall cliffs above the sea. The island, which is 320 kilometers from Alaska, is only 6.4 kilometers (about 4 miles) wide and 51 kilometers long.

World War II started to wind down before any reindeer were shot. The Coast Guard pulled out, leaving behind some seabirds, arctic foxes, and voles—and a herd of healthy reindeer with nothing big enough to hunt them down.

In 1957 a biologist with the U.S. Fish and Wildlife Service, who later became a University of Alaska professor, visited St. Matthew. On a hike from one end of the island to the other, David Klein counted 1,350 well-fed reindeer. He noticed lichens that had been overgrazed and trampled.

Six years after that, Klein and three other biologists returned to the small island. They counted 6,000 reindeer. They could not help but notice the profusion of reindeer tracks and feces, and a lot of pummeled lichens.

Klein did not return to St. Matthew until the summer of 1966. Bleached-out reindeer bones littered the island. Forty-two reindeer were still alive. Only one was a male, and it had abnormally shaped antlers. There were no fawns. The population had plummeted to 1 percent of the founding herd! Apparently, thousands had starved to death during the winter of Klein's previous visit. By the time the 1980s rolled around, there were no reindeer at all.

St. Matthew Island is small, with clear boundaries, so it is easy to draw a lesson from this unintended experiment in population ecology: *A population's growth depends on environmental resources, few of which are unlimited.*

Does the lesson apply to other populations, in other places? Another case, on another isolated island, suggests that it does. Remember the Chapter 27 introduction? That human population, too, plummeted in size after growing beyond the capacity of the environment to sustain it.

What if the environment for a population is as big as a continent or a sea? Do resources still run out? Consider this: There are more whitetail deer (*Odocoileus virginianus*) in North America than there were five centuries ago. There are an estimated 20 million to 33 million. The nation's forests no longer can sustain them. Deer now overbrowse ground vegetation. They strip trees of leaves and bark. By eating so many acorns and seedlings, they have stopped the self-renewal of many oak forests. Forests remain in good shape only when there are no more than twenty deer per square mile. Today, densities are often greater than seventy deer per square mile.

Figure 45.1 What happens when you import a small herd of herbivores to a remote island where there are no natural predators and then forget about them?

IMPACTS, ISSUES

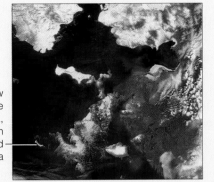

Saint Matthew Island in the Bering Sea, between Alaska and Siberia

The rampant browsing of the whitetail deer population also is endangering nesting birds, wildflowers, and other species. Deer spill into human habitats as well. They cause highway accidents that kill 200 or so people and cost about a billion dollars' worth of property damage annually. Each year, farmers lose about 400 million dollars' worth of crops to deer. Many of us know firsthand what even a few hungry deer can do to gardens.

Fewer people are hunting deer for sport, and hunting for the commercial market is banned. Animal rights groups are pleased. They want to control local populations with birth control or other nonlethal methods. However, it is difficult and expensive to track down and treat deer with injectable birth control drugs. Efforts to introduce birth control drugs into the deer's food sources could harm other herbivores.

The point is, *certain principles govern the growth and sustainability of all populations over time.* These principles are the bedrock of **ecology**—the systematic study of how organisms interact with one another and with the physical and chemical environment. Ecological interactions start within and between populations and extend on through communities, ecosystems, and the biosphere. They are the focus of this last unit of the book. After presenting the basic principles, this chapter invites you to apply them to the past, present, and future of the human species.

How Would You Vote?

Some people oppose any deer hunting, while others see hunters as a logical substitute for an absence of natural predators. Do you support encouraging hunting in areas where the presence of too many deer is harming the habitat? See BiologyNow for details, then vote online.

Key Concepts

WHAT ARE THE DEMOGRAPHICS?

Ecological principles govern the growth and sustainability of all populations. Genetic factors and a population's size, density, distribution, and the number of individuals in its various age categories influence patterns of growth. Sections 45.1, 45.2

EXPONENTIAL RATES OF GROWTH

Any population that is growing at a rate proportional to its size is showing exponential growth. Depending on the size of the population's reproductive base, exponential growth may be slow or fast. Section 45.3

LIMITS ON INCREASES IN SIZE

In time, exponential growth typically overshoots the carrying capacity, which is the maximum number of individuals of a population that environmental resources can sustain indefinitely. Some populations stabilize after a crash. Others never do recover. Section 45.4

PATTERNS OF SURVIVAL AND REPRODUCTION

Competition, disease, predation, and other factors that control population growth vary among species and help shape their life history patterns. Sections 45.5, 45.6

THE HUMAN POPULATION

Historically, expansion into new habitats around the world, cultural interventions, and technological innovations have allowed human populations to postpone abiotic and biotic limits to growth. However, the operative word is "postpone." Sections 45.7–45.10

Links to Earlier Concepts

Earlier chapters defined the population, a unit of biological organization that undergoes evolution (Sections 1.4, 17.3, 19.1–19.3). They introduced you to certain morphological, physiological, and behavioral traits that help characterize populations in general (17.3, 18.1, 18.8). With this chapter we turn to demographics—population size and other vital statistics—and the factors that limit increases in size.

45.1 Characteristics of Populations

LINKS TO
SECTIONS
17.3, 18.1, 18.8

By this point in the book, you know that a population is a group of individuals of the same species. Ecological interactions begin with characteristics of populations. We call these vital statistics **demographics.**

Each population has a gene pool and an evolutionary history, as explained in Chapters 17 and 18. It also has a characteristic size, density, distribution, and number of individuals in its various age categories.

Population size is the number of individuals that actually or potentially contribute to the gene pool. The **age structure** is the number of individuals in each of several age categories. For instance, individuals often are grouped by *pre-reproductive*, *reproductive*, and *post-reproductive* ages. Those in the first category have the capacity to produce offspring when mature. Together with individuals in the second category, they make up the population's **reproductive base.**

Population density is the number of individuals in some specified area or volume of a habitat. A *habitat*, remember, is the type of place where a species lives. We characterize a habitat by its physical and chemical features and its particular array of species. **Population distribution** is the pattern in which the individuals are dispersed in a specified area.

Crude density is a measured number of individuals in a specified area. It does not reveal how much of the habitat is actually being used for living space. Even areas that seem rather uniform, such as a long, sandy shoreline, are more like tapestries of light, moisture, temperature, composition, and many other variables. Only one portion of the habitat might be suitable for a given population. It might be suitable all of the time or only some of the time, as in summer versus winter.

Different species occupying the same area typically compete for energy, nutrients, living space, and other resources. Such *interspecific* interactions influence each population's density and dispersion through a habitat.

Theoretically, populations show a clumped, nearly uniform, or random distribution pattern (Figure 45.2). Clumping is the most common, for several reasons. First, each species is adapted to particular conditions and resources, which often are not uniform through a habitat. Some animals cluster by a water hole, seeds sprout only in moist soil, and so on. Second, animals may live in social groups, which offer more protection and mating opportunities. A school of fish is like this. Third, many plant seedlings and the offspring of many animals cannot disperse far from their parents.

With nearly uniform distribution, individuals are more evenly spaced than we would expect on the basis of chance alone. Uniform distribution is relatively rare in nature. It sometimes occurs when competition for resources or territory is fierce, as in a nesting colony of seabirds. Figures 45.2 and 49.16 show examples.

We observe random dispersion only when habitat conditions are nearly uniform, resource availability is fairly steady, and individuals of a population or pairs of them neither attract nor avoid one another. Each wolf spider does not hunt far from its burrow, which can be almost anywhere in forest soil (Figure 45.2).

clumped

nearly uniform

random

Each population has characteristic demographics, such as size, density, distribution pattern, and age structure.

Environmental conditions and species interactions shape these characteristics, which may change over time.

Figure 45.2 Three patterns of population distribution: clumped, as in squirrelfish schools; more or less uniform, as in a royal penguin nesting colony; and random, as when wolf spiders live in randomly located burrows in forest soil.

Ecologists go into the field to test theories about species interactions and population dynamics, and to monitor the health of threatened or endangered populations.

As the chapter introduction indicated, deer are all around us in forests, grasslands, golf courses, and gardens. How would you go about counting the ones living near you?

A full count would be a measure of absolute population density. Census takers supposedly make such a count of human populations every ten years, although not everyone answers the door. Ecologists make counts of large species in small areas, such as birds in a forest, northern fur seals at their breeding grounds, and sea stars in a tidepool. More often, however, a full count is impractical, so ecologists sample part of a population and estimate its total density.

For instance, you could divide a map of your county into small plots, or quadrats. **Quadrats** are sampling areas of the same size and shape, such as rectangles, squares, and hexagons. You could count individual deer in several plots and, from that, extrapolate the average number for the county as a whole. Ecologists often conduct such counts for plants and other species that stay put (Figure 45.3). Some counts in small areas also help them estimate the population sizes of migrating animals.

Deer are among the animals that do not stay put. How can ecologists be sure that the individuals being counted in a given plot are not the same ones counted earlier in a different plot? **Capture–recapture methods** are one way to sample a population of mobile animals. Such individuals are captured and marked in some way. Deer get collars, squirrels get tattoos, salmon get tags, birds get leg rings, butterflies get wing markers, and so on (Figure 45.4). The marked animals are released at time 1. At time 2, traps are reset. When all goes well, the proportion of marked animals in the second sample is representative of the proportion marked in the whole population:

$$\frac{\text{Marked individuals in sampling at time 2}}{\text{Total captured in sampling 2}} = \frac{\text{Marked individuals in sampling at time 1}}{\text{Total population size}}$$

Ideally, both marked and unmarked individuals of the population are captured at random, none of the marked animals is overlooked, and none dies or otherwise departs during the study interval.

In the real world, recapturing marked individuals might not be random. For example, squirrels that were marked after being attracted to bait in boxes might now be trap-happy or trap-shy. Such individuals may overrepresent or underrepresent their population. Other examples: Instead of mailing tags of marked fish to ecologists, a fisherman may keep them as good-luck charms. Birds lose leg rings.

Your estimate also depends on the time of year when you make a sampling. Population distribution varies with time, as during migrations in response to environmental rhythms. Few places yield abundant resources all year long, so many populations move between habitats as seasons change. Canada geese are like this. So are deer. In such cases, capture–recapture methods might be used more than once a year, for several years.

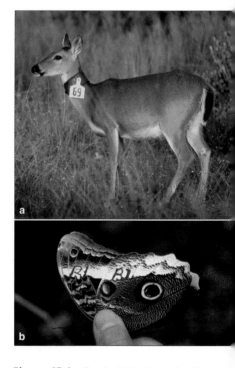

Figure 45.3 Near the eastern base of the Sierra Nevada, a population of creosote bushes showing nearly uniform distribution. The plants compete for scarce water in this desert climate zone, which has extremely hot, dry summers and mild winters.

Figure 45.4 Two individuals marked for population studies. (**a**) Florida Key deer and (**b**) Costa Rican owl butterfly (*Caligo*).

45.3 Population Size and Exponential Growth

LINK TO
SECTION
18.8

Populations are dynamic units of nature. Depending on the species, they may add or lose individuals every minute of every day, season, or year. Sometimes they glut portions of their habitat with individuals. Other times, individuals are scarce. Populations even drive themselves or are driven to extinction.

GAINS AND LOSSES IN POPULATION SIZE

You can measure change in population size in terms of birth rates, death rates, and how many individuals are entering and leaving during a specified interval.

Population size increases as a result of births and **immigration**, the arrival of new residents from other populations of the same species. Its size decreases as a result of deaths and **emigration**, the departure of individuals that take up permanent residence in some other place. As one example, Arnold Schwarzenegger emigrated from Austria to the United States, where he became a celebrated immigrant. His permanent move decreased the Austrian population by 1 and increased the United States population by 1.

For many species, population size changes during seasonal or daily migrations. However, **migration** is a recurring round trip between two distinct regions, so we need not consider its transient effects in this initial look at the nature of increases in population size.

FROM ZERO TO EXPONENTIAL GROWTH

To keep things simple, assume that immigration and emigration balance each other over time so that you can ignore the effects of both on population size. By doing so, you can define **zero population growth** as an interval in which the number of births is balanced by the number of deaths. During such an interval, the population size remains stable, with no net increase or decrease in the number of individuals.

Births, deaths, and the other variables that might change population size can be measured in terms of **per capita** rates, or rates per individual. *Capita* means head, as in head counts.

Visualize 2,000 mice living in a cornfield. Twenty or so days after their eggs get fertilized, the females give birth to a litter, then they nurse the offspring for a while. Then they get pregnant again. Suppose 1,000 mice are born in one month. The birth rate is 0.5 per mouse per month (1,000 births/2,000 mice). If 200 of the 2,000 die during that interval, the death rate will be 200/2,000 = 0.1 per mouse per month.

Assume further that the birth rate and death rate remain constant. By doing so, you can combine both variables into a single variable—the **net reproduction per individual per unit time**, or r for short. For our mice, r is 0.5 – 0.1 = 0.4 per mouse per month.

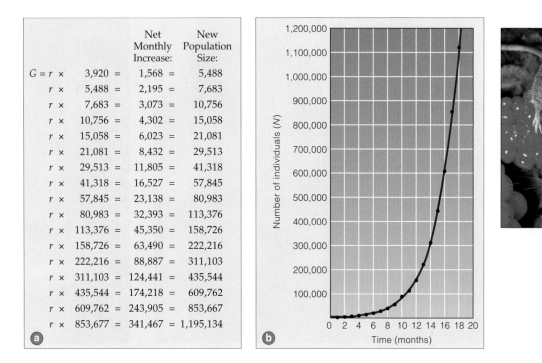

		Net Monthly Increase:		New Population Size:
$G = r \times$	3,920 =	1,568	=	5,488
$r \times$	5,488 =	2,195	=	7,683
$r \times$	7,683 =	3,073	=	10,756
$r \times$	10,756 =	4,302	=	15,058
$r \times$	15,058 =	6,023	=	21,081
$r \times$	21,081 =	8,432	=	29,513
$r \times$	29,513 =	11,805	=	41,318
$r \times$	41,318 =	16,527	=	57,845
$r \times$	57,845 =	23,138	=	80,983
$r \times$	80,983 =	32,393	=	113,376
$r \times$	113,376 =	45,350	=	158,726
$r \times$	158,726 =	63,490	=	222,216
$r \times$	222,216 =	88,887	=	311,103
$r \times$	311,103 =	124,441	=	435,544
$r \times$	435,544 =	174,218	=	609,762
$r \times$	609,762 =	243,905	=	853,667
$r \times$	853,677 =	341,467	=	1,195,134

Figure 45.5 *Animated!* (a) Net monthly increases in a population of field mice living in a cornfield. Start to finish, the list shows a pattern typical of exponential growth. **(b)** Graph the numerical data and you end up with a J-shaped growth curve.

Figure 45.6 Effect of deaths on the rate of increase in two hypothetical populations of bacteria. Plot population growth for bacterial cells that reproduce every half hour and you get growth curve *1*. Plot the growth of a population of cells that divide every half hour, with 25 percent dying between divisions, and you get growth curve *2*. Deaths do slow the rate of increase, but as long as birth rate exceeds the death rate, exponential growth will continue.

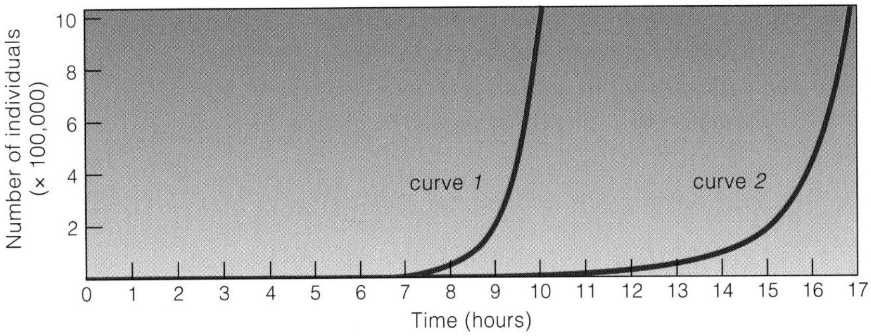

Thus the hypothetical mice in a cornfield give us a simple way to represent population growth as:

$$\begin{array}{l}\text{population}\\\text{growth per}\\\text{unit time}\end{array} = \begin{array}{l}\text{net population}\\\text{growth rate}\\\text{per individual}\\\text{per unit time}\end{array} \times \begin{array}{l}\text{number of}\\\text{individuals}\end{array}$$

or, more simply, $G = rN$.

When the next month starts, 2,800 mice are living in the cornfield. The net increase of 800 fertile mice means the reproductive base has become larger. All of these mice reproduce, so the population size expands, for a net increase of $0.4 \times 2,800 = 1,120$. Population size is now 3,920. Suppose all mice in the population reproduce month after month. As Figure 45.5*a* shows, within two years, the number of mice in the cornfield will increase from 2,000 to more than 1 million!

Plot the monthly increases against time and you end up with a graph line in the shape of a "J," as in Figure 45.5*b*. When the growth of any population over time plots out as a J-shaped curve, you know that you are tracking exponential growth.

Exponential growth refers to any quantity that is growing at a rate proportional to its size. For instance, a population that is growing by a fixed percentage every day, month, or some other specified interval is growing exponentially. This does not necessarily mean eyepoppingly fast increases; a small population can grow exponentially, too, but at a slow rate. Even so, as long as *r* remains constant, the rate of growth will be proportional to the number of individuals that make up the reproductive base. A population that has 6,000 successfully reproducing individuals will grow three times faster in a given year than a population of 2,000. *The larger a population's reproductive base, the greater will be the rate of growth in a specified interval.*

Now look at other aspects of exponential growth. Start by supplying one bacterium in a culture flask with all the nutrients required for growth. After thirty minutes, the cell divides in two. Its two daughter cells divide, and so on every thirty minutes. Assume none of the cells dies between divisions. The population size doubles in each interval—from 1 to 2, then 4, 8,

16, 32, and so on. The time it takes for a population to double in size is its **doubling time**.

After 9–1/2 hours, or nineteen doublings, there are more than 500,000 cells. Ten hours (twenty doublings) later, there are more than a million. Curve *1* in Figure 45.6 is a plot of this outcome.

Will deaths put the brakes on exponential growth? Suppose that 25 percent of the descendant cells die every half hour. It now takes about seventeen hours, not ten, for the population to reach 1 million. *Deaths slowed the rate of increase but did not stop exponential growth* (curve 2 in Figure 45.6). As long as birth rates exceed death rates, exponential growth will continue.

WHAT IS THE BIOTIC POTENTIAL?

Now visualize a population in a habitat where living conditions are ideal. Every individual has sufficient shelter, food, and other vital resources. No predators, pathogens, or pollutants lurk anywhere in the habitat. The population may well display its **biotic potential**. This term refers to the maximum rate of increase per individual for any population that is growing under ideal conditions.

Each species has a characteristic maximum rate of increase. For many bacteria, it is 100 percent each half hour or so. For humans and other large mammals, the estimated biotic potential is 2 to 5 percent per year. The *actual* rate depends on how old individuals are at the onset of reproduction, how often they reproduce, and how many offspring they produce over a lifetime. The human population is not now displaying its full biotic potential, but it still is growing exponentially.

During a specified interval, population size is generally an outcome of births, deaths, immigration, and emigration.

A population that is growing at a rate proportional to the size of its reproductive base in a given interval is showing exponential growth.

As long as the per capita birth rate remains above the per capita death rate, a population will grow exponentially.

45.4 Limits on the Growth of Populations

LINKS TO
SECTIONS
17.3, 21.8

Many complex interactions take place within and between populations in nature, and it is not always easy to identify all the factors that can restrict population growth.

DENSITY-DEPENDENT LIMITING FACTORS

Most of the time, environmental circumstances keep a population from fulfilling its biotic potential. That is why sea stars—the females of which could produce 2,500,000 eggs each year—do not fill the oceans.

To get a sense of what some of the constraints may be, start again with a bacterial cell in a culture flask, where you can control the variables. First you enrich the culture medium with glucose and other nutrients necessary for bacterial growth. Then you sit back and let bacterial cells reproduce for many generations.

At first, growth may be exponential. Then it slows, and population size remains relatively stable. After a stable period, population size plummets until all of the bacterial cells are dead. *What happened?* The larger population required more nutrients. In time, nutrient levels declined, which acted as an environmental cue for cells to stop dividing. Even when growth stopped, the population continued to absorb nutrients until no nutrients were left, so the population died out.

Any essential resource that is in short supply is a **limiting factor** on population growth. Food, mineral ions, refuge from predators, living space, and absence of pollutants are common examples (Figure 45.7). The number of limiting factors can be extensive, and their effects vary. Even so, one factor alone is often enough to put the brakes on population growth.

Suppose you kept freshening the nutrient supply. After growing exponentially, the population collapsed anyway. Like every other organism, bacteria generate metabolic wastes. The population produced so many wastes that it drastically altered the living conditions inside the culture flask. Collectively, the bacterial cells polluted the experimentally designed habitat and put a stop to further exponential growth.

CARRYING CAPACITY AND LOGISTIC GROWTH

Think of a small population of individuals dispersed through the habitat. As it increases in size, more and more individuals must compete for nutrients, living quarters, and other resources. The share available to each diminishes, fewer offspring are born, and more die from starvation or nutrient deficiencies. Now the population's growth rate declines until the births are balanced or outnumbered by deaths.

Ultimately, the *sustainable* supply of resources will determine population size. **Carrying capacity** is the maximum number of individuals of a population that a given environment can sustain indefinitely.

We can use the pattern of **logistic growth** to show how carrying capacity may affect population size. By this pattern, a small population starts growing slowly in size, then it grows rapidly, and finally its size levels off once the carrying capacity is reached. This pattern plots out as an S-shaped curve, as in Figure 45.8 (time A to C). We also may represent the pattern of logistic growth by the following equation:

$$\begin{matrix} \text{population} \\ \text{growth} \\ \text{per unit} \\ \text{time} \end{matrix} = \begin{matrix} \text{maximum net} \\ \text{population} \\ \text{growth rate} \\ \text{per individual} \\ \text{per unit time} \end{matrix} \times \begin{matrix} \text{number} \\ \text{of} \\ \text{individuals} \end{matrix} \times \begin{matrix} \text{proportion} \\ \text{of resources} \\ \text{not yet} \\ \text{used} \end{matrix}$$

The S-shaped curve is only an approximation of what goes on in nature. For instance, a population that grows too fast may drastically overshoot the carrying capacity. The death rate skyrockets and the birth rate plummets, which may well drive the population far below the carrying capacity. That is what happened to the St. Matthew reindeer (Figure 45.9).

Figure 45.7 Response to a scarcity of nesting sites, a limiting factor for weaver bird populations in Africa.

(**a**) African weavers construct densely woven, cup-shaped nests that are only wide and deep enough for a hen and her nestlings. Many nests hang from the same spindly limbs. (**b**) This nest in Namibia is like an apartment house in a place where few trees are available. Between 100 and 300 pairs of sparrow weavers occupy their own flask-shaped nests, each with its own tubular entrance.

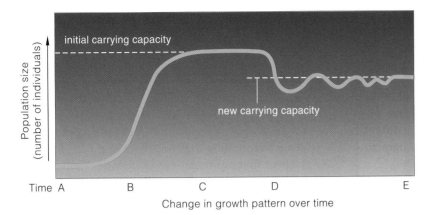

Figure 45.8 *Animated!* An idealized S-shaped curve, characteristic of logistic growth. Population growth slows after a phase of rapid increase (time A to C). The growth curve flattens as the carrying capacity is reached (time C to D). An S-shaped growth curve can show variations, as when changes in the environment lower the carrying capacity (time D to E).

Fluctuations in the pattern emerged after the bubonic plague swept through Europe's crowded cities in the fourteenth century. Remember the introduction to Chapter 18? One pandemic claimed 25 million lives.

Food and other essential resources are not the only factors that come into play when populations become too dense. When either exponential or logistic growth leads to overcrowding, many abiotic and biotic factors function as **density-dependent controls**, which means that they reduce the odds for individual survival.

For instance, interactions with predators or other species can drive the number of individuals below the maximum sustainable level. Predators, parasites, and pathogens have more intense effects in overcrowded populations of prey or host populations. In most cases, they thin out the population and thereby remove the very conditions that invited their controlling effect on population growth, as the next chapter explains.

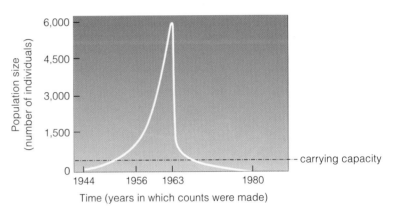

Figure 45.9 Growth curve for the reindeer herd barged over to St. Matthew Island in 1944, as described in the chapter introduction. This population did not recover after overshooting the carrying capacity.

DENSITY-INDEPENDENT LIMITING FACTORS

Density-independent factors can cause more deaths or fewer births regardless of population density. For instance, each year, millions of monarch butterflies fly from Canada to Mexico's forested mountains, where they spend the winter (Figure 45.10). Deforestation is going on in those mountains. In 2002 a sudden freeze —made worse by deforestation—killed off millions of them. That freeze would have killed them regardless of the density of the butterfly population.

Carrying capacity is the maximum number of individuals of a population that can be sustained indefinitely by the resources in a given environment.

With logistic growth, population growth is fast during times of low density, then it slows as the population approaches carrying capacity, where the numbers may level off.

Density-dependent factors, such the availability of a vital resource, exert control after populations become too dense as a result of exponential or logistic growth. Other factors exert control independently of population density.

Figure 45.10 Monarch butterflies at their wintering ground in Central Mexico. Each year these migratory insects travel hundreds of kilometers south to places that normally have been cool and humid in winter. If they were to stay in their northern breeding grounds, monarchs would risk being killed by more severe climatic conditions. Even so, deforestation and winter freezes in Mexico are now killing millions of them annually.

45.6 Natural Selection and Life Histories

LINKS TO
SECTIONS
1.4, 18.3, 26.2

Earlier you read that jaws evolved among certain fishes during the Cambrian. This key innovation led to adaptive radiations among predators and diverse defenses among prey. No one witnessed that coevolutionary arms race. However, experimental studies show that predators are still acting as selective agents, and prey are still evolving.

Several years ago, with fishnets in hand and drenched in sweat, two evolutionary biologists conducted fieldwork in the mountains of Trinidad, an island in the southern Caribbean Sea. They wanted to capture small fishes that live in the shallow freshwater streams (Figure 45.12). The fish were guppies (*Poecilia reticulata*). David Reznick and John Endler were starting an eleven-year study of the variables that help shape guppy life history patterns.

Male guppies are generally smaller and have brighter fish scales than females of the same age. The colors serve as visual signals for mating during the guppy's complex courtship rituals. Females are drab colored. Unlike males, they continue to grow after they reach sexual maturity.

Reznick and Endler were interested in the effects of predation on guppy evolution. They chose their study site because different predators act on different populations of guppies in the mountain streams of Trinidad. Different predators are present even along the length of the same

a *Right*, guppy that shared a stream with killifishes (*below*).

b *Right*, guppy that shared a stream with pike-cichlids (*below*).

Figure 45.12 (**a,b**) Two guppies (*Poecilia reticulata*) and two guppy eaters. (**c**) Biologist David Reznick contemplating interactions among guppies and their predators in a freshwater stream in Trinidad.

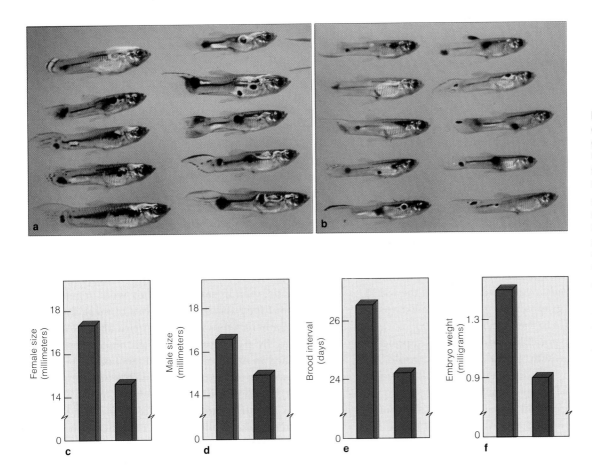

Figure 45.13 Typical size differences between (**a**) guppies that were preyed upon by killifishes and (**b**) guppies that were preyed upon by pike-cichlids.

(**c–f**) Graphs of some of the experimental evidence of natural selection among guppy populations that experimenters subjected to different predation pressures.

Compared to the guppies raised with killifish (*red* bars), guppies raised with pike-cichlids (*green* bars) differed in body size and length of time between broods. Killifish are small and prey on smaller guppies. Pike-cichlids are large fish and prey on larger guppies. The two predators select for differences in guppy life histories.

stream, because waterfalls prevent them from moving upstream or downstream from their habitat (Figure 45.12c).

Two major predators of guppies are killifishes (*Rivulus hartii*) and pike-cichlids (*Crenicichla*), as shown in Figure 45.12a,b. A killifish is not a very big fish. It preys efficiently on small, immature guppies but not on the larger adults. Pike-cichlids live in other streams. They prey on larger and sexually mature guppies and tend to ignore small ones.

Reznick and Endler hypothesized that predation is a selective agent that has shaped guppy life history patterns. As they knew, guppies in pike-cichlid streams grow faster and are smaller at maturity, compared with the guppies in killifish streams (Figure 45.13a,b). Also, guppies hunted by pike-cichlids reproduce earlier and more often, and they have more offspring per brood (Figure 45.13c–f).

Were these differences genetic, or did other variables influence life history patterns in killifish and pike-cichlid streams? To find out, the researchers shipped live guppies from each stream back to their laboratory in the United States. They allowed the guppies to reproduce in separate, predator-free aquariums for two generations. All other physical and chemical conditions in the artificial habitats were identical for the different experimental groups.

As it turned out, the experimental guppy populations displayed the same differences that the researchers saw in natural populations. The conclusion? *Differences between guppies preyed upon by different predators have a genetic foundation and therefore are subject to natural selection.*

What would happen if the selective pressure on a guppy population were to change? Reznick and Endler answered the question with sets of field experiments. For one set, they introduced guppies upstream from a small waterfall. Before the experiment, the waterfall had barred guppies and large pike-cichlids from emigrating upstream, where killifish were in the stream. Guppies introduced to the upstream experimental site were taken from a population that had evolved with pike-cichlids downstream from the waterfall.

Eleven years later, after thirty to sixty generations of guppies had been born, researchers revisited the stream. The experimental population had evolved. Guppies now had traits like those of guppies that had been living with killifishes for a longer time. The difference in the type of predator had influenced guppy body size, the frequency of reproduction, and other aspects of the life history patterns. Laboratory experiments with two generations of guppies confirmed that the differences have a genetic basis.

Reznick and Endler showed that life history traits, like other characteristics, can be inherited. They demonstrated that these traits can evolve. Traits that affect life history can be altered in a surprisingly short time in response to particular selection pressures.

45.7 Human Population Growth

LINKS TO
SECTIONS
21.8, 26.15

Human population size surpassed 6.4 billion in 2005. Take a look now at what the number means.

THE HUMAN POPULATION TODAY

Worldwide, the average rate of increase for the human population in 2004 was 1.3 percent. As long as birth rates continue to exceed death rates, annual additions to the population will drive a *larger* absolute increase each year into the foreseeable future.

Human population size is expanding even though more than 1 billion people are already confronted by limits to growth. They are malnourished or starving (Figure 45.14). They do not have clean drinking water, adequate shelter, access to health care systems, and sewage treatment facilities. Most of the population is expanding in already overcrowded parts of the world. Figure 45.15 is a graphic clue to what the expansion means with respect to the carrying capacity.

Even if it were possible to double food supplies to keep pace with growth, living conditions would still be marginal for most people. At least 10 million would continue to die each year from starvation.

For a time, it will be like the Red Queen's garden in Lewis Carroll's *Through the Looking Glass*, where one is forced to run as fast as one can to stay in the same place. What happens when our population doubles again? Can you brush the doubling aside as being too far in the future? *It is no farther removed from you than the sons and daughters of the next generation.*

EXTRAORDINARY FOUNDATIONS FOR GROWTH

How did we get into this predicament? For most of its history, the human population grew slowly. Things started to pick up about 10,000 years ago, and in the past two centuries, growth rates skyrocketed. Three trends promoted the rates of increase. First, humans gradually developed the capacity to expand into new habitats and climate zones. Second, humans increased the carrying capacity of their existing habitats. Third, human populations sidestepped limiting factors that tend to restrain the growth of other species.

Reflect on the first point. Early humans evolved in woodlands, then in savannas. They were vegetarians, mostly, but they also scavenged bits of meat. Bands of hunter–gatherers moved out of Africa about 2 million years ago. By 40,000 years ago, their descendants were established in much of the world (Section 26.15).

Few species can expand into such a broad range of habitats. Having a truly complex brain, early humans drew on learning and memory to figure out how to build fires, make shelters, make clothing, make tools, and cooperate in hunts. With the advent of language, knowledge did not die with the individual. It spread quickly among groups. *The human population expanded into diverse environments far more rapidly compared to the long-term geographic dispersals of other species.*

Reflect on the second point. Starting about 11,000 years ago, many bands of hunter–gatherers shifted to agriculture. Instead of simply following the migratory game herds, they settled in fertile valleys and other regions that favored seasonal harvesting of fruits and grains. In this way, they developed a more dependable basis for life. A pivotal factor was the domestication of wild grasses, including species ancestral to modern wheat and rice. People harvested, stored, and planted seeds all in one place. They domesticated animals for food and pulling plows. They dug irrigation ditches and diverted water to croplands.

Agricultural productivity was a basis for increases in population growth rates. Towns and cities formed. Later in time, food supplies increased again, and yet again, by the use of chemical fertilizers, herbicides, and pesticides. Transportation improved, as did food distribution. *Thus, even at its simplest, management of food supplies through agriculture increased the carrying capacity for the human population.*

What about sidestepping of limiting factors? Until about 300 years ago, poor hygiene, malnutrition, and

Figure 45.14 Far from well-fed humans who live in the highly developed countries, an Ethiopian showing some morphological outcomes of starvation. Ethiopia is one of the poorest developing countries, with an annual per capita income of 120 dollars. Average caloric intake is more than 25 percent below the minimum required to maintain good health. The population has one of the highest annual rates of increase—about 2.7 percent in 2003. In that year Ethiopia's total fertility rate was 5.9 children per woman. In addition, Ethiopia is being torn apart by prolonged civil war. Even so, in 2003, its population of 74 million was expected to double in less than twenty-five years.

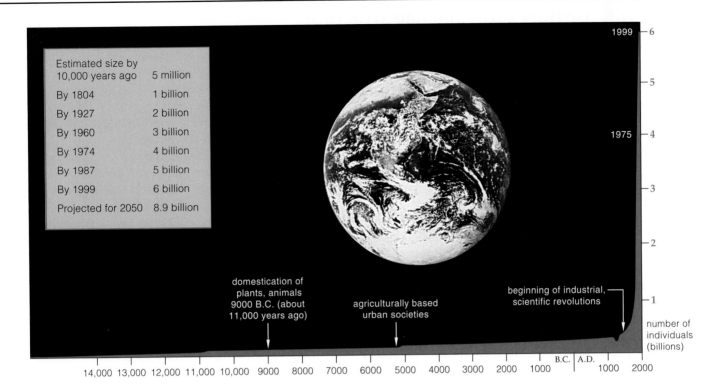

Estimated size by 10,000 years ago	5 million
By 1804	1 billion
By 1927	2 billion
By 1960	3 billion
By 1974	4 billion
By 1987	5 billion
By 1999	6 billion
Projected for 2050	8.9 billion

domestication of plants, animals 9000 B.C. (about 11,000 years ago)

agriculturally based urban societies

beginning of industrial, scientific revolutions

number of individuals (billions)

14,000 13,000 12,000 11,000 10,000 9000 8000 7000 6000 5000 4000 3000 2000 1000 | B.C. A.D. | 1000 2000

Figure 45.15 Growth curve (*red*) for the world human population. The *blue* box indicates how long it took for the human population to increase from 5 million to 6 billion. The dip between years 1347 and 1351 marks the time when 60 million people died during a bubonic plague, as explained in the Chapter 18 introduction.

infectious diseases kept death rates high enough to more or less balance birth rates. Infectious diseases became density-dependent controls. Epidemics swept through overcrowded settlements and cities that were infested with fleas and rodents. Then came plumbing and new methods of sewage treatment. Over time, vaccines, antibiotics, and other drugs were developed as weapons against many pathogens. The death rates dropped sharply. Births began to exceed deaths—and rapid population growth was under way.

In the industrial revolution of the mid-eighteenth century, people discovered how to harness the energy stored in fossil fuels, starting with coal. Within a few decades, large industrialized societies began to form in western Europe and North America. The urgency of World War I sparked the development of even more technologies. After the war, factories mass-produced cars, tractors, and other affordable goods. Advances in agriculture meant that fewer farmers were required to support a larger population.

In sum, by controlling disease agents and tapping into fossil fuels—forms of energy already concentrated for the taking—the human population has managed to sidestep big factors that had previously limited its rate of increase.

Where have the far-flung dispersals and stunning advances in agriculture, industrialization, and health care taken us? Starting with *Homo habilis*, it took about 2.5 million years for human population size to reach 1 billion. As Figure 45.15 shows, it took just 123 years to reach 2 billion, 33 more to reach 3 billion, 14 more to reach 4 billion, and then 13 more to get to 5 billion. It

took only 12 more years to arrive at 6 billion! Given the principles governing population growth, we may expect the rate of increase to decline as birth rates fall or death rates rise. Alternatively, the rates of increase may continue to rise if breakthroughs in technology expand the carrying capacity. *Even so, continued growth cannot be sustained indefinitely.*

Why? Continuing increases in population size are invitations to certain density-dependent controls. For instance, globe-hopping travelers introduce pathogens to dense urban areas all around the world in a matter of weeks (Section 21.8). Also, emigration away from economic hardship and civil strife have put 50 million individuals on the move within and between nations. Will relocations of so many individuals be peaceable? How much food, clean water, and other resources will become available to them, wherever they end up?

Through expansion into new habitats, cultural interventions, and technological innovations, the human population has temporarily skirted environmental resistance to growth.

As population increases, density-dependent controls, such as disease and competition for resources, may slow growth.

45.8 Fertility Rates and Age Structure

LINK TO
SECTION
39.10

Acknowledgment of the risks posed by rising populations has resulted in increased family planning in almost every region. Putting the brakes on population growth is not easy, and numbers are expected to continue increasing.

Most governments recognize that population growth, resource depletion, pollution, and the quality of life are interconnected. Most are working to lower long-term birth rates, as with family planning programs. Details vary among countries, but most are offering information on available methods of fertility control.

The attempts are having impact. Birth rates are now slowing worldwide. Death rates are declining, mainly because improved diets and health care are lowering infant mortality rates: the number of infants per 1,000 who die in their first year. However, AIDS sent death rates soaring in some African countries (Section 39.10).

We still expect the world population to peak at 8.9 billion by 2050 and possibly to decline near the end of the century. Think about all the resources that will be required. We will have to boost food production and find more sources of energy and fresh water to meet even basic needs, something that still eludes close to half of the population. The large-scale manipulations of resources will intensify pollution.

We expect to see the most growth in India, China, Pakistan, Nigeria, Bangladesh, and Indonesia, in that order. China (with 1.3 billion people) and India (with 1.09 billion) dwarf other countries; together, they make up 38 percent of the world population. Next in line is the United States, with 294 million.

The **total fertility rate** (TFR) is the average number of children born to the women of a population during their reproductive years. TFR estimates are based on current age-specific rates. In 1950, the worldwide TFR averaged 6.5. Currently it is 2.8, which is still far above the replacement level of 2.1—or the average number of children a couple must bear to replace themselves.

These numbers are averages. TFRs are at or below replacement levels in many developed countries; the developing countries in western Asia and Africa have the highest. Figure 45.16 has some examples of the disparities in demographic indicators.

Comparing the age structure diagrams for different populations is revealing. In Figure 45.17, focus on the reproductive age category for the next fifteen years. The average range for childbearing years is 15–49. We can expect populations with a broad base to increase in size at a faster rate. The United States population has a relatively narrow base and is undergoing slow growth, which has implications for 78 million *baby-boomers* (Figure 45.18). This cohort started forming in 1946 when American soldiers came home after World War II and began to raise families. The cohort is big, and the workforce must sustain their retirement.

Even if every couple decides to bear no more than two children, world population growth will not slow for sixty years, because 1.9 billion are about to enter the reproductive age bracket. *More than one-third of the world population is in the broad pre-reproductive base.*

China has the most wide-reaching family planning program. Its government discourages premarital sex; it urges people to delay marriage and limit families to one or two children. It offers abortions, contraceptives, and sterilization at no cost to married couples. Even in remote rural areas, paramedics and mobile units offer access to these measures. Couples who follow these guidelines receive more food, free medical care, better housing, and salary bonuses. Their offspring receive free tuition and preferential treatment when they are old enough to enter the job market. Parents who have more than two children lose government benefits and pay more taxes.

Although the policy might sound harsh, it works. Since 1972, China's TFR has fallen sharply, from 5.7 to 1.8. An unintended consequence has been a shift in the country's sex ratio. Traditional cultural preference

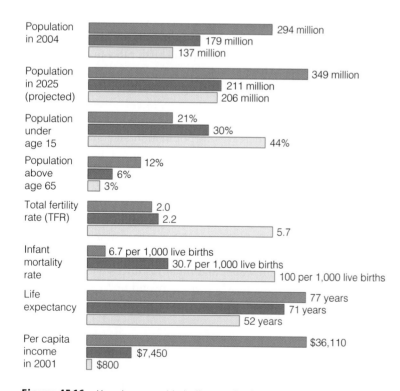

Figure 45.16 Key demographic indicators for three countries, mainly in 2004. The United States (*brown* bar) is highly developed, Brazil (*red* bar) is moderately developed, and Nigeria (*gold* bar) is less developed.

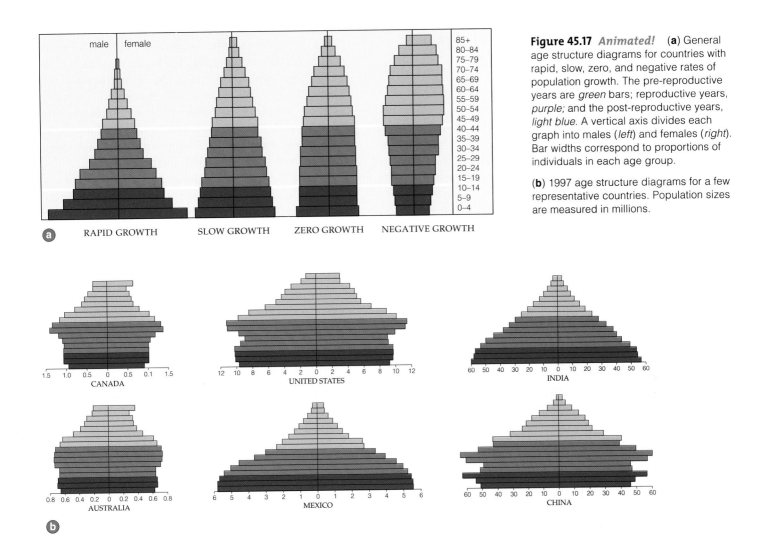

male | female

85+
80–84
75–79
70–74
65–69
60–64
55–59
50–54
45–49
40–44
35–39
30–34
25–29
20–24
15–19
10–14
5–9
0–4

a RAPID GROWTH   SLOW GROWTH   ZERO GROWTH   NEGATIVE GROWTH

Figure 45.17 *Animated!* (**a**) General age structure diagrams for countries with rapid, slow, zero, and negative rates of population growth. The pre-reproductive years are *green* bars; reproductive years, *purple;* and the post-reproductive years, *light blue.* A vertical axis divides each graph into males (*left*) and females (*right*). Bar widths correspond to proportions of individuals in each age group.

(**b**) 1997 age structure diagrams for a few representative countries. Population sizes are measured in millions.

1.5 1.0 0.5 0 0.5 0.1 1.5
CANADA

12 10 8 6 4 2 0 2 4 6 8 10 12
UNITED STATES

60 50 40 30 20 10 0 10 20 30 40 50 60
INDIA

0.8 0.6 0.4 0.2 0 0.2 0.4 0.6 0.8
AUSTRALIA

6 5 4 3 2 1 0 1 2 3 4 5 6
MEXICO

60 50 40 30 20 10 0 10 20 30 40 50 60
CHINA

b

for sons, especially in the rural areas, has led some parents to abort developing females or even commit infanticide. Worldwide, 1.06 boys are born for every girl, but in China the latest census reports 1.19 boys per girl. Also, more than 100,000 girls are abandoned each year. The government now is offering additional cash and tax incentives to the parents of girls. In the meantime, China's population time bomb continues to tick. About 150 million Chinese girls are now in the pre-reproductive age category.

The worldwide total fertility rate has been dropping, but it is still above the replacement level that would move the population growth rate close to zero.

Most countries support family planning programs of some sort. Even with the slowdowns, the human population will continue to increase; its pre-reproductive base is immense.

At present, more than one-third of the human population is in a very broad pre-reproductive base.

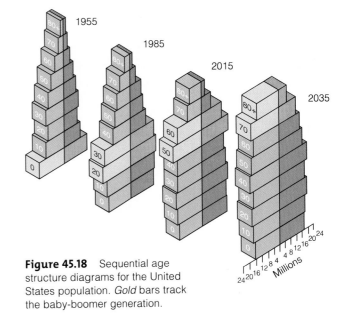

Figure 45.18 Sequential age structure diagrams for the United States population. *Gold* bars track the baby-boomer generation.

45.9 Population Growth and Economic Effects

LINK TO
SECTION
39.10

The most highly developed countries have the slowest growth rates and use the most resources.

DEMOGRAPHIC TRANSITIONS

Changes in population growth rates often correlate with four stages of economic development, the heart of the **demographic transition model**. By this model, living conditions are harshest in a *preindustrial* stage, before technology and medical advances spread. Birth and death rates are high, so the rate of growth is low. In the *transitional* stage, industrialization begins. Food production and health care improve and death rates slow. Birth rates stay high in the agricultural societies, where big families provide help in the fields. Annual growth rates are between 2.5 and 3 percent. As living conditions improve and birth rates begin to decline, growth generally starts to level off (Figure 45.19).

In the *industrial* stage, industrialization is in full swing and growth slows. People move to cities, and couples often want small families. As they accumulate goods, many decide the time and cost of raising more than a few children conflict with their goals.

In the *postindustrial* stage, population growth rates become negative. The birth rate falls below the death rate, and population size slowly decreases.

The United States, Canada, Australia, and most of western Europe, Japan, and much of the former Soviet Union are in the industrial stage. Most developing countries, such as Mexico, are now in the transitional stage, without enough skilled workers to complete the transition to a fully industrial economy.

By some projections, many developing countries will make the demographic transition to an industrial stage in the next few decades. However, there also are signs that the still-rapid population growth in many of those countries will overwhelm economic growth, food production, and health care systems. If that were to happen, they would be trapped demographically, unable to pass through the transition stage.

Africa and some other developing countries already are caught in the demographic trap. Here you might wish to reflect again on the magnitude of the AIDS pandemic that is wreaking havoc on populations, as in sub-Saharan Africa (Section 39.10). One outcome is that many African populations are being driven back to the lowest stage of economic development.

The demographic transition model might not apply to many developing countries, because the conditions on which it is based no longer prevail in some places. For instance, how many can compete in a new global economy without a base of high-tech workers? How many have funds for fast economic growth? How much of what they do have is used to pay interest on debts already owed? Recognizing the problem, in 2004, the world's richest nations agreed to write off 40 billion dollars owed to them by the poorest nations.

A QUESTION OF RESOURCE CONSUMPTION

The industrialized nations use the most resources. For example, the United States has about 4.6 percent of the world's population and produces about 21 percent of all goods and services. Yet it requires thirty-five times

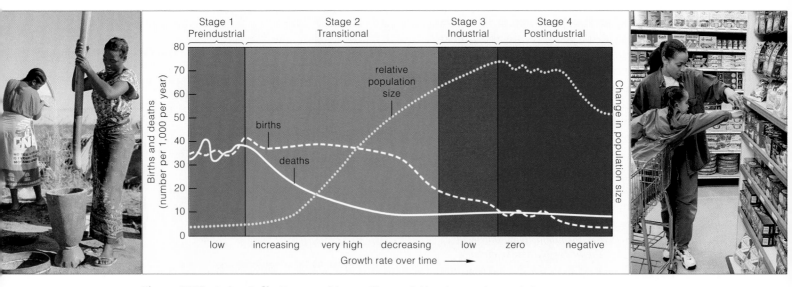

Figure 45.19 *Animated!* Demographic transition model for changes in population growth rates and sizes, correlated with long-term changes in the economy.

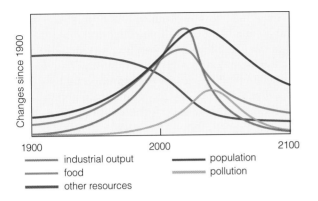

Figure 45.20 Computer-based projection of what might happen if human population size continues to skyrocket without dramatic policy changes and technological innovation. The assumptions were that the population has already overshot the carrying capacity and current trends will continue unchanged.

Chart legend:
— industrial output
— food
— other resources
— population
— pollution

1900   2000   2100

Changes since 1900

more goods and services than India. It uses 25 percent or so of the world's processed minerals and much of the available energy supplies. The United States is not alone in this. China and India are now demanding an ever increasing share of the economic pie.

G. Tyler Miller once estimated that it would take 12.9 billion impoverished individuals living in India to have as much impact on the environment as 284 million people living at the time in the United States. He pointed out that the projected increase in human population growth rates raises serious questions. Will there be enough food, energy, water, and other basic resources to sustain so many people? Will regional governments be able to provide adequate education, housing, medical care, and other social services for them? Computer models suggest not (Figure 45.20). Even so, some analysts claim we can adapt politically and socially to a more crowded world if innovative technologies improve harvests, if food resources are distributed more equitably, and if dietary preferences are shifted away from animal products.

There are no easy answers. If you have not been doing so, start following the arguments in the media. It is a good idea to become an informed participant in a debate that will have impact on your future.

Differences in population growth and resource consumption among countries can be correlated with levels of economic development. Growth rates are typically greatest during the transition to industrialization.

Global conditions have so changed that the demographic transition model may no longer apply to many nations.

45.10 Social Impact of No Growth

For humans, as for all species, the biological implications of exponential population growth are sobering. However, so are the social implications of what would happen if human population growth were to approach zero.

In a growing population, most of the individuals are in lower age brackets. When living conditions favor moderate growth, then the age distribution should guarantee a future workforce. But the distribution has social implications. Why? It takes a large workforce to support individuals in the higher age brackets.

In the United States, most seniors who have retired expect the government to subsidize their medical care and provide low-cost housing as well as many other social programs. Remember the baby-boomers?

However, as one outcome of better medicine and hygiene, people in these brackets are living far longer than seniors did when the nationwide social security program was established. Cash benefits now exceed contributions that individuals made to the program when *they* were younger.

If the human population does reach and maintain zero growth, a larger proportion of individuals will end up in higher age brackets. Even slower growth poses problems. Will these people continue to receive goods and services as the workforce carries more and more of the economic burden? Put it to yourself. How much economic hardship are you willing to bear for the sake of your parents? Your grandparents? How much will your children bear for you?

We have arrived at a turning point, not only in our biological evolution but in our cultural evolution as well. The decisions awaiting us are among the most pressing and difficult we will ever have to make.

All species face limits to growth. We might think we are different from the rest, and in some respects we are. The uniquely human capacity to undergo rapid cultural evolution has helped us postpone the action of most factors that limit growth. However, the crucial word is *postpone*. On the basis of all the models that are available to us, we can be fairly sure of this: Much of the human population will not escape the impact of limiting factors in the environment.

The human population has sidestepped a number of the constraints that typically restrict the population growth of other species.

By doing so, staggering numbers of individuals have now become more vulnerable to the laws of nature that cannot be repealed.

Summary

Sections 45.1, 45.2 A population is a group of individuals of the same species in a specified area. It is characterized in part by size, density, distribution, and age structure, which are measurable. Most populations in nature have a clumped distribution pattern.

Counting the number of individuals in quadrats is one way to estimate the density of a population in a specified area. Using capture–mark–recapture methods is a way to estimate the density for mobile animals.

Biology ⊜ Now
Learn how to estimate population size with the interaction on BiologyNow.

Section 45.3 The growth rate for a population in a specified interval depends on the rates of birth, death, immigration, and emigration. By putting aside the effects of immigration and emigration, we may then represent population growth (G) as

$$G = rN$$

where r is the net reproduction per individual per unit time and N is the number of individuals.

In cases of exponential growth, a population's rate of growth is proportional to its size. The reproductive base, and population size, increases at a fixed rate in a given interval. This trend plots out as a J-shaped growth curve. The rate of growth may be slow or rapid. As long as the population's per capita birth rate remains above its per capita death rate, it shows exponential growth.

Biology ⊜ Now
Observe a pattern of exponential growth with the animation on BiologyNow.

Section 45.4 The maximum number of individuals of a population that can be sustained indefinitely by the resources in their environment is called the carrying capacity. Food and other essential resources, disease, competition, and predation are examples of density-dependent factors that can limit population growth. Density-independent factors affect growth regardless of how crowded the individuals are.

Unlike exponential growth, a logistic growth pattern plots out as an S-shaped curve. As one example, a small population increases slowly in size, then rapidly, then levels off once the carrying capacity is reached.

Biology ⊜ Now
Learn about logistic growth on BiologyNow.

Sections 45.5. 45.6 Each species has a life history pattern characterized by the age at first reproduction, number of offspring per generation, life span, and other traits. Three general types of survivorship curves are common: a high death rate late in life, a constant rate at all ages, or a high rate early in life. Many aspects of life histories have a genetic basis, are subject to natural selection, and differ among populations and species.

Section 45.7 The human population has surpassed 6.4 billion. Its rapid growth in the past two centuries

occurred through expansion into many diverse habitats and through agricultural, medical, and technological developments that raised carrying capacity.

Section 45.8 Family planning refers to societal efforts to slow population growth. The total fertility rate (TFR) is the average number of children born to women of a population during their reproductive years. The global TFR is declining. Even so, the pre-reproductive base of the world population is so large that human population size will continue to increase for at least sixty years.

Biology ⊜ Now
Compare age structure diagrams with the interaction on BiologyNow.

Section 45.9 The demographic transition model correlates industrial and economic development with changes in population growth rates, although global conditions have changed so much that the model may no longer apply to many of the developing nations. Per capita consumption of resources in developed nations is far higher than it is in developing nations.

Biology ⊜ Now
Learn about the demographic transition model with the animation on BiologyNow.

Section 45.10 Zero population growth has social repercussions, as when a population's age structure is such that the number of older individuals exceeds the number of young workers that must support them.

Self-Quiz   *Answers in Appendix II*

1. The rate at which a population grows or declines depends on the rate of _____ .
   a. births   c. immigration   e. a and b
   b. deaths   d. emigration   f. all of the above

2. Populations grow exponentially when _____ .
   a. its rate of increase is proportional to the size of its reproductive base in a given interval
   b. the size of a low-density population increases slowly, then quickly, then levels off once the carrying capacity is reached
   c. a and b are characteristics of exponential growth

3. For a given species, the maximum rate of increase per individual under ideal conditions is its _____ .
   a. biotic potential   c. environmental resistance
   b. carrying capacity   d. density control

4. Resource competition, disease, and predation are _____ controls on population growth rates.
   a. density-independent   c. age-specific
   b. population-sustaining   d. density-dependent

5. A life history pattern for a population is a set of adaptations that influence the individual's _____ .
   a. longevity   c. age at reproductive maturity
   b. fertility   d. all of the above

6. In 2004, the worldwide average rate of increase for the human population at midyear was _____ percent.
   a. 0   c. 1.3   e. 3.8
   b. 0.5   d. 2.7   f. 4.6

Figure 45.21 Saguaros (*Canegiea gigantea*) growing very slowly in Arizona's Sonoran desert.

Figure 45.22 A young Malian, with a 10 percent chance of becoming a mother by age fifteen, and a 50 percent chance before nineteen. Mali's TFR, about 7, is one of the world's highest.

7. Match each term with its most suitable description.

____ carrying capacity	a. maximum rate of increase per individual under ideal conditions
____ exponential growth	b. population growth plots out as an S-shaped curve
____ biotic potential	c. maximum number of individuals sustainable by the resources in a given environment
____ limiting factor	d. population growth plots out as a J-shaped curve
____ logistic growth	e. essential resource that restricts population growth when scarce

Additional questions are available on Biology⬛Now™

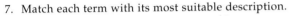

Critical Thinking

1. If house cats that have not been neutered or spayed live up to their biotic potential, two can be the start of many kittens—12 the first year, 72 the second year, 429 the third, 2,574 the fourth, 15,416 the fifth, 92,332 the sixth, 553,019 the seventh, 3,312,280 the eighth, and 19,838,741 kittens the ninth year. Is this a case of logistic growth? Exponential growth? Irresponsible cat owners?

2. Reflect on Section 45.6. When researchers moved guppies from populations that were prey of pike-cichlids to a habitat with killifish, the life histories of the transplanted guppies evolved. They came to resemble those of guppy populations that had been preyed upon by killifish. The age of first reproduction increased, as did body size. The males became gaudier. Some of their scales formed larger, more colorful spots. How could a decrease in predation pressure on sexually mature fish influence male guppy coloration?

3. Each summer, a giant saguaro cactus produces tens of thousands of tiny black seeds. Most die, but a few land in a sheltered spot and sprout the following spring. The saguaro is a slow-growing CAM plant (Section 7.7). After fifteen years, it may be only knee high, and it will not flower for another fifteen years. It may live for 200 years. Saguaros share their habitat with annuals, such as poppies, that sprout, form seeds, and die in just a few weeks (Figure 45.21). Speculate on how these different life histories can both be adaptive in the same desert environment.

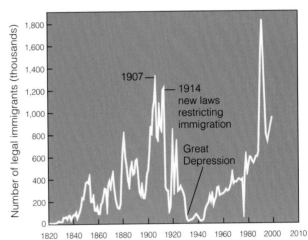

Figure 45.23 Chart of legal immigration to the United States between 1820 and 2000.

4. A third of the world population is younger than fifteen (Figure 45.22). Describe the effect this age distribution will have on the human population's growth rate. If you suspect it will have severe impact, what humane recommendations would you make to encourage individuals of this age group to limit their family size? What are some social, economic, and environmental factors that might prevent them from following the recommendations?

5. Figure 45.23 charts the legal immigration to the United States between 1820 and 2000. The greatest increase came after the Immigration Reform and Control Act of 1986 gave legal status to undocumented immigrants who proved they had lived in the country for years. Economic downturns in the 1980s and 1990s fanned resentment against immigrants.

Many people in the United States would like to limit legal immigration to 300,000–450,000 per year and to deport undocumented individuals. Others would like to have open borders; they say a rigidly enforced documentation policy would discriminate against legal immigrants of the same ethnic background. Do some research, then write an essay on the arguments on both sides of this volatile issue, which has social and economic ramifications.

6. Write a short essay about a population having one of the age structures signified by the diagrams at right. Speculate on that population's current economic status and the social and economic problems it may face in the future.

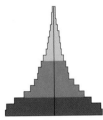

46 COMMUNITY STRUCTURE AND BIODIVERSITY

Fire Ants in the Pants

Solenopsis richteri and *S. invicta* entered the United States in the 1930s, probably as stowaways on cargo ships. These two species of Argentine fire ants infiltrated communities throughout the Northern Hemisphere, starting with the southeastern states. *S. invicta*, the imported red fire ant, recently colonized Southern California and New Mexico.

Disturb a fire ant that crawled onto your skin and it will bite down even as it pumps venom into you through a stinger. Searing pain follows, then a pus-filled bump forms where you were bitten (Figure 46.1). At one time or another, about half of all the Americans who live where fire ants are common have been stung. More than eighty people have died from the attacks.

Imported fire ants menace more than people. These insects attack just about anything that disturbs them, including livestock, pets, and wildlife. They also are more competitive than native ant species and other animals that feed on insects. The imports may be contributing to the declines of some native wildlife species.

To give an example, the Texas horned lizard (*Phrynosoma cornutum*) vanished from most of its home range when red fire ants moved in and displaced the native ants. To

the Texas horned lizard, native ants are the food of choice, and it cannot tolerate eating the invaders.

The young of other lizard species are worse off. So are the hatchlings of quail and other ground-nesting birds. Fire ants swarm all over them and kill them directly.

Invicta means "invincible" in Latin. So far, *S. invicta* is living up to its species name. Pesticides have not slowed its invasions of new habitats. To the contrary, they might even be facilitating the invasions by wiping out most of the native ant populations.

Ecologists are enlisting biological controls. Two phorid fly species attack *S. invicta* in its native habitat. Both are parasitoids, a specialized type of parasite that kills its host in a rather gruesome way. A female fly pierces the cuticle of an adult ant, then lays an egg in the ant's soft tissues. The egg hatches into a larva, which grows and then eats

Figure 46.1 Fire ant mounds in west Texas, and agitated fire ants swarming over a leather boot. *Facing page*, skin eruptions that typically follow a concerted attack by these exotic imports.

IMPACTS, ISSUES

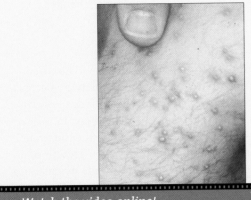

Watch the video online!

its way through tissues to the ant's head. After it gets big enough, the larva secretes an enzyme that makes the head fall off. The larva, sheltered inside the detached, cuticle-covered head, undergoes metamorphosis into an adult.

The flies are choosy about where they lay their eggs. Native ants are not candidates. Knowing this, ecologists released one of the parasitic fly species in Florida in 1997. They released a second species in other southern states in 2001. It is too soon to know whether these biological controls are working.

As ecologists wait for results, they are exploring other options. One idea is to use imported, pathogenic fungi or protists that will infect *S. invicta* but not the native ants. Another idea is to introduce a parasitic ant species that invades *S. invicta* colonies and decapitates the queens.

This example invites you into the sometimes rough-and-tumble aspects of **community structure**, or patterns in the number of species and their relative abundances. As you will see, species interactions and disturbances to the habitat shift community structure in small and large ways—some predictable, others unexpected.

✓ How Would You Vote?

Currently, only a fraction of the crates being imported into the United States are inspected for the inadvertent or deliberate presence of exotic species. Would the cost of added inspections be worth it? See BiologyNow for details, then vote online.

Key Concepts

COMMUNITY CHARACTERISTICS

A community consists of all species in a habitat. Each species has a niche—the sum of activities and relationships in which its individuals take part as they secure and use vital resources. The habitat's history and characteristics, resource availability over time, and the history, adaptations, and interactions of its array of species shape community structure.   Section 46.1

FORMS OF SPECIES INTERACTIONS

Commensalism, mutualism, competition, predation, and parasitism are forms of symbiotic interactions that directly involve two or more species.   Sections 46.2–46.7

COMMUNITY STABILITY AND CHANGE

By an older model, a predictable succession of species in a habitat stabilizes as a climax community, which thereafter does not change much. It now appears that abiotic and human-created disturbances are more significant factors in shaping community structure.   Sections 46.8–46.10

GLOBAL PATTERNS IN COMMUNITY STRUCTURE

Biogeographers identify patterns in species richness of mainland and island communities around the world. Two of the most striking patterns correlate with distance from the equator and from colonizing sources.   Section 46.11

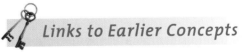

Links to Earlier Concepts

In this chapter, you will see how studies from diverse fields of inquiry often converge to explain big patterns in life. You will draw on the Unit IV survey of biodiversity. You will deepen your sense of the challenges facing conservation biology, as by species introductions (Chapter 27). You will revisit biogeography and take a closer look at global patterns in species richness (Sections 17.1, 17.3). You also will see how microevolutionary processes and population dynamics can influence community structure (18.4, 18.7, 45.1, 45.5). You will come across modern expressions of evolutionary arms races that started in the Cambrian seas (26.2).

46.1 Which Factors Shape Community Structure?

*The type of place where each organism normally lives is its **habitat**. All species that directly or indirectly associate with one another in a habitat represent a **community**.*

Each community has a characteristic structure, which we define by **species richness**—the number of species—and their relative abundances. That structure arises largely in response to these abiotic and biotic factors:

1. The physical and chemical conditions that prevail in the habitat, such as temperature, rainfall, soil type, size, and annual incoming solar radiation.

2. The type, amount, and seasonal availability of food and other resources, as in Figure 46.2.

3. The evolutionary history of the habitat and of each resident species.

4. The morphological, physiological, and behavioral traits that help species survive in the habitat.

5. Interactions among species.

6. Natural and human-induced physical disturbances that vary unpredictably in magnitude and frequency.

It will take more than one chapter to survey these factors. Chapter 47, for example, focuses on energy flow and nutrient cycling among species. Here we start with the niche of each species in the community.

THE NICHE

All species of a community share the same habitat—the same "address"—but each has a "profession" that sets it apart. It has a distinct **niche**, the sum of its activities and interactions as it goes about acquiring and using the resources it must have to survive and reproduce. Its *fundamental* niche would prevail even in the absence of competition or any other factors that might limit how individuals get and use resources.

However, constraining factors come into play, and they tend to bring about a more limited, *realized* niche. The realized niche is dynamic. It shifts over time, in small or large ways, in response to a mosaic of changes.

CATEGORIES OF SPECIES INTERACTIONS

Even in the simplest communities, dozens to hundreds of species interact. Interactions between two species may have indirect effects on others, but focus now on five forms of *symbiosis*, or close associations between two or more species during part or all of the life cycle. Each can promote or suppress population growth of a participating species. Let's simplify things by casting the definitions in terms of two-species interactions.

Commensalism directly helps one species but affects the other little, if at all. A bird may get a roosting site from a tree, which gets no benefit but is not harmed. In **mutualism**, both species benefit. Don't think of this as cozy cooperation; the benefits flow from a two-way exploitation. In **interspecific competition**, one species wins or loses with respect to access to some resource. **Predation** and **parasitism** directly benefit one species. Predators typically kill and eat prey. Parasites live in or on hosts and weaken but rarely kill them outright.

A habitat is the type of place where individuals of a species normally live. All of its species form a community. The community's structure arises from a habitat's physical and chemical features, resource availability over time, adaptive traits of its species, how its species interact, and the history of the habitat and its occupants.

A niche is the sum of all activities and relationships in which individuals of a species engage as they secure and use the resources necessary to survive and reproduce.

Commensalism, mutualism, competition, predation, and parasitism are all forms of symbiotic interactions.

Figure 46.2 Three of twelve fruit-eating pigeon species in Papua New Guinea's tropical rain forests. *Left to right*, the tiny pied imperial pigeon, the superb crowned fruit pigeon, and the turkey-sized Victoria crowned pigeon. The forest's trees differ in the size of fruit and fruit-bearing branches. The big pigeons eat big fruit. Smaller ones, with smaller bills, cannot peck open big, thick-skinned fruit. They eat small, soft fruit on branches too spindly to hold big pigeons.

Trees feed the birds, which help the trees. Seeds in fruit resist digestion in the bird gut. Flying pigeons disperse seed-rich droppings, often some distance from tall, mature trees that are established competitors for water, minerals, and sunlight. With dispersal, some seedlings have a better chance to take hold.

46.2 Mutualism

In a mutualistic interaction, two species take advantage of their partner in ways that benefit both, as when one withdraws nutrients from the other while sheltering it.

Interactions in which positive benefits flow both ways abound in nature. Remember Section 31.2? Flowering plants and the insects, birds, bats, and other animals that pollinate them are vivid examples. Similarly, rain forest trees give pigeons food, and pigeons disperse seeds from the trees to new sites (Figure 46.2).

In *facultative* mutualism, the interaction is helpful but not vital. Ants and aphids get along well without each other, but ants do protect aphids as they feed on sugar droplets exuding from aphids (Figure 31.14).

With *obligatory* mutualism, each species must have access to the other in order to complete its life cycle and reproduce. Yucca plants and the yucca moths that pollinate them are obligatory mutualists (Figure 46.3). So are fungi that interact with a photobiont in lichens or with plant roots in mycorrhizae (Section 24.6). In a mycorrhiza, fungal hyphae penetrate root cells or form a dense, velvety mat around them. The plant pilfers mineral ions from the fungus, which absorbs far more ions than the plant could do on its own. The fungus pilfers a few photosynthetic products—sugars—from the plant. The fungus depends on the plant mutualist for its reproductive success. It will stop making spores if the plant stops photosynthesizing.

Anemone fishes can hide out among the tentacles of one or more species of sea anemones; a thick coat of mucus makes them impervious to nematocysts. In one mutualistic interaction, the sea anemone benefits, too. An aggressive anemone fish chases off another kind of fish that likes to eat the tentacles (Figure 46.4).

Or reflect on the apparent endosymbiotic origin of eukaryotes (Section 20.4). Long ago, phagocytes were engulfing aerobic bacterial cells—but some resisted digestion, tapped into host nutrients, and then kept reproducing independently of the host cell body. In time, the hosts came to depend on the ATP produced by the guests—which evolved into mitochondria and chloroplasts. If those ancient prokaryotic cells had not coevolved as mutualists, you and all other eukaryotic species would not be around today.

Mutualism is a common form of symbiosis. Each species benefits as it exploits a partner in some way that helps assure its own reproductive success. In cases of obligatory mutualism, one or both partners cannot complete its life cycle in the absence of the interaction.

LINKS TO
SECTIONS 20.4,
24.6, 28.5, 31.2

Figure 46.3 Mutualism on a rocky slope of the high desert in Colorado.

Only one yucca moth species pollinates plants of each *Yucca* species; it cannot complete its life cycle with any other plant. The moth matures when yucca flowers blossom. The female has specialized mouthparts that collect and roll sticky pollen into a ball. She flies to another flower and pierces its ovary, where seeds will form and develop, and lays eggs inside. As she crawls out, she pushes a ball of pollen onto the flower's pollen-receiving platform.

After pollen grains germinate, they give rise to pollen tubes, which grow through the ovary tissues and deliver sperm to the plant's eggs. Seeds develop after fertilization.

Meanwhile, moth eggs develop into larvae that eat a few seeds, then gnaw their way out of the ovary. Seeds that larvae do not eat give rise to new yucca plants.

Figure 46.4 The sea anemone *Heteractis magnifica*, which shelters about a dozen fish species. It has a mutualistic association with the pink anemone fish (*Amphiprion perideraion*). This tiny but aggressive fish chases away predatory butterfly fishes that bite off the tips of its partner's tentacles. In return, the fish and its eggs get protection and shelter—scarce commodities on tropical reefs (Section 27.5).

46.3 Competitive Interactions

LINKS TO
SECTIONS
29.5, 45.4

Where you come across limited supplies of energy, nutrients, living space, and other natural resources, there you are likely to find organisms competing for a share of them.

Competition for resources is typically intense between individuals of the same species (Chapters 45 and 49). At the community level, competition between species usually is not as intense. Why not? *The requirements of two species may be similar but are never as close as they are among individuals of the same species.* Let's consider two forms of interspecific competition.

In **interference competition**, one species controls or blocks access of another species to some resource, regardless of its abundance. The leaves of some plant species exude aromatic compounds that taint soil and prevent potential competitors from taking root. A few aggressive chipmunk species keep others out of their habitats (Figure 46.5). In spring and early summer, a male broadtailed hummingbird evicts birds of its own species from its richly flowered territory in the Rocky Mountains. In August, however, *rufous* hummingbirds migrate through the Rockies on their way to Mexico. Until they fly on, the stronger, more aggressive rufous males force the male broadtails to give up territory.

In **exploitative competition**, different species have equal access to a resource, but one is better at using it. In one experiment, a large and a small species of water

flea (*Daphnia*) that feed on the same alga were grown together in an alga-enriched culture flask. The larger species increased in body mass. Also, its population expanded. The smaller species lost body mass, and its population shrank—which leads us to a theory.

THEORY OF COMPETITIVE EXCLUSION

Any two species differ to a greater or lesser extent in their capacity to secure and use resources. The more they overlap in these respects, the less likely they are to coexist in the same habitat.

Years ago, G. Gause found evidence of this when he grew two species of *Paramecium* separately and then together (Figure 46.6). Both of these ciliated protozoans hunt the same prey—bacteria—and compete intensely for it. Gause's species, which use identical resources, could not coexist indefinitely. Later experiments with water fleas and many other species yielded the same results, in support of what ecologists now call the theory of **competitive exclusion**.

Gause also studied two other *Paramecium* species that did not overlap much in requirements. He grew them together. One species tended to feed on bacteria suspended in culture tube liquid. The other ate yeast cells near the bottom of the tube. Population growth rates slowed for both species—but the overlap in use

Figure 46.5 Example of interspecific competition in nature. On the slopes of the Sierra Nevada, competition helps keep nine species of chipmunks (*Tamias*) in different habitats.

The alpine chipmunk (**a**) lives in the alpine zone, the highest elevation. Below it are the lodgepole pine, piñon pine, and then sagebrush habitat zones. Lodgepole pine chipmunks (**b**), least chipmunks (**c**), and other species live in the forest zones. Merriam's chipmunk (**d**) lives at the base of the mountains, in sagebrush. Its traits would allow it to move up into the pines, but the aggressively competitive behavior of forest-dwelling chipmunks won't let it. Food preferences keep the pine forest chipmunks out of the sagebrush habitat.

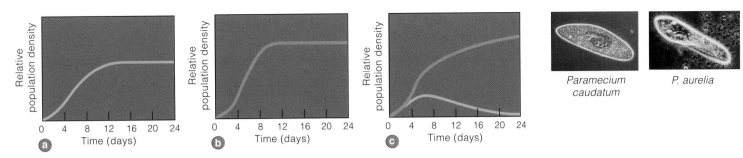

Figure 46.6 *Animated!* Results of competitive exclusion between two protozoan species that compete for the same food. (**a**) *Paramecium caudatum* and (**b**) *P. aurelia* were grown in separate culture flasks and established stable populations. The S-shaped graph curves indicate logistic growth and stability.

(**c**) Then the two species were grown together. *P. aurelia* (*brown* curve) drove *P. caudatum* toward extinction (*green* curve in **c**). This experiment and others suggest that two species cannot coexist indefinitely in the same habitat *when they require identical resources*. If their requirements do not overlap much, one might influence the population growth rate of the other, but they may still coexist.

Figure 46.7 Two coexisting species of salamanders: (**a**) *Plethodon glutinosus* complex and (**b**) *P. jordani*.

of resources was not enough for one species to fully exclude the other. The two continued to coexist.

Field experiments also reveal effects of competition. For instance, N. Hairston studied salamanders in the Balsam Mountains and Great Smoky Mountains. One species, *Plethodon glutinosus*, lives at lower elevations than its relative *P. jordani*, but the home ranges overlap in some areas (Figure 46.7). Hairston removed one or the other species from test plots in the overlap areas. He left some plots untouched as controls. Five years later, nothing had changed in those control plots; the species were coexisting. Population sizes in test plots were growing. Plots cleared of *P. jordani* had a greater proportion of *P. glutinosus*. In addition, plots cleared of *P. glutinosus* had a greater proportion of *P. jordani*.

Hairston concluded that, where populations of the two salamander species coexist in nature, competitive interactions suppress the growth rate of both.

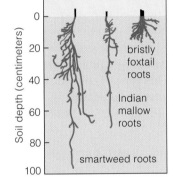

Figure 46.8 Resource partitioning among three annual plant species in an abandoned field. The plants differ in how they are adapted to secure soil water and mineral ions. The roots of each species tap into different depths of soil.

bristly foxtail

Indian mallow   smartweed

RESOURCE PARTITIONING

Think back on those fruit-eating pigeon species. They all use the same resource: fruit. Yet they overlap only a bit in their use of it, because each prefers fruits of a certain size. They are a case of **resource partitioning** —a *subdividing* of some category of similar resources, which allows competing species to coexist.

Similarly, three annual plant species live in the same plowed, abandoned field. All require sunlight, water, and minerals. Each exploits a slightly different part of the habitat (Figure 46.8). Bristly foxtail grasses have a shallow, fibrous root system that absorbs water fast during rains. They grow where moisture shifts daily, and are drought-tolerant. Indian mallow has a taproot

system in deeper soil that is moist early in spring and drier later. The taproot system of smartweed branches in topsoil and soil below the roots of other species. It grows where soil is perpetually moist (Section 29.5).

In some competitive interactions, one species controls or blocks access to a resource, regardless of whether it is scarce or abundant. In other interactions, one is better than another at exploiting a shared resource.

When two species overlap too much in their requirements, they cannot coexist in the same habitat unless they share required resources in different ways or at different times.

46.4 Predator–Prey Interactions

LINKS TO
SECTIONS
1.5, 1.6, 26.2

Predators are consumers that obtain energy and nutrients from living organisms—their prey—which they generally capture and kill. The quantity and types of prey species affect predator diversity and abundances, and the types of predators and their numbers do the same for prey.

COEVOLUTION OF PREDATORS AND PREY

Coevolution influences predator and prey interactions. The term refers to species that evolve jointly as their close ecological interaction exerts selection pressure on each other over the generations. If a gene mutation in a prey organism leads to a more effective defense against predators, the mutant allele will increase in frequency in the prey population. Its bearers and their offspring will tend to survive in greater numbers. If a gene mutation in a predator leads to a better way to overcome the novel prey defense, its bearers and their offspring will eat better; they will tend to survive and leave more descendants in the predator population.

Thus, over time, the predators are selective agents that favor improved prey defenses in prey. The prey with better defenses are selective agents that favor more effective predators. This type of coevolutionary arms race started among vertebrate predators and their prey when jawed fishes emerged (Section 26.2).

MODELS FOR PREDATOR–PREY INTERACTIONS

The extent to which predators limit numbers of prey depends on several factors. A key factor is the response of individual predators to increases or decreases in prey density. Figure 46.9*a* is an overview of the three general patterns of functional responses.

By the type I model, a predator removes a constant proportion of prey over time, regardless of levels of prey abundance. The number of prey killed in a given interval depends only on the prey density. This model applies to passive predators, such as web spiders. The more flies there are, the more get caught in webs.

By the type II model, the capacity of predators to consume and digest prey determines how many prey they capture. When prey density rises, the proportion captured rises steeply at first, then slows as predators are exposed to more prey than they can deal with at one time. Figure 46.9*b* offers an example. A wolf that just killed a caribou will not hunt another until it has eaten and digested the first one.

By the type III model, predator response is lowest when prey density is low. It is highest at intermediate prey densities, then levels off. This type of response is observed for predators that can switch to other prey when individuals of a prey species are scarce and hard to find. Predators that can make the type I and type II responses can limit prey at a stable equilibrium point.

Other factors besides individual predator response to prey density are at work. For example, predator and prey reproductive rates affect the interaction. So do hiding places for prey, the presence of other prey or predator species, and carrying capacities.

THE CANADIAN LYNX AND SNOWSHOE HARE

In some cases, shifts in environmental conditions can cause predator and prey densities to oscillate. At the lowest level, predation will strongly depress the prey density. At the highest level, predation is absent and the prey population nears the carrying capacity.

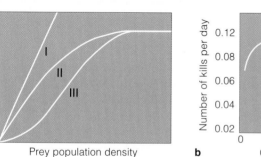

a Prey population density

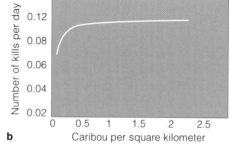

b Caribou per square kilometer

Figure 46.9 *Animated!* (**a**) Three models for responses of predators to prey density. Type I: Prey consumption rises linearly as prey density rises. Type II: Prey consumption is high at first, then levels off as predator bellies stay full. Type III: When prey density is low, it takes longer to hunt prey, so the predator response is low. (**b**) A type II response in nature. For one winter month in Alaska, B. W. Dale and his coworkers observed four wolf packs (*Canis lupus*) feeding on caribou (*Rangifer tarandus*). The interaction fit the type II model for the functional response of predators to the prey density.

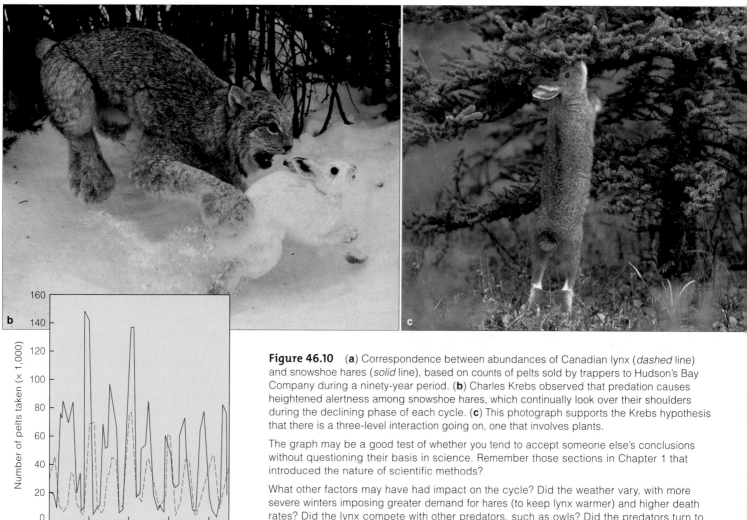

Figure 46.10 (**a**) Correspondence between abundances of Canadian lynx (*dashed* line) and snowshoe hares (*solid* line), based on counts of pelts sold by trappers to Hudson's Bay Company during a ninety-year period. (**b**) Charles Krebs observed that predation causes heightened alertness among snowshoe hares, which continually look over their shoulders during the declining phase of each cycle. (**c**) This photograph supports the Krebs hypothesis that there is a three-level interaction going on, one that involves plants.

The graph may be a good test of whether you tend to accept someone else's conclusions without questioning their basis in science. Remember those sections in Chapter 1 that introduced the nature of scientific methods?

What other factors may have had impact on the cycle? Did the weather vary, with more severe winters imposing greater demand for hares (to keep lynx warmer) and higher death rates? Did the lynx compete with other predators, such as owls? Did the predators turn to alternative prey during low points of the hare cycle? When fur prices rose in Europe, did the trapping increase? When the pelt supply outstripped the demand, did trapping decline?

Consider a ten-year oscillation in populations of a predator, the Canadian lynx, and the snowshoe hare that is its main prey (Figure 46.10). To identify the causes of this pattern, the ecologist Charles Krebs and his coworkers tracked hare population densities for ten years in Alaska, in the Yukon River Valley. They set up 1-square-kilometer control plots and experimental plots. Electric fences kept predatory mammals out of some plots. Extra food or fertilizers that fanned plant growth were placed in other plots. The team captured and released more than a thousand snowshoe hares, lynx, and other animals, giving each a radio collar.

In predator-free plots, the hare density doubled. In plots with extra food, it tripled. In plots having extra food and fewer predators, it increased elevenfold.

The experimental manipulations delayed the cyclic declines in population density but did not stop them.

Why not? Owls and other raptors flew over the fences. Only 9 percent of the collared hares starved to death; predators devoured most of the rest. Krebs concluded that a simple predator–prey or plant–herbivore model cannot fully explain his Yukon River Valley results. For the Canadian lynx and snowshoe hare cycle, other variables are at work, during multilevel interactions.

Predator and prey populations tend to exert coevolutionary pressures on one another.

Predators may affect prey density. There are three general patterns of response to changes in prey density. Population levels of prey may also show periodic oscillations.

Predator and prey numbers often vary in complex ways that reflect the multiple levels of interaction in a community.

46.5 An Evolutionary Arms Race

LINKS TO
SECTIONS 1.6,
18.4, 26.8, 33.1

As explained in the preceding section, predators and prey exert selective pressure on one another. One defends itself and the other must overcome defenses. Such interactions are often evidence of a coevolutionary arms race.

PREY DEFENSES

Camouflage Many heritable traits help an organism hide in the open; they function in **camouflaging**. Body form, patterning, color, behavior, or some combination of these blend with the surroundings and help the organism avoid detection. Consider Figure 46.11. Some nesting birds thrust their beak upward and sway slightly, like the plants around them. A caterpillar with special color patterns passes itself off as a bird dropping. When a certain desert plant (*Lithops*) is not flowering, it looks like a rock. It flowers only during a brief rainy season, when herbivores are more likely to be distracted by the profuse growth of other plants. Section 18.4 explains the genetic basis for camouflage among rock pocket mice as part of an example of natural selection.

Mimicry Many prey species closely resemble a hard-to-catch, dangerous, or unpalatable species. **Mimicry** is the name for an ecological association between one species that is a *model* for deception and a different species—a *mimic*, which very closely resembles it in form, behavior, or both. Predators often avoid a model species because of a repellent taste, toxic secretion, or painful bite or sting, and so they tend also to avoid the mimic. Section 1.6 offers an experimental test of mimicry. Here, Figure 46.12 shows the deceptive look of three tasty but weaponless mimics. All strongly resemble a very aggressive wasp that can sting repeatedly, with painful results.

Chemical Defenses The leaves, flowers, and seeds of many plants contain bitter, hard-to-digest, or dangerous repellents. Peach, apricot, and rose seeds are loaded with cyanide. Remember the Chapter 14 introduction? The castor bean plant did not develop its capacity to make the lethal chemical ricin in an evolutionary vacuum. Ricin protects this plant from herbivores that would otherwise eat it.

Many prey species that taste bad or that make toxins announce their unpalatability with **warning coloration**. They have conspicuous patterns and colors that predators learn to recognize as avoidance signals. For instance, a young, inexperienced bird might eat an orange-and-black patterned monarch butterfly once. It quickly learns to associate the butterfly's coloration and patterning with "Eat me and you will vomit foul-tasting toxins."

Figure 46.11 Prey camouflage. (**a**) What bird??? When a predator approaches its nest, the least bittern stretches its neck (which is colored like the surrounding withered reeds), points its bill upward, and sways like reeds in the wind. (**b**) An inedible bird dropping? No. This caterpillar's body coloration and its capacity to hold its body in a rigid position help camouflage it from predatory birds. (**c**) Find the plants (*Lithops*) hiding in the open from herbivores with the help of their stonelike form, pattern, and coloration.

a A dangerous model   **b** One of its edible mimics   **c** Another edible mimic   **d** And another edible mimic

Figure 46.12 An example of mimicry. Edible insect species often resemble toxic or unpalatable species that are not at all closely related. (**a**) A yellowjacket can deliver a painful sting. It might be the model for nonstinging wasps (**b**), beetles (**c**), and flies (**d**) of strikingly similar appearance.

Truly dangerous or repugnant species often make little or no attempt to conceal themselves. Remember the vividly colored and poisonous frogs (Section 33.1)? Or think about skunks, which spray one of the most odious repellents.

Moment-of-Truth Defenses When luck runs out and an animal is cornered or under attack, survival may turn on a last-chance trick. Many animals try to startle predators. For instance, some hiss, puff up, flash big eye-shaped spots on their body, bare sharp teeth, or flare neck ruffs (Figure 26.17*d*). Opossum and hognose snakes make a big show of pretending to be dead. Many cornered animals, including hognose snakes and certain beetles, secrete or squirt out irritating chemical repellents or toxins (Figure 46.13*a*).

ADAPTIVE RESPONSES OF PREDATORS

Again, predators tend to counter prey defenses with their own adaptations. Stealth, camouflage, and ingenious ways of avoiding repellents are some countermeasures. Consider the edible beetles that direct sprays of noxious chemicals at attackers. A grasshopper mouse grabs the beetle and plunges the sprayer end into the ground, and then chews on the tasty, unprotected head (Figure 46.13*a,b*).

Some prey can outrun even cheetahs when they get a head start. But the cheetah is the world's fastest land animal. One was clocked at 114 kilometers (70 miles) per hour. Compared to other big cats, the cheetah has longer legs relative to its body size and nonretractable claws that act like cleats to increase traction. Thomson's gazelle, its main prey, can run longer but not as fast (80 kilometers per hour). Without a head start, it is toast, so to speak.

Camouflaging helps predators as well as prey. Think of white polar bears stalking seals over ice, striped tigers crouched in tall-stalked, golden grasses, and scorpionfish hidden on the seafloor (Figure 46.13*c*). Camouflage is often stunning among predatory insects (Figure 46.13*d*). With camouflaging, predators select for enhanced sensory systems in prey. By one theory, primate color vision may have evolved in part to enhance detection of predators.

Figure 46.13 Predator responses to prey defenses. (**a**) Some beetles spray noxious chemicals at attackers, which deters them some of the time. (**b**) At other times, grasshopper mice plunge the chemical-spraying tail end of their beetle prey into the ground and feast on the head end. (**c**) Find the scorpionfish, a venomous predator with camouflaging fleshy flaps, multiple colors, and profuse spines. (**d**) Where do the pink flowers end and the pink praying mantis begin?

46.6 Parasite–Host Interactions

LINKS TO
SECTIONS 21.8,
25.5, 25.10, 25.16

Parasites spend all or part of their life cycle in or on other living organisms, from which they draw nutrients. They weaken a host but usually do not kill it outright. Different kinds complete their life cycle in one or more host species.

PARASITES AND PARASITOIDS

Parasites have pervasive impacts on populations. By draining nutrients from hosts, they alter the amount of energy and nutrients the host population demands from a habitat. Also, weakened hosts are usually more vulnerable to predation and less attractive to potential mates. Some parasite infections cause sterility. Others shift the ratio of host males to females. In such ways, parasitic infections lower birth rates, raise death rates, and affect intraspecific and interspecific competition.

Sometimes the gradual drain of nutrients during a parasitic infection indirectly leads to death. The host becomes so weakened that it can't fight off secondary infections. Nevertheless, in evolutionary terms, killing a host too quickly is bad for a parasite's reproductive success. A parasitic infection must last long enough to give the parasite time to produce some offspring. The longer it lives in the host, the more offspring. We may therefore expect selective agents to favor parasites that have less-than-fatal effects on hosts (Section 21.8).

Usually, death occurs only when a parasite attacks a novel host—one with no coevolved defenses against it—or when too many parasitic individuals attack at the same time and collectively overwhelm the body.

You looked at many parasites in the diversity unit, especially in Chapters 21, 24, and 25. You saw how some species require a single host and how others are free-living some of the time or residents of different hosts at different times. Many types ride inside insects and other arthropods, which are vectors between one host organism and the next (Section 25.16).

All viruses and some bacteria, protists, and fungi are parasites. Figure 46.14 shows a young trout that was parasitized by *Myxobolus cerebralis*, a protist.

Even a few plants are parasitic. Nonphotosynthetic types, such as dodders, obtain energy and nutrients from other plants (Figure 46.15). Other types carry out photosynthesis but still tap into the nutrients and water in tissues of a host plant. Mistletoe is like this; its modified roots invade the sapwood of host trees.

Many tapeworms, flukes, and certain roundworms are well-known invertebrate parasites (Figure 46.16). So are ticks, many insects, and many crustaceans.

You already read about **parasitoids**. An immature stage of these insects matures in a different insect's body, which they devour from the inside out. Unlike parasites, parasitoids always kill their hosts directly. About 15 percent of all insects may be parasitoids.

Social parasites are animals that take advantage of the social behavior of a host as a way to complete the life cycle. The cuckoos and North American cowbirds, described shortly, are like this.

Figure 46.14 (**a**) A young trout with a twisted spine and darkened tail caused by whirling disease, which damages cartilage and nerves. Jaw deformities and whirling movements are other symptoms. (**b**) Spores of *Myxobolus cerebralis*, the introduced protist that causes the disease. It is now in many lakes and streams in Western and Northeastern states.

Figure 46.15 Dodder (*Cuscuta*), also known as strangleweed or devil's hair. This parasitic flowering plant's sporophytes have no chlorophylls. They wind around a host plant during growth. Modified roots penetrate the host's vascular tissues and absorb water and nutrients from them.

Figure 46.16 Adult roundworms (*Ascaris*), an endoparasite, packed inside the small intestine from a host pig. Sections 25.5 and 25.10 give more examples of parasitic worms.

Figure 46.17 Biological control agent: a commercially raised parasitoid wasp about to deposit an egg in an aphid. This wasp reduces aphid populations. It stops the aphid from laying eggs even before the wasp egg develops into a larva that will eat it.

USES AS BIOLOGICAL CONTROLS

Parasites and parasitoids are commercially raised and released in target areas as *biological controls*. They are promoted as a workable alternative to pesticides. The chapter introduction and Figure 46.17 give examples.

Effective biological controls display five attributes. The agents are adapted to a specific host species and to its habitat; they are good at locating hosts; their population growth rate is high compared to the host's; their offspring are good at dispersing; and they make a type III functional response to prey, without much lag time after shifts occur in the host population size.

Biological control is not without risks of its own. Releasing more than one kind of biological control agent in an area may invite competition among them, which can lower their effectiveness against an intended target. In addition, an introduced parasite sometimes parasitizes nontargeted species as well as—or instead of—the species they were expected to control.

In Hawaii, the introduction of several parasitoids to control an imported stink bug resulted in the decline of the koa bug, the state's largest native bug. Few koa bugs have been collected since 1978. Apparently the koa bugs, which congregate in big groups, were more tempting to parasitoids. Also, introduced parasitoids have been implicated in the ongoing decline of many native Hawaiian butterflies and moths.

> *Natural selection favors parasitic species that temper their attacks in ways that ensure an adequate supply of hosts.*
>
> *Parasitic species belong to many groups, including bacteria, protists, invertebrates, and plants. Parasitoids are insects that feed on and kill other insects. Social parasites use the social behavior of another species to their own benefit.*

*The brown-headed cowbird's genus name (*Molothrus*) means "intruder" in Latin. This bird intrudes, sneakily, into the life cycle of other species. Let us ask: Why?*

Brown-headed cowbirds (*Molothrus ater*) evolved in the Great Plains of North America. They lived as commensalists with bison. Great herds of these hefty ungulates stirred up plenty of insects as they migrated through the grasslands, and, being insect-eaters, the cowbirds wandered around with them (Figure 46.18a).

A vagabond way of life did not lend itself to nesting in any one place. However it happened, cowbirds learned to lay eggs in nests constructed by other species, then leave them and move on with the herds. Many species became "hosts"; they did not have the neural wiring to recognize the differences between cowbird eggs and their own eggs. Concurrently, cowbird hatchlings became innately wired for hostile takeovers. Even before hatchlings open their eyes, they shove the owner's eggs out of the nest and demand to be fed as rightful occupants (Figure 46.18b). Thus, for thousands of years, cowbirds have perpetuated their genes by way of parasitic chutzpah.

When American pioneers moved west, many cleared swaths of woodlands for pastures. Cowbirds now moved in the other direction. They adapted easily to a life with new ungulates—cattle—in the manmade grasslands; hence their name. They started to penetrate adjacent woodlands and exploit novel species. Today, brown-headed cowbirds parasitize at least fifteen species of native North American birds. Some of those birds are threatened or endangered.

Besides being successful opportunists, cowbirds are big-time reproducers. A female can lay an egg a day for ten days, give her ovaries a rest, do the same again, and then again in one season. As many as thirty eggs in thirty nests—that is a lot of cowbirds.

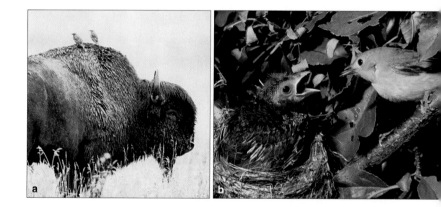

Figure 46.18 Oh give me a home, where the buffalo roam—brown-headed cowbirds (*Molothrus ater*) originally evolved as commensalists with bison and as social parasites of other bird species of the North American Great Plains. When conditions changed, they expanded their range. They became nest usurpers in woodlands as well as grasslands in much of the United States.

46.8 Ecological Succession

LINKS TO
SECTIONS
17.4, 23.11

*By an older model for **ecological succession**, a community comes into being through competition and other species interactions and in time stabilizes into a predictable array of species. However, abiotic forces, including fire, storms, and human-created disturbances, may be more important in shaping community structure.*

SUCCESSIONAL CHANGE

A concept of "nature in balance" once guided studies in community ecology. Researchers knew that **pioneer species** are the start of community structure. These are opportunistic colonizers of new or newly vacated habitats. They have high dispersal rates, they grow and mature quickly, and they produce many offspring. In time, more competitive species replace them. Then the replacements are replaced.

Primary succession is a process that begins when pioneer species colonize a barren habitat, such as a new volcanic island and land exposed when a glacier retreats (Figure 46.19). Pioneers include lichens and plants, such as club mosses, that are small, have short life cycles, and can survive intense sunlight, extreme temperature changes, and nutrient-poor soil. Early on, hardy annual flowering plants put out many small seeds, which are quickly dispersed.

Established pioneers often improve soil and other conditions. In doing so, they typically set the stage for their own replacement. Many of the new arrivals are mutualists with nitrogen-fixing bacteria, so they can grow in nitrogen-poor habitats. Seeds of later species find shelter inside mats of the pioneers, which do not grow high enough to shade out the new seedlings.

Organic wastes and remains accumulate over time, which add volume and nutrients to soil, which favors invasions by other species. Later successional species crowd out earlier ones, whose spores and seeds travel as fugitives on wind and water—destined, perhaps, for another new but temporary habitat.

In **secondary succession**, a disturbed area within a community recovers. If improved soil is still present, secondary succession can be fast. It commonly occurs in abandoned fields, burned forests, and tracts of land cleared by volcanic eruptions.

INTERMEDIATE DISTURBANCE HYPOTHESIS

In a traditional view, a predictable array of species in the habitat stabilizes as the *climax* community, after which not much changes. The community is adapted to many factors, such as topography, climate, soil, and species interactions, and it may show some variation

Figure 46.19 In Alaska's Glacier Bay region, one pathway of primary succession. (**a**) As a glacier retreats from the sea, meltwater leaches minerals from the glacial till. (**b**) Lichens, horsetails, mosses, fireweed, and mountain avens are pioneer species; some are mutualists with nitrogen-fixing microbes. Within twenty years, alder, cottonwood, and willow seedlings take hold. Alders have nitrogen-fixing symbionts. (**c**) Within fifty years, they form dense, mature thickets in which cottonwood, hemlock, and a few evergreen spruce grow fast. (**d**) After eighty years, western hemlock and spruce crowd out mature alders. (**e**) In areas deglaciated for more than a century, forests of Sitka spruce dominate.

Figure 46.20 A natural laboratory for succession after the 1980 Mount Saint Helens eruption (**a**). The community at the base of this Cascade volcano was destroyed. (**b**) In less than a decade, pioneer species took hold. (**c**) Twelve years later, seedlings of the dominant species, Douglas firs, were taking hold.

along gradients of environmental conditions. In this view, even after a disturbance, the community reverts to a climax state. Later, Henry Gleason proposed that most communities are *not* stable, that unpredictable disturbances can alter the direction of succession.

It turned out that the magnitude and frequency of disturbances may be more important than interactions among species in defining the community. According to the **intermediate disturbance hypothesis**, species richness of a community becomes greatest in between disturbances of moderate intensity or frequency. There is enough time for many colonizing species to enter the habitat but not enough time for many species to be competitively excluded from it:

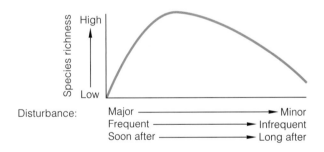

Tolerance, inhibition, and facilitation may be three successional mechanisms. In cases of *tolerance*, an early colonizer has no effect on which species will colonize the habitat after it. In *inhibition*, the early colonizer changes conditions of the habitat in specific ways that bar colonization by later species. In *facilitation*, early colonizers improve conditions for later ones.

Ecologists documented inhibition and facilitation in intertidal zone succession. After Wayne Sousa cleared algae from an intertidal zone in Southern California, different algae moved in. After Teresa Turner removed attached algae from plots in an Oregon intertidal zone, surf grass could not move in, because they use algae as anchoring sites. Experimental studies in old fields, temperate forests, and other land regions where major disturbances have occurred also give evidence of these three mechanisms of succession.

For instance, after Washington state's Mount Saint Helens erupted in 1980, the blast wave, superheated mudslides, and floods obliterated approximately 600 square kilometers of forests (Figure 46.20). Afterward, ecologists moved in to monitor succession first-hand. They observed and recorded in detail natural patterns of colonization. They also manipulated plots inside the blast zone. William Morris and David Wood showed that facilitation and inhibition were factors in plant succession. By adding seeds of certain plant species to some plots and keeping some other plots barren, they demonstrated that early colonizers helped several other species of colonizing plants move in. They also found that earlier colonizers kept some plant species out.

A community develops through a succession of stages, starting with pioneer species that are replaced by others. Biotic (biological) and abiotic (physical and chemical) factors affect community structure.

Disturbances are unpredictable and vary in magnitude and frequency. By an intermediate disturbance hypothesis, species richness is greatest between moderate disturbances.

46.9 Species Interactions and Community Instability

LINKS TO
SECTIONS
17.4, 23.11

The loss or addition of even one species may destabilize the number and relative abundances of species in a community.

As you read earlier, short-term physical disturbances can knock a community out of equilibrium. Long-term changes in climate or another environmental variable also have destabilizing effects. Besides this, a shift in species interactions also can tip a community out of its uneasy balance. Remember, resources are sustained as long as populations do not flirt dangerously with the carrying capacity. Predators and their prey coexist as long as neither wins. Competitors have no sense of fair play. Mutualists are stingy, as when plants make as little nectar as necessary to attract pollinators and the pollinators take as much nectar as they can for the least possible effort.

Whether biotic or abiotic, a disturbance sometimes causes the number and relative abundances of species to shift irrevocably. For instance, if some occupants of the habitat happen to be rare or do not compete well with the others, they might be driven to extinction.

THE ROLE OF KEYSTONE SPECIES

The uneasy balancing of forces in a community comes into focus when we observe the effects of a **keystone species**. Such a species has a disproportionately large effect on a community relative to its abundance. Robert Paine was the first to describe the role of a keystone species after his experiments on the rocky shores of California's coast. Species in this rocky intertidal zone survive by clinging to rocks, and access to spaces to cling to is a limiting factor. Paine set up control plots with the sea star *Pisaster ochraceus* and its main prey— chitons, limpets, barnacles, and mussels. He removed all sea stars from his experimental plots.

Mussels (*Mytilus*) happen to be the prey of choice for sea stars. In the absence of sea stars, they took over Paine's experimental plots; they became the strongest competitors and crowded out seven other species of invertebrates. In this intertidal zone, predation by sea stars normally keeps the number of prey species high because it restricts competitive exclusion by mussels.

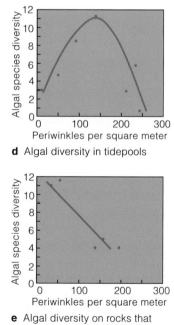

d Algal diversity in tidepools

e Algal diversity on rocks that become exposed at high tide

Figure 46.21 Effect of competition and predation in an intertidal zone. (**a**) Grazing periwinkles (*Littorina littorea*) affect the number of algal species in different ways in different marine habitats. (**b**) *Chondrus* and (**c**) *Enteromorpha*, two kinds of algae in their natural habitats. (**d**) By grazing on the dominant alga in tidepools (*Enteromorpha*), the periwinkles promote the survival of less competitive algal species that would otherwise be overgrown. (**e**) *Enteromorpha* doesn't grow on rocks. Here, *Chondrus* is dominant. Periwinkles find *Chondrus* tough and dine instead on less competitive algal species. By doing so, periwinkles decrease the algal diversity on the rocks.

Table 46.1 Adverse Effects of Some Species Introduced Into the United States

Species Introduced	Origin	Mode of Introduction	Outcome
Water hyacinth	South America	Intentionally introduced (1884)	Clogged waterways; other plants shaded out
Dutch elm disease:			
Ophiostoma ulmi (fungus)	Asia (by way	Accidental; on infected elm timber (1930)	Millions of mature elms destroyed
Bark beetle (vector)	of Europe)	Accidental; on unbarked elm timber (1909)	
Chestnut blight fungus	Asia	Accidental; on nursery plants (1900)	Nearly all eastern American chestnuts killed
Zebra mussel	Russia	Accidental; in ballast water of ship (1985)	Clog pipes and water intake valves of power plants; displacing native Great Lake bivalves
Japanese beetle	Japan	Accidental; on irises or azaleas (1911)	Close to 300 plant species (e.g., citrus) defoliated
Sea lamprey	North Atlantic	Ship hulls, through canals (1860s, 1921)	Trout, other fish species destroyed in Great Lakes
European starling	Europe	Intentional release, New York City (1890)	Outcompete native cavity-nesting birds; crop damage; swine disease vector
Nutria	South America	Accidental release of captive animals being raised for fur (1930)	Crop damage, destruction of levees, overgrazing of marsh habitat

Remove all the sea stars, and the community shrinks from fifteen species to eight.

The impact of a keystone species can vary between habitats that differ in their species arrays. Periwinkles (*Littorina littorea*) are alga-eating snails of intertidal zones. Jane Lubchenco showed that their removal can increase *or* decrease the diversity of algal species in different habitats (Figure 46.21).

In tidepools, the periwinkles prefer to eat the alga *Enteromorpha*, which can outgrow other algal species. By keeping *Enteromorpha* in check, periwinkles help less competitive algal species survive. However, on exposed rocks in the lower intertidal zone, they avoid *Chondrus* and other tough, unpalatable red algae that persist as the dominant species. Periwinkles on these rocks graze on competitively weaker algal species. In short, they help *maintain* the number of algal species in tidepools but *reduce* it on exposed rock surfaces.

HOW SPECIES INTRODUCTIONS TIP THE BALANCE

Instabilities also are set in motion when residents of established communities move out from their home range and successfully take up residence elsewhere. This type of directional movement, called **geographic dispersal**, happens in three ways.

First, over a number of generations, a population might expand its home range by slowly moving into outlying regions that prove hospitable. Second, some individuals might be rapidly transported across great distances, an event called *jump* dispersal. This often takes individuals across regions where they could not survive on their own, as when insects travel from the mainland to Maui in a ship's cargo hold. Third, some population might be moved away from a home range by continental drift, at an almost imperceptibly slow pace over long spans of time.

Successful dispersal and colonization of a vacant adaptive zone can be remarkably rapid. Consider one of Amy Schoener's experiments in the Bahamas. She set out plastic sponges on barren sand at the bottom of Bimini Lagoon. How fast did aquatic species take up residence on or in the artificial habitats? Schoener recorded occupancy by 220 species within thirty days.

When you hear someone bubbling enthusiastically about an exotic species, you can safely bet the speaker isn't an ecologist. An **exotic species** is a resident of an established community that dispersed from its home range and became established elsewhere. Unlike most imports, which never do take hold outside the home range, an exotic species permanently insinuates itself into a new community.

Following jump dispersal, more than 4,500 exotic species have become established in the United States. We put some of the new arrivals, including soybeans, rice, wheat, corn, and potatoes, to use as food crops.

Accidental imports also alter community structure. You learned about imported fire ants in the chapter introduction. Table 46.1 lists others, and the section to follow describes the unintended impact of a few more.

A keystone species is one that has a major effect on species richness and relative abundances in particular habitats.

Species introductions and other biotic disturbances can permanently alter community structure.

46.10 Exotic Invaders

LINKS TO
SECTIONS
17.4, 23.11

Nonnative species are on the loose in communities on every continent. They can alter habitats; they often outcompete and displace native species.

THE ALGA TRIUMPHANT

They looked so perfect in saltwater aquariums, those long, green, feathery branches of *Caulerpa taxifolia*. So Stuttgart Aquarium researchers in Germany developed a hybrid, sterile strain of this green alga and magnanimously shared it with other marine institutions. Was it from Monaco's Oceanographic Museum that the hybrid strain escaped into the wild? Some say yes, Monaco says no.

The aquarium strain grows asexually by runners, just a few centimeters a day, but boat propellers and fishing nets dispersed it. Between 1984 and 2000, this alga blanketed over 30,000 hectares of seafloor near the Mediterranean coast (Figure 46.22a). Scuba divers found it growing off the Southern California coast. Someone might have drained water from a home aquarium into a storm drain or into the lagoon itself. Governmental and private groups sprang into action. They tarped over the area to shut out sunlight, pumped chlorine into the mud to poison the alga, and used welders to boil it. So far, eradication and surveillance programs have worked, but they have cost more than 3.4 million dollars.

It is now illegal to import the harmful strain into the United States. Interstate sale also is prohibited. Some still slip into the country because the aquarium industry has successfully lobbied against a ban on all *Caulerpa* species,

and it is difficult to distinguish the invasive strain without genetic analysis.

Just how bad is it? The aquarium strain of *C. taxifolia* thrives on sandy or rocky shores and in mud. It can live ten days after being discarded in meadows. Unlike its tropical parents, it survives in cool water and polluted water. It also displaces endemic algae. Its toxin poisons invertebrates and fishes, including herbivorous types that might keep it in check. It has the potential to overgrow reefs and destroy marine food webs. Can you sense why this algal strain has been nominated as one of the 100 worst exotic invaders?

THE PLANTS THAT ATE GEORGIA

One more of the infamous 100: In 1876, kudzu (*Pueraria montana*) from Japan was introduced to the United States. In its native habitat—temperate regions of Asia—this vine is a well-behaved legume with a strong root system. It *seemed* like a good idea to use it for forage and to control erosion. But kudzu grew faster in the Southeast, where herbivores, pathogens, and less competitive plants posed no serious threat to it.

With nothing to stop it, kudzu shoots grow sixty meters per year. Its vines now blanket streambanks, trees, telephone poles, houses, and almost everything else in their path (Figure 46.22b). It withstands burning, and its deep roots resist being dug up. Grazing goats and herbicides help. But goats eat most other plants along with it, and herbicides taint water supplies. Kudzu invasions now stretch from Connecticut down to Florida and are reported in Arkansas. It has crossed

Figure 46.22 (**a**) Aquarium strain of *Caulerpa taxifolia* suffocating yet another richly diverse marine ecosystem.

(**b**) Kudzu (*Pueraria montana*) taking over part of Lyman, South Carolina. This vine has become invasive in many states from coast to coast. Ruth Duncan of Alabama, who makes 200 kudzu vine baskets a year, just can't keep up.

Figure 46.23 Rabbit-proof fence? Not quite. This is part of a fence built to hold back the 200 million to 300 million rabbits that are wreaking havoc with the vegetation in Australia. It didn't work.

the Mississippi River into Texas, and thanks to jump dispersal, it is now an invasive species in Oregon.

On the bright side, Asians use a starch extracted from kudzu in drinks, herbal medicines, and candy. A kudzu processing plant in Alabama may export this starch to Asia, where the demand currently exceeds the supply. Also, kudzu may help save trees; it can be an alternative source for paper. Today, about 90 percent of Asian wallpaper is kudzu-based.

THE RABBITS THAT ATE AUSTRALIA

During the 1800s, British settlers in Australia just couldn't bond with koalas and kangaroos, and so they imported familiar animals from home. In 1859, in what would be the start of a major disaster, a landowner in northern Australia imported and then released two dozen European rabbits (*Oryctolagus cuniculus*). Good food and sport hunting—that was the idea. An ideal rabbit habitat with no natural predators—that was the reality.

Six years later, the landowner had killed 20,000 rabbits and was besieged by 20,000 more. The rabbits displaced livestock and caused the decline of native wildlife. Now 200 to 300 million are hippity-hopping through the southern half of the country. They graze on grasses in good times and strip bark from shrubs and trees during droughts. Thumping hordes turn shrublands as well as grasslands into eroded deserts. Their burrows undermine the soil and set the stage for widespread erosion.

Rabbit warrens have been shot at, fumigated, plowed under, and dynamited. The first all-out assaults killed 70 percent of them, but the rabbits rebounded in less than a year. When a fence 2,000 miles long was built to protect western Australia, rabbits made it from one side to the other before workers could finish the job (Figure 46.23).

In 1951, the government introduced a myxoma virus that normally infects South American rabbits. The virus causes *myxomatosis*. This disease has mild effects on its coevolved host but nearly always kills *O. cuniculus*. Mosquitoes and fleas transmit the virus to new host. Having no coevolved defenses against the import, European rabbits died in droves. But natural selection has since favored a rise in rabbit populations resistant to the imported virus.

In 1991, on an uninhabited island in Australia's Spencer Gulf, researchers released rabbits that were injected with a calicivirus. The rabbits died from blood clots in their lungs, heart, and kidneys. The test virus escaped from the island in 1995, perhaps on insect vectors.

By 2001, the rabbit population sizes were staying 80 to 85 percent below their peak values. Grasses, nonwoody shrubs, and woody shrubs are rebounding. Different kinds of herbivores are increasing in density.

The rabbit calicivirus was discovered in China in 1984 and is now found in Europe and other countries as well. To date, tests on more than forty animal species indicate that it replicates in rabbits alone. However, other caliciviruses can and do cross species barriers. The jury is still out on the long-term impact of the viral releases.

As you might have deduced, *O. cuniculus* is another one of the 100 worst exotic invaders. Also on the list are two *Anopheles* species, the vectors for malaria. So is the cane toad (*Bufo marinus*). It was introduced as a biological control of pests in fields of sugarcane and other crops all over the world, but it eats almost everything. Despite its catchy name, the banana bunchy top virus is another one of the worst. So is the house cat (*Felis catus*) turned feral. Finally, the house mouse (*Mus musculus*) probably has a greater distribution than any other mammal except humans. Populations of this prolific breeder destroy crops and consume or contaminate much of our food supplies. They are implicated in the extinction of many species. Interested in learning more? Go to http://www.issg.org/ for some eye-openers.

46.11 Biogeographic Patterns in Community Structure

LINKS TO
SECTIONS 17.1,
17.3, 18.7, 26.15,
45.3, CHAPTER 27

The richness and relative abundances of species differ from one habitat or one world province to another. Often these differences correspond to predictable patterns that have biogeographic and historical foundations.

Unit IV gave you a sense of the sweep of biodiversity, and Chapter 27 placed it in evolutionary perspective. Starting with Alfred Wallace and other naturalists of the 1800s, it became apparent that communities show *patterns* in biodiversity, as measured by the richness and relative abundances of species. Certain patterns follow environmental gradients in sunlight intensity, temperature, rainfall, and other factors that differ by latitude, elevation, and depth. Other patterns have

their roots in the history of a habitat and its species, which vary in their resource requirements, physiology, capacity for dispersal, and the specific ways in which they interact with one another.

MAINLAND AND MARINE PATTERNS

Perhaps the most striking pattern of species richness corresponds with distance from the equator. *For most groups of plants and animals, the number of coexisting species on land and in the seas is greatest in the tropics, and it systematically declines from the equator to the poles.* Figure 46.24 shows two clear examples of this pattern. Consider just a few factors that help bring about such a pattern and maintain it.

First, for reasons explained in Section 46.1, tropical latitudes intercept more intense sunlight and receive more rainfall, and their growing season is longer. As one outcome, resource availability tends to be greater and more reliable in the tropics than elsewhere. This favors a degree of specialized interrelationships not possible where species are active for shorter periods.

Second, tropical communities have been evolving for a longer time than temperate ones, some of which did not start forming until the end of the last ice age.

Third, species richness may be self-reinforcing. The number of species of trees in tropical forests is much greater than in comparable forests at higher latitudes. When more plant species compete and coexist, so will more species of herbivores, partly because no single herbivore species can overcome all chemical defenses of all plants. Also, more predatory and parasitic species evolve in response to more kinds of prey and hosts. The same effect applies to the number of species on tropical reefs.

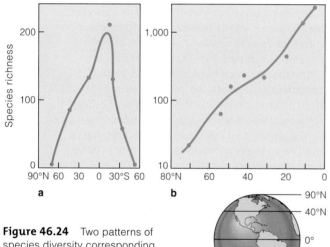

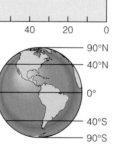

Figure 46.24 Two patterns of species diversity corresponding to latitude. The number of ant species (**a**) and breeding birds (**b**) in the Americas.

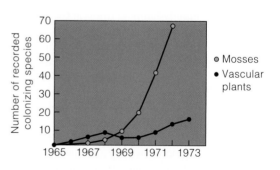

Figure 46.25 Surtsey, a volcanic island, at the time of its formation. Newly formed, isolated islands are natural laboratories for ecologists. The chart gives the number of colonizing species between 1965 and 1973.

Figure 46.26 **(a)** Two island biodiversity patterns. Distance effect: Species richness on islands of a specified size declines with increasing distance from a source of colonizing species. *Green* circles signify islands less than 300 kilometers from the colonizing source. *Orange* triangles signify islands more than 300 kilometers from the source areas. Area effect: Among islands the same distance from a source of colonizing species, the larger ones support more species.

(b) Wandering albatross, one travel agent for jump dispersals. Seabirds that island-hop long distances often have seeds stuck to their feathers. Seeds that successfully germinate in a new island community may give rise to a population of new immigrants.

ISLAND PATTERNS

As you saw in Chapter 45, islands are laboratories for population studies. They also have been laboratories for community studies. For instance, a 1965 volcanic eruption quickly formed Surtsey, an island southwest of Iceland. Within six months, bacteria, fungi, seeds, flies, and seabirds were established on it. A vascular plant appeared two years after the island formed; the first mosses came along two years after that (Figure 46.25). As the island soil became enriched, more and more plant species began to take hold.

As is the case for other islands, the number of new species on Surtsey will not increase indefinitely. Why not? Models based on studies of island communities around the world suggest some answers.

First, larger islands tend to support more species than smaller ones the same distance from a colonizing source. This is the **area effect** (Figure 46.26a). Larger islands generally have more varied habitats, and more of them. Most have complex topography and higher elevations. Such variations promote species richness. Also, being bigger, the larger islands intercept more of the accidental tourists that winds and ocean currents move from the mainland but offer no way back.

Second, islands that are far away from a source of potential colonists receive fewer colonizing species. The few that do arrive naturally are adapted for long-distance dispersal (Figure 46.26a). This is the **distance effect**. Remember the nature of individual extinctions (Section 27.1)? Extinctions are more prevalent on the small islands. Because immigration rates are low and extinction rates are high, small islands support fewer species once the balance is struck. Island populations are far more vulnerable to famine, storms, droughts, disease and genetic drift. Remember the account of St. Matthew Island that opened Chapter 45?

One more island pattern: Remember the miniature *Homo* species that was discovered on the Indonesian island Flores (Section 26.15)? There is a trend, among new arrivals, for the big to get smaller and the small to get bigger. They adapt to fewer or different resources than in the place left behind. Biogeographers know more about patterns of diversity and the disruptions of them. If you wish to learn more, David Quammen's *Song of the Dodo* is a good place to start.

Species richness shows global patterns, as when it correlates with environmental gradients in latitude, elevation, and depth. Microenvironments along these gradients often introduce variations in the overall patterns.

Species richness in a given area also is an outcome of the evolutionary history of each species, its requirements for resources, its physiology, its capacity for dispersal, and its rates of birth, death, immigration, and emigration.

Generally, species richness is highest in the tropics and lowest at the poles. The number of species on an island also depends on its size and distance from a colonizing source.

Summary

Section 46.1 A habitat is the type of place where individuals of a species normally live. A community is an association of all populations of species that occupy a habitat. Each species in a community has a niche, the sum of all of the activities and relationships in which its individuals engage as they secure and use the resources required for their survival and reproduction.

Community structure arises from a habitat's physical and chemical features, resource availability over time, adaptive traits of its species, how its species interact, and the history of the habitat and its occupants.

Direct symbiotic interactions help shape community structure. They include commensalism, mutualism, competition, predation, and parasitism.

Section 46.2 Mutualism is a species interaction that benefits both participants. Some mutualists cannot complete their life cycle without the interaction.

Section 46.3 By the competitive exclusion theory, when two (or more) species require identical resources, they cannot coexist indefinitely. Species may coexist when they differ in their use of a resource, share it in different ways, or share it at different times.

Biology Now
Learn about competitive interactions with the animation on BiologyNow.

Sections 46.4, 46.5 Predators and prey exert selection pressure on each other. Densities of predator and prey populations often oscillate. The carrying capacity, density dependencies, refuges, predator efficiency, and often alternative prey sources affect the cycles. Threat displays, chemical weapons, camouflage, stealth, and mimicry may be outcomes of coevolution between predators and their prey.

Biology Now
Compare the three alternative models for predator responses to prey density with the animation on BiologyNow.

Read the InfoTrac article "How the Pufferfish Got Its Puff," Carl Zimmer, Discover, September 1997.

Sections 46.6, 46.7 Parasites live in or on other living hosts and withdraw nutrients from host tissues for part of their life cycle. Hosts may or may not die as a result. Parasitoids kill their hosts, and social parasites take over some aspect of a host's life cycle.

Section 46.8 By a model for ecological succession, a community develops in predictable sequence, from its pioneer species to a climax community—a stable, self-perpetuating array of species that are in equilibrium with one another and the environment. However, abiotic and biotic disturbances have destabilizing effects. They are unpredictable and vary in magnitude and frequency. By an intermediate disturbance hypothesis, species richness is greatest between moderate disturbances.

Sections 46.9, 46.10 Community structure reflects an uneasy balance between biotic as well as abiotic forces, including predation and competition, that can shift over time. Species introductions can change the structure.

Section 46.11 Many studies of mainland and island communities reveal global patterns in species richness.

Biology Now
Learn about the area effect and distance effect with the interaction on BiologyNow.

Read the InfoTrac article "Island Biogeography's Lasting Impact," Fred Powledge, Bioscience, November 2003.

Self-Quiz
Answers in Appendix II

1. A habitat _____ .
   a. has distinguishing physical and chemical features
   b. is where individuals of a species normally live
   c. is occupied by various species
   d. all of the above

2. A niche is _____ .
   a. the sum of activities and relationships by which individuals of a species secure and use resources
   b. unvarying for a given species
   c. something that shifts in large and small ways
   d. both a and c

3. Two species may coexist indefinitely in some habitat when they _____ .
   a. differ in their use of resources
   b. share the same resource in different ways
   c. use the same resource at different times
   d. all of the above

4. A predator population and prey population _____ .
   a. always coexist at relatively stable levels
   b. may undergo cyclic or irregular changes in density
   c. cannot coexist indefinitely in the same habitat
   d. both b and c

5. Parasites _____ .
   a. weaken their hosts   c. feed on host tissues
   b. can kill novel hosts   d. all of the above

6. By a currently favored hypothesis, species richness of a community is greatest between physical disturbances of _____ intensity or frequency.
   a. low   b. intermediate   c. high   d. variable

7. Match the terms with the most suitable descriptions.
   ____ geographic dispersal
   ____ area effect
   ____ pioneer species
   ____ climax community
   ____ keystone species
   ____ exotic species
   ____ resource partitioning

   a. opportunistic colonizer of barren or disturbed habitat
   b. greatly affects other species
   c. individuals leave home range, become established elsewhere
   d. more species on large islands than small ones at same distance from the source of colonists
   e. array of species at the end of successional stages in a habitat
   f. allows competitors to coexist
   g. often outcompete, displace native species of established community

Additional questions are available on Biology Now™

Figure 46.27 Phasmids. (**a**) South African stick insect. (**b**) Leaf insect from Java. (**c**) Phasmid eggs often look like seeds.

Figure 46.28 One of the nominations for the worst 100 invaders: water hyacinths (*Eichhornia crassipes*) choking a Florida waterway.

Critical Thinking

1. With antibiotic resistance rising, researchers are looking for ways to reduce use of antibiotics. Cattle were once fed antibiotic-laced food but now get *probiotic feeds* that contain cultured bacteria that can establish or bolster populations of helpful bacteria in the animal's gut. The idea is that if a large population of beneficial bacteria is in place, then the harmful bacteria cannot become established or thrive. Which ecological theory is guiding this research?

2. Most phasmids resemble sticks or leaves (Figure 46.27). All are herbivorous insects. Most are motionless in the day, and move and feed only at night. If disturbed, a phasmid will fall to the ground, as if dead. Speculate on the selective pressures that may have shaped phasmid morphology and behavior. Suggest an experiment with one species to test whether its appearance and behavior may be adaptive.

3. The water hyacinth (*Eichhornia crassipes*) is an aquatic plant native to South America. Today, this plant lives in nutrient-rich waters from Florida to San Francisco. It has displaced many native species, and choked rivers and canals (Figure 46.28). Research and write a brief account of how it got from one continent to another.

4. Answering this question well should earn you big points. Long ago, Alfred Wallace puzzled over an odd pattern in the distribution of organisms in the islands of Indonesia. Deep

water separates Bali and Lombok and, farther north, the larger islands of Borneo and Sulawesi (Figure 46.29). Most major groups on the Asian mainland had representative species on Borneo and Bali—but few or none on Lombok and Sulawesi. The boundary he had identified came to be called *Wallace's Line*, and his explanation for it is still valid.

In Wallace's time, geologists had already discovered evidence of past ice ages, when much of the ocean's waters became locked up in vast ice sheets. Wallace's line marks the boundary of the Asian continent when the sea level fell 75 fathoms (450 feet). All of the shallow seas and straits from the Asian mainland to Borneo and Bali became dry land. Wallace inferred that many species dispersed to the east. When the sea level rose again, they became cut off from the mainland. Some survived; others vanished.

Even during the ice ages, Lombok and Sulawesi never were connected to the Asian mainland. If a plant or animal could not fly, swim, or be blown or rafted across an expanse of deep water, then they never got across Wallace's line. An expanse of deep water also separates Lombok and Sulawesi from Australia and Papua New Guinea.

Sulawesi is famous for its remarkably high percentage of endemic bird species. About one-third are endemic or close to it. By comparison, Borneo is home to relatively few endemic species of birds. Section 19.2 presents a model for speciation on island archipelagos. Review this section, and then formulate a hypothesis to explain why there are more endemic bird species on Sulawesi than on Borneo.

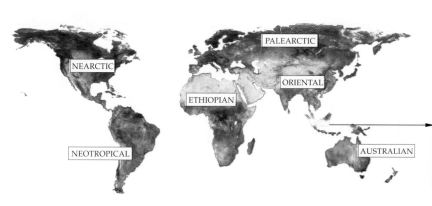

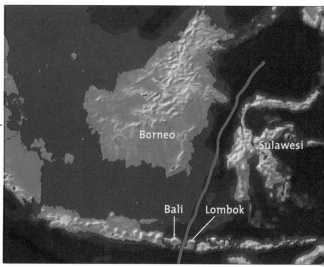

Figure 46.29 Wallace's Line (*red*), which helped nineteenth-century naturalists mark a boundary between two biogeographic realms (Oriental and Australian). The other realms shown were identified later. They have since become subdivided into biomes and then into ecoregions, which include the water provinces.

Bye-Bye, Blue Bayou

Each Labor Day, the coastal Louisiana town of Morgan City celebrates the region's economic mainstays with the Louisiana Shrimp and Petroleum Festival. The state is the nation's third-largest petroleum producer and the leader in shrimp harvesting. But the petroleum industry's success may be contributing indirectly to the possible disappearance of the state's fisheries.

The global air temperature is rising, and fossil fuel burning is a contributing factor. Warmer air heats water near the sea surface, heated water expands, and so the sea level is rising. Warmer air also is melting ancient glaciers and ice caps, and meltwater is adding to the sea volume.

Since the 1940s, Louisiana has lost an area the size of Rhode Island to the sea. Low elevations along the United States coastline—including *14,720,000 acres* next to the Gulf of Mexico and Atlantic Ocean—may be one to three feet under water within fifty years.

Given that it has more than 40 percent of the nation's saltwater marshes, Louisiana has the most to lose (Figure 47.1). Its wetlands are already sinking, because extensive dams and levees interfere with the deposition of sediments that could replace those washed out to sea. In time, a rise in sea level will make 70 percent of the nation's wetlands *really* wet, with no land at all.

Are ecological and economic disasters now unfolding? What will happen to the livelihoods of people who harvest more than 3 billion dollars' worth of shellfish and fish from Louisiana's wetlands each year? What will happen to the more than 5 million birds—about 40 percent of North America's migratory ducks—that overwinter here? What will happen to villages, cities, and natural ecosystems at the low inland elevations, which Louisiana's wetlands buffer from storm surges and hurricanes?

More bad news: Warmer water may promote algal blooms and huge fish kills. Also, populations of many pathogenic bacteria increase in warmer water, so more people might get sick after swimming in contaminated water or eating contaminated shellfish.

Inland, heat waves and wildfires will become more intense. Deaths related to heat stroke will climb. Warmer temperatures will permit mosquitoes to extend their inland ranges. Some mosquitoes are vectors for agents of malaria, West Nile virus, and other diseases.

For some time, researchers have been predicting that global warming will raise evaporation rates, alter weather patterns, and cause prolonged drought for some regions and severe flooding for others, including Louisiana. They worry that 3 billion people may run out of fresh drinking

Figure 47.1
Cypress swamp in Louisiana. Inland saltwater intrusions threatening these trees, which are actually adapted to freshwater habitats. *Facing page*, Dawn on the bayou.

Watch the video online!

water within twelve years. Reflect on Katrina, the category 5 hurricane that made a direct hit on Gulf Coast lowlands in 2005. Reflect on the devastation, flooding, contaminated freshwater sources, displaced populations, and impact on the nation's economy. Are predictions becoming reality?

This chapter can get you thinking about energy flow through ecosystems, starting with energy inputs from the sun. It will show how ecosystems depend on inputs, cycling, and outputs of nutrients—and how nutrients are cycled on a global scale.

The chapter also can get you thinking more about a related concept of equal importance. We have become players in the global flows of energy and nutrients even before we fully comprehend how the game plans work. Decisions we make today about global warming and other environmental issues may affect the quality of human life and the environment far into the future.

 How Would You Vote?

Emissions from motor vehicles are a major source of greenhouse gases. Many people buy large vehicles that use more fuel but are viewed as safer and more useful. Should such vehicles be additionally taxed to discourage sales and offset their environmental costs? Can we expect better fuels as well as more of the fuel-efficient, larger vehicles that are becoming available? See BiologyNow for details, then vote online.

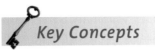 Key Concepts

ORGANIZATION OF ECOSYSTEMS

An ecosystem is a community and its physical environment. It is maintained by a one-way flow of energy and a cycling of materials through its interacting participants. It is an open system, with inputs, internal transfers, and outputs of both energy and nutrients. Section 47.1

FOOD WEBS

Food chains are linear sequences of feeding relationships, from producers through consumers, decomposers, and detritivores. The chains cross-connect, as food webs.

Most of the energy that enters a food web returns to the environment, mainly as metabolic heat. Most of the nutrients are cycled, but some are lost to the environment.

Biological magnification is the increasing concentration of a substance in the tissues of organisms as it moves up food chains. Sections 47.2, 47.3

PRIMARY PRODUCTIVITY

An ecosystem's primary productivity is the rate at which its producers capture and store energy in their tissues during a given interval. The amount stored depends on the number of producers and on the balance between photosynthesis and aerobic respiration. Section 47.4

CYCLING OF WATER AND NUTRIENTS

Primary productivity is influenced by the availability of water, carbon, nitrogen, phosphorus, and other substances, the ions or molecules of which move slowly from environmental reservoirs, among organisms of food webs, then back to the reservoirs. Human activities intervene in these cycles in measurable ways. Sections 47.5–47.12

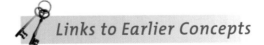 Links to Earlier Concepts

This chapter takes a closer look at the main participants in ecosystems, especially the autotrophs (Sections 1.2, 7.8). It builds on your understanding of how the one-way flow of energy in nature shapes the organization of life (6.1, 6.2). You will place nitrogen-fixing microbes as well as soil erosion and leaching in the context of a global nitrogen cycle (30.1, 30.2). You will revisit pesticides (32.2), algal blooms (22.6), and methane hydrates (Chapter 3 introduction). You will come across more effects of deforestation (23.10).

47.1 The Nature of Ecosystems

LINKS TO
SECTIONS
1.2, 6.1, 7.8, 46.1

In the preceding chapter, you focused on the dynamic nature of community structure. Turn now to the ways in which the energy and raw materials available in the physical environment help organize the interactions among species in that community. By identifying these interactions, ecologists can make predictions about whether they will remain stable and how they might change over the long term.

OVERVIEW OF THE PARTICIPANTS

Diverse natural systems abound on Earth's surface. In climate, landforms, soil, vegetation, animal life, and other features, deserts differ from hardwood forests, which differ from tundra and the prairies. Reefs differ from the open ocean, which differs from streams and lakes. *Even so, despite their differences, all of the systems are alike in many aspects of their structure and function.*

These systems run on energy that autotrophs—the self-feeders—capture. The most familiar autotrophs, remember, are plants and phytoplankton (Sections 7.8 and 22.6). As you know, both convert energy from the sun to chemical bond energy and use it to synthesize organic compounds from simple inorganic materials. These photoautotrophs are the **primary producers** for the system (Figure 47.2).

All other organisms in the system are **consumers**. They are different kinds of heterotrophs that feed on the tissues, products, and remains of other organisms. We may describe consumers by their diets. *Herbivores* eat plants. *Carnivores* eat flesh. *Parasites* live in or on a host and feed on its tissues. Earthworms, crabs, and other **detritivores** eat particles of decomposing organic matter (detritus), such as decaying bits of fallen leaves. **Decomposers** break down organic remains and wastes of all organisms. Hundreds of thousands of species of bacteria, protists, and fungi are decomposers.

Bear in mind, we cannot place some consumers in simple categories. Different kinds are *omnivores*, which may feed on animals, plants, fungi, protists, and even bacteria. A red fox is an example (Figure 47.3). It also scavenges when the opportunity presents itself. In the natural world, a full-time *scavenger* is a type of animal that feeds on the flesh of dead and decaying animals. Vultures are full-time scavengers. Hyenas hunt to kill but, like foxes, are opportunistic scavengers.

How does a system cycle nutrients? First, primary producers get hydrogen, oxygen, and carbon atoms from water and carbon dioxide in their environment. They also take up minerals, such as phosphorus and nitrogen. These are materials for biosynthesis. Later on, decomposition of organic wastes and remains releases

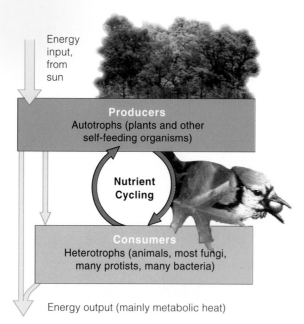

Figure 47.2 *Animated!* Model for ecosystems. Energy flows one way: into an ecosystem and out from it. Nutrients are cycled among autotrophs and heterotrophs. In nearly all ecosystems, energy flow starts with autotrophs that capture energy from the sun.

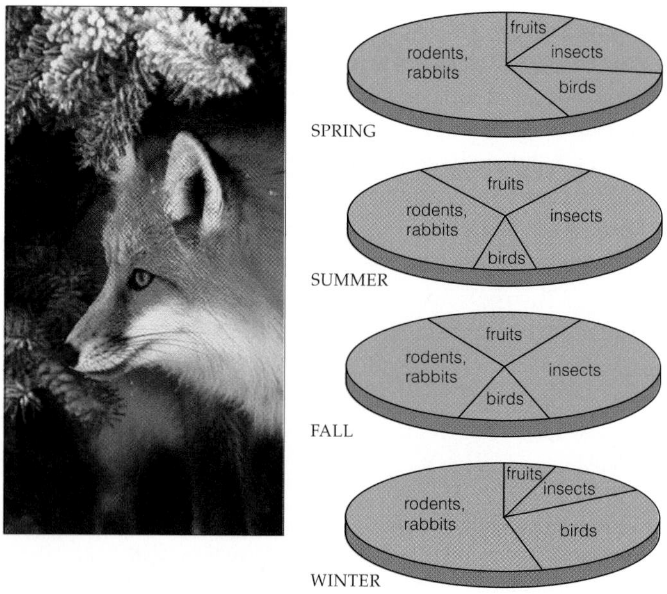

Figure 47.3 Red fox, an omnivore. Its diet shifts with seasonal changes in available food. Rodents, rabbits, and some birds make up the bulk of its diet in spring and winter. It eats more fruits and insects in the summer and fall.

nutrients back into the environment. Unless released substances move out of the system, as when mineral ions become dissolved in a stream that flows from a meadow, producers usually take them up again.

What we have just outlined is the **ecosystem**. We define each ecosystem as an array of organisms and their physical environment, all interacting through a one-way flow of energy and a cycling of the materials required to sustain life. It is an open system, in that it cannot sustain itself.

Energy inputs to most ecosystems are in the form of sunlight. There may be *nutrient inputs,* as from a creek delivering dissolved minerals to a lake. There are also *energy outputs* and *nutrient outputs.*

Energy transfers, remember, cannot be 100 percent efficient (Section 6.1). Over time, the energy originally harnessed by producers escapes to the environment, mainly as metabolically generated heat.

STRUCTURE OF ECOSYSTEMS

We can classify all organisms of an ecosystem by their functional roles in a hierarchy of feeding relationships called **trophic levels** (*troph,* nourishment). "Who eats whom?" we might ask. If organism B eats organism A, energy is transferred from A to B. All organisms at a given trophic level are the same number of transfer steps away from the energy input into an ecosystem.

As one example, think about some organisms of a tallgrass prairie ecosystem. The flowering plants and other producers that tap energy from the sun are at the first trophic level. Plants are eaten by herbivores, such as cutworms, which are at the next trophic level. Cutworms are one of the primary consumers that are eaten by carnivores at the third trophic level, and so on up through tiers of trophic levels.

At each trophic level, organisms interact with the same sets of predators, prey, or both. Omnivores feed at several levels, so we would partition them among different levels or assign them to a level of their own.

A **food chain** is a straight-line sequence of steps by which energy originally stored in autotroph tissues moves to higher trophic levels. In one tallgrass prairie food chain, for instance, energy from a plant flows to a cutworm that eats its juicy parts and on to a garter snake that eats the cutworm, to a crow that eats snake eggs and hatchlings, and finally to a marsh hawk that eats crow eggs and hatchlings (Figure 47.4).

Identifying a food chain is a simple way to start thinking about who eats whom in ecosystems. Bear in mind, many different species are usually competing for food in complex ways. Tallgrass prairie producers

fifth trophic level
top carnivore
(fourth-level consumer)

fourth trophic level
carnivore
(third-level consumer)

third trophic level
carnivore
(second-level consumer)

second trophic level
herbivore
(primary consumer)

first trophic level
autotroph
(primary producer)

Figure 47.4 *Animated!* Example of a simple food chain and its corresponding trophic levels in a tallgrass prairie.

(mainly flowering plants) feed grazing mammals and herbivorous insects. But many more species interact in the tallgrass prairie and nearly all other ecosystems, particularly at the lower trophic levels. A number of food chains *cross-connect* with one another—as **food webs**—and that is the topic of the next section.

An ecosystem is a community of organisms that interconnect with one another and with their physical environment by a one-way energy flow and a cycling of materials.

Autotrophs tap into an environmental energy source and make their own organic compounds from inorganic raw materials. They are the ecosystem's primary producers.

Autotrophs are at the first trophic level of a food chain, a straight-line sequence of feeding relationships that proceeds through one or more levels of heterotrophs, or consumers.

In ecosystems, food chains cross-connect, as food webs.

47.2 The Nature of Food Webs

LINKS TO
SECTIONS
6.1, 6.2

Food chains cross-connect with one another in food webs. By untangling the chains of many food webs, ecologists discovered patterns of organization. Those patterns reflect environmental constraints and the inefficiency of energy transfers from one trophic level to the next.

Recall, from Section 6.1, that energy concentrated in one place tends to spread out, or disperse, on its own. The collective strength of chemical bonds resists this spontaneous direction of energy flow. Every organism in an ecosystem must tap into a concentrated energy source and use it to build complex molecules even as they continually lose energy, as metabolic heat.

Plants capture energy that is concentrated in rays from the sun. They use some of it to drive metabolism, store about half of it in new plant tissues, and lose the rest as heat. Consumers tap into energy that became stored in plant tissues, remains, and wastes. They too, lose metabolic heat. *Taken together, all of the heat losses represent a one-way flow of energy out of the ecosystem.*

HOW MANY TRANSFERS?

When ecologists compared food chains in different kinds of food webs, a pattern emerged. In most cases, energy initially captured by producers passes through no more than four or five trophic levels. Even the rich ecosystems with complex food webs, such as the one in Figure 47.5, do not have lengthy food chains. The inefficiency of energy transfers may limit the sequence.

higher trophic levels

Complex array of carnivores, omnivores, parasites, detritivores, decomposers, and other consumers. Many feed at more than one trophic level all the time, seasonally, or whenever an opportunity presents itself.

second trophic level

Primary consumers (e.g., herbivores, detritivores, and decomposers)

first trophic level

Primary producers

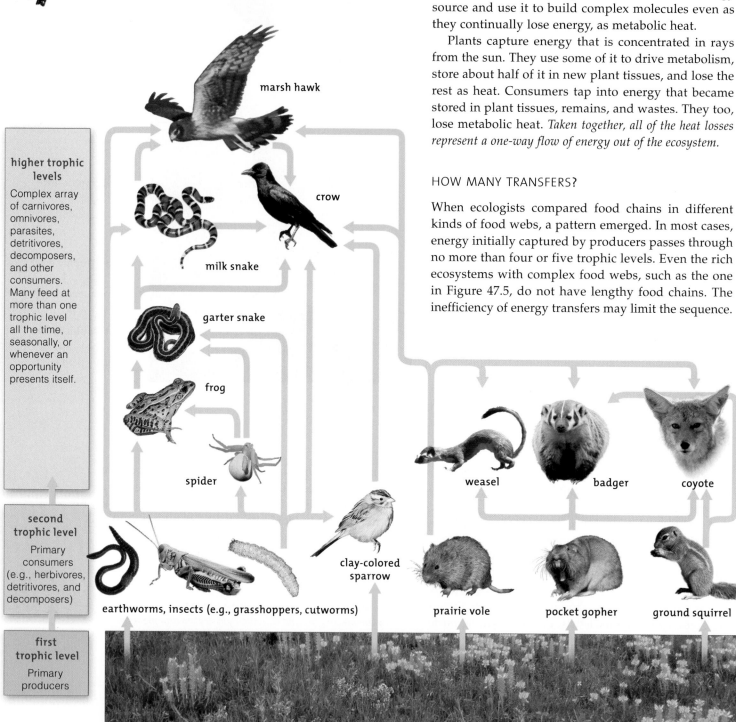

marsh hawk

crow

milk snake

garter snake

frog

spider

clay-colored sparrow

earthworms, insects (e.g., grasshoppers, cutworms)

weasel

badger

coyote

prairie vole

pocket gopher

ground squirrel

grasses, composites

Figure 47.5 *Animated!* A small sampling of the organisms at successively higher trophic levels for a tallgrass prairie food web.

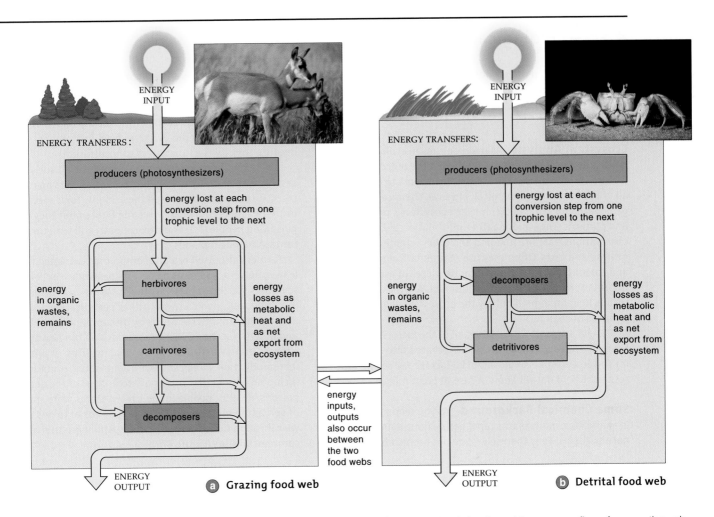

a Grazing food web   **b** Detrital food web

Figure 47.6 Generalized sketches of the one-way flow of energy through the participants of (**a**) a grazing food web and (**b**) a detrital food web.

Field studies and computer simulations of aquatic and land food webs reveal more patterns. Chains in food webs tend to be shortest where environmental conditions usually vary widely over time. Food chains tend to be longer in habitats that are more stable, such as ocean depths. The most complex webs tend to have many herbivorous species, as happens in grasslands. By comparison, the food webs with fewer connections tend to have more carnivores.

TWO CATEGORIES OF FOOD WEBS

Energy from producers—the organisms closest to a primary source—flows in one direction through two kinds of webs. In a **grazing food web**, energy flows mostly into herbivores, carnivores, then decomposers. In a **detrital food web**, energy from producers flows mainly into detritivores and decomposers. Figure 47.6 summarizes the flow through these food webs.

In nearly all ecosystems, both kinds of webs cross-connect. For example, in a rocky intertidal ecosystem, energy captured by algae flows to snails, which are eaten by herring gulls as part of a grazing food web. However, gulls also hunt crabs, which are among the primary consumers in the detrital food web.

The amount of energy that moves through the two kinds of food webs differs among ecosystems, and it often varies with the seasons. In most cases, however, most of the energy stored in producer tissues moves through *detrital* food webs. Think of cattle that graze heavily in a pasture. About half the energy stored in the grass plants enters the grazers. But cattle cannot access all of the stored energy. A lot is still present in undigested plant parts and in feces, and decomposers and detritivores go to work. Similarly, in marshes, most of the energy initially stored in the marsh grass tissues enters detrital food webs when the plants die.

The inherent inefficiency in energy transfers between trophic levels limits the length of food chains.

Tissues of living photosynthetic organisms are the basis for grazing food webs. Remains and wastes of these organisms are the basis for detrital food webs. In nearly all ecosystems, both types of food webs prevail and interconnect.

47.3 Biological Magnification in Food Webs

LINKS TO
SECTIONS 27.3,
28.4, 32.2, 46.5

We turn now to a premise that opened this chapter—that disturbances to one part of an ecosystem often can have unexpected effects on other, seemingly unrelated parts.

Ecosystem Analysis Many programs in ecology devise models as a way to monitor and predict the outcome of disturbances to ecosystems. Researchers work to identify all of the interacting biological, physical, chemical, and geologic factors that determine an ecosystem's processes and patterns. They might gather information by direct observations, satellite imaging and other remote sensing devices, and tests. Often they use mathematical models and computer programs to integrate pieces of available information on how the factors interact. Analysis of the results help them predict how the ecosystem will react to forces of change.

Results are most useful when all of the factors have been identified and accurately incorporated into a model for the ecosystem. The most crucial factor may be one that researchers do not yet know. A case in point follows.

Some Chemical Background As you read in Sections 28.4 and 46.5, many plants repel herbivorous animals with natural toxins. They themselves are not harmed by these organic compounds, but the chemical effects may repel or kill individuals of a different species. We encounter traces of natural plant toxins, even in such familiar foods as hot peppers, potatoes, figs, celery, rhubarb, and alfalfa sprouts. We do not get sick or die in droves from hot peppers, often because toxicity is a function of concentration.

Just a few thousand years ago, farmers used sulfur, lead, arsenic, and mercury to help protect crop plants against insects. They freely dispensed these highly toxic metals until the late 1920s, when someone figured out they were poisoning people. Traces of toxic metals still turn up in contaminated croplands.

Farmers also used organic compounds extracted from leaves, flowers, and roots as natural pesticides. In 1945, scientists started to make synthetic toxins and to identify mechanisms by which toxins attack pests. *Herbicides*, such as synthetic auxins, kill weeds by disrupting metabolism and growth (Section 32.2). *Insecticides* clog the airways of a target insect, disrupt its nerves and muscles, or prevent its reproduction. *Fungicides* work against harmful fungi, including a mold that makes aflatoxin, one of the deadliest poisons. By 1995, people in the United States were spraying or spreading more than 1.25 billion pounds of toxins each year through fields, gardens, homes, and industrial and commercial sites (Figure 47.7).

a *2,4-D* (2,4-dichlorophenoxyacetic acid), a synthetic auxin widely used as a herbicide. Enzymes of weeds and microbes cannot easily degrade 2,4-D, compared to natural auxins.

b *Atrazine*, the best-selling herbicide, kills weeds within a few days, as do glyphosate (Roundup), alachlor, (Lasso), and daminozide (Alar). It now appears that atrazine causes abnormal sexual development in frogs, even in trace amounts below the level allowed in drinking water.

c Dichlorodiphenyltrichloroethane, or *DDT*. It takes two to fifteen years for this nerve cell poison to break down. Chlordane, another type of insecticide, also persists for a long time in the environment.

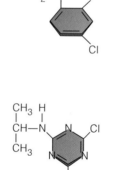

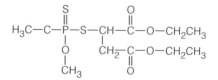

d *Malathion*. Like other organophosphates, it is cheap, breaks down faster than chlorinated hydrocarbons, and is more toxic. Organophosphates represent half of all insecticides used in the United States. Some are now banned for crops; application of others must end at least three weeks before harvest. Farmers who contest this policy want the Environmental Protection Agency to consider economic and trade issues as well as human health.

Figure 47.7 A few pesticides, some more toxic than others. The photograph shows one of the crop dusters that intervene in the competition for nutrients between crop plants and pests, including weeds.

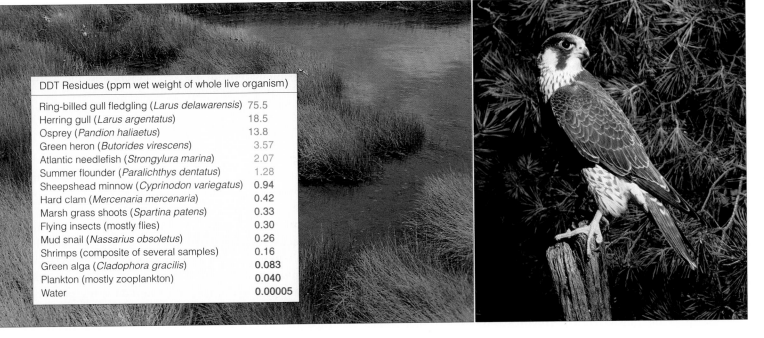

DDT Residues (ppm wet weight of whole live organism)	
Ring-billed gull fledgling (*Larus delawarensis*)	75.5
Herring gull (*Larus argentatus*)	18.5
Osprey (*Pandion haliaetus*)	13.8
Green heron (*Butorides virescens*)	3.57
Atlantic needlefish (*Strongylura marina*)	2.07
Summer flounder (*Paralichthys dentatus*)	1.28
Sheepshead minnow (*Cyprinodon variegatus*)	0.94
Hard clam (*Mercenaria mercenaria*)	0.42
Marsh grass shoots (*Spartina patens*)	0.33
Flying insects (mostly flies)	0.30
Mud snail (*Nassarius obsoletus*)	0.26
Shrimps (composite of several samples)	0.16
Green alga (*Cladophora gracilis*)	0.083
Plankton (mostly zooplankton)	0.040
Water	0.00005

Figure 47.8 Biological magnification in an estuary on the south shore of Long Island, New York, as reported in 1967 by George Woodwell, Charles Wurster, and Peter Isaacson. The researchers knew of broad correlations between the extent of DDT exposure and mortality. For instance, residues in birds known to have died from DDT poisoning were 30–295 ppm, and they were 1–26 ppm in several fish species. Some DDT concentrations measured during this study were below lethal thresholds but were still high enough to interfere with reproductive success.

Figure 47.9 Peregrine falcon, a top carnivore in some food webs. This raptor almost became extinct as a result of biological magnification of DDT. A wildlife management program successfully brought back its population sizes. Peregrine falcons were reintroduced into wild habitats. They have adapted to cities. There, they hunt pigeons, large populations of which are a messy nuisance.

DDT in Food Webs The DDT molecule highlighted in Figure 47.7c is a fairly stable hydrocarbon that is nearly insoluble in water. Therefore, you might think—as many others did—that it would exert its toxic effects only where it was applied. However, winds can easily disperse DDT in vapor form, and water can disperse fine particles of it.

Given its molecular properties, DDT is highly soluble in fats, and so it can accumulate in the tissues of organisms. That is why DDT can show **biological magnification**. By this occurrence, a substance that degrades slowly or not at all becomes ever more concentrated in tissues of organisms at higher trophic levels of a food web.

Most of the DDT that becomes concentrated in all of the organisms that a consumer eats during its lifetime ends up in the consumer's own tissues. DDT and its modified forms disrupt metabolic activities and are toxic to many aquatic and terrestrial animals.

Several decades ago, DDT started to infiltrate food webs and exert its effects on diverse organisms in ways that no one had predicted. Where people sprayed DDT to control Dutch elm disease, songbirds died (Section 27.3). In forests where DDT was sprayed to kill budworm larvae, fish in the forest streams died. In fields sprayed to control one kind of pest, new pests moved in. *DDT was indiscriminately killing the natural predators that keep pest populations in check.*

Then side effects of biological magnification started to show up in habitats far removed from where the DDT had been applied—*and much later in time.* Most vulnerable were brown pelicans, bald eagles, peregrine falcons, and other top carnivores of some food webs (Figures 47.8 and 47.9). Why? A product of DDT breakdown interferes with some physiological processes. As one outcome, bird eggs developed brittle shells; many chick embryos did not even hatch. Some species were facing extinction.

In the United States, DDT has been banned since the 1970s except where necessary to protect public health. Many species hit hardest have recovered. Some birds still lay thin-shelled eggs because they pick up DDT at their winter ranges in Latin America. As late as 1990, a fishery near Los Angeles was closed. DDT from industrial waste discharges that had stopped twenty years earlier was still contaminating that ecosystem.

Today, ecologists are monitoring more than pesticides in ecosystems. Radiosotopes and heavy metals, including copper, zinc, lead, and mercury, also can become ever more concentrated in organisms. For example, in fields near heavily trafficked highways, ecologists found out that the soil concentration of lead can be as high as 1,200 parts per million (ppm)—and it gets magnified as it moves up food chains. The longer the chain, the greater the magnification.

47.4 Studying Energy Flow Through Ecosystems

LINKS TO
SECTIONS
7.8, 8.7

Ecologists measure the amount of energy and nutrients entering an ecosystem, how much is captured, and the proportion stored in each trophic level.

WHAT IS PRIMARY PRODUCTIVITY?

The rate at which producers capture and store energy in their tissues during a given interval is the **primary productivity** of an ecosystem. How much energy gets stored depends on (1) how many producers there are and (2) the balance between photosynthesis (energy trapped) and aerobic respiration (energy used). *Gross primary production* is all energy initially trapped by the producers. *Net* primary production is the fraction of trapped energy that producers funnel into growth and reproduction. **Net ecosystem production** is the gross primary production *minus* the energy used by the producers and soil detritivores and decomposers. That amount of energy is subtracted because it cannot be transferred to herbivores at the next trophic level.

On land and in the water provinces, many factors impact net production, its seasonal patterns, and its distribution through a habitat (Section 7.8 and Figure 47.10). For instance, the size and form of the primary producers, the temperature range, the availability of mineral ions, and the amount of sunlight and rainfall in each growing season affect energy acquisition and storage. The harsher the conditions are, the less new growth plants add in a given season, and the lower the primary productivity.

ECOLOGICAL PYRAMIDS

Ecologists often represent the trophic structure of an ecosystem in the form of an ecological pyramid. In such pyramids, all primary producers form a base for successive tiers of consumers above them.

A **biomass pyramid** depicts the dry weight of all of an ecosystem's organisms at each tier. Figure 47.11 shows a biomass pyramid for one aquatic ecosystem. The amounts measured are grams per square meter at some specified time. Most commonly, the primary producers have most of the biomass in pyramids like this, and top carnivores are few. But some biomass pyramids are "upside-down," in that the smallest tier is on the bottom. This happens in springtime blooms of phytoplankton, which grow and reproduce quickly. The primary producers of these aquatic communities support a larger biomass of zooplankton, which eat them about as fast as they can reproduce.

An **energy pyramid** illustrates how the amount of usable energy diminishes as it is transferred through an ecosystem. Sunlight energy is captured at the base (first trophic level) and declines through successive levels to its tip (the top carnivores). Energy pyramids have a large energy base at the bottom, so they are always "right-side up." Such pyramids can provide a clear picture of energy flow from an outside source

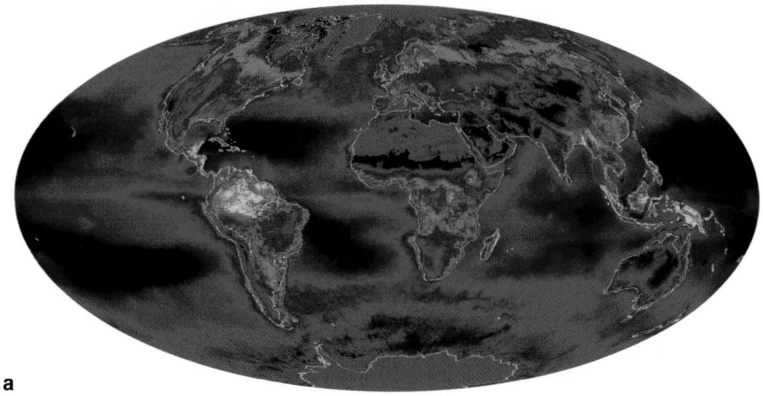

a

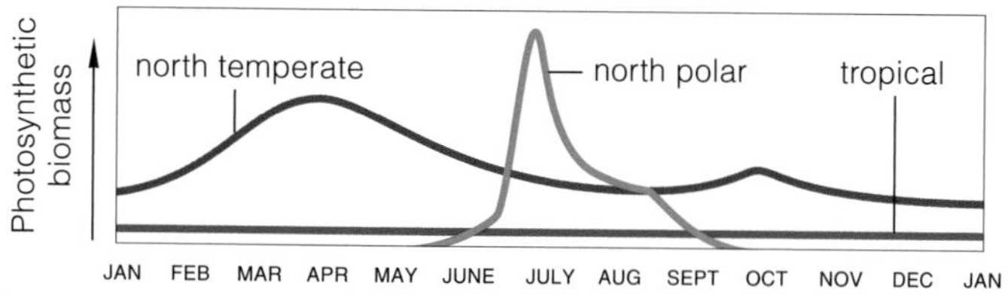

b

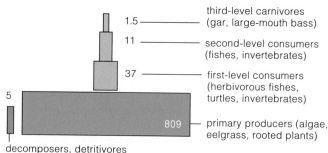

Figure 47.10 (a) Summary of satellite data on net primary productivity during 2002. Productivity is coded as *red* (highest) down through *orange*, *yellow, green, blue,* and *purple* (lowest). Although average productivity per unit of sea surface is lower than it is on land, total productivity on land and in seas is about equal, because most of Earth's surface is covered by water. (b) Examples of seasonal shifts in net primary productivity for three categories of ocean ecosystems.

Figure 47.11 Biomass pyramid for Silver Springs, a small aquatic ecosystem in which biomass decreases in successively higher tiers. In different ecosystems, autotrophs are eaten almost as fast as they grow and reproduce. In such cases, biomass accumulates faster in consumers, so the biomass pyramid would be upside down.

Figure 47.12 *Animated!* Breakdown of the annual energy flow through Silver Springs, Florida, as measured in kilocalories/square meter/year. Most of the primary producers in this small spring are aquatic plants. Most carnivores are insects and small fishes; the top carnivores are larger fishes. The original energy source, sunlight, is available all year. Detritivores and decomposers cycle organic compounds from the other trophic levels.

Producers trapped 1.2 percent of the incoming solar energy, and only a little more than a third of that became fixed in new plant biomass. The producers used more than 63 percent of the fixed energy for their own metabolism.

About 16 percent of the fixed energy was transferred to herbivores. Most was used for metabolism or transferred to detritivores and decomposers. Of the energy that transferred to herbivores, only 11.4 percent reached the next trophic level (carnivores). About 5.5 percent of the energy in the lower-level carnivores flowed to the top carnivores.

By the end of the specified interval, all 20,810 kilocalories of energy that flowed through the system appeared as metabolically generated heat.

and on through its departure, mainly by losses of the metabolic heat that each organism generates.

ENERGY FLOW THROUGH SILVER SPRINGS

Visualize yourself with ecologists who are gathering data to construct an energy pyramid for a freshwater spring over the course of one year. They measure how much energy one individual of each species takes in, loses as metabolic heat, stores in its body tissues, and then loses as wastes. They multiply the energy per individual by population size, then calculate energy inputs and outputs. Then they express the energy flow per unit of water (or land) per unit of time. Figure 47.12 was constructed from data that were gathered this way during a long-term study of a grazing food web in this type of aquatic ecosystem. It shows some calculations that ecologists used to depict the energy flow in pyramid form in Figure 47.13.

Based on many such studies, ecologists arrived at this generalization: Given the metabolic demands of organisms and the amount of energy lost in organic wastes, only 6 to 16 percent of the energy entering one trophic level is available for organisms at the next.

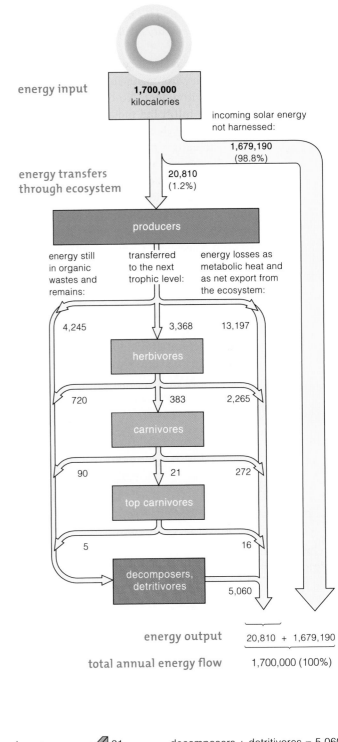

Gross primary productivity is an ecosystem's total rate of photosynthesis during a specified interval. The net amount is the rate at which primary producers store energy in tissues in excess of their rate of aerobic respiration. Heterotrophic consumption affects the rate of energy storage.

The trophic structure of an ecosystem may be represented by an ecological pyramid. Biomass pyramids may be top- or bottom-heavy depending on the ecosystem. In contrast, an energy pyramid always has the largest tier on the bottom.

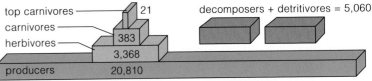

Figure 47.13 Pyramid of energy flow through Silver Springs, in kilocalories/square meter/year. This is a summary of the data used to construct Figure 47.12. Compare Figure 47.11, the biomass pyramid for this same ecosystem.

47.5 Overview of Biogeochemical Cycles

Without water and the nutrients dissolved in it, there would be no primary productivity, and no life.

In a **biogeochemical cycle**, an essential element moves from the environment, through ecosystems, then back to the environment. No other element can directly or indirectly fulfill the metabolic role of such elements, or **nutrients**, which is why we call them essential. As you read earlier, oxygen, hydrogen, carbon, nitrogen, and phosphorus are among them.

Figure 47.14 is one model for these cycles. Transfers to and from environmental reservoirs are usually far slower than rates of exchange among organisms of an ecosystem. Water is the main source for hydrogen and oxygen. Gaseous or ionized forms of other elements are dissolved in it. Solid forms of elements are tied up in rocks or sediments.

Nutrients move into and out of ecosystems by way of natural geologic processes. Weathering of rocks is a common source of nutrient inputs into an ecosystem. Erosion and runoff put nutrients into streams that carry them away. Most often, the quantity of a nutrient being cycled through an ecosystem each year is greater than the amount entering and leaving.

Decomposers help cycle the nutrients in ecosystems. Various prokaryotic species help transform solids and ions into gases, then back again. Through their action, they convert some elements that function as nutrients to forms that primary producers can take up.

In three types of biogeochemical cycles, portions of the environment are reservoirs for specific elements. In the *hydrologic* cycle, oxygen and hydrogen move, on a grand scale, in molecules of water. In *atmospheric* cycles, some gaseous form of the nutrient is the one available to ecosystems. Carbon and nitrogen cycles are examples. Phosphorus and other solid nutrients that have no gaseous form move in *sedimentary* cycles. They accumulate on the seafloor and eventually return to land through geological uplifting, which typically has taken millions of years. Earth's crust is the biggest reservoir for nutrients that have sedimentary cycles.

> *Primary productivity depends on water and nutrients that become dissolved in it. In a biogeochemical cycle, a nutrient moves slowly through the environment, then rapidly among organisms, and back to environmental reservoirs.*

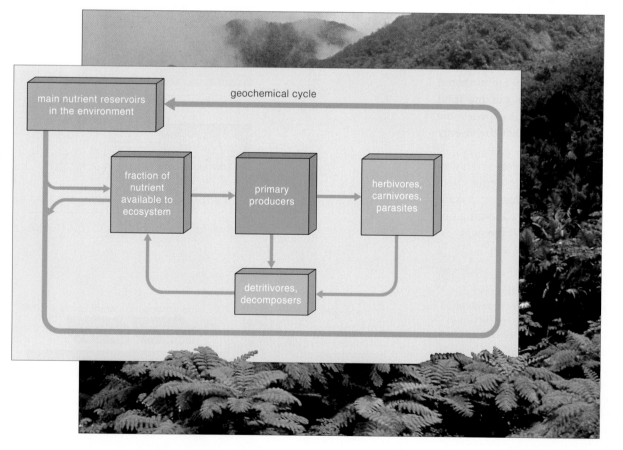

Figure 47.14 One generalized model of nutrient flow through an ecosystem on land. The overall movement of nutrients from the physical environment, through organisms, and then back to the environment is a biogeochemical cycle.

47.6 Hydrologic Cycle

On land, the availability of water, and nutrients dissolved in it, is not plentiful all of the time in all ecosystems. The variation affects primary productivity.

Driven by solar energy, Earth's waters slowly move from the ocean into the atmosphere, to land, and back to the ocean—the main reservoir. Figure 47.15 shows this **hydrologic cycle**. Water that evaporates into the lower atmosphere stays aloft as vapor, clouds, and ice crystals, then falls mainly as rain and snow. Ocean circulation and wind patterns influence the cycle.

Water moves nutrients into and out of ecosystems. A **watershed** is any region where precipitation flows into a single stream or river. Watersheds may be as small as the area that drains into a stream or as vast as the Amazon River or Mississippi River basin. Most water entering a watershed seeps into soil or joins surface runoff into streams. Plants take up water from soil and lose it by transpiration (Figure 47.16).

In the hydrologic cycle, water slowly moves on a global scale from the world ocean—the main reservoir—through the atmosphere, onto land, then back to the ocean.

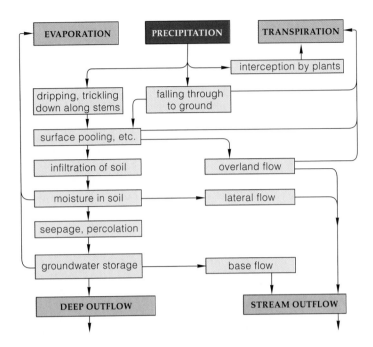

Figure 47.16 Model for how water moves through watersheds in general. *Dark blue* box is the input to the watershed; *light blue*, the distribution within it; and *medium blue*, outputs from it. Transpiration is the name for evaporation of water from leaves and other plant parts exposed to air (Section 30.3). Plants absorb water from soil and groundwater stores.

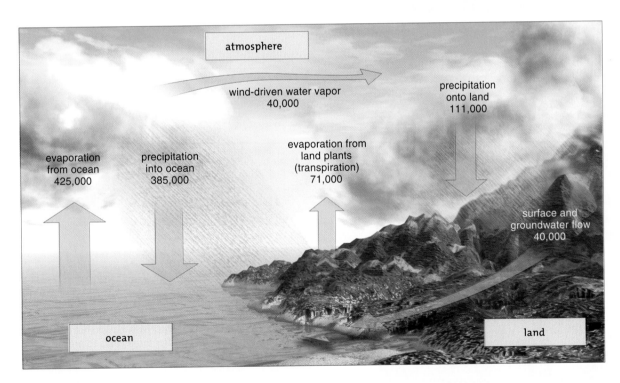

Figure 47.15 *Animated!* Hydrologic cycle. Water moves from the ocean to the atmosphere, land, and back. Arrows identify processes that move water, as measured in cubic kilometers per year. With 1,370,000,000 cubic kilometers, the ocean is the main reservoir. The next largest, polar ice and glaciers, locks up 29,000,000. Groundwater makes up only 4,000,000, lakes and rivers only 241,000, soil 67,000, and the atmosphere 14,000 cubic kilometers.

47.7 Watershed Experiments

LINKS TO
SECTIONS
23.10, 30.1

Water is vital for all organisms. It also is a transport medium; it moves nutrients into and out of ecosystems. Its role in moving nutrients became clear in long-term studies of watersheds.

A watershed, again, is any region in which the precipitation becomes funneled into just one stream or one river. Figure 47.17*a* shows part of an experimental forest in the Hubbard Brook Valley of New Hampshire. The watersheds in this forest have a surface area of 14.6 hectares (36 acres), on average. Over the years, ecologists have painstakingly measured nutrient inputs and outputs in this forest. Such measurements have many practical applications.

For instance, cities that draw from a watershed's supply of surface water can adjust their usage in compliance with seasonal shifts in the volume of water. Measurements also reveal the extent to which vegetation cover influences the movement of nutrients through the ecosystem phase of biogeochemical cycles.

For example, you might think that surface runoff in a watershed would swiftly leach out calcium ions and other minerals dissolved in soil water. (Here you may wish to review the Section 30.1 explanation of leaching.) However, in the young, undisturbed forests of the Hubbard Brook watersheds, each hectare lost only eight kilograms or so of its calcium. The weathering of rocks and rainfall were replacing the lost calcium. In addition, tree roots were "mining" the soil by absorbing dissolved mineral ions, so calcium was being stored in a growing biomass of tree tissues.

In experimental watersheds in the Hubbard Brook Valley, deforestation caused a shift in nutrient outputs. The results, summarized in Figure 47.17*d*, are sobering. Calcium and other nutrients cycle very slowly, which means that deforestation may disrupt the availability of nutrients for an entire ecosystem. This is especially the case for forests that cannot regenerate themselves over the short term because of their soil properties and other characteristics. As you will see in the next chapter, the northern coniferous forests and tropical rain forests require long recovery times.

a

b

c

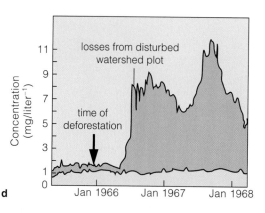

d

Figure 47.17 (**a**) Experimental deforestation in the Hubbard Brook watershed of New Hampshire. Researchers monitor surface runoff flowing over concrete catchments (**b**) on its way to a stream below. For some experiments, they stripped vegetation from forest plots but did not disturb the soil. They applied herbicides for three years to prevent regrowth. (**c**) They compared concentrations of calcium ions and other minerals in runoff over a control catchment against concentrations in runoff from an undisturbed area.

(**d**) Calcium losses were *six times* greater in deforested plots. Removing all vegetation from such forests clearly alters nutrient outputs in ways that can disrupt nutrient availability for an entire ecosystem.

47.8 A Global Water Crisis

Most of Earth's water is too salty to drink or use for agriculture. The skyrocketing increases in human population size make this a big problem.

Two-thirds of the fresh water that humans use goes directly into irrigating fields (Figure 47.18). Ironically, irrigation makes the land less suitable for agriculture. Piped-in water commonly has high concentrations of mineral salts. Where soil drains poorly, evaporation results in **salinization**, or a build-up of salt in soil that stunts crop plants and decreases yields.

Soil and aquifers hold **groundwater**. About half of the United States population taps into groundwater as a source of drinking water. Chemicals leached from landfills, hazardous waste dumps, and underground tanks that store gasoline, oil, and some solvents often contaminate it. Unlike flowing streams, which recover fast, polluted groundwater is difficult and expensive to clean up.

Groundwater overdrafts, or the amount that nature has not replenished, are high in many areas. Figure 47.19 shows some regions of aquifer depletion in the United States. Overdrafts have now depleted half of the great Ogallala aquifer, which supplies irrigation water for 20 percent of the Midwest's croplands.

Inputs of sewage, animal wastes, and many toxic chemicals from power-generating plants and factories make water unfit to drink. Sediments and pesticides run off from fields into water, along with phosphates and other nutrients that promote algal blooms. The pollutants accumulate in lakes, rivers, and bays before reaching the ocean. Many cities all over the world are still dumping untreated sewage into coastal waters.

If current rates of human population growth and water depletion continue, the amount of fresh water available for everyone will soon be 55 to 66 percent less than it was in 1976. In this past decade, thirty-three nations have already engaged in conflicts over reductions in water flow, pollution, and silt buildup in aquifers, rivers, and lakes. Among the squabblers are the United States and Mexico, Pakistan and India, and Israel and the Palestinian territories.

Could we meet our water needs by **desalinization**, or removal of salt from seawater? Salt can be removed by distillation or pushing water through membranes. The processes require fossil fuels, which makes them more feasible in Saudi Arabia and other countries with small populations and big fuel reserves. Most likely, desalinization will not be cost-effective for large-scale agriculture. It also produces mountains of salts.

We may be in for upheavals and wars over water rights. Does this sound far-fetched? Consider the new

Figure 47.18 If the world has so little fresh water, why are we irrigating deserts?

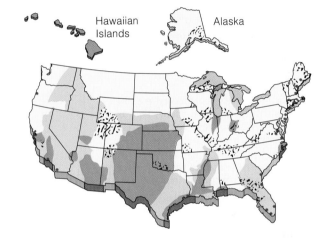

Figure 47.19 Aquifer depletion, seawater intrusion, and groundwater contamination in the United States. *Green* signifies high overdrafts, *gold*, moderate overdrafts, and *pale yellow*, insignificant withdrawals. Shaded areas are sites of major groundwater pollution. *Blue* squares indicate saltwater intrusion from nearby seas.

dam across the Euphrates River. By building immense irrigation systems and dams at the headwaters of the Tigris and Euphrates rivers, Turkey can, in the view of one of its dam site managers, shut off water flow into Syria and Iraq for as long as eight months "to regulate their political behavior." One might say that regional, national, and global planning is overdue.

Agriculture accounts for about two-thirds of the human population's use of freshwater.

Aquifers that supply much of the world's drinking water are becoming polluted and depleted. Regional conflicts over access to clean, drinkable water are likely to increase.

LINKS TO
SECTIONS
7.8, 23.10,
INTRODUCTIONS
CHAPTERS 3, 22

47.9 Carbon Cycle

Most of the world's carbon is locked in ocean sediments and rocks. It moves into and out of ecosystems in gaseous form, so its movement is said to be an atmospheric cycle.

In the **carbon cycle**, carbon moves through the lower atmosphere and all food webs on its way to and from its large reservoirs (Figure 47.20). Earth's crust holds the most carbon—66 million to 100 million gigatons, of which 4,000 are present in fossil fuels. (One gigaton is a billion tons.) Remember Chapter 22? Many single-celled organisms of ancient aquatic habitats, including foraminiferans and coccolithophores, formed shells of calcium carbonate. Uncountable numbers of cells died, sank, and were buried in seafloor sediments. Carbon in their remains have been cycled exceedingly slowly, after geologic forces uplift part of the seafloor. Such cycling cannot be measured in years, obviously.

Most of the annual cycling takes place between the ocean and atmosphere. The ocean holds 38,000–40,000 gigatons of dissolved carbon, primarily in the form of bicarbonate and carbonate ions. The atmosphere holds about 766 gigatons of carbon, mainly combined with oxygen in the form of carbon dioxide (CO_2).

Detritus in soil holds another 1,500–1,600 gigatons of carbon atoms. Another 540–610 gigatons is present in biomass. Methane hydrates form a huge reservoir that was, oddly, overlooked in the past. Between 10,000 and 11,000 gigatons are sequestered off the coasts of continents and in permafrost. As explained in Section 48.10, permafrost consists of perpetually frozen peat bogs that are sometimes more than 500 meters thick. You read about methane hydrates in the introduction to Chapter 3. Also, on page 865 of this chapter, you are invited to consider how unstable deposits on the seafloor can have big impact on the carbon cycle.

Why doesn't all of the CO_2 dissolved in warm sea surface waters escape into the atmosphere? Driven by winds and regional differences in water density, ocean water makes a gigantic loop from the surface of the Pacific and Atlantic oceans down to the Atlantic and

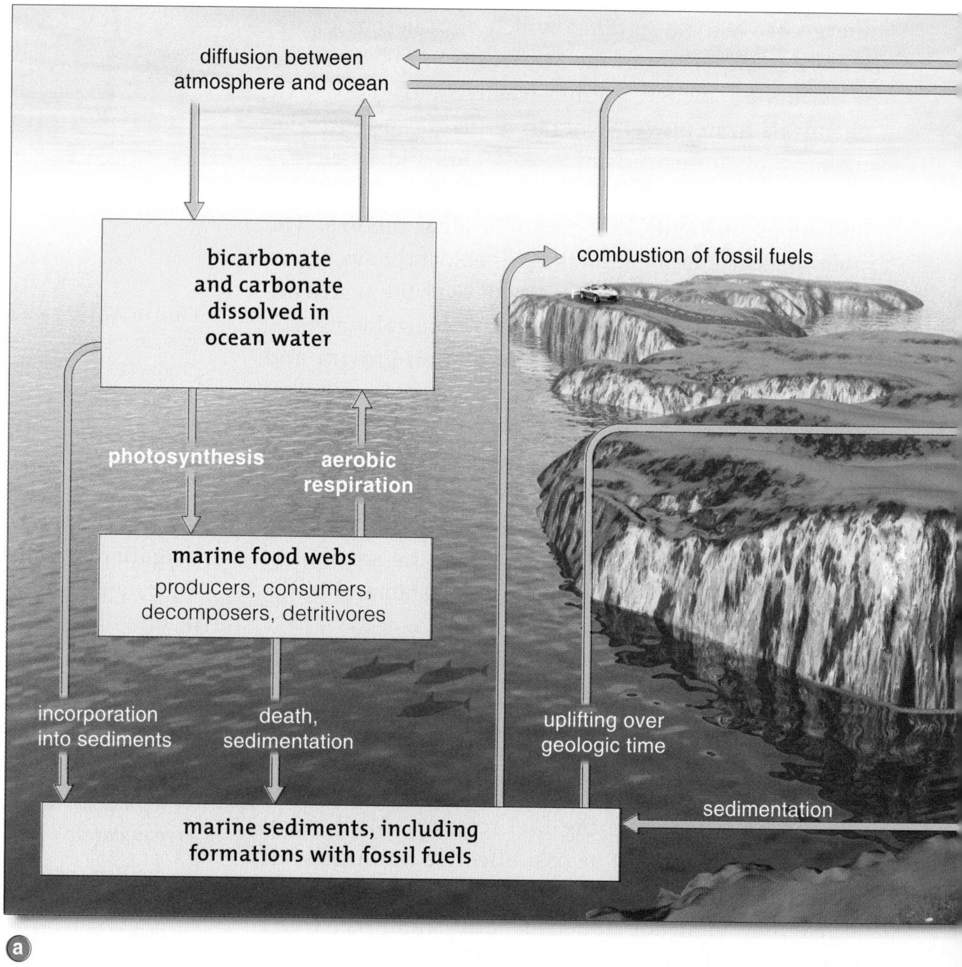

Figure 47.20 *Animated!* Global carbon cycle through typical marine ecosystems (**a**) and land ecosystems (**b**). *Gold* boxes show the main carbon reservoirs. The vast majority of carbon atoms are in sediments and rocks, followed by ever lesser amounts in ocean water, soil, the atmosphere, and biomass. Here are typical annual fluxes in the global distribution of carbon, in gigatons:

From atmosphere to plants by carbon fixation	120
From atmosphere to ocean	107
To atmosphere from ocean	105
To atmosphere from plants	60
To atmosphere from soil	60
To atmosphere from fossil fuel burning	5
To atmosphere from net destruction of plants	2
To ocean from runoff	0.4
Burial in ocean sediments	0.1

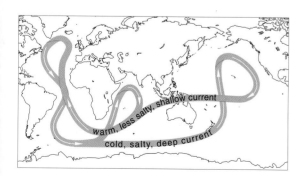

Figure 47.21 Ocean loop that moves carbon dioxide to carbon's deep ocean reservoir. It sinks in the cold, salty North Atlantic and rises in the warmer Pacific.

diffusion between
atmosphere and ocean

bicarbonate and carbonate dissolved in ocean water

combustion of fossil fuels

photosynthesis   aerobic respiration

marine food webs
producers, consumers, decomposers, detritivores

incorporation into sediments

death, sedimentation

uplifting over geologic time

sedimentation

marine sediments, including formations with fossil fuels

warm, less salty, shallow current

cold, salty, deep current

a

Antarctic seafloors. The CO_2 moves into deep storage reservoirs before water loops back up (Figure 47.21). The loop effectively mediates the annual fluxes in the global distribution of carbon.

As you know, CO_2 is the key source of carbon for autotrophs, and for food webs in ecosystems on land and in the seas. That is why biologists often refer to the glocal cycling of carbon as the *carbon–oxygen cycle*. When photosynthetic autotrophs fix carbon, they lock up billions of metric tons of carbon atoms in organic compounds annually (Section 7.8). When aerobic cells engage in aerobic respiration, they release CO_2. More CO_2 is released when fossil fuels or forests burn and when volcanoes erupt.

The average time that an ecosystem holds a given carbon atom varies. As examples, organic wastes and remains decompose so fast in tropical rain forests that carbon does not build up at the soil surface. Bogs and other anaerobic habitats do not favor decomposition, so the organic material is not degraded to smaller bits and carbon accumulates in peat. Humans withdraw 4 to 5 gigatons from fossil fuel reservoirs every year. At the same time, human activities put about 6 gigatons more carbon in the atmosphere than can be cycled to the ocean reservoirs by natural processes.

Only about 2 percent of the excess carbon entering the atmosphere will become dissolved in ocean water. Most researchers now suspect that the carbon build-up in the atmosphere is amplifying the greenhouse effect. In other words, the increase might be contributing to global warming. The next section takes a look at this possibility and its environmental implications.

> Earth's crust holds the vast majority of carbon. The ocean is the next largest reservoir. Most of the annual cycling of carbon occurs between the ocean and atmosphere.
>
> Carbon moves into and out of ecosystems mainly when combined with oxygen, as in carbon dioxide, bicarbonate, and carbonate. We refer to this as the carbon–oxygen cycle.

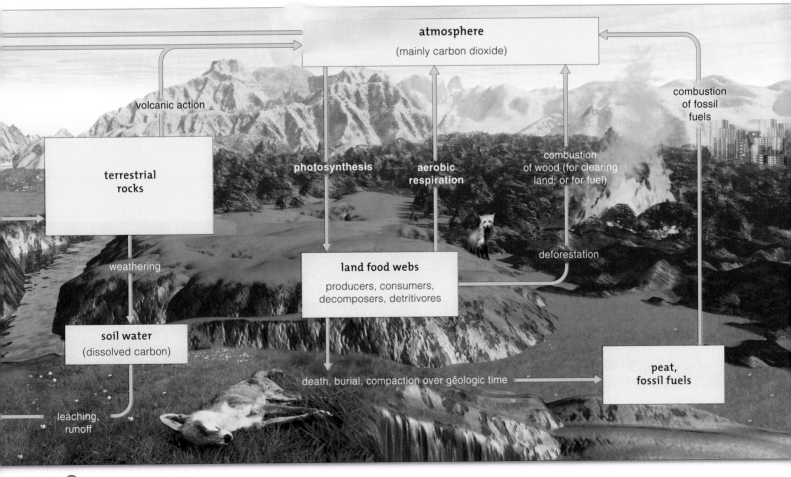

atmosphere
(mainly carbon dioxide)

volcanic action

combustion of fossil fuels

terrestrial rocks

photosynthesis

aerobic respiration

combustion of wood (for clearing land; or for fuel)

weathering

deforestation

land food webs
producers, consumers, decomposers, detritivores

soil water
(dissolved carbon)

death, burial, compaction over geologic time

peat, fossil fuels

leaching, runoff

b

47.10 Greenhouse Gases, Global Warming

LINKS TO
SECTIONS
7.1, 21.5, 23.10

The atmospheric concentrations of gaseous molecules help determine the average temperature near Earth's surface. Human activities are contributing to increases that may cause dramatic climate change.

Concentrations of a variety of gaseous molecules in Earth's atmosphere profoundly influence the average temperature near its surface. Temperature, in turn, has far-reaching effects on global and regional climates.

Atmospheric molecules of carbon dioxide, water, nitrous oxide, methane, and chlorofluorocarbons are among the main players in interactions that affect global temperature. Collectively, these gases function like the panes of glass in a greenhouse—hence their name, "greenhouse gases." The wavelengths of visible light pass through these gases to Earth's surface, which absorbs them and then emits longer, infrared wavelengths—heat. Greenhouse gases impede the escape of heat energy from Earth into space. How? The gaseous molecules absorb the longer wavelengths, then radiate much of it back toward Earth (Figure 47.22).

Constant reradiation of heat by greenhouse gases occurs lockstep with the constant bombardment and absorption of wavelengths from the sun. As heat builds up in the lower atmosphere, the air temperature near Earth's surface rises. The warming action is known as the **greenhouse effect**. Without it, Earth's surface would be so cold that it could not support life.

In the 1950s, laboratory researchers on Hawaii's highest volcano set out to measure the atmospheric concentrations of greenhouse gases. That remote site is almost free of local airborne contamination; it is also representative of overall atmospheric conditions for the Northern Hemisphere. What did they find? Briefly, carbon dioxide concentrations follow annual cycles of primary production. They drop during the summer, when photosynthesis rates are highest. They rise

Figure 47.23 *Facing page*, graphs of recent increases in four categories of atmospheric greenhouse gases. A key factor is the sheer number of gasoline-burning vehicles in large cities. *Above*, Mexico City on a smoggy morning. With 10 million residents, it is the world's largest city.

in winter, when photosynthesis rates decline but aerobic respiration is still going on.

Alternating troughs and peaks along the graph line in Figure 47.23*a* are annual lows and highs of global carbon dioxide concentrations. For the first time, we could see the integrated effects of carbon balances for an entire hemisphere. Notice the midline of the troughs and peaks in the cycle. It shows that carbon dioxide concentration is steadily increasing—as are the concentrations of other major greenhouse gases.

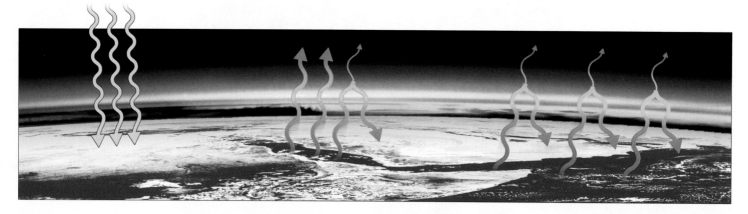

a Wavelengths in rays from the sun penetrate the lower atmosphere, and they warm the Earth's surface.

b The surface radiates heat (infrared wavelengths) to the atmosphere. Some heat escapes into space. But greenhouse gases and water vapor absorb some infrared energy and radiate a portion of it back toward Earth.

c Increased concentrations of greenhouse gases trap more heat near Earth's surface. Sea surface temperatures rise, so more water evaporates into the atmosphere. Earth's surface temperature rises.

Figure 47.22 *Animated!* The greenhouse effect.

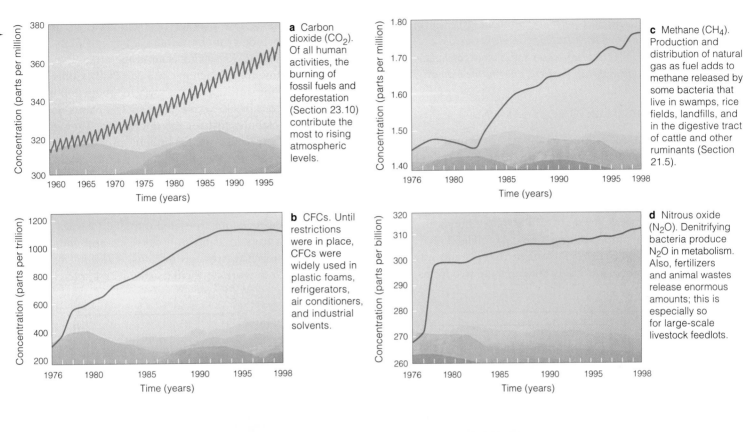

a Carbon dioxide (CO_2). Of all human activities, the burning of fossil fuels and deforestation (Section 23.10) contribute the most to rising atmospheric levels.

c Methane (CH_4). Production and distribution of natural gas as fuel adds to methane released by some bacteria that live in swamps, rice fields, landfills, and in the digestive tract of cattle and other ruminants (Section 21.5).

b CFCs. Until restrictions were in place, CFCs were widely used in plastic foams, refrigerators, air conditioners, and industrial solvents.

d Nitrous oxide (N_2O). Denitrifying bacteria produce N_2O in metabolism. Also, fertilizers and animal wastes release enormous amounts; this is especially so for large-scale livestock feedlots.

Figure 47.24 Recorded changes in global temperature between 1880 and 2000. At this writing, the hottest year on record was 1998.

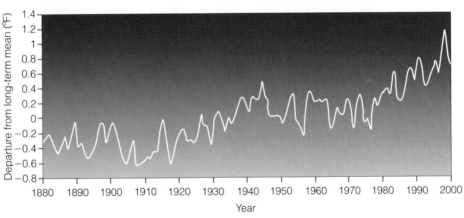

Atmospheric levels of greenhouse gases are far higher than they were in most of the past. Carbon dioxide may be at its highest level since 420,000 years ago, and possibly since 20 million years ago. There is a growing consensus that the rise in greenhouse gases is caused by some human activities, mainly burning of fossil fuels. The big worry is that the increase may have far-reaching environmental consequences.

The increase in greenhouse gases may be a factor in **global warming**, a long-term increase in temperature near Earth's surface. Since direct atmospheric readings started in 1861, the lower atmosphere's temperature has risen by more than 1°F, mostly since 1946 (Figure 47.24). Also since then, nine of the ten hottest years on record occurred between 1990 and the present. Data from satellites, weather stations and balloons, research ships, and supercomputer programs suggest that irreversible climate changes are already under way. Polar ice is melting; glaciers are retreating. This past century, the sea level may have risen as much as twenty centimeters (eight inches).

We can expect continued temperature increases to have drastic effects on climate. As evaporation increases, so will global precipitation. Intense rains and flooding are expected to become more frequent in some regions.

It bears repeating: As investigations continue, a key research goal is to investigate all of the variables in play. With respect to the consequences of global warming, the most crucial variable may be the one we do not know.

47.11 Nitrogen Cycle

LINKS TO
SECTIONS 21.4,
22.6, 24.6, 30.2

Gaseous nitrogen makes up about 80 percent of the lower atmosphere. Successively smaller reservoirs are seafloor sediments, ocean water, soil, biomass on land, nitrous oxide in the atmosphere, and marine biomass.

INPUTS INTO ECOSYSTEMS

Gaseous nitrogen (N_2) travels in an atmospheric cycle called the **nitrogen cycle**. Triple covalent bonds join its two atoms ($N\equiv N$). Volcanic action and lightning convert some N_2 into forms that enter food webs. Far more enters by **nitrogen fixation**. With this metabolic process, bacteria split all three bonds in N_2 and use the atoms to form ammonia (NH_3). Later, ammonia is converted to ammonium (NH_4^+) and nitrate (NO_3^-). Most plants easily take up these two forms of nitrogen.

Figure 47.25 shows the nitrogen cycle. Its nitrogen fixers include cyanobacteria in aquatic habitats and in many lichens. *Rhizobium* is a nitrogen fixer in nodules on legume roots (Sections 24.6 and 30.2). Collectively, these photoautotrophic bacteria fix about 200 million metric tons of nitrogen each year. The plants do pay a high metabolic cost for the interaction. They give up sugars and other photosynthetic products that take large investments of ATP and NADPH. Such plants have a competitive edge in nitrogen-poor soil. Other plants that do not pay the metabolic price commonly displace them in nitrogen-rich soil.

The nitrogen incorporated into plant tissues moves through trophic levels of ecosystems and ends up in nitrogen-rich wastes and remains, where bacteria and fungi go to work on them (Sections 21.4 and 24.6). By

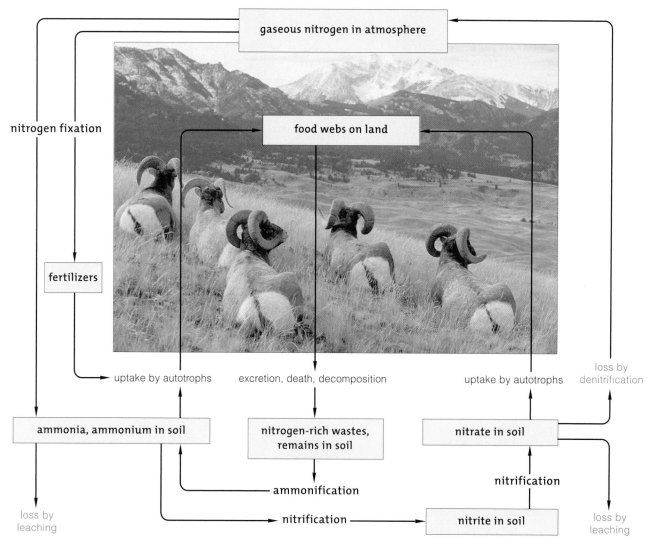

Figure 47.25 *Animated!* The nitrogen cycle in an ecosystem on land. Activities of nitrogen-fixing bacteria make nitrogen available to plants. Other bacterial species cycle nitrogen to plants. They break down organic wastes to ammonium and nitrates.

the process of **ammonification**, these microbes break down nitrogenous materials, and ammonium forms. They use some of the ammonium and release the rest to soil. Plants take it up, as do some nitrifying bacteria. In the first step of **nitrification**, certain bacteria cause nitrite (NO_2^-) to form when they strip electrons from ammonium. Different nitrifying bacteria use the nitrite in reactions that form nitrate (NO_3^-).

NATURAL LOSSES FROM ECOSYSTEMS

Ecosystems lose nitrogen through **denitrification**. By this process, denitrifying bacteria convert nitrate or nitrite to gaseous nitrogen or to nitrogen oxide (NO_2). Most denitrifying bacteria are anaerobic; they live in waterlogged soils and aquatic sediments.

Ammonium, nitrite, and nitrate are also lost from a land ecosystem in runoff and by leaching, the removal of some nutrients as water percolates through the soil (Section 30.1). Leaching removes nitrogen from land ecosystems and adds it to aquatic ones.

DISRUPTIONS BY HUMAN ACTIVITIES

Deforestation and grassland conversion for agriculture also cause big nitrogen losses. With each clearing and harvest, nitrogen in plant tissues is removed. Soil also becomes more vulnerable to erosion and leaching.

Many farmers counter nitrogen losses by rotating crops, as by alternating wheat with legumes. Rotation can help keep soil stable and productive. In developed countries, farmers also spread nitrogen-rich fertilizers. They even select new strains of crop plants that have a greater capacity to take up fertilizers from soil. By such practices, crop yields per hectare have doubled and sometimes quadrupled over the past forty years.

High temperature and pressure converts nitrogen and hydrogen gases to ammonia fertilizers. The use of these manufactured fertilizers greatly increases crop yields. It also can alter soil chemistry by disrupting a pH-dependent process called **ion exchange**. By this process, ions dissociate from soil particles, and then other ions in soil water replace them (Section 30.2). The most abundant exchangeable ions are calcium and magnesium. The hydrogen ions in nitrogen fertilizers makes soil water more acidic, and they displace other ions from binding sites on soil particles. Too many calcium and magnesium ions, which also are required for plant growth, trickle away in soil water.

Deposition of nitrogen in acid rains can have the same effect as overfertilization. Fossil fuel burned in power plants and vehicles releases nitrogen oxides,

Figure 47.26 Dead and dying trees in Smoky Mountain National Park. Forests are among the casualties of nitrogen oxides and other forms of air pollution.

which contribute to global warming and to acid rain. Winds often carry these air pollutants far from their sources (Figure 47.26). By some estimates, pollutants are putting ten times the normal amounts of nitrogen into certain forests in eastern Europe.

Different plant species respond in different ways to high nitrogen levels. Increases in nitrogen can disrupt the balance among competing species in a community (Section 46.1), and diversity may decline. The impact can be pronounced in forests at high elevations and high latitudes, which have nitrogen-poor soils.

Some human activities disrupt aquatic ecosystems through nitrogen enrichment. Crop plants cannot take up all of the nitrogen in fertilizers. About half of the nitrogen applied to fields runs off into rivers, lakes, and estuaries. Sewage from cities and animal wastes puts even more nitrogen into the water provinces. As one outcome, nitrogen inputs promote algal blooms. So does the phosphorus in fertilizers, as explained in Sections 22.6 and 47.12.

The ecosystem phase of the nitrogen cycle starts with nitrogen fixation. Bacteria convert gaseous nitrogen in the air to ammonia and then to ammonium, which is a form that plants easily take up.

By ammonification, bacteria and fungi make additional ammonium available to plants when they break down nitrogen-rich organic wastes and remains.

By nitrification, bacteria convert nitrites in soil to nitrate, which also is a form that plants easily take up.

The ecosystem loses nitrogen when denitrifying bacteria convert nitrite and nitrate back to gaseous nitrogen, and when nitrogen is leached from soil.

47.12 Sedimentary Cycles

LINKS TO
SECTIONS
3.8, 22.6

Unlike carbon and nitrogen, phosphorus does not cycle into and out of ecosystems in gaseous form. Like nitrogen, phosphorus can be taken up by plants only in ionized form, and it, too, is often a limiting factor on plant growth.

In the **phosphorus cycle**, phosphorus passes quickly through food webs as it moves from land to ocean sediments, then slowly back to dry land. Earth's crust is the largest reservoir of phosphorus.

The phosphorus in rock formations is mainly in the form of phosphate (PO_4^{3-}). Weathering and erosion deliver these ions to streams and rivers, which move them onward to the sea (Figure 47.27). The phosphates gradually accumulate and form insoluble deposits on submerged continental shelves. After many millions of years, crustal movements might uplift part of the seafloor and expose the phosphates on land surfaces. There, weathering and erosion will release phosphates from exposed rocks and start the cycle over again.

Phosphates are required building blocks for ATP, phospholipids, nucleic acids, and other compounds. Plants take up dissolved phosphates from soil water. Herbivores get them by eating plants; carnivores get them by eating herbivores. Animals lose phosphate in urine and in feces. Bacterial and fungal decomposers release phosphate from organic wastes and remains, then plants take them up again.

The hydrologic cycle helps move phosphorus and other minerals through ecosystems. Water evaporates from the ocean and falls on land. As it flows back to the ocean, it transports silt and dissolved phosphates that the primary producers require for growth.

Of all minerals, phosphorus is often the limiting factor in ecosystems. Only newly weathered, young soils are high in phosphorus. In aquatic habitats, most phosphorus is locked up in sediments. Not much is in gaseous form, so little is lost to the atmosphere.

Phosphorus is being lost from many tropical and subtropical ecosystems, many of which already have phosphorus-poor soils. Phosphorus stored in biomass and released from decomposing organic matter can sustain undisturbed forests or grasslands. As trees are harvested or land is cleared, phosphorus is lost. Crop yields start out low and soon are nonexistent. After fields are abandoned, natural regrowth is sparse. In developing countries especially, 1 to 2 billion hectares may be already depleted of phosphorus.

What about the developed countries? After years of fertilizer applications, many soils have phosphorus overloads. It is concentrated in eroded sediments and

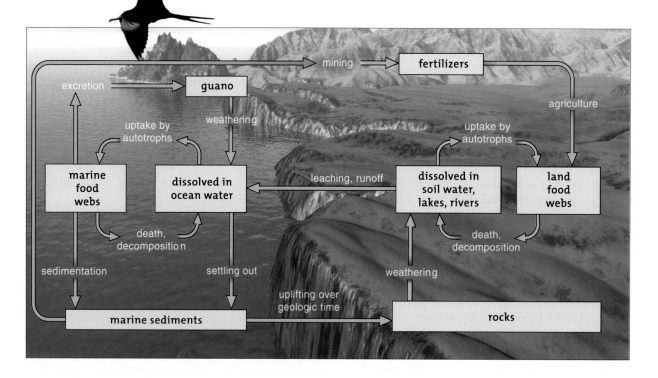

Figure 47.27 *Animated!* Phosphorus cycle. In this sedimentary cycle, phosphorus moves mainly in the form of phosphate ions (PO_4^{3-}) to the ocean. It moves through phytoplankton of marine food webs, then to fishes that eat plankton. Seabirds eat the fishes, and their droppings (guano) accumulate on islands. Humans collect and use guano as a phosphate-rich fertilizer.

runoff from agricultural fields. Phosphorus also is a waste in the outflows from sewage treatment plants and factories, and in the runoff from fields. Dissolved phosphorus that gets into streams, rivers, lakes, and estuaries can promote destructive algal blooms. Like plants, all photosynthetic algae require phosphorus, nitrogen, and other ions to grow. In many freshwater ecosystems, nitrogen-fixing bacteria keep the nitrogen levels high, so phosphorus becomes the limiting factor. When phosphate-rich pollutants pour in, populations of algae soar. As aerobic decomposers break down the remains of the algae, the water becomes depleted of the oxygen that fishes and other organisms require.

Eutrophication refers to the nutrient enrichment of any ecosystem that is otherwise low in nutrients. It is a process of natural succession. Phosphorus inputs as from agriculture can accelerate it, as the experiment shown in Figure 47.28 demonstrated.

Sedimentary cycles, in combination with the hydrologic cycle, move most mineral elements, such as phosphorus, through terrestrial and aquatic ecosystems.

Agriculture, deforestation, and other human activities upset the nutrient balances of ecosystems.

Figure 47.28 One of the eutrophication experiments. Researchers put a plastic curtain across a channel between two basins of a natural lake. They added nitrogen, carbon, and phosphorus on one side of the curtain (here, the *lower* part of the lake) and added nitrogen and carbon on the other side. Within months, the phosphorus-rich basin was eutrophic, with a dense algal bloom (green) covering its surface.

Summary

Section 47.1 An ecosystem consists of an array of organisms together with their physical and chemical environment. There is a one-way flow of energy into and out of an ecosystem, and a cycling of materials among the organisms. All ecosystems have inputs and outputs of energy and nutrients.

Sunlight is the initial energy source for almost all ecosystems. Primary producers capture energy from the sun. They also assimilate nutrients that they, and all consumers, require. Consumers include herbivores, carnivores, omnivores, decomposers, and detritivores.

Organisms in an ecosystem are classified by trophic levels. Those at the same level are the same number of steps away from the energy input into the ecosystem.

Biology⑤Now
Learn about energy flow and material cycling with the animation on BiologyNow.

Section 47.2 Linear sequences by which energy and nutrients move through ever higher trophic levels are food chains, which interconnect as food webs. The efficiency of energy transfers is always low, so most ecosystems support no more than four or five trophic levels away from an original energy source. In a grazing food web, energy captured by producers flows directly to consumers. In a detrital food web, it flows directly to detritivores and decomposers. In nearly all ecosystems, both types of food webs interconnect.

Biology⑤Now
Explore a food web with the animation on BiologyNow.

Section 47.3 By biological magnification, some chemical substance is passed from organisms at one trophic level to those above and becomes increasingly concentrated in body tissues. DDT is an example.

Section 47.4 A system's primary productivity is the rate at which producers capture and store energy in their tissues. It varies with climate, season, nutrient availability, and other factors.

Ecologists construct energy pyramids and biomass pyramids to show how energy and organic compounds are distributed in an ecosystem. Energy pyramids are largest at their base. The lowest trophic level has the greatest proportion of the energy in an ecosystem.

Long-term studies of the Silver Springs ecosystem in Florida illustrate the inefficiency of energy transfers. At each trophic level, far more energy was lost to the environment or in wastes and remains than was passed on to the next trophic level.

Biology⑤Now
See how energy flows through one ecosystem with the animation on BiologyNow.

Section 47.5 In a biogeochemical cycle, water or a nutrient moves through the environment, then through organisms, then back to an environmental reservoir.

Sections 47.6–47.8 In the hydrologic cycle, water moves from the ocean into the atmosphere, to land, and back to the ocean—the main reservoir. Human actions are disrupting the cycle in ways that result in shortages and pollution of water.

Biology Now
Learn about the hydrologic cycle with the animation on BiologyNow.

Section 47.9 The carbon cycle moves carbon from its main reservoirs in rocks and seawater, through its gaseous form (carbon dioxide) in the atmosphere, and then through ecosystems. Deforestation and the burning of wood and fossil fuels are adding more carbon dioxide to the atmosphere than the oceans can absorb.

Biology Now
Observe the flow of carbon through its global cycle with the animation on BiologyNow.

Section 47.10 Collectively, greenhouse gases trap heat in the lower atmosphere, which helps make Earth's surface warm enough to support life. Natural processes and human activities are adding more greenhouse gases, including carbon dioxide, CFCs, methane, and nitrous oxide, to the atmosphere. The rise correlates with a rise in global temperatures and other climate changes.

Biology Now
Explore the causes of the greenhouse effect and global warming with the animation on BiologyNow.

Section 47.11 The atmosphere is the main reservoir for N_2, a gaseous form of nitrogen that plants cannot use. In nitrogen fixation, some soil bacteria degrade N_2 and assimilate the two nitrogen atoms into ammonia. Other reactions convert ammonia to ammonium and nitrate, which plants are able to take up. Some nitrogen is lost to the atmosphere by the action of denitrifying bacteria.

Human activities add nitrogen to ecosystems; for example, through fertilizer applications and fossil fuel burning, which releases nitrogen oxides.

Biology Now
Learn how nitrogen is cycled with the animation on BiologyNow.

Section 47.12 The phosphorus cycle is the main sedimentary cycle. Earth's crust is the largest reservoir. Phosphorus is often the limiting factor on the population growth of producers. Excess inputs of phosphorus to aquatic ecosystems contribute to eutrophication.

Biology Now
Learn how phosphorus is cycled with the animation on BiologyNow.

Self-Quiz

Answers in Appendix II

1. Ecosystems have _____ .
   a. energy inputs and outputs
   b. one trophic level
   c. no nutrient outputs; all nutrients are cycled
   d. a and b

2. Organisms at the lowest trophic level in a tallgrass prairie are all _____ .
   a. at the first step away from the original energy input
   b. autotrophs   d. both a and b
   c. heterotrophs   e. both a and c

3. Decomposers are commonly _____ .
   a. fungi   b. animals   c. bacteria   d. a and c

4. Trophic levels are _____ .
   a. structured feeding relationships
   b. a case of who eats whom in an ecosystem
   c. a hierarchy of energy transfers
   d. all of the above

5. Primary productivity on land is affected by _____ .
   a. nutrient availability   c. temperature
   b. amount of sunlight   d. all of the above

6. If biological magnification occurs, the _____ will have the highest levels of toxins in their systems.
   a. producers   c. primary carnivores
   b. herbivores   d. top carnivores

7. Disruption of the _____ cycle is depleting aquifers.
   a. hydrologic   c. nitrogen
   b. carbon   d. phosphorus

8. Earth's largest carbon reservoir is _____ .
   a. the atmosphere   c. seawater
   b. sediments and rocks   d. living organisms

9. The _____ cycle is a sedimentary cycle.
   a. hydrologic   c. nitrogen
   b. carbon   d. phosphorus

10. _____ is often a limiting factor for plant growth.
    a. Nitrogen   d. both a and c
    b. Carbon   e. all of the above
    c. Phosphorus

11. Nitrogen fixation converts _____ to _____ .
    a. nitrogen gas; ammonia   d. ammonia; nitrates
    b. nitrates; nitrites   e. nitrites; nitrogen oxides
    c. ammonia; nitrogen gas

12. Match the terms with suitable descriptions.
    ____ producers   a. feed on plants
    ____ herbivores   b. feed on small bits of
    ____ decomposers   organic matter
    ____ detritivores   c. degrade organic
                             wastes and remains to
                             inorganic forms
                          d. capture sunlight energy

Additional questions are available on Biology Now™

Critical Thinking

1. Visualize and then describe an extreme situation in which you are a participant in a food chain rather than a food web.

2. Marguerite is growing a vegetable garden in Maine. Eduardo is growing arugula in Spain. What are some of the variables that influence primary production in each of these locations?

3. Look around you and name all of the objects, natural or manufactured, that might be contributing to amplification of the greenhouse effect.

4. Polar ice shelves are vast, thickened sheets of ice that float on seawater. In March 2002, 3,200 square kilometers (1,410 square miles) of Antarctica's largest ice shelf broke free from the continent and shattered into thousands of icebergs (Figure 47.29). Scientists knew the ice shelf was shrinking and breaking up, but this was the single largest loss ever observed at one time. Why should this concern people who live in more temperate climates?

5. Fishes are a fine source of protein and of *omega-3 fatty acids*, which are necessary for the normal development of the nervous system. This would seem to make fish a good choice for pregnant women. But coal-burning power plants put mercury into the environment, and some of it ends up in fish. Eating mercury-tainted fish during pregnancy can adversely affect development of a fetal nervous system.

Tissues of predatory marine fishes, such as swordfishes, tunas, marlins, and sharks, have especially high levels of mercury. The Environmental Protection Agency has issued health advisories to pregnant women, suggesting that they limit their consumption of fish species most likely to be tainted with mercury. Although sardines are harvested from the same ocean, they have lower mercury levels and are not on the warning list. Explain why two species of fishes that live in the same place can have very different levels of mercury in their tissues.

6. Methane, remember, is a gaseous molecule of one carbon atom to which four hydrogens are attached (Section 3.1). *Methane hydrate* is a methane molecule surrounded by an icelike lattice of water molecules, and it forms only at low temperatures, high pressures, and high concentrations of methane. Such conditions prevail beneath the seafloor and in the arctic tundra, where vast deposits of methane hydrate have formed in frozen peat bogs. The deposits typically are hundreds of meters thick.

Over millions of years, ancient organisms that died and sank to the ocean floor were buried in sediments. Their carbon-rich remains are food for anaerobic archaeans living far beneath the seafloor. The archaeans produce methane, which bubbles up to the seafloor. There, the high pressures and low temperatures freeze the methane into solid blocks of methane hydrates (Figure 47.30).

Again, ocean deposits of methane hydrate hold 10,000 to 11,000 gigatons of methane. The next largest deposit, in arctic regions, only amounts to hundreds of gigatons. Significant amounts of carbon continually enter and leave the deposits. For instance, bacterial activity continually converts some of the methane to carbon dioxide, which helps control the amount of methane that can escape into the atmosphere.

The oceanic reservoir of methane hydrates may contain more carbon than all of the known reserves of oil, coal, and natural gas. However, the deposits are highly unstable; ice fills spaces in sediments, so increases in pressure or ocean water temperature can trigger catastrophic landslides and tsunamis. Reflect again on the global carbon cycle, then list some of the consequences of a catastrophic release of methane in terms of global temperature, climate, glaciation and sea level changes, and the composition of ocean water.

7. Reflect once more on Figure 47.30. The methane-eating and sulfate-eating bacteria near seeps on the seafloor are the start of food webs. Do some research and identify the producers and consumers in these deep-sea communities. Before you start, would you expect the food chains to be short or long? Make a list that organizes the participants by trophic levels.

Figure 47.29 Antarctica's Larson B ice shelf in (**a**) January and (**b**) March 2002. About 720 billion tons of ice broke from the shelf, forming thousands of icebergs. (**c**) These are the just the tips of the icebergs, projecting twenty-five meters (eighty-two feet) above the sea surface.

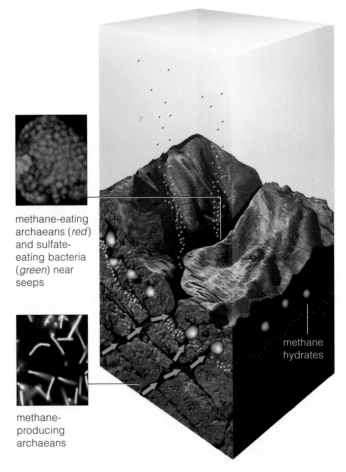

methane-eating archaeans (*red*) and sulfate-eating bacteria (*green*) near seeps

methane-producing archaeans

methane hydrates

Figure 47.30 Methane cycling on the seafloor. Archaeans beneath the seafloor produce methane that seeps out from seafloor sediments. There, methane-eating microbes release carbon dioxide as well as hydrogen sulfide as metabolic products that become the basis of deep-sea food webs.

Surfers, Seals, and the Sea

The stormy winter of 1997–1998 was a very good time for surfers on the lookout for the biggest waves, and a very bad time for seals and sea lions. As Ken Bradshaw rode a monster wave (Figure 48.8), half of the population of sea lions on the Galápagos Islands were dying, including nine of every ten sea lion pups. In California, the number of liveborn Northern fur seals plummeted. Most of the ones that made it through birth were dead within a few months. Diverse forms of life in distant parts of the world were connected by the full fury of El Niño—a recurring event that ushers in an often spectacular seesaw in the world climate.

That winter, a massive volume of warm water from the southwestern Pacific moved east. It piled into coasts from California down through Peru and displaced currents that otherwise would have churned up tons of nutrients from the deep. Without a replenishment of nutrients, primary producers for marine food webs declined in numbers. The scarcity of producers, in combination with the warmer water temperature, drove away consumers. The great populations of anchovies and some other fishes dispersed elsewhere. Fishes and squids that could not do so starved to death. So did many seals and sea lions, because fishes and squids are the mainstay of their diet.

Marine mammals did not dominate the headlines, because the 1997–1998 El Niño gave humans plenty of other things to worry about. It battered Pacific coasts with fierce winds and torrential rains that resulted in massive flooding and landslides. As another rippling effect, an ice storm in New York, New England, and central and eastern Canada crippled regional electrical grids. Three weeks later, 700,000 people still had no electricity. Meanwhile, the global seesaw caused drought-driven crop failures and raging wildfires in Australia and Indonesia.

Figure 48.1 Ken Bradshaw surfing a monster wave, more than twelve meters high, during the most powerful El Niño of the past century. In January 1998, a storm formed off the Siberian coast. It generated an ocean swell that, at the time, was the biggest wave known to hit Hawaii.

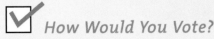

All told, that one El Niño episode killed thousands of people and drained away tens of billions of dollars from economies around the world.

The impact of El Niño on diverse communities invites us back to **biogeography**—the study of factors that give rise to patterns in the distribution of organisms through the environment. This chapter invites you to reflect on how climate, topography, human activities, and other factors influence that distribution in the space of the biosphere.

The **biosphere**, again, is the sum of all places where we find life on Earth. Organisms live in the *hydrosphere*—the ocean, ice caps, and other bodies of water, liquid and frozen. They live on and within sediments and soils of the *lithosphere*—Earth's outer, rocky layer. Many lift off into the lower region of the *atmosphere*—the gases and airborne particles that envelop Earth.

How Would You Vote?

We cannot stop an El Niño from happening, but we might be able to minimize its environmental, social, and economic impacts. Would you support the use of taxpayer dollars to fund research into the causes and effects of El Niño? See BiologyNow for details, then vote online.

Key Concepts

AIR CIRCULATION PATTERNS

Indirectly, energy inputs from the sun are the starting point for the global distribution of communities and ecosystems. Its continual inputs of heat energy warm the atmosphere and drive Earth's weather systems. Sections 48.1, 48.2

OCEAN CIRCULATION PATTERNS

Atmospheric and ocean circulation patterns interact with topography and cause regional variations in temperature and rainfall. The interactions influence soils and sediments, which in turn affect the population growth and distribution of primary producers. Section 48.3

LAND PROVINCES

A biome is a vast region of land characterized partly by its distinctive, dominant vegetation. The global distribution of biomes is an outcome of Earth's physical history as well as climate, topography, soil type, and species interactions. Sections 48.4–48.11

WATER PROVINCES

Water provinces cover more than 71 percent of Earth's surface. Freshwater and marine ecosystems have gradients in light availability, temperature, and dissolved gases that vary daily and seasonally. Primary productivity shifts with the variations. Sections 48.12–48.15

APPLYING THE CONCEPTS

Understanding the interactions among the atmosphere, ocean, and land can lead to discoveries about specific events—in this case, recurring cholera epidemics—that impact human life. Section 48.16

Links to Earlier Concepts

With this chapter, you have arrived at the highest level of organization in nature (Section 1.1). You will draw on earlier topics, from acid rain (2.6) to eutrophication (47.12), the ozone layer (7.1), soils (30.1), the carbon cycle (47.9), and ecosystem productivity (7.8, 47.4, 47.6). You will revisit agricultural practices (45.7), deforestation (23.10, 27.7, 47.7), coral reefs (27.5), and hydrothermal vents (20.2, 21.5). You will deepen your knowledge of biogeography (17.1, 27.6, 46.1, 46.11). The chapter ends with an example of the power of a scientific approach to problem solving (1.5, 1.6).

48.1 Global Air Circulation Patterns

LINKS TO
SECTIONS
7.1, 23.5

The biosphere encompasses ecosystems that range from continent-straddling forests to rainwater pools in cup-shaped clusters of leaves. Except for a few ecosystems at hydrothermal vents, climate influences all of them.

CLIMATE AND TEMPERATURE ZONES

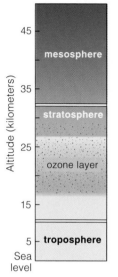

Climate refers to average weather conditions, such as cloud cover, temperature, humidity, and wind speed, over time. It arises from variations in solar radiation, Earth's daily rotation and annual path around the sun, the distribution of land masses and seas, and land elevations. Interactions among these factors are the source of prevailing winds and ocean currents, even the composition of soils.

First, sunlight warms air near Earth's surface (Figure 48.2). The sun's rays are spread out over a greater area near the poles (which intercept them at an angle) than they are at the equator (which intercepts them head-on). The intense light at the equator warms air a lot more. Warm air rises, then spreads north and south toward cooler regions—and so begins a global pattern of air circulation.

Earth's rotation and its curvature alter the initial air circulation pattern. Air masses moving north or south are not attached to Earth—which, like any ball, spins fastest at its equator during each spin around its axis. Air masses moving north or south from a given point *seem* to be deflected east or west relative to the curve of the surface spinning under them. This is the basis of prevailing east and west winds (Figure 48.3).

Also, land absorbs and gives up heat faster than the ocean, so air parcels above it sink and rise faster. Warm air expands, so its pressure is lower where it rises and greater where it cools and sinks. The uneven distribution of land and water between regions causes pressure differences. Regional winds arise and disrupt the overall flow of air from the equator to the poles.

In short, *latitudinal variations in solar heating cause a north–south pattern of air circulation, and then east–west deflections and air pressure differences over land and water introduce variations in the pattern.*

Differences in rainfall by latitude accompany the air circulation patterns. Warm air holds more moisture than cool air does. At the equator, warm air masses pick up moisture from the ocean, then rise and cool.

Figure 48.2 *Above,* Earth's atmosphere. Air circulates mainly in the lower atmosphere, or troposphere, where temperatures cool rapidly with increases in altitude. An ozone layer between 17 and 27 kilometers above sea level absorbs most of the ultraviolet (UV) wavelengths in the sun's rays. Ozone and oxygen in the atmosphere absorb most of the incoming UV light.

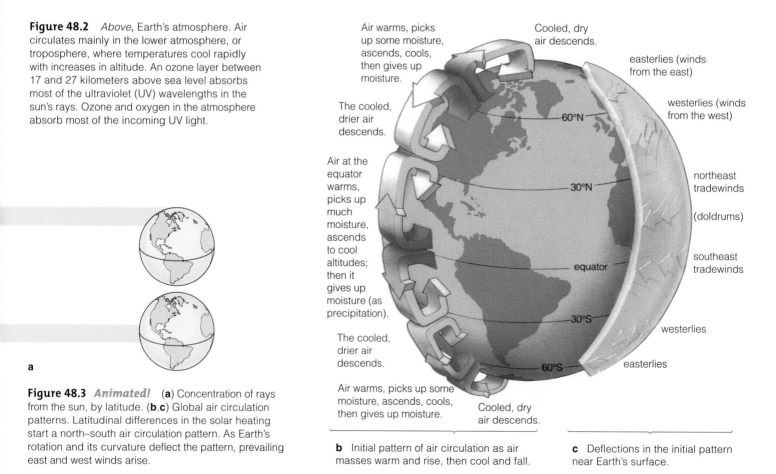

Figure 48.3 *Animated!* (a) Concentration of rays from the sun, by latitude. (b,c) Global air circulation patterns. Latitudinal differences in the solar heating start a north–south air circulation pattern. As Earth's rotation and its curvature deflect the pattern, prevailing east and west winds arise.

b Initial pattern of air circulation as air masses warm and rise, then cool and fall.

c Deflections in the initial pattern near Earth's surface.

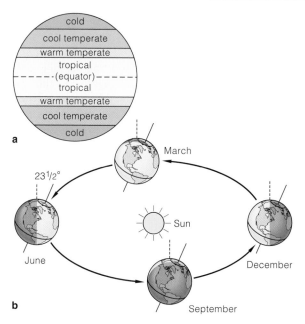

a

b

Figure 48.4 (**a**) World temperature zones. (**b**) Incoming solar radiation varies annually. The northern end of Earth's fixed axis tilts toward the sun in June and away from it in December, changing the equator's position relative to the day–night boundary of illumination. Variations in sunlight intensity and daylength cause seasonal temperature shifts.

Figure 48.5 (**a**) Large arrays of electricity-producing photovoltaic cells in panels that collect solar energy. (**b**) In California, a field of turbines harvesting wind energy.

They give up some moisture as rain, which supports forest ecosystems. The now-drier air moves north or south. It warms and loses more moisture as it descends at latitudes 30°, where deserts typically form. Farther north and south, air picks up moisture, ascends, and gives up rain near latitudes 60°. It descends in polar regions, where cold temperatures and a near-absence of precipitation result in cold, dry polar deserts.

We divide the temperature gradients from Earth's equator to the poles into **temperature zones** (Figure 48.4*a*). Temperatures shift in the zones as Earth's orbit puts it closer to and farther from the sun during each annual cycle (Figure 48.4*b*). Temperatures, the hours of daylight, and winds change with the seasons. The changes are far greater inland (away from the ocean's moderating effects) and farther from the equator. The more pronounced the change, the more that primary productivity rises and falls on land and in the seas.

HARNESSING THE SUN AND WIND

Paralleling the human population's J-shaped growth curve is a steep rise in its total and per capita energy consumption. You may think we have an abundance of energy, but there is a big difference between total and net amounts. *Net* is what is left after subtracting the energy it takes to find, extract, transport, store, and deliver energy to consumers. Some sources, such as coal, are not renewable (Section 23.5), but solar and wind energy are another matter. Every year, incoming solar energy surpasses, by about ten times, the energy that is stored in all known fossil fuel reserves.

We already know how to get *solar–hydrogen energy*. Photovoltaic cells (Figure 48.5*a*) hold electrodes that, when exposed to the sun's rays, generate an electric current that splits water molecules into oxygen and hydrogen gas, which can be stored efficiently. The gas can directly fuel cars. It can heat and cool buildings. Water is the only waste. It also costs less to distribute hydrogen gas than electricity. Space satellites run on it.

Unlike fossil fuels, the sun's energy and seawater are unlimited. A solar–hydrogen age might end smog, oil spills, acid rain, and a reliance on nuclear energy, and it might help reduce global warming.

Solar energy also is converted into the mechanical energy of winds. Turbines at "wind farms" exploit the wind patterns that arise from latitudinal variations in sunlight intensity. Wind power provides California with 1 percent of its electricity. Winds of North and South Dakota might meet 80 percent of the current energy needs of the United States. Winds do not blow constantly, but when they do, wind energy can be fed into utility grids. Wind power also has potential for islands and other areas remote from utility grids.

Major temperature zones and climates start with global patterns of circulation, which arise through interacting factors: latitudinal variations in incoming solar radiation, Earth's rotation and annual path around the sun, and the distribution of land masses and seas.

48.2 Circulating Airborne Pollutants

LINKS TO
SECTIONS 2.6,
7.1, 42.9, 47.9

*Through activities that pollute the air, human populations interact with global air circulation patterns in unexpected ways, with unintended consequences. **Pollutants** are any natural or synthetic substances that have accumulated in harmful or disruptive amounts because organisms have had no prior evolutionary experience with them.*

A Fence of Wind and Ozone Thinning The ozone layer (Figure 48.2) is nearly twice as high above sea level as Mount Everest. From September to October, it thins above both poles. Seasonal **ozone thinning** is so vast that it was once called an "ozone hole" in the stratosphere. Figure 48.6 has an example. With a decline in the ozone concentration, far more UV radiation reaches Earth's surface. That is why the declining concentration correlates with increasing cases of skin cancers, cataracts, and weakened immunity. It also alters the atmosphere's composition indirectly. UV radiation kills phytoplankton, which results in drastic declines in their oxygen-releasing activity (Section 7.8).

Chlorofluorocarbons (CFCs) are major ozone destroyers. These odorless gases have been used as propellants in aerosol cans, coolants in refrigerators and air conditioners, and in solvents and plastic foam. They slowly seep into the air, and they resist breakdown. A free CFC molecule gives up a chlorine atom when it absorbs UV light. Reaction of this atom with ozone yields oxygen and chlorine monoxide. Chloride monoxide in turn reacts with free oxygen and releases another chlorine. Each chlorine atom can break apart more than 10,000 ozone molecules!

Chlorine monoxide concentrations above polar regions are 100 to 500 times higher than at midlatitudes. Why? Like a dynamic fence, winds rotate around the poles for most of the winter. Chlorine compounds are split apart on ice crystals in the clouds. After the almost perpetually dark polar winter ends, UV light in the sun's rays invites chlorine to start destroying ozone. Methyl bromide is worse for the ozone layer. Each year in the United States, farmers spray about 60 million pounds of this fumigant over croplands to kill insects, nematodes, and other pests.

Developed countries phased out CFC production. The developing countries may phase it out by 2010. Methyl bromide production is expected to end at that time as well. Even so, it will be one or two centuries before the ozone layer fully recovers.

In 2004 the destruction of ozone reached near-record levels, with close to a 50 percent loss in some parts of the stratosphere. However, in 2005, it became apparent that global forces can mediate the impact. Stratospheric wind patterns shifted and transported ozone-rich air northward from the midlatitudes. Ozone levels in polar regions were restored to near-normal levels.

No Wind, Lots of Pollutants, and Smog Certain weather conditions trap a layer of cool, dense air under a warm air layer. This is a *thermal inversion*. It intensifies **smog**, an atmospheric condition in which winds cannot disperse air pollutants that have accumulated and are trapped under a thermal inversion layer (Figure 48.7). Thermal inversions in certain parts of the world have contributed to some of the worst air pollution disasters.

Where winters are cold and wet, *industrial* smog forms as a gray haze over cities that burn a lot of coal and other fossil fuels. Burning releases smoke, soot, ashes, asbestos, oil, particles of lead and other heavy metals, and sulfur oxides. If winds and rain do not disperse them, these air

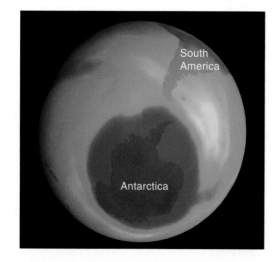

Figure 48.6 Seasonal ozone thinning above Antarctica during 2001. *Darkest blue* indicates the area with the lowest ozone level, at that time the largest recorded.

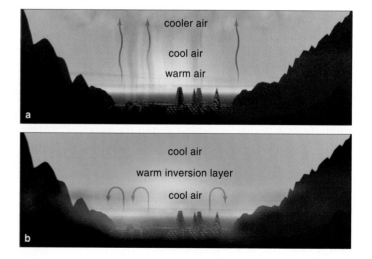

Figure 48.7 (**a**) Normal pattern of air circulation in smog-forming regions. (**b**) Air pollutants are trapped under a thermal inversion layer.

pollutants can reach lethal concentrations. Most industrial smog now forms in China, India, and eastern Europe.

Big cities in warm climate zones are bathed in a brown haze called *photochemical* smog. This type of smog is most dense when the city is in a natural topographic basin. Los Angeles and Mexico City are two classic cases (Figure 47.23). Exhaust fumes from vehicles contain nitric oxide, a major pollutant that forms nitrogen dioxide by combining with oxygen. When sunlight strikes it, nitrogen dioxide reacts with hydrocarbon gases to form photochemical oxidants. Most hydrocarbon gases are released from spilled or partly burned gasoline.

Winds and Acid Rain   Coal-burning power plants, smelters, and factories emit sulfur dioxides. Vehicles, power plants that burn gas and oil, and nitrogen-rich fertilizers all emit nitrogen oxides. In dry weather, airborne oxides fall as dry acid deposition. In moist air, they form nitric acid vapor, sulfuric acid droplets, and sulfate and nitrate salts. Winds typically disperse them far from their source. They fall to Earth in rain and snow. We call this wet acid deposition, or **acid rain**.

The pH of typical rainwater is 5 or so (Section 2.6). Acid rain can be 0 to 100 times more acidic, as potent as lemon juice! It corrodes metals, marble, rubber, plastics, nylon stockings, and other materials. It harms organisms, and it can alter the chemistry of ecosystems.

Depending on the soil type and vegetation cover, some regions are more sensitive to acid rain (Figure 48.8). Highly alkaline soil neutralizes acids before they enter streams and lakes. Also, highly alkaline water can neutralize the acid inputs. But many of the watersheds of northern Europe, southeastern Canada, and regions throughout the United States have thin soil layers on top of solid granite. These soils cannot buffer much of the acidic inputs.

Rain in much of eastern North America is thirty to forty times more acidic than it was even a few decades ago. Crop yields are declining. Fish populations have disappeared from more than 200 lakes in the Adirondack Mountains of New York. Air pollutants from distant industrial regions are changing the acidity of rainfall. The change is contributing to the decline of forest trees and mycorrhizae that support new growth.

As Harvard and Brigham Young University researchers report, living with airborne particles of dust, soot, smoke, or acid droplets shortens the human life span by a year or so. Smaller airborne particles damage lung tissues. High levels of ultrafine particles can increase the risk of lung cancer (Figure 48.9).

At one time, the world's tallest smokestack, in the Canadian province of Ontario, produced 1 percent by weight of the world's annual emissions of sulfur dioxide. Today, however, Canada gets more acid deposition from the midwestern United States than it sends across its southern border. *Prevailing winds—hence air pollutants—do not stop at national boundaries.*

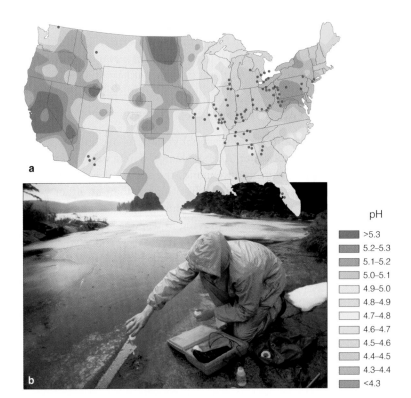

Figure 48.8   (**a**) The average 1998 precipitation acidities in the United States. *Red* dots mark large coal-burning power and industrial plants. (**b**) Biologist measuring the pH of New York's Woods Lake during a spring melt of acidified snow. Acid rain has already altered the lake.

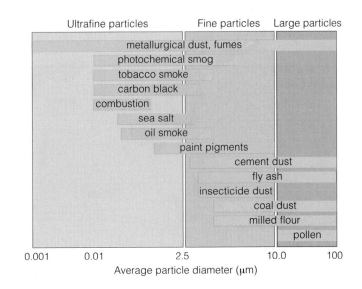

Figure 48.9   Suspended particulate matter. These solids and liquid droplets are small enough to stay aloft for variable intervals.

Ultrafine particles contribute to respiratory disorders. Carbon black is a powdered form of carbon used in paints, tires, and other goods. About 6 million tons are manufactured annually. About 45 million tons of fly ash are generated annually by coal combustion.

48.3 The Ocean, Landforms, and Climates

LINKS TO
SECTIONS
26.13, 47.10

*The **ocean** is a continuous body of water that covers more than 71 percent of Earth's surface. Driven by solar heat and wind friction, its upper 10 percent moves in currents that distribute nutrients through marine ecosystems.*

OCEAN CURRENTS AND THEIR EFFECTS

Latitudinal and seasonal variations in sunlight warm and cool water. At the equator, where vast volumes of water warm and expand, the sea level is about eight centimeters (three inches) higher than at either pole. The volume of water in this "slope" is enough to get sea surface water moving in response to gravity, most often toward the poles. The moving water warms air parcels above it. At midlatitudes it transfers *10 million billion* calories of heat energy per second to the air!

Ocean currents are large volumes of water flowing in response to the tug of trade winds and westerlies. Their direction and properties are outcomes of Earth's rotation and topography. They circulate clockwise in the Northern Hemisphere and counterclockwise in the Southern Hemisphere (Figure 48.10).

Swift, deep, and narrow currents of nutrient-poor water parallel the east coast of continents. Along the eastern coast of North America, warm water moves northward, as the Gulf Stream. Slower, shallow, broad currents of cold water paralleling the western coast of continents flow toward the equator.

These currents affect climate zones. Why are Pacific Northwest coasts cool and foggy in summer? The cold California Current is giving up moisture; warm winds by the coast are transferring heat to it. Why are Boston and Baltimore muggy in summer? Air masses pick up heat and moisture from the warm Gulf Stream, then southerly and easterly winds put them over the cities. Why are the winters milder in London and Edinburgh than they are in Ontario and central Canada—which are at the same latitude? Warm water from the Gulf Stream flows into the North Atlantic Current (Figure 48.10). That current gives up heat to prevailing winds, which warm northwestern Europe.

The patterns changed in the past. They may change again if global warming disrupts ocean currents and climate zones (Sections 26.13 and 47.10).

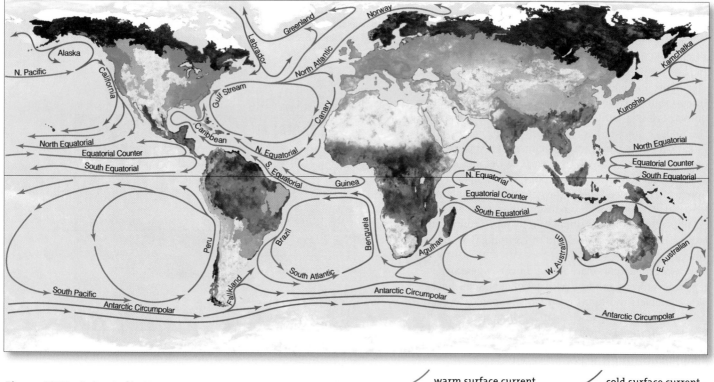

Figure 48.10 *Animated!* Major climate zones correlated with surface currents and surface drifts of the world ocean. The warm surface currents start moving from the equator toward the poles, but prevailing winds, Earth's rotation, gravity, the shape of ocean basins, and land masses all influence the direction of flow. Water temperatures, which differ with latitude and depth, contribute to differences in air temperature and in rainfall patterns.

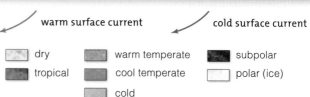

warm surface current   cold surface current

dry   warm temperate   subpolar
tropical   cool temperate   polar (ice)
cold

(a) Prevailing winds move moisture inland from the Pacific

(b) Clouds pile up and rain forms on side of mountain range facing prevailing winds

(c) Rain shadow on side facing away from the prevailing winds makes arid conditions

4,000/ 75
3,000/ 85
1,800/ 125
1,000/ 85
15/ 25
moist habitats
2,000/ 25
1,000/ 25

Figure 48.11 *Animated!* Rain shadow effect and the Sierra Nevada. On the side of mountains facing away from the prevailing winds, rainfall is light. *Black* numbers signify annual precipitation, in centimeters, averaged on both sides of this mountain range. The *white* numbers signify elevations, in meters.

RAIN SHADOWS AND MONSOONS

Mountains, valleys, and other topographical features affect regional climates. *Topography* refers to a region's surface features, such as elevation. Track a warm air mass after it picks up moisture off California's coast. It moves inland, as wind from the west, and piles up against the Sierra Nevada. This high mountain range parallels the distant coast. The air cools as it rises in altitude and loses moisture as rain (Figure 48.11). The result is a **rain shadow**—a semiarid or arid region of sparse rainfall on the leeward side of high mountains. *Leeward* is the side facing away from the wind. The Himalayas, Andes, Rockies, and other great mountain ranges cause vast rain shadows on continents.

Belts of vegetation at different elevations relate to differences in temperature and moisture. Grasslands form at the western base of the Sierra Nevada. Higher up, in the cooler air, deciduous and evergreen species reign. Higher still are a few evergreen species that are adapted to the cold habitat. Above the subalpine belt, hardy, dwarfed plants are all that can withstand even more extreme temperatures.

There are rain shadows on the leeward side of the Andes, Rockies, Himalayas, and other great mountain ranges. On the side of Hawaii's high volcanic peaks facing into the wind (*windward*), lush tropical forests flourish. Desert conditions prevail on the other side.

Air circulation patterns called **monsoons** influence continents that lie north or south of warm oceans. The land near these oceans heats intensely in summer, and vast, low-pressure air parcels form above it. The low pressure draws moisture-laden air from the ocean, as from the Bay of Bengal into Bangladesh. Trade winds converge here. Together with the equatorial sun, they invite intense heating and heavy rainfall. Air moving northward and southward across the continents gives rise to alternating dry and wet seasons, to alternating drought conditions and flooding.

Recurring coastal breezes are like mini-monsoons. Water and coastal land have different heat capacities. In the morning, water does not warm as fast as land. When warm air above land rises, cooler marine air moves in. After sunset, the land loses heat faster than water, so land breezes flow in the reverse direction.

Surface ocean currents influence regional climates and help distribute nutrients in marine ecosystems.

Air circulation patterns, ocean currents, and landforms interact in ways that influence regional temperatures and moisture levels. Thus they also influence the distribution and dominant features of ecosystems.

48.4 Biogeographic Realms

LINKS TO
SECTIONS 17.1,
27.6, 46.11

Differences in physical and chemical properties, and in evolutionary history, help explain why deserts, grasslands, forests, and tundra form where they do, and why some water provinces are richer than others in biodiversity.

Suppose you live in the coastal hills of California and decide to tour the Mediterranean coast, the southern tip of Africa, and central Chile. In each region, you see highly branched, tough-leafed woody plants that look a lot like the highly branched, tough-leafed chaparral plants back home. Vast geographic and evolutionary distances separate the plants. Why are they alike?

You decide to compare their locations on a global map. You discover that American and African desert plants live about the same distance from the equator. Chaparral plants and their distant look-alikes all grow along the western and southern coasts of continents positioned between latitudes 30° and 40°. As Alfred

Wallace, Charles Darwin, and other naturalists did so long ago, you have just stumbled onto one of many biogeographic patterns.

Those naturalists divided Earth's land masses into six **biogeographic realms**—vast expanses where they could expect to find communities of certain types of plants and animals, such as palm trees and camels in the Ethiopian realm (Section 46.11 and Figure 46.29). In time, the classic realms shown in Figure 48.12 were subdivided, as when Hawaii, parts of Indonesia and Japan, and Polynesia, Micronesia, Papua New Guinea, and New Zealand became the Oceania realm.

Biomes are finer subdivisions of the great realms, but they are still identifiable on a global scale. Their distribution has been shaped partly by evolutionary processes and events, as when evolving species were slowly rafted about on different land masses after the splitting of Pangea (Chapter 17). They also owe their

- desert
- dry shrubland, dry woodland
- warm grassland (e.g., savanna)
- temperate grassland
- mountain grassland
- tropical broadleaf forest
- temperate deciduous forest
- tropical coniferous forest
- temperate coniferous forest (e.g., rain forest)
- northern coniferous forest (boreal forest)
- tropical dry forest
- tundra
- mountains, complex zonation
- mangrove swamp
- perpetual ice cover
- marine ecoregions

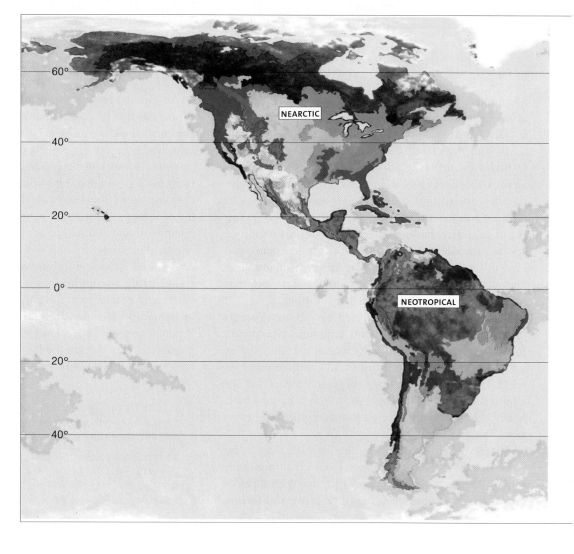

Figure 48.12 *Animated!* Global distribution of major categories of biomes and marine ecoregions.

distribution to topography, climate, and interactions among species. Some biomes extend across more than one continent, and their communities, although similar in many ways, consist of unique species. Notice, in Figure 48.12, the distribution of dry shrublands and woodlands (coded dark brown). Chaparral is part of that biome. Similarly, the temperate grassland biome occurs in North America (the prairie), South America (pampa), southern Africa (veld), and Eurasia (steppe). In each case, the community is dominated by grasses that can tolerate strong winds and drought.

In general, communities that characterize a biome are arrays of species adapted to specific temperatures, patterns of rainfall, soil type, and one another. The community's consumers are adapted to the dominant vegetation, as when tall forest trees invite competitive birds to partition resources between the forest floor and the canopy (Sections 46.1 and 46.3). Adaptations,

recall, extend from the form, function, and behavior of each species to its life-history pattern.

Evolutionary events and environmental conditions also underlie the distribution of communities in the seas. Figure 48.12 shows the main marine ecoregions.

Remember Section 27.6? It sketches out the sweep of conservation biology. Conservationists are working to locate, inventory, and protect the world's **hot spots**, portions of biomes and ecoregions that are richest in biodiversity and the most vulnerable to species losses. There are 150 hot spots in North America alone.

Biomes are vast expanses of land dominated by distinct kinds of communities. Marine ecoregions are realms of biodiversity in the seas. Their distribution is an outcome of evolutionary history, topography, climate, and species interactions, which vary in different parts of the world.

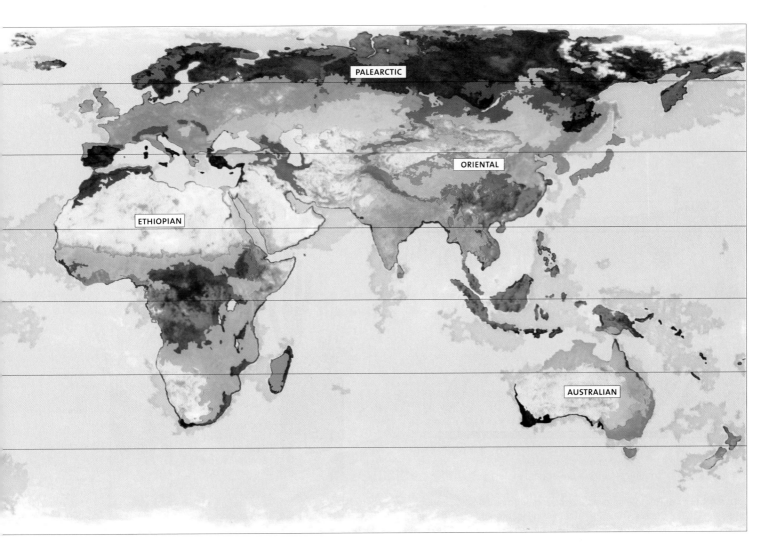

48.5 Availabililty of Sunlight, Soils, and Moisture

LINKS TO
SECTIONS 23.10,
24.6, 30.1, 47.11

Biomes differ in their soil profiles. They differ in their daily and seasonal supplies of water and other resources, and therefore in their primary productivity.

Soils are mixtures of mineral particles and varying amounts of the decomposing organic material we call humus. When rocks are weathered and broken down, they form coarse-grained gravel, then sand, silt, and finely grained clay. Water and air fill spaces between soil particles. As explained in Section 30.1, the types, proportions, and compaction of particles differ within and between regions. Clay is richest in minerals, but its fine, closely packed particles drain poorly. It also does not have enough air spaces for roots to take up oxygen. Gravelly or sandy soils invite leaching, which depletes them of water and mineral ions.

Biome soils differ in their layered structure, or *soil profile*, which reflects how they formed (Figure 48.13). Natural grasslands and then deciduous forests tend to have the most topsoil. Topsoil is richest in humus but is easily eroded. It is less than 1 centimeter thick on steep slopes. It can be more than 1 meter thick in grasslands, which is why former grasslands are the top choice for agriculture. Tropical forests also are cleared for agriculture, but little topsoil accumulates above their poorly draining layers. Clearing exposes topsoil to rains that leach nutrients, so crops perform poorly unless heavily fertilized, as Section 47.11 explains.

Reflect on the map in Figure 48.13*a*. Where there is plenty of moisture, competition for sunlight influences community structure. That is why the sun-drenched deciduous forests have an obvious vertical structure. As you will see, dense stands of trees form a canopy high above the ground. Shorter, understory trees and shrubs compete for light filtering through the canopy.

Where desert biomes have formed, you seldom find competition for sunlight. Water is the key limiting factor in deserts. As you saw in Section 45.2, widely spaced desert plants are evidence of root systems that compete for this scarce resource.

Some parts of desert biomes are monotonously low in biodiversity. Figure 48.14*a* shows part of hundreds of square kilometers of the Sonoran Desert in Arizona, where you see little more than widely spaced creosote bushes. Add some moisture and a higher elevation to the mix, and you see more diversity, as in this desert's wetter uplands (Figure 48.14*b*). The caption to Figure 48.15 offers a glimpse into how desert soils form.

> *A biome's soil profile and its proportions of sand, silt, clay, gravel, and humus influence its primary productivity.*
>
> *Even within the boundaries of the same biome, differences in water availability and temperature can bring about big differences in species richness.*

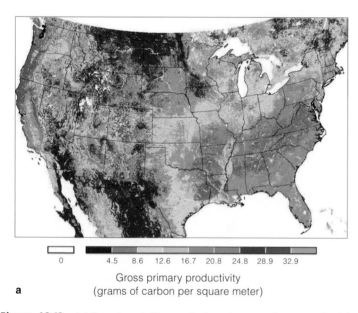

Gross primary productivity
(grams of carbon per square meter)

a

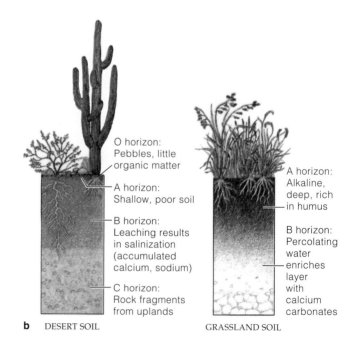

b DESERT SOIL   GRASSLAND SOIL

Figure 48.13 (**a**) Remote satellite monitoring of gross primary productivity across the United States. The differences roughly correspond with variations in soil types and moisture. (**b**) Soil profiles from a few representative biomes.

a   **b**

Figure 48.14 Two views of a single biome—the Sonoran Desert of Arizona. Sunlight is equally intense in the lowlands (**a**) and uplands (**b**), but differences in water availability, temperature, and soil types influence plant growth.

Compared to wetter biomes, deserts are low in primary productivity. All desert soils are mineral-rich, but they are in an early stage of development. Particles in the upper horizons are loose and not yet stabilized. Little organic matter accumulates at the surface, so decomposers cannot form much humus. The soils cannot hold on to water, which falls only in intense pulses during a brief rainy season. Also, evaporative water loss from desert soils is high, so dissolved salts rise to the soil surface. Extensive saltpans form, and they support little to no plant growth.

The uplands rarely frost over. Also, they get heavy pulses of rain in summer and light but prolonged rain in winter. Annuals and perennials that bloom in summer and winter are not as water-stressed in the uplands. In hotter, drier, flat valley floors, temperatures are extremely high, rainfall is sparse, and woody plants are widely spaced. Creosote bush (*Larrea*), sometimes mixed with bursage (*Ambrosia*), dominates.

Figure 48.15 How desert soils start. Over millions of years, rain, wind, freezes and thaws, and chemical processing erode mountain ranges. Runoff moves small, loose rocks and particles toward the floor of adjoining valleys. Gravel, sand and silt collect in lowland deserts. *Cryptobiotic crusts* may form. These desert encrustations are mainly lichens, which slowly enrich the soil (Section 24.6). Where a thin humus layer forms, as in the Sonoran uplands, the soil stabilizes and holds some water. When dune buggies race across "empty" deserts, they destroy the crusts.

The photograph shows debris in a gully at the base of some mountains in the Chihuahuan Desert in Texas. The gully channels water draining from the slopes, and accumulated debris has stabilized enough to support woody shrubs and herbaceous plants. Plants are sparse on the gully's right slope. They are less water-stressed on the left side, which is shaded during the hottest time of day.

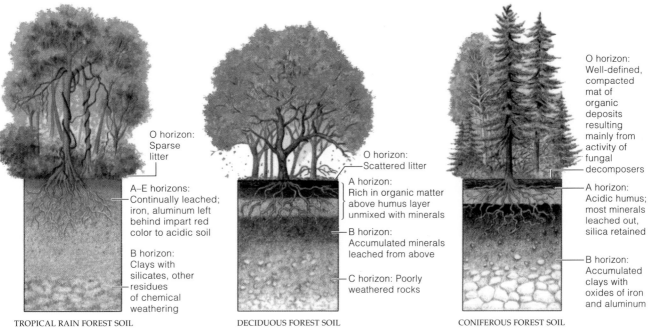

O horizon: Sparse litter

A–E horizons: Continually leached; iron, aluminum left behind impart red color to acidic soil

B horizon: Clays with silicates, other residues of chemical weathering

TROPICAL RAIN FOREST SOIL

O horizon: Scattered litter

A horizon: Rich in organic matter above humus layer unmixed with minerals

B horizon: Accumulated minerals leached from above

C horizon: Poorly weathered rocks

DECIDUOUS FOREST SOIL

O horizon: Well-defined, compacted mat of organic deposits resulting mainly from activity of fungal decomposers

A horizon: Acidic humus; most minerals leached out, silica retained

B horizon: Accumulated clays with oxides of iron and aluminum

CONIFEROUS FOREST SOIL

48.6 Moisture-Challenged Biomes

LINK TO
SECTION
27.4

At certain latitudes with sparse rainfall and high rates of evaporation, drought-tolerant plants predominate.

DESERTS

Deserts tend to form near latitudes 30° north and south. Annual rainfall is less than ten centimeters or so, and evaporation rates are high. Brief, infrequent pulses of rain swiftly erode the topsoil. Humidity is so low that the sun's rays easily penetrate air masses. The ground heats fast, then cools fast at night. The arid or semiarid conditions do not support large, leafy plants, but, as the preceding section explained, the species richness is often high where moisture is available in more than one season. Figure 48.16 shows another example.

DRY SHRUBLANDS, WOODLANDS, AND GRASSLANDS

Dry shrublands receive less than 25 to 60 centimeters of rain a year. They dominate Mediterranean, South African, and California regions that are vulnerable to lightning-sparked, wind-driven firestorms in summer.

DESERTS

Figure 48.16 From the Mojave Desert of California, cactus (cholla), golden poppies, and—in the distance—tall saguaro cacti. The inset shows flowers of a claret cup cactus.

The shrubs have highly flammable leaves and quickly burn to the ground. They are adapted to episodes of fire and resprout from the root crowns. Trees do not fare as well during firestorms. Shrubs that withstand the fires have a competitive edge (Figure 48.17).

Dry woodlands prevail where annual rainfall is 40 to 100 centimeters. Trees are often tall, but they do not provide a continuous canopy. Eucalyptus-dominated areas of southwestern Australia and oak woodlands of California and Oregon are examples.

Grasslands form in the interior of continents in the zones between deserts and temperate forests (Figure 48.18). Summers are warm; winters are cold. Annual rainfall of 25 to 100 centimeters prevents deserts from forming, but this is not enough to support forests. The primary producers tolerate strong winds, sparse and infrequent rain, rapid evaporation, and drought. The dominant animals are grazing and burrowing species. Their activities, combined with infrequent fires, help stop shrublands and forests from encroaching on the fringes of many grasslands.

Between the tropical forests and the hot deserts of Africa, South America, and Australia are savannas—broad belts of grasslands with a few shrubs and trees (Figure 48.18a). Rainfall averages 90–150 centimeters per year. Droughts are seasonal. Where rainfall is low, fast-growing grasses prevail. The savannas grade into tropical woodlands, where low trees and shrubs, and coarse grasses, die and often burn in the dry season.

On flat or rolling land in North America, native shortgrass and tallgrass prairie once prevailed (Figure 48.18b,c). Roots of perennial grasses extended deep into rich topsoil. In the 1930s, shortgrass prairie in the Great Plains was overgrazed. It also was plowed to grow wheat. Prolonged droughts killed this crop, and strong winds stripped away the topsoil. The region became "the Dust Bowl." John Steinbeck's novel *The Grapes of Wrath* is a glimpse into the human tragedies that unfolded because of this environmental disaster.

Grasses two meters tall gave the tallgrass prairie its name. Legumes and composites, including daisies, thrived amidst the grasses in the continent's eastern interior. Nearly all of the original tallgrass prairie has now been converted to cereal croplands (Section 27.4). A few patches that did escape the plow are protected reserves; some others are being restored.

Deserts, dry shrublands, dry woodlands, and grasslands form in regions of sparse rainfall, dry air, strong winds, and recurring episodes of drought and fire.

Figure 48.17 *Above*, two views of California chaparral. The dominant plants are multibranched, woody, and typically only a few meters tall. They are adapted to periodic fires; some produce seeds that germinate only after a fire. This fire raced through chaparral-choked canyons above Malibu. California has 2.4 million hectares (6 million acres) of this dry shrubland.

Figure 48.18 *Right*, (**a**) scarce water in the African savanna, attracting a big herd of wildebeest. Other herbivores of this immense grassland include Cape buffalos, zebras, impalas, giraffes, and Thomson's gazelles.

(**b**) Shortgrass prairie, east of the Rocky Mountains, that once sustained an estimated 60 million bison.

(**c**) A rare patch of natural tallgrass prairie in eastern Kansas.

48.7 More Rain, Broadleaf Forests

LINKS TO
SECTIONS
23.10, 30.1

In forest biomes, tall trees grow close together and form a fairly continuous canopy over a broad region. Rainfall and distance from the equator influence which trees dominate.

TROPICAL RAIN FORESTS

Evergreen broadleaf forests form between latitudes 10° north and south—equatorial zones of Africa, the East Indies, Malaysia, Southeast Asia, South America, and Central America. Rainfall is a whopping 130 to 200 centimeters per year. Regular rains, an annual mean temperature of 25°C, and at least 80 percent humidity support tropical rain forests of the sort shown on the facing page. In structure and diversity, these biomes are the most complex. Some trees are 30.5 meters (100 feet) tall. Many form a closed canopy that stops much of the sunlight from reaching the forest floor. Hence we see a profusion of epiphytes, vines, and climbers.

Decomposition and mineral cycling are rapid in these forests, so litter does not accumulate. The soils are highly weathered, heavily leached, and very poor nutrient reservoirs. You read about the consequences in earlier chapters (Sections 23.10 and 30.1).

SEMI-EVERGREEN AND DECIDUOUS BROADLEAF FORESTS

Away from the equatorial forests, between latitudes 10° to 25°, the dry season grows longer, and broadleaf forests are less and less complex. In humid, tropical regions of Southeast Asia and India, decomposition is slow, and *semi-evergreen forests* form. Where less than 2.5 centimeters (1 inch) of rain falls in the dry season, *tropical deciduous broadleaf* forests form.

Broadleaf deciduous forests form in warm or humid subtropical parts of eastern North America, western and central Europe, and eastern Asia, including Japan. There is a six-month growing season. About 50–150 centimeters (about 20–60 inches) of precipitation fall throughout the year. Oak, hickory, maple, beech, elm, basswood, sweet gum, chestnut, and walnut trees are dominant. Rhododendrons, azaleas, mountain laurel, lichens, club mosses, and mosses thrive in the forests.

Particularly in undisturbed forests of eastern North America, leaves turn almost fluorescent red, orange, and yellow before dropping in autumn (Figure 48.19). The leaf drop is considerable from these fast-growing trees, and it starts decaying into rich humus in spring. Soils of temperate deciduous forests in the Northern Hemisphere rank among the most fertile. That is why vast tracts have been cleared for agriculture. In some parts of the world, the cleared tracts have been under cultivation for thousands of years.

> *Conditions in broadleaf forest biomes favor dense stands of trees that form a continuous canopy over broad regions. As in other forest biomes, tree structure and growth patterns are adapted to patterns in rainfall and temperature.*

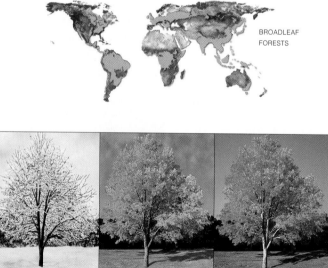

BROADLEAF FORESTS

Figure 48.19 North American temperate deciduous forest. The series above shows changes in one deciduous tree's foliage from fall (*far left*) through winter, spring, and summer.

48.8 You and the Tropical Forests

You read about economic aspects of the destruction of tropical rain forests in earlier chapters. We invite you to reflect on a few more consequences of deforestation.

Developing nations in Latin America, Southeast Asia, and Africa have the fastest-growing populations but not enough food, fuel, and lumber. Of necessity, they turn to their forests for growth-sustaining resources. Most of the forests may disappear within your lifetime. That possibility elicits the most outcries from concerned groups in highly developed nations—which happen to use most of the world's resources, including forest products.

For purely ethical reasons, many people condemn the destruction of so much biodiversity. Tropical rain forests have the greatest variety and numbers of insects, and the world's largest ones. They are homes to the most species of birds and to plants with the largest flowers (*Rafflesia*). The forest canopy and understory support monkeys, tapirs, and jaguars in South America and apes, okapi, and leopards in Africa. Massive vines twist around trees. Orchids, mosses, lichens, and other organisms grow on branches, absorbing minerals that rains deliver to them. Entire communities of microbes, insects, spiders, and amphibians live, breed, and die in small pools of water that collect in furled leaves.

Their disappearance will have repercussions on your life. The few strains of crop plants and livestock that sustain most human populations are vulnerable to ever evolving pathogens. Tissue-culture specialists and genetic engineers use genes of forest species to develop new or hybrid strains that can make our food base less vulnerable. Geneticists use them to develop better antibiotics and vaccines. Aspirin, the most widely used painkiller, is based on a chemical blueprint of an extract from tropical willow leaves. Many ornamental plants, spices, and foods, including cinnamon, cocoa, and coffee, originated in tropical forests. So did the latex, gums, resins, dyes, waxes, and oils used in tires, shoes, toothpaste, ice cream, shampoo, compact discs, condoms, and perfumes. And think about this: The burning of tropical forests around the world is releasing enough air pollutants to change the air you breathe and possibly to help overheat the planet in your lifetime.

And so conservation biologists rightly decry the mass extinction, the assaults on species diversity, the depletion of much of the world's genetic reservoir. Yet something else is going on here. Too many of us grow uneasy when we pass through obliterated forests in our own country. Is it because we are losing the comfort of our heritage—a connection with our evolutionary past? Many millions of years ago, our earliest primate ancestors moved into the trees of tropical forests. Through countless generations, their nervous and sensory systems evolved and became highly responsive to information-rich, arboreal worlds.

Does our neural wiring still resonate with rustling leaves, with shafts of light and mosaic shadows? Are we innately attuned to the forests of Eden—or have time and change buried recognition of home?

Rafflesia

48.9 Wet Summers, Cold Winters, and Conifers

LINKS TO
CHAPTER 23
INTRODUCTION
AND SECTION 23.10

Conifers—cone-bearing trees—are primary producers of *coniferous* forests. Most have thick cuticles, needle-shaped leaves, and recessed stomata, adaptations that help conserve water through droughts and snows. They dominate the boreal forests, montane coniferous forests, temperate rain forests, and pine barrens.

Boreal forests of mostly pine, fir, and spruce stretch across northern Europe and Asia, and North America. They are known as *taigas*, meaning "swamp forests." Most form in glaciated regions with cold lakes and streams (Figure 48.20). Most rain falls in the summer, and little water evaporates into the cool summer air. Winters are cold, dry, and far more severe in eastern parts of these biomes than in the west, where ocean winds moderate climate. Spruce and fir dominate the boreal forests of North America. Forests of pine, birch, and aspen form in burned or logged areas. In poorly drained soil, acidic bogs prevail. Farther to the north, boreal forests thin and grade into arctic tundra.

Coniferous forests also form in high mountains. At high northern elevations, too, spruce and fir dominate.

The montane coniferous forests give way to pine forests in the south and at lower elevations, where the winters are milder. Coniferous forests also flourish in some temperate lowlands and near the Pacific coast from Alaska into northern California. The tallest trees in the world are in these forests—Sitka spruce to the north and coast redwoods to the south. Large tracts have been logged over (Chapter 23).

Coniferous forests also endure in warmer regions. For instance, forests of pine and scrub oak thrive in New Jersey. Southern pine forests dominate coastal plains of the southern Atlantic and Gulf states. In the Deep South, palmettos grow beneath loblolly pines. Many species are adapted to dry, sandy, nutrient-poor soil and to periodic, lightning-sparked fires.

Coniferous forests prevail across the Northern Hemisphere in regions where wet summers alternate with cold, dry winters. They also prevail at high elevations.

CONIFEROUS FORESTS

Figure 48.20 (**a**) Taiga in Alberta, Canada. (**b**) Montane coniferous forest near Mount Ranier, Washington.

48.10 Brief Summers and Long, Icy Winters

Tundra lies between the polar ice cap and belts of boreal forests in the Northern Hemisphere. It is the youngest biome, having first evolved about 10,000 years ago.

The Northern Hemisphere's treeless plain, the *arctic tundra*, is extremely cold. Annual snowmelt and rain are usually less than 25 centimeters. Plants grow fast in the nearly continuous sunlight of a brief growing season (Figure 48.21).

Not much more than the surface soil thaws during summer. Just below is **permafrost**, a frozen layer 500 meters thick in some places. It prevents drainage, so the soil above is perpetually waterlogged (page 865). The cool, anaerobic conditions do not favor nutrient cycling. Plant remains accumulate in soggy masses and decay slowly. The accumulation of undecayed organic matter in permafrost makes the arctic tundra one of Earth's greatest stores of carbon. However, the frozen layer is starting to thaw. Global warming is melting ice and snow, which reflect sunlight. Newly exposed soil is absorbing heat from the sun's rays.

Alpine tundra is a similar biome, but it develops in high mountains throughout the world (Figure 48.22). At night, the below-freezing temperatures make it too difficult for trees to grow. Even in summer, shaded patches of snow persist in this biome, but there is no permafrost. Alpine soil is thin and well drained, but it is nutrient-poor. As a result, primary productivity is low. Grasses, heaths, and small-leafed shrubs grow in patches where better soil has formed. These low plant species are adapted to withstand strong winds.

Figure 48.22 Compact, low-growing, hardy plants typical of alpine tundra in the Washington Cascade range.

Arctic tundra prevails at high latitudes, where short, cold summers alternate with long, cold winters. Alpine tundra prevails in high, cold mountains regardless of seasonal differences in latitude.

Figure 48.21 (a) Arctic tundra in the summer. Hardy lichens and shallow-rooted, low-growing plants are a base for food webs that include voles, arctic hares, caribou, arctic foxes, wolves, and polar bears. Great numbers of migratory birds nest here in summer, when the air is thick with mosquitoes and other kinds of flying insects.

(b) Arctic tundra makes up about 4 percent of Earth's land mass. It is blanketed with snow for as long as nine months of the year. Most occurs in northern Russia and Canada, followed by Alaska and Scandinavia. Bands of humans have herded reindeer, hunted, and fished in these sparsely populated regions for hundreds of thousands of years. More people, and machines, are moving in to extract mineral and fossil fuels. If the operations alter the vegetation and soils, it might take decades for a region to recover. Why? Tundra plants grow very slowly, and their seasonal growth is limited to just a few months of each year.

48.11 Converting Marginal Land for Agriculture

LINKS TO
SECTIONS 23.10,
30.1, 32.9, 45.8

*Desertification has become one of the most far-reaching
environmental challenges facing us today. If it continues
unchecked, gains in human well-being in marginal lands
will be arrested and, in some cases, reversed.*

What Is Desertification?

The conversion of large
tracts of land to a more desertlike condition is known as
desertification. The term applies as well when a similar
conversion of rain-fed or irrigated croplands results in
a 10 percent or greater decline in agricultural productivity.
Over the past fifty years, about 9 million square kilometers
worldwide have become desertified.

Each year, on all continents except Antarctica, natural
processes and human activities are converting at least
200,000 square kilometers to deserts. Humans contribute
to desertification by cutting dryland shrubs for firewood,
allowing livestock to overgraze on marginal rangelands,
and by other destructive agricultural practices. Prolonged
droughts accelerate the process, just as one did years ago
in the Great Plains of the American Midwest (Section 30.1).

This is a huge problem, because human populations
already use nearly 21 percent of Earth's land surfaces for
cropland or for grazing. Another 28 percent of the surface
is said to be potentially suitable for agriculture, but the
primary productivity would be so low that the conversion
may not be worth the cost (Figure 48.23). As you know
from Section 45.8, the human population size is far from
stabilized; it is expected to increase to 8.9 billion within
the next fifty years. You do the math.

Impact on Human Populations

Severe, recurring
food shortages are common in heavily populated regions
of Asia, Central and South America, Africa, and the Middle
East. In these regions, people already have 80 percent of
the productive land under cultivation, and pressure is on
to expand into marginal drylands.

More than 500 million of the most impoverished people
live on marginal lands in the Sahel and at high watershed
elevations in the Andes and Himalayas. Under current policies
and conditions, that number is expected to rise to 800 million
by 2020. Their lives are directly and acutely impacted by the
deterioration of natural resources.

About 1.5 billion people around the world are now living
in deserts, dry shrublands, and dry woodland. In China, nearly
all of the 65 million people who are officially recognized as
income-poor live in remote mountains with poor soils.

Scientists continue to make valiant efforts to improve
production on existing cropland. Under the banner of the
green revolution, their research has been directed toward
(1) improving the genetic traits of crop plants for higher
yields and (2) exporting modern agricultural practices and
equipment to the developing countries. Here you may wish
to refer to the Section 32.9 case study on quinoa cultivation.

Many of these countries rely on subsistence agriculture,
which runs on energy inputs from sunlight and human
labor. They also rely heavily on animal-assisted agriculture,
with energy inputs from oxen and other draft animals. By
contrast, mechanized agriculture requires massive inputs
of fertilizers, pesticides, and ample irrigation to sustain
high-yield crops. It requires fossil fuel energy to drive farm
machines. Crop yields are four times as high. However, the
modern practices require a hundred times more energy
and soil nutrients. Also, there are signs that limiting factors
may be about to come into play. If so, they will slow down
additional increases in crop yields.

Impact on Biodiversity

As Figure 48.24 indicates,
effects of desertification are not confined to local regions.
Desertification invites immense dust storms, as in Figure
48.25. The loss of vegetation cover results in downstream
flooding and sedimentation. It also disrupts the global
cycling of carbon. Such factors have adverse impact on
the stability of communities—and species richness—
throughout the world.

With a little ingenuity, it may be possible to maintain
biodiversity and meet human needs. In Africa, imported
livestock—cattle, sheep, and goats—are a major factor in
desertification. These domesticated animals evolved in
areas where water is plentiful and so require more water
than native African herbivores do. The introduced species
continually move back and forth between their grazing
areas and watering holes. As they do, they trample grasses
and compact the soil surface (Figure 48.26). By contrast,

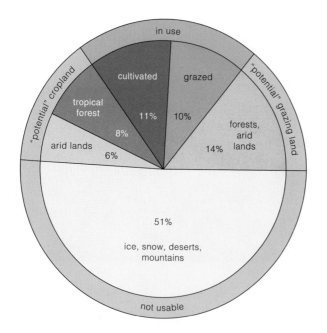

Figure 48.23 Classification of land in terms of its suitability for
agriculture. Theoretically, clearing vast tracts of tropical forests
and irrigating marginal lands could more than double the world's
cropland. Doing so would destroy valuable forest resources,
damage the environment, cause severe losses in biodiversity,
and possibly cost more than it is worth.

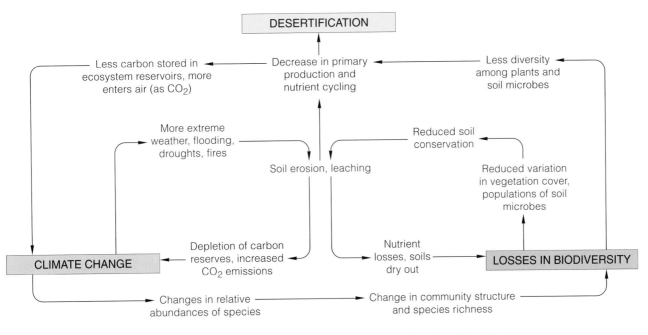

Figure 48.24 Model for the connections, including feedback loops, among desertification, global climate change, and losses in biodiversity. After Millenium Ecosystem Assessment's Desertification Synthesis.

gazelles and other wild herbivores get most (if not all) of their water from plants. They also are better at conserving water; they lose little in feces, compared to imported livestock.

In 1976 a biologist, David Hopcraft, began rearing gazelles, antelopes, giraffes, wildebeasts, ostriches, and other native herbivores on a ranch in Kenya. In one study, he compared the meat yield and evironmental impact of raising cattle versus gazelles in two adjacent study areas. Compared to

cattle, the gazelles left much more of the vegetation intact, which helped maintain plant biodiversity. In addition, the gazelles yielded almost twice the amount of meat per acre.

Similar ranches are now being set up in other parts of Africa, but there are obstacles. Such ranches are more expensive to set up and they require specialized butchering, packing, and marketing efforts. Also, most Africans prefer the taste of imported cattle to that of native animals, which tends to be very lean.

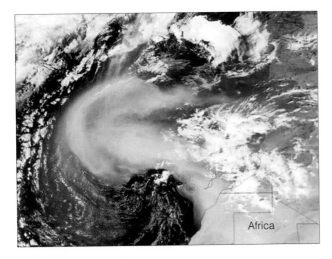

Figure 48.25 Satellite image of an immense dust storm that formed in the hot, dry Sahara Desert, then continued past the west coast of Africa, and is well on its way across the Atlantic Ocean.

Figure 48.26 Desertification in the Sahel, a region of West Africa that forms a belt between the hot, dry Sahara Desert and tropical forests. This savanna area is undergoing rapid desertification due to overgrazing, overfarming, and prolonged drought.

48.12 Standing Freshwater Ecosystems

LINKS TO
SECTIONS
47.11, 47.12

Freshwater and saltwater provinces cover more of Earth's surface than all biomes combined. They include the world ocean, lakes, ponds, wetlands, and coral reefs. There are no "typical" regions. Ponds are shallow; Siberia's Lake Baikal is 1.7 kilometers deep. All aquatic ecosystems have gradients in light penetration, temperature, and dissolved gases, but values differ greatly. All we can do here is sample the diversity, starting with freshwater provinces called lakes.

A lake is a body of standing freshwater, as in Figure 48.27. Over time, erosion and sedimentation alter its dimensions; it usually ends up filled in or drained. A young lake has littoral, limnetic, and profundal zones (Figure 48.28*a*). Its littoral zone extends all around the shore to the depth where rooted aquatic plants cannot grow. The well-lit, shallow, warm water supports high biodiversity. The limnetic zone encompasses the sunlit, open water away from shore to depths where light and photosynthesis are limited. Its *phyto*plankton includes cyanobacteria, green algae, and diatoms. *Zoo*plankton includes rotifers and copepods. The profundal zone is all open water at depths that are impenetrable to the most efficient wavelengths for photosynthesis. Below it, sediments house communities of decomposers.

SEASONAL CHANGES IN LAKES

In temperate regions, many lakes undergo seasonal changes in density and temperature gradients, from surface to bottom. In midwinter, a surface layer of ice often forms. Water under the ice is near the freezing point and is the least dense. Water at 4°C is the most dense; it forms deeper layers that are a bit warmer.

In spring, there are more daylight hours and the air warms. The ice melts, the temperature of the surface layer rises to 4°C, and temperature gradients vanish. Winds blowing across the lake surface cause a **spring overturn**, in which strong vertical movements deliver dissolved oxygen from surface waters to the depths. At the same time, nutrients released by decomposers of the lake bottom sediments move to the surface.

In summer, the lake develops a *thermocline*, which means it becomes thermally stratified in between the surface and its depths. The midlayer cools, which stops the vertical mixing between the warmer, oxygen-rich surface layer and the cold, oxygen-poor layer below it (Figure 48.28*b*). Decomposers now deplete the oxygen dissolved near the lake bottom.

In autumn, the upper layer cools and gets denser. It sinks and the thermocline vanishes. During this **fall overturn**, water mixes vertically, and again dissolved oxygen moves down and nutrients move up.

Primary productivity is seasonal. After the spring overturn, longer daylengths and the cycled nutrients

Figure 48.27 Lake in Chile's Torres del Paine National Park.

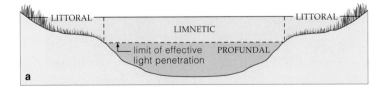

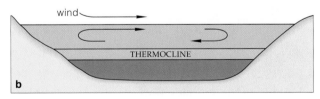

Figure 48.28 (**a**) Lake zonation. A lake's littoral zone extends all around the shore to a depth where aquatic plants stop growing. Its profundal zone is all water below the depth of light penetration. Above the profundal are the open, sunlit waters of a limnetic zone. (**b**) In temperate regions, thermal layering occurs in many lakes during the summer.

Figure 48.29 Crater Lake, Oregon, a collapsed volcanic cone filled with rainwater and snowmelt. Like the rest of the volcanoes of the Cascade Range, it started forming at the dawn of the Cenozoic.

The chart compares the main characteristics of oligotrophic and eutrophic lakes. Some lakes fall in a category that is intermediate between these two groups; they are called mesotrophic lakes. In which category would you place Crater Lake?

Oligotrophic Lake	Eutrophic Lake
Deep, steeply banked	Shallow with broad littoral
Large deep-water volume relative to surface-water volume	Small deep-water volume relative to surface-water volume
Highly transparent	Limited transparency
Water blue or green	Water green to yellow- or brownish-green
Low nutrient content	High nutrient content
Oxygen abundant through all levels throughout year	Oxygen depleted in deep water during summer
Not much phytoplankton; green algae and diatoms dominant	Abundant, thick masses of phytoplankton; and cyanobacteria dominant
Abundant aerobic decomposers favored in profundal zone	Anaerobic decomposers
Low biomass in profundal	High biomass in profundal

favor increased primary productivity. Phytoplankton and aquatic plants take up phosphorus, nitrogen, and other nutrients. During the growing season, vertical thermocline mixing ends. Nutrients do not move up, and photosynthesis slows. By late summer, shortages of nutrients limit growth. The fall overturn does cycle nutrients back to the surface, and there is a sudden burst of primary productivity. But a sustained burst is not possible, given the fewer daylight hours. Primary productivity will not rise again until spring.

TROPHIC NATURE OF LAKES

Each lake's topography, climate, and geologic history affect the number and relative abundances of species, how they are dispersed through the lake, and how they cycle nutrients. Soils of the region and sediments in the lake basin contribute to the type and amount of nutrients available. Oxygen, phosphorus, and nitrogen dissolved in the water influence primary production (Sections 47.11 and 47.12). So does the water volume and how long it has been standing in the lake.

Interplays among climate, soil, basin shape, and metabolic activities of the lake's residents contribute to a lake's trophic status. *Oligotrophic* lakes typically are newly formed, deep, clear, and nutrient-poor, with low primary productivity. *Eutrophic* lakes are older, shallower, nutrient-rich, and higher in their primary productivity (Figure 48.29). As a lake ages, sediments accumulate, water becomes opaque and shallow, and phytoplankton come to dominate the community. A filled-in basin is the final successional stage.

In Section 47.12, you read about an experiment in **eutrophication**. The term refers to natural or artificial processes that enrich a body of water with nutrients. Here is an another example:

Human activities caused eutrophication of Seattle's Lake Washington. From 1941 to 1963, phosphate-rich sewage drained into the lake. This made nitrogen the major limiting factor for the photosynthetic species. Cyanobacteria dominate when nitrogen is scarce; they can fix gaseous nitrogen (N_2). They became dominant and formed slimy mats in summer, which made the lake completely useless for recreation. Sewage inputs were stopped. By 1975, the lake neared full recovery.

Water provinces are far more extensive than the biomes. Like the other aquatic ecosystems, lakes show gradients in light penetration, temperature, and dissolved gases.

48.13 Flowing Freshwater Ecosystems

The composition of streams, rivulets, and rivers becomes altered as water drains away from watersheds on its journey to the sea.

STREAM ECOSYSTEMS

Flowing-water ecosystems called **streams** start out as freshwater springs or seeps. As they flow downslope, they grow and merge, then often converge as a river. Between a river's headwaters and end, we find riffles, pools, and runs (Figure 48.30). The *riffles* are shallow, turbulent stretches where the water flows swiftly over a rough, sandy and rocky bottom. In *pools*, deep water flows slowly over a smooth, sandy, or muddy bottom. *Runs* are smooth-surfaced, rapidly flowing stretches above bedrock or rock and sand.

Rainfall, snowmelt, geography, altitude, even the shade cast by plants influence a stream's average flow volume and its temperature. Streambed composition, along with agricultural, industrial, and urban wastes, shape the solute concentrations in the water.

A stream imports organic matter into food webs, especially in forest ecosystems. Trees cast shade and thus hamper photosynthesis, but their litter is the start of detrital food webs (Section 47.2). Aquatic species take up and then release nutrients as the water flows downstream. Nutrients move upstream only inside the tissues of migratory fishes and other animals. They spiral between aquatic organisms and the water as it flows on a one-way course to the sea.

RIVER SYSTEMS

More than 250,000 rivers drain the watersheds of the United States. The Mississippi River drains the largest watershed. Each second, its flow volume into the Gulf of Mexico is about 16,792 cubic meters (593,000 cubic feet). It is not the longest river in the United States. From its headwaters high in the Rocky Mountains to its convergence with the Mississippi, the Missouri River flows for more than 4,000 kilometers (2,500 miles). The shortest river in the United States is Oregon's Devil River. It moves water 36.5 meters (120 or so feet) from Devil's Lake to the Pacific Ocean.

More and more sediments accumulate in the water of long rivers during the journey to the sea. Figure 48.31 shows the Mississippi River Delta, formed from sediments as far away as the Missouri's headwaters.

Water flows downhill. By the time it has moved as rivulets and streams to a river, its composition reflects that of the watersheds it drained. Nutrients in fallen leaves alter the composition. So do mudslides, heavy rains, debris from fires, and other natural events. So do human activities.

WATER POLLUTION

Ever since cities formed, streams and rivers have been sewers for the wastes from agriculture, industry, and cities. Runoff from poorly managed agricultural fields contains silt, animal wastes, pesticides, and nutrients such as phosphates, which promote algal blooms that choke streams. Power-generating plants and factories add toxic chemicals to the water.

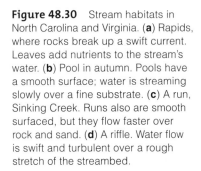

Figure 48.30 Stream habitats in North Carolina and Virginia. (**a**) Rapids, where rocks break up a swift current. Leaves add nutrients to the stream's water. (**b**) Pool in autumn. Pools have a smooth surface; water is streaming slowly over a fine substrate. (**c**) A run, Sinking Creek. Runs also are smooth surfaced, but they flow faster over rock and sand. (**d**) A riffle. Water flow is swift and turbulent over a rough stretch of the streambed.

Unlike lakes, streams are resilient. They can recover fairly quickly when pollution is controlled. However, pollutants make streamwater unfit to drink.

Water pollutants collect in lakes, rivers, and bays before water gets to the sea. Many cities throughout the world dump untreated sewage into their coastal waters. The cities along rivers and harbors maintain shipping channels by dredging the polluted muck and barging it out to sea. They also barge out sewage sludge—coarse, settled solids enriched with bacteria, viruses, and toxic metals.

In the United States, 15,000 or so facilities partially treat liquid wastes from 70 percent of the population and 87,000 industries. Wastes from suburban and rural populations are treated in septic tanks or lagoons, but much is discharged, untreated, into waterways.

There are three levels of *wastewater treatment*. With primary treatment, screens and settling tanks remove sludge, which is dried, burned, dumped in landfills, or treated further. Chlorine kills most pathogens but forms carcinogens by reacting with some chemicals. In secondary treatment, microbial populations break down organic matter after the primary treatment but before water is chlorinated. The two treatments get rid of most of the solids and oxygen-demanding wastes —but not all nitrogen, phosphorus, toxins, and heavy metals. Tertiary treatment cleanses the water but is still largely experimental and expensive. Only 5 percent of the nation's wastewater gets this level of treatment.

Thus, most wastewater is not treated adequately. A pattern gets repeated thousands of times along the waterways. The water for drinking is drawn upstream from a city, and the wastes from industry and sewage treatment are discharged downstream. That is how pollution intensifies as rivers flow down to the sea. In Louisiana, waters drained from the central states flow toward the Gulf of Mexico. Its high pollution levels threaten public health as well as ecosystems.

This picture might be bleak to most of us, but not to biologist John Todd. He constructed experimental wastewater treatment facilities in several greenhouses and artificial lagoons (Figure 48.32). When it works properly, his solar–aquatic treatment system produces water fit to drink. Such natural alternatives cannot work for the large urban areas. They are an attractive alternative for small towns and rural areas.

You may wish to investigate where the drinking water for your own city comes from, and what it has picked up from its surroundings.

Figure 48.31 Satellite image of the Mississippi River Delta, where the biggest river in North America empties into the Gulf of Mexico. The Mississippi is not the world's largest river. The flow volume at the mouth of the Amazon River is far greater; it is close to 220,000 cubic meters (7 million cubic feet) per second.

Figure 48.32 An experimental wastewater treatment facility in Rhode Island. Treatment begins when sewage flows into rows of large water tanks in which water hyacinths, cattails, and other aquatic plants are growing. Decomposers inside the tanks degrade wastes—which contain nutrients that promote plant growth. Heat from incoming sunlight speeds the decomposition.

From these tanks, water flows through an artificial marsh of sand, gravel, and bulrushes that filter out algae and organic wastes. Then it flows into aquarium tanks where zooplankton and snails consume microorganisms suspended in the water. There also, zooplankton become food for crayfishes and fishes, such as tilapia, that can be sold as bait. After ten days, the now-clear water flows into a second artificial marsh for final filtering and cleansing.

48.14 Life at Land's End

LINKS TO
SECTIONS
22.8, 25.17, 27.5

Near the coasts of continents, around islands and reefs, concentrations of nutrients support some of the world's most productive ecosystems.

WETLANDS AND THE INTERTIDAL ZONE

Like freshwater ecosystems, estuaries and mangrove wetlands have distinct physical and chemical features that include depth, water temperature, salinity, and the light penetration. **Estuaries** are partly enclosed coast regions where seawater mixes with nutrient-rich fresh water from rivers, streams, and runoff (Figure 48.33a). The confined region, slow mixing of water, and tidal action combine to trap dissolved nutrients. Water flow continually replenishes nutrients, which is one reason estuaries can support highly productive ecosystems.

Primary producers are phytoplankton, plants that tolerate submergence at high tide, and algae. Detrital food webs are common. So many larval and juvenile stages of invertebrates and some fishes develop here that estuaries are sometimes called marine nurseries. Many migratory birds use the estuaries as rest stops.

Estuaries range from broad, shallow Chesapeake Bay, Mobile Bay, and San Francisco Bay to the narrow, deep fjords of Norway and others like them in Alaska and British Columbia. Many estuaries are in decline because fresh water is being diverted upstream, and agricultural runoff and wastes are flowing in.

In tidal flats at tropical latitudes, we find nutrient-rich *mangrove wetlands*. "Mangrove" refers to forests of salt-tolerant plants in sheltered areas along tropical coasts. These plants have shallow or branching prop roots that extend from the trunk (Figure 48.33b). Many

have root extensions that take up oxygen. Mangrove trees germinate while still attached to the parent tree. Seedlings drop to the water and float upright until the currents deposit them in the mud of shallow water.

Net primary productivity of a mangrove wetland depends partly on the tidal volume and flow rate, as well as on salinity and nutrient availability. But tidal circulation and nutrient input combined can support notable biomass. Especially along the Gulf of Mexico and the Malaysian peninsula, currents carry nutrient-rich detritus away from mangrove wetlands and into the neighboring estuarine ecosystems.

ROCKY AND SANDY COASTLINES

Rocky and sandy coastlines support ecosystems of the intertidal zone, which is not renowned for creature comforts. Waves batter its residents; tides alternately submerge and expose them. The higher up they are, the more they dry out, freeze in winter, and bake in summer, and the less food comes their way. The lower they are, the more they compete in limited spaces. At low tides, birds, rats, and raccoons move in and feed on them. High tides bring the predatory fishes.

Generalizing about coastlines is not easy, for waves and tides continually resculpt them. One feature that rocky and sandy shores share is vertical zonation.

Rocky shores have three zones. Their *upper* littoral zone is submerged only at the highest tide of the lunar cycle and is sparsely populated (Figure 48.34a). The *mid*littoral is submerged at the highest regular tide and exposed at the lowest. Its tidepools hold algae, fishes, hermit crabs, nudibranchs, sea urchins, and sea

Figure 48.33 (a) South Carolina salt marsh. A marsh grass (*Spartina*) is a major producer.
(b) In the Florida Everglades, a mangrove wetland lined with red mangroves (*Rhizophora*).

upper littoral of intertidal zone; submerged only at highest tide of lunar cycle

midlittoral; submerged at each highest regular tide and exposed at lowest tide

lower littoral; exposed only at lowest tide of lunar cycle

a   b

Figure 48.34 (**a**) Tidepool on a rocky shore of the Pacific Northwest, with algae and invertebrates. (**b**) In this coastal region, you are likely to come across pronounced vertical zonation, as shown here.

Around the world, the difference between the high and low tide marks ranges from a few centimeters in the Mediterranean Sea to about fifteen meters in the Bay of Fundy, near Nova Scotia.

Figure 48.35 Coral fragment washed up on the sandy shore of Heron Island, part of Australia's Great Barrier Reef.

Figure 48.36 Two inhabitants of coral reefs. *Left*, clownfish with sea anemone. Compare Section 47.2. *Right*, sea fan (*Gorgonia*), one of the corals.

stars (Sections 22.8 and 25.17). The *lower* littoral zone is exposed only during the lunar cycle's lowest tide. It has the greatest diversity. In all three zones, erosion is so swift that detritus cannot build up, so grazing food webs prevail (Figure 48.34*b*).

Waves and currents continually rearrange the loose sediments of *sandy* and *muddy* shores. Few big plants can grow in unstable places, so you will not discover grazing food webs here. Detrital food webs start with inputs from land or offshore (Figure 48.35).

CORAL REEFS

As Section 27.5 explains, *coral reefs* develop in clear, warm waters near coasts or around volcanic islands, mainly between latitudes 25° north and south. Each is a wave-resistant formation of the slowly accumulated remains of marine organisms. Hard corals as well as mineral-hardened cell walls of red algae, cemented together, formed the reef spine. Figure 48.36 shows two more examples of the wealth of warning colors, tentacles, and stealth of reef species—signs of danger and fierce competition for resources by individuals that must interact in a limited space.

Life thrives where land meets the sea. Wetlands and coral reefs show high primary productivity. Rocky and sandy shores are not renowned for their creature comforts.

48.15 The Open Ocean

LINKS TO
SECTIONS 7.8,
20.2, 20.3, 25.4

The world ocean consists of two vast provinces (Figure 48.37). Its benthic (bottom) province starts at continental shelves and extends to deep-sea trenches. Its pelagic province is the full volume of ocean water. The neritic zone is the volume above continental shelves, and the oceanic zone is the volume above the ocean basins.

SURPRISING DIVERSITY

Photosynthesis is seasonal and intense near the ocean surface. Drifting in seawater are phytoplankton, the "pastures" that feed copepods, krill, whales, squids, fishes, and other members of marine food webs. Also near the surface, photoautotrophs account for most of the ocean's primary productivity. They range in size from bacterial cells less than five micrometers across (ultraplankton) to coccolithophores and other types of cells as much as fifty micrometers (nannoplankton).

Deeper ocean water is too dark for photosynthesis. There, food webs start with **marine snow**. These tiny bits of organic matter drift down from communities above. They are the base for staggering biodiversity; midoceanic water may be home to 10 million species!

Also, in what might be the greatest of all circadian migrations, a number of species rise thousands of feet to feed in upper waters at night and move down the next morning. Carnivores at the top of the food webs range from familiar types, including sharks and giant squids, to the visually jarring deep-sea angler fishes and immense siphonophores.

The benthic province includes largely unexplored ecosystems on seamounts and at hydrothermal vents (Figure 48.38). *Seamounts* are extinct volcanoes that rise at least 1,000 meters from the seafloor but are far below the ocean surface. At *hydrothermal vents*, near-freezing water seeps into fissures in the seafloor and becomes superheated. As it spews back out, it leaches mineral ions from rocks. Dissolved in the outpouring are iron, zinc, copper sulfides, and sulfates of calcium and magnesium. The minerals settle out and form rich deposits, which are an energy source for bacterial and archaean chemoautotrophs. These prokaryotes are the primary producers for rich food webs, which include a variety of fishes, crustaceans, clams, and tube worms, such as those shown in Figure 48.38b.

Many biologists suspect that life originated in such hot, nutrient-rich places on the seafloor. Sections 20.2 and 20.3 offer some evidence of this possibility.

UPWELLING AND DOWNWELLING

As you read earlier, prevailing winds that parallel the western coasts of continents tug on the ocean surface. Wind friction gets the surface waters moving. Earth's rotational force deflects masses of slow-moving water away from the coasts. Cold, deep, often nutrient-rich

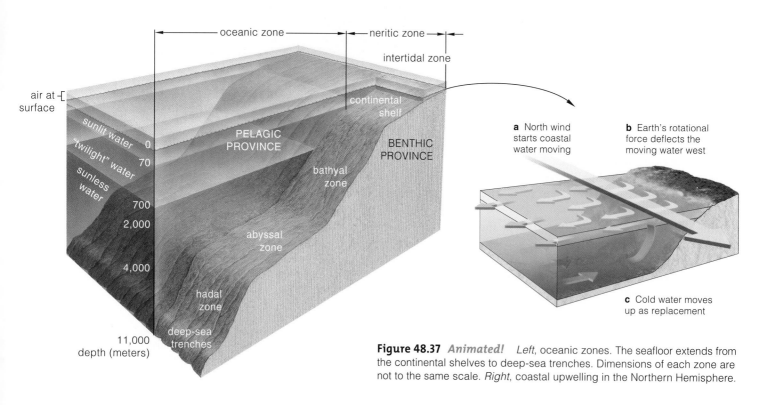

Figure 48.37 *Animated!* *Left*, oceanic zones. The seafloor extends from the continental shelves to deep-sea trenches. Dimensions of each zone are not to the same scale. *Right*, coastal upwelling in the Northern Hemisphere.

water moves in vertically in its place (Figure 48.37). Cold, deeper water moving up this way is called an **upwelling**. Upwelling occurs in equatorial currents and along continents, and it cools air masses above it. When thick fogbanks form near California's coast, an upwelling of cold water is interacting with warm air.

In the Southern Hemisphere, commercial fisheries depend on wind-induced upwelling along Peru and Chile. Prevailing coastal winds blow from the south and southeast, tugging surface water away from the shore. Cold, deeper water carried to the continental shelf by the Humboldt Current moves up near the surface. Large quantities of nitrate and phosphate are pulled up and carried north by the cold Peru Current. The nutrients sustain phytoplankton that are the basis of one of the world's richest fisheries.

Every three to seven years, warm surface waters of the western equatorial Pacific Ocean move eastward. This massive displacement of warm water acts on the prevailing wind direction. The eastward flow speeds up so much that it hampers the vertical movement of water along the coasts of Central and South America.

Surface water piling into a coast is forced down and flows away from it. Near Peru's coast, prolonged **downwelling** of nutrient-poor water displaces cooler waters of the Humboldt Current and puts a stop to upwelling. The warm current typically arrives around Christmas. Fishermen in Peru named it **El Niño** ("the little one," in reference to the baby Jesus). The name became incorporated into a more inclusive, scientific explanation called the El Niño Southern Oscillation, or ENSO. The next section takes a closer look at some of the consequences of this recurring event.

OCEAN AS GARBAGE DUMP

The North Pacific is the largest ocean realm, about the size of Africa, in the Northern Hemisphere. Circular winds above it drag the surface in an ever tightening spiral, the north central Pacific gyre. This gyre is the most stable feature of Earth's weather systems. It also has become a self-perpetuating garbage dump.

Anything that remains afloat on the North Pacific ocean ends up here. Organic debris breaks down, but plastics do not. Bacterial decomposers cannot degrade the synthetic compounds. Sunlight breaks them down into individual molecules. By one estimate, there may be 2.7 kilograms (6 pounds) of plastic debris for every .45 kilograms (1 pound) of plankton in this continent-sized stretch of floating plastic sand. To field biologist Shawn Farry, the sight of floating plastic, light bulbs, shoes, bottles, and discarded or lost fishing lines and

Figure 48.38 What lies beneath—a vast, largely unexplored world of marine life. (**a**) Flytrap anemone on Davidson seamount, off the California coast. There are an estimated 30,000 seamounts. They may be home to rich marine ecosystems that also may be resting stops for hammerhead sharks and other marine vertebrates that migrate across the open ocean.

(**b**) Tube worms, part of a hydrothermal vent ecosystem on the ocean floor.

(**c**) *Praya dubia*, a relative of the Portuguese man-of-war. This stinging, bioluminescent siphonophore (a type of cnidarian) is one of the longest existing animals. Some specimens have measured fifty meters, from a mouthless, pulsating swimming bell to the tip of a narrow stem to which reproductive medusae, tentacles, and feeding polyps connect. *P. dubia* moves vertically in a circadian migration. (**d**) Deep-sea angler fish.

nets floating around even the most remote islands is nothing new. What was shocking to him was the vast concentration anywhere he dove in the middle of the ocean. Even at depths of about 30.5 meters (100 feet), researchers aboard the research vessel *Alguita* watched invertebrate filter-feeders consuming debris and also becoming entangled in it. Think about that the next time you see a piece of Styrofoam bobbing along.

From the ocean's coral reefs down to hydrothermal vents, throughout the pelagic province, we find astounding levels of primary productivity and biodiversity.

48.16 Rita in the Time of Cholera

LINKS TO
SECTIONS
7.8, 22.8. 25.13

We turn now to an application that reinforces a unifying ecological concept. Events in the atmosphere and ocean, and on land, interconnect in ways that can profoundly influence the world of life.

An El Niño Southern Oscillation, or **ENSO**, is defined by changes in sea surface temperatures and in the air circulation patterns. "Southern oscillation" refers to a seesawing of the atmospheric pressure in the western equatorial Pacific—the world's greatest reservoir of warm water and warm air. It is the source of heavy rainfall, which releases enough heat energy to drive global air circulation.

Between ENSOs, the warm waters and heavy rains move westward (Figure 48.39a). *During* an ENSO, the prevailing surface winds over the western equatorial Pacific pick up speed and "drag" surface waters east (Figure 48.39b). As they do, the westward transport of water slows down. Sea surface temperatures rise, evaporation accelerates, and air pressure falls. These changes have global repercussions.

El Niño episodes usually persist for six to eighteen months. Then another oscillation called **La Niña** starts up, and the weather seesaws again.

As you read in the chapter opening, 1997 ushered in the most powerful ENSO event of the century. The average sea surface temperatures in the eastern Pacific rose 9°F (about 5°C). The warmer water extended 9,660 kilometers west from the coast of Peru.

The 1997–1998 El Niño/La Niña rollercoaster had record-breaking impact on primary productivity in the equatorial Pacific. With the massive eastward flow of nutrient-poor warm water, photoautotrophs were almost undetectable (Figure 48.40a).

During the La Niña rebound, cooler, nutrient-rich water welled up to the sea surface and was displaced westward all along the equator. As satellite images clearly revealed, the upwelling had sustained a vast algal bloom, one that stretched across the equatorial Pacific (Figure 48.40b; also compare Section 7.8).

During the 1997–1998 El Niño event, 30,000 cases of *cholera* were reported in Peru alone, compared to only 60 cases from January to August in 1997. People knew that water contaminated by *Vibrio cholerae* causes epidemics of cholera (Figure 48.41). The disease agent triggers severe diarrhea, and it thereby enters water supplies in feces. Individuals who are forced to use the tainted water become infected.

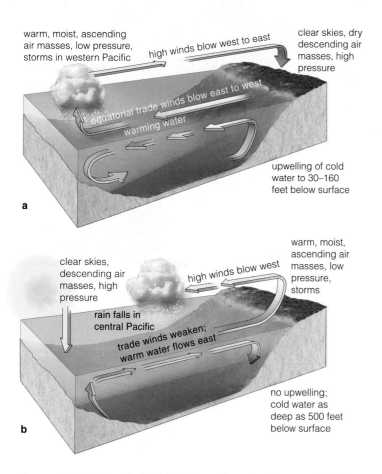

Figure 48.39 *Animated!* (**a**) Westward flow of cold, equatorial surface water between ENSOs. (**b**) Eastward dislocation of warm water during an ENSO.

a Near-absence of phytoplankton in the equatorial Pacific during an El Niño.

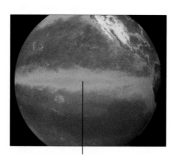

b Huge algal bloom in the equatorial Pacific in the La Niña rebound event.

Figure 48.40 Satellite data on primary productivity in the equatorial Pacific Ocean. The concentration of chlorophyll was used as the measure. (**a**) During the 1997–1998 El Niño episode, a massive amount of nutrient-poor water moved to the east, and so photosynthetic activity was negligible. (**b**) During a subsequent La Niña episode, massive upwelling and westward displacement of nutrient-rich water led to a vast algal bloom that stretched all the way to the coast of Peru.

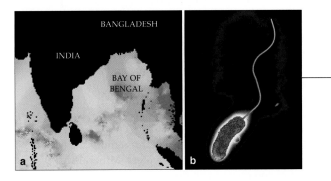

What people did *not* know was where *V. cholerae* remained between cholera outbreaks. It could not be found in humans or in water supplies. Even so, the pathogen would show up simultaneously in places that could be far apart—usually coastal cities where the urban poor draw water from rivers to the sea.

Marine biologist Rita Colwell had been thinking about the fact that humans are not the host between outbreaks. Was there an environmental reservoir for the pathogen? Maybe. But nobody had detected it in water samples subjected to standard culturing.

Then Colwell had a flash of insight: What if no one could find the pathogen because it changes its form and enters a dormant stage between outbreaks?

During one cholera outbreak in Louisiana, Colwell realized that she could use an antibody-based test to detect a protein unique to *V. cholerae*'s surface. Later, tests in Bangladesh revealed bacteria in fifty-one of fifty-two samples of water. Standard culture methods had missed it in all but seven samples.

V. cholerae lives in rivers, estuaries, and seas. As Colwell knew, plankton also thrive in these aquatic environments. She decided to restrict her search for the unknown host to warm waters near Bangladesh, where outbreaks of cholera occur seasonally (Figure 48.41c). It was here that she discovered the dormant *V. cholerae* stage inside copepods. These tiny marine crustaceans graze on algae and other phytoplankton species (Figure 48.41b). The number of copepods— and of the *V. cholerae* cells inside them—rises and falls with shifts in phytoplankton abundances.

Colwell already knew about seasonal variations in sea surface temperatures. Remember the old saying, *Chance favors the prepared mind*? In one sense, she was prepared to recognize a connection between cholera cases and seasonal temperature peaks in the Bay of Bengal. She compared data from the 1990–1991 and 1997–1998 El Niño episodes. Her correlation held. Four to six weeks after the sea surface temperatures go up, so do cases of cholera!

Figure 48.41 (**a**) Satellite data on rising sea surface temperatures in the Bay of Bengal correlated with cholera cases in the region's hospitals. *Red* signifies warmest summer temperatures. (**b**) *Vibrio cholerae*, agent of cholera. Copepods host a dormant stage of this bacterium that waits out adverse environmental conditions that do not favor its growth and reproduction. (**c**) A typical Bangladesh waterway from which water samples were drawn for analysis.

(**d**) In Bangladesh, Rita Colwell comparing samples of unfiltered and filtered drinking water.

Today, Colwell and Anwarul Huq, a Bangladeshi scientist, are investigating salinity and other factors that may influence outbreaks. Their goal is to design a model for predicting where cholera will break out next. They have advised women in Bangladesh to use sari cloth as a filter to remove *V. cholerae* cells from the water (Figure 48.41d). Copepod hosts are too big to pass through the thin cloths, which can be rinsed in clean water, sun-dried, and used again and again. This simple and inexpensive method has cut cholera outbreaks by half.

Combine knowledge about life with knowledge about the physical and chemical aspects of the biosphere, and who knows what you may discover.

Summary

Sections 48.1, 48.2 Global air circulation patterns affect climate and the distribution of communities. The patterns start with latitudinal variations in incoming solar radiation. The basic patterns are influenced by Earth's daily rotation and annual path around the sun, the distribution of continents and seas, and elevations of land masses. Solar energy, and the winds it drives, are renewable, clean sources of energy.

Human activities alter the atmosphere. The use of CFCs and methyl bromide depletes ozone in the upper atmosphere. With seasonal ozone thinning, more UV radiation reaches Earth's surface.

Smog, a form of air pollution, arises in areas where large amounts of fossil fuels are burned. Coal-burning power plants are also the main contributors to acid rain, which alters habitats and kills many organisms.

Biology⊗Now
Learn how sunlight energy drives global patterns of air circulation with the interaction on BiologyNow.

Section 48.3 Latitudinal and seasonal variations in sunlight warm ocean water and set currents in motion. The currents distribute heat energy around the seas and affect weather patterns. Ocean currents, air currents, and topography interact to shape global climate zones.

Biology⊗Now
See the patterns of major ocean currents with the animation on BiologyNow.

Section 48.4 The world's land masses are realms of biodiversity, each with an evolutionary history and a tapestry of physical and chemical conditions. Biomes are vast expanses characterized by specific arrays of species, mainly plants and animals. Regional variations in climate, landforms, and soils influence them. Marine ecoregions are comparable realms of biodiversity.

Biology⊗Now
Examine the distribution of biomes with the animation on BiologyNow.

Section 48.5 Soil characteristics vary among biomes and help determine their primary productivity.

Sections 48.6–48.10 Deserts form near latitudes 30° north and south if annual rainfall is sparse. Slightly moister southern or western coastal regions support dry woodlands and shrublands. In the interior of midlatitude continents, vast deserts or grasslands form.

From the equator to latitudes 10° north and south, evergreen tropical forests grow in regions of high rainfall, high humidity, and mild temperatures. Semi-evergreen and deciduous broadleaf forests form between latitudes 10°–25°, depending on how much of the annual rainfall occurs in a prolonged dry season.

Where a cold, dry season alternates with a cold, rainy season, coniferous forests dominate.

Low-growing, hardy plants of the tundra dominate at high latitudes and high altitudes.

Section 48.11 Desertification, the conversion of marginally productive lands to desertlike conditions, is one of the major current threats to biodiversity.

Sections 48.12–48.14 Lakes, streams, and other aquatic ecosystems show gradients in penetration of sunlight, water temperature, salinity, and dissolved gases. These factors vary over time and affect primary productivity. Coastal zones and tropical reefs support diverse ecosystems. Primary productivity is high in coastal wetlands and on coral reefs.

Section 48.15 Life persists throughout the ocean. Diversity is highest in sunlit waters. Mineral-rich waters support communities at deep-sea hydrothermal vents. Upwelling is an upward movement of deep, cool, often nutrient-rich ocean water, typically along the coasts of continents. An El Niño event disrupts upwelling, and it triggers massive, reversible changes in rainfall as well as other weather patterns around the world.

Biology⊗Now
Learn about the oceanic zones with the animation on BiologyNow.

Section 48.16 Drawing on knowledge of microbial ecology as well as biogeographic patterns, Rita Colwell found a crucial bit of information that led to effective countermeasures against cholera outbreaks.

Biology⊗Now
Observe how an El Niño event affects ocean currents and upwelling with the animation on BiologyNow.

Self-Quiz   Answers in Appendix II

1. Solar radiation drives the distribution of weather systems and so influences _____ .
   a. temperature zones   c. seasonal variations
   b. rainfall distribution   d. all of the above

2. _____ shields life against the sun's UV wavelengths.
   a. A thermal inversion   c. The ozone layer
   b. Acid precipitation   d. The greenhouse effect

3. Regional variations in the global patterns of rainfall and temperature depend on _____ .
   a. global air circulation   c. topography
   b. ocean currents   d. all of the above

4. A rain shadow is a reduction in rainfall _____ .
   a. on the leeward side of a mountain range
   b. during an El Niño event
   c. that occurs seasonally in the tropics

5. Acid rain is one outcome of _____ .
   a. coal burning   c. nitrogen-rich fertilizers
   b. gas and oil burning   d. all of the above

6. Biomes are _____ .
   a. water provinces   d. partly characterized
   b. water and land zones   by dominant plants
   c. vast expanses of land   e. both c and d

7. Biome distribution depends on _____ .
   a. climate   c. soils
   b. topography   d. all of the above

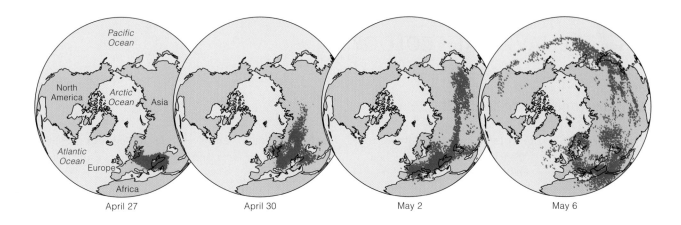

April 27   April 30   May 2   May 6

8. Grasslands most often predominate _____ .
   a. near the equator   c. in interior of continents
   b. at high altitudes   d. b and c

9. During _____ , deeper, often nutrient-rich water moves to the surface of a body of water.
   a. spring overturns   c. upwellings
   b. fall overturns   d. all of the above

10. Match the terms with the most suitable description.
   _____ tundra   a. equatorial broadleaf forest
   _____ chaparral   b. partly enclosed by land; where
   _____ desert   freshwater and seawater mix
   _____ savanna   c. type of grassland with trees
   _____ estuary   d. has low-growing plants at
   _____ boreal forest   high latitudes or elevations
   _____ tropical rain   e. at latitudes 30° north and south
         forest   f. mineral-rich, superheated water
   _____ hydrothermal   supports communities here
         vents   g. conifers dominate
                             h. dry shrubland

Additional questions are available on **Biology ⊛Now**™

Critical Thinking

1. On April 26, 1986, in Ukraine, a meltdown occurred at the Chernobyl nuclear power plant. Nuclear fuel burned for nearly ten days and released 400 times more radioactive material than the atomic bomb that dropped on Hiroshima. Winds carried radioactive fallout around the globe (Figure 48.42). Thirty-one people died right after the meltdown. Thousands more are still likely to die from cancers and other harmful effects of radiation.

The Chernobyl accident stiffened opposition to *nuclear power* in the United States, but recent developments have some people reconsidering. Increasing the use of nuclear energy would diminish the country's dependence on oil from the politically unstable Middle East. Nuclear power does not contribute to global warming, acid rain, or smog. It does produce highly radioactive wastes. Investigate the pros and cons of nuclear power, and decide if you think the environmental benefits outweigh the risks. Would you feel differently if a nuclear power plant were about to be built ten kilometers upwind from your home?

2. Use of off-road recreational vehicles may double over the next twenty years. Many off-road enthusiasts would like increased access to government-owned desert areas.

Figure 48.42 Global distribution of radioactive fallout after the 1986 meltdown of the Chernobyl nuclear power plant in Ukraine. The meltdown put 300 million to 400 million people at risk for leukemia and other radiation-induced disorders. By 1998, the rate of thyroid abnormalities in children living downwind from the site was nearly seven times as high as for those upwind; their thyroid gland concentrated the iodine radioisotopes.

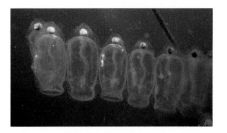

Figure 48.43 Chain of salps (*Thalia democratica*). Like you, salps have a nerve cord. The cord's anterior end develops into a rudimentary brain, with a light-sensitive eyespot.

Some argue that it's just the perfect place for off-roaders because "There's nothing there." Do you agree? If not, how would you counter this argument?

3. Write a short description of how global warming may affect spring overturn and thermocline formation in a Minnesota lake. What would be some ecological effects?

4. *Thalia democratica*, a salp, is one of our remote chordate relatives (Figure 48.43). This urochordate is part of marine plankton, usually in warm and temperate seas but also in cold, deep water. Salps swim, separately or in loose chains, in vast numbers. Salps range in size from 1.5 centimeters to a reported tubular specimen (*Pyrostremma*) that was 20 meters long and wide enough for a scuba diver to swim through. Refer to Section 25.1, on the feeding mode of urochordates. Then formulate a hypothesis on how the deep curtain of plastic debris in the ocean may adversely affect the vast salp populations and, through them, marine food webs.

5. *Southern pine forests* dominate the coastal plains of the southern Atlantic and Gulf states. Many pine species are adapted to the periodic, lightning-sparked fires, but fires are suppressed where buildings are encroaching on the forests. Suppression results in an accumulation of dry undergrowth that can fuel uncontrollable wildfires, such as the one shown at right. Do some research on whether and how Florida and other vulnerable states carry out controlled burns. How many acres are at risk?

My Pheromones Made Me Do It

A few years ago, as Toha Bererub walked down a street near her Las Vegas home, she felt a sharp pain above her right eye. Then another, and then another. Seconds later, hundreds of stinging bees covered the upper half of her body. Firefighters in protective gear rescued her, but not before she had been stung more than 500 times.

Bererub's tiny attackers were Africanized honeybees, a hybrid between the mild-tempered European honeybee and an African strain that is easy to provoke (Figure 49.1). Breeders had imported African bees to Brazil in the 1950s. They thought cross-breeding experiments would result in a mild-tempered but zippier pollinator for commercial orchards. However, some of the captive imports escaped and started mating with the locals.

Then, in a grand example of geographic dispersal, some descendant bees buzzed all the way from Brazil to Mexico and on into the United States. So far, the Africanized bees have established themselves in Texas, New Mexico, Nevada, California to the west, and Alabama, Virginia, and Florida to the east.

Honeybees sting only once. All species make the same kind of venom. Africanized bees make a bit less venom, but they get riled up faster and mount collective attacks. One squadron reportedly chased a perceived threat for a quarter of a mile.

The Africanized bees became known as "killer bees," although they rarely kill their target. Their stings are extremely painful, but adults in good health usually can survive a collective attack. Bererub was seventy years old when attacked, and she recovered fully after spending a week in the hospital.

What makes the Africanized bees so testy? Isopentyl acetate. This chemical, which smells like bananas, is a key component of honeybee alarm pheromone.

A **pheromone**, remember, is a chemical signal released by one individual that may cause another individual of the species to alter its behavior. A honeybee releases an alarm pheromone when it recognizes and stings a perceived threat. The signaling molecules diffuse through air and form a concentration gradient, which guides other bees to the individual sounding the alarm.

Researchers once studied hundreds of colonies of Africanized honeybees and European honeybees to compare their responses to alarm pheromone. They positioned a tiny target in front of each colony and then released a small quantity of an artificial pheromone. The Africanized bees flew out of the colony and zeroed in on the perceived threat much faster. They also plunged six to eight times as many stingers into it.

Figure 49.1 Good bee, bad bee. At left, a European honeybee about to pollinate a flower. *Facing page*, two of its aggressive relatives, Africanized honeybees, that stand guard at the entrance to their hive. If a potential intruder appears, they will release an alarm pheromone that stimulates hivemates to join an attack.

The two kinds of honeybees show other differences in their behavior. Compared to European bees, Africanized bees are less picky about where they establish a colony. They are more likely to abandon their colony after being disturbed. Of more concern to beekeepers, they are less interested in stashing large amounts of honey.

Such differences among honeybees lead us into the world of **animal behavior**, to the coordinated responses that animal species make to stimuli. We invite you to reflect on the genetic basis of behavior before turning to the instinctive and learned mechanisms that arise from it. Along the way, we also will consider the adaptive value of behavior.

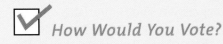

How Would You Vote?

Africanized bees are slowly expanding their range in North America. Some think the more we know about them, the better we will be able to protect ourselves. Should we fund more research into the genetic basis of their behavior? See BiologyNow for details, then vote online.

Key Concepts

FOUNDATIONS FOR BEHAVIOR

An individual's behavior starts with interactions among gene products, such as hormones and pheromones. Most behavior has innate components but can be modified by environmental factors.

Behavioral traits that have a heritable basis and that enhance the individual's reproductive success can evolve by natural selection. Sections 49.1–49.3

CUES FOR SOCIAL BEHAVIOR

Evolved modes of communication underlie social behavior. Communication signals hold clear meaning for both the sender and the receiver of signals. Section 49.4

COSTS AND BENEFITS OF BEHAVIOR

Life in social groups has reproductive benefits and costs. Not every environment favors the evolution of such groups. Self-sacrificing behavior has evolved among a few kinds of animals that live in large family groups. Sections 49.5–49.7

FOUNDATIONS FOR HUMAN SOCIAL BEHAVIOR

The social behavior of all primates, including humans, has evolved in complex ways. Only humans consistently make moral choices about their behavior. Sections 49.8, 49.9

Links to Earlier Concepts

This chapter builds on your understanding of the nervous system and sensory systems (Sections 34.1, 35.1). You will consider the functions of neurotransmitters, hormones, and pheromones in behavior (36.1, 36.2). You will revisit the social parasites (46.7).

Be sure you understand the concepts of directional selection (18.4), sexual selection (18.6), and adaptation (18.9). You will see how predation (46.4) can exert selection pressure on the evolution of behavior. You may wish to review the sections on the evolution of primates and modern humans (26.12, 26.15).

49.1 Behavior's Heritable Basis

LINKS TO
SECTIONS
34.1, 35.1, 36.1

The nervous and endocrine systems govern behavioral responses to stimuli. Because genes specify the substances required for constructing and operating those systems, they are the heritable foundation for animal behavior.

GENES AND BEHAVIOR

Before an animal is even born or hatched, the nervous system becomes prewired to detect, interpret, and then issue commands for response to stimuli. A **stimulus**, recall, is a piece of information about the external or internal environment that a specific type of sensory receptor has detected. It takes gene products to build and operate sensory receptors, nerves and, in most

Figure 49.2 (**a**) Banana slug, food for (**b**) an adult garter snake of coastal California. (**c**) Newborn garter snake from a coastal population, tongue-flicking at a cotton swab drenched with tissue fluids from a banana slug.

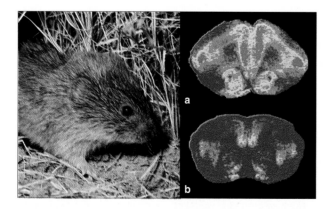

Figure 49.3 Distribution of oxytocin receptors (*red*) inside the brain of (**a**) a mate-for-life prairie vole and (**b**) a promiscuous mountain vole.

species, a brain. Gene products also affect behavioral responses to stimuli.

Stevan Arnold found experimental evidence of the genetic basis for behavior in the feeding preferences of coastal and inland snake populations in California. Garter snakes living near the coast hunt banana slugs (Figure 49.2*a*). Snakes living inland hunt tadpoles and fishes. Offer them a banana slug and they ignore it.

Arnold offered captive newborn snakes a bit of slug as a first meal. Newborn coastal snakes usually ate it and flicked their tongue at cotton swabs drenched in essence of slug. (Snakes "smell" by tongue-flicking, which pulls odors into the mouth.) Newborn inland snakes ignored the swabs and rarely ate bits of slug. Here was a big difference between captive snakes that had no prior experience with slugs. These snakes are programmed before hatching to accept or reject slugs; they did not learn feeding preferences by taste trials.

Did allelic differences influence how odor-detecting mechanisms form in the garter snake embryo? Arnold crossed coastal with inland snakes. He predicted that hybrid offspring would make an *intermediate* response to slug chunks and odors, and they did. Many hybrid baby snakes tongue-flicked at slug-perfumed swabs more often than newborn inland snakes did—but not as often as newborn coastal snakes did (Figure 49.2*c*).

Some genes have been linked to specific behaviors. The *fruitless* gene controls male fruit fly courtship. A female will not mate unless a male waves his wings and licks and taps her body. Male flies that researchers induced to make the female version of the fruitless protein became more attracted to males than females. Female flies induced to make the male version of the protein waved their wings at females. These altered flies had no interest in ordinary males, but they did court males who had been made to smell like females.

This gene product is a master switch in the nervous system, with far-reaching effects on complex behavior.

HORMONES AND BEHAVIOR

Some hormones are behavior-guiding gene products. For instance, all mammals make and secrete oxytocin, which affects more than labor and lactation (Section 44.14). In many species, it guides social behavior, such as pair bonding, aggression, and territoriality.

In prairie voles (*Microtus ochrogaster*), oxytocin is the hormonal key that unlocks the female's heart. The female of these small rodents bonds with a male after a night of repeated matings, and she mates for life. To test oxytocin's impact, researchers kept a female vole with a male for a few hours but blocked mating. They

Figure 49.4 Instinctive behavior of the European cuckoo. (**a**) This social parasite lays eggs in the nests of other birds. Even before a cuckoo hatchling opens its eyes, it reacts instinctually to anything round—typically the host bird's egg—and shoves it from the nest. (**b**) The clueless foster parents instinctually respond to a gaping mouth, not to a usurper's size or other traits that differ from their own species.

Figure 49.5 Instinctive behavior of a human baby who is imitating an adult's facial expression.

injected oxytocin into the female, and the pair bonded without the normally required sex act. By contrast, pair-bonded female prairie voles that were injected with an oxytocin blocker immediately dumped their former partners.

Whether a vole species is monogamous depends on the number and distribution of receptors for oxytocin. Monogamous prairie voles have more receptors than highly promiscuous mountain voles (Figure 49.3).

Monogamous voles also have more receptors in the brain for antidiuretic hormone (ADH). As you know, kidney cells have receptors for this hormone, but so do cells in the brain. Researchers isolated the gene for an ADH receptor in monogamous prairie voles. They transferred copies of the gene into some forebrain cells of male meadow voles (*M. pennsylvanicus*). Afterward, the males of this more promiscuous species showed an increased tendency to partner with one female only.

Male meadow voles used as a control group also got copies of the gene, but in a brain region not known to be involved in pair-bonding. Unlike the experimental group, the males retained their promiscuous ways.

INSTINCTIVE BEHAVIOR

Like many other animals, slug-loving garter snakes, wing-waving fruit flies, and pair-bonding voles offer us evidence of **instinctive behavior**—they perform a behavior without having first learned it through actual experience in the environment. They are prewired to recognize sign stimuli before being born or hatched. *Sign stimuli* are one or two simple, well-defined cues that trigger a suitable response. For garter snakes, the cue is a specific slug scent that calls for a *fixed action pattern*—a stereotyped motor program of coordinated muscle activity that runs to completion independently of feedback from the environment. The baby snake is compelled to strike, capture, and eat a slug.

Or consider cuckoos, a type of social parasite. Like cowbirds, the females lay eggs in nests of other birds.

A newly hatched cuckoo is blind; skin covers its eyes. But contact with an egg (or any round object) triggers a fixed action pattern. That hatchling maneuvers the egg onto its back, then pushes it from the nest (Figure 49.4*a*). Its behavior helps the hatchling get undivided attention. Its "foster parents" are oblivious to the odd color and size of the usurper. They respond only to one sign stimulus—the gaping mouth of a chick—and continue with their parenting (Figure 49.4*b*).

Humans, too, display instinctive behavior. Three days after birth, a human infant already displays a capacity to mimic facial expressions of an adult who comes close to it (Figure 49.5). The infant cannot see its own face, nor can it feel which facial muscles the adult is using. Somehow it is able to open its mouth, protrude its tongue, or rotate its head the same way as the adult. Infants will also respond to a simplified stimulus—a flat, face-sized mask with two dark spots for eyes. One "eye" won't do the trick.

Behavior, or coordinated responses to stimuli, starts with genes. Some gene products construct and operate the nervous system, which governs behavior. Other products, such as hormones, help control the mechanisms required for specific forms of behavior.

Animals start out life neurally wired to recognize vital cues and to make an instinctively suitable response, one that has not been learned through actual experience.

Many animals execute a fixed action pattern, a stereotyped program of coordinated muscle activity in response to one or two simple, well-defined environmental cues.

49.2 Learned Behavior

*With **learned behavior**, an individual draws from past experiences and varies or changes its response to stimuli. A classic example, imprinting, occurs early in life, during a genetically determined period.*

Animals process information about experiences and then use it to change or adjust responses to stimuli. Learned behavior arises as the environment directly or indirectly influences gene expression. Sensory input and good or bad nutrition are typical factors that lead to alterations in how and what an animal learns.

Birdsong (Figure 49.6) is an instinctive behavior. Even so, songbirds can learn variations, or dialects, of the species song in different habitats. As Peter Marler demonstrated, many male birds learn the full song ten to fifty days after hatching by listening to other birds sing it. The male nervous system is prewired to recognize the species song; a learning mechanism is primed to select and respond to acoustical input. But what the male bird hears during a sensitive period shapes his *rendition* of the song.

In one study, Marler raised white-crown nestlings to maturity in soundproof chambers so they could not hear adult males. Their songs did not have the exact structure of a typical adult song. Marler also isolated captive nestlings and let them hear recorded songs of white-crown sparrows *and* song sparrows. When the captives matured, they sang just the white-crown song and mimicked the species dialect of the unseen tutor.

In another experiment, Marler did not use taped songs. He let young, hand-reared male white-crowns interact with a "social tutor" of the different species. The males tended to learn the tutor's song.

Results from many such experiments support this hypothesis: Birdsong starts with the genetically based capacity to learn from acoustical cues.

Imprinting is a classic case of learned behavior. This time-dependent form of learning is triggered by exposure to a sign stimulus. Exposure normally takes place during a sensitive period when the animal is young. Imprinting of baby geese on their mother is a favorite example of animal behaviorists (Figure 49.7).

Animals process information about experiences in the environment and use it to change or vary their response to a stimulus. This learned behavior involves interactions between gene products and environmental inputs.

Figure 49.6 Male marsh wren belting out a territorial song that has been likened to a loud gurgle. Males of this species start to imitate their species song when they are about fifteen days old. Unlike many species, marsh wrens continue to learn songs throughout their life.

Figure 49.7 No one can tell these imprinted baby geese that Konrad Lorenz is not Mother Goose!

(**a**) In response to a moving object and probably acoustical cues, baby geese imprint on the mother goose and follow her during a short, sensitive period right after hatching. They are neurally wired to learn crucial information—the identity of the one individual that will be most likely to protect them in the months ahead. Usually that will be their mother.

(**b**) Konrad Lorenz, one of the early investigators of animal behavior, presented these baby geese with sign stimuli that made them form an attachment to him.

The Adaptive Value of Behavior

If forms of behavior have a genetic basis, then they may evolve in various ways through natural selection. Alleles that encode the most adaptive versions of a trait tend to to increase in frequency in a population, and alternative alleles do not. In time, genetic changes in behavior that yield greater reproductive success are favored.

Natural selection theory helps us develop and test explanations of why some behavior persists and how it offers reproductive benefits that offset reproductive costs (disadvantages) associated with it. If a behavior is adaptive, it promotes the *individual's* production of offspring. Here are five definitions to keep in mind:

1. **Reproductive success.** An individual reproduces, and at least some offspring survive.

2. **Adaptive behavior.** A form of behavior that helps perpetuate the individual's genes. Its frequency in a population is maintained or increases over time.

3. **Social behavior.** Behavior expressed in the context of interactions among individuals of the same species.

4. **Selfish behavior.** Form of behavior that improves an individual's chance to produce or protect its own offspring regardless of the impact on the population.

5. **Altruism.** Self-sacrificing behavior. An individual behaves in a way that helps others in the population but reduces its own chance of producing offspring.

When biologists speak of selfish or altruistic behavior, they do not mean that the individual is consciously aware of some behavior or its reproductive goal. A hungry lion does not have to know that eating zebras is good for reproductive success. Its nervous system simply calls for HUNTING BEHAVIOR! when that lion sees a zebra. Hunting behavior persists in lion populations because genes for neural mechanisms that command hunting behavior are persisting.

To assess the adaptive value of any behavior, look for how it might promote reproductive success. For example, starlings (*Sturnus vulgaris*) nest in cavities of trees and decorate the nest bowl with sprigs of fresh leaves of pungent plants, such as wild carrot (*Daucus carota*). Larry Clark and Russell Mason hypothesized that nest decorating behavior suppresses populations of mites that infest nests and parasitize birds. Even a few mites produce thousands of descendants. In large numbers, mites suck enough blood from a nestling to weaken it and affect its growth and survival.

Clark and Mason tested their hypothesis with a set of experimental nests, some with fresh-cut wild carrot leaves and some without. They removed natural nests that starlings were using. Half of the nesting starling

Figure 49.8 Experimental test of the adaptive value of starling nest-decorating behavior (**a**). Nests designated *A* did not have fresh sprigs of wild carrot (**b**) and other plants that make aromatic compounds. Nests designated *B* had fresh sprigs added every seven days. (**c**) Twenty-one days after the experiment started, the chicks left and researchers made counts of the mites (*Ornithonyssus sylviarum*) infesting each nest. The test results supported the hypothesis that aromatic compounds suppress development of juvenile mites into (**d**) adult mites.

pairs got new nests with wild carrot sprigs. Swapped nests for the other pairs were sprigless. Figure 49.8*c* gives the results. Sprig-free nests had more mites than sprig-festooned nests. At the end of one experiment, sprig-free nests teemed with an average of 750,000 mites. Nests with sprigs had 8,000 mites. Why? Wild carrot sprigs contain an aromatic steroid compound that repels herbivores and helps plants survive. By coincidence, it prevents mites from maturing sexually.

Selection theory continues to guide experiments on this behavior. For instance, other researchers did not yield the same evidence that greenery reduces mite populations. Rather, their test results indicate that the decorating behavior may be adaptive either because it deters different kinds of parasites or because it boosts the immune function of the nestlings.

A genetically determined behavior may persist or increase in frequency in a population when it is adaptive. A behavior is adaptive when it increases the number of descendants that an individual successfully produces.

49.4 Communication Signals

LINKS TO
SECTIONS
18.6. 35.1, 36.1

Competing for food, defending territory, alerting others to danger, advertising sexual readiness, forming bonds with a mate, caring for the offspring—such intraspecific behaviors require unambiguous forms of communication.

THE NATURE OF COMMUNICATION SIGNALS

Communication signals are unambiguous cues sent and received among individuals of a species, and they involve instinctive and learned forms of behavior. Information-laden cues from a signaler are meant to change the behavior of receivers. Chemical, acoustical, and visual cues are among the most common.

Pheromones, again, are communication signals. The *signaling* pheromones induce the receiver to respond fast. They include chemical alarms, such as honeybee calls to action against potential threats. They include sex attractants. Bombykol is one of them. Bombykol molecules released by a female silk moth can attract males that are kilometers away. *Priming* pheromones bring about physiological (not behavioral) responses. As one example, a volatile odor in the urine of certain male mice can trigger and enhance estrus in female mice of the same species.

Acoustical signals are common. Male birds, frogs, grasshoppers, whales, and many other animals make sounds that attract females. Prairie dogs bark alarms. Wolves howl and kangaroo rats drum their feet on the ground when advertising possession of territory.

Some signals never vary. Zebra ears pressed flat to the head convey hostility; ears pointing up convey its absence. Different signals convey the intensity of the message. A zebra with ears laid back is not too riled up as long as its mouth is open only a bit. When the ears are laid back and its mouth gapes, watch out. That combination is a type of *composite* signal. Such signals have information encoded in two or more cues.

Signals often take on different meaning in different contexts. A lion emits a spine-tingling roar to keep in touch with its pride *or* to threaten rivals. Also, a signal can convey information about signals to follow. Dogs and wolves solicit play behavior with their play bow, as in Figure 49.9a. Without the bow, the signal receiver may construe the behaviors that follow as aggressive, sexual, or even exploratory—but not playful.

Signals evolve or persist in a population when they promote reproductive success of both the sender *and* receiver. If a signal is harmful, natural selection will favor individuals that don't send it or respond to it.

COMMUNICATION DISPLAYS

The play bow is a *communication* display, a pattern of behavior that is a social signal. The *threat* display is another common pattern. It announces that a signaler is prepared to attack a signal receiver. If a rival for a receptive female confronts a dominant male baboon, the dominant animal will roll his eyes upward and "yawn," which exposes sharp canines (Figure 49.9b). The signaler can benefit when the rival backs down, because he can control access to the female without having to fight. The signal receiver benefits because he can avoid a serious beating, infected wounds, and possibly death.

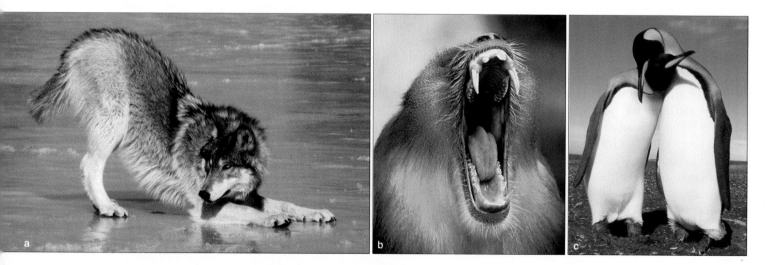

Figure 49.9 Communication displays. (**a**) Play bow of a young male wolf soliciting a romp. (**b**) Part of a male baboon's threat display: exposed canines. (**c**) Courtship display of Adelé penguins.

Figure 49.10 *Animated!* Honeybee dances, a classic example of a tactile display. (**a**) Honeybees that have visited a source of food close to their hive return and perform a *round* dance on the hive's honeycomb. Worker bees that maintain contact with the foraging bee throughout the dance will fly out and search for food near the hive.

(**b**) A bee that visits a feeding station more than 100 meters distant from the hive performs a *waggle* dance. During the dance, it makes a straight run and waggles its abdomen. A waggle dancer also varies the dance speed to convey more information about distance to a food source. For example, when food is 150 meters away, a bee dances much faster, and with more waggles per straight run, compared to a dance about a food source that is 500 meters away.

(**c**) As Karl von Frisch discovered, a *straight* run's orientation varies, depending on the direction in which a food source is located. He put one dish of honey on a direct line between a hive and the sun. Foragers that located it returned to the hive and oriented their straight runs right up the honeycomb. He put another dish of honey at right angles to a line between the hive and the sun. Foraging bees made their straight runs 90 degrees to vertical on the honeycomb. Thus, a honeybee "recruited" into foraging orients its flight *with respect to the sun and the hive.* By doing so, it wastes less time and energy during its food-gathering expedition.

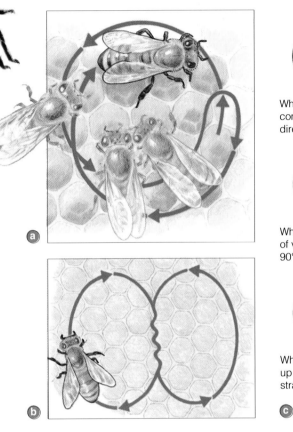

When bee moves straight down comb, recruits fly to source directly away from the sun.

When bee moves to right of vertical, recruits fly at 90° angle to right of the sun.

When bee moves straight up comb, recruits fly straight toward the sun.

Such displays are ritualized, with intended changes in the function of common behavior patterns. Normal movements might be exaggerated or frozen. Feathers, manes, claws, and other body parts are often notably enlarged, patterned, and colored. Ritualization is well developed in *courtship* displays—the steps that must precede pair formation. Courtship displays are well developed among birds (Section 18.6 and Figure 49.9c).

With *tactile* displays, a signaler touches the receiver in ritualized ways. After locating a source of pollen or nectar, a foraging honeybee returns to its colony (the hive) and performs a complex dance. It moves in a defined pattern, jostling a crowd of workers that stay in close physical contact with it. Its signals give other bees information about the general location, distance, and direction of a food source (Figure 49.10).

ILLEGITIMATE SIGNALERS AND RECEIVERS

Unintended recipients can intercept communication signals. Male tungara frogs make two kinds of calls—one simple, the other complex. The calls mean "come on over" to female frogs, but they mean "dinner over here" to fringe-lipped bats. Complex calls are more inviting to females but make it easier for bats to find the caller. When bats are near, male frogs vocalize less and are more likely to make the simpler call.

There are illegitimate signalers, too. Some assassin bugs can borrow the scent of their prey—termites— by hooking a dead termite on their back. By signaling that they "belong" to a termite colony, they can more easily hunt termites. As another example, if a female of certain predatory firefly species sees a flash from a male of a different species, she will flash back. If she lures him into attack range, she will capture and eat him. Getting eaten is an evolutionary cost of having an otherwise useful response to a come-hither signal.

A communication signal transfers information from one individual to another individual of the same species. Such signals benefit both the signaler and the receiver.

Some individuals of a different species act illegitimately as a communication signaler or receiver.

49.5 Mates, Offspring, and Reproductive Success

LINKS TO
SECTIONS
18.6, 26.10

For reasons we need not explore here, many people find mating and parenting behaviors of animals fascinating. How useful is selection theory in helping us interpret such behavior? Take a look.

SEXUAL SELECTION AND MATING BEHAVIOR

Competition among members of one sex for access to mates is common. So is choosiness in selecting a mate. Such activities, recall, are forms of **sexual selection**. This microevolutionary process favors traits that give the individual a competitive advantage in attracting and often holding on to mates (Section 18.6).

But *whose* reproductive success is it—the male's or the female's? Male animals, remember, produce many tiny sperm, and females produce far larger but fewer eggs. For the male, success generally depends on how many eggs he can fertilize. For the female, it depends more on how many eggs she produces or how many offspring she can raise. Usually, the most important factor in a female's sexual preference is the quality of the mate, not the quantity of partners.

Female hangingflies (*Harpobittacus apicalis*) provide an instructive example. They choose males that offer superior food. A male hunts and kills a moth or some other insect. Then he releases a sex pheromone, which attracts females to him and his "nuptial gift" (Figure 49.11*a*). A female tends to select the male that offers a large calorie-rich gift. Only after the female has been eating the gift for five minutes or so does she start to accept sperm from her partner. She lets the male continue inseminating her—but only for as long as it takes for her to devour the gift.

Before twenty minutes are up, a female hangingfly can break off the mating at any point. If she does, she might well mate with a different male hangingfly and accept his sperm. Doing so dilutes the reproductive success of her first partner.

Females of different species shop around for males with the best burrows. Consider the fiddler crabs that live along muddy shores from Massachusetts down to Florida. In males, one of the two claws is enlarged. Some of those claws are enlarged enough to make up more than half the body weight (Figure 49.11*b*). When spring tides are favorable, the male crabs build their elaborate mating burrows in the same area. Each male stands beside his burrow, waving his oversized claw. Females stroll by, checking out details of the burrows. When a female likes what she sees, she will follow the male into his burrow and engage in sex.

Many female birds are choosy. Male sage grouse (*Centrocercus urophasianus*) converge at a **lek**, a type of communal display ground. Each male stakes out a few square meters. With tail feathers erect, males use their large, puffed-out neck pouches to emit booming calls (Figure 49.11*d*). As they do, they stamp about on

Figure 49.11 (**a**) Male hangingfly dangling a moth as a nuptial gift for a potential mate. Females of some hangingfly species choose sexual partners that offer the largest gift to them. By waving his enlarged claw, a male fiddler crab (**b**) may attract the eye of a female fiddler crab (**c**). A male sage grouse (**d**) showing off as he competes for female attention at a communal display ground.

their patch of prairie, a bit like wind-up toys. Females tend to select and mate with one male sage grouse. Afterward, they go off to nest and raise the young by themselves. Many females often select the same male, so most of the males never do mate.

In another behavioral pattern, sexually receptive females of some species cluster in defendable groups. Where you come across such a group, you are likely to observe males competing for access to clusters. The competition for ready-made harems has resulted in combative male lions, sheep, elk, elephant seals, and bison, to name a few types of animals (Figure 49.12).

PARENTAL CARE

When females fight for males, then we can expect that the males provide more than sperm delivery. Some help with parenting. Midwife toads are an example. A male wraps strings of fertilized eggs around his legs until the eggs hatch (Figure 49.13a). With her eggs being cared for, a female can mate with other males, if she can find some that are not already caring for eggs. Late in the breeding season, unencumbered males are rare, and female toads fight for access to them. The females even attempt to pry mating pairs apart.

Parental behavior uses up time and energy, which parents otherwise might spend on living long enough to reproduce again. However, for many species, the benefit of immediate reproductive success outweighs the cost of parenting. Reproductive success might be more chancy later on.

For amphibians and reptiles, parenting is rare once the young are hatched. Crocodilians are an exception. Crocodilian parents construct a nest, as birds do. Their young call out when they are ready to hatch. Parents dig up the young and care for them for some time.

Most birds are monogamous, and both parents often care for the young (Figure 49.13b). In mammals, males typically leave after mating. Females raise the young alone, and males attempt to mate again or conserve energy for the next breeding season (Figure 49.13c). Mammalian species in which males do help care for the young tend to be monogamous. About 5 percent of all mammals fall into this category.

> *Researchers use selection theory to explain some aspects of mating behavior.*
>
> *Male or female preferences for certain behavioral traits can provide the individual with a competitive edge and promote its reproductive success.*

Figure 49.12 Male bison locked in combat during the breeding season.

Figure 49.13 (**a**) Male midwife toad with developing eggs wrapped around his legs. (**b**) Male and female Caspian terns cooperate in the care of their chick. (**c**) A female grizzly will care for her cub for as long as two years. The male takes no part in its upbringing.

49.6 Costs and Benefits of Social Groups

LINKS TO
SECTIONS
45.1, 46.4

*Survey the animal kingdom and you will find a range
of social groups, with evolutionary costs and benefits.*

COOPERATIVE PREDATOR AVOIDANCE

Cooperative responses to predators help some groups reduce the net risk to all. Vulnerable individuals, too, can be on the alert for predators, join a counterattack, or engage in more effective defenses (Figure 49.14).

Vervet monkeys, meerkats, prairie dogs, and many other mammals cooperate with their alarm calls, as in Figure 49.14a. A prairie dog makes a particular bark when it sights an eagle and a different signal when it sights a coyote. Others dive into burrows to escape an eagle's attack or they stand erect, the better to scan the horizon and zero in on the threat.

Ecologist Birgitta Sillén-Tullberg observed group benefits for Australian sawfly caterpillars that live in clumps on branches (Figure 49.14b). When disturbed, individuals collectively rear back, writhe, and vomit partly digested eucalyptus leaves, which are toxic to songbirds and other animals that prey on them.

As Sillén-Tullberg hypothesized, individual sawfly caterpillars benefit from their coordinated repulsion of predatory birds. She used her hypothesis to predict that birds are more likely to eat a lone caterpillar. She tested her prediction with young hand-reared birds. Birds that were offered one caterpillar at a time ate an average of 5.6 caterpillars. Birds that were offered a clump of caterpillars ate an average of 4.1. Individuals were safer in a group, as predicted.

THE SELFISH HERD

Simply by their physical position in the group, some individuals form a living shield against predation on others. They belong to a **selfish herd**, a simple society that benefits their reproductive self-interest. Selfish-herd behavior has been studied in bluegill sunfishes. Male sunfishes build adjacent nests on the bottom of a lake. Then females deposit their eggs where males have used their fins to scoop out depressions in mud.

If a colony of bluegill males is a selfish herd, then we can predict competition for the "safe" sites—at the center of a colony. Compared to eggs at the periphery, eggs in nests at the center are less likely to be eaten by snails and largemouth bass. Competition does indeed occur. The largest, most powerful males tend to claim centermost locations. Other, smaller males assemble around them and bear the brunt of predatory attacks. Even so, they are better off in the group than on their own, fending off a bass single-handedly, so to speak.

COOPERATIVE HUNTING

Many predatory mammals, including wolves, lions, and wild dogs, live in social groups and cooperate in hunts (Figure 49.15). Are group hunts more successful than solitary hunts? Often they are not. Researchers observed a solitary lion that captured prey about 15 percent of the time. Two lions hunting together did capture prey twice as often, but they had to share it, so the number of successful hunts per lion balanced

Figure 49.14 Group defenses. (**a**) Black-tailed prairie dogs bark an alarm call that warns others of predators. Does this put the caller at risk? Not much. Prairie dogs usually act as sentries only if they are done feeding and are standing beside their burrows. (**b**) Australian sawfly caterpillars form clumps and collectively regurgitate a fluid (the yellow blobs) that is toxic to most predators. (**c**) Musk oxen adults (*Ovibos moshatus*) form a ring of horns, often around the young.

Figure 49.15 Members of a wolf pack (*Canis lupus*). Wolves cooperate in hunting, caring for the young, and defending a territory. Benefits are not distributed equally. Only the highest ranking individuals, the alpha male and alpha female, breed.

out. When more lions joined the hunt, the success rate per lion fell. Wolves show a similar pattern. Among many cooperative hunters, hunting success in itself might not explain group living. Individuals do hunt together, but they also may fend off scavengers, care for one another's young, and protect territory.

DOMINANCE HIERARCHIES

Many social groups share resources unequally among some individuals that are subordinate to others. Most wolf packs, for instance, have one dominant male that breeds with just one dominant female. Other wolves in the pack are nonbreeding brothers and sisters, or aunts and uncles. They all hunt and bring food to the individuals that guard the young in their den.

Baboons live in large troops. A female stays with the group into which she was born and inherits social standing from her mother. Dominant females get more food, water, and grooming. Their young grow and mature faster than those of lower-ranking females.

Why would a subordinate give up resources and, often, breeding privileges? It might get injured or die if it challenges a strong individual. It might not be able to survive on its own. A subordinate might even get a chance to reproduce if it lives long enough or if its dominant peers are taken out by a predator or old age. Some subordinate wolves and baboons do move up the social ladder when the opportunity arises.

REGARDING THE COSTS

If social behavior is advantageous, *then why are there so few social species?* In most habitats, the costs outweigh benefits. For instance, packed-together individuals do compete more for a share of resources (Section 45.1). Cormorants, puffins, and many other seabirds form dense breeding colonies, as in Figure 49.16. All must compete for a share of the same ecological pie.

Figure 49.16 Nearly uniform spacing in a crowded cormorant colony.

Large social groups also attract more predators. If individuals are crowded together, they invite parasites and contagious diseases that jump from host to host. The individuals may also be at risk of being killed or exploited by others. Given the opportunity, breeding pairs of herring gulls cannibalize a neighbor's eggs and any chicks that wander away from their nest.

> Living in a social group can provide benefits, as through cooperative defenses or shielding against predators.
>
> Group living has costs, in terms of increased competition, increased vulnerability to infections, and exploitation by others of the group.

49.7 Why Sacrifice Yourself?

LINK TO
CHAPTER 46
INTRODUCTION

Extreme cases of sterility and self-sacrifice have evolved in only two groups of insects and one group of mammals. How are genes of the nonreproducers perpetuated?

SOCIAL INSECTS

Honeybees and fire ants (Chapter 46) are among the true social (eusocial) insects. Like termites, they stay together for generations in a group that has a division of labor. Many permanently sterile individuals care cooperatively for the offspring of just a few breeding individuals. Often they are highly specialized in form and function (Figure 49.17).

Consider a honeybee hive. The only fertile female, a queen, secretes a pheromone that other female bees distribute through the hive. This signaling molecule suppresses the development of ovaries in all the other females, which makes them sterile. The queen bee is larger than worker bees partly because of her enlarged egg-producing ovaries (Figure 49.18a).

About 30,000 to 50,000 female workers feed larvae, clean and maintain the hive, and build honeycomb from waxy secretions. Adult worker bees live for about six weeks in the spring and summer. When foragers return to the hive after finding a rich source of nectar or pollen, they engage others in a dance. This tactile display recruits more foragers (Figure 49.10). Workers also cooperate through the transfer of food from one to another. They guard the entrance to the hive and will sacrifice themselves to repel intruders.

Males, the stingless drones, develop only in spring and in summer. They have no part in the day-to-day work and subsist on food gathered by their worker sisters. Drones live for sex. Each day, they fly out in search of a mate. If one is lucky, he will find a virgin queen on her single flight away from her colony. The sole function of her flight is to meet up with and mate with a drone. A drone dies right after he inseminates a virgin queen, which then founds a new colony. She will store and use his sperm for years, perpetuating his genes and those of his original colony.

Like honeybees, termites live in enormous family groups with a queen specialized for producing eggs (Figure 49.18c). Unlike a honeybee hive, each termite colony holds sterile individuals of both sexes. A king supplies the female with sperm. Winged reproductive termites of both sexes develop seasonally.

SOCIAL MOLE-RATS

Vertebrates are not known for sterility and extreme self-sacrifice. The only eusocial mammals are African mole-rats. The best studied is *Heterocephalus glaber*, the naked mole-rat. Clans of this nearly hairless rodent build and occupy burrows in arid parts of East Africa.

A reproducing female dominates the clan, and she mates with one to three males (Figure 49.18b). Other, nonbreeding members live just to protect and care for the "queen" and "king" (or kings) and their offspring. The sterile diggers excavate subterranean tunnels and chambers that are living rooms or dumps for wastes. When a digger comes across a tasty tuber or root, it hauls some back to the main chamber, where it emits a series of chirps. Its chirps recruit others, which help carry the tuber back to the chamber. In this way, the queen, her retinue of males, and her offspring get fed. Digger mole-rats also deliver food to other helpers that seem to loaf about, shoulder to shoulder and belly to back, with the reproductive royals. These "loafers" actually spring to action when a snake or some other enemy threatens the clan. Collectively, and at great risk, they chase away or attack and kill the predator.

Figure 49.17 Specialized ways of serving and defending the colony. (**a**) An Australian honeypot ant worker. This sterile female is a living container for her colony's food reserves. (**b**) Army ant soldier (*Eciton burchelli*) with formidable mandibles. (**c**) Eyeless soldier termite (*Nasutitermes*). It bombards intruders with a stream of sticky goo from its nozzle-shaped head.

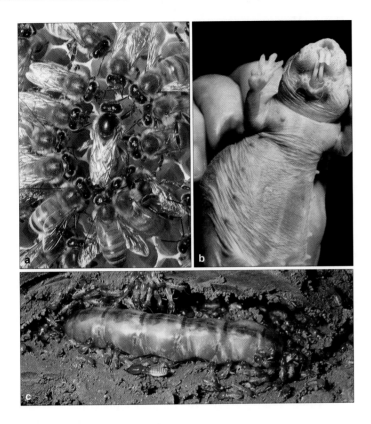

Figure 49.18 Three queens. (**a**) A queen honeybee with her court of sterile worker daughters. (**b**) This queen naked mole-rat has twelve mammary glands, the better to feed her many offspring. In a laboratory colony at Cornell University, one female produced a litter of twenty-eight pups. She gave birth to more than 900 offspring during her lifetime. (**c**) A termite queen (*Macrotermes*) dwarfs her offspring and her mate. Her body pumps out thousands of eggs a day.

Figure 49.19 Damaraland mole-rats in a burrow. Like their relatives, the naked mole-rats, they live in colonies having nonbreeding workers. Unlike naked mole-rats, these fuzzy burrowers are not highly inbred.

INDIRECT SELECTION FOR ALTRUISM

None of the altruistic individuals of a honeybee hive, termite colony, or naked mole-rat clan directly passes genes to the next generation. So how are genes that underlie altruistic behavior perpetuated? According to William Hamilton's theory of **inclusive fitness**, genes associated with altruism can be favored by selection if they lead to behavior that will increase the number of offspring produced by an altruist's closest relatives.

A sexually reproducing, diploid parent caring for offspring is not helping exact genetic copies of itself. Each of its gametes, and each of its offspring, inherits one-half of its genes. Other individuals of the social group that have the same ancestors also share genes with their parents. Two siblings (brothers or sisters) are as genetically similar as a parent and its offspring. Nephews and nieces share about one-fourth of their uncle's genes.

Sterile workers may be indirectly promoting genes for "self-sacrifice" through altruistic behavior that will benefit their close relatives. All of the individuals in honeybee, termite, and ant colonies are members of a great extended family. Nonbreeding family members support siblings, a few of which are future kings and queens. Although a guard bee dies after driving her stinger into a bear, siblings in the hive will perpetuate some of her genes.

Does close kinship explain why naked mole-rats are the only eusocial mammals? DNA fingerprinting studies of one naked mole-rat clan revealed that all of the individuals are *very* close relatives and genetically different from individuals of other clans. Each clan is highly inbred after many generations of brother–sister, mother–son, and father–daughter matings.

However, inbreeding might not even be necessary for mole-rat eusociality. The social organization of the Damaraland mole-rat (*Cryptomys damarensis*) resembles that of *H. glaber* (Figure 49.19). Nonbreeding members of both sexes cooperatively assist one breeding pair. Even so, breeding pairs of wild Damaraland mole-rat colonies usually are unrelated.

Researchers are now searching for other factors that select for eusocial behavior in mole-rats. According to one hypothesis, arid habitats and patchy food sources favor mole-rat genes that give rise to cooperation in digging burrows, searching for food, and fending off competitors of other species for resources.

Altruistic behavior may persist when individuals pass on genes indirectly, by helping relatives survive and reproduce.

By the theory of inclusive fitness, genes associated with altruistic behavior that is directed toward relatives may spread through a population in certain situations.

49.8 A Look at Primate Social Behavior

LINKS TO
SECTIONS
26.12, 26.15

Primates, especially chimpanzees and bonobos, live in groups. Their social environment is a significant factor in determining an individual's reproductive success.

In the 1960s Jane Goodall, a young primatologist, set out on her lifelong study of chimpanzees in Tanzania. One of her earliest discoveries was the chimpanzee's capacity to make and use simple tools—"fishing sticks" —by stripping leaves from branches. The long, flexible sticks are inserted into a termite mound, as shown in Figure 49.20a, which agitates the termites. The stick is carefully withdrawn after termites swarm on it, and the chimpanzee gets a high-protein snack. Thicker sticks are used to make holes in the mound, then the fishing sticks are inserted into the holes.

Different chimpanzee groups use slightly different tool-shaping and termite-fishing methods. Youngsters of each group learn by imitating the adults.

Male chimpanzees spend their lives in the group in which they are born and form strong social bonds. The females are often unrelated and interact little with one another. A female's status is dictated mostly by how she gets along with the males. Before the rainy season, mature females that are entering their fertile cycle go through hormone-driven physiological and behavioral changes. Their external genitalia become swollen and vivid pink. The swellings are strong visual signals to males. They are flags for sexual jamborees—for great gatherings of highly stimulated chimpanzees in which any males present may have a turn at copulating with the same female.

Male chimpanzees cooperatively hunt for monkeys, small pigs, and antelopes. They may also cooperate in attacks on neighboring groups. Males sometimes even kill infants. By one hypothesis, infanticidal behavior of males may exert selection pressure for promiscuity in females. A female who mates with many males might protect her offspring by obscuring their paternity. A male would be expected to avoid killing an infant that might carry his genes.

Comparative studies of the closely related bonobo reveal contrasting sexual and social behavior (Figure 49.20b). As with chimpanzees, adult males are related and females are not. Yet bonobo females form strong social bonds. Unlike female chimpanzees, they can be receptive to sex at any time, not just during the fertile cycle. Male bonobos display less social cohesion than male chimpanzees do. They do not hunt together, and no one has come across an infanticidal male.

What explains the differences? Does a higher level of interaction help female bonobos deter potentially infanticidal males? Does unlimited access to sexually receptive females interfere with male–male bonding or diffuse male aggression? We do not know. Hormones that affect pair bonding may play a role. Like prairie voles and mountain voles, chimpanzees and bonobos differ in a regulatory region near a gene that encodes one ADH receptor. In voles, a longer sequence in this region correlates with more family-oriented behavior. Interestingly, the bonobo sequence for this region is about 360 bases longer than the chimpanzee sequence. What about humans? Our sequence in this region is nearly identical to that of the bonobos—and with this in mind, we turn briefly to human behavior.

Chimpanzees and bonobos both live in social groups. They differ in the details of social organization, degree of female cooperation, and extent of male aggression.

Figure 49.20 (**a**) Chimpanzees (*Pan troglodytes*) using sticks as tools for extracting tasty termites from a nest.

(**b**) Female bonobo (*Pan paniscus*) with her offspring. Like humans, bonobos are bipedal; they often walk on two legs. Also like humans, and unlike chimpanzees, the bonobo females have sexual organs that allow them to copulate facing their partner, and they can be sexually receptive at any time of year. They use sex as a means of strengthening social bonds.

49.9 An Evolutionary View of Human Social Behavior

Evolutionary forces shaped animal behavior—but humans alone consistently make moral choices about their behavior.

LINK TO
CHAPTER 43
INTRODUCTION

EXAMPLES OF BEHAVIORAL CUES

Is it possible that molecular cues help humans form social attachments, as they do in other animals? Think of how oxytocin and ADH help control pair bonding in voles. Now think about *autism*. Someone affected by this behavioral disorder cannot enter normal social relationships. An autistic child has significantly low levels of oxytocin in blood. Also, a control sequence near the gene for one ADH receptor is shorter than normal. This same sequence is shorter in chimpanzees than in bonobos and humans.

Researchers are studying how hormones influence mother–infant bonding and romantic bonds. Nursing stimulates oxytocin secretion. So does orgasm, even a friendly massage. Does the brief increase in oxytocin or other hormonal responses contribute to what we perceive as love? That is an open question.

Also, human pheromones may be present in sweat or other secretions. When females live in proximity, as they do in college dormitories, their menstrual cycles typically become synchronized. Martha McClintock and Kathleen Stern demonstrated that one woman's menstrual cycle will lengthen or shorten after she has become exposed to sweat secreted by a woman who was in a different phase of the cycle.

Men and women secrete different chemicals and respond differently to them. PET scans reveal that one chemical component in male sweat activates certain brain areas in women but not in most men. Similarly, a chemical component of female urine activates brain areas in most males more than it does in females. Intriguingly, male homosexuals show the same brain response to male sweat as women do.

In most mammals, pheromones bind to receptors in a vomeronasal organ, or VNO. Neurons connect it to parts of the brain that control behavior. In humans, the VNO is a tiny, ductlike structure on the septum, a tissue that divides the nose into two nostrils. Many scientists hypothesize that human VNOs are vestigial structures—no longer functional. Others suspect that the human VNO does connect to the brain, by way of some pathway that has not yet been discovered.

EVOLUTIONARY QUESTIONS

If we are comfortable with studying the evolutionary basis of the behavior of termites, naked mole-rats, and other animals, why do so many people resist the idea of analyzing human behavior in the same way? Often they fear that attempts to identify the adaptive value of some human trait will be used to define its morality. However, there is a clear difference between trying to explain behavior in terms of its evolutionary history and attempting to justify it. To a biologist, "adaptive" does not mean "morally right." It simply means useful in perpetuating an individual's genes.

An example: Infanticide is morally repugnant. Is it unnatural? No. It happens in many animal groups and all human cultures. Male lions often kill the offspring of other males when they take over a pride. Doing so frees up the lionesses to breed with them, which can increase the infanticidal male's reproductive success.

Biologists may predict that unrelated human males are a threat to infants, and evidence supports this. The absence of a biological father and the presence of an unrelated male increases risk of death for an American child under age two by seventy times.

What about parents who kill their own offspring? In her book on maternal behavior, primatologist Sarah Blaffer Hrdy cites a study of a village in Papua New Guinea in which about 40 percent of the newborns were killed by parents. She argues that when resources or social support are hard to come by, a mother might increase her fitness by killing a newborn. She can then allocate child-rearing energy to her other offspring or save it for children she may have in the future.

Do most of us find such behavior appalling? Yes. Does such behavior warrant attention? Think about all you have learned in this book, then decide.

A behavior that might be adaptive in the evolutionary sense may still be judged by society to be morally wrong.

Summary

Section 49.1 Animal behavior starts with genes that specify products required for development of the nervous, endocrine, and muscular systems. Hormones are among the gene products that affect behavior.

Instinctive behavior is performed without having been learned by experience in the environment. It is a prewired response to one or two simple, well-defined environmental cues.

Section 49.2 An animal learns when it processes and integrates information from experiences, then uses that information to vary or change responses to stimuli. Imprinting is one form of learning that happens only during a sensitive period early in life.

Section 49.3 A behavior that has a genetic basis is subject to evolution by natural selection. Adaptive forms of behavior evolved as a result of individual differences in reproductive success in past generations. A behavior persists when its reproductive benefits exceed the reproductive costs.

Section 49.4 Communication signals are meant to change the behavior of individuals of the same species. Pheromones are signaling molecules that have roles in social communication.

Visual signals are key components of courtship displays and threat displays. Acoustical signals are sounds that have precise, species-specific information. Tactile signals are specific forms of physical contact between a signaler and a receiver.

Biology ⓔ Now
Explore the honeybee dance language with the animation on BiologyNow.

Section 49.5 Sexual selection favors traits that give an individual a competitive edge in attracting and often holding on to mates. Females of many species select for males that have traits or engage in behaviors they find attractive. When large numbers of females cluster in defensible areas, males may compete with one another to control the areas.

Parental care has reproductive costs in terms of future reproduction and survival. It is adaptive when benefits to a present set of offspring offset the costs.

Biology ⓔ Now
Read the InfoTrac article "Something Fishy in the Nest," Bryan Neff, Natural History, February 2004.

Section 49.6 Animals that live in social groups may benefit by cooperating in predator detection, defense, and rearing the young. Benefits of group living are often distributed unequally. Species that live in large groups incur costs, including increased disease and parasitism, and increased competition for resources.

Biology ⓔ Now
Read the InfoTrac article "Caterpillars as Social Insects," James Costa, American Scientist, March–April 1997.

Section 49.7 Ants, termites, and some other insects as well as two species of mole-rats are eusocial. They live in colonies with overlapping generations and have a reproductive division of labor. Most colony members do not reproduce; they assist their relatives and rear their offspring.

According to the theory of inclusive fitness, such extreme altruism is perpetuated because altruistic individuals have some number of genes in common with their reproducing relatives. Altruistic individuals in the social group pass on "by proxy" the genes that underlie this behavior.

Sections 49.8, 49.9 Researchers are identifying the mechanisms and adaptive significance of primate social behavior. With respect to humans, a behavior that is adaptive in the evolutionary sense may still be judged by society to be morally wrong.

Self-Quiz

Answers in Appendix II

1. Genes affect the behavior of individuals by _____ .
   a. influencing the development of nervous systems
   b. affecting the kinds of hormones in individuals
   c. governing development of muscles and skeletons
   d. all of the above

2. A behavior is defined as adaptive if it _____ .
   a. varies among individuals of a population
   b. occurs without prior learning
   c. increases an individual's reproductive success
   d. is widespread across a species
   e. benefits unrelated members of the species

3. Steven Arnold offered slug meat to newborn garter snakes from different populations to test his hypothesis that the snakes' response to slugs _____ .
   a. was shaped by indirect selection
   b. is an instinctive behavior
   c. is based on pheromones
   d. is adaptive

4. Generally, living in a social group costs the individual, in terms of _____ .
   a. competition for food, other resources
   b. vulnerability to contagious diseases
   c. competition for mates
   d. all of the above

5. Social behavior evolves because _____ .
   a. social animals are more advanced than solitary ones
   b. under some conditions, the costs of social life to individuals are offset by benefits to the species
   c. under some conditions, the benefits of social life to an individual offset the costs to that individual
   d. under most conditions, social life has no costs to an individual.

6. Eusocial insects _____ .
   a. live in extended family groups
   b. are found among almost all insect orders
   c. show a reproductive division of labor
   d. a and c
   e. all of the above

7. Helping other individuals at a reproductive cost to oneself might be adaptive if those helped are _____ .
   a. members of another species
   b. competitors for mates
   c. close relatives
   d. illegitimate signalers

8. Match the terms with their most suitable description.
   ____ fixed action pattern
   ____ altruism
   ____ basis of instinctive and learned behavior
   ____ imprinting
   ____ pheromone

   a. time-dependent form of learning requiring exposure to key stimulus
   b. genes plus actual experience
   c. stereotyped motor program that runs to completion independently of feedback from environment
   d. assisting another individual at one's own expense
   e. one communication signal

Additional questions are available on **Biology ⓔ Now™**

Figure 49.21 Behaviorally confused rooster.

Critical Thinking

1. Sexual imprinting is common in birds. During a short sensitive period in early life, the bird learns features that it will seek later, when ready to mate. Figure 49.21 shows an amorous rooster wading into the water after ducks. Speculate on what might have caused this behavior.

2. Nazca boobies (*Sula granti*) lay two eggs, several days apart. No matter how much food is available, only one chick survives to adulthood (Figure 49.22). The first chick to hatch pushes its younger sibling from the nest, and that sibling dies of starvation and neglect. Formulate a hypothesis on how it might be adaptive for parents of this species to lay two eggs if one of the hatchlings tends to kill the other. Design an experiment to test your hypothesis.

3. In 2002 Svante Paabo proposed how one gene, *FOXP2*, might have been pivotal in the evolution of *language*. Humans who have a base-pair substitution in this gene cannot speak intelligibly, understand complex sentences, or make certain movements of the mouth and face.

   All mammals have the *FOXP2* gene, which has a 715 base-pair sequence. The gene has mutated very little over evolutionary time. The one in chimpanzees differs from the one in mice by a single base pair. But two more base pairs mutated after the ancestors of humans diverged from the lineage that led to chimpanzees. This altered version of the gene became fixed in the lineage that led to modern humans. Why? By one hypothesis, language-related traits that arose from the two recent substitutions were favored by directional selection. Speculate on how a capacity to make and comprehend more complex auditory signals shaped the social behavior of the forerunners of humans.

4. A cheetah scent-marks plants in its territory with certain exocrine gland secretions. What evidence would you require to demonstrate that the cheetah's action is an evolved communication signal?

5. Among primates, differences in sexual behavior tend to be related to the size of a male's gonads. Gorillas have relatively tiny testicles. In a 450-pound male, they may weigh about an ounce. Gorillas live in groups consisting of a male, a few females, and offspring. This is the most

Figure 49.22 A Nazca booby attends to its single surviving chick.

typical kind of primate social group. When a female is ready to mate, there usually is only one adult male around to inseminate her.

In contrast, a female chimpanzee advertises her fertile period and mates with many males (Section 49.8). A 100-pound chimpanzee male has testicles about four times as weighty as a gorilla's. By making far more sperm, a male chimpanzee increases the odds that his sperm, not a rival's, will fertilize a female's egg.

An adult human male is larger than a chimpanzee, but his testicles are only about half the weight. What might this suggest about female promiscuity and male competition to fertilize eggs in the lineage that led to humans?

6. In moths and many other insects, potential mates find one another with the help of species-specific pheromones. The pheromones are usually mixes of chemicals derived from fatty acids. Explain how a mutation could result in a change in the mix of chemicals in a moth pheromone. How might such mutations encourage speciation?

Epilogue

BIOLOGICAL PRINCIPLES AND THE HUMAN IMPERATIVE

Molecules, single cells, tissues, organs, organ systems, multicelled organisms, populations, communities, ecosystems, and the biosphere. These are architectural systems of life, assembled in increasingly complex ways over the past 3.8 billion years. We are latecomers to this immense biological building program. And yet, within the relatively short span of 10,000 years, many of our activities have been changing the character of the land, ocean, and atmosphere, even the genetic character of species.

It would be presumptuous to think that we alone have had profound impact on the world of life. As long ago as the Proterozoic, photosynthetic organisms were irrevocably changing the course of biological evolution by enriching the atmosphere with oxygen. During the past as well as the present, competitive adaptations led to the rise of some groups, whose dominance assured the decline of others. Change is nothing new. What *is* new is the capacity of one species to comprehend what might be going on.

We now have the population size, technology, and cultural inclination to use up energy and modify the environment at rapid rates. Where will this end? Will feedback controls operate as they do, for instance, when population growth exceeds carrying capacity? In other words, will negative feedback controls come into play and keep things from getting too far out of hand?

Feedback control will not be enough, for it does not get under way until the deviation has reached a critical

threshold. Our patterns of resource consumption and our population growth are founded on an illusion of unlimited resources and a forgiving environment. A prolonged, global shortage of food or the passing of a critical threshold for the global climate can come too fast to be corrected; in which case the impact of the deviation may be too great to be reversed.

What about feedforward mechanisms, which might serve as early warning systems? For example, when sensory receptors near the surface of skin detect a drop in outside air temperature, each sends messages to the nervous system. That system responds by triggering mechanisms that raise the body's core temperature before the body itself becomes dangerously chilled. Extrapolating from this, if we develop feedforward control mechanisms, would it not be possible to start corrective measures before we do too much harm?

Feedforward controls alone will not work, for they operate after change is under way. Think of the DEW line—the Distant Early Warning system. It is like a vast sensory receptor for detecting missiles launched against North America. By the time it does what it is supposed to, it may be too late to stop widespread destruction.

It would be naive to assume we can ever reverse who we are at this point in evolutionary time, to de-evolve ourselves culturally and biologically into becoming less complex in the hope of averting disaster. Yet there is reason to believe we can avert disaster by using a third kind of control mechanism—a capacity to anticipate events even before they happen. We are not locked into responding only after irreversible change has begun. We have the capacity to anticipate the future—it is the essence of our visions of utopia and hell. *We all have the capacity to adapt to a future that we can partly shape.*

For instance, we can stop trying to "beat nature" and learn to work with it. Individually and collectively, we can work to develop long-term policies that take into account biotic and abiotic limits on population growth. Far from being a surrender, this would be one of the most intelligent behaviors of which we are capable.

Having a capacity to adapt and using it are not the same thing. We have already put the world of life on dangerous ground because we have not yet mobilized ourselves as a species to work toward self-control.

Our survival depends on predicting possible futures. It depends on preserving, restoring, and constructing ecosystems that fit with our definition of basic human values and available biological models. Human values can change; our expectations can and must be adapted to biological reality. *For the principles of energy flow and resource utilization, which govern the survival of all systems of life, do not change.*

It is our biological and cultural imperative that we come to terms with these principles, and ask ourselves this: What will be our long-term contribution to the world of life?

Appendix I. Classification System

This revised classification scheme is a composite of several that microbiologists, botanists, and zoologists use. The major groupings are agreed upon, more or less. However, there is not always agreement on what to name a particular grouping or where it might fit within the overall hierarchy. There are several reasons why full consensus is not possible at this time.

First, the fossil record varies in its completeness and quality. Therefore, the phylogenetic relationship of one group to other groups is sometimes open to interpretation. Today, comparative studies at the molecular level are firming up the picture, but the work is still under way. Also, molecular comparisons do not always provide definitive answers to questions about phylogeny. Comparisons based on one set of genes may conflict with those comparing a different part of the genome. Or comparisons with one member of a group may conflict with comparisons based on other group members.

Second, ever since the time of Linnaeus, systems of classification have been based on the perceived morphological similarities and differences among organisms. Although some original interpretations are now open to question, we are so used to thinking about organisms in certain ways that reclassification often proceeds slowly.

A few examples: Traditionally, birds and reptiles were grouped in separate classes (Reptilia and Aves); yet there are compelling arguments for grouping the lizards and snakes in one group and the crocodilians, dinosaurs, and birds in another. Many biologists still favor a six-kingdom system of classification (archaea, bacteria, protists, plants, fungi, and animals). Others advocate a switch to the more recently proposed three-domain system (archaea, bacteria, and eukarya).

Third, researchers in microbiology, mycology, botany, zoology, and other fields of inquiry inherited a wealth of literature, based on classification systems that have been developed over time in each field of inquiry. Many are reluctant to give up established terminology that offers access to the past.

For example, botanists and microbiologists often use *division*, and zoologists *phylum*, for taxa that are equivalent in hierarchies of classification.

Why bother with classification frameworks if we know they only imperfectly reflect the evolutionary history of life? We do so for the same reasons that a writer might break up a history of civilization into several volumes, each with a number of chapters. Both are efforts to impart structure to an enormous body of knowledge and to facilitate retrieval of information from it. More importantly, to the extent that modern classification schemes accurately reflect evolutionary relationships, they provide the basis for comparative biological studies, which link all fields of biology.

Bear in mind that we include this appendix for your reference purposes only. Besides being open to revision, it is not meant to be complete. Names shown in "quotes" are polyphyletic or paraphyletic groups that are undergoing revision. For example, "reptiles" comprise at least three and possibly more lineages.

The most recently discovered species, as from the mid-ocean province, are not listed. Many existing and extinct species of the more obscure phyla are also not represented. Our strategy is to focus primarily on the organisms mentioned in the text or familiar to most students. We delve more deeply into flowering plants than into bryophytes, and into chordates than annelids.

PROKARYOTES AND EUKARYOTES COMPARED

As a general frame of reference, note that almost all bacteria and archaea are microscopic in size. Their DNA is concentrated in a nucleoid (a region of cytoplasm), not in a membrane-bound nucleus. All are single cells or simple associations of cells. They reproduce by prokaryotic fission or budding; they transfer genes by bacterial conjugation.

Table A lists representative types of autotrophic and heterotrophic prokaryotes. The authoritative reference, *Bergey's Manual of Systematic Bacteriology*, has called this a time of taxonomic transition. It references groups mainly by numerical taxonomy (Section 19.1) rather than by phylogeny. Our classification system does reflect evidence of evolutionary relationships for at least some bacterial groups.

The first life forms were prokaryotic. Similarities between Bacteria and Archaea have more ancient origins relative to the traits of eukaryotes.

Unlike the prokaryotes, all eukaryotic cells start out life with a DNA-enclosing nucleus and other membrane-bound organelles. Their chromosomes have many histones and other proteins attached. They include spectacularly diverse single-celled and multicelled species, which can reproduce by way of meiosis, mitosis, or both.

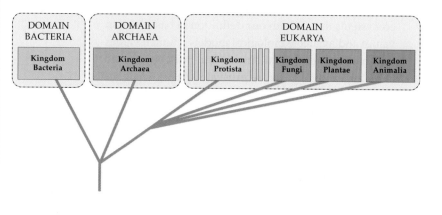

DOMAIN BACTERIA — Kingdom Bacteria
DOMAIN ARCHAEA — Kingdom Archaea
DOMAIN EUKARYA — Kingdom Protista, Kingdom Fungi, Kingdom Plantae, Kingdom Animalia

DOMAIN OF BACTERIA

KINGDOM BACTERIA The largest, and most diverse group of prokaryotic cells. Includes photosynthetic autotrophs, chemosynthetic autotrophs, and heterotrophs. All prokaryotic pathogens of vertebrates are bacteria.

PHYLUM AQIFACAE Most ancient branch of the bacterial tree. Gram-negative, mostly aerobic chemoautotrophs, mainly of volcanic hot springs. *Aquifex.*

PHYLUM DEINOCOCCUS-THERMUS Gram-positive, heat-loving chemoautotrophs. *Deinococcus* is the most radiation resistant organism known. *Thermus* occurs in hot springs and near hydrothermal vents.

PHYLUM CHLOROFLEXI Green nonsulfur bacteria. Gram-negative bacteria of hot springs, freshwater lakes, and marine habitats. Act as nonoxygen-producing photoautotrophs or aerobic chemoheterotrophs. *Chloroflexus.*

PHYLUM ACTINOBACTERIA Gram-positive, mostly aerobic heterotrophs in soil, freshwater and marine habitats, and on mammalian skin. *Propionibacterium, Actinomyces, Streptomyces.*

PHYLUM CYANOBACTERIA Gram-negative, oxygen-releasing photoautotrophs mainly in aquatic habitats. They have chlorophyll *a* and photosystem I. Includes many nitrogen-fixing genera. *Anabaena, Nostoc, Oscillatoria.*

PHYLUM CHLOROBIUM Green sulfur bacteria. Gram-negative nonoxygen-producing photosynthesizers, mainly in freshwater sediments. *Chlorobium.*

PHYLUM FIRMICUTES Gram-positive walled cells and the cell wall-less mycoplasmas. All are heterotrophs. Some survive in soil, hot springs, lakes, or oceans. Others live on or in animals. *Bacillus, Clostridium, Heliobacterium, Lactobacillus, Listeria, Mycobacterium, Mycoplasma, Streptococcus.*

PHYLUM CHLAMYDIAE Gram-negative intracellular parasites of birds and mammals. *Chlamydia.*

PHYLUM SPIROCHETES Free-living, parasitic, and mutualistic gram-negative spring-shaped bacteria. *Borelia, Pillotina, Spirillum, Treponema.*

PHYLUM PROTEOBACTERIA The largest bacterial group. Includes photoautotrophs, chemoautotrophs, and heterotrophs; free-living, parasitic, and colonial groups. All are gram-negative.

Class Alphaproteobacteria. *Agrobacterium, Azospirillum, Nitrobacter, Rickettsia, Rhizobium.*

Class Betaproteobacteria. *Neisseria.*

Class Gammaproteobacteria. *Chromatium, Escherichia, Haemopilius, Pseudomonas, Salmonella, Shigella, Thiomargarita, Vibrio, Yersinia.*

Class Deltaproteobacteria. *Azotobacter, Myxococcus.*

Class Epsilonproteobacteria. *Campylobacter, Helicobacter.*

DOMAIN OF ARCHAEA

KINGDOM ARCHAEA Prokaryotes that are evolutionarily between eukaryotic cells and the bacteria. Most are anaerobes. None are photosynthetic. Originally discovered in extreme habitats, they are now known to be widely dispersed. Compared with bacteria, the archaea have a distinctive cell wall structure and unique membrane lipids, ribosomes, and RNA sequences. Some are symbiotic with animals, but none are known to be animal pathogens.

PHYLUM EURYARCHAEOTA Largest archean group. Includes extreme thermophiles, halophiles, and methanogens. Others are abundant in the upper waters of the ocean and other more moderate habitats. *Methanocaldococcus, Nanoarchaeum.*

PHYLUM CRENARCHAEOTA Includes extreme theromophiles, as well as species that survive in Antarctic waters, and in more moderate habitats. *Sulfolobus, Ignicoccus.*

PHYLUM KORARCHAEOTA Known only from DNA isolated from hydrothermal pools. As of this writing, none have been cultured and no species have been named.

DOMAIN OF EUKARYOTES

KINGDOM "PROTISTA" A collection of single-celled and multicelled lineages, which does not constitute a monophyletic group. Some biologists consider the groups listed below to be kingdoms in their own right.

PARABASALIA Parabasalids. Flagellated, single-celled anaerobic heterotrophs with a cytoskeletal "backbone" that runs the length of the cell. There are no mitochondria, but a hydrogenosome serves a similar function. *Trichomonas, Trichonympha.*

DIPLOMONADIDA Diplomonads. Flagellated, anaerobic single-celled heterotrophs that do not have mitochondria or Golgi bodies and do not form a bipolar spindle at mitosis. May be one of the most ancient lineages. *Giardia.*

EUGLENOZOA Euglenoids and kinetoplastids. Free-living and parasitic flagellates. All with one or more mitochondria. Some photosynthetic euglenoids with chloroplasts, others heterotrophic. *Euglena, Trypanosoma, Leishmania.*

RHIZARIA Formaminiferans and radiolarians. Free-living, heterotrophic amoeboid cells that are enclosed in shells. Most live in ocean waters or sediments. *Pterocorys, Stylosphaera.*

ALVEOLATA Single cells having a unique array of membrane-bound sacs (alveoli) just beneath the plasma membrane.

Ciliata. Ciliated protozoans. Heterotrophic protists with many cilia. *Paramecium, Didinium.*

Dinoflagellates. Diverse heterotrophic and photosynthetic flagellated cells that deposit cellulose in their alveoli. *Gonyaulax, Gymnodinium, Karenia, Noctiluca.*

Apicomplexans. Single-celled parasites of animals. A unique microtubular device is used to attach to and penetrate a host cell. *Plasmodium.*

STRAMENOPHILA Stramenophiles. Single-celled and multicelled forms; flagella with tinsel-like filaments.

Oomycotes. Water molds. Heterotrophs. Decomposers, some parasites. *Saprolegnia, Phytophthora, Plasmopara.*

Chrysophytes. Golden algae, yellow-green algae, diatoms, coccolithophores. Photosynthetic. *Emiliania, Mischococcus.*

Phaeophytes. Brown algae. Photosynthetic; nearly all live in temperate marine waters. All are multicellular. *Macrocystis, Laminaria, Sargassum, Postelsia.*

RHODOPHYTA Red algae. Mostly photosynthetic, some parasitic. Nearly all marine, some in freshwater habitats. Most multicellular. *Porphyra, Antithamion.*

CHLOROPHYTA Green algae. Mostly photosynthetic, some parasitic. Most freshwater, some marine or terrestrial. Single-celled, colonial, and multicellular forms. Some biologists place the chlorophytes and charophytes with the land plants in a kingdom called the Viridiplantae. *Acetabularia, Chlamydomonas, Chlorella, Codium, Udotea, Ulva, Volvox.*

CHAROPHYTA Photosynthetic. Closest living relatives of plants. Include both single-celled and multicelled forms. Desmids, stoneworts. *Micrasterias, Chara, Spirogyra.*

AMOEBOZOA True amoebas and slime molds. Heterotrophs that spend all or part of the life cycle as a single cell that uses pseudopods to capture food. *Amoeba, Entoamoeba* (amoebas), *Dictyostelium* (cellular slime mold), *Physarum* (plasmodial slime mold).

KINGDOM FUNGI

Nearly all multicelled eukaryotic species with chitin-containing cell walls. Heterotrophs, mostly saprobic decomposers, some parasites. Nutrition based upon extracellular digestion of organic matter and absorption of nutrients by individual cells. Multicelled species form absorptive mycelia and reproductive structures that produce asexual spores (and sometimes sexual spores).

PHYLUM CHYTRIDIOMYCOTA Chytrids. Primarily aquatic; saprobic decomposers or parasites that produce flagellated spores. *Chytridium.*

PHYLUM ZYGOMYCOTA Zygomycetes. Producers of zygospores (zygotes inside thick wall) by way of sexual reproduction. Bread molds, related forms. *Rhizopus, Philobolus.*

PHYLUM ASCOMYCOTA Ascomycetes. Sac fungi. Sac-shaped cells form sexual spores (ascospores). Most yeasts and molds, morels, truffles. *Saccharomyces, Morchella, Neurospora, Claviceps, Candida, Aspergillus, Penicillium.*

PHYLUM BASIDIOMYCOTA Basidiomycetes. Club fungi. Most diverse group. Produce basidiospores inside club-shaped structures. Mushrooms, shelf fungi, stinkhorns. *Agaricus, Amanita, Craterellus, Gymnophilus, Puccinia, Ustilago.*

"IMPERFECT FUNGI" Sexual spores absent or undetected. The group has no formal taxonomic status. If better understood, a given species might be grouped with sac fungi or club fungi. *Arthobotrys, Histoplasma, Microsporum, Verticillium.*

"LICHENS" Mutualistic interactions between fungal species and a cyanobacterium, green alga, or both. *Lobaria, Usnea.*

KINGDOM PLANTAE

Most photosynthetic with chlorophylls *a* and *b*. Some parasitic. Nearly all live on land. Sexual reproduction predominates.

BRYOPHYTES (NONVASCULAR PLANTS)

Small flattened haploid gametophyte dominates the life cycle; sporophyte remains attached to it. Sperm are flagellated; require water to swim to eggs for fertilization.

PHYLUM HEPATOPHYTA Liverworts. *Marchantia.*

PHYLUM ANTHOCEROPHYTA Hornworts.

PHYLUM BRYOPHYTA Mosses. *Polytrichum, Sphagnum.*

SEEDLESS VASCULAR PLANTS

Diploid sporophyte dominates, free-living gametophytes, flagellated sperm require water for fertilization.

PHYLUM LYCOPHYTA Lycophytes, club mosses. Small single-veined leaves, branching rhizomes. *Lycopodium, Selaginella.*

PHYLUM MONILOPHYTA

Subphylum Psilophyta. Whisk ferns. No obvious roots or leaves on sporophyte, very reduced. *Psilotum.*

Subphylum Sphenophyta. Horsetails. Reduced scalelike leaves. Some stems photosynthetic, others spore-producing. *Calamites* (extinct), *Equisetum.*

Subphylum Pterophyta. Ferns. Large leaves, usually with sori. Largest group of seedless vascular plants (12,000 species), mainly tropical, temperate habitats. *Pteris, Trichomanes, Cyathea* (tree ferns), *Polystichum.*

SEED-BEARING VASCULAR PLANTS

PHYLUM CYCADOPHYTA Cycads. Group of gymnosperms (vascular, bear "naked" seeds). Tropical, subtropical. Compound leaves, simple cones on male and female plants. Plants usually palm-like. Motile sperm. *Zamia, Cycas.*

PHYLUM GINKGOPHYTA Ginkgo (maidenhair tree). Type of gymnosperm. Motile sperm. Seeds with fleshy layer. *Ginkgo.*

PHYLUM GNETOPHYTA Gnetophytes. Only gymnosperms with vessels in xylem and double fertilization (but endosperm does not form). *Ephedra, Welwitchia, Gnetum.*

PHYLUM CONIFEROPHYTA Conifers. Most common and familiar gymnosperms. Generally cone-bearing species with needle-like or scale-like leaves. Includes pines (*Pinus*), redwoods (*Sequoia*), yews (*Taxus*).

PHYLUM ANTHOPHYTA Angiosperms (the flowering plants). Largest, most diverse group of vascular seed-bearing plants. Only organisms that produce flowers, fruits. Some families from several representative orders are listed:

BASAL FAMILIES

Family Amborellaceae. *Amborella.*
Family Nymphaeaceae. Water lilies.
Family Illiciaceae. Star anise.

MAGNOLIIDS

Family Magnoliaceae. Magnolias.
Family Lauraceae. Cinnamon, sassafras, avocados.
Family Piperaceae. Black pepper, white pepper.

EUDICOTS

Family Papaveraceae. Poppies.
Family Cactaceae. Cacti.
Family Euphorbiaceae. Spurges, poinsettia.
Family Salicaceae. Willows, poplars.
Family Fabaceae. Peas, beans, lupines, mesquite.
Family Rosaceae. Roses, apples, almonds, strawberries.
Family Moraceae. Figs, mulberries.
Family Cucurbitaceae. Squashes, melons, cucumbers.
Family Fagaceae. Oaks, chestnuts, beeches.
Family Brassicaceae. Mustards, cabbages, radishes.
Family Malvaceae. Mallows, okra, cotton, hibiscus, cocoa.
Family Sapindaceae. Soapberry, litchi, maples.
Family Ericaceae. Heaths, blueberries, azaleas.
Family Rubiaceae. Coffee.
Family Lamiaceae. Mints.
Family Solanaceae. Potatoes, eggplant, petunias.
Family Apiaceae. Parsleys, carrots, poison hemlock.
Family Asteraceae. Composites. Chrysanthemums, sunflowers, lettuces, dandelions.

MONOCOTS

Family Araceae. Anthuriums, calla lily, philodendrons.
Family Liliaceae. Lilies, tulips.
Family Alliaceae. Onions, garlic.
Family Iridaceae. Irises, gladioli, crocuses.
Family Orchidaceae. Orchids.
Family Arecaceae. Date palms, coconut palms.
Family Bromeliaceae. Bromeliads, pineapples.
Family Cyperaceae. Sedges.
Family Poaceae. Grasses, bamboos, corn, wheat, sugarcane.
Family Zingiberaceae. Gingers.

KINGDOM ANIMALIA

Multicelled heterotrophs, nearly all with tissues and organs, and organ systems, that are motile during part of the life cycle. Sexual reproduction occurs in most, but some also reproduce asexually. Embryos develop through a series of stages.

PHYLUM PORIFERA Sponges. No symmetry, tissues.

PHYLUM PLACOZOA Marine. Simplest known animal. Two cell layers, no mouth, no organs. *Trichoplax.*

PHYLUM CNIDARIA Radial symmetry, tissues, nematocysts.
Class Hydrozoa. Hydrozoans. *Hydra, Obelia, Physalia, Prya.*
Class Scyphozoa. Jellyfishes. *Aurelia.*
Class Anthozoa. Sea anemones, corals. *Telesto.*

PHYLUM PLATYHELMINTHES Flatworms. Bilateral, cephalized; simplest animals with organ systems. Saclike gut.

Class Turbellaria. Triclads (planarians), polyclads. *Dugesia.*
Class Trematoda. Flukes. *Clonorchis, Schistosoma.*
Class Cestoda. Tapeworms. *Diphyllobothrium, Taenia.*

PHYLUM ROTIFERA Rotifers. *Asplancha, Philodina.*

PHYLUM MOLLUSCA Mollusks.

Class Polyplacophora. Chitons. *Cryptochiton, Tonicella.*

Class Gastropoda. Snails, sea slugs, land slugs. *Aplysia, Ariolimax, Cypraea, Haliotis, Helix, Liguus, Limax, Littorina.*

Class Bivalvia. Clams, mussels, scallops, cockles, oysters, shipworms. *Ensis, Chlamys, Mytelus, Patinopectin.*

Class Cephalopoda. Squids, octopuses, cuttlefish, nautiluses. *Dosidiscus, Loligo, Nautilus, Octopus, Sepia.*

PHYLUM ANNELIDA Segmented worms.

Class Polychaeta. Mostly marine worms. *Eunice, Neanthes.*

Class Oligochaeta. Mostly freshwater and terrestrial worms, many marine. *Lumbricus* (earthworms), *Tubifex.*

Class Hirudinea. Leeches. *Hirudo, Placobdella.*

PHYLUM NEMATODA Roundworms. *Ascaris, Caenorhabditis elegans, Necator* (hookworms), *Trichinella.*

PHYLUM ARTHROPODA

Subphylum Chelicerata. Chelicerates. Horseshoe crabs, spiders, scorpions, ticks, mites.

Subphylum Crustacea. Shrimps, crayfishes, lobsters, crabs, barnacles, copepods, isopods (sowbugs).

Subphylum Myriapoda. Centipedes, millipedes.

Subphylum Hexapoda. Insects and sprintails.

PHYLUM ECHINODERMATA Echinoderms.

Class Asteroidea. Sea stars. *Asterias.*
Class Ophiuroidea. Brittle stars.
Class Echinoidea. Sea urchins, heart urchins, sand dollars.
Class Holothuroidea. Sea cucumbers.
Class Crinoidea. Feather stars, sea lilies.
Class Concentricycloidea. Sea daisies.

PHYLUM CHORDATA Chordates.

Subphylum Urochordata. Tunicates, related forms.
Subphylum Cephalochordata. Lancelets.

CRANIATES

Class Myxini. Hagfishes.

VERTEBRATES (SUBGROUP OF CRANIATES)

Class Cephalaspidomorphi. Lampreys.

Class Chondrichthyes. Cartilaginous fishes (sharks, rays, skates, chimaeras).

Class "Osteichthyes." Bony fishes. Not monophyletic (sturgeons, paddlefish, herrings, carps, cods, trout, seahorses, tunas, lungfishes, and coelocanths).

TETRAPODS (SUBGROUP OF VERTEBRATES)

Class Amphibia. Amphibians. Require water to reproduce.
Order Caudata. Salamanders and newts.
Order Anura. Frogs, toads.
Order Apoda. Apodans (caecilians).

AMNIOTES (SUBGROUP OF TETRAPODS)

Class "Reptilia." Skin with scales, embryo protected and nutritionally supported by extraembryonic membranes.

Subclass Anapsida. Turtles, tortoises.
Subclass Lepidosaura. *Sphenodon*, lizards, snakes.
Subclass Archosaura. Crocodiles, alligators.

Class Aves. Birds. In some classifications birds are grouped in the archosaurs.

Order Struthioniformes. Ostriches.
Order Sphenisciformes. Penguins.
Order Procellariiformes. Albatrosses, petrels.
Order Ciconiiformes. Herons, bitterns, storks, flamingoes.
Order Anseriformes. Swans, geese, ducks.
Order Falconiformes. Eagles, hawks, vultures, falcons.
Order Galliformes. Ptarmigan, turkeys, domestic fowl.
Order Columbiformes. Pigeons, doves.
Order Strigiformes. Owls.
Order Apodiformes. Swifts, hummingbirds.
Order Passeriformes. Sparrows, jays, finches, crows, robins, starlings, wrens.
Order Piciformes. Woodpeckers, toucans.
Order Psittaciformes. Parrots, cockatoos, macaws.

Class Mammalia. Skin with hair; young nourished by milk-secreting mammary glands of adult.

Subclass Prototheria. Egg-laying mammals (monotremes; duckbilled platypus, spiny anteaters).

Subclass Metatheria. Pouched mammals or marsupials (opossums, kangaroos, wombats, Tasmanian devils).

Subclass Eutheria. Placental mammals.

Order Edentata. Anteaters, tree sloths, armadillos.
Order Insectivora. Tree shrews, moles, hedgehogs.
Order Chiroptera. Bats.
Order Scandentia. Insectivorous tree shrews.
Order Primates.

Suborder Strepsirhini (prosimians). Lemurs, lorises.
Suborder Haplorhini (tarsioids and anthropoids).

Infraorder Tarsiiformes. Tarsiers.
Infraorder Platyrrhini (New World monkeys).

Family Cebidae. Spider monkeys, howler monkeys, capuchin.

Infraorder Catarrhini (Old World monkeys and hominoids).

Superfamily Cercopithecoidea. Baboons, macaques, langurs.
Superfamily Hominoidea. Apes and humans.

Family Hylobatidae. Gibbon.
Family "Pongidae." Chimpanzees, gorillas, orangutans.
Family Hominidae. Existing and extinct human species (*Homo*) and humanlike species, including the australopiths.

Order Lagomorpha. Rabbits, hares, pikas.
Order Rodentia. Most gnawing animals (squirrels, rats, mice, guinea pigs, porcupines, beavers, etc.).
Order Carnivora. Carnivores (wolves, cats, bears, etc.).
Order Pinnipedia. Seals, walruses, sea lions.
Order Proboscidea. Elephants, mammoths (extinct).
Order Sirenia. Sea cows (manatees, dugongs).
Order Perissodactyla. Odd-toed ungulates (horses, tapirs, rhinos).
Order Tubulidentata. African aardvarks.
Order Artiodactyla. Even-toed ungulates (camels, deer, bison, sheep, goats, antelopes, giraffes, etc.).
Order Cetacea. Whales, porpoises.

Appendix II. Answers to Self-Quizzes

CHAPTER 1
1. cell — *1.1*
2. energy — *1.2*
3. homeostasis — *1.2*
4. domains — *1.3*
5. d — *1.2, 1.4*
6. d — *1.2*
7. mutation — *1.4*
8. adaptive — *1.4*
9. a — *1.6*
10. c — *1.1*
 e — *1.4*
 d — *1.5*
 b — *1.5*
 a — *1.5*

CHAPTER 2
1. False — *2.1*
2. b — *2.1*
3. d — *2.2*
4. c — *2.4*
5. a — *2.4*
6. e — *2.5*
7. f — *2.6*
8. acid, base — *2.6*
9. c — *2.6*
10. e — *Introduction*
 d — *2.6*
 b — *2.4*
 c — *2.5*
 a — *2.1*

CHAPTER 3
1. complex carbo-hydrates:
 simple sugars — *3.3*
 lipids; fatty acids or sterol rings — *3.4*
 proteins; amino acids — *3.5*
 nucleic acids; nucleotides — *3.7*
2. d — *3.1*
3. c — *3.2*
4. f — *3.3*
5. b — *3.4*
6. b — *3.4*
7. e — *3.4*
8. d — *3.5, 3.7*
9. d — *3.6*
10. d — *3.7*
11. b — *3.7*
12. c — *3.5*
 e — *3.7*
 b — *3.4*
 d — *3.7*
 a — *3.3*

CHAPTER 4
1. c — *4.1*
2. See Figure 4.15
3. d — *4.4*
4. d — *4.9*
5. False — *4.9*
6. c — *4.3*
7. e — *4.7*
 d — *4.8*
 a — *4.1, 4.4*
 b — *4.6*
 c — *4.6*

CHAPTER 5
1. c — *5.1*
2. c — *5.1*
3. a — *5.1*
4. d — *5.2*
5. d — *5.3*
6. d — *5.3*
7. b — *5.4*
8. a — *5.5*
9. c — *5.6*
10. d — *5.6*
 g — *5.4*
 a — *5.2*
 e — *5.4*
 c — *5.1*
 b — *5.3*
 f — *5.2*

CHAPTER 6
1. c — *6.1*
2. d — *6.1*
3. b — *6.1*
4. d — *6.3*
5. d — *6.4*
6. b — *6.5*
7. c — *6.2*
 g — *6.2*
 a — *6.2*
 d — *6.2*
 e — *6.2*
 b — *6.3*
 f — *6.2*

CHAPTER 7
1. carbon dioxide, sunlight — *Introduction*
2. b — *7.1*
3. a — *7.4*
4. b — *7.4*
5. c — *7.4*
6. d — *7.4*
7. c — *7.6*
8. b — *7.6*
9. d — *7.6*
10. c — *7.5*
 a — *7.6*
 b — *7.6*

CHAPTER 8
1. d — *8.1*
2. c — *8.2*
3. b — *8.4*
4. c — *8.4*
5. See Figure 8.3
6. c — *8.5*
7. b — *8.5*
8. b — *8.6*
9. b — *8.2*
 c — *8.5*
 a — *8.3*
 d — *8.4*

CHAPTER 9
1. d — *9.1*
2. b — *9.1*
3. c — *9.1*
4. d — *9.2*
5. a — *9.2*
6. c — *9.2*
7. a — *9.3*
8. b — *9.3*
9. d — *9.3*
 b — *9.3*
 c — *9.3*
 a — *9.3*

CHAPTER 10
1. c — *10.1*
2. b — *10.2*
3. a — *10.2*
4. d — *10.1*
5. d — *10.2*
6. b — *10.2*
7. d — *10.3*
8. Sister chromatids remain attached. — *10.3*
9. d — *10.3*
10. e — *10.4*
11. d — *10.2*
 a — *10.1*
 c — *10.3*
 b — *10.3*

CHAPTER 11
1. a — *11.1*
2. b — *11.1*
3. a — *11.1*
4. b — *11.1*
5. c — *11.2*
6. a — *11.2*
7. d — *11.3*
8. c — *11.5*
9. a — *11.5*
10. b — *11.3*
 d — *11.2*
 a — *11.1*
 c — *11.1*

CHAPTER 12
1. d — *12.2*
2. c — *12.5*
3. b — *12.3*
4. b — *12.3*
5. b — *12.3*
6. False — *12.7*
7. d — *12.7*
8. e — *12.8*
9. d — *12.9*
10. False — *12.9*
11. c — *12.9*
12. a — *12.10*
13. c — *12.9*
 e — *12.8*
 d — *12.9*
 b — *12.8*
 a — *12.2*
 f — *12.9*

CHAPTER 13
1. c — *13.2*
2. d — *13.2*
3. c — *13.2*
4. a — *13.3*
5. d — *13.3*
6. b — *13.2*
7. b — *13.4*
8. c — *13.1*
 e — *13.4*
 a — *13.2*
 b — *13.3*
 d — *13.3*
 f — *13.3*

CHAPTER 14
1. c — *14.1*
2. b — *14.1*
3. c — *14.1*
4. c — *14.1*
5. d — *14.2*
6. a — *14.3*
7. f — *14.5*
8. e — *14.5*
 c — *14.4*
 a — *14.1*
 f — *14.2*
 d — *14.3*
 g — *14.1*
 b — *14.2*

CHAPTER 15
1. d — *15.1*
2. d — *15.1*
3. d — *15.1, 15.4*
4. d — *15.1*
5. h — *15.1*
6. d — *15.1*
7. d — *15.1*
8. d — *15.2*
9. c — *15.2*
10. b — *15.3*
11. b — *15.3*
12. b — *15.4*
13. e — *15.2*
 a — *15.2*
 b — *15.4*
 d — *15.2*
 c — *15.1*
 f — *15.1*

CHAPTER 16
1. c — *16.1*
2. plasmid — *16.1*
3. b — *16.1*
4. a — *16.2*
5. d — *16.3*
6. b — *16.3*
7. b — *16.5*
8. b — *16.5*
9. d — *16.4*
 c — *16.7*
 b — *Introduction*
 e — *16.2*
 a — *16.10*

CHAPTER 17
1. d — *17.1*
2. d — *17.4*
3. a — *17.5*
4. Gondwana — *17.6*
5. b — *17.7*
6. d — *17.7*
7. d — *17.8*
8. c — *17.8*
 g — *17.4*
 a — *17.4*
 f — *17.8*
 e — *17.5*
 c — *17.7*
 b — *17.2*
 d — *17.7*

CHAPTER 18
1. populations — *18.1*
2. b — *Issues, Impacts*
3. a — *18.1*
4. c — *18.1, 18.3*
5. b — *18.4*
6. c — *18.5*
7. c — *18.6*
8. e — *18.6*
9. a — *18.7*
10. c — *18.8*
 d — *18.1, 18.3*
 a — *18.1*
 b — *18.7*

CHAPTER 19
1. d — *19.1*
2. d — *19.1*
3. a — *19.2*
4. c — *19.4*
5. c — *19.4*
6. c — *19.5*
7. d — *19.5*
8. c — *19.6*
9. d — *19.5*
10. e — *19.5*
 d — *19.4*
 a — *19.5*
 f — *19.5*
 b — *19.5, 19.6*
 c — *19.4*

CHAPTER 20
1. c — *20.1*
2. c — *20.2*
3. c — *20.3*
4. b — *20.4*
5. d — *20.4*
6. f — *20.1*
 c — *20.2*
 d — *20.3*
 a — *20.3*
 b — *20.5*
 e — *20.5*

CHAPTER 21
1. d — *21.1*
2. c — *21.2*
3. c — *21.1*
4. d — *21.4*
5. b — *21.4*
6. c — *21.4*
7. c — *21.2*
8. b — *21.6*
9. false — *21.6*
10. d — *21.7*
11. d — *21.5*
 e — *21.4*
 b — *21.6*
 f — *21.1*
 g — *21.5*
 a — *21.8*
 c — *21.1*

CHAPTER 22
1. f — *22.3*
2. a — *22.4*
3. d — *22.5*
4. a — *22.8*
5. b — *22.8*
6. d — *22.12*
7. e — *22.3*
 a — *22.3*
 b — *22.9*
 c — *22.6*
 f — *22.4*
 d — *22.11*

CHAPTER 23
1. c — *23.1*
2. a — *23.2*
3. b — *23.3*
4. c — *23.4*
5. a — *23.5*
6. e — *23.3, 23.4*
7. b — *23.6*
8. c — *23.2*
9. c — *23.7*
 e — *23.2*
 g — *23.4*
 h — *23.8*
 f — *23.3*
 a — *23.2*
 b — *23.2*
 d — *23.8*

CHAPTER 24
1. b — *24.6*
2. a — *24.1*
3. a — *24.1*
4. c — *24.1*
5. c — *24.1, 24.3*
6. e — *24.2*
 c — *24.6*
 d — *24.1*
 g — *24.1, 24.3*
 b — *24.4*
 f — *24.4*
 a — *24.1*

CHAPTER 25
1. a — *25.1*
2. a — *25.2*
3. a — *25.4*
4. a — *25.5*
5. b — *25.15*
6. c — *25.5, 25.6*
7. b — *25.17*
8. i — *25.1*
 c — *25.3*
 h — *25.4*
 b — *25.5*
 a — *25.10*
 f — *25.6*
 d — *25.11*
 e — *25.8*
 g — *25.17*

CHAPTER 26

1.	b	26.1
2.	a	26.2
3.	d	26.3
4.	b	26.6
5.	f	26.6
6.	a	26.8
7.	c	26.9
8.	f	26.10, 26.12
9.	c	26.15
10.	g	26.3
	a	26.4
	e	26.8
	b	26.9
	c	26.10
	f	26.10
	d	26.12

CHAPTER 27

1.	b	27.1
2.	f	27.4
3.	d	27.4
4.	a	27.4
5.	d	27.6
6.	d	27.6
7.	d	27.7

CHAPTER 28

1.	a	28.1
2.	c	28.1
3.	d	28.2
4.	d	28.5
5.	b	28.4
	c	28.3
	a	28.5
	d	28.3

CHAPTER 29

1. Eudicot above, with blue vascular bundles separating ground tissue into outer cortex and inner pith. Dicot, below, with vascular bundles dispersed in ground tissue. 29.1, 29.3

2.	a	29.1
3.	d	29.1, 29.6
4.	eudicot, monocot	29.1
5.	c	29.2
6.	c	29.4
7.	b	29.2
8.	b	29.2
9.	d	29.7
10.	b	29.1
	d	29.1
	e	29.2
	c	29.2
	f	29.5
	a	29.6

CHAPTER 30

1.	a	30.1
2.	c	30.2
3.	b	30.2
4.	c	30.3
5.	d	30.3
6.	a	30.4
7.	b	30.4
8.	c	30.4
	g	30.1
	e	30.5
	b	30.2
	d	30.3
	a	30.3
	f	30.5

CHAPTER 31

1.	a	31.1
2.	d	31.1, 31.3
3.	b	31.1
4.	c	31.4
5.	b	31.3
6.	b	31.3
7.	a	31.4
8.	b	31.4
9.	d	31.4
10.	d	31.4
11.	c	31.7
12.	c	31.1
	f	31.1
	a	31.3
	e	31.1
	d	31.1
	b	31.3

CHAPTER 32

1.	c	32.1
2.	c	32.2
3.	e	32.2
4.	d	32.1
5.	d	32.4
6.	a	32.5, 32.6
7.	c	32.6
8.	a	32.7
9.	d	32.6
	e	32.3
	a	32.1
	b	32.4
	c	32.2

CHAPTER 33

1. (a) epithelium, sheetlike with one free surface 33.1; (b) skeletal muscle, striated contractile cells 33.3; (c) loose connective tissue, scattered cells and fibers in a extracelluar matrix of their own secretions 33.2; (d) adipose tissue, cells swollen with stored fat, nuclei pushed to the side 33.2.

2.	a	33.1
3.	c	33.1
4.	a	33.1
5.	b	33.1, 33.2
6.	b	33.2
7.	c	33.2
8.	c	33.3
9.	d	33.3
10.	d	33.4
11.	b	33.1
	g	33.1
	a	33.2
	c	33.5
	d	33.3
	f	33.2
	e	33.1

CHAPTER 34

1.	a	34.1
2.	d	34.2
3.	d	34.3
4.	a	34.4
5.	c	34.6
6.	c	34.8
7.	b	34.8
8.	a	34.5
9.	a	34.9
10.	f	34.7
	d	34.4, 34.5
	g	34.11
	b	34.9
	h	34.10
	a	34.9
	e	34.6
	i	34.8
	c	34.9

CHAPTER 35

1.	a	35.1
2.	c	35.1
3.	c	35.2
4.	e	35.3
5.	a	35.2
6.	b	35.4
7.	b	35.5
8.	See Figure 35.17	
9.	d	35.8
	g	35.5
	f	35.6, 35.7
	a	35.4, 35.5
	c	35.8
	e	35.3
	b	35.4
	h	35.2

CHAPTER 36

1.	f	36.1
2.	b	36.3
3.	a	36.3
4.	e	36.2
5.	b	36.6
6.	b	36.4
7.	b	36.6
8.	d	36.8
	f	36.4
	c	36.4
	e	36.6
	a	36.10
	b	36.1

CHAPTER 37

1.	d	37.4
2.	b	37.5
3.	b	37.6
4.	b	37.6
5.	d	37.6
6.	d	37.8
7.	e	37.4
	f	37.9
	g	37.9
	h	37.4
	c	37.6
	c	37.4
	b	37.3
	i	37.6
	d	37.9

CHAPTER 38

1.	c	38.1
2.	b	38.1
3.	d	38.2
4.	b	38.4
5.	d	38.2
6.	b	38.6
7.	c	38.7
8.	a	38.7
9.	c	38.8
10.	d	38.10
11.	f	38.8
	a	38.10
	e	38.2
	g	38.7
	b	38.6
	c	38.8
	d	38.5

12. See Figure 38.13
13. artery, high speed transport away from the heart; vein; transport to the heart and blood reservoir; arteriole, adjusts blood distribution; capillary, diffusion zone

CHAPTER 39

1.	f	39.2
2.	e	39.3
3.	d	39.4
4.	d	39.5
5.	e	39.4
6.	a	39.5
7.	b	39.7
8.	c	39.3
	b	39.5, 39.6
	a	39.1
	e	39.6
	d	39.9

CHAPTER 40

1.	a	40.1
2.	d	40.1, 40.5
3.	c	40.2
4.	a	40.3
5.	c	40.4, 40.5
6.	d	40.6
7.	a	40.6
8.	a	40.5
9.	d	40.4
	h	40.4
	f	40.4
	e	40.1
	g	40.4
	c	40.4
	b	40.4
	a	40.6

CHAPTER 41

1.	d	41.1
2.	b	41.4
3.	c	41.5
4.	b	41.4, 41.5
5.	a	41.5
6.	c	41.6
7.	b	41.9
8.	f	41.4
`	b	41.6
	a	41.4
	d	41.5
	e	41.4
	c	41.2, 41.4

CHAPTER 42

1.	c	42.1
2.	a	42.2
3.	b	42.2
4.	See Figures 42.5 and 42.6	
5.	d	42.3
6.	b	42.3
7.	a	42.4
8.	a	42.4
9.	c	42.2
	a	42.2
	b	42.2
	e	42.4
	d	42.4
10.	d	42.8
11.	d	42.8
12.	False	42.4
13.	b	42.7
	a	42.7
	d	42.7
	c	42.7
	e	42.7

CHAPTER 43

1.	d	43.1
2.	c	43.2
3.	c	43.2
4.	b	43.3
5.	c	43.4
6.	d	43.4
7.	c	43.5
8.	d	43.5
9.	b	43.5
10.	a	43.3
11.	c	43.2
	d	43.2
	a	43.2, 43.3
	b	43.2, 43.4
	f	43.4
	e	43.4

CHAPTER 44

1. See Figure 44.4
2. Figure 44.11

3.	d	44.2
4.	c	44.4
5.	c	44.4
6.	e	44.8
7.	c	44.9
8.	a	44.9, 44.12
9.	c	44.6
10.	c	44.9
11.	c	44.11
12.	f	44.10
13.	e	44.12
14.	c	44.2
	h	44.3
	a	44.11
	g	44.14
	d	44.4
	e	44.6
	b	44.1
	f	44.3

CHAPTER 45

1.	f	45.3
2.	a	45.3
3.	a	45.3
4.	d	45.4
5.	d	45.5
6.	c	45.7
7.	c	45.4
	d	45.3
	a	45.3
	e	45.4
	b	45.4

CHAPTER 46

1.	d	46.1
2.	d	46.1
3.	d	46.3
4.	b	46.4
5.	d	46.6
6.	b	46.8
7.	c	46.9
	d	46.11
	a	46.8
	e	46.8
	b	46.9
	g	46.9
	f	46.3

CHAPTER 47

1.	a	47.1
2.	d	47.1
3.	d	47.1
4.	d	47.1
5.	d	47.4
6.	d	47.3
7.	a	47.6, 47.8
8.	b	47.9
9.	d	47.12
10.	b	47.11, 47.12
11.	a	47.11
12.	d	47.1
	a	47.1
	c	47.1
	b	47.1

CHAPTER 48

1.	c	48.1
2.	c	48.2
3.	d	48.1, 48.3
4.	a	48.3
5.	d	48.2
6.	d	48.4
7.	d	48.4
8.	c	48.6
9.	d	48.12, 48.15
10.	d	48.10
	h	48.4
	e	48.6
	c	48.6
	b	48.14
	g	48.9
	a	48.7
	f	48.15

CHAPTER 49

1.	d	49.1
2.	c	49.3
3.	b	49.1
4.	d	49.6
5.	b	49.6
6.	d	49.7
7.	c	49.7
8.	c	49.1
	d	49.3, 49.7
	b	49.1
	a	49.2
	e	49.4

Appendix III. Answers to Genetics Problems

CHAPTER 11

1. a. Both parents are heterozygotes (*Aa*). Their children may be albino (*aa*) or unaffected (*AA* or *Aa*).

   b. All are homozygous recessive (*aa*).

   c. Homozygous recessive (*aa*) father, and heterozygous (*Aa*) mother. The albino child is *aa*, the unaffected children *Aa*.

2. Possible outcomes of an experimental cross between F_1 rose plants heterozygous for height (*Aa*):

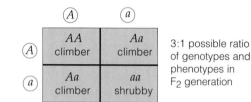

3:1 possible ratio of genotypes and phenotypes in F_2 generation

Possible outcomes of a testcross between an F_1 rose plant heterozygous for height and a shrubby rose plant:

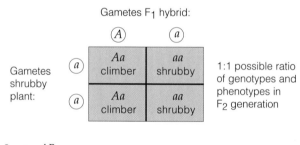

1:1 possible ratio of genotypes and phenotypes in F_2 generation

3. a. *AB*
   b. *AB, aB*
   c. *Ab, ab*
   d. *AB, Ab, aB, ab*

4. a. All offspring will be *AaBB*.
   b. 1/4 *AABB* (25% each genotype)
       1/4 *AABb*
       1/4 *AaBB*
       1/4 *AaBb*
   c. 1/4 *AaBb* (25% each genotype)
       1/4 *Aabb*
       1/4 *aaBb*
       1/4 *aabb*
   d. 1/16 *AABB* (6.25% of genotype)
       1/8 *AaBB* (12.5%)
       1/16 *aaBB* (6.25%)
       1/8 *AABb* (12.5%)
       1/4 *AaBb* (25%)
       1/8 *aaBb* (12.5%)
       1/16 *AAbb* (6.25%)
       1/8 *Aabb* (12.5%)
       1/16 *aabb* (6.25%)

5. a. *ABC*
   b. *ABC, aBC*
   c. *ABC, aBC, ABc, aBc*
   d. *ABC*
       aBC
       AbC
       abC
       ABc
       aBc
       Abc
       abc

6. A mating of two M^L cats yields 1/4 *MM*, 1/2 M^LM, and 1/4 M^LM^L. Because M^LM^L is lethal, the probability that any one kitten among the survivors will be heterozygous is 2/3.

7. Yellow is recessive. Because F_1 plants have a green phenotype and must be heterozygous, green must be dominant over the recessive yellow.

8. a. *RR* and *rr*
   b. all *Rr*

9. Because all F_1 plants of this dihybrid cross had to be heterozygous for both genes, then 1/4 (25%) of the F_2 plants will be heterozygous for both genes.

10. A mating between a mouse from a true-breeding, white-furred strain and a mouse from a true-breeding, brown-furred strain would provide you with the most direct evidence. Because true-breeding strains of organisms typically are homozygous for a trait being studied, all F_1 offspring from this mating should be heterozygous. Record the phenotype of each F_1 mouse, then let them mate with one another. Assuming only one gene locus is involved, these are possible outcomes for the F_1 offspring:

   a. All F_1 mice are brown, and their F_2 offspring segregate: 3 brown : 1 white. *Conclusion*: Brown is dominant to white.

   b. All F_1 mice are white, and their F_2 offspring segregate: 3 white : 1 brown. *Conclusion*: White is dominant to brown.

   c. All F_1 mice are tan, and the F_2 offspring segregate: 1 brown : 2 tan : 1 white. *Conclusion*: The alleles at this locus show incomplete dominance.

11. The data reveal that these genes do not assort independently because the observed ratio is very far from the 9:3:3:1 ratio expected with independent assortment. Instead, the results can be explained if the genes are located close to each other on the same chromosome, which is called linkage.

12. Fred could use a testcross to find out if his pet's genotype is *WW* or *Ww*. He can let his black guinea pig mate with a white guinea pig having the genotype *ww*.

   If any F_1 offspring are white, then the genotype of his pet is *Ww*. If the two guinea pig parents are allowed to mate repeatedly and all the offspring of the matings are black, then there is a high probability that his pet guinea pig is *WW*.

(For instance, if ten offspring are all black, then the probability that the male is *WW* is about 99.9 percent. The greater the number of offspring, the more confident Fred can be of his conclusion.)

13. a. _1/2_ red _1/2_ pink _____ white
    b. _____ red _All_ pink _____ white
    c. _1/4_ red _1/2_ pink _1/4_ white
    d. _____ red _1/2_ pink _1/2_ white

14. 9/16 walnut
    3/16 rose
    3/16 pea
    1/16 single

15. Because both parents are heterozygotes (Hb^AHb^S), the following are the probabilities for each child:

    a. 1/4 Hb^SHb^S
    b. 1/4 Hb^AHb^A
    c. 1/2 Hb^AHb^S

16. 2/3

17. The smooth rind/furrowed rind ratio is 230:230, which is exactly 1:1. The nonexplosive rind/explosive rind ratio is 227:233, which is close to a 1:1 ratio.

The overall ratio is close to a 1:1 ratio, which indicates that the genes are assorting independently.

18. See the percentages in the graph below. The varied colors in wheat kernels are due to the combined effects of incomplete dominance of alleles of two genes that influence the same phenotype.

CHAPTER 12

1. a. Human males (XY) inherit their X chromosome from their mother.

b. A male can produce two kinds of gametes. Half carry an X chromosome and half carry a Y chromosome. All the gametes that carry the X chromosome carry the same X-linked allele.

c. A female homozygous for an X-linked allele produces only one kind of gamete.

d. Fifty percent of the gametes of a female who is heterozygous for an X-linked allele carry one of the two alleles at that locus; the other fifty percent carry its partner allele for that locus.

2. Because Marfan syndrome is a case of autosomal dominant inheritance and because one parent bears the allele, the probability that any child of theirs will inherit the mutant allele is 50 percent.

3. a. Nondisjunction might occur during anaphase I or anaphase II of meiosis.

b. As a result of translocation, chromosome 21 may get attached to the end of chromosome 14. The new individual's chromosome number would still be 46, but its somatic cells would have the translocated chromosome 21 in addition to two normal chromosomes 21.

4. A daughter could develop this muscular dystrophy only if she inherited two X-linked recessive alleles— one from each parent. Males who carry the allele are unlikely to father children because they develop the disorder and die early in life.

5. In the mother, a crossover between the two genes at meiosis generates an X chromosome that carries neither mutant allele.

6. The phenotype appeared in every generation shown in the diagram, so this must be a pattern of autosomal dominant inheritance.

7. There is no scientific answer to this question, which simply invites you to reflect on the difference between a scientific and a subjective interpretation of this individual's condition.

Genotype	Phenotype	Number Displaying the Trait	Percent of Population
$A^1A^1B^1B^1$	Dark red	181	9.05
$A^1A^1B^1B^2$ or $A^1A^2B^1B^1$	Red	360	18.00
$A^1A^2B^1B^2$ or $A^1A^1B^2B^2$ or $A^2A^2B^1B^1$	Salmon	922	46.10
$A^1A^2B^2B^2$ or $A^2A^2B^1B^2$	Pink	358	17.90
$A^2A^2B^2B^2$	White	179	8.95
	Totals	2,000	100

Appendix IV. Periodic Table of the Elements

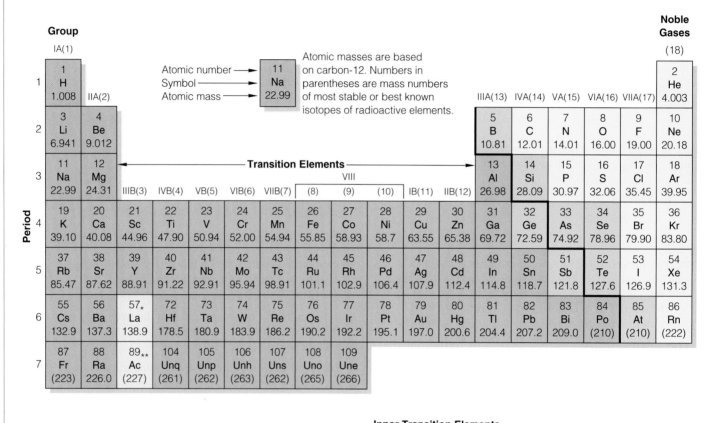

Group

Noble Gases
(18)

Atomic masses are based on carbon-12. Numbers in parentheses are mass numbers of most stable or best known isotopes of radioactive elements.

Atomic number → 11
Symbol → Na
Atomic mass → 22.99

Transition Elements

Inner Transition Elements

	58 Ce 140.1	59 Pr 140.9	60 Nd 144.2	61 Pm (145)	62 Sm 150.4	63 Eu 152.0	64 Gd 157.3	65 Tb 158.9	66 Dy 162.5	67 Ho 164.9	68 Er 167.3	69 Tm 168.9	70 Yb 173.0	71 Lu 175.0
Lanthanide Series 6														
Actinide Series 7	90 Th 232.0	91 Pa 231.0	92 U 238.0	93 Np 237.0	94 Pu (244)	95 Am (243)	96 Cm (247)	97 Bk (247)	98 Cf (251)	99 Es (252)	100 Fm (257)	101 Md (258)	102 No (259)	103 Lr (260)

Appendix V. The Amino Acids

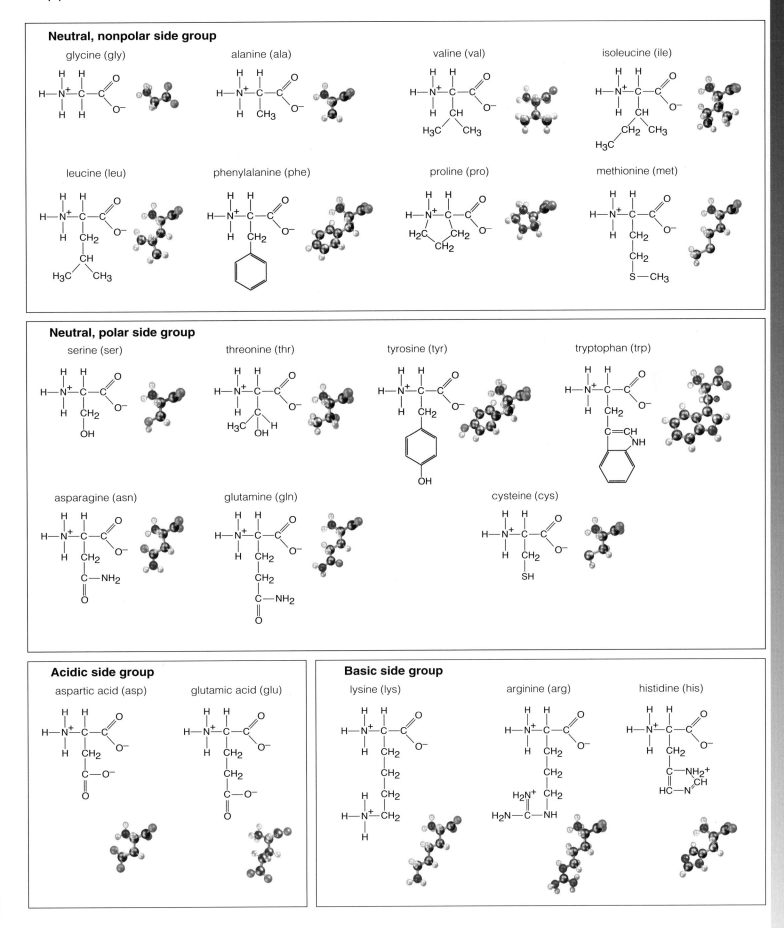

Neutral, nonpolar side group

glycine (gly)

alanine (ala)

valine (val)

isoleucine (ile)

leucine (leu)

phenylalanine (phe)

proline (pro)

methionine (met)

Neutral, polar side group

serine (ser)

threonine (thr)

tyrosine (tyr)

tryptophan (trp)

asparagine (asn)

glutamine (gln)

cysteine (cys)

Acidic side group

aspartic acid (asp)

glutamic acid (glu)

Basic side group

lysine (lys)

arginine (arg)

histidine (his)

Appendix VI. Units of Measure

Metric-English Conversions

Length

English		Metric
inch	=	2.54 centimeters
foot	=	0.30 meter
yard	=	0.91 meter
mile (5,280 feet)	=	1.61 kilometer

To convert	multiply by	to obtain
inches	2.54	centimeters
feet	30.00	centimeters
centimeters	0.39	inches
millimeters	0.039	inches

Weight

English		Metric
grain	=	64.80 milligrams
ounce	=	28.35 grams
pound	=	453.60 grams
ton (short) (2,000 pounds)	=	0.91 metric ton

To convert	multiply by	to obtain
ounces	28.3	grams
pounds	453.6	grams
pounds	0.45	kilograms
grams	0.035	ounces
kilograms	2.2	pounds

Volume

English		Metric
cubic inch	=	16.39 cubic centimeters
cubic foot	=	0.03 cubic meter
cubic yard	=	0.765 cubic meters
ounce	=	0.03 liter
pint	=	0.47 liter
quart	=	0.95 liter
gallon	=	3.79 liters

To convert	multiply by	to obtain
fluid ounces	30.00	milliliters
quart	0.95	liters
milliliters	0.03	fluid ounces
liters	1.06	quarts

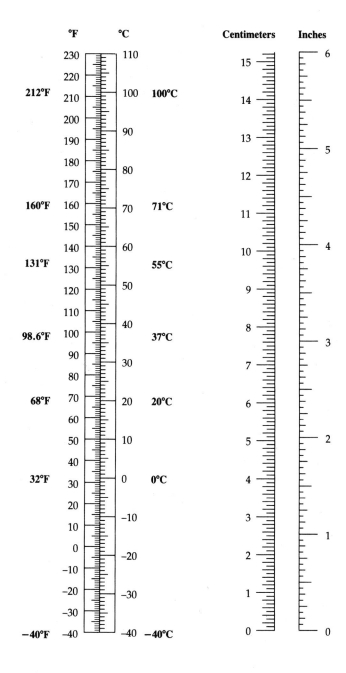

Appendix VII. Closer Look at Some Major Metabolic Pathways

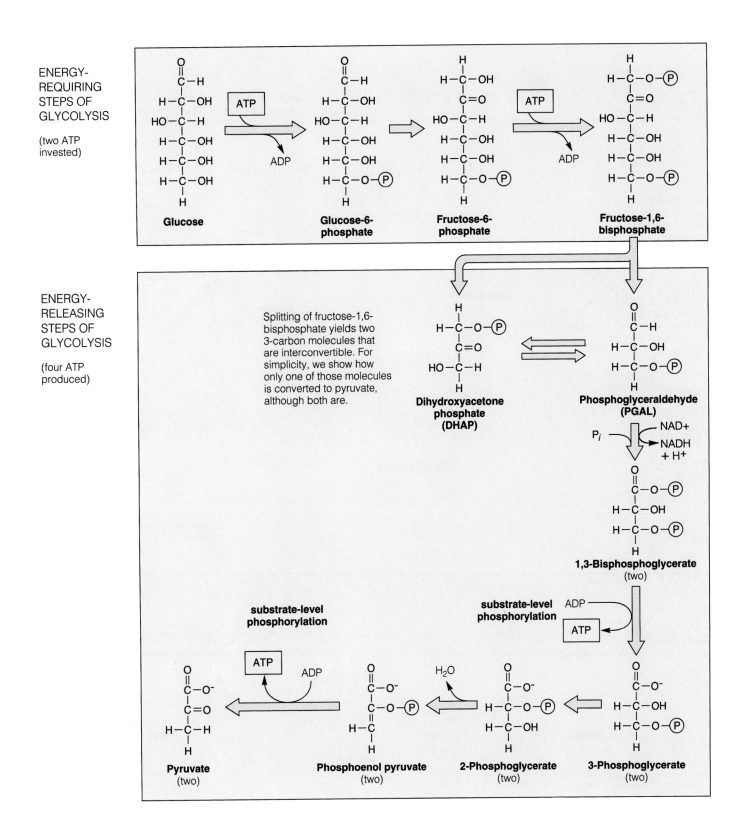

Figure A Glycolysis, ending with two 3-carbon pyruvate molecules for each 6-carbon glucose molecule entering the reactions. The *net* energy yield is two ATP molecules (two invested, four produced).

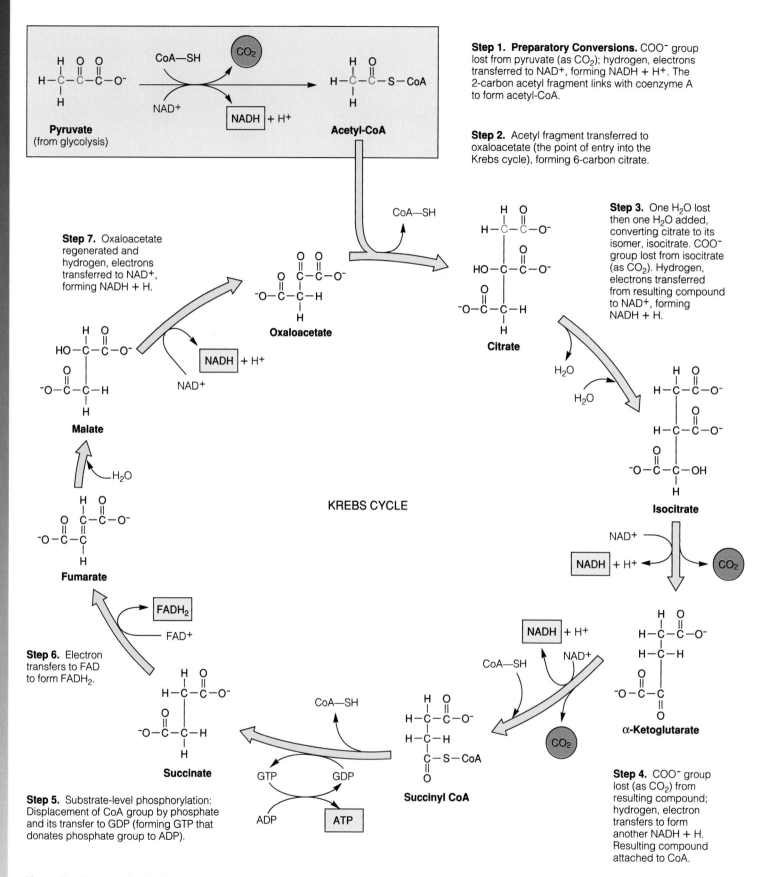

Figure B Krebs cycle, also known as the citric acid cycle. *Red* identifies carbon atoms entering the cyclic pathway (by way of acetyl-CoA) and leaving (by way of carbon dioxide). These cyclic reactions run twice for each glucose molecule that has been degraded to two pyruvate molecules.

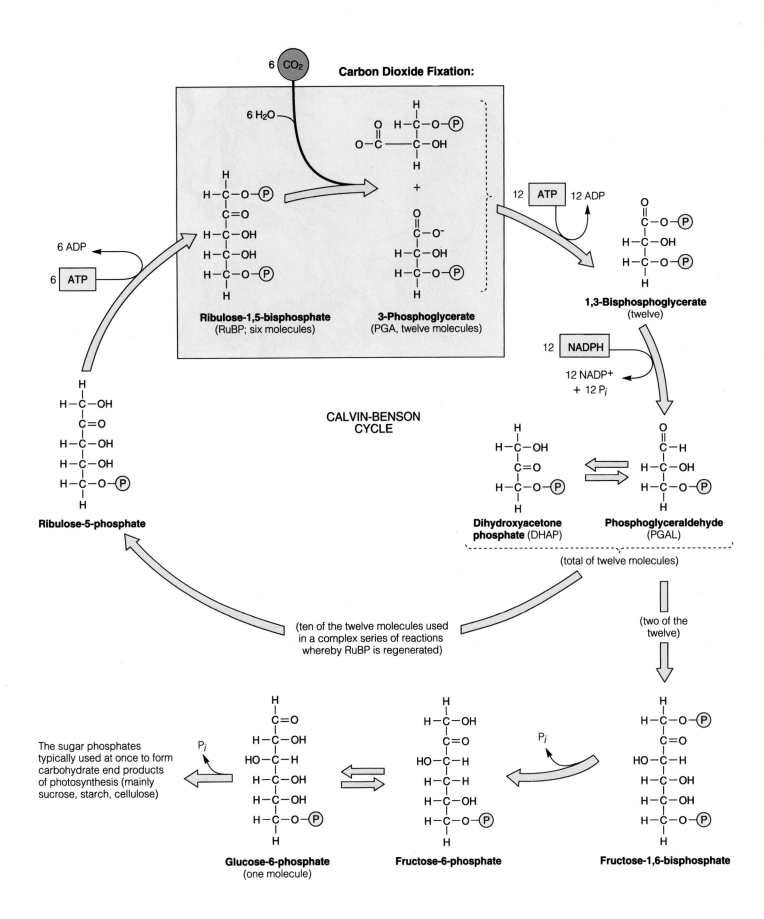

Figure C Calvin–Benson cycle of the light-independent reactions of photosynthesis.

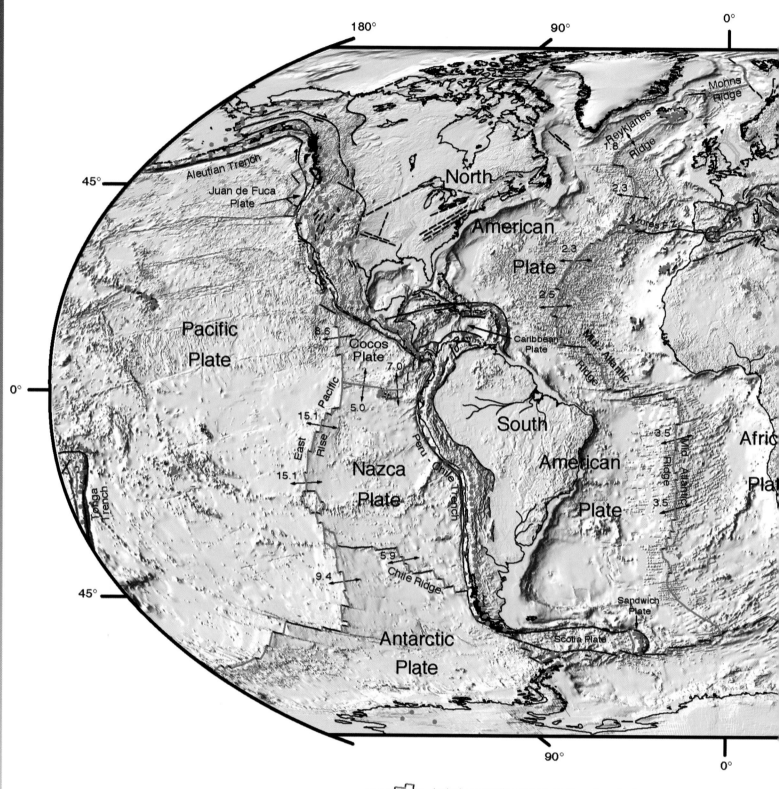

180° **90°** **0°**

45°

Aleutian Trench

Juan de Fuca
Plate

North

American

Plate

Mohns
Ridge

Reykjanes
Ridge

7.8

2.3

Azores F.Z.

2.3

2.5

Pacific

Plate

8.6

Cocos
Plate

7.0

Caribbean
Plate

Mid-Atlantic
Ridge

0°

5.0

15.1

East
Pacific
Rise

Nazca

Plate

South

American

Plate

3.5

Mid-Atlantic
Ridge

Afric

Pla

15.1

Peru-Chile Trench

Chile Ridge

3.5

Tonga
Trench

5.9

9.4

45°

Antarctic

Plate

Sandwich
Plate

Scotia Plate

90° **0°**

Appendix VIII.
Restless Earth—Life's Changing
Geologic Stage

This NASA map summarizes the tectonic and volcanic
activity of Earth during the past 1 million years. The
reconstructions at far right indicate positions of Earth's
major land masses through time.

Actively-spreading ridges and transform faults

1.4   Total spreading rate, cm/year

Major active fault or fault zone; dashed where nature,
location, or activity uncertain

Normal fault or rift; hachures on downthrown side

Reverse fault (overthrust, subduction zones); generalized;
barbs on upthrown side

Volcanic centers active within the last one million years;
generalized. Minor basaltic centers and seamounts omitted.

90°
180°
45°
0°
45°
90°

ian Plate

Baikal
Rift

Altyn Tagh F.

Tan-Lu F.

Arabian
Plate

Indian
Plate

Somalia
Plate

Philippines
Plate

Pacific
Plate

Marina Trench

Caroline
Plate

Ontong-Java
Plateau

Java Trench

Australian
Plate

Darling F.

S.E. Indian
Ocean Ridge

7.2

7.5

S.W. Indian
Ocean Ridge

Alpine F.

Antarctic Plate

P.D.L.

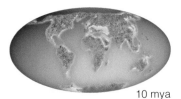

10 mya

Middle Miocene. Polar regions again iced over, as in Cambrian. All land masses are assuming their current distribution

65 mya

Cretaceous into Tertiary. Extinction of dinosaurs; rise of mammals

Pangea

240 mya

Permian into Triassic. Vast swamp forests (eventual coal source); seed plants evolve

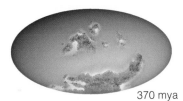

370 mya

Devonian. Jawed fishes evolve, diversify; ancestors of amphibians invade land

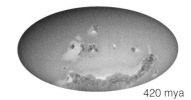

420 mya

Silurian. Sea level rises, diverse marine life; plants, invertebrates invade land

540 mya

Cambrian. Fragments of Rodinia, the first supercontinent. Major adaptive radiations in equatorial seas; icy polar regions

Appendix IX. Annotations to A Journal Article

This journal article reports on the movements of a female wolf during the summer of 2002 in northwestern Canada. It also reports on a scientific process of inquiry, observation and interpretation to learn where, how and why the wolf traveled as she did. In some ways, this article reflects the story of "how to do science" told in section 1.5 of this textbook. These notes are intended to help you read and understand how scientists work and how they report on their work.

(1) ARCTIC

(2) VOL. 57, NO. 2 (JUNE 2004) P. 196–203

(3) Long Foraging Movement of a Denning Tundra Wolf

(4) Paul F. Frame,[1,2] David S. Hik,[1] H. Dean Cluff,[3] and Paul C. Paquet[4]

(5) *(Received 3 September 2003; accepted in revised form 16 January 2004)*

(6) ABSTRACT Wolves (*Canis lupus*) on the Canadian barrens are intimately linked to migrating herds of barren-ground caribou (*Rangifer tarandus*). We deployed a Global Positioning System (GPS) radio collar on an adult female wolf to record her movements in response to changing caribou densities near her den during summer. This wolf and two other females were observed nursing a group of 11 pups. She traveled a minimum of 341 km during a 14-day excursion. The straight-line distance from the den to the farthest location was 103 km, and the overall minimum rate of travel was 3.1 km/h. The distance between the wolf and the radio-collared caribou decreased from 242 km one week before the excursion to 8 km four days into the excursion. We discuss several possible explanations for the long foraging bout.

(7) *Key words:* wolf, GPS tracking, movements, *Canis lupus*, foraging, caribou, Northwest Territories

(8) RÉSUMÉ Les loups (*Canis lupus*) dans la toundra canadienne sont étroitement liés aux hardes de caribous des toundras (*Rangifer tarandus*). On a équipé une louve adulte d'un collier émetteur muni d'un système de positionnement mondial (GPS) afin d'enregistrer ses déplacements en réponse au changement de densité du caribou près de sa tanière durant l'été. On a observé cette louve ainsi que deux autres en train d'allaiter un groupe de 11 louveteaux. Elle a parcouru un minimum de 341 km durant une sortie de 14 jours. La distance en ligne droite de la tanière à l'endroit le plus éloigné était de 103 km, et la vitesse minimum durant tout le voyage était de 3,1 km/h. La distance entre la louve et le caribou muni du collier émetteur a diminué de 242 km une semaine avant la sortie à 8 km quatre jours après la sortie. On commente diverses explications possibles pour ce long épisode de recherche de nourriture.

Mots clés: loup, repérage GPS, déplacements, *Canis lupus*, recherche de nourriture, caribou, Territoires du Nord-Ouest

Traduit pour la revue *Arctic* par Nésida Loyer.

(9) Introduction

Wolves (*Canis lupus*) that den on the central barrens of mainland Canada follow the seasonal movements of their main prey, migratory barren-ground caribou (*Rangifer tarandus*) (Kuyt, 1962; Kelsall, 1968; Walton et al., 2001). However, most wolves do not den near caribou calving grounds, but select sites farther south, closer to the tree line (Heard and Williams, 1992). Most caribou migrate beyond primary wolf denning areas by mid-June and do not return until mid-to-late July (Heard et al., 1996; Gunn et al., 2001). Conse-

quently, caribou density near dens is low for part of the summer.

During this period of spatial separation from the main caribou herds, wolves must either search near the homesite for scarce caribou or alternative prey (or both), travel to where prey are abundant, or use a combination of these strategies.

Walton et al. (2001) postulated that the travel of tundra wolves outside their normal summer ranges is a response to low caribou availability rather than a pre-dispersal exploration like that observed in terri torial wolves (Fritts and Mech, 1981; Messier, 1985). The authors postulated this because most such travel was directed toward caribou calving grounds. We report details of such a long-distance excursion by a breeding female tundra wolf wearing a GPS radio collar. We discuss the relationship of the excursion to movements of satellite-collared caribou (Gunn et al., 2001), supporting the hypothesis that tundra wolves make directional, rapid, long-distance movements in response to seasonal prey availability.

[1] Department of Biological Sciences, University of Alberta, Edmonton, Alberta T6G 2E9, Canada
[2] Corresponding author: pframe@ualberta.ca
[3] Department of Resources, Wildlife, and Economic Development, North Slave Region, Government of the Northwest Territories, P.O. Box 2668, 3803 Bretzlaff Dr., Yellowknife, Northwest Territories X1A 2P9, Canada; Dean_Cluff@gov.nt.ca
[4] Faculty of Environmental Design, University of Calgary, Calgary, Alberta T2N 1N4, Canada; current address: P.O. Box 150, Meacham, Saskatchewan S0K 2V0, Canada

196

---

1 Title of the journal, which reports on science taking place in Arctic regions.

2 Volume number, issue number and date of the journal, and page numbers of the article.

3 Title of the article: a concise but specific description of the subject of study—one episode of long-range travel by a wolf hunting for food on the Arctic tundra.

4 Authors of the article: scientists working at the institutions listed in the footnotes below. Note #2 indicates that P. F. Frame is the *corresponding author*—the person to contact with questions or comments. His email address is provided.

5 Date on which a draft of the article was received by the journal editor, followed by date one which a revised draft was accepted for publication. Between these dates, the article was reviewed and critiqued by other scientists, a process called peer review. The authors revised the article to make it clearer, according to those reviews.

6 ABSTRACT: A brief description of the study containing all basic elements of this report. First sentence summarizes the *background* material. Second sentence encapsulates the *methods* used. The rest of the paragraph sums up the *results*. Authors introduce the main *subject* of the study—a female wolf (#388) with pups in a den—and refer to later *discussion* of possible explanations for her behavior.

7 Key words are listed to help researchers using computer databases. Searching the databases using these key words will yield a list of studies related to this one.

8 RÉSUMÉ: The French translation of the abstract and key words. Many researchers in this field are French Canadian. Some journals provide such translations in French or in other languages.

9 INTRODUCTION: Gives the background for this wolf study. This paragraph tells of known or suspected wolf behavior that is important for this study. Note that (a) major species mentioned are always accompanied by scientific names, and (b) statements of fact or *postulations* (claims or assumptions about what is likely to be true) are followed by references to studies that established those facts or supported the postulations.

10 This paragraph focuses directly on the wolf behaviors that were studied here.

11 This paragraph starts with a statement of the *hypothesis* being tested, one that originated in other studies and is supported by this one. The hypothesis is restated more succinctly in the last sentence of this paragraph. This is the *inquiry* part of the scientific process—asking questions and suggesting possible answers.

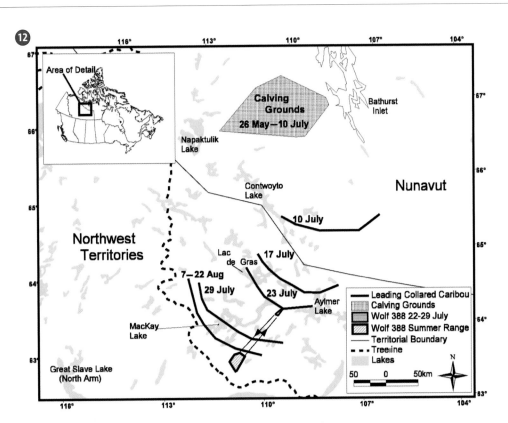

Figure 1. Map showing the movements of satellite radio-collared caribou with respect to female wolf 388's summer range and long foraging movement, in summer 2002.

12 This map shows the study area and depicts wolf and caribou locations and movements during one summer. Some of this information is explained below.

13 STUDY AREA: This section sets the stage for the study, locating it precisely with latitude and longitude coordinates and describing the area (illustrated by the map in Figure 1).

14 Here begins the story of how prey (caribou) and predators (wolves) interact on the tundra. Authors describe movements of these nomadic animals throughout the year.

15 We focus on the denning season (summer) and learn how wolves locate their dens and travel according to the movements of caribou herds.

13 Study Area

Our study took place in the northern boreal forest–low Arctic tundra transition zone (63° 30′ N, 110° 00′ W; Figure 1; Timoney et al., 1992). Permafrost in the area changes from discontinuous to continuous (Harris, 1986). Patches of spruce (*Picea mariana, P. glauca*) occur in the southern portion and give way to open tundra to the northeast. Eskers, kames, and other glacial deposits are scattered throughout the study area. Standing water and exposed bedrock are characteristic of the area.

14 *Details of the Caribou-Wolf System*

The Bathurst caribou herd uses this study area. Most caribou cows have begun migrating by late April, reaching calving grounds by June (Gunn et al., 2001;

Figure 1). Calving peaks by 15 June (Gunn et al., 2001), and calves begin to travel with the herd by one week of age (Kelsall, 1968). The movement patterns of bulls are less known, but bulls frequent areas near calving grounds by mid-June (Heard et al., 1996; Gunn et al., 2001). In summer, Bathurst caribou cows generally travel south from their calving grounds and then, parallel to the tree line, to the northwest. The rut usually takes place at the tree line in October (Gunn et al., 2001). The winter range of the Bathurst herd varies among years, ranging through the taiga and along the tree line from south of Great Bear Lake to southeast of Great Slave Lake. Some caribou spend the winter on the tundra (Gunn et al., 2001; Thorpe et al., 2001).

15 In winter, wolves that prey on Bathurst caribou do not behave territorially. Instead, they follow the herd throughout its winter range (Walton et al., 2001; Musiani, 2003). However, during denning (May–

Foraging Movement of A Tundra Wolf **197**

16 Other variables are considered—prey other than caribou and their relative abundance in 2002.

17 METHODS: There is no one scientific method. Procedures for each and every study must be explained carefully.

18 Authors explain when and how they tracked caribou and wolves, including tools used and the exact procedures followed.

19 This important subsection explains what data were calculated (average distance ...) and how, including the software used and where it came from. (The calculations are listed in Table 1.) Note that the behavior measured (traveling) is carefully defined.

20 RESULTS: The heart of the report and the *observation* part of the scientific process. This section is organized parallel to the Methods section.

21 This subsection is broken down by periods of observation. Pre-excursion period covers the time between 388's capture and the start of her long-distance travel. The investigators used visual observations as well as telemetry (measurements taken using the global positioning system (GPS)) to gather data. They looked at how 388 cared for her pups, interacted with other adults, and moved about the den area.

Table 1. Daily distances from wolf 388 and the den to the nearest radio-collared caribou during a long excursion in summer 2002.

Date (2002)	Mean distance from caribou to wolf (km)	Daily distance from closest caribou to den
12 July	242	241
13 July	210	209
14 July	200	199
15 July	186	180
16 July	163	162
17 July	151	148
18 July	144	137
19 July[1]	126	124
20 July	103	130
21 July	73	130
22 July	40	110
23 July[2]	9	104
29 July[3]	16	43
30 July	32	43
31 July	28	44
1 August	29	46
2 August[4]	54	52
3 August	53	53
4 August	74	74
5 August	75	75
6 August	74	75
7 August	72	75
8 August	76	75
9 August	79	79

[1] Excursion starts.
[2] Wolf closest to collared caribou.
[3] Previous five days' caribou locations not available.
[4] Excursion ends.

August, parturition late May to mid-June), wolf movements are limited by the need to return food to the den. To maximize access to migrating caribou, many wolves select den sites closer to the tree line than to caribou calving grounds (Heard and Williams, 1992). Because of caribou movement patterns, tundra denning wolves are separated from the main caribou herds by several hundred kilometers at some time during summer (Williams, 1990:19; Figure 1; Table 1).

16 Muskoxen do not occur in the study area (Fournier and Gunn, 1998), and there are few moose there (H.D. Cluff, pers. obs.). Therefore, alternative prey for wolves includes waterfowl, other ground-nesting birds, their eggs, rodents, and hares (Kuyt, 1972; Williams, 1990:16; H.D. Cluff and P.F. Frame, unpubl. data). During 56 hours of den observations, we saw no ground squirrels or hares, only birds. It appears that the abundance of alternative prey was relatively low in 2002.

17 Methods

Wolf Monitoring

18 We captured female wolf 388 near her den on 22 June 2002, using a helicopter net-gun (Walton et al., 2001). She was fitted with a releasable GPS radio collar (Merrill et al., 1998) programmed to acquire locations at 30-

minute intervals. The collar was electronically released (e.g., Mech and Gese, 1992) on 20 August 2002. From 27 June to 3 July 2002, we observed 388's den with a 78 mm spotting scope at a distance of 390 m.

Caribou Monitoring

In spring of 2002, ten female caribou were captured by helicopter net-gun and fitted with satellite radio collars, bringing the total number of collared Bathurst cows to 19. Eight of these spent the summer of 2002 south of Queen Maud Gulf, well east of normal Bathurst caribou range. Therefore, we used 11 caribou for this analysis. The collars provided one location per day during our study, except for five days from 24 to 28 July. Locations of satellite collars were obtained from Service Argos, Inc. (Landover, Maryland).

Data Analysis

Location data were analyzed by ArcView GIS software (Environmental Systems Research Institute Inc., Redlands, California). We calculated the average distance from the nearest collared caribou to the wolf and the den for each day of the study. **19**

Wolf foraging bouts were calculated from the time 388 exited a buffer zone (500 m radius around the den) until she re-entered it. We considered her to be traveling when two consecutive locations were spatially separated by more than 100 m. Minimum distance traveled was the sum of distances between each location and the next during the excursion.

We compared pre- and post-excursion data using Analysis of Variance (ANOVA; Zar, 1999). We first tested for homogeneity of variances with Levene's test (Brown and Forsythe, 1974). No transformations of these data were required.

Results **20**

Wolf Monitoring

Pre-Excursion Period: Wolf 388 was lactating when **21** captured on 22 June. We observed her and two other females nursing a group of 11 pups between 27 June and 3 July. During our observations, the pack consisted of at least four adults (3 females and 1 male) and 11 pups. On 30 June, three pups were moved to a location 310 m from the other eight and cared for by an uncollared female. The male was not seen at the den after the evening of 30 June.

Before the excursion, telemetry indicated 18 foraging bouts. The mean distance traveled during these bouts was 25.29 km (± 4.5 SE, range 3.1–82.5 km). Mean greatest distance from the den on foraging

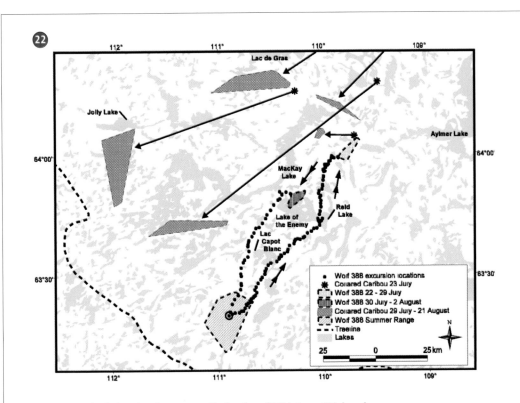

22

Figure 2. Details of a long foraging movement by female wolf 388 between 19 July and 2 August 2002. Also shown are locations and movements of three satellite radio-collared caribou from 23 July to 21 August 2002. On 23 July, the wolf was 8 km from a collared caribou. The farthest point from the den (103 km distant) was recorded on 27 July. Arrows indicate direction of travel.

22 The key in the lower right-hand corner of the map shows areas (shaded) within which the wolves and caribou moved, and the dotted trail of 388 during her excursion. From the results depicted on this map, the investigators tried to determine when and where 388 might have encountered caribou and how their locations affected her traveling behavior.

23 The wolf's excursion (her long trip away from the den area) is the focus of this study. These paragraphs present detailed measurements of daily movements during her two-week trip—how far she traveled, how far she was from collared caribou, her time spent traveling and resting, and her rate of speed. Authors use the phrase "minimum distance traveled" to acknowledge they couldn't track every step but were measuring samples of her movements. They knew that she went at least as far as they measured. This shows how scientists try to be exact when reporting results. Results of this study are depicted graphically in the map in Figure 2.

bouts was 7.1 km (± 0.9 SE, range 1.7–17.0 km). The average duration of foraging bouts for the period was 20.9 h (± 4.5 SE, range 1–71 h).

The average daily distance between the wolf and the nearest collared caribou decreased from 242 km on 12 July, one week before the excursion period, to 126 km on 19 July, the day the excursion began (Table 1).

23 **Excursion Period:** On 19 July at 2203, after spending 14 h at the den, 388 began moving to the northeast and did not return for 336 h (14 d; Figure 2). Whether she traveled alone or with other wolves is unknown. During the excursion, 476 (71%) of 672 possible locations were recorded. The wolf crossed the southeast end of Lac Capot Blanc on a small land bridge, where she paused for 4.5 h after traveling for 19.5 h (37.5 km). Following this rest, she traveled for 9 h (26.3 km) onto a peninsula in Reid Lake, where she spent 2 h before backtracking and stopping for 8 h just off the peninsula. Her next period of travel lasted 16.5 h (32.7 km), terminating in a pause of 9.5 h just 3.8 km from a concentration of locations at the far end of her excursion, where we presume she encountered caribou. The mean duration of these three movement periods was 15.7 h (± 2.5 SE), and that of the pauses, 7.3 h (± 1.5). The wolf required 72.5 h (3.0 d) to travel a minimum of 95 km from her den to this area near caribou (Figure 2). She remained there (35.5 km2) for 151.5 h (6.3 d) and then moved south to Lake of the Enemy, where she stayed (31.9 km²) for 74 h (3.1 d) before returning to her den. Her greatest distance from the den, 103 km, was recorded 174.5 h (7.3 d) after the excursion

24 Post-excursion measurements of 388's movements were made to compare with those of the pre-excursion period. In order to compare, scientists often use *means*, or averages, of a series of measurements—mean distances, mean duration, etc.

25 In the comparison, authors used statistical calculations (F and df) to determine that the differences between pre- and post-excursion measurements were *statistically insignificant*, or close enough to be considered essentially the same or similar.

26 As with wolf 388, the investigators measured the movements of caribou during the study period. The areas within which the caribou moved are shown in Figure 2 by shaded polygons mentioned in the second paragraph of this subsection.

27 This subsection summarizes how distances separating predators and prey varied during the study period.

28 DISCUSSION: This section is the *interpretation* part of the scientific process.

29 This subsection reviews observations from other studies and suggests that this study fits with patterns of those observations.

30 Authors discuss a prevailing *theory* (CBFT) which might explain why a wolf would travel far to meet her own energy needs while taking food caught closer to the den back to her pups. The results of this study seem to fit that pattern.

---

began, at 0433 on 27 July. She was 8 km from a collared caribou on 23 July, four days after the excursion began (Table 1).

The return trip began at 0403 on 2 August, 318 h (13.2 d) after leaving the den. She followed a relatively direct path for 18 h back to the den, a distance of 75 km.

The minimum distance traveled during the excursion was 339 km. The estimated overall minimum travel rate was 3.1 km/h, 2.6 km/h away from the den and 4.2 km/h on the return trip.

24 **Post-Excursion Period:** We saw three pups when recovering the collar on 20 August, but others may have been hiding in vegetation.

Telemetry recorded 13 foraging bouts in the post-excursion period. The mean distance traveled during these bouts was 18.3 km (+ 2.7 SE, range 1.2–47.7 km), and mean greatest distance from the den was 7.1 km (+ 0.7 SE, range 1.1–11.0 km). The mean duration of these post-excursion foraging bouts was 10.9 h (+ 2.4 SE, range 1–33 h).

When 388 reached her den on 2 August, the distance to the nearest collared caribou was 54 km. On 9 August, one week after she returned, the distance was 79 km (Table 1).

Pre- and Post-Excursion Comparison

25 We found no differences in the mean distance of foraging bouts before and after the excursion period (F = 1.5, df = 1, 29, p = 0.24). Likewise, the mean greatest distance from the den was similar pre- and post-excursion (F = 0.004, df = 1, 29, p = 0.95). However, the mean duration of 388's foraging bouts decreased by 10.0 h after her long excursion (F = 3.1, df = 1, 29, p = 0.09).

26 *Caribou Monitoring*

Summer Movements: On 10 July, 5 of 11 collared caribou were dispersed over a distance of 10 km, 140 km south of their calving grounds (Figure 1). On the same day, three caribou were still on the calving grounds, two were between the calving grounds and the leaders, and one was missing. One week later (17 July), the leading radio-collared cows were 100 km farther south (Figure 1). Two were within 5 km of each other in front of the rest, who were more dispersed. All radio-collared cows had left the calving grounds by this time. On 23 July, the leading radio-collared caribou had moved 35 km farther south, and all of them were more widely dispersed. The two cows closest to the leader were 26 km and 33 km away, with 37 km between them. On the next location (29 July), the most southerly caribou were 60 km

farther south. All of the caribou were now in the areas where they remained for the duration of the study (Figure 2).

A Minimum Convex Polygon (Mohr and Stumpf, 1966) around all caribou locations acquired during the study encompassed 85 119 km^2.

Relative to the Wolf Den: The distance from the **27** nearest collared caribou to the den decreased from 241 km one week before the excursion to 124 km the day it began. The nearest a collared caribou came to the den was 43 km away, on 29 and 30 July. During the study, four collared caribou were located within 100 km of the den. Each of these four was closest to the wolf on at least one day during the period reported.

28 Discussion

Prey Abundance

Caribou are the single most important prey of tundra **29** wolves (Clark, 1971; Kuyt, 1972; Stephenson and James, 1982; Williams, 1990). Caribou range over vast areas, and for part of the summer, they are scarce or absent in wolf home ranges (Heard et al., 1996). Both the long distance between radio-collared caribou and the den the week before the excursion and the increased time spent foraging by wolf 388 indicate that caribou availability near the den was low. Observations of the pups' being left alone for up to 18 h, presumably while adults were searching for food, provide additional support for low caribou availability locally. Mean foraging bout duration decreased by 10.0 h after the excursion, when collared caribou were closer to the den, suggesting an increase in caribou availability nearby.

Foraging Excursion

One aspect of central place foraging theory (CPFT) **30** deals with the optimality of returning different-sized food loads from varying distances to dependents at a central place (i.e., the den) (Orians and Pearson, 1979). Carlson (1985) tested CPFT and found that the predator usually consumed prey captured far from the central place, while feeding prey captured nearby to dependants. Wolf 388 spent 7.2 days in one area near caribou before moving to a location 23 km back towards the den, where she spent an additional 3.1 days, likely hunting caribou. She began her return trip from this closer location, traveling directly to the den. While away, she may have made one or more successful kills and spent time meeting her own energetic needs before returning to the den. Alternatively, it may have taken several attempts to make a kill, which she then fed on before beginning her return trip. We do not know if she returned food to the pups, but such behavior would be supported by CPFT.

31 Other workers have reported wolves' making long round trips and referred to them as "extraterritorial" or "pre-dispersal" forays (Fritts and Mech, 1981; Messier, 1985; Ballard et al., 1997; Merrill and Mech, 2000). These movements are most often made by young wolves (1–3 years old), in areas where annual territories are maintained and prey are relatively sedentary (Fritts and Mech, 1981; Messier, 1985). The long excursion of 388 differs in that tundra wolves do not maintain annual territories (Walton et al., 2001), and the main prey migrate over vast areas (Gunn et al., 2001).

Another difference between 388's excursion and those reported earlier is that she is a mature, breeding female. No study of territorial wolves has reported reproductive adults making extraterritorial movements in summer (Fritts and Mech, 1981; Messier, 1985; Ballard et al., 1997; Merrill and Mech, 2001). However, Walton et al. (2001) also report that breeding female tundra wolves made excursions.

Direction of Movement

32 Possible explanations for the relatively direct route 388 took to the caribou include landscape influence and experience. Considering the timing of 388's trip and the locations of caribou, had the wolf moved northwest, she might have missed the caribou entirely, or the encounter might have been delayed.

A reasonable possibility is that the land directed 388's route. The barrens are crisscrossed with trails worn into the tundra over centuries by hundreds of thousands of caribou and other animals (Kelsall, 1968; Thorpe et al., 2001). At river crossings, lakes, or narrow peninsulas, trails converge and funnel towards and away from caribou calving grounds and summer range. Wolves use trails for travel (Paquet et al., 1996; Mech and Boitani, 2003; P. Frame, pers. observation). Thus, the landscape may direct an animal's movements and lead it to where cues, such as the odor of caribou on the wind or scent marks of other wolves, may lead it to caribou.

33 Another possibility is that 388 knew where to find caribou in summer. Sexually immature tundra wolves sometimes follow caribou to calving grounds (D. Heard, unpubl. data). Possibly, 388 had made such journeys in previous years and killed caribou. If this were the case, then in times of local prey scarcity she might travel to areas where she had hunted successfully before. Continued monitoring of tundra wolves may answer questions about how their food needs are met in times of low caribou abundance near dens.

34 Caribou often form large groups while moving south to the tree line (Kelsall, 1968). After a large aggregation of caribou moves through an area, its scent can linger for weeks (Thorpe et al., 2001:104). It is conceivable that 388 detected caribou scent on the wind, which was blowing from the northeast on 19–21 July (Environment Canada, 2003), at the same time her excursion began. Many factors, such as odor strength and wind direction and strength, make systematic study of scent detection in wolves difficult under field conditions (Harrington and Asa, 2003). However, humans are able to smell odors such as forest fires or oil refineries more than 100 km away. The olfactory capabilities of dogs, which are similar to wolves, are thought to be 100 to 1 million times that of humans (Harrington and Asa, 2003). Therefore, it is reasonable to think that under the right wind conditions, the scent of many caribou traveling together could be detected by wolves from great distances, thus triggering a long foraging bout.

Rate of Travel

35 Mech (1994) reported the rate of travel of Arctic wolves on barren ground was 8.7 km/h during regular travel and 10.0 km/h when returning to the den, a difference of 1.3 km/h. These rates are based on direct observation and exclude periods when wolves moved slowly or not at all. Our calculated travel rates are assumed to include periods of slow movement or no movement. However, the pattern we report is similar to that reported by Mech (1994), in that homeward travel was faster than regular travel by 1.6 km/h. The faster rate on return may be explained by the need to return food to the den. Pup survival can increase with the number of adults in a pack available to deliver food to pups (Harrington et al., 1983). Therefore, an increased rate of travel on homeward trips could improve a wolf's reproductive fitness by getting food to pups more quickly.

Fate of 388's Pups

36 Wolf 388 was caring for pups during den observations. The pups were estimated to be six weeks old, and were seen ranging as far as 800 m from the den. They received some regurgitated food from two of the females, but were unattended for long periods. The excursion started 16 days after our observations, and it is improbable that the pups could have traveled the distance that 388 moved. If the pups died, this would have removed parental responsibility, allowing the long movement.

Our observations and the locations of radio-collared caribou indicate that prey became scarce in

31 Here our authors note other possible explanations for wolves' excursions presented by other investigators, but this study does not seem to support those ideas.

32 Authors discuss possible reasons for why 388 traveled directly to where caribou were located. They take what they learned from earlier studies and apply it to this case, suggesting that the lay of the land played a role. Note that their description paints a clear picture of the landscape.

33 Authors suggest that 388 may have learned in traveling during previous summers where the caribou were. The last two sentences suggest ideas for future studies.

34 Or maybe 388 followed the scent of the caribou. Authors acknowledge difficulties of proving this, but they suggest another area where future studies might be done.

35 Authors suggest that results of this study support previous studies about how fast wolves travel to and from the den. In the last sentence, they speculate on how these observed patterns would fit into the theory of evolution.

36 Authors also speculate on the fate of 388's pups while she was traveling. This leads to . . .

37 Discussion of cooperative rearing of pups and, in turn, to speculation on how this study and what is known about cooperative rearing might fit into the animal's strategies for survival of the species. Again, the authors approach the broader theory of evolution and how it might explain some of their results.

38 And again, they suggest that this study points to several areas where further study will shed some light.

39 In conclusion, the authors suggest that their study supports the hypothesis being tested here. And they touch on the implications of increased human activity on the tundra predicted by their results.

40 ACKNOWLEDGEMENTS: Authors note the support of institutions, companies and individuals. They thank their reviewers ad list permits under which their research was carried on.

41 REFERENCES: List of all studies cited in the report. This may seem tedious, but is a vitally important part of scientific reporting. It is a record of the sources of information on which this study is based. It provides readers with a wealth of resources for further reading on this topic. Much of it will form the foundation of future scientific studies like this one.

---

the area of the den as summer progressed. Wolf 388 may have abandoned her pups to seek food for herself. However, she returned to the den after the excursion, where she was seen near pups. In fact, she foraged in a similar pattern before and after the excursion, suggesting that she again was providing for pups after her return to the den.

37 A more likely possibility is that one or both of the other lactating females cared for the pups during 388's absence. The three females at this den were not seen with the pups at the same time. However, two weeks earlier, at a different den, we observed three females cooperatively caring for a group of six pups. At that den, the three lactating females were observed providing food for each other and trading places while nursing pups. Such a situation at the den of 388 could have created conditions that allowed one or more of the lactating females to range far from the den for a period, returning to her parental duties afterwards. However, the pups would have been weaned by eight weeks of age (Packard et al., 1992), so nonlactating adults could also have cared for them, as often happens in wolf packs (Packard et al., 1992; Mech et al., 1999).

Cooperative rearing of multiple litters by a pack could create opportunities for long-distance foraging movements by some reproductive wolves during summer periods of local food scarcity. We have recorded multiple litters at one or more tundra wolf dens per year since 1997. This reproductive strategy may be an adaptation to temporally and **38** spatially unpredictable food resources. All of these possibilities require further study, but emphasize both the adaptability of wolves living on the barrens and their dependence on caribou.

Long-range wolf movement in response to caribou **39** availability has been suggested by other researchers (Kuyt, 1972; Walton et al., 2001) and traditional ecological knowledge (Thorpe et al., 2001). Our report demonstrates the rapid and extreme response of wolves to caribou distribution and movements in summer. Increased human activity on the tundra (mining, road building, pipelines, ecotourism) may influence caribou movement patterns and change the interactions between wolves and caribou in the region. Continued monitoring of both species will help us to assess whether the association is being affected adversely by anthropogenic change.

40 Acknowledgements

This research was supported by the Department of Resources, Wildlife, and Economic Development, Government of the Northwest Territories; the Department of Biological Sciences at the University of Alberta; the Natural Sciences and Engineering Research Council of Canada; the Department of Indian and Northern Affairs Canada; the Canadian Circumpolar Institute; and DeBeers Canada, Ltd. Lorna Ruechel assisted with den observations. A. Gunn provided caribou location data. We thank Dave Mech for the use of GPS collars. M. Nelson, A. Gunn, and three anonymous reviewers made helpful comments on earlier drafts of the manuscript. This work was done under Wildlife Research Permit – WL002948 issued by the Government of the Northwest Territories, Department of Resources, Wildlife, and Economic Development.

41 References

BALLARD, W.B., AYRES, L.A., KRAUSMAN, P.R., REED, D.J., and FANCY, S.G. 1997. Ecology of wolves in relation to a migratory caribou herd in northwest Alaska. Wildlife Monographs 135. 47 p.

BROWN, M.B., and FORSYTHE, A.B. 1974. Robust tests for the equality of variances. Journal of the American Statistical Association 69:364–367.

CARLSON, A. 1985. Central place foraging in the red-backed shrike (*Lanius collurio* L.): Allocation of prey between forager and sedentary consumer. Animal Behaviour 33:664–666.

CLARK, K.R.F. 1971. Food habits and behavior of the tundra wolf on central Baffin Island. Ph.D. Thesis, University of Toronto, Ontario, Canada.

ENVIRONMENT CANADA. 2003. National climate data information archive. Available online: http://www.climate.weatheroffice.ec.gc.ca/Welcome_e.html

FOURNIER, B., and GUNN, A. 1998. Musk ox numbers and distribution in the NWT, 1997. File Report No. 121. Yellowknife: Department of Resources, Wildlife, and Economic Development, Government of the Northwest Territories. 55 p.

FRITTS, S.H., and MECH, L.D. 1981. Dynamics, movements, and feeding ecology of a newly protected wolf population in northwestern Minnesota. Wildlife Monographs 80. 79 p.

GUNN, A., DRAGON, J., and BOULANGER, J. 2001. Seasonal movements of satellite-collared caribou from the Bathurst herd. Final Report to the West Kitikmeot Slave Study Society, Yellowknife, NWT. 80 p. Available online: http://www.wkss.nt.ca/HTML/08_ProjectsReports/PDF/SeasonalMovementsFinal.pdf

HARRINGTON, F.H., and ASA, C.S. 2003. Wolf communication. In: Mech, L.D., and Boitani, L., eds. Wolves: Behavior, ecology, and conservation. Chicago: University of Chicago Press. 66–103.

HARRINGTON, F.H., MECH, L.D., and FRITTS, S.H. 1983. Pack size and wolf pup survival: Their relationship under varying ecological conditions. Behavioral Ecology and Sociobiology 13:19–26.

HARRIS, S.A. 1986. Permafrost distribution, zonation and stability along the eastern ranges of the cordillera of North America. Arctic 39(1):29–38.

HEARD, D.C., and WILLIAMS, T.M. 1992. Distribution of wolf dens on migratory caribou ranges in the Northwest

Territories, Canada. Canadian Journal of Zoology 70:1504–1510.

HEARD, D.C., WILLIAMS, T.M., and MELTON, D.A. 1996. The relationship between food intake and predation risk in migratory caribou and implication to caribou and wolf population dynamics. Rangifer Special Issue No. 2:37–44.

KELSALL, J.P. 1968. The migratory barren-ground caribou of Canada. Canadian Wildlife Service Monograph Series 3. Ottawa: Queen's Printer. 340 p.

KUYT, E. 1962. Movements of young wolves in the Northwest Territories of Canada. Journal of Mammalogy 43:270–271.

———. 1972. Food habits and ecology of wolves on barren-ground caribou range in the Northwest Territories. Canadian Wildlife Service Report Series 21. Ottawa: Information Canada. 36 p.

MECH, L.D. 1994. Regular and homeward travel speeds of Arctic wolves. Journal of Mammalogy 75:741–742.

MECH, L.D., and BOITANI, L. 2003. Wolf social ecology. In: Mech, L.D., and Boitani, L., eds. Wolves: Behavior, ecology, and conservation. Chicago: University of Chicago Press. 1–34.

MECH, L.D., and GESE, E.M. 1992. Field testing the Wildlink capture collar on wolves. Wildlife Society Bulletin 20:249–256.

MECH, L.D., WOLFE, P., and PACKARD, J.M. 1999. Regurgitative food transfer among wild wolves. Canadian Journal of Zoology 77:1192–1195.

MERRILL, S.B., and MECH, L.D. 2000. Details of extensive movements by Minnesota wolves (Canis lupus). American Midland Naturalist 144:428–433.

MERRILL, S.B., ADAMS, L.G., NELSON, M.E., and MECH, L.D. 1998. Testing releasable GPS radiocollars on wolves and white-tailed deer. Wildlife Society Bulletin 26:830–835.

MESSIER, F. 1985. Solitary living and extraterritorial movements of wolves in relation to social status and prey abundance. Canadian Journal of Zoology 63:239–245.

MOHR, C.O., and STUMPF, W.A. 1966. Comparison of methods for calculating areas of animal activity. Journal of Wildlife Management 30:293–304.

MUSIANI, M. 2003. Conservation biology and management of wolves and wolf-human conflicts in western North America. Ph.D. Thesis, University of Calgary, Calgary, Alberta, Canada.

ORIANS, G.H., and PEARSON, N.E. 1979. On the theory of central place foraging. In: Mitchell, R.D., and Stairs, G.F., eds. Analysis of ecological systems. Columbus: Ohio State University Press. 154–177.

PACKARD, J.M., MECH, L.D., and REAM, R.R. 1992. Weaning in an arctic wolf pack: Behavioral mechanisms. Canadian Journal of Zoology 70:1269–1275.

PAQUET, P.C., WIERZCHOWSKI, J., and CALLAGHAN, C. 1996. Summary report on the effects of human activity on gray wolves in the Bow River Valley, Banff National Park, Alberta. In: Green, J., Pacas, C., Bayley, S., and Cornwell, L., eds. A cumulative effects assessment and futures outlook for the Banff Bow Valley. Prepared for the Banff Bow Valley Study. Ottawa: Department of Canadian Heritage.

STEPHENSON, R.O., and JAMES, D. 1982. Wolf movements and food habits in northwest Alaska. In: Harrington, F.H., and Paquet, P.C., eds. Wolves of the world. New Jersey: Noyes Publications. 223–237.

THORPE, N., EYEGETOK, S., HAKONGAK, N., and QITIR-MIUT ELDERS. 2001. The Tuktu and Nogak Project: A caribou chronicle. Final Report to the West Kitikmeot/Slave Study Society, Ikaluktuuttiak, NWT. 160 p.

TIMONEY, K.P., LA ROI, G.H., ZOLTAI, S.C., and ROBINSON, A.L. 1992. The high subarctic forest-tundra of northwestern Canada: Position, width, and vegetation gradients in relation to climate. Arctic 45(1):1–9.

WALTON, L.R., CLUFF, H.D., PAQUET, P.C., and RAMSAY, M.A. 2001. Movement patterns of barren-ground wolves in the central Canadian Arctic. Journal of Mammalogy 82:867–876.

WILLIAMS, T.M. 1990. Summer diet and behavior of wolves denning on barren-ground caribou range in the Northwest Territories, Canada. M.Sc. Thesis, University of Alberta, Edmonton, Alberta, Canada.

ZAR, J.H. 1999. Biostatistical analysis. 4th ed. New Jersey: Prentice Hall. 663 p.

Glossary of Biological Terms

ABC model Model for the genetic basis of flower formation; products of three master genes (*A*, *B*, *C*) control the development of sepals, petals, and stamens and carpels from meristematic tissue.

ABC transporter One of a distinct class of membrane proteins, each a channel or pump for a specific hydrophobic substance (e.g., ions, sugars, amino acids).

ABO blood typing Method of identifying which self-recognition proteins of types A and B are at the surface of an individual's red blood cells; the absence of either type is designated O.

abortion Premature expulsion of embryo or fetus from the uterus; called miscarriage when spontaneous rather than induced.

abscisic acid Plant hormone; stimulates stomatal closure in response to water stress, protein storage in seeds, maturation of embryo sporophyte; possibly induces and maintains dormancy in some plants.

abscission (ab-SIH-zhun) The dropping of leaves, flowers, fruits, or other parts from a plant in response to seasonal change, drought, injury, or nutrient deficiencies.

absorption Of pigments, interception of photon energy. Of cells, uptake of water and solutes from the surroundings. Of digestion, uptake of water and solutes into the internal environment.

absorption spectrum Range of wavelengths that a given type of pigment can absorb.

accessory pigment A pigment that absorbs and transfers light energy to a photosystem; extends the range of light wavelengths for photosynthesis; e.g., a carotenoid.

acclimatization Processes by which the body adjusts in physiology and behavior to a new environment; e.g., after moving from sea level to a high-altitude habitat.

acetylation The attachment of an acetyl group (CH3CO) to an organic compound.

acetylcholine (ACh) Neurotransmitter with stimulatory or inhibitory effects in the brain, spinal cord, glands, and muscles.

acetyl–CoA Coenzyme A bound to a two-carbon fragment from pyruvate, which it transfers to oxaloacetate for Krebs cycle.

acid [L. *acidus*, sour] Any water-soluble substance that releases hydrogen ions (H^+) in water, yielding a pH below 7.0.

acid–base balance Outcome of control over solute concentrations; extracellular fluid is neither too acidic nor too basic (alkaline).

acidity A solution with a pH of less than 7.

acid rain Acidic precipitation; rain or snow with high levels of sulfur and nitrogen oxides.

acoelomate animal Any invertebrate with no fluid-filled cavity between its gut and body wall; e.g., a flatworm.

acoustical receptor Any mechanoreceptor sensitive to sound.

acoustical signal A sound or sounds used in intraspecific communication.

actin Protein monomer of microfilaments that functions in contraction, cell division, and reinforcing or reconfiguring the shape of a cell or its contents.

action potential Of excitable cells, a self-propagating, abrupt reversal in the voltage difference across the plasma membrane.

activation energy Minimum amount of energy required to start a reaction; enzyme action lowers this energy barrier. Reactions differ in the amount required.

activator A regulatory molecule that binds to specific sequences in DNA and thereby promotes transcription.

active site Chemically stable crevice in an enzyme where substrates bind and a reaction can be catalyzed repeatedly.

active transport Pumping of a specific solute across a cell membrane against its concentration gradient, through the interior of a transport protein. Requires energy input, as from ATP.

adaptation, evolutionary [L. *adaptare*, to fit] Any long-term, heritable aspect of form, function, or behavior that improves an individual's chances of surviving and reproducing; outcome of natural selection and other microevolutionary processes.

adaptation, sensory Of sensory neurons, a decline or cessation of action potentials when a stimulus of constant strength does not end.

adaptive immunity The mechanisms that defend the vertebrate body against specific threats to health, as characterized by self/nonself recognition, specificity, diversity, and memory. All antibody-mediated and cell-mediated responses to antigen.

adaptive radiation A macroevolutionary pattern. A burst of genetic divergences from a lineage that gives rise to many species, each able to use a novel resource or to move into a new, or newly vacated, habitat.

adaptive trait An aspect of form, function, or behavior that helps an individual survive and reproduce under prevailing conditions.

adaptive zone A set of different niches that become be filled by a group of species.

adenine One of four nitrogen-containing bases in nucleotide monomers of DNA or RNA; also refers to a nucleotide having an adenine base component.

ADH Antidiuretic hormone. Hypothalamic hormone released by the posterior pituitary that induces water conservation by kidneys.

adhering junction Complex of adhesion proteins that anchors cells to each other and to extracellular matrixes.

adhesion protein Of multicelled species, a plasma membrane protein that helps cells stick together in tissues and to extracellular matrixes such as basement membrane.

adipose tissue Type of connective tissue having an abundance of fat-storing cells.

ADP Adenosine diphosphate (ah-DEN-uh-seen die-FOSS-fate). A nucleotide with an adenine base and two phosphate groups.

adrenal cortex Outer zone of the adrenal gland; secretes cortisol and aldosterone.

adrenal gland Endocrine gland located on top of the kidney; regulates stress reponses and influences glucose metabolism.

adrenal medulla Inner zone of adrenal gland that secretes epinephrine and norepinephrine.

aerobic respiration (air-OH-bik) [Gk. *aer*, air, + *bios*, life] Oxygen-requiring pathway of ATP formation in mitochondria: from glycolysis, to Krebs cycle and electron transport phosphorylation. Typical net energy yield: 36 ATP per glucose molecule.

African emergence model Model for the origin of modern humans; *Homo sapiens* is said to have originated in Africa, then replaced archaic *Homo* populations in different parts of the world.

age structure Of a population, the number of individuals in each age category.

agglutination (ah-glue-tin-AY-shun) A vertebrate defense response; antibodies bind antigen and form insoluble clumps that attract phagocytes.

aging Of complex multicelled organisms, a time-dependent progressive deterioration of molecules, cells, tissues, and organs that weakens the body's capacity to function.

AIDS Acquired immune deficiency syndrome. A set of chronic disorders that develops after prolonged infection by HIV has weakened the immune system.

alcohol Organic compound having one or more hydroxyl groups that dissolves easily in water; e.g., ethanol.

alcoholic fermentation An anaerobic ATP-forming pathway using pyruvate and NADH from glycolysis. NADH transfers electrons to an intermediate, acetaldehyde, forming ethanol. Occurs in cytoplasm only. The net yield is 2 ATP from glycolysis; the steps remaining only regenerate NAD^+.

aldosterone (al-DOSS-tuh-rohn) Adrenal cortex hormone; acts in kidneys to promote sodium retention.

alga, plural **algae** Informal term for groups of single-celled or multicelled eukaryotic photoautotrophs, mostly aquatic; e.g., kelps.

algal bloom Rapid, huge increases in algal population sizes after nutrient enrichment of an aquatic habitat.

alkylation Of a molecule, replacement of a hydrogen atom by a hydrocarbon group.

allantois (ah-LAN-twahz) [Gk. *allas*, sausage] One of four extraembryonic membranes of amniote eggs. In reptiles, birds, and some mammals, it exchanges gases and stores metabolic wastes; in humans, it helps form placental blood vessels, urinary bladder.

allele One of two or more molecular forms of a gene at a given locus; alleles arise by mutation and encode slightly different versions of the same trait.

allele frequency Abundance of one allele relative to others at a gene locus among individuals of a population.

allergen A normally harmless substance that can provoke immune responses.

allergy Hypersensitivity to an allergen.

allopatric speciation [Gk. *allos*, different, + L. *patria*, native land] Speciation model. A physical barrier arises and separates populations or subpopulations of a species, ends gene flow, and so favors divergences that result in new species.

alternation of generations The alternation of haploid (gamete-producing) and diploid (spore-producing) phases in the life cycle of an organism.

alternative splicing Event by which the same gene can specify two or more slightly different proteins. All exons in a pre-mRNA transcript of the gene are retained or some are removed and the rest spliced in various combinations for the mature transcript.

altruism (AL-true-IZ-um) Social behavior that decreases the individual's chance of reproductive success while improving the chances for others of its species.

alveolate A type of single-celled eukaryote that has many tiny, membrane-bound sacs just beneath the plasma membrane; e.g., a ciliate, apicomplexan, or dinoflagellate.

alveolus (ahl-VEE-uh-lus), plural **alveoli** At the endings of respiratory bronchioles, one of great numbers of thin-walled, cup-shaped outpouching where air in the lungs exchange gases with blood.

amino acid (uh-MEE-no) A small organic compound with a carboxylic acid group, an amino group, and a characteristic side group (R); monomer of polypeptide chains.

ammonification (uh-moan-ih-fih-KAY-shun) Part of the nitrogen cycle; soil fungi and bacteria decompose nitrogen-containing compounds, the result being ammonia and ammonium ions that plant roots can absorb.

amnion One of the four extraembryonic membranes of amniote eggs; the boundary layer of a fluid-filled, cushioning sac in which the embryo develops.

amniote Type of tetrapod that produces amniote eggs. Major groups are synapsids (mammals and early mammal-like reptiles) and sauropsids ("reptiles" and birds).

amniote egg An egg, often shelled, with four extraembryonic membranes (amnion, chorion, yolk sac, allantois). Pivotal factor in the early evolution of reptiles, birds, and mammals in habitats on land.

amoeba A single-celled amoebozoan that moves on pseudopods. All are predatory or parasitic; none forms colonies.

amphibian A tetrapod, or a descendant of one, having a body plan and reproductive mode between fishes and reptiles.

anaerobic electron transfer (an-uh-ROW-bik) [Gk. *an*, without, + *aer*, air] Of some bacteria and archaeans, ATP formation by way of a flow of electrons through transfer chains in the plasma membrane to a final electron acceptor that is not oxygen.

anaerobic pathway Any set of metabolic reactions for which a substance other than oxygen is the final acceptor of electrons stripped from substrates.

anagenesis A major pattern of speciation. Directional changes in allele frequencies and morphology are confined within a single lineage, and in time a new type differs so much from the ancestral type that it is classified as a separate species.

analogous structures (an-AL-uh-gus) [Gk. *analogos*, similar to one another] Dissimilar body parts that have become similar in structure, function, or both in lineages that are not closely related but were subjected to similar pressures.

anaphase, meiosis Nuclear division stage. In anaphase I, each chromosome and its homologue move to opposite poles; both are still duplicated. In anaphase II, sister chromatids of each chromosome separate from each other, move to opposite poles.

anaphase, mitosis (AN-uh-faze) Nuclear division stage. Sister chromatids of each chromosome are separated from each other and move to opposite spindle poles.

anatomy Study of the internal parts of an organism to ascertain their structure and positions relative to one another. Now conducted at levels of organization from molecules through organ systems.

anemia Type of disorder resulting from having too few functional red blood cells.

aneuploidy (AN-yoo-ploy-dee) A type of chromosome abnormality in which body (somatic) cells have one extra or one less chromosome relative to the parental chromosome number.

angiosperm [Gk. *angeion*, vessel + *spermia*, seed] A flowering plant; its egg-containing ovules mature into seeds within closed, protected chambers called ovaries.

angiotensin II Plasma protein that helps constrict arterioles and stimulates ADH and aldosterone secretion.

animal Any multicelled heterotroph that ingests other organisms or their tissues, develops through a series of embryonic stages, and is motile during part or all of the life cyle. Most species have epithelial tissues and extracellular matrixes.

animal behavior A coordinated response to stimuli that involves motor, neural, and endocrine components. Animal behavior has a genetic basis, it can evolve, and it can be modified by learning.

Animalia Kingdom of animals.

annelid A bilateral invertebrate having a highly segmented body; major groups are polychaetes, oligochaetes, and leeches. Except in leeches, segments have clusters of chitin-reinforced bristles.

annual A plant having a life cycle that starts and ends in one growing season.

anthocyanin One of a class of accessory pigments that reflect red to blue light.

anthropoid One of a group of primates; a monkey, ape, or human.

antibiotic [Gk. *anti*, against] In nature, a metabolic product of certain bacteria and fungi in soil that is toxic to their microbial competitors for nutrients.

antibody Antigen-binding glycoprotein made and secreted only by B cells; during adaptive immunity, activates complement, neutralizes toxins, enhances phagocytosis, immobilizes internal pathogens or parasites.

antibody-mediated immune response One of two arms of adaptive immunity in which antibodies are produced in response to a specific antigen; mediated by B cells.

anticodon Series of three nucleotide bases in tRNA that can base-pair with mRNA codons.

antigen (AN-tih-jen) Any molecular pattern that triggers an immune response.

antigen–MHC complex Fragment(s) of antigen bound to MHC markers at a cell's plasma membrane; recognized by T cells.

antigen-presenting cell Lymphocyte that binds and processes antigen to present to T cells as antigen–MHC complexes; secretes cytokines that stimulate proliferation and differentiation of lymphocytes.

antioxidant Any enzyme or cofactor that helps neutralize free radicals before they damage tissues.

anus The terminal opening of a complete digestive system.

aorta (ay-OR-tah) Of vertebrates, the main artery of systemic circulation.

apical dominance (AY-pih-kul) Growth-inhibiting effect on lateral (axillary) buds, caused by auxin diffusing down a shoot tip from the terminal bud.

apical meristem (MARE-ih-stem) [L. *apex*, top, + Gk. *meristos*, divisible] Mass of dividing cells at root tips and shoot tips.

apicomplexan A parasitic alveolate having a unique microtubular device at its anterior end; the device attaches to and penetrates a host cell.

apoptosis (APP-oh-TOE-sis) Programmed cell death. A cell is induced to commit suicide as part of growth, development, and maintenance of a multicelled body.

appendicular skeleton (ap-en-DIK-yoo-lahr) Bones of limbs, hips, and shoulders.

appendix Small, narrow outpouching from the cecum, vulnerable to infection.

aquaporin A type of passive transporter that assists diffusion of water molecules across the plasma membrane.

Archaea Domain of prokaryotic species; one of two lineages that evolved shortly after life originated. Archaeans have many unique molecular and biochemical traits but also share some traits with bacteria and other traits with eukaryotic species.

Archean Eon extending from the time that life originated, 3.8 billion years ago, to 2.5 billion years ago.

archipelago A chain or cluster of islands, often of volcanic origin in the open ocean.

area effect Biogeographical pattern; larger islands support more species than smaller ones at equivalent distances from sources of colonizer species.

arteriole (ar-TEER-ee-ole) The type of blood vessel in between arteries and capillaries. Selectively distributing more of the total blood volume to different organs in a given interval requires controls over dilation and constriction of the diameter of arterioles.

arteriosclerosis Chronic disease in which the wall of arteries thickens abnormally, hardens, and loses its elasticity.

artery A thick-walled, muscular, rapid-transport vessel that smooths out pulses of pressure generated by heartbeats.

arthropod Type of invertebrate having a hardened exoskeleton and specialized segments with jointed appendages; e.g., millipedes, spiders, lobsters, insects.

artificial selection Manipulation of the reproduction of a species as by breeding practices. Only individuals of a captive population that display a valued trait are allowed to reproduce, the goal being to increase the trait's magnitude and frequency over the generations.

ascospore Sexual spore of sac fungi.

asexual reproduction Any reproductive mode by which offspring arise from one parent and inherit that parent's genes only; e.g., prokaryotic fission, transverse fission, budding, vegetative propagation.

atherosclerosis A medical condition in which arteries narrow in diameter as lipids and other deposits accumulate in their wall.

atmosphere, Earth's The volume of gases, water vapor, and airborne particles that envelopes Earth's surface.

atom The smallest unit of an element that still retains the element's properties.

atomic number The number of protons in the nucleus of atoms of a given element.

ATP Adenosine triphosphate (ah-DEN-uh-seen try-FOSS-fate). A type of nucleotide that functions as the main energy carrier between reaction sites in cells. Consists of the base adenine, the five-carbon sugar ribose, and three phosphate groups.

ATP/ADP cycle How a cell regenerates its ATP supply. ADP forms when ATP gives up a phosphate group, then ATP forms as ADP binds to inorganic phosphate or a phosphate group split from a molecule.

ATP synthase A type of membrane-bound active transport protein that also catalyzes the formation of ATP.

australopith (OHSS-trah-low-pith) [L. *australis*, southern, + Gk. *pithekos*, ape] One of the early hominids of Africa.

autoimmune response Inappropriate lymphocyte attack on normal body cells.

automated DNA sequencing Extremely rapid, robotic method of identifying the nucleotide sequence of a region of DNA. Gel electrophoresis and laser detection of fluorescent tracers are part of method.

autonomic nerve (AH-toe-NOM-ik) One of the nerves from the central nervous system that helps control smooth muscle, cardiac muscle, and glands of viscera.

autosome Of a sexually reproducing species, any chromosome of a type that is the same in both males and females.

autotroph [Gk. *auto*, self, and *trophos*, feeder] An organism that synthesizes its own food from simple inorganic compounds in its environment with energy captured from the sun or from oxidizing inorganic substances; e.g., a photoautotroph or chemoautotroph.

auxin (OX-in) A type of plant hormone that stimulates lengthening of shoots and coleoptiles, vascular cambium activity, and vascular tissue differentiation; also inhibits lateral bud formation, abscission, and fruit formation; responsive to gravity and light.

axial skeleton (AX-ee-uhl) Skull, backbone, ribs, and breastbone (sternum).

axon A neuron's signal-conducting zone; action potentials typically self-propagate away from the cell body on this slender, typically long process.

B lymphocyte B cell. Type of white blood cell central to immune responses; only cell that makes antibodies.

bacillus Rod-shaped prokaryotic cell.

Bacteria Domain of prokaryotic species; the first kinds of cells that formed after life originated. Collectively, bacteria are the most metabolically diverse organisms. Most kinds are chemoheterotrophs.

bacterial chromosome A circular, double-stranded molecule of prokaryotic DNA.

bacteriophage (bak-TEER-ee-oh-fahj) One of a class of viruses that infects bacteria.

balanced polymorphism An outcome of natural selection against homozygotes, so that two or more alleles for a trait are being maintained in the population.

bark Of woody plants, all tissues that are external to the vascular cambium.

Barr body Of the two X chromosomes in the somatic cells of female mammals, the one that has been condensed.

basal body An organelle that started out as a centriole, the source of a 9+2 array of microtubules in a cilium or flagellum. It remains below the finished array.

base Any water-soluble substance that releases hydroxyl ions in water to yield a pH greater than 7.0. Also the nitrogen-containing component of a nucleotide.

base-pair substitution Mutation in which one nucleotide is wrongly substituted for another during DNA replication.

base sequence Linear order of nucleotides that compose a DNA or RNA strand.

basidiospore Sexual spore of club fungi.

basophil White blood cell circulating in blood that secretes histamine and other substances with roles in inflammation.

bell curve Idealized statistical distribution of the continuous variation in a population for a trait of interest.

biennial (bi-EN-yul) A flowering plant that requires two growing seasons to complete its life cycle; flowers and fruits form in the second season.

big bang Model for the origin of universe, by a nearly instantaneous distribution of all matter and energy through all of space.

bilateral symmetry Body plan in which the main axis divides the body into two halves that are mirror images of one another.

bile Mix of salts, cholesterol, and pigments made by the liver and used in fat digestion.

binary fission Asexual reproductive mode of certain invertebrates; the body splits spontaneously, then both parts grow what is missing. *See also* Prokaryotic fission.

binding energy Energy released as weak bonds form between a substrate, enzyme, and any cofactor.

binomial system A scientific system of naming species, whereby each kind of organism receives a two-part name. The first part indicates the genus; the second part is the species epithet (descriptor).

biodiversity [Gk. *bios*, life] Biological diversity in an environment, as measured by the number of species and their relative abundances.

biofilm Large microbial populations that anchored themselves to epithelium, rocks, or other surfaces by their own secretions.

biogeochemical cycle Slow movement of an element from environmental reservoirs, through food webs, then back.

biogeographic realm [Gk. *bios*, life, + *geographein*, to describe Earth's surface] One of many vast expanses of land where one can expect to find communities of certain types of plants and animals.

biogeography Scientific study of patterns in the geographic distribution of species and communities.

biological clock Internal time-measuring mechanism by which individuals adjust their activities seasonally, daily, or both in response to environmental cues.

biological magnification Ever increasing concentration of a slowly degradable or nondegradable substance in body tissues as it is passed along food chains.

biological species concept Definition of a sexually reproducing species as one or more populations of individuals that interbreed under natural conditions, produce fertile offspring, and are reproductively isolated from other such populations.

biology The scientific study of life.

bioluminescence Fluorescent light formed when certain organisms convert chemical bond energy to photon energy.

biomass Of an ecosystem, the combined weight of all organisms at a trophic level.

biomass pyramid Chart in which the size of successive tiers depicts the measured

biomass (dry weight) of an ecosystem's producers, consumers, and decomposers.

biome One of the finer subdivisions of a biogeographic realm.

biosphere [Gk. *bios*, life, + *sphaira*, globe] All regions of Earth's waters, crust, and atmosphere in which organisms live.

biosynthetic pathway Any metabolic pathway by which one or more organic compounds are synthesized.

biotic potential The maximum rate of increase per individual for any population that is growing under ideal conditions.

bipedalism Habitually walking upright on two feet, as by ostriches and hominids.

bipolar spindle Of eukaryotic cells, a dynamic array of microtubules that moves chromosomes with respect to its two poles during mitosis or meiosis.

bird A warm-blooded, feathered amniote classified as a sauropsid; the dinosaurs and crocodilians are its closest relatives.

blastocyst (BLASS-tuh-sist) A type of blastula with a surface layer of blastomeres, a cavity filled with their secretions, and an inner cell mass; e.g., a human blastocyst.

blastomere One of the small, nucleated cells that forms during the cleavage stage of animal development.

blastula A ball of blastomeres and a cavity filled with their own secretions; outcome of the cleavage stage of animal development.

blood A fluid connective tissue that is the transport medium of circulatory systems. Mostly water in which ions, molecules, blood cells, and platelets are dissolved.

blood–brain barrier Specialized blood capillaries that protect the brain and spinal cord by exerting some control over which solutes enter the cerebrospinal fluid.

blood pressure Fluid pressure generated by heartbeats that causes blood circulation.

bone Type of vertebrate organ consisting of mineral-hardened connective tissue. Functions in movement, mineral storage, and protection; blood cells form in some.

bone remodeling Ongoing depositions and withdrawals of mineral ions from bone by certain bone cells; their antagonistic actions adjust bone strength as well as calcium and phosphorus levels in blood.

bone tissue Of vertebrate skeletons, a tissue of osteoblast secretions, which have become hardened with mineral ions.

bony fish Aquatic vertebrate that has an endoskeleton, including a cranium, mostly of bone tissue.

boreal forest *Taiga.* One of the extensive northern coniferous forests of Europe, Asia, and North America.

bottleneck Severe reduction in the size of a population, brought about by intense selection pressure or a natural calamity.

Bowman's capsule First part of a nephron, where its wall balloons back on itself in the shape of a double-layer cup; under the force of blood pressure, water and solutes are filtered out of blood and into the cup.

brain Of most nervous systems, a major integrating center that receives, processes, and often stores sensory input, and issues coordinated commands for responses.

brain stem The most ancient nerve tissue in all three divisions of a vertebrate brain.

bronchiole Finely branched airway that is part of the bronchial tree inside a lung.

bronchus, plural **bronchi** (BRONG-cuss, BRONG-kee) [Gk. *bronchos*, windpipe] A tubular airway that starts at the trachea and becomes the main branch into a lung.

brown alga A stramenopile; a multicelled marine photoautotroph with an abundance of the pigment fucoxanthin; e.g., kelps.

bryophyte Nonvascular land plant. The haploid stage dominates its life cycle, and its sperm require standing water to reach eggs. A moss, liverwort, or hornwort.

bud A dormant shoot, mostly meristematic tissue and often sheathed in small, young leaves. A lateral (axillary) bud forms in a leaf axil; a terminal bud forms at a shoot tip and is the main zone of primary growth.

buffer system A weak acid and the salt that forms when it dissolves. The two work as a pair to counter slight shifts in pH.

bulk Of the vertebrate gut, the volume of undigested material in the small intestine that cannot be decreased by absorption.

bulk flow The mass movement of one or more substances in the same direction, most often in response to pressure.

C3 plant Type of plant in which three-carbon PGA is the first stable intermediate to form after carbon fixation.

C4 plant Type of plant in which four-carbon oxaloacetate is the first stable intermediate to form after initial carbon fixation; in these plants, carbon is fixed twice, in two different types of photosynthetic cells.

calcium pump Active transport protein; pumps calcium ions across a cell membrane against their concentration gradient.

Calvin–Benson cycle Cyclic reactions that form sugar and regenerate RuBP in the second stage of photosynthesis. The reactions require carbon (from carbon dioxide). They use energy from ATP and hydrogens and electrons from NADPH, both of which form in the first stage.

CAM plant Type of plant that conserves water by opening stomata only at night, when it fixes carbon by repeated turns of the C4 pathway; stands for crassulacean acid metabolism.

camera eye A camera-like eyeball having an inner darkened chamber, one opening for light, and a retina (like film) on which visual stimuli are focused. Cephalopods and vertebrates have camera eyes.

camouflage Body coloration, patterning, form, or behavior that helps predators or prey blend with the surroundings and possibly escape detection.

cancer A malignant neoplasm; a mass of abnormally dividing cells that can leave their home tissue and invade and form new masses in other parts of the body.

capillary, blood Smallest diameter blood vessel; the exchanges between interstitial fluid and blood occur across its wall, which is only one cell thick.

capillary bed One of the diffusion zones for circulatory systems; great numbers of blood capillaries exchange substances with interstitial fluid.

capture–recapture method Individuals of a mobile species are captured (or selected) at random, marked, then released so they can mix with unmarked individuals. One or more samples are taken. The ratio of marked to unmarked individuals is used for estimating the size of the population.

carbamino hemoglobin $HbCO_2$, the form in which 30 percent or so of the CO_2 in blood is transported.

carbohydrate Any molecule of carbon, hydrogen, and oxygen typically in a 1:2:1 ratio. Main kinds are monosaccharides, oligosaccharides, and polysaccharides. They serve as structural materials, energy stores, and transportable energy forms.

carbon cycle Atmospheric cycle. Carbon moves from its environmental reservoirs (sediments, rocks, the ocean), through the atmosphere (mostly as CO_2), food webs, and back to the reservoirs.

carbon fixation Process by which any autotrophic cell incorporates carbon atoms into a stable organic compound. Different cells get carbon dioxide from the air or dissolved in water.

carbonic anhydrase Enzyme in red blood cells that speeds formation of bicarbonate from CO_2 and water.

carcinogen (kar-SIN-uh-jen) Any agent or substance that can cause cancer.

cardiac conduction system [Gk. *kardia*, heart, + *kyklos*, circle] A set of specialized cardiac muscle cells that initiate and send signals that make regular cardiac muscle cells contract. The SA node, AV node, and junctional fibers that link them.

cardiac cycle A recurring sequence of muscle contraction and relaxation that corresponds to one heartbeat.

cardiac muscle tissue A contractile tissue present only in the heart wall.

cardiac pacemaker Sinoatrial (SA) node; a cluster of self-excitatory cardiac muscle cells that spontaneously started contracting first; they continue to set the normal rate of heartbeat.

carnivore An animal that eats primarily the flesh of other animals.

carnivorous plant A plant that traps and digests insects and other small animals, and absorbs the released nutrients, which are otherwise scarce in its habitat.

carotenoid One of a class of accessory pigments in photosynthesis that reflect red, orange, and yellow light. One kind, beta-carotene, is a precursor of vitamin A.

carpel (KAR-pul) Female reproductive parts of flowers; a sticky or hairlike stigma, often stalked, above a chamber (ovary) in which one or more ovules mature into seeds.

carrying capacity Maximum number of individuals in a population or species that a given environment can sustain indefinitely.

cartilage Connective tissue consisting of fine collagen fibers packed in a secreted, rubbery matrix that resists compression.

cartilaginous fish Jawed fish having an endoskeleton, including a cranium, of cartilage and large fins; e.g., sharks.

Casparian strip A waxy, impermeable band that seals abutting cell walls of endodermis (and exodermis) in roots; forces water and solutes to pass through cells, which helps control the type and amount of solutes that enter the vascular cylinder.

catastrophism Idea that abrupt changes in the geologic and fossil records are evidence of divinely invoked catastrophes.

cDNA DNA synthesized from an mRNA transcript through the use of the enzyme reverse transcriptase.

cell Smallest unit that still displays the properties of life; it has the capacity to survive and reproduce on its own.

cell communication How free-living cells or cells of a multicelled species coordinate activities; they send, receive, transduce, and respond to signaling molecules.

cell cortex A dynamic mesh of crosslinked cytoskeletal elements just underneath the plasma membrane and attached to it.

cell count The number of cells of a given type present in one microliter of blood.

cell cycle Of eukaryotic cells, a series of events from the time a cell forms until it reproduces. A cycle consists of interphase, mitosis, and cytoplasmic division.

cell differentiation In developing embryos of multicelled organisms, the process by which different cell lineages selectively express a different fraction of their genome and thereby become specialized in their composition, structure, and function.

cell junction Of a tissue, any molecular structure that connects adjoining cells physically, chemically, or both at their plasma membranes.

cell-mediated immune response Actions of sensitized phagocytes and cytotoxic T cells that directly destroy infected or cancerous body cells.

cell plate formation The mechanism of cytoplasmic division in plant cells. After nuclear division, vesicles derived from Golgi bodies deposit the material for a cross-wall that cuts through the cytoplasm and connects to the parent cell wall.

cell theory All organisms consist of one or more cells, the cell is the smallest unit of organization still displaying the properties of life, and life's continuity arises directly from growth and division of single cells.

cell wall Of many cells (not animal cells), a semirigid but permeable structure that surrounds the plasma membrane; helps a cell retain its shape and resist rupturing.

Cenozoic The modern geologic era, from 65 million years ago to the present.

centipede Venomous predatory arthropod with many segments having paired legs.

central nervous system Of vertebrates, the brain and spinal cord.

central vacuole In many mature, living plant cells, an organelle that stores amino acids, sugars, and some wastes; when it enlarges during growth, it forces the cell to enlarge and increase its surface area.

centriole A barrel-shaped structure that arises from a centrosome and organizes newly forming microtubules into a 9+2 array inside a cilium or flagellum.

centromere Of a eukaryotic chromosome, a constricted region having binding sites (kinetochores) for spindle microtubules.

centrosome Dense mass of material in the cytoplasm of eukaryotic cells from which microtubules start to grow.

cephalization (SEF-ah-lah-ZAY-shun) [Gk. *kephalikos*, head] During the evolution of most kinds of animals, the increasing concentration of sensory structures and nerve cells at the anterior end of the body.

cephalochordate Lancelet. A filter-feeding invertebrate chordate with tapered ends.

cephalopod Soft-bodied mollusk with a closed circulatory system. Moves by jet propulsion of water from a siphon; e.g., squids, octopuses, chambered nautilus.

cerebellum (ser-ah-BELL-um) Hindbrain region with reflex centers that maintain posture and smoothing limb movements.

cerebral cortex Thin surface layer of two cerebral hemispheres; its interneurons receive, integrate, and store sensory information and coordinate responses.

cerebrospinal fluid Clear extracellular fluid that bathes and protects the brain and spinal cord; contained in a system of canals and chambers.

cerebrum (suh-REE-bruhm) A forebrain region concerned with olfactory input and motor responses. In mammals, it evolved into the most complex integrating center.

charophyte A type of photoautotroph once grouped with other green algae but now known to be more closely related to land plants; e.g., desmids, stoneworts.

checkpoint gene A gene that encodes a protein that can help delay, advance, or block the cell cycle when something goes wrong during DNA replication or repair.

chemical bond A union between the electron structures of two or more atoms.

chemical energy Potential energy in the bonds between atoms in molecules.

chemical equilibrium No net change in concentrations of reactants and products in a reversible chemical reaction.

chemical synapse (SIN-aps) A cleft between a presynaptic neuron and postsynaptic cell that neurotransmitters diffuse across.

chemiosmotic theory Well-supported but confusingly named theory of ATP formation by way of an electrochemical gradient. The action of electron transfer chains causes H^+ to accumulate inside a membrane-bound compartment. ATP forms as H^+ follows the resulting concentration/electric gradients across the membrane through ATP synthases.

chemoautotroph (KEE-moe-AH-toe-trofe) Any prokaryotic cell that makes its own food by oxidizing inorganic substances.

chemoreceptor Sensory receptor; detects dissolved ions or molecules in fluid.

chlamydia A group of bacteria; all are intracellular parasites that cannot make ATP; they pilfer it from animal cells.

chlorofluorocarbon (KLORE-oh-FLOOR-oh-car-bun) CFC; an organic compound that contains chlorine and fluorine; contributes to ozone thinning in the atmosphere.

chlorophyll *a* (KLOR-uh-fill) [Gk. *chloros*, green, + *phyllon*, leaf] In plants and algae, a pigment that is a receptor for the photon energy required to start photosynthesis; it absorbs mainly violet and red light and reflects or transmits green light.

chlorophyll *b* An accessory pigment that absorbs mainly blue and orange light.

chlorophyte Formal taxon for most species of green algae. Single-celled, multicelled, and colonial photoautotrophs of mostly aquatic habitats, closely related to land plants; e.g., *Ulva, Chlamydomonas, Chlorella.*

chloroplast Organelle of photosynthesis in plants and algae. Two outer membranes enclose a semifluid interior, the stroma. A third membrane forms a compartment inside that functions in ATP and NADPH formation; sugars form in the stroma.

choanoflagellate Single-celled eukaryote having a microvilli collar around a single flagellum at their anterior end. A sister taxon of animals and fungi.

cholecystokinin (CCK) A hormone that stimulates gallbladder contractions and the secretion of pancreatic enzymes; also may suppress appetite.

chordate Animal with a notochord, dorsal hollow nerve cord, pharynx, gill slits in the pharynx wall, and a tail that extends past the anus. These traits develop in embryos but some may not persist in the adult form.

chorion (CORE-ee-on) An extraembryonic membrane of amniote eggs; becomes part of the placenta. Villi form at its surface and facilitate the exchange of substances between the embryo and mother.

chromatid One of the two DNA molecules of a duplicated chromosome that remain attached to each other at the centromere region during nuclear division.

chromatin All of the DNA molecules and associated proteins in a nucleus.

chromosome In eukaryotic cells, a linear DNA double helix with many histones and other proteins attached. *See also* Bacterial chromosome.

chromosome number The sum of all of the chromosomes in cells of a given type.

chrysophyte One of the stramenopiles; most are single-celled photoautotrophs and components of phytoplankton; e.g., coccolithophores, diatoms.

chyme Semidigested food in the gut.

chytrid A flagellated intracellular parasite of a fungal group, the microsporidians.

ciliate An alveolate having arrays of cilia at its surface; traditionally called a ciliated protozoan. Most are free-living predators.

cilium, plural **cilia** A motile structure with a 9+2 array of microtubules that projects from the plasma membrane of certain eukaryotic cells. Modified cilia, such as those of hair cells, have sensory functions.

circadian rhythm (ser-KAYD-ee-un) [L. *circa*, about, + *dies*, day] Any biological activity repeated in cycles, each about twenty-four hours long, independently of any shifts in environmental conditions.

circulatory system Organ system that rapidly transports substances to and from cells; typically consists of a heart, blood vessels, and blood. Helps stabilize body temperature and pH in some animals.

clade [Gk. *klados-*, branch] All species that share a unique trait, being descended from an ancestral species in which the trait first evolved.

cladogenesis One speciation pattern. A lineage branches when one or more of its populations or subpopulations become reproductively isolated, and then genetic divergences result in new species.

cladogram Evolutionary tree diagram that depicts relative relatedness among groups. Each branch is monopyletic; it includes only an ancestral species in which a unique trait first evolved and all of its descendants.

classification system A way to organize and retrieve information about species.

cleavage Early stage of development in animals. Mitotic cell divisions cut up a fertilized egg into many smaller, nucleated cells (blastomeres); the original volume of egg cytoplasm does not increase.

climate Prevailing weather conditions of a region; e.g., temperature, cloud cover, wind speed, rainfall, and humidity.

climax community An array of species that develops by ecological succession. By a traditional model, it has stabilized under prevailing habitat conditions.

cloaca (kloe-AY-kuh) Last gut chamber or duct of some animals, roles in excretion, reproduction, and sometimes respiration.

clone A genetically identical copy of DNA, a cell, or a multicelled organism.

cloning A method used in recombinant DNA technology to make multiple copies of a DNA fragment. Also, manipulated reproductive interventions that bypass sexual reproduction; e.g., embryo cloning, adult cloning, and therapeutic cloning.

cloning vector Any DNA molecule that can accept foreign DNA and that can be replicated inside a host cell.

club fungus Fungus that produces sexual spores in a club-shaped cell, a basidium.

cnidarian (nye-DAR-ee-un) A type of radial invertebrate having epithelial tissues and a saclike gut. The only animal that makes nematocysts.

coal A nonrenewable energy source that formed more than 280 million years ago from submerged, undecayed, and slowly compacted plant remains.

coccolithophore A single-celled marine autotroph having calcium carbonate plates; one of the chrysophytes that are abundant in phytoplankton and the leading source of calcium deposits on the seafloor.

coccus A spherical prokaryotic cell.

cochlea [Gk. *koklias,* snail] A fluid-filled, coiled structure in the inner ear; transduces sound waves into action potentials.

codominance A condition in which a pair of nonidentical alleles that influence two different phenotypes are expressed at the same time in heterozygotes.

codon A base triplet; a linear sequence of three nucleotides in an mRNA transcript; a code for an amino acid or a termination signal that gets translated during protein synthesis.

coelom (SEE-lum) Between the gut and body wall, a cavity lined with peritoneum.

coenzyme An organic molecule that is a necessary participant in some enzymatic reactions; helps catalysis by donating or accepting electrons or functional groups; e.g., a vitamin, ATP, NAD$^+$.

coevolution The joint evolution of two species interacting so closely that a change in the structure, function, or behavior of one exerts selection pressure on the other over the generations.

cofactor A metal ion or a coenzyme that assists an enzyme in catalysis by accepting or donating electrons or functional groups.

cohesion A capacity to resist rupturing when placed under tension (stretched).

cohesion–tension theory Explanation of how water is transported from roots to leaves in plants. Water evaporation from leaves pulls water up in xylem by creating a continuous negative pressure (tension) that extends to roots.

cohort Group of individuals of the same age.

coitus Sexual intercourse.

coleoptile A thin tissue sheath that forms in embryo sporophytes of grasses; protects a primary shoot while growth is pushing it through soil after a seed germinates.

collar cell A sponge cell having a ring of food-trapping villi around a flagellum.

collecting duct A small tube into which as many as eight nephrons drain and that in turn drains into the renal pelvis.

collenchyma (coll-ENG-kih-mah) One of three simple plant tissues. Flexibly supports rapidly growing plant parts. Its elongated cells, alive at maturity, have a pectin-rich primary wall that is thickened where three or more collenchyma cells abut.

colon (CO-lun) Large intestine.

commensalism An ecological interaction in which one species benefits directly and one or more others are affected little, if at all.

communication display A social signal, often ritualized with intended changes in functions of common patterns.

communication protein A membrane protein that helps form an open channel between the cytoplasm of adjoining cells.

communication signal A social cue that is encoded in stimuli, such as the body's surface coloration or patterning, odors, sounds, and postures.

community All populations of all species in a habitat.

community structure The number of species and their relative abundances in a habitat, which shift over time.

companion cell Specialized parenchyma cell that helps load sugars into conducting cells of phloem.

comparative morphology [Gk. *morph,* form] Scientific study of comparable external body parts of embryonic stages and adult forms of major lineages.

compartmentalization In some plants, a defense response to attack, including secretion of sticky resins and toxins.

competition, exploitative An ecological interaction in which different species have equal access to a resource but one is better at exploiting it.

competition, interference An ecological interaction in which one species restricts or blocks access of another species to a resource regardless of its abundance.

competition, interspecific An ecological interaction in which the individuals of different species are competing for a share of resources.

competition, intraspecific An ecological interaction in which individuals of a population compete for resources.

competitive exclusion Theory that two or more species that require identical resources cannot coexist indefinitely.

complement system Set of proteins that circulate in inactive form in blood. Active kinds attract phagocytes and enhance their binding to antigen, promote inflammation, and induce lysis of pathogens during both innate and adaptive immune responses.

complete digestive system A tubular digestive system having a mouth at one end and an anus at the other.

composite signal A communication signal with more than one information-laden cue.

compound Molecule consisting of two or more elements in proportions that do not vary, as they can in mixtures.

compound eye Crustacean or insect eye having multiple rodlike units, each of which samples part of the visual field.

concentration gradient Difference in the number of molecules or ions of any one substance between two adjoining regions.

condensation reaction Type of chemical reaction in which two molecules become covalently bonded as a larger molecule; water often forms as a by-product.

conduction, heat Of two objects in contact with each other, an exchange of heat as a result of a thermal gradient between them.

cone Reproductive structure of certain seed-bearing plants; has clusters of scales with exposed ovules on their surface.

cone cell A vertebrate photoreceptor that responds to intense light and contributes to sharp daytime vision and color perception.

conifer A type of gymnosperm adapted to conserve water through droughts and cold winters. Cone-producing woody trees or shrubs with thickly cuticled needlelike or scalelike leaves.

conjugation Among prokaryotic species, a mode of gene transfer that is possible when one of the cells has an F plasmid. Also a sexual reproductive mode among some single-celled eukaryotes.

connective tissue Most abundant type of animal tissue. Soft connective tissues differ in the amounts and arrangements of fibroblasts, fibers, ground substance. Adipose tissue, cartilage, bone tissue, and blood are specialized types.

conservation biology An international field of inquiry. Biodiversity, as measured by species richness and abundances, is being surveyed and its evolutionary and ecological origins are being identified. Methods to maintain and use biodiversity for the benefit of the human population are being identified.

conservation of mass, law of The total mass of all substances entering a reaction equals the total mass of all products.

consumer Type of heterotroph that feeds on the tissues of other organisms as its source of carbon and energy.

continuous variation Of individuals of a population, a range of small differences in the phenotypic expression of a trait.

contractile cell A cell having cytoskeletal elements that help it contract (shorten) in response to stimulation and then lengthen (relax) and return to a resting position.

contractile ring mechanism Mechanism of cytoplasmic division of animal cells. Just beneath the plasma membrane, a thin band of contractile filaments around the cell midsection contracts and pinches the cytoplasm in two.

contractile vacuole [L. *contractus,* to draw together] Organelle in some single-celled eukaryotes that collects excess water in the cell body and then expels it.

control group In experimental tests, a group used as a standard for comparison against one or more experimental groups.

convection Air or water movement driven by a temperature gradient; also the transfer of heat by moving molecules.

convergent evolution In response to similar environmental pressures, evolutionarily distant lineages slowly evolve in similar ways and end up alike in some aspect of biochemistry, morphology, or behavior.

coral reef A formation of accumulated hard parts of corals and other organisms in warm, clear waters between latitudes 25° north and south.

core temperature The internal temperature of a large-bodied animal.

cork Tissue component of bark with many suberized layers; waterproofs, insulates, and protects woody stem and root surfaces.

cork cambium A lateral meristem, the descendants of which replace epidermis with cork on woody plant parts.

cornea Transparent, light-admitting part of the outer layer of a vertebrate eyeball.

corpus luteum (CORE-pus LOO-tee-um) A glandular structure that forms from cells of a ruptured follicle after ovulation; its progesterone and estrogen secretions help thicken the endometrium in preparation for a pregnancy.

cortex Generally, a rindlike layer. In cells, a mesh of cytoskeletal elements beneath the plasma membrane. In vascular plants, a ground tissue, mostly parenchyma, that supports parts and stores food.

cortisol A hormone that helps maintain the blood level of glucose between meals; its level rises when the body is stressed.

cotyledon (KOT-uhl-EE-dun) Seed leaf; part of a flowering plant embryo. In eudicots, two cotyledons absorb nutrients from endosperm, emerge aboveground as the seed germinates, and transfer nutrients that sustain early growth; photosynthetic before true leaves form. In most monocots, one small cotyledon helps transfer nutrients from endosperm to the embryo, but it remains underground when a seed germinates and is never photosynthetic.

countercurrent exchange In fish gills, the exchange of gases between blood flowing one way in blood vessels and water that is flowing over the vessels in the opposite direction. At membrane proteins of the ascending and descending limbs of the loop of Henle of a nephron, the pumping of specific solutes into and out of interstitial fluid in ways that alter its composition.

countercurrent flow Any movement of two fluids in opposing directions.

courtship display A pattern of ritualized social behavior between potential mates.

covalent bond (koe-VAY-lunt) [L. *con*, together, + *valere*, to be strong] A sharing of one or more electrons between two atoms. In a polar covalent bond, the atoms share electrons unequally; in a nonpolar covalent bond, each atom gets an equal share of the electrons.

craniate A vertebrate that has its brain protected inside a cranium; all modern fishes, amphibians, reptiles, birds, and mammals are craniates.

creatine phosphate Organic compound that can transfer phosphate to ADP in a fast, short-term, ATP-generating pathway.

CRH Corticotropin-releasing hormone; a hypothalamic releaser that stimulates the secretion of ACTH from the adrenal gland.

cross-bridge formation In sarcomeres of muscle fibers, a reversible, ATP-driven interaction between an actin filament and myosin head that results in a short power stroke; the basis of muscle contraction.

crossing over At prophase I of meiosis, reciprocal exchange of segments between two nonsister chromatids of a pair of homologous chromosomes. Puts novel combinations of alleles in gametes.

crust, Earth's Outer zone of low-density rocks resting on Earth's mantle.

crustacean One of the abundant "insects of the seas," mostly marine arthropods having a hardened, flexible exoskeleton and pairs of jointed appendages.

culture Sum of behavior patterns of a social group, passed between generations by learning and symbolic behavior.

cuticle (KEW-tih-kull) Of plants, a cover of transparent waxes and cutin on the outer wall of epidermal cells. Of annelids, a thin, flexible coat. Of arthropods, a lightweight exoskeleton hardened with chitin.

cutin Lipid polymer synthesized by land plants and deposited in cell walls and on the outer surface of epidermal cells.

cyanobacterium A type of single-celled photoautotroph; the first to use a noncyclic pathway of photosynthesis, which slowly enriched the early atmosphere with oxygen.

cycad A gymnosperm of subtropical or tropical habitats; pollen-bearing and seed-bearing strobili on separate plants.

cyclic AMP (SIK-lik) A nucleotide that is often a second messenger; relays a signal from outside through the cytoplasm.

cyclic pathway of ATP formation Oldest photosynthetic pathway. Photon energy forces electrons out of membrane-bound photosystems to transfer systems, which return them to the photosystems. Electron flow across the membrane sets up H^+ gradients that drive ATP formation.

cyst Of many microbes, a resting stage with a secreted cover. Also an abnormal, fluid-filled sac in skin with no opening.

cytochrome (SIGH-toe-krome) An iron-containing protein molecule of electron transfer systems.

cytokinesis (SIGH-toe-kih-NEE-sis) [Gk. *kinesis*, motion] Cytoplasmic division.

cytokinin (SIGH-toe-KYE-nin) A type of plant hormone; promotes cell division and leaf expansion, and retards leaf aging.

cytoplasm (SIGH-toe-plaz-um) All cell parts, particles, and semifluid substances between the plasma membrane and the nucleus or nucleoid.

cytoplasmic division Cytokinesis. After nuclear division, a splitting of the parent cell cytoplasm that completes formation of daughter cells.

cytoplasmic localization The accumulation of different kinds of proteins or RNAs in specific regions of the egg cytoplasm. These "maternal messages" are partitioned into different blastomeres at the cleavage stage of animal development.

cytosine (SIGH-toe-seen) One of the four nitrogen-containing bases in nucleotide monomers of DNA or RNA; also applies to a nucleotide that contains a cysteine base.

cytoskeleton In a eukaryotic cell, the dynamic framework of diverse protein filaments that structurally support, organize, and move the cell and internal structures. Prokaryotic cells have a few similar protein filaments.

cytotoxic T cell T lymphocyte that acts in adaptive immune responses; touch-kills infected or cancerous body cells.

day-neutral plant A plant that flowers when mature, independently of seasons.

decomposer [L. *dis*–, to pieces] One of the prokaryotic or fungal heterotrophs that obtains carbon and energy by breaking down wastes or remains of organisms. The collective action of decomposers helps cycle nutrients to producers in ecosystems.

deforestation Removal of all trees from a large tract of land.

degradative pathway Any of the stepwise series of metabolic reactions that break down organic compounds.

deletion Loss of a chromosome segment; often leads to genetic disorders. Also the loss of one or more nucleotide bases from a DNA molecule.

demographics The vital statistics of a population; e.g., size, age structure.

demographic transition model Model that correlates changes in population growth with stages of economic development; may no longer apply to developing countries, which now compete in a global market.

denaturation (deh-NAY-chur-AY-shun) Disruption of hydrogen bonds and other interactions holding a molecule in its three-dimensional shape, which thereby changes. Increases in temperature, shifts in pH, and detergents can cause it.

dendrite (DEN-drite) [Gk. *dendron*, tree] Short, slender extension from cell body of a neuron; commonly a signal input zone.

dendritic cell Phagocytic white blood cell; mainly presents antigen to naive T cells.

denitrification (DEE-nite-rih-fih-KAY-shun) Conversion of nitrate or nitrite to gaseous nitrogen (N_2) or nitrogen oxide (NO_2) by metabolic activity of certain bacteria in soil.

dense, irregular connective tissue A type of animal tissue with fibroblasts and many fibers asymmetrically arrayed in a matrix. In skin and some capsules around organs.

dense, regular connective tissue A type of animal tissue with rows of fibroblasts between parallel bundles of fibers. In tendons, elastic ligaments.

density-dependent control Any factor that comes into play in an overcrowded population; reduces birth rate or raises death and dispersal rates, as by intensified predation, parasitism, disease, competition.

density-independent factor Any factor that causes fewer births or more deaths in a population regardless of its density; e.g., a severe storm or flood.

dentition (den-TIH-shun) The type, size, and number of an animal's teeth.

deoxyribonucleic acid *See* DNA.

derived trait A novel feature shared only by descendants of an ancestral species in which it originated.

dermal tissue system Tissues that cover and protect all exposed plant surfaces.

dermis Skin layer beneath the epidermis; mostly dense connective tissue.

desalinization Removal of salt from water.

desert Biome of areas where the potential for evaporation greatly exceeds rainfall, where soil is thin and vegetation sparse.

desertification (dez-urt-ih-fih-KAY-shun) Conversion of grassland or irrigated or rain-fed cropland to desertlike conditions.

detrital food web (dih-TRY-tul) Cross-connecting food chains in which energy flows mainly from plants through arrays of detritivores and decomposers.

detritivore Any animal that feeds on decomposing particles of organic matter; e.g., a crab, earthworm, or roundworm.

deuterostome (DUE-ter-oh-stome) [Gk. *deuteros*, second, + *stoma*, mouth] A bilateral animal of a lineage characterized in part by events of embryonic development, as when the second indentation to appear on the early embryo's surface becomes the mouth; e.g., an echinoderm or a chordate.

development Of complex multicelled species, a series of stages from formation of gametes, then fertilization, and on through embryonic and adult forms.

diaphragm [Gk. *diaphragma*, to partition] A muscular partition between the thoracic and abdominal cavities. Also a fertility control device inserted into the vagina to prevent sperm from entering uterus.

diatom A single-celled photoautotroph with a perforated silica shell, which has two overlapping parts that fit together like a pillbox; one of the chrysophytes.

dicot *See* eudicot.

diffusion Net movement of like ions or molecules from a region where they are most concentrated to an adjoining region where they are less concentrated; they move down their concentration gradient.

digestive system Body sac or tube, often having specialized regions where food is ingested, digested, and absorbed, and undigested residues expelled. Incomplete systems have one opening; the complete systems have two (mouth and anus).

dihybrid experiment Type of experiment that starts with a cross between two true-breeding, homozygous parents that differ in two traits governed by alleles of two genes. The actual experiment is a cross between two of their F_1 offspring that are identically heterozygous for alleles of the two genes; e.g., *AaBb* x *AaBb*.

dimorphism Persistence of two forms of the same trait in a population.

dinoflagellate One of the alveolates that deposits cellulose in alveoli, often as thick protective plates. Predators, parasites and photoautotrophs; some cause red tides.

dinosaur One of a group of reptiles that arose in the Triassic and became dominant land vertebrates for 125 million years.

diploid chromosome number (DIP-loyd) Of many sexually reproducing species, having two chromosomes of each type, or pairs of homologues, in somatic cells.

diplomonad A flagellated heterotroph with no mitochondria and three flagella at its anterior end and one at its trailing end. Belongs to one of earliest lineages of single-celled eukaryotes; e.g., *Giardia*.

directional selection Mode of natural selection by which forms at one end of a range of phenotypic variation are favored.

disaccharide (die-SAK-uh-ride) [Gk. *di*, two, + *sakcharon*, sugar] A carbohydrate composed of two sugar monomers.

disease Condition that arises when the body's defenses cannot overcome infection and activities of the pathogen or parasite interfere with normal body functions.

disruptive selection Mode of natural selection that favors different forms of a trait at both ends of a range of variation; intermediate forms are selected against.

distal tubule Tubular part of nephron where water and sodium reabsorption are adjusted by hormonal controls.

distance effect A major biogeographic pattern. Only species adapted for long-distance dispersal are potential colonists of islands far from their home range.

diversity of life Sum of all variations in form, function, and behavior in all lineages, from life's origin to the present.

division of labor Of multicelled species, a splitting up of tasks among different types of cells, tissues, and often organs, and organ systems, which collectively help the whole organism survive. Also a splitting up of tasks among different stages of the life cycle, as in insects.

DNA Deoxyribonucleic acid (dee-OX-ee-RYE-bow-new-CLAY-ik). Double-stranded nucleic acid twisted into a helical shape; its base sequence encodes the primary hereditary information for all living organisms and many viruses.

DNA chip Microarray of thousands of gene sequences that represents a large subset of a genome; stamped onto a glass plate and used to study gene expression.

DNA clone Fragment of DNA inserted into a vector such as a plasmid and introduced into a host organism; used to make many copies of a particular segment of DNA.

DNA fingerprinting A way to distinguish one individual from all others based on unique differences in parts of their DNA; fragments cut from an individual's DNA (RFLPs) have a unique pattern of sizes.

DNA ligase (LYE-gaze) Type of enzyme that catalyzes the sealing of short stretches of DNA into a continuous strand during replication; also seals strand breaks.

DNA polymerase Type of enzyme that catalyzes the addition of free nucleotides to new DNA strands during replication; also proofreads and corrects mismatches.

DNA proofreading mechanism Any enzyme-mediated process that fixes DNA replication errors or strand breaks.

DNA replication Process by which a cell duplicates its DNA molecules before it divides into daughter cells.

domain Of protein structure, part or all of a polypeptide chain that is a structurally stable, functional unit. Of one classification system, the most inclusive taxon.

dominance hierarchy Social organization in which some individuals of the group have a subordinate status to others.

dominant allele Of diploid cells, an allele that masks the phenotypic effect of any recessive allele paired with it.

dopamine Neurotransmitter that affects fine motor control, pleasure-seeking behavior.

dormancy [L. *dormire*, to sleep] Of many spores, cysts, seeds, perennials, and some animals, a predictable time of metabolic inactivity during the life cycle.

dosage compensation A gene control mechanism in female mammals in which most genes on one of two X chromosomes in somatic cells are inactivated; ensures that X chromosome genes are expressed at the same levels as in males (XY).

double-blind study A study in which neither the subjects nor the experimenter know if any particular subject is in the experimental group or the control group; minimizes bias.

double fertilization Of flowering plants only, fusion of a sperm and egg nucleus, and the fusion of another sperm nucleus with nuclei of a cell that gives rise to the endosperm, a nutritive tissue in seeds.

doubling time The time it takes for a population to double in size.

downwelling Water forced down and away from a coast after winds shift and make a surface current pile into the coast.

drug addiction Dependence on a drug, which assumes an "essential" biochemical role following habituation and tolerance.

dry acid deposition Airborne oxides of sulfur, nitrogen fall during dry weather.

dry shrubland Biome of areas that get less than 25 to 60 centimeters of rain; short, multibranched woody shrubs dominate.

dry woodland Biome of areas that get about 40 to 100 centimeters of rain may have many tall trees but no dense canopy.

duplication Base sequence in DNA that has been repeated two or more times.

ecdysone Hormone of many insect life cycles; roles in metamorphosis, molting.

echinoderm One of the protostomes; a radial invertebrate with some bilateral features and calcified spines or plates on the body wall; e.g., sea stars.

echolocation Use of echoes from self-generated ultrasounds as a navigational mechanism, as by bats and dolphins.

ecological succession Traditional view that a community arises by species interactions and in time forms a stable array of species, a climax community. Primary succession starts with pioneer species that colonize a barren habitat; secondary succession is the recovery of a disturbed climax community. *See* Intermediate disturbance hypothesis.

ecology [Gk. *oikos*, home, + *logos*, reason] Scientific study of how organisms interact with one another and the environment.

ecoregion Broad land or ocean province influenced by abiotic and biotic factors.

ecosystem Array of organisms, together with their environment, interacting by a flow of energy and cycling of materials.

ecosystem modeling Analytical method; computer programs and models predict effects of disturbances to an ecosystem.

ectoderm [Gk. *ecto*, outside, + *derma*, skin] First-formed, outer primary tissue layer of animal embryos; gives rise to nervous tissues and outer layer of the integument.

ectotherm An animal that can stay warm mainly by absorbing environmental heat, as from the sun's rays.

Ediacaran One of a diverse collection of tiny multicelled precambrian species having a highly flattened body, sometimes with many unspecialized segments.

effector Muscle (or gland); helps bring about movement (or chemical change) in response to neural or endocrine signals.

effector cell Antigen-sensitized B cell or T cell that carries out adaptive immunity.

egg Mature female gamete, or ovum.

El Niño Massive eastward flow of warm surface waters of the western equatorial Pacific that displaces cool water off South America. Recurs, disrupts global climates.

electric gradient A difference in electric charge between adjoining regions.

electromagnetic spectrum All wavelengths of photon energy from gamma rays less than 10^{-5} nanometers long to radio waves more than 10 kilometers long.

electron Negatively charged subatomic particle. Electrons occupy orbitals around the atomic nucleus.

electron transfer chain Array of enzymes and other molecules in a cell membrane that accept and give up electrons in sequence; operation of chain releases the energy of the electrons in small, usable increments.

electron transfer phosphorylation Final stage of aerobic respiration; electron flow through electron transfer chains in inner mitochondrial membrane sets up H^+ concentration and electric gradients that drives ATP formation. Oxygen accepts electrons at the end of the chain.

element Fundamental form of matter that cannot be degraded to a simpler form by ordinary means. All atoms of an element have the same atomic number.

embryo Of animals, a new individual that forms by cleavage, gastrulation, and other early stages of development. Of plants, a young sporophyte until germination.

embryonic induction A change in the composition, structure, or both of cells in an embryo by exposure to signals from cells of nearby tissues; basis of pattern formation.

emergent property With respect to life's levels of organization, a new property that emerges through interactions of entities at lower levels, none of which displays the property; e.g., living cells that emerge from "lifeless" molecules.

emerging pathogen A newly mutated or opportunistic strain of a deadly pathogen.

emigration Permanent move of one or more individuals out of a population.

emulsification In the gut, the coating of fat droplets with bile salts so that fats remain suspended in chyme.

encapsulated receptor A type of skin mechanoreceptor that detects pressure, temperature, and low vibrations.

endangered species A species endemic (native) to a habitat, found nowhere else, and highly vulnerable to extinction.

endergonic reaction (en-dur-GONE-ik) A chemical reaction that requires a net energy input and converts more stable reactants into less stable products; not spontaneous.

endocrine gland A ductless gland that secretes hormone molecules, which typically travel in blood to target cells.

endocrine system Control system of cells, tissues, and organs that interacts intimately with the nervous system; secretes hormones and other signaling molecules.

endocytosis (EN-doe-sigh-TOE-sis) Cell uptake of substances by forming vesicles from patches of plasma membrane. Three modes are receptor-mediated endocytosis, phagocytosis, and the bulk transport of extracellular fluid.

endoderm Inner primary tissue layer of animal embryos; source of the inner gut lining and organs derived from it.

endodermis Cylindrical, sheetlike cell layer around the root vascular cylinder; helps control water and solute uptake.

endomembrane system Endoplasmic reticulum, Golgi bodies, and transport vesicles concerned with modification of many new proteins, lipid assembly, and their transport within the cytoplasm or to the plasma membrane for export.

endometrium [Gk. *metrios*, of the womb] Inner lining of the uterus.

endophytic fungus A fungal symbiont in leaves and stems of most plants. Helpful, neutral, or sometimes harmful effects.

endoplasmic reticulum ER. Organelle that extends from the nuclear envelope through cytoplasm. Ribosomes coat the cytoplasmic side of rough ER, which modifes many new polypeptide chains in its lumen. Membrane lipids are assembled, fatty acids are broken down, and some toxins are inactivated in the lumen of smooth ER.

endorphin One of the neuromodulators that is a natural painkiller.

endoskeleton Of chordates, an internal framework consisting of cartilage, bone, or both; works with skeletal muscle to position, support, and move the body.

endosperm Nutritive tissue in the seeds of flowering plants only.

endospore Of certain bacteria, a resting structure enclosing a bit of cytoplasm and the DNA; resists heat, irradiation, drying, acids, disinfectants, and boiling water. It germinates when conditions favor growth and a bacterium emerges from it.

endosymbiosis [*Endo–*, within + *symbiosis*, living together] An intimate, permanent ecological interaction in which one species lives and reproduces in the other's body to the benefit of one or both.

endotherm An animal that can stay warm mainly by metabolically generated heat.

energy A capacity to do work.

energy carrier A molecule that delivers chemical energy from one reaction site to another; mainly ATP.

energy pyramid Diagram that depicts the energy stored in the tissues of organisms at each trophic level in an ecosystem. Lowest tier of the pyramid, consisting of primary producers, is always the largest.

enhancer A small sequence in DNA that binds transcription-regulating molecules; enhances transcription rates.

enkephalin A neuromodulator that is a natural painkiller.

ENSO El Niño Southern Oscillation. A recurring seesaw of atmospheric pressure in the western equatorial Pacific that has global repercussions on climates.

entropy Measure of how much and how far a concentrated form of energy has been dispersed after an energy change.

enzyme A type of protein that catalyzes (speeds) a chemical reaction. Some RNAs also show catalytic activity.

eosinophil A white blood cell that, during inflammatory responses, secretes enzymes and toxins that target extracellular parasites too large for phagocytosis.

epidemic Rapid spread, then subsidence, of a disease within a population.

epidermis Outermost tissue layer of plants and nearly all animals.

epiglottis Flaplike structure between the pharynx and larynx; controlled positional changes direct air into the trachea or food into the esophagus.

epinephrine (ep-ih-NEF-rin) A signaling molecule of the adrenal medulla that acts as a hormone or neurotransmitter on different targets; affects metabolism, heart function; works with norepinephrine in the fight–flight response. Also known as adrenaline.

epistasis (eh-PISS-tah-sis) An interaction among products of two or more gene pairs that influence the same trait.

epitheliomuscular cell Of structurally simple invertebrates, an epithelial cell with elongated extensions that contain parallel arrays of contractile filaments. Muscle fibers evolved from such cells.

epithelium (EP-ih-THEE-lee-um) Animal tissue covering external and internal body surfaces. A key innovation that favored larger, more complex bodies; cells started interacting as functional units.

EPSP Excitatory postsynaptic potential. Graded potential that drives an excitable cell's membrane toward threshold.

equilibrium model of island biogeography A model describing the number of species expected to inhabit a habitat island of a particular size and distance from colonists.

ER *See* endoplasmic reticulum.

erosion, soil Wearing away of land surface by wind, running water, and ice.

erythropoietin Kidney hormone; induces stem cells in bone marrow to give rise to red blood cells.

esophagus (ee-SOF-uh-gus) A muscular tube between the pharynx and stomach.

essential amino acid Any amino acid that an organism cannot synthesize for itself and must obtain from food.

essential fatty acid Any fatty acid that an organism cannot synthesize for itself and must obtain from food.

estrogen A female sex hormone. It helps oocytes mature and prime the endometrium for pregnancy; affects growth, development, and female secondary sexual traits.

estuary Partly enclosed coastal region where seawater mixes with fresh water and runoff from land, as in rivers.

ethylene Plant hormone; promotes fruit ripening and leaf, flower, fruit abscission.

eudicot (YOO-dih-kot) Flowering plant characterized by having embryos with two cotyledons; net-veined leaves; and floral parts in fours, fives, or multiples of these.

euglenoid A single eukaryotic cell with a crystalline rod reinforcing a thick flagellum. Different kinds are colorless heterotrophs or photoautotrophs of aquatic habitats.

Eukarya Domain of eukaryotic species; all "protists," plants, fungi, and animals.

eukaryotic cell Type of cell that starts life with a nucleus and other membrane-bound organelles.

eutherian Placental mammal.

eutrophication Nutrient enrichment of a body of water that promotes population growth of phytoplankton and opacity.

evaporation Process of conversion of a liquid to a gas; requires energy input.

evolution, biological [L. *evolutio*, an unrolling] Genetic change in a line of descent by microevolutionary events (gene mutation, natural selection, genetic drift, and gene flow); basis of large-scale patterns, rates, and trends in the history of life.

evolutionary tree A treelike diagram in which each branch point represents a divergence from a shared ancestor; each branch is a separate line of descent.

excretion Removal of excess water and solutes by urinary system or glands.

exercise Increased contractile activity.

exergonic reaction (EX-ur-GONE-ik) Any chemical reaction with a net energy loss.

exocrine gland Glandular structure that secretes products, usually through ducts or tubes, to a free epithelial surface.

exocytosis Fusion of a cytoplasmic vesicle with the plasma membrane; as it becomes part of the membrane, its contents are released to extracellular fluid.

exodermis Cylindrical sheet of cells close to root epidermis of most flowering plants; helps control uptake of water and solutes.

exon A base sequence in eukaryotic DNA that is part or all of a protein-encoding gene; may or may not be excised from a pre-mRNA during transcript processing.

exoskeleton [Gk. *sklēros*, hard, stiff] An external skeleton; e.g., a hardened cuticle.

exotic species Species that has become established in a new community after dispersing from its home range.

experiment, scientific A test that simplifies observation in nature or the laboratory by manipulating and controlling conditions under which observations are made.

experimental group A group of objects or individuals that display or are exposed to the variable under investigation. Test results for this group are compared against the results for a control group.

exponential growth (EX-po-NEN-shul) Any quantity that is growing at a rate proportional to its size. For populations, it plots out as a J-shaped curve.

external ear The sound-collecting flap of cartilage-reinforced skin of many ears.

extinction Irrevocable loss of a species.

extracellular digestion, absorption Mode of nutrition; the organism grows in or on organic matter, digests it with secreted enzymes, and absorbs digested bits.

extracellular fluid All fluid not in cells; e.g., blood's plasma and interstitial fluid.

extracellular matrix Secretions and other deposits on or between cells of a tissue.

extreme halophile Bacterium or archaean adapted to an extremely salty habitat.

extreme thermophile Bacterium or archaean adapted to a hot aquatic habitat; e.g., a hot spring or hydrothermal vent.

eye Sensory organ that incorporates a dense array of photoreceptors.

F₁, F₂ The first and second generation offspring of experimental crosses.

FAD Flavin adenine dinucleotide. A type of nucleotide coenzyme; transfers electrons and H$^+$ from one reaction site to another.

fall overturn Vertical mixing of a body of water in fall. Its upper oxygenated layer cools, gets dense, and sinks; nutrient-rich water from the bottom moves up.

fat Type of lipid with one, two, or three fatty acid tails attached to a glycerol head.

fate map Surface diagram of certain early embryos (e.g., *Drosophila*) showing where differentiated cells of the adult originate.

fatty acid Organic compound having a carboxyl group and a backbone of as many as thirty-six carbon atoms; saturated types have single bonds only; unsaturated types include one or more double covalent bonds.

feather Of birds, lightweight structures used in flight, as body insulation, and often in courtship displays.

feedback inhibition Mechanism by which a change that results from some cellular activity triggers responses that decrease or shut down the activity.

fermentation *See* alcoholic fermentation, lactate fermentation.

fern A seedless vascular plant having fronds that often are divided in leaflets.

fertilization Fusion of a sperm nucleus and an egg nucleus, the result being a single-celled zygote.

fetus In mammalian development, the stage after all major organ systems have formed until time of birth.

fever An internally induced rise in core body temperature above a set point in the hypothalamic temperature control center.

fibrous, irregular connective tissue One of the soft connecive tissues; its matrix is packed with fibroblasts and collagen fibers oriented in all directions.

fibrous, regular connective tissue One of the soft connective tissues, with orderly rows of fibroblasts in between parallel, tightly packed bundles of fibers.

fibrous root system Lateral branchings of adventitious roots arising from a new stem.

Fick's law The rate at which a gas will diffuse across a respiratory surface is proportional to its partial pressure and the surface area.

filter feeder Animal that filters food from a current of water that flows through pores or slits of some body structure.

filtration *See* Glomerular filtration and Ultrafiltration.

fin An appendage that helps stabilize, orient, and propel most fishes in water.

first law of thermodynamics Energy cannot be created or destroyed.

fish An aquatic animal of the oldest and most diverse vertebrate lineage; a jawless, jawed cartilaginous, or jawed bony fish.

fitness The degree of adaptation to the environment, as measured by the relative genetic contribution to future generations.

fixation Of a population, the loss of all alleles but one at a gene locus; all individuals have become homozygous for the allele.

fixed action pattern Instinctual program of coordinated, stereotyped movements that runs its course independently of any feedback from environment.

flagellum, plural **flagella** Of many eukaryotic cells, a long, whip-like motile structure with an inner 9+2 array of microtubules. Prokaryotic flagella do not have this array and are not whiplike; they rotate like a propeller.

flatworm One of the simplest existing animals with organ systems that form from three primary tissue layers.

flavoprotein A plant pigment that absorbs blue light and can induce phototropism.

flower A reproductive structure of fertile parts (stamens, carpels) nonfertile parts (sepals, petals), and a receptacle (modified base of floral shoot).

flowering plant A magnoliid, eudicot, or monocot. The most successful group of plants; most coevolved with pollinators.

fluid mosaic model A cell membrane has a mixed composition (mosaic) of lipids and proteins, the interactions and motions of which impart fluidity to it.

fluorescence Light that may become visible after a molecule has absorbed a photon, then emits another photon of lower energy.

folate Water-soluble B vitamin especially essential for embryonic development.

follicle (FOLL-ih-kul) Small sac, pit, or cavity, as around a hair; a mammalian oocyte with its surrounding layer of cells.

food chain Linear sequence of steps by which energy stored in autotroph tissues enters higher trophic levels.

food pyramid Chart of a purportedly well-balanced diet; continually updated.

food web Cross-connecting food chains consisting of producers, consumers, and decomposers, detritivores, or both.

foramen magnum Opening in the skull where the spinal cord and brain connect. Its position relative to the skull's base helps researchers determine whether a fossilized animal was bipedal or a tetrapod.

foraminiferan Single-celled, predatory eukaryote with a richly perforated shell through which thin pseudopods project.

forebrain Part of vertebrate brain that includes the cerebrum, olfactory lobes, and hypothalamus.

forest A community in which tall trees grow close enough together to form a fairly continuous canopy.

fossil Recognizable, physical evidence of an organism that lived in the distant past.

fossil fuel Coal, petroleum, or natural gas; nonrenewable energy source that formed long ago from remains of swamp forests.

fossilization How fossils form over time. An organism or evidence of it gets buried in sediments or volcanic ash; water slowly infiltrates the remains, and metal ions and other inorganic compounds dissolved in it replace the minerals in bones and other hardened tissues.

founder effect A form of bottlenecking. By chance, a few individuals that establish a new population differ in allele frequencies relative to the original population.

free nerve ending One of the simplest sensory receptors in skin, internal tissues; unmyelinated or thinly myelinated.

free radical Any unbound molecular fragment with an unpaired electron.

fruit [L. after *frui*, to enjoy] Mature ovary, often with accessory parts, from a flower.

fruiting body Spore-bearing structures formed by some bacteria, fungi.

FSH Follicle-stimulating hormone of the anterior lobe of pituitary gland; has reproductive roles in both sexes.

functional group An atom or a group of atoms with characteristic properties that is covalently bonded to the carbon backbone of an organic compound.

functional-group transfer One molecule donates a functional group to another.

Fungi Kingdom of fungi.

fungus, plural **fungi** Type of eukaryotic heterotroph that obtains nutrients by extracellular digestion and absorption; fungi are notable for being prolific spore producers and major decomposers.

GABA Gamma amino butyric acid. A neuromodulator that blocks neurotransmitter release by other neurons in the brain.

gallbladder Organ that stores bile from the liver; its duct connects to the small intestine.

gamete (GAM-eet) Haploid cell formed by meiotic cell division of a reproductive cell; required for sexual reproduction.

gamete formation Formation of cells used in sexual reproduction; e.g., sperm or eggs.

gametophyte (gam-EET-oh-fite) [Gk. *phyton*, plant] A haploid multicelled body in which haploid gametes form during the life cycle of plants and some algae.

ganglion (GANG-lee-on), plural **ganglia** Distinct cluster of cell bodies of neurons.

gap junction Cylindrical arrays of proteins in the plasma membrane of adjoining cells; they pair up as open channels for rapid flows of ions and small molecules.

gastric fluid Extremely acidic mixture of secretions from the stomach lining.

gastrodermis Glandular epithelium that lines the gut of many invertebrates.

gastrula Early animal embryo with two or three primary tissue layers (germ layers).

gastrulation (gas-tru-LAY-shun) Stage of animal development; the reorganization of embryonic cells that formed by cleavage into two or three primary tissue layers.

gel electrophoresis Method of separating DNA molecules according to length, or protein molecules according to size and charge. The molecules move apart while migrating through a gel matrix in response to a weak electric current.

gene Unit of heritable information in DNA, transmissable from parents to offspring.

gene control A molecular mechanism that governs if, when, or how a specific gene is transcribed or translated.

gene expression Conversion of heritable information in a gene into a product; e.g., from DNA to mRNA to a structural or functional protein.

gene flow Microevolutionary process; alleles enter and leave a population by immigration and emigration. Counters mutation, natural selection, and genetic drift, hence reproductive isolation.

gene library Collection of host cells that contain different cloned DNA fragments representing all or most of a genome.

gene locus A gene's location along the length of a chromosome.

gene mutation Small-scale change in the nucleotide sequence of a gene; can result in an altered protein product.

gene pair Two alleles at the same locus on a pair of homologous chromosomes.

gene pool All genotypes in a population; a pool of genetic resources.

gene therapy Generally, a transfer of one or more normal genes into an organism to correct or minimize a genetic disorder.

genetic abnormality A less common or rare version of a heritable trait.

genetic code Correspondence between triplets of nucleotides in DNA and mRNA, and specific sequences of amino acids in a polypeptide chain; near-universal language of protein synthesis; mitochondria and a few species have a few variant code words.

genetic disorder An inherited condition causing mild to severe medical problems.

genetic divergence An accumulation of differences in the gene pools of two or more populations or subpopulations of a species after gene flow stops entirely; mutation, natural selection, and genetic drift operate independently in each one.

genetic drift Change in allele frequencies over generations due to chance alone. Most pronounced effects in small populations.

genetic engineering Manipulation of an organism's DNA, usually to alter at least one aspect of phenotype.

genetic equilibrium In theory, a state in which a population is not evolving with respect to a specified gene locus. *Compare* Hardy–Weinberg rule.

genetic recombination Outcome of any process that puts new genetic information in a DNA molecule; e.g., by crossing over.

genome All DNA in a haploid number of chromosomes for a species.

genomics The study of genes and gene function in humans and other organisms.

genotype (JEEN-oh-type) Genetic makeup of an individual; a single gene pair or the sum total of an individual's genes.

genus, plural **genera** (JEEN-US, JEN-er-ah) [L. *genus*, race or origin] A grouping of species more closely related to one another in morphology, ecology, and history than to others at the same taxonomic level.

geographic dispersal A movement of individuals out of their home range and their integration in a new community.

geologic time scale Time scale for Earth's history; major subdivisions correspond to mass extinctions. Dates are now absolute as a result of radiometrically dating.

germ cell Animal cell set aside for sexual reproduction; gives rise to gametes.

germination (jur-mih-NAY-shun) Of a seed or a spore, the resumption of growth after dormancy, dispersal, or both.

ghrelin Hormone secreted from cells of the stomach lining that stimulates an appetite control center.

gibberellin (JIB-er-ELL-un) Plant hormone; induces stem elongation, helps seeds break dormancy, role in flowering in some species.

gill Respiratory organ with a thin, moist, vascularized layer for gas exchange.

gill slit One of the openings in a thin-walled pharynx that functions in food-trapping, respiration, or both.

ginkgo A deciduous gymnosperm; its ancestors were diverse in dinosaur times.

gland A saclike, secretory organ that opens onto a free epithelial surface. Hormone-secreting endocrine glands have ducts; exocrine glands are ductless.

gland cell A cell that secretes products unrelated to its own metabolism.

global broiling hypothesis An asteroid impact caused the K–T mass extinction, the debris from which raised the global air temperature by thousands of degrees.

global warming Long-term increase in temperature of Earth's lower atmosphere.

glomerular capillary (glow-MARE-you-lar) [L. *glomus*, ball] A set of blood capillaries in Bowman's capsule of nephron.

glomerular filtration First step in urine formation, when blood pressure forces water out of glomerular capillaries.

glomerulus Bowman's capsule and the glomerular capillaries that it cups around.

glottis Opening between the vocal cords.

glucagon Pancreatic hormone; stimulates conversion of glycogen and amino acids to glucose when blood glucose levels fall.

glyceride (GLISS-er-ide) Molecule of one, two, or three fatty acid tails attached to a glycerol backbone; one of the fats or oils.

glycerol Three-carbon compound having three hydroxyl groups; in fats and oils.

glycocalyx Sticky meshlike capsule or slime layer around a prokaryotic cell wall.

glycogen (GLY-kuh-jen) Highly branched polysaccharide of glucose monomers; the main storage carbohydrate in animals.

glycolysis Breakdown of glucose or another organic compound to two pyruvates. First stage of aerobic respiration, fermentation, or anaerobic electron transfer. Oxygen has no role in glycolysis, which takes place in the cytoplasm of all cells. Two NADH form. Net yield: 2 ATP per glucose molecule.

glycoprotein Protein with linear or branched oligosaccharides covalently bonded to it.

gnetophyte A type of woody, vinelike or shrubby gymnosperm.

GnRH Gonadotropin-releasing hormone, induces the anterior pituitary to release LH and FSH.

golden alga A chrysophyte with silica scales or other hard parts and fucoxanthin.

Golgi body Organelle of endomembrane system; its enzymes modify many new polypeptide chains, assemble lipids, and package both inside vesicles for secretion or for use inside cell.

gonad (GO-nad) Primary reproductive organ in animals; produces gametes.

Gondwana Paleozoic supercontinent that later became part of Pangea.

graded potential Of excitable cells, a change in the resting membrane potential at an input zone; can vary in magnitude.

gradual model, speciation Addresses the rate of speciation and cites fossil evidence that morphological changes accumulate slowly over great time spans.

Gram-positives Informal name for mostly chemoheterotrophic bacteria that have a multilayered wall; not a monophyletic group.

Gram stain Microbiology diagnostic tool. Cells are exposed to purple dye, iodine, an alcohol wash, and then counterstain. Cell walls of Gram-positive species stay purple; Gram-negative species turn pink.

granum, plural grana Of chloroplasts, one portion of the thylakoid membrane in the shape of a stack of flattened disks.

grassland Biome in flat or rolling interiors of continents with warm summers, 25–100 centimeters of rain, and recurring natural fires that regenerate the dominant plants.

gravitropism (GRAV-ih-TROPE-izm) Growth in a direction influenced by gravity.

gray crescent Of amphibian eggs, partially pigmented region of cell cortex; establishes the embryo's anterior–posterior axis.

gray matter Areas of neuron cell bodies, dendrites, and unmyelinated axons, plus neuroglia, in the brain and spinal cord.

grazing food web Cross-connecting food chains in which energy flows from plants to an array of herbivores, then carnivores.

green alga *See* charophyte; chlorophyte.

green revolution Use of improved crop strains and modern equipment to increase crop yields in the developing countries.

greenhouse effect Trapping of heat near Earth's surface by the action of atmospheric gases. The gases absorb infrared wavelengths (heat) from the sun-warmed surface, and then radiate some wavelengths downward.

ground tissue system Parenchyma and other tissues; the bulk of a plant body.

groundwater Water in soil and aquifers.

growth Of multicelled species, increases in the number, size, and volume of cells. Of single-celled prokaryotes, increases in the number of cells of a population.

growth factor A protein that stimulates increases in size; e.g., by inducing mitosis.

growth ring One of the alternating bands of early and late wood; a "tree ring."

growth, tissue specialization Stage of animal development; new organs enlarge and assume specialized functions. This stage continues into adulthood.

guanine One of four nitrogen-containing bases in nucleotide monomers of DNA or RNA; also may refer to a nucleotide that contains a guanine base.

guard cell One of two cells that define a stoma across leaf or stem epidermis.

gut A sac or tube in which food is digested. Also the gastrointestinal tract from the stomach onward.

gymnosperm (JIM-noe-sperm) [Gk. *gymnos*, naked, + *sperma*, seed] A vascular plant that forms seeds on exposed surfaces of spore-producing structures; e.g., conifers, cycads.

habitat [L. *habitare*, to live in] The place where an organism or species normally lives, characterized by its physical and chemical features and its array of species.

habitat fragmentation The break-up of a habitat into patches too small to support successful breeding; may make species more vulnerable to losses.

habitat island An area of endemic species that is surrounded by a "sea" of habitats unsuitable for sustaining the species; e.g., a lake surrounded by a forest.

habitat loss Reduction in suitable living space and closure of part of a habitat as an outcome of chemical pollution.

hair A flexible structure rooted in skin, with a shaft above the skin's surface.

hair cell Hairlike mechanoreceptor; it is activated when sufficiently bent or tilted.

half-life The unvarying time it takes for half of a quantity of any radioisotope to decay into a more stable form.

haploid chromosome number The sum of all chromosomes in cells with one of each type of chromosome characteristic of the species; e.g., in a gamete.

hardwood Strong, dense wood with many vessels, tracheids, and fibers in xylem.

Hardy–Weinberg rule Theoretical baseline for tracking changes in allele frequencies over the generations. Frequencies do not change as long as there is no mutation, the population is infinitely large and isolated from other populations, and all individuals are reproducing equally and randomly.

HCG Human chorionic gonadotropin; secreted by blastocyst; helps maintain the endometrium until placenta secretes it, about eleven weeks later.

HDL A high-density lipoprotein in blood; it transports dietary cholesterol to the liver, which metabolizes it.

hearing Perception of sound.

heart Muscular pump; its contractions circulate blood through the animal body.

heartwood Dense, dry tissue at the core of aging tree stems and roots; helps trees defy gravity and store metabolic wastes.

heat A transfer of thermal energy.

HeLa cell Cancer cell of a lineage used in research laboratories around the world.

helicase Type of enzyme that catalyzes breaking of hydrogen bonds during DNA replication so the two strands of double helix can unwind from each other.

helper T cell CD4 lymphocyte. Central to adaptive immunity; induces responsive T and B cells to form antigen-sensitive armies.

heme Oxygen-transporting cofactor of many enzymes and pigments; one iron atom at the center of an organic ring structure.

hemoglobin (HEEM-oh-glow-bin) [Gk. *haima*, blood, + L. *globus*, ball] A heme-containing protein produced by red blood cells; carries most of the oxygen in blood.

hemostasis (HEE-mow-STAY-sis) [Gk. *stasis*, standing] Process that stops blood loss from a damaged blood vessel by coagulation, spasm, and other mechanisms.

herbicide Natural or synthetic toxin that can kill or inhibit growth of target plants.

herbivore [L. *herba*, grass, + *vovare*, to devour] Plant-eating animal.

hermaphrodite (her-MAH-froe-dyte) An individual with male and female gonads.

heterocyst (HET-er-oh-sist) Self-modified cyanobacterial cell; synthesizes nitrogen-fixing enzyme when nitrogen is scarce.

heterotherm An animal that maintains its core temperature by controlling metabolic activity some of the time and allowing it to rise or fall at other times.

heterotroph (HET-er-oh-trofe) [Gk. *heteros*, other, + *trophos*, feeder] Organism that cannot make its own food; feeds on other organisms, their wastes, or their remains.

heterozygous condition (HET-er-oh-ZYE-guss) [Gk. *zygoun*, join together] Having nonidentical alleles at a given gene locus on a pair of homologous chromosomes.

higher taxon, plural **taxa** One of ever more inclusive groupings of species; e.g., family, order, class, phylum, kingdom.

hindbrain Medulla oblongata, cerebellum, and pons. Its reflex centers are vital for respiration and other basic functions, and for coordinating motor responses.

histamine Local signaling molecule that stimulates inflammation; makes arterioles vasodilate and capillaries more permeable.

histone Type of structural protein that helps organize and condense eukaryotic chromosomes and control access to genes during interphase.

HLA Human leukocyte antigens. *See* MHC.

homeostasis (HOE-me-oh-STAY-sis) [Gk. *homo*, same, + *stasis*, standing] State in which physical and chemical aspects of internal environment (blood, interstitial fluid) are being maintained within ranges that are tolerable for cell activities.

homeotic gene One of a class of master genes; helps determine identity of body parts during embryonic development.

hominid [L. *homo*, man] All humanlike and human species.

hominoid Apes, humans, and their most recent ancestors.

homologous chromosome (huh-MOLL-uh-gus) [Gk. *homologia*, correspondence] One of a pair of chromosomes in body cells of diploid organisms; except for a pairing of nonidentical sex chromosomes, a pair has the same size, shape, and gene sequence.

homozygous dominant condition Having a pair of dominant alleles at a gene locus on homologous chromosomes; e.g., *AA*.

homozygous recessive condition Having a pair of recessive alleles at a gene locus on homologous chromosomes; e.g., *Aa*.

hormone [Gk. *hormon*, stir up] Signaling molecule secreted by one cell that can alter activities of any cell with receptors for it.

hornwort Bryophyte having a horn-shaped sporophyte attached to a flat gametophyte.

horsetail Seedless vascular plant having rhizomes, scale-like leaves, and hollow stems with silica-reinforced ribs.

host, of parasite Living organism in or on which a parasite must complete its life cycle. A definitive host harbors the mature stage. Intermediate hosts harbor immature stages.

host, of symbiont The larger, stronger, or more dominant partner of two symbionts.

hot spot A region where human activities are driving many species to extinction. Also

a region where superplumes have ruptured the crust; e.g., at the Hawaiian Archipelago.

human Primate of species *Homo sapiens*.

human gene therapy The transfer of one or more normal or modified genes into a person to correct a genetic defect, or to boost resistance to a disease.

humus Variably thick layer of organic matter that is decomposing in soil.

hybrid Individual having a nonidentical pair of alleles for a trait being studied.

hydrocarbon Organic compound with only hydrogen bonded to its carbon backbone.

hydrogen bond A weak attraction that has formed between a covalently bonded hydrogen atom and an electronegative atom taking part in another covalent bond.

hydrogen ion Free (or unbound) proton; one hydrogen atom that lost its electron and now bears a positive charge (H^+).

hydrologic cycle Biogeochemical cycle driven by solar energy; water moves through atmosphere, on or through land, to the ocean, and back to the atmosphere.

hydrolysis (high-DRAWL-ih-sis) [L. *hydro*, water, + Gk. *lysis*, loosening] A cleavage reaction; an enzyme splits a molecule, then the components of water (—OH and —H) are attached to the fragments.

hydrophilic substance [Gk. *philos*, loving] A polar molecule that dissolves easily in water; e.g., glucose.

hydrophobic substance [Gk. *phobos*, dreading] A nonpolar molecule that resists dissolving in water; e.g., oil.

hydrostatic pressure Pressure exerted by a volume of fluid against a cell wall, membrane, or some other structure that contains it; also called turgor pressure.

hydrostatic skeleton A fluid-filled cavity against which a contractile force can act.

hydrothermal vent A steaming fissure on the ocean floor; has unique ecosystems.

hypertonic solution Of two fluids, the one with the higher solute concentration.

hypha (HIGH-fuh), plural **hyphae** Fungal filament having chitin-reinforced walls; component of a mycelium.

hypothalamus [Gk. *hypo*, under, + *thalamos*, inner chamber] Forebrain region; a center of homeostatic control of internal environment (e.g., salt–water balance, core temperature); influences hunger, thirst, sex, other viscera-related behaviors, and emotions.

hypothesis, scientific An explanation of a phenomenon, one that has the potential to be proven false by experimental tests.

hypotonic solution Of two fluids, the one with the lower solute concentration.

imbibition Water molecules move into a seed, attracted by hydrophilic groups of proteins; assists in germination.

immigration One or more individuals move and take up residence in another population of its species.

immune system White blood cells and signaling molecules of vertebrate adaptive immunity; they recognize self and nonself and have the potential to target a billion specific antigens, with lasting effect.

immunity The body's ability to resist and combat infections.

immunization A process that promotes immunity from disease; e.g., vaccination.

immunoglobulin One of five classes of antibodies, each with antigen-binding and class-specific structural components.

implantation A process in pregnancy; a blastocyst burrows into the endometrium; it establishes connections by which the embryo that forms from its inner cell mass will exchange substances with the mother.

imprinting A form of learning triggered by exposure to sign stimuli; time-dependent, often in a young animal's sensitive period.

in vitro fertilization Conception outside the body, "in glass" petri dishes or tubes.

inbreeding Nonrandom mating among very close relatives that share many identical alleles; may fix harmful alleles.

inclusive fitness theory Idea that genes associated with caring for relatives may be favored in some situations.

incomplete digestive system Saclike gut in which both food intake and waste output occur through a single opening.

incomplete dominance Condition in which one allele of a pair is not fully dominant; the heterozygous phenotype is somewhere between both homozygous phenotypes.

independent assortment An outcome of random alignments at metaphase I of meiosis. Each homologous chromosome and its partner—and the genes they carry —are assorted into different gametes independently of the other pairs. Crossing over can affect the outcome.

indicator species Any species which, by its abundance or scarcity, is a measure of the health or degradation of its habitat.

induced-fit model Explanation of how some enzymes work; their shape changes and fits a bound substrate more closely, and the tension destabilizes substrate bonds so that they can break.

infection Invasion and multiplication of a pathogen or parasite in a host. Disease follows if defenses are not mobilized fast enough against the tissue disruptions.

inflammation, acute Rapid, nonspecific response to tissue invasion or injury. White blood cells release the cytokines that attract phagocytes and cause local vasodilation. Signs include redness, heat, swelling, pain.

inheritance Transmission, from parents to offspring, of genes that underlie the traits characteristic of their species.

inhibin Hormone that inhibits secretion of GnRH and FSH from the hypothalamus and pituitary gland.

inhibiting hormone Signaling molecule from the hypothalamus that suppresses release of an anterior pituitary hormone.

inhibitor Any substance that binds to a molecule and interferes with its function.

innate immunity Immediate, off-the-shelf set of responses to tissue invasion that rid the body of most pathogens. Recognition of a fixed set of conserved pathogen-associated molecular patterns triggers phagocytosis, inflammation, and complement activation.

inner ear Of vertebrates, the primary organ of equilibrium and hearing; includes a vestibular apparatus and cochlea.

insertion A mutation by which one or more bases are introduced into a DNA strand. Also a movable attachment of muscle to bone.

instinctive behavior Behavior performed without having first been learned through actual experience in the environment.

insulin Pancreatic hormone. Its actions lower the blood level of glucose.

integrator A control center that receives, processes, and stores sensory input, and coordinates the responses; e.g., a brain.

integument Of animals, a protective body covering; e.g., skin. Of seed-bearing plants, one of the layers around an ovule that mature into a seed coat.

integumentary exchange Of some animals, gas exchange across thin, moistened skin or some other external body surface.

interleukin A type of cytokine; a signaling molecule of cells of immune system.

intermediate A substance formed between the start and end of a metabolic pathway.

intermediate disturbance hypothesis An explanation of community structure; holds that species richness is greatest in between disturbances that are moderate in intensity, frequency, or both.

intermediate filament Cytoskeletal element that mechanically strengthens some cells.

internal environment The body fluid *not* inside cells, or extracellular fluid; in most animals, blood and interstitial fluid.

interneuron Any of the diverse neurons of the brain or spinal cord.

internode Plant stem between two nodes.

interphase In a eukaryotic cell cycle, the interval between mitotic divisions when a cell grows in mass, roughly doubles the number of its cytoplasmic components, and replicates its DNA.

interstitial fluid (IN-ter-STISH-ul) All fluid in spaces between cells of all tissues except the connective tissue called blood.

intertidal zone Between low and high water marks of a rocky or sandy shore.

intervertebral disk In between vertebrae, a cartilaginous flex point and shock absorber.

intron One of the noncoding sequences in eukaryotic genes; it is excised from the pre-mRNA transcripts before translation.

inversion A chromosomal alteration; part of the DNA sequence gets oriented in the reverse direction, with no molecular loss.

invertebrate Animal without a backbone.

ion Atom having an unequal number of protons and electrons; it carries a positive or negative electric charge.

ion exchange pH-dependent process; ions dissociate from soil particles, then other ions dissolved in soil water replace them.

ionic bond Ions interacting through the attraction of their opposite charges.

ionizing radiation Form of radiation with enough energy to eject electrons from atoms.

IPSP An inhibitory postsynaptic potential; a type of graded potential at an input zone of an excitable cell that drives its membrane away from threshold.

isotonic solution Any fluid having the same solute concentration as another fluid to which it is being compared.

isotope One of two or more atoms of the same element (same number of protons) that differ in their number of neutrons.

jaw Paired, hinged cartilaginous or bony feeding structures of most chordates. Many invertebrates have hinged feeding structures but not of bone or cartilage.

joint Area of contact between bones.

J-shaped curve Diagrammatic curve that emerges as unrestricted exponential growth of a population is plotted against time.

juvenile Post-embryonic stage of certain animals; it changes in size and proportion before adulthood, with no metamorphosis.

karyotype Preparation of an individual's metaphase chromosomes arranged by length, centromere location, and shape.

keratinocyte Skin cell that makes keratin, a tough, water-insoluble protein.

key innovation A chance modification in some body structure or function that gives a species the opportunity to exploit the environment more efficiently or in a novel way; e.g., modifications of the forelimbs of amniotes into diverse legs and wings during radiations into adaptive zones.

keystone species A species that influences community structure in disproportionally large ways relative to its abundance.

kidney One of a pair of vertebrate organs that filter ions and other substances from blood; it controls the amounts returned to help maintain the internal environment.

kilocalorie 1,000 calories of heat energy; amount needed to raise the temperature of 1 kilogram of water by 1°C. Standard unit of measure for food's caloric content.

kinase Type of enzyme that transfers a phosphate-group to an organic molecule.

kinetic energy Energy of motion.

kinetochore A mass of protein and DNA in the centromere to which microtubules of the spindle attach.

kinetoplastid A colorless flagellate; the only eukaryote with mitochondrial DNA massed inside a mitochondrion almost as long as the cell; e.g., *Trypanosoma*.

knockout experiment An experiment in which a living organism is engineered so that one of its genes does not function.

Krebs cycle The second stage of aerobic respiration in which many coenzymes form as pyruvate from glycolysis is fully broken down to CO_2 and H_2O. Two ATP also form. Occurs only in mitochondria.

K–T asteroid impact theory A massive asteroid struck Earth 65 million years ago and caused a mass extinction; casualties included the last of the dinosaurs.

K–T boundary The boundary between the Cretaceous and Tertiary periods.

La Niña Cooler climatic event between ENSOs; disrupts global climates.

labor Time of childbirth.

lactate fermentation One of the anaerobic pathways of ATP formation. NADH from glycolysis donates hydrogen and electrons to pyruvate, converting it to three-carbon lactate, and regenerating NAD^+. The net energy yield is 2 ATP (from glycolysis).

lactation Milk production and secretion by hormone-primed mammary glands.

lake A body of standing fresh water in a basin characterized by light penetration, temperature gradients, and other features.

Langerhans cell Antigen-presenting cell in skin; engulfs viruses and bacteria.

large intestine Colon. The bacteria-rich region of the vertebrate gut that absorbs water and mineral ions and also compacts undigested food residues for elimination.

larva, plural **larvae** An immature stage between the embryo and adult in the life cycle of many animals.

larynx (LARE-inks) Tubular airway leading to lungs; has vocal cords in some animals.

lateral bud Axillary bud. A dormant shoot that forms in a leaf axil.

lateral meristem Vascular cambium or cork cambium. A sheetlike cylinder of meristem inside older stems and roots.

lateral root Outward branching from the first (primary) root of a taproot system.

LDL Low-density lipoprotein; transports cholesterol; excess amounts in blood may contribute to atherosclerosis.

leaching Removal of some nutrients from soil as water percolates through it.

leaf Chlorophyll-rich plant organ of sunlight interception and photosynthesis.

learned behavior Enduring modification of a behavior as an outcome of experience in the environment.

lek A communal courtship display ground.

lens Of camera eyes, a transparent body that bends light rays so they all converge suitably onto photoreceptors of a retina.

leptin An appetite-suppressing hormone produced mainly by adipose tissue.

lethal mutation Mutation having drastic effects on phenotype; usually causes death.

Leydig cell Testosterone-secreting cell in mammalian testes.

LH Luteinizing hormone. An anterior pituitary hormone; roles in reproductive function of male and female mammals.

lichen (LY-kun) Mutualism between a fungus and one or more photoautotrophs.

life cycle A series of stages through which an individual passes from the time it forms by way of a mode of sexual or asexual reproduction until its own reproduction. Among complex species, the individual typically ages and dies at some time after it has reproduced.

life history pattern Of many species, the pattern of when and how many offspring are produced during a typical lifetime.

ligament A strap of dense connective tissue that bridges a skeletal joint.

light-dependent reactions First stage of photosynthesis. Pigments trap photon energy, which is transduced to ATP chemical energy. In a noncyclic pathway, a reduced coenzyme, NADPH, also forms.

light-independent reactions Second stage of photosynthesis. Involves carbon fixation and cyclic reactions that form sugars and regenerate an organic compound that is the cycle's entry point. ATP from the first stage delivers energy that drives the reactions. NADPH from the first stage donates electrons and hydrogen building blocks. The carbon and nitrogen come from CO_2.

lignin Gluelike polymer deposited in secondary cell walls; makes some plant parts stronger, more waterproof, and less vulnerable to attacks.

limbic system Centers in cerebrum that govern emotions; roles in memory.

limiting factor Any essential resource that limits population growth when scarce.

lineage (LIN-ee-edge) Line of descent.

linkage group All genes on a chromosome.

lipid One of the nonpolar hydrocarbons; e.g., a fat, oil, wax, sterol, phospholipid, or glycolipid. Cells use as storage forms of energy and building blocks.

lipid bilayer Structural basis of all cell membranes; mainly phospholipids arranged tail-to-tail in two layers, with hydrophilic heads of one dissolved in cytoplasmic fluid and heads of the other in extracellular fluid.

lipoprotein A protein complexed with cholesterol, triglycerides, or phospholipids that were absorbed from the small intestine.

liver A large gland that stores, converts, and helps maintain blood levels of organic compounds; also inactivates most hormone molecules after signaling ends as well as compounds that are toxic at high levels.

liverwort One of the bryophytes.

loam Soil best for plant growth; roughly the same proportions of sand, silt, clay.

lobe-finned fish Only bony fish having ventral fins with fleshy extensions and internal skeletal elements.

local signaling molecule One of many cell secretions into extracellular fluid. All have potent effects but are inactivated so fast that the signal is confined to local tissues; e.g., prostaglandins.

logistic growth (low-JISS-tik) Population growth pattern. A low-density population slowly increases in size, enters a phase of rapid growth, then levels off in size once the carrying capacity has been reached.

long-day plant A plant that flowers in spring, when nights are shorter (and days longer) than some critical value.

loop of Henle Hairpin-shaped, tubular part of a nephron where water and solutes are reabsorbed from interstitial fluid.

loose connective tissue Animal tissue with fibers and fibroblasts loosely arrayed in a semifluid matrix of cell secretions.

lung One of a pair of internal sac-shaped respiratory surfaces that originated in oxygen-poor aquatic habitats. Exclusive organs of respiration in birds, reptiles, and mammals; supplements respiration in some fishes and most amphibians.

lungfish A type of bony fish having both gills and one or two lung-like outpouchings of the gut wall that assist in respiration.

lycophyte A type of seedless vascular plant, typically with true leaves, roots, and stems; e.g., a club moss.

lymph Interstitial fluid that has entered vessels of the lymphatic system.

lymph node Lymphoid organ that is a key site for immune responses, as executed by its organized arrays of lymphocytes.

lymph vascular system The portion of the lymphatic system that takes up and conducts excess tissue fluid, absorbed fats, and reclaimable solutes to blood.

lymphatic system Organ system with vessels that return excess interstitial fluid and reclaimable solutes to blood and with lymphoid organs that function in defense.

lymphocyte A class of white blood cells. *See* B lymphocyte, T lymphocyte, NK cell.

lysis Gross damage to a cell wall, plasma membrane, or both that lets cytoplasm leak out; causes cell death.

lysogenic pathway A latent period that extends many viral replication cycles. Viral genes are integrated into host chromosome and may remain inactivated through many host cell divisions before being replicated.

lysosome Vesicle filled with enzymes that functions in intracellular digestion.

lysozyme Infection-fighting enzyme in mucous membranes; e.g., of mouth.

lytic pathway A rapid viral replication pathway that ends with lysis of host cell.

macroevolution Large-scale patterns, rates of change, and trends among lineages.

macrophage Phagocytic white blood cell; in vertebrates, it takes part in nonspecific defenses and adaptive immunity.

magnoliid One of three major flowering plant groups; e.g., magnolias, avocados.

malpighian tubule One of many small tubes that help insects on land dispose of toxic wastes without losing body water.

mammal Only amniote that makes hair and nourishes offspring with milk from the female's mammary glands.

mangrove wetland Tidal flat community at tropical latitudes; rich in nutrients.

mantle Of mollusks, a tissue draped over the visceral mass. Of Earth, a rocky zone of intermediate density under the crust.

marine snow Organic matter drifting down from ultraplankton to mid-oceanic water; supports food webs and marine biodiversity only now being explored.

marsupial Pouched mammal.

mass extinction Catastrophic event or phase in geologic time when families or other major groups are lost.

mass number Sum of protons and neutrons in the nucleus of an element's atoms.

mast cell White blood cell in connective tissue; secretes most of the cytokines during an innate immune response.

master gene One of the genes encoding products that map out the body plan in developing embryos. Gene products form gradients by diffusing from a source tissue. Cells along the gradient differentiate in ways that give rise to tissues and organs in expected places.

mechanoreceptor Sensory cell that detects mechanical energy (a change in pressure, position, or acceleration).

medulla oblongata Hindbrain region. Its reflex centers control respiration and other basic tasks; coordinate motor responses with complex reflexes; e.g., coughing.

medusa (meh-DOO-sah) [Gk. *Medousa*, one of three sisters in Greek mythology with snake-entwined hair] Of cnidarian life cycles, a free-swimming, bell-shaped stage, often with oral lobes and tentacles.

megaspore Haploid meiotic spore in ovary of seed-bearing plants; gives rise to a female gametophyte with egg cell.

meiosis (my-OH-sis) [Gk. *meioun*, to diminish] A nuclear division process that halves the parental chromosome number, to a haploid (*n*) number. Prerequisite to the formation of gametes and sexual spores.

melanin A brownish-black pigment.

melanocyte A skin cell that produces and releases melanin to keratinocytes.

memory A capacity to store and retrieve information about sensory experiences.

memory cell A sensitized B or T cell that forms in a primary immune response but is reserved for recurrence of the same antigen.

menstrual cycle Recurring cycle in adult human females and some other primates. A secondary oocyte is released from an ovary and the uterine lining is primed for pregnancy, all under hormonal control.

meristem [Gk. *meristos*, divisible] One of the localized zones where dividing cells gives rise to differentiated cell lineages that form all mature plant tissues.

mesoderm (MEH-zoe-derm) Primary tissue layers gives rise to many internal organs and part of the integument; pivotal in the evolution of large, complex animals.

mesoglea Of cnidarians, a gelatinous matrix with scattered cells between the epidermis and gastrodermis; functions as a buoyant, deformable skeleton.

mesophyll (MEH-zoe-fill) Photosynthetic parenchyma with many air spaces.

Mesozoic Era of spectacular expansion in the range of global diversity; lasted from 240 million to 65 million years ago.

messenger RNA mRNA. A single strand of ribonucleotides transcribed from DNA; the only type of RNA that carries protein-building information to ribosomes.

metabolic pathway A stepwise sequence of enzyme-mediated reactions.

metabolism (meh-TAB-oh-lizm) All the controlled, enzyme-mediated chemical reactions by which cells acquire and use energy as they synthesize, store, degrade, and eliminate substances.

metamorphosis (me-tuh-MOR-foe-sis) [Gk. *meta*, change, + *morphe*, form] Major changes in body form of certain animals. Hormonally controlled growth, tissue reorganization, and remodeling of body parts leads to adult form.

metaphase Of meiosis I, stage when all pairs of homologues are positioned at the equator of a bipolar spindle. Of mitosis or meiosis II, the stage when all duplicated chromosomes are positioned at the equator.

metastasis Abnormal migration of cancer cells that break away from home tissues and may start colonies in other tissues.

methanogen Any bacterium or archaean that produces methane gas as by-product of anaerobic reactions.

methylation Attachment of a methyl group to an organic compound; also a common gene control mechanism.

MHC molecule Also called HLA. Type of proteins at the surface of body cells that T cells recognize as self-markers. Sounds the immune alarm when it becomes complexed with antigen fragments.

micelle formation The combining of bile salts with fatty acids into tiny droplets.

microevolution Of a population, a small-scale change in allele frequencies resulting from mutation, genetic drift, gene flow, natural selection, or a combination of them.

microfilament The thinnest cytoskeletal element; consists of actin subunits that function in cell contraction, movement, and structural support.

micrograph Photograph of an image formed with the aid of a microscope.

microorganism Microbe. Any organism, usually single celled, that is too small to be observed without a microscope.

microspore Type of walled haploid spore of gymnosperms and angiosperms that gives rise to pollen grains.

microsporidian Intracellular fungal parasite of aquatic habitats that forms flagellated spores; belongs to one of the most ancient eukaryotic lineages.

microtubular spindle *See* Bipolar spindle.

microtubule Largest cytoskeletal element; a filament of tubulin subunits. Contributes to cell shape, growth, and motion.

microtubule organizing center MTOC. Of eukaryotic cells, a mass of cytoplasmic substances, the number, type, and location of which dictate how microtubules will be organized and oriented in a given type of cell; e.g., a centrosome.

microvillus (MY-crow-VILL-us) [L. *villus*, shaggy hair] Slender extension from free surface of certain cells; arrays of many microvilli greatly increase the absorptive or secretory surface area of a cell.

midbrain A vertebrate brain region with centers for coordinating reflex responses to visual and auditory input; also relays signals to forebrain.

middle ear The eardrum and ear bones that transmit air waves to the inner ear.

migration Of many animals, a recurring pattern of movement between two or more regions in response to seasonal change or other environmental rhythms.

millipede An arthropod with a great many unspecialized segments and paired legs; scavenges decaying plant material.

mimicry (MIM-ik-ree) A case of one species (the mimic) closely resembling another (its model) in form, behavior, or both.

mineral Element or inorganic compound required for normal cell functioning.

mitochondrion (MY-toe-KON-dree-on) Double-membraned organelle of ATP formation; only site of the second and third stages of aerobic respiration.

mitosis (my-TOE-sis) [Gk. *mitos*, thread] Type of nuclear division that maintains the parental chromosome number. The basis of growth in size, tissue repair, and often asexual reproduction for eukaryotes.

mixture Two or more types of molecules intermingled in proportions that can and usually do vary.

model Theoretical explanation of any object or event that has not been or cannot be directly observed.

molar Tooth with cusps that crush, grind, and shear food; one of the cheek teeth.

molecular clock Model used to calculate the time of origin of one lineage relative to others; assumes that a group of genes accumulates mutations at a constant rate, measurable as a series of predictable ticks back through time. The last tick stops close to the time the lineage originated.

molecule Two or more covalently bonded atoms of the same or different elements.

mollusk Only invertebrate with a mantle draped over a soft, fleshy visceral mass; most have an external or internal shell; e.g., gastropods, bivalves, cephalopods.

molting Periodic shedding of worn-out or too-small body structures. Permits an animal to grow in size or renew parts.

monocot (MON-oh-kot) Monocotyledon; flowering plant characterized by embryo sporophytes having one cotyledon; floral parts usually in threes (or multiples of three); and often parallel-veined leaves.

monohybrid experiment An experiment that starts with a cross between two true-breeding, homozygous parents that differ in a trait governed by alleles of one gene. The experiment is a cross between two F_1 offspring that are identically heterozygous for the two genes; e.g., *Aa* x *Aa*.

monomer Any small molecule that is a repeating subunit in a polymer; e.g., the sugar monomers of starch.

monophyletic group A set of species that share a derived trait, a novel feature that evolved in one species and is present only in its descendants; all of the evolutionary branchings from a single stem.

monosaccharide [Gk. *monos*, alone, single, + *sakcharon*, sugar] A simple sugar.

monotreme Egg-laying mammal.

monsoon Air circulation pattern; moves moisture-laden air above warm oceans to continents north or south of them.

morphogen An inducer molecule. Diffuses through embryonic tissues; the resulting gradient sequentially activates master genes.

morphogenesis (MORE-foe-JEN-ih-sis) [Gk. *morphe*, form, + *genesis*, origin] Orderly, genetically programmed changes in size, proportion, and shape of body parts of an animal embryo through which specialized tissues and organs form.

morphological convergence A pattern of macroevolution. In response to similar environmental pressures, body parts of evolutionarily distant lineages slowly evolve in similar ways and end up being alike in function, appearance, or both.

morphological divergence Pattern of macroevolution. One or more body parts of genetically diverging lineages undergo structural and functional changes from the parts in the common ancestor.

mosaic tissue effect In female mammals, an outcome of random X chromosome inactivation; different patches of tissue are expressing different X-linked alleles.

mosaicism Two or more genetically distinct cell lineages in an individual.

moss Most common kind of bryophyte.

motor neuron Neuron that relays signals from the brain or spinal cord to muscle cells or gland cells.

motor protein A type of accessory protein that interacts with microfilaments or with microtubules to move cell structures or the whole cell; e.g., myosin.

motor unit One motor neuron and all muscle cells that form junctions with its axon endings.

multicelled organism Organism that consists of many cells that, at the least, have formed layers; most have many differentiated cells that have formed true tissues, organs, and organ systems.

multiple allele system Three or more slightly different molecular forms of a gene that persists among the individuals of a population.

multiregional model Idea that modern humans evolved from *Homo erectus* groups that spread through much of the world by about 1 million years ago and evolved into regionally distinctive "races."

muscle fatigue Decline in muscle tension when tetanic contraction is continuous.

muscle fiber A group of muscle cells in parallel array. Each large, multinucleated fiber of skeletal and cardiac muscles formed when a group of undifferentiated muscle cells fused during embryonic development. Smooth muscle fibers are shorter, and each cell has retained its own nucleus.

muscle spindle A sensory organ that detects muscle stretching; its input zones are enclosed in a sheath that runs parallel with the muscle.

muscle tension Mechanical force exerted by a contracting muscle; resists opposing forces; e.g., weight of an object being lifted.

muscle tissue Tissue with muscle fibers arranged in parallel to bring about the directional contraction of a body part.

muscle twitch A sequence of muscle contraction and relaxation in response to a brief stimulus.

mushroom Aboveground reproductive structure produced by many club fungi.

mutation [L. *mutatus*, a change, + *-ion*, act, result, or process] Heritable change in DNA's molecular structure. Original source of new alleles and life's diversity.

mutation rate Of a given gene locus, the probability that a spontaneous mutation will happen in a specified interval.

mutualism [L. *mutuus*, reciprocal] A type of symbiotic interaction that benefits both participants.

mycelium (my-SEE-lee-um), plural **mycelia** [Gk. *mykes*, fungus] Underground mesh of tiny, branching filaments (hyphae); the food-absorbing portion of most fungi.

mycorrhiza (MY-coe-RIZE-uh) "Fungus-root." A form of mutualism between a fungus and young plant roots. Hyphae withdraw some carbohydrates from the plant, which withdraws some absorbed mineral ions from hyphae.

myelin sheath Lipid-rich wrappings of oligodendrocytes around axons of many sensory and motor neurons; enhances long-distance propagation of action potentials.

myofibril (MY-oh-FY-brill) One of many long, thin structures divided into contractile units that run parallel with the long axis of a muscle fiber.

myoglobin A pigment that is structurally similar to a hemoglobin chain but stores oxygen; abundant in some muscle fibers.

myosin (MY-uh-sin) An ATP-energized motor protein that moves cell components on cytoskeletal tracks. Interacts with actin in sarcomeres to bring about contraction.

NAD+ Nicotinamide adenine dinucleotide. A nucleotide coenzyme; after it accepts electrons and H^+, abbreviated as NADH.

NADP+ Nicotinamide adenine dinucleotide phosphate. A phosphorylated nucleotide coenzyme; after it accepts electrons and H^+, abbreviated NADPH$_2$.

nannoplankton Coccolithophores and other marine photoautotrophs from five to fifty micrometers across.

natural selection Microevolutionary process; the outcome of differences in survival and reproduction among individuals of a population that differ in the details of their heritable traits.

necrosis (neh-CROW-sis) Passive death of many cells after severe tissue damage.

nectar Dilute, sucrose-rich fluid secreted from a nectary that connects to phloem; attracts pollinators.

negative control Control mechanism by which one or more regulatory proteins slow down a cell activity.

negative feedback mechanism A main homeostatic mechanism by which some activity changes conditions in a cell or multicelled organism and thereby triggers a response that reverses the change.

nematocyst (NEM-at-uh-sist) [Gk. *nema*, thread + *kystis*, pouch] A fluid-filled, jack-in-the-box capsule housed in one of three types of sensory–effector cells in cnidarians. It has a mechanoreceptor projecting above the cell surface and a dischargeable, tubular thread, often with barbs or toxin-drenched. Only cnidarians make nematocysts.

neoplasm Mass of cells (tumor) that lost control over the cell cycle.

nephridium, plural **nephridia** Of some invertebrates, one of many water-regulating units that help control the composition and volume of tissue fluid.

nephron (NEFF-ron) [Gk. *nephros*, kidney] One of millions of tubules in kidneys; it filters water and solutes from blood, then reabsorbs adjusted amounts of both.

nerve A sheathed, cordlike communication line that holds bundled fibers of sensory neurons, motor neurons, or both.

nerve cell A neuron.

nerve cord Of bilateral animals, a line of communication, usually paired, that runs parallel with the anterior–posterior axis. In large or long invertebrates, it often has one or more large axons. In chordates, it develops as a hollow, neural tube that gives rise to the spinal cord and brain.

nerve net Nervous system of cnidarians and some other invertebrate groups; an asymmetrical mesh of sensory and motor neurons that controls simple movements. It activates epitheliomuscular cells arrayed as sheets or rings in the body wall.

nervous system Organ system of neurons and, in many animals, neuroglia. Detects, distributes, processes, and issues signals for responses to sensory information; also stores information in complex species.

nervous tissue Tissue consisting of neurons and often neuroglia.

net ecosystem production All the energy the primary producers have accumulated during growth, reproduction in a specified interval (net primary production), *minus* energy that producers, and decomposers, have used.

neural tube Embryonic and evolutionary forerunner of the brain and spinal cord.

neuroglia (NUR-oh-GLEE-uh) Collectively, cells that structurally and metabolically support neurons; about half the volume of nervous tissue in vertebrates.

neuromodulator Any signaling molecule that reduces or magnifies the influence of a neurotransmitter on target cells.

neuromuscular junction A chemical synapse between a motor neuron's axon endings and a muscle fiber.

neuron (NUR-on) A nerve cell; the basic communication unit in nervous systems.

neurotransmitter Any of a diverse class of signaling molecules that are secreted by neurons. It acts in a synaptic cleft, then is rapidly degraded or recycled.

neutral mutation A mutation with no effect on phenotype; natural selection thus cannot change its frequency in a population.

neutron Type of subatomic particle in the nucleus of all atoms except hydrogen; has mass but no charge.

neutrophil Abundant circulating white blood cell; mainly phagocytic in innate immunity; its enzymes kill extracellular microbes and stimulate inflammation.

niche (NITCH) [L. *nidas*, nest] Sum total of all activities and relationships in which individuals of a species engage as they secure and use the resources required to survive and reproduce.

nitrification (nye-trih-fih-KAY-shun) One stage of the nitrogen cycle. Soil bacteria break down ammonia or ammonium to nitrite, then other bacteria break down nitrite to nitrate, which plants can absorb.

nitrogen cycle An atmospheric cycle. Nitrogen moves from its largest reservoir (atmosphere), then through the ocean, ocean sediments, soils, and food webs, then back to the atmosphere.

nitrogen fixation One stage of the nitrogen cycle process. Bacteria convert gaseous nitrogen to ammonia, which dissolves in their cytoplasm to form ammonium for use in biosynthesis.

NK cell Natural killer cell. One of the cytotoxic lymphocytes of innate and adaptive immunity; touch-kills tumor cells and virus-infected cells.

node A location along the length of a stem where one or more leaves form.

noncyclic pathway of ATP formation (non-SIK-lik) [L. *non*, not, + Gk. *kylos*, circle] The light-dependent reactions of photosynthesis that produce both ATP and NADPH; its oxygen by-product is the basis of Earth's oxygen-rich atmosphere.

nondisjunction Failure of sister chromatids or homologous chromosomes to move apart in meiosis or mitosis. Daughter cells get too many or too few chromosomes.

nonionizing radiation Form of radiation that carries enough energy to boost electrons to higher energy levels but not enough to eject them from an atom.

nonshivering heat production Increase in metabolically generated heat in response to prolonged or severe cold exposure.

norepinephrine (NOR-epih-NEF-rin) A stress hormone released from the adrenal medulla and released as a neurotransmitter from sympathetic neurons and the brain. Affects metabolic rates, heart function; acts with epinephrine in the fight–flight response.

notochord (KNOW-toe-kord) A rod of stiffened tissue, neither cartilage nor bone, that develops in chordate embryos and that may or may not persist as a supporting structure for the adult body.

nuclear envelope A double membrane that is the outer boundary of the nucleus.

nucleic acid Single-stranded or double-stranded molecule of nucleotides joined at phosphate groups; e.g., DNA, RNA.

nucleic acid hybridization Any base-pairing between DNA or RNA strands from different sources.

nucleoid (NEW-KLEE-oid) The portion of a prokaryotic cell where DNA is physically organized but not enclosed in a membrane.

nucleolus (new-KLEE-oh-lus) [L. *nucleolus*, tiny kernel] In an interphase nucleus, a mass of material from which RNA and proteins are assembled into the subunits of ribosomes.

nucleosome Small stretch of eukaryotic DNA wound twice around a spool of proteins called histones.

nucleotide Small organic compound with a five-carbon sugar, a nitrogen-containing base, and a phosphate group. Functions as coenzymes or monomers of nucleic acids.

nucleus Large organelle with an outer envelope of two pore-ridden lipid bilayers that separates eukaryotic chromosomes from the cytoplasm.

numerical taxonomy In microbiology, a method of classifying an unidentified microbe by comparing it with a known group on the basis of shape, wall staining attributes, and other observable traits; the more traits shared, the closer is the inferred relatedness.

nutrient Any element having a direct or indirect role in metabolism that no other element can fulfill.

nutrition Collectively, processes by which an organism takes in, digests, absorbs, and converts food into organic compounds.

nymph Immature, post-embryonic stage of some insect life cycles.

obesity Having an excessive amount of fat in adipose tissue; caloric intake has exceeded the body's energy output.

ocean A continuous body of water that covers more than 71 percent of Earth; its currents distribute nutrients in marine ecosystems and affect regional climates.

olfactory receptor Chemoreceptor for a water-soluble or volatile substance.

oligosaccharide (oh-LIG-oh-SAC-uh-rid) Short-chain carbohydrate of two or more covalently bonded sugar monomers; e.g., sucrose and other disaccharides.

omnivore [L. *omnis*, all, + *vovare*, to devour] A type of animal that eats other organisms at more than one trophic level.

oncogene (ON-koe-jeen) A gene which, when mutated or expressed at abnormal levels, is associated with cancer.

oocyte A type of immature egg.

oomycote "Egg fungus." Heterotrophic stramenopile, once wrongly grouped with fungi. Many are pathogens of plants; e.g., water molds, downy mildews.

operator Part of an operon; a DNA binding site for a regulatory protein.

operon Group of bacterial genes together with a promoter–operator DNA sequence that controls their transcription.

organ Body structure with definite form and function made of more than one tissue.

organ formation Developmental stage in which primary tissue layers give rise to differentiated cell lineages, the descendants of which form all organs of the adult.

organ system A set of organs that are interacting chemically, physically, or both in a common task.

organelle One of the membrane-bound compartments that carry out specialized metabolic functions in eukaryotic cells; e.g., a nucleus, mitochondria.

organic compound Any carbon-based molecule that also incorporates atoms of hydrogen and, often, oxygen, nitrogen, and other elements; e.g., fats, proteins.

osmoreceptor Type of sensory receptor that detects shifts in water volume.

osmosis Diffusion of water across a selectively permeable membrane from a region where the water concentration is higher to a region where it is lower.

osmotic pressure The amount of pressure which, when applied to a hypertonic fluid, will stop osmosis from occurring across a semipermeable membrane.

osteoblast Bone-forming cell; it secretes organic substances that get mineralized.

osteoclast Bone-digesting cell; it secretes enzymes that digest bone's organic matrix, which releases calcium and phosphorus for uptake by blood when metabolism requires more of these ions.

osteocyte A mature bone cell, imprisoned in its own secretions.

ostracoderm An early craniate; a filter-feeding bottom-dwelling jawless fish that became extinct after jawed fishes evolved.

ovary (OH-vuh-ree) Of animals, a female gonad. Of flowering plants, the enlarged base of a carpel in which one or more ovules develop into seeds.

oviduct (OH-vih-dukt) Duct between the ovary and uterus where fertilization most often occurs. Also called a Fallopian tube.

ovulation (OHV-you-LAY-shun) Release of a secondary oocyte from an ovary.

ovule (OHV-youl) [L. *ovum*, egg] Of seed-bearing plants, an egg-containing female gametophyte surrounded by tissue layers; a mature ovule is a seed.

ovum Mature secondary oocyte.

oxaloacetate (ox-AL-oh-ASS-ih-tate) A four-carbon compound with roles in metabolism; e.g., the point of entry into the Krebs cycle.

oxidation–reduction reaction Transfer of electrons between reactant molecules.

oxidized molecule A molecule that has lost one or more electrons.

oxygen debt A lower O_2 level in blood after muscle cells use up more ATP than they have formed by aerobic respiration.

oxyhemoglobin In red blood cells only, oxygen bound to hemoglobin; HbO_2.

oxytocin Animal hormone with roles in labor, lactation, and recovery of uterus after pregnancy. In many animals, also guides social behavior; e.g., pair bonding.

ozone thinning Pronounced seasonal thinning of the atmosphere's ozone layer.

P_i Abbreviation for inorganic phosphate.

pain Perception of injury to a body region.

pain receptor A nociceptor; a sensory receptor that detects tissue damage.

Paleozoic Era from 544 million to 248 million years ago; Cambrian through Permian.

PAN Peroxylacyl nitrate. An oxidant in photochemical smog.

pancreas (PAN-cree-us) Glandular organ. Secretes enzymes and bicarbonate that help digestion in the small intestine; secretes insulin and glucagon that have central roles in organic metabolism, glucose especially.

pancreatic islet Any of 2 million or so clusters of endocrine cells of the pancreas.

pandemic An epidemic that breaks out in several countries at the same time.

Pangea Paleozoic supercontinent; the first land plants and animals evolved on it.

parabasalid A flagellated heterotroph with bundled microtubules as long as the cell and giving rise to four to thousands of flagella; one of the earliest lineages of single-celled eukaryotes; e.g., *Trichomonas*.

parapatric speciation A speciation model. Populations in contact along a common border evolve into new species; hybrids that form in the contact zone are less fit than individuals on either side of it and thereby act as a reproductive isolating mechanism.

parasite [Gk. *para*, alongside, + *sitos*, food] Organism that withdraws nutrients from a living host, which it usually does not kill outright.

parasitism Symbiotic interaction in which a parasitic species benefits as it exploits and harms (but usually does not kill) the host.

parasitoid A type of insect that, in a larval stage, grows inside a host (usually another insect), feeds on its soft tissues, and kills it.

parasympathetic neuron A neuron of the autonomic nervous system. Its signals slow overall activities and divert energy to basic tasks; also works in opposition with sympathetic neurons to make small ongoing adjustments in the activities of internal organs that they both innervate.

parathyroid gland One of four small glands embedded in the back of the thyroid gland; their secretions trigger increases in blood calcium levels.

parenchyma (par-EN-kih-mah) One of the simple plant tissues; makes up the bulk of the plant. Its living cells have roles in photosynthesis, storage, and other tasks.

parthenogenesis (par-THEN-oh-GEN-uh-sis) The development of an embryo from an unfertilized egg.

partial pressure Contribution of one gas to the total pressure of a mixture of gases.

passive transport Diffusion of a solute across a cell membrane, through the interior of a transport protein.

pathogen [Gk. *pathos*, suffering, + *genēs*, origin] A virus, bacterium, fungus, protist, or parasitic worm that infects an organism and multiplies in it, thus causing disease.

pattern formation In animal embryonic development, the sculpting of specialized tissues and organs from clumps of cells in the proper places, in the proper order by way of embryonic induction.

PCR Polymerase chain reaction. A method to rapidly copy DNA fragments.

peat bog Compressed, soggy, acidic mat of accumulated remains of peat mosses.

pedigree Chart of connections among individuals related by descent.

pellicle A thin, flexible, protein-rich body covering of some single-celled eukaryotes.

peptide hormone A hormone that binds to a membrane receptor, which activates enzymes and often a second messenger in the cytoplasm.

per capita [L. *capita*, head] A term used in head counts of a population.

perception Understanding of a stimulus.

perennial [L. *per-*, throughout, + *annus*, year] Plant having a life cycle that extends through three or more growing seasons.

pericarp The fleshy part of a fruit; the endocarp, mesocarp, and exocarp.

pericycle (PARE-ih-sigh-kul) [Gk. *peri-*, around, + *kyklos*, circle] One or more cell layers inside the endodermis; gives rise to lateral roots and also contributes to secondary growth.

periderm Protective cover that replaces plant epidermis on older stems and roots.

periodic table of the elements Tabular arrangement of elements in order of their increasing atomic number.

peripheral nervous system (per-IF-ur-uhl) [Gk. *peripherein*, to carry around] All nerves leading into and out of the spinal cord and brain, plus their ganglia.

peripheral vasoconstriction Diameters of arterioles constrict, decreasing blood's delivery of heat to body surface.

peripheral vasodilation Blood vessels in the skin dilate; more blood flows to skin, which then dissipates excess body heat.

peristalsis (pare-ih-STAL-sis) Recurring waves of contraction of muscles in the wall of a tubular or saclike organ.

peritoneum (pare-ih-tuh-NEE-um) The membrane that lines the coelom.

peritubular capillaries A set of blood capillaries around the tubular parts of a nephron that reabsorbs water and solutes; and that excretes excess H^+, other solutes.

permafrost An impermeable, perpetually frozen layer, sometimes 500 meters thick, that underlies arctic tundra.

peroxisome Enzyme-filled vesicle that breaks down amino acids, fatty acids, and toxic substances such as ethanol.

PGA Phosphoglycerate. During glycolysis, the intermediate that results after ATP has formed by substrate-level phosphorylation; also the first stable intermediate of the Calvin–Benson cycle of photosynthesis.

PGAL Phosphoglyceraldehyde. During glycolysis, the intermediate that gives up electrons and hydrogen to form NADH. During turns of the Calvin–Benson cycle, two PGALs form one sugar; rearrangements of ten others regenerate a compound that is the entry point for the cycle.

pH scale Measure of the H^+ concentration of a solution. pH 7 is neutral.

phagocytosis [Gk. *phagein*, to eat] "Cell eating," a common endocytic pathway by which various cells engulf food bits, microbes, and cellular debris.

pharynx A muscular tube. Invertebrate chordates use theirs in filter-feeding and respiration. In land vertebrates, it is the entrance to the esophagus and trachea.

phenotype (FEE-no-type) [Gk. *phainein*, to show + *typos*, image] Observable trait or traits of an individual.

pheromone Nearly odorless exocrine gland secretion. A hormone-like signaling molecule between individuals of the same species that integrates social behavior.

phloem (FLOW-um) Plant vascular tissue that distributes photosynthetic products through the plant body. Its conducting tubes are interconnecting, living cells assisted by companion cells that help load solutes into the tubes.

phospholipid A lipid with a phosphate group in its hydrophilic head. The main constituent of cell membranes.

phosphorus cycle A sedimentary cycle. Phosphorus (mainly phosphate) moves from land, through food webs, to ocean sediments, then back to land.

phosphorylation Enzyme-mediated transfer of a phosphate group to an organic compound.

photoautotroph Any photosynthetic autotroph; e.g., nearly all plants, most algae, and a few bacteria.

photolysis (foe-TALL-ih-sis) [Gk. *photos*, light, + *-lysis*, breaking apart] Reactions that split water molecules, which release electrons for the noncyclic pathway of photosynthesis; oxygen is a by-product.

photon Unit of electromagnetic energy; has wave-like and particle-like properties.

photoperiodism Biological response to change in the relative lengths of daylight and darkness.

photoreceptor A light-sensitive sensory cell of invertebrates and vertebrates.

photosynthesis The process by which photoautotrophs capture sunlight energy and use it in the formation of ATP and NADPH, then in the formation of sugars from carbon dioxide and water. ATP gives up energy that drives the sugar-building reactions, and NADPH donates electrons and hydrogen building blocks.

photosystem In photosynthetic cells, a cluster of membrane-bound pigments and other molecules; it converts light energy to chemical energy.

phototropism Change in the direction of cell movement or growth in response to a light source.

photovoltaic cell Unit in a device that converts sunlight energy into electricity.

phycobilin One of a class of accessory pigments in cyanobacteria and red algae that reflects red to blue light.

phylogeny Evolutionary relationships among species.

physiology Study of how the multicelled body functions in its environment; more specifically, of the mechanisms by which its component parts grow, develop, and are maintained and reproduced.

phytochrome A light-sensitive pigment that helps set plant circadian rhythms based on length of night. Influences stem lengthening and branching, leaf expansion, and often flowering.

phytoplankton (FIE-toe-PLANK-tun) [Gk. *phyton*, plant, + *planktos*, wandering] An aquatic community of floating or weakly swimming photoautotrophs.

pigment Any light-absorbing molecule.

pilomotor response Formation of a layer of still air next to skin as hairs or feathers become erect.

pilus Among prokaryotic cells, a short, filamentous protein that projects above the cell wall and can adhere to surfaces; a sex pilus functions in conjugation.

pineal gland Light-sensitive, melatonin-secreting endocrine gland. Seasonal change in melatonin levels affect biological clocks, overall activity, and reproductive cycles.

pioneer species An opportunistic colonizer of barren or disturbed habitats. Adapted for rapid growth and dispersal.

pith Of most eudicot stems, ground tissue inside the ring of vascular bundles.

pituitary gland Vertebrate endocrine gland; interacts with the hypothalamus to control many physiological functions, including activity of many other glands. Its posterior lobe stores and secretes hormones from the hypothalamus; its anterior lobe produces and secretes its own hormones.

placenta (plah-SEN-tuh) Of pregnant female placental mammals, a blood-engorged organ that forms from endometrial tissue and extraembryonic membranes. Lets a mother exchange substances with a fetus but keeps their blood circulation separate.

placoderm An early jawed craniate with paired fins, armor plates on head; extinct.

placozoan An asymmetric, soft-bodied animal with two simple tissues around a thin, inner matrix.

plankton Aquatic community of mostly microscopic autotrophs and heterotrophs.

plant A multicelled photoautotroph, most with well-developed roots and shoots (e.g., stems, leaves), as well as photosynthetic cells that include starch grains as well as chlorophylls *a* and *b*, and polysaccharides such as cellulose, pectin, and lignin in cell walls. The primary producers on land.

Plantae Kingdom of plants.

planula Of cnidarians, a type of swimming or creeping larva, usually with a ciliated epidermis.

plasma (PLAZ-muh) Liquid portion of blood; mainly water and dissolved ions, proteins, sugars, gases, and other solutes.

plasma membrane Outer cell membrane; the structural and functional boundary between cytoplasm and extracellular fluid.

plasmid A small, circular bacterial DNA molecule having a few genes; replicated independently of the bacterial chromosome.

plasmodesma (PLAZ-moe-DEZ-muh), plural **plasmodesmata** A plant cell junction that connects the cytoplasm of adjoining cells.

plasmodium Of plasmodial slime molds, a multinucleated mass that forms when a single diploid cell undergoes rounds of mitosis without cytoplasmic division.

plate tectonics Theory that great slabs or plates of Earth's outer layer float on a hot, semi-molten mantle. All plates are moving slowly and have rafted continents to new positions over time.

platelet A megakaryocyte fragment; it releases substances that help form clots.

pleiotropy A case of alleles at a single gene locus having positive or negative impact on two or more traits.

polar body One of four cells that form by meiotic cell division of an oocyte but that does not become the ovum.

pollen grain [L. *pollen*, fine dust] A tiny structure that forms from microspores; consists of a sturdy wall around a few cells that will develop into a mature, sperm-bearing, male gametophyte.

pollination Arrival of pollen on a carpel's stigma in a flower of the same species.

pollinator Any agent that delivers pollen grains to the egg-containing structures in flowers of the same species; e.g., wind, water, or birds, bats, and other animals.

pollutant Natural or synthetic substance of types or in amounts that are novel in the history of an ecosystem, so there is no evolved mechanism that can prevent the substance from accumulating to harmful or disruptive levels.

polygenic inheritance Inheritance of multiple genes that affect the same trait.

polymer Large molecule of multiple linked monomers.

polymorphism (poly-MORE-fizz-um) [Gk. *polus*, many, + *morphe*, form] Persistence of two or more qualitatively different forms of a trait, or morphs, in a population.

polyp (POH-lip) Vase-shaped, sedentary stage of cnidarian life cycles.

polypeptide chain Three or more amino acids linked by peptide bonds.

polyploidy A case of somatic cells having three or more of each type of chromosome characteristic of the species.

polysaccharide [Gk. *polus*, many, + *sakcharon*, sugar] Straight or branched chain of covalently bonded monomers of the same or different kinds of sugars; e.g., cellulose, starch, and glycogen.

polysome A series of ribosomes that are all translating the same mRNA molecule at the same time.

polytene chromosome Of some insects, a chromosome consisting of many parallel copies of the same DNA molecule.

pons Hindbrain traffic center for signals between the cerebellum and forebrain.

population All individuals of the same species living in a specified area.

population density Count of individuals of a population in a specified area or volume of a habitat.

population distribution The pattern in which individuals of a population are dispersed through their habitat.

population size The total number of individuals that make up a population.

positive control Use of regulatory proteins to promote gene expression.

positive feedback mechanism Major form of homeostatic control; initiates a chain of events that intensify a change in conditions; e.g., a complement cascade.

potential energy A object's capacity to do work owing to its position in space or the arrangement of its parts.

predation Ecological interaction in which a predator feeds on a prey organism.

predator [L. *prehendere*, to grasp, seize] A heterotroph that eats other living organisms (its prey), does not live in or on them, and most often kills them.

prediction A statement, based on a hypothesis, about what you expect to observe in nature; the "if-then process."

pressure flow theory In vascular plants, organic compounds flow through phloem in response to pressure and concentration gradients between sources (e.g., leaves) and sinks (use or storage in growing parts).

pressure gradient Difference in pressure between two adjoining regions.

prey Any organism that another organism captures as a food source.

primary growth Plant growth originating at root tips and shoot tips.

primary oocyte Of human females, an immature egg that is arrested in prophase I of meiosis until eight to ten hours before being released from an ovary.

primary producer An autotroph at the first trophic level of an ecosystem.

primary productivity The rate at which an ecosystem's primary producers secure and store energy in tissues in a given interval.

primary root First root of a seed plant.

primary succession *See* ecological succession.

primary wall The first thin, pliable wall of young plant cells.

primate A type of mammal; a prosimian, a tarsioid, or an anthropoid.

primer Short nucleotide sequence that researchers design as an initiation site for synthesis of a DNA strand on a DNA or RNA template.

prion A type of protein particle normally in vertebrate nervous systems that turns infectious when its shape changes.

probability The odds that each outcome of an event will occur is proportional to the total number of ways in which that outcome can be reached.

probe Short nucleotide sequence that has been labeled with a tracer; designed to hybridize with part of a gene or mRNA.

producer An autotrophic organism.

product A substance remaining at the end of a reaction.

progesterone (pro-JESS-tuh-rown) One of the sex hormones; ovaries and the corpus luteum secrete it.

proglottid One of many tapeworm body units that bud behind the scolex.

progymnosperm Among earliest plants to produce seedlike structures or seeds.

prokaryotic cell [L. *pro*, before, + Gk. *karyon*, kernel] A single-celled organism, often walled, that does not have the organelles characteristic of eukaryotic cells. Only bacteria and archaeans are prokaryotic.

prokaryotic fission Cell reproduction mechanism of prokaryotic cells only.

prolactin Hormone that induces synthesis of enzymes used in milk production.

promoter Short stretch of DNA to which RNA polymerase binds. Transcription then begins at the gene closest to the promoter.

prophase, meiosis In prophase I in a germ cell, all duplicated chromosomes condense, typically undergo crossing over with their homologue, then get tethered to a spindle and move to its equator. In prophase II, one member of each pair of homologous chromosomes is tethered to the opposite spindle pole and moved to the equator.

prophase, mitosis All of the duplicated chromosomes in a cell condense and get attached to a newly forming spindle.

protein Organic compound consisting of one or more polypeptide chains. Diverse kinds have structural, functional, and regulatory roles in all organisms.

proteobacteria A group of Gram-negative bacteria; the most diverse monophyletic group of prokaryotic cells.

Proterozoic Era between 2.5 million to 544 million years ago. An oxygen-rich early atmosphere formed, sparking the Cambrian explosion of biodiversity.

"protist" Informal name for all structurally simple eukaryotes, which are now being classified as monophyletic groups.

proto-cell Presumed stage of chemical evolution that preceded living cells.

proton Positively charged subatomic particle in the nucleus of all atoms.

protostome (PRO-toe-stome) [Gk. *proto*, first, + *stoma*, mouth] A bilateral animal of a branching lineage characterized partly by events in embryonic development, as when the first indentation to form on the early embryo's surface becomes a mouth; e.g., mollusks, annelids, arthropods.

protozoan Traditional name for one of the motile predatory or parasitic species of single-celled eukaryotes.

proximal tubule Tubular portion of a nephron closest to Bowman's capsule.

pseudocoel False coelom; a main body cavity incompletely lined with tissue derived from mesoderm.

pseudopod A dynamic lobe of membrane-enclosed cytoplasm; functions in motility and phagocytosis by amoebas, amoeboid cells, and many white blood cells.

puberty Of humans, the post-embryonic stage when gametes start to mature and secondary sexual traits emerge.

pulmonary circuit Cardiovascular route in which oxygen-poor blood flows to lungs from the heart, gets oxygenated, then flows back to the heart.

punctuation model, speciation Addresses the rate of speciation; cites fossil evidence that morphological changes required for reproductive isolation evolve in a relatively brief time span, within the tens to hundreds of thousands of years when two or more populations are diverging from each other.

Punnett-square method A simple way to predict the probable outcomes of a genetic cross by constructing and filling in a diagram of all possible combinations of genotypes, phenotypes, or both.

pupa, plural **pupae** An immature, post-embryonic stage of many insect life cycles.

purine A nucleotide base with a double ring structure; e.g., adenine or guanine.

pyrimidine A nucleotide base with a single ring structure; e.g., cytosine, thymine, uracil.

pyruvate Three-carbon compound that forms as an end product of glycolysis.

quadrat One of a number of sampling areas of the same size and shape used to estimate population size.

r Net reproduction per individual per unit time; a variable in population growth equations for which birth and death rates are assumed to remain constant.

radial symmetry Animal body plan with four or more roughly equivalent parts around an anterior–posterior axis.

radiation Any form of radiant energy.

radioactive decay Natural, inevitable process by which an atom emits energy as subatomic particles and x-rays as its unstable nucleus spontaneously breaks apart; transforms one element into another in a predictable time span.

radioisotope Any isotope that has an unstable nucleus.

radiolarian A single-celled predatory eukaryote that has pseudopods projecting from a perforated shell and a cell cortex with buoyancy-imparting vacuoles.

radiometric dating Method of measuring proportions of a radioisotope in a mineral trapped long ago in newly formed rock and a daughter isotope that formed from it by radioactive decay in the same rock. Used to assign absolute dates to fossil-containing rocks and to the geologic time scale.

rain shadow Reduction in rainfall on the leeward side of a high mountain range that results in arid or semiarid conditions.

ray-finned fish A bony fish having fin supports derived from skin, a swim bladder, and thin, flexible scales.

reabsorption At a capillary bed, osmotic movement of some interstitial fluid into plasma. *See also* tubular reabsorption.

reactant Substance that enters a reaction.

reaction center At a photosystem's center, a special pair of chlorophyll *a* molecules; the center loses electrons on absorption of photon energy, thereby initiating the light-dependent reactions of photosynthesis.

rearrangement, molecular Conversion of one organic compound to another through changes in its internal bonds.

receptor, molecular A protein or some other molecule with a binding site for a specific signaling molecule.

receptor, sensory Sensory cell or a specialized ending of one that detects a particular kind of stimulus.

recessive allele Allele whose expression in heterozygotes is fully or partially masked by expression of a dominant partner allele. It is fully expressed only in homozygous recessives.

reciprocal cross A paired cross that may identify the role of parental sex on the inheritance of a trait. In the second cross, a trait characteristic of each sex is reversed compared to the original cross.

recognition protein One of a class of glycoproteins or glycolipids that project above the plasma membrane and that identify a cell as *nonself* (foreign) or *self* (belonging to one's own body tissue).

recombinant DNA A DNA molecule that contains genetic material from more than one organism of the same species or from different species.

recombinant DNA technology Techniques by which DNA molecules from different species can be cut into fragments, spliced together into cloning vectors, and then amplified to useful quantities.

recombination, genetic Introduction of nonparental combinations of alleles in chromosomes, as by crossing over.

rectum Last part of the mammalian gut that briefly stores feces before their expulsion.

red alga An aquatic, mostly multicelled photoautotroph having an abundance of phycobilins that masks its chlorophyll *a*.

red blood cell Erythrocyte; functions in the efficient transport of oxygen in blood.

red marrow Site of blood cell formation in the spongy tissue of many bones.

red tide An algal blood that turns the water near coasts rust-red or brown.

reduced molecule A molecule to which one or more electrons were transferred.

reflex [L. *reflectere*, to bend back] Simple, stereotyped movement in response to a stimulus; sensory neurons synapse on motor neurons in the simplest reflex arcs.

regulatory protein Part of mechanisms that control transcription, translation, and gene products by interacting with DNA, RNA, new polypeptide chains, or proteins such as enzymes.

releaser Hypothalamic signaling molecule that enhances or slows the secretion of a specific anterior pituitary hormone.

renal corpuscle Bowman's capsule and the glomerular capillaries it cups around.

renal failure Condition in which nephrons of both kidneys no longer function.

repair enzyme Type of enzymes that repairs nucleotide mismatches in a DNA strand.

repressor Type of protein that can block transcription of a prokaryotic gene by binding to an operator.

reproduction Any asexual or sexual process by which a parent cell or organism produces offspring.

reproductive base The number of actually and potentially reproducing individuals of a population.

reproductive isolating mechanism Any heritable feature of body form, function, or behavior that prevents interbreeding between two or more populations; sets the stage for genetic divergences.

reproductive success Of individuals, the production of viable, fertile offspring.

"reptile" No longer a formal taxon; not a monophyletic group. The name persists as a means to refer to sauropsid lineages other than birds that show basic amniote features but not derived traits that define birds or mammals; e.g., a turtle, crocodile.

resource partitioning The sharing of a resource in different ways or at different times that permits two or more species to coexist in a habitat.

respiration [L. *respirare*, to breathe] The sum of physiological processes that move O_2 from the surroundings to metabolically active tissues in the animal body and CO_2 from tissues to the outside.

respiratory cycle One in-and-out breath.

respiratory membrane Fused-together alveolar and blood capillary epithelial and the basement membrane in between; a respiratory surface in the human lung.

respiratory pigment A protein complexed with one or more metal ions that binds O_2 in oxygen-rich animal tissues and gives it up where O_2 levels are lowest.

respiratory surface Any thin, moist body surface that functions in gas exchange.

respiratory system Animal organ system that takes in O_2 for aerobic respiration and rids the body of its CO_2 wastes.

resting membrane potential The voltage difference across the plasma membrane of a neuron or other excitable cell that is not receiving outside stimulation.

restoration ecology Work to reestablish biodiversity in ecosystems severely altered by mining, agriculture, other disturbances.

restriction enzyme One of hundreds of proteins that recognize and cut specific base sequences in double-stranded DNA.

reticular formation Mesh of interneurons that is a low-level pathway of information flow through the upper spinal cord, brain stem, and cerebral cortex.

retina Of vertebrate and many invertebrate eyes, a tissue packed with photoreceptors and interwoven with sensory cells.

reverse transcriptase A viral enzyme that catalyzes the assembly of free nucleotides into a strand of DNA on an RNA template.

Rh blood typing Method of determining whether Rh^+, a type of surface recognition protein, is present on an individual's red blood cells; if absent, the cell is Rh^-.

rhizoid A rootlike absorptive structure.

rhizome A short absorptive stem that grows underground in a horizontally branching pattern, most often.

rhyniophyte The first seedless vascular plants; originated in Gondwana lowlands.

ribosomal RNA rRNA. A class of RNA that becomes complexed with proteins to form ribosomes; some catalyze assembly of polypeptide chains.

ribosome The site of polypeptide chain synthesis in all cells. An intact ribosome has two subunits of rRNA and proteins.

riparian zone The narrow corridor of vegetation along a stream or river.

RNA Ribonucleic acid. Any of a class of single-stranded nucleic acids involved in gene transcription and translation; some RNAs show enzyme activity.

RNA polymerase Enzyme that catalyzes transcription of DNA into RNA.

RNA world Model for a time prior to the evolution of DNA; a self-replicating system chemically evolved in which RNA strands were templates for protein synthesis.

rod cell Vertebrate photoreceptor that detects very dim light; contributes to the coarse perception of movement.

root Typically belowground plant part. It absorbs water and dissolved minerals, often anchors aboveground parts and stores food.

root hair Hairlike, absorptive extension of a young, specialized root epidermal cell.

root nodule Mutualistic association of nitrogen-fixing bacteria and roots of some legumes and other plants; infection leads to a localized tissue swelling.

root system Underground vascular plant structures that absorb water, mineral ions.

rotifer Bilateral, cephalized animal with a false coelom and a crown of cilia.

roundworm Bilateral invertebrate with a false coelom and complete digestive system in a cylindrical body. Most are decomposers; many are parasites.

rubisco RuBP carboxylase. Carbon-fixing enzyme of the C3 photosynthesis pathway.

RuBP Ribulose bisphosphate. A five-carbon organic compound; the entry point for the Calvin–Benson cycle, which regenerates it.

ruminant Hoofed, herbivorous mammal that has multiple stomach chambers.

sac fungus Fungus that produces sexual spores in sac-shaped cells; e.g., truffles.

salinization Salt buildup in soil by poor drainage, evaporation, or heavy irrigation.

saliva Salivary gland secretion into the mouth that starts starch breakdown.

salt Any compound that releases ions other than H^+ and OH^- in solution.

saltatory conduction Of a myelinated neuron, a rapid form of action potential propagation. Excitation hops node to node between jellyrolled membranes of neuroglial cells of the myelin sheath.

sampling error Using a sample or subset of a population, an event, or some other aspect of nature as an experimental group that is not large enough to be representative of the whole.

saprobe Heterotroph that extracts energy and carbon from nonliving organic matter and so causes its decay.

sapwood Of an older stem or root, the moist secondary growth between the vascular cambium and heartwood.

sarcomere (SAR-koe-meer) One of many basic units of contraction, defined by Z lines, along the length of a muscle fiber. It shortens by ATP-driven interactions between its parallel arrays of actin and myosin components.

sarcoplasmic reticulum Specialized ER that forms flattened, membrane-bound chambers around muscle fibers; takes up, stores, and releases Ca^{++} for contraction.

sauropsid A "reptile" or a bird.

savanna Broad belt of warm grassland with a smattering of shrubs and trees.

scale Of a fish, one of a number of small, bony plates that protect the body without weighing it down.

Schwann cell Type of neuroglial cell that myelinates many axons.

scientific theory *See* Theory, scientific.

sclerenchyma (skler-ENG-kih-mah) One of three simple plant tissues; supports mature parts and often protects seeds. Lignin often thickens and reinforces its cell walls.

second law of thermodynamics Energy tends to flow from concentrated to less concentrated forms.

second messenger Molecule in a cell that relays a hormonal signal; e.g., cyclic AMP.

secondary growth A thickening of older stems and roots; wood when extensive.

secondary oocyte A haploid cell which, with a first polar body, is produced by the first meiotic division of a primary oocyte; the cell released from the ovary of a female vertebrate at ovulation.

secondary sexual trait A trait associated with maleness or femaleness but with no direct role in reproduction (e.g., body hair distribution). The primary sexual trait is the presence of male or female gonads.

secondary succession *See* Ecological succession.

secondary wall A rigid, permeable wall inside the primary wall of many plant cells; forms after the first growing season.

secretion Release of a substance from a cell or gland to its surroundings.

sedimentary cycle Any biogeochemical cycle in which an element having no gaseous phase moves from land, through food webs, to the seafloor, then back to land through long-term uplifting.

seed A mature ovule.

seed bank A storage facility where genes of diverse plant lineages are preserved.

seed fern One of the earliest plants to make seedlike structures, or seeds.

segmentation Of animal body plans, a series of units that may or may not be similar in appearance. Of tubular organs, an oscillating movement produced by rings of circular muscle in the tube wall.

segregation, theory of Mendelian theory that two genes of a pair on homologous chromosomes are separated from each other at meiosis, eventually to end up in different gametes.

selective gene expression Outcome of controls over which gene products a cell makes or activates in a specified interval. Basis of cell differentiation.

selective permeability Built-in capacity of a cell membrane to prevent or allow specific substances from crossing it at certain times, in certain amounts.

selfish behavior An individual increases its own chance to reproduce regardless of the biological costs to its social group.

selfish herd Social group held together by reproductive self-interest.

semen (SEE-mun) Sperm-bearing fluid expelled from a penis during sex.

semiconservative replication [Gk. *semi–*, half, + L. *conservare*, to keep] Mechanism by which a DNA molecule is duplicated. The double helix unzips along its length, exposed bases of each strand are a template upon which a new strand is assembled, then each conserved strand and its new partner wind up in a double helix. Two double helixes, each with a parental strand and new strand of DNA, result.

seminiferous tubule One of the coiled tubes in testes where sperm start forming.

senescence (sen-ESS-cents) [L. *senescere*, to grow old] Of differentiated multicelled organisms, the phase in a life cycle from maturity until death; also applies to death of parts, such as plant leaves.

sensation Conscious awareness of a stimulus.

sensory neuron Type of neuron that detects a stimulus and relays information about it toward an integrating center.

sensory system Collectively, all sensory cells of a nervous system that detect and report information about external and internal stimuli to integrating centers.

serotonin A neurotransmitter that affects mood, memory, and sleep behavior.

Sertoli cell A type of cell in seminiferous tubules with FSH receptors; helps nourish and support developing sperm.

sessile animal (SESS-ihl) Animal that is attached to a substrate during part of the life cycle; e.g., an adult barnacle.

sex chromosome One of two kinds of homologous chromosomes that, in certain combinations, dictate the gender of the new individual. Also has genes unrelated to sexual traits.

sexual dimorphism A notable difference between female and male phenotypes of a population.

sexual reproduction Production of genetically variable offspring by meiosis, gamete formation, and fertilization.

sexual selection A category of natural selection; an outcome of differences in success at attracting mates and reproducing among individuals of a population.

shell model Model for how electrons are distributed in an atom; all of the orbitals are shown as a nested series of shells.

shifting cultivation A practice of cutting and burning trees, then tilling ashes into the soil of a small plot of land. Once called slash-and-burn agriculture.

shivering response Rhythmic tremors in response to cold; raises heat production.

shoot system Aboveground plant parts; e.g., stems, leaves, flowers.

short-day plant Plant that flowers in late summer or early fall, when night length is longer than a critical value.

sieve tube A conducting tube in phloem.

sieve-tube member A living cell that helps form a conducting tube in phloem.

sign stimulus Simple environmental cue that triggers a response to a stimulus; the nervous system is prewired to recognize it.

signal reception Activation of a molecular or sensory receptor when a hormone or another signaling molecule binds to it.

signal transduction Conversion of an extracellular signal into a molecular or chemical form that causes a change in some activity inside a target cell.

signaling molecule Any secretion from one cell type that can alter the behavior of a different cell that bears a receptor for it; a means of cell communication.

sister chromatid (CROW-mah-tid) One of the two attached members of a duplicated eukaryotic chromosome.

six-kingdom classification system The grouping of all organisms into kingdoms Bacteria, Archaea, Protista, Fungi, Plantae, and Animalia.

skeletal muscle Organ of many muscle fibers bundled inside a connective tissue sheath and attached to bone by tendons.

skeletal muscle tissue Contractile tissue that is the functional partner of bone.

skin Vertebrate integument and diverse structures derived from it.

sliding-filament model Model for how the sarcomeres of muscle fibers contract. ATP-activated myosin heads repeatedly bind actin filaments (tethered to Z lines) and tilt in short power strokes that slide the actin toward the sarcomere's center.

slime mold An amoebozoan; one of the free-living, amoebalike cells that also cluster into a migrating mass, differentiate, and form reproductive structures.

small intestine Part of the vertebrate gut in which digestion is completed and from which most dietary nutrients are absorbed.

smog Atmospheric condition in which winds cannot disperse airborne pollutants that have become trapped under a thermal inversion.

smooth muscle tissue Contractile tissue in the wall of soft internal organs.

social behavior Interacting individuals of a species that display, send, and respond to shared forms of communication.

social parasite Species that completes its life cycle by taking advantage of the social behavior of a host species, thus harming it.

sodium–potassium pump Cotransporter that, when energized, actively transports sodium out of a cell and helps potassium passively diffuse into it at the same time.

softwood Wood with tracheids, no fibers or vessels; less dense than hardwood.

soil Mix of mineral particles of variable sizes, decomposing organic material, and air and water in spaces between particles.

solar-hydrogen energy Sunlight energy is used to convert water to H_2 as a fuel source.

solar tracking A photoperiodic response to sun's changing angle through the day.

solute (SOL-yoot) [L. *solvere*, to loosen] Any substance dissolved in a solution.

solvent Any fluid (e.g., water) in which one or more substances are dissolved.

somatic cell (so-MAT-ik) [Gk. *soma–*, body] Any body cell that is not a germ cell.

somatic nervous system The nerves that connect the vertebrate central nervous system and skeletal muscles.

somatic sensation Perception of touch, pain, pressure, temperature, motion, or positional changes of body parts.

somatosensory cortex Part of the outer gray matter of the cerebral hemispheres.

somite One of many paired segments in a vertebrate embryo that gives rise to most bones, skeletal muscles of the head and trunk, and the dermis.

special sense Vision, hearing, olfaction, or another sensation involving receptors that are restricted to certain body parts.

speciation (spee-see-AY-shun) One of the macroevolutionary processes; formation of daughter species from a population or subpopulation of a parent species; the routes vary in their details and duration.

species (SPEE-sheez) [L. *species*, a kind] Of sexually reproducing species, one or more natural populations of individuals that successfully interbreed and are isolated reproductively from other such groups. By a cladistic definition, one or more natural populations of individuals with at least one unique trait derived a common ancestor that occurs in no other groups.

specific epithet The last part of a two-part species name; the genus is first.

sperm Mature male gamete.

sphere of hydration A clustering of water molecules around molecules or ions of a solute by positive and negative interactions.

sphincter A ring of muscles that alternately contract and relax, which closes and opens a passageway between two organs.

spinal cord The part of a central nervous system inside a vertebral canal. Basic reflex centers, and tracts to and from the brain.

spindle, microtubular *See* Bipolar spindle.

spirillum A spiral-shaped prokaryotic cell.

spirochaete A motile, parasitic or symbiotic bacterium that looks like a stretched spring.

spleen The largest lymphoid organ, with phagocytic white blood cells and B cells; filters antigen and used-up platelets and worn-out or dead red blood cells. In embryos only, a site of red blood cell formation.

sponge Structurally, the simplest existing animal. Its asymmetrical body has a spicule-reinforced matrix in two cell layers (not epithelium). Its phagocytic collar cells trap food from water flowing through pores in its wall.

spore A structure of one or a few cells, often walled or coated, that protects and/or disperses a new sexual or asexual generation. Many bacteria as well as apicomplexans, fungi, and plants form spores.

sporophyte [Gk. *phyton*, plant] A spore-producing vegetative body of a plant or multicelled alga that grows by mitotic cell divisions from a zygote.

sporozoan *See* Apicomplexan.

spring overturn Of large bodies of water, a downward movement of oxygenated surface water and an upward movement of nutrient-rich water from below during spring; fans primary productivity.

S-shaped curve Type of diagrammatic curve that emerges when plotting logistic population growth against time.

stabilizing selection Mode of natural selection; intermediate phenotypes are favored over extremes at both ends of the range of variation.

stamen (STAY-mun) An anther, typically raised on a stalked filament.

statolith A cluster of particles that acts as a gravity-sensing mechanism.

STD A sexually transmitted disease.

stem cell Self-perpetuating, undifferentiated animal cell. A portion of its daughter cells becomes specialized; e.g., red blood cells from stem cells in bone marrow.

steroid hormone Cholesterol-derived, lipid-soluble hormone.

sterol Any lipid consisting of a rigid backbone of four fused carbon rings.

stigma Sticky or hairy surface tissue on the top of a carpel or fused carpels; captures pollen and promotes its germination.

stimulus [L. *stimulus*, goad] A specific form of energy that activates a sensory receptor able to detect it; e.g., pressure.

stoma, plural **stomata** A gap between two plumped guard cells that lets water vapor and gases diffuse across the epidermis of a leaf or primary stem; diffusion stops when the cells lose water and collapse.

stomach Muscular, stretchable sac; mixes and stores ingested food and helps break it apart mechanically and chemically.

strain A type of organism which, when compared against an organism of known type, has differences that are too minor to classify it as a separate species.

stramenopile A single-celled or multicelled eukaryote with four outer membranes and thin tinsel-like filaments projecting from one of two flagella; e.g., a photosynthetic chrysophyte or a colorless oomycote.

stratification Stacks of sedimentary rock layers, built up by deposition of silt and other materials over time.

stream A flowing-water ecosystem that starts out as a freshwater spring or seep.

strip logging A way to minimize erosion from deforestation; a narrow corridor that parallels contours of sloped land is cleared; the upper part is used as a log-hauling road then is reseeded from intact forest above it.

strobilus Of certain nonflowering plants, a cluster of spore-producing structures.

stroma The semifluid matrix between the thylakoid membrane system and two outer membranes of a chloroplast where sucrose, starch, cellulose, and other end products of photosynthesis are built.

stromatolite Fossilized remains of dome-shaped mats of shallow-water communities, cyanobacterial species especially, that were infiltrated with dissolved minerals and fine sediments. Some are 3 billion years old.

substance P A neuromodulator that enhances pain perception.

substrate A reactant molecule that is specifically acted upon by an enzyme.

substrate-level phosphorylation Direct, enzyme-mediated transfer of a phosphate group from a substrate to another molecule.

succession *See* ecological succession.

suppressor T cell Type of lymphocyte that helps end an immune response.

surface-to-volume ratio A relationship in which the volume of an object increases with the cube of the diameter, but the surface area increases with the square.

survivorship curve Plot of age-specific survival of a cohort, from the time of birth until the last individual dies.

swim bladder Adjustable flotation sac that helps many fishes maintain neutral buoyancy in water; its volume changes as it exchanges gases with blood.

symbiosis [Gk. *sym*, together, + *bios*, life, mode of life] An ecological interaction in which one or more individuals interact closely with individuals of a different species for some or all of the life cycle; e.g., mutualism, predation, parasitism.

sympathetic neuron A neuron of the autonomic nervous system. Its signals cause increases in overall activities in times of stress or heightened awareness. Also works in opposition with sympathetic neurons to make small ongoing adjustments in activities of internal organs they both innervate.

sympatric speciation [Gk. *sym*, together, + *patria*, native land] A speciation model. Occurs inside the home range of a species in the absence of a physical barrier; e.g., by way of polyploidy in flowering plants.

synapsid An amniote lineage of early mammal-like reptiles and mammals.

synaptic integration (sin-AP-tik) The summation of excitatory and inhibitory signals that are arriving at an excitable cell's input zone at the same time.

syndrome The set of symptoms that characterize a medical condition.

system acquired resistance Of many plants, a mechanism that induces cells to produce and release compounds that will protect tissues from attack.

systemic circuit Cardiovascular route in which oxygenated blood flows from the heart through the rest of the body, where it gives up oxygen and takes up carbon dioxide, then flows back to the heart.

T lymphocyte T cell. White blood cell that regulates vertebrate immune responses by way of cytokines; cytotoxic T cells carry out cell-mediated immunity.

tactile display A type of ritualized social interaction involving physical contact.

tandem repeat One of many copies of short base sequences positioned one after another on a chromosome; used in DNA fingerprinting.

taproot system A primary root and all of its lateral branchings.

target cell Any cell that has molecular receptors for a signaling molecule.

taste receptor A type of chemoreceptor that detects solutes in the fluid bathing it.

taxon, plural **taxon** A set of organisms of a given type.

taxonomy Field of biology that identifies, names, and classifies species.

TCR Antigen-binding receptor of T cells.

tectum Midbrain's roof. In fishes and amphibians, coordinates most sensory inputs and initiates motor responses. In most vertebrates (not mammals), a reflex center; relays sensory input to forebrain.

telomere A cap of repetitive DNA sequences on the end of a chromosome. Each nuclear division, enzymes digest a bit of it; cells stop dividing when only a nubbin remains.

telophase (TEE-low-faze) Of meiosis I, a stage when one member of each pair of homologous chromosomes has arrived at a spindle pole. Of mitosis and of meiosis II, the stage when chromosomes typically decondense into threadlike structures and two daughter nuclei form.

temperature Measure of molecular motion.

temperature zone Globe-spanning bands of temperature defined by latitude; e.g., cool temperate, equatorial.

tendon A cord or strap of dense connective tissue that attaches a muscle to bone.

terminal bud *See* Bud.

territory An area that an animal defends against competitors for food, water, living space, mates, and other resources.

test, scientific Any standardized or innovative means by which a prediction based on a hypothesis might be disproved; often requires designing and conducting experiments, making observations, or developing models.

testcross A cross that might reveal the (unknown) genotype of an individual showing dominance for a trait; the individual is crossed with a known homozygous recessive individual.

testis, plural **testes** A type of gonad where male gametes and sex hormones form.

testosterone (tess-TOSS-tuh-rown) A sex hormone necessary for the development and functioning of the male reproductive system of vertebrates.

tetanus (TET-uh-nuss) A large muscle contraction. Repeated stimulation of a motor unit causes muscle twitches to run together. In the disease tetanus, muscles cannot be released from contraction.

tetrapod A vertebrate that is a four-legged walker or a descendant of one.

thalamus (THAL-uh-muss) Forebrain region; a coordinating center for sensory input and a relay station for signals to the cerebrum.

theory, scientific A time-tested, widely accepted intellectual framework used to interpret a broad range of observations and data about some aspect of nature. Tested rigorously but is still open to tests, revision, and tentative acceptance or rejection.

thermal inversion A layer of dense, cool air trapped beneath a layer of warm air.

thermal radiation Emission of radiant energy (heat) from any object.

thermocline Thermal stratification in a large body of water; a cool midlayer stops vertical mixing between warm surface water above it and cold water below it.

thermoreceptor Type of sensory cell that detects radiant energy (heat).

thigmotropism (thig-MOE-truh-pizm) [Gk. *thigm*, touch] Redirected growth in response to physical contact with a solid object; e.g., a vine curling around a post.

thirst center Part of the hypothalamus; promotes water-seeking behavior when osmoreceptors in the brain detect a rise in the blood level of sodium.

threat display Ritualized intraspecific signal conveying intent to attack.

three-domain system A classification system that groups all organisms into domains Bacteria, Archaea, and Eukarya.

thylakoid membrane A chloroplast's inner membrane system, often folded as flattened sacs, that forms a continuous compartment in the stroma. In the first stage of photosynthesis, pigments and enzymes in the membrane function in the formation of ATP and NADPH.

thymine (THY-meen) One of four nitrogen-containing bases in nucleotide monomers of DNA; also applies to a nucleotide with a thymine base component.

thymus gland Lymphoid organ; secretes hormones that influences the maturation of T cells that circulate to this gland right after they have formed in bone marrow.

thyroid gland Endocrine gland; secretes hormones that influence overall growth, development, and rates of metabolism.

tidal volume Volume of air flowing in and out of lungs in one respiratory cycle.

tight junction An array of many strands of fibrous proteins collectively joining the sides of cells that make up an epithelium; the array prevents solutes from leaking between the cells.

tissue Of multicelled organisms, a group of cells and matrixes interacting in the performance of one or more tasks.

tissue culture propagation Inducing the vegetative growth of a plant fragment or cell in a culture medium.

titin Elastic protein that keeps myosin filaments centered in a sarcomere and lets relaxed muscles passively resist stretching.

tongue A vertebrate organ of membrane-covered skeletal muscles used to position food and swallow, also to make sounds.

tonicity (toe-NISS-ih-TEE) Relative solute concentrations of two fluids.

tooth A hardened appendage used to cut, shred, pierce, or pummel food.

topsoil Uppermost soil layer with the most nutrients for plant growth.

torsion A drastic twisting of the body, including the visceral mass, as certain molluscan embryos develop.

total fertility rate TFR. Of humans, the average number of children born to females during their reproductive years.

touch-killing Mechanism by which a cytotoxic T cell kills target cells; it directly releases perforins and toxins onto them.

toxin Normal metabolic product that can damage or kill cells of a different species.

trace element Any element making up less than 0.01 percent of body weight.

tracer Any substance with a radioisotope attached; researchers can track it after delivering it into a cell, a multicelled body, ecosystem, or some other system.

trachea (TRAY-kee-uh), plural **tracheae** An air-conducting tube used in respiration. Of land vertebrates, the windpipe.

tracheal system Finely branching tubes for respiration that start at openings across the integument and dead-end in body tissues of arthropods; e.g., grasshoppers.

tracheid (TRAY-kid) A type of cell in xylem that conducts water and mineral ions.

tract Cordlike bundle of axons of sensory neurons, motor neurons, or both in the brain or spinal cord.

transcription [L. *trans*, across, + *scribere*, to write] First stage of protein synthesis. An RNA strand is assembled from nucleotides using a gene region in DNA as a template.

transfer RNA tRNA. One of a class of small RNA molecules that delivers amino acids to a ribosome. Its anticodon pairs with an mRNA codon during translation.

transition state A fleeting point when a chemical reaction can run to product or back to reactant.

translation Second stage of protein synthesis. At ribosomes, information encoded in an mRNA transcript guides the synthesis of a new polypeptide chain from amino acids.

translocation Attachment of a piece of a broken chromosome to another chromosome. Also, a mechanism by which organic compounds are conducted in phloem.

transpiration Evaporative water loss from a plant's aboveground parts.

transport protein Membrane protein that passively or actively assists specific ions or molecules into or out of a cell. The solutes move through the protein's interior.

transposon Transposable element. A stretch of DNA that jumps spontaneously and randomly to a different location in the genome and may mutate a gene.

triglyceride A lipid with three fatty acid tails attached to a glycerol backbone.

trisomy Having one extra chromosome in somatic cells; e.g., trisomy 21 (2n + 1).

trophic level (TROE-fik) All organisms the same number of transfer steps away from the energy input into an ecosystem.

tropical rain forest A biome in regions of regular, heavy rainfall, an annual mean temperature of 25°C, and humidity greater than 80 percent. Rich in biodiversity but low in topsoil; decomposition is too fast.

tropism (TROE-pizm) Directional growth response to an environmental factor; e.g., a shoot bending toward a light source.

true breeding lineage A group consisting of parents and their offspring in which only one version of a trait persists over time.

tubular reabsorption A process by which peritubular capillaries reclaim water and solutes that leak or are pumped out of a nephron's tubular regions.

tubular secretion Transport of H$^+$, urea, other solutes out of peritubular capillaries and into nephrons for excretion.

tumor Tissue mass of cells dividing at an abnormally high rate. Benign tumor cells stay in their home tissue; malignant ones metastasize, or slip away and invade other places in the body, where they may start new tumors. *See also* neoplasm.

tundra Biome of high-latitudes or high elevations with poor drainage, very little decomposition, very low temperatures, and a short growing season.

tunicate One of the baglike, filter-feeding urochordates.

turgor pressure (TUR-gore) Hydrostatic pressure. The pressure that any volume of fluid exerts against a wall, membrane, or some other structure containing it.

ultrafiltration Bulk flow of some protein-free plasma out of a blood capillary when outward-directed blood pressure exceeds the inward-directed osmotic movement of interstitial fluid.

ultraplankton Photosynthetic bacteria less than five micrometers across that help form "pastures of the seas."

uniformity theory Theory that Earth's surface has changed in slow, uniformly repetitive ways except for expected annual catastrophes, such as big floods. Changed Darwin's view of evolution; has since been discredited by plate tectonics theory.

upwelling Upward movement of deep, often nutrient-rich water near a coastline; replaces a mass of surface ocean water forced away by prevailing winds.

uracil (YUR-uh-sill) One of four nitrogen-containing bases in nucleotide monomers of RNA; also applies to a nucleotide with a uracil base component. Like thymine, uracil can base-pair with adenine.

urea Waste product formed in the liver from ammonia (derived from protein breakdown) and CO_2; excreted in urine.

ureter A urine-conducting tube from each kidney to the urinary bladder.

urethra A tube that drains the urinary bladder and opens at the body surface.

urinary bladder Distensible sac in which urine is stored before being excreted.

urinary excretion Mechanism by which excess water and solutes are removed from the body by the urinary system.

urinary system Vertebrate organ system that adjusts blood's volume and composition; helps maintain extracellular fluid.

urine Fluid consisting of excess water, wastes, and solutes that forms in kidneys by filtration, reabsorption, and secretion.

urochordate A bag-shaped chordate with larvae that have a firm, flexible notochord extending through a tail; e.g., a tunicate.

uterus (YOU-tur-us) [L. *uterus*, womb] Of a female placental mammal, a muscular, pear-shaped organ in which embryos are housed and nurtured during pregnancy.

vaccination Immunization procedure against a specific pathogen.

vaccine A type of antigen-containing preparation introduced into the body to prime the immune system to recognize the threat before actual infection.

vagina Of female mammals, the organ that receives sperm, forms part of birth canal, and channels menstrual flow.

variable Of experimental tests, a specific aspect of an object or event of interest that may differ over time and among individuals. A single variable is directly manipulated in an experimental group.

vascular bundle Multistranded, sheathed bundle of primary xylem and phloem in the ground tissue system of a stem or leaf.

vascular cambium A lateral meristem that forms in older stems or roots.

vascular cylinder Multistranded, sheathed, cylindrical array of primary xylem and phloem inside a root.

vascular plant Plant with xylem, phloem, and usually well-developed roots, stems, and leaves.

vascular tissue system All xylem and phloem in plants that are structurally more complex than bryophytes.

vasoconstriction A decrease in blood vessel diameter, arterioles especially.

vasodilation An increase in blood vessel diameter, arterioles especially.

vegetative growth Growth of a new plant from an extension or fragment of another.

vein Of a cardiovascular system, any of the large-diameter vessels that lead back to the heart. Of leaves, a vascular bundle threading through photosynthetic tissue.

venule A small blood vessel that connects several capillaries to a vein.

vernalization Stimulation of flowering in spring by low temperature in the season preceding it.

vertebra, plural **vertebrae** One of a series of hard bones that protects the spinal cord and forms the structural backbone for the anterior–posterior body axis.

vertebrate Animal having a backbone.

vesicle A small, membrane-bound sac in the cytoplasm; different sacs transport or store substances or hold enzymes that digest their contents.

vessel member Type of cell in xylem, dead at maturity; its wall becomes part of a water-conducting vessel.

vestibular apparatus Of vertebrates, an organ of equilibrium.

vestigial (ves-TIDJ-ul) A small body part, tissue, or organ that developed abnormally or degenerated over the generations and is unable to function as it normally might; e.g., vestigial wings of mutant fruit flies and human "tail bones."

villus (VIL-us), plural **villi** A fingerlike absorptive structure projecting from the free surface of some epithelia; e.g., the profusion of intestinal villi.

viroid Infectious particle of short, tightly folded strands or circles of RNA.

virus A noncellular infectious agent of DNA or RNA, a protein coat and, in some types, an outer lipid envelope; it can be replicated only after its genetic material enters a host cell and subverts the host's metabolic machinery.

viscera All soft organs inside an animal body; e.g., heart, lungs, and stomach.

vision Perception of visual stimuli based on light focused on a retina and image formation in the brain.

visual accommodation Light-focusing adjustments in a lens position or shape.

visual field The portion of the outside world that an animal sees.

visual signal An observable action or cue that functions as a communication signal.

vital capacity Air volume leaving lungs in one breath after maximum inhalation.

vitamin Any organic substance that an organism requires in trace amounts for metabolism but that it generally cannot synthesize for itself. Many coenzymes function as vitamins.

vitamin D A fat-soluble vitamin; helps the body absorb dietary calcium.

vocal cord One of the thick, muscular folds of the larynx that help some animals produce sound waves for vocalization.

warning coloration Of many toxic species and their mimics, strong colors, patterns, and other signals that predators learn to recognize and avoid.

wastewater treatment Removal of toxins, sludge, organic matter from liquid wastes.

water mold A stramenopile; most are saprobic decomposers or opportunistic parasites, of aquatic habitats.

water table Upper limit at which ground in a region is fully saturated with water.

watershed A region of any specified size in which all precipitation drains into one stream or river.

water–vascular system Of echinoderms, a system of tube feet connected to canals, through which controlled water flow can extend the feet in coordinated ways.

wavelength The distance between the crests of two successive wavelike forms of energy in motion.

wax A lipid with long-chain fatty acids attached to an alcohol other than glycerol.

whisk fern Seedless vascular plant having a branching form and no true roots; e.g., *Psilotum*.

white blood cell Leukocyte. A participant in innate or adaptive immunity, or both; e.g., an eosinophil, neutrophil, basophil, macrophage, T cell, B cell.

white matter Tracts of myelinated axons in the brain and spinal cord.

wild-type allele Of a given gene locus, the allele that occurs normally or with the greatest frequency among individuals of a population.

wind farm Collection of turbines used to convert mechanical energy into electricity.

wing A body part that functions in flight, as among birds, bats, and many insects.

X chromosome A type of sex chromosome that influences sex determination; e.g., XX mammalian embryo becomes female; an XY pairing causes it to develop into a male.

X chromosome inactivation In a female mammalian embryo, the programmed painting of special RNAs over most of one of the two X chromosomes, which cuts off access to the majority of its genes. Which X chromosome gets painted in each cell is a random event, so tissues of adult female mammals are a mosaic of traits. *See also* Dosage compensation.

xanthophyll One of a class of accessory pigments in photosynthesis that reflects yellow to orange light.

xenotransplantation Surgical transfer of an organ from one species to another.

X-linked gene Any gene on an X chromosome.

X-linked recessive inheritance Recessive condition in which the responsible, mutated gene is on the X chromosome.

x-ray diffraction image Film image of x-rays scattered by a crystalline sample; the resulting pattern of streaks and dots can be used to calculate the spacing between the atoms in the crystal lattice.

xylem (ZYE-lum) [Gk. *xylon*, wood] Of vascular plants, a complex tissue that conducts water and solutes through tubes of interconnected walls of cells that are dead at maturity.

Y chromosome Distinctive chromosome in males or females of many species (not both); e.g., human males XY, females, XX.

yellow-green alga A chrysophyte that does not make fucoxanthin; species are common in salt marshes.

yellow marrow Of most mature bones, a fatty tissue that produces red blood cells when blood loss from the body is severe.

Y-linked gene Gene on a Y chromosome.

yolk Protein- and lipid-rich substance that nourishes embryos in animal eggs.

yolk sac Extraembryonic membrane. In most shelled eggs, it holds nutritive yolk; in humans, part becomes a blood cell formation site, and some cells give rise to forerunners of gametes.

zero population growth No net increase or decrease in population size during a specified interval.

zooplankton A community of mostly microscopic heterotrophs suspended or weakly swimming in an aquatic habitat.

zygomycetes Type of parasitic or saprobic fungus in which diploid zygotes develop into zygospores, a type of thick-walled sexual spore in a thin, clear covering.

zygospore Sexual spore of zygomycetes.

zygote (ZYE-goat) A fertilized egg.

pp. 17248 (1992). **13.9** (1–3) © James King-Holmes/SPL/Photo Researchers, Inc.; (4) © McLeod Murdo/CORBIS Sygma. **13.10** Photos by Victor Fisher, courtesy Genetic Savings & Clone. **13.12** © SPL/Photo Researchers, Inc. **13.13** Right, Shahbaz A. Janjua, MD/Dermatlas; http://www.dermatlas.org.

CHAPTER 14 **14.1** Left, © Vaughan Fleming/SPL/Photo Researchers, Inc.; right, PDB ID: 2AAI; Rutenber, E., Katzin, B.J., Ernst, S., Collins, E.J., Mlsna, D., Ready, M. P., Robertus, J. D., Crystallographic Refinement of ricin to 2.5 A. Proteins 10pp.240 (1991). **14.7** Above, tRNA model by Dr. David B. Goodin, The Scripps Research Institute. **14.8** (a) Courtesy of Thomas A. Steitz from *Science*. **14.11** Left, Nik Kleinberg; right, P. J. Maughan. **14.12** © John W. Gofman and Arthur R. Tamplin. From *Poisoned Power: The Case Against Nuclear Power Plants Before and After Three Mile Island*, Rodale Press, PA, 1979. **14.14** © Dr. M. A. Ansary/SPL/Photo Researchers, Inc.

CHAPTER 15 **15.1** From the archives of www.breastpath.com, courtesy of J. B. Askew, Jr., M.D., P. A. Reprinted with permission, copyright 2004 Breastpath.com. **Page 231** Courtesy of Robin Shoulla and Young Survival Coalition. **15.2** (b) From the collection of Jamos Werner and John T. Lis. **15.4** (a) Dr. Karen Dyer Montgomery. **15.5** Jack Carey. **15.6** Above, Juergen Berger, Max Planck Institute for Developmental Biology–Tuebingen, Germany; (a) below, © Jose Luis Riechmann; (b) below, © Jose Luis Riechmann. **15.7** (a) Left, © Visuals Unlimited; (a) right, UCSF Computer Graphics Laboratory, National Institutes, NCRR Grant 01081; (c) from left, © Walter J. Ghering/University of Basel, Switzerland; © Carolina Biological/Visuals Unlimited; © Carolina Biological/Visuals Unlimited; © Carolina Biological/Visuals Unlimited; Courtesy of Edward B. Lewis, California Institute of Technology **15.8** (a) Palay/Beaubois after Robert F. Weaver and Philip W. Hedrick, *Genetics*. © 1989 W. C. Brown Publishers; (b–c) © Jim Langeland, Jim Williams, Julie Gates, Kathy Vorwerk, Steve Paddock, and Sean Carroll, HHMI, University of Wisconsin-Madison. **15.9** PDB ID: 1CJG; Spronk, C. A. E. M., Bonvin, A. M. J. J., Radha, P. K., Melacini, G., Boelens, R., Kaptein, R.: The Solution Structure of Lac Repressor Headpiece 62 Complexed to a Symmetrical Lac Operator. Structure (London) 7 pp. 1483 (1999). Also PDB ID: 1LBI; Lewis, M., Chang, G., Horton, N. C., Kercher, M. A., Pace, H. C., Schumacher, M. A., Brennan, R. G., Lu, P.: Crystal structure of the lactose operon repressor and its complexes with DNA and inducer. *Science* 271 pp. 1247 (1996); lactose pdb files from the Hetero-Compound Information Centre - Uppsala (HIC-Up). **15.11** © Jim Langeland, Jim Williams, Julie Gates, Kathy Vorwerk, Steve Paddock, and Sean Carroll, HHMI, University of Wisconsin-Madison.

CHAPTER 16 **16.1** (a) Per Hardestam/www.hardestam.se; (b) Dr. Jorge Mayer, Golden Rice Project. **Page 243** ScienceUV/Visuals Unlimited. **16.3** (a) Dr. Huntington Potter and Dr. David Dressler; (b) with permission of © QIAGEN, Inc. **Page 248** © TEK IMAGE/Photo Researchers, Inc. **16.9** Left, © David Parker/SPL/Photo Researchers, Inc.; right, Cellmark Diagnostics, Abingdon, UK. **16.10** Right, © Volker Steger/SPL/Photo Researchers, Inc. **16.11** Courtesy of Joseph DeRisa. From *Science*, 1997 Oct. 24; 278 (5338) 680–686. **Page 252** © Professor Stanley Cohen/SPL/Photo Researchers, Inc. **16.12** (a) Courtesy Calgene, LLC; (b) Dr. Vincent Chiang, School of Forestry and Wood Products, Michigan Technology

University. **16.13** (d) © Lowell Georgis/CORBIS; (e) Keith V. Wood. **16.14** (a) © Adi Nes, Dvir Gallery Ltd.; (b) Transgenic goat produced using nuclear transfer at GTC Biotherapeutics. Photo used with permission; (c) © Matt Gentry/Roanoke Times. **16.14** R. Brinster, R. E. Hammer, School of Veterinary Medicine, University of Pennsylvania. **Page 256** © Jeans for Gene Appeal.

Page 259 UNIT III © Wolfgang Kaehler/Corbis.

CHAPTER 17 **17.1** Left, NASA Galileo Imaging Team; Right © David A. Kring, NASA/Univ. Arizona Space Imagery Center. **17.2** Art by Don Davis. **17.3** (a, c) © Wolfgang Kaehler/Corbis; (b) © Earl & Nazima Kowall/Corbis; (d, e) Edward S. Ross. **17.4** right, © Bruce J. Mohn; (inset) Phillip Gingerich, Director, University of Michigan. Museum of Paleontology. **17.5** © Jonathan Blair/Corbis. **17.6** (a) Courtesy George P. Darwin, Darwin Museum, Down House; (b) Christopher Ralling; (e) Dieter & Mary Plage/Survival Anglia. **Page 265** Heather Angel. **17.7** (a) © Joe McDonald/Corbis; (b) © Karen Carr Studio/www.karencarr.com. **17.8** (a) © Gerra and Sommazzi/www.justbirds.org; (b) © Kevin Schafer/Corbis; (c) Alan Root/Bruce Coleman Ltd. **17.9** Down House and The Royal College of Surgeons of England. **17.10** Left, H. P. Banks; Right, Jonathan Blair. **17.11** © Jonathan Blair/Corbis. **17.12** (b) © 2001 Photodisc, Inc. **17.14** © Corbis. **17.15** (a) NASA/GSFC. **17.16** (a-d) After A.M. Ziegler, C.R. Scotese, and S.F. Barrett, "Mesozoic and Cenozoic Paleogeographic Maps," and J. Krohn and J. Sundermann (Eds.), Tidal Frictions and the Earth's Rotation II, Springer-Verlag, 1983; (e) © Martin Land/Photo Researchers, Inc.; (f) © John Sibbick. **17.18** (a) J. Scott Altenbach, University of New Mexico, computer enhanced by Lisa Starr; (b) Frans Lanting/Minden Pictures; computer enhanced by Lisa Starr; (c) above, © Stephen Dalton/Photo Researchers, Inc.); (c bottom) Natural History Collection, Royal BC Museum. **17.19** (a, b) Courtesy of Professor Richard Amasino, University of Wisconsin-Madison; (c) Juergen Berger, Max Planck Institute for Developmental Biology—Tuebingen, Germany; (d) Courtesy of Professor Martin F. Yanofsky, UCSD. **17.21** (a) above, Tait/Sunnucks Peripatus Research; (a) below, © Jennifer Grenier, Grace Boekhoff-Falk and Sean Carroll, HMI, University of Wisconsin-Madison; (b) above, Herve Chaumeton/Agence Nature; (b) below, © Jennifer Grenier, Grace Boekhoff-Falk and Sean Carroll, HMI, University of Wisconsin-Madison; (c) above, © Peter Skinner/Photo Researchers, Inc.; (c) below, Courtesy of Dr. Giovanni Levi; (d) Dr. Chip Clark. **Page 278** © TEK IMAGE/Photo Researchers, Inc. **17.23** Left, Kjell B. Sandved/Visuals Unlimited; Center, Jeffrey Sylvester/FPG/Getty Images; Right, Thomas D. Mangelsen/Images of Nature. **17.24** John Klausmeyer, University of Michigan Exhibit of Natural History.

CHAPER 18 **18.1** Above, © Bettmann/Corbis; Below, © Reuters NewMedia, Inc./Corbis. **Page 283** © St. Bartholomew's Hospital/Science Photo Library/Photo Researchers, Inc. **18.2** Clockwise from left, © Peter Bowater/Photo Researchers, Inc.; © Owen Franken/Corbis; © Sam Kleinman/Corbis; Alan Solem; © Christopher Briscoe/Photo Researchers, Inc.; © Jim Cornfield/Corbis. **18.3** © 2002/Photodisc/Getty Images. **18.6** J. A. Bishop, L. M. Cook. **18.7** Courtesy of Hopi Hoekstra, University of California, San Diego. **18.10** © Peter Chadwick/Science Photo Library/Photo Researchers, Inc.. **18.11** Thomas Bates Smith. **18.12** Bruce Beehler. **18.13** (a–b) After Ayala and others; (c) © Michael Freeman/Corbis. **18.14** Adapted from S. S. Rich, A. E. Bell, and S. P. Wilson, "Genetic

drift in small populations of Tribolium," *Evolution* 33:579-584, Fig. 1, p. 580, 1979. Used by permission of the publisher. **Page 294** © Steve Bronstein/The Image Bank/Getty Images. **18.15** Frans Lanting/Minden Pictures (computer-modified by Lisa Starr). **18.16** Left, David Neal Parks; right W. Carter Johnson. **18.17** (b) John W. Merck, Jr., University of Maryland. **18.18** Left, © Thomas Mangelsen; right © Theo Allofs/Corbis. **18.19** Left, © Francois Gohier/Photo Researchers, Inc.; right © David Parker/SPL/Photo Researchers, Inc.. **18.20** Elliot Erwitt/Magnum Photos, Inc..

CHAPTER 19 **19.1** Left, © Jack Jeffrey Photography; Right, Image courtesy of the Image Analysis Laboratory, NASA/Johnson Space Center. **19.2** Bill Sparklin/Ashley Dayer. **19.4** (a) John Alcock, Arizona State University; (b) © Alvin E. Staffan/Photo Researchers, Inc.; (c) G. Ziesler/ZEFA. **19.5** Left © Digital Vision/PictureQuest; (a) © Joe McDonald/Corbis. **19.6** (a) © Graham Neden/Corbis; (b) © Kevin Schafer/ Corbis; Center, © Ron Blakey, Northern Arizona University; (c) © Rick Rosen/Corbis SABA. **19.7** Po' ouli, Bill Sparklin/Ashley Dayer; All others, © Jack Jeffrey Photography. **19.8** Above, Steve Gartlan; Below, © Below Water Photography/www.belowwater.com. **19.9** Jean-Claude Carton/Bruce Coleman, Inc. **19.10** After W. Jensen and F. B. Salisbury, Botany: An Ecological Approach, Wadsworth, 1972. **19.11** Courtesy of Dr. Robert Mesibov. **19.13** Courtesy of Daniel C. Kelley, Anthony J. Arnold, and William C. Parker, Florida State University Department of Geological Science. **19.14** © Carnegie Museum of Natural History. **19.15** From left, © Science Photo Library/Photo Researchers, Inc.; © Galen Rowell/Corbis; © Kevin Schafer/Corbis; Courtesy of Department of Entomology, University of Nebraska-Lincoln; Bruce Coleman, Ltd. **19.18** From left, © Hans Reinhard/Bruce Coleman, Inc; © Phillip Colla Photography; © Randy Wells/Corbis; © Cousteau Society/The Image Bank/Getty Images; © Robert Dowling/Corbis. **19.20** Left, Courtesy of Department of Library Services, American Museum of Natural History (Neg. #K10273); Right, Photo by Lisa Starr. **19.21** © Gulf News, Dubai, UAE.

CHAPTER 20 **20.1** Left, Courtesy of Agriculture Canada; Right, © Raymond Gehman/Corbis. **20.2** © Philippa Uwins/The University of Queensland. **20.3** Jeff Hester and Paul Scowen, Arizona State University, and NASA. **20.4** Left, Painting by William K. Hartmann; Right, Painting by Chesley Bonestell. **20.6** (a) Eiichi Kurasawa/Photo Researchers, Inc.; (b) © Dr. Ken MacDonald/SPL/Photo Researchers, Inc.; (c) © Micheal J. Russell, Scottish Universities Environmental Research Centre. **20.7** (a) Sidney W. Fox; (b) From Hanczyc, Fujikawa, and Szostak, Experimental Models of Primitive Cellular Compartments: Encapsulation, Growth, and Division; www.sciencemag.org, *Science* 24 October 2003; 302;529, Figure 2, page 619. Reprinted with Permission of the authors and AAAS. **20.8** (a) Stanley M. Awramik; (b-c) © Bruce Runnegar, NASA Astrobiology Institute; (d) © N. J. Butterfield, University of Cambridge. **20.9** (a) © Chase Studios/Photo Researchers, Inc.; (b) © John Reader/SPL/Photo Researchers, Inc.; (c) © Sinclair Stammers/SPL/Photo Researchers, Inc. **20.10** (a) © CNRI/Photo Researchers, Inc.; (b) © Robert Trench, Professor Emeritus, University of British Columbia.

Page 331 UNIT IV © Layne Kennedy/Corbis.

CHAPTER 21 **21.1** © David Lees/Getty Images. **Page 333** © R. Sorensen/J. Olsen/Photo Researchers, Inc. **21.3** (a) P. Hawtin, University of Southampton/SPL/Photo Researchers, Inc.; (b) ©

Dr. Manfred Schloesser, Max Planck Institute for Marine Microbiology; (c) CNRI/SPL/Photo Researchers, Inc.; (d) © Dr. Dennis Kunkel/Visuals Unlimited. **21.4** L. J. LeBeau, University of Illinois Hospital/BPS. **21.5** Micrograph L. Santo. **21.7** Electron micrograph of Aquifex pyrophilus, platinum shadowed. Bar: 1 micrometer. Image by R. Rachel and K. O. Stetter, University Regensburg, Germany. **21.8** (a) P. W. Johnson and J. MeN. Sieburth, Univ. Rhode Island/BPS; (b) © Dr. Jeremy Burgess/SPL/ Photo Researchers, Inc.; (c) © Stem Jems/Photo Researchers, Inc.; (d) Dr. Terry J. Beveridge, Department of Microbiology, University of Guelph, Ontario, Canada. **21.9** Richard Blakemore. **21.11** (a) Courtesy Jack Jones, *Archives of Microbiology*, Vol. 136, 1983, pp. 254-261. Reprinted by permission of Springer-Verlag; (b) © Dr. John Brackenbury/ Science Photo Library/Photo Researchers, Inc. **21.12** (a) © Martin Miller/Visuals Unlimited; (b) © Alan L. Detrick, Science Source/Photo Researchers, Inc.; (c) © Dr. Harald Huber, Dr. Michael Hohn, Prof. Dr. K.O.Stetter, University of Regensburg, Germany. **21.13** After Stephen L. Wolfe. **21.14** (a) © CAMR/A. B. Dowsett/SPL/Photo Researchers, Inc.; (b) © Dr. Linda Stannard, UCT/SPL/Photo Researchers, Inc.; (c-d) Kenneth M. Corbett. **21.15** © Science Photo Library/Photo Researchers, Inc. **Page 346** Left, above, © CAMR, Barry Dowsett/ Science Photo Library/Photo Researchers, Inc.; below © Sercomi/Photo Researchers, Inc.; right, © Camr/B. Dowsett/SPL/Photo Researchers, Inc. **21.17** (a) © Lily Echeverria/Miami Herald; (b) © APHIS photo by Dr. Al Jenny; (bottom) PDB ID: 1QLX; Zahn, R., Liu, A., Luhrs, T., Riek, R., Von Schroetter, C., Garcia, F.L., Billeter, M., Calzolai, L., Wider, G., Wuthrich, K.: NMR Solution Structure of the Human Prion Protein, Proc. Nat. Acad. Sci. USA 97 pp. 145 (2000). **21.18** right, E. A. Zottola, University of Minnesota.

CHAPTER 22 **22.1** Wim van Egmond/Visuals Unlimited. **Page 351** Above, © Adam Woolfitt/ Corbis: below, © Ric Ergenbright/Corbis. **22.3** (a) © Dr. Dennis Kunkel/Visuals Unlimited; (b-c) Dr. Stan Erlandsen, University of Minnesota. **22.4** (a) P. L. Walne and J. H. Arnott, Planta, 77:325-354, 1967; (b) Photo by Stephen Durr. **22.5** (a) After Prescott et al., Microbiology, third edition; (b) © Oliver Meckes/Photo Researchers, Inc. **22.6** (a) Courtesy of Allen W. H. Bé and David A. Caron; (b) © Wim van Egmond; (c) © G. Shih and R. Kessel/Visuals Unlimited. **22.7** (a) Above, Courtesy Professor Steve Beck, Nassau Community College; (a) Below, Courtesy James Evarts; (b) Gary W. Grimes and Steven L'Hernault; (c) Redrawn from V. & M. Pearse and M. & R. Buchsbaum, Living Invertebrates, The Boxwood Press, 1987. Used by permission. **22.8** © John Walsh/SPL/Photo Researchers, Inc. **22.9** (a) © Wim van Egmond/Micropolitan Museum; (b) © Dr. David Phillips/Visuals Unlimited. **22.10** © Lexey Swall/Staff from article, " Deep Trouble: Bad Blooms" October 3, 2003 by Eric Staats. **22.11** Left (mosquito), © Sinclair Stammers/Photo Researchers, Inc.; top right, © London School of Hygiene & Tropical Medicine/Photo Researchers, Inc.; bottom right © Moredum Animal Health, Ltd./ Photo Researchers, Inc.; center bottom (gametocyte), Micrograph Steven L'Hernault. **22.12** (a-b) Ron Hoham, Dept. of Biology, Colgate University; (c) Emiliania huxleyi. Photograph by Vita Pariente. Scanning electron micrograph taken on a Jeol T330A instrument at the Texas A & M University Electron Microscopy Center; (d) Greta Fryxell, University of Texas, Austin. **22.13** Left (background) Steven C. Wilson/Entheos; (a) © Jeffrey Levinton, State University of New York, Stony Brook; (b) © Lewis Trusty/Animals Animals; (c) From T. Garrison, Oceanography: An Invitation to Marine Science, Brooks/Cole, 1993. **22.14** Claude Taylor and the

University of Wisconsin Dept. of Botany. **22.15** (a) Heather Angel; (b) International Potato Center, Lima, Peru. **22.16** Left, © Susan Frankel, USDA-FS; right, Dr. Pavel Svihra. **22.17** © Photodisc/Getty Images. **22.18** © Wim van Egmond. **22.19** (a) © Lawson Wood/Corbis; (b) Monterey Bay Aquarium; (c) Courtesy of Professor Astrid Saugestad; (d) Courtesy of Knut Norstog and the Botanical Society of America; (e) D. S. Littler. **22.20** (b) Courtesy of Professeur Michel Cavalla. **22.21** © Astrid Hanns-Frieder Michler/SPL/Photo Researchers, Inc. **22.22** (a) Edward S. Ross; (b) Courtesy of www.hiddenforest.co.nz. **22.23** (b) M. Claviez, G. Gerish, and R. Guggenheim; (c-e) Carolina Biological Supply Company; (f) Courtesy Robert R. Kay from R. R. Kay, et al., Development, 1989 Supplement, pp. 81-90, (c) The Company of Biologists Ltd., 1989. **Page 351** Above right, Nature Publishing Group, www.nature.com. 1: Figure, Number 1 from Nature, Vol. 410, pp. 430, *Asexual reproduction: 'Midwives' assist dividing aboebae* by David Biron, Pazit Libros, Dror Sagi, David Mirelman, Ellisha Moses, et al. **22.24** © W. P. Armstrong; inset, Courtesy Brian Duval. **22.25** Lauren and Homer, Photography by Gary Head. **23.1** © T. Kerasote/Photo Researchers, Inc. **Page 371** © Craig Allikas/www.orchidworks.com.

CHAPTER 23 **23.3** (a) © Wim van Egmond; (b) Courtesy Microbial Culture Collection, National Institute for Environmental Studies, Japan. **23.4** (a) © Reprinted with permission from Elsevier; (b) Patricia G. Gensel; (c) Illustration by Zdenek Burian, © Jeri Hochman and Martin Hochman; (d) © Karen Carr Studio/www.karencarr.com. **23.6** (b) After E.O. Dodson and P. Dodson, Evolution: Process and Product, Third Ed., p. 401, PWS. **23.7** Robert Potts, California Academy of Sciences. **23.8** (a) Craig Wood/Visuals Unlimited; (b) Jane Burton/Bruce Coleman Ltd. **23.9** (a) Fred Bavendam/Peter Arnold, Inc.; (b) John D. Cunningham/Visuals Unlimited. **23.10** (a) © University of Wisconsin-Madison, Department of Biology, Anthoceros CD; (b) Left, National Park Services, Paul Stehr-Green; (b) Right, National Park Services, Martin Hutten; (c) © Wayne P. Armstrong, Professor of Biology and Botany, Palomar College, San Marcos, California. **23.11** (a) Gerald D. Carr; (b) Ed Reschke/Peter Arnold, Inc.; (c) © Colin Bates, www.coastalimage-works.com; (d) Left, Photo by A. Murray, University of Florida, Center for Aquatic and Invasive Plants. Used with permission; (e) Derrick Ditlchburn/ Visuals Unlimited. **23.12** (a) Above right, A. & E. Bomford/Ardea, London; (b) © Klein, Hubert/ Peter Arnold, Inc. **23.13** Above, Brian Parker/Tom Stack & Associates; Below Field Museum of Natural History, Chicago (Neg. #7500C). **23.14** (a) Ralph Pleasant/FPG/Getty Images; (b) © Earl Roberge/ Photo Researchers, Inc.; (c) George Loun/Visuals Unlimited; (d) Courtesy of Water Research Commission, South Africa. **Page 381** Above right, © George J. Wilder/Visuals Unlimited, computer enhanced by Lachina Publishing Services, Inc. **23.15** (a) © Dave Cavagnaro/Peter Arnold, Inc.; (b) © Robert & Linda Mitchell Photography; (c) © E. Webber/Visuals Unlimited; (d) © Michael P. Gadomski/Photo Researchers, Inc.; (e) © Sinclair Stammers/Photo Researchers, Inc.; (f) Courtesy of Wayside Gardens/www.waysidegardens.com; (g) © Gerald & Buff Corsi/Visuals Unlimited; (h) © Fletcher and Baylis/Photo Researchers, Inc. **23.16** Left, Robert Potts, California Academy of Sciences; (a) Robert & Linda Mitchell Photography; (b) © R. J. Erwin/Photo Researchers, Inc. **23.17** From top, Ed Reschke; Lee Casebere; Robert & Linda Mitchell Photography; Runk & Schoenberger/Grant Heilman, Inc. **23.18** (a) © Michelle Garrett/Corbis; (b) © Sanford/Agliolo/Corbis; (c) © Gregory G. Dimijian/Photo Researchers, Inc.; (d) © Darrell

Gulin/Corbis; (e) Photo provided by DLN/ Permission by Dr. Daniel L. Nickrent. **23.21** Gerry Ellis/The Wildlife Collection. **23.22** © William Campbell/TimePix/Getty Images. **23.23** Left, © 1989 Clinton Webb.

CHAPTER 24 **24.1** © Charles Lewallen. **Page 391** © Jacques Langevin/Corbis Sygma. **Page 392** Micrograph Garry T. Cole, University of Texas, Austin/BPS. **24.3** (a-e) Robert C. Simpson/Nature Stock. **24.4** (a) Courtesy of Ken Nemuras; (b) CDC. **24.5** N. Allin and G. L. Barron. **24.6** (a-b) Micrographs Ed Reschke; Below right, Micrograph J. D. Cunningham/Visuals Unlimited. **24.7** Left, Micrograph Garry T. Cole, University of Texas, Austin/BPS; right, © Michael W. Clayton/ University of Wisconsin-Madison, Department of Biology; art, After T. Rost, et al., *Botany*, Wiley 1979. **24.9** (a) Above, © North Carolina State University, Department of Plant Pathology; (a) below, © Michael Wood/mykob.com; (b) © Fred Stevens/ Mykoweb.com; (c) © Dennis Kunkel Microscopy, Inc.; (d) Garry T. Cole, University of Texas, Austin/ BPS. **24.10** (a) © Dr. P. Marazzi/SPL/Photo Researchers, Inc.; (b) Eric Crichton/Bruce Coleman; (c) © Harry Regin. **24.11** (a) © Mark Mattock/Planet Earth Pictures; (c) © Mark E. Gibson/Visuals Unlimited; (d) After Raven, Evert, and Eichhorn, *Biology of Plants*, 4th ed., Worth Publishers, New York, 1986. **24.12** (a) Prof. DJ Read, University of Sheffield; (b) © 1990 Gary Braasch; (c) F. B. Reeves. **24.13** John Hodgin. **24.14** Robert C. Simpson/ Nature Stock. **24.15** (a) Jane Burton/Bruce Coleman, Ltd.; (b) © Chris Worden.

CHAPTER 25 **25.1** © K.S. Matz. **Page 403** © Callum Roberts, University of York. **25.2** © David Aubrey/Corbis. **25.4** David Patterson, courtesy micro*scope/http://microscope.mbl.edu. **25.8** (a-b) Neville Pledge/South Australian Museum; (c) Dr. Chip Clark. **25.9** (a) Marty Snyderman/Planet Earth Pictures; (b) Bruce Hall; (c) © David Aubrey/ Corbis; (d) Don W. Fawcett/Visuals Unlimited. **25.11** (a) © Robert Brons/livingreefimages.com; (b left) After Laszlo Meszoly, in L. Margulis, *Early Life*, Jones and Bartlett, 1982. **25.12** (c) © Brandon D. Cole/Corbis; (d) © Jeffrey L. Rotman/Corbis. **25.13** (a) After Eugene Kozloff; (b) Courtesy of Dr. William H. Hamner. **25.14** (a) Kim Taylor/Bruce Coleman, Ltd.; (b) © A.N.T./Photo Researchers, Inc. **25.15** (Above) After T. Storer, et al., *General Zoology*, Sixth Edition; Right, © Wim van Egmond/ Micropolitan Museum. **25.17** (a) © James Marshall/ Corbis. **25.18** Right © Andrew Syred/SPL/Photo Researchers, Inc. **25.19** (a) J. Solliday/BPS; (b) Jon Kenfield/Bruce Coleman Ltd.; (c-d) Adapted from Rasmussen, "Ophelia," Vol. 11, in Eugene Kozloff, Invertebrates, 1990. **25.20** J. A. L. Cooke/Oxford Scientific Films. **25.21** (g) © Cabisco/Visuals Unlimited; (h) © Science Photo Library/Photo Researchers, Inc. **Page 416** Danielle C. Zacherl with John McNulty. **25.24** (a) © B. Borrell Casals/Frank Lane Picture Agency/Corbis; (b) Jeff Foott/Tom Stack & Associates; (c) © Joe McDonald/Corbis; (d) © Reinhard Dirscherl/Visuals Unlimited; (e) Alex Kirstitch. **25.25** (a) Herve Chaumeton/Agence Nature. **25.26** (a) Illustrations by Zdenek Burian, © Jeri Hochman and Martin Hochman; (c) Alex Kirstitch; (d) Bob Cranston. **25.27** Below, Micrograph, J. Sulston, MRC Laboratory of Molecular Biology. **25.28** (a) © Sinclair Stammers/ SPL/Photo Researchers, Inc.; (b) © L. Jensen/ Visuals Unlimited; (c) Dianora Niccolin. **25.29** Jane Burton/Bruce Coleman, Ltd. **25.30** (a) © Angelo Giampiccolo/FPG/Getty Images; (b) © Frans Lemmens/The Image bank/Getty Images. **25.31** (a) Redrawn from Living Invertebrates, V. & J. Pearse/M. & R. Buchsbaum, The Boxwood Press, 1987. Used by permission; (b) © Corbis; (c) ©

Andrew Syred/Photo Researchers, Inc. **25.32** (a) © Jeff Hunter/The Image Bank/Getty Images; (b) © Peter Parks/Imagequestmarine.com; (c) © Science Photo Library/Photo Researchers, Inc.; d-e) After D.H. Milne, Marine Life and the Sea, Wadsworth, 1995. **25.34** (a) © Michael & Patricia Fogden/Corbis; (b) Steve Martin/Tom Stack & Associates. **25.37** (a) David Maitland/Seaphot Limited/Planet Earth Pictures; (b-g) Edward S. Ross; (h) © Mark Moffett/Minden Pictures; (i) Courtesy of Karen Swain, North Carolina Museum of Natural Sciences; (j) Chris Anderson/Darklight Imagery; (k) Joseph L. Spencer. **25.38** (a) John H. Gerard; (b) © D. Suzio/Photo Researchers, Inc. **25.39** (a) © Stem Jems/Photo Researchers, Inc.; (b) © California Department of Health Services; (c) © Bernard Cohen, M.D., Dermatlas; http://www.dermatalas.org. **25.40** William Dow/Corbis. **25.41** (a) Marlin E. Rice, Iowa State University; (b-c) John Obermeyer, Purdue University, Dept of Entomology. **25.42** Photo by James Gathany, Centers for Disease Control. **25.44** (a) Herve Chaumeton/Agence Nature; (b) © Fred Bavendam/Minden Pictures; (c) © George Perina, www.seapix.com; (d) Jan Haaga, Kodiak Lab, AFSC/NMFS; (e) Herve Chaumeton/Agence Nature; (g) Jane Burton/Bruce Coleman, Ltd. **25.45** (a) Walter Deas/Seaphot Limited/Planet Earth Pictures; (b) © E. Webber/Visuals Unlimited. **25.46** Jane Burton/Bruce Coleman, Ltd. **26.1** © Karen Carr Studio/www.karencarr.com. **Page 433** P. Morris/Ardea London.

CHAPTER 26 **26.3** (a) © 2002 Gary Bell/Taxi/Getty Images; (b-c) Redrawn from Living Invertebrates, V. & J. Pearse and M. & R. Buchsbaum. The Boxwood Press, 1987. Used by permission. **26.5** © Brandon D. Cole/Corbis. **26.6** (a) © John and Bridgette Sibbick; (b-c) © Jenna Hellack, Department of Biology, Univerisy of Central Oklahoma. (a–c) Adapted from A.S. Romer and T.S. Parsons, The Vertebrate Body, Sixth Edition, Saunders, 1986; Left, Photo by Lisa Starr; right, Courtesy of John McNamara, www.paleodirect.com. **26.9** (a) © Jonathan Bird/Oceanic Research Group, Inc.; (b) © Gido Braase/Deep Blue Productions; (c) © Tom McHugh/Photo Researchers, Inc.; (d) Robert & Linda Mitchell Photography; (e) © Ivor Fulcher/Corbis; (f) Patrice Ceisel/© 1986 John G. Shedd Aquarium. **26.10** (a) © Norbert Wu/Peter Arnold, Inc.; (b) Wernher Krutein/photovault.com; (c) © Alfred Kamajian. **26.11** (left) Adapted from A.S. Romer and T.S. Parsons, The Vertebrate Body, Sixth Edition, Saunders, 1986. (a) © Bill M. Campbell, MD; (b) © Stephen Dalton/Photo Researchers, Inc.; (c) John Serraro/Visuals Unlimited. **26.12** Juan M. Renjifo/Animals Animals. **26.13** (a) Pieter Johnson; (b) Stanley Sessions/Hartwick College. **26.14** © 1989 D. Braginetz. **26.15** © Karen Carr Studio/www.karencarr.com. **Page 443** Bottom right, © Julian Baum/SPL/Photo Researchers, Inc. **Page 444** Left, S. Blair Hedges. **26.17** (a) © Kevin Schafer/Corbis; (c) © Joe McDonald/Corbis; (d) © David A. Northcott/Corbis; (e) © Pete & Judy Morrin/Ardea London; (f) © Stephen Dalton/Photo Researchers, Inc.; (g) Kevin Schafer/Tom Stack & Associates. **26.18** © James Reece, Nature Focus, Australian Museum. **26.19** (a) From The Life of Birds, fourth edition, L. Baptista and J.C. Whelty, 1988, Saunders; (b–c) Courtesy of Dr. M. Guinan, University of California-Davis, Anatomy, Physiology and Cell Biology, School of Veterinary Medicine. **26.21** (a) Gerard Lacz/ANT Photolibrary; (b) © Kevin Schafer/Corbis. **26.23** (a) Sandy Roessler/FPG/Getty Images; (b) After M. Weiss and A. Mann, Human Biology and Behavior, 5th Edition, HarperCollins, 1990 **26.24** Painting © Ely Kish. **26.25** (a) Painting © Ely Kish; (b) © Karen Carr Studio/www.karencarr.com. **26.26** (e) D. & V. Blagden/ANT Photo Library; (f) © Nigel J. Dennis/

Gallo Images/Corbis; (g) © Tom Ulrich/Visuals Unlimited. **26.27** (a) © Tom McHugh/Photo Researchers, Inc.; (b) Corbis Images/PictureQuest; (c) Mike Jagoe/Talune Wildlife Park, Tasmania, Australia. **26.28** (b) © Mike Johnson. All rights reserved, www.earthwindow.com; (c) © Marine Themes Stock Photo Library; (d) © Douglas Faulkner/Photo Researchers, Inc.; (e) © David Parker/SPL/Photo Researchers, Inc.; (f) Bryan and Cherry Alexander Photography; (g) © Stephen Dalton/Photo Researchers, Inc.; (h) © Merlin D. Tuttle/Bat Conservation International; (i) Alan and Sandy Carey. **26.30** (a) Larry Burrows/Aspect Photolibrary; (c) Allen Gathman, Biology department, Southeast Missouri State University; (d) © Dallas Zoo, Robert Cabello; (e) Bone Clones®, www.boneclones.com; (g) Courtesy of Dr. Takeshi Furuichi, Biology, Meiji-Gakuin University-Yokohama. **26.31** (a) © Utah's Hogle Zoo; left, Gerry Ellis/The Wildlife Collection. **26.32** After National Geographic, February 1997, p. 82. **26.33** Left, MPFT/Corbis Sygma; (all others) National Museum of Ethiopia, Addis Ababa. © 1985 David L. Brill. **26.34** (a) Dr. Donald Johanson, Institute of Human Origins; (b) Kenneth Garrett/National Geographic Image Collection; (c) Louise M. Robbins; (d) Kenneth Garrett/National Geographic Image Collection. **26.35** Kenneth Garrett/National Geographic Image Collection. **26.36** Left, Jean Paul Tibbles; right, National Museum of Ethiopia, Addis Ababa. © 1985 David L. Brill. **26.37** Left, © Elizabeth Delaney/Visuals Unlimited; right, John Reader © 1981. **26.39** National Museum of Ethiopia, Addis Ababa. © 1985 David L. Brill. **26.41** Left, National Museum of Ethiopia, Addis Ababa. © 1985 David L. Brill; right, NASA. **26.43** Left, Sandak/FPG/Getty Images; right, Douglas Mazonowicz/Gallery of Prehistoric Art. **26.44** © California Academy of Sciences. **26.45** © Jean Phillipe Varin/Jacana/Photo Researchers, Inc. **26.47** Z. Leszczynski/Animals Animals.

CHAPTER 27 **27.1** © David Nunuk/PhotoResearchers, Inc. **27.3** (a) Painting by Charles Knight/American Museum of Natural History; (b) Mansell Collection/Time, Inc./Getty Images. **27.4** Photo, Gary Head; Topographical maps, U.S. Geological Survey. **27.5** Eric Hartmann/Magnum Photos. **27.6** James Marshall/Corbis. **27.7** © Greenpeace/Cunningham. **27.8** © Bagla Pallava/Corbis Sygma; inset, © A. Bannister/Photo Researchers, Inc. (b) From T. Garrison, Oceanography: An Invitation to Marine Science, Third Edition, Brooks/Cole, 2000. All rights reserved. **27.10** (a) C. B. & D. W. Frith/Bruce Coleman, Ltd.; (b) © Douglas Faulkner/Photo Researchers, Inc.; (c) Douglas Faulkner/Sally Faulkner Collection; (d) Sea Studios/Peter Arnold, Inc.; (e) From left, top Douglas Faulkner/Sally Faulkner Collection; Peter Scoones; center, Jeff Rotman; Alex Kirstitch; Douglas Faulkner/Sally Faulkner Collection; bottom (all), Douglas Faulkner/Sally Faulkner Collection. **27.11** © Greenpeace/Grace. **27.14** Hans Renner. **27.15** Above, © R. Bieregaard/Photo Researchers, Inc.; below, photo © 2000 Photodisc, Inc. **27.16** Bureau of Land Management. **27.17** Left, Peter Scoones; right, NASA **27.18** Courtesy of Eternal Reefs, Inc., www.eternalreefs.com.

CHAPTER 28 **28.1** Star Tribune/Minneapolis-St. Paul. **Page 479** (Left) © VVG/Science Photo Library/Photo Researchers, Inc.; (right) © Michael Davidson/Mortimer Abramowitz Gallery of Photomicrography/www.olympusmicro.com **28.2** Top Courtesy of Charles Lewallen; Center & bottom, © Bruce Iverson. **28.3** Left (Art by Lisa Starr with) ©

2000 Photodisc, Inc; Right, above, © CNRI/SPL/Photo Researchers, Inc.; below, Dr. Robert Wagner/University of Delaware, www.udel.edu/Biology/Wags. **28.4** Left © Pat Johnson Studios Photography; Right © Darrell Gulin/The Image Bank/Getty Images. **28.5** (a) © Cory Gray; (b) © Photodisc/Getty Images; (c) Heather Angel; (d) © Biophoto Associates/Photo Researchers, Inc. **28.6** (a) © Geoff Tompkinson/SPL/Photo Researchers, Inc.; (b) © John Beatty/SPL/Photo Researchers, Inc. **28.8** © VVG/Science Photo Library/Photo Researchers, Inc. **28.9** Galen Rowell/Peter Arnold, Inc. **28.10** © Niall Benvie/Corbis. **28.11** Left © Kennan Ward/Corbis; Right G. J. McKenzie (MGS). **28.12** Frank B. Salisbury. **28.14** (a) Courtesy of Dr. Kathleen K. Sulik, Bowles Center for Alcohol Studies, the University of North Carolina at Chapel Hill. **28.15** © John DaSiai, MD/Custom Medical Stock Photo. **28.16** (a) Courtesy of Dr. Consuelo M. De Moraes; (b–d) © Andrei Sourakov and Consuelo M. De Moraes.

CHAPTER 29 **29.1** (a) © Michael Westmoreland/Corbis; (b) © Charles O'Rear/Corbis. **Page 493** © Reuters/Corbis. **29.4** (a) From top, © Bruce Iverson; © Ernest Manewal/Index Stock Imagery; © Simon Fraser/Photo Researchers, Inc.; © Andrew Syred/Photo Researchers, Inc.; (b) From top, © Mike Clayton/University of Wisconsin Department of Botany; © Darrell Gulin/Corbis; Gary Head; © Andrew Syred/Photo Researchers, Inc. **29.6** Left, © Donald L. Rubbelke/Lakeland Community College; right, © Andrew Syred/Photo Researchers, Inc. **29.7** (a) © Dr. Dale M. Benham, Nebraska Wesleyan University; (b) D. E. Akin and I. L. Risgby, Richard B. Russel Agricultural Research Center, Agricultural Research Service, U.S. Dept. Agriculture, Athens, GA; (c) Kingsley R. Stern. **29.8** Below, © Andrew Syred/Photo Researchers, Inc. **29.9** George S. Ellmore. **29.10** (a) Below, © Dr. Dale M. Benham, Nebraska Wesleyan University. **29.11** (a) Left, Ray F. Evert; (a) right, James W. Perry; (b) Left, Carolina Biological Supply Company; (b) right, James W. Perry. **29.13** David Cavagnaro/Peter Arnold, Inc. **Page 500** Right, © 2001 Photodisc, Inc. **29.14** (a) © N. Cattlin/Photo Researchers, Inc.; (c) C. E. Jeffree, et al., Planta, 172(1):20-37, 1987. Reprinted by permission of C. E. Jeffree and Springer-Verlag; (d) Jeremy Burgess/SPL/Photo Researchers, Inc. **29.15** (a) © Simon Fraser/Photo Researchers, Inc. **29.16** (a) Left, John Limbaugh/Ripon Microslides, Inc.; (a right) After Salisbury and Ross, Plant Physiology, Fourth Edition, Wadsworth; (b) © Mike Clayton/University of Wisconsin Department of Botany. **29.17** Carolina Biological Supply Company. **29.18** © Omikron/Photo Researchers, Inc.. **29.21** Alison W. Roberts, University of Rhode Island. **29.23** John Lotter Gurling/Tom Stack & Associates. **29.24** (b) H. A. Core, W. A. Cote, and A. C. Day, Wood Structure and Identification, 2nd Ed., Syracuse University Press, 1979. **29.25** (a) © Peter Ryan/SPL/Photo Researchers, Inc.; (b) © Jon Pilcher; (c) © George Bernard/SPL/Photo Researchers, Inc. **29.26** Edward S. Ross. **29.27** (a) © NOAA; (b) © USDA/Forestry Service; (c) © David W. Stahle, Department of Geosciences, University of Arkansas.

CHAPTER 30 **30.1** (a) © OPSEC Control Number #4 077-A-4; (b) © Billy Wrobel, 2004; (c) © Keith Weller/USDA-ARS. **30.2** William Ferguson. **30.3** (a) © Robert Frerch/Stone/Getty Images; (b) Courtesy of NOAA. **30.4** (a) Mark E. Dudley and Sharon R. Long; (b) Adrian P. Davies/Bruce Coleman, Ltd; (c) NifTAL Project, Univ. of Hawaii, Maui. **30.5** Left, © Andrew Syred/Photo Researchers, Inc.; right, Courtesy of Mark Holland,

Salisbury University. **30.6** (b) Micrograph Chuck Brown. **30.7** Left, Alison W. Roberts, University of Rhode Island; center and right, H. A. Core, W. A. Cote and A. C. Day, *Wood Structure and Identification*, 2nd Ed., Syracuse University Press, 1979. **30.8** The Ohio Historical Society, Natural History Collections. **30.9** © Claude Nuridsany & Marie Perennou/ Science Photo Library/Photo Researchers, Inc. **30.10** Micrograph by Bruce Iverson, computer-enhanced by Lisa Starr. **30.13** (a) Don Hopey/ Pittsburgh Post-Gazette, 2002, all rights reserved. Reprinted with permission; (b) Jeremy Burgess/ SPL/Photo Researchers, Inc.; (c) © Jeremy Burgess/ SPL/Photo Researchers, Inc. **30.14** (a) © James D. Mauseth, MCDB; (b) © J.C. Revy/ISM/Phototake. **30.15** Martin Zimmerman, Science, 1961, 133:73-79, © AAAS. **30.18** (a–b) Robert & Linda Mitchell Photography; (c) John N. A. Lott, Scanning Electron Microscope Study of Green Plants, St. Louis: C. V. Mosby Company, 1976; (d) Robert C. Simpson/ Nature Stock. **31.1** Left, Courtesy of Caroline Ford, School of Plant Sciences, University of Reading, UK; right © James L. Amos/Corbis.

CHAPTER 31 **31.2** Left, © John McAnulty/Corbis; right © Robert Essel NYC/Corbis. **31.3** (b) Robert E. Bayse Rose Breeding and Genetics Research Program, Department of Horticulture, Texas A&M University. **31.4** (a) David M. Phillips/Visuals Unlimited; (b) © Dr. Jeremy Burgess/SPL/Photo Researchers, Inc.; (c) David Scharf/Peter Arnold, Inc. **31.5** (a) Thomas Eisner, Cornell University; (b) Robert A. Tyrrell. **31.6** Left, John Alcock/Arizona State University; right, Merlin D. Tuttle, Bat Conservation International. **31.7** Photo by Marcel Lecoufle. **31.8** © Dr. John Hilty. **31.10** Center, Dr. Charles Good, Ohio State University–Lima; All others, Michael Clayton, University of Wisconsin. **31.11** (a–c) Janet Jones; (e) Dr. Dan Legard, University of Florida GCREC, 2000; (f) Richard H. Gross; (g) © Andrew Syred/SPL/Photo Researchers, Inc.; (i) Mark Rieger. **31.12** (a) R. Carr; (b) © Gregory K. Scott/Photo Researchers, Inc.; (c) © Robert H. Mohlenbrock © USDA-NRCS PLANTS Database/USDA SCS. 1989. "Midwest wetland flora; field office illustrated guide to plant species." Midwest National Technical Center, Lincoln, NE. **31.13** John Alcock/Arizona State University. **31.14** © Darrell Gulin/Corbis. **31.15** © Professor Dr. Hans Hanks-Ulrich Koop. **31.16** Edward S. Ross.

CHAPTER 32 **32.1** (a–b) R. Lyons/Visuals Unlimited; (c) Michael A. Keller/FPG/Getty Images. **32.2** © Dr. John D. Cunningham/Visuals Unlimited **32.3** Right, above, Barry L. Runk/Grant Heilman, Inc.; below From Mauseth. **32.4** Right, Herve Chaumeton/Agence Nature. **32.5** Sylvan H. Wittwer/Visuals Unlimited. **32.6** © Eric B. Brennan. **32.12** (a–b) Michael Clayton, University of Wisconsin; (c–d) © Muday, GK and P. Haworth (1994) "Tomato root growth, gravitropism, and lateral development: Correlations with auxin transport." "Plant Physiology and Biochemistry 32, 193-203" with permission from Elsevier Science. **32.13** Micrographs courtesy of Randy Moore from "How Roots Respond to Gravity," M. L. Evans, R. Moore, and K. Hasenstein, *Scientific American*, December 1986. **32.14** (c) Cathlyn Melloan/Stone/Getty Images. **32.16** Cary Mitchell. **32.19** (a) © Clay Perry/Corbis; (b) © Eric Chrichton/Corbis. **32.20** (a) Ray Evert, University of Wisconsin; (b) © Clay Perry/Corbis; (c) © Eric Chrichton/Corbis. **32.21** Eric Welzel/Fox Hill Nursery, Freeport, Maine. **32.22** Left, © Roger Wilmshurst, Frank Lane Picture Agency/Corbis; Right © Dr. Jeremy Burgess/Photo Researchers, Inc.. **32.23** Larry D. Nooden. **32.24** R. J. Downs. **32.25** Grant Heilman Photography, Inc. **32.26** Dan Fairbanks. **32.27** Jan Zeevaart.

CHAPTER 33 **33.1** Left, © Don W. Fawcett/Photo Researchers, Inc.; Right © Science Photo Library/ Photo Researchers, Inc. **33.2** Left, © Jerry Ohlinger/ Corbis Sygma; Right © Ron Sachs/Corbis Sygma. **33.3** (a) Left, Manfred Kage/Bruce Coleman, Ltd.; Right, Focus on Sports; (b) Left, © Ray Simmons/ Photo Researchers, Inc.; Center, Ed Reschke/Peter Arnold, Inc.; Right, © Don W. Fawcett. **33.4** © Gregory Dimijian/Photo Researchers, Inc.; right) Adapted from C.P. Hickman, Jr., L.S. Roberts, and A. Larson, *Integrated Principles of Zoology*, ninth Edition, Wm. C. Brown, 1995. **33.6** Above, (a) © John Cunningham/Visuals Unlimited; (b–c) Ed Reschke; (d) © Science Photo Library/Photo Researchers, Inc.; (e) © Michael Abbey/Photo Researchers, Inc.; (f) © University of Cincinnati, Raymond Walters College, Biology. **33.7** Roger K. Burnard. **33.8** © Science Photo Library/Photo Researchers, Inc.. **33.9** Above, © Tony McConnell/Science Photo Library/ Photo Researchers, Inc.; (a–b) Ed Reschke; (c) © Biophoto Associates/Photo Researchers, Inc.. **33.10** © Triarch/Visuals Unlimited. **33.11** Kim Taylor/ Bruce Coleman, Ltd. **33.14** © John D. Cunningham/ Visuals Unlimited. **33.15** (b) © Frank Trapper/ Corbis Sygma; (c) © AFP/Getty Images. **33.16** Michael Keller/FPG. **Page 570** Self-Quiz, (a) Ed Reschke/ Peter Arnold, Inc.; (b–d) Ed Reschke. **33.18** © Sean Sprague/Stock, Boston. **33.19** David Macdonald. **33.20** Dr. Preston Maxim and Dr. Stephen Bretz, Department of Emergency Services, San Francisco General Hospital.

CHAPTER 34 **34.1** © Jamie Baker/Getty Images. **Page 573** Left, © PA Photos; Right © Manni Mason's Pictures. **34.3** (d) After Eugene Kozloff. **34.6** (d) SEM of typical motor neuron. **34.9** (c) © Jeff Greenberg/Index Stock Imagery. **34.10** (b) Micrograph, Dr. Constantino Sotelo from International Cell Biology, p. 83, 1977. Used by copyright permission of the Rockefeller University Press. **34.11** Micrograph by Don Fawcett, *Bloom and Fawcett*, 11th edition, after J. Desaki and Y. Uehara/ Photo Researchers, Inc. **34.13** (a) AP/Wide World Photos; (b–c) from Neuro Via Clinical Research Program, Minneapolis VA Medical Center. **34.14** © Nancy Kedersha/UCLA/Photo Researchers, Inc. **34.18** Washington University/www.thalamus .wustl.edu. **34.21** (a) © Colin Chumbley/Science Source/Photo Researchers, Inc.; (b) C. Yokochi and J. Rohen, Photographic Anatomy of the Human Body, 2nd Ed., Igaku-Shoin, Ltd., 1979. **34.22** (a) After Penfield and Rasmussen, *The Cerebral Cortex of Man*, © 1950 Macmillan Library Reference. Renewed 1978 by Theodore Rasmussen; (b) © Colin Chumbley/Science Source/Photo Researchers, Inc. **34.23** (b) Marcus Raichle, Washington Univ. School of Medicine. **34.30** PET scans from E. D. London, et al., *Archives of General Psychiatry*, 47:567-574, 1990. **34.31** Herve Chaumeton/Agence Nature.

CHAPTER 35 **35.1** © Phillip Colla, OceanLight.com. All Rights Reserved Worldwide. **35.2** (a) © David Turnley/Corbis. **35.3** From Hensel and Bowman, *Journal of Physiology*, 23:564-568, 1960. **35.4** Left, After Penfield and Rasmussen, *The Cerebral Cortex of Man*, © 1950 Macmillan Library Reference. Renewed 1978 by Theodore Rasmussen; Right, © Colin Chumbley/Science Source/Photo Researchers, Inc. **Page 605** Left, © AFP Photo/Timothy A. Clary/ Corbis. **35.11** (a) Right, Fabian/Corbis Sygma; (d) Medtronic Xomed; (e) above, Micrograph by Dr. Thomas R. Van De Water, University of Miami Ear Institute. **35.12** Robert E. Preston, courtesy Joseph E. Hawkins, Kresge Hearing Research Institute, University of Michigan Medical School. **35.13** (d) Frank Park/ANT Photo Library; (e) Keith Gillett/

Tom Stack & Associates. **35.14** (a, below) After M. Gardiner, *The Biology of Vertebrates*, McGraw-Hill, 197; (a, above) © E. R. Degginger; (b) G. A. Mazohkin-Porshnykov (1958). Reprinted with permission from Insect Vision © 1969 Plenum Press. **35.15** Chris Newbert. **35.16** Chase Swift. **35.19** Micrograph Lennart Nilsson © Boehringer Ingelheim International GmbH. **35.20** www.2 .gasou.edu/psychology/courses/muchinsky and www.occipita.cfa.cmu.edu. **35.21** After S. Kuffler and J. Nicholls, *From Neuron to Brain*, Sinauer, 1977. **35.22** © Ophthalmoscopic image from Webvision http://webvision.med.utah.edu/. **35.23** © Lydia V. Kibiuk. **35.24** Photos by Gary Ellis/The Wildlife Collection. **35.25** (a) Eric A. Newman; (b) Merlin D. Tuttle, Bat Conservation International. **35.26** © F. Spoor using Voxel-man. **35.27** Edward W. Bower/ © 1991 TIB.

CHAPTER 36 **36.1** Left, © David Ryan/ SuperStock; Right, © Catherine Ledner Photography. **Page 619** © David Aubrey/Corbis. **36.5** Left, photograph by Lisa Starr; Right, Courtesy of Dr. Erica Eugster. **36.8** (a) © Scott Camazine/ Photo Researchers, Inc.; (b) © Biophoto Associates/ SPL/Photo Researchers, Inc. **36.9** The Stover Group/D. J. Fort. **36.10** (top) PDB ID: 1MSO; Smith, G.D.; Pangborn, W.A.; Blessing, R.H.: *The Structure of T6 Human Insulin at 1.0 A Resolution Acta Crystallogr, Sect D* 59, pp. 474 (2003). **36.11** Left, Ralph Pleasant/FPG/Getty Images; Right, © Yoav Levy/Phototake. **36.13** Mitch Reardon/Photo Researchers, Inc. **36.14** John S. Dunning/Ardea, London. **36.15** Frans Lanting/Bruce Coleman, Ltd. **36.16** (a) © Dr. Carlos J. Bourdony; (b) Courtesy of G. Baumann, MD, Northwestern University. **36.17** © Kevin Fleming/Corbis.

CHAPTER 37 **37.1** Michael Neveux. **Page 639** Ed Reschke. **37.2** Linda Pitkin/Planet Earth Pictures. **37.3** © Stephen Dalton/Photo Researchers, Inc. **37.4** © Stephen Dalton/Photo Researchers, Inc. **37.5** Herve Chaumeton/Agence Nature. **37.6** Bone Clones®, www.boneclones.com. **37.7** Yokochi and J. Rohen, Photographic Anatomy of the Human Body, 2nd Ed., Igaku-Shoin, Ltd., 1979. **37.8** (a) Micrograph Ed Reschke. **37.10** (a–b) © Professor P. Motta/Department of Anatomy/La Sapienza, Rome/SPL/Photo Researchers, Inc. **37.12** N.H.P.A./ANT Photolibrary. **37.14** (a, below), Dance Theatre of Harlem, by Frank Capri; (b–c) © Don Fawcett/Visuals Unlimited, from D. W. Fawcett, The Cell, Philadelphia; W. B. Saunders Co., 1966. **37.21** Painting by Sir Charles Bell, 1809, courtesy of Royal College of Surgeons, Edinburgh. **37.22** (a) © Paul Sponseller, MD/Johns Hopkins Medical Center; (b) Courtesy of the family of Tiffany Manning. **37.23** John Brandenberg/Minden Pictures.

CHAPTER 38 **38.1** (a) From A. D. Waller Physiology: The Servant of Medicine, Hitchcock Lectures, University of London Press, 1910; (b) Courtesy of The New York Academy of Medicine Library. **Page 657** © Mark Thomas/Science Photo Library/Photo Researchers, Inc. **38.2** Left, above, © Darlyne A Murawski/Getty Images; left, below © Gay Bumgarner/Getty Images. **38.3** (d) After Labarbera and S. Vogel, *American Scientist*, 1982, 70:54-60. **38.4** Above, © National Cancer Institute/ Photo Researchers, Inc. **38.5** Left, © 2001 EyeWire; (Art) After *Bloodline Image Atlas*, University of Nebraska-Omaha, and Sherri Wicks, *Human Physiology and Anatomy*, University of Wisconsin Web Education System, and others. **38.6** From: Maslak P. Blast Crisis of Chronic Myelogenous Leukemia (posted online December 5, 2001). ASH Image Bank. Copyright American Society of Hematology, used with permission. **38.7** Lester V.

46.9 (a-b) After Rickleffs & Miller, *Ecology*, Fourth Edition, page 459 (Fig. 23.13a) and page 461 (Fig. 23.14); photo, © W. Perry Conway/Corbis. **46.10** (b) © Ed Cesar/Photo Researchers, Inc.; (c) © Kennan Ward. **46.11** (a) © JH Pete Carmichael; (b) Edward S. Ross; (c) W. M. Laetsch; (d) © Nigel Jones. **46.13** (a–b) Thomas Eisner, Cornell University; (c) © Jeffrey Rotman Photography; (d) © Bob Jensen Photography. **46.14** (a) MSU News Service, photo by Montana Water Center; (b) Karl Andree. **46.15** Left, © The Samuel Roberts Noble Foundation, Inc.; right, Courtesy of Colin Purrington, Swarthmore College. **46.16** © C. James Webb/Phototake USA. **46.17** © Peter J. Bryant/Biological Photo Service. **46.18** (a) © Richard Price/Getty Images; (b) © E.R. Degginger/Photo Researchers, Inc. **46.19** (a) © Doug Peebles/Corbis; (b) © Pat O'Hara/Corbis; (c–d) © Tom Bean/Corbis; (e) © Duncan Murrell/Taxi/Getty Images. **46.20** (a) R. Barrick/USGS; (b–c) © 1980 Gary Braasch. **46.21** (a, c) Jane Burton/Bruce Coleman, Ltd.; (b) Heather Angel; (d–e) Based on Jane Lubchenco, *American Naturalist*, 112:23-19, © 1978 University of Chicago Press. Used with permission. **46.22** Above left and (b) © Angelina Lax/Photo Researchers, Inc.; (a) © Dr. Alexande Meinesz, University of Nice-Sophia Antipolis; right, © The University of Alabama Center for Public TV. **46.23** Peter Bird/Australian Picture Library/Westlight. **46.24** After W. Dansgaard et al., *Nature*, 364:218-220, July 15 1993; D. Raymond et al., *Science*, 259:926-933, February 1993; W. Post, *American Scientist*, 78:310-326, July-August 1990. **46.25** © Pierre Vauthey/Corbis Sygma; art, After S. Fridriksson, *Evolution of Life on a Volcanic Island*, Butterworth, London 1975. **46.26** (a) © Susan G. Drinker/Corbis; (b) Frans Lanting/Minden Pictures. **46.27** (a) © Anthony Bannister, Gallo Images/Corbis; (b) © Bob Jensen Photography. **46.28** Heather Angel/Biofotos. **46.29** Photograph NASA.

CHAPTER 47 **47.1** LA Wildlife & Fisheries, Natural Heritage Program, Patti Faulkner. **Page 843** NOAA, photo by Commander Grady Tuell. **47.2** Above, Photodisc, Inc.; below, David Neal Parks. **47.3** Photograph Alan and Sandy Carey; (right) After R.L. Smith, *Ecology and Field Biology*, Fifth Edition. **47.4** Field of flowers, © Frank Oberle/Stone/Getty Images; marsh hawk, © J. Lichter/Photo Researchers, Inc.; crow, © Ed Reschke; garter snake, Michael Jeffords; cutworm, © Nigel Cattlin/Holt Studios International/Photo Researchers, Inc. **47.5** Field of flowers, © Frank Oberle/Stone/Getty Images; marsh hawk, © J. Lichter/Photo Researchers, Inc.; crow, © Ed Reschke; milk snake, Mike Pingleton/www.pingleton.com; garter snake and spider, Michael Jeffords; frog, John H. Gerard; Earthworm, grasshopper, badger, coyote, and ground squirrel, Photodisc/Getty Images; cutworm, © Nigel Cattlin/Holt Studios International/Photo Researchers, Inc.; clay-colored sparrow, © Rod Planck/Photo Researchers, Inc.; weasel, Courtesy of Biology Department, Loyola Marymount University; prairie vole and pocket gopher, © Tom McHugh/Photo Researchers, Inc. **47.6** Left, D. Robert Franz; right, Frans Lanting/Bruce Coleman, Ltd.; art, After Paul Hertz. **47.7** Right, Inga Spence/Tom Stack & Associates. **47.9** © David T. Grewcock/Corbis. **47.10** (a) NASA's Earth Observatory. **47.14** Gerry Ellis/The Wildlife Collection. **47.17** (a) www.hubbardbrook.org; (b) © Hubbard Brook Experimental Forest/www.hubbardbrook.org; (c) Gene E. Likens from G. E. Likens, et. al., *Ecology Monograph*, 40(1):23-47, 1970; (d) After G.E. Likens and F. H. Bormann, "An Experimental Approach to New England Landscapes," In A. D. Hasler (ed.), *Coupling of Land and Water Systems*, Chapman & Hall, 1975. **47.18** © Craig Aurness/Corbis. **47.20**

(wolf) © 2000 Photodisc, Inc. **47.22** NASA photograph from JSC Digital Image Collection. **47.23** © Yann Arthus-Bertrand/Corbis. **47.24** (a) Compilation of data from Mauna Loa Observatory, Keeling and Whorf, Scripps Institute of Oceanography; (b) Compilation of data from Prinn, et al., CDIAC, Oak Ridge National Laboratory, World Resources Institute; Khalil and Rasmussen, Oregon Graduate Institute of Science and Technology, CDIAC DB-1010; Liefer and Chan, CDIAC DB-1019; (c,d) Compilation of data from World Resources Institute; Law Dome ice core samples, Etheridge, Pearman, and Fraser, Commonwealth Scientific and Industrial Research Organization; Prinn et al., CDIAC, Oak Ridge National Laboratory; Leifer and Chan, CDIAC DB-1019. **47.25** © Jeff Vanuga/Corbis. **47.26** © Frederica Georgia/Photo Researchers, Inc. **47.27** Above, © 2000 Photodisc, Inc. **47.28** Fisheries & Oceans Canada, Experimental Lakes Area. **47.29** (a–b) Courtesy of NASA's Terra satellite, supplied by Ted Scambos, National Snow and Ice Data Center, University of Colorado, Boulder; (c) Courtesy of Keith Nicholls, British Antarctic Survey. **47.30** Left, above, © Boetius et al. 2000, *Nature* 407, 623-626; below, Courtesy of K. O. Stetter & R. Rachel, University of Regensburg.

CHAPTER 48 **48.01** © Hank Fotos Photography. **Page 867** NASA. **48.5** Alex MacLean/Landslides. **48.06, 48.10** NASA. **48.8** (a) Adapted from *Living in the Environment* by G. Tyler Miller, Jr., p. 428. © 2002 by Brooks/Cole, a division of Thomson Learning; (b) © Ted Spiegel/Corbis. **48.11** Left, © Sally A. Morgan, Ecoscene/Corbis; right, © Bob Rowan, Progressive Image/Corbis. **48.12** NASA. **48.13** NASA's Earth Observatory; (Art) After Whittaker, Bland, and Tilman. **48.14, 48.15** Courtesy of Jim Deacon, The University of Edinburgh. **48.16** © George H. Huey/Corbis; inset, © John M. Roberts/Corbis. **48.17** Left, © John C. Cunningham/Visuals Unlimited; right, AP/Wide World Photos. **48.18** (a) Jonathan Scott/Planet Earth Pictures; (b) © Tom Bean Photography; (c) Ray Wagner/Save the Tall Grass Prairie, Inc. **48.19** Left © James Randklev/Corbis; trees, © Randy Wells/Corbis. **Page 867** Above, Gerry Ellis/The Wildlife Collection; below © 1991 Gary Braasch; inset Edward S. Ross. **48.20** (a) © Raymond Gehman/Corbis; (b) © Thomas Wiewandt/ChromoSohm Media, Inc./Photo Researchers, Inc. **48.21** (a) © Darrell Gulin/Corbis; (b) © Paul A. Souders/Corbis. **48.22** © Pat O'Hara/Corbis. **48.23** Based on data from G.T. Miller, Jr. **48.25** © Orbimage Imagery. **48.26** U.S. Agency for International Development. **48.27** © Onne van der Wal/Corbis. **48.28** (b) After E.S. Deevy, Jr., *Scientific American*, October 1951. **48.29** Jack Carey. **48.30** (a) Bruce M. Herman/Photo Researchers, Inc.; (b–d) E. F. Benfield, Virginia Tech. **48.31** Image courtesy of the Image Analysis Laboratory, NASA Johnson Space Center. **48.32** Ocean Arks International. **48.33** (a) © Annie Griffiths Belt/Corbis; (b) © Douglas Peebles/Corbis. **48.34** (a) © Nancy Sefton; (b) Courtesy of J. L. Sumich, Biology of Marine Life, 7th ed., W. C. Brown, 1999. **48.35** © Paul A. Souders/Corbis. **48.36** Left © Amos Nachoum/Corbis; right, © Corbis. **48.38** (a) Image courtesy of NOAA and MBARI; (b) Robert Vrijenhoek, MBARI; (c) Steven Haddock, MBARI; (d) © Peter David/FPG/Getty Images. **48.40** (a) NASA–Goddard Space Flight Center Scientific Visualization Studio. **48.41** (a) CHAART, at NASA Ames Research Center; (b) © Eye of Science/Photo Researchers, Inc.; (c) Courtesy of Dr. Anwar Huq and Dr. Rita Colwell, University of Maryland; (d) Raghu Rai/Magnum Photos. **48.42** After M. H. Dickerson, "ARAC: Modeling an Ill Wind," in *Energy and Technology Review*, August 1987. Used by permission of University of California Lawrence Livermore National Laboratory and U.S.

Dept. of Energy. **48.43** © Lawson Wood/Corbis. **Page 987** Right, © Nigel Cook/Dayton Beach News Journal/Corbis Sygma.

CHAPTER 49 **49.1** © Stephen Dalton/Photo Researchers, Inc. **Page 899** © Scott Camazine. **49.2** (a) Eugene Kozloff; (b–c) Stevan Arnold. **49.3** (a) © Robert M. Timm & Barbara L. Clauson, University of Kansas; (b) Reprinted from Trends in Neuroscience, Vol. 21, Issue 2, 1998, L. J. Young, W. Zuoxin, T. R. Insel, "Neuroendocrine bases of monogamy", Pages 71 – 75, ©1998, with permission from Elsevier Science. **49.4** (a) Eric Hosking; (b) © Stephen Dalton/Photo Researchers, Inc. **49.5** © Jennie Woodcock, Reflections Photolibrary/Corbis. **49.6** © James Zipp/Photo Researchers, Inc. **49.7** © Nina Leen/TimePix/Getty Images; inset, © Robert Semeniuk/Corbis. **49.8** (a) Robert Maier/Animals Animals; (b) © John Bova/Photo Researchers, Inc.; (d) Jack Clark/Comstock, Inc. **49.9** (a) © Monty Sloan/www.wolfphotography.com; (b) Tom and Pat Leeson, leesonphoto.com; (c) © Kevin Schafer/Corbis. **49.10** © Stephen Dalton/Photo Researchers, Inc. **49.11** (a) John Alcock, Arizona State University; (b–c) © Pam Gardner/Frank Lane Picture Agency/Corbis; (d) © D. Robert Franz/Corbis. **49.12** Michael Francis/The Wildlife Collection. **49.13** (a) © B. Borrell Casals/Frank Lane Picture Agency/Corbis; (b) © Steve Kaufman/Corbis; (c) © John Conrad/Corbis. **49.14** (a) Tom and Pat Leeson, leesonphoto.com; (b) John Alcock, Arizona State University; (c) © Paul Nicklen/National Geographic/Getty Images. **49.15** © Jeff Vanuga/Corbis. **49.16** © Eric and David Hosking/Corbis. **49.17** (a) © Australian Picture Library/Corbis; (b) © Alexander Wild; (c) © Professor Louis De Vos. **49.18** (a) Kenneth Lorenzen; (b) © Nicola Kountoupes/Cornell University; (c) © Peter Johnson/Corbis. **49.19** © Dr. Tim Jackson, University of Pretoria. **49.20** (a) Steve Bloom Images/www.stevebloom.com; (b) © Gallo Images/Corbis. **Page 913** © Matthew Alan/Corbis. **49.22** F. Schutz. **49.23** © Brad Bergstrom. **Page 916** © Joseph Sohm, Visions of America/Corbis.